Recommended Dietary Allowances (RDAs) and Adequate Intakes (AIs) for Vitamins

Life-Stage Group	Thiamin RDA (mg/day)	Riboflavin RDA (mg/day)	Niacin RDA (mg/day)a	Biotin AI (µg/day)	Pantothenic acid AI (mg/day)	Vitamin B6 RDA (mg/day)	Folate RDA (µg/day)b	Vitamin B12 RDA (µg/day)	Choline AI (mg/day)	Vitamin C RDA (mg/day)	Vitamin A RDA (µg/day)c	Vitamin D RDA (µg/day)d	Vitamin E RDA (mg/day)e	Vitamin K AI (µg/day)
Infants														
0–6 mo	0.2	0.3	2	5	1.7	0.1	65	0.4	125	40	400	10	4	2.0
7–12 mo	0.3	0.4	4	6	1.8	0.3	80	0.5	150	50	500	10	5	2.5
Children														
1–3 y	0.5	0.5	6	8	2	0.5	150	0.9	200	15	300	15	6	30
4–8 y	0.6	0.6	8	12	3	0.6	200	1.2	250	25	400	15	7	55
Males														
9–13 y	0.9	0.9	12	20	4	1.0	300	1.8	375	45	600	15	11	60
14–18 y	1.2	1.3	16	25	5	1.3	400	2.4	550	75	900	15	15	75
19–30 y	1.2	1.3	16	30	5	1.3	400	2.4	550	90	900	15	15	120
31–50 y	1.2	1.3	16	30	5	1.3	400	2.4	550	90	900	15	15	120
51–70 y	1.2	1.3	16	30	5	1.7	400	2.4	550	90	900	15	15	120
>70 y	1.2	1.3	16	30	5	1.7	400	2.4	550	90	900	20	15	120
Females														
9–13 y	0.9	0.9	12	20	4	1.0	300	1.8	375	45	600	15	11	60
14–18 y	1.0	1.0	14	25	5	1.2	400	2.4	400	65	700	15	15	75
19–30 y	1.1	1.1	14	30	5	1.3	400	2.4	425	75	700	15	15	90
31–50 y	1.1	1.1	14	30	5	1.3	400	2.4	425	75	700	15	15	90
51–70 y	1.1	1.1	14	30	5	1.5	400	2.4	425	75	700	15	15	90
>70 y	1.1	1.1	14	30	5	1.5	400	2.4	425	75	700	20	15	90
Pregnancy														
14–18 y	1.4	1.4	18	30	6	1.9	600	2.6	450	80	750	15	15	75
19–30 y	1.4	1.4	18	30	6	1.9	600	2.6	450	85	770	15	15	90
31–50 y	1.4	1.4	18	30	6	1.9	600	2.6	450	85	770	15	15	90
Lactation														
14–18 y	1.4	1.6	17	35	7	2.0	500	2.8	550	115	1200	15	19	75
19–30 y	1.4	1.6	17	35	7	2.0	500	2.8	550	120	1300	15	19	90
31–50 y	1.4	1.6	17	35	7	2.0	500	2.8	550	120	1300	15	19	90

Note: For all nutrients, values for infants are AIs.
aNiacin recommendations are expressed as niacin equivalents (NE), except for recommendations for infants younger than 6 months, which are expressed as preformed niacin.
bFolate recommendations are expressed as dietary folate equivalents (DFE).
cVitamin A recommendations are expressed as retinol activity equivalents (RAE).
dVitamin D recommendations are expressed as cholecalciferol.
eVitamin E recommendations are expressed as α-tocopherol.

Recommended Dietary Allowances (RDAs) and Adequate Intakes (AIs) for Minerals

Life-Stage Group	Sodium AI (mg/day)	Chloride AI (mg/day)	Potassium AI (mg/day)	Calcium RDA (mg/day)	Phosphorus RDA (mg/day)	Magnesium RDA (mg/day)	Iron RDA (mg/day)	Zinc RDA (mg/day)	Iodine RDA (µg/day)	Selenium RDA (µg/day)	Copper RDA (µg/day)	Manganese AI (mg/day)	Fluoride AI (mg/day)	Chromium AI (µg/day)	Molybdenum RDA (µg/day)
Infants															
0–6 mo	120	180	400	200	100	30	0.27	2	110	15	200	0.003	0.01	0.2	2
7–12 mo	370	570	700	260	275	75	11	3	130	20	220	0.6	0.5	5.5	3
Children															
1–3 y	1000	1500	3000	700	460	80	7	3	90	20	340	1.2	0.7	11	17
4–8 y	1200	1900	3800	1000	500	130	10	5	90	30	440	1.5	1	15	22
Males															
9–13 y	1500	2300	4500	1300	1250	240	8	8	120	40	700	1.9	2	25	34
14–18 y	1500	2300	4700	1300	1250	410	11	11	150	55	890	2.2	3	35	43
19–30 y	1500	2300	4700	1000	700	400	8	11	150	55	900	2.3	4	35	45
31–50 y	1500	2300	4700	1000	700	420	8	11	150	55	900	2.3	4	35	45
51–70 y	1300	2000	4700	1000	700	420	8	11	150	55	900	2.3	4	30	45
>70 y	1200	1800	4700	1200	700	420	8	11	150	55	900	2.3	4	30	45
Females															
9–13 y	1500	2300	4500	1300	1250	240	8	8	120	40	700	1.6	2	21	34
14–18 y	1500	2300	4700	1300	1250	360	15	9	150	55	890	1.6	3	24	43
19–30 y	1500	2300	4700	1000	700	310	18	8	150	55	900	1.8	3	25	45
31–50 y	1500	2300	4700	1000	700	320	18	8	150	55	900	1.8	3	25	45
51–70 y	1300	2000	4700	1200	700	320	8	8	150	55	900	1.8	3	20	45
>70 y	1200	1800	4700	1200	700	320	8	8	150	55	900	1.8	3	20	45
Pregnancy															
14–18 y	1500	2300	4700	1300	1250	400	27	12	220	60	1000	2.0	3	29	50
19–30 y	1500	2300	4700	1000	700	350	27	11	220	60	1000	2.0	3	30	50
31–50 y	1500	2300	4700	1000	700	360	27	11	220	60	1000	2.0	3	30	50
Lactation															
14–18 y	1500	2300	5100	1300	1250	360	10	13	290	70	1300	2.6	3	44	50
19–30 y	1500	2300	5100	1000	700	310	9	12	290	70	1300	2.6	3	45	50
31–50 y	1500	2300	5100	1000	700	320	9	12	290	70	1300	2.6	3	45	50

Tolerable Upper Intake Levels (ULs) for Vitamins

Life-Stage Group	Niacin (mg/day)[a]	Vitamin B6 (mg/day)	Folate (µg/day)[a]	Choline (mg/day)	Vitamin C (mg/day)	Vitamin A (µg/day)[b]	Vitamin D (µg/day)	Vitamin E (mg/day)[c]
Infants								
0–6 mo	—	—	—	—	—	600	25	—
7–12 mo	—	—	—	—	—	600	38	—
Children								
1–3 y	10	30	300	1000	400	600	63	200
4–8 y	15	40	400	1000	650	900	75	300
Adolescents								
9–13 y	20	60	600	2000	1200	1700	100	600
14–18 y	30	80	800	3000	1800	2800	100	800
Adults								
19–70 y	35	100	1000	3500	2000	3000	100	1000
>70 y	35	100	1000	3500	2000	3000	100	1000
Pregnancy								
14–18 y	30	80	800	3000	1800	2800	100	800
19–50 y	35	100	1000	3500	2000	3000	100	1000
Lactation								
14–18 y	30	80	800	3000	1800	2800	100	800
19–50 y	35	100	1000	3500	2000	3000	100	1000

[a]The UL for niacin and folate apply to synthetic forms obtained from supplements, fortified foods, or a combination of the two.

[b]The UL for vitamin A applies to the preformed vitamin only.
[c]The UL for vitamin E applies to any form of supplemental α-tocopherol, fortified foods, or a combination of the two.

Tolerable Upper Intake Levels (ULs) for Minerals

Life-Stage Group	Sodium (mg/day)	Chloride (mg/day)	Calcium (mg/day)	Phosphorus (mg/day)	Magnesium (mg/day)[d]	Iron (mg/day)	Zinc (mg/day)	Iodine (µg/day)	Selenium (µg/day)	Copper (µg/day)	Manganese (mg/day)	Fluoride (mg/day)	Molybdenum (µg/day)	Boron (mg/day)	Nickel (mg/day)
Infants															
0–6 mo	—[e]	—[e]	1000	—	—	40	4	—	45	—	—	0.7	—	—	—
7–12 mo	—[e]	—[e]	1500	—	—	40	5	—	60	—	—	0.9	—	—	—
Children															
1–3 y	1500	2300	2500	3000	65	40	7	200	90	1000	2	1.3	300	3	0.2
4–8 y	1900	2900	2500	3000	110	40	12	300	150	3000	3	2.2	600	6	0.3
Adolescents															
9–13 y	2200	3400	3000	4000	350	40	23	600	280	5000	6	10	1100	11	0.6
14–18 y	2300	3600	3000	4000	350	45	34	900	400	8000	9	10	1700	17	1.0
Adults															
19–70 y	2300	3600	2500[f]	4000	350	45	40	1100	400	10,000	11	10	2000	20	1.0
>70 y	2300	3600	2000	3000	350	45	40	1100	400	10,000	11	10	2000	20	1.0
Pregnancy															
14–18 y	2300	3600	3000	3500	350	45	34	900	400	8000	9	10	1700	17	1.0
19–50 y	2300	3600	2500	3500	350	45	40	1100	400	10,000	11	10	2000	20	1.0
Lactation															
14–18 y	2300	3600	3000	4000	350	45	34	900	400	8000	9	10	1700	17	1.0
19–50 y	2300	3600	2500	4000	350	45	40	1100	400	10,000	11	10	2000	20	1.0

[d]The UL for magnesium applies to synthetic forms obtained from supplements or drugs only.
[e]Source of intake should be from human milk (or formula) and food only.
[f]The UL for calcium for 19–50 y is 2500 mg; the UL for calcium is reduced to 2000 mg for 51–70 y.

Note: An Upper Limit was not established for vitamins and minerals not listed and for those age groups listed with a dash (—) because of a lack of data, not because these nutrients are safe to consume at any level of intake. All nutrients can have adverse effects when intakes are excessive.

SOURCE: Adapted with permission from the *Dietary Reference Intakes series*, National Academies Press. Copyright 1997, 1998, 2000, 2001, 2011, by the National Academy of Sciences. Courtesy of the National Academies Press, Washington, D.C.

Acceptable Macronutrient Distribution Ranges (AMDRs)

Macronutrient	Range (percent of energy)		
	Children, 1–3 years	Children, 4–18 years	Adults 19+ years
Fat	30–40	25–35	20–35
ω-6 polyunsaturated acids[a] (linoleic acid)	5–10	5–10	5–10
ω-3 polyunsaturated fatty acids[a] (linolenic acid)	0.6–1.2	0.6–1.2	0.6–1.2
Carbohydrate	45–65	45–65	45–65
Protein	5–20	10–30	10–35

[a]Approximately 10% of the total can come from longer-chain ω-3 or ω-6 fatty acids.

SOURCE: Adapted from Institute of Medicine, *Dietary Reference Intakes for Energy, Carbohydrate, Fiber, Fat, Fatty Acids, Cholesterol, Protein, and Amino Acids,* National Academies Press; Washington, D.C., 2005.

Nutritional Sciences

FROM FUNDAMENTALS TO FOOD

Michelle McGuire, PhD
Washington State University

Kathy A. Beerman, PhD
Washington State University

WADSWORTH
CENGAGE Learning

Australia • Brazil • Japan • Korea • Mexico • Singapore • Spain • United Kingdom • United States

WADSWORTH
CENGAGE Learning·

Nutritional Sciences: From Fundamentals to Food, Third Edition
Michelle McGuire and Kathy A. Beerman

Publisher: Yolanda Cossio

Senior Acquisitions Editor: Peggy Williams

Developmental Editor: Suzannah Alexander

Assistant Editor: Elesha Feldman

Editorial Assistant: Sean Cronin

Media Editor: Miriam Myers

Senior Marketing Manager: Laura McGinn

Senior Marketing Communications
    Manager: Linda Yip

Marketing Coordinator: Jing Hu

Senior Content Project Manager:
    Carol Samet

Creative Director: Rob Hugel

Art Director: John Walker

Print Buyer: Karen Hunt

Rights Acquisitions Specialist:
    Tom McDonough

Production Service: Lynn Lustberg,
    MPS Limited, a Macmillan Company

Text Designer: Ellen Pettengell

Photo Researcher: Josh Garvin,
    Bill Smith Group

Copy Editor: Laurene Sorensen

Cover Designer: Ellen Pettengell

Cover Image: Gettyimages

Compositor: MPS Limited, a Macmillan
    Company

For product information and technology assistance, contact us at
Cengage Learning Customer & Sales Support, 1-800-354-9706.

For permission to use material from this text or product,
submit all requests online at **www.cengage.com/permissions**.
Further permissions questions can be e-mailed to
permissionrequest@cengage.com.

Library of Congress Control Number: 2011925356

ISBN-13: 978-0-8400-5820-1

ISBN-10: 0-8400-5820-9

Wadsworth
20 Davis Drive
Belmont, CA 94002-3098
USA

Cengage Learning is a leading provider of customized learning solutions with office locations around the globe, including Singapore, the United Kingdom, Australia, Mexico, Brazil, and Japan. Locate your local office at **www.cengage.com/global**.

Cengage Learning products are represented in Canada by Nelson Education, Ltd.

To learn more about Wadsworth, visit **www.cengage.com/wadsworth**

Purchase any of our products at your local college store or at our preferred online store **www.cengagebrain.com**.

Printed in the United States of America
2 3 4 5 18 17 16 15

CONTENTS IN BRIEF

CONTENTS

CHAPTER 3 Chemical, Biological, and Physiological Aspects of Nutrition 69

CHAPTER 4 Carbohydrates 113

CHAPTER 5 Protein 161

CHAPTER 6   Lipids   **217**

CHAPTER 7 Energy Metabolism   269

CHAPTER 7 Energy Metabolism 269

CHAPTER 6   Lipids   217

CHAPTER 10 Water-Soluble Vitamins 417

CHAPTER 11 Fat-Soluble Vitamins 459

CHAPTER 12 The Major Minerals and Water   505

CHAPTER 13 The Trace Minerals 553

LIST OF *FOCUS ON* BOXES

EVERYBODY HAS A STORY: Michelle "Shelley" McGuire

Courtesy of The Dannon Institute

Dr. Michelle "Shelley" McGuire got her academic start in the small town of Polo, Illinois, where her mother-in-law-to-be (a high school English teacher) taught her how to write, and her family and community instilled within her a strong Midwestern work ethic and love of learning. Shelley earned a bachelor's degree in biology from the University of Illinois, a master's degree in nutritional sciences from the University of Illinois, and a doctorate in human nutrition (with minors in physiology and international nutrition) from Cornell University. She is now a faculty member at Washington State University, where she conducts research related to nutrition and lactation. Recent research, conducted in collaboration with her husband Dr. Mark McGuire (University of Idaho) and funded primarily by the National Institutes of Health and the Bill and Melinda Gates Foundation, has focused on the discovery that human milk naturally contains "healthy" bacteria that likely benefit both mothers and infants. Shelley is an active member of the American Society for Nutrition, for which she is a National Spokesperson and a member of various committees. She is also a member of the International Society for Research in Human Milk and Lactation, for which she is secretary/treasurer. Dr. McGuire has been enthusiastically teaching Washington State University's basic nutrition course for mixed majors for over 15 years and has a strong appreciation for its goals and objectives, as well as the wide variety of student needs associated with the course. She also teaches an upper-level science communications course for biology majors. These experiences, coupled with her strong writing skills, research background, and personal connections within the international nutrition research community, have helped her create an exciting and refreshing text for the introductory nutrition course. When Shelley is not writing, teaching, or overseeing research, she enjoys being with family and friends, traveling, running, practicing yoga, cooking, gardening, and performing classical music.

With continued gratitude to Mark, Emily, Grace, and Keith for providing my "equilibrium" each and every day.

This book is dedicated to my dear "academic mothers" and mentors— Drs. Mary Frances Picciano and Kathleen Rasmussen. Mary Frances, thank you for gently yet securely diverting me from a path leading straight to medical school to one paving the way to academia. You taught me that a woman can be both a serious scientist and a devoted mother/wife. You also taught me that husband/wife academic teams can be incredibly productive, loving, and long-lasting. I miss your advice and wit on a daily basis. Kathy, your guidance and continued support have been invaluable. From experimental design to statistical analysis; from critical evaluation of the literature to application of current science to public health recommendations; from listening to me to providing leadership opportunities. I cannot adequately express my appreciation for all you have done and continue to do for me. You both believed in and encouraged me when I was weary and unsure of how a sensitive female could possibly cope in the sometimes brutal world of rigorous science. Thank you for teaching me that critical evaluation and continued questioning of dogma require humility, humor, courage, persistence, and (especially) a sturdy backbone.

Shelley

EVERYBODY HAS A STORY: Kathy Beerman

I express my deepest gratitude to those who supported me through the many twists and turns of this endeavor. I am very fortunate to have such wonderful friends, and you have each contributed to this book in many ways. I am especially grateful to my husband, children, brother, and sister-in-law who provided endless and unwavering love each step of the way. My greatest inspiration comes from my father, Morris Beerman, whose profound wisdom has guided me throughout life. It is to him that I dedicate this book.

Kathy

Dr. Kathy Beerman was born and raised in Buffalo, New York. She earned her bachelor's and master's degrees from the State University of New York at Brockport, Department of Health Sciences. Although her urban, East Coast roots run deep, she relocated to Corvallis, Oregon, to attend graduate school at Oregon State University, where she earned her PhD. After finishing her doctorate, Kathy and her husband (Steven) moved to Moscow, Idaho. Now a professor of nutrition, Kathy teaches in the School of Biological Sciences at Washington State University. The author of several published articles, she specializes in research focusing on dietary practices of college students and the effects of isoflavones on health parameters (immune response, thyroid function, and memory) in postmenopausal women. A collaborative clinical intervention study currently underway involves the assessment of L-arginine and L-citrulline oral supplementation on blood pressure in hypertensive men. She is a member of the American Society for Nutrition. Dr. Beerman teaches the introductory nutrition course for health-related majors, as well as courses in nutrition education, biology of women, and life cycle nutrition. She also provides educational training to graduate students in preparation for instructional roles as teaching assistants. Dr. Beerman will also be co-teaching a course with her husband in Puntarenas, Costa Rica, that explores the connectivity between agriculture, environmental quality, human nutrition, and health. Since joining the faculty at Washington State University in 1989, she has taught more than 15,000 students, and has been the recipient of several college and university teaching awards including the Burlington Northern Faculty Meritorious Achievement in Teaching Award, the R. M. Wade Foundation Award for Excellence in Teaching, and the WSU Mortar Board (honor society of college seniors) Distinguished Professor Award. In 2008, Dr. Beerman was awarded the highest teaching honor offered at Washington State University—the Sahlin Faculty Excellence Award for Instruction. In addition, Dr. Beerman was inducted into the President's Teaching Academy at Washington State University, which is composed of faculty members who provide leadership to strengthen undergraduate and graduate teaching and learning. Dr. Beerman's years of teaching experience combined with her wide knowledge base in nutrition and health sciences have helped her create this innovative introductory nutrition textbook. At home, Kathy enjoys spending time with her husband, Steven, and children, Anna and Michael. She also enjoys traveling, gardening, running, and being a part of the Moscow community.

PREFACE

Like you, we teach introductory nutrition at the college level, and our experiences in the classroom have taught us many things. We would like to share two primary insights that shaped the inception of this textbook and have guided us throughout the writing of each edition.

- First and foremost, an introductory nutrition textbook must explain nutrition concepts accurately, clearly, and completely in a way that all students, regardless of level or background, can understand. To do this most effectively, it is essential to provide students with fundamental scientific concepts. Only then can these concepts be applied to the science of nutrition in a meaningful way.
- For some students, scientific concepts are abstract and difficult to grasp. For this reason, the use of well-designed figures greatly enhances learning by providing integrated, contextually-based illustrative examples.

These two straightforward observations formed the genesis of this textbook. It is our hope that this book will allow all students, *even those with little or no science background*, to find the science of nutrition approachable, understandable, and—perhaps most imperative—useful in their lives.

Critically important to the evolution of this textbook is our commitment to getting the facts straight, updating the text as new knowledge unfolds, and presenting information in an objective, unbiased manner. As such, the revision process is an integral part of all stages of textbook development, writing, and implementation. When necessary, subject matter experts are consulted to clarify concepts, discuss implications of a new study, provide commentary on new scientific discoveries, or even help us define terms. If you are one of these people we have reached out to, thank you! We also attend national meetings, participate in conferences, and seek out nutrition-related events to ensure that we are on the cutting edge of the science. In addition, and perhaps most importantly, we use this textbook in our classrooms and solicit feedback from our students. Sometimes we learn as much from them as they learn from us.

All of these approaches made us even more eager to roll up our sleeves and dive into this new edition. If you used previous editions of this book, you will immediately notice that many changes have been made to every chapter. These changes are described in more detail later in this preface. Although we are confident that this book is one of the most accurate, clearly written, current, beautiful, and student-friendly nutrition books available, we will have already started working on the next edition by the time you are reading this page. Please do not hesitate to let us know whether these changes address your needs and those of your students, as well as what you would like us to address in subsequent editions.

The Fundamentals Are Important

We think you will agree that students need a strong knowledge base to master nutrition concepts. Without knowledge of fundamental scientific principles and associated vocabulary, many capable students in introductory nutrition courses are lost from the start. For example, when teaching nutrition we often refer to chemical bonds—how else do we distinguish between saturated and unsaturated fatty acids? Similarly, when teaching the importance of ATP, we discuss the breaking of phosphate bonds. But how can a student grasp these concepts without understanding what a chemical bond is? The solid scientific foundation provided by this book will enable students to navigate through the most difficult of concepts, regardless of their background. For example, Chapter 3 not only introduces the basic principles of chemistry, biology, and physiology, but also applies these concepts to the study of nutrition. To our knowledge, this is the first introductory nutrition textbook to present such a "primer" on nutritional biochemistry and physiology at the level appropriate for both science and nonscience students. In fact, because we believe that the fundamentals are so crucially important, we worked hard to define and describe each new science-based term and concept when it is introduced. You will see evidence of this in every chapter.

An Integrated, Yet Systematic, Approach

As the field of nutrition grows, so does the amount of information we must teach. Consequently, it can be difficult to cover the necessary basic science and applied aspects of nutrition in a single semester. For instance, it is important to first learn about *all* the related micronutrients and macronutrients before launching into a full discussion of nutrition and bone health. Therefore, our approach has been to "lead" with the basics, followed by application whenever possible. To accomplish this, we have organized the text in a methodical, integrated manner. First, the main chapters of the book are organized in the traditional way—starting with the macronutrients and then discussing the micronutrients. However, applied concepts are integrated throughout the book. For example, active and passive transport mechanisms are carefully described in Chapter 3, but then referred to in subsequent chapters throughout the book. Another pedagogical tool, called **Connections,** is conveniently placed on pages to provide students with a quick reminder of important terms and concepts. Similarly, a structural organization of headings and subheadings provides students with a framework that presents a broad view, which then progresses to a more detailed examination of information. By posing first-level headings as questions, students are able to self-check to ensure they are extracting critical information.

Second, we have continued to use a feature entitled **Focus on...,** which highlights issues related to scientific innovation, diet and health, food, clinical applications, life cycle nutrition, and sports nutrition. These features assimilate topics that are of great interest to students within appropriate chapters, and are especially useful to instructors who do not have time to more extensively cover them during the semester. For example, a Focus on... discussion of high-fructose corn syrup accompanies the section on simple sugars, and a Focus on... discussion of nutrient–drug interactions accompanies the section on vitamin K. Most chapters have several Focus on... features, which students will enjoy reading and from which they will learn a great deal. In fact, these features received high praise from our students, who found the topics interesting and provocative.

Third, to improve the integration of important nutrients related to the maintenance of health or the risk for chronic disease, we include segments called **Nutrition Matters,** which conclude most of the chapters. These "minichapters" deal with important nutrition-related issues—such as food safety, nutrition and cancer, and dysfunctional eating. They are up-to-date and comprehensive, yet clear and student-friendly. Because these pieces conclude most chapters, instructors can easily choose to assign them in any order that works for their class, or not assign them at all if time does not allow.

Putting a *Personal* Face on Nutritional Science

We all know that stories can greatly enhance learning. Without a doubt, teachers double as storytellers. If you are like us, your "stories" are often about people with nutrition-related challenges in their lives. And you would probably agree that everybody has such a story to tell. A special feature called **Everybody Has a Story** made its first debut in the second edition of the textbook. As in the second edition, each chapter opens with a story about a person with a real nutrition-related challenge. In this new edition, we are pleased to introduce several new people, all of whom generously shared their captivating stories. It is our hope that students can relate to these stories and refer back to them as they read the chapter. For example, Chapter 2 (Nutritional Assessment and Dietary Planning) features Emily, a student studying at the University of Cincinnati. This story was selected to get students to think about how easily overlooked medical conditions—such as celiac disease—can cascade into a host of nutrient-related concerns, especially in college students facing a multitude of changes and stresses in their lives. Chapter 8 (Energy Balance and Body Weight Regulation) features August, who describes her lifelong struggle with obesity and her decision to have gastric bypass surgery after repeated failed weight-loss attempts. This story presents a thought-provoking perspective to students about the physical, psychological, and sociocultural aspects of obesity. Most importantly, it challenges students to examine their beliefs about this sensitive topic. We are hopeful that these personal stories will help students connect the fundamentals of nutritional science to everyday living and health.

This book also provides numerous opportunities for students to apply their knowledge to the art and science of making their own good food choices on a daily basis. The **Food Matters** sections, for example, help transform the 2010 Dietary Guidelines for Americans (which are described in detail in Chapter 2) into practical ideas for choosing and preparing foods in the most health-conscious manner. These pieces can be found in most chapters, helping integrate the fundamental concepts into personal food-related decisions.

Book Length

Most introductory nutrition courses are taught over one quarter or semester. Yet textbooks written for these courses typically contain substantially more information than can be covered within a 10- to 15-week period. Addressing concepts in a concise manner has enabled us to create a slimmer, trimmer textbook that can be more easily covered in a single semester. Additional resources that might otherwise be found in an appendix are conveniently provided to students via the book's website or can be bundled as supplements to the textbook. An example is the food composition table, which can be provided free as a separate booklet with every new text.

Pedagogical Tools

As previously described, we are convinced that outstanding illustrations enhance learning. All students benefit from seeing scientific concepts articulated in clear, well-organized illustrations. The figures, tables, and graphs in this book were designed using a unique captioning system, and the consistent use of blue text boxes and lines quickly identifies the key points of each figure. Many of our visual summaries take students step by step through complex processes, from the whole-body "big picture" to the details. We affectionately refer to these as "you are here" illustrations and have noticed other textbook writers adopting this idea and using it in their books as well.

Additionally, and to help with recognition, extra effort has been made to ensure that components found in many figures are displayed consistently throughout the book. For example, the anatomical drawings are rendered in the same way so that various organs such as the liver, pancreas, and stomach have a consistent appearance; glucose is always colored blue; and phospholipids are always drawn and colored the same way.

In addition to beautiful illustrations, each chapter contains a number of helpful tools that assist students in learning the material. New to this edition, tear-out **Study Cards** are gathered on perforated card stock at the back of the book. These cards include not only chapter summaries that identify key concepts, arranged by major subject headings, but also review questions that enable students to test their knowledge. Question types include multiple choice, essay, and practice calculations (specific mathematics-related problems encountered in nutrition). Using these cards, students can easily identify content that requires further review and locate where the information is presented in the chapter. Students can also carry these handy summaries and questions instead of the book to conveniently review key concepts and definitions for exams. In addition, **Diet Analysis PLUS Activities** are included at the end of most chapters. These activities are designed to relate the chapter concepts to the dietary analysis software that can accompany this text.

New to the Third Edition

Perhaps most exciting to us are the many subtle and not-so-subtle changes we have made to transform this book into its third edition. Many of these changes were made in response to excellent feedback that we received from both colleagues and students, and we thank you all for helping us make this edition even better. First, you can look forward to reading new challenges in the **Everybody Has a Story** features, and we welcome your feedback on how these helped in the classroom. In this edition, we also worked to strengthen the writing by using a more direct, inclusive voice. We hope that this style will resonate with students, and help them grasp important concepts more readily.

And, of course, throughout the book we have updated all of the nutrition-related guidelines (such as the 2010 Dietary Guidelines for Americans and MyPlate food guidance system) and programs (such as the National School Breakfast and Lunch Programs) discussed, as well as health statistics. Importantly, the timing of this edition nicely coincided with the releases of the 2010 Dietary Guidelines for Americans and the MyPlate food guidance system—cogent events that serve as the cornerstones of nutritional guidance. We are indeed fortunate, therefore, that they are aptly described and incorporated into this newly released book. Another of the most important considerations during the revision process was to avoid layering new material on top of old. Therefore, sections that no longer seemed relevant or supported by the literature were removed.

Perhaps most important to any rigorous college-level textbook, each chapter now includes new information that reflects changes to our knowledge that have surfaced since the last edition was released. Some of these are highlighted here.

Chapter 1 (The Science of Nutrition)

- Added new Focus on... concerning "Industrialization, Population Growth, and the Nutrition Transition."

Chapter 2 (Nutritional Assessment and Dietary Planning)

- Replaced previous Everybody Has a Story piece with one describing Emily, a college student who learns she has secondary iron deficiency caused by celiac disease.
- Updated discussion of DRIs to include new values for calcium and vitamin D released in 2011.
- Replaced information related to the 2005 Dietary Guidelines for Americans and MyPyramid food guidance system with that reflecting the 2010 Dietary Guidelines for Americans and MyPlate food guidance system.
- Updated descriptions of Healthy People to reflect newest edition (2010), and introduced the relevance of the Healthy, Hunger-Free Kids Act of 2010 to our nation's youth.

Chapter 3 (Chemical, Biological, and Physiological Aspects of Nutrition)

- While this chapter continues to provide a strong foundation in physiological functions, we have provided more examples that demonstrate their applicability to the study of nutrition.
- Added a new Everybody Has a Story chapter opener featuring a high school student who was experiencing an unspecified illness that was eventually diagnosed as Crohn's disease.

Chapter 4 (Carbohydrates; Nutrition and Diabetes)

- Added a new Everybody Has a Story chapter opener featuring an Ironman® competitor who faces the challenges of training while also dealing with the complications of having type 1 diabetes.
- Updated information regarding the controversy on high-fructose corn syrup and obesity.
- Updated information based on the 2010 Dietary Guidelines for Americans regarding added sugar consumption in the United States.
- Updated information based on the 2010 Dietary Guidelines for Americans regarding emphasis on fiber, fruit, and vegetable intake.
- Relocated information on metabolic syndrome from Chapter 4 to Chapter 8 in order to broaden the discussion of its relevance to obesity.
- Updated CDC trend maps showing recent regional estimates of adults with diagnosed diabetes.
- Incorporated 2010 Dietary Guidelines for Americans and MyPlate food guidance system where appropriate.
- Included American Diabetes Association's 2011 guidelines for carbohydrate intake.

Chapter 5 (Protein; Food Safety)

- Updated and expanded upon concepts of *epigenetics* and *nutrigenomics*, and discussion concerning personalized dietary prescriptions.
- Revised protein food intake recommendations based on the 2010 Dietary Guidelines for Americans, MyPlate food guidance system, and American College of Sports Medicine guidelines (for athletes).
- Revised recommendations regarding safe internal temperatures for meats based on new FightBAC® guidelines.
- Updated information on federal regulations related to mad cow disease, and included brief description of the FDA Food Safety Modernization Act passed in 2011.
- Updated many of the foodborne illness cases in the Food Safety Nutrition Matters to reflect outbreaks that occurred between 2009 and 2011.

Chapter 6 (Lipids; Nutrition and Cardiovascular Health)
- Replaced previous Everybody Has a Story piece with one describing Nancy, a women who has her gallbladder removed and endures somewhat troubling complications.
- Revised food intake recommendations (especially those related to omega-3 fatty acids) based on the 2010 Dietary Guidelines for Americans and MyPlate food guidance system.
- Expanded section on Tangier disease to reflect current knowledge.
- Revised information concerning nutritional essentiality of docosahexaenoic acid (DHA) based on current knowledge.
- Revised statistics related to incidence and prevalence of cardiovascular disease.

Chapter 7 (Energy Metabolism; Alcohol, Health, and Disease)
- Included new Everybody Has a Story chapter opener describing how dietary management of an inherited metabolic disease impacts daily food choices.
- Revised figure illustrating the process of enzymatic catalysis.
- Added new Focus on... feature concerning the use of therapeutic ketogenic diets to help control seizures in some children with epilepsy.
- Added a "You are here" illustration to provide students with an integrative view of the various pathways involved in energy metabolism.
- Added application of basic concepts of energy metabolism to topics such as alcohol metabolism and sports nutrition.

Chapter 8 (Energy Balance and Body Weight Regulation; Disordered Eating)
- Expanded discussion pertaining to food cravings and aversions.
- Updated information regarding the regulatory role of ghrelin in hunger and satiety.
- Included overview of the National Association for Sports and Physical Education guidelines for physical education in schools.
- Expanded presentation of 2008 Physical Activity Guidelines for Americans.
- Updated information regarding obesity, inflammation, and chronic disease.
- Included new, updated tables summarizing feelings and behaviors associated with anorexia nervosa and bulimia nervosa.
- Expanded discussion of other dysfunctional eating patterns such as food neophobia, muscle dysmorphia, and night eating syndrome.
- Updated information and statistics regarding U.S. obesity trends.

Chapter 9 (Physical Activity and Health)
- Included new Everybody Has a Story chapter opener describing the transformation of a sedentary middle-aged man into an Ironman® contender.
- Expanded text related to 2008 Physical Activity Guidelines for Americans and recommendations by the American College of Sports Medicine.
- Expanded and updated discussion of nutritional guidelines for athletes.

Chapter 10 (Water-Soluble Vitamins)
- Updated information on vitamin and phytochemical function based on current knowledge.
- Expanded and updated discussion of folate fortification, its reported benefits, and its possible unintended consequences.
- Revised intake recommendations for water-soluble, vitamin-rich foods based on the 2010 Dietary Guidelines for Americans and MyPlate food guidance system.

Chapter 11 (Fat-Soluble Vitamins; Nutrition and Cancer)
- Updated information concerning recommended intakes of vitamin D based on new DRI values.
- Revised intake recommendations for fat-soluble vitamin-rich foods based on the 2010 Dietary Guidelines for Americans and MyPlate food guidance system.
- Updated statistics related to cancer morbidity and mortality.

Chapter 12 (Major Minerals and Water; Nutrition and Bone Health)
- Updated information concerning recommended intakes of calcium based on new DRI values.
- Revised intake recommendations for major mineral-rich foods based on the 2010 Dietary Guidelines for Americans and MyPlate food guidance system.
- Updated discussion concerning dairy products, calcium, and weight maintenance to reflect current knowledge.
- Updated statistics related to osteoporosis morbidity and mortality.

Chapter 13 (Trace Minerals)
- Introduced and described importance of hepcidin to iron homeostasis.
- Revised section on hereditary hemochromatosis to reflect current knowledge related to hepcidin production.
- Updated dietary intake recommendations related to mineral-rich whole-grain products based on the 2010 Dietary Guidelines for Americans and MyPlate food guidance system.

Chapter 14 (Life Cycle Nutrition; Food Security, Hunger, and Malnutrition)
- Included new discussion related to pregnancy-induced hypertension, gestational diabetes, and dietary management of common pregnancy-related discomforts.
- Updated information pertaining to domestic nutrition-related legislation where appropriate.
- Updated DRI values for calcium and vitamin D throughout the life cycle.
- Introduced several important contemporary concepts such as food deserts and ready-to-use therapeutic foods.
- Updated information and statistics reflecting national and international food security and hunger.
- Replaced previously used CDC growth charts with currently recommended WHO growth charts for infants.
- Included data from Health, United States, 2010, which presents national trends in health statistics related to pregnancy outcomes.

Supplements

- **Diet Analysis PLUS™,** a comprehensive diet assessment program, features the latest Dietary Reference Intakes and a database with over 20,000 foods that can be personalized with recipes. The program allows students to create personal profiles based on height, weight, age, gender, and activity level. A dynamic interface makes it easy for students to track the types and serving sizes of the foods they consume, from one day to 365 days. The program also allows students to generate reports that analyze their diets and see the health implications of their eating habits. New labs and assignments promote critical thinking.
- **WebTutor™:** Provides customizable, text-specific content that allows instructors to edit, reorganize, or delete content to meet their course needs. WebTutor offers quizzing, videos, animations, Pop-up Tutors, and test bank materials along with direct access to Diet Analysis PLUS, Global Nutrition Watch, and an interactive eBook.

- **CourseMate™:** Interested in a simple way to complement your text and course content with study and practice materials? Cengage Learning's *Nutrition* CourseMate brings course concepts to life with interactive learning, study, and exam preparation tools that support the printed textbook. Access an eBook, chapter-specific learning tools, including flashcards, quizzes, videos, and more in your *Nutrition* CourseMate, accessed through **www.cengagebrain.com**.
- The student **Study Guide,** which has been thoroughly updated for the third edition, provides a thorough review of each chapter and Nutrition Matters sections through practice tests, fill-in-the-blank summaries, key term matching sets, discussion questions, word problems, and figure identification exercises.
- **PowerLecture™ DVD:** This one-stop course preparation and presentation resource makes it easy for you to assemble, edit, publish, and present custom lectures for your course, using PowerPoint®. The PowerLecture includes PowerPoint® with stepped art, animations, BBC® video clips, the instructor's manual, the test bank, "clicker" content, and ExamView computerized testing.
- The **Instructor's Manual with Test Bank** features a robust assortment of knowledge- and application-level test items, the majority of which are multiple-choice format and organized by chapter section to facilitate selection. New to this edition are several food-focused class activities (with accompanying PowerPoint presentations) designed to engage students in making more healthful choices. Also provided are student assignment materials, enrichment activity suggestions, answer keys to the Everybody Has a Story questions, instructor guides to the in-text Diet Analysis PLUS Activities, and lecture outlines.
- The **Correlation Guide for Transparency Acetates** allows the transparencies to be easily mapped to the third edition of the text.

ACKNOWLEDGMENTS

Revising a textbook is much like the process of spring cleaning. That is, to stay fresh, organized, and reinvigorated one must be insightful about what has proven worthy, and what can be discarded. Only when these decisions have been made and the "closet emptied" is there room for new elements to be woven into the existing tapestry of the book's pages. That is exactly what we tried to do in the third edition of this book, and we certainly could not have done it alone. We are indeed fortunate to have teamed up with the experienced and skillful professionals at Wadsworth | Cengage Learning. Their guidance was invaluable, enabling us to transform our evolving vision into this quality textbook. We will forever be grateful for their encouragement and continued support throughout this entire process. Our special thanks are extended to our visionary leader Peggy Williams, who makes herself readily available to the entire team. Peggy not only understands the world of textbook publishing, but also has the necessary insights to approach projects from multiple perspectives—including those of students and faculty. Her tough-yet-tender tact has been perfect and appreciated. In the world of publishing, she is known for taking bold and progressive steps, and we are fortunate and honored to have her serve as captain of our team. Much appreciation also goes to Suzannah Alexander for her steadfast guidance throughout the entire revision process. From beginning to end, she was keenly watchful and aware of what everyone was doing and kept us moving forward in an orderly and synchronized fashion. Under her direction and guidance, we all arrived at the finish line together with a book of which we are all so proud. A special thank you is also extended to Carol Samet, who closely monitored every step of the production process to transform the various pieces of this book into a beautiful and meaningful whole. We are also grateful to Lynn Lustberg, for her endless and needed patience and assiduity; to the talented team of artists who continued to take our sometimes-sketchy ideas and transform them into artful illustrations; and to our creative photo researcher Josh Garvin who presented us with numerous excellent photo options from which to select. We think you will agree that these photos add meaning and instructional depth to the many concepts related to nutritional science. We also appreciate his finding such a wonderful assortment of stunning photographs that illuminate the opening of each chapter. These visuals have greatly improved the book, making each page lively and engaging. Special thanks also go to Elesha Feldman, who again skillfully developed the instructor's manual, test bank, and student study guide that accompany this text; Sean Cronin, for providing his support and assistance; Laura McGinn, for her continued efforts in coordinating the marketing and promotion of the third edition; and Miriam Myers who has created the media supplements.

We are also incredibly grateful to the professors who provided their expertise in preparing various supplements: thank you to Jennifer Bueche and Carolyn J. Haessig La Potin of State University of New York College at Oneonta for preparing the test bank and creating new activities for the instructor's manual; to Susan E. Helm of Pepperdine University for writing the study guide and contributing to the instructor's manual; and to Lisa Esposito, consulting dietitian, and Prithiva Chanmugam and Judy Myhand of Louisiana State University for their skilled contributions to the instructor's manual. Finally, many thanks to Mithia Mukutmoni of Sierra College for preparing the Diet Analysis PLUS Activities that appear at the end of most chapters and to Judy Myhand for her suggestions.

We are also thankful for the many expert reviewers who took the time to read each chapter and provide invaluable feedback in terms of content and clarity. We certainly could not have produced an up-to-date textbook without this critical input. And we would like to extend a heartfelt thank you to those who allowed us to share their interesting life stories with the readers of this textbook. These stories, as featured in the new Everybody Has a Story pieces, would not be possible without these people's willingness to discuss the challenges they face on a daily basis. We all have a great deal to learn from their experiences and triumphs. Last, to all our friends, families, and colleagues who helped in more ways than they will ever know: Thank you for being fellow travelers on this ever-important journey.

Reviewers of the Third Edition

Carmen Boyd
Missouri State University

Georgia R. Brown
Southern University

Margaret Craig-Schmidt
Auburn University

Karen Gabrielsen
Everett Community College

Janice Grover
Truckee Meadows Community College

Margaret L. Gunther
Palomar Community College

Joe S. Hughes
California State University, San Bernardino

Allen W. Knehans
University of Oklahoma

Shaynee Roper
University of Houston

Reviewers of the Second Edition

Jaclyn M. Abbott
Rutgers University

Jonathan Allen
North Carolina State University

Samuel Besong
Delaware State University

Mark Blegen
Springfield College

Carmen Boyd
Missouri State University

Jennifer Bueche
State University of New York College at Oneonta

Prithiva Chanmugam
Louisiana State University

Mary Anne Cukr
Oakland University

Shawn Dunnagan
Montana State University

Kelly K. Eichmann
Fresno City College

George F. Estabrook
University of Michigan

Karon Felten
University of Nevada, Reno

Arthur F. Fishkin
Creighton University School of Medicine

Betty Forbes
West Virginia University

Bernard Frye
University of Texas, Arlington

Naomi K. Fukagawa
University of Vermont College of Medicine

Leonard E. Gerber
University of Rhode Island

Ellen R. Glovsky
Northeastern University

Shelby Goldberg
Pima Community College

Heather Graham
Truckee Meadows Community College

Jessica Grieger
Penn State University

Margaret Gunther
Palomar College

Susan Helm
Pepperdine University

Helen L. Henry
University of California, Riverside

Paula Inserra
Virginia State University

Carol Jones
Central Connecticut State University

Coleen Kaiser
Montana State University, Bozeman

Robert D. Lee
Central Michigan University

Charles McCormick
Cornell University

Mark S. Meskin
California State Polytechnic University, Pomona

Molly Michelman
University of Nevada, Las Vegas

Donna Mueller
Drexel University

Anna M. Page
Johnson County Community College

Janet T. Peterson
Linfield College

Elizabeth Quintana
West Virginia University

Sudha Raj
Syracuse University

Kathleen M. Rasmussen
Cornell University

Robert Rucker
University of California, Davis

Kevin Schalinske
Iowa State University

Adria Sherman
Rutgers University

Ann C. Skulas-Ray
Penn State University

Joy E. Swanson
Cornell University

Kathy Timmons
Murray State

Trinh T. Tran
City College of San Francisco

Gabrielle Turner-McGrievy
University of Alabama

Dhiraj Vattem
Texas State University, San Marcos

Priya Venkatesan
Pasadena City College

Weiqun Wang
Kansas State University

Reviewers of the First Edition

Jonathan P. Allen
North Carolina State University

Janet Anderson*
Utah State University

Susan S. Appelbaum*
St. Louis Community College, Florissant

Lynn Ausman
Tufts University

Georgina Awipi[†]
University of Tennessee, Martin

Susan E. Bascone-Dalrymple*
Miami University

Carolyn M. Bednar*
Texas Woman's University

Bonnie Beezhold[†]
Arizona State University

Beverly A. Benes*
University of Nebraska, Lincoln

Samuel Besong[†]
Delaware State University

Margaret Ann Bock
New Mexico State University

Carmen Boyd[†]
Missouri State University

Kenneth S. Broughton*
University of Wyoming

Blakely D. Brown*
University of Montana

Eston Brown*
University of Akron

Jennifer Bueche*[†]
State University of New York College at Oneonta

Jay Burgess*
Purdue University

*Denotes editorial board participation.
[†]Denotes class test participation.

Judith Burns Lowe
Ball State University

Melanie Tracy Burns
Eastern Illinois University

Aleda F. Carpenter
Grayson County College

Timothy P. Carr
University of Nebraska

Prithiva Chanmugum*
Louisiana State University

Alana D. Cline*
University of Northern Colorado

Carla E. Cox
University of Montana

Margaret Craig-Schmidt
Aubum University

Annette C. Crawford-Harris
Hudson Valley Community College

Sylvia Crixell*
Texas State University, San Marcos

Sharon Croft†
Clayton College and State University

Robert Cullen*
Illinois State University

Sharron Dalton
New York University

Mary DeBusman
San Francisco State University

Richard P. Dowdy
University of Missouri, Columbia

Carol O. Eady
University of Memphis

Denise Eagan
Marshall University

Liz Emery*
Drexel University

Stephanie England
University of Tennessee, Chattanooga

George F. Estabrook
University of Michigan

Elizabet Fiddler
University of Washington

Janet Fisher
Clark Atlanta University

Pam Fletcher
Albuquerque TVI Community College

Mary M. Flynn†
Brown University

Betty Forbes*†
West Virginia University

Joey Kathleen Freeman
Seattle Pacific University

Bernard L. Frye
University of Texas, Arlington

Candance Gabel
University of Missouri, Columbia

Susan Gaumont†
Arizona State University

Ellen Glovsky
Northeastern University

Cynthia Gonzalez*
Santa Monica City College

Nanci Grayson
University of Colorado, Boulder

Guy Groblewski
University of Wisconsin

Sandra M. Gross
West Chester University of Pennsylvania

Margaret Gunther
Palomar Community College

Evette M. Hackman
Seattle Pacific University

Carolyn J. Haessig*
State University of New York College at Oneonta

Vivian Haley-Zitlin
Clemson University

Eric E. Hall
Elon University

Shelley R. Hancock
University of Alabama

Donna Handley*
University of Rhode Island

James Hargrove*
University of Georgia

Edward Hart
Bridgewater State College

Cynthia Heiss
California State University, Northridge

Susan Edgar Helm*†
Pepperdine University

Nancy R. Hudson
University of California, Davis

Karen Israel
Anne Arundel Community College

Thunder Jalili*
University of Utah

Barbara L. Jendrysik
Manatee Community College

Mary Beth Kavanagh
Case Western Reserve University

Michael Kennan*†
Louisiana State University

Mark Kern
San Diego State University

Lon Kilgore
Midwestern State University

Elizabeth M. Kitchin
University of Alabama, Birmingham

Melinda S. Kreisberg
West Liberty State College

Laura Kruskall*
University of Nevada at Las Vegas

Anda Lam*
Pasadena City College

Dale A. Larson
Johnson County Community College

Christina Lengyel
University of North Carolina, Greensboro

Edith Lerner
Case Western Reserve University

Cathy Levenson
Florida State University

George Liepa
Eastern Michigan University

Susan C. Linnenkohl
Marshall University

Ingrid Lofgren
University of New Hampshire

Linda Johnston Lolkus*
Indiana University–Purdue University, Fort Wayne

Clara Lowden
Riverside Community College

Myrtle McCulloch
Georgetown University

Glen F. McNeil
Fort Hays State University

Mary Mead
University of California, Berkeley

Kathleen J. Melanson
University of Rhode Island

Juliet Mevi-Shiflett
Diablo Valley College

Anahita Mistry*
Eastern Michigan University

Huanbia Mo
Texas Woman's University

Gaile Moe
Seattle Pacific University

Mohey Mowafy
Northern Michigan University

Kathy D. Munoz
Humboldt State University

Judy Myhand*†
Louisiana State University

Steven Nizielski
Grand Valley State University

Amy Ozier*
Northern Illionois University, DeKalb

Anna M. Page*†
Johnson County Community College

Kathleen Page
Bucknell University

Eleanor B. Pella
Harrisburg Area Community College, Gettysburg

Susan Polasek
University of Texas, Austin

Mercy Popoola†
Clayton College and State University

William R. Proulx
State University of New York College at Oneonta

Elizabeth Quintana
West Virginia University

Jenice Rankins*†
Florida State University

Maureen Redenauer
Camden County College

Robert D. Reynolds
University of Illinois, Chicago

Judy Richman†
Northeastern University

Chris Roberts*
University of California, Los Angeles

Carmen R. Roman-Shriver*
Texas Tech University

Andrew Rorschach*†
University of Houston

Connie S. Ruiz
Lamar University

Ross Santell*
Alcorn State University

Linda Sartor
University of Pennsylvania

Peter Schaefer*
Hudson Valley Community College

Kevin Schalinske*
Iowa State University

Claudia Schopper
West Virginia University

Neil Shay
University of Notre Dame

Adria Sherman
Rutgers University

Sandra S. Short
Cedarville University

Sarah Short
Syracuse University

Brent Shriver*
Texas Tech University

Deborah Silverman
Eastern Michigan University

Joanne Slavin*
University of Minnesota

Margaret K. Snooks
University of Houston, Clear Lake

LuAnn Soliah
Baylor University

Arlene Spark*
Hunter College

Diana-Marie Spillman*†
Miami University of Ohio

Catrinel Stanciu
Louisiana State University

Carol Stinson
University of Louisville

Jon A. Story*
Purdue University

Maria Sun
Southwest Tennessee Community College

Melanie Taylor
University of Utah

Forrest Thye*
Virginia Tech University

Carol Turner
New Mexico State University

Dhiraj Vattem*†
Texas State University

Priya Venkatesan*
Pasadena City Colege

Eric Vlahov
University of Tampa

Janella Walter
Baylor University

Dana Wassmer
Cosumnes River College

M. K. (Suzy) Weems
Baylor University

Lavern Whisenton-Davidson*
Millersville University

Mary W. Wilson
Eastern Kentucky University

Stacie L. Wing-Gaia
University of Utah

Ira Wolinsky
University of Houston

Jane E. Ziegler
Cedar Crest College

Donna L. Zoss
Purdue University

Nutritional Sciences

FROM FUNDAMENTALS TO FOOD

The Science of Nutrition

CHAPTER 1

Let thy food be thy medicine and thy medicine be thy food.

—Hippocrates (460–364 B.C.)

Life would not be possible without the nourishment of food. Indeed, our quality of life depends greatly on which foods we choose to eat, and experts unanimously agree that the foods we eat can greatly influence both our immediate health and our risk of disease as we age. Hopefully, you are reading this textbook because you are interested in making sure your diet is as healthful as possible. The information you learn in this course—if you apply it to your life—will be beneficial.

Perhaps the first questions you should ask yourself as you embark upon this journey of learning are "*Do I choose my foods wisely?*" and "*What changes might I make to ensure optimal nutrition and a long and healthy life?*" To answer these questions, you must first understand what nutrients are and how your body uses them to maintain and fuel all of its complex physiologic processes. In other words, you must understand the science of nutrition.

Science is powerful. It helps explain our world, makes it a better place to live, and contributes to good health. Yet scientific progress almost always generates considerable debate. Not surprisingly, nutritional discoveries that have helped prevent and cure diseases are often met with both excitement and skepticism. For example, one day you may read in a newspaper, "Vitamin A decreases risk of heart disease." Later, another headline claims, "Vitamin A increases risk for cancer." Likewise, an article that says, "You should eat more fish" may be followed by one that asserts, "Fish contains dangerous heavy metals." While nutritional science offers great hope for improving health, you are likely well aware that it also generates controversy.

There is no argument, however, that nutrition and its impact on human health are of crucial importance. Nutritional deficiencies have always posed major health challenges worldwide, but today nutritional abundance and imbalance also contribute to many of our health problems.[1] Poor dietary choices you make now very well might play a major role in predisposing you to obesity, cancer, heart disease, osteoporosis, and type 2 diabetes later in your life.[2] This chapter discusses the fundamental concepts necessary to understand how good nutrition is basic to your health. You will also learn how scientists study nutrition. With this knowledge, you will be able to make sound decisions about selecting a healthy diet—based on scientific reason, not rumor—for years to come.

Steven Nilsson/Getty Images

Choosing Nutrition as a Career Path

Katherine has always enjoyed good food and the pleasures of eating. But she never imagined that she would pursue a career in nutrition. However, when she recalls her experiences as an overweight child competing at the local swimming club, she realizes that she has long understood the importance of a healthy diet for both physical and mental health. Science was not one of Katherine's passions as she went through her high school years. Instead, she was much more interested in music and tennis, and Katherine was emphatic that she would do nothing related to science, partly because she found her high school science laboratories "contrived and not useful to her daily experience." She instead entered the University of Idaho as an elementary education major and planned to pursue a career as a teacher in the public schools. To fulfill her general education requirement in biology, however, Katherine enrolled in an introductory microbiology class during her first year of studies—not because she was at all interested in the subject, but because her friends reassured her that the professor was engaging and fair. Much to her surprise, Katherine rapidly learned that college-level science was very different from what she had experienced in high school. She understood, for the first time, that many scientific disciplines were incredibly applicable to her own life as well as the health and well-being of society. Of exceptional interest in her microbiology course was the section on foodborne illness because it emphasized how a person's dietary choices can immediately and profoundly impact his or her health. Within a few months, Katherine had switched her major to dietetics and was working in a nutrition research laboratory on campus. She soon discovered that the science of nutrition is not only compelling, challenging, and useful, but that she thoroughly enjoyed the basic work required to carry out nutrition experiments and analyze their results. Katherine is now a graduate student working on her doctorate in nutrition at the University of Idaho. Her graduate research focuses on how a breastfeeding woman's diet influences the composition of her milk, and she has presented her work both nationally and internationally. For Katherine, the discovery that science—in particular, nutritional science—was her career calling came somewhat late in her undergraduate career. And it came as a complete surprise. Indeed, keeping one's options open during college and exploring various disciplines of study are some of the most important tasks of any college student. Sometimes one class can change your life.

Courtesy Katherine Hunt

Critical Thinking: Katherine's Story

Why are you taking this nutrition class, and what are your personal goals for what you will learn? Are there specific issues that are of special interest to you as you embark upon your study of nutrition? Why are you especially interested in these topics? Has nutrition been an interest of yours for many years, or has some situation or experience made you particularly interested in learning more about it?

What Do We Mean by "Nutrition"?

Perhaps the first question you may have is what the term *nutrition* actually means. The term **nutrition** refers to the science of how living organisms obtain and use food to support all the processes required for their existence, and the study of nutrition incorporates a wide variety of scientific disciplines. Some nutritionists, for example, are interested in food production and availability, whereas others conduct research on why people choose to eat certain foods. Still others investigate the relationships between diet and heart disease, how nutrition can influence athletic performance, or whether the composition of our meals can influence weight management. Indeed, the field of nutrition encompasses a broad array of important scientific and cultural aspects.

As such, scientists who study nutrition, called **nutritional scientists,** can be found in many disciplines, including immunology, medicine, genetics, biology, physiology, biochemistry, education, psychology, and sociology, as well as nutrition. **Dietitians** are nutrition professionals who help people make dietary changes and food choices to support a healthy lifestyle. A dietitian has the credential of "RD," which stands for *r*egistered *d*ietitian. Many dietitians are also involved in research. Thus the science of nutrition, collectively called the **nutritional sciences,** reflects a broad spectrum of academic and social disciplines.

NUTRIENTS SUPPORT ALL WE DO

But what are nutrients, and why do we need them? **Nutrients** have traditionally been defined as substances in foods required or used by the body for at least one of the following: energy, structure, or regulation of chemical reactions. For example, carbohydrates supply energy to fuel your body's activities, calcium and phosphorus are important building blocks of your teeth and bones, and many of the vitamins are essential for chemical reactions such as those needed to protect your cells from the damaging effects of excessive sunlight and pollution. There are also many other substances present in food that appear to have health benefits such as decreasing risks for cancer and heart disease. Scores of these compounds have only recently been discovered and are therefore less understood than the "traditional" nutrients. Clearly, the definition of what is a nutrient is evolving, and the list of established nutrients will likely expand as researchers learn more about how the thousands of substances found in foods can promote health and well-being.

The dietary choices you make today can influence your health for years to come.

FOODS CONTAIN NUTRIENTS AND NONNUTRIENTS

You may be surprised to learn that not all compounds in food are nutrients. To convince yourself of this, you need only examine almost any food label. Many of these compounds found in foods, such as artificial colors, are not nutrients because they are not needed to support basic functions in your body. In general, scientists classify nutrients into six categories based on their chemical structure and composition: carbohydrates, proteins, lipids, water, minerals, and vitamins. Scientists also categorize nutrients and foods in other ways, and some of these classifications are described next.

Essential, Nonessential, and Conditionally Essential Nutrients Although our bodies can use all the nutrients in foods, we only need to consume some of them. These nutrients are referred to as the **essential nutrients.** Essential nutrients

nutrition The science of how living organisms obtain and use food to support processes required for life.

nutritional scientist A person who conducts and/or evaluates nutrition-related research.

dietitian A nutritionist who helps people make healthy dietary choices.

nutritional sciences A broad spectrum of academic and social disciplines related to nutrition.

nutrient A substance in foods used by the body for energy, maintenance of body structures, or regulation of chemical processes.

essential nutrient A substance that must be obtained from the diet, because the body needs it and cannot make it in required amounts.

must be obtained from your diet, because your body needs them and either cannot make them at all or cannot make them in adequate amounts. **Nonessential nutrients** are those your body can make in amounts needed to satisfy its physiological requirements. Hence, you do not actually need to consume nonessential nutrients. Most foods contain a mixture of essential and nonessential nutrients. For example, milk contains a variety of essential vitamins and minerals (such as vitamin A and calcium) as well as nonessential nutrients (such as cholesterol).

However, there are situations when a normally nonessential nutrient can become essential. During these times, the nutrient is called a **conditionally essential nutrient.** For example, older children and adults must obtain two essential lipids through the diet, whereas babies are thought to require at least four, which they are unable to make. The additional lipids are therefore "conditionally essential" during early life. Certain diseases also cause normally nonessential nutrients to become conditionally essential. You will learn about some of these in later chapters.

Macronutrients versus Micronutrients Nutrients can also be classified on the basis of how much of them we require from *our* diet. (Figure 1.1). Water, carbohydrates, proteins, and lipids are called **macronutrients,** because they are needed in large quantities (over a gram each day). Vitamins and minerals are called **micronutrients,** because we need only very small amounts of them (often micrograms or milligrams each day). For example, a typical adult requires about 2,726 pounds (1,239 kilograms) of the macronutrient protein over the course of a lifetime but only about 0.3 pounds (0.14 kilograms) of the micronutrient iron.

ORGANIC NUTRIENTS ARE DIFFERENT FROM ORGANIC FOODS

We can also classify nutrients as being organic or inorganic. By definition, molecules that contain carbon atoms bonded to hydrogen atoms or other carbon atoms are called **organic compounds.** Carbohydrates, proteins, lipids, and vitamins are chemically organic nutrients. Water and minerals are **inorganic** because they do not contain carbon–carbon or carbon–hydrogen bonds. In this way, all foods are considered organic—at least in the chemical sense of the term. However, the term *organic* also has an additional and very different meaning when it is used to describe how a food (plant or animal) is grown and harvested. When a food is labeled "**Certified Organic,**" it has been grown and processed according to U.S. Department of Agriculture (USDA) national organic standards. For example, a farmer cannot use conventional pesticides and herbicides on organically grown crops. You can find out what percentage of organic ingredients a product has by reading its food label (Figure 1.2) and learn more about organic foods in the Focus on Food feature.

"Certified organic" foods can be identified by this seal.

nonessential nutrient A substance found in food and used by the body to promote health but not required to be consumed in the diet.

conditionally essential nutrient Normally nonessential nutrient that, under certain circumstances, becomes essential.

macronutrients Nutrients that we need to consume in relatively large quantities (>1 gram/day).

micronutrients Nutrients that we need to consume in relatively small quantities (<1 gram/day).

organic compound A substance that contains carbon–carbon bonds or carbon–hydrogen bonds.

inorganic compound A substance that does not contain carbon–carbon bonds or carbon–hydrogen bonds.

certified organic foods Plant and animal foods that have been grown, harvested, and processed without conventional pesticides, fertilizers, growth promoters, bioengineering, or ionizing radiation.

FIGURE 1.1 Micronutrients versus Macronutrients Vitamins and minerals are micronutrients, whereas water, carbohydrates, proteins, and lipids are macronutrients.

We need only small amounts, <1 gram/day, of the micronutrients.

Micronutrients
• Vitamins
• Minerals

Macronutrients
• Water
• Carbohydrates
• Proteins
• Lipids (fats and oils)

We need larger amounts, >1 gram/day, of the macronutrients.

©2007 Cengage Learning

A chemist understands the term organic to generally mean a carbon-containing compound. However, the food industry uses the term *organic* to mean something quite different. In 1992, the U.S. federal government established the National Organic Standards Board (NOSB) to help develop standards for substances to be used (or not used) in organic food production. To learn more about these standards, you can visit the USDA's National Organic Program website at http://www.ams.usda.gov/AMSv1.0/nop.htm. The NOSB developed the following definition: *"organic agriculture is an ecological production management system that promotes and enhances biodiversity, biological cycles and soil biological activity … based on management practices that restore, maintain, and enhance ecological harmony."* As such, an organic food must be produced, grown, and harvested without the use of most conventional pesticides, fertilizers made with synthetic ingredients, bio-engineering, or ionizing radiation. Furthermore, organic meat, eggs, and dairy products must come from livestock raised without the use of growth-promoting hormones and antibiotics.

Foods with the USDA organic seal labeled as being "100% organic" must have at least 95% organically produced ingredients. Foods labeled as being "organic" must have at least 70% organic ingredients. Products with less than 70% organic ingredients may list specific organically produced ingredients on the side panel of the package but may not make any organic claims on the front of the package (see Figure 1.2).

The USDA makes no claims that organically produced food is safer or more nutritious than conventionally produced food, and the labeling of foods as "organic" is not meant to suggest enhanced nutritional quality or food safety. In fact, there is mixed evidence that organic foods are nutritionally superior to conventional foods.[3] Rather, the difference between organic foods and conventionally produced foods largely involves the methods used to grow, handle, and process them. Whether these alternative agricultural practices promote enhanced environmental integrity and balance is an area of active debate.

PHYTOCHEMICALS, ZOONUTRIENTS, AND FUNCTIONAL FOODS

As scientists learn more about the relationship between diet and health, they are discovering that, in addition to the traditional or established macronutrients and micronutrients, foods also contain other substances that influence our health. When health-promoting compounds such as these are found in plants, they are called phytochemicals. Others, called zoonutrients, are found in animal foods. Although phytochemicals and zoonutrients are not considered to be nutrients by many nutritionists, researchers think that many are

FIGURE 1.2 Understanding Food Labels of Organic Products

Must have 95–100% certified organic ingredients.

Must have at least 70% certified organic ingredients.

Organic ingredients can be listed on side panel.

No organic claim is being made.

beneficial to health. As scientists learn more about these compounds, some of them may be reclassified as nutrients in the future.

Phytochemicals: Beneficial Substances from Plant Foods Although they are not nutrients, **phytochemicals** (also called phytonutrients) are substances found in plants that may help reduce the risk for developing certain diseases.[4] In fact, many "health claims" on food packaging labels refer not to traditional nutrients but instead to phytochemicals. For example, consuming phytochemicals found in tomatoes and garlic may decrease your risk of cancer. Grapes and wine contain phytochemicals that may reduce the risk of heart disease. You will learn more about these and other phytochemicals throughout this book.

Zoonutrients: Beneficial Substances from Animal Foods Like phytochemicals, which are found in plants, **zoonutrients** (also called zoochemicals) are compounds present in animal foods that provide health benefits beyond the provision of traditional nutrients and energy.[5] Examples of zoonutrients include a variety of nonessential lipids, found in fish and dairy products, that are thought to decrease your risk for heart disease. Another example of a zoonutrient is found in the larval jelly produced by honeybees. This substance is antimicrobial and may reduce the risk of infection.[6]

Functional Foods May Offer Important Health Benefits **Functional foods** are those that may promote optimal health, above and beyond simply helping the body meet its basic nutritional needs.[7] Functional foods contain (1) enhanced amounts of traditional nutrients, (2) phytochemicals, and/or (3) zoonutrients. For example, soy milk is considered a functional food because it contains phytochemicals thought to decrease risk for some cancers. Other examples are conventional cow's milk, which has been shown to be rich in zoonutrients that may lower your risks of cancer and high blood pressure. Although consuming functional foods may improve your health, the mechanisms by which this occurs are often poorly understood.

Why might scientists consider this plate of spaghetti and glass of wine functional foods?

What Are the Major Nutrient Classes?

Nutrients are needed by your body to provide structure, regulate chemical reactions (metabolism), and supply energy. Protein, for example, is important for providing the basic structure of muscles, many vitamins help regulate the hundreds of chemical reactions that occur in your body, and dietary fats provide an important source of energy needed to power your body's activities. Each class of nutrients consists of many different compounds and contributes to most of these functions in one way or another. You will learn more about each of the nutrient classes in upcoming chapters.

CARBOHYDRATES ARE VITAL FOR ENERGY AND REGULATORY ROLES

Carbohydrates consist of carbon, hydrogen, and oxygen atoms and serve a variety of functions in the body. There are many different types of carbohydrates; for example, those found in starchy foods like rice and pasta are quite different from those found in fruits and sweet desserts. Of the various carbohydrates that exist, perhaps the most important is glucose. Indeed, most cells use glucose as their primary source of energy. Your body uses carbohydrates for many other purposes as well. For instance, some are needed to make the genetic material (DNA) in cells. Other carbohydrates such as

phytochemical (also called phytonutrient) (phy – to – CHEM – i – cal) A substance found in plants and thought to benefit human health above and beyond the provision of essential nutrients and energy.

zoonutrient (zo – o – NU – tri – ent) A substance found in animal foods and thought to benefit human health above and beyond the provision of essential nutrients and energy.

functional food A food that contains enhanced levels of an essential nutrient, phytochemical, or zoonutrient and thought to benefit human health.

dietary fiber play roles in maintaining the health of your digestive system and may help decrease your risk of certain conditions, including heart disease and type 2 diabetes. Carbohydrates are also important structural and regulatory components of the membranes that surround the millions of cells in your body.

PROTEINS MAKE UP MUSCLES AND ARE IMPORTANT FOR ENERGY AND REGULATION

Protein is abundant in many foods, including meat, legumes (such as dried peas), and some cereal products. Although most proteins consist primarily of carbon, oxygen, nitrogen, and hydrogen atoms, some also contain sulfur or selenium atoms. The thousands of proteins in your body have numerous roles in addition to serving as a source of energy. Proteins also comprise the major structural material in various parts of your body, including muscle, bone, and skin. Proteins allow us to move, support our complex internal communication systems, keep us healthy by their roles in the immune system (which protects against infection and disease), and regulate many of the chemical reactions needed for life.

LIPIDS ARE MORE THAN ABUNDANT ENERGY SOURCES

Lipids, which include a variety of oils and fats found in foods and the body, generally consist of carbon, oxygen, and hydrogen atoms. They provide large amounts of energy, are important for the structure of cell membranes, and are needed for your nervous and reproductive systems to function properly. Lipids also regulate a variety of cellular processes. Many foods contain lipids, although the types found in plant-based products such as corn oil and nuts are typically quite different from those found in animal-based foods such as meat, fish, eggs, and milk.

WATER IS THE ESSENCE OF LIFE ITSELF

Without water, there would be no life. Indeed, water—which is made of oxygen and hydrogen atoms—makes up approximately 60% of your total body weight. Without exception, you consume water every day, be it in beverages or in the foods you eat. The functions of water are varied and vital, including transport of nutrients, gases, and waste products; serving as a medium in which chemical reactions occur; and involvement in many chemical reactions. Water is also important in regulating body temperature and protecting your internal organs from damage.

VITAMINS REGULATE REACTIONS AND PROMOTE GROWTH AND DEVELOPMENT

Vitamins have a variety of chemical structures and are abundant in most naturally occurring foods—especially fruits, vegetables, and grains. Although they all contain carbon, oxygen, and hydrogen atoms, some vitamins also

Grains and cereals provide most of the carbohydrates in the diet.

There are many good sources of protein in the diet, including milk.

Lipids found in olives are thought to impart important health benefits.

Most fruits and vegetables are excellent sources of vitamins.

contain substances such as phosphorus and sulfur atoms. Your body needs vitamins to regulate its hundreds of chemical reactions as well as promote growth and development. Some vitamins, called antioxidants, also protect your body from the damaging effects of toxic compounds such as those found in air pollution.

Unlike carbohydrates, proteins, and lipids, vitamins are not used directly for structure or energy. However, they play important roles in the chemical processes required for building and maintaining tissue as well as in using the energy contained in the macronutrients. Vitamins can be classified, based on how they interact with water, as either water soluble (vitamin C and the B vitamins) or fat soluble (vitamins A, D, E, and K). Much research today is focused on the roles of vitamins in preventing and managing diseases such as heart disease and cancer.

MINERALS PROVIDE STRUCTURE AND ASSIST WITH REGULATION

Technically speaking, minerals (such as iron, selenium, and sodium) are inorganic substances that occur naturally in the earth. At least 16 minerals are essential nutrients, each serving its own specific purpose. For example, calcium is abundant in dairy products and provides the matrix for various structural components in your body (e.g., bone). Other minerals, such as the sodium you add to foods when you salt them, help regulate a variety of body processes (e.g., water balance). Still other minerals, such as selenium, which is abundant in many seeds and nuts, facilitate chemical reactions. Like vitamins, minerals are not used directly for energy, although many are involved in energy-producing reactions. Scientists are still discovering the many ways that minerals can prevent and perhaps even treat various diseases.

How Do Foods Provide Energy?

We all know that we need to get energy from food, but how food provides this commodity may be less clear. **Energy** is defined as the capacity of a physical system to do work. Thus, if something "has energy," it can cause something else to happen. In terms of nutrition, carbohydrates, proteins, and lipids all contain chemical energy. Cells in your body can transfer the chemical energy from these nutrients into a special substance called **adenosine triphosphate (ATP)**, which stores energy somewhat like a "molecular battery." Your body can then use the energy in ATP to power its many processes. For example, the energy in ATP allows your muscles to move, drives nutrient digestion, and keeps you warm. Carbohydrates, proteins, and lipids are called **energy-yielding nutrients** because their energy can be used to generate ATP. It is also important to note that many vitamins and minerals are involved in "energy production," because they help regulate the many chemical reactions required to transform the energy in energy-yielding nutrients to ATP.

Your ability to use or harvest the energy in nutrients can be compared with the use of wind power to light a room. Wind has mechanical energy. This is evident during a windstorm, when the energy moves and breaks things. In its natural form, however, the wind's energy cannot run an electrical appliance. Instead, wind's mechanical energy must first be transformed into electrical energy. This is the basic underlying principle behind how windmills can be used to "produce" electrical energy. You may have learned, however, that energy can be neither created nor destroyed—just transferred from one form to

energy The capacity to do work.

adenosine triphosphate (ATP) (a – DEN – o – sine tri – PHOS – phate) A chemical used by the body to perform work.

energy-yielding nutrient A nutrient that the body can use to produce ATP.

another. As such, the chemical energy in energy-yielding nutrients must first be transformed into ATP before the body can use it. Note that energy *per se* is not a nutrient. Instead, the body uses energy in foods to grow, develop, move, and fuel the many chemical reactions required for life.

ENERGY IN FOOD IS MEASURED IN UNITS CALLED CALORIES

The amount of energy in foods varies and is measured in units called **calories**— the more calories a food has, the more ATP the body can make from it. Because 1 calorie represents a very small amount of energy, the energy content of foods is typically expressed in units of 1,000 calories, or **kilocalories.** This is often abbreviated as "kcalories" or "kcal." In addition, a kilocalorie is sometimes referred to as a Calorie (note the capital "C"), as on food labels. Therefore, 1 Calorie is equivalent to 1,000 calories or 1 kilocalorie.

Measuring the Caloric Content of Macronutrients and Foods Scientists can measure the caloric content (or energy content) of a food in the laboratory using a device called a **bomb calorimeter.** As shown in Figure 1.3, this is done by placing the food in an airtight chamber surrounded by water, pumping oxygen into the chamber, and combusting the food. The energy contained in the food's chemical bonds is released as heat, causing the temperature of the surrounding water to increase. The change in temperature is directly related to the amount of energy originally in the food. A calorie is defined as the amount of heat required to raise the temperature of 1 g of water 1 degree Celsius. A kilocalorie is the amount of heat required to raise the temperature of 1 kilogram (kg) of water 1 degree Celsius.

Estimating Food Calories Using Mathematical Equations Obviously, most people do not use a bomb calorimeter to determine the caloric content of their food. Instead, we typically estimate it mathematically. Although these mathematical computations were historically done by hand, they are now more commonly computed using dietary software. Carbohydrates and proteins provide approximately 4 kcal/gram, whereas lipids provide approximately 9 kcal/gram.

calorie A unit of measure used to express the amount of energy in a food.

kilocalorie (kcal or Calorie) 1,000 calories.

bomb calorimeter (cal – o – RIM – e – ter) A device used to measure the amount of energy in a food.

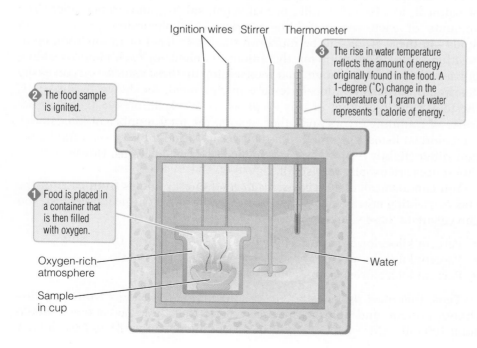

Ignition wires   Stirrer   Thermometer

3 The rise in water temperature reflects the amount of energy originally found in the food. A 1-degree (°C) change in the temperature of 1 gram of water represents 1 calorie of energy.

2 The food sample is ignited.

1 Food is placed in a container that is then filled with oxygen.

Oxygen-rich atmosphere

Sample in cup

Water

FIGURE 1.3 Bomb Calorimeter
Bomb calorimetry is used to determine the energy content of a food.

TABLE 1.1 Calculating the Caloric Content of a Typical Breakfast

Food	Kilocalories from Energy-Yielding Nutrients			Total Kilocalories (kcal)
	Carbohydrates (4 kcal/g)	Protein (4 kcal/g)	Lipids (9 kcal/g)	

Oatmeal, 1 cup
  Carbohydrates: 25 g
  Protein: 6 g
  Lipids: 2 g
	25 × 4 = 100	6 × 4 = 24	2 × 9 = 18	142

Milk, 1 cup
  Carbohydrates: 12 g
  Protein: 8 g
  Lipids: 2 g
	12 × 4 = 48	8 × 4 = 32	2 × 9 = 18	98

Brown sugar,
2 tablespoons
  Carbohydrates: 24 g
  Protein: 0 g
  Lipids: 0 g
	24 × 4 = 96	0 × 4 = 0	0 × 9 = 0	96

Raisins, 1/2 ounce
  Carbohydrates: 11 g
  Protein: 0 g
  Lipids: 0 g
	11 × 4 = 44	0 × 4 = 0	0 × 9 = 0	44

Orange juice, 1 cup
  Carbohydrates: 27 g
  Protein: 2 g
  Lipids: 0 g
	27 × 4 = 108	2 × 4 = 8	0 × 9 = 0	116

Totals	396	64	36	496

Thus, 10 g of a pure carbohydrate or protein would contain 40 (4 × 10) kcal, whereas 10 g of a pure lipid would contain 90 (9 × 10) kcal. Although alcohol is not considered a nutrient, it provides 7 kcal per gram.

For example, try estimating the caloric content of a breakfast consisting of oatmeal, low-fat (1%) milk, brown sugar, raisins, and orange juice. The amounts of each energy-yielding nutrient—carbohydrates, proteins, and lipids—in these foods can be found on the food's label or in any food composition table. By multiplying the amount (weight) of each energy-yielding nutrient by its caloric content and then adding up these values, you can easily determine how many kilocalories are in this meal. As shown in Table 1.1, the total caloric content of this breakfast is 496 kcal, or 496 Calories. Because of rounding errors and other factors, the total number of kilocalories (or Calories) listed for a food in a food composition table or on a food label may differ slightly from the value obtained from calculations. However, these differences are usually very small.

You can also calculate what *percentage* of energy comes from each of the energy-yielding nutrient classes. In our example of the oatmeal breakfast, you can calculate these values as follows.

- Percent kilocalories from carbohydrates: (396 ÷ 496) × 100 = 80%
- Percent kilocalories from protein: (64 ÷ 496) × 100 = 13%
- Percent kilocalories from lipids: (36 ÷ 496) × 100 = 7%

Thus, this meal provides 80%, 13%, and 7% of its energy from carbohydrates, protein, and lipids, respectively. Note that these percentages should total 100 (80 + 13 + 7). Currently, it is recommended that 45 to 65% of your energy come from carbohydrates, 10 to 35% from proteins, and 20 to 35% from lipids. Being able to do these calculations is also sometimes important in disease states (such as kidney disease) that may require a person to consume a diet low in a particular macronutrient (such as protein).

Few things are more familiar and important to us than food. As you have seen, food is also an exciting frontier of scientific research, with many questions still to be answered about how food interacts with our bodies to influence health. But how do scientists carry out this research, and how can you distinguish reliable information about food and health from false and exaggerated claims?

How Is Nutrition Research Conducted?

Scientists test theories (including those related to nutrition) in many ways, most of which involve a series of steps collectively called the **scientific method.** There are three steps involved in the scientific method: making an observation, proposing a hypothesis, and testing the hypothesis.[8] You will learn about each of these steps next.

STEP 1: THE OBSERVATION MUST BE ACCURATE

Making an appropriate and accurate observation about an event serves as both the framework and foundation for the rest of the scientific method. If the observation is flawed, the resulting conclusion will likely be flawed as well. For example, consider the observation that there has been an alarming rise in childhood obesity during the past few decades.[9] Before moving on to develop an explanation for this observation, you should first ask yourself several questions. For instance, are girls more likely to be obese than boys? And at what age do the rates of obesity increase? Answering these questions helps ensure that the observation is complete and accurate.

A careful researcher interested in studying childhood obesity using the scientific method must consider these and other questions before moving on to the next step of developing an explanation for the observation. This is because the scientist's likely explanation depends on the answers to these questions. Although it is tempting to think that an observation is simply a statement of "the facts," scientists must carefully consider whether they have taken into account all the available data to make sure that their observation is complete and correct.

STEP 2: A HYPOTHESIS MAKES SENSE OF AN OBSERVATION

Once the scientist has made an observation and understands the details associated with it, the next step is to explain why the event occurred. Scientists propose **hypotheses** to explain their observations. For example, someone might hypothesize that the increase in childhood obesity is due to a lack of exercise. Another hypothesis might be that childhood obesity is caused by consuming too much fat. Or someone could hypothesize that childhood obesity is caused by both a lack of exercise and too much fat. Regardless of which is correct, all three are reasonable hypotheses (or explanations) as to why childhood obesity has increased.

Two General Types of Hypotheses: Causative and Correlative Scientists can make two general types of hypotheses: those that predict cause-and-effect relationships (also called causal relationships) and those that predict

scientific method Steps used by scientists to explain observations.

hypothesis (hy – PO – the – sis) A prediction about the relationship between variables.

Although television watching is *related* to risk for obesity in children, it is other factors (like snacking and lack of exercise) that actually *cause* weight gain.

cause-and-effect relationship (also called causal relationship) When an alteration in one variable causes a change in another variable.

correlation (also called association) When a change in one variable is related to a change in another variable.

positive correlation An association between factors in which a change in one is related to a similar change in the other.

negative correlation (also called inverse correlation) An association between factors in which a change in one is related to a change in the other in the opposite direction.

simple relationship A relationship between two factors that is not influenced or modified by another factor.

complex relationship A relationship that involves one or more interactions.

interaction When the relationship between two factors is influenced or modified by another factor.

lifestyle factor Behavioral component of our lives over which we may or may not have control (such as diet and tobacco use).

environmental factor An element or variable in our surroundings over which we may or may not have control (such as pollution and temperature).

genetic factor An inherited element or variable in our lives that cannot be altered.

correlations (also called associations). It is important to understand the difference between these kinds of hypotheses because they require different kinds of studies to test them. A **cause-and-effect relationship** describes a situation in which a change in one factor *causes* a change in another factor. In other words, factor A (such as excess energy intake) causes factor B (such as weight gain). When the relationship between A and B is causal, we can say that a change in A will generally cause a change in B.

When two factors are simply correlated (or associated) with each other, we can only say that a change in one is *related to* a change in the other. This relationship is referred to as a **correlation.** The term *correlation* is used when two or more events occur simultaneously, but one does not necessarily cause the other to happen. For example, if your alarm clock goes off every day close to the time the sun rises, the two events are correlated, but neither causes the other. A correlation can be positive or negative. In **positive correlations,** the variables change in the same direction—as one increases or decreases, so does the other. For example, television watching is *positively* correlated with the prevalence of obesity in children, because as one increases, the other tends to do the same. In **negative correlations,** the variables change in opposite directions—as one increases, the other decreases. A negative correlation is also called an inverse correlation.

Understanding the difference between causal relationships and correlations is important in all scientific disciplines, including nutrition. Although many studies are designed to test for correlations, their results are unfortunately interpreted or reported as proving causal relationships. Thus, it is important to remember that *correlation does not necessarily mean causation*.

Simple versus Complex Relationships It is also helpful for scientists to determine whether they expect a relationship between two variables to be simple or complex. A **simple relationship** between two variables is one that cannot be altered by other factors. For example, we know that consuming inadequate iron for a long period of time results in iron-deficiency anemia. This relationship is ultimately true for all people and is not dependent on other factors such as a person's age, activity level, sex, or other dietary factors.

However, most relationships between diet and health are not this simple; they are instead **complex relationships** because they involve one or more interactions. An **interaction** exists when one factor can alter or modify a relationship between two other factors. For example, it is generally true that fat consumption is related to risk for heart disease. This, however, is a complex relationship because the association between excessive fat intake and heart disease can be altered or modified by other factors such as exercise and overall caloric intake. Consequently, scientists often conclude that fat intake, other lifestyle and environmental factors, and genetics *interact* to influence risk for heart disease. The difference between a simple and complex relationship is shown in Figure 1.4.

Lifestyle, Environment, and Genetics—A Complex Interaction There are numerous examples of how **lifestyle factors** (such as diet and exercise), **environmental factors** (such as exposure to pollutants), and **genetic factors** (such as inherited differences that influence physiology) interact to influence health. Typically, scientific research first focuses on simple relationships and then moves on to explore the many interactions that exist among factors. An example is the long-held advice that all people should limit their intake of salt to prevent hypertension (high blood pressure). However, scientists now know that high salt intake increases blood pressure only in some people.[10] High salt intake does not influence blood pressure in other people, and some current recommendations have been adjusted to reflect this information. It is important

FIGURE 1.4 Simple Relationships versus Complex Relationships Most relationships between nutrition and health are not simple but instead involve interactions with other factors.

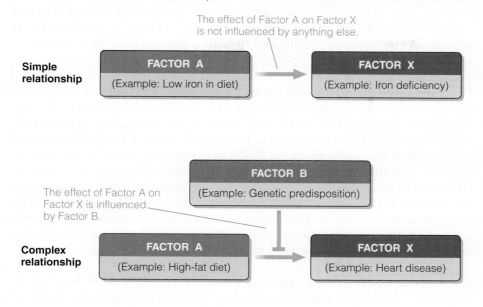

The effect of Factor A on Factor X is not influenced by anything else.

Simple relationship

FACTOR A
(Example: Low iron in diet)

FACTOR X
(Example: Iron deficiency)

FACTOR B
(Example: Genetic predisposition)

The effect of Factor A on Factor X is influenced by Factor B.

Complex relationship

FACTOR A
(Example: High-fat diet)

FACTOR X
(Example: Heart disease)

to remember that relationships initially thought simple often turn out to be more complex than originally believed. This process can lead to changes in dietary recommendations over time.

STEP 3: DATA ARE COLLECTED TO TEST THE HYPOTHESIS

Although making an accurate observation (Step 1) and developing an appropriate hypothesis (Step 2) are critical to the scientific method, these are not nearly as complex as the final step—data collection. Without supporting data, a hypothesis is simply an unproven conjecture—not a scientific finding. Scientific studies require the researcher to design an appropriate study, carefully conduct the study, and interpret the data correctly. If the study design is flawed, an important observation and hypothesis can be completely wasted.

A basic understanding of experimental practices can help you discern which nutrition claims are unfounded and which ones are valid. A thorough discussion of experimental design is beyond the scope of this book, but a general understanding of the appropriate use of study designs can help you differentiate between nutrition fact and fiction.

Epidemiologic Studies: Testing for Correlations When an observation suggests that one factor is correlated or associated with another factor, scientists often conduct an **epidemiologic study** to test their hypothesis. In these studies, researchers measure the relationship between factors in a group of people by "simply" making observations and recording information. In epidemiologic studies people are not asked to change their behaviors, alter their food intake patterns, or undergo any sort of treatment.

Epidemiologic studies should not be used to test hypotheses predicting causal relationships—only correlations. For example, if the hypothesis predicts, "Dietary fat intake *is related to* obesity in American children," researchers could conduct an epidemiologic study to estimate fat intake and body weight in a group of American children. In this type of study, the scientists would not try to control the amount of fat eaten but simply document what the children choose to eat and then measure body weight. Data would then be analyzed to determine the strength of the relationship between the two

epidemiologic study (e – pi – de – mi – o – LO – gic) A study in which data are collected from a group of individuals who are not asked to change their behaviors in any way.

Courtesy of Shelley McGuire

Lifestyle factors, such as exercise and nutrition, interact with genetics to determine how healthy we are.

factors of interest (fat intake and obesity). Remember, however, that the results of this type of study should never be used to make a causal conclusion. Unfortunately, this is sometimes done in nutrition research. Two of the most famous nutrition-related epidemiologic studies (The Framingham Heart Study and The National Health and Nutrition Examination Survey) are described next.

The Framingham Heart Study—A Classic Epidemiologic Investigation Nutritional scientists often use epidemiologic studies when they do not know enough to put forth a causal hypothesis, or when conducting another type of study would be impractical or unethical. An example is the **Framingham Heart Study,** which was initiated in the 1940s to investigate the complex relationship between lifestyle choices and heart health.[11] In fact, the Framingham Heart Study provided the first convincing evidence that what a person eats is related to his or her risk for heart disease. Hundreds of scientific articles resulting from the original Framingham Heart Study have since provided important information about how genetic, lifestyle, and environmental factors interactively influence long-term health.

The National Health and Nutrition Examination Survey (NHANES) Another example of an important nutrition-related epidemiologic study is the **National Health and Nutrition Examination Survey (NHANES),** which has been ongoing for more than four decades. By the late 1960s, researchers were beginning to discover links between dietary habits and disease. In response, the **National Center for Health Statistics (NCHS)** initiated a large, ongoing epidemiologic study to simultaneously monitor nutrition and health in the U.S. population. This was the beginning of the NHANES. These important surveys continue to monitor the many factors—especially nutrition—that are important in fostering good health and long life in the U.S. population. You can learn more about the NHANES surveys by visiting the NCHS website at http://www.cdc.gov/nchs/nhanes.htm.

Advantages and Limitations of Epidemiologic Studies As with all study designs, there are both positive aspects and limitations to epidemiologic studies. One of their strengths is that they can be used to explore the complex interactions among genetic, environmental, and lifestyle factors before the underlying causal nature of the relationship is understood. For example, in the Framingham Heart Study, scientists uncovered a relationship between nutrition and heart disease but did not know what happens in the body to connect the two. In addition, epidemiologic studies often do not require extensive technical and laboratory training of the research personnel.

As previously discussed, a limitation of epidemiologic studies is that, although they can show an association between two factors, they do not prove that one *causes* the other. It is also important to recognize that subjects studied in an epidemiologic study may not be representative of the entire population. These considerations are important for you to think about when determining whether the study's findings are valid.

INTERVENTION STUDIES TEST FOR CAUSALITY

When a hypothesis suggests a causal relationship, the scientist has several options for what type of experiment to conduct. Because epidemiologic studies cannot prove causal relationships—they can only assess associations—they are typically not used. Instead, the researcher usually conducts an intervention study either with humans, with experimental animals, or with cell culture systems.

Framingham Heart Study A large epidemiologic study begun in the 1940s designed to assess the relationship between lifestyle factors and risk for heart disease.

National Health and Nutrition Examination Survey (NHANES) A federally funded epidemiologic study begun in the 1970s to assess trends in diet and health in the U.S. population.

National Center for Health Statistics (NCHS) A component of the U.S. Public Health Service whose mission is to compile statistical information to be used in improving the health of Americans.

Use of a Control Group In contrast to epidemiologic studies, **intervention studies** require "participants"—regardless of whether they are humans, animals, or cells—to undergo a treatment or intervention. Usually, some participants receive the treatment, while others do not. Participants who do not receive the treatment or intervention are said to be in the **control group.** A control group is needed to determine whether the effects seen in the treatment group are actually caused by the treatment, by chance, or by some other aspect of the study. For example, researchers interested in understanding the factors that might be contributing to the rise in childhood obesity could test the hypothesis that nutrition education can decrease obesity in children. To do this, they might have some children attend a nutrition education class (the intervention group), whereas others would not (the control group). The researchers could then measure whether the children receiving the intervention (education) gained less weight than children in the control group.

Avoiding Hawthorne Effect, Placebo Effect, and Researcher Bias In addition to using a control group, researchers must also try to minimize factors that might bias the results of the study. For example, just being in a study can influence a person's behaviors, which in turn can influence the results of the study. This phenomenon is called the **Hawthorne effect.** For instance, consider the previously mentioned study designed to test the effect of nutrition education on weight gain. In this case, just knowing that they were in a "body weight study" might cause some children to eat fewer sweets. This might affect their weight over the period of the study—regardless of whether they were in the control or intervention group. Obviously, this change in behavior would alter the study's outcome.

Another commonly discussed phenomenon, called the **placebo effect,** occurs when an observable effect of the treatment seems to arise just because the individual *expects* or *believes* the treatment will work. For example, many studies have shown that the simple act of taking a sugar pill can actually influence blood pressure in some people.[12] How the placebo effect works remains mysterious to scientists, but it clearly illustrates a strong mind–body connection.

In addition, a scientist conducting the study can inadvertently influence the outcome by knowing which subjects are receiving the treatment and which ones are not. This is called **researcher bias.** For example, knowing that certain subjects received a treatment thought to improve memory, a researcher might inadvertently score a memory test more favorably for these individuals. This type of bias may affect the final results and should be avoided.

Blinded and Placebo-Controlled Studies An important technique that minimizes the Hawthorne effect, placebo effect, and researcher bias is the "blinding" of the experimenter and/or participant so that neither knows to which group participants have been assigned. When the researchers, but not the participants, know who is in the treatment and placebo groups, the study is said to be a **single-blind study.** When neither knows, the study is called a **double-blind study.**

In addition, many study designs require the participants in the control group to consume or experience something that looks, smells, tastes, and/or seems just like the real treatment. This "fake" or imitation treatment is called a **placebo.** For example, a sugar pill could be given instead of a vitamin supplement, or a nonnutrition lecture might be given instead of a nutrition education class. If simply taking a pill or attending a lecture causes either the placebo or Hawthorne effect, then this effect will be seen in both the intervention and control groups. As such, any *additional* effects seen in the intervention group compared with the control group can then confidently be attributed to the actual intervention and not to placebo or Hawthorne effects.

intervention study An experiment in which something is altered or changed to determine its effect on something else.

control group A group of people, animals, or cells in an intervention study that does not receive the experimental treatment.

Hawthorne effect Phenomenon in which study results are influenced by an unintentional alteration of a behavior by the study participants.

placebo effect (pla – CE – bo) The phenomenon in which there is an apparent effect of the treatment because the individual expects or believes that it will work.

researcher bias When the researcher influences the results of a study.

single-blind study A human experiment in which the participants do not know to which group they have been assigned.

double-blind study A human experiment in which neither the participants nor the scientists know to which group the participants have been assigned.

placebo A "fake" treatment, given to the control group, that cannot be distinguished from the actual treatment.

Random Assignment and Controlling for Confounding Variables Another way to avoid bias is by **random assignment** of participants to the treatment and control groups. This is important, because it distributes possible confounding variables equally among study groups. **Confounding variables** are factors other than the ones of interest that might influence the outcome of the study. For example, consider our study designed to test whether nutrition education influences childhood obesity. Random assignment of children to either treatment or control groups would help ensure that confounding variables (such as usual intake of sweets or exercise) are equally distributed between study groups. All reports of intervention studies should state whether participants were randomly assigned to study groups. When evaluating a nutrition intervention study, it is important for you to determine whether randomization was used. As you might imagine, randomized, double-blind, placebo-controlled intervention studies are considered to have the ideal experimental design to test a hypothesis about causal relationships (Figure 1.5).

The unintentional influence of confounding variables can also be minimized by excluding certain participants from being in the experiment. For example, heavy smoking can influence the birth weight of a baby. Therefore, investigators wanting to study the relationship between caffeine intake during pregnancy and birth weight might want to exclude smokers from the study. Controlling for confounding factors is especially important in nutrition studies because of the many interactions among genetic, lifestyle, and environmental factors.

Advantages and Limitations of Human Intervention Studies Well-designed human intervention studies are scientifically powerful, because their results can provide evidence that the relationship between two factors is causal in nature. For example, epidemiologic studies have long suggested a relationship between low maternal folate (a B vitamin) intake and increased risk for certain types of birth defects. However, it was not until intervention studies were conducted that this was confirmed.[13] Results from these intervention studies led the United States and several other countries to add folate to their national food supplies.[14] Another major strength of a human intervention study is that the results can be directly applied to humans—which is typically not the case for animal and cell culture studies.

There are, however, limitations to human intervention trials as well. For example, they are often quite costly and time consuming. In addition, as with epidemiologic studies, it is sometimes difficult to control for confounding

random assignment When study participants have equal chance of being assigned to each experimental group.

confounding variable A factor, other than the one of interest, that might influence the outcome of an experiment.

FIGURE 1.5 The Ideal Nutrition Intervention Study Randomized, double-blind, placebo-controlled intervention studies are considered the "gold standard" in nutrition research.

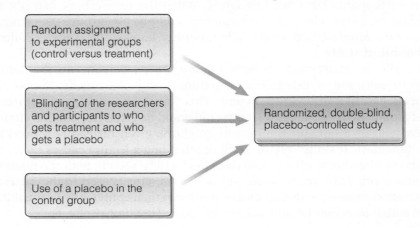

factors, and before extending the conclusions to the general population, you must be sure the study participants are representative of the population of interest.

Animal and Cell Culture Studies **Animal studies** can provide important information concerning nutrition and health, especially when, for ethical or practical reasons, it is not possible to have human subjects participate in the experiment. Using animals, such as mice, rats, or primates, in nutrition research has many advantages. For instance the genetic variability among laboratory animals is much less than that among people. In addition, scientists can study interactions in animals more easily because researchers can control more aspects of an animal's environment than a human's environment.

Although animal studies have advantages, they also have limitations. For example, one must always question whether the data collected from these studies are applicable to humans. Nonetheless, a series of convincing animal studies can lay the foundation for scientists to move forward and conduct appropriate follow-up human intervention trials.

Sometimes scientists want to study causal relationships at the cellular level—an example would be how vitamin A might influence the production of a particular protein in liver cells. In these situations, researchers can use **cell culture systems,** which are specific types of cells that can be grown in laboratory conditions. Cell culture systems are examples of what scientists call *in vitro* systems, meaning that the researchers are studying natural phenomena in an environment outside of a living organism. Note that the term *in vitro* literally means *in glass*. Of course, modern laboratories now use both glass and plastic for their *in vitro* research. In contrast, human and animal studies are *in vivo* systems, meaning they involve the study of natural phenomena within a living organism (as opposed to within glass or plastic). Examples of cell culture systems include those composed of breast, adipose (fat), skin, and muscle cells. By using cultured cells that can be shared among laboratories or purchased from general supply companies, scientists worldwide can use identical *in vitro* systems.

Using cell culture systems provides researchers with a powerful tool to study what is happening inside a cell in response to a particular treatment. However, applying data from cell culture systems to human health has limitations. First, cells that grow readily in the laboratory are usually not representative of normal, healthy cells. In addition, cells within the body interact with other cells; these interactions cannot be completely duplicated *in vitro*.

Courtesy Katherine Hunt

Critical Thinking: Katherine's Story In her research, Katherine observed that breastfeeding women with higher body fat levels produced milk that had higher fat content. Can you propose two hypotheses related to dietary intake as to why you think this might be true? Based on whether these hypotheses suggest correlative or causal relationships, would you conduct epidemiologic or intervention studies to test them?

animal study The use of experimental animals such as mice, rats, or primates in a study.

cell culture system Specific type of cells that can be grown in the laboratory and used for research purposes.

in vitro Involving the use of cells or environments that are not part of a living organism.

in vivo Involving the study of natural phenomena in a living organism.

Are All Nutrition Claims Believable?

As you now know, nutrition research takes many shapes and forms, and careful use of the scientific method helps ensure that conclusions are valid and appropriate. Because nutrition research continues to reveal new findings, our

Determining fact from fiction when it comes to nutrition claims can sometimes be difficult.

understanding about the relationship between diet and health is continually changing. Without a doubt, it is the nature of science to be a long and winding road of discoveries. Therefore, you should not be surprised when dietary recommendations change over time. This is expected, and it is important for you to be able to evaluate nutritional claims as they are made. To help in this quest, you should keep the following important issues in mind when you make nutrition-related decisions.

DETERMINE THE SOURCE OF THE INFORMATION

When determining whether a nutrition claim is reputable, it is important to consider the source of the information. As you can imagine, not all sources of information have the same credibility. This often makes it difficult to judge the validity of nutrition-related claims when you first see or hear them. Instead, you must determine the **primary source** of the information—in other words, where it was *first* reported or published. In general, primary sources of information you can trust are **peer-reviewed journals** such as the *Journal of Nutrition*. Having a paper published in a peer-reviewed journal means it was read and "approved" by a group of scientists (peers) knowledgeable in that area of study. Many private and governmental organizations, such as the National Institutes of Health, also publish credible information concerning diet and health. These agencies are considered some of the most reliable and unbiased sources of information available to the public and to the scientific community. Table 1.2 provides a list of some reputable peer-reviewed journals and other organizations that publish nutrition-related articles. As a rule, you should question nutrition claims that have not been first published in a peer-reviewed journal or other reputable publication. Often you need not go any further than this step to determine if a nutrition claim is even worth considering.

primary (information) source The location (e.g., scientific journal) in which a scientific finding is first published.

peer-reviewed journal A publication that requires a group of scientists to read and approve a study before it is accepted.

TABLE 1.2 Some Reliable Sources of Nutrition Information

Peer-Reviewed Journals	Government and Private Agencies
American Journal of Clinical Nutrition	American Cancer Society (http://www.cancer.org)
Annals of Nutrition and Metabolism	American Diabetes Association (http://www.diabetes.org)
Annual Review of Nutrition	American Dietetic Association (http://www.eatright.org)
Appetite	American Heart Association (http://www.americanheart.org)
British Journal of Nutrition	American Institute for Cancer Research (http://www.aicr.org)
Clinical Nutrition	American Medical Association (http://www.ama-assn.org)
European Journal of Nutrition	American Society for Nutrition (http://www.nutrition.org)
Journal of Human Nutrition and Dietetics	Centers for Disease Control and Prevention (http://www.cdc.gov)
Journal of Nutrition	Institute of Medicine (http://www.iom.edu)
Journal of the American College of Nutrition	Mayo Clinic (http://www.mayoclinic.org)
Journal of the American Dietetic Association	National Academy of Sciences (http://www.nas.edu)
Journal of the American Medical Association (JAMA)	National Institutes of Health (http://www.nih.gov)
Journal of Pediatric Gastroenterology and Nutrition	NIH Office of Dietary Supplements (http://ods.od.nih.gov)
Lancet	NIH National Center for Complementary and Alternative Medicine
Nature	(http://nccam.nih.gov)
New England Journal of Medicine	U.S. Department of Agriculture and its Food and Nutrition Information Center
Nutrition	(http://www.usda.gov and http://fnic.nal.usda.gov)
Nutrition Research	U.S. Food and Drug Administration (http://www.fda.gov)
Public Health Nutrition	
Science	
Scientific American	

CREDIBILITY OF THE RESEARCHERS IS IMPORTANT

Next you should ask, *"Who conducted the research?"* In general, most reliable nutrition research is conducted by scientists at universities or medical schools. Researchers at a variety of private and public institutions and organizations conduct sound nutrition research as well. It is important that the individuals conducting the research are qualified and knowledgeable, and finding out where they work and what their qualifications are can help you make this determination.

WHO PAID FOR THE RESEARCH?

It is also important to consider who paid for the research. There are many ways that scientists receive money to fund their research, but most involve applying for grants from private companies or state or federal agencies. Researchers must take all necessary steps to ensure that the funding sources do not bias or influence the outcomes of their studies, even if the funding agencies have something to gain or lose by the studies' results.[15] For example, consider a study designed to assess the effect of milk consumption on weight gain in children. The dairy industry might be interested in funding such a study. This arrangement poses no ethical problems whatsoever, as long as the results are not biased by what the dairy industry would like the researchers to conclude. It is very difficult to determine if a funding agency has influenced a study's outcome, but it is something you should keep in mind when evaluating the validity of a nutrition claim.

EVALUATE THE EXPERIMENTAL DESIGN

Once you have determined that a nutrition claim (1) has been published in a reputable journal or report, (2) was conducted by a qualified researcher, and (3) was likely not biased by the funding source, you are ready to consider the research itself. In other words, was the research conducted in a way that was appropriate to test the hypothesis? And do the conclusions fit the study design?

One of the best resources for finding the details about a study is the U.S. National Library of Medicine, which supports a searchable biomedical database called **PubMed.*** This information can help you answer important questions about a study. For example, was it an epidemiologic study or a human intervention trial? Do the results suggest an association or causal relationship? Was an appropriate control group used? Was the study double-blinded? Was a placebo used in the control group? With your knowledge of experimental design, you can evaluate the research and determine whether the nutrition claim is likely to be valid.

DO PUBLIC HEALTH ORGANIZATIONS CONCUR?

Because even the best experiment does not always provide conclusive evidence that a particular nutrient influences health in a certain way, public health experts usually wait for the results of several studies before they begin to make overall claims about the effect of a certain nutrient on health. Thus,

PubMed A computerized database that allows access to approximately 11 million biomedical journal citations.

*The PubMed database can be found at http://www.ncbi.nlm.nih.gov/PubMed.

before believing it to be true, it is advisable for you to determine whether a claim is supported by major public health organizations. For example, you could check whether the American Heart Association supports a claim concerning nutrition and heart disease or whether the American Cancer Society supports a claim concerning nutrition and cancer.

Nutrition and Health: What Is the Connection?

As you have just learned, nutritional scientists continually study the importance of nutrition on health because experts agree that nutrition plays a powerful role in health and disease. Indeed, consuming either too little or too much of a nutrient can cause illness. But, what do we really know about nutrition and health? And how do scientists and organizations track the overall health of our nation? Although extensive answers to these questions are beyond the scope of this chapter, it is important for you to understand some basic concepts related to how the relationship between nutrition and health is assessed on a national level and how this relationship is ever evolving.

PUBLIC HEALTH AGENCIES ASSESS THE HEALTH OF THE NATION

To understand the complex relationship between nutrition and health, you first need to understand how societal (rather than individual) "health" is assessed. In other words, what do scientists and other health professionals measure when they want to determine whether a population is becoming more or less healthy?

In the United States, the organization responsible for monitoring health trends and compiling health-related statistics is the **Centers for Disease Control and Prevention (CDC).**[†] Established in 1946 to fight malaria, the CDC now provides a system of national and international health surveillance to monitor and prevent disease outbreaks, implement disease prevention strategies, and maintain national health statistics.

MORTALITY AND MORBIDITY RATES MEASURE DEATH AND ILLNESS OVER TIME

The CDC monitors many aspects of societal health, but the most frequently used are morbidity and mortality rates. A **rate** is a measure of some event, disease, or condition affecting a specified group of people within a specific time span. One example of a health-related rate is **mortality rate,** which assesses the number of deaths that occur in a certain population group during a period of time (usually one year).

A more specific example of a health-related mortality rate is **infant mortality rate,** defined as the number of infant deaths (<1 year of age) per 1,000 live births in a given year. Infant mortality rates are often used to assess the health and well-being of a society, as they reflect a complex web of environmental,

U.S. Centers for Disease Control and Prevention (CDC) A governmental agency that monitors the nation's health in order to prevent and control disease outbreaks.

rate A measure of the occurrence of a certain type of event within a specific period of time.

mortality rate (mor – TAL – i – ty) The number of deaths in a given period of time.

infant mortality rate The number of infant deaths (<1 year of age) per 1,000 live births in a given year.

[†]Originally, the abbreviation "CDC" stood for Communicable Disease Center.

FIGURE 1.6 Changes in Life Expectancy and Infant Mortality Rate Since 1900, life expectancy has increased while the infant mortality rate has decreased. These shifts indicate improved societal health.

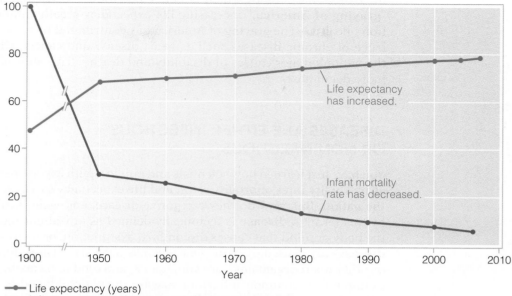

- Life expectancy (years)
- Infant mortality rate (deaths per 1,000 births per year)

SOURCES: National Center for Health Statistics (NCHS). Health, United States, 2010. Available from: http://www.cdc.gov/nchs/data/hus/hus10.pdf. Centers for Disease Control and Prevention. Achievements in public health 1990–1999; healthier mothers and babies. Morbidity and Mortality Weekly Report. 1999:48:849–58.

social, economic, medical, and technological factors that interact to influence overall health. As shown in Figure 1.6, mortality rates in the United States have dropped dramatically in the past century.[16] Many factors influence infant mortality rate, including genetic factors, access to medical care, substance abuse, adequate maternal nutrition, and weight gain during pregnancy.

Mortality rates can also be assessed for various diseases, such as heart disease, cancer, stroke, and diabetes. These rates are typically expressed as the number of deaths from a particular disease per 100,000 people in a given year. Because of a combination of factors, U.S. mortality rates from various diseases have shifted dramatically over the past century.[17]

Whereas mortality rates assess the number of deaths in a given period of time, **morbidity rates** reflect illness in a given period of time. Morbidity rates can be expressed as either incidence or prevalence. The **incidence** of a disease is the number of people who are *newly diagnosed* with the disease in a given period of time, whereas the **prevalence** is the *total number of people* with a particular disease in a given period of time. Like mortality rates, morbidity rates (incidence and prevalence) for certain diseases in the United States have changed drastically during the past few decades. For example, tuberculosis incidence has declined over the past century, whereas prevalence of type 2 diabetes and acquired immunodeficiency syndrome (AIDS) has increased.

morbidity rate (mor – BID – i – ty) The number of illnesses in a given period of time.

incidence The number of people who are newly diagnosed with a condition in a given period of time.

prevalence The total number of people who have a condition in a given period of time.

life expectancy A statistical prediction of the average number of years of life remaining to a person at a specific age.

LIFE EXPECTANCY HAS INCREASED DRAMATICALLY

Another indicator of societal health is **life expectancy,** which is defined as the average number of years of life remaining to a person *at a particular age*—for example, at 25 years of age. As you might have guessed, average life expectancy in the United States has increased dramatically over the past century (see Figure 1.6).[18] There are many reasons for this increase in life expectancy, one of which is better nutrition.[19] Often referred to as the **"graying of America,"** increasing life expectancy greatly influences our nation's health.[20] The graying of America has contributed to a rise in the prevalence of chronic diseases such as heart disease and cancer, which are now the nation's major causes of disability and death.[17] This shift is described in more detail next.

DISEASES ARE EITHER INFECTIOUS OR NONINFECTIOUS

You have just learned how scientists and public health experts use indices such as morbidity rates, mortality rates, and life expectancy to assess the health of the nation. But, how do they categorize disease, and what is actually meant by this term? A **disease** is technically defined as any abnormal condition of the body or mind that causes discomfort, dysfunction, or distress. **Infectious diseases** are caused by pathogens (such as bacteria, viruses, fungi, parasites, or other microorganisms), are contagious, and tend to be acute in nature. An example of a common infectious disease during childhood is chicken pox, which is caused by the varicella virus. Conversely, **noninfectious diseases** are not spread from one person to another, do not involve an infectious agent, and tend to be chronic in nature. An example of a noninfectious disease is type 2 diabetes. The cause of a disease is often called its **etiology.** As you will soon learn, noninfectious diseases have replaced infectious diseases as the most common causes of death in the United States. As such, understanding the complex etiologies of noninfectious diseases is becoming increasingly critical.

Categories of Noninfectious Diseases One type of noninfectious disease that is commanding the attention of today's medical community is referred to as autoimmune disease. **Autoimmune diseases** occur when the immune system—which typically rids the body of pathogens or diseased tissue—instead attacks its own healthy tissues. Examples of autoimmune diseases include some forms of thyroid disease and type 1 diabetes. Other forms of noninfectious diseases are the nutrient deficiency diseases that result from insufficient consumption of nutrients. A third category of noninfectious disease is **chronic degenerative diseases,** which develop slowly and persist for a long time, including heart disease, osteoporosis, and cancer.

Fortunately, although chronic degenerative diseases are among the most common and costly health problems we face today, they are also among the most preventable. Adopting healthy behaviors such as eating nutritious foods, being physically active, and avoiding tobacco use can prevent or control the devastating effects of these diseases. Indeed, scientists believe that most chronic degenerative diseases are caused by a combination of genetic and lifestyle factors—in particular poor nutrition. You will learn much more about this throughout this book.

CHRONIC DISEASES ARE THE LEADING CAUSES OF DEATH

The evolving relationship between nutrition and health that has occurred in the past century is paralleled by a shift in disease prevalence over time. For example, it may surprise you that the major causes of death in the United States during

graying of America The phenomenon occurring in the United States in which the proportion of elderly individuals in the population is increasing with time.

disease A condition that causes physiological or psychological discomfort, dysfunction, or distress.

infectious disease A contagious illness caused by a pathogen such as a bacteria, virus, or parasite.

noninfectious disease An illness that is not contagious.

etiology The cause or origin of a disease.

autoimmune disease A condition in which the immune system attacks an otherwise healthy part of the body.

chronic degenerative disease A noninfectious disease that develops slowly and persists over time.

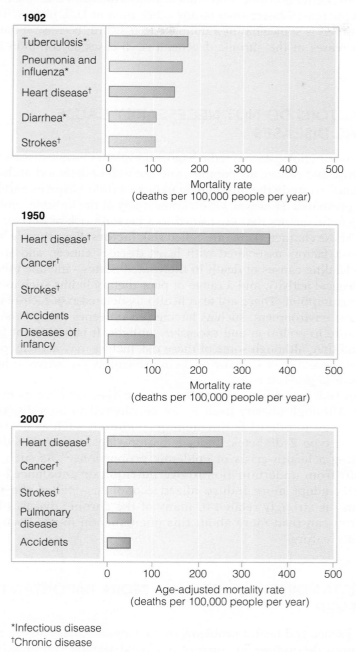

FIGURE 1.7 Five Leading Causes of Death in 1902, 1950, and 2007 Chronic diseases such as heart disease, cancer, and stroke have replaced infectious diseases such as tuberculosis, pneumonia, and diarrhea as our leading causes of death.

*Infectious disease
†Chronic disease

SOURCES: Data for 1902 and 1950: Centers for Disease Control and Prevention. Leading causes of death, 1900–1998. Available from: http://www.cdc.gov/nchs/data/dvs/lead1990_98.pdf. Data for 2007: Centers for Disease Control and Prevention. Health, United States, 2010. Available from: http://www.cdc.gov/nchs/data/hus/hus10.pdf.

the early 1900s were infectious diseases, whereas the leading causes of death now are chronic degenerative diseases (Figure 1.7).[21] Infectious diseases such as pneumonia, tuberculosis, and influenza were rampant during the early part of the 20th century, accounting for one-third of the nation's deaths. In addition to infectious disease, nutritional deficiencies contributed to the high morbidity and mortality rates in the early 1900s.[22] Public health efforts such as installation of water treatment and sewage disposal facilities, the development of antibiotics,

Today, people can expect to live longer than ever before.

and implementation of childhood vaccination programs helped to reduce the incidence of many infectious diseases.[16] As a combined result of decreased infectious disease, better nutrition, and other advances in health care, life expectancy rates have increased faster than at any other time in U.S. history. This aging trend, however, is somewhat of a double-edged sword, because it has brought with it increases in the chronic degenerative diseases so common in today's society.

RISK FACTORS DO NOT NECESSARILY CAUSE CHRONIC DISEASES

The causes—or etiologies—of today's most prevalent chronic diseases, such as heart disease and cancer, are complex and not well understood. Indeed, scientists continue to study the physiologic nature of these diseases and how they might be prevented. However, we do know many of the lifestyle, environmental, and genetic factors that are *related* to a person's risk of developing these diseases. These characteristics are called **risk factors.** For example, the major lifestyle risk factors associated with heart disease, cancer, and stroke—the top three leading causes of death in the United States—include tobacco use, lack of physical activity, and a range of poor dietary habits such as excessive energy consumption. There are also health-associated risk factors related to genetics and environment, such as having certain genes that promote cancer and exposure to pollution and excessive sunlight. It is important for you to understand that, although some of these risk factors may actually play a role in *causing* these diseases, many of them are simply *predictive* in helping us know who is at greatest risk.

A major risk factor for many of the chronic diseases is being overweight or obese. Although obesity itself is not considered to be a disease, it can predispose you to other life-threatening conditions such as heart disease, stroke, and type 2 diabetes. Indeed, nationwide and worldwide obesity is becoming a health crisis of epidemic proportions.[23] As such, the dramatic shift from undernutrition to overnutrition or unbalanced nutrition as societies adopt more industrialized economies—called the **nutrition transition**—is strongly related to many of the chronic diseases facing us today.[24] You can read more about this phenomenon in the Focus on Diet and Health feature.

UNDERSTANDING NUTRITION IS MORE IMPORTANT THAN EVER

Chronic disease and health problems in the United States caused, at least in part, by poor diet represent some of our most serious and pressing public health issues. Researchers estimate that 70% of adults older than 20 years of age are either overweight or obese and that more than 280,000 deaths each year are due to obesity alone.[18] More than 64 million Americans now have cardiovascular disease (the leading cause of death), 50 million have high blood pressure, and more than 1 million have type 2 diabetes.[18] In addition, cancer accounts for 25% of all deaths in the United States annually.[18] Fortunately, consuming a healthy balance of nutrients, phytochemicals, and zoonutrients can help you decrease your risk of developing all these conditions. Indeed, as the incidence and prevalence of these diseases increase, it is becoming even more important for you to pay attention to what you eat throughout your entire life.

risk factor A lifestyle, environmental, or genetic factor related to a person's chances of developing a disease.

nutrition transition The shift from undernutrition to overnutrition or unbalanced nutrition that often occurs simultaneously with the industrialization of a society.

FOCUS ON DIET AND HEALTH
Industrialization, Population Growth, and the Nutrition Transition

Until recent times, being very poor was typically associated with being underfed and underweight. In contrast, being overweight or obese was a sign of wealth. This is because food has traditionally represented a costly and difficult-to-obtain item, especially in regions where poverty was common. Over the past 20 years, however, the prevalence of obesity has been on the rise worldwide—even in economically challenged and developing countries. Moreover, increasing rates of obesity have been accompanied by unwelcome and dangerous increases in other chronic diseases such as type 2 diabetes and hypertension. For example, scientists estimate that the number of people with type 2 diabetes worldwide will increase from 171 million in the year 2000 to 366 million by 2030 (see map), with some of the largest increases in poor regions such as sub-Saharan Africa and India.[25] The World Health Organization (WHO) similarly expects the number of people with type 2 diabetes in developing countries to increase from 84 million to 228 million in the next few decades.

Although the reason for the increasing rates of obesity and its related diseases even in poor nations may not be immediately obvious, it appears that they are due to a complex interaction among the impacts of population growth, decreased physical activity, and overconsumption of readily available, inexpensive, high-energy foods. This phenomenon, called the nutrition transition, is a commonly observed characteristic of nations as they begin to adopt a Western lifestyle. However, it appears to be less common in developing regions of the world where some important aspects of the traditional lifestyles (such as high levels of physical activity and consumption of a moderate and balanced diet) have been preserved during the transition from a rural to an industrialized, urban society. International leaders in public health are hopeful that nationwide initiatives to encourage people to remain physically active and consume reasonable amounts of traditional foods will help prevent this unhealthy trend. This is especially important in regions of the world, such as many Latin American and African countries, that are currently adopting aspects of the Western lifestyle while gaining economic strength and striving to better their standards of living.

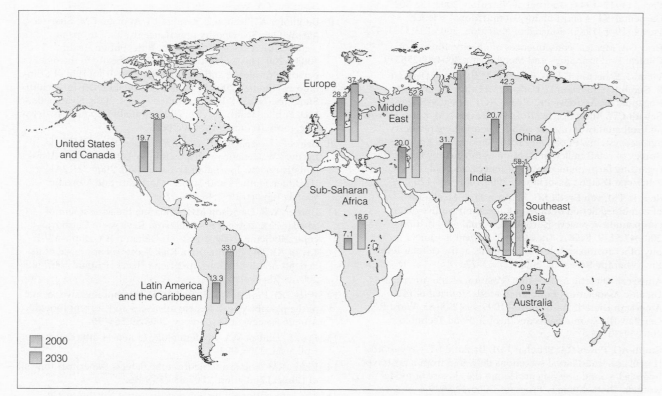

The prevalence of type 2 diabetes is expected to increase in the next few decades—especially in developing countries. Data represent millions of people.

SOURCE: From Hossain, Kawar, Nahas, *Obesity and Diabetes in the Developing World—A Growing Challenge*, NEJM 356; 3, 2007. Copyright © 2007 Massachusetts Medical Society. All rights reserved.

As the saying goes, *"An ounce of prevention is worth a pound of cure."* We could not agree more with this age-old adage, and hope the information and insight you gain from studying nutrition will help you lead a longer, more healthful, and more enjoyable life. Perhaps stated most succinctly and accurately by Sir Francis Bacon more than 400 years ago, *"Knowledge is power."*

Critical Thinking: Katherine's Story Recall that Katherine, the student featured at the beginning of this chapter, changed her major from elementary education to nutrition soon after beginning her college studies. Now is also a good time for you to determine which field of study best suits you. What other majors (aside from the one you are in) might be of interest to you? How would you find out about their requirements? Is it possible that nutrition, or a related field, might be right for you? If so, with whom could you talk about this on your campus?

Notes

1. Carpenter KJ. A short history of nutritional science: Part 1 (1785–1885). Journal of Nutrition. 2003;133:638–45. Carpenter KJ. A short history of nutritional science: Part 2 (1885–1912). Journal of Nutrition. 2003;133:975–84. Carpenter KJ. A short history of nutritional science: Part 3 (1912–1944). Journal of Nutrition. 2003;133:3023–32. Carpenter KJ. A short history of nutritional science: Part 4 (1945–1985). Journal of Nutrition. 2003;133:3331–42.

2. Bray GA. Medical consequences of obesity. Journal of Clinical Endocrinology and Metabolism. 2004;89:2583–9.

3. Vicini J, Etherton T, Kris-Etherton P, Ballam J, Denham S, Staub R, Goldstein D, Cady R, McGrath M, Lucy, M. Reganold JP, Andrews PK, Reeve JR, Carpenter-Boggs L, Schadt CW, Alldredge JR, Ross CF, Davies NM, Zhou J. Fruit and soil quality of organic and conventional strawberry agroecosystems. PLoS One. 2010. Sept. 1;5(9). ppi:e12346. Survey of retail milk composition as affected by label claims regarding farm-management practices. Journal of the American Dietetic Association. 2008;108:1198–1203.

4. de Kok TM, van Breda SG, Manson MM. Mechanisms of combined action of different chemopreventive dietary compounds: A review. European Journal of Nutrition. 2008;47:51–9. Kale A, Gawande S, Kotwal S. Cancer phytotherapeutics: Role for flavonoids at the cellular level. Phytotherapy Research. 2008;22:567–77.

5. American Dietetic Association. Position of the American Dietetic Association: Functional foods. Journal of the American Dietetic Association. 1999;99:1278–85. Ward RE, German JB. Zoonutrients and health. Food Technology. 2003;57:30–36.

6. Smith AG, Powis RA, Pritchard DI, Britland ST. Greenbottle (Lucilia sericata) larval secretions delivered from a prototype hydrogel wound dressing accelerate the closure of model wounds. Biotechnology Progress. 2006;22:1690–6.

7. Bansal T, Garg S. Probiotics: From functional foods to pharmaceutical products. Current Pharmacology and Biotechnology. 2008;9:267–87. Olmedilla-Alonso B, Granado-Lorencio F, Herrero-Barbudo C, Blanco-Navarro I, Blázquez-García S, Pérez-Sacristán B. Consumption of restructured meat products with added walnuts has a cholesterol-lowering effect in subjects at high cardio-vascular risk: A randomised, crossover, placebo-controlled study. Journal of the American College of Nutrition. 2008;27:342–8.

8. Carey SS. A beginner's guide to scientific method, 3rd ed. Belmont, CA: Wadsworth/Thomson Learning; 2004.

9. Berghöfer A, Pischon T, Reinhold T, Apovian CM, Sharma AM, Willich SN. Obesity prevalence from a European perspective: A systematic review. BMC Public Health. 2008;8:200. Haslam D. Understanding obesity in the older person: Prevalence and risk factors. British Journal of Community Nursing. 2008;13:115–6. National Center for Health Statistics (NCHS). Health, United States, 2008. Hyattsville, MD: Public Health Service; 2008. Available from: http://www.cdc.gov/nchs/data/hus/hus08.pdf.

10. Beeks E, Kessels AG, Kroon AA, van der Klauw MM, de Leeuw PW. Genetic predisposition to salt-sensitivity: A systematic review. Journal of Hypertension. 2004;22:1243–9. Healthier mothers and babies. Morbidity and Mortality Weekly Report. 1999;48:849–58.

11. Borden WB, Davidson MH. Updating the assessment of cardiac risk: beyond Framingham. Reviews in Cardiovascular Medicine. 2009. 10:63–71. Hemann BA, Bimson WF, Taylor AJ. The Framingham Risk Score: an appraisal of its benefits and limitations. American Heart Hospital Journal. 2007, 5:91–6.

12. Price DD, Finniss DG, Benedetti F. A comprehensive review of the placebo effect: Recent advances and current thought. Annual Review of Psychology. 2008;59:565–90.

13. Frey L, Hauser WA. Epidemiology of neural tube defects. Epilepsia. 2003;44:4–13.

14. Pitkin RM. Folate and neural tube defects. American Journal of Clinical Nutrition. 2007;85:285S–8S.

15. The International Life Sciences Institute North America Working Group on Guiding. Funding food science and nutrition research: financial conflicts and scientific integrity. Journal of Nutrition. 2009;77:264–72.

16. Armstrong BL, Conn LA, Pinner RW. Trends in infectious disease mortality in the United States during the 20th century.

Journal of the American Medical Association. 1999;281:61–6. Centers for Disease Control and Prevention. Achievements in public health, 1900–1999: Control of infectious diseases. Morbidity and Mortality Weekly Report. 1999;48:621–9.

17. Centers for Disease Control and Prevention. Leading causes of death, 1900–1998. Available from: http://www.cdc.gov/nchs/data/dvs/lead1900_98.pdf. National Center for Health Statistics (NCHS). Health, United States, 2008. Hyattsville, MD: Public Health Service; 2008. Available from: http://www.cdc.gov/nchs/data/hus/hus08.pdf.

18. National Center for Health Statistics (NCHS). Health, United States, 2010. Available from: http://www.cdc.gov/nchs/data/hus/hus10.pdf.

19. Perls T, Terry D. Understanding the determinants of exceptional longevity. Annals of Internal Medicine. 2003;139:445–9.

20. Anderson RE, Smith RD, Benson ES. The accelerated graying of American pathology. Human Pathology, 1991;22:210–4.

21. Centers for Disease Control and Prevention. Achievements in public health, 1900–1999: Control of infectious diseases. Morbidity and Mortality Weekly Report. 1999;48:621–9. Armstrong GL, Conn LA, Pinner RW. Trends in infectious disease mortality in the United States during the 20th century. Journal of the American Medical Association. 1999;281:61–6.

22. Carpenter KJ. A short history of nutritional science: Part 2 (1885–1912). Journal of Nutrition. 2003;133:975–84.

23. Wild S, Roglic G, Green A, Sicree R, King H. Global prevalence of diabetes: Estimates for the year 2000 and projections for 2030. Diabetes Care. 2004;27:1047–53.

24. Hedley AA, Ogden CL, Johnson CL, Carroll MD, Curtin LR, Flegal KN. Prevalence of overweight and obesity among US children, adolescents, and adults, 1999–2002. Journal of the American Medical Association. 2004;291:2847–50. National Center for Health Statistics (NCHS). Health, United States, 2008. Available from: http://www.cdc.gov/nchs/data/hus/hus08.pdf.

25. Khan NC, Khoi HH. Double burden of malnutrition: The Vietnamese perspective. Asia Pacific Journal of Clinical Nutrition. 2008;17:116–8. Haslam DW, James WP. Obesity. Lancet. 2005;366:1197–1209. Wild S, Roglic G, Green A, Sicree R, King H. Global prevalence of diabetes: Estimates for the year 2000 and projections for 2030. Diabetes Care. 2004;27:1047–53. Hossain P, Kawar B, El Nahas M. Obesity and diabetes in the developing world—A growing challenge. New England Journal of Medicine. 2007;356:213–15.

Nutritional Assessment and Dietary Planning

As you undoubtedly realize, food and eating make up a large part of our culture. For most of us, food is plentiful and associated with good times—family gatherings, social occasions, and special events. For others, food is limited, and getting enough to eat is an ongoing concern. In either case, food and the rituals associated with eating are integral to the fabric of our lives.

But there is much more to food than just how it contributes to our social, cultural, and family activities. On the most basic level, food is a vehicle for nutrients that provide the building blocks and energy for all the body's structures and functions. Because health can deteriorate when we consume too little or too much of certain nutrients, getting appropriate amounts of each essential nutrient is vital for life itself. Consequently, the decisions you make about which foods to eat are important, and you need to consider your choices thoughtfully. But how do you know if you are consuming enough, but not too much, of all the essential nutrients? And how can you better choose foods to meet your nutritional needs?

To answer these questions today and for years to come, you need to understand several important concepts related to the nutrients in foods and your body. These concepts include such things as nutritional adequacy, nutritional status, dietary assessment, and diet planning. In this chapter, you will learn the fundamentals of these concepts and how to apply them to your food choices. You will also learn about important dietary regulations and guidelines that have been developed to help in this quest. Collectively, this basic information and these dietary planning tools will help you determine which foods—and how much of them—to choose to help optimize your health.

Nutrient Deficiencies—Primary or Secondary?

Emily always maintained an active lifestyle, excellent grades, and busy work and volunteer schedules. But when she began her studies in architecture at the University of Cincinnati, there were many changes in her life, including increased stress and lack of sleep. Therefore, when she started to experience gastrointestinal upset and chronic exhaustion, she assumed a combination of school, stress, and lack of sleep was the cause. That is, until she started feeling so tired that she could not stay awake and was so weak that walking across campus became a major effort. Finally, Emily made an appointment to see a physician. As expected, the doctor took Emily's weight, asked questions about her lifestyle and eating habits, and had her blood drawn. Responding to the facts that Emily had recently lost five pounds, she was profoundly fatigued, and her blood had very low levels of iron, her doctor concluded she was iron deficient. He sent her back to her dormitory with a high-dose iron supplement and asked her to come back for another evaluation in four weeks. Emily was meticulous about taking her iron supplements, got plenty of rest, and within a couple of weeks started feeling better. So when she returned to the student health center several weeks later, she was surprised that her iron levels were only slightly higher than they had been. Even four weeks later, she was still iron deficient. Puzzled by how long it was taking to become healthy, Emily finished her spring term and returned home for the summer.

Like many students, Emily looked forward to going home and enjoying family meals—a welcome break from eating in the dining halls. She also enjoyed getting plenty of sleep at home. But several weeks into her summer, Emily started to wonder why she was again experiencing serious stomach upset and fatigue. She began reading books and searching the Internet for clues, and to her surprise, all signs pointed to a condition called celiac disease. People with celiac disease react to the protein gluten, found in wheat and other grains, causing diarrhea and bloating. Celiac patients also frequently experience multiple nutritional deficiencies—not because they are not eating well (primary nutrient deficiency), but rather because their damaged intestines cannot absorb and use the nutrients they are consuming (secondary nutrient deficiency). Emily went to her family doctor and asked to be tested for celiac disease, and found that her suspicion was correct. She immediately eliminated all gluten-containing foods from her diet and, within days, noticed an immense difference in her health. The mystery had been solved. Although it is sometimes tricky to make sure she does not eat gluten-containing food, Emily has been very successful in remaining gluten-free since she was diagnosed—even after returning to college. Most importantly, her iron levels are now normal and she is feeling great. In fact, feeling healthier than she ever has, Emily just completed her first half marathon.

Courtesy Megan Philibin

Critical Thinking: Emily's Story

Emily's story is not unique. Can you recall changes to your health that occurred when you first went to college? Do you think any of these changes might have been due to altered eating habits? Can you list three nutritious foods that you ate before college but now rarely consume? Would it be possible to add these food items back to your dietary pattern?

What Do We Mean by "Nutritional Status"?

Most people know that a balanced diet is needed for optimal physiologic function and long-term health. Indeed, consuming too little of a nutrient, a situation called **undernutrition,** can cause nutritional deficiency, which can be serious and sometimes fatal. It is equally important that we do not eat too much of certain nutrients and foods. For example, consuming too many fatty foods can lead to obesity and its related health consequences, and overconsumption of some vitamins and minerals (**nutritional toxicity**) can be fatal. Undernutrition and overnutrition make up the extreme ends of what is called the **nutritional status** continuum, and both are examples of **malnutrition.** Malnutrition is defined as a state of poor nutrition due to an imbalance between the body's nutrient requirements and nutrient consumption. Figure 2.1 shows the relationships among nutrient availability (e.g., food intake), physiologic function or health, and nutritional status. Both undernutrition and overnutrition can lead to suboptimal health. An example of this is what happens with vitamin A malnutrition: deficiency causes poor growth and night blindness, whereas toxicity can lead to blurred vision, weak bones, and even birth defects. Thus, a person with optimal nutritional status is in the "center" of the nutritional status continuum.

PRIMARY AND SECONDARY MALNUTRITION CAN LEAD TO POOR NUTRITIONAL STATUS

Many underlying factors can contribute to poor nutritional status—whether it be due to under- or overnutrition. **Primary malnutrition** is due to inadequate or excess food intake, while **secondary malnutrition** is caused by other factors. For example, a person may be deficient in one of the B vitamins because his or her diet is lacking vitamin-rich fruits and vegetables (this is primary malnutrition) or because an illness interferes with vitamin B absorption (secondary malnutrition). It is important to know whether primary or secondary malnutrition

Both undernutrition and overnutrition are forms of malnutrition and can lead to poor health.

FIGURE 2.1 Nutrient Intake Largely Determines Nutritional Status and Contributes to Health
Both under- and overnutrition can cause poor nutritional status, which leads to suboptimal health.

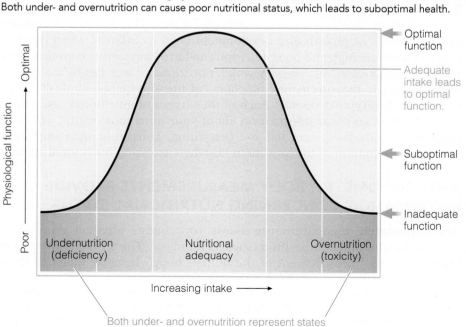

Optimal function

Adequate intake leads to optimal function.

Suboptimal function

Inadequate function

Physiological function — Optimal / Poor

Undernutrition (deficiency)

Nutritional adequacy

Overnutrition (toxicity)

Increasing intake →

Both under- and overnutrition represent states of malnutrition or poor nutritional status.

undernutrition (or nutritional deficiency) Inadequate intake of one or more nutrients and/or energy.

nutritional toxicity Overconsumption of a nutrient resulting in dangerous (toxic) effects.

nutritional status The health of a person as it relates to how well his or her diet meets that person's individual nutrient requirements.

malnutrition Poor nutritional status caused by either undernutrition or overnutrition.

primary malnutrition Poor nutritional status caused strictly by inadequate diet.

secondary malnutrition Poor nutritional status caused by factors such as illness.

is causing poor nutritional status, because the treatments are very different. Whereas deficiencies caused by primary malnutrition can be treated directly by making changes in a person's diet, in the case of secondary malnutrition the underlying cause must first be addressed. An example of this is what happened to Emily, the college student featured in this chapter's Everybody Has a Story. Although Emily's severe iron deficiency was first thought to be a result of inadequate iron intake (primary iron deficiency), it was later found to be caused by celiac disease (secondary iron deficiency). Once her celiac disease was properly treated, her iron status returned to normal.

ADEQUATE NUTRIENT INTAKE CAN BE DIFFERENT AMONG INDIVIDUALS

How can a health care professional know if a person is malnourished? And how can you know the amount of each essential nutrient you need to be healthy? The answers to these questions are complex because different people need different amounts of nutrients for optimal health. Indeed, each of us has our own nutrient needs, determined by a host of factors such as sex, age, physical activity, and genetics. For instance, a college-age male who is a member of the soccer team is likely to have higher nutritional requirements than a comparable classmate who is relatively inactive. Nonetheless, both these individuals need to consume sufficient amounts of the various nutrients to achieve **nutritional adequacy**—when nutrient intake is optimal.

There are several ways that health professionals such as dietitians can determine whether your nutritional status is optimal or perhaps could benefit from different dietary choices. You can use many of these same tools to assess your own nutritional status and dietary adequacy. It is important that you gain a basic understanding of these various methods and tools because they will be referred to frequently as you continue your study of the essential nutrients and overall nutritional health. Some of these tools are described in the next two sections.

How Is Nutritional Status Assessed?

Because adequate nutrition is required for optimal health, health care providers routinely assess a person's nutritional status. This is especially critical during periods of growth and development such as infancy, when nutrient requirements are high and poor nutrition can have long-lasting consequences. In general, there are four ways by which nutritional status can be assessed—anthropometric measurements, biochemical measurements, clinical assessment, and dietary assessment.* Each of these types of nutritional assessment methods provides some information about your nutritional status, although each one—by itself—cannot tell you everything. Thus, it is often important to use a combination of the four methods when assessing nutritional status.

ANTHROPOMETRY: BODY MEASUREMENTS PROVIDE INFORMATION CONCERNING NUTRITIONAL STATUS

Anthropometric measurements assess your body's physical dimensions (such as height) and composition (such as fat mass). The term *anthropometry* literally means "measurement of the human body." Because many of these

---

nutritional adequacy The situation in which a person consumes the required amount of a nutrient to meet physiological needs.

anthropometric measurements Measurements or estimates of physical aspects of the body such as height, weight, circumferences, and body composition.

---

*These tools are sometimes referred to as the "ABCD" methods of nutritional assessment (*a*nthropometric, *b*iochemical, *c*linical, and *d*ietary). Sometimes an "E" is added to this list of nutritional status assessment methods, referring to the importance of *e*conomic status when assessing nutritional risk.

measurements are easy and inexpensive to obtain, anthropometry is routinely used in clinical and research settings. In fact, you are taking an anthropometric measurement whenever you use a bathroom scale to weigh yourself. However, anthropometric measurements alone are not considered diagnostic and must be supported by other measures of nutritional status. For example, poor growth cannot tell you whether you are deficient in any particular essential nutrient. But it can provide clinicians with important clues concerning a person's nutritional status. The most commonly used anthropometric measurements are described next.

Physical Dimensions: Height, Weight, and Circumferences Because being overweight or obese can increase a person's risk for certain chronic degenerative diseases, assessing body weight relative to height can be important. Changes in body weight and height can also provide information regarding the progression of certain diseases. For example, loss of height in an elderly person might indicate a decline in bone density. Significant loss of body weight in a college student might indicate an eating disorder. Height and weight are also commonly used to assess nutritional status throughout infancy, childhood, and pregnancy. This is why they are typically measured throughout childhood, during infancy, and throughout pregnancy.

Other physical dimensions that are sometimes measured include various circumferences such as those of the waist, hips, and head. Waist and hip circumferences provide an indication of body fat distribution, and head circumference is frequently measured to monitor brain growth during infancy. You will learn more about how these measurements are used to assess nutritional status and health throughout this book.

Body Composition—What You Are Made Of Nutritional assessments can also include estimates of **body composition**—the proportions of fat, water, muscle, and mineral (bone) mass that make up your tissues. The distribution of these body components is an important indicator of nutritional status and overall health. For instance, adequate hydration status (water content) is important for optimal athletic performance, loss of lean mass (muscle) can indicate advanced disease in cancer patients, too much body fat can lead to cardiovascular disease, and loss of bone mass is a major risk factor for osteoporosis. Body composition measurements, therefore, are often used along with other anthropometric measurements to provide more detailed information concerning nutritional status. Many campus recreation centers and health clinics offer free body composition testing. You might want to see if this type of testing is available at your university because knowing this information about yourself will be useful as you learn more about nutrition and health.

LABORATORY TESTS ARE IMPORTANT BIOCHEMICAL INDICATORS OF NUTRITIONAL STATUS

To further assess health and nutritional status beyond anthropometric measurements, **biochemical measurements** are often used. These involve laboratory analysis of a biological sample, such as blood or urine. In some cases, the sample is analyzed for a specific nutrient. For example, blood iron levels can be measured to determine iron status, as was done in the case of Emily (featured earlier in the chapter). In other laboratory tests, the sample is analyzed for an indicator that reflects the nutrient's function, or what is called a **biological marker** or biomarker. For example, the hemoglobin content of blood is often measured as a biological marker of iron status. This is because hemoglobin levels decrease during iron deficiency.

body composition Components of the body such as fat, lean mass (muscle), water, and minerals.

biochemical measurement Laboratory analysis of biological samples, such as blood and urine, used in nutritional assessment.

biological marker (biomarker) A measurement in a biological sample, such as blood or urine, that reflects a nutrient's function.

Biochemical measurements are powerful because they can help diagnose a *specific* nutrient deficiency or excess. Note that this is not true of anthropometric measurements, which are only general indicators of *overall* nutritional status. Because collecting and analyzing biological samples often require technical expertise and costly procedures, biochemical analyses are often done only when malnutrition is already suspected.

CLINICAL EVALUATIONS ASSESS SIGNS AND SYMPTOMS OF DISEASE

Another method of assessing nutritional status is conducting a face-to-face clinical assessment. Clinical assessments usually involve a series of activities. For example, the clinician or researcher may ask questions about previous diseases, unusual weight loss and/or weight gain, surgeries, or medications and ask about other relevant information such as family history. This is called "taking a **medical history**" and can be helpful in determining a person's overall health and risk for disease. During a clinical examination, the health care worker will likely note any visible **signs** of illness, which are those objective outcomes of disease that can be seen or assessed by someone else. An example of a sign of iron deficiency would be pale skin or shortness of breath. Other observable signs that may indicate poor nutritional status include skin rashes, swollen ankles (edema), and bleeding gums. These signs can suggest vitamin B, protein, and vitamin C deficiency, respectively. Anthropometric assessments are often done during clinical evaluations as well.

During a clinical evaluation, you may also be asked about whether you are experiencing anything unusual in your health—in other words, whether you have any **symptoms** of disease or malnutrition. Symptoms differ from signs because symptoms are subjective and cannot be observed or noticed by someone else—they must be reported or revealed by the patient. For example, lack of "energy" is a symptom commonly associated with iron deficiency, and loss of appetite is a symptom of zinc toxicity. It is important for clinicians—such as clinical dietitians, nurses, and doctors—to know the various signs and symptoms associated with nutrient deficiencies so that they can make the appropriate observations and ask the right questions.

Clinical assessment has many advantages over other forms of nutritional assessment. For example, it allows health care providers to determine if there are symptoms of malnutrition. Furthermore, because malnutrition is often associated with distinct and recognizable signs, observing them can make clinical diagnosis of a particular nutrient deficiency or toxicity quite accurate. Because the signs and symptoms of many nutrient deficiencies are not apparent until they become severe, however, clinical assessment may not be able to detect malnutrition in its early stages.

ANALYSIS OF YOUR DIET CAN ALSO BE HELPFUL

Although anthropometric, biochemical, and clinical assessments can be used to evaluate some aspects of your nutritional status, it is also very useful to examine the adequacy of your diet. In general, there are two types of **dietary assessment** methods: (1) **retrospective dietary assessment,** which requires you to remember foods consumed in the past, and (2) **prospective dietary assessment,** which requires you to keep track of which and how much food you consume during a specified period of time.

Retrospective Methods: Recalls and Questionnaires The two main retrospective dietary assessment methods are the 24-hour recall and the food frequency questionnaire. In the **24-hour recall** method, you must record everything

Clinical assessments are valuable in nutritional assessment because they can uncover signs and symptoms of malnutrition.

medical history Questions asked to assess overall health.

sign Physical indicator of disease that can be seen by others, such as pale skin and skin rashes.

symptom Subjective manifestation of disease, such as stomach pain or loss of appetite, that cannot be seen by others.

dietary assessment The evaluation of a person's dietary intake.

retrospective dietary assessment Type of dietary assessment that assesses previously consumed foods and beverages.

prospective dietary assessment Type of dietary assessment that evaluates adequacy of food and beverage intake.

24-hour recall A retrospective dietary assessment method that analyzes each food and drink consumed over the previous 24 hours.

you have eaten or drunk in the previous 24 hours and then analyze the information to estimate nutrient intake. Because 24-hour recalls are based on a single day and require a person to remember information, they often do not represent usual food intake. Another retrospective method is the **food frequency questionnaire,** which typically asks for information on food intake *patterns* over an extended period of time. For this method, you must complete a questionnaire that lists certain foods of interest. For example, a food frequency questionnaire used to determine overall fruit intake would ask questions about which fruits you typically eat and how much you normally consume in a serving. Because it does not assess nutrient intake (only food intake patterns), information from a food frequency questionnaire is limited in accuracy and completeness.

Prospective Methods: Diet Records Although retrospective methods of dietary assessment are relatively simple, their usefulness depends on a person's memory. To more accurately assess your diet, it is better to record foods and beverages *as they are consumed* in what is called a **diet record,** or food record. To keep a diet record, you must either estimate portion sizes using standard household measurements (such as tablespoons or cups) or weigh the food before you eat it. Ideally, food records should be kept for three days, one of which should be a weekend day. This information is then analyzed to estimate your nutrient intake. Clinicians consider the diet record to be one of the most accurate methods of dietary assessment.

Note that a diet record may be conducted by a medical professional, but it can also be part of a self-assessment whereby you assess your own nutritional adequacy. Dietary self-assessment is particularly appropriate if you are generally healthy and want to make sure you are following a diet that can promote and maintain health. Individuals who suspect they have a disease or are malnourished should instead consult with a dietitian or other qualified health care provider.

Courtesy Megan Philibin

Critical Thinking: Emily's Story After suffering with somewhat debilitating symptoms for months, Emily learned that she had secondary iron deficiency due to celiac disease. Why do you think this possibility was not explored when she first visited her campus health care provider? What forms of nutritional assessment did her doctor use when examining her? Why might this misdiagnosis have been avoided if she had completed a dietary assessment?

FOOD COMPOSITION TABLES AND DIETARY ANALYSIS SOFTWARE ARE IMPORTANT TOOLS

After completing a dietary record, the next step is to determine the micronutrient, macronutrient, and energy (calorie) contents of your diet. For example, if you are interested in your vitamin C status, you will need to determine how much vitamin C you consumed. There are basically two ways to find information concerning the nutrient composition of foods: **food composition tables** and **computerized nutrient databases**. Food composition tables, such as the booklet that may have accompanied this textbook, can be purchased or accessed free-of-charge from the U.S. Department of Agriculture (USDA).[†] Using printed food composition tables to calculate nutrient intake can be time consuming and tedious. Fortunately, easy-to-use, computerized nutrient databases are also available. For example, the USDA offers an online dietary

[†] You can access the USDA's food composition tables at http://www.ars.usda.gov/nutrientdata.

food frequency questionnaire A retrospective dietary assessment method that assesses food selection patterns over an extended period of time.

diet record A prospective dietary assessment method that requires the individual to write down detailed information about foods and drinks consumed over a specified period of time.

food composition table Tabulated information concerning the nutrient and energy contents of foods.

computerized nutrient database Software that provides information concerning the nutrient and energy contents of foods.

analysis tool that accompanies its MyPlate food guidance system. This is described in more detail later in this chapter.

As you now know, there are many ways to assess your nutritional status, and these should be components of your routine health care plan. Although scrutinizing your diet should always be used in conjunction with anthropometric, biochemical, and clinical assessment, it is one of the most effective methods of nutritional assessment. But even after you have determined the specifics of your dietary intake—for example, what your vitamin C intake is—how can you know if your intake is adequate? The answer to this question lies in being able to refer to a variety of dietary intake reference standards and recommendations, several of which are described next.

How Much of a Nutrient Is Adequate?

Now that you have learned how dietary assessments can help assess nutritional adequacy, you may be wondering how to apply this information to your food choices. In other words, it is not enough to simply quantify the nutrients and energy you consume. You also need to know how to interpret these values. Indeed, just knowing how much vitamin C has been consumed does not tell you if this amount is adequate for you.

Dietary assessment must go one step further than estimating how much of each nutrient is consumed. You must also determine what level of nutrient intake is adequate *for you*. Recall that different people require different amounts of nutrients depending on their sex, age, etc. Thus, there is no simple way to know what your nutrient needs actually are. To this end, the Institute of Medicine has developed a set of nutritional standards to help both medical professionals and interested individuals assess dietary adequacy. These standards can help you judge whether your typical dietary intake is likely to provide too little, too much, or just about the right amount of all the essential micronutrients, macronutrients, and energy.

DIETARY REFERENCE INTAKES (DRIs) PROVIDE REFERENCE STANDARDS

Given the fact that all individuals have different dietary requirements, establishing nutrient intake standards represented quite a challenge to the scientific community. This enormous task required the input and analysis of an impressive assembly of researchers organized by the Institute of Medicine, a division of the National Academy of Sciences. The National Academy of Sciences is a private (nongovernmental) organization of distinguished scholars dedicated to furthering science and general welfare and has had a long-standing interest in nutrition and nutrient recommendations. In 1994 this organization established a set of dietary reference standards called the **Dietary Reference Intakes (DRIs),** which comprise four main sets of reference values for nutrient consumption.[1] These four sets of standards are outlined in Table 2.1, illustrated in Figure 2.2, and described below. Before examining each set of the DRIs in detail, however, it is helpful to know a little about the rich history of nutrition recommendations and standards in the United States.

Dietary Reference Intakes (DRIs) A set of four types of nutrient intake reference standards used to assess and plan dietary intake; these include the Estimated Average Requirements (EARs), Recommended Dietary Allowances (RDAs), Adequate Intake levels (AIs), and the Tolerable Upper Intake Levels (ULs).

A Historical Perspective on Nutrition Recommendations Health experts in the United States have for decades been interested in providing guidance as to optimal nutrient intake, and one of our nation's first official sets of dietary standards was published in 1943 by the National Academy of Sciences. These standards were called the Recommended Dietary Allowances (RDAs). The

TABLE 2.1 Available Dietary Reference Intake (DRI) Standards

Nutrient	EAR	RDA	AI	UL
Macronutrients[a]				
Water			■	
Linoleic acid			■	
Linolenic acid			■	
Carbohydrates	■	■		
Fats*			■	
Protein	■	■		
Vitamins				
Thiamin	■	■		
Riboflavin	■	■		
Niacin	■	■		■
Biotin			■	
Pantothenic Acid			■	
Vitamin B$_6$	■	■		■
Folate	■	■		■
Vitamin B$_{12}$	■	■		
Vitamin C	■	■		■
Vitamin A	■	■		■
Vitamin D	■	■		■
Vitamin E	■	■		■
Vitamin K			■	
Minerals				
Sodium				■
Chloride				■
Potassium			■	
Calcium	■	■		■
Phosphorus	■	■		■
Magnesium	■	■		■
Iron	■	■		■
Zinc	■	■		■
Iodine	■	■		■
Selenium	■	■		■
Copper	■	■		■
Manganese			■	■
Fluoride			■	■
Chromium			■	■
Molybdenum	■	■		■

[a]Note that there are no DRIs for energy, *per se*. Instead, you can estimate your caloric needs by using the Estimated Energy Requirement (EER) calculations.

*AI values have only been set for infants, not any other life-stage group.

RDAs, in one form or another, have been used ever since as goals for good nutrition. When the first RDAs were released, malnutrition in the United States was generally due to *under*nutrition, and nutritional deficiencies were common. Thus, the RDAs were initially developed to address these health issues. However, nutritional issues changed, knowledge advanced, and it became apparent that this single set of nutrient intake standards had important

FIGURE 2.2 Dietary Reference Intake (DRI) Standards There are four sets of DRI reference values: EARs, RDAs, AIs, and ULs. Each of these standards is used for a different purpose.

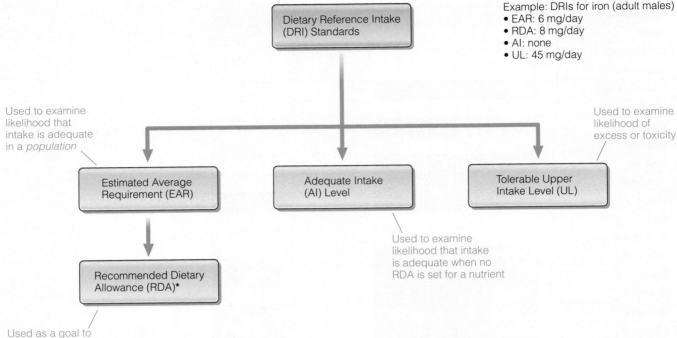

Example: DRIs for iron (adult males)
• EAR: 6 mg/day
• RDA: 8 mg/day
• AI: none
• UL: 45 mg/day

Used to examine likelihood that intake is adequate in a *population*

Used to examine likelihood of excess or toxicity

Used to examine likelihood that intake is adequate when no RDA is set for a nutrient

Used as a goal to help ensure adequate intake in an *individual*

*Note that in the case of energy, an Estimated Energy Requirement (EER) is provided instead.

limitations. Researchers were also learning that optimal nutrition went far beyond just preventing deficiencies—it was also important for decreasing risk for chronic degenerative diseases such as heart disease and cancer. In response, the DRIs were developed. Note that both the United States and Canada recognize the DRIs as their official set of dietary reference standards.

DRI VALUES DEPEND ON MANY FACTORS

Using all four types of DRI values allows health professionals to comprehensively assess the nutritional adequacy of individuals as well as populations. To do this effectively, it is critical to understand some fundamental concepts underlying the DRIs. For example, nutrient requirements among individuals differ by sex and life stage and the DRIs provide different values depending on these variables. In this context, a life stage typically refers to an age group and/or a physiologic state such as pregnancy or lactation. In all, there are 16 "life-stage groups" for females and 10 "life-stage groups" for males. One must be careful to use the correct values for his or her appropriate sex and life-stage group, because using the incorrect ones may result in faulty conclusions.

It is also important to understand the term **nutrient requirement,** which is the amount of a nutrient that a person must consume to promote optimal health. You have already learned that we each have our own unique nutrient requirements. In general, nutrient requirements of all the individuals in a population are distributed in a bell-shaped manner, meaning the vast majority of people have requirements at some mid-level amount, with some requiring much less and others requiring much more. Although the DRIs take into account the influence of sex and life stage on nutrient and energy requirements, there are other factors such as genetics, medications, lifestyle choices, and environmental influences that can also affect your nutrient requirements. For instance, exposure to tobacco smoke can increase your

nutrient requirement The lowest intake level of a nutrient that supports basic physiological functions and promotes optimal health.

vitamin C requirement. As such, it is virtually impossible to know how much of each essential nutrient you really need, and the DRIs are only *estimates* of nutrient requirements and intake goals in a healthy population. Your personal level may actually be less or more.

Nonetheless, the DRI values can provide powerful tools in assessing nutritional status and planning a healthy diet. But how can you use the DRIs for these purposes? The answer to this question requires a basic understanding of the various sets of standards that comprise the DRIs. In the following sections, you will learn about each of these four main dietary reference standards in detail and how to use them to determine if you are getting too little or too much of each important nutrient.

Because of an array of genetic, lifestyle, and environmental factors, each of us has different nutritional requirements.

ESTIMATED AVERAGE REQUIREMENTS (EARs) REFLECT A *POPULATION'S* AVERAGE NEED

For each essential micronutrient (vitamins and minerals) and macronutrient (carbohydrate, protein, fat, and water), the DRI committees first attempted to establish an **Estimated Average Requirement (EAR)** for each life-stage group. Note that, because energy is not a nutrient, EARs for it were not established. Instead, Estimated Energy Requirements (EER) are provided, and this reference standard is described later in this chapter.

EAR values represent the intakes thought to meet the requirements of half the healthy individuals in each particular life stage and sex (Figure 2.3). For example, the EAR for iron in breastfeeding women is 6.5 mg/day. This means that about half of the women in this life-stage group require less than 6.5 mg/day iron, and the other half needs more. It is important to remember that the EAR meets or exceeds the true nutritional needs of half the population but is inadequate for the other half.

Estimated Average Requirement (EAR) The amount of a nutrient that meets the physiological requirements of half the healthy population of similar individuals.

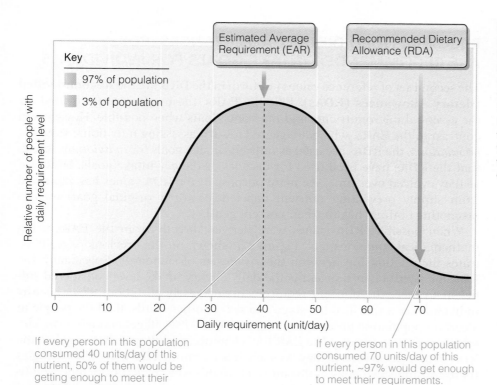

FIGURE 2.3 Estimated Average Requirements (EARs) Compared with Recommended Dietary Allowances (RDAs) EARs, which are not used as dietary intake goals, represent intake values needed by about half of a healthy population. RDAs, which are used as dietary intake goals, represent the intake values needed by about 97% of a healthy population. RDA values are always higher than their EAR counterparts.

To help understand how EARs are used, it is helpful to know how they were developed. As an example, consider the EARs for iron. These values were determined by a panel of iron experts who first reviewed the functions of iron in the body as well as the potential outcomes of poor iron status throughout the life cycle. Then they selected the best measure that could be used to assess iron status during each life stage. When appropriate, they also distinguished between males and females. For example, they estimated the average iron requirements for older infants by taking into account increasing blood volume and increasing iron storage, which are important during this period of life. For women of childbearing age, they took into account iron loss associated with menstruation. Then, using mathematical equations, they established EARs that predicted the amount of iron needed to meet the requirements of half the population in each of the life-stage groups. Note that there are not EAR values for all the nutrients—only the ones for which there is sufficient high-quality published research.

The EARs are very useful in research and public health settings to evaluate whether a *group* of people is likely to be consuming an adequate amount of a nutrient. However, it is inappropriate to use the EAR values as recommended dietary intake goals for *individuals* because doing so would result in half the population not consuming adequate amounts. For instance, the American Cancer Society might want to conduct a study to determine whether American men are consuming sufficient amounts of selenium, a mineral that protects cells from oxidative damage that can lead to cancer. The EAR for selenium in adult men is 45 µg/day. If the results of their study show that the average selenium intake in men is 50 µg/day, they could conclude that, in general, this population is likely consuming enough selenium. This is because the EARs are meant to be used exactly for this purpose—to assess the likelihood that a population is consuming enough of a nutrient. However, it would be inappropriate for the American Cancer Society to recommend that adult males aim to consume 45 µg/day of this nutrient because this amount is less than what is needed by 50% of this group.

RECOMMENDED DIETARY ALLOWANCES (RDAs) ARE RECOMMENDED INTAKE GOALS FOR *INDIVIDUALS*

The second set of reference values put forth in the DRIs are the **Recommended Dietary Allowances (RDAs).** Like the EARs, these values are provided for the essential micronutrients and macronutrients when possible. However, in contrast to the EARs, which are geared toward assessing nutritional status of *populations*, the RDAs are used as nutrient-intake goals for *individuals*. Recall that the RDAs have been used for decades as dietary intake goals. However, as they evolved over time, the main purpose of the RDA values has expanded from simply preventing nutrient deficiencies (their original goal) to also promoting optimal health (their current goal).

When possible, RDA values were derived directly from the EARs using mathematical equations, and Figure 2.3 (shown on the previous page) illustrates the relationship between these two sets of standards. Specifically, the RDA for a particular nutrient is the daily dietary intake level considered sufficient to meet the nutrient requirements of nearly all (about 97%) healthy individuals of a specified life stage and sex. In other words, if all the people in a certain population group consume the RDA, 97% will get enough. Consider selenium as an example. The EAR for selenium in adult men is 45 µg/day; the RDA in this group is 55 µg/day. As such, it is recommended that all adult males try to consume 55 µg/day of this nutrient. In this way, the RDAs include a built-in safety margin to help ensure adequate nutrient intake at an individual level.

Recommended Dietary Allowance (RDA)
The average intake of a nutrient thought to meet the nutrient requirements of nearly all (97%) healthy people in a specified life stage and sex.

It is important to understand that, unless specifically noted, RDAs do not distinguish between whether the nutrient is found naturally in foods, is added to foods, or is consumed in supplement form. Also, because RDA values could only be established for nutrients that had EAR values, not all nutrients have RDAs. For the other nutrients, another set of standards—called the Adequate Intake (AI) levels—was established to provide provisional intake goals until more rigorous scientific studies allow EARs and RDAs to be developed.

ADEQUATE INTAKE (AI) LEVELS WERE SET WHEN DATA WERE LACKING FOR EARs

When scientific evidence was insufficient to establish an EAR and thus accurately set an RDA, the DRI committee derived an **Adequate Intake (AI) level** instead. As such, the existence of an AI instead of an RDA for a nutrient tells you that more research is needed. Like the RDAs, AIs are meant to be used as nutrient intake goals for *individuals*. But, unlike the RDAs, which were based on rigorous scientific studies, the AIs were based on intake levels that seem to maintain adequate nutritional status in healthy people. An example of a nutrient with an AI instead of an RDA is sodium. This may surprise you, because scientists and health care professionals have long known that sodium is essential for maintaining healthy fluid volume within the body. However, more research is required before EARs and RDAs can be established for this mineral. Like the RDAs, AI levels are thought to exceed true nutrient requirements for most people. However, in the case of the AIs, the extent to which the value exceeds the actual mean requirement is not known.

Because rigorous studies cannot ethically be done on young infants (0 to 6 months of age), there are no RDAs for this life-stage group. Instead, there are only AIs, and these were based on the average daily nutrient intakes by breastfed babies. Breastfeeding is considered the ideal way to feed a baby, so these nutrient intakes are assumed to be adequate. You can see which nutrients have AI versus RDA values in each life-stage group by examining the DRI tables inside the front cover of this book as well as Table 2.1.

Together, EAR, RDA, and AI values set goals to help us consume nutrients in sufficient quantities to support health. Conversely, another set of values—the Tolerable Upper Intake Levels (UL)—helps us avoid consuming nutrients in such large quantities that they actually do harm. These are described next.

TOLERABLE UPPER INTAKE LEVELS (ULs) REFLECT SAFE MAXIMAL INTAKES

The RDA and AI values have been established to prevent deficiencies and promote optimal health. However, avoiding nutrient overconsumption and toxicity is also important. To help prevent such excesses, **Tolerable Upper Intake Levels (ULs)** have been established as the highest level of usual daily nutrient intake likely to be safe. For instance, the UL for selenium in men is 400 µg/day. This means that, for a typical college-age male, consuming 400 µg/day of this mineral is not likely to cause problems, although exceeding this amount might lead to selenium toxicity.

The ULs are not to be used as goals for dietary intake. Instead, they were developed to provide limits for those who take supplements or consume large amounts of fortified foods. Such limits are necessary because some nutrients are harmful at very high levels. Note that scientific data are insufficient to allow the establishment of UL values for all nutrients. However, the lack of a UL for a particular nutrient does not automatically mean that high intakes are safe but instead indicates a need for continued caution when consuming high levels.

Adequate Intake (AI) level Nutrient intake of healthy populations that appears to support adequate nutritional status; established when RDAs cannot be determined.

Tolerable Upper Intake Level (UL) The highest level of chronic intake of a nutrient thought to be not detrimental to health.

FIGURE 2.4 Using EARs, RDAs, and ULs to Assess Dietary Adequacy In general, dietary intake is considered adequate if it is between the RDA and the UL. If only AI values (not shown here) are available, you can assume your intake is probably adequate if it exceeds the AI value.

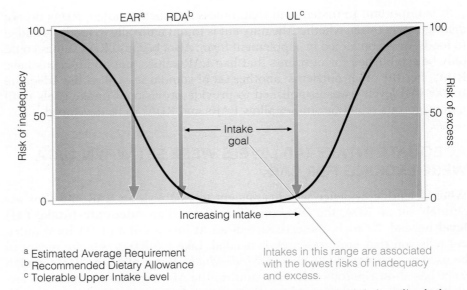

a Estimated Average Requirement
b Recommended Dietary Allowance
c Tolerable Upper Intake Level

Intakes in this range are associated with the lowest risks of inadequacy and excess.

SOURCE: Adapted from the Institute of Medicine. Dietary reference intakes for energy, carbohydrates, fiber, fat, fatty acids, cholesterol, protein, and amino acids. Washington, DC: National Academies Press; 2005.

Using EARs, RDAs, AIs, and ULs to Assess Your Nutrient Intake Now that you understand the four sets of standards included in the DRIs, you can use this information to assess your nutrient intake. Luckily, there are some relatively simple guidelines you can use to make inferences about your diet by comparing the results of your dietary assessment with your collective EAR, RDA, AI, and UL values. These concepts are illustrated in Figure 2.4 and are listed below.

When EARs, RDAs, and ULs have been established

- If your intake of a nutrient is much less than your EAR, then your intake is likely to be inadequate—increasing your risk of nutrient deficiency.
- If your intake is between the EAR and the RDA, then you should probably increase your intake.
- If your intake is between the RDA and the UL, then your intake is probably adequate.
- If your intake is above the UL, then your intake is probably too high.

When only AIs are available

- If your intake of a nutrient falls between the AI and UL, then your intake is probably adequate.
- If your intake falls below your AI, no conclusion can be made concerning the adequacy of your diet.

For example, consider a 20-year-old female who, upon completing a food record and dietary assessment, learns that her vitamin A intake is 1,500 μg/day. Because this value falls between the RDA (700 μg/day) and the UL (3,000 μg/day), she should conclude that her vitamin A intake is probably adequate. Had her vitamin A intake been 600 μg/day, it would have been between her EAR (500 μg/day) and the RDA (700 μg/day), and she should probably consume more. Had it been 3,500 μg/day (above her UL), then she should have concluded that her intake was too high.

ENERGY INTAKE CAN ALSO BE EVALUATED

So far you have learned about how to use the DRIs to assess your nutritional status in terms of the micronutrients and macronutrients. However, it is also important to determine whether you are consuming the right amount of calories and whether these calories are coming from the right mix of foods. To address this issue, the DRI publications—in addition to including the EAR, RDA, AI, and UL values—also include two types of standards and goals that you can use to assess your energy intake: the Estimated Energy Requirements (EERs) and the Acceptable Macronutrient Distribution Ranges (AMDRs).[2]

The **Estimated Energy Requirements (EERs)** are similar in theory and application to the EARs and can be thought of in the same way. As such, these values represent the average energy intakes needed to maintain weight (or grow, if appropriate for the life-stage group) in a healthy person of a particular age, sex, weight, height, and physical activity level. Note that this is different from the other DRI reference values, which only consider life stage and sex.

EERs are calculated using mathematical equations, and you can see EER equations for adult men and women of healthy weight below. Additional EER equations for other life-stage groups can be found in Appendix B.

- Adult man: EER $= 662 - [9.53 \times \text{age (y)}] + \text{PA} \times [15.91 \times \text{wt (kg)} + 539.6 \times \text{ht (m)}]$
- Adult woman: EER $= 354 - [6.91 \times \text{age (y)}] + \text{PA} \times [9.36 \times \text{wt (kg)} + 726 \times \text{ht (m)}]$

In this equation, "PA" refers to physical activity, which is categorized as sedentary, low active, active, or very active. The lower your PA value, the less active you are, and the lower your EER. Table 2.2 provides examples of these activity categories and their corresponding PA values, and Figure 2.5 illustrates how EERs are influenced by age and activity level. As you can see, at every age active individuals need more energy than do their sedentary counterparts.

Calculating your EER is not difficult, as it requires only that you insert the appropriate information concerning your age, PA value, weight, and height into the correct equation. Here is an example. *Kyung-Soon is a 38-year-old woman who weighs 115 pounds, is 5 feet 4 inches (5.3 feet) tall, and has a low activity level.* Remember that you can convert pounds to kilograms by dividing by 2.2 and feet to meters by dividing by 3.3. Kyung-Soon's EER is calculated as follows.

- EER $= 354 - [6.91 \times \text{age (y)}] + \text{PA} \times [9.36 \times \text{wt (kg)} + 726 \times \text{ht (m)}]$
  $= 354 - (6.91 \times 38) + 1.12 \times (9.36 \times 52.3 + 726 \times 1.6)$
  $= 354 - 262.6 + 1.12 \times (489.5 + 1161.6)$
  $= 91.4 + 1.12 \times 1651.1$
  $= 1,941 \text{ kcal}$

Estimated Energy Requirement (EER)
Average energy intake required to maintain energy balance in healthy individuals based on sex, age, physical activity level, weight, and height.

TABLE 2.2 Physical Activity (PA) Categories and Values[a]

Activity Level Category	Physical Activity (PA) Value		Description
	Men	Women	
Sedentary	1.00	1.00	No physical activity aside from that needed for independent living
Low active	1.11	1.12	1.5 to 3 miles/day at 2 to 4 miles/hour in addition to the light activity associated with typical day-to-day life
Active	1.25	1.27	3 to 10 miles/day at 2 to 4 miles/hour in addition to the light activity associated with typical day-to-day life
Very active	1.48	1.45	10 or more miles/day at 2 to 4 miles/hour in addition to the light activity associated with typical day-to-day life

[a] Please note that these values only apply to normal weight, nonpregnant, nonlactating adults. Values for children, pregnant or lactating women, and overweight or obese individuals are different.

SOURCE: Institute of Medicine. Dietary reference intakes for energy, carbohydrate, fiber, fat, fatty acids, cholesterol, protein, and amino acids. Washington, DC: National Academies Press; 2005.

FIGURE 2.5 Effects of Age and Activity Level on Estimated Energy Requirements (EERs) In both males and females, choosing an active lifestyle leads to increased energy requirements throughout the life cycle.

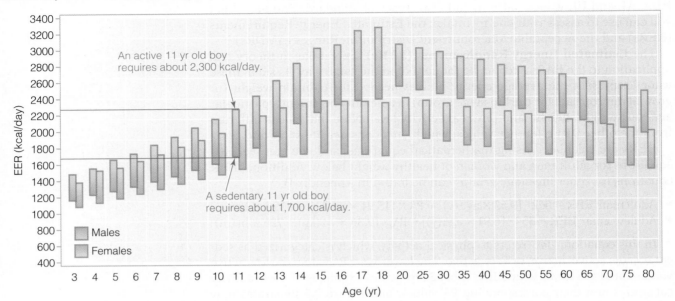

We strongly encourage you to take the time right now to determine your own EER. This should only take a few minutes. Knowing how much energy you require on a daily basis will help you immediately apply much of what you will learn from this course to your own life.

Acceptable Macronutrient Distribution Ranges (AMDRs) Aside from knowing if you are consuming the right *amount* of total calories, it is also important to know whether your *distribution* of energy sources—carbohydrates, proteins, and fats—is healthy. To answer this question, the **Acceptable Macronutrient Distribution Ranges (AMDRs)** were published along with the EERs. The AMDRs reflect the ranges of intakes for each class of energy-yielding nutrient associated with reduced risk for chronic disease while providing adequate intakes of essential micronutrients. The AMDRs are expressed as percentages of total energy intake, and they are listed next as well as inside the front cover of your book for easy future reference.

Acceptable Macronutrient Distribution Ranges (AMDRs)
- Carbohydrates: 45 to 65% of total energy
- Protein: 10 to 35% of total energy
- Fat: 20 to 35% of total energy

For example, if your EER is 2,400 kcal/day, you should be getting 1,080 (0.45 × 2,400) to 1,560 (0.65 × 2,400) kcal/day from the carbohydrates in your diet. Again, we encourage you to determine how many calories you should be getting from each group of energy-yielding macronutrients, as this information is helpful as you continue to study the science of nutrition and how it applies to your own health.

How Can You Easily Assess and Plan Your Diet?

Completing a dietary assessment using a food record along with DRI, EER, and AMDR values can help you determine whether your nutrient and energy intakes are likely adequate. However, this process can be rather cumbersome and is therefore usually only done in research settings or when nutrient inadequacies

Acceptable Macronutrient Distribution Ranges (AMDRs) Recommendations concerning the distribution or percentages of energy from each of the macronutrient groups.

are suspected. Luckily, there are numerous tools available for simplifying this process. In addition, many reputable organizations have streamlined dietary assessment even further in that they have summarized what a "healthy diet" generally looks like. In other words, they have attempted to answer the question *"What kinds of foods should I eat, and how much of them should I eat?"* by formulating various types of nutritional guidelines. These include the Dietary Guidelines for Americans and the MyPlate food guidance system. This section discusses these important dietary tools and shows you how to use them to determine more easily if the foods you eat are likely optimizing your health.

FOOD GUIDANCE SYSTEMS HAVE BEEN PART OF DIETARY PLANNING FOR DECADES

Several government agencies are involved in ensuring the nation's health and in fact have been providing information concerning nutrition and health for more than a century. One such agency is the USDA. As far back as 1894, the USDA published its first set of nutritional recommendations for Americans. Since that time, the USDA has continued to make recommendations to the U.S. public, and some of these publications are shown in Figure 2.6.[3] Although these first recommendations were limited and referred mostly to total calorie and protein intake, they provided the groundwork from which all other nutrient guidelines would later be developed.

Since 1894, there has been a succession of federally supported recommendations designed to provide guidelines for dietary planning.[4] Generally known as the **USDA Food Guides** (now called the USDA Food Patterns), these dietary recommendations combine nutritionally similar foods into "food groups" and make suggestions as to how much we should eat from each group. For example, poultry, meat, seafood, and legumes (e.g., dried beans) have been generally grouped together because they are all excellent sources of protein. Similarly, dairy products are grouped together partly because they are good sources of calcium. The composition and number of "food groups" has changed over the years as science has evolved and socioeconomic times have changed. For instance, as you can see in Figure 2.6, during the Depression,

USDA Food Guides and Food Patterns
Dietary recommendations developed by the USDA based on categorizing foods into "food groups."

FIGURE 2.6 Dietary Guidance Has Evolved The U.S. Department of Agriculture (USDA) has published many types of food-related recommendations during the past century.

1902

Depression era

Basic (1956–1970s)

SOURCE: USDA, at http://www.usda.gov

food was scarce and malnutrition was common. In response, the focus of the USDA's Food Guide was to recommend relatively inexpensive, high-fat foods such as milk, peanuts, and cheese that would provide maximal energy and nutrients at minimal cost. In contrast, in 1956, when the "Basic 4" food groups were unveiled, the USDA's Food Guide included a bread/cereal group and a vegetable/fruit group. This reflected both the positive economic situation and the growing evidence that these types of foods provided important vitamins and minerals to the diet. As you will soon learn, the current food guide has five food groups, each contributing an important set of nutrients.

In 1980 the U.S. Department of Health and Human Services (DHHS) and the USDA introduced an expanded form of dietary recommendations called the **Dietary Guidelines for Americans,** which provided specific advice about how good dietary habits can promote health and reduce risk for major chronic disease. These guidelines are now revised every five years to reflect new scientific information. Note that, in addition to giving overarching advice concerning healthy eating, the Dietary Guidelines for Americans now contains the USDA's Food Guide, recently renamed "Food Patterns." The current version of the Dietary Guidelines and its associated Food Patterns were published in 2010. Indeed, these guidelines are designed to evolve as our understanding of nutrition and health promotion advances, and you can follow these changes on the USDA website (http://www.usda.gov).

To help convey the major messages in the Dietary Guidelines and Food Patterns to the public, the USDA also provides an image and website designed to illustrate and demonstrate their important components. Currently, this graphic is called **MyPlate,** which provides visual, hands-on, practical, and user-friendly information helping to answer the important questions *"Is my diet adequate?"* and *"What can I do to make it better?"* In the next section, you will learn about the current Dietary Guidelines, Food Patterns, and MyPlate and how to use them to both assess and plan your diet.

2010 Dietary Guidelines for Americans: Our Current Recommendations

The 2010 Dietary Guidelines for Americans were developed by a highly esteemed committee of nutrition scientists who systematically evaluated the published scientific literature and formulated recommendations regarding dietary and physical activity patterns to optimize the health of the U.S. population. To do this, they had to keep in mind two important and sometimes competing facts:

- Poor diet and physical inactivity are the most important factors contributing to an epidemic of overweight and obesity in all segments of the society, and
- Nearly 15% of American households are currently unable to acquire adequate food to meet their dietary needs.[5]

In addition, the committee recognized that, while obesity has become the nation's leading health threat, many Americans (even those who are overweight or obese) have less-than-optimal intake of certain nutrients. This is true even among those with adequate financial resources.

The overarching goals of the 2010 Dietary Guidelines for Americans were to help individuals (1) maintain energy balance over time to achieve and sustain a healthy weight, and (2) focus on consuming nutrient-dense foods and beverages. Underlying these broad goals was the premise that nutrient needs should be met primarily through consuming foods (not supplements). The 2010 Dietary Guidelines for Americans encompass four groups of "key recommendations," which are listed in Table 2.3 and discussed next.

Dietary Guidelines for Americans Dietary recommendations that give specific nutritional guidance to individuals as well as advice about physical activity, alcohol intake, and food safety.

MyPlate Graphic representation of the major concepts of the Dietary Guidelines for Americans; interactive website available at www.choosemyplate.gov.

TABLE 2.3 2010 Dietary Guidelines for Americans—Major Concepts and Key Recommendations

Major Concepts	Key Recommendations
Balance Calories to Manage Weight	• Prevent and/or reduce overweight and obesity through improved eating and physical activity behaviors. • Control total calorie intake to manage body weight. For people who are overweight or obese, this will mean consuming fewer calories from foods and beverages. • Increase physical activity and reduce time spent in sedentary behaviors. • Maintain appropriate calorie balance during each stage of life.
Reduce Certain Foods and Food Components	• Reduce daily sodium intake to less than 2,300 mg/day and further reduce intake to 1,500 mg/day among individuals with increased risk for hypertension. • Consume less than 10% of calories from saturated fats by replacing them with monounsaturated and polyunsaturated fats. • Consume less than 300 mg/day of dietary cholesterol. • Keep *trans* fatty acid consumption as low as possible. • Reduce intake of calories from solid fats (saturated and *trans* fats) and added sugars. • Limit consumption of foods that contain refined grains. • If you consume alcohol, consume it in moderation—up to one drink per day for women and two drinks per day for men; only adults of legal drinking age should drink alcoholic beverages.
Increase Certain Foods and Nutrients[1]	• Increase vegetable and fruit intake. • Eat a variety of vegetables, especially peas and beans and vegetables that are dark green, red, or orange. • Consume at least half of all grains as whole grains. • Increase intake of fat-free or low-fat milk and milk products or fortified soy beverages. • Choose a variety of protein foods, including seafood, lean meat and poultry, eggs, beans and peas, soy products, and unsalted nuts and seeds. • Replace protein foods high in solid fats with choices that are lower in solid fats and calories. • Choose foods that provide more potassium, dietary fiber, calcium, and vitamin D, such as vegetables, fruits, whole grains, and dairy products.
Build Healthy Eating Patterns	• Select an eating pattern that meets nutrient needs over time at an appropriate calorie level. • Account for all foods and beverages consumed and assess how they fit within a total healthy eating pattern. • Follow food safety recommendations when preparing and eating foods to reduce the risk of foodborne illness.

[1] Additional recommendations are also made for specific population groups. For example, women capable of becoming pregnant are advised to choose foods that supply heme iron, additional iron sources, and enhancers of iron absorption (e.g., vitamin C). It is also recommended that they consume synthetic folic acid (from foods or supplements) in addition to foods high in folate. Women who are pregnant or breastfeeding are advised to consume 8 to 12 oz seafood per week, being careful to avoid those types known to be high in mercury. Individuals who are 50 years or older should consume foods fortified with vitamin B_{12} or a dietary supplement.

SOURCE: U.S. Department of Agriculture and U.S. Department of Health and Human Services. Dietary guidelines for Americans, 2010. 7th edition, Washington, DC: US Government Printing Office; December 2010.

BALANCE CALORIES TO MANAGE WEIGHT

Achieving and sustaining appropriate body weight throughout life is vital to maintaining good health. Simply put, it is important to balance the number of calories you consume with those you expend—otherwise, weight gain and its associated health risks will surely ensue, especially as you age. Unfortunately, because of a multitude of lifestyle and environmental factors, energy intake often exceeds energy expenditure—resulting in what has been called an "obesity epidemic." In response, several recommendations related to calorie balance (energy consumed relative to energy spent) are put forth in the 2010 Dietary Guidelines. For instance, recognizing that all people—regardless of age—can become obese, the Dietary Guidelines recommend that we maintain appropriate calorie balance *during each stage of life*. To help actualize these guidelines, the following strategies are suggested.

• Focus especially on *total number* of calories consumed.
• Monitor food intake, body weight, and physical activity so that changes that might lead to poor health can be detected.
• When eating out, choose smaller portions or lower-calorie options.
• Prepare, serve, and consume smaller portions of foods and beverages, especially those high in calories.

- Eat a nutrient-dense breakfast—in other words, one that has maximal nutrients for each calorie.
- Limit "screen time," such as television watching and computer gaming, to no more than one to two hours each day.

Know Your Energy Requirements Clearly, a main focus of the 2010 Dietary Guidelines is to reverse current obesity trends. To determine personal energy requirements, calorie consumption must be compared with calorie needs. The Estimated Energy Requirement (EER) equations, published as part of the DRIs, are designed perfectly for this purpose. Once energy requirements have been determined using the appropriate EER equation, a person can either (1) aim to eat that amount to maintain weight, or (2) consume fewer calories (e.g., 500 kcal/day less) to lose weight. The Dietary Guidelines recommend that individuals control their calorie intake by increasing intake of whole grains, vegetables, and fruits; reducing intake of sugar-sweetened beverages; and monitoring calorie intake from alcoholic beverages.

REDUCE CERTAIN FOODS AND FOOD COMPONENTS

Whereas the first group of recommendations put forth in the 2010 Dietary Guidelines is related to weight maintenance, the second focuses on reducing intakes of foods and food components known to increase risk for chronic degenerative disease. Following these recommendations will also increase the likelihood of meeting nutritional needs without exceeding energy requirements. In particular, the Dietary Guidelines recommend that we reduce our intakes of sodium, saturated fats, cholesterol, *trans* fatty acids, solid fats (saturated fats + *trans* fats), added sugars, and refined grains. Suggested strategies to meet these goals are listed next.

- Read Nutrition Facts labels for information on the sodium content of foods and purchase foods that are low in sodium.
- Consume more fresh foods and fewer processed foods that are high in sodium.
- Eat more home-prepared foods and use little or no salt when cooking or eating.
- When eating at restaurants, ask that salt not be added to your food or order low-sodium options.
- Focus on eating the most nutrient-dense forms of foods from all food groups.
- Limit the amount of solid fats and added sugars when cooking or eating out; for example, trim fat from meat, use less butter and margarine, and use less table sugar.
- Consume fewer and smaller portions of foods and beverages that contain solid fats or added sugars, such as grain-based desserts, sodas, and other sugar-sweetened beverages.

It is worth noting that several of these guidelines (such as those for sodium) are drawn directly from the DRI values; in this case, the sodium recommendations are based on the UL values. This is a great example of how national nutrition recommendations and guidelines frequently overlap, with each one drawing on the others.

INCREASE CERTAIN FOODS AND NUTRIENTS

The Dietary Guidelines also identify several "**nutrients of concern**" that are known to be somewhat lacking in the American diet. These include potassium, dietary fiber, calcium, and vitamin D. Intakes of iron, folate, and vitamin B_{12} are also of concern for specific populations. Similarly, there are some especially wholesome and healthy foods (such as seafood, beans and

nutrients of concern Those nutrients, identified in the 2010 Dietary Guidelines for Americans, that are somewhat lacking in the typical U.S. diet.

Working Toward the Goal: Maximizing Nutrient Intake by Increasing Fruit and Vegetable Intake

The 2010 Dietary Guidelines recommend that we maximize our nutrient intake by choosing a variety of fruits and vegetables each day, while not consuming too many calories. The following selection and preparation tips will help you meet these goals. For students who live in dormitories, remember that you can always ask your food service provider to consider helping you meet these goals.

- Buy a variety of fruits and vegetables that are easy to prepare. For example, one week buy packages of precut melon. The next, purchase precut green peppers for a healthy snack in seconds. In other words, mix it up, and make it convenient!

- Consider frozen juice bars (100% juice) or concentrate as healthy alternatives to high-fat snacks and sugary drinks, and make sure you do not always buy the same variety.

- Make a Waldorf salad—which contains raisins, apples, celery, walnuts, and dressing—or combine fruits with low-fat vanilla yogurt for a nice alternative to a lettuce salad.

- Add an interesting assortment of fruits and vegetables such as pineapples, peaches, onions, and eggplants to kabobs as part of a barbecue meal.

- Keep a package of dried fruit or vegetable chips in your desk or bag.

- As a snack, spread peanut butter on apple slices or top frozen yogurt with berries or slices of kiwi fruit.

- Plan some meals around a mixed-vegetable main dish, such as a vegetable stir-fry or hearty vegetable soup.

- Shred carrots or zucchini into meatloaf, casseroles, quick breads, and muffins.

- Include a combination of chopped vegetables in pasta sauce or lasagna.

- Add diced fruits, such as apples, pears, and blood oranges, to a lettuce salad to add both variety and additional nutrients.

- Keep a bowl of cut-up vegetables and a bowl of fruit in a transparent container in the refrigerator, and alternate what types of vegetables and fruit you have on hand.

peas, whole-grains, and low-fat dairy products) of which Americans would benefit from eating more. In response, the Dietary Guidelines include several key recommendations aimed at increasing the nation's consumption of these important foods and food components. Of course, we must always remember that these goals should be met as part of a healthy eating pattern while staying within our calorie needs.

BUILD HEALTHY EATING PATTERNS

The 2010 Dietary Guidelines also emphasize the important concept that there are many ways that individuals and families can incorporate nutrient-related recommendations into their lives. In other words, there is more than one healthful eating pattern—especially when considering the incredible diversity of cultural, ethnic, traditional, and personal preferences in the United States. Food costs and availability are also critical pieces of the puzzle for many people when determining what they will (and will not) eat. To help consumers of all ethnicities, traditions, preferences, and socioeconomic levels choose a healthy dietary pattern, the Dietary Guidelines put forth three key recommendations related to overall meal planning and preparation (Table 2.3). In general, these recommendations emphasize the importance of considering each food and beverage that you consume in terms of balancing nutrient and energy consumption. Keeping foods safe to eat is also highlighted.

Establishing the Current Food Patterns It is helpful for people to have information about what types of overall food intake *patterns* tend to be healthy. As such, the 2010 Dietary Guidelines describe an array of eating patterns that have been scientifically shown to impart optimal health, supply all of the essential nutrients, and limit calories. Importantly, the Dietary Guidelines provide specific recommendations about what kinds of foods you should consume and

Consuming a nutrient-dense diet allows you to have more flexibility in eating high-fat foods without gaining weight.

in what proportions you should consume them. For instance, one of the key recommendations is to "eat a variety of vegetables, especially peas and beans and vegetables that are dark green, red, or orange."

But how many servings of these different types of vegetables do you actually need? Answering this question forms the basis of the accompanying Food Patterns list, which currently includes five food groups: vegetables, fruits, grains, dairy products, and protein foods. There are also subcategories within some of the food groups. For example, the following list represents the important vegetable subgroups: dark green (e.g., broccoli, spinach); red and orange (e.g., tomatoes, pumpkin); beans and peas (e.g., kidney beans, lentils); starchy (e.g., white potatoes, corn); and "other" (e.g., iceberg lettuce, onions). The food patterns also include an allowance for oils and limits on the maximum number of calories that should be consumed from solid fats and added sugars.

There are 12 different Food Patterns based on caloric needs, and as you might guess, the more calories you need, the more food you should eat from each of the food groups (Table 2.4). For instance, people requiring 2,000 kcal/day are urged to eat 2½ cups of vegetables each day, whereas those requiring 2,400 kcal/day should strive to eat 3 cups daily.

Focus on Nutrient Density The Food Patterns emphasize selection of most foods in nutrient-dense forms—that is, with little or no solid fats and added sugars. **Nutrient-dense** foods and beverages provide maximal amounts of nutrients for minimal number of calories. In other words, nutrient-dense foods are the opposite of "junk foods," which provide calories but few nutrients. Junk foods are also sometimes said to provide "empty calories." Ideally, nutrient-dense foods are those that retain their naturally occurring components, such as dietary fiber. Unless they are altered during preparation, all vegetables, fruits, whole grains, seafood, eggs, beans and peas, unsalted nuts and seeds, fat-free and low-fat milk and milk products, and lean meats and poultry are considered to be nutrient-dense foods. A comparison of high- and low-nutrient–dense foods can be found in Figure 2.7.

Remember that Beverages Count The current Dietary Guidelines pay special attention to reminding us that we need to consider everything we eat *and drink* when we are considering how closely our diets match the recommended

nutrient density The relative ratio of nutrients in a food in comparison to total calories.

TABLE 2.4 Amounts of Each Food Group Recommended by the USDA Food Guide and MyPlate

Food Category	Amounts Recommended (per day)[a]	Dietary Significance
Grains	3–10 oz (or equivalent)	Major sources of B vitamins, iron, magnesium, selenium, energy, and dietary fiber
Vegetables	1–4 cups	Rich sources of potassium; vitamins A, E, and C; folate; and dietary fiber
Fruits	1–2½ cups	Rich sources of folate, vitamins A and C, potassium, and fiber
Dairy products	2–3 cups	Major sources of calcium, potassium, vitamin D, and protein
Protein foods	2–7 oz (or equivalent)	Rich sources of protein, magnesium, iron, zinc, B vitamins, vitamin D, energy, and potassium

[a] Amounts depend on age, sex, and physical activity level. Personalized recommendations can be generated by going to the MyPlate website (http://www.choosemyplate.gov).

FIGURE 2.7 Examples of the Calories in Related Nutrient-Dense and Non–Nutrient-Dense Food Choices Many foods come in a variety of forms—for instance, milk can be fat free or whole. Choosing the lower-fat or lower-sugar option will increase the nutrient density of your diet.

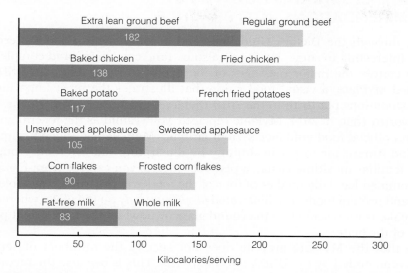

■ Calories in nutrient-dense form of the food ■ Additional calories in food with lower nutrient densities

SOURCE: Adapted from U.S. Department of Agriculture and U.S. Department of Health and Human Services. Dietary guidelines for Americans, 2010. 7th edition. Washington, DC: US Government Printing Office; December 2010.

dietary patterns. Beverages contribute substantially to calorie intake for most Americans. However, although they provide water and some are good sources of important nutrients, many are not nutrient dense—they often supply "empty calories" and added sugars. The beverages most commonly consumed by U.S. adults are regular soda, energy and sports drinks, and alcoholic varieties. These low-nutrient–dense drinks supply almost 120 kcal to the diet every day while contributing little or no additional nutritional value in terms of vitamins, minerals, or other important substances. Similar trends are found in children. The Dietary Guidelines recommend that individuals drink water and other beverages with few or no calories, in addition to recommended amounts of low-fat or fat-free milk and 100% fruit juices.

HELPING AMERICANS MAKE HEALTHY CHOICES

In addition to putting forth key recommendations as to how *individuals* should choose their foods for optimal health, the 2010 Dietary Guidelines go one step further by acknowledging that "everyone has a role in the movement to make Americans healthy." Specifically, we are reminded that by working together through policies, programs, and partnerships, our nation can improve the health of the current generation and take responsibility for ensuring better health for future generations. Although we all make our own food and physical activity choices, it is important for us all to have access to the best choices possible. As such, the 2010 Dietary Guidelines' "Call to Action" includes the following guiding principles.

- Ensure that all Americans have access to nutritious foods and opportunities for physical activity.
- Facilitate individual behavior change through environmental strategies.
- Set the stage for lifelong healthy eating, physical activity, and weight management behaviors.

For these principles to be realized, individuals will need to work cooperatively with local, state, and federal groups to make sure that healthy foods and activity options are both available *and* desirable to all. An example of the strategies put forth to meet these goals includes provision of comprehensive health, nutrition, and physical education programs in schools.

MYPLATE ILLUSTRATES HOW TO PUT RECOMMENDATIONS INTO PRACTICE

Wading through the Dietary Guidelines and accompanying Food Patterns may be interesting to some, but it can also be time consuming and complex. To help convey the major messages of its reports to the public, the USDA developed MyPlate, a visual food guide that illustrates the most important recommendations put forth in the 2010 Dietary Guidelines for Americans. It is noteworthy that, in 2011, MyPlate replaced MyPyramid as the federal government's official food guidance system graphic. MyPlate is designed simply to remind Americans to eat healthfully by illustrating the five food groups using a familiar mealtime visual: a place setting. As you can see from Figure 2.8, recommended daily intakes of four of the food groups (fruits, vegetables, grains, and protein foods) are illustrated *proportionately* on the plate, whereas dairy intake is represented by the round glass or bowl on the upper-right periphery of the plate.

Note that the MyPlate graphic does not specify the numbers of servings recommended by the USDA Food Patterns. This is because the recommended amount of each food group depends on your age, sex, and physical activity level, and the creators of MyPlate wanted to encourage each person to determine how much food is needed on an individual basis. To do this, you must log on to the MyPlate website (http://www.choosemyplate. gov) and provide personal information (age, sex, and physical activity level) under the section called "Daily Food Plan" found within the interactive tools. Once you have provided your own personal information, the website will generate specific recommendations on food-intake patterns, serving sizes, and menu selection. In all, there are 12 different sets of recommended dietary patterns for the general population. This corresponds

FIGURE 2.8 USDA MyPlate Icon MyPlate graphically illustrates the USDA's recommended food consumption pattern as a consumer-friendly icon.

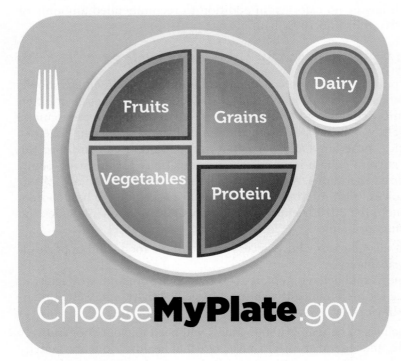

SOURCE: USDA, http://www.usda.gov

My Daily Food Plan

Based on the information you provided, this is your daily recommended amount for each food group.

FIGURE 2.9 An Example of a Personalized Food Plan There are 12 different food plans in the MyPlate food guidance system, each based on a particular energy requirement. That shown here is for an 1,800-kcal/day diet.

GRAINS 6 ounces	VEGETABLES 2 1/2 cups	FRUITS 1 1/2 cups	DAIRY 3 cups	PROTEIN FOODS 5 ounces
Make half your grains whole Aim for at least **3 ounces** of whole grains a day	**Vary your veggies** Aim for these amounts each week: **Dark green veggies** = 1 1/2 cups **Red & orange veggies** = 5 1/2 cups **Beans & peas** = 1 1/2 cups **Starchy veggies** = 5 cups **Other veggies** = 4 cups	**Focus on fruits** Eat a variety of fruit Choose whole or cut-up fruits more often than fruit juice	**Get your calcium-rich foods** Drink fat-free or low-fat (1%) milk, for the same amount of calcium and other nutrients as whole milk, but less fat and Calories Select fat-free or low-fat yogurt and cheese, or try calcium-fortified soy products	**Go lean with protein** Twice a week, make seafood the protein on your plate Vary your protein routine— choose beans, peas, nuts, and seeds more often Keep meat and poultry portions small and lean

Find your balance between food and physical activity

Be physically active for at least **150 minutes** each week.

Know your limits on fats, sugars, and sodium

Your allowance for oils is **5 teaspoons** a day.
Limit Calories from solid fats and added sugars to **160 Calories** a day.
Reduce sodium intake to less than **2300 mg** a day.

Your results are based on a 1800 Calorie pattern.   Name: _____

This Calorie level is only an estimate of your needs. Monitor your body weight to see if you need to adjust your Calorie intake.

SOURCE: USDA, http://www.usda.gov

to the 12 different energy-intake levels listed in the Food Patterns on which MyPlate was based. An example of a personalized food plan, which was created for a 20-year-old woman with low physical activity, is provided in Figure 2.9. Food plans can also be generated for pregnant and breastfeeding women by clicking on "For Moms."

The MyPlate website also provides in-depth descriptions of types and amounts of foods that fit into each food group, as well as specific information about meal planning and special subgroups (such as vegetarians). For example, you can see how a meal or entire day's menu contributes to your MyPlate recommendations by clicking on the "Menu Planner" option. Nutrient composition and serving sizes of individual foods can be found in "MyFood-a-pedia." Information on meeting your physical activity recommendations and Let's Move!, a federally funded initiative designed to combat childhood obesity, is also available on the MyPlate website.[§]

Key Themes and Messages of the MyPlate Symbol There are four basic themes of the MyPlate website and its components. These themes are to (1) build a healthy plate, (2) cut back on foods high in solid fats, added sugars, and salt, (3) eat the right amount of calories for you, and (4) be physically active "your way." As such, MyPlate is much like its predecessor MyPyramid, which emphasized similar overarching goals. However, unlike previous food guidance icons, MyPlate is part of a much larger communications initiative to help U.S. consumers make better food choices. This initiative involves strategic partnering between the USDA and dozens of public and private-sector groups to help promote and amplify seven key consumer messages based on MyPlate's four basic themes. These user-friendly messages are listed below.

- Make half your plate fruits and vegetables.
- Enjoy your food, but eat less.
- Drink water instead of sugary drinks.

[§]It is noteworthy that, at the time this textbook was revised, the USDA had not fully updated its MyPlate website and was planning to do so in the fall or winter of 2011. Consequently, there may be slight differences between what is written in this book and details of the MyPlate website.

- Make at least half your grains whole grains.
- Avoid oversized portions.
- Compare sodium in foods and choose the foods with lower numbers.
- Switch to fat-free or low-fat (1%) milk.

For instance, during the spring and summer of 2012, you may notice numerous advertisements and other forms of communications encouraging people to drink water instead of sugary drinks. In the winter of 2013, the coast-to-coast focus will be on choosing fat-free or low-fat milk. Throughout this nationwide initiative, the overarching theme of "be physically active your way" will be integrated within each key message. It is the hope of the USDA that this highly orchestrated and targeted communications campaign will systematically translate the basic information illustrated within the MyPlate icon onto the American dinner plate.

Using MyPlate Food Tracker for Conducting a Dietary Self-Assessment The MyPlate website also offers an excellent opportunity for you to conduct a self-assessment of your own dietary intake using the **Food Tracker** option, which allows you to assess whether your diet meets your recommended USDA Food Pattern (for example, whether you are consuming enough servings of milk) as well as recommendations put forth in the DRIs (for example, whether you are consuming your RDA for calcium). To use Food Tracker (click on "Analyze my diet" on the MyPlate homepage), you must keep track of everything you eat and drink for at least one day and enter the information into the website. Ideally, however, you should keep a food record for three days—one of which should be a weekend day. Other tips for keeping an accurate food record are listed next.

- *Detail is important.* It is important to include as much detail, such as brand names and preparation methods, as you can for all foods and beverages that are consumed. For mixed dishes (also called composite foods), estimate the amount of each ingredient. An example is "salad," which might include lettuce, tomatoes, eggs, and so on. The more detail that is included in the diet record, the more accurate the analysis can be.
- *Estimate or measure serving sizes accurately.* Ideally, weigh or measure foods using standard household devices such as measuring cups or spoons. If this is not possible, estimate serving sizes as carefully as possible. Some restaurants provide detailed information concerning the amounts of food served, and such detail can be very helpful.
- *Choose representative, "normal" days.* Choose days that are representative of typical eating patterns, and avoid "special" days such as holidays and birthdays. Because being sick or under unusual stress can influence food preferences and overall intake, it is best to avoid these times as well.
- *Do not change your normal eating patterns.* When keeping a diet record, people often alter their eating patterns to be more convenient or healthful. Resist this temptation, as it is especially important that diet records reflect normal intake.

When these guidelines are followed, Food Tracker can be a powerful tool in helping you determine whether the foods and beverages you typically choose are likely to be contributing to or deterring you from being well.

HEALTHY PEOPLE 2020 OUTLINES OUR NATION'S GOALS FOR HEALTHY LIVING

Whereas the Dietary Guidelines and MyPlate are meant to help you choose foods that promote optimal health on a personal level, there are guidelines

Food Tracker A component of the MyPlate website that allows individuals to conduct a dietary self-assessment.

that have been developed to optimize "national health." For instance, the U.S. Department of Health and Human Services has developed a comprehensive document called **Healthy People 2020,** which outlines a set of overall health objectives for the nation to achieve by the year 2020.[6] Since 1990, Healthy People documents have been published every 10 years. The goals and objectives put forth in Healthy People 2020 are meant to be used by government agencies, communities, and professional organizations to develop programs that will help improve the health of our nation. For example, the National School Lunch Program and the Supplemental Nutrition Assistance Program (SNAP)** were both altered, in response to goals put forth by the previously published Healthy People 2010 document.[7]

Briefly, Healthy People 2020 was designed to achieve four overarching health-related goals for the United States.

The goals outlined in Healthy People 2010 are used by federal and state programs such as the National School Lunch Program to help ensure the health of the nation.

- Attain high-quality, longer lives free of preventable disease, disability, injury, and premature death.
- Achieve health equity, eliminate disparities, and improve the health of all groups.
- Create social and physical environments that promote good health for all.
- Promote quality of life, healthy development, and healthy behaviors across all life stages.

Within these broad goals, there are 39 different topic areas, such as cancer; diabetes; food safety; maternal, infant, and child health; and nutrition/weight status. A corresponding goal is then identified for each of these topic areas. For example, the goal for "nutrition/weight status" is the *promotion of health and reduction of chronic disease risk through consumption of healthful diets and achievement and maintenance of healthy body weights*. These goals are then broken down further into achievable outcomes; for instance, it is hoped that the proportion of adults who are at a healthy weight will increase by 10% over the next decade.

Although it is beyond the scope of this book to discuss each topic area and goal, it is worth noting that many of them are consistent with the recommendations put forth in the Dietary Guidelines for Americans. This is no coincidence. In fact, Healthy People 2020 often refers to the Dietary Guidelines for specific detail concerning nutrition recommendations. In this way, the Dietary Guidelines help *individuals* meet their specific nutrition goals, whereas Healthy People helps policy makers assess whether our *national* health goals are met. As new Guidelines are developed, so will updated Healthy People documents. You can stay abreast of new Healthy People publications by visiting their website (www.healthypeople.gov).

The Dietary Guidelines for Americans, MyPlate, and Healthy People documents have all been developed and updated to translate the basic science underlying the DRIs into user-friendly recommendations for how Americans should choose their foods. How these guidance systems are related is illustrated in Figure 2.10. Understanding that many of our federal health programs interact to provide health and wellness information to our population is important.

There are also many federal guidelines concerning what information can or must be included on food labels. The goal of these regulations is to provide accurate, easily accessible, and understandable information to the American consumer. You will learn how you can best use food labels to choose your foods in the next section.

** Previously referred to as the Food Stamp Program.

Healthy People 2020 A federally developed document that provides overall health objectives for the nation.

FIGURE 2.10 Many Private and Federal Groups Work Together to Keep Us Healthy Development of the Dietary Guidelines for Americans relies heavily on the DRI, AMDR, and EER values as well as evolving nutrition research. The MyPlate graphic and website as well as Healthy People 2020, in turn, rely on the recommendations made in the Dietary Guidelines.

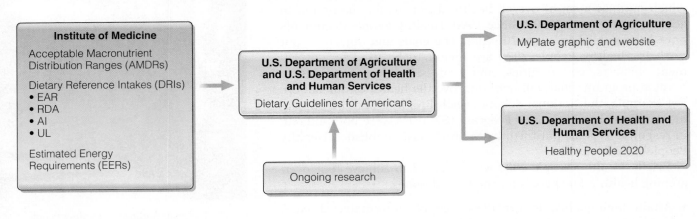

How Can You Use Food Labels to Plan a Healthy Diet?

Understanding the recommendations put forth by the Dietary Guidelines and MyPlate provides an excellent starting place for healthful eating. However, it is important to be aware of other tools that are available to help you choose the appropriate amounts of nutritious foods on a daily basis. These tools are intended to help you optimize your health by making the best food-related decisions daily. Although there are many ways you can do this, perhaps one of the most important is understanding the vast amount of information found on food packaging. Indeed, learning how to read a food label and then apply that knowledge to making healthful food choices can greatly contribute to your well-being.

UNDERSTANDING NUTRITION FACTS PANELS

Reading food labels can help you choose the most nutritious products.

You probably know that most foods have food labels, but did you know that the information on them was not always regulated? In fact, consistent nutrition labeling for foods was not established until 1973, when the U.S. Food and Drug Administration (FDA) implemented a series of rules to help consumers become aware of a food's nutrient content. Since that time, the FDA has required that most packaged foods with more than one ingredient have certain information printed on their labels. Required components of a food label include the following information.

- Product name and place of business
- Product net weight
- Product ingredient content (from most abundant to least abundant ingredient)
- Company name and address
- Country of origin
- Product code (UPC bar code)
- Product dating, if applicable
- Religious symbols, if applicable (such as kosher)

- Safe-handling instructions, if applicable (such as for raw meats)
- Special warning instructions, if applicable (such as for aspartame and peanuts)
- Nutrition Facts panel outlining specified nutrient information

The most recent addition to this list is mandating that the "country of origin" be included on food labels. This became a requirement in late 2008 and applies to most seafood, beef, veal, pork, lamb, poultry, fresh and frozen fruits and vegetables, and peanuts. Having this information assists consumers who want to avoid foods produced in certain countries or only purchase foods from nearby locations. You can also see that some of the information on a food label is not required on every food; the information is only required if it is applicable to the particular product. For example, foods are not required to state whether they are "kosher" or not. Instead only manufacturers that want to market their products as being kosher need to label their foods accordingly. You can find out more about what it means for a food to be kosher in the Focus on Food feature.

Kosher foods have been part of the Jewish heritage for centuries.

Of special interest to many people is the **Nutrition Facts panel,** which is required on most food labels. An example of a food label—including a Nutrition Facts panel—is shown in Figure 2.11. The FDA requires several elements on a Nutrition Facts panel. First, the manufacturer must include the serving size of the food. Serving sizes have been standardized so that nutrient contents of similar foods can be easily compared. For example, you can compare the amount of iron in a single serving of two similar breakfast cereals (like unsweetened wheat flakes) by simply comparing the amounts listed on their Nutrition Facts panels. In addition, the label must list the total energy (listed as Calories), total carbohydrates (including dietary fiber), sugar, and protein.

Nutrition Facts panels must also provide information concerning specific nutrients and food components that the Dietary Guidelines suggest limiting. Recall that these include total fat, saturated fat, *trans* fat, cholesterol, and sodium. Conversely, the FDA requires that information concerning dietary fiber, vitamins A and C, and the minerals iron and calcium be included on Nutrition Facts panels. This is because these are nutrients that we typically need to consume more of. In this way, food labels can help consumers choose foods that specifically meet the goals of the Dietary Guidelines.[††]

Food labels can also contain information not regulated or specified by the FDA. For example, you may have noticed that some foods are labeled as having been "produced locally." Although the FDA has not established a definition for what it means to be "produced locally," some consumers prefer to purchase foods that have not been grown and shipped from across the nation or even another country. The new requirement for including country of origin on food labels will help consumers choose only products produced in the United States. However, this will not assist the small but growing segment of the population who wishes to only purchase foods produced and manufactured in even a smaller radius. In response to this public demand, some manufacturers and retailers label locally grown or manufactured products as such. You can learn more about this trend in the Focus on Food feature.

Some people prefer to buy locally grown and prepared foods because they believe this practice to be more nutritionally, environmentally, and economically sound.

Nutrition Facts panel A required component of food packaging that contains information about the nutrient content of the food.

[††] You can learn more details about food labels by visiting the FDA website at http://www.cfsan.fda.gov/~dms/foodlab.html.

There are many traditions related to food preparation and meals that, for centuries, have helped define families, communities, and nations. Indeed, food traditions play a critical role in the lives of both individuals and cultures. One such food tradition is that of kosher foods. But why do so many people choose to eat kosher foods, and what makes a food kosher? A **kosher food** is one that has been prepared in accordance with Jewish dietary laws. These laws originated in the Old Testament of the Bible, although they have been reinterpreted over the centuries in response to changes in the food industry, the Jewish people, and world culture. In general, a kosher food satisfies Jewish food standards, some of which are described here.

- If properly prepared, meat from ruminant, cloven-hoofed mammals such as beef, deer, sheep, and goats can be kosher. However, food made from pigs or rabbits can never be kosher.

- Only foods made from certain birds can be considered kosher in the United States. This includes chickens, ducks, geese, and turkeys. Whereas some fish can be kosher, reptiles and shellfish cannot.

- Allowed meats, fish, and poultry must be slaughtered in a prescribed manner to be kosher. Only people who are trained and qualified are allowed to slaughter kosher animals.

- To be considered kosher, meat and dairy products cannot be prepared or served together.

- Processed foods (including pickles) must be prepared in a facility overseen and/or inspected by a rabbi or other qualified person.

You may wonder why Jewish people believe that they should follow these food laws. Although there is active debate about the exact reasons for these food laws, some Jewish scholars believe that these food preparation and consumption "rules" were established to help ensure ethical treatment of animals as well as food safety. However, regardless of *why* people of the Jewish faith are supposed to follow these food laws, their use indicates observance to a long-standing and rich religious history. In other words: tradition.

Products that have been certified as kosher are usually labeled with kosher symbols, some of which are shown here. Unlike the FDA-approved symbol used to indicate that a food is organic, kosher symbols are registered trademarks of various private kosher certification organizations and cannot be placed on a food label without the organization's permission.

Today millions of people from various religious and ethnic backgrounds eat kosher food for religious, cultural, health, and quality reasons. For example, Muslims may buy kosher food products because they fit the Quran's dietary laws. And people who are health-conscious may purchase a kosher food because they believe it is healthier and safer as a result of the extra supervision during its preparation. Regardless of the reason, consumption of certified kosher foods represents a commitment to a deep-rooted food tradition that has been part of human existence for centuries.

Kosher foods are often labeled with symbols representing a certification organization.

Daily Values (DVs) Although Nutrition Facts panels provide an impressive amount of nutrition information, using this information for diet planning can be somewhat challenging. For instance, a Nutrition Facts panel on a cereal box may tell you how much vitamin C is in a serving of the cereal, but how do you know if that amount is a little or a lot? To help answer this question, Nutrition Facts panels include **Daily Values (DVs),** which were created to give consumers a benchmark for knowing whether a food is a good source of a nutrient and allowing them to easily compare one food with another.

kosher food A type of food that has been prepared and served in ways that are observant of Jewish law.

Daily Value (DV) Recommended intake of a nutrient based on either a 2,000- or 2,500-kcal diet.

FIGURE 2.11 Understanding Food Labels and Nutrition Facts Panels The FDA requires food labels to have specific information such as manufacturer's name, net contents, and nutrition information. Nutrition Facts panels provide a variety of information useful in dietary planning.

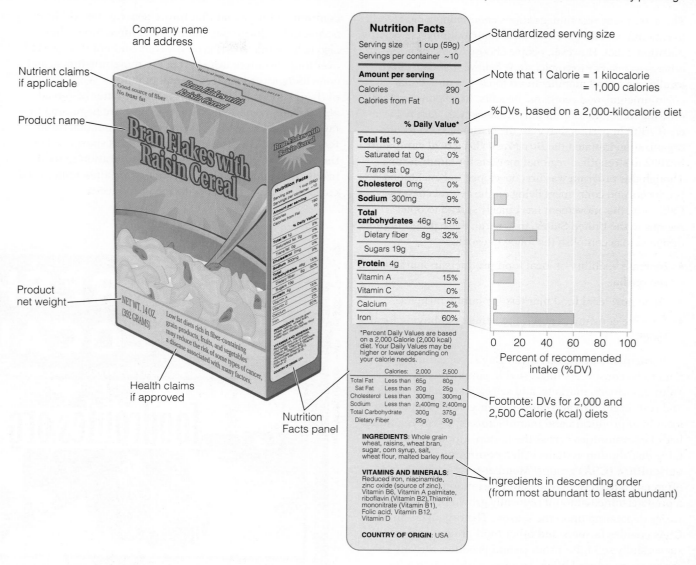

There are two basic "types" of DVs. The first type, such as that used for selected vitamins and minerals, represents recommended intake goals or reference points for people who require approximately 2,000 kcal/day. For example, the DV for vitamin C is 60 mg/day. Thus, a cereal providing 30 mg of vitamin C per serving would contain half of the DV—in other words, the food would have a %DV of 50%. You can see an example of this in Figure 2.11. By reading the Nutrition Facts label on this box of cereal, you can see that one serving provides 15, 0, 2, and 60% of an "average" person's DVs for vitamin A, vitamin C, calcium, and iron, respectively.

The other type of DV represents "upper limits" and gives daily amounts that you should try *not* to exceed. This type of DV is used for total fat, saturated fat, cholesterol, sodium, and total carbohydrates. Note from the figure that one serving of cereal would provide 2, 0, 0, 9, and 15% of what would be considered "upper limits" for these substances, respectively, for a person requiring 2,000 kcal/day. If the food's package is big enough, the actual DV values as well as %DVs for these nutrients are provided in a footnote. As you can see in the lower portion of the food label, the DV (or upper limit) for saturated fat for someone requiring 2,000 kcal is 20 g/day; this increases to 25 g/day if you need 2,500 kcal/day.

There are many recommendations concerning which foods to eat (and not eat) based on their contributions to a healthy, nutritional diet. However, people choose foods for many other reasons. For example, some believe that consuming locally grown and manufactured foods has nutritional, environmental, and economic benefits. In response to this growing trend, the Community Alliance with Family Farmers (CAFF) in partnership with the National Food Routes organization initiated the ***Buy Fresh Buy Local* campaign** in 2002 to strengthen regional markets for family farms. Though this program was first developed to benefit food producers and consumers living on the central coast of California, this movement has quickly spread to other regions in the United States. In general, these programs are designed to accomplish the following goals.

- Increase visibility of local food products in markets and the media.
- Open new local food markets for family farmers.
- Educate consumers about the benefits of buying local foods.

Participants in the *Buy Fresh Buy Local* campaigns tend to be consumers interested in supporting local agriculture, farmers, markets, and selected schools, retailers, and restaurants. To accomplish their goal of supporting local food producers and manufacturers, a growing number of communities across the nation support farm-to-table distribution systems called **community-supported agriculture (CSA)** groups. Member-supporters of CSAs often pay a fixed weekly or monthly amount in return for a predetermined amount of locally grown produce that varies depending upon the season. The organization of CSAs enables farmers and other food producers to more successfully sell their produce and products directly to the consumer. In addition, in areas with an active *Buy Fresh Buy Local* campaign, locally produced foods are often identified as such in local grocery stores and food cooperatives via logos like the one shown here.

Supporters of this "back-to-basics" food distribution system believe that it helps ensure that family farms will continue to thrive and that future generations will have access to locally grown and manufactured food. They also believe that foods grown locally and not transported over long distances taste better, are more nutritious, and contribute less to environmental pollution. Whether any of these beliefs are true, however, is a matter of considerable debate. Nonetheless, many people are beginning to favor the purchase of locally grown and manufactured foods. Regardless of whether this trend is short lived or the beginning of a more long-lasting alteration in food production and purchase patterns, it certainly unites food consumers more closely with food producers.

© Food Routes

The *Buy Fresh Buy Local* campaign encourages consumers to purchase locally produced and manufactured foods.

Buy Fresh Buy Local campaign A grass-roots movement that emphasizes purchasing foods grown, processed, and distributed locally.

community-supported agriculture (CSA) A system that connects local food growers to local consumers such that food "bundles" are purchased on a weekly or monthly basis.

In addition to helping you know what percent of your daily requirement or upper limit of nutrients a food provides, you can also use the %DVs to determine if a food is or is not a good source of a particular nutrient. A food with a %DV of less than 5% for a nutrient is considered a poor source of that nutrient. Conversely, one with a %DV of at least 20% for a nutrient is a good source of that nutrient. Thus, this cereal would be considered high in vitamin A and low in saturated fat. Note that the DVs and %DVs are based on a combination of older RDA values and previously established guidelines for macronutrient intake. It is likely that these references will be updated using the current DRI and AMDR values in the near future. Because not all foods have Nutrition Facts panels—for example, fresh fruits and vegetables—you have to use food composition tables and DRI values to determine whether they are good sources of specific nutrients.

NUTRIENT CONTENT CLAIMS, STRUCTURE/FUNCTION CLAIMS, AND HEALTH CLAIMS

Food packages can also contain additional nutrition-related information that can help you plan a healthy diet. For example, **nutrient content claims** describe in a very consumer-friendly way how much of a nutrient (or its "content") is in a food. These include phrases such as "sugar free," "low sodium," and "good source of." Thus, if you wanted to increase your fiber intake, you could consistently choose foods labeled as "good sources of fiber." The use of these terms is regulated by the FDA, and some of the approved definitions are provided in Table 2.5. Another type of information that can be included on food packaging is a **structure/function claim,** which provides consumers with information about how a particular nutrient or other dietary ingredient might affect normal structure or function in humans. For instance, a label might state that "calcium builds strong bones" or "fiber maintains bowel regularity." Unlike nutrient content claims, structure/function claims are not pre-approved by the FDA but are expected to be truthful. If a dietary supplement label includes such a claim, it must state in a disclaimer that the FDA has not evaluated the claim and that the product is not intended to "diagnose, treat, cure, or prevent any disease."

Some food manufacturers also include information about potential health benefits you might get by consuming their products. These are called

nutrient content claim FDA-regulated phrase or words that can be included on a food's packaging to describe its nutrient content.

structure/function claim A statement that can be placed on food packaging stating the relationship between a nutrient or other dietary ingredient and health; not FDA approved.

TABLE 2.5 FDA-Approved Nutrient Content Claims

Wording	Description
"Light" or "Lite"	If 50% or more of the calories are from fat, fat must be reduced by at least 50% as compared with a regular product. If less than 50% of calories are from fat, fat must be reduced by at least 50% or calories reduced by at least one-third as compared with a regular product.
"Reduced Calories"	At least 25% fewer calories per serving compared with a regular product.
"Calorie Free"	Less than 5 kcal (Calories) per serving.
"Fat Free"	Less than 0.5 g fat per labeled serving.
"Low Fat"	3 g fat or less per serving.
"Saturated-Fat Free"	Less than 0.5 g saturated fat and less than 0.5 g *trans* fatty acids per serving.
"Low in Saturated Fat"	1 g saturated fat per serving and containing 15% or less of calories from saturated fat.
"Cholesterol Free"	Less than 2 mg cholesterol per serving. Note that cholesterol claims are only allowed when food contains 2 g or less saturated fat per serving.
"Low in Cholesterol"	20 mg cholesterol or less per serving.
"Sodium Free"	Less than 5 mg sodium per serving.
"Low in Sodium"	140 mg or less sodium per serving.
"Sugar Free"	Less than 0.5 g sugars per serving. This does not include sugar alcohols.
"High," "Rich in," or "Excellent Source of"	Contains 20% or more of the Daily Value (DV); used to describe protein, vitamins, minerals, dietary fiber, or potassium per serving.
"Good Source of," "Contains," or "Provides"	10–19% of the Daily Value (DV) per serving.
"More," "Added," "Extra," or "Plus"	10% or more of the Daily Value (DV) per serving. May only be used for vitamins, minerals, protein, dietary fiber, and potassium.
"Fresh"	A raw food that has not been frozen, heat processed, or otherwise preserved.
"Fresh Frozen"	Food that was quickly frozen while still fresh.

SOURCE: Adapted from U.S. Department of Health and Human Services and U.S. Food and Drug Administration. A food labeling guide—Appendix B. Available from: http://www.cfsan.fda.gov/~dms/flg-6b.html.

health claims. For example, a manufacturer of oatmeal can claim that consumption of its product decreases your risk for heart disease. Health claims are quite different from nutrient content claims and structure/function claims, which simply state that a food contains or does not contain a particular nutrient or that a nutrient can influence health in a particular way. Rather, health claims about a food's beneficial qualities must be supported by sufficient scientific evidence. Like other parts of a food's package, health claims must be approved by the FDA.

Manufacturers can make two kinds of health claims: **regular health claims** and **qualified health claims.** Both are claims concerning the relationship between a specific food component or whole food and a health-related condition. However, regular health claims are supported by considerable research, while qualified health claims have less scientific backing and must be accompanied by a disclaimer (or qualifying statement). You can usually tell if a health claim is qualified, because it contains a statement such as *"However, the FDA has determined that this evidence is limited and not conclusive."* As new research emerges, new health claims might be approved, while existing health claims might be disapproved.

Can You Put These Concepts into Action?

You now have the information needed to assess your own diet and begin to choose foods to improve it. Remember, however, that dietary assessment is only one component of conducting a complete nutritional assessment. To really assess your nutritional adequacy, you should visit a health professional for complete clinical, anthropometric, and biochemical assessments as well. Nonetheless, an example follows of how one college student (Jodi) used the same basic knowledge you have just acquired to assess her diet and make changes that positively affected her health.

STEP 1: SET THE STAGE AND SET YOUR GOALS

Consider a 21-year-old college student named Jodi who is interested in making sure that her diet is adequate. Jodi is generally in good health. Lately, however, she has felt tired and is wondering if her diet might be responsible. As such, she would like to know whether her diet might be lacking adequate amounts of iron or other nutrients.

STEP 2: ASSESS YOUR NUTRITIONAL STATUS

First, Jodi examines her anthropometric measurements to determine whether there is any evidence of overall malnutrition. Given that Jodi is 5 feet, 3 inches tall (1.6 m) and weighs 135 pounds (61.4 kg), she is likely at a healthy body weight and probably not consuming too few or too many calories. However, because weight and height are not good indicators of overall nutritional adequacy, Jodi decides to conduct a dietary self-assessment using a three-day diet record and MyPlate's Food Tracker dietary assessment program. She carefully records everything she eats and drinks for three days, paying close attention to portion sizes and describing all components of more complex foods such as the lasagna served in the cafeteria.

Next Jodi logs on to the MyPlate website and enters all her information into its database. Using this free software, she is able to find out how her dietary intake compares with the DRI values for all the required vitamins, minerals, and macronutrients as well as energy. In addition, she is able to compare her

health claim FDA-approved claim that can be included on a food's packaging to describe a specific health benefit.

regular health claim Statement concerning scientifically backed health benefit associated with a food or food component.

qualified health claim Statement concerning less well established health benefits that have been ascribed to a particular food or food component.

dietary intake of certain food groups with that recommended in the Dietary Guidelines and MyPlate.

The results of her dietary analysis indicate that almost all of the necessary nutrients are supplied in her diet at levels that fall above their AI or between the RDAs and ULs. In addition, her total energy intake is acceptable. However, the percentage of calories coming from fat is 40%—higher than recommended—and her fiber intake is below the recommended 20 to 25 g/day. Furthermore, her intake of iron is only 13 mg/day (72% of her RDA), and her vitamin B_{12} intake is only 1.9 micrograms (µg)/day (79% of her RDA). This suggests she may have inadequate intakes of fiber, iron, and vitamin B_{12}. Jodi does some research on why the body requires iron and vitamin B_{12} and learns that both are needed for energy (ATP) production. This might explain why she has been so tired. As such, Jodi thinks she should increase intakes of both these nutrients, try to decrease her fat intake, and increase her fiber consumption.

STEP 3: SET THE TABLE TO MEET YOUR GOALS

Using nutrient databases on the USDA website, information on the MyPlate website, and a food composition table, Jodi discovers that lean meats and fortified breakfast cereals are good sources of iron but do not supply high amounts of fat. She also learns that meat and dairy products supply vitamin B_{12} and that fiber is found in whole-grain products, peas and lentils, and some fruits and vegetables. Jodi begins to specifically choose these foods while limiting her total fat intake.

She also begins reading Nutrition Facts panels on packaged foods and, when possible, chooses foods with lower total fat contents. In the cafeteria, Jodi looks at cereal box labels and begins to eat high-fiber, fortified breakfast cereals instead of low-fiber, unfortified ones.[§§] In addition, she selects foods containing at least 20% of the DV for iron, looks for products that contain relatively large amounts of whole-grain components, and makes sure she eats enough lean meat and low-fat dairy products.

STEP 4: COMPARE YOUR PLAN WITH YOUR ASSESSMENT: DID YOU SUCCEED?

Jodi wants to know whether the changes she made to her diet have improved her nutrient intake. Therefore, she does another three-day dietary self-assessment and learns that these simple dietary changes have resulted in her consuming adequate intakes of iron, vitamin B_{12}, and fiber while simultaneously decreasing her total fat intake. Jodi also seems to be less tired and is able to concentrate on her studies for longer periods of time. It appears that she has succeeded. However, just to make sure that she has not overlooked something, Jodi makes an appointment with her campus health care provider to make sure her overall health is good.

This example of using dietary self-assessment and planning techniques can be easily applied to any person's life—including yours. Indeed, you now have all the tools needed to do each of these steps and the knowledge to make real changes in your diet and health. Some introductory nutrition classes require students to

Doing what you can right now to improve your health will help you live a longer and healthier life. This is especially important as life expectancy increases.

[§§] Fortified foods are those to which nutrients have been added during manufacturing.

undertake this exercise as a class assignment. If this is the case for you, then you will get firsthand experience at these procedures. If not, you might want to consider doing this on your own. Whether required or optional, a personal dietary self-assessment will make the rest of this course more meaningful because you will be able to apply the knowledge you gain to your own diet and overall health.

THERE IS NO TIME LIKE THE PRESENT

Now that you know how to assess your nutritional status—for example, by completing a dietary assessment—you can better use the many reference standards (such as DRI values), dietary recommendations (such as the Dietary Guidelines), and nutrition tools (such as Nutrition Facts panels) that are available to help you choose a diet that fits your personal nutritional needs.

Experts agree that food habits established early can influence your eating patterns and health for years to come. Therefore, we encourage you to take the time right now to set yourself up for success by eating right and being as healthy as possible.

Courtesy Megan Philbin

Critical Thinking: Emily's Story Recall Emily, the college student with secondary iron deficiency whom you met at the beginning of the chapter. What tools might Emily use on a daily basis to make sure she is consuming only gluten-free foods? How can she continue to monitor her intake of iron so that her iron status is maintained? What might be barriers to accomplishing this in a college setting, and how might university foodservice personnel help?

Diet Analysis PLUS ✚ Activity

Understanding Food Labels

"White" Wheat Bread vs. Whole-Grain Wheat Bread

Part I. Using a Nutrition Facts panel from any brand of **refined (white) wheat bread** (in which refined flour is listed as the first ingredient), fill in the following information concerning number of grams of each of the energy-yielding macronutrients.

Grams of **total** carbohydrate per serving _____

    Grams of sugar per serving _____

    Grams of fiber per serving _____

Grams of **total** fat per serving _____

    Grams of saturated fat per serving _____

    Grams of *trans* fat per serving _____

Grams of protein per serving _____

A. Using the values determined above for refined wheat bread, calculate the number of kilocalories from carbohydrates, fats, and protein in a single serving. Then calculate the *total* number of kilocalories in a serving.

    _____ grams of total carbohydrates × 4 kcal/g

    = _____ kcal

_____ grams of sugar × 4 kcal/g

= _____ kcal

_____ grams of fiber × 4 kcal/g

= _____ kcal

_____ grams of total fat × 9 kcal/g

= _____ kcal

_____ grams of saturated fat × 9 kcal/g

= _____ kcal

_____ grams of *trans* fat × 9 kcal/g

= _____ kcal

_____ grams of protein × 4 kcal/g

= _____ kcal

Total kilocalories/serving = (kilocalories from carbohydrates) + (kilocalories from fat) + (kilocalories from protein) = _____ kcal/serving

Why were the kilocalories from sugar, fiber, saturated fat, and *trans* fats not individually included when calculating total number of kilocalories/serving in this product?

B. To determine the percentage of kilocalories from each energy-yielding macronutrient, perform the following calculations.

Carbohydrates: _____ kcal/serving from carbohydrates ÷ _____ total kcal/serving × 100 = _____ % kcal from carbohydrates

Fat: _____ kcal from fat/serving ÷ _____ total kcal/serving × 100 = _____ % kcal from fat

Protein: _____ kcal from protein/serving ÷ _____ total kcal/serving × 100 = _____ % kcal from protein

- What is the purpose of calculating the percentage of kilocalories from each energy-yielding macronutrient?
- According to the 2010 Dietary Guidelines for Americans, what is the recommended distribution of kilocalories from each macronutrient class?
- How do these values compare to the Institute of Medicine's Acceptable Macronutrient Distribution Ranges (AMDRs)?

C. To determine how a serving of this particular food contributes to your overall daily nutrient requirements (assuming you consume 2,000 kcal/day), calculate the percent Daily Value (%DV) for fiber and fat.

Fiber: _____ grams of fiber/serving ÷ 25 grams of fiber recommended per day × 100 = _____ %DV

What percentage is listed next to fiber on your food label (if any)? _____ %DV

Are these similar? If not, why do you think they differ?

Fat: _____ grams of fat/serving ÷ 65 grams of fat allowed per day × 100 = _____ %DV

What percentage is listed next to total fat on your food label? _____ %DV

Are these similar? If not, why do you think they differ?

Part II. Now, determine the same information (as above) for any brand of **whole-grain wheat bread** (in which whole-wheat flour is the first ingredient).

- What do you notice about the caloric difference between refined wheat and whole-wheat breads?
- How do the sugar and fiber contents compare?
- How do the percentages of kilocalories from each macronutrient compare?
- Which type of bread provides more dietary fiber?
- How do the %DVs for fat compare between the refined wheat and whole-wheat breads?
- Compare the entire Nutrition Facts panels on the two types of bread. Which type of bread is more nutritious in terms of vitamins and minerals?

Notes

1. Institute of Medicine. Dietary reference intakes: Application in dietary assessment. Washington, DC: National Academies Press; 2000.

2. Institute of Medicine. Dietary reference intakes for energy, carbohydrate, fiber, fat, fatty acids, cholesterol, protein, and amino acids. Washington, DC: National Academies Press; 2005.

3. Atwater WO. Foods: Nutritive value and cost. Farmers' Bulletin. No. 23. Washington, DC: U.S. Government Printing Office; 1894.

4. U.S. Department of Agriculture, Center for Nutrition Policy and Promotion. A brief history of USDA food guides. June 2011. Available at http://www.choosemyplate.gov/downloads/MyPlate/ABriefHistoryOfUSDAFoodGuides.pdf.

5. U.S. Department of Agriculture and U.S. Department of Health and Human Services. Dietary guidelines for Americans, 2010. 7th Edition, Washington, DC: U.S. Government Printing Office, December 2010.

6. U.S. Department of Health and Human Services. Office of Disease Prevention and Health Promotion. Healthy People 2020. ODPHP Publication No. B0132. November 2010. Available at www.healthypeople.gov.

7. Food and nutrition act of 2008. Public Law 110-246. Available at http://www.fns.usda.gov/snap/rules/Legislation/pdfs/PL_110-246.pdf; Healthy, hunger-free kids act of 2010. Public Law 111-296. Available at http://www.gpo.gov/fdsys/pkg/PLAW-111publ296/pdf/

Chemical, Biological, and Physiological Aspects of Nutrition

CHAPTER **3**

Your body is made of carbon, hydrogen, oxygen, nitrogen, and a few other assorted elements. When joined together, elements can form large, life-sustaining molecules, such as proteins, carbohydrates, lipids, and nucleic acids. The building blocks for these biological substances come from the foods you eat. Without proper nourishment, cells—the basic units of all living organisms—die. To satisfy its nutritional needs, your body extracts and utilizes nutrients from the multitude of complex foods you eat. The first step in this process takes place in the gastrointestinal tract, where food is physically and chemically broken down into its most basic nutrient components. Once absorbed, these nutrients circulate through the extensive network of arteries and veins that make up your vascular system. Nutrients taken up by cells then undergo a series of chemical transformations that often produce metabolic waste products. The kidneys, lungs, and skin assist in eliminating these potentially harmful substances from the body. To ensure that these activities take place under optimal conditions, your endocrine and nervous systems maintain a nonstop communication network. In this chapter, you will learn about the biochemical and physiological events that take place every time you eat, and you will gain an appreciation for the intricate and varied tasks required to nourish your body.

Living with Crohn's Disease

Paige is a typical high school student who enjoys school, good friends, and an active social life. An accomplished athlete, Paige plays volleyball and is on the high school swim team. These sports are physically demanding, so it was common for Paige to feel tired at the end of the day. However, the fatigue she experienced during her junior year of high school was unlike anything she had experienced before. Having little energy, she found it increasingly more difficult to make it through the day without feeling exhausted. Next came weight loss, stomachaches, and diarrhea. Everyone was quick with advice—you are probably anemic, you are having another growth spurt, your body is telling you that you need to get more sleep. Even her doctors were quick to dismiss Paige's symptoms as nothing out of the ordinary. But, what was going on inside Paige's body was anything but ordinary—especially for an active teenage girl.

Nobody took Paige's health concerns seriously, but she was convinced something was terribly wrong. The weight loss continued, her appetite diminished further, and the stomachaches only got worse. Her doctor first suspected a parasite, but testing eliminated that as a possible cause of Paige's health problems. More doctors, more specialists, and one by one other possible causes were ruled out. If it was not a food intolerance, celiac disease, food allergy, or irritable bowel syndrome, what could it be? After a colonoscopy, the answer became alarmingly clear—Paige had Crohn's disease.

Like many people, Paige and her family had never heard of Crohn's disease. They quickly learned that Crohn's disease causes inflammation of the small intestine and/or colon, and is often recurrent. The next piece of news, that Paige would require surgery, was equally frightening. Not only was her colon swollen and inflamed, but an abscess had developed, and it needed to be removed. To prepare for surgery, Paige endured 10 days on a clear liquid diet. Once her condition was stabilized, Paige underwent a surgical procedure during which 30 centimeters (about 12 inches) of her colon were removed. Recovery came quickly, and within a few days Paige was eating solid foods again. She was released from the hospital and was back in school the following week. Paige is hopeful that she will not experience a recurrence of Crohn's disease. Although no further treatment is needed at this time, Paige takes dietary supplements to make sure she is getting all the nutrients her body needs. She has resumed her active lifestyle, and the only thing that reminds her of this long, painful ordeal is a 3-inch scar. Even that is starting to fade away, along with the memory of how sick she had been.

muzsy/Shutterstock.com

Critical Thinking: Paige's Story

Having a chronic disease can impact almost every aspect of a person's life. If you are familiar with Crohn's disease, you are already aware of the many challenges a person may face. This is particularly true for Paige, who was 16 years old when she was diagnosed. After reading Paige's story, how would you respond to this diagnosis? How would it impact your daily routines and lifestyle?

How Does Chemistry Apply to the Study of Nutrition?

Chemistry is fundamental to the study of nutrition. The organization of atoms into simple molecules, simple molecules into complex molecules, complex molecules into cells, cells into tissues, tissues into organs, and organs into organ systems is indeed remarkable (Figure 3.1). This entire circuit is made of and fueled by the nutrients contained in food. To appreciate these life-sustaining functions, it is important that you first have a basic understanding of chemistry—the science that deals with matter.

FIGURE 3.1 Levels of Organization in the Body

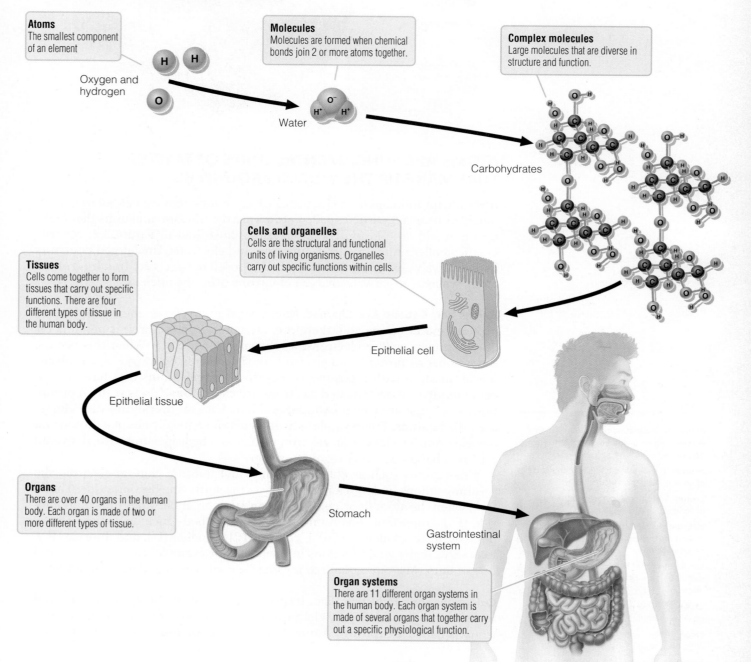

Atoms
The smallest component of an element

Oxygen and hydrogen

Molecules
Molecules are formed when chemical bonds join 2 or more atoms together.

Water

Complex molecules
Large molecules that are diverse in structure and function.

Carbohydrates

Cells and organelles
Cells are the structural and functional units of living organisms. Organelles carry out specific functions within cells.

Epithelial cell

Tissues
Cells come together to form tissues that carry out specific functions. There are four different types of tissue in the human body.

Epithelial tissue

Organs
There are over 40 organs in the human body. Each organ is made of two or more different types of tissue.

Stomach

Gastrointestinal system

Organ systems
There are 11 different organ systems in the human body. Each organ system is made of several organs that together carry out a specific physiological function.

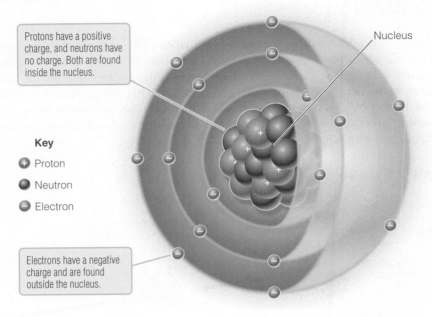

FIGURE 3.2 **A Model of an Atom** Atoms consist of subatomic particles—protons, neutrons, and electrons.

Protons have a positive charge, and neutrons have no charge. Both are found inside the nucleus.

Nucleus

Key

⊕ Proton

● Neutron

⊖ Electron

Electrons have a negative charge and are found outside the nucleus.

ATOMS ARE FUNDAMENTAL UNITS OF MATTER THAT MAKE UP THE WORLD AROUND US

It is difficult to imagine the existence of something that we cannot see, taste, touch, or hear. Yet particles called **atoms** are the fundamental units that make up the world around us. The atom itself, illustrated in Figure 3.2, consists of still smaller units—neutrons, protons, and electrons. Uncharged **neutrons** and positively charged **protons** are both housed in the center (*nucleus*) of the atom, whereas negatively charged **electrons** orbit the nucleus.

Anions and Cations Are Charged Atoms Most atoms have equal numbers of protons and electrons and, therefore, are neutral. However, it is possible for atoms to gain or lose electrons, as shown in Figure 3.3. When this occurs, the number of protons and electrons in the atom is no longer equal, resulting in an atom with a positive or negative charge. Atoms that have an unequal number of protons and electrons are called **ions.*** Ions with an overall positive charge are called **cations,** and ions with an overall negative charge are called **anions.** For example, a hydrogen atom typically has one proton and one electron. The loss of an electron results in a hydrogen ion with an overall positive charge—in other words, a cation.

When atoms such as chlorine, iodine, and fluorine gain an electron, the resulting anions undergo a name change; the suffix -ine becomes -ide. Therefore, when the atom fluor*ine* (F) gains an electron, it becomes the anion fluor*ide* (F$^-$). Important ions found in the human body include sodium (Na$^+$), potassium (K$^+$), calcium (Ca^{2+}), chloride (Cl$^-$), iodide (I$^-$), and fluoride (F$^-$). Ions serve many vital functions in the body. For example, calcium (Ca^{2+}) and magnesium (Mg^{2+}) are required for muscle contraction and nerve function.

Oxidation and Reduction The transfer of electrons between atoms and molecules is an important chemical event. The loss of one or more electrons is called **oxidation.** Atoms that lose one or more electrons are said

*Note that molecules such as hydroxide (OH$^-$) can also be ions.

atom The smallest portion into which an element can be divided into and still retain its properties.

neutron A subatomic particle, in the nucleus of an atom, with no electrical charge.

proton A subatomic particle, in the nucleus of an atom, that carries a positive charge.

electron A subatomic particle that orbits around the nucleus of an atom, and carries a negative charge.

ion An atom that has acquired an electrical charge by gaining or losing one or more electrons.

cation (CAT – i – on) An ion with a net positive charge.

anion (AN – i – on) An ion with a net negative charge.

oxidation (ox – i – DA – tion) The loss of one or more electrons.

FIGURE 3.3 Formation of Cations and Anions

Atoms with unequal numbers of protons and electrons are ions. An ion can have a positive charge (cation) or a negative charge (anion).

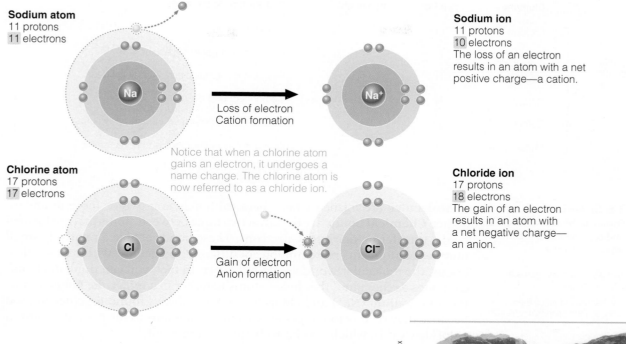

Sodium atom
11 protons
11 electrons

Sodium ion
11 protons
10 electrons
The loss of an electron results in an atom with a net positive charge—a cation.

Loss of electron
Cation formation

Notice that when a chlorine atom gains an electron, it undergoes a name change. The chlorine atom is now referred to as a chloride ion.

Chlorine atom
17 protons
17 electrons

Chloride ion
17 protons
18 electrons
The gain of an electron results in an atom with a net negative charge—an anion.

Gain of electron
Anion formation

to be oxidized and become more positively charged. For example, when iron (Fe) loses two electrons, the overall charge changes to Fe^{2+}. When Fe is fully oxidized, it has lost three electrons, becoming Fe^{3+}. Therefore, the oxidation of iron increases the net positive charge (Fe to Fe^{3+}), reflecting the loss of three negatively charged electrons. In fact, when iron-containing proteins in red meat are exposed to air for a prolonged period of time, they become oxidized, causing the meat to turn brown.

Conversely, the gain of one or more electrons is called **reduction,** and atoms become more negative during this process. **Reduction–oxidation** (or **redox**) **reactions** often take place simultaneously and are said to be coupled reactions (Figure 3.4). That is, the loss of an electron by one atom (oxidation) results in the gain of an electron by another (reduction). There are many important redox reactions in the body that involve the transfer of electrons between molecules.

CHEMICAL BONDS ENABLE ATOMS TO FORM MILLIONS OF DIFFERENT MOLECULES

An **element** is defined as a pure substance made up of only one type of atom. There are approximately 92 naturally occurring elements, 20 of which are essential for human health. In fact, just 6 elements—carbon, oxygen, hydrogen, nitrogen, calcium, and phosphorus—account for 99% of your total body weight. These important elements (Figure 3.5) provide the raw materials needed to form large, complex molecules found in living systems, such as proteins, carbohydrates, lipids, and nucleic acids.

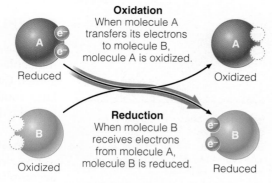

Tony Cenicola/The New York Times/Redux

When red meat is exposed to the air too long, iron-containing proteins become oxidized, causing the surface of the meat to turn a grayish-brown color.

FIGURE 3.4 Electron Transfer in Reduction–Oxidation (Redox) Reactions

Redox reactions are coupled reactions in which one or more electrons are transferred between molecules.

Oxidation
When molecule A transfers its electrons to molecule B, molecule A is oxidized.

A — Reduced
A — Oxidized

Reduction
When molecule B receives electrons from molecule A, molecule B is reduced.

B — Oxidized
B — Reduced

reduction The gain of one or more electrons.

reduction–oxidation (redox) reactions Chemical reactions that take place simultaneously whereby one molecule gives up one or more electrons (is oxidized) while the other molecule receives one or more electrons (is reduced).

element A pure substance made of only one type of atom.

FIGURE 3.5 The Most Abundant Elements in the Human Body

Just 6 elements account for 99% of body weight in humans.

Element	Atomic symbol	% of human body by weight	Distribution in the body
Oxygen	O	65	Found in water
Carbon	C	18	Found in all organic molecules
Hydrogen	H	10	Found in most molecules, including water
Nitrogen	N	3	Component of proteins
Calcium	Ca	2	Component of bones, teeth, and body fluids
Phosphorus	P	1	Found in cell membranes and bone matrix

FIGURE 3.6 Understanding Molecular Formulas Molecular formulas are used to indicate the number and types of atoms in a molecule.

Some molecules such as hydrogen are made of only 1 type of atom.

The subscript indicates there are 2 hydrogen atoms in this molecule.

H_2

Some molecules are made of 2 or more types of atoms.

Glucose is made of 6 carbon, 12 hydrogen, and 6 oxygen atoms.

$C_6H_{12}O_6$

A number in front of the molecule indicates the number of molecules.

There are 3 water molecules. Each molecule is made of 2 hydrogen atoms and 1 oxygen atom.

$3H_2O$

molecule A substance held together by chemical bonds.

chemical bonds The attractive force between atoms formed by the transfer, sharing, or interaction of electrons.

molecular formula (mo – LEC – u – lar) Indicates the number and types of atoms in a molecule.

compound A molecule made of two or more different types of atoms.

condensation A chemical reaction that results in the formation of water.

hydrolysis (hy – DRO – ly – sis) A chemical reaction whereby compounds react with water and are split apart.

Molecules are formed when chemical bonds join two or more atoms together. **Chemical bonds** are the attractive force between atoms that are formed by the transfer or sharing of electrons. These bonds enable a relatively small number of atoms to create millions of different molecules. Without chemical bonds, the molecular world would fall apart. In fact, you can think of chemical bonds as the "glue" that holds atoms together in a molecule. For example, the ions sodium (Na^+) and chloride (Cl^-) are always found together in food because they readily join together via chemical bonds to form a salt—sodium chloride (NaCl), which you know better as "table salt."

Understanding a Molecular Formula A **molecular formula,** such as H_2O, describes the number and type of atoms present in a molecule. For example, glucose, an important source of energy in your body, has a molecular formula of $C_6H_{12}O_6$. These numbers and letters tell you that one molecule of glucose consists of 6 carbon, 12 hydrogen, and 6 oxygen atoms. When more than one molecule of a substance is present, a number is placed before the molecular formula. For example, three molecules of water is written as $3H_2O$. Although some molecules, such as oxygen (O_2), consist of only one type of atom, most are made of several types of atoms. Molecules composed of two or more different types of atoms, such as water (H_2O) and glucose ($C_6H_{12}O_6$), are called **compounds** (Figure 3.6).

COMPLEX MOLECULES ARE VITAL TO CELL FUNCTION

So far, we have talked about molecules that are small and simple. However, molecules can also be very large, consisting of thousands of atoms bonded together. Carbohydrates, lipids, proteins, and nucleic acids (DNA and RNA) are large molecules and are vital to the functions of cells. The raw materials used to make these complex molecules come from the nutrients in foods that we eat.

Condensation and Hydrolysis: Make-and-Break Reactions Complex molecules such as proteins and lipids are often assembled and disassembled within cells. This is accomplished by chemical reactions that either form or break chemical bonds. One type of chemical reaction that joins molecules together is **condensation.** For example, a condensation reaction takes place when a chemical bond forms between two glucose molecules. This reaction results in the formation and release of a water molecule. The opposite type of reaction, called **hydrolysis,** can split molecules apart. During hydrolysis, a molecule of water is used to break chemical bonds. You can think of condensation and hydrolysis as opposite "make-and-break" reactions. Condensation and hydrolysis play important roles in digestion and metabolism, and they are illustrated in Figure 3.7.

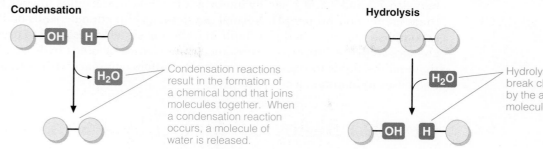

FIGURE 3.7 Condensation and Hydrolysis Condensation and hydrolysis are "make-and-break" reactions.

Condensation

Condensation reactions result in the formation of a chemical bond that joins molecules together. When a condensation reaction occurs, a molecule of water is released.

Hydrolysis

Hydrolysis reactions break chemical bonds by the addition of a molecule of water.

ACID–BASE CHEMISTRY IS IMPORTANT TO THE STUDY OF NUTRITION

Grapefruits and lemons taste sour, baking soda tastes bitter, and pure water has no taste at all. These taste differences are attributed, in part, to the level of acidity, ranging from acidic (such as citrus) to neutral (such as water) to alkaline (such as baking soda). Acid–base chemistry is important to the study of nutrition. In fact, diseases that disrupt the acid–base balance of the body can cause serious health problems. To better understand acid–base chemistry, it is best to begin with an understanding of water, the medium in which chemical reactions take place.

Ionization of Water Molecules: The Basis of the pH Scale Although the chemical bonds holding together the hydrogen and oxygen atoms in a water molecule are quite strong, water molecules can also separate, or dissociate, into their charged (ionic) components. Water molecules dissociate to form hydrogen (H^+) and hydroxide (OH^-) ions, which can reform back into water molecules (H_2O).[†] If you analyzed a sample of pure water, you would find that the numbers of hydrogen and hydroxide ions are extremely small. That is, in a sample of pure water, only a small fraction of the water molecules are in an ionized state. You would also find that the concentration of hydrogen ions equals that of hydroxide ions. Thus, water is neither acidic nor basic, but neutral.

The ionization of water molecules is the basis for the **pH scale,** which ranges from 0 to 14. Because pure water has equal amounts of hydrogen (H^+) and hydroxide (OH^-) ions, it is neutral and has a pH of 7. However, when an acid is added to water, hydrogen ions (H^+) are released. This increases the concentration of hydrogen ions (H^+) in the solution, causing it to become more **acidic.** Fluids that contain a higher concentration of hydrogen ions (H^+) than hydroxide ions (OH^-) have a pH less than 7. As a solution becomes more acidic, its pH decreases even further.

When a base is added to water, hydroxide ions (OH^-) are released. Fluids that contain more hydroxide ions (OH^-) than hydrogen ions (H^+) have a pH greater than 7, making them **basic** or **alkaline.** The pH value of a solution increases as it becomes more basic.

Each consecutive number on the pH scale represents a 10-fold increase or decrease in the concentration of hydrogen ions (H^+). Thus, a fluid with a pH of 3 is 10 times more acidic than a fluid with a pH of 4. The various fluids in your body all have specific pH ranges. For example, the pH of urine

pH scale A scale, ranging from 0 to 14, that signifies the acidity or alkalinity of a solution.

acidic Having a pH less than 7.

basic (also called alkaline) Having a pH greater than 7.

[†]Technically, this reaction involves two water molecules and forms a hydronium ion (H_3O^+), and the ionization reaction is written as $2H_2O \leftrightarrow H_3O^+ + OH^-$; however, for simplicity we will describe water ionization as forming H^+ and OH^-.

typically ranges between 5.5 and 7.5, whereas the pH of blood is usually between 7.3 and 7.5. It is vitally important for the various fluids in your body to maintain their proper pH. Normal physiological function depends on it. To accomplish this, your body has built-in buffering systems designed to prevent changes in pH. A **buffer** is a substance that releases or binds hydrogen ions that enables fluids to resist changes in pH. Buffers can react with both acids and bases to maintain a constant pH.

How Do Biological Molecules Form Cells, Tissues, Organs, and Organ Systems?

Biological molecules such as carbohydrates, proteins, lipids, water, and nucleic acids are basic to life. However, these molecules by themselves cannot function in useful ways. Rather, some of these building blocks are used to form structural and functional units called cells. Cells make up tissues, which in turn organize to form organs. Last, organs work together as part of an organ system. The human body has 11 organ systems, all of which are pertinent to the study of nutrition.

SUBSTANCES CROSS CELL MEMBRANES BY PASSIVE AND ACTIVE TRANSPORT

Cells are like microscopic cities, full of activity. These activities are carried out by structures called **organelles,** which are distributed in a gel-like matrix called the **cytoplasm** (also called the cytosol). Cells are surrounded by a protective cell membrane that provides a boundary between the **extracellular** (outside the cell) and **intracellular** (within the cell) environments. Thus, it is the cell membrane that regulates the movement of substances into and out of cells.

Transport across Cell Membranes Cell membranes are selectively permeable, meaning that they allow some substances to cross them more readily than others. The movement of nutrients and other substances across a cell membrane occurs through a variety of processes, which are referred to collectively as *transport mechanisms*. Broadly speaking, a process that does not require energy (ATP) is called a **passive transport mechanism,** whereas one that does require energy (ATP) is called an **active transport mechanism.** Both passive and active transport mechanisms are utilized for the passage of nutrients into and out of all the cells in your body, including those that line the small intestine.

Passive Transport Mechanisms The three main types of passive transport mechanisms are simple diffusion, facilitated diffusion, and osmosis (Figure 3.8). **Simple diffusion** enables substances to cross cell membranes from a region of higher concentration to a region of lower concentration without using energy (ATP) in the process. When this occurs, the substance is said to move passively "down its concentration gradient," similar to a floating raft moving downstream. Once the concentration of the substance is equal on both sides of the cell membrane, a state of *equilibrium* has been reached. **Facilitated diffusion** also involves the passive movement of a substance down its concentration gradient (high concentration to low concentration) but differs from simple diffusion in that it requires the assistance of a membrane-bound transport protein that "escorts" materials across cell membranes.

Another type of passive transport is called **osmosis,** which is defined as the movement of water molecules across a selectively permeable membrane, such as those that surround cells. However, because cells rupture if they contain too

buffer A substance that releases or binds hydrogen ions in order to resist changes in pH.

organelles (or – gan – ELLES) Cellular structures that have a particular function.

cytoplasm (also called cytosol) The gel-like matrix inside cells.

extracellular (ex – tra – CEL – lu – lar) Situated outside of a cell.

intracellular (in – tra – CEL – lu – lar) Situated within a cell.

passive transport mechanism Transport mechanism that enables substances to cross cell membranes without expenditure of energy (ATP).

active transport mechanism Transport mechanism that enables substances to cross cell membranes, requiring the expenditure of energy (ATP).

simple diffusion A passive transport mechanism whereby substances cross cell membranes from a region of higher concentration to a region of lower concentration without using energy (ATP).

facilitated diffusion A passive transport mechanism whereby substances cross cell membranes from a region of higher concentration to a region of lower concentration with the assistance of a transport protein.

osmosis Movement of water molecules from a region of lower solute concentration to that of a higher solute concentration, until equilibrium is reached.

FIGURE 3.8 Passive Transport Passive transport mechanisms—such as simple diffusion, facilitated diffusion, and osmosis—do not require energy (ATP).

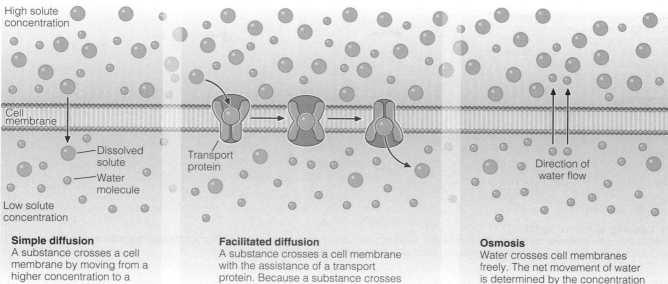

Simple diffusion
A substance crosses a cell membrane by moving from a higher concentration to a lower concentration. Energy is not required.

Facilitated diffusion
A substance crosses a cell membrane with the assistance of a transport protein. Because a substance crosses a cell membrane by moving from a higher concentration to a lower concentration, energy is not required.

Osmosis
Water crosses cell membranes freely. The net movement of water is determined by the concentration of solutes. Water moves by osmosis from a lower solute concentration to a higher solute concentration, until the solute concentration is the same on both sides of the membrane. Energy is not required.

much water and collapse if they contain too little, osmosis is carefully regulated. Fluids in your body contain different types of dissolved substances, such as proteins and electrolytes. Although the terms *ions* and *electrolytes* are often used interchangeably, **electrolytes** are substances such as salt that dissolve or dissociate into ions when put in water. Substances that are dissolved in fluids are called **solutes,** which, when uniformly dispersed, form a **solution.** The concentration of these dissolved substances inside and outside of the cell determines the direction and the amount of water that crosses cell membranes. A solution with a low solute concentration contains relatively more water (and fewer solutes) than a solution with a high solute concentration. Therefore, water moves from a region of lower solute concentration to that of a higher solute concentration, until equilibrium is reached.

Active Transport Mechanisms Similar to swimming upstream, some substances must cross cell membranes against the prevailing concentration gradient, moving from a region of lower concentration to that of a higher concentration. In cells, this uphill journey is accomplished by active transport mechanisms, which include carrier-mediated active transport and vesicular active transport (Figure 3.9). **Carrier-mediated active transport** requires both energy (ATP) and the assistance of a transport protein. Energy is used to "pump" molecules across cell membranes against their concentration gradients.

Another type of active transport used to move molecules into and out of cells is **vesicular active transport.** There are two types of vesicular active transport mechanisms: endocytosis and exocytosis. **Endocytosis** moves substances from the extracellular to the intracellular environment. In endocytosis, a portion of the cell membrane surrounds an extracellular particle, enclosing it in a saclike structure called a vesicle. The contents of the vesicle are then released to the inside of the cell. The reverse process, **exocytosis,** enables substances to leave cells by packaging them in vesicles, which are then released into the surrounding extracellular fluid.

electrolytes Substances such as salt that dissolve or dissociate into ions when put in water.

solute (SOL – ute) A substance that dissolves in a solvent.

solution A mixture of two or more substances that are uniformly dispersed.

carrier-mediated active transport An energy-requiring mechanism whereby a substance moves from a region of lower concentration to a region of higher concentration, requiring the assistance of a carrier protein.

vesicular active transport (ve – SIC – u – lar) An energy-requiring mechanism whereby large molecules move into or out of cells by an enclosed vesicle.

endocytosis (en – do – cy – TO – sis) A form of vesicular active transport whereby the cell membrane surrounds extracellular substances and releases them to the cytoplasm.

exocytosis (ex – o – cy – TO – sis) A form of vesicular active transport whereby intracelllular cell products are enclosed in a vesicle and the contents of the vesicle are released to the outside of the cell.

FIGURE 3.9 Active Transport Active transport mechanisms—such as carrier-mediated active transport and vesicular active transport—require energy (ATP).

A. Carrier-mediated active transport
A solute crosses a cell membrane with the assistance of a transport protein. Energy (ATP) is required, because the substance moves against its concentration gradient, moving from a lower concentration to a higher concentration.

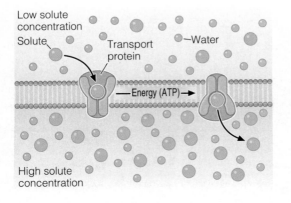

B. Vesicular active transport
Part of the cell membrane surrounds the substance, forming a vesicle. The vesicle then moves across the cell membrane and the substances are released inside (endocytosis) or outside (exocytosis) the cell.

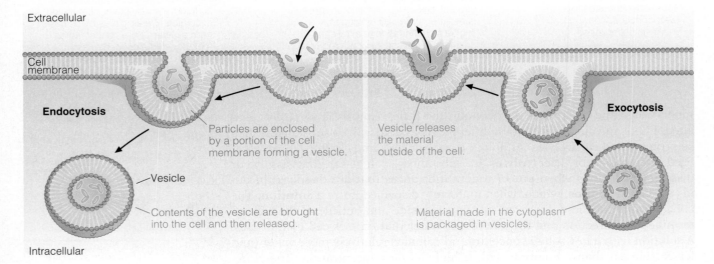

CELL ORGANELLES CARRY OUT SPECIALIZED FUNCTIONS CRITICAL FOR LIFE

Within the cell, small membrane-bound structures called organelles carry out specialized functions that are critical for life. Each type of organelle is responsible for a specific function. Some organelles produce substances needed for cellular activities, whereas others function as waste-disposal systems that assist with degrading and recycling worn-out cellular components. For example, organelles called **mitochondria** serve as power stations, converting the chemical energy in energy-yielding nutrients (glucose, fatty acids, and amino acids) into a form of energy (ATP) that is used by cells. Another organelle, called the **nucleus,** houses the genetic material DNA, which provides the "blueprint" for protein synthesis. Figure 3.10 provides an overview of cellular organelles and their functions.

mitochondria (mi – to – CHON – dri – a) Cellular organelles involved in generating energy (ATP).

nucleus A membrane-enclosed organelle that contains the genetic material DNA.

FIGURE 3.10 A Typical Cell

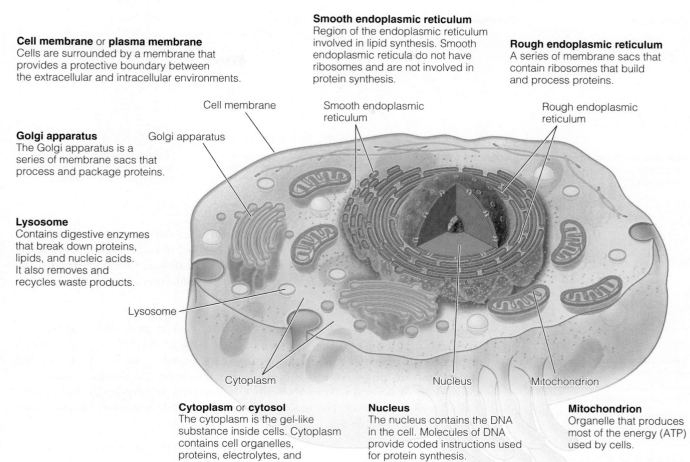

Cell membrane or **plasma membrane**
Cells are surrounded by a membrane that provides a protective boundary between the extracellular and intracellular environments.

Smooth endoplasmic reticulum
Region of the endoplasmic reticulum involved in lipid synthesis. Smooth endoplasmic reticula do not have ribosomes and are not involved in protein synthesis.

Rough endoplasmic reticulum
A series of membrane sacs that contain ribosomes that build and process proteins.

Golgi apparatus
The Golgi apparatus is a series of membrane sacs that process and package proteins.

Lysosome
Contains digestive enzymes that break down proteins, lipids, and nucleic acids. It also removes and recycles waste products.

Cell membrane

Golgi apparatus

Smooth endoplasmic reticulum

Rough endoplasmic reticulum

Lysosome

Cytoplasm

Nucleus

Mitochondrion

Cytoplasm or **cytosol**
The cytoplasm is the gel-like substance inside cells. Cytoplasm contains cell organelles, proteins, electrolytes, and other molecules.

Nucleus
The nucleus contains the DNA in the cell. Molecules of DNA provide coded instructions used for protein synthesis.

Mitochondrion
Organelle that produces most of the energy (ATP) used by cells.

GROUPS OF CELLS MAKE UP TISSUES, TISSUES MAKE UP ORGANS, AND ORGANS MAKE UP ORGAN SYSTEMS

So far, you have learned that atoms make up molecules, molecules make up cells, and cells carry out the basic functions of life. The next level of complexity is the tissue, consisting of cells that carry out specialized functions. In the human body, four types of tissue make up more than 40 organs, which in turn make up 11 unique organ systems—all of which are relevant to the study of nutrition.

Four Different Types of Tissue **Tissue** is formed when a large number of cells with similar structure and function group together. The human body contains four different types of tissue: (1) epithelial, (2) connective, (3) muscle, and (4) neural, shown in Figure 3.11. **Epithelial tissue** (or epithelium) provides a protective layer on body surfaces (skin) as well as lining internal organs, ducts, and cavities. **Connective tissue** is the "glue" that holds the body together. Tendons, cartilage, and some parts of bones are examples of connective tissue. Blood is also a type of connective tissue, consisting of cells and platelets in a liquid called **plasma.** The body contains three major types of **muscle tissue**—skeletal muscle, smooth muscle, and cardiac muscle. Large muscles in the body consist of skeletal muscle tissue, which is needed for voluntary movement. Smooth muscle tissue is found in the lining of organs such as the esophagus, stomach, and small intestine, and is even found in the interior lining of

tissue (TIS – sue) An aggregation of specialized cells that are similar in form and function.

epithelial tissue (ep – i – THE – li – al) Tissue that forms a protective layer on bodily surfaces and lines internal organs, ducts, and cavities.

connective tissue (con – NEC – tive) Tissue that supports, connects, and anchors body structures.

plasma (PLAS – ma) The fluid component of blood.

muscle tissue Tissue that specializes in movement.

FIGURE 3.11 Four Basic Types of Tissue Epithelial, connective, neural, and muscle tissue make up all the organs in the human body.

Epithelial tissue

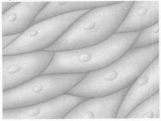

Epithelial tissue covers and lines body surfaces, organs, and cavities.

Connective tissue

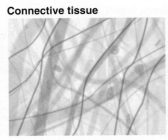

Connective tissue provides structure to the body by binding and anchoring body parts.

Muscle tissue

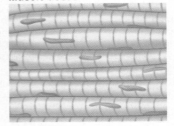

Muscle tissue contracts and shortens when stimulated, playing an important role in movement.

Neural tissue

Neural tissue plays a role in communication by receiving and responding to stimuli.

blood vessels. Because smooth muscles are not under our conscious control, they are referred to as involuntary muscles. Cardiac muscle is found exclusively in the heart. The fourth type of tissue is **neural tissue,** which makes up the brain, spinal cord, and nerves. Neural tissue plays an important communicative role in the body.

Organs Form Organ Systems **Organs** consist of two or more different types of tissue, functioning together to perform a variety of related tasks. An **organ system** is formed when several organs work together, each organ carrying out important physiological functions. For example, the digestive system is composed of several organs that work collectively to physically and chemically break down food. The major organ systems and their basic functions are summarized in Table 3.1.

neural tissue Tissue that specializes in communication via nerves.

organ A group of tissues that combine to carry out coordinated functions.

organ system Organs that work collectively to carry out related functions.

TABLE 3.1 Organ Systems and Related Major Functions

System	Major Organs and Structures	Major Function
Integumentary	Skin, hair, nails, and sweat glands	Protects against pathogens and helps regulate body temperature.
Skeletal	Bones, cartilage, and joints	Provides support and structure to the body. The bone marrow of some bones produces blood cells. Also provides a storage site for certain minerals.
Muscular	Smooth, cardiac, and skeletal muscle	Assists in voluntary and involuntary body movements.
Nervous	Brain, spinal cord, nerves, and sensory receptors	Interprets and responds to information. Controls the basic senses, movement, and intellectual functions.
Endocrine	Endocrine glands	Produces and releases hormones that control physiological functions such as reproduction, hunger, satiety, blood glucose regulation, metabolism, and stress response.
Respiratory	Lungs, nose, mouth, throat, and trachea	Governs gas exchange between the blood and air. Also assists in regulating blood acid–base (pH) balance.
Circulatory	Heart, blood vessels, blood, lymph vessels, lymph nodes, and lymph organs	Transports nutrients, waste products, gases, and hormones. Also plays a role in regulating body temperature. Helps remove foreign substances and plays a role in immunity.
Digestive	Mouth, esophagus, stomach, small intestine, large intestine, liver, gallbladder, pancreas, and salivary glands	Governs the physical and chemical breakdown of food into a form that can be absorbed into the circulatory system. Eliminates solid wastes.
Reproductive	Gonads and genitals	Carries out reproductive functions and is associated with sexual characteristics, sexual function, and sexual behaviors.
Urinary	Kidneys, bladder, and ureters	Removes metabolic waste products from the blood; governs nutrient reabsorption, acid–base balance, and regulates water balance.
Immune	White blood cells, lymph vessels, bone marrow, and lymphatic tissue	Provides a defense against foreign bodies, such as bacteria and viruses, and unregulated cell growth.

Organ Systems Work Together to Maintain Homeostasis The ability of organ systems to work together to carry out common functions requires constant communication. In other words, the right hand must know what the left hand is doing. The body has two well-developed communication systems that coordinate physiologic processes—the nervous and endocrine systems. The nervous system receives and transmits information via electrical impulses and **neurotransmitters** between nerve cells, whereas the endocrine system communicates via chemical messengers, called **hormones,** in the blood. Hormones are released from glands or cells in response to various stimuli, and exert their effects by binding to receptors on specific tissues. When this occurs, tissues initiate an appropriate response to the initial stimulus. Together, the nervous and endocrine systems continuously monitor our internal environment, responding to change and restoring balance. These mechanisms allow complex organisms, such as humans, to adapt in an ever-changing environment, a process known as **homeostasis.**

Homeostasis, a state of equilibrium or balance, is an important concept in physiology and nutrition. **Negative feedback systems** play vital roles in homeostasis by opposing changes in the internal environment and by initiating corrective responses that restore balance. An example of a negative feedback system is the regulation of blood glucose. In response to a carbohydrate-rich meal, blood glucose levels rise (a change in the internal environment). The pancreas detects this change and initiates a response—the release of the hormone insulin. Insulin then binds to specific receptors on cell membranes, which facilitates the uptake of glucose. As a result, blood glucose levels are restored. This is why diseases that disrupt homeostatic responses can have serious health consequences.

How Does the Digestive System Break Down Food into Absorbable Components?

To nourish your body, your digestive system methodically disassembles the complex molecules in food into simpler basic components. This arduous task requires many organ systems, but primarily it is the digestive system that gets the job done. Your digestive system is made up of the digestive tract and accessory organs. The digestive tract, more commonly known as the **gastrointestinal (GI) tract** or alimentary tract, can be thought of as a hollow tube that runs from the mouth to the anus (Figure 3.12). Organs that make up the GI tract include the mouth, esophagus, stomach, small intestine, and large intestine.

The accessory organs, which participate in digestion but are not part of the GI tract, include the salivary glands, pancreas, and biliary system (liver and gallbladder). The accessory organs release secretions needed for the process of digestion into ducts, which empty into the **lumen,** the inner cavity that spans the entire length of the GI tract. Together, the GI tract and accessory organs carry out three important functions: (1) **digestion,** the physical and chemical breakdown of food; (2) **absorption,** the transfer of nutrients from the GI tract into the blood or lymphatic circulatory systems; and (3) **egestion,** the process whereby solid waste (feces) is expelled from the body.

THE GI TRACT HAS FOUR TISSUE LAYERS THAT CONTRIBUTE TO THE PROCESS OF DIGESTION

As shown in Figure 3.13, the digestive tract contains four major tissue layers—the mucosa, submucosa, muscularis, and serosa. Each tissue layer contributes to the overall function of the GI tract by providing secretions, movement, communication, and protection.

neurotransmitters Chemical messengers released from nerve cells that transmit information.

hormones Substances released from glands or cells in response to various stimuli that exert their effect by binding to receptors on specific tissues.

homeostasis (ho – me – o – STA – sis) A state of balance or equilibrium.

negative feedback systems Corrective responses that oppose change and restore homeostasis.

gastrointestinal (GI) tract A tubular passage that runs from the mouth to the anus that includes several organs that participate in the process of digestion; also called the digestive tract.

lumen (LU – men) The cavity inside a tubular structure in the body.

digestion The physical and chemical breakdown of food by the digestive system into a form that allows nutrients to be absorbed.

absorption The passage of nutrients through the lining of the GI tract into the blood or lymphatic circulation.

egestion The process whereby solid waste (feces) is expelled from the body.

FIGURE 3.12 Organs of the Digestive System The digestive system consists of the gastrointestinal tract and the accessory organs.

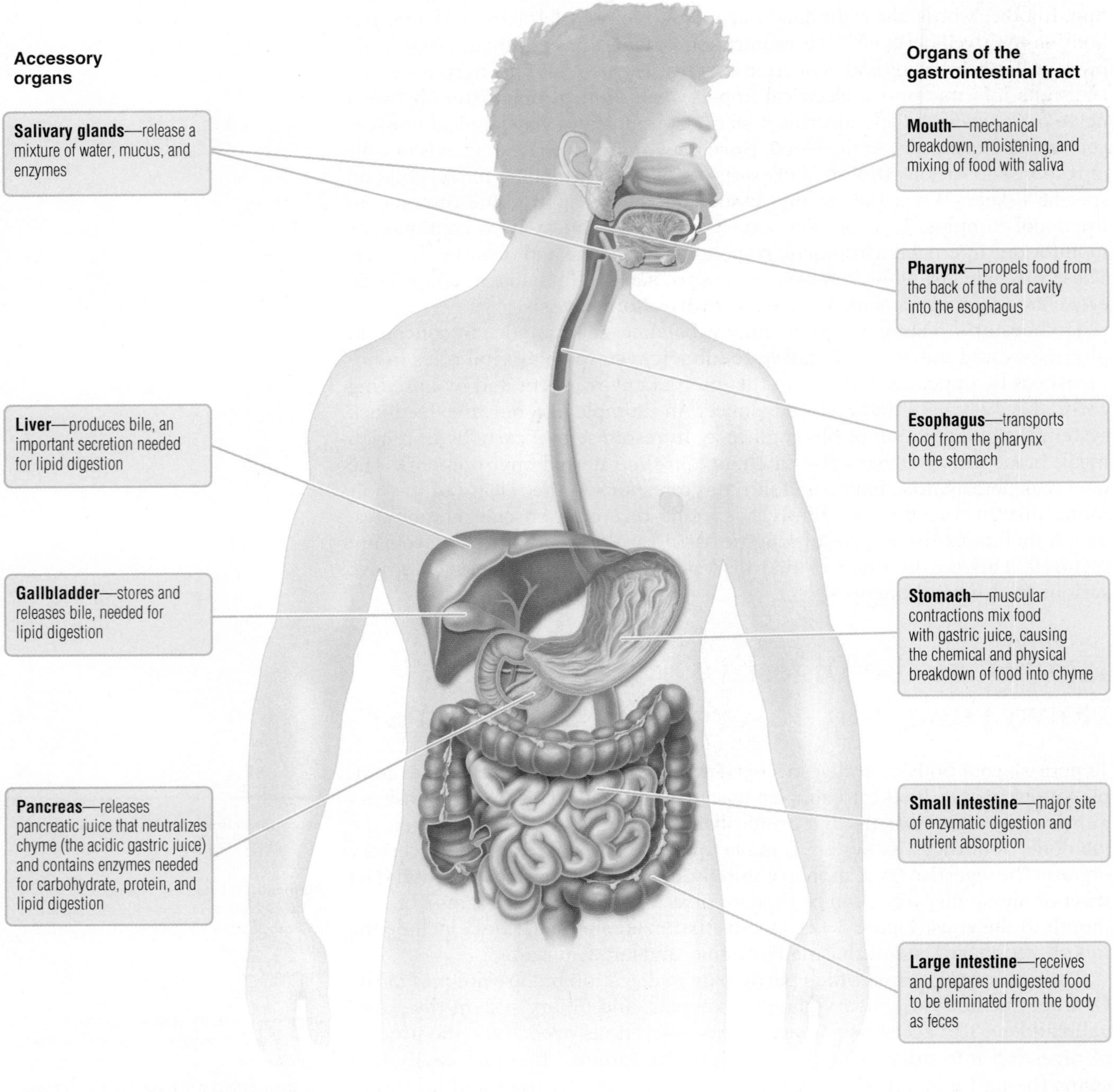

Accessory organs

Salivary glands—release a mixture of water, mucus, and enzymes

Liver—produces bile, an important secretion needed for lipid digestion

Gallbladder—stores and releases bile, needed for lipid digestion

Pancreas—releases pancreatic juice that neutralizes chyme (the acidic gastric juice) and contains enzymes needed for carbohydrate, protein, and lipid digestion

Organs of the gastrointestinal tract

Mouth—mechanical breakdown, moistening, and mixing of food with saliva

Pharynx—propels food from the back of the oral cavity into the esophagus

Esophagus—transports food from the pharynx to the stomach

Stomach—muscular contractions mix food with gastric juice, causing the chemical and physical breakdown of food into chyme

Small intestine—major site of enzymatic digestion and nutrient absorption

Large intestine—receives and prepares undigested food to be eliminated from the body as feces

mucosa (mu – CO – sa) The lining of the gastrointestinal tract that is made up of epithelial cells; also called mucosal lining.

GI secretions Substances released by organs that make up the digestive system that facilitate the process of digestion; also called digestive juices.

The Mucosa: Source of GI Secretions The innermost lining of the digestive tract, called the **mucosa,** consists mainly of epithelial tissue, and carries out a variety of digestive functions. The mucosa, often called the mucosal epithelial lining, produces a variety of secretions such as enzymes and hormones, referred to collectively as **GI secretions,** that facilitate the chemical breakdown of food in the GI tract. Because the cells that make up the mucosal lining are continuously exposed to harsh digestive secretions within the GI tract, their lifespan is a mere two to five days. Once the mucosal epithelial cells wear out, they slough off and are replaced with new cells. Worn-out cells

FIGURE 3.13 The Layers of the Gastrointestinal Tract The gastrointestinal tract consists of four tissue layers: the mucosa, submucosa, muscularis, and serosa. Each layer carries out specific functions—communication, secretion, movement, and protection—all of which facilitate digestion and nutrient absorption.

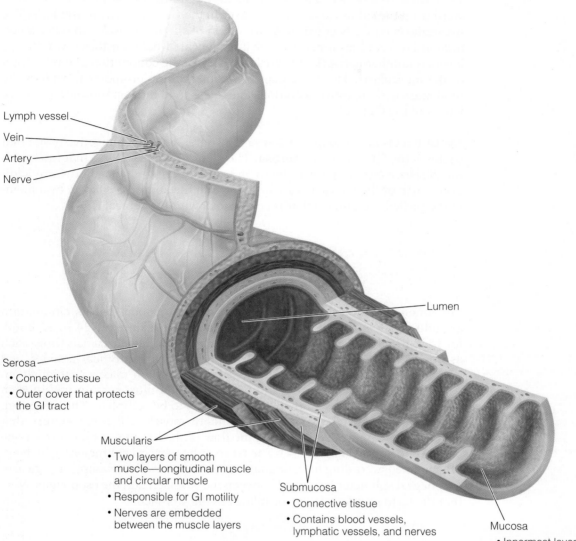

Lymph vessel
Vein
Artery
Nerve

Lumen

Serosa
• Connective tissue
• Outer cover that protects the GI tract

Muscularis
• Two layers of smooth muscle—longitudinal muscle and circular muscle
• Responsible for GI motility
• Nerves are embedded between the muscle layers

Submucosa
• Connective tissue
• Contains blood vessels, lymphatic vessels, and nerves

Mucosa
• Innermost layer of epithelial tissue
• Produces and releases secretions needed for digestion

from the mucosal lining are eliminated from the body in the feces. To support this rapid rate of cell turnover, the mucosa has high nutrient requirements. Nutrient deficiencies can profoundly affect the ability to maintain the mucosal lining, impairing digestion and nutrient absorption.

Blood Vessels and Nerves Embedded in the Submucosa A layer of connective tissue called the **submucosa** surrounds the mucosal layer. The submucosa contains a rich supply of blood vessels, delivering nutrients and oxygen to the inner mucosal layer and the next outward layer, called the muscularis. These blood vessels also circulate most of the nutrients absorbed from the small intestine away from the GI tract. In addition to blood vessels, the submucosa contains lymphatic vessels that are filled with a fluid called **lymph.** Lymph aids in the circulation of water-insoluble substances such as dietary fat away

submucosa (SUB – mu – co – sa) A layer of tissue that lies between the mucosa and muscularis tissue layers.

lymph A fluid found in lymphatic vessels.

How Does the Digestive System Break Down Food into Absorbable Components?   **83**

from the GI tract. The submucosa also contains a network of nerves, which regulate the release of GI secretions from cells making up the mucosal lining.

The Muscularis Enables Food to Mix and Move through the GI Tract Moving outward from the submucosa, the next layer in the GI tract is the muscularis. The **muscularis** typically consists of two layers of smooth muscle: an outer *longitudinal* layer and an inner *circular* layer. Located between these two muscle layers is another network of nerves that control the contraction and relaxation of the muscularis. The movement of the muscularis promotes mixing of the food mass with digestive secretions and keeps food moving through the entire length of the GI tract.

The GI Tract Is Enclosed by the Serosa The outer layer of connective tissue that encloses the GI tract is the **serosa.** The serosa secretes a fluid that lubricates the digestive organs, preventing them from adhering to one another. In addition, much of the GI tract is anchored within the abdominal cavity by a membrane (called mesentery) that is continuous with the serosa.

How Do Gastrointestinal Motility and Secretions Facilitate Digestion?

The amount of time between the consumption of food and its elimination as solid waste is called **transit time.** It takes approximately 24 to 72 hours for food to pass from mouth to anus. Many factors affect transit time, such as composition of diet, illness, certain medications, physical activity, and emotions. Because each organ in the digestive system makes a unique contribution to the overall process of digestion, food must remain within each region long enough for all the digestive events to be complete. This is accomplished in part by circular bands of smooth muscle called **sphincters** that act like one-way valves, regulating the flow of the luminal contents from one organ to the next (Figure 3.14). The GI tract has several sphincters, which are often named according to their anatomical locations. For example, the gastroesophageal sphincter is located between the stomach and the esophagus. Note that the term *gastric* pertains or relates to the stomach.

muscularis (mus – cu – LAR – is) The layer of tissue in the gastrointestinal tract that consists of at least two layers of smooth muscle.

serosa (se – RO – sa) Connective tissue that encloses the gastrointestinal tract.

transit time Amount of time between the consumption of food and its elimination as solid waste.

sphincter (SFINK-ter) A muscular band that narrows an opening between organs in the GI tract.

FIGURE 3.14 Sphincters Regulate the Flow of Food Sphincters are circular bands of muscle located between organs that regulate the flow of material through the gastrointestinal tract.

The gastroesophageal sphincter, located between the esophagus and the stomach, relaxes briefly to allow food to enter the stomach.

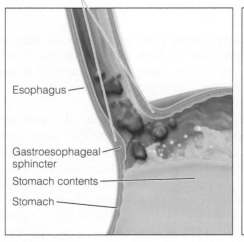

After the food passes into the stomach, the gastroesophageal sphincter closes to prevent the stomach contents from re-entering the esophagus.

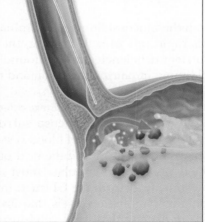

Esophagus

Gastroesophageal sphincter

Stomach contents

Stomach

GASTROINTESTINAL (GI) MOTILITY MIXES AND PROPELS FOOD IN THE GI TRACT

The term **GI motility** refers to the mixing and propulsion of material by muscular contractions in the GI tract. These vigorous movements result from the contraction and relaxation of muscles that make up the muscularis. There are two types of movement in the GI tract: segmentation and peristalsis. **Segmentation** both mixes and slowly propels food, whereas **peristalsis** involves more vigorous propulsive movements. These movements serve different functions, as shown in Figure 3.15. For example, when segmentation occurs in the small intestine, circular muscles move the food mass back and forth in a bidirectional fashion. These mixing and propulsive movements increase the contact between food particles and digestive secretions, giving the intestine an appearance of a chain of sausages. In contrast, peristalsis involves rhythmic, wavelike muscle contractions that propel food along the entire length of the GI tract. The contraction of circular muscles behind the food mass causes the longitudinal muscles to shorten. When the longitudinal muscles subsequently lengthen, the food is

GI motility Mixing and propulsive movements of the gastrointestinal tract caused by contraction and relaxation of the muscularis.

segmentation A muscular movement in the gastrointestinal tract that moves the contents back and forth within a small region.

peristalsis (per – i – STAL – sis) Waves of muscular contractions that move materials in the GI tract in a forward direction.

FIGURE 3.15 Segmentation and Peristalsis GI motility involves both mixing and propulsive movements. Segmentation (A) both mixes and slowly propels food, whereas peristalsis (B) involves more vigorous propulsive movements.

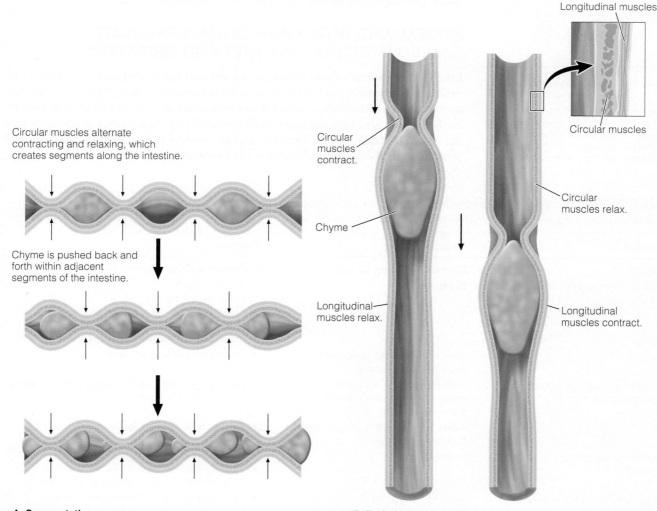

Circular muscles alternate contracting and relaxing, which creates segments along the intestine.

Chyme is pushed back and forth within adjacent segments of the intestine.

Longitudinal muscles

Circular muscles

Circular muscles contract.

Chyme

Longitudinal muscles relax.

Circular muscles relax.

Longitudinal muscles contract.

A. Segmentation
Segmentation both mixes and slowly moves food through the GI tract. The circular muscles contract and relax, creating a "chopping" motion.

B. Peristalsis
Peristalsis consists of a series of wavelike rhythmic contractions and relaxation involving both the circular and longitudinal muscles. This action propels food forward through the GI tract.

propelled forward. Thus, peristalsis is similar to the motion generated when a crowd at a sporting event does "the wave."

GASTROINTESTINAL (GI) SECRETIONS AID DIGESTION AND PROTECT THE GI TRACT

The organs that make up the digestive system release a variety of secretions—water, acid, electrolytes, mucus, salts, enzymes, bile, bicarbonate, and other substances—that are important for digestion and the protection of the GI tract. For example, **mucus** forms a protective coating that lubricates the mucosal lining. **Digestive enzymes** are biological catalysts that facilitate chemical reactions that break down complex food particles. Specifically, digestive enzymes catalyze hydrolytic chemical reactions (hydrolysis) that break chemical bonds by the addition of water. As a result, complex molecules such as starch and protein are broken down into smaller, simpler molecules.

Organs that release digestive secretions include the salivary glands, stomach, pancreas, gallbladder, small intestine, and large intestine. In fact, approximately 7 liters of secretions, most of which is water, are released daily into the lumen of your GI tract. Fortunately, the body has an elaborate recycling system that enables much of this water to be reclaimed. The major GI secretions and their related functions are summarized in Table 3.2.

NEURAL AND HORMONAL SIGNALS REGULATE GASTROINTESTINAL MOTILITY AND SECRETIONS

mucus (MU – cus) A substance that coats and protects mucous membranes.

digestive enzymes Biological catalysts that facilitate chemical reactions that break chemical bonds by the addition of water (hydrolysis), resulting in the breakdown of large molecules into smaller components.

Like the conductor of an orchestra, neural and hormonal signals precisely coordinate GI motility and the release of GI secretions. These involuntary regulatory activities ensure that complex food particles are physically and chemically broken down and that the food mass moves along the GI tract at the appropriate rate. The GI tract has three regulatory control mechanisms—two of which provide neural control and the other hormonal.

TABLE 3.2 Summary of Major Secretions Produced and Released by the Gastrointestinal Tract and Accessory Organs and Their Related Functions

Secretion	Source	Function
Mucus	Mucosal cells of the GI tract	• Protects and lubricates the GI tract
Saliva (contains water, mucus, and enzymes)	Salivary glands	• Moistens foods • Helps form the bolus • Facilitates taste • Aids swallowing • Chemically breaks down food via enzymes
Enzymes	Salivary glands, stomach, small intestine, and pancreas	• Chemically break down foods
GI hormones (e.g., gastrin, secretin, cholecystokinin)	Endocrine cells in the gastric pits, and endocrine cells in the mucosal lining of the small intestine	• Facilitate communication, and regulate GI motility and release of GI secretions
Bile	Made in the liver; stored and released from the gallbladder	• Enables lipid globules to disperse in water
Pancreatic juice (contains bicarbonate and enzymes)	Pancreas	• Neutralizes chyme • Provides enzymes needed for the chemical breakdown of carbohydrates, proteins, and lipids
Gastric juice (contains hydrochloric acid, enzymes, water, intrinsic factor)	Exocrine cells of the gastric pits (mucus-secreting cells, parietal cells, and chief cells)	• Provides enzymes needed for the chemical breakdown of some nutrients • Hydrochloric acid is needed for forming chyme and activating some enzymes • Intrinsic factor is needed for absorption of vitamin B_{12}

Neural Regulation of Digestive Functions The GI tract has its own "local" nervous system that consists of neural networks that are embedded in the submucosa and muscularis layers. Collectively, these networks of nerves are referred to as the **enteric nervous system.** Although the term *enteric* typically pertains to the intestine, the enteric nervous system actually spans the entire length of the GI tract. The enteric nervous system receives information from sensory receptors located within the GI tract. **Sensory receptors** monitor conditions and changes related to digestive activities. For example, **chemoreceptors** detect changes in the chemical composition of the luminal contents, whereas **mechanoreceptors** detect stretching or distension in the walls of the GI tract. As you might expect, the presence of food in the GI tract can stimulate both chemo- and mechanoreceptors. Information from sensory receptors is relayed to the enteric nervous system, which responds by communicating with the muscles and hormone-producing cells of the GI tract. In response, muscles and glands carry out the appropriate response to help digest food, such as increasing peristalsis or releasing secretions.

While the enteric nervous system controls digestive functions on the local level, the GI tract also communicates with the **central nervous system,** which consists of the brain and spinal cord. The neural network connecting the central and enteric nervous systems keeps the GI tract and the brain in close communication. Because the central nervous system can also initiate neural communication with the GI tract, sensory and emotional stimuli can affect GI function. For example, sensory stimuli such as the sight, smell, and thought of food stimulate GI motility and the release of GI secretions. Similarly, emotional factors such as sadness, anger, and anxiety can disrupt digestive functions, causing GI distress such as an upset stomach.

Hormonal Regulation of Digestive Functions Recall that the mucosal lining of the GI tract contains hormone-producing endocrine cells. **GI hormones** play an important communicative role in the process of digestion by acting as chemical messengers. Released into the blood in response to chemical and physical changes in the GI tract, these hormones alert other organs to the impending arrival of food. Like neural signals, hormones also influence transit time and the release of secretions that aid digestion. In addition, some GI hormones communicate with appetite centers in the brain—influencing the desire to eat. For example, release of the hormone ghrelin increases in response to fasting, stimulating hunger and food intake. The major roles of various GI hormones are summarized in Table 3.3 and discussed in more detail in subsequent chapters.

enteric nervous system (en – TER – ic) Neurons located within the submucosa and muscularis layers of the digestive tract.

sensory receptors Receptors that monitor conditions and changes in the GI tract.

chemoreceptor (CHE – mo – re – cep – tor) A type of sensory receptor that responds to a chemical stimulus.

mechanoreceptor (mech – A – no – re – cep – tor) A type of sensory receptor that responds to pressure, stretching, or mechanical stimulus.

central nervous system The part of the nervous system consisting of the brain and spinal cord.

GI hormones Hormones secreted by the mucosal lining of the GI tract that regulate GI motility and secretion.

TABLE 3.3 The Major GI Hormones and Their Related Functions

Hormone	Site of Production	Stimuli for Release	Major Activities
Gastrin	Stomach	• Food in the stomach • Stretching of the stomach walls • Alcohol • Caffeine • Cephalic stimuli (smell, taste)	• Stimulates gastric motility • Stimulates secretion of gastric juice • Increases gastric emptying
Secretin	Small intestine	• Arrival of chyme into the duodenum	• Inhibits gastric motility • Inhibits secretion of gastric juice • Stimulates release of pancreatic juice containing sodium bicarbonate and enzymes
Cholecystokinin (CCK)	Small intestine	• Arrival of partially digested protein and fat into the duodenum	• Stimulates gallbladder to contract and release bile • Stimulates release of pancreatic enzymes
Ghrelin	Stomach (and other tissues)	• Not well understood, but release is greater during fasting	• Stimulates hunger

How Does the Gastrointestinal Tract Coordinate Functions to Optimize Digestion and Nutrient Absorption?

The process of digestion begins with the intake, or ingestion, of food. Over the next 24 to 72 hours, the food is physically and chemically transformed. To optimize digestion and nutrient absorption, the intensity of GI motility and the timing of the release of digestive secretions must be synchronized with the arrival of food. For this reason, digestion is often divided into three phases: the cephalic, gastric, and intestinal phases.

The **cephalic phase** begins even before food enters your mouth. During this phase, the thought, smell, and sight of food stimulate the central nervous system, which in turn stimulates GI motility and the release of digestive secretions. This response serves as a "wake-up" call to your GI tract, preparing it to receive and digest food. The **gastric phase** of digestion begins with the arrival of food in the stomach. During this phase, muscular contractions (gastric motility) become more forceful and the release of gastric secretions increases, which prepare the stomach for its role in the digestive process. By the time food reaches the small intestine, it has undergone considerable physical and chemical change, and no longer resembles the food that you consumed. Yet the process of digestion is not complete. In preparation for the next phase of digestion, the **intestinal phase,** hormonal signals from the small intestine slow the churning action of the stomach (motility), decreasing the rate at which material passes out of the stomach and into the small intestine. As food enters the small intestine, hormonal responses alert the accessory organs (pancreas and gallbladder) of the digestive tasks that lie ahead, signifying that the intestinal phase of digestion is under way.

DIGESTION BEGINS IN THE MOUTH WITH CHEWING AND MIXING FOOD

Whereas the cephalic phase prepares the GI tract to receive and digest food, digestion actually begins when food enters the mouth (also called the oral cavity). The forceful grinding action of your teeth breaks food into manageable pieces. This process, called **mastication,** results in the physical (mechanical) breakdown of food. The presence of food in the mouth stimulates the salivary glands to release **saliva**—as much as a quart per day. Saliva consists of water, mucus, digestive enzymes, and antibacterial agents. As food is broken apart, it mixes with saliva, becoming moist and easier to swallow.

Saliva is also a necessary factor in taste sensation, because some food components must first be dissolved before they can be detected by taste buds. Taste buds, located on specific regions of the tongue, were initially thought to discriminate only four basic tastes: salty, sour, sweet, and bitter. More recently, a fifth taste called **umami** has been identified, and is often described as a meat-like taste sensation. When food is consumed, gustatory (taste) cells and olfactory (smell) cells are stimulated, sending neural signals to the brain. Together, these signals enable the brain to distinguish the thousands of different flavors we enjoy in our foods. Olfactory cells in particular have a profound effect on our ability to taste food, accounting for approximately 80% of taste. Blocked nasal passages make it difficult for the aroma molecules emitted by food to stimulate olfactory cells, which is

cephalic phase (ce – PHAL – ic) The response of the central nervous system to sensory stimuli, such as smell, sight, and taste, that occurs before food enters the GI tract; characterized by increased GI motility and release of GI secretions.

gastric phase The phase of digestion stimulated by the arrival of food into the stomach; characterized by increased GI motility and release of GI secretions.

intestinal phase The phase of digestion in which chyme enters the small intestine; characterized by both a decrease in gastric motility and secretion of gastric juice.

mastication Chewing and grinding of food by the teeth to prepare for swallowing.

saliva A secretion released into the mouth by the salivary glands; moistens food and starts the process of digestion.

umami (u – MAM – i) A taste, in addition to the four basic taste components, that imparts a savory or meat-like flavor.

why it is difficult to taste food when your nasal passages are congested.

Swallowing Moves Food from the Mouth to the Esophagus The tongue, made primarily of muscle, assists in chewing and swallowing. As food mixes with saliva, the tongue manipulates the food mass, pushing it up against the hard, bony palate making up the roof of the mouth. Infants born with an anatomical birth defect called a cleft palate have difficulty swallowing because the bones that form the hard palate are not fused properly. This results in an opening between the oral and nasal cavities. Fortunately, this is a treatable condition, and corrective surgery is typically performed within a few months of birth. Until then, special care must be given to ensure that the infant is not experiencing feeding problems and is adequately nourished.

Figure 3.16 shows that swallowing takes place in two phases. As you prepare to swallow, your tongue directs the soft, moist mass of food, now referred to as a **bolus,** toward the back of your mouth, an area known as the **pharynx.** The pharynx is the shared space between the oral and nasal cavities. This phase of swallowing is under voluntary control, but once

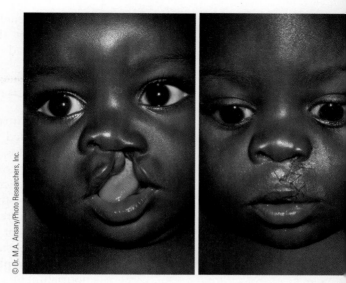

A cleft palate occurs when the bones that make up the hard palate do not completely fuse together. This birth defect can be corrected surgically, often leaving minimal visible signs.

bolus (BO – lus) A soft, moist mass of chewed food.

pharynx (PHAR – nyx) Region toward the back of the mouth that is the shared space between the oral and nasal cavities.

FIGURE 3.16 Voluntary and Involuntary Phases of Swallowing The first phase of swallowing is under our control (voluntary), whereas the second phase is not (involuntary).

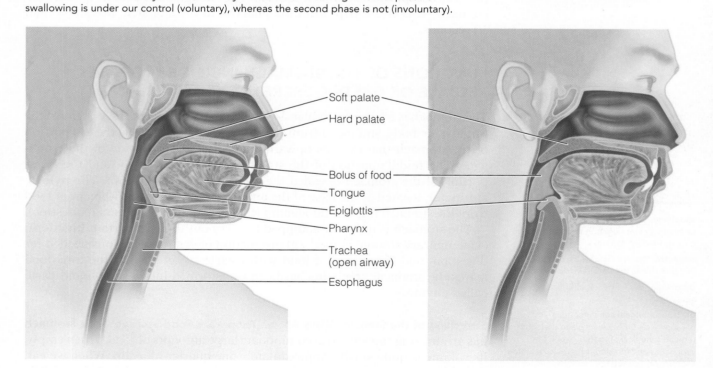

- Soft palate
- Hard palate
- Bolus of food
- Tongue
- Epiglottis
- Pharynx
- Trachea (open airway)
- Esophagus

A. Voluntary phase
The tongue pushes the bolus of food against the hard palate. Next, the tongue pushes the bolus against the soft palate, triggering the swallowing response.

B. Involuntary phase
The soft palate rises, in turn preventing the bolus from entering the nasal cavity. The epiglottis covers the trachea, blocking the opening to the lungs. The bolus enters the esophagus and is propelled toward the stomach by peristalsis.

the bolus reaches the pharynx the involuntary phase of swallowing begins. At this point, the bolus is ready to enter the **esophagus,** a narrow muscular tube that leads to the stomach.

THE ESOPHAGUS DELIVERS FOOD TO THE STOMACH

During the involuntary phase of swallowing, the upper, back portion of your mouth (called the soft palate) rises, blocking the entrance to the nasal cavity. This helps guide the bolus into the correct passageway—the esophagus. Swallowing is also facilitated by the movement of the soft palate that pulls the larynx (vocal cords) upward, causing the **epiglottis** (a cartilage flap) to cover the trachea—the airway leading to the lungs. Once the bolus moves past this dangerous intersection, the muscles relax and prepare for the next swallow. Disorders that affect skeletal muscles or nerves, such as Parkinson's disease and strokes, can affect our ability to swallow. Impaired swallowing, or what is called **dysphagia,** can make it difficult for someone to obtain adequate nourishment.

The esophagus is lubricated and protected by a thin layer of mucus, which facilitates the passage of food. Peristalsis propels the food toward the stomach, where it encounters the first of several sphincters in the GI tract. The **gastroesophageal sphincter** (also called the lower esophageal sphincter or the cardiac sphincter) forms a juncture between the esophagus and the stomach. To prevent the contents of the stomach from re-entering the esophagus, the gastroesophageal sphincter remains closed. However, as the bolus travels toward the stomach, stretch receptors trigger a neural response that signals the gastroesophageal sphincter to relax long enough for the bolus to enter the stomach. This entire trip, from the pharynx to the stomach, takes less than 10 seconds.

FUNCTIONS OF THE STOMACH INCLUDE STORAGE, RELEASE OF GASTRIC SECRETIONS, AND MIXING

The stomach is a large, muscular, J-shaped sac composed of three regions: the fundus, the body, and the antrum (Figure 3.17). The fundus is the top portion of the stomach that extends upward, above the lower portion of the esophagus. The middle portion of the stomach is called the body, and the lower portion of the stomach is called the pyloric region or antrum. The **pyloric sphincter,** located at the base of the antrum, regulates the movement of food from the stomach into the duodenum (the first portion of the small intestine).

The stomach is uniquely equipped to carry out three important functions: (1) temporary storage of food, (2) production of gastric secretions needed for digestion, and (3) mixing of food with gastric secretions. By the time food leaves the stomach, the bolus has been transformed into a semi-liquid paste called **chyme.**

Stretching of the Stomach Walls Allows Temporary Food Storage Your stomach has an amazing capacity to accommodate large amounts of food. When empty, its volume is quite small—approximately one quarter of a cup. When we eat, the walls of the stomach expand to increase its capacity to 1 to 2 quarts. The ability to expand to this extent is due to the special interior lining of the stomach, which is folded into convoluted pleats called **rugae.** Like an accordion, the rugae unfold and flatten, allowing the stomach walls to stretch as it fills with food. This stretching triggers a neural response (via mechanoreceptors), signaling the brain that the stomach is becoming full. Shortly thereafter, hunger diminishes, causing a person to stop eating.

esophagus (e – SOPH – a – gus) The passageway that begins at the pharynx and ends at the stomach.

epiglottis (ep – i – GLOT – tis) A cartilage flap that covers the trachea while swallowing.

dysphagia (dys – PHA – gia) Difficulty swallowing.

gastroesophageal sphincter (gas – tro – e – soph – a – GEAL) A circular muscle that regulates the flow of food between the esophagus and the stomach; also called lower esophageal sphincter or cardiac sphincter.

pyloric sphincter (py – LOR – ic) A circular muscle that regulates the flow of food between the stomach and the duodenum.

chyme The thick fluid resulting from the mixing of food with gastric secretions in the stomach.

rugae (RU – gae) Folds that line the inner stomach wall.

FIGURE 3.17 Anatomy of the Stomach and Its Role in Digestion The stomach is divided into three regions: the fundus, the body, and the antrum.

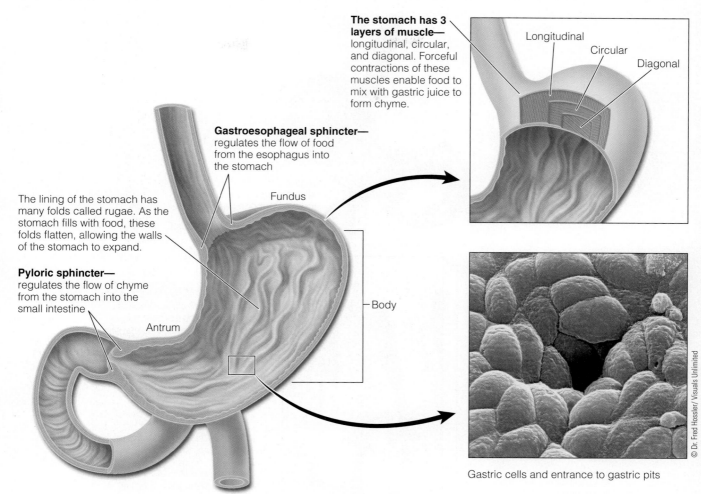

The stomach has 3 layers of muscle—longitudinal, circular, and diagonal. Forceful contractions of these muscles enable food to mix with gastric juice to form chyme.

Longitudinal

Circular

Diagonal

Gastroesophageal sphincter—regulates the flow of food from the esophagus into the stomach

Fundus

The lining of the stomach has many folds called rugae. As the stomach fills with food, these folds flatten, allowing the walls of the stomach to expand.

Pyloric sphincter—regulates the flow of chyme from the stomach into the small intestine

Antrum

Body

Gastric cells and entrance to gastric pits

© Dr. Fred Hossler/ Visuals Unlimited

Gastric Secretions Help Liquefy Solid Food At first glance, the stomach lining appears to be covered with numerous small holes. When the holes are magnified, you can see that they penetrate deep into the mucosal layer and form structures called **gastric pits** (Figure 3.18). The stomach contains several million gastric pits, that are formed by cells that produce and release a variety of gastric secretions. Some of these cells are **endocrine** cells that release their secretions (hormones) into the blood. Others, called **exocrine** cells, release their secretions into ducts that empty directly into the cavity of the gastric pit. These secretions, collectively called **gastric juice,** consist mainly of water, hydrochloric acid, digestive enzymes, mucus, and intrinsic factor—a substance needed for vitamin B_{12} absorption. Your stomach produces more than 2 liters (roughly 2 quarts) of gastric juice daily.

The presence of food in the stomach causes the endocrine cells in the gastric pits to release the hormone **gastrin.** Even the thought, smell, taste, and anticipation of food can trigger its release. Gastrin stimulates exocrine cells to release hydrochloric acid (HCl) and intrinsic factor from **parietal cells** and digestive enzymes from neighboring **chief cells.** HCl, a major component of gastric juice, dissolves food particles, destroys bacteria that may be present in food, and provides an optimal acidic environment (pH 2) for digestive enzymes produced in the stomach to function.

gastric pits Invaginations of the mucosal lining of the stomach that contain specialized endocrine and exocrine cells.

endocrine cells Those that produce and release hormones into the blood.

exocrine cells Those that produce and release their secretions into ducts.

gastric juice Digestive secretions produced by exocrine cells that make up gastric pits.

gastrin (GAS – trin) A hormone secreted by endocrine cells that stimulates the production and release of gastric juice.

parietal cells (pa – RI – e – tal) Exocrine cells within the gastric mucosa that secrete hydrochloric acid and intrinsic factor.

chief cells Exocrine cells in the gastric mucosa that produce digestive enzymes.

FIGURE 3.18 **Gastric Pits** The mucosal lining of the stomach is made up of exocrine and endocrine cells. The exocrine cells (mucus-secreting cells, chief cells, and parietal cells) secrete gastric juices, whereas the endocrine cells secrete hormones.

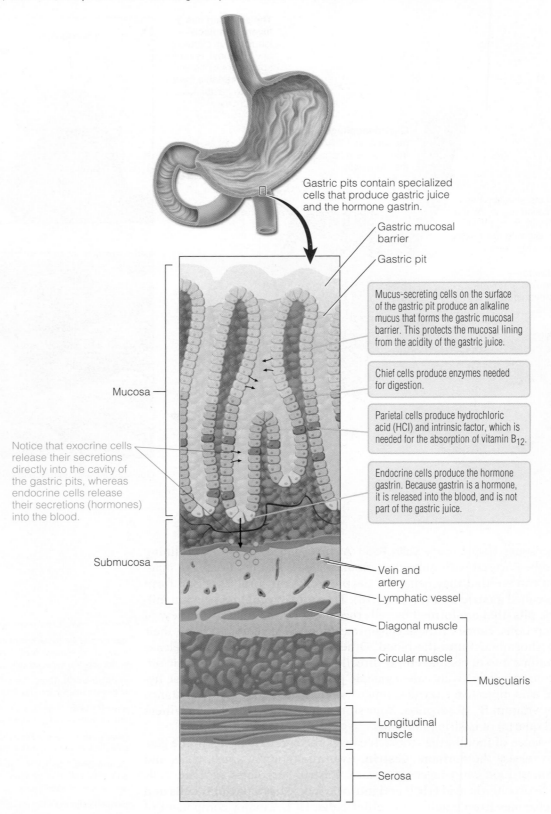

Gastric pits contain specialized cells that produce gastric juice and the hormone gastrin.

Gastric mucosal barrier

Gastric pit

Mucus-secreting cells on the surface of the gastric pit produce an alkaline mucus that forms the gastric mucosal barrier. This protects the mucosal lining from the acidity of the gastric juice.

Chief cells produce enzymes needed for digestion.

Parietal cells produce hydrochloric acid (HCl) and intrinsic factor, which is needed for the absorption of vitamin B_{12}.

Endocrine cells produce the hormone gastrin. Because gastrin is a hormone, it is released into the blood, and is not part of the gastric juice.

Mucosa

Notice that exocrine cells release their secretions directly into the cavity of the gastric pits, whereas endocrine cells release their secretions (hormones) into the blood.

Submucosa

Vein and artery

Lymphatic vessel

Diagonal muscle

Circular muscle

Muscularis

Longitudinal muscle

Serosa

Located near the entrance of the gastric pits are numerous secretory cells that release a thin, watery mucus. The mucus forms a protective layer called the **gastric mucosal barrier,** which prevents the acidic gastric juice from damaging the delicate lining of the stomach. Without this layer, the mucosal lining could not withstand its harsh environment, resulting in inflammation and the formation of sores or ulcers. Similarly, a condition called gastroesophageal reflux disease results when the unprotected lining of the esophagus is repeatedly exposed to gastric juice.[1]

Gastric Ulcers and Gastroesophageal Reflux Disease Although the mucosal barrier is usually successful in protecting the stomach and esophagus from the harsh gastric juices, this protection sometimes fails, resulting in two common GI disorders: gastric ulcers and gastroesophageal reflux disease (GERD). As you can see from Figure 3.19, an ulcer looks similar to an open canker sore in the mouth and, if left untreated, can erode through the various tissue layers. This can result in severe complications such as bleeding and infection in the abdominal cavity. Ulcers can occur when gastric juice erodes areas of the mucosal lining in the esophagus (esophageal ulcer), stomach (**gastric ulcer**), or duodenum (duodenal ulcer). Collectively, GI ulcers (stomach, duodenum, or esophagus) are referred to as **peptic ulcers.**

The primary cause of GERD is the relaxation of the gastroesophageal sphincter, which enables the stomach contents (chyme) to move from the stomach back into the esophagus, or what is referred to as reflux (Figure 3.20). Repeated exposure to the acidic gastric juice can cause the delicate lining of the esophagus to become inflamed. You can read more about these two common GI disorders (peptic ulcers and GERD) in the Focus on Clinical Applications feature.

gastric mucosal barrier A thick layer of mucus that protects mucosal lining of the stomach from the acidic gastric juice.

gastric ulcer A sore in the lining of the stomach.

peptic ulcer An irritation or erosion of the mucosal lining in the stomach, duodenum, or esophagus.

FIGURE 3.19 Peptic Ulcers
Peptic ulcers are erosions that occur in the mucosal lining of the esophagus, stomach, or duodenum.

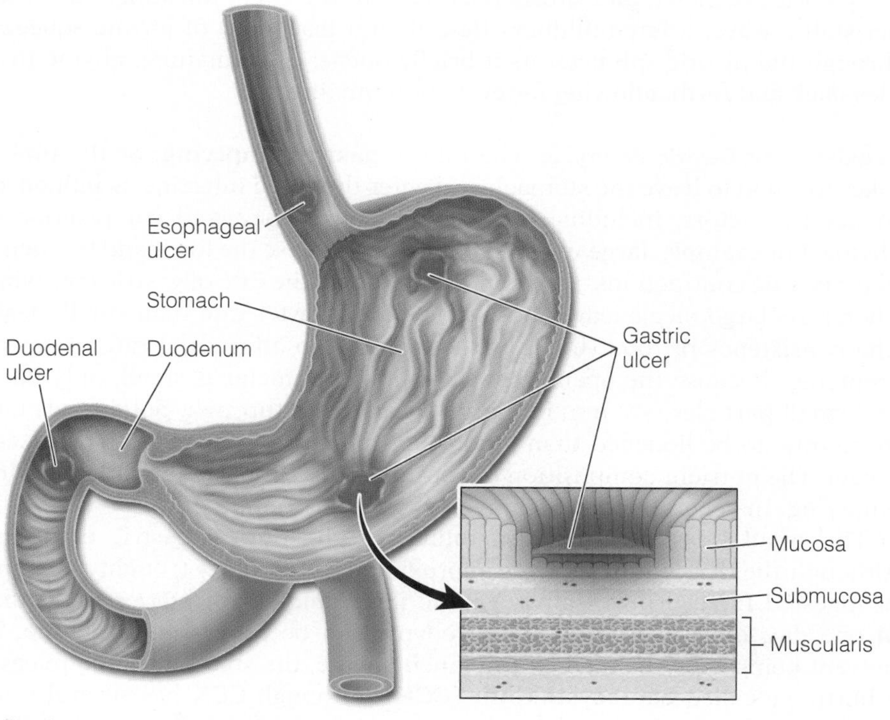

Peptic ulcers occur when the mucosal lining of the esophagus, stomach, or duodenum becomes eroded. If left untreated, an ulcer can penetrate through the layers of the GI tract.

FIGURE 3.20 Gastroesophageal Reflux Disease (GERD) GERD is caused by dysfunction of the gastroesophageal sphincter.

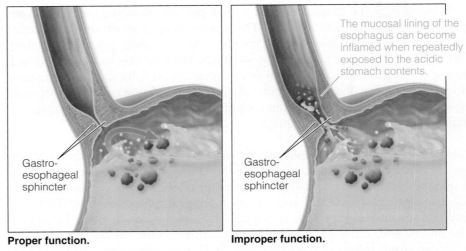

The mucosal lining of the esophagus can become inflamed when repeatedly exposed to the acidic stomach contents.

Gastro-esophageal sphincter

Gastro-esophageal sphincter

Proper function.

To prevent food from flowing back into the esophagus, the gastroesophageal sphincter normally remains contracted.

Improper function.

If the gastroesophageal sphincter weakens, the stomach contents flow back into the esophagus. The reflux of stomach contents into the esophagus is called gastroesophageal reflux disease.

Gastric Mixing Helps Form Chyme In addition to storing food and producing secretions, the stomach also mixes the food with the gastric juices to form chyme. Unlike the rest of the GI tract, your stomach has a third (diagonal) layer of smooth muscle that generates a forceful churning action, much like the kneading of bread. The strength of these contractions increases under the influence of the hormone gastrin. Within three to five hours after ingestion, the partially digested food is thoroughly mixed with the gastric juices. The resulting chyme, now the consistency of a soupy paste, is pushed toward the pyloric sphincter, and slowly released into the small intestine. With each peristaltic wave, a few milliliters (less than a teaspoon) of chyme squeezes through the pyloric sphincter as it briefly opens; the remaining chyme tumbles back and forth, allowing for even more mixing.

Regulation of Gastric Emptying The rate of **gastric emptying,** or the time it takes for food to leave the stomach and enter the small intestine, is influenced by several factors, including the volume, consistency, and composition of chyme. For example, large volumes of chyme increase the force and frequency of peristaltic contractions, which in turn increase the rate of gastric emptying. Therefore, large meals leave your stomach at a faster rate than small meals. The consistency of food (liquid versus solid) also affects the rate of gastric emptying. Because the opening of the pyloric sphincter is small, only fluids and small particles (<2 mm in diameter) can pass through. Solid foods take more time to be liquefied than fluids and, therefore, remain in the stomach longer. The nutrient composition of your meal also impacts the rate of gastric emptying. In general, foods high in fat slow gastric emptying.[2]

The small intestine itself also influences the rate of gastric emptying. Although the duodenum receives approximately 10 quarts (roughly 40 cups) of ingested food, drinks, saliva, gastric juice, and so forth over the course of one day, only small amounts of chyme can be processed at a time. To prevent being overwhelmed by too much chyme, the small intestine releases a hormone called **cholecystokinin (CCK).** Although CCK has several functions, one of its most important functions is to slow gastric emptying. This delay enables the small intestine to properly digest the chyme that it receives. When the small intestine is ready to receive more chyme, the release of CCK decreases.

gastric emptying The process by which food leaves the stomach and enters the small intestine.

cholecystokinin (CCK) (CHO – le – cys – to – KI – nin) A hormone, produced by the small intestine, that stimulates the release of enzymes from the pancreas and contraction of the gallbladder.

***Helicobacter pylori* (*H. pylori*)** A bacterium residing in the GI tract that causes peptic ulcers.

gastroesophageal reflux disease (GERD) A condition caused by the weakening of the gastroesophageal sphincter, which enables gastric juices to reflux into the esophagus, causing irritation to the mucosal lining.

FOCUS ON CLINICAL APPLICATIONS
Peptic Ulcers and Gastroesophageal Reflux Disease

Millions of Americans describe the symptoms as "fire" in the belly—that burning sensation often mistaken for indigestion. However, peptic ulcers and gastroesophageal reflux disease (GERD) are more than just a little indigestion. These terms both refer to painful conditions that affect the upper GI tract.

Peptic Ulcers

Contrary to popular belief, the majority of peptic ulcers are not caused by stress or by eating spicy foods. Rather, most ulcers (80%) are caused by a small spiral-shaped bacterium called **Helicobacter pylori** (*H. pylori*).[3] This discovery revolutionized standard medical treatment that traditionally consisted of a diet of bland food, milk, stress reduction, and rest. However, the idea that a bacterium could cause ulcers was slow to gain acceptance in the medical community. It seemed highly doubtful that bacteria could survive the acidic environment of the stomach. This changed in 1982 when Drs. Barry Marshall and Robin Warren demonstrated that *H. pylori* could burrow into the thick protective mucus layer, exposing the sensitive underlying stomach layers to acidic gastric juice.

To prove their theory that *H. pylori* could survive the acidic environment of the stomach, Dr. Marshall willingly swallowed the bacterium.[4] Ten days later, he developed acute gastritis (inflammation of the stomach lining). The presence of *H. pylori* was later confirmed by examining a sample of his gastric mucosal lining. Because of Marshall and Warren's scientific conviction, the *H. pylori* bacterium is now widely accepted as the primary cause of ulcers. In 2005, the Nobel Prize in Physiology or Medicine was awarded to Marshall and Warren for their discovery that peptic ulcers and gastritis could be caused by a bacterium.

The most common symptom associated with ulcers is dull, gnawing pain in the stomach that is often relieved by eating. Other symptoms include intermittent pain in the abdominal region, weight loss, loss of appetite, and vomiting. Today, doctors treat most people diagnosed with ulcers with a combination of therapies, including antibiotics to address the underlying bacterial infection and acid-blocking medications to help promote healing. Although most ulcers are caused by *H. pylori*, irritants such as nonsteroidal anti-inflammatory agents (aspirin and ibuprofen) and alcohol can cause them as well.

Gastroesophageal Reflux Disease (GERD)

Gastroesophageal reflux disease (GERD) is a persistent disorder that affects a striking percentage of the U.S. population. According to the American College of Gastroenterology, more than 15 million Americans experience GERD symptoms daily. It occurs when there is reverse flow (reflux) of the stomach contents into the esophagus. Since the esophagus does not have a thick protective mucus layer, repeated exposure to acidic chyme can irritate its lining.

This causes a burning sensation in the upper chest, the most common symptom of GERD. Because of the location of the pain, people with GERD often complain of what is

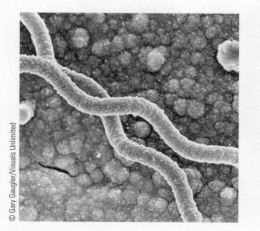

In 1982, Drs. Robin Warren (left) and Barry Marshall (right) proposed that this small spiral-shaped bacterium called *Helicobacter pylori* (*H. pylori*) was the cause of most peptic ulcers. In 2005, these researchers were awarded the Nobel Prize in Physiology or Medicine for their discovery.

(continued)

How Does the Gastrointestinal Tract Coordinate Functions to Optimize Digestion and Nutrient Absorption?　**95**

frequently called heartburn. Other physical complaints associated with GERD include a dry cough, asthma, and difficulty swallowing.

The primary cause of GERD is the inappropriate relaxation of the gastroesophageal sphincter, which normally prevents the stomach contents (chyme) from moving back into the esophagus. Dietary habits that contribute to this include eating large portions of foods and lying down soon after eating. GERD is also associated with consuming certain foods and beverages such as onions, chocolate, mint, high-fat foods, spicy foods, citrus juices, alcohol, and caffeinated beverages. Other factors associated with GERD include being overweight, smoking, and wearing tight-fitting clothes. Pregnancy-related hormonal and anatomical changes can also contribute to GERD.

Making the appropriate lifestyle changes is the first step in preventing and managing GERD. However, when lifestyle changes are not enough, over-the-counter or prescription medications may be necessary. Fortunately, many people are able to manage GERD effectively by making lifestyle changes such as eating smaller portions of foods, avoiding foods that trigger reflux, and remaining upright after eating.

GERD can lead to chronic inflammation of the esophagus that, left untreated, is a risk factor for esophageal cancer. A procedure called **endoscopy** is used to examine the lining of the esophagus for damage and early precancerous changes. For this reason, it is important for all people experiencing GERD to seek medical attention and treatment.

---

endoscopy A procedure used to examine the lining of the GI tract

duodenum (du – o – DE – num) The first segment of the small intestine.

jejunum (je – JU – num) The midsection of the small intestine, located between the duodenum and the ileum.

ileum (IL – e – um) The last segment of the small intestine that comes after the jejunum.

THE SMALL INTESTINE IS THE PRIMARY SITE OF CHEMICAL DIGESTION AND NUTRIENT ABSORPTION

The small intestine is the primary site of chemical digestion and nutrient absorption. This 20-foot-long narrow tube has a diameter of about 1 inch and is well-suited to carry out these functions. The small intestine consists of three regions—the **duodenum,** the **jejunum,** and the **ileum** (Figure 3.21). Chyme

FIGURE 3.21 Overview of the Small Intestine and Accessory Organs

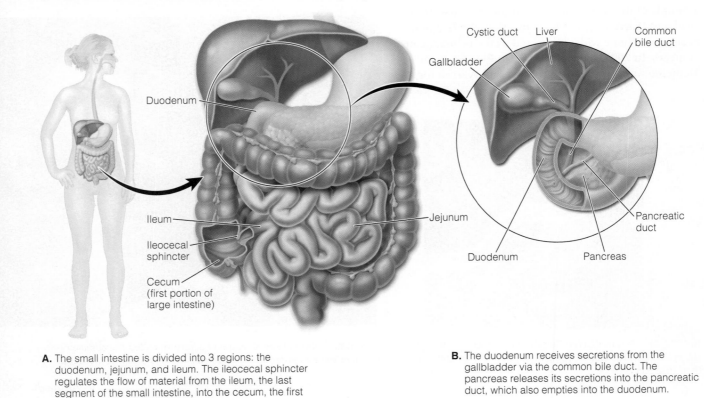

A. The small intestine is divided into 3 regions: the duodenum, jejunum, and ileum. The ileocecal sphincter regulates the flow of material from the ileum, the last segment of the small intestine, into the cecum, the first portion of the large intestine.

B. The duodenum receives secretions from the gallbladder via the common bile duct. The pancreas releases its secretions into the pancreatic duct, which also empties into the duodenum.

first passes into the duodenum, the receiving end of the small intestine. In addition to chyme, the duodenum receives secretions from the gallbladder (bile) and the pancreas (pancreatic juice). When the gallbladder contracts, bile is forced into the cystic duct, which ultimately joins the common bile duct and empties into the duodenum. Similarly, pancreatic juice is released into the pancreatic duct, which also empties into the duodenum.

The lining of the small intestine has a large surface area, which is aptly suited for the process of digestion and nutrient absorption. As illustrated in Figure 3.22, the inner lining of the small intestine (the mucosa and underlying submucosa) is arranged in large, circular folds called **plica circulares.** These circular folds face inward, toward the lumen of the small intestine, and

plica circulares Circular folds in the mucosal lining of the small intestine.

FIGURE 3.22 Absorptive Surface of the Small Intestine

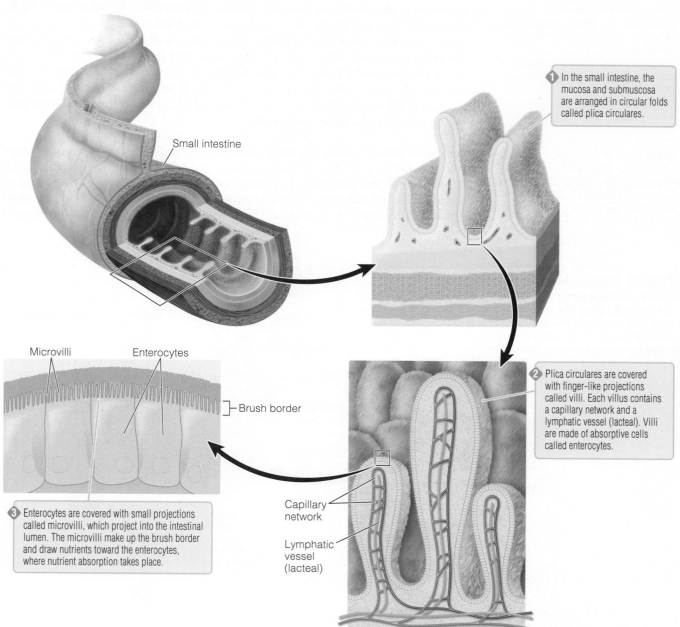

Small intestine

1 In the small intestine, the mucosa and submuscosa are arranged in circular folds called plica circulares.

2 Plica circulares are covered with finger-like projections called villi. Each villus contains a capillary network and a lymphatic vessel (lacteal). Villi are made of absorptive cells called enterocytes.

Microvilli   Enterocytes

Brush border

3 Enterocytes are covered with small projections called microvilli, which project into the intestinal lumen. The microvilli make up the brush border and draw nutrients toward the enterocytes, where nutrient absorption takes place.

Capillary network

Lymphatic vessel (lacteal)

are covered with tiny finger-like projections called **villi** (plural form of *villus*). Another way to think of the inner lining of the small intestine is to imagine a looped-style bathroom rug folded like an accordion. The folds in the rug represent the plica circulares, whereas each tiny loop that covers the surface of the rug represents a villus.

Each villus is made up of hundreds of absorptive epithelial cells called **enterocytes.** The surface of the enterocyte that faces the intestinal lumen is covered with thousands of minute projections called **microvilli.** The microvilli comprise the absorptive surface of the small intestine, or what is called the **brush border.** This vast surface area is approximately the size of the playing surface on a standard tennis court. The brush border of the small intestine is where the final stages of nutrient digestion and absorption take place. Each villus contains a network of blood capillaries and a lymph-containing lymphatic vessel called a **lacteal,** both of which circulate absorbed nutrients away from the small intestine.

Digestion in the Small Intestine Is Regulated by Hormones In addition to CCK, the small intestine produces a hormone called **secretin.** With great precision, these hormones help regulate the process of digestion by coordinating the release of secretions from the pancreas and gallbladder, the relaxation of sphincters, and GI motility. The actions of secretin and CCK ensure that nutrient digestion and absorption in the small intestine are rapid and efficient. Indeed, within 30 minutes of the arrival of chyme in the small intestine, the final stages of digestion are complete.

Pancreatic Juice Protects the Small Intestine The pancreas plays an important role in protecting the small intestine from the acidity of chyme (Figure 3.23). Recall that chyme has a pH of approximately 2 and is potentially damaging to the unprotected lining of the small intestine. The arrival of chyme in the small intestine stimulates the release of the hormone secretin, which in turn signals

villi (plural of *villus*) (VI – li, VI – lus) Small, finger-like projections that cover the inner surface of the small intestine.

enterocytes (en – TER – o – cytes) Epithelial cells that make up the lumenal surface of each villus.

microvilli (MI – cro – vi – li) Hairlike projections on the luminal surface of enterocytes.

brush border The absorptive surface of the small intestine made up of thousands of microvilli that cover the luminal surface of enterocytes.

lacteal (LAC – te – al) A lymphatic vessel found in an intestinal villus.

secretin (se – CRE – tin) A hormone, secreted by the small intestine, that stimulates the release of sodium bicarbonate and enzymes from the pancreas.

FIGURE 3.23 The Pancreas The pancreas releases pancreatic juice into the pancreatic duct. The pancreatic juice helps neutralize chyme when it enters the small intestine and contains enzymes needed for nutrient digestion.

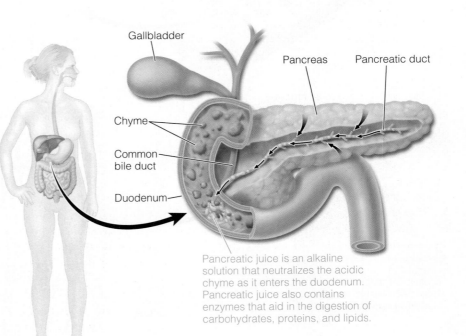

Gallbladder

Pancreas   Pancreatic duct

Chyme

Common bile duct

Duodenum

Pancreatic juice is an alkaline solution that neutralizes the acidic chyme as it enters the duodenum. Pancreatic juice also contains enzymes that aid in the digestion of carbohydrates, proteins, and lipids.

FIGURE 3.24 The Role of the Liver and Gallbladder in Digestion The liver produces bile, which is stored in the gallbladder. The gallbladder releases bile into the common bile duct, which empties into the small intestine. Bile plays an important role in lipid digestion.

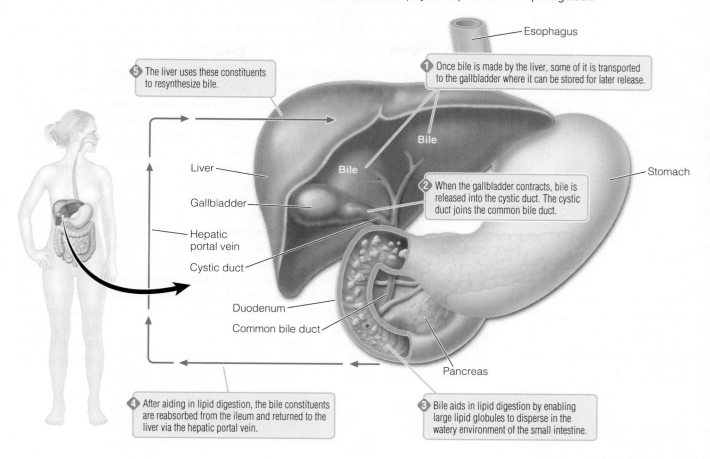

① Once bile is made by the liver, some of it is transported to the gallbladder where it can be stored for later release.

⑤ The liver uses these constituents to resynthesize bile.

② When the gallbladder contracts, bile is released into the cystic duct. The cystic duct joins the common bile duct.

④ After aiding in lipid digestion, the bile constituents are reabsorbed from the ileum and returned to the liver via the hepatic portal vein.

③ Bile aids in lipid digestion by enabling large lipid globules to disperse in the watery environment of the small intestine.

Esophagus
Bile
Bile
Liver
Stomach
Gallbladder
Hepatic portal vein
Cystic duct
Duodenum
Common bile duct
Pancreas

the pancreas to release **pancreatic juice.** This alkaline solution consists of water, sodium bicarbonate, and various enzymes needed for digestion. The sodium bicarbonate in pancreatic juice quickly neutralizes chyme as it enters the duodenum.

Bile Is Needed for Fat Digestion A substance called **bile** also plays an important role in digestion, especially when you consume fatty foods (Figure 3.24). Bile, which is made in the liver, is a watery solution that consists primarily of cholesterol, bile acids, and bilirubin—a pigment that gives bile its characteristic yellowish-green color. Once bile is formed, it is transported to the gallbladder, where some of it is stored and the rest is released into the small intestine.

Fats are not soluble in the watery environment of the small intestine and are therefore more difficult to digest and absorb than other food components. To counteract this, the presence of fat-containing chyme in the small intestine signals the release of CCK, which causes the gallbladder to contract, emptying its contents into the duodenum. The bile acids and cholesterol in bile act like detergents, dispersing large globules of fat into smaller droplets that are easier for the enzymes to digest. Once lipid digestion is complete, bile is reabsorbed through the ileum and returned to the liver via the **hepatic portal vein.** This process, called **enterohepatic circulation,** enables the liver to recycle many of the constituents that make up bile. In fact, only 5% of the bile escapes into the large intestine and is lost in the feces.

pancreatic juice Pancreatic secretions that contain bicarbonate and enzymes needed for digestion.

bile A fluid, made by the liver and stored in and released from the gallbladder, that contains bile salts, cholesterol, water, and bile pigments.

hepatic portal vein A blood vessel that circulates blood to the liver from the GI tract.

enterohepatic circulation (en – ter – O – he – PA – tic) Circulation between the small intestine and the liver used to recycle compounds such as bile.

Digestion in the Small Intestine Is Facilitated by Enzymes Both the small intestine and pancreas provide enzymes needed to chemically break down nutrients even further. Pancreatic enzymes are released into the duodenum in response to the hormone CCK. Intestinal enzymes are made in the brush border epithelial cells, where the final steps of enzymatic digestion take place.

Nutrient Absorption When the process of digestion is complete, nutrients are ready to be absorbed. The transfer of nutrients from the GI tract to the circulatory system, or what is referred to as **nutrient absorption,** takes place by passive and active transport mechanisms: simple diffusion, facilitated diffusion, carrier-mediated active transport, and to a lesser extent endocytosis. The stomach has a minor role in nutrient absorption—only water and alcohol are absorbed there to any significant extent. By far, the majority of nutrients are absorbed along the brush border surface of the small intestine. Once absorbed, nutrients enter either the blood or lymphatic circulatory systems.

The vast surface area and unique structure of the small intestine make nutrient absorption very efficient. The sweeping action of the microvilli traps and pulls nutrients toward the enterocytes. However, the transfer of nutrients from the lumen of the small intestine into the enterocyte is really only the first step in nutrient absorption. To enter the blood or lymph, nutrients must also cross the **basolateral membrane,** the cell membrane of the enterocyte that faces away from the lumen toward the submucosa. Thus, nutrient absorption includes both entry into and exit out of the enterocyte, as illustrated in Figure 3.25. Disease states that affect the absorptive surface of the small intestine, such as celiac disease, can lead to nutritional deficiencies. You can read more about celiac disease in the Focus on Clinical Applications feature.

⟨CONNECTIONS⟩ Recall that an autoimmune disease develops when the immune system produces antibodies that attack and destroy cells in the body (Chapter 1, page 24).

nutrient absorption The transfer of nutrients from the lumen of the GI tract to the circulatory system.

basolateral membrane The cell membrane that faces away from the lumen of the GI tract and toward the submucosa.

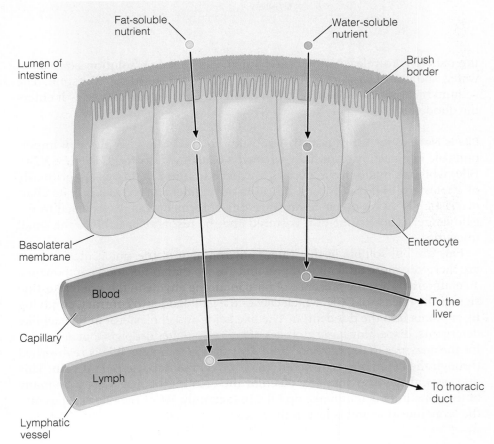

FIGURE 3.25 Nutrient Absorption and Circulation Nutrient absorption includes both entry into and exit out of enterocytes. Once nutrients cross the basolateral membrane, they are circulated away from the intestine by either blood or lymph. Water-soluble nutrients are circulated in blood, whereas fat-soluble nutrients are circulated in lymph.

The extent to which a nutrient or other food component is absorbed is called its **bioavailability.** The bioavailability of a particular nutrient can be influenced by physiological conditions, other dietary components, and certain medications. For example, the body absorbs only the amount of iron it needs and excretes the excess in feces. This regulatory step helps protect us from iron toxicity. It is also well known that some nutrients can markedly affect the bioavailability of other nutrients. For example, vitamin C can enhance the absorption of certain forms of iron, and vitamin D can increase the bioavailability of calcium. This is why it is important to be aware of the impact that certain nutrients and drugs may have on nutrient bioavailability.

How Does the Body Circulate Nutrients and Eliminate Cellular Waste Products?

Once nutrient digestion and absorption are complete, the next task is to transport the nutrients throughout the body. This is accomplished by the body's extensive circulatory system, which is made up of veins, arteries, and lymphatic vessels. In addition to delivering nutrients and oxygen, the circulatory system aids in the elimination of cellular waste products.

NUTRIENTS ABSORBED FROM THE SMALL INTESTINE ARE CIRCULATED TO THE LIVER

Upon absorption, water-soluble nutrients enter the bloodstream through the capillaries contained within each villus. (As we will discuss later in this chapter, fat-soluble nutrients instead enter lymphatic vessels.) Once water-soluble nutrients enter the bloodstream, they circulate directly to the liver. This arrangement gives the liver first access to the nutrient-rich blood leaving the small intestine. Nutrients taken up by the liver can then be stored, undergo metabolic changes, or be released into the systemic circulation, which in turn delivers the nutrients to other parts of the body.

THE CARDIOVASCULAR SYSTEM CIRCULATES NUTRIENTS, OXYGEN, AND OTHER SUBSTANCES

In addition to having a continuous supply of essential nutrients and oxygen, cells must also have a way to rid themselves of metabolic waste products. The cardiovascular system, which consists of the heart and an elaborate vascular network, helps meet these needs. The cardiovascular system consists of two separate loops: (1) the systemic circulation and (2) the pulmonary circulation. As illustrated in Figure 3.26 on p. 103, each of these circulatory loops delivers blood to specific regions within the body.

The Systemic Circulation Delivers Blood to Body Organs Substances are transported to and from cells in the blood, which flows through vessels called **arteries** and **veins** in a continuous loop that begins and ends at the heart. This part of the cardiovascular system, called the **systemic circulation,** delivers blood to all the body's organs except the lungs. Oxygenated blood leaves the heart through the **aorta,** which then branches into an intricate maze of arteries. Arteries circulating blood away from the heart divide, subdivide, and eventually form beds of microscopic vessels called **capillaries** situated around organs and tissues. Capillaries have thin walls

bioavailability The extent to which nutrients are absorbed into the blood or lymphatic system.

artery A blood vessel that carries blood away from the heart.

vein A blood vessel that carries blood toward the heart.

systemic circulation The division of the cardiovascular system that begins and ends at the heart and delivers blood to all the organs except the lungs.

aorta (a – OR – ta) The main artery that initially carries blood from the heart to all areas of the body except the lungs.

capillaries (CAP – il – lar – ies) Blood vessels with thin walls, which allow for the exchange of materials between blood and tissues.

It has only been in the last 50 years that researchers have begun to understand celiac disease and how to treat it. In 1888, Dr. Samuel Gee recognized that, when celiac patients avoided foods that contained starch, they "suffered" far less.[5] Many years later, Dr. Gee's observations were confirmed. Indeed, **celiac disease** is an inflammatory response to a specific protein called **gluten** found in a variety of cereal grains such as wheat, rye, barley, and possibly oats. When people with celiac disease consume gluten-containing foods, the lining of the small intestine becomes damaged and eventually causes severe symptoms. Celiac disease is far more common than previously estimated, and its prevalence may be as high as 3 million—1 out of 133 Americans.[6]

Researchers now know that people with celiac disease (also called gluten-sensitive enteropathy) experience an immunological response to gluten. More specifically, the consumption of gluten-containing foods triggers the production of specific antibodies that attack the intestinal microvilli, causing them to flatten.[7] As a result, nutrient absorption is impaired. Like many other types of auto-immune diseases, celiac disease runs in families.

Because of the progressive damage to the absorptive surface of the small intestine, people with celiac disease experience diarrhea, weight loss, and malnutrition. In fact, children with celiac disease often experience slow growth and tend to be small for their age.[8] More recently, migraine headaches, osteoporosis, miscarriages, and infertility have also been attributed to celiac disease.[9] Celiac disease is often misdiagnosed because the signs and symptoms are similar to other common GI disorders. When celiac disease is suspected, a blood test may be performed to screen for the presence of antibodies made in response to gluten. Although antibody testing is important, a definitive diagnosis can only be made by biopsy, a procedure that requires taking a small piece of tissue from the intestinal lining.

Fortunately, once diagnosed, many people with celiac disease manage to live symptom-free by eliminating gluten from their diet. This was certainly the case with Emily, who shared her story in Chapter 2. Recall that it took several months before she was accurately diagnosed with celiac disease. Once she realized that her iron deficiency was due to celiac disease and not insufficient dietary iron, her health quickly returned to normal. However, given the numerous food products made with wheat and other cereal grains (such as breads, crackers, cookies, cakes, and pasta), adherence to a gluten-free diet is easier advised than done. In addition to occurring naturally in many cereal-based products, gluten is often added to many processed foods. People with celiac disease should be vigilant when reading food labels because gluten may be present in less obvious foods such as meats, soups, candies, soy sauce, malt beverages, and even in some medications. In addition, foods made with modified food starch, hydrolyzed vegetable protein, and binders may contain gluten. For this reason, foods claiming to be wheat free are not always gluten free.

For people with celiac disease, it is important to buy foods that are gluten free. A variety of gluten-free products are available in most grocery stores.

---

celiac disease (CE – li – ac) An autoimmune response to the protein gluten that damages the absorptive surface of the small intestine; also called gluten-sensitive enteropathy.

gluten (GLU – ten) A protein found in cereal grains such as wheat, rye, barley, and possibly oats.

interstitial fluid (in – ter – STI – tial) Fluid that surrounds cells.

and narrow diameters, making them well suited for their primary function—the exchange of materials, nutrients, and gases between the blood and tissues. Tiny pores in the capillary walls allow materials such as nutrients and oxygen to pass from the blood into the **interstitial fluid** that surrounds cells (Figure 3.27 on p. 104). At the same time, cellular waste products (such as carbon dioxide) can be taken up from the interstitial fluid and carried away by the blood.

A capillary network marks the end of the arterial blood flow to the cell and the beginning of the venous blood flow away from the cell and back to the

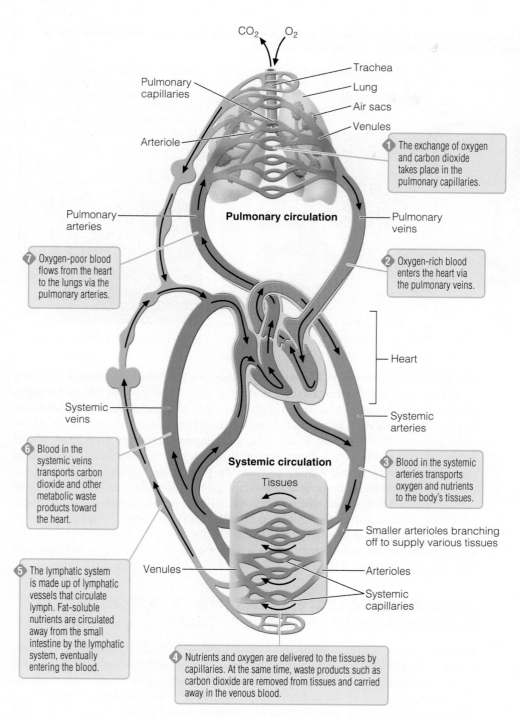

CO₂   O₂

Trachea
Pulmonary capillaries
Lung
Air sacs
Venules
Arteriole

❶ The exchange of oxygen and carbon dioxide takes place in the pulmonary capillaries.

Pulmonary circulation

Pulmonary arteries
Pulmonary veins

❼ Oxygen-poor blood flows from the heart to the lungs via the pulmonary arteries.

❷ Oxygen-rich blood enters the heart via the pulmonary veins.

Heart

Systemic veins
Systemic arteries

❻ Blood in the systemic veins transports carbon dioxide and other metabolic waste products toward the heart.

Systemic circulation

Tissues

❸ Blood in the systemic arteries transports oxygen and nutrients to the body's tissues.

Smaller arterioles branching off to supply various tissues

❺ The lymphatic system is made up of lymphatic vessels that circulate lymph. Fat-soluble nutrients are circulated away from the small intestine by the lymphatic system, eventually entering the blood.

Venules
Arterioles
Systemic capillaries

❹ Nutrients and oxygen are delivered to the tissues by capillaries. At the same time, waste products such as carbon dioxide are removed from tissues and carried away in the venous blood.

heart. Although the arterial and venous vascular systems have many similarities, they differ in several ways.

- Oxygen-rich arterial blood flows toward capillaries, whereas the oxygen-poor venous blood flows away from them.
- Arteries leading to the capillaries become progressively smaller (arteries → **arterioles** → capillaries), whereas veins leading away from capillaries become progressively larger (capillaries → **venules** → veins).
- Arterial circulation flows away from the heart, whereas the venous blood flows toward the heart.
- Arterial blood delivers nutrients and oxygen to cells, whereas the venous blood carries metabolic waste products away from cells.

arteriole Small blood vessel that branches off from arteries.

venule Small blood vessel that branches off from veins.

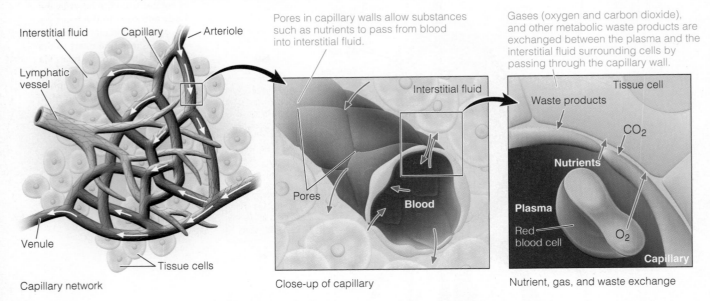

FIGURE 3.27 Nutrient and Gas Exchange across the Capillary Wall The exchange of nutrients, gases (oxygen and carbon dioxide), and other cellular waste products takes place between the plasma and the interstitial fluid.

Interstitial fluid   Capillary   Arteriole

Lymphatic vessel

Venule

Tissue cells

Capillary network

Pores in capillary walls allow substances such as nutrients to pass from blood into interstitial fluid.

Interstitial fluid

Pores

Blood

Close-up of capillary

Gases (oxygen and carbon dioxide), and other metabolic waste products are exchanged between the plasma and the interstitial fluid surrounding cells by passing through the capillary wall.

Tissue cell

Waste products

CO_2

Nutrients

Plasma

Red blood cell

O_2

Capillary

Nutrient, gas, and waste exchange

The Pulmonary Circulation Moves Blood between the Heart and Lungs Another component of the cardiovascular system involves the flow of blood between the heart and lungs. This circuit, referred to as the **pulmonary circulation,** begins with the arrival of partially deoxygenated (oxygen-poor, carbon dioxide–rich) venous blood to the heart. The **pulmonary arteries** transport blood from the right side of the heart to the lungs, where the exchange of carbon dioxide and oxygen takes place across the pulmonary (lung) capillaries. By exhaling air through our nose and mouth, carbon dioxide that has crossed out of the capillaries and into the air sacs of the lungs is eliminated from the body. Likewise, during inhalation, oxygen is taken into the lungs, where it crosses into the capillaries and enters the blood. The oxygen-rich blood returns to the heart through the **pulmonary veins** and is pumped out of the heart through the aorta to the rest of the body.

THE LYMPHATIC SYSTEM TRANSPORTS FAT-SOLUBLE NUTRIENTS AWAY FROM THE GI TRACT

Another major component of the circulatory system, the **lymphatic system,** plays an important role in circulating fat-soluble nutrients (mostly lipids and some vitamins) away from the GI tract and eventually delivering them to the cardiovascular system. Each villus contains a lacteal that connects to a network of lymphatic vessels that circulate a clear liquid called lymph. This circulatory route initially bypasses the liver, eventually emptying into the blood. At this point, fat-soluble nutrients can circulate in the bloodstream where they can be taken up and used by cells.

THE KIDNEYS PLAY AN IMPORTANT ROLE IN EXCRETING CELLULAR WASTE PRODUCTS

Nutrients taken up by cells undergo considerable metabolic change. That is, they are transformed, broken down, or used to synthesize other materials. These processes result in the formation of a variety of cellular waste products such as carbon dioxide, water, and urea—a nitrogen-containing compound resulting from the breakdown of protein. Because the accumulation of waste

pulmonary circulation (PUL – mo – nar – y) The division of the cardiovascular system that circulates deoxygenated blood from the heart to the lungs, and oxygenated blood from the lungs back to the heart.

pulmonary artery Blood vessel that transports oxygen-poor blood from the right side of the heart to the lungs.

pulmonary vein Blood vessel that transports oxygen-rich blood from the lungs to the heart.

lymphatic system (lym – PHAT – ic) A component of the circulatory system made up of lymphatic vessels and lymph that flows from organs and tissues, drains excess fluid from spaces that surround cells, and picks up dietary fats from the digestive tract.

FIGURE 3.28 Overview of the Urinary System The urinary system serves three important functions—filtration, reabsorption, and excretion.

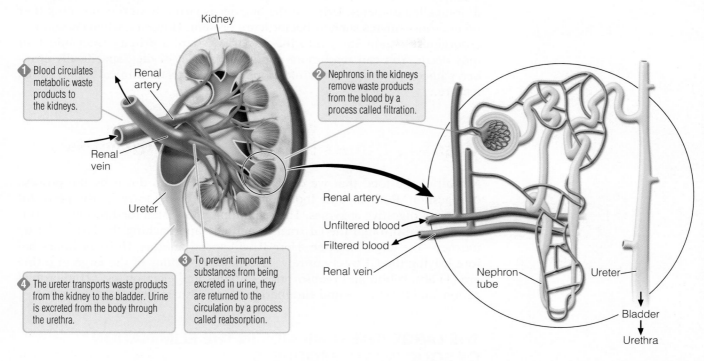

1 Blood circulates metabolic waste products to the kidneys.

Kidney

Renal artery

Renal vein

Ureter

2 Nephrons in the kidneys remove waste products from the blood by a process called filtration.

Renal artery

Unfiltered blood

Filtered blood

Renal vein

Nephron tube

Ureter

Bladder

Urethra

3 To prevent important substances from being excreted in urine, they are returned to the circulation by a process called reabsorption.

4 The ureter transports waste products from the kidney to the bladder. Urine is excreted from the body through the urethra.

products can be toxic, it is important that they be eliminated from the body. Whereas the respiratory system removes carbon dioxide, the kidneys have the responsibility of ridding the body of other cellular wastes, via the urine.

Urinary Excretion of Cellular Waste Products As shown in Figure 3.28, the kidneys play an important role in removing metabolic wastes from the plasma and delivering them for excretion in the urine. Blood flows to the kidneys at a rate of about 1,200 mL (approximately 1.5 quarts) per minute. **Nephrons,** which are the functional units of the kidney, perform the important functions of filtration and reabsorption. **Filtration** initially removes substances such as urea, excess water, electrolytes, salts, and minerals from the blood as it flows through the nephrons, most of which are then excreted from the body in the urine. Because the body can reuse some of the materials that have been filtered out of the blood, the kidneys carry out another important function called **reabsorption,** which means to "absorb again." Substances that are reabsorbed are returned to the blood, enabling the body to reclaim compounds such as amino acids, glucose, and other important nutrients that would otherwise be excreted in the urine.

The inability of our kidneys to perform these important functions can cause toxic waste products to accumulate in the blood. Although some forms of kidney disease are associated with hereditary factors, diabetes and high blood pressure are the two most common causes of impaired kidney function. People with impaired kidney function often need to adhere to strict dietary restrictions so that their kidneys do not become overburdened. However, if kidney function becomes severely impaired, a person may require hemodialysis. **Hemodialysis** is a medical procedure that uses a machine to filter waste products from the blood and to restore proper fluid balance.

The Formation and Excretion of Urine Urine contains a variety of substances including water, salts, and urea. In fact, the composition of urine is so well

nephron (NEPH – ron) Functional unit of the kidney that filters waste materials from the blood that are later excreted in the urine.

filtration The process of selective removal of metabolic waste products from the blood.

reabsorption The return of previously removed materials to the blood.

hemodialysis A medical procedure that uses a machine to filter waste products from the blood and to restore proper fluid balance.

defined that urine analysis can be used to diagnose certain diseases. For example, finding large amounts of glucose in the urine could indicate a condition called diabetes. Urine in the bladder is virtually sterile, meaning that no microorganisms such as bacteria are present. However, when certain microorganisms gain entry into the urinary system, a urinary tract infection may occur. You can read more about the impact of diet, specifically cranberry juice, on urinary tract infections in the Focus on Clinical Applications feature.

What Is the Role of the Large Intestine?

Not all of the food that we eat is completely broken down by the process of digestion. Rather, some food components are able to resist the powerful actions of digestive enzymes. The remaining undigested food residue continues to move through the GI tract, eventually approaching the last leg of its journey—the large intestine. Here it will spend another 10 to 24 hours before leaving the GI tract entirely. The major functions of the large intestine are (1) absorption and reabsorption of fluids and electrolytes, (2) microbial action, and (3) storage and elimination of solid waste (feces).

THE LARGE INTESTINE AIDS IN THE ELIMINATION OF SOLID WASTE PRODUCTS

cecum (CE – cum) The first portion of the large intestine.

The large intestine has four general regions: the cecum, colon, rectum, and anal canal (Figure 3.29). The **cecum,** the first portion of the large intestine, is a short, saclike structure with an attached appendage, consisting of

FIGURE 3.29 Overview of the Large Intestine The large intestine has four general regions: the cecum, colon, rectum, and anal canal.

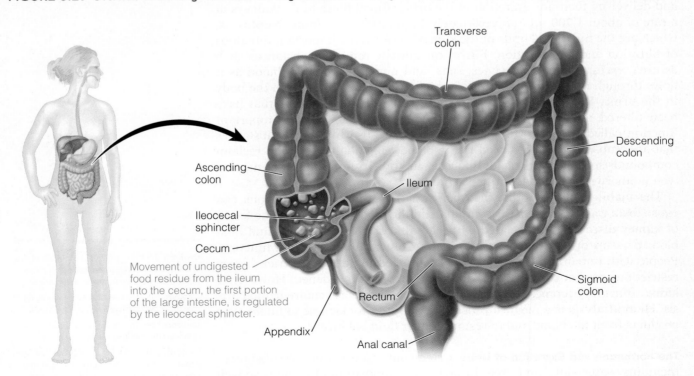

Transverse colon

Descending colon

Ascending colon

Ileum

Ileocecal sphincter

Cecum

Movement of undigested food residue from the ileum into the cecum, the first portion of the large intestine, is regulated by the ileocecal sphincter.

Sigmoid colon

Rectum

Appendix

Anal canal

Hype or reality: is there evidence to support the claim that cranberry juice helps prevent urinary tract infections (UTIs)? For years, the evidence was anecdotal, but several clinical intervention studies now provide some evidence that cranberry juice may protect against the recurrence of UTIs.[10] UTIs, which are more common in women than men, result from bacteria adhering to the lining of the bladder and urinary tract. The bacterium *Escherichia coli (E. coli)* is the most common cause of UTIs. More than 11 million prescriptions are written yearly for antibiotics to treat these infections.

Typically, a person with a UTI experiences urgent, frequent, and painful sensations associated with urination. How does drinking cranberry juice help? In a word—proanthocyanidins. It may be a mouthful to pronounce, but proanthocyanidins are phytochemicals, found in both cranberry and blueberry juices, which may help prevent UTIs. These compounds, which are the pigments responsible for the deep red or purple color of these juices, appear to prevent *E. coli* from adhering to the bladder lining. A clinical intervention study published in the *British Medical Journal* reported a significant reduction in the incidence of UTIs among women who consumed daily a concentrate consisting mainly of cranberry juice.[11] The rate of UTI recurrence was 16% for women given cranberry juice and 36% for women given a placebo. This and several other studies provide considerable support that cranberry juice helps prevent recurrent UTIs.[12] Although more studies are needed to clarify effective doses, as little as 10 ounces of cranberry juice daily may be enough to help ward off UTIs. It is important to emphasize, however, that cranberry juice is not a treatment for an established

UTI. Instead, drinking cranberry juice on a daily basis may offer protection from the recurrence of UTIs and is an example of how a food can offer both medicinal and nutritional benefits. This is one reason why cranberry juice, in addition to being nutritious, is considered by some to be a functional food.

sarsmis/Shutterstock.com

Proanthocyanidins are phytochemicals found in cranberries that may help protect against the recurrence of urinary tract infections.

lymphatic tissue, called the appendix. On occasion, trapped material can cause the appendix to become inflamed, which can necessitate an appendectomy—its surgical removal. The function of the appendix remains unclear, and people do not experience ill effects related to its removal. The **ileocecal sphincter** regulates the intermittent flow of material from the ileum into the cecum.

The **colon,** which makes up most of the large intestine, is shaped like an inverted letter U (∩). The right side of the colon is called the ascending colon, whereas the portion spanning right to left across the abdomen is the transverse colon. From there, the descending colon continues downward, on the left side of the body. Following the colon is the **rectum,** which terminates at the anal canal, the segment of the large intestine that leads outside of the body. A thickening of smooth muscle around the anal canal forms the internal and external **anal sphincters.**

ileocecal sphincter (il – e – o – CE – cal) The sphincter that separates the ileum from the cecum and regulates the flow of material between the small and large intestines.

colon The portion of the large intestine that carries material from the cecum to the rectum.

rectum The lower portion of the large intestine between the sigmoid colon and the anal canal.

anal sphincters Internal and external sphincters that regulate the passage of feces through the anal canal.

FLUIDS AND ELECTROLYTES ARE ABSORBED AND REABSORBED IN THE LARGE INTESTINE

Material entering the large intestine consists mostly of undigested fibrous material from plants, water, bile, cellular debris, and electrolytes. In the large intestine, slow, churning segmentation movements called **haustral contractions** help expose the undigested food residue to the absorptive (mucosal) lining of the colon. Some water and electrolytes are absorbed for the first time from the colon. However, most are actually being reabsorbed because they were released into the colon as GI secretions. As material passes through the various regions of the colon, water and electrolytes are extracted and circulated away from the GI tract in the blood for reuse by the body. This is another example of the body's ability to reclaim its important resources. Between intervals of haustral contractions, peristalsis slowly propels the material forward for further processing.

The consistency of the remaining material, now called **feces,** reflects how much water was reabsorbed. For example, diarrhea, characterized by loose, watery fecal matter, can result when material moves too quickly through the colon, not allowing sufficient time for water removal. Prolonged diarrhea can result in excessive loss of fluids and electrolytes from the body, which can lead to serious complications such as dehydration. Conversely, slow colonic movements can cause too much water to be removed, resulting in hard, dry fecal matter, a condition known as constipation. Constipation can make elimination difficult and put excessive strain on the colonic muscles. Two other conditions that can cause intestinal discomfort are irritable bowel syndrome and inflammatory bowel disease, which are discussed further in the Focus on Clinical Applications feature.

MICROBIAL ACTION IN THE LARGE INTESTINE BREAKS DOWN UNDIGESTED FOOD RESIDUE

Although bacteria reside throughout the GI tract, the large intestine provides the most suitable environment for them. The colon's optimal pH, sluggish haustral contractions, and lack of antimicrobial secretions present ideal conditions for bacteria to grow and flourish. The number and variety of bacteria residing in the large intestine is astronomical. In fact, more than 400 species of bacteria can be found in the large intestine, contributing to nearly one-third of the dry weight of feces. This natural microbial population, also referred to as the **intestinal microbiota** (or **microbiome**), is important for a healthy colonic ecosystem. First, these bacteria break down some of the undigested food residue, which consists mostly of fibrous plant material. Intestinal bacteria also produce vitamin K, limited amounts of certain B vitamins, and some lipids. Nutrients and other substances produced by the intestinal bacteria are absorbed into the blood. Perhaps most important, the natural microbiota help protect us from infection by competing with pathogenic bacteria for limited resources (nutrients and space) in the large intestine. You can read more about how to establish a healthy intestinal microbiota in the next Focus on Clinical Applications feature.

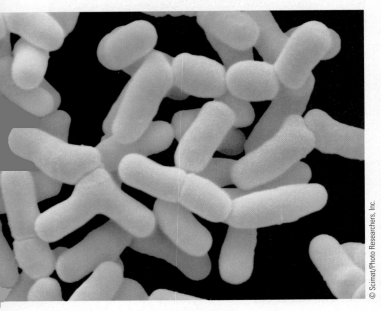

Bifidobacterium cells on colon epithelium.

© Scimat/Photo Researchers, Inc.

haustral contractions (HAU – stral) Slow muscular movements that move the colonic contents back and forth, helping to compact the feces.

feces Waste matter consisting mostly of undigested food residue that is eliminated from the body through the anus.

intestinal microbiota (MI – cro – bi – O – ta) Bacteria that reside in the large intestine (also called intestinal microbiome).

After we eat and enjoy our food, the GI tract dutifully takes over without us having to give it further thought. That is, unless something goes wrong. Health conditions that affect GI function can seriously impair the ability of the GI tract to digest food and absorb nutrients. This is certainly the case with **inflammatory bowel disease (IBD)** such as ulcerative colitis or Crohn's disease. Another GI disorder is irritable bowel syndrome (IBS), a poorly understood condition that affects up to 20% of Americans. While irritable bowel syndrome and inflammatory bowel disease (IBD) may sound similar, they are very different. Nonetheless, both these GI disturbances can have serious implications in terms of nutritional health.

Because **irritable bowel syndrome (IBS)** is not associated with any known structural abnormalities, it is considered to be a functional disorder in that normal bowel activity is disrupted. People with IBS experience bouts of abdominal discomfort such as cramping, bloating, diarrhea, and constipation. Although the underlying cause of IBS has yet to be determined, some clinicians believe that it is a psychological manifestation. While emotional stress itself is not a cause of IBS, it can be a contributing factor. There is also some speculation that IBS may occur when the colon overreacts to stimuli, causing it to spasm. Typically, the diagnosis of IBS is based on ruling out other, better-defined intestinal disorders such as ulcerative colitis or Crohn's disease. It is important for people with IBS to identify and avoid foods that trigger IBS episodes and seek out those that bring comfort and relief. IBS is sometimes treated with antispasmodic medication. Fortunately, IBS does not progress to other, more serious illnesses.

Ulcerative colitis and **Crohn's disease** are both considered forms of IBD, and are characterized by inflammation of the lining of the GI tract.[13] IBD is classified as an autoimmune disease, although this is somewhat speculative.

Many researchers believe that IBD can develop when the intestinal lining is exposed to a foreign protein, or what is called an antigen. Possible immune triggers (antigens) include proteins found in food and exposure to viruses or bacteria. In response to the antigen, a person's immune system produces antibodies, which triggers inflammation. Because the inflammatory response is often prolonged and excessive, the intestinal lining can become damaged. For reasons that are not clear, the occurrence of IBD has dramatically increased over that past 30 years.[14] Currently, it is estimated that more than 1.5 million Americans have IBD, 25–30% of whom are children or adolescents.[15]

Crohn's disease and ulcerative colitis share many of the same clinical signs and symptoms. However, Crohn's disease tends to affect the lower portion of the small intestine (ileum), although it can occur anywhere along the GI tract. In contrast, ulcerative colitis tends to occur along the inner lining of the colon. IBD is typically diagnosed by a procedure called colonoscopy, which involves the insertion of a small scope into the anus. This scope is threaded through the rectum, allowing the physician to inspect the intestinal wall for signs of inflammation or ulcers. A biopsy (tissue sample) can also be taken at this time. Because IBD increases a person's risk of developing colon cancer, it is important for people with this disease to have regular medical exams.

IBD flare-ups can cause diarrhea, fatigue, weight loss, abdominal pain, diminished appetite, and, on occasion, rectal bleeding. Although dietary practices do not cause IBD, nutritional support is important. Nutrient malabsorption and loss of appetite can cause significant weight loss and a variety of nutritional problems. Poor nutritional status can even make IBD worse because adequate nutrient and energy intake are needed to repair the damaged tissue. This is why the right assistance by a qualified team of health care professionals can help prevent further complications associated with IBD.

THE LARGE INTESTINE STORES AND ELIMINATES SOLID WASTE PRODUCTS FROM THE BODY

Egestion refers to the process whereby solid waste (feces) is eliminated from the body. As solid waste moves through the colon, it eventually collects in the rectum, which serves as a holding chamber for feces. The accumulation of

Critical Thinking: Paige's Story Now that you understand the function of the large intestine, think back to Paige's story. Paige was slowly losing function in her large intestine. As a result, she was experiencing weight loss, diarrhea, and severe abdominal pain. However, Paige's condition was not quickly diagnosed. In fact, it was first suggested that she was lactose intolerant or had celiac disease or irritable bowel syndrome. Why could any of these conditions cause the same signs and symptoms as Crohn's disease?

inflammatory bowel disease (IBD) Chronic conditions such as ulcerative colitis and Crohn's disease that cause inflammation of the lower GI tract.

irritable bowel syndrome (IBS) A condition that typically affects the lower GI tract, causing abdominal pain, muscle spasms, diarrhea, and constipation.

ulcerative colitis (co – LI – tis) A type of inflammatory bowel disease (IBD) that causes chronic inflammation of the colon.

Crohn's disease A chronic inflammatory condition that usually affects the ileum and first portion of the large intestine.

The role that intestinal bacteria play in GI function and disease prevention has only recently become appreciated. It has been estimated that several million bacteria reside in the human GI tract, a number far greater than previously believed. Bacteria begin to colonize the GI tract shortly after birth. Age-related changes in the gut, antibiotic therapy, and dietary choices can disrupt this balance. In fact, by 65 years of age, the number of "friendly" microbes residing in a human's colon declines a thousand fold.[16] For these and other reasons, there has been considerable interest in fully understanding factors that help establish and maintain a healthy microbiota.

There is now substantial evidence that certain types of food contribute to a healthy intestinal microbiota.[17] These foods are referred to as probiotic and prebiotic foods. **Probiotic foods** contain live bacterial cultures that colonize the colon and have health-promoting benefits. For example, microorganisms associated with probiotic foods help inhibit the colonization of pathogenic bacteria by adhering to the intestinal lining. Other potential health benefits include immune-enhancing effects and protection against cancer, allergies, and autoimmune diseases. Dietary supplements are one source of probiotic bacteria, as are some "cultured" dairy products such as yogurt, buttermilk, sour cream, and cottage cheese. Some yogurts produced in the United States are made by adding the probiotic bacteria *Streptococcus thermophilus* to milk. Foods with live bacteria in them are often labeled as such.

Scientists have also found that the addition of probiotic bacteria to soy milk improved the bioavailabilty of bioactive compounds (isoflavones) and calcium, effects that are beneficial to health.[18] Thus, probiotic foods may have the added advantage of enhancing the body's ability to absorb certain components in food. Studies are also investigating a potential therapeutic role of various probiotics in managing disease states such as IBS and IBD.[19] Researchers have also discovered that the millions of bacteria that populate our intestinal tract may play a role in body weight regulation.[20] Although information linking body weight and human microbiota is limited, it has been hypothesized that intestinal bacteria may use energy from food to maintain a stable microbial "community." Future treatments for weight management may someday include populating the intestine with certain bacteria through ingestion of probiotic foods. Although probiotic foods have been available for years, the American public is just beginning to recognize their potential as a natural approach to staying healthy. In fact, probiotics are considered part of the functional foods trend.

Another way to increase colonization of beneficial bacteria is to consume **prebiotic foods,** which are typically fiber-rich and selectively promote the growth of nonpathogenic bacteria. Because dietary fiber can resist digestion, it passes into the colon and provides a source of nourishment for the microbiota. Dietary fiber is found in whole grains, cereals, fruits, vegetables, and legumes and is discussed in detail in Chapter 4. Together, consumption of probiotic and prebiotic foods helps maintain a well-colonized microbial population in the large intestine, providing an important defense against pathogenic bacteria. In addition, the microbiota produce many substances that are likely beneficial to our health.

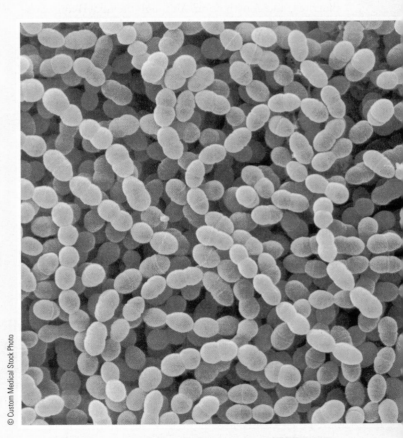

© Custom Medical Stock Photo

Streptococcus thermophilus is a bacteria used to produce yogurt. It is beneficial to the gastrointestinal tract, and therefore considered a probiotic bacteria.

Critical Thinking: Paige's Story As a result of Crohn's disease, Paige had a portion of her colon surgically removed. How might this impact normal physiological functions in the colon? Why do you think her physician recommended she take nutrient supplements and probiotics?

feces causes the walls of the rectum to stretch, signaling the need to **defecate.** However, the external anal sphincter, which is under conscious control, enables us to determine whether the time is right for waste elimination. By keeping the external anal sphincter contracted, a person can delay defecation. When the internal anal sphincter relaxes, the feces move into the anal canal and then are expelled from the body, via the process of defecation.

probiotic (PRO – bi – o – tic) Food or dietary supplement that contains beneficial live bacteria.

prebiotic food (PRE – bi – o – tic) Food that stimulates the growth of beneficial bacteria that naturally reside in the large intestine.

defecation The expulsion of feces from the body through the rectum and anal canal.

Notes

1. Spechler SJ. Clinical manifestations and esophageal complications of GERD. American Journal of Medical Sciences. 2003;326:279–84.

2. Meyer JH. Gastric emptying of ordinary food: effect of antrum on particle size. American Journal of Physiology. 1980;239:G133–5. Hunt JN. Mechanisms and disorders of gastric emptying. Annual Review of Medicine. 1983;34:219–29.

3. DeCross AJ, Marshall BJ. The role of *Helicobacter pylori* in acid-peptic disease. American Journal of Medical Sciences. 1993;306:381–92.

4. Blaser MJ. The bacteria behind ulcers. Scientific American. 1996;274:104–7. Meuler DA. Helicobacter pylori and the bacterial theory of ulcers. 2010 National Center for Case Study Teaching in Science. University of Buffalo. Available from: http://sciencecases.lib.buffalo.edu/cs/files/peptic_ulcer.pdf.

5. Impact. A publication of the University of Chicago Celiac Disease Center. A brief history of celiac disease. 2007;7:1–2. Available from: http://www.celiacdisease.net/assets/pdf/SU07CeliacCtr.News.pdf.

6. Fasano A, Berti I, Gerarduzzi T, Not T, Colletti RB, Drago S, Elitsur Y, Green P, Guandalini S, Hill ID, Pietzak M, Ventura A, Thorpe M, Kryszak D, Fornaroli F, Wasserman SS, Murray JA, Horvath K. Prevalence of Celiac Disease in At-Risk and Not-At-Risk Groups in the United States. A Large Multicenter Study. Archives of Internal Medicine. 2003;163:286–92.

7. Nelsen DA. Gluten-sensitive enteropathy (celiac disease): More common than you think. American Family Physician. 2002;66:2259–66.

8. Mearin ML. Celiac disease among children and adolescents. Current Problems in Pediatrics and Adolescent Health Care. 2007;37:86–105.

9. Hernandez L, Green PH. Extraintestinal manifestations of celiac disease. Current Gastroenterology Reports. 2006;8:383–9.

10. McMurdo ME, Bissett LY, Price RJ, Phillips G, Crombie IK. Does ingestion of cranberry juice reduce symptomatic urinary tract infections in older people in hospital? A double-blind, placebo-controlled trial. Age and Ageing. 2005;34:256–61. Howell AB. Cranberry proanthocyanidins and the maintenance of urinary tract health. Critical Reviews in Food Science and Nutrition. 2002;42:273–8.

11. Kontiokari T, Sundqvist K, Nuutinen M, Pokka T, Koskela M, Uhari M. Randomised trial of cranberry-lingonberry juice and Lactobacillus GG drink for the prevention of urinary tract infections in women. British Medical Journal. 2001;30;322:1571.

12. L Strothers. A randomized trial to evaluate effectiveness and cost effectiveness of naturopathic cranberry products as prophylaxis against urinary tract infection in women. Canadian Journal of Urology 2002;9:1558–62.

13. Xavier RJ, Podolsky DK. Unravelling the pathogenesis of inflammatory bowel disease. Nature. 2007;448:427–34.

14. Lakatos PL. World recent trends in the epidemiology of inflammatory bowel diseases: Up or down? Journal of Gastroenterology. 2006;12:6102–8.

15. Abraham C, Cho JH. Inflammatory bowel disease. New England Journal of Medicine. 2009;361:2066–78.

16. Kolida S, Saulnier DM, Gibson GR. Gastrointestinal microflora: Probiotics. Advances in Applied Microbiology. 2006;59:187–219.

17. Rastall RA. Bacteria in the gut: Friends and foes and how to alter the balance. Journal of Nutrition. 2004;134:2022S–6S.

18. Pham TT, Shah NP. Biotransformation of isoflavone glycosides by bifidobacterium animalis in soymilk supplemented with skim milk powder. Journal of Food Science. 2007;72:316–24.

19. Guslandi MJ. Probiotic agents in the treatment of irritable bowel syndrome. Journal of International Medical Research. 2007;35:583–9. Hedin C, Whelan K, Lindsay JO. Evidence for the use of probiotics and prebiotics in inflammatory bowel disease: A review of clinical trials. Proceedings of the Nutrition Society. 2007;66:307–15.

20. DiBaise JK, Zhang H, Crowell MD, Krajmalnik-Brown R, Decker GA, Rittmann BE. Gut microbiota and its possible relationship with obesity. Mayo Clinic Proceedings. 2008;83:460–9. Kalliomäki M, Collado MC, Salminen S, Isolauri E. Early differences in fecal microbiota composition in children may predict overweight. American Journal of Clinical Nutrition. 2008;87:534–8. Cani PD, Delzenne NM. Gut microflora as a target for energy and metabolic homeostasis. Current Opinion in Clinical Nutrition and Metabolic Care. 2007;10:729–34.

Carbohydrates

Carbohydrates comprise a diverse group of compounds produced primarily by plants. The sugars in fruit, the fibers in celery, and the starch in potatoes are all different forms of carbohydrate. Although all carbohydrates are made of sugar molecules, they differ from each other in a variety of ways. For example, some carbohydrates contain only one or two sugar molecules, whereas others are made of hundreds of sugar molecules. As the number of sugar molecules in a carbohydrate increases, so does its size and complexity. Carbohydrates are plentiful in a variety of foods and serve many essential functions within the body. For example, carbohydrates provide cells with a vital source of energy and are components of ribonucleic and deoxyribonucleic acids (RNA and DNA, respectively). As such, they are an important part of a healthy, well-balanced diet. In this chapter, we will look at carbohydrate chemistry, dietary sources of carbohydrates, functions of carbohydrates in the body, and guidelines for carbohydrate intake.

What It Takes to Stay in the Race

After completing over 20 marathons, Laura was ready for her next challenge—the Ironman® competition. This ultra-endurance event is one of the most rigorous athletic competitions: a 2.4-mile swim, a 112-mile bike ride, and a marathon (26.2 miles). After six months of relentless training, Laura finally began the final stages of planning and preparation. As she packed her bag, she put the two most essential items in last—her insulin pump and her glucometer. Yes, Laura has type 1 diabetes. In addition to worrying about flat tires, bike crashes, and consuming enough energy to make it through the grueling 14-hour competition, Laura has an additional worry that most Ironman competitors do not have—monitoring and regulating her blood glucose levels throughout the competition. For Laura, plummeting blood glucose levels could easily mean the end of the race.

Like most competitive athletes, Laura maintains a strict training regimen. In addition to recording her distances and times, she also keeps a detailed food diary. Laura understands that rigorous exercise can cause unexpected highs and lows in blood glucose, and determining what and how much food to eat can be a real challenge.

For those with type 1 diabetes, testing blood glucose throughout the day using a glucometer becomes routine. This requires obtaining a small drop of blood using a lancet device, placing the blood onto the test strip, and then inserting the test strip into the glucometer. How to manage this while biking and running was an obvious challenge for Laura. After a few near-crashes on her bike she decided to try a continuous glucose-monitoring device. The tiny glucose sensor inserted under the skin allows Laura to more easily monitor and record her blood glucose levels throughout the day and night. The information from the sensor is transmitted to a wireless monitor. Continuous glucose monitoring is not intended for long-term use, but it enables Laura to determine the right combination and amount of food she needs to compete.

Training for the Ironman® competition is hard for anyone, but it is especially hard for someone with diabetes.

But Laura's motto has always been "Be ready for anything." This means being able to detect and respond to sudden and unexpected fluctuations in blood glucose. Knowing when to eat, what to eat, and when it is time for insulin is critical. Watches, alarms, needles, pumps, and carefully selected food keep Laura on pace and in the race. You may be wondering why someone with type 1 diabetes, or anyone for that matter, competes in such an intense and exhausting competition. In addition to family and work, much of Laura's inspiration comes from a nonprofit organization called Triabetes. This online network for triathletes with type 1 diabetes provides support and strategies to navigate these largely uncharted waters. As the finisher's medal was placed around Laura's neck, she thought to herself, "What an amazing experience."

Laura Benson

Critical Thinking: Laura's Story
Type 1 diabetes is a condition that is not easily ignored. In fact, a person with this disease must be vigilant when it comes to blood glucose control. What insights did you gain from Laura's story about what it is like to have type 1 diabetes? What do you think your biggest challenges would be if you had this condition?

What Are Simple Carbohydrates?

A carbohydrate is an organic compound made up of one or more sugar molecules (Figure 4.1). Most people think of sugar as a substance used to sweeten their foods. Although this is true, sugars encompass far more than this. For example, cells use a special type of sugar, glucose, for an important source of energy. **Carbohydrates** are abundant in a wide variety of foods, and there are many different types. A carbohydrate consisting of a single sugar molecule is called a **monosaccharide;** a carbohydrate made of two sugar molecules is a **disaccharide.** Because of their small size, monosaccharides and disaccharides are called **simple carbohydrates** or simple sugars.

MONOSACCHARIDES ARE SINGLE SUGAR MOLECULES

Monosaccharides are single-sugar molecules that are made up of carbon, hydrogen, and oxygen atoms in the ratio of 1:2:1. For example, if a sugar has 6 carbon atoms, it also has 12 hydrogen and 6 oxygen atoms (written $C_6H_{12}O_6$). Because the number of carbon atoms and the arrangement of atoms can vary, monosaccharides typically have different shapes and sizes. There are hundreds of different naturally occurring monosaccharides, but the three most plentiful in food are glucose, fructose, and galactose. Although the structures of these sugars differ, they all have one thing in common: each contains six carbon atoms. Therefore these monosaccharides are referred to as **hexose** (*hexa*, meaning six) sugars.

Although glucose, fructose, and galactose have the same molecular formula ($C_6H_{12}O_6$), each has a different arrangement of atoms (Figure 4.2). Notice that each monosaccharide exists as a ring structure and that each carbon atom in the ring is assigned a number. This numbering system provides a way to describe important structural differences in monosaccharides. As you will soon learn, it also provides the basis for describing different types of chemical bonds that attach one sugar molecule to the next.

Glucose Is the Most Abundant Sugar in Blood **Glucose,** the most abundant monosaccharide in the body, is produced when chlorophyll-containing plants combine carbon dioxide and water in the presence of sunlight (Figure 4.3). This process, called **photosynthesis,** provides an important energy source

carbohydrate Organic compound made of varying numbers of monosaccharides.

monosaccharide (mo – no – SAC – cha – ride) (*mono-*, one; *-saccharide*, sugar) Carbohydrate consisting of a single sugar.

disaccharide (di – SAC – cha – ride) (*di-*, two) Carbohydrate consisting of two monosaccharides bonded together.

simple carbohydrate, or simple sugar Category of carbohydrates consisting of mono- and disaccharides.

hexose Monosaccharide made of six carbon atoms.

glucose (GLU – kose) A six-carbon monosaccharide produced by photosynthesis in plants.

photosynthesis (*photo-*, light; *-synthesis*, product) Process whereby plants use energy from the sun to produce glucose from carbon dioxide and water.

FIGURE 4.1 Classification of Carbohydrates Carbohydrates are classified as simple or complex. These categories are further subdivided into mono-, di-, oligo-, and polysaccharides, depending on the number of monosaccharides (sugars) they contain.

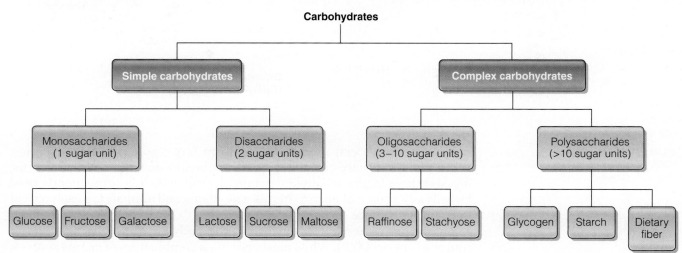

FIGURE 4.2 The Structure of Monosaccharides Glucose, galactose, and fructose are monosaccharides that contain six carbon atoms.

Monosaccharides that have 6 carbon atoms are called hexose sugars.
All hexose sugars have 6 carbon, 12 hydrogen, and 6 oxygen atoms (1:2:1).

Fructose has a 5-sided ring structure, whereas glucose and galactose have 6-sided ring structures.

Glucose

Galactose

Fructose

Notice that glucose and galactose have a similar chemical structure with the exception of the hydroxyl groups (–OH), which face in opposite directions.

FIGURE 4.3 Photosynthesis The process of photosynthesis combines carbon dioxide and water in the presence of sunlight to produce glucose. When we consume plants, the glucose provides us with a source of energy.

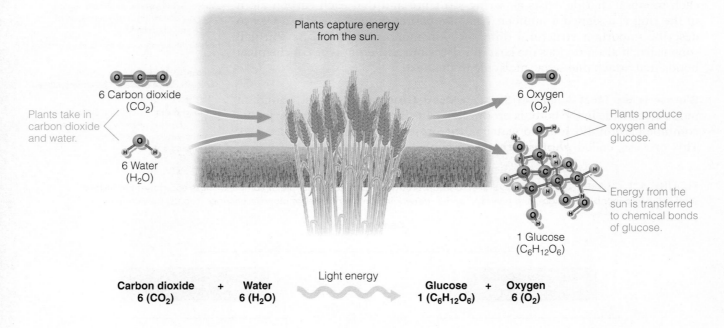

Plants capture energy from the sun.

Plants take in carbon dioxide and water.

6 Carbon dioxide (CO_2)

6 Water (H_2O)

6 Oxygen (O_2)

Plants produce oxygen and glucose.

Energy from the sun is transferred to chemical bonds of glucose.

1 Glucose ($C_6H_{12}O_6$)

Carbon dioxide	+	Water	Light energy	Glucose	+	Oxygen
6 (CO_2)		6 (H_2O)		1 ($C_6H_{12}O_6$)		6 (O_2)

⟨**CONNECTIONS**⟩ The nervous system is made up of the brain, spinal cord, and nerves (Chapter 3, page 82).

for plants. Plants can use glucose to form large, complex carbohydrates. When plant foods are consumed, the body breaks down (or in other words, digests) these large carbohydrates into glucose, which is subsequently used by cells as a source of energy. In this way, plants and animals are each part of the delicate balance that exists in nature.

The primary function of glucose in the body is to provide cells with a source of energy (ATP). While most cells use a combination of energy sources, glucose is the preferred energy source for the nervous system and the sole source of energy for red blood cells. Glucose is also used to synthesize other compounds in the body. For example, it can be converted to some amino acids (the building blocks of proteins) and fat for long-term energy storage. In addition, the body can store small amounts of glucose as a compound called glycogen.

Fructose Is the Most Abundant Sugar in Fruits and Vegetables **Fructose** is a naturally occurring monosaccharide found primarily in honey, fruits, and vegetables. Even though fructose has a five-sided ring structure, it is still classified as a hexose sugar because it contains a total of six carbons. While fructose is abundant in fruits and vegetables, the majority of fructose in the Western diet comes from foods made with high-fructose corn syrup. **High-fructose corn syrup** (HFCS) is a widely used sweetener found in soft drinks, fruit juice beverages, and a variety of other foods. Derived from corn, HFCS consists of almost equal amounts of fructose and glucose. It is used so extensively by food manufacturers that it now accounts for approximately 7% of total energy intake in the United States.[1] The U.S. Department of Agriculture estimates that the typical American consumes approximately 35 pounds of HFCS per year.[2] In fact, consumption of foods and beverages sweetened with HFCS now exceeds that of those sweetened with table sugar (Figure 4.4). Some scientists speculate that excess consumption of HFCS may be fueling the American obesity epidemic.[3] This controversy is discussed further in the Focus on Diet and Health feature.

Galactose Is Found in Dairy Products At first glance, the monosaccharides glucose and **galactose** look rather similar. However, notice that the hydroxyl (–OH) groups on carbon 4 actually face in opposite directions. Even this minor structural variation results in important differences in physiological functions of galactose and glucose. Few foods contain galactose in its free

fructose (FRUC – tose) A six-carbon monosaccharide found in fruits and vegetables; also called levulose.

high-fructose corn syrup A substance derived from corn that is used to sweeten foods and beverages.

galactose (ga – LAC – tose) A six-carbon monosaccharide found mainly bonded with glucose to form the milk sugar lactose.

FIGURE 4.4 Trends (1970–2010) in Consumption of Foods and Beverages Sweetened with High-Fructose Corn Syrup and Table Sugar

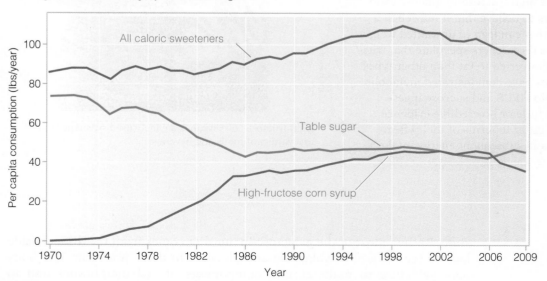

SOURCES: U.S. Department of Agriculture, Economic Research Service. 2011. Table 51—Refined cane and beet sugar: estimated number of per capita calories consumed daily, by calendar year. Table 52—High fructose corn syrup: estimated number of per capita calories consumed daily, by calendar year. Available from: http://www.ers.usda.gov/Briefing/Sugar/Data.htm.

Fructose is a naturally occurring sugar that makes fruits deliciously sweet. However, most of the fructose we consume in the United States does not come from fruit but instead comes from foods and beverages sweetened with high-fructose corn syrup (HFCS).

HFCS is produced when corn is treated with enzymes, resulting in a glucose-rich syrup. During this process, some of the glucose molecules are converted to fructose. While sucrose consists of glucose and fructose in roughly equal amounts, the ratio of fructose to glucose in HFCS can be considerably higher. For example, HFCS used in the production of soft drinks contains approximately 55% fructose and 45% glucose. According to the National Soft Drink Association, HFCS works better than refined sugar in beverage production because it blends more readily with liquids. In addition, the cost to make products with HFCS is substantially less than that of using sucrose.

Recently, scientists observed that an increase in HFCS consumption has paralleled an increasing rate of obesity.[3] Coincidence or not, some claim that HFCS has unique properties that may be cause for concern. Although researchers have raised questions about a link between HFCS consumption and obesity, as yet studies have not shown a definite relationship.[4] Clearly, no single cause is responsible for the obesity epidemic. And it should come as no surprise that weight gain is likely to ensue when any energy-dense foods and beverages are consumed in excess. So what is it about HFCS, aside from its caloric content, that has sounded the alarm?

Both sucrose and HFCS, when consumed in excess, can provide considerable amounts of fructose. However, the fact that HFCS has a high fructose-to-glucose ratio has led some researchers to believe that it may elicit a physiological response that can lead to unwanted weight gain. For example, some researchers speculate that fructose is more efficiently converted to fat than other types of sugars. Yet few studies have demonstrated metabolic differences in response to HFCS and sucrose ingestion.[5] Similarly, there is little or no credible evidence that HFCS and sucrose differ in terms of their effect on satiety, food intake, and subsequent energy intake.[6]

Until scientists learn more about the effects of HFCS on health, we can all agree that consuming too many calories in the form of sugar and HFCS can contribute to unwanted weight gain. Undoubtedly, the relationship between high intakes of HFCS and obesity is likely due to excess consumption of energy-dense foods and beverages and not HFCS *per se*.

© David Hancock/Alamy

High-fructose corn syrup is a widely used sweetener found in a variety of foods and beverages.

state. Rather, most of the galactose in our diet comes from the disaccharide lactose (comprised of galactose and glucose) in dairy products. The body uses galactose to make certain components of cell membranes and to synthesize lactose, an important sugar in breast milk. However, the majority of galactose in the body is converted to glucose and used as a source of energy (ATP).

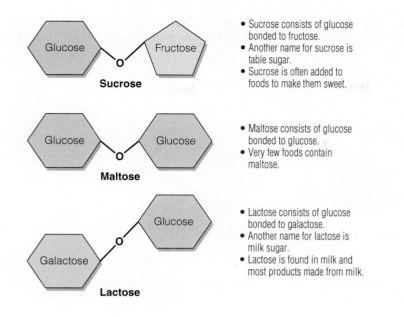

- Sucrose consists of glucose bonded to fructose.
- Another name for sucrose is table sugar.
- Sucrose is often added to foods to make them sweet.

Sucrose

- Maltose consists of glucose bonded to glucose.
- Very few foods contain maltose.

Maltose

- Lactose consists of glucose bonded to galactose.
- Another name for lactose is milk sugar.
- Lactose is found in milk and most products made from milk.

Lactose

FIGURE 4.5 Disaccharides The disaccharides sucrose, maltose, and lactose consist of two monosaccharides bonded together. Notice that each disaccharide has at least one glucose molecule.

DISACCHARIDES CONSIST OF TWO MONOSACCHARIDES

Disaccharides consist of two monosaccharides bonded together. The most common disaccharides are **lactose** (galactose and glucose), **maltose** (glucose and glucose), and **sucrose** (fructose and glucose). In all of these disaccharides, at least one of the monosaccharides in the pair is glucose (Figure 4.5). A condensation reaction chemically joins monosaccharides together by a glycosidic bond. As shown in Figure 4.6, a **glycosidic bond** is formed when the hydroxyl group (–OH) from one monosaccharide interacts with a hydrogen group (–H) from another monosaccharide, resulting in the loss of one molecule of water (H_2O). Disaccharides have a molecular formula of $C_{12}H_{22}O_{11}$, which reflects the loss of the water molecule ($2C_6H_{12}O_6 \rightarrow C_{12}H_{22}O_{11} + H_2O$).

In addition to consisting of different pairs of monosaccharides, disaccharides can also have different types of glycosidic bonds. Numbers are used to designate which carbon atoms form the glycosidic bond. For example, a

lactose Disaccharide consisting of glucose and galactose; produced by mammary glands.

maltose (MAL – tose) Disaccharide consisting of two glucose molecules bonded together; formed during the chemical breakdown of starch.

sucrose (SU – crose) Disaccharide consisting of glucose and fructose; found primarily in fruits and vegetables.

glycosidic bond (gly – co – SI – dic) A type of chemical bond that forms between two monosaccharides.

⟨CONNECTIONS⟩ A condensation reaction occurs when molecules are bonded together with the release of a molecule of water (Chapter 3, page 74).

FIGURE 4.6 Formation of Glycosidic Bonds A glycosidic bond is formed by a condensation reaction between two monosaccharides. In this example, two glucose molecules join to form water and the disaccharide maltose.

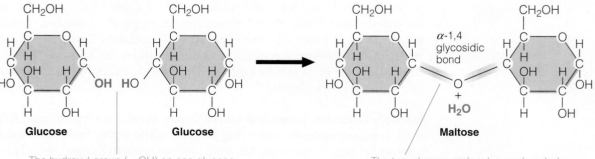

The hydroxyl group (—OH) on one glucose molecule bonds with a hydrogen atom (H) from the other glucose molecule.

The two glucose molecules are bonded together by an α-1,4 glycosidic bond, forming maltose and a molecule of water.

FIGURE 4.7 The Naming of Glycosidic Bonds A glycosidic bond joins two monosaccharides. The numbers in a glycosidic bond refer to the carbon atoms that form the bond. If the glycosidic bond is facing down, it is called an alpha (α) glycosidic bond. A glycosidic bond facing up is called a beta (β) glycosidic bond.

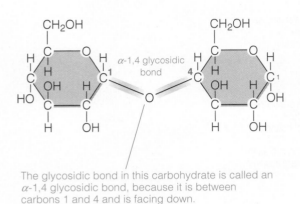

The glycosidic bond in this carbohydrate is called an α-1,4 glycosidic bond, because it is between carbons 1 and 4 and is facing down.

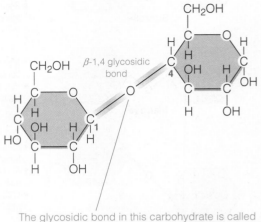

The glycosidic bond in this carbohydrate is called a β-1,4 glycosidic bond, because it is between carbons 1 and 4 and is facing up.

alpha (α) glycosidic bond A downward-facing type of glycosidic bond between two monosaccharides.

beta (β) glycosidic bond An upward-facing type of glycosidic bond between two monosaccharides.

glycosidic bond between carbon 1 on the glucose molecule and carbon 4 on the other monosaccharide is referred to as a 1,4 glycosidic bond. In addition, the bond may be an **alpha (α) bond,** which faces down, or a **beta (β) bond,** which faces up (Figure 4.7). This system offers a convenient way to describe important structural features of carbohydrates. Understanding these basic differences between glycosidic bonds is also important because they can determine whether a carbohydrate is digestible or indigestible.

Lactose, Maltose, and Sucrose Are Disaccharides The disaccharides lactose, maltose, and sucrose are found in a wide variety of foods. **Lactose** is the most abundant carbohydrate in milk, and is the only disaccharide that has a beta glycosidic bond between the monosaccharides. During lactation, enzymes in the mammary glands (breasts) combine glucose and galactose to produce lactose. Thus, milk and milk products (such as yogurt, cheese, and ice cream) contain lactose, although its concentrations can vary.

The disaccharide maltose is formed during starch digestion. Maltose is not found in many foods, but the breakdown of starch to maltose is an important step in the production of beer. During the brewing process, enzymes convert cereal starches such as those found in barley to maltose. Maltose is then fermented by bacteria to produce the alcohol found in beer.

Sucrose is found in many plants, and is especially abundant in sugar cane and sugar beets. These plants can be crushed to produce a juice that is processed to make a brown liquid called molasses. Further treatment and purification forms pure crystallized sucrose, otherwise known as refined table sugar. Because most people enjoy the intense sweetness of sucrose, it is often added to processed foods.

Naturally Occurring Sugars and Added Sugars Foods like fruits, vegetables, and milk contain naturally occurring sugars, while many processed foods contain added sugar or syrups. On food labels, the term "sugar" usually refers to both added sugar and those that occur naturally. As you can see in Table 4.1, many types of sugars and syrups can be added to foods, including white sugar, brown sugar, corn syrup, honey, and molasses. Sweetness is not the only reason sugar is added to foods. For example, sugars and syrups can be used to thicken and to alter the texture of foods.

Although added sugars are chemically identical to naturally occurring sugars, foods with naturally occurring sugars are usually nutrient dense, providing high amounts of nutrients relative to the amount of calories. For example, besides sucrose, fruits are naturally rich in vitamins, minerals, and fiber. In contrast, foods with large amounts of added sugars, such as soft drinks, cakes, and candy, often have little nutritional value beyond the calories they contain. Some people believe that sweeteners such as honey are healthier alternatives to sweeteners such as refined white sugar. However, like refined sugars, natural sweeteners also have limited nutritional value other than energy.

In recent years, consumption of added sugars has increased throughout the United States. Today, the average American consumes about 89 grams (22 teaspoons) of added sugars—a total of 355 kcal—every day.[7] Although this may not sound like much, the calories can add up quickly. Nearly half (46%) of the added sugars in the American diet are from sugar-sweetened soft drinks.[8] Moderate evidence suggests that greater consumption of sugar-sweetened beverages is associated with increased body weight in adults.[9] This is not surprising when you consider that the average 12-oz can of sweetened soft drink contains 8 teaspoons of added sugar—the equivalent of 130 kcal. To lower the caloric content of foods, food manufacturers often use nonsugar, low-calorie sweeteners. Because sugar-laden foods are associated with tooth decay, nonsugar substitutes may provide a healthier sweet alternative. You can learn more about these products in the Focus on Food feature.

TABLE 4.1 Selected Terms Used on Food Labels for Added Sweeteners

Brown rice syrup

Brown sugar

Concentrated fruit juice sweetener

Confectioner's sugar

Corn syrup

Dextrose

Fructose

Glucose

Granulated sugar

High-fructose corn syrup

Honey

Invert sugar

Lactose

Levulose

Maltose

Maple sugar

Molasses

Natural sweeteners

Raw sugar

Sucrose

Turbinado sugar

White sugar

What Are Complex Carbohydrates?

In contrast to the simple carbohydrates that contain one or two monosaccharides, **complex carbohydrates** consist of many monosaccharides bonded together in a variety of arrangements. Complex carbohydrates include **oligosaccharides,** made of 3 to 10 monosaccharides, and **polysaccharides,** which consist of more than 10 monosaccharides. However, most polysaccharides are made of hundreds of monosaccharides bonded together. You can refer back to Figure 4.1 to see how oligosaccharides and polysaccharides fit in the overall scheme of carbohydrate classification.

OLIGOSACCHARIDES ARE COMPONENTS OF CELL MEMBRANES

Oligosaccharides are present in a variety of foods, including dried beans, soybeans, peas, and lentils. Raffinose and stachyose are the two most common oligosaccharides. However, because humans lack the enzymes needed to digest these two carbohydrates, they pass undigested into the large intestine, where bacteria break them down. As a result, some people experience abdominal discomfort (bloating and cramps) and flatulence (gas). Commercial products such as Beano® supply enzymes needed to break down oligosaccharides, making them more digestible and therefore less available to intestinal bacteria.

In the body, oligosaccharides are components of cell membranes, and allow cells to recognize and interact with one another. Oligosaccharides are

complex carbohydrates Category of carbohydrate that includes oligosaccharides and polysaccharides.

oligosaccharide (o – li – go – SAC – cha – ride) (*oligo-*, few; *-saccharide*, sweet) Carbohydrate made of relatively few (3 to 10) monosaccharides.

polysaccharide (po – li – SAC – cha – ride) (*poly-*, many; *-saccharide*, sweet) Complex carbohydrate made of many monosaccharides.

Nutritive sweeteners are naturally occurring, digestible carbohydrates such as sucrose. However, food manufacturers can also use nonsugar alternatives to sweeten foods, including saccharin, aspartame, acesulfame K, and sugar alcohols. In addition to these alternative sweeteners, a sweet-tasting herbal dietary product called *stevia* has recently been approved by the Food and Drug Administration (FDA). This long-awaited decision means that food companies can now use stevia to sweeten foods and beverages.

Choosing which nonsugar sweetener to use is not as simple as you might think. For example, some lose their sweetness when heated and therefore are not recommended for baking or cooking. Others are not chemically stable and become bitter over time. In addition, ongoing debates questioning the safety of some nonsugar substitutes have raised health concerns among consumers.

- **Saccharin.** Saccharin was one of the first nonsugar *artificial* sweeteners to be widely used in the United States. During World War I, the use of saccharin increased because of sugar rationing. Although there was concern that saccharin might cause cancer, recent studies support its safety for human consumption. Most experts agree that the use of saccharin in moderation poses no health risks.[10] Saccharin is extremely sweet, very stable, and inexpensive to produce. Commercial products with saccharin include Sweet'N Low® and Sugar Twin®.

- **Aspartame.** Aspartame, another nonsugar sweetener, sold by its trade names NutraSweet® and Equal®, consists of two amino acids (phenylalanine and aspartic acid) bonded together. Although aspartame has the same energy content as sucrose (4 kcal per gram), it is almost 200 times as sweet. The food industry uses aspartame in sugar-free beverages. However, it is not heat stable and therefore cannot be used in products that require cooking. Beverages made with aspartame must be clearly labeled because of a potential risk to people with a genetic condition called phenylketonuria (PKU). Individuals with PKU cannot metabolize the amino acid phenylalanine, a component of this sweetener. Although the FDA has judged aspartame safe, some people claim it has adverse effects, including headaches, dizziness, nausea, and seizures. While several studies have reported that feeding laboratory animals aspartame increased the occurrence of cancer, hundreds of studies have failed to find similar effects in humans.[11]

- **Acesulfame K.** Another nonsugar sweetener, called acesulfame K, is actually a salt that contains potassium (K is the symbol for potassium). Acesulfame K has been used extensively in Europe and was approved in 1998 for use in the United States, where it is sold under the trade name of Sweet One®. Unlike aspartame, this artificial sweetener is heat stable and can be used in a wide variety of commercial products. Acesulfame K has been approved for use in refrigerated and frozen desserts, yogurt, dry dessert mixes, candies, gum, syrups, and alcoholic beverages.

- **Sucralose.** Sucralose (trade name Splenda®), another low-calorie sweetener, was approved by the FDA in 1990. The production of sucralose starts with sucrose, which then undergoes considerable processing. The end result is a sugar molecule with chlorine atoms attached. Sucralose is about 600 times sweeter than sucrose but provides minimal calories because it is difficult for the body to digest and absorb. Because it is water soluble and stable, sucralose is used in a broad range of foods and beverages.

- **Sugar Alcohols.** Sugar alcohols, another group of alternative sweeteners, are neither sugars nor alcohols. Rather, they are "polyols," meaning that the sugar molecule has multiple alcohol groups attached. Sugar alcohols occur naturally in plants, particularly fruits, and have half the sweetness and calories of sucrose (only 2 to 3 kcal per gram). Sorbitol, mannitol, and xylitol, the most common sugar alcohols, are often found in "sugar-free" products such as chewing gums, breath mints, candies, toothpastes, mouthwashes, and cough syrups. One advantage of sugar alcohols is that, unlike sucrose, they do not readily promote tooth decay. However, when eaten in excessive amounts, sugar alcohols can have a laxative effect, leading to diarrhea.

- **Stevia.** Stevia is sometimes referred to as the "natural sugar substitute" because it is derived from sunflowers that grow freely in tropical regions of Central and South America. The native people of these regions have long known about the sweet properties of stevia. Stevia is essentially calorie free and is considerably sweeter than table sugar. Now that the FDA no longer objects to the use of stevia, product development by several large food manufacturers is underway. Stevia has multiple trade names including Only Sweet™, PureVia™, Reb-A™, Rebiana™, SweetLeaf™, and Truvia®.

Although foods sweetened with nonsugar alternatives may appear to offer sweet indulgences with little caloric cost, this is not always the case. It is important to understand that foods sweetened with low-energy sweeteners are not necessarily healthy food choices. Many of these foods offer little in terms of nutritional value. Furthermore, replacing caloric sweeteners with low-energy sugar substitutes is a common practice among those trying to lose weight. Substituting nonsugar sweeteners for sugar sweeteners alone will not facilitate weight loss.[12] As obesity rates continue to climb, so does the use of low-energy sugar substitutes.

also made in the breasts, where they are incorporated into human milk. These oligosaccharides are part of a complex system that helps protect infants from disease-causing pathogens, and they are one of the many reasons why mothers are encouraged to breastfeed.

POLYSACCHARIDES DIFFER IN THE TYPES AND ARRANGEMENTS OF SUGAR MOLECULES

Polysaccharides are made of many monosaccharides bonded together by glycosidic bonds. The types and arrangements of the sugar molecules determine the shape and form of the polysaccharide. For example, some polysaccharides have an orderly linear appearance, whereas others are shaped like branches on a tree. The three most common polysaccharides—starch, glycogen, and dietary fiber—are discussed next.

Amylose and Amylopectin Are Types of Starch Recall that plants synthesize glucose by the process of photosynthesis. To store this important source of energy, plants convert the glucose to starch. There are two forms of starch, **amylose** and **amylopectin,** both of which consist entirely of glucose molecules. However, what distinguishes amylose from amylopectin is the

amylose (A – my – lose) A type of starch consisting of a linear (unbranched) chain of glucose molecules.

amylopectin (a – my – lo – PEK – tin) A type of starch consisting of a highly branched arrangement of glucose molecules.

FIGURE 4.8 Structure of Starch Plants store glucose in the form of starch. Both forms of starch (amylose and amylopectin) consist of glucose molecules bonded together.

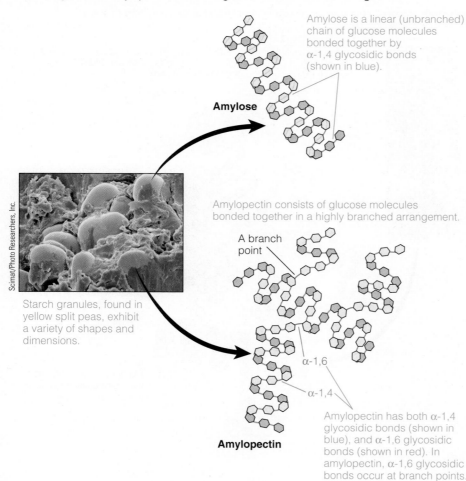

Amylose is a linear (unbranched) chain of glucose molecules bonded together by α-1,4 glycosidic bonds (shown in blue).

Amylose

Amylopectin consists of glucose molecules bonded together in a highly branched arrangement.

A branch point

α-1,6

α-1,4

Amylopectin

Amylopectin has both α-1,4 glycosidic bonds (shown in blue), and α-1,6 glycosidic bonds (shown in red). In amylopectin, α-1,6 glycosidic bonds occur at branch points.

Scimat/Photo Researchers, Inc.

Starch granules, found in yellow split peas, exhibit a variety of shapes and dimensions.

arrangement of glucose molecules (Figure 4.8). Whereas amylose is a linear (unbranched) chain of glucose molecules held together entirely by α-1,4 glycosidic bonds, amylopectin is a highly branched arrangement of glucose molecules. The linear portions of amylopectin contain α-1,4 glycosidic bonds, whereas α-1,6 glycosidic bonds occur at branch points.

Plants typically contain a mixture of both types of starch—amylose and amylopectin. Examples of starchy foods include grains (such as corn, rice, and wheat), products made from them (such as pasta and bread), and legumes (such as lentils and split peas). Potatoes and winter (hard) squashes are also sources of starch. Starch is often added to food to enhance its texture and stability. For instance, cornstarch forms a thick gel when heated, which is why it is used to thicken sauces and gravies. Food scientists have also developed ways to chemically modify starch to improve its functionality. This is why modified food starch is an ingredient commonly used in food production.

The Body Stores Glucose as Glycogen The body stores small amounts of glucose in the form of **glycogen**. Although many tissues store small amounts of glycogen, the majority is found in liver and skeletal muscles. Like amylopectin, glycogen is a highly branched arrangement of glucose molecules, consisting of both α-1,4 (linear portions) and α-1,6 (branch points) glycosidic bonds (Figure 4.9). The numerous branch points in glycogen provide a physiological advantage because enzymes can hydrolyze multiple glycosidic bonds simultaneously. As a result, glycogen can be broken down quickly when energy is needed.

glycogen (GLY – co – gen) Polysaccharide consisting of a highly branched arrangement of glucose molecules; found primarily in liver and skeletal muscle.

FIGURE 4.9 Structure of Glycogen Animals store glucose in the form of glycogen, a polysaccharide consisting of many glucose molecules.

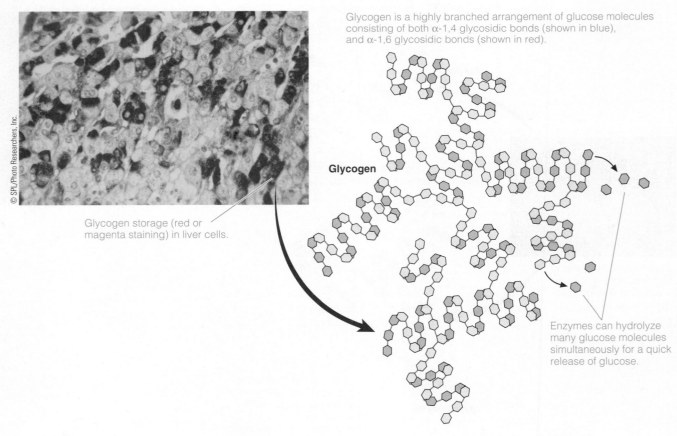

Glycogen is a highly branched arrangement of glucose molecules consisting of both α-1,4 glycosidic bonds (shown in blue), and α-1,6 glycosidic bonds (shown in red).

Glycogen

© SPL/Photo Researchers, Inc.

Glycogen storage (red or magenta staining) in liver cells.

Enzymes can hydrolyze many glucose molecules simultaneously for a quick release of glucose.

FIGURE 4.10 Humans Are Unable to Digest Dietary Fiber Unlike starch (which can be fully broken down into glucose molecules), dietary fiber remains intact in the gastrointestinal tract. This is because humans lack digestive enzymes that can break the glycosidic bonds in fiber.

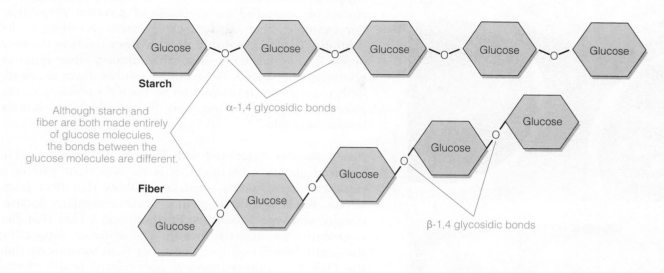

Although starch and fiber are both made entirely of glucose molecules, the bonds between the glucose molecules are different.

Because glucose is an important energy source for cells, the body turns to glycogen when glucose availability is low, such as during fasting and strenuous or prolonged exercise. When liver glycogen is broken down, the glucose can be released directly into the blood. Therefore, liver glycogen plays an important role in blood glucose regulation. However, muscle glycogen serves a somewhat different role. Unlike the liver, muscle lacks the enzyme needed to release glucose into the blood. In this case, the glucose that results from the breakdown of muscle glycogen is used to fuel physical activity. Some athletes try to increase their muscle glycogen stores by a technique called **carbohydrate loading.** This is described in greater detail in Chapter 9.

Humans Are Unable to Digest Fiber The term **fiber** generally refers to a diverse group of plant polysaccharides that, unlike starch, cannot be digested in the human small intestine. This is because fiber contains β-glycosidic bonds that are resistant to digestive enzymes (Figure 4.10). Undigested fiber passes from the small to the large intestine relatively intact. Intestinal bacteria then begin to break down the fiber, producing gas, lipids, and other substances. Although gas production (flatulence) by intestinal bacteria may be an annoyance, other substances made by these microorganisms serve useful purposes such as nourishing cells that line the colon. In addition, dietary fiber promotes the selective growth of beneficial intestinal bacteria, which in turn help inhibit the growth of other, disease-causing (pathogenic) bacteria.[13]

Classification of Fiber Types Our diets contain many kinds of fiber. **Dietary fiber** occurs naturally in plants, whereas **functional fiber** refers to fiber added to food as an ingredient. Functional fiber is typically derived from natural fibrous plant sources and used to manufacture foods or other types of products. It can also be synthetically manufactured. For a fiber to be called functional fiber, it must have demonstrated beneficial physiological effects. Functional fiber can increase the total fiber content of certain foods, which in turn provides health benefits. Thus, the term **total fiber** refers to the combination of dietary fiber that exists naturally in a food plus any functional fiber that is added during manufacturing.

carbohydrate loading A technique used to increase the body's glycogen stores.

fiber Polysaccharide found in plants that is not digested or absorbed in the human small intestine.

dietary fiber Fiber that naturally occurs in plants.

functional fiber Fiber that is added to food to provide beneficial physiological effects.

total fiber The combination of dietary fiber and functional fiber.

Fiber is an important part of a healthy diet. Soluble fiber retains water, and can help lower blood cholesterol. Insoluble fiber adds bulk, and can aid in weight management. Most fiber-rich foods such as dried beans and peanuts provide a combination of both soluble and insoluble fiber.

〈CONNECTIONS〉 Recall that prebiotics promote the growth of healthy bacteria in the colon (Chapter 3, page 110).

soluble fiber Dietary fiber that dissolves in water.

insoluble fiber Dietary fiber that does not dissolve in water.

Dietary fiber is found in a variety of plant foods such as whole grains, legumes, vegetables, and fruits. Different foods contain different types of dietary fiber, which is commonly classified on the basis of physical properties. For example, fiber is often categorized according to its solubility in water. **Soluble dietary fiber** tends to dissolve or swell in water, while **insoluble dietary fiber** remains relatively unchanged. Gel-forming soluble fibers are readily broken down (fermentable) by bacteria residing in the colon, as opposed to insoluble fibers, which are largely nonfermentable.

Health Benefits Associated with Fiber While fiber is not a required dietary component, it is an important part of a healthful diet. Studies consistently show that fiber from whole foods protects against cardiovascular disease, obesity, and type 2 diabetes. Furthermore, a fiber-rich diet is essential for digestive health. The evidence supporting the health benefits of fiber-rich foods is so substantive that the FDA has approved several fiber-related health claims (Table 4.2).[14] Though fiber is not a magic bullet, its credentials are impressive.

Foods such as oats, barley, legumes, rice, bran, psyllium seeds, soy, and some fruits contain mostly soluble, viscous dietary fibers that "absorb" water and swell. Examples of fibers with these physical properties include pectin, gums, and β-glucan. The sponge-like effect of these types of fiber can help soften fecal matter, reducing strain and making elimination easier. Consumption of soluble, viscous fiber can also help reduce blood cholesterol levels in some people.[15] Research indicates that the viscous fiber may bind with dietary fat and cholesterol in the GI tract, making it less likely to be absorbed. Similarly, eating foods with soluble, viscous fiber can delay gastric emptying, which can help promote satiety. Delayed gastric emptying may also help lower blood glucose levels.[16] In addition to these health-promoting benefits, soluble, fermentable fibers can also function as a prebiotic, promoting the growth of "friendly" bacteria in the colon.

Cellulose is the most abundant insoluble dietary fiber in food. Examples of foods high in cellulose include whole-grain flour, wheat bran, whole-grain breakfast cereals, seeds, and many vegetables including carrots, broccoli, celery, peppers, and cabbage. In addition to cellulose, insoluble dietary

TABLE 4.2 List of FDA-Approved Health Claims Concerning Fiber

- "Diets low in fat and rich in high-fiber foods may reduce the risk of certain cancers."
- "Diets low in fat and rich in soluble fiber may reduce risk of heart disease."
- "Diets low in fat and rich in fruits and vegetables may reduce the risk of certain cancers."
- "Diets low in fat and rich in whole oats and psyllium seed husk can help reduce the risk of heart disease."
- "Diets high in whole-grain foods and other plant foods and low in total fat, saturated fat, and cholesterol may help reduce the risk of heart disease and certain cancers."

SOURCE: U.S. Food and Drug Administration Health Claims Meeting Significant Scientific Agreement (SSA). Available from http://www.fda.gov/Food/LabelingNutrition/LabelClaims/HealthClaimsMeetingSignificantScientificAgreement SSA/default.htm

TABLE 4.3 Common Dietary Fibers, Food Sources, and Potential Physiological Effects

Dietary Fiber	Description	Food Sources	Potential Physiological Effects
Cellulose	Insoluble fiber consisting of glucose molecules with β-glycosidic bonds; the main structural component of plant cell walls.	Whole grains Bran Cereals Broccoli	Increases stool weight; may decrease transit time
Hemicellulose	Insoluble fiber consisting of a variety of different monosaccharide molecules (e.g., glucose, arabinose, mannose, and xylose).	Cabbage Legumes Apples Root vegetables	
Pectin	Soluble fiber found in the skin of ripe fruits; consists of a variety of monosaccharide molecules; is used commercially to make jams and jellies.	Apples Citrus fruits Strawberries Raspberries	Some evidence exists that it lowers cholesterol by increased excretion of bile acids and cholesterol; significantly reduces glycemic response
β-Glucan	A nonstarch polysaccharide composed of branched chains of glucose molecules.	Mushrooms Barley Oats	Oat bran increases stool weight because of viscosity and fermentation by bacteria; reduces cholesterol; may reduce blood glucose levels
Gums	Highly soluble and viscous nonstarch polysaccharides used to thicken foods.	Oats Legumes	Reduces blood cholesterol and blood glucose
Psyllium	Insoluble nonpolysaccharide dietary fiber, consisting of numerous alcohol units, found within the woody portion of plants.	Berries Wheat	May decrease lipid absorption

SOURCE: Institute of Medicine. Dietary reference intakes for energy, carbohydrate, fiber, fat, fatty acids, cholesterol, protein, and amino acids. Washington, DC: National Academies Press; 2005.

fiber includes hemicellulose and lignin, both of which are found in wheat and some green, leafy vegetables (Table 4.3). Because insoluble fibers do not readily dissolve in water, they do not form viscous gels. Nor do bacteria in the colon readily ferment them. Instead, insoluble fiber passes intact through the GI tract, which helps to increase fecal weight and volume. In general, large amounts of fecal mass move through the colon quickly by stimulating peristaltic contractions in the colon, propelling the material forward. When consumed with sufficient amounts of fluid, insoluble fiber can help prevent and alleviate constipation.

The production of hard, dry feces, characteristic of insufficient fiber, not only makes elimination more difficult but may also contribute to a condition called **diverticular disease** (Figure 4.11). When it is chronic, constipation can cause areas along the colon wall to become weak. As a result, protruding pouches, called diverticula, can form. It is not uncommon for older adults, especially those with low intakes of fibrous foods, to develop diverticula. By 70 years of age, approximately 50% of adults have diverticulosis, 10 to 20% of whom develop complications such as diverticulitis.[17] **Diverticulitis** occurs when the diverticula become infected or inflamed (note that the suffix –*itis* refers to inflammation). Symptoms include cramping, diarrhea, fever, and on occasion bleeding from the rectum. Scientists think that dietary fiber helps prevent the formation of diverticula by increasing fecal mass, making bowel movements easier. Therefore, a diet high in fiber with plenty of fluids may help protect against diverticular disease.

Whole-Grain Foods There are many ways to ensure adequate fiber intake. In addition to eating a variety of fruits and vegetables every day, it is important to select foods made from whole grains. The nutritional value of grains is greatest when all three components of the grain—bran, germ, and endosperm—are

diverticular disease, or **diverticulosis** (di – ver – TI – cu – lar) (di – ver – ti – cu – LO – sis) Condition in the large intestine; characterized by the presence of pouches that form along the intestinal wall.

diverticulitis (di – ver – ti – cu – LI – tis) Inflammation of diverticula (pouches) in the lining of the large intestine.

FIGURE 4.11 Diverticular Disease
Diverticular disease is most common in older adults who consume low intakes of fibrous foods. Diverticulitis is the inflammation of diverticula.

Diverticulosis results when small, protruding pouches called diverticula form along the wall of the large intestine.

Diverticular disease is characterized by the formation of outpouchings along the wall of the large intestine.

Du Cane Medical Imaging Ltd./Photo Researchers, Inc.

bran The outer layer of a grain; contains most of the fiber.

germ The portion of a grain that contains most of its vitamins and minerals.

endosperm The portion of a grain that contains mostly starch.

whole-grain foods Cereal grains that contain bran, endosperm, and germ in the same relative proportion as they exist naturally.

present (Figure 4.12). Whereas the **bran** contains most of the fiber, the **germ** supplies much of the vitamins and minerals. The **endosperm** is mostly starch. Because milling removes the bran and germ, products made with refined flour often contain very little fiber. Although some of the lost nutritive value is restored when food manufacturers fortify their products with vitamins and minerals, many important nutrients, as well as fiber, may still be lacking.

When reading food labels, look for the words "whole-grain cereal" and "whole-wheat flour." Foods made with "wheat flour" are not necessarily high in fiber. To make it easier for consumers to select whole-grain products, the FDA has published a tentative definition: **whole-grain foods** should contain

FIGURE 4.12 Anatomy of a Wheat Kernel
Foods made with whole-wheat flour are typically more nutritious than foods made with refined wheat flour because the components of the wheat kernel—the bran, the germ, and the endosperm—have not been removed. Each component contributes important nutrients needed for good health.

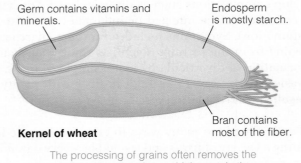

Germ contains vitamins and minerals.

Endosperm is mostly starch.

Kernel of wheat

Bran contains most of the fiber.

The processing of grains often removes the germ and bran portions, which contain the majority of the vitamins, minerals, and fiber.

TABLE 4.4 The Dietary Fiber Content of Selected Foods

Food	Serving Size	Insoluble Fiber (g)	Soluble Fiber (g)	Total Dietary Fiber (g)
Fruits				
Apple	1 medium	2.0	0.9	2.9
Orange	1 medium	0.7	1.3	2.0
Banana	1 medium	1.4	0.6	2.0
Vegetables				
Broccoli	1 stalk	1.4	1.3	2.7
Carrot	1 large	1.6	1.3	2.9
Tomato	1 small	0.7	0.1	0.8
Potato	1 medium	0.8	1.0	1.8
Corn	²/₃ cup	1.4	0.2	1.6
Grains				
All-Bran® cereal	¹/₂ cup	7.6	1.4	9.0
Oat bran	¹/₂ cup	2.2	2.2	4.4
Cornflakes® cereal	1 cup	0.5	0	0.5
Rolled oats	³/₄ cup	1.7	1.3	3.0
Whole-wheat bread	1 slice	1.1	0.3	1.4
White bread	1 slice	0.1	0.3	0.4
Macaroni	1 cup cooked	0.3	0.5	0.8
Legumes				
Green peas	²/₃ cup	3.3	0.6	3.9
Kidney beans	¹/₂ cup	4.9	1.6	6.5
Pinto beans	¹/₂ cup	4.7	1.2	5.9
Lentils	²/₃ cup	3.9	0.6	4.5

SOURCE: Anderson JW, Bridges SR. Dietary fiber content of selected foods. American Journal of Clinical Nutrition. 1988; 47: 440–7.

the three key ingredients of cereal grains—bran, endosperm, and germ—in the same relative proportion as the naturally-occurring grain.[18]

Based on USDA food-labeling guidelines, a food labeled as an "excellent source of fiber" must contain 20% or more of the Daily Value of fiber or at least 5 grams of fiber per serving. Foods with labels claiming to be "good" sources of fiber have 2.5 to 4.9 g of fiber per serving. The fiber content of selected foods is shown in Table 4.4.

How Are Carbohydrates Digested, Absorbed, and Circulated in the Body?

Because carbohydrates are structurally diverse, the steps involved in the digestive process vary. During digestion, carbohydrates undergo extensive chemical transformations as they move through the GI tract. With the help of digestive enzymes, the glycosidic bonds that hold disaccharides and starches together are broken. The ultimate goal of carbohydrate digestion is to break down large, complex molecules such as starches into small, absorbable monosaccharides. This process requires a series of enzymes produced in the salivary glands, pancreas, and small intestine.

The nutritional value is greater in whole-grain foods than in refined-grain foods because all three components of the grain—bran, germ, and endosperm—are present.

⟨CONNECTIONS⟩ The brush border is the absorptive surface of the small intestine that is made up of minute projections called microvilli, which line enterocytes. (Chapter 3, page 98).

⟨CONNECTIONS⟩ Hydrolytic reactions break chemical bonds by the addition of water (Chapter 3, page 74).

salivary α-amylase (A – my – lase) Enzyme, produced by the salivary glands, which digests starch by hydrolyzing α-1,4 glycosidic bonds.

dextrin A partial breakdown product formed during starch digestion, consisting of varying numbers of glucose units.

pancreatic α-amylase (pan – cre – A – tic Al – pha A – my – lase) Enzyme, produced by the pancreas, which digests starch by hydrolyzing α-1,4 glycosidic bonds.

maltase (MAL – tase) Brush border enzyme that hydrolyzes maltose into two glucose molecules.

brush border enzyme Enzyme, produced by enterocytes, which aids in the final steps of digestion.

limit dextrin (DEX – trin) A partial breakdown product formed during amylopectin digestion that contains three to four glucose molecules and an α-1,6 glycosidic bond.

α-dextrinase (DEX – stri – nase) Brush border enzyme that hydrolyzes α-1,6 glycosidic bonds.

disaccharidase (di – SAC – cha – ri – dase) Brush border enzyme that hydrolyzes glycosidic bonds in disaccharides.

sucrase Brush border enzyme that hydrolyzes sucrose into glucose and fructose.

lactase Brush border enzyme that hydrolyzes lactose into glucose and galactose.

lactose intolerance Inability to digest the milk sugar lactose; caused by a lack of the enzyme lactase.

STARCH DIGESTION BEGINS IN THE MOUTH

Recall that the two basic forms of starch, amylose and amylopectin, are both made entirely of glucose molecules, although the types of bonds are slightly different. Amylose has only α-1,4 glycosidic bonds, while amylopectin contains both α-1,4 and α-1,6 glycosidic bonds. Nonetheless, many of the same enzymes are involved in digesting both amylose and amylopectin (Figure 4.13).

The chemical digestion of starch begins in the mouth when the salivary glands release the enzyme salivary α-amylase. **Salivary α-amylase** hydrolyzes the α-1,4 glycosidic bonds in both amylose and amylopectin, resulting in shorter polysaccharide chains of varying lengths called **dextrins.**

However, because food stays in the mouth only a short time, very little starch digestion actually takes place there. Once dextrins enter the stomach, the acidic environment stops the enzymatic activity of salivary α-amylase. Dextrins then pass unchanged from the stomach into the small intestine, where they encounter pancreatic α-amylase, an enzyme produced by the pancreas. Like salivary α-amylase, **pancreatic α-amylase** hydrolyzes α-1,4 glycosidic bonds, transforming dextrins into the disaccharide maltose. Last, **maltase**, a **brush border enzyme** produced by enterocytes, finishes the job of starch digestion by hydrolyzing the last remaining chemical bond in maltose, resulting in two free (unbound) glucose molecules.

The combined efforts of salivary α-amylase, pancreatic α-amylase, and maltase complete the chemical transformation of amylose into multiple glucose molecules. However, an additional step, the hydrolysis of α-1,6 glycosidic bonds, is needed to complete the digestion of amylopectin. Because salivary and pancreatic α-amylases only hydrolyze α-1,4 glycosidic bonds, partial breakdown products called limit dextrins form during amylopectin digestion. **Limit dextrins** consist of three to four glucose molecules and contain α-1,6 glycosidic bonds that were located at branch points in the original amylopectin molecules. The enzyme **α-dextrinase,** also a brush border enzyme, accomplishes the hydrolysis of α-1,6 glycosidic bonds, completing the digestion of amylopectin. Thus, amylose and amylopectin digestion results in the production of numerous glucose molecules that are now ready to be transported into the enterocytes.

DISACCHARIDES ARE DIGESTED IN THE SMALL INTESTINE

The digestion of disaccharides (maltose, sucrose, and lactose) takes place entirely in the small intestine (Figure 4.14). The enterocytes provide the enzymes needed for disaccharide digestion, and are collectively referred to as the **disaccharidases.** Each disaccharide requires a specific disaccharidase. For example, **sucrase** hydrolyzes sucrose into glucose and fructose, maltase hydrolyzes maltose into two glucose molecules, and **lactase** hydrolyzes lactose into glucose and galactose. Once disaccharides have been digested into their component monosaccharides, they can be transported into the enterocytes.

Lactose Intolerance Although most babies produce enough lactase to digest the high amounts of lactose found in milk, some people produce very little or none of this enzyme, so they have difficulty digesting lactose—a condition called **lactose intolerance**. When people with lactose intolerance consume lactose-containing foods, much of the lactose enters the large intestine undigested. Bacteria in the large intestine break down (ferment) the lactose, producing several by-products, including gas. Symptoms such as abdominal cramping,

FIGURE 4.13 Digestion of Amylose and Amylopectin Amylose and amylopectin digestion require many of the same enzymes. Through the process of digestion, both amylose and amylopectin are broken down into molecules of glucose. Most starch digestion takes place in the small intestine.

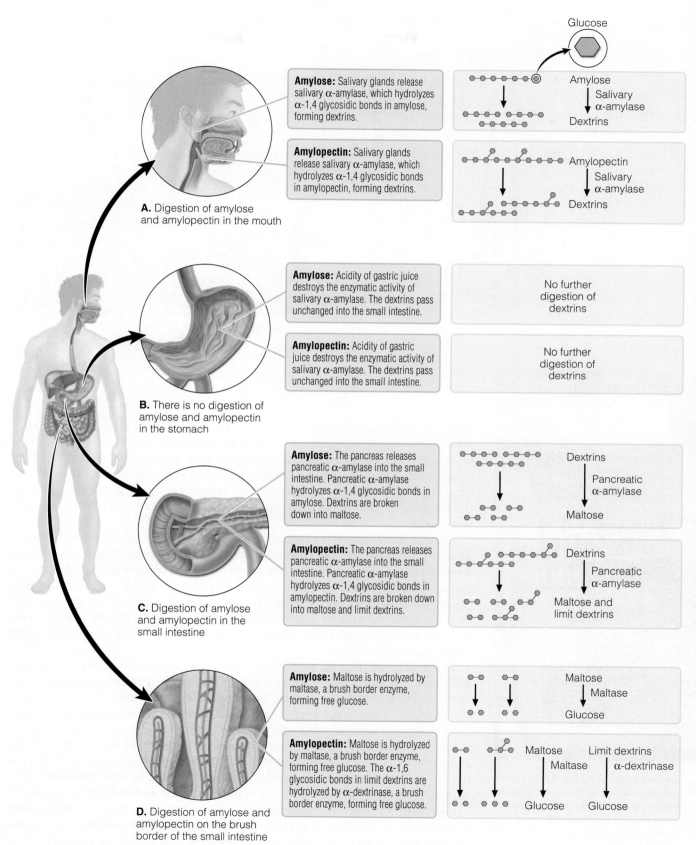

Glucose

Amylose: Salivary glands release salivary α-amylase, which hydrolyzes α-1,4 glycosidic bonds in amylose, forming dextrins.

Amylopectin: Salivary glands release salivary α-amylase, which hydrolyzes α-1,4 glycosidic bonds in amylopectin, forming dextrins.

A. Digestion of amylose and amylopectin in the mouth

Amylose — Salivary α-amylase → Dextrins

Amylopectin — Salivary α-amylase → Dextrins

Amylose: Acidity of gastric juice destroys the enzymatic activity of salivary α-amylase. The dextrins pass unchanged into the small intestine.

Amylopectin: Acidity of gastric juice destroys the enzymatic activity of salivary α-amylase. The dextrins pass unchanged into the small intestine.

B. There is no digestion of amylose and amylopectin in the stomach

No further digestion of dextrins

No further digestion of dextrins

Amylose: The pancreas releases pancreatic α-amylase into the small intestine. Pancreatic α-amylase hydrolyzes α-1,4 glycosidic bonds in amylose. Dextrins are broken down into maltose.

Amylopectin: The pancreas releases pancreatic α-amylase into the small intestine. Pancreatic α-amylase hydrolyzes α-1,4 glycosidic bonds in amylopectin. Dextrins are broken down into maltose and limit dextrins.

C. Digestion of amylose and amylopectin in the small intestine

Dextrins — Pancreatic α-amylase → Maltose

Dextrins — Pancreatic α-amylase → Maltose and limit dextrins

Amylose: Maltose is hydrolyzed by maltase, a brush border enzyme, forming free glucose.

Amylopectin: Maltose is hydrolyzed by maltase, a brush border enzyme, forming free glucose. The α-1,6 glycosidic bonds in limit dextrins are hydrolyzed by α-dextrinase, a brush border enzyme, forming free glucose.

D. Digestion of amylose and amylopectin on the brush border of the small intestine

Maltose — Maltase → Glucose

Maltose — Maltase → Glucose   Limit dextrins — α-dextrinase → Glucose

FIGURE 4.14 Digestion of Disaccharides Disaccharides are digested along the brush border by intestinal enzymes, collectively known as disaccharidases.

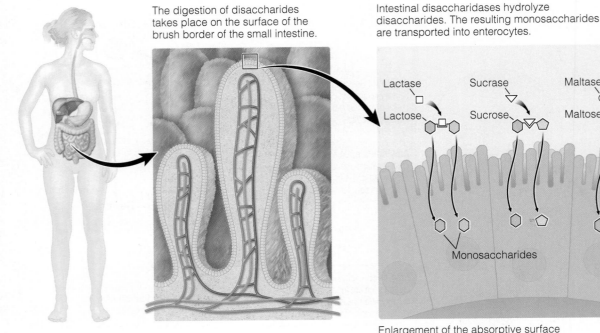

The digestion of disaccharides takes place on the surface of the brush border of the small intestine.

Intestinal disaccharidases hydrolyze disaccharides. The resulting monosaccharides are transported into enterocytes.

Lactase · Sucrase · Maltase

Lactose · Sucrose · Maltose

Monosaccharides

Enlargement of the absorptive surface

⬡ Glucose · ⬡ Galactose · ⬠ Fructose

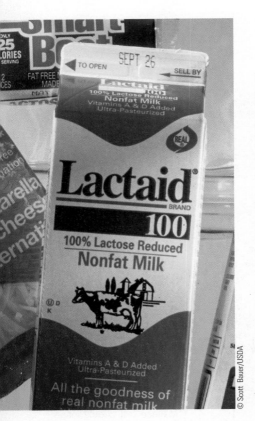

© Scott Bauer/USDA

Lactose-reduced milk and other dairy products are widely available at most grocery stores. These dairy products contain all of the nutrients found in regular milk products, with the exception of lactose.

bloating, flatulence, and diarrhea can occur within 30 to 60 minutes. The severity of symptoms depends on how much lactose is consumed and how little lactase the person produces. Most people with lactose intolerance can consume small amounts of dairy products without experiencing any discomfort. Some dairy products such as yogurt and cheese are easier to tolerate because the bacteria used to make these dairy products convert some of the lactose to lactic acid.

Lactose intolerance is especially common among certain ethnic groups; more than half of Asian Americans, Native Americans, and African Americans are reportedly lactose intolerant.[19] In comparison, only 12% of people of northern European descent are lactose intolerant. Certain diseases, medications, and surgery that damage the intestinal mucosa can also increase the risk of lactose intolerance, although this is usually temporary; lactase production returns once the underlying condition is no longer present.

In most cases, lactose intolerance does not pose a serious health threat. Although it may be annoying and sometimes inconvenient, people can easily live with it. Products such as lactose-free milk and over-the-counter lactase enzyme tablets have made it easier for lactose-intolerant people to enjoy dairy products and consume sufficient calcium. These lactose-reduced dairy products contain all of the nutrients found in regular milk products, with the exception of lactose. However, if these calcium-rich, lactose-reduced alternatives are not available or acceptable, it is important for lactose-intolerant individuals to choose alternate sources of calcium such as fortified soy products, fish with edible bones, fortified orange juice, and calcium-rich vegetables such as Swiss chard and rhubarb. They may also need calcium supplements to ensure the body is receiving adequate amounts of this important mineral.

FIGURE 4.15 Absorption and Circulation of Monosaccharides Glucose and galactose are absorbed into the enterocytes by carrier-dependent, energy-requiring active transport. Fructose is absorbed by facilitated diffusion. Once absorbed, monosaccharides are circulated to the liver via the hepatic portal system.

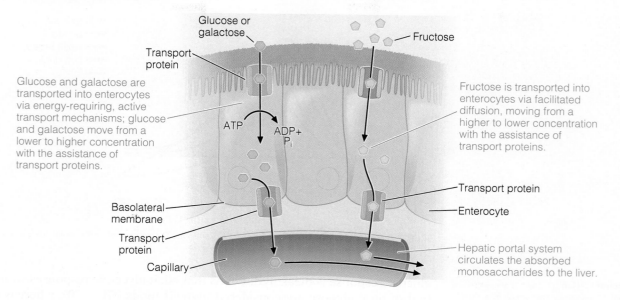

Glucose and galactose are transported into enterocytes via energy-requiring, active transport mechanisms; glucose and galactose move from a lower to higher concentration with the assistance of transport proteins.

Fructose is transported into enterocytes via facilitated diffusion, moving from a higher to lower concentration with the assistance of transport proteins.

Hepatic portal system circulates the absorbed monosaccharides to the liver.

MONOSACCHARIDES ARE READILY ABSORBED FROM THE SMALL INTESTINE

Once disaccharide and starch digestion is complete, the resulting monosaccharides (glucose, galactose, and fructose) are readily absorbed from the small intestine into the blood. As shown in Figure 4.15, glucose and galactose are absorbed across the lumenal membrane of the enterocytes via carrier-mediated active transport, whereas fructose is absorbed through facilitated diffusion. After crossing the basolateral membrane (via facilitated diffusion), monosaccharides then circulate in the blood directly to the liver via the hepatic portal system.

The Glycemic Response A rise in blood glucose levels can be detected shortly after we eat carbohydrate-rich foods. However, not all carbohydrates have the same effect on blood glucose levels. Some foods cause blood glucose levels to rise quickly and remain elevated, while others elicit a more gradual increase. The change in blood glucose following the ingestion of a specific food is called the **glycemic response** (Figure 4.16).

Scientists have long believed that simple carbohydrates elicit higher glycemic responses than complex carbohydrates. However, this is not always the case. In fact, blood glucose response to starchy foods—such as potatoes, refined cereal products, white bread, some whole-grain breads, and white rice—are sometimes higher than foods rich in simple sugars, such as soft drinks. Scientists now recognize that, although carbohydrate complexity may influence the glycemic response elicited by a food, other factors such as the presence of fat or protein, viscosity, processing, ripeness, and food additives are also important.

The **glycemic index (GI)** is a rating system based on a scale of 0 to 100 that can be used to compare the glycemic responses of various foods. In this system, blood glucose response to the consumption of 50 grams of pure glucose is used as a reference point represented by the highest GI value, 100. The GI value of a food is determined experimentally: subjects consume an amount of food that contains 50 grams of carbohydrate, and their blood

⟨**CONNECTIONS**⟩ Carrier-mediated active transport requires a transport protein and energy (ATP) to transport material across cell membranes moving from a lower to a higher concentration. Facilitated diffusion requires only a transport protein moving from a higher to a lower concentration (Chapter 3, page 77).

glycemic response (*glyc-*, sugar) The change in blood glucose following the ingestion of a specific food.

glycemic index (GI) A rating system used to categorize foods according to the relative glycemic responses they elicit.

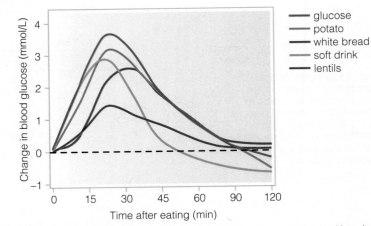

FIGURE 4.16 Glycemic Response Some foods cause blood glucose levels to rise quickly and remain elevated (high glycemic response), while others elicit a more gradual increase (low glycemic response).

SOURCE: Brand-Miller J, McMillan-Price J, Steinbeck K, Carterson I. Dietary glycemic index: Health implications. Journal of the American College of Nutrition. 2009;28:446S–9S.

⟨**CONNECTIONS**⟩ The hepatic portal system is the vascular connection from the small intestine to the liver that circulates absorbed nutrients (Chapter 3, page 99).

glucose response is measured. Foods that elicit glycemic responses similar to that of pure glucose are considered high-GI foods (GI ≥ 70), whereas those that cause a lower or more gradual rise in blood glucose are considered low-GI foods (GI ≤ 55).

One limitation of using GI values to compare the effects of foods on glycemic response is that the amount of carbohydrate found in a typical serving of a food is not taken into account. Rather, GI values are based on a standard *amount* of carbohydrate (50 grams), which may or may not represent the *amount* a person would normally eat. For example, to consume 50 g of carbohydrates from carrots, a person would need to eat more than a pound of carrots, which is unrealistic. To counter this problem, another rating system is sometimes used to evaluate the glycemic response to various foods. In this system, foods are assigned a value called its **glycemic load (GL).** The GL of a food is different from its GI in that the GL takes into account the typical portion of food consumed.

Glycemic load (GL) is calculated by dividing the glycemic index (GI) of a food by 100 and then multiplying by the grams of carbohydrate in one serving. For example, the GL of carrots that have a GI of 16 and contain approximately 8 g carbohydrates per serving is 1.3 (16 ÷ 100 × 8). By comparison, the GL of a 12-oz soft drink that has a GI of 63 and 39 g carbohydrate is approximately 25 (63 ÷ 100 × 39). These values are probably closer to what you may have predicted. You can compare the GI and GL values of selected foods in Table 4.5.

The impact of glycemic response on long-term health remains controversial. While some studies have demonstrated health benefits associated with low-GL diets, others are less clear.[20] Until we know more, it is important to make healthy food choices by increasing our consumption of whole grains, fruits, and vegetables, while decreasing our consumption of foods made with refined flour and added sugar.

MONOSACCHARIDES HAVE SEVERAL FUNCTIONS IN THE BODY

Once absorbed, monosaccharides circulate directly to the liver via the hepatic portal system, where the majority of galactose and fructose is converted into other compounds—most notably glucose. However, some

glycemic load (GL) A rating system used to categorize the body's glycemic response to foods that takes into account the glycemic index as well as the amount of carbohydrate typically found in a single serving of the food.

TABLE 4.5 Glycemic Index (GI) and Glycemic Load (GL) of Selected Foods Based on Glucose as a Reference Food

Value	Glycemic Index (GI)	Glycemic Load (GL)
High	≥70	≥20
Medium	56–69	11–19
Low	≤55	≤10

Food (serving size)	GI	Carbohydrates per Serving (g)	GL	Food	GI	Carbohydrates per Serving (g)	GL
Pastas, Grains, Legumes, Breads, Starchy Vegetables, and Cereals							
Baked potato (150 g)	85	30	26	Corn Chex™ (30 g)	83	30	25
Waffles (35 g)	76	13	10	Corn (80 g)	48	16	8
French fries (150 g)	75	29	22	Popcorn (20 g)	54	11	6
Bagel (70 g)	72	35	25	Cracked wheat (150 g)	48	26	12
Oat bran bread (30 g)	44	18	8	Pancakes (80 g)	67	58	39
White rice (150 g)	56	42	24	Apple muffins (60 g)	44	29	13
Angel food cake (50 g)	67	29	19	Lentils (50 g)	30	40	12
Raisin bran cereal (30 g)	61	19	12				
Fruits, Beverages, and Snack Foods							
Apple (22 g)	38	22	8	Grapes (120 g)	43	17	7
Raisins (60 g)	64	44	28	Apple juice (250 mL)	39	25	10
Banana (120 g)	48	25	12	Tomato juice (250 mL)	38	9	3
Potato chips (28 g)	54	15	8	Plums (120 g)	24	14	3
Jelly beans (28 g)	80	26	21	Chocolate (28 g)	49	18	9
Cherries (120 g)	22	12	3	Sucrose (5 g)	65	5	3

NOTES: Glycemic load (GL) of a food is the glycemic index (GI) divided by 100 and multiplied by the amount of carbohydrate in a single serving (in grams). Therefore, a food's GL is numerically lower than its GI.

SOURCE: Foster-Powell K, Holt SH, Brand-Miller JC. International table of glycemic index and glycemic load values: 2002. American Journal of Clinical Nutrition. 2002; 76: 5–56.

monosaccharides are converted to ribose, a constituent of many vital compounds including ATP, RNA, and DNA. While monosaccharides serve numerous functions within the body, the ability to transform the energy contained in glucose into ATP is probably their most important role. Details related to glucose and energy metabolism are presented in Chapter 7.

How Do Hormones Regulate Blood Glucose and Energy Storage?

The concentration of glucose in your blood fluctuates throughout the day. After you have gone several hours without eating, your blood glucose decreases. Conversely, blood glucose increases after you eat a meal rich in carbohydrates. Blood glucose levels are the lowest in the morning after an overnight fast, returning to normal shortly after breakfast. Because our cells need energy 24 hours a day, the pancreatic hormones insulin and glucagon work vigilantly to maintain blood glucose levels within an acceptable range at all times.

THE HORMONES INSULIN AND GLUCAGON ARE PRODUCED BY THE PANCREAS

The pancreas, which is partially composed of hormone-secreting cells, plays a major role in glucose homeostasis. These endocrine cells, collectively known as the islets of Langerhans, consist mainly of beta (β) cells that produce the hormone **insulin** and alpha (α) cells that produce the hormone **glucagon** (Figure 4.17). The pancreas, which has a rich blood supply, releases these hormones directly into the circulation in response to fluctuations in blood glucose.

Insulin and glucagon assist in blood glucose regulation and energy storage. When blood glucose levels increase, the pancreas takes action by releasing more insulin. Insulin in turn lowers blood glucose by facilitating the uptake of glucose into many kinds of cells. In addition, when meals provide more glucose than we require, insulin stimulates its storage as glycogen. Once muscles and the liver reach their glycogen storage capacity, excess glucose is converted to fat, which is stored primarily in adipose tissue. The hormonal balance shifts toward glucagon when blood glucose levels decrease. To increase glucose availability, glucagon stimulates the breakdown of glycogen stores in the liver. It is important to recognize that the release of insulin and glucagon is not an all-or-nothing situation. That is, the relative concentrations of these two hormones in the blood determine the shift between energy storage and mobilization (Figure 4.18). We will now take a closer look at these two important hormones and how each contributes to blood glucose regulation.

⟨**CONNECTIONS**⟩ Homeostasis is a state of balance or equilibrium (Chapter 3, page 81).

insulin Hormone secreted by the pancreatic β-cells in response to increased blood glucose.

glucagon (GLU – ca – gon) Hormone secreted by the pancreatic α-cells in response to decreased blood glucose.

FIGURE 4.17 Release of Insulin and Glucagon from the Pancreas Insulin is made and released by the pancreatic β-cells, and glucagon is made and released by the pancreatic α-cells. Both hormones play an important role in blood glucose regulation.

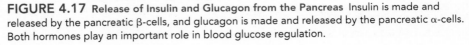

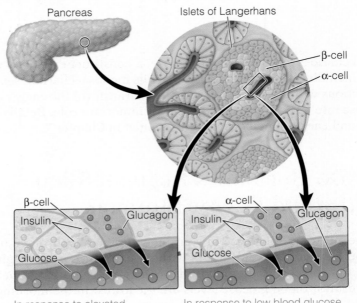

In response to elevated blood glucose levels, the pancreas increases its release of the hormone insulin.

In response to low blood glucose levels, the pancreas increases its release of the hormone glucagon.

FIGURE 4.18 Hormonal Regulation of Blood Glucose Insulin increases when the level of glucose in the blood increases. Glucagon increases when the level of glucose in the blood decreases.

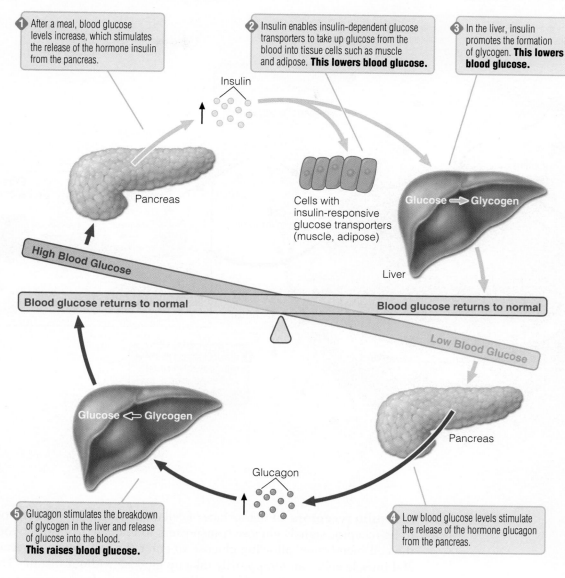

1. After a meal, blood glucose levels increase, which stimulates the release of the hormone insulin from the pancreas.

2. Insulin enables insulin-dependent glucose transporters to take up glucose from the blood into tissue cells such as muscle and adipose. **This lowers blood glucose.**

3. In the liver, insulin promotes the formation of glycogen. **This lowers blood glucose.**

Insulin

Pancreas

Cells with insulin-responsive glucose transporters (muscle, adipose)

Glucose ⇌ Glycogen

Liver

High Blood Glucose

Blood glucose returns to normal

Blood glucose returns to normal

Low Blood Glucose

Glucose ⇌ Glycogen

Pancreas

Glucagon

5. Glucagon stimulates the breakdown of glycogen in the liver and release of glucose into the blood. **This raises blood glucose.**

4. Low blood glucose levels stimulate the release of the hormone glucagon from the pancreas.

INSULIN LOWERS BLOOD GLUCOSE AND PROMOTES ENERGY STORAGE

Glucose enters cells via facilitated diffusion, mediated by carrier proteins known as **glucose transporters** (Figure 4.19). Some glucose transporters require insulin to transport glucose across the cell membrane, whereas others do not. In fact, it is the type of glucose transporter in a cell that determines if insulin is needed for the uptake of glucose. Examples of tissues that have **insulin-responsive glucose transporters** are skeletal muscle and adipose tissue. Brain and liver tissue have glucose transporters that do not require insulin for the uptake of glucose.

After a person eats carbohydrate-containing foods, blood glucose levels quickly rise, causing the pancreas to increase its release of insulin. When insulin encounters cells with insulin-responsive glucose transporters, it binds

glucose transporters Proteins that assist in the transport of glucose molecules across cell membranes.

insulin-responsive glucose transporters Glucose transporters that require insulin to function.

FIGURE 4.19 **Role of Insulin in Cellular Uptake of Glucose** Binding of insulin to its receptor causes insulin-responsive glucose transporters to relocate from the cytoplasm to the cell membrane, facilitating the movement of glucose molecules across the cell membrane.

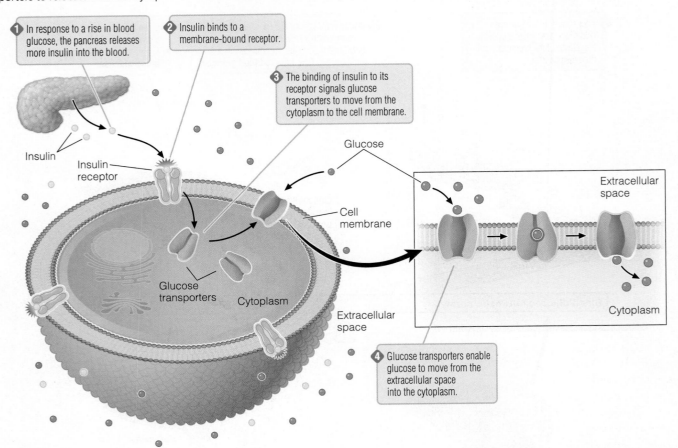

1 In response to a rise in blood glucose, the pancreas releases more insulin into the blood.

2 Insulin binds to a membrane-bound receptor.

3 The binding of insulin to its receptor signals glucose transporters to move from the cytoplasm to the cell membrane.

Insulin

Insulin receptor

Glucose

Cell membrane

Glucose transporters

Cytoplasm

Extracellular space

Extracellular space

Cytoplasm

4 Glucose transporters enable glucose to move from the extracellular space into the cytoplasm.

to **insulin receptors** on the surface of cell membranes. The binding of insulin to its receptor signals glucose transporters to relocate from the cytoplasm to the cell membrane, allowing glucose to enter the cell. During exercise, skeletal muscle cells can temporarily take up glucose without insulin. This is because muscle contractions, like insulin, can activate the movement of glucose transporters from the cytoplasm to the cell membrane, allowing for glucose uptake. The insulin-responsive glucose transporters return to their insulin-requiring state within a few hours after exercise stops.

Sometimes insulin-responsive glucose transporters have difficulty transporting glucose across cell membranes, causing glucose to accumulate in the blood, a condition called **hyperglycemia.** When this occurs, a person is said to have **impaired glucose regulation** or, in more serious cases, diabetes. Impaired blood glucose regulation and diabetes are serious health conditions that can lead to long-term complications. Diabetes prevention and management are fully addressed in the Nutrition Matters following this chapter.

Insulin Promotes Energy Storage After a meal more glucose may be available than is needed. When this occurs, the body stores the excess energy contained in glucose for later use. The hormone insulin promotes the storage of excess energy from glucose in the form of glycogen and body fat (Figure 4.20). In addition, insulin stimulates protein synthesis and inhibits the breakdown of muscle.

insulin receptors Proteins, located on the surface of certain cell membranes, that bind insulin.

hyperglycemia (hi – per – gly – CE – mi – a) (*hyper-*, excessive) Abnormally high level of glucose in the blood.

impaired glucose regulation Condition characterized by elevated levels of glucose in the blood.

FIGURE 4.20 Insulin Promotes Energy Storage The pancreas increases its release of the hormone insulin in response to high blood glucose. Insulin stimulates glucose transport into cells and promotes energy storage.

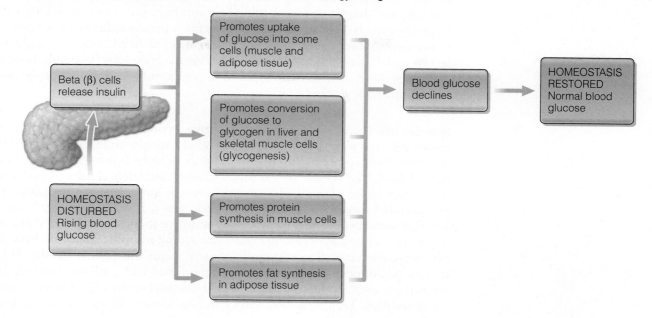

To store excess glucose as glycogen, insulin stimulates a process called **glycogenesis,** which occurs mainly in the liver and skeletal muscle. However, the body can only store a limited amount of glycogen. Once the limit is reached, insulin promotes the uptake of excess glucose by adipose tissue, where the glucose is converted to fatty acids (a type of lipid). Unlike what occurs with glycogen storage, the body has a seemingly endless capacity to store excess energy from glucose as body fat. It is important to note that the conversion of glucose into a fatty acid is irreversible. That is, once glucose is transformed into a fatty acid, the fatty acid cannot be converted back into glucose. This is very different from the reversible metabolic transformation between glucose and glycogen.

Critical Thinking: Laura's Story Think back to Laura's story about her experience with type 1 diabetes, a condition characterized by the body's inability to produce insulin. Based on what you have read about the role of insulin in blood glucose regulation and energy storage, why is it so important for Laura to monitor her blood glucose while exercising? How might low insulin and high blood glucose levels, common in people with diabetes, influence Laura's ability to train for and compete in the Ironman® competition?

GLUCAGON HELPS INCREASE BLOOD GLUCOSE

Clearly insulin is instrumental in helping the body use glucose after a meal and to promote energy storage. However, to maintain homeostasis, the body also requires another hormone that predominates during periods of low blood glucose. Several hours after a person eats, blood glucose levels begin to fall. Unless glucose is replenished by food, the body must begin to break down liver glycogen to maintain adequate levels of blood glucose. The hormone glucagon provides the signal for this to occur.

Because the brain and other components of the nervous system cannot store glucose, they are dependent on circulating glucose for energy. The brain is very sensitive to low blood glucose levels, and even a relatively small decrease

glycogenesis (gly – co – GE – ne – sis) (-*genesis,* coming into being) Formation of glycogen.

in blood glucose, a condition referred to as **hypoglycemia,** can make some people feel nauseated, dizzy, anxious, lethargic, and irritable. This is one reason why it is hard to concentrate when you have not eaten for a long time. Hypoglycemia has many causes. For example, a person with diabetes can experience hypoglycemia if too much insulin is injected. In addition, blood glucose levels can fall in response to prolonged or vigorous exercise. In nondiabetic individuals, there are two main types of hypoglycemia—reactive hypoglycemia and fasting hypoglycemia.

Reactive and Fasting Hypoglycemia **Reactive hypoglycemia** (now more commonly referred to as idiopathic postprandial hypoglycemia) occurs when the pancreas "over-responds" to high blood glucose levels following the consumption of food. The release of too much insulin ultimately results in extraordinarily low blood glucose. Researchers are not sure what causes idiopathic postprandial (reactive) hypoglycemia and even if it really exists, but speculate that some people may be especially sensitive to increases in blood glucose levels after a meal. People who experience symptoms associated with hypoglycemia are encouraged to eat small, frequent meals throughout the day. Other dietary recommendations include avoiding foods with caffeine, alcohol, and foods that cause a rapid and/or elevated glycemic response. Until blood glucose levels stabilize, keeping a food diary can help identify foods that trigger reactive hypoglycemia.

Unlike reactive hypoglycemia, **fasting hypoglycemia** is not associated with eating. Instead, it occurs when the pancreas releases too much insulin, even when food has not been consumed. This form of hypoglycemia is typically caused by pancreatic tumors, medications, hormonal imbalances, and certain illnesses, and it requires medical attention.

The Breakdown of Glycogen Glucagon stimulates the breakdown of glycogen in the liver and the release of glucose into the blood. This metabolic process, called **glycogenolysis,** literally means the breakdown of glycogen. Liver glycogen can supply glucose for approximately 24 hours before being depleted. Although glycogenolysis also occurs in skeletal muscle, the process differs in two important ways from that in the liver. First, muscle tissue is not responsive to the hormone glucagon; in other words, it is glucagon insensitive. Instead, the hormone **epinephrine,** which is released from the adrenal glands, is an important regulator of glycogenolysis in skeletal muscle. Another important difference is that muscles lack the enzyme needed to release glucose garnered from glycogenolysis into the blood. Rather, glucose released from glycogen during muscle glycogenolysis is used by muscle cells themselves and is not made available to other tissues. Therefore, muscle glycogen does not play a role in blood glucose regulation. The role of glucagon in blood glucose regulation and in the breakdown of energy stores is summarized in Figure 4.21.

Gluconeogenesis Increases Glucose Availability The breakdown of liver glycogen is an effective short-term solution for providing cells with glucose. However, because this energy reserve can be quickly depleted, the body must find an alternative source of glucose. Because fatty acids cannot be converted to glucose, the body uses amino acids from muscle protein to make glucose when glycogen stores are no longer available. The synthesis of glucose from noncarbohydrate sources such as amino acids is called **gluconeogenesis.** The same hormones that stimulate glycogenolysis—glucagon and epinephrine—also stimulate gluconeogenesis. Not surprisingly, insulin inhibits gluconeogenesis.

hypoglycemia (*hypo-*, under or below) Abnormally low level of glucose in the blood.

reactive hypoglycemia (also called idiopathic postprandial hypoglycemia) Low blood glucose that occurs when the pancreas releases too much insulin in response to eating carbohydrate-rich foods.

fasting hypoglycemia Low blood glucose that occurs when the pancreas releases excess insulin during periods of low food intake.

glycogenolysis (gly – co – ge – NO – ly – sis) (*-lysis*, to break apart) The breakdown of liver and muscle glycogen into glucose.

epinephrine (e – pi – NEPH – rine) Hormone released from the adrenal glands in response to stress; helps increase blood glucose levels by promoting glycogenolysis.

gluconeogenesis (glu – co – ne – o – GE – ne – sis) (*neo-*, new; *-genesis*, bringing forth) Synthesis of glucose from noncarbohydrate sources.

FIGURE 4.21 Glucagon Promotes Mobilization of Stored Energy The pancreas increases its release of the hormone glucagon in response to low blood glucose. Glucagon promotes the breakdown of liver glycogen, fat (adipose tissue), and protein (muscle). Glucagon also promotes the synthesis of ketones and the production of glucose from noncarbohydrate sources.

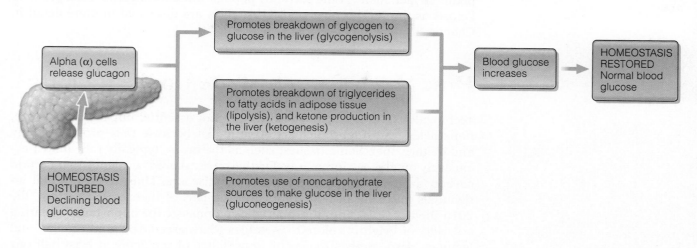

FIGHT-OR-FLIGHT RESPONSE PROVIDES AN IMMEDIATE ENERGY SOURCE

Whereas glucagon is involved in the day-to-day homeostatic regulation of blood glucose, other hormones dominate when there is a need for immediate energy. For example, the adrenal glands release the hormones epinephrine and **cortisol** when the body needs a sudden surge in blood glucose. These hormones have a variety of metabolic effects. For example, epinephrine increases glucose availability by stimulating glycogenolysis and gluconeogenesis. In addition to stimulating gluconeogenesis, cortisol also decreases glucose uptake by tissues other than the brain. These hormonal responses to the body's need for immediate energy, components of what is sometimes called the fight-or-flight response, ensure glucose availability under extreme circumstances.

KETONES ARE THE BODY'S ALTERNATIVE ENERGY SOURCE

Although gluconeogenesis increases glucose availability, it can also have negative consequences. This is because the breakdown of muscle protein to provide a source of glucose can have damaging effects, especially if it involves organs such as the heart. To minimize the loss of muscle, the body reduces its dependency on glucose by using an alternative energy source called **ketones.** These organic compounds form when fatty acids are broken down in a relative absence of glucose. This process, called **ketogenesis,** occurs in the liver and is stimulated by the hormone glucagon. Once formed, ketones are released into the blood and used for energy by the brain, heart, skeletal muscle, and kidneys. This glucose-sparing response helps minimize the loss of muscle protein by lessening the body's demand for glucose. The importance of ketone use in overall energy metabolism is discussed in Chapter 7.

Ketone synthesis is not without its own consequences, however. A condition called **ketosis** occurs when ketone production exceeds the rate of ketone use, resulting in the accumulation of ketones in the blood. This happens when

〈**CONNECTIONS**〉 Dietary Reference Intakes (DRIs) are reference values for nutrient intake. When adequate information is available, Recommended Dietary Allowances (RDAs) are established, but when less information is available, Adequate Intake (AI) values are provided. Tolerable Upper Intake Levels (ULs) are intake levels that should not be exceeded (Chapter 2, page 38).

cortisol (COR – ti – sol) Hormone secreted by the adrenal glands in response to stress; helps increase blood glucose availability via gluconeogenesis and glycogenolysis.

ketone (KE – tone) Organic compound used as an energy source during starvation, fasting, low-carbohydrate diets, or uncontrolled diabetes.

ketogenesis (ke – to – GE – ne – sis) Metabolic pathway that leads to the production of ketones.

ketosis (ke – TO – sis) Condition resulting from excessive ketones in the blood.

⟨**CONNECTIONS**⟩ The daily value (DV) is the recommended intake of a nutrient based on either a 2,000- or 2,500-kcal diet (Chapter 2, page 59).

energy intake is very low or when the diet provides insufficient amounts of carbohydrate. Ketosis causes a variety of complications, including loss of appetite. In fact, many of the currently popular low-carbohydrate diets promote ketosis as the "fat-burning key." These diets are discussed in more detail in Chapter 8.

How Much Carbohydrate Do We Require?

Technically speaking, carbohydrates are not essential nutrients. They are, nonetheless, an important part of your diet because they provide energy and dietary fiber. Some tissues, such as the brain, typically rely almost exclusively on glucose for energy. Furthermore, there is evidence that certain carbohydrates may help prevent chronic diseases. Thus, to ensure that individuals make "healthy" carbohydrates a part of their eating pattern, the 2010 Dietary Guidelines for Americans emphasize the importance of eating a variety of brightly colored vegetables (dark green, red, and orange) and legumes. Also emphasized is the importance of replacing at least half our intake of refined grains with whole-grain foods. This recommendation is particularly important because refined-grain foods often contain solid fats, added sugars, and excess sodium. By making this one dietary change, overall health can be improved in many ways. The recommendations discussed in the next section—developed to minimize risk for chronic disease and promote optimal health—point out that it is important to pay attention to the amount as well as to the source of dietary carbohydrates.

DIETARY REFERENCE INTAKES FOR CARBOHYDRATES

The Institute of Medicine's Dietary Reference Intakes (DRIs) established a Recommended Dietary Allowance (RDA) for carbohydrate of 130 g per day for adults; this represents the minimum amount of glucose utilized by the brain. However, during pregnancy and lactation, glucose requirements increase to 175 g and 210 g per day, respectively. The Institute of Medicine does not provide any special recommendations about carbohydrate intake for athletes. However, people who are physically active and work out regularly are often advised to consume more carbohydrate to prevent fatigue and replenish glycogen stores. DRI values for carbohydrates are provided inside the front cover of this book.

There is no Tolerable Upper Intake Level (UL) for any individual carbohydrate class or for total carbohydrates. However, the DRI committee recognizes that there is considerable evidence that overconsumption of added sugars is associated with various adverse health effects, such as dental caries and obesity. The Institute of Medicine recommends that added sugars contribute no more than 25% of total calories.

To meet the body's need for energy and decrease the risk for chronic diseases, the Institute of Medicine suggests an Acceptable Macronutrient Distribution Range (AMDR) of 45 to 65% of total energy from carbohydrates. As there are 4 kcal per gram of carbohydrate, this means a person needing 2,000 kcal daily should consume between 225 and 325 g of carbohydrate (900 and 1,300 kcal, respectively) each day. You can easily determine how much carbohydrate is recommended for you by following the steps outlined in Table 4.6.

Recommendations for Dietary Fiber Intake Based on the DRI recommendations, the 2010 Dietary Guidelines for Americans recommend that adults consume

TABLE 4.6 Calculating Recommended Total Carbohydrate Intake

Total Energy Requirement (kcal/day)	Recommended % kcal from Carbohydrate (AMDR)	Calculation	Recommended Total Carbohydrate Intake (g/day)
1,500	45–65	1,500 × 0.45/4 1,500 × 0.65/4	169–244
2,000	45–65	2,000 × 0.45/4 2,000 × 0.65/4	225–325
2,500	45–65	2,500 × 0.45/4 2,500 × 0.65/4	281–406

NOTE: Intake goals are based on the Acceptable Macronutrient Distribution Ranges (AMDRs).

14 g of dietary fiber per 1,000 kcal. Adequate Intake (AI) values for dietary fiber range from 21 to 38 g per day for adults. The average daily intake of fiber in the United States is about half this amount (approximately 15 g).[21] One reason for this is the rise in popularity of low-carbohydrate weight-loss diet plans, some of which recommend minimal intakes of fruits, vegetables, and grains. Encouraging people to consume more fruits, vegetables, and whole-grain foods is a crucial step in helping increase fiber consumption.

Although UL values for dietary fiber and functional fiber have not been established, some people may experience indigestion and other gastrointestinal disturbances when they consume large amounts of fiber. While some fibers have a tendency to bind minerals such as calcium, zinc, iron, and magnesium, it is doubtful that consuming dietary fiber in recommended amounts affects mineral status in healthy adults. Most experts agree that high-fiber diets are beneficial and have minimal adverse effects.

DIETARY GUIDELINES AND MYPLATE EMPHASIZE WHOLE GRAINS, FRUITS, AND VEGETABLES

When it comes to carbohydrate-rich foods, there are many from which to choose. While the choices can be overwhelming, health experts agree that the best strategy is to maximize your intake of foods that are nutrient dense and consume plenty of fiber, vitamins, and minerals. Conversely, it is important to minimize your intake of foods high in fat and sugar. Following the 2010 Dietary Guidelines for Americans can help eliminate much of the guesswork when it comes to determining which carbohydrate-rich foods to choose. One general guideline is to consume more fruits, vegetables, and whole-grain foods; these foods provide naturally occurring sugars, fiber, and plenty of vitamins and minerals. The MyPlate food guidance system, which emphasizes whole grains, fruit, and vegetables as the foundation of a healthy diet, is another useful tool for evaluating your carbohydrate food choices. The current recommendation for adults is to consume an equivalent of 5 to 10 ounces of grain-based foods daily, half of which should be whole grain.

Another general guideline is to minimize highly processed foods that contain added sugars. Although there is no consensus as to how much total added sugar to consume, health experts generally agree that one's consumption of foods with high amounts of added sugar (such as cookies, soda, sugary cereals, and heavy syrups) should be minimized. Table 4.7 lists the average amounts of added sugar in selected foods and beverages. This chapter's Food Matters feature provides suggestions for reducing your intake of added sugars.

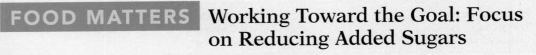

Working Toward the Goal: Focus on Reducing Added Sugars

A healthy eating pattern means selecting foods that are nutrient dense—high in nutrients relative to the amount of calories. To achieve this goal, the 2010 Dietary Guidelines for Americans suggest that we reduce the amount of added sugars in our diet. In many cases, this can be easily achieved by replacing highly processed foods with fresh fruits, vegetables, and whole grains. The following suggestions can help you reduce the amount of added sugars in your diet.

- The Nutrition Facts panel on food labels lists both total carbohydrates and total sugars. In highly processed foods, most of the sugar is likely to be added sugar. Compare the number of grams of sugar with the number of grams of carbohydrate. If the difference in these two numbers is small, the food is likely to be high in added sugar and should only be consumed in moderation.

- Limit intake of soft drinks, fruit punch, and other sweetened beverages. Water is the best way to quench your thirst. If you do drink these beverages, avoid large sizes.

- When you want something sweet, choose fresh fruit rather than candy or cookies. Also, when buying canned fruit, choose brands that are packed in natural juices rather than sweetened varieties.

- Avoid presweetened breakfast cereals. Small amounts of sweeteners can be added to oatmeal and other cereals if desired.

- Add fruit to plain yogurt rather than buying sweetened varieties with fruit added.

TABLE 4.7 Added Sugars in Your Diet

Food	Serving Size	Added Sugar (g)*
Soft drink	12 oz	43
Milkshake	10 oz	36
Fruit punch	8 oz	38
Chocolate	1.5 oz	24
Sweetened breakfast cereal (flakes)	1 cup	15
Yogurt with fruit	1 cup	33
Ice cream (vanilla)	1 cup	28
Cake with frosting	1 slice	28
Cookies (chocolate chip)	2 (medium)	14
Jam or jelly	1 tbsp	2

*1 teaspoon equals 4.75 grams

SOURCE: United States Department of Agriculture Database for the Added Sugars Content of Selected Foods, Release 1 (2006). Available from www.nal.usda.gov/fnic/foodcomp/Data/add_sug/addsug01.pdf. Krebs-Smith SM. Choose beverages and foods to moderate your intake of sugars: Measurement requires quantification. The Journal of Nutrition. 2001;131:527S–5S. Available from: http://jn.nutrition.org/content/131/2/527S.full.

Laura Benson

Critical Thinking: Laura's Story Think back to Laura's story about her experience with type 1 diabetes. Based on what you have read about how foods affect blood glucose, are there foods that Laura should avoid or limit? Review the recommendations and guidelines for carbohydrate intake and identify some ways these might be different for an athlete with type 1 diabetes compared to an athlete without diabetes.

Diet Analysis PLUS ✚ Activity

PART A. Total Carbohydrates

1. Using your "Profile DRI goals" report from your Diet Analysis Plus printouts, how many total kilocalories (kcals) should you be consuming per day? _____

2. Next to your recommendation for total carbohydrates in italics, what percentage range of total kcals is recommended to come from carbohydrates? _____

3. Using this information, calculate the recommended number of grams (g) of carbohydrates you should be eating.

   Lower end of range:

   AMDR total kcals × 0.45 = kcal of carbohydrates ÷ 4 kcal/gram = grams of carbohydrates

   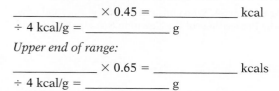

   _____ × 0.45 = _____ kcal
   ÷ 4 kcal/g = _____ g

   Upper end of range:

   _____ × 0.65 = _____ kcals
   ÷ 4 kcal/g = _____ g

4. Compare your "Profile DRI goals" with your printed and calculated DRI range for total carbohydrates.

   DRI recommended range for total carbohdyrates

   _____

   Calculated range for number of grams of carbohydrates _____

   Compare these values to determine if your diet is providing the recommended amounts (g) and recommended range of total kilocalories for carbohydrates.

PART B. Fiber

1. Refer to your "Intake and DRI Goals Compared" report. How much fiber did you actually consume? _____ g

2. Does the amount you actually consumed meet your AI for fiber? _____

3. Look at your individual days' "Intake Spreadsheets." List the one food consumed each day that contains the most fiber.

4. If you answered no to Question 2, what foods could you eat more of to meet your AI for fiber?

PART C. Added Sugar

1. The DRI suggests that added sugars should account for no more than 25% of the day's total energy intake. To determine the percentage of sugar calories you actually consumed, fill in the blanks below. Use your "Intake and DRI Goals Compared" report to locate the information you need.

   Grams of sugar consumed _____ × 4 kcal/g
   = _____ kcal of sugar consumed
   _____ (kcal from sugar) ÷ _____ (total kcal consumed) × 100 = _____% of calories as sugars

2. What is your assessment of your sugar intake?

3. Looking at your individual days' "Intake Spreadsheets," identify the food that contributed the most sugar each day.

Conclusions

Do you need to improve or change your overall carbohydrate intake? If so, what changes need to be made?

Notes

1. Bray GA, Nielsen SJ, Popkin BM. Consumption of high-fructose corn syrup in beverages may play a role in the epidemic of obesity. American Journal of Clinical Nutrition. 2004;9:537–43. Gaby AR. Adverse effects of dietary fructose. Alternative Medicine Review. 2005;10:294–306.

2. U.S. Department of Agriculture, Economic Research Service. 2011. Table 51—Refined cane and beet sugar: estimated number of per capita calories consumed daily, by calendar year. Table 52—High fructose corn syrup: estimated number of per capita calories consumed daily, by calendar year. Available from: http://www.ers.usda.gov/Briefing/Sugar/Data.htm.

3. Jacobson MF. High-fructose corn syrup and the obesity epidemic. American Journal of Clinical Nutrition. 2004;60:1081–2.

4. Forshee RA, Storey ML, Allison DB, Glinsmann WH, Hein GL, Lineback DR, Miller SA, Nicklas TA, Weaver GA, White JS. A critical examination of the evidence relating high fructose corn syrup and weight gain. Critical Reviews in Food Science and Nutrition. 2007;47:561–82.

5. Melanson KJ, Zukley L, Lowndes J, Nguyen V, Angelopoulos TJ, Rippe JM. Effects of high-fructose corn syrup and sucrose consumption on circulating glucose, insulin, leptin, and ghrelin and on appetite in normal-weight women. Nutrition. 2007;23:103–12.

6. Monsivais P, Perrigue MM, Drewnowski A. Sugars and satiety: Does the type of sweetener make a difference? American Journal of Clinical Nutrition. 2007;86:116–23.

7. Johnson R, Appel LJ, Brands M, Howard BV, Lefevre M, Lustig RH, Sacks F, Steffen LM, Wylie-Rosett J. Dietary sugars intake and cardiovascular health: A scientific statement from the American Heart Association. Circulation. 2009;120:1011–20.

8. National Cancer Institute. Sources of added sugars in the diets of the U.S. population ages 2 years and older, NHANES 2005–2006. Risk Factor Monitoring and Methods. Cancer Control and Population Sciences. http://riskfactor.cancer.gov/diet/foodsources/added_sugars/table5a. html.

9. U.S. Department of Agriculture and U.S. Department of Health and Human Services. Dietary Guidelines for Americans, 2010, 7th edition. Washington, DC: U.S. Government Printing Office, December 2010.

10. Whitehouse CR, Boullata J, McCauley LA. The potential toxicity of artificial sweeteners. American Association of Occupational Health Nurses. 2008;56:251–9.

11. Soffritti M, Belpoggi F, Tibaldi E, Esposti DD, Lauriola M. Life-span exposure to low doses of aspartame beginning during prenatal life increases cancer effects in rats. Environmental Health Perspectives. 2007;11:1293–7. Magnuson BA, Burdock GA, Doull J, Kroes RM, Marsh GM, Pariza MW, Spencer PS, Waddell WJ, Walker R, Williams GM. Aspartame: A safety evaluation based on current use levels, regulations, and toxicological and epidemiological studies. Critical Reviews in Toxicology. 2007;37:629–727.

12. Swithers SE, Davidson TL. A role for sweet taste: calorie predictive relations in energy regulation by rats. Behavioral Neuroscience. 2008;122:161–73.

13. Slavin J. Why whole grains are protective: Biological mechanisms. Proceedings of the Nutrition Society. 2003;62:129–34.

14. Marquart L, Wiemer KL, Jones JM, Jacob B. Whole grains health claims in the USA and other efforts to increase whole-grain consumption. Proceedings of the Nutrition Society. 2003;62:151–60.

15. Jenkins DJ, Kendall CW, Axelsen M, Augustin LS, Vuksan V. Viscous and nonviscous fibres, nonabsorbable and low glycaemic index carbohydrates, blood lipids and coronary heart disease. Current Opinion in Lipidology. 2000;11:49–56.

16. Anderson JW. Whole grains protect against atherosclerotic cardiovascular disease. Proceedings of the Nutrition Society. 2003;62:135–42.

17. Stollman N, Raskin JB. Diverticular disease of the colon. Lancet. 2004;363:631–9.

18. U.S. Department of Health and Human Services, Food and Drug Administration. Whole grain label statements. 2006 Available from: http://www.cfsan.fda.gov/-dms/flgragui.html.

19. Vesa TH, Marteau P, Korpela R. Lactose intolerance. Journal of the American College of Nutrition. 2000;19:165S–75S.

20. Foster-Powell K, Holt SH, Brand-Miller JC. International table of glycemic index and glycemic load values: 2002. American Journal of Clinical Nutrition. 2002;76:5–56.

21. Lang R, Jebb SA. Who consumes whole grains, and how much? Proceedings of the Nutrition Society. 2003;62:123–7.

Nutrition and Diabetes

Eat a healthy diet, exercise, and *watch your weight* are all important health recommendations that you have undoubtedly heard before. Yet there is a convincing reason for us to take this advice seriously—in a word, *diabetes.* In the past 40 years, the prevalence of diabetes has dramatically increased. In fact, diabetes has now reached epidemic proportions. According to the Centers for Disease Control and Prevention (CDC), 11% of the U.S. adult population (aged 20 years or older) has diabetes (18.8 million diagnosed and 7 million undiagnosed), which adds up to approximately 25.8 million Americans. Furthermore, an additional 79 million Americans have a prediabetic condition characterized by blood glucose levels that are elevated but not high enough to be considered diabetes.[1] These numbers are staggering—but not surprising to many health experts.

Although some types of diabetes cannot be prevented, many professionals believe the most common form of diabetes, type 2, is largely attributable to lifestyle. While genetics plays a role, physical inactivity, obesity, and unhealthy dietary practices have paved the way for this modern epidemic. The good news is that diet and other lifestyle practices can play an equally powerful role in managing and preventing type 2 diabetes.[2] Although some aspects of diabetes are not within our control, action is the key to diabetes prevention, treatment, and management.

What Is Diabetes?

Diabetes mellitus is a metabolic disorder characterized by elevated levels of glucose in the blood, otherwise known as hyperglycemia. Diabetes was first described more than 2,000 years ago as an affliction that caused excessive thirst, weight loss, and honey-sweet urine that attracted ants and flies. By the 16th century, physicians began treating diabetic patients with special diets. For example, some physicians recommended diets consisting of milk, barley, water, and bread to replace the sugar lost in the urine. Physicians later tried diets of meat and fat. Regardless of diet, patients died within a few months of diagnosis. With nothing else to offer, physicians attempted to keep patients alive by restricting their food intake. Until the discovery of insulin, there was little hope for people with diabetes.

THE DISCOVERY OF INSULIN

More than 100 years ago, two physiologists made the unexpected discovery that the surgical removal of the pancreas in dogs caused diabetes. This medical finding captured the interest of Frederick Banting, who was certain it would lead to a cure for diabetes. Banting—along with associates Charles Best, John Macleod, and James Collip—were the first to successfully control blood glucose in diabetic dogs by injecting them with secretions from the pancreas. Shortly thereafter, Leonard Thompson, a 14-year-old boy who was dying of diabetes, was the first human to be injected with Banting and Best's pancreatic extract. Amazingly, the young boy gained weight and seemed to recover from the debilitating effects of diabetes. The pancreatic extract, which was later given the name insulin, was obtained from the pancreas of cattle and pigs. Today, human insulin is produced by genetic engineering, and is used successfully by millions of people to manage their diabetes.

DIABETES IS CLASSIFIED BY ITS UNDERLYING CAUSE

Since the discovery of insulin, major advances have been made in understanding diabetes. By 1960, it became clear that there were different types of diabetes, which over the years were classified in a variety of ways. For example, diabetes was once categorized according to typical age of onset: juvenile-onset diabetes versus adult-onset diabetes. This classification scheme was

---

diabetes mellitus (di – a – BE – tes MEL – lit – tus) (*diabetes*, to siphon; *mellitus*, sugar) Medical condition characterized by a lack of insulin or impaired insulin utilization that results in elevated blood glucose levels.

Courtesy of Clendening History of Medicine Library, University of Kansas

Billy Leroy, one of the first patients to receive insulin. Pictures were taken of Billy as an infant before diabetes, with diabetes, and after several months of treatment with insulin.

What Is Type 1 Diabetes?

Type 1 diabetes occurs when the pancreas is no longer able to produce insulin. Without insulin, most cells cannot take up glucose, causing blood glucose levels to become dangerously high (Figure 1). This is the reason why individuals with type 1 diabetes require daily insulin injections to control blood glucose. It is also the reason why it was previously called insulin-dependent diabetes mellitus. Approximately 5 to 10% of all people with diabetes have type 1.[4] Although type 1 diabetes can develop at any age, it most often occurs during childhood and early adolescence, typically around 12 to 14 years of age. This is why type 1 diabetes was once called juvenile-onset diabetes.

confusing because the various forms of diabetes can occur at any age. Diabetes was later classified on the basis of whether insulin was required as a treatment: insulin-dependent diabetes versus non–insulin-dependent diabetes. The problem with this classification system was that some people diagnosed with non–insulin-dependent diabetes require insulin to control hyperglycemia. In 1997 the American Diabetes Association developed a new system of diabetes classification based on etiology, or underlying cause.[3] Today diabetes is categorized as type 1 diabetes, type 2 diabetes, gestational diabetes, or secondary diabetes. These are summarized in Table 1.

type 1 diabetes Previously known as juvenile-onset diabetes and as insulin-dependent diabetes mellitus, this form of diabetes results when the pancreas is no longer able to produce insulin due to a loss of insulin-producing β-cells.

TABLE 1 Classification of Types of Diabetes

Category	Typical Age of Onset	Underlying Cause	Description
Type 1 diabetes; formerly called insulin-dependent diabetes mellitus (IDDM) or juvenile-onset diabetes	Childhood and adolescence; can develop in adults	Lack of insulin production by the pancreatic ß-cells	An autoimmune disorder that destroys the insulin-producing ß-cells of the pancreas, resulting in little or no insulin production. Person requires insulin delivered via daily injections or insulin pump.
Type 2 diabetes; formerly called non–insulin-dependent diabetes mellitus (NIDDM) or adult-onset diabetes mellitus	Middle-aged and older adults; increasing incidence in childhood and adolescence	Insulin resistance	Skeletal muscle and adipose tissue develop insulin resistance, resulting in blood glucose levels that are above normal. Genetics, obesity, and physical inactivity play major roles in insulin resistance. Can often be managed by diet, weight loss, and exercise; may require glucose-lowering medication or insulin injections.
Gestational diabetes	Occurs in pregnant women	Insulin resistance	A temporary form of diabetes that develops in 4 to 7% of pregnant women. Characterized by insulin resistance brought on by hormonal changes that take place during pregnancy. Women who develop gestational diabetes are at increased risk for developing type 2 diabetes.
Secondary diabetes	Varies	Varies	Brought on by other diseases, medical conditions, and medications.

FIGURE 1

FIGURE 1 Type 1 Diabetes Blood glucose regulation depends on the release of insulin from the pancreas in response to elevated blood glucose. In the case of type 1 diabetes, the pancreas is not able to produce insulin.

1 The insulin producing β-cells of the pancreas have been destroyed, and are unable to produce insulin.

2 Without insulin, glucose transporters do not receive a signal to relocate to the cell membrane. As a result, glucose is unable to move from outside the cell (extracellular space) into the inside (intracellular space) of the cell.

3 Unable to cross the cell membrane, glucose accumulates in the blood, resulting in hyperglycemia.

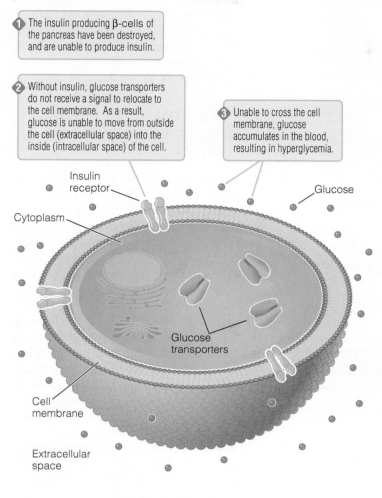

Insulin receptor

Glucose

Cytoplasm

Glucose transporters

Cell membrane

Extracellular space

TYPE 1 DIABETES IS CAUSED BY A LACK OF INSULIN PRODUCTION

Type 1 diabetes is caused by complex interactions among genetics, environmental factors, and the immune system that lead to an inability of the pancreas to produce insulin. Scientists have long recognized that type 1 diabetes seems to be more common in some families and in certain ethnic groups, suggesting a genetic component.[5] Type 1 diabetes, which is classified as an autoimmune disease, is triggered by something in the environment. Recall that autoimmune diseases are disorders caused by the production of antibodies that attack and destroy tissues in the body. Many researchers believe that, in genetically susceptible people, a viral infection stimulates the immune system to produce antibodies.[6] Type 1 diabetes results when the antibodies attack and destroy the insulin-producing cells (β-cells) of the pancreas.[7] Why this occurs is not fully understood.

The destruction of pancreatic β-cells is a gradual process. With little or no insulin-producing ability left, the pancreas cannot keep up with the body's need for insulin, resulting in severe hyperglycemia (Figure 2). Symptoms tend to develop rapidly and, because of their severity, are not easily overlooked.

METABOLIC DISTURBANCES RESULT FROM TYPE 1 DIABETES

The lack of insulin causes the rapid breakdown of fat stores and muscle. This is why people with untreated type 1 diabetes experience profound weight loss. Furthermore, without insulin, most cells cannot take up glucose, an energy source on which they normally depend.

Cells compensate for the lack of glucose by metabolizing fat and protein (muscle) for energy. However, this can have serious consequences. Using large amounts of fat for energy with a relative absence of glucose results in ketone formation. The accumulation of acidic ketones in the blood brings on flu-like symptoms such as abdominal pain, nausea, and vomiting. If the situation is not corrected, a life-threatening condition called **diabetic ketoacidosis** can occur. Diabetic ketoacidosis can lead to coma or death and requires immediate medical attention.

In addition to rapid weight loss, other symptoms associated with type 1 diabetes include extreme thirst, hunger, and frequent urination (Table 2). These symptoms are caused by high levels of blood glucose. For example, the kidneys are not able to remove the excess glucose from the blood, resulting in the "spilling" of glucose into the urine. However, before the glucose can be excreted in the

⟨**CONNECTIONS**⟩ Recall that autoimmune disorders occur when the immune system produces antibodies that attack and destroy tissues in the body. Antibodies are proteins produced by the immune system and typically respond to the presence of foreign proteins in the body (Chapter 1, page 24).

diabetic ketoacidosis (ke – to – a – ci – DO – sis) Severe metabolic condition resulting from the accumulation of ketones in the blood.

FIGURE 2 Cause of Type 1 Diabetes Type 1 diabetes is caused by an autoimmune disorder in which antibodies destroy the insulin-producing cells of the pancreas.

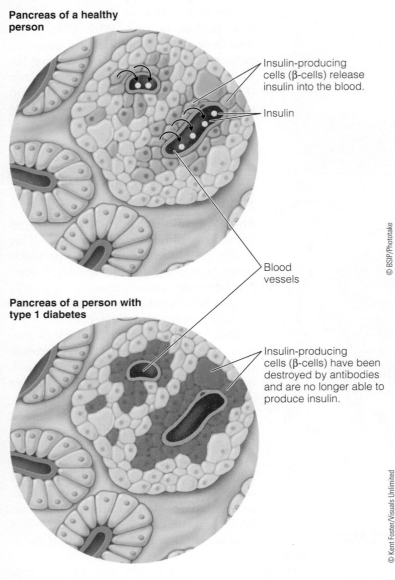

Pancreas of a healthy person

Insulin-producing cells (β-cells) release insulin into the blood.

Insulin

Blood vessels

Pancreas of a person with type 1 diabetes

Insulin-producing cells (β-cells) have been destroyed by antibodies and are no longer able to produce insulin.

© BSIP/Phototake

© Kent Foster/Visuals Unlimited

urine, it must first be diluted. This is accomplished by drawing water out of the blood. As a result, people urinate frequently, resulting in dehydration and thirst. Type 1 diabetes also causes a person to feel hungry and weak because cells are starved for energy. In fact, diabetes is often described as "starvation in the midst of plenty."

TYPE 1 DIABETES REQUIRES INSULIN INJECTIONS OR A PUMP

Insulin cannot be taken orally because it is a protein and therefore would be destroyed by enzymes in the digestive tract. Therefore, people with type 1 diabetes need multiple daily injections of insulin to control blood glucose. However, some people prefer to use an insulin pump, which looks like a small pager and is worn on a belt or on the waistband of clothing. Work is also under way to develop an artificial pancreas that can monitor blood glucose levels and automatically dispense the right amount of insulin into the blood. Hopefully this new technology will make diabetes care and blood glucose management easier for those who currently depend on a daily, labor-intensive regimen to maintain control of blood glucose.

To know how much insulin to administer, it is important for people with type 1 diabetes to monitor their blood glucose using a medical device called a **glucometer.** Glucometers are able to measure the approximate concentration of glucose in the blood using only a small drop of blood. Most glucometers have memory chips that can display a readout of average blood glucose measures over time.

The most successful care plans for type 1 diabetes focus on individual needs, preferences, and cultural practices. There is no such thing as "one-plan-fits-all" when it comes to treating and managing type 1 diabetes. However, the primary focus for people with type 1 diabetes is to balance insulin injections with a healthy diet and physical activity.

glucometer A medical device used to measure the concentration of glucose in the blood.

TABLE 2 Symptoms Associated with Type 1 and Type 2 Diabetes

Type 1 Diabetes	Type 2 Diabetes
Frequent urination	Frequent urination
Excessive thirst	Excessive thirst
Fatigue	Fatigue
Unusual weight loss	Frequent infections
Extreme hunger	Blurred vision
Ketosis	Vaginal itching
	Cuts or bruises that are slow to heal
	Tingling or numbness in hands or feet
	Frequent urinary tract infections

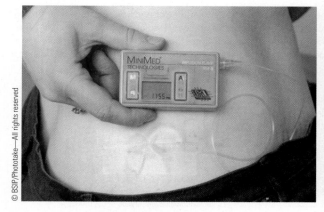

Individuals with type 1 diabetes must either inject insulin or use an insulin pump. Insulin pumps deliver insulin directly under the skin through narrow tubing.

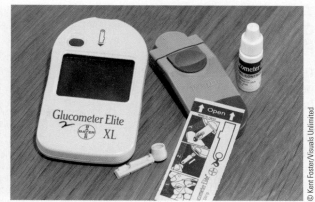

Glucometers are an important part of diabetes management. It is important for people who have diabetes to monitor blood glucose levels several times daily.

Type 1 Diabetes Can Lead to Serious Health Complications It is important for people to know that type 1 diabetes can lead to other serious health problems. For example, diabetes increases the risk for having heart attacks and strokes. It can also lead to blindness, limb amputation, impaired kidney function, and a loss of feeling in the feet and hands. These long-term complications are largely attributed to the harmful effects of excess glucose on blood vessels and nerves.

For many years, it was unclear if the occurrence of long-term complications associated with type 1 diabetes could be delayed or reduced by controlling blood glucose levels as close to normal as possible. However, a large clinical study called the Diabetes Control and Complications Trial (DCCT) was conducted to answer this important question.[8] This study showed that maintaining blood glucose levels as close to normal as possible can prevent many of the long-term complications associated with type 1 diabetes.

What Is Type 2 Diabetes?

Type 2 diabetes is by far the most common form of diabetes, with 90 to 95% of people with diabetes falling into this category.[1] In the United States, type 2 diabetes has become so widespread that an estimated one out of four people have or will have this disease or have a family member with it.[9] Because of the high prevalence of type 2 diabetes, the American Diabetes Association recommends that all adults be considered for screening beginning at 45 years of age.[10] Individuals who have one or more risk factors for type 2 diabetes (Table 3) should be screened at a younger age. Screening for type 2 diabetes involves an overnight fast and a blood draw. This screening method, called a fasting plasma glucose, is relatively simple and inexpensive.

Although type 2 diabetes can occur at any age, it most frequently develops in middle-aged and older adults. For this reason, type 2 diabetes was once called adult-onset diabetes. However, the growing number of children and teens diagnosed with type 2 diabetes is alarming.[11] Nearly 2 million adults were newly diagnosed with type 2 diabetes in 2010. Perhaps more alarming is the fact that between 2002 and 2005, approximately 3,500 children and adolescents were newly diagnosed with type 2 diabetes annually.[1] Therefore, type 2 diabetes can no longer be thought of as a condition that affects adults only.

TABLE 3 Risk Factors for Type 2 Diabetes

- Family history of diabetes (parent, brother, or sister with diabetes)
- History of gestational diabetes or delivery of a baby weighing more than nine pounds at birth
- Sedentary lifestyle (exercise fewer than three times per week)
- Overweight or obese (BMI ≥ 25 kg/m²)
- History of vascular disease
- Being over 45 years of age
- Being of African, Hispanic, Native American, or Pacific Island descent
- Having polycystic ovary syndrome (females)
- Having high blood pressure (≥140/90 mm Hg)
- Having low HDL cholesterol (≤35 mg/dL)
- Having high triglycerides (≥250 mg/dL)

SOURCE: Adapted from American Diabetes Association. Position statement from the American Diabetes Association: Screening for type 2 diabetes. Diabetes Care. 2004;27:s11–s14. Available from: http://care.diabetesjournals.org/content/27/suppl_1/s11.full.

TYPE 2 DIABETES IS CAUSED BY INSULIN RESISTANCE

Unlike type 1 diabetes, most people with type 2 diabetes have normal or even elevated levels of insulin in their blood. Because type 2 diabetes is often managed without insulin treatments, it was once called non–insulin-dependent diabetes mellitus (NIDDM). **Type 2 diabetes** is caused by **insulin resistance,** which means that insulin receptors have difficulty recognizing or responding to insulin (Figure 3).[12] As discussed in Chapter 4, insulin acts like a cellular key by binding to insulin receptors and enabling glucose to enter the cell. Because insulin is not able to do its job, the transport of glucose across the cell membrane is impaired, which results in the accumulation of glucose in the blood.

Elevated blood glucose levels cause the pancreas to release even more insulin into the blood. In many cases, this response keeps blood glucose levels within a relatively normal range. It is not clear why some people with insulin resistance develop type 2 diabetes whereas others do not. Perhaps after many years of working overtime to produce extra insulin, the pancreas becomes worn out and can no longer produce insulin in amounts needed to lower blood glucose. When blood glucose levels remain elevated over time, a person begins to experience symptoms associated with type 2 diabetes. Blood glucose values used to diagnose diabetes are shown in Table 4.

GENETIC AND LIFESTYLE FACTORS INCREASE THE RISK OF DEVELOPING TYPE 2 DIABETES

Many risk factors are associated with the development of type 2 diabetes (see Table 3, shown previously). For example, the prevalence of type 2

FIGURE 3 Type 2 Diabetes Blood glucose regulation depends on the release of insulin from the pancreas in response to elevated blood glucose. Insulin binds to insulin receptors, allowing glucose to be taken up by the cell. Type 2 diabetes occurs when insulin receptors are unable to respond to insulin, resulting in hyperglycemia. This is called "insulin resistance."

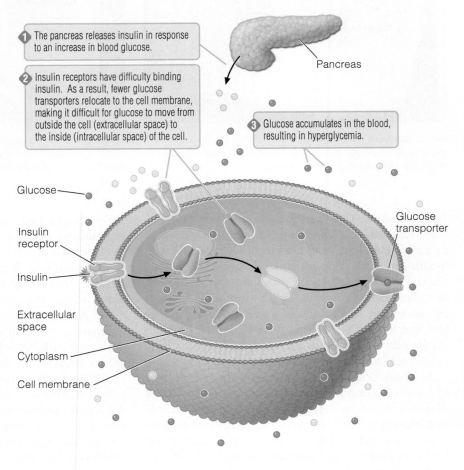

1. The pancreas releases insulin in response to an increase in blood glucose.

2. Insulin receptors have difficulty binding insulin. As a result, fewer glucose transporters relocate to the cell membrane, making it difficult for glucose to move from outside the cell (extracellular space) to the inside (intracellular space) of the cell.

3. Glucose accumulates in the blood, resulting in hyperglycemia.

Pancreas

Glucose

Glucose transporter

Insulin receptor

Insulin

Extracellular space

Cytoplasm

Cell membrane

TABLE 4 Diagnostic Values for Diabetes

Diagnosis	Fasting Plasma Glucose[a]	Two-Hour Oral Glucose-Tolerance Test[b]
Normal	<100 mg/dL	<140 mg/dL
Diabetes	≥126 mg/dL	≥200 mg/dL

[a] Person must fast for at least 8 hours prior to testing.
[b] Person must fast for at least 8 hours prior to testing. After an initial fasting blood sample is taken, a 75-g glucose beverage is consumed. A second blood sample is taken two hours later.

SOURCE: American Diabetes Association. Standards of medical care in diabetes—2011. Diabetes Care. 2011;34:s11–s61.

type 2 diabetes Previously known as adult-onset diabetes and non–insulin-dependent diabetes mellitus, this form of diabetes results when insulin-requiring cells have difficulty responding to insulin.

insulin resistance Condition characterized by the inability of insulin receptors to respond to the hormone insulin.

diabetes among the Caucasian population is approximately 10%.[1] However, in the African American population, it increases to about 19%.[1] The Pima Indians have the highest rates of type 2 diabetes in the United States. Aside from ethnicity and genetics, other risk factors associated with type 2 diabetes are obesity and the nutritional environment *in utero* and during infancy.

Obesity and Type 2 Diabetes The risk of developing type 2 diabetes is influenced, in part, by genetics.[13] However, this risk dramatically increases when accompanied by obesity. Overweight adults are more likely to develop type 2 diabetes than are lean individuals.[14] Approximately 80% of people diagnosed with type 2 diabetes are overweight. In addition, having a particular distribution of body fat can influence risk for type 2 diabetes. Body fat stored in the abdominal region of the body poses a greater risk than does that stored in the lower body regions of the body.[15] As the prevalence of obesity in the United States continues to climb, so does the prevalence of type 2 diabetes (Figure 4). If these trends continue, researchers estimate that 18 million Americans will have type 2 diabetes by the year 2020.[16]

The mechanisms that link obesity and insulin resistance remain unclear. However, a recent discovery that adipose tissue secretes a variety of hormones that trigger insulin resistance has generated new insights.[17] Although there is more to learn about these hormones, the link between obesity and insulin resistance is indisputable.

Birth Weight, Early Growth, and Type 2 Diabetes In addition to genetics and obesity, there is evidence that a poor nutritional environment *in utero* and during infancy can influence a person's risk for developing type 2 diabetes later in life.[18] A U-shaped relationship between birth weight and type 2 diabetes indicates that both accelerated (high birth weight) and slow (low birth weight) fetal growth can increase a person's risk of type 2 diabetes later in life (Figure 5).[19] This relationship is most evident when low birth weight is followed by rapid postnatal weight gain. Although the mechanisms remain unclear, scientists believe that epigenetic modification of DNA may be the means whereby adverse nutritional conditions early in life can increase health risks as adults.[20] You will learn more about epigenetics in Chapter 5.

SIGNS AND SYMPTOMS OF TYPE 2 DIABETES ARE OFTEN IGNORED

Because symptoms associated with type 2 diabetes tend to develop gradually, they are easily ignored. This is why type 2 diabetes can go undiagnosed

FIGURE 4 2008 Age-Adjusted Estimates of the Percentages of Adults with Diagnosed Diabetes

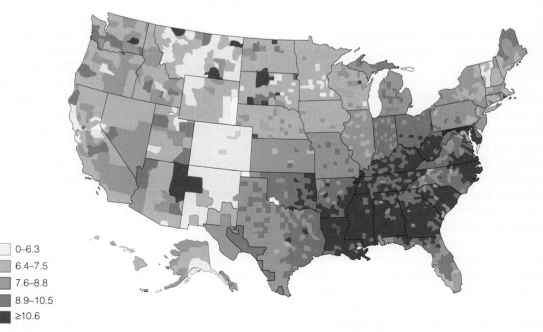

0–6.3
6.4–7.5
7.6–8.8
8.9–10.5
≥10.6

SOURCE: Centers for Disease Control and Prevention, Behavioral Risk Factor Surveillance System (BRFSS).

FIGURE 5 Relationship Between Birth Weight and Relative Risk of Developing Type 2 Diabetes Later in Life The U-shaped curve indicates that both low (<5.5 lb) and high birth weight (>9.9 lb) are associated with a higher occurrence of type 2 diabetes later in life.

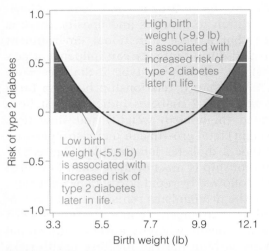

SOURCE: Harder T, Rodekamp E, Schellong K, Dudenhausen JW, Plageman A. Birth weight and subsequent risk of type 2 diabetes: A Meta-Analysis. American Journal of Epidemiology. 2007;165:849–57.

for many years. Some of the early symptoms associated with type 2 diabetes include fatigue, frequent urination, and excessive thirst (see Table 2, shown previously). However, if type 2 diabetes is left untreated, it eventually causes more noticeable symptoms, such as blurred vision, frequent urinary tract infections, slow wound healing, and vaginal itching.

MANAGING TYPE 2 DIABETES CAN HELP PREVENT LONG-TERM COMPLICATIONS

Like type 1 diabetes, people with type 2 diabetes are at increased risk of developing long-term complications such as heart disease, blindness, and impaired kidney function. However, the good news is that maintaining blood glucose within near-normal ranges can often prevent these complications. In the case of type 2 diabetes, this can often be accomplished through weight management, regular exercise, and a healthy diet. For example, losing as little as 10 to 15 pounds or taking a 30-minute walk daily has been shown to lower blood glucose. Similar to the Diabetes Control and Complications Trial, the United Kingdom Prospective Diabetes Study revealed that controlling blood glucose within near-normal ranges can help

reduce long-term complications associated with type 2 diabetes.[21]

While many people with type 2 diabetes are able to control blood glucose by making lifestyle changes, others may require medication that lowers blood glucose by making cells more responsive to insulin. Even though the pancreas produces insulin, some people with type 2 diabetes need additional insulin to lower blood glucose. This is why some people with type 2 diabetes require insulin injections. However, even if medication is used to manage blood glucose, diet remains a critical component of diabetes management.

Diet and Type 2 Diabetes Nutritional guidelines developed by the American Diabetes Association stress the importance of diet in the management of type 2 diabetes.[22] Most of these guidelines, which are summarized in Table 5, can be applied by anyone who wants to eat a healthy diet. To help control type 2 diabetes, the American Diabetes Association recommends consuming carbohydrates mainly from fruits, vegetables, whole grains, legumes, and low-fat milk. Strategies such as the the diabetic exchange system and carbohydrate counting can be used to help monitor total carbohydrate intake (see Appendix D). Although people with diabetes are often told to avoid foods high in sugar, there is no need to totally eliminate sugars from the diet. Still, whole grains, fruits, and vegetables are preferred over high-sugar foods because they are good sources of micronutrients, phytochemicals, and dietary fiber.

Other dietary interventions for controlling type 2 diabetes and other health-related complications include reducing energy intake in order to reduce body weight and limiting foods high in saturated and *trans* fatty acids, cholesterol, and sodium. Consuming two or more servings of fish per week is also recommended to ensure adequate intakes of omega-3 fatty acids, which are important for good health. At this time there is insufficient evidence to recommend the use of vitamin or mineral supplements to treat diabetes. Rather, nutritionists recommend that a well-balanced diet be the source of these important nutrients.

Physical Activity and Type 2 Diabetes For people with type 2 diabetes, the benefits of regular exercise and physical activity are substantial. When combined with an energy-controlled meal plan, exercise can facilitate weight loss and can help in

TABLE 5 American Diabetes Association Dietary Recommendations for Type 2 Diabetes

- Overweight or obese individuals are encouraged to lose weight by decreasing energy intake and increasing physical activity.
- Carbohydrates from fruits, vegetables, whole grains, legumes, and low-fat milk should be emphasized.
- A diet that provides a variety of fiber-containing foods is encouraged.
- There is no need to totally eliminate sugars from the diet, but whole grains, fruits, and vegetables are recommended over high-sugar foods.
- Limit dietary cholesterol to <200 mg/day, saturated fats to <7% of total calories, and minimize consumption of *trans* fatty acids.
- Two or more servings of fish per week should be consumed to provide omega-3 fatty acids.
- There is not sufficient evidence to recommend the use of vitamin and mineral supplements to prevent or treat diabetes.
- Moderate alcohol consumption, defined as one drink/day for adult women and two drinks/day for adult men, poses no significant health concern for people with diabetes.

SOURCE: American Diabetes Association. Standards of Medical Care in Diabetes. Diabetes Care. 2011;34:S11–S61.

long-term maintenance of weight loss.[23] Regular exercise has many other health benefits including improved insulin sensitivity and blood glucose regulation. In addition, exercise can also help reduce blood pressure, improve cardiovascular fitness, reduce stress, and decrease lipid levels in the blood. For these reasons, exercise and physical activity should always be encouraged. Exercise ought to be thought of as a vital component of a type 2 diabetes management plan, and as little as 30 minutes of moderate physical activity on most days of the week can benefit health.

LIFESTYLE PRACTICES CAN INFLUENCE RISK OF DEVELOPING TYPE 2 DIABETES

We cannot do much about our genetics; however, there is much we can do in terms of lifestyle choices to help prevent type 2 diabetes. Studies show that even individuals who are at high risk for developing type 2 diabetes can significantly reduce their risk by eating a variety of healthy foods, staying physically fit, and maintaining a recommended body weight.[24] The story of the Pima Indians, presented in the Focus on the Process of Science feature, provides an example of how alterations in diet and physical activity can help prevent type 2 diabetes, even in a genetically susceptible population.

Dietary Recommendations for Preventing Type 2 Diabetes Individuals at high risk for developing type 2 diabetes are encouraged to eat a healthy diet that balances energy intake with energy expenditure.

Diets that provide whole grains and adequate amounts of dietary fiber have been shown to be beneficial in preventing diabetes.[25] Currently, there is not enough evidence to suggest that eating foods with low-glycemic loads can help prevent diabetes.[26] Although moderate alcohol consumption poses no significant health concern for people with type 2 diabetes and some studies suggest that moderate alcohol intake may actually reduce the risk for diabetes, the American Diabetes Association does not recommend alcohol consumption for diabetes prevention.[27] To help prevent type 2 diabetes in children and adolescents, a healthy, well-balanced diet that adequately supports normal growth and development is especially important.

What Are Secondary Diabetes and Gestational Diabetes?

The majority of people with diabetes have type 1 or type 2. However, a small percentage of people have **secondary diabetes,** which is hyperglycemia that develops as a result of certain diseases, medical conditions, or medications. For example, some medications may affect how the body uses or produces insulin. Once the underlying cause of secondary diabetes is treated, normal blood glucose control is often restored. Another type of diabetes, called gestational diabetes, is brought on by pregnancy.

secondary diabetes Diabetes that results from other diseases, medical conditions, or medication.

FOCUS ON THE PROCESS OF SCIENCE
The Story of the Pima Indians

The nature-versus-nurture issue is an ongoing and hotly debated topic. Are chronic diseases determined by our genes or by our lifestyle? Is it a combination of both, and, if so, does one have more influence than the other? The Pima Indians have provided researchers with a unique opportunity to study these important questions.[28]

For thousands of years the Pima Indians, originally from the Sierra Madre Mountains in Mexico, farmed, hunted, and fished as a way of life. The Pima, living in a hot, dry climate, developed ingenious farming practices and an elaborate irrigation system that allowed their crops to flourish. For some Pima, this way of life changed about 100 years ago, when new settlers to the region began farming upstream from the Pima, diverting the water supply for their own crops. Unable to farm, many Pima migrated north and relocated to the state of Arizona. For these people, the traditional lifestyle was gone. Not only did their diet change, but so did their way of life. The physical demands of farming gave way to a sedentary lifestyle. Instead of traditional foods, they consumed diets rich in sugar, white flour, and lard.

For the past 30 years researchers from the National Institutes of Health have been studying the Pima Indians.[29] The high rates of obesity and type 2 diabetes among this group have been of particular concern. This is because the prevalence of type 2 diabetes is an astonishing 34% of men and 41% of women.[30] Many Pima develop type 2 diabetes at a remarkably young age, and 95% of those with type 2 diabetes are obese. To complete their investigation, the research team also studied a group of Pima Indians still living in a remote area in Mexico. The diet and lifestyle of these traditional Mexican Pima were similar to those of 100 years ago. Although the two groups of Pima Indians (those living in Arizona and those living in Mexico) were genetically similar, the Mexican Pima had very low rates of obesity and type 2 diabetes.

These divergent lifestyles of the U.S. and Mexican Pima clearly show that lifestyle can influence the development of obesity and type 2 diabetes. Although both groups of Pima have similar genetics, only those living in a permissive environment of abundant food and reduced physical activity developed diabetes. Further studies have shown that lifestyle changes, such as increased exercise and improved diets, can significantly reduce the high rates of obesity and type 2 diabetes among the Pima living in Arizona.[31] The good news for all of us is that lifestyle practices can sometimes override the influence of genetics on development of obesity and type 2 diabetes.

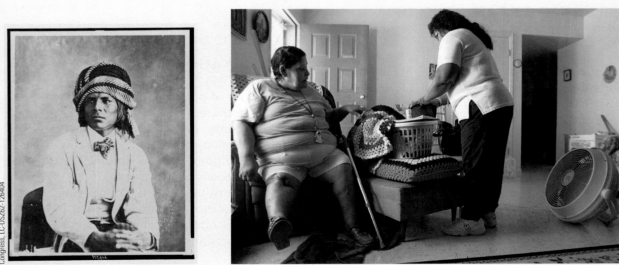

National Photo Company Collection, Prints & Photographs Division, Library of Congress, LC-USZ62-126404

© AP Photo/J. Pat Carter

Genetics and environment both play a role in the development of type 2 diabetes. Research conducted on the Pima Indians has provided evidence that obesity is a major risk factor in the development of diabetes. The prevalence of diabetes is only 8% among Pima Indians living a traditional lifestyle in Mexico, while it is greater than 34% among Pima Indians living in the United States.

SOME PREGNANT WOMEN DEVELOP GESTATIONAL DIABETES

During pregnancy, most women experience insulin resistance to some extent. This normal and healthy response to pregnancy helps make glucose more available to the fetus. However, approximately 18% of all pregnant women develop a more severe form of insulin resistance called **gestational diabetes.**[1] This temporary form of diabetes typically develops after the 24th week of pregnancy, disappearing within 6 weeks after delivery. Gestational diabetes is more common in obese women and those who have a family history of type 2 diabetes.[32] Risk factors associated with gestational diabetes are listed in Table 6.

Effects of Gestational Diabetes Although maternal insulin does not cross the placenta, glucose does. As a result, the fetus of a woman with gestational diabetes is exposed to high levels of glucose. This causes the fetus's pancreas to increase its production of insulin, which can lead to increased glucose uptake and the deposit of extra fat in the growing baby. For this reason, infants born to women with gestational diabetes tend to be large, which can make delivery difficult. Babies born to mothers with poorly controlled gestational diabetes can weigh as much as 10 to 12 pounds.

Women who develop gestational diabetes tend to have a high recurrence of this condition in subsequent pregnancies. In addition, having gestational diabetes increases a woman's risk for developing type 2 diabetes by up to 60% in the next 10–20 years.[1] Infants born to women with gestational diabetes may also be at increased risk for developing type 2 diabetes, especially if there is a family history of diabetes.[33]

TABLE 6 Risk Factors Associated with Gestational Diabetes

- Overweight or very overweight
- History of abnormal glucose tolerance
- Over 25 years of age
- Being of African, Hispanic, Native American, or Pacific Islander descent
- Family history of type 2 diabetes or gestational diabetes
- Previous delivery of a stillborn or a very large baby
- Having gestational diabetes with previous pregnancy

SOURCE: U.S. Department of Health and Human Services, National Institutes of Health, and the National Institute of Child Health and Human Development. NIH Pub. No. 00–4818; 2005. Available from: http://www.nichd.nih.gov/publications/pubs/upload/gest_diabetes_risk_2005.pdf.

Diagnosing and Managing Gestational Diabetes Unless a woman is considered low risk for gestational diabetes (younger than 25 years with a body mass index less than 27 kg/m²), the American Diabetes Association recommends routine screening for gestational diabetes.[34] This is typically done around the 24th week of pregnancy and involves measuring blood glucose levels. To help achieve and maintain optimal blood glucose control, it is important for women with gestational diabetes to receive counseling on diet and exercise. The American Diabetes Association recommends a meal plan that provides adequate nutrients and energy to meet the needs of pregnancy but restricts carbohydrate intake to 35 to 40% of daily calories.[35] Another important component of gestational diabetes management is exercise. A physically active lifestyle is an important part of a healthy pregnancy, and this can help lower maternal glucose concentrations, reduce stress, and prevent excess weight gain. However, every pregnant woman should first consult with her physician before beginning an exercise program.

Managing Diabetes Today Can Help Prevent Health Problems Tomorrow

While the prevalence of diabetes continues to increase in the United States and around the world, so do the related medical expenditures. Indeed, having diabetes can be costly both emotionally and monetarily. It has been estimated that people with diabetes, on average, have medical expenditures that are approximately 2.3 times higher than those without diabetes.[36] Making dietary and other lifestyle changes is not easy and often requires assistance from specially trained health care professionals. The day-to-day management of these medical conditions can be demanding, and the emotional side of this task is often overlooked. It is not uncommon for people with diabetes to follow "all the rules" and still have difficulties controlling blood glucose, creating a feeling of frustration. However, people who participate actively as members of their health care team by asking questions and making good choices can avoid complications and achieve positive effects on their health and well-being.

gestational diabetes (ges – TA – tion – al) Type of diabetes characterized by insulin resistance that develops in pregnancy.

Key Points

What Is Diabetes?

- The four types of diabetes are categorized by the underlying cause.

What Is Type 1 Diabetes?

- Type 1 diabetes is classified as an autoimmune disease and is caused by an interaction among genetics, environment, and the immune system.

- In the case of type 1 diabetes, antibodies produced in response to something in the environment attack and destroy the insulin-producing cells of the pancreas.

- Without insulin, cells are unable to take up glucose, resulting in hyperglycemia, rapid weight loss, accumulation of ketones in the blood, extreme thirst, hunger, frequent urination, and fatigue.

- Treatment of type 1 diabetes focuses on balancing insulin injections with a healthy diet and physical activity.

- Controlling blood glucose reduces the complications associated with diabetes.

- Chronic hyperglycemia damages blood vessels and nerves, leading to heart attacks, strokes, blindness, limb amputation, impaired kidney function, and loss of feeling in the feet and hands.

What Is Type 2 Diabetes?

- Type 2 diabetes is caused by insulin resistance, meaning that insulin-requiring cells do not respond to insulin in a normal way.

- Most people with diabetes have type 2, which is strongly associated with obesity.

- Symptoms associated with type 2 diabetes include fatigue, frequent urination, excessive thirst, chronic hyperglycemia, blurred vision, frequent urinary tract infections, slow wound healing, and vaginal itching.

- The treatment focus of type 2 diabetes is on maintaining near-normal blood glucose through a healthy diet, weight management, and physical activity.

What Are Secondary Diabetes and Gestational Diabetes?

- Diabetes that develops as a result of other diseases, medical conditions, or medications is called secondary diabetes.

- Gestational diabetes is a form of diabetes that can develop during the later stages of pregnancy. It is more common in obese women and those with a family history of type 2 diabetes.

- Infants born to mothers with gestational diabetes are often large because the fetal pancreas increases its production of insulin in response to being exposed to high blood glucose levels from the mother.

- Women who develop gestational diabetes are at high risk for recurrence, and both the infant and mother are at increased risk for developing type 2 diabetes later in life.

Managing Diabetes Today Can Help Prevent Health Problems Tomorrow

- A team made up of health care professionals is important for management of diabetes and metabolic syndrome.

- A healthy lifestyle is a critical component of all management plans.

Notes

1. Centers for Disease Control and Prevention. National diabetes fact sheet: national estimates and general information on diabetes and prediabetes in the United States, 2011. Atlanta, GA: U.S. Department of Health and Human Services, Centers for Disease Control and Prevention, 2011. Available from: http://www.cdc.gov/diabetes/pubs/pdf/ndfs_2011.pdf.

2. Lindstrom J, Peltonen M, Tuomilehto J. Lifestyle strategies for weight control: Experience from the Finnish diabetes prevention study. Proceedings of the Nutrition Society. 2005;64:81–8.

3. Expert Committee on the Diagnosis and Classification of Diabetes Mellitus. Report of the Expert Committee on the diagnosis and classification of diabetes mellitus. Diabetes Care. 1997;20:1183–97.

4. Sperling MA, editor. Type 1 diabetes: Etiology and treatment. Totowa, NJ: Human Press; 2003.

5. Eisenbarth GS. Type 1 diabetes: Molecular, cellular and clinical immunology. Advances in Experimental Medicine and Biology. 2004;552:306–10. Hirschhorn JN. Genetic epidemiology of type 1 diabetes. Pediatrics and Diabetes. 2003;4:87–100.

6. Hintermann E, Christen U. Viral infection—a cure for type 1 diabetes? Current Medicinal Chemistry. 2007;14:2048–52.

7. Knip M. Environmental triggers and determinants of beta-cell autoimmunity and type 1 diabetes. Endocrine and Metabolic Disorders. 2003;4:213–23.

8. Diabetes Control and Complications Trial Research Group. The effect of intensive treatment of diabetes on the development and progression of long-term complications in insulin-dependent diabetes mellitus. New England Journal of Medicine. 1993;329:977–86.

9. American Diabetes Association. Position statement from the American Diabetes Association: The prevention or delay of type 2 diabetes. Diabetes Care. 2002;25:742–9.

10. American Diabetes Association. Position statement from the American Diabetes Association: Screening for type 2 diabetes. Diabetes Care. 2004;27:s11–s14.

11. Draznin MB. Type 2 diabetes. Adolescence Medicine State of the Art Review. 2008;3:498–506.

12. Leahy JL. Pathogenesis of type 2 diabetes mellitus. Archives of Medical Research. 2005;36:197–209.

13. Malecki MT. Genetics of type 2 diabetes mellitus. Diabetes Research and Clinical Practice. 2005;68:S10–21. Tusie Luna MT. Genes and type 2 diabetes mellitus. Archives of Medical Research. 2005;36:210–22.

14. van Dam RM, Rimm EB, Willett WC, Stampfer MJ, Hu FB. Dietary patterns and risk for type 2 diabetes mellitus in U.S. men. Annals of Internal Medicine. 2002;136:201–9.

15. Solomon CG, Manson JE. Obesity and mortality: A review of the epidemiologic data. American Journal of Clinical Nutrition. 1997;66:1044S–50S.

16. Green A, Christian Hirsch N, Pramming SK. The changing world demography of type 2 diabetes. Diabetes/Metabolism Research and Reviews. 2003;19:3–7.

17. Fischer-Posovszky P, Wabitsch M, Hochberg Z. Endocrinology of adipose tissue—an update. Hormone and Metabolic Research. 2007;39:314–21.

18. Mathers JC. Early nutrition: Impact on epigenetics. Forum of Nutrition. 2007;60:42–8.

19. Dunger DB, Salgin B, Ong KK. Session 7: Early nutrition and later health early developmental pathways of obesity and diabetes risk. The Proceedings of the Nutrition Society. 2007;66:451–7.

20. Nobili V, Alisi A, Panera N, Agostoni C. Low birth weight and catch-up-growth associated with metabolic syndrome: a ten year systematic review. Pediatric Endocrinology Reviews. 2008;2:241–7.

21. UK Prospective Diabetes Study (UKPDS) Group. Intensive blood-glucose control with sulphonylureas or insulin compared with conventional treatment and risk of complications in patients with type 2 diabetes (UKPDS 33). Lancet. 1998;352:837–53. Anderson JW, Kendall CW, Jenkins DJA. Importance of weight management in type 2 diabetes: Review with meta-analysis of clinical studies. Journal of the American College of Nutrition. 2003;22:331–9.

22. American Diabetes Association. Position statement from the American Diabetes Association: Nutrition recommendations and interventions for diabetes. Diabetes Care. 2008;31:S61–78.

23. Sigal RJ, Kenny GP, Wasserman DH, Castaneda-Sceppa C. Physical activity, exercise, and type 2 diabetes. Diabetes Care. 2005;27:2518–39.

24. Wareham NJ. Epidemiological studies of physical activity and diabetes risk, and implications for diabetes prevention. Applied Physiology, Nutrition, and Metabolism. 2007;32:778–82.

25. Kelley DE. Sugars and starch in the nutritional management of diabetes mellitus. American Journal of Clinical Nutrition. 2003;78:858S–64S.

26. Thomas D, Elliott EJ. Low glycaemic index, or low glycaemic load, diets for diabetes mellitus.Cochrane Database Systematic Reviews. 2009;1:CD006296.

27. Katsilambros N, Liatis S, Makrilakis K. Critical review of the international guidelines: What is agreed upon—what is not? Nestle Nutrition Workshop Series. Clinical Performance and Programme. 2006;11:207–18.

28. Bennett PH. Type 2 diabetes among the Pima Indians of Arizona: An epidemic attributable to environmental change? Nutrition Reviews. 1999;57:S51–4.

29. Baier LJ, Hanson RL. Genetic studies of the etiology of type 2 diabetes in Pima Indians: Hunting for pieces to a complicated puzzle. Diabetes. 2004;53:1181–6.

30. Schulz LO, Bennett PH, Ravussin E, Kidd JR. Effects of traditional and western environments on prevalence of Type 2 diabetes in Pima Indians in Mexico and the U.S. Diabetes Care. 2006;29:1866–71.

31. Pavkov ME, Hanson RL, Knowler WC, Bennett PH, Krakoff J, Nelson RG. Changing patterns of type 2 diabetes incidence among Pima Indians. Diabetes Care. 2007;30:1758–63.

32. American Diabetes Association. Position statement from the American Diabetes Association: Gestational diabetes mellitus. Diabetes Care. 2008;31:s12–s54.

33. Stocker CJ, Arch JR, Cawthorne MA. Fetal origins of insulin resistance and obesity. Proceedings of the Nutrition Society. 2005;64:143–51.

34. The American College of Obstetricians and Gynecologists. Screening for gestational diabetes mellitus: Recommendations and rationale. Obstetrics & Gynecology. 2003;101:393–5.

35. American Diabetes Association. Position statement from the American Diabetes Association: Standards of medical care in diabetes—2011. Diabetes Care. 2011;34:s11–s61.

36. Economic costs of diabetes in the U.S. in 2007. Diabetes Care. 2008;31:596–615.

Protein

In Chapter 4, you learned about carbohydrates, which are a critical source of energy and serve as building blocks for complex molecules such as deoxyribonucleic acid (DNA) and ribonucleic acid (RNA). However, the importance of dietary carbohydrates to health is arguably rivaled by the next class of macronutrient: protein. The term *protein* was derived more than 170 years ago from the Greek *prota*, meaning "of first importance."[1] There is no debate that the proteins your body makes (or does not make) can be critical to your health. And because you need to eat protein to make protein, obtaining enough of the right kinds of protein from your diet is vitally important.

Proteins constitute the most abundant organic substance in your body, making up at least 50% of your dry weight.[2] But why exactly do you need these proteins, and how can the proteins you eat influence the proteins you make? The answers to these questions are somewhat complex, and understanding some fundamental concepts related to protein nutrition will help you answer them.

In this chapter, you will learn what proteins are, what foods are good sources of them, how you digest and absorb dietary proteins, and how your body uses these substances to maintain health. You will also learn about vegetarianism; how certain proteins in foods cause allergic reactions; and how dietary choices, genetic makeup, and environment can interact to influence your risk of various diseases. With this information, you can make more informed decisions concerning which protein sources to choose and how much is needed to optimize long-term health and well-being.

Living with Peanut Allergy

Tyler is an adorable 10-year-old boy who loves to camp and play ice hockey, and you might think he does not have a care in the world. However, that assumption is far from the truth. When Tyler was two years old, he and his mother, Christie, were camping. After eating a peanut butter sandwich, Tyler began to feel sick and complained that his stomach hurt. When he would not stop crying, his mother took him to the emergency room. At that point, life dramatically changed, as Tyler was diagnosed with having a severe allergic response to peanuts. In fact, he was allergic to all legumes.

Nothing is easy when you have a food allergy. For example, Tyler's mom once gave him a bowl of vegetable soup that contained a couple of lima beans, a type of legume. She was horrified when his tongue swelled and he began having difficulty breathing. Once again, a trip to the emergency room was necessary. Clearly, living with food allergies is extremely scary for Tyler and his family, and Christie told us that she lives with the fear that she "will lose Tyler because someone just was not being careful."

When Tyler was a toddler, it was not difficult to limit him to "safe" foods. However, as he got older, this became increasingly difficult. "Peanut-Free Zone" signs were posted in his classroom and at a special table in the lunchroom. Notes were sent home requesting that only peanut-free foods be brought for birthdays and other celebrations. It was also very important for Tyler to be aware of which foods were "safe" and "unsafe." But even these precautions were not enough. One day Tyler traded a bag of carrots for a chocolate cookie with a friend at lunch. Both kids thought the cookie was safe, but it had been in a bag with a peanut butter sandwich. When hives developed on Tyler's neck and his breathing became labored, the school called Christie, and she once again rushed him to the emergency room.

Along with Christie, Tyler's teachers and school nurse continue to be vigilant in keeping him safe. The school nurse told us that "handling these sometimes life-threatening conditions can be very confusing, but the best advice is continued education and awareness." Although Christie says her goal has always been to prevent peanut allergy from affecting Tyler's social life, striking a balance between being overprotective and not hurting someone's feelings is difficult. For example, Tyler is not allowed to eat with many of his friends at school—instead, he must eat at the designated "peanut-free table."

Clearly, living with peanut allergy is challenging. We asked Tyler what it is like, and he told us the following: "It's really scary. If I could have one wish in the world, it would be to be like my friends and not have peanut allergy."

© Shelley McGuire

Critical Thinking: Tyler's Story

Do you or anyone you know have a food allergy? If so, how has this affected your (or his or her) life choices? How might having a food allergy be especially difficult in college, and what might you do differently if you learned that your roommate had a severe allergy to some type of food?

What Are Proteins?

Perhaps the first questions you might ask when beginning your study of proteins are *"What are proteins, how does one protein differ from another, and what makes proteins different from other nutrients?"* **Proteins** (also called **peptides**) are complex molecules made from smaller subunits called **amino acids** that are joined together by **peptide bonds.** The formation of a peptide bond is an example of a condensation reaction in which a hydroxyl group ($-OH$) from one amino acid is joined with a hydrogen atom ($-H$) from another amino acid, releasing a molecule of water (H_2O) in the process. Proteins contain appreciable amounts of nitrogen, since each amino acid contains one or more nitrogen atoms. This makes proteins chemically distinct from the other macronutrients (carbohydrates and lipids).

Protein size is quite variable, depending on the number of amino acids present. Whereas some are very simple, containing only a few amino acids, others contain thousands of these building blocks. Most proteins, however, have 250 to 300 amino acids. This is why proteins are often classified on the basis of their number of amino acids: dipeptides have two amino acids, tripeptides have three, and so forth. Those with 2 to 12 amino acids are collectively called oligopeptides, and proteins that have more than 12 amino acids are called **polypeptides.**

AMINO ACIDS ARE THE BUILDING BLOCKS OF PROTEINS

The numerous proteins in the body are amazingly diverse. But what makes one protein different from another? The key to this diversity lies in the number and types of amino acids they contain as well as the sequence in which they are linked together. To understand protein diversity, you must first understand the basic components of the amino acids and what makes each one unique.

Most amino acids have three common parts: (1) a central carbon atom bonded to a hydrogen atom, (2) a nitrogen-containing **amino group** ($-NH_2$), and (3) a carboxylic acid group ($-COOH$). In the body, the amino and carboxylic acid groups almost always exist in "charged" states. Specifically, the amino group has a positive charge, and the carboxylic acid has a negative charge. Each amino acid also contains a side-chain, called an **R-group.** The structure of the R-group makes each amino acid uniquely different from the others. The R-group can be as simple as a hydrogen atom or as complex as a ring structure. The chemical and physical properties of amino acids depend on subtle differences in the R-groups. For example, some R-groups are negatively charged, some are positively charged, and some have no charge at all. The three charged components of amino acids—the amino group, the carboxylic acid group, and the R-group—cause proteins to bend and take on unique and complex shapes important for their functions. A "generic" amino acid and some examples of R-groups are shown in Figure 5.1.

AMINO ACIDS ARE CLASSIFIED AS ESSENTIAL, NONESSENTIAL, OR CONDITIONALLY ESSENTIAL

The body needs 20 different amino acids to make all the proteins it requires, and these amino acids can be categorized as essential, nonessential, or conditionally essential (Table 5.1).*[3] The nine essential amino acids are those you

---

*Essential amino acids are also called indispensable amino acids, nonessential amino acids are called dispensable amino acids, and conditionally essential amino acids are called conditionally indispensable amino acids.

---

protein (peptide) Nitrogen-containing macronutrient made from amino acids.

amino acid Nutrient composed of a central carbon bonded to an amino group, carboxylic acid group, and a side-chain group (R-group).

peptide bond A chemical bond that joins amino acids.

polypeptide A string of more than 12 amino acids held together via peptide bonds.

amino group ($-NH_2$) The nitrogen-containing component of an amino acid.

R-group The portion of an amino acid's structure that distinguishes it from other amino acids.

FIGURE 5.1 The Main Components of an Amino Acid
Amino acids have four main components: a central carbon, an amino group, a carboxylic acid group, and an R-group.

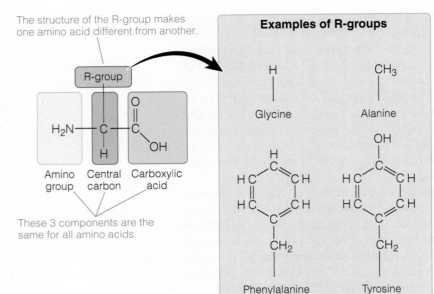

The structure of the R-group makes one amino acid different from another.

R-group

Amino group | Central carbon | Carboxylic acid

These 3 components are the same for all amino acids.

Examples of R-groups

Glycine

Alanine

Phenylalanine

Tyrosine

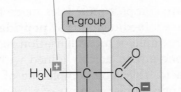

This nitrogen is positively charged, because it has gained a hydrogen atom.

R-group

This oxygen is negatively charged, because it has lost a hydrogen atom.

must consume in your diet because the body cannot make them or cannot make them in required amounts. The remaining 11 amino acids are nutritionally nonessential because the body can make them from other compounds. To synthesize a nonessential amino acid, the body transfers an amino group from one amino acid to another compound called an **α-keto acid,** which is basically an amino acid without its amino group. This process, called **transamination,** results in the synthesis of a new amino acid.

Under some conditions, however, the body may be unable to synthesize one or more of the nonessential amino acids. For example, some infants born prematurely cannot make several of the traditionally nonessential amino acids. When this happens, some nonessential amino acids become conditionally essential because they must be obtained from the diet until the baby matures.[4] Fortunately, infants fed human milk and/or infant formula receive sufficient amounts of these conditionally essential amino acids. For premature infants, however, it is sometimes necessary to fortify milk with conditionally essential amino acids to ensure optimal growth and development.[5]

α-keto acid A compound that accepts an amino group from an amino acid in the process of transamination.

transamination The process whereby an amino group is formed via the transfer of an amino group from one amino acid to another organic compound (an α-keto acid).

TABLE 5.1 Essential, Nonessential, and Conditionally Essential Amino Acids

Essential	Nonessential	Conditionally Essential
Histidine	Alanine	Arginine
Isoleucine	Asparagine	Cysteine
Leucine	Aspartic acid	Glutamine
Lysine	Glutamic acid	Glycine
Methionine	Serine	Proline
Phenylalanine		Tyrosine
Threonine		
Tryptophan		
Valine		

SOURCE: Institute of Medicine. Dietary Reference Intakes for energy, carbohydrate, fiber, fat, fatty acids, cholesterol, protein, and amino acids. Washington, DC: National Academies Press; 2005.

Diseases, such as **phenylketonuria (PKU),** can also cause traditionally nonessential amino acids to become conditionally essential. People with PKU do not make one of the enzymes required to convert the essential amino acid phenylalanine to the normally nonessential amino acid tyrosine. Thus, tyrosine becomes conditionally essential. In all, six amino acids are considered to be conditionally essential, and these are listed in Table 5.1.

Are All Food Proteins Equal?

It is not surprising that some foods contain more protein than others. For example, meat, poultry, fish, eggs, dairy products, and nuts provide more protein (per gram) than grains, fruits, and vegetables. An exception to this is legumes such as dried beans, lentils, peas, and peanuts. Leguminous plants are unique in that their roots are associated with bacteria that can take nitrogen from the air and incorporate it into amino acids. These amino acids are then used by the associated plant. This is why legumes tend to have higher protein content than most other plants.

Moreover, foods with the same amount of total protein can contain different combinations of amino acids. For example, both a cup of cottage cheese and a cup of cooked lima beans provide about 15 g of protein. However, the amino acids in these two foods are quite different. Some foods provide balances of amino acids that are more nutritionally useful for the body than others. This is somewhat analogous to having two piggy banks—one containing a pound of coins made up of ½ pennies and ½ nickels and the other containing a pound of coins made up of ⅓ each of nickels, dimes, and quarters. In this analogy, you can think about the pennies and nickels as nonessential amino acids and the dimes and quarters as essential amino acids. Although both of these banks contain 1 pound of coins, that containing the dimes and quarters is more valuable. Similarly, some food proteins are "worth" more than others, with animal foods generally being more "valuable" than plant foods. Take our comparison of cottage cheese versus lima beans. Foods of animal origin like cottage cheese tend to be more "valuable" than do plant-derived foods like lima beans because the former generally have greater amounts of *essential* amino acids than do the latter.

phenylketonuria (PKU) An inherited disease in which the body cannot convert phenylalanine into tyrosine.

complete protein source A food that contains all the essential amino acids in relative amounts needed by the body.

incomplete protein source A food that lacks or contains very low amounts of one or more essential amino acids.

limiting amino acid The essential amino acid in the lowest concentration in an incomplete protein source.

COMPLETE AND INCOMPLETE PROTEINS

Food proteins can be categorized based on the balance of amino acids they contain. Those containing adequate amounts of all essential amino acids are said to be **complete protein sources,** whereas those supplying low amounts of one or more of the essential amino acids are **incomplete protein sources.** The amino acids that are missing or in low amounts in an incomplete protein source are called **limiting amino acids.** When a limiting amino acid is missing from the diet, we cannot make proteins that contain that particular amino acid, even if we have all of the other essential amino acids.

In general, meat, poultry, eggs, and dairy products are complete protein sources, whereas plant products are incomplete protein sources. For example, corn protein is generally low in lysine and tryptophan—these are considered two of its limiting amino acids. Another incomplete protein source is wheat, which provides only small amounts of lysine.

Animal-derived foods such as meat, milk, fish, and eggs are excellent sources of complete protein.

Protein complementation involves consumption of a variety of incomplete protein sources so that all essential amino acids are consumed.

PROTEIN COMPLEMENTATION

You may be wondering how people who eat only plant foods (vegetarians) get all of their essential amino acids if these types of foods only contain incomplete proteins. The answer to this question is that combining a variety of foods with incomplete proteins can provide adequate amounts of all the essential amino acids. This dietary practice, called **protein complementation,** is customary around the world, especially in regions that traditionally rely heavily on plant sources for protein.[6] Examples of commonly consumed foods whose proteins "complement" each other are rice and beans as well as corn and beans. Both rice and corn have several limiting amino acids (for example, lysine) but generally provide adequate methionine. Beans and other legumes tend to be limiting in methionine but provide adequate lysine. In general, protein complementation allows diets containing a variety of plant protein sources to provide all of the needed essential amino acids.

PROTEIN QUALITY

The "quality" of a food protein, however, is more complex than simply which amino acids it contains. Factors that determine whether food proteins are good sources of essential amino acids include whether the protein is complete or incomplete and the body's ability to absorb the amino acids the protein contains (i.e., its bioavailability).[7] If a food is a complete protein source and it provides easily absorbed amino acids (has high bioavailability), it is a **high-quality protein source.** In general, animal-derived foods are sources of high-quality protein, and foods containing incomplete proteins and/or those in which the protein has low bioavailability are **low-quality protein sources.**

For example, protein from processed (polished) rice is of low quality because it has low amounts of an essential amino acid (lysine) and has low bioavailability. Thus, people who rely solely on rice for their amino acid requirements become lysine deficient unless they consume extremely large amounts of rice. A person with lysine deficiency cannot make enough of *any* of the proteins that require this amino acid. Thus, lysine deficiency negatively affects many aspects of health, including oxygen transport (insufficient hemoglobin), muscle synthesis (inadequate muscle proteins), and digestive processes (lack of digestive enzymes).

Traditionally, the protein quality of a plant was something that we could not influence, making protein complementation vital in cultures consuming little or no animal-derived food. However, scientists are now able to alter the amino acid composition of some plants with the ultimate goal of transforming low-quality protein sources into high-quality protein sources. You can read more about this in the Focus on Food feature.

How Are Proteins Made?

You now know the fundamental concepts related to what makes up a protein, how proteins differ from each other, and why some foods are considered better sources of protein than others. But to really understand why proteins are essential in the diet, you must also understand what your body does with protein once you eat it. In other words, how does your body convert the proteins you eat into the exact and functional proteins you need? And how do these new proteins function? To understand this you must first know how the body makes its own proteins. This is because proteins acquire their distinct structures and functions as they are made in your cells, and it is the shape

protein complementation Combining incomplete protein sources to provide all of the essential amino acids in relatively adequate amounts.

high-quality protein source A complete protein source with high amino acid bioavailability.

low-quality protein source A food that is either an incomplete protein source or one that has low amino acid bioavailability.

Which specific proteins a particular plant or animal produces is determined by its genetic code. Therefore, alterations in the genetic material (DNA) can influence the proteins a plant makes and ultimately affect the protein quality of foods that we make from the plant. Although purposefully altering the DNA of an organism was in the realm of science fiction only decades ago, this is no longer the case. Indeed, the DNA and protein quality of a food can now be enhanced by the production of a genetically modified plant or animal (also called a **genetically modified organism,** or GMO).[8] Genetic modification involves manipulating the genetic material of an organism to "force" it to produce different or altered proteins. Sometimes this is done by simply modifying the DNA that the organism has. Other times, DNA from another organism is inserted in the nucleus. In this way, a plant with one or more limiting amino acids can be enriched with those amino acids and, as a result, be transformed into a complete protein source.

An example of a GMO is the Opaque-2 corn plant.[9] To produce this modified corn, scientists altered the corn's genetic code so that it produces proteins with more lysine and tryptophan—the two limiting amino acids in unmodified corn. The availability of this genetically modified corn may be especially important for people living in areas of the world that rely heavily on corn for their protein intake, because its consumption decreases the risk of amino acid deficiency in these at-risk populations.

However, the development and use of GMOs is somewhat controversial. For example, some people worry that GMOs may increase risk of food allergies and antibiotic resistance and result in unintended transfer of genetic materials to other organisms. Others have expressed concern that it is unethical to tamper with nature by mixing genes among species. This is currently an area of active debate. In response, the U.S. Food and Drug Administration (FDA) has developed policies and guidelines that must be followed in developing new foods made with GMOs. For example, genetic modification must not negatively alter the nutrient quality of the food. The FDA does not require foods made with GMOs to be so labeled, although those that do not contain GMOs can be labeled as such. You can find out more about GMOs and the FDA's rules and regulations related to them on its website (http://www.fda.gov).

that determines a protein's function. The process of protein synthesis, which involves three basic steps—cell signaling, transcription, and translation—is discussed next and illustrated in Figure 5.2.

STEP 1: CELL SIGNALING INITIATES PROTEIN SYNTHESIS

Nearly all cells in the body make proteins. However, different cells need different proteins, and the amounts needed are highly variable. As a result, protein synthesis within a cell is not random; instead, it is regulated by what the individual cell and the entire body need. To initiate this process, the cell must first be told to make a particular protein. This process, called **cell signaling,** is initiated by proteins (like hormones) or other substances (like vitamins) associated with the cell. Cell signaling communicates environmental conditions or cellular needs to the nucleus of the cell, just as an "indoor/outdoor" thermometer can be used to monitor the outside temperature from inside a house.

The "turning on" of protein synthesis by cell signaling is called **up-regulation.** Conversely, sometimes cells are instructed to "turn off" the synthesis of a certain protein. This type of cell signaling is called **down-regulation.** Nutrients are involved in both of these processes. The body's ability to orchestrate the up- and down-regulation of protein synthesis via cell signaling allows it to produce only the proteins it needs—and to stop producing proteins that are not needed. This involves a cell's genes: the genetic material found in its nucleus. For example, cell signaling of the genes needed for calcium absorption in your small intestine is up-regulated when your blood calcium level is low and down-regulated when it is elevated. In this way, blood calcium level is maintained in its optimal range.

genetically modified organism (GMO) An organism (plant or animal) made by genetic engineering.

cell signaling The first step in protein synthesis, in which the cell receives a signal to produce a protein. Note that this term is also used for a variety of other processes (aside from protein synthesis) within the cell.

up-regulation In the context of protein synthesis, increased expression of a gene.

down-regulation In the context of protein synthesis, decreased expression of a gene.

FIGURE 5.2 **The Steps of Protein Synthesis** Protein synthesis involves three basic steps: cell signaling, transcription, and translation.

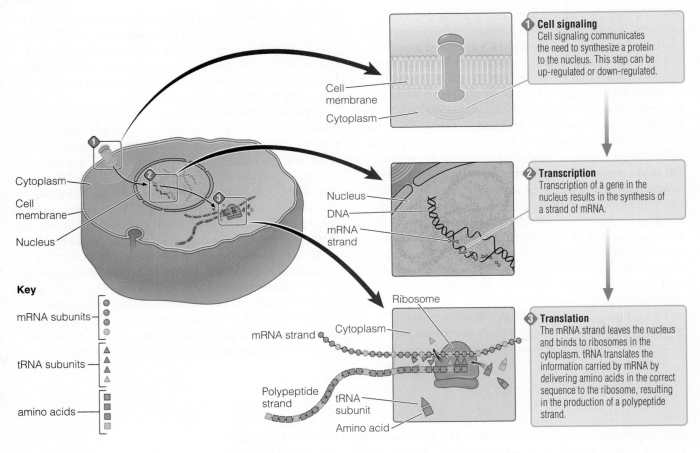

1 Cell signaling
Cell signaling communicates the need to synthesize a protein to the nucleus. This step can be up-regulated or down-regulated.

2 Transcription
Transcription of a gene in the nucleus results in the synthesis of a strand of mRNA.

3 Translation
The mRNA strand leaves the nucleus and binds to ribosomes in the cytoplasm. tRNA translates the information carried by mRNA by delivering amino acids in the correct sequence to the ribosome, resulting in the production of a polypeptide strand.

Key

mRNA subunits

tRNA subunits

amino acids

STEP 2: TRANSCRIPTION TRANSFERS GENETIC INFORMATION TO mRNA

Cell signaling (or more specifically, the up-regulation of protein synthesis) initiates the second step of protein synthesis, **transcription.** One way to think of transcription is to compare it to reading an instruction manual describing how to put together a newly purchased product. In protein synthesis, the "written" instruction manual is in the nucleus of each cell—coded in strands of deoxyribonucleic acid, or DNA. Found in a cell's nucleus, coiled strands of DNA combine with special proteins to form **chromosomes.** Each chromosome is subdivided into thousands of units called **genes.** Each gene provides the instructions needed to make a peptide. In other words, a gene tells a cell which amino acids are needed and in what order they must be arranged to synthesize a particular protein (or portion thereof).

However, proteins are not made directly from the genes comprising your chromosomes. Rather, the gene's DNA code must first be communicated to protein-synthesizing organelles found outside the cell's nucleus. This is accomplished by enzymes that "unzip" the DNA. Next, the genetic information contained in the DNA is "decoded" or transcribed by a molecule called **messenger ribonucleic acid (mRNA).** In the nucleus, a series of mRNA subunits bind to the up-regulated gene. These mRNA subunits then join, forming a strand of mRNA. The newly formed mRNA strand separates from the DNA, exits the nucleus, and enters the cytoplasm, where the mRNA participates in the next step of protein synthesis, called translation. As such, the information

transcription The process whereby mRNA is made using DNA as a template.

chromosome A strand of DNA and associated proteins in a cell's nucleus.

gene A portion of a chromosome that codes for the primary structure of a protein.

messenger ribonucleic acid (mRNA) A form of RNA involved in gene transcription.

originally contained in the DNA sequence of a gene is transcribed into an mRNA sequence, which can be understood by the protein-synthesizing organelles housed in the cytoplasm.

STEP 3: TRANSLATION PRODUCES A NEW PEPTIDE CHAIN

The third step of protein synthesis, called **translation,** begins when the mRNA strand binds to ribosomes. **Ribosomes** are found in the cytoplasm and assist in translating the information contained in mRNA into a peptide. This process requires yet another type of RNA called **transfer ribonucleic acid (tRNA).** The function of tRNA is to transport amino acids to the ribosomes for protein synthesis. For translation to proceed, the ribosome moves along the mRNA strand "reading its sequence." The sequence of the mRNA, in turn, instructs specific tRNAs to transfer the amino acids they are carrying to the ribosome. One by one, amino acids join together via peptide bonds to form a growing peptide chain. When translation is complete, the newly formed peptide separates from the ribosome. However, this is not the final step in the formation of a new protein. The final structure and shape of the protein is yet to be determined, and how this happens is described next.

How Do Proteins Get Their Shapes?

When genetic material has been transcribed and translated, a peptide chain is released from the ribosome. However, to form the complex shape of the final protein the peptide must fold and, in some cases, combine with other peptide chains or substances. Because the final shape must be exact for the protein to function correctly, it is important to consider the many levels of protein structure and what can happen when something goes wrong.

PRIMARY STRUCTURE DICTATES A PROTEIN'S BASIC IDENTITY

There are several levels of protein structure, but the most basic is determined by the number and sequence of amino acids in a single peptide chain. This is called a protein's **primary structure** or primary sequence (Figure 5.3) and is determined by the DNA code. Each peptide chain has a unique primary structure and is therefore a unique molecule. Although understanding the concept of primary structure is relatively simple, take a moment to contemplate the enormous number of primary structures that can be made from just 20 amino acids. Consider, as an analogy, the English alphabet, which has a somewhat similar number of letters—26 in all. As you know, the number and variety of words that can be made with only 26 letters is astonishing. Some are short; some are long. Some contain just a few different letters; others contain many different letters. The same holds true for proteins: some are short, some are long. Some contain a handful of different amino acids, whereas others may contain all 20 amino acids. The possibilities are seemingly endless.

A Protein's Primary Structure Is Critical The primary structure of a protein is critical to its function because it determines the protein's most basic chemical and physical characteristics. In other words, the primary structure represents the basic identity of the protein. Changes in a protein's primary structure can profoundly affect the ability of the protein to do its job. Some

translation The process whereby amino acids are linked together via peptide bonds on ribosomes, using mRNA and tRNA.

ribosome An organelle, associated with the endoplasmic reticulum in the cytoplasm, involved in gene translation.

transfer ribonucleic acid (tRNA) A form of RNA in the cytoplasm involved in gene translation.

primary structure The sequence of amino acids that make up a single peptide chain.

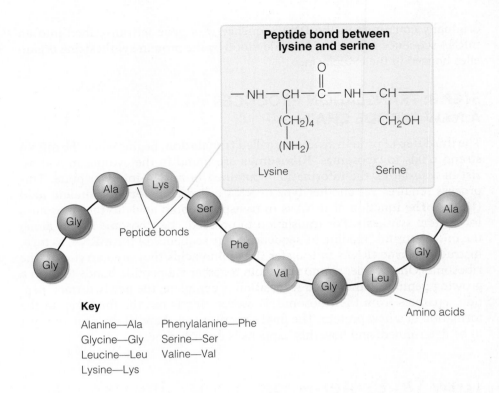

FIGURE 5.3 The Primary Structure of a Protein The primary structure of a protein is the sequence and number of the amino acids in its peptide chain.

Peptide bond between lysine and serine

Lysine   Serine

Peptide bonds

Amino acids

Key

Alanine—Ala   Phenylalanine—Phe
Glycine—Gly   Serine—Ser
Leucine—Leu   Valine—Val
Lysine—Lys

alterations in a protein's primary structure can be caused by inherited genetic variations in the DNA sequence. **Sickle cell anemia** (also called sickle cell disease) is an example of a disease caused by an inherited genetic variation. The complications associated with sickle cell anemia, such as fatigue and increased risk for infection, are due to a small "error" in the DNA that ultimately results in the production of defective molecules of the protein hemoglobin within red blood cells.[10] Because hemoglobin is responsible for circulating oxygen and carbon dioxide in the blood, these complications can sometimes be serious. Additional detail on sickle cell anemia is provided in the Focus on Clinical Applications feature.

SECONDARY STRUCTURE FOLDS AND TWISTS A PEPTIDE CHAIN

In addition to each peptide's unique primary structure, most proteins have a three-dimensional shape that is more complex than a linear chain of amino acids. Recall that the backbone of the peptide chain is made of a series of positively charged amino acids and negatively charged carboxylic acid groups. These charges attract and repel each other like magnets, folding portions of the peptide into organized and predictable patterns. These interactions form weak bonds[†] that twist and fold the primary structure into a three-dimensional pattern. This level of folding is called the **secondary structure** of a protein. The two most common folding patterns are the **α-helix** and **β-folded sheets,** illustrated in Figure 5.4. You can think of an α-helix as being like a spiral staircase and a β-folded sheet as being similar to a folded paper fan.

TERTIARY STRUCTURE ADDS COMPLEXITY

The next level of protein complexity is called **tertiary structure,** which is additional folding due to interactions among the R-groups. This causes the

[†]Technically, these are hydrogen bonds.

sickle cell anemia A disease in which a small change in the amino acid sequence of hemoglobin causes red blood cells to become misshapen and decreases the ability of the blood to carry oxygen and carbon dioxide.

secondary structure Folding of a protein because of weak bonds that form between elements of the amino acid backbone (not R-groups).

α-helix (AL – pha – he – lix) A common configuration that makes up many proteins' secondary structures.

β-folded sheet (BAY – ta – fold – ed) A common configuration that makes up many proteins' secondary structures.

tertiary structure (TER – ti – a – ry) Folding of a polypeptide chain because of interactions among the R-groups of the amino acids.

Sickle cell anemia is a disease caused by a single error in the amino acid sequence of the protein hemoglobin. More specifically, people with sickle cell anemia have an alteration, or mutation, in the gene that codes for one of the polypeptides in hemoglobin. Hemoglobin is a large and complex protein, found in red blood cells, that carries oxygen from the lungs to the body's tissues and carbon dioxide back to the lungs for removal. The sickle cell mutation results in the insertion of an incorrect amino acid (valine) for the correct amino acid (glutamic acid) during translation. As a result of this seemingly small error in its primary structure, the shape of the hemoglobin molecule is completely altered, causing it to function improperly.

Red blood cells with normal hemoglobin are smooth and glide through blood vessels. Red blood cells with "sickle cell" hemoglobin are rigid, sticky, and shaped like a sickle or crescent. This alteration in shape causes the cells to form clumps and get stuck in blood vessels. The clumps of sickle-shaped cells can block blood flow in the vessels that lead to the limbs and organs, resulting in pain, serious infections, and organ damage. Signs and symptoms of this disease include anemia (the inability of the blood to carry oxygen); pain in the chest, abdomen, and joints; severe fatigue; swollen hands and feet; frequent infections; stunted growth; and vision problems.

Scientists hypothesize that, hundreds of years ago, the genetic alteration responsible for sickle cell anemia somehow protected people from the serious and sometimes deadly disease malaria.[11] During this time, a malaria epidemic killed tens of thousands of people in parts of Africa, the Mediterranean, the Middle East, and India. People with the sickle cell mutation, however, are thought to have survived this outbreak better than those who did not have the mutated gene. As a result, survivors were able to reproduce and pass on their genetic code to their offspring. Today, millions of people all over the world have the gene for sickle cell anemia, especially those with African, Mediterranean, Middle Eastern, or Indian ancestry. In the United States, sickle cell anemia is most common in people of black African heritage; 1 in 500 African Americans has the disease.[11] Note that in order for a person to have

sickle cell anemia he or she must have received a faulty gene from *both* parents. If a sickle cell gene is inherited from only one parent, the person will not have serious signs or symptoms of sickle cell anemia but could pass the gene on to the next generation. Such individuals are said to have the sickle cell "trait." Approximately 1 in 12 African Americans has sickle cell trait.[11]

Although there is currently no cure for sickle cell anemia, scientists hope that they might soon have the technology to "correct" the defective DNA.[12] This technology, called **gene therapy,** may be able to cure this and other genetically based diseases. In the meantime, complications of this disease are treated with analgesic drugs to reduce pain, antibiotics to treat infection, blood transfusions, supplemental oxygen, folic acid supplementation, and bone marrow transplants.

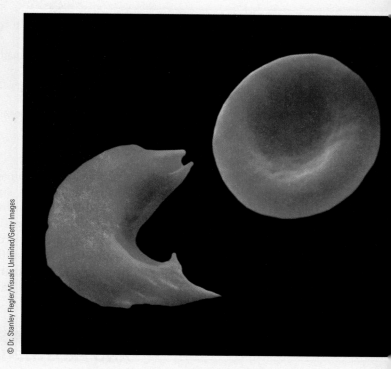

Normal red blood cells (right) are disc shaped, whereas those of people with sickle cell anemia (left) are crescent shaped.

entire protein to have an even more complex, three-dimensional structure. Imagine, for example, what would happen to a folded paper fan if you were to gently crumple it in your hand. The fan's original folds (analogous to a protein's secondary structure) would remain, but they might be further bent and twisted in some regions (analogous to a protein's tertiary structure). This is what happens in a protein to form its tertiary structure.

gene therapy The use of altered genes to enhance health.

FIGURE 5.4 The Secondary Structure of a Protein Weak chemical bonds fold proteins into α-helix and β-folded sheet patterns, resulting in a secondary structure.

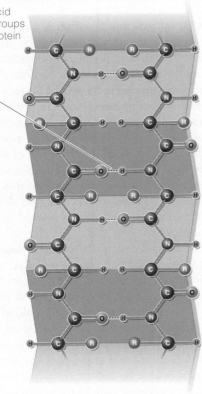

Weak bonds between carboxylic acid and amino groups cause the protein to fold into a secondary structure.

α-helix

β-folded sheet

Some of the strongest interactions between R-groups occur between amino acids that contain sulfur atoms (for example, cysteine), which react with other sulfur-containing amino acids to form sulfur–sulfur (disulfide) bonds. These bonds are particularly stable, and anything that disturbs them can severely disrupt the protein's tertiary structure. In fact, it is the number and arrangement of disulfide bonds in hair that determine if it is straight or wavy. Breaking and reforming these bonds is the basis for some "permanent wave" treatments used in hairstyling.

SOME PROTEINS HAVE QUATERNARY STRUCTURE AND PROSTHETIC GROUPS

The fourth level of protein structure is called **quaternary structure.** Quaternary structure exists when two or more peptide chains come together, as shown in Figure 5.5. This level of complexity would be somewhat like putting two or three crumpled paper fans together. Not all proteins have a quaternary structure—only those made from more than one peptide chain.

To function properly, some proteins have precisely positioned nonprotein components called **prosthetic groups,** which often contain minerals. Hemoglobin is an example of a protein with quaternary structure and prosthetic groups. Specifically, it is made from four separate polypeptide chains (hemoglobin's quaternary structure), each of which combines with an iron-containing heme (hemoglobin's prosthetic group). Heme is the portion of hemoglobin that actually transports the oxygen and carbon dioxide gases in the blood.

quaternary structure (quat – ER – nar – y) The combining of peptide chains with other peptide chains in a protein.

prosthetic group A nonprotein component of a protein that is part of the quaternary structure.

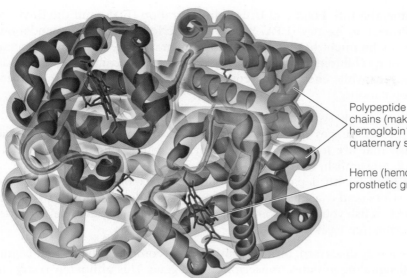

Polypeptide chains (making up hemoglobin's quaternary structure)

Heme (hemoglobin's prosthetic group)

DENATURING AGENTS ALTER A PROTEIN'S SHAPE AND FUNCTION

A protein's primary, secondary, tertiary, and quaternary structures determine its final shape. This shape is critical to allowing the protein to carry out its function. However, there are many conditions that can disrupt a protein's shape. One example is **denaturation.** Denaturation occurs when a protein unfolds. Note, however, that only the secondary, tertiary, and quaternary structures are affected by denaturation; the primary structure remains intact. This would be akin to flattening out one of the pieces of paper making up your folded paper fan "protein." Flattening out the fan would result in a fan that does not work, just like denaturation can cause a protein to lose its function. Compounds and conditions that cause denaturation are called denaturing agents and include physical agitation (e.g., shaking), heat, detergents, acids, bases, salts, alcohol, and heavy metals (e.g., lead and mercury).

A familiar example of denaturation occurs when an egg white is heated; the proteins unfold, and the egg white changes from a thick, clear liquid to an opaque solid. Another example of a denaturing agent is mercury, which can disrupt disulfide bonds and thus tertiary structure. Because of its denaturing effects on proteins involved in neural function, high levels of mercury exposure can cause numbness, hearing loss, visual problems, difficulty walking, and severe emotional and cognitive difficulties. In fact, we now know that a condition termed "mad hatter's disease" that was widespread in the 19th century was due to mercury poisoning.[13] During that period, hat makers used large amounts of mercury-containing compounds to treat felt and fur and suffered the consequences in that they went "mad."

Because of the denaturing effects of heavy metals such as lead and mercury, efforts are being made to decrease the concentrations of these compounds in the environment. For example, there are concerns that certain types of fish contain dangerously high amounts of mercury. In fact, although most fish are safe

© Lisa Romerein/Taxi/Getty Images

It is important to limit our consumption to certain types of fish because some contain mercury, which can cause protein denaturation in the body.

denaturation The alteration of a protein's three-dimensional structure by heat, acid, chemicals, enzymes, or agitation.

to consume, the U.S. Food and Drug Administration (FDA) and the Environmental Protection Agency (EPA) have made the following recommendations for women who might become pregnant, women who are pregnant or lactating, and young children.[14] These recommendations are meant to decrease mercury exposure while encouraging consumption of low-mercury fish in these at-risk populations.

- Do not eat shark, swordfish, king mackerel, or tilefish.
- Eat up to 12 ounces (two average meals) a week of a variety of fish and shellfish that are lower in mercury: shrimp, canned light tuna, salmon, pollock, and catfish.
- Check local health departments about the safety of fish caught in local lakes, rivers, and coastal areas. If no advice is available, eat up to 6 ounces per week of fish you catch from local waters, but do not consume any other fish during that week.

As previously discussed, protein synthesis is a complex process that begins when cell signaling initiates transcription of a gene. This ultimately results in the body producing all of the many proteins that it needs. But have you ever considered why some cells make different proteins than other cells, and how your nutritional status may affect this process? We will now look at how genes interact with nutrition to determine the particular proteins you make, in turn influencing your health and well-being.

Genetics, Epigenetics, Nutrition, and Nutrigenomics

The genetic material (DNA) in a cell's nucleus provides the cell with instructions to make all the proteins it needs, and the exact DNA sequences individuals have are determined by what DNA their parents passed on to them. When a sperm cell fertilizes an egg cell, the chromosomes in each combine to become the genetic makeup of the offspring. This is how an infant inherits his or her parents' **genetic makeup** or genotype. Except for identical twins, no two individuals have the same genotype, which in part explains the vast diversity among humans.

Although the overall sequences of genes coding for specific proteins are often identical among people, some genes have modifications that are unique. These genetic differences give us each our individual physical characteristics, such as eye color and hair color. These characteristics do not really affect our health. However, alterations in other genes can have more important consequences.

GENETIC ALTERATIONS: MUTATIONS AND POLYMORPHISMS

Sometimes a modification in a gene results in a protein with an altered amino acid sequence. When this is due to a chance genetic modification, it is called a **mutation.** While some mutations have no measurable effect on health, others do. For example, you have learned that sickle cell anemia (which negatively impacts the function of red blood cells) is a sometimes serious disease caused by a mutation. Other mutations, such as the one involved in PKU, can influence metabolism. Still others alter protein synthesis, causing cells to experience uncontrollable growth, which, in turn, can lead to cancer. Alterations (mutations) in the DNA of egg and sperm cells can be passed on to the offspring and are therefore inherited. PKU and sickle cell anemia are the result of inherited DNA mutations.

genetic makeup (genotype) The particular DNA contained in a person's cells.

mutation The alteration of a gene.

Some mutations are more common than others, and when a particular genetic mutation is present in at least 1% of the population, it is called a **polymorphism.** Polymorphisms can impact a person's health and risk for disease. They can also influence nutrient requirements. Thus, nutritional scientists are very interested in understanding common genetic polymorphisms and how they interact with nutrition to influence health.

EXPERTS BELIEVE THAT NUTRITION MAY BE RELATED TO EPIGENETICS

Although a person's genetic makeup (genotype) has long been thought to reflect only the *sequence* of the DNA making up his or her chromosomes, scientists have recently learned that the connection between genes and physiology is actually much more complex. For example, the DNA and its accompanying organelles can be altered in a variety of ways (not related to DNA sequence) that regulate whether a particular gene is expressed or not. The term **epigenetics** (the prefix *epi-* is derived from the Greek word meaning "over" or "above") refers to these types of modifications. As such, two people can have exactly the same genetic sequences in their DNA but produce different amounts and combinations of proteins. Like mutations in DNA, some of these epigenetic differences can be passed on to the next generation; in other words, they are inheritable.

Epigenetic modifications are responsible for establishing and maintaining the diverse patterns of gene expression that distinguish different cell and tissue types.[15] For example, although bone and nerve cells contain exactly the same DNA, they synthesize very different proteins in large part because of different epigenetics. Scientists now think that epigenetic modifications may also play important roles in the development of many chronic degenerative diseases such as cancer, type 2 diabetes, and cardiovascular disease.[16] For instance, epigenetic changes can influence kidney function and therefore have an effect on blood pressure regulation.

Of great importance to the field of nutrition is growing evidence that nutritional status in very early life may affect long-term health via epigenetic modifications. For example, babies who are malnourished during fetal life but then experience accelerated growth in infancy or childhood may be at increased risk for cardiovascular disease and type 2 diabetes later in life, partly due to differences in epigenetic patterns.[17] Whether later alterations in a person's environment, such as better nutrition, can reverse this epigenetic effect remains to be discovered. Clearly epigenetics is an exciting new area of nutrition research.

THE HUMAN GENOME PROJECT HAS OPENED THE DOOR TO NUTRIGENOMICS

Until about 20 years ago, scientists knew very little about the genes that make up the human genome. However, the Human Genome Project was carried out in the 1990s to describe all the genes within our chromosomes. The Human Genome Project, described in more detail in the Focus on the Process of Science feature, has allowed scientists to better understand how genetic variations influence protein synthesis, overall health, and risk for disease. In addition, it has opened up the field of nutrition research referred to as nutrigenomics.[18] **Nutrigenomics** is the study of how nutrition and genetics interact to influence health. Nutritional scientists and other health professionals hope that, by understanding your individual nutritional needs, nutritional status, *and* genetics, they will someday be able to optimize your health and

polymorphism An alternation in a gene that is present in at least 1% of the population.

epigenetics (epi – ge – NE – tics) Alterations in gene expression that do not involve changes in the DNA sequence.

nutrigenomics (nu – tri – gen – O – mics) The science of how genetics and nutrition together influence health.

Scientists, philosophers, and psychologists have long been interested in what makes one person different from the next. Although this question may represent the ultimate mystery, researchers continue to try to understand the physiological and chemical reasons for why we all differ. In the 1800s, Gregor Mendel discovered that many physical characteristics could be passed on to offspring, launching the field of modern genetics. Genetics has, however, progressed significantly since that time.

In 1990, the U.S. government and an international team of scientists initiated the Human Genome Project, which was designed to decode all the genes making up human chromosomes. This information has become invaluable for our understanding of genetic diversity and how it influences health and well-being.

Nutritional scientists can now directly test what they have long thought: nutrition interacts with genetics to influence health. For example, it is now much easier to study how dietary factors influence whether a gene is turned on or turned off. Additionally, emerging studies suggest that nutrition likely determines some epigenetic alterations.

Indeed, many scientists and health professionals hope that someday each of us will be able to inexpensively and noninvasively find out what type of diet is best for our personal genetic makeup. This may be a relatively simple process only requiring that you provide a swab taken from the inside of your cheek. The sample could then be mailed to a laboratory, which would analyze it and make dietary recommendations about specific amounts of foods, nutrients, supplements, and exercise to best help prevent illness.

For example, if your genes suggest that your risk for heart disease is high, your list of recommendations would likely stress a diet low in fat, saturated fat, *trans* fatty acids, and sodium but high in cardioprotective foods. In other words, you would receive a *personalized* dietary prescription. Today a handful of commercial laboratories claim to be able to make these personal diet recommendations for anywhere from $500 to $2,000. However, their validity is the subject of *significant* debate. Nonetheless, this amazing concept is no longer science fiction but is clearly the direction in which modern nutritional science is moving.

well-being by precisely matching dietary recommendations with your unique genetic—and perhaps epigenetic—makeup.

You now know how proteins are made in the body and how alterations in genetics and epigenetics can influence this process. However, to make proteins, the body must first obtain their basic building blocks—the amino acids—from the diet. As such, dietary proteins must be digested, absorbed, and circulated to the cells that need them. How the body completes these tasks is described next.

How Are Dietary Proteins Digested, Absorbed, and Circulated?

The process of digestion disassembles food proteins into amino acids that are then absorbed and circulated to cells where the amino acids are reassembled into the proteins that the body needs. This is somewhat like disassembling someone else's house and then using the materials to rebuild another house that perfectly fits your own exact needs. In addition to using *dietary* proteins, the body efficiently and systematically breaks down and recycles its own proteins when they become old and nonfunctional. In this way, intestinal enzymes, digestive secretions, and degraded cells can contribute more than once to the amino acids available for protein synthesis. In fact, you can think of the body as having its own "protein recycling center." The stages of protein digestion, absorption, and circulation are shown in Figure 5.6 and described next.

⟨CONNECTIONS⟩ Recall that hydrochloric acid (HCl) is released from parietal cells, whereas pepsinogen is released from chief cells (Chapter 3, page 91).

PROTEIN DIGESTION BEGINS IN THE STOMACH

Proteins in the foods you eat must be broken down into their component amino acids; only then can they be used by the body. Although a small amount of

FIGURE 5.6 Overview of Protein Digestion Protein digestion occurs in both the stomach and small intestine.

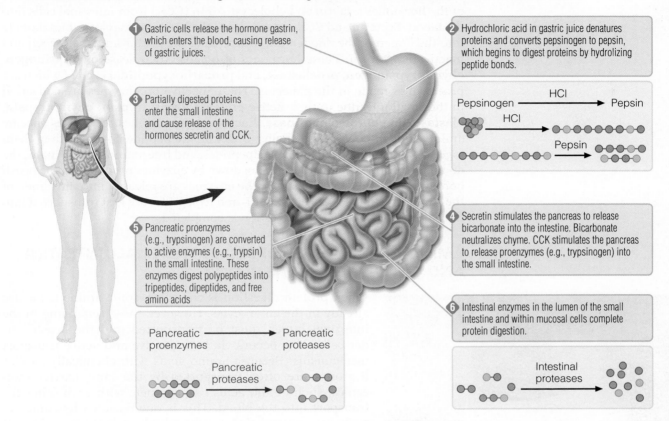

1 Gastric cells release the hormone gastrin, which enters the blood, causing release of gastric juices.

2 Hydrochloric acid in gastric juice denatures proteins and converts pepsinogen to pepsin, which begins to digest proteins by hydrolizing peptide bonds.

Pepsinogen $\xrightarrow{\text{HCl}}$ Pepsin

3 Partially digested proteins enter the small intestine and cause release of the hormones secretin and CCK.

4 Secretin stimulates the pancreas to release bicarbonate into the intestine. Bicarbonate neutralizes chyme. CCK stimulates the pancreas to release proenzymes (e.g., trypsinogen) into the small intestine.

5 Pancreatic proenzymes (e.g., trypsinogen) are converted to active enzymes (e.g., trypsin) in the small intestine. These enzymes digest polypeptides into tripeptides, dipeptides, and free amino acids

6 Intestinal enzymes in the lumen of the small intestine and within mucosal cells complete protein digestion.

Pancreatic proenzymes $\longrightarrow$ Pancreatic proteases

Pancreatic proteases

Intestinal proteases

mechanical digestion of food protein occurs as you chew, chemical digestion of protein begins when it enters the stomach. The presence of food in your stomach causes some gastric cells to release a hormone called **gastrin** which is released from endocrine cells found deep in the gastric pits. Gastrin triggers the release of hydrochloric acid (HCl), **pepsinogen,** mucus, and substances from other stomach cells that make up the gastric pits. Remember these substances are collectively referred to as *gastric juices*. Note that pepsinogen is an example of a **proenzyme** (also called a zymogen), which is an inactive form of an enzyme.

It is important to understand how all these substances (HCl and enzymes) work together to contribute to protein digestion. First, HCl disrupts the chemical bonds responsible for the protein's secondary, tertiary, and quaternary structures. This process of unfolding, or denaturation, straightens out the complex protein structure so that the peptide bonds can be exposed to digestive enzymes. Next, HCl converts pepsinogen (the proenzyme form) into its active form, called **pepsin.** Pepsin is an example of a **protease.** Protease enzymes hydrolyze or break peptide bonds between amino acids. Note that the stomach does not produce the active protease enzyme—in this case, pepsin. Instead, it produces and stores the inactive or "safe" proenzyme (pepsinogen), thus protecting itself from the enzyme's protein-digesting function until it is needed.

As a result of the actions of the gastric juices, proteins that you eat are partially digested to shorter peptides and some free amino acids. The partially broken down proteins are then ready to leave the stomach and enter the small intestine to be digested further.

PROTEIN DIGESTION CONTINUES IN THE SMALL INTESTINE

Protein digestion in the small intestine takes place both in the lumen and within the enterocytes that line it. Initiating this cascade of digestive events, amino

⟨CONNECTIONS⟩ Recall that hydrolysis is a chemical reaction in which chemical bonds are split by the addition of a water molecule (Chapter 3, page 74).

gastrin (GAS – trin) A hormone, secreted by endocrine cells in the stomach, which stimulates the production and release of gastric juice.

pepsinogen (pep – SIN – o – gen) The inactive form (proenzyme) of pepsin, produced in the stomach.

proenzyme (zymogen) An inactive precursor of an enzyme.

pepsin (PEP – sin) (*peptein*, to digest) An enzyme needed for protein digestion.

protease An enzyme that cleaves peptide bonds.

acids and smaller polypeptides coming from the stomach stimulate the release of the hormones secretin and cholecystokinin (CCK) from intestinal cells into the blood. Secretin and CCK in turn signal the pancreas to release bicarbonate into the lumen of the duodenum, neutralizing the acid from the stomach and inactivating pepsin. These hormones also cause the release of **trypsinogen, chymotrypsinogen, proelastase,** and **procarboxypeptidase,** all of which are proenzymes made in the pancreas. These inactive proenzymes are converted in the small intestine to their active protease forms: **trypsin, chymotrypsin, elastase,** and **carboxypeptidase.** Each of these enzymes recognizes specific amino acids in polypeptide chains, breaking the peptide bonds holding them together and forming tripeptides, dipeptides, and free amino acids. Finally, the di- and tripeptides are further broken down by enzymes produced in the brush border enterocytes. In some cases, the enzymes are released into the lumen of the small intestine. However, most di- and tripeptides are first transported into the enterocytes, where their digestion is completed.

AMINO ACIDS ARE ABSORBED IN THE SMALL INTESTINE AND CIRCULATED IN THE BLOOD

People who have peanut allergy must be very careful to not eat foods, like peanut butter, that contain peanuts.

When protein digestion is complete, some amino acids are already in the enterocytes. However, those remaining in the intestinal lumen must be transported into the brush border cells. This occurs via both passive and active transport mechanisms. Because amino acids with chemically similar R-groups are often transported by the same carrier molecules, such amino acids can compete with each other for transport into the blood. This is one reason why some experts recommend that you avoid taking large quantities of certain amino acid supplements. Most amino acids are absorbed in your duodenum, where they enter your blood and circulate to the liver via the hepatic portal system.

What Are Food Allergies and Intolerances? The breakdown of proteins into amino acids is usually quite complete and typically results in the absorption of amino acids (not proteins) into the circulation. Sometimes, however, larger peptide chains are absorbed. When this happens, the body's immune system may respond as if these peptides were dangerous. In such cases, the person is said to have an "allergic response," or what is more commonly called a **food allergy.**[19] The majority of food allergies are caused by proteins present in eggs, milk, peanuts, soy, and wheat. Researchers estimate that approximately 2% of adults and 5% of infants and young children in the United States have food allergies. Recall Tyler, the boy with a severe food allergy featured at the beginning of this chapter. Tyler is allergic to peanuts and other legumes. This is because relatively large portions of intact legume proteins are absorbed into his bloodstream, and his immune system responds as if they were dangerous—mounting a strong and sometimes dangerous inflammatory response.

Note, however, that not all adverse reactions to foods are true food allergies. A nonimmunological reaction to a substance in a food is called a **food intolerance** (or food sensitivity). An example of a food intolerance is lactose intolerance, which was discussed in Chapter 4.

Signs and Symptoms of Food Allergies For some people, an allergic reaction to a particular food protein may cause mild physical reactions such as skin rashes or gastrointestinal (GI) distress. For others, like Tyler, an allergic food reaction

trypsinogen, chymotrypsinogen, proelastase, and **procarboxypeptidase** Inactive proenzymes produced in the pancreas and released into the small intestine in response to CCK.

trypsin, chymotrypsin, elastase, and **carboxypeptidase** Active enzymes (proteases) involved in protein digestion in the small intestine.

food allergy A condition in which the body's immune system reacts against a protein in food.

food intolerance A condition in which the body reacts negatively to a food or food component but does not mount an immune response.

can be frightening and even life threatening. Signs and symptoms of a food allergy usually develop within a few minutes to an hour after eating the food and are dependent upon what type of food allergy a person has and how his or her immune system reacts to it. The most common signs and symptoms are listed here.

- Tingling in the mouth
- Hives, itching, or eczema
- Swelling of the lips, face, tongue, and throat or other parts of the body
- Wheezing, nasal congestion, or trouble breathing
- Abdominal pain, diarrhea, nausea, or vomiting
- Dizziness, lightheadedness, or fainting

In a severe allergic reaction to food, a person may have more extreme symptoms. For example, **anaphylaxis** might result. Anaphylaxis is a rapid immune response that causes a sudden drop in blood pressure, rapid pulse, dizziness, and a narrowing of the airways. This can, almost immediately, block normal breathing. When this occurs emergency treatment is critical. Anaphylaxis due to food allergies is responsible for thousands of emergency room visits and as many as 200 deaths in the United States each year.[20]

What to Do if You Have a Food Allergy If you have a food allergy, the best way to prevent an allergic reaction is to know which foods to avoid. It is especially important to read food labels carefully, and wearing a medical alert bracelet or necklace may be advantageous. The U.S. Food and Drug Administration (FDA) requires that all foods containing the most common allergens be labeled as such. In the case of children, parents should talk with members of their families as well as friends, child care providers, and school personnel so that they can help avoid exposing the child to offending foods when he or she is under their supervision. It is also important to make sure that the child knows what foods to avoid and to ask for help if needed.

Now that you understand better what causes food allergies and how to avoid them, you may want to reconsider the questions posed at the beginning of this chapter concerning the issue of food allergies on a college campus. For example, what actions can you take to help protect a roommate or friend who has a food allergy?

© Shelley McGuire

Critical Thinking: Tyler's Story Explain, at a basic physiologic level, what causes the allergic responses Tyler experiences when he eats peanuts and other legumes. Specifically, how is this related to protein nutrition, digestion, and absorption? Can you imagine ways that food manufacturers might treat or process peanut-containing foods to make them safe for people with peanut allergy?

What Are the Major Functions of Proteins and Amino Acids in the Body?

Once amino acids are circulated away from the GI tract, the body uses them to make the thousands of proteins it needs via the protein synthetic reactions previously described. Using the previous analogy, this is the stage at which you would use all of the disassembled materials from someone else's house to build one of your own. However, there are many different types of things you can build even from a single material like wood, such as walls, cabinets, stairways,

anaphylaxis A severe and potentially life-threatening allergic reaction.

TABLE 5.2 The Major Functions of Proteins in the Body

Function	Description	Example(s)
Structure	Proteins making up the basic structure of tissues such as bones, teeth, skin	• Hydroxyapatite in bones • Collagen in skin, teeth, ligaments, and tendons • Keratin in hair and fingernails
Catalysis	Enzymes	• Lingual lipase digests lipids in mouth • Pancreatic amylase digests carbohydrates in intestine • Pepsin digests proteins in intestine
Movement	Proteins found in muscles, ligaments, and tendons	• Actin and myosin in muscle
Transport	Proteins involved in the movement of substances across cell membranes and within the circulatory system	• Glucose and sodium transporters in cell membranes • Lipoproteins and hormone transport proteins in blood
Communication	Protein hormones and cell-signaling proteins	• Insulin and glucagon regulate blood glucose • CCK helps regulate digestion in the small intestine • Cell-signaling proteins initiate protein synthesis
Protection	Skin proteins and immune proteins	• Collagen in skin • Fibrinogen helps blood clot • Antibodies fight off infection
Regulation of fluid balance	Proteins that—via the process of osmosis—regulate the distribution of fluid in the body's various compartments	• Albumin is a major regulator of fluid balance in the circulatory system
Regulation of pH	Proteins that readily take up and release hydrogen ions (H^+) to maintain pH of the body	• Hemoglobin is an important regulator of blood pH

and furniture. Each of these household items has its own function. Similarly, the proteins the body makes can be classified into general categories related to their functions (Table 5.2). For example, some proteins, like those in the muscles, are needed for movement. Others, like the hormone insulin, are used to communicate blood glucose levels. Proteins can also be broken down for energy (ATP), and some amino acids can be used to make glucose. In this section, you will learn more about the various types of proteins you need as well as other ways your body uses proteins and amino acids to function and stay healthy.

PROTEINS PROVIDE STRUCTURE

Proteins provide most of the structural materials in the body, being constituents of the muscles, skin, bones, hair, and fingernails. For instance, collagen is a structural protein that forms a supporting matrix in bones, teeth, ligaments, and tendons. Proteins are also important structural components of cell membranes and organelles. The synthesis of structural proteins is especially important during periods of active growth and development such as infancy and adolescence.

ENZYMES ARE PROTEINS THAT CATALYZE CHEMICAL REACTIONS

Molecules called "enzymes" (most of which are proteins) function as biological catalysts, driving the myriad chemical reactions that occur in the body. A **catalyst** is a substance that speeds up a chemical reaction but is not consumed or altered in the process. Without the catalytic functions of protein enzymes, the thousands of chemical reactions needed by the body would simply not occur or, at best, occur at very slow rates.

As an analogy, consider everything it takes to prepare a meal. Although this clearly requires the availability of all the ingredients, just having them in your kitchen will never result in their being prepared and served. Instead, there must also be a person who is willing and able to slice, dice, cook, and season the foods appropriately. In the same way, chemical reactions (like meals) will not occur readily without enzymes (like chefs) to arrange the molecules in the correct positions and supply the needed expertise to facilitate the appropriate chemical changes. Examples of protein enzymes you have already learned about are amylase and pepsin, which catalyze reactions needed to digest carbohydrates and proteins, respectively.

MUSCLE PROTEINS FACILITATE MOVEMENT

Protein is also needed for movement, which results from the contraction and relaxation of the many muscles in the body. There are three distinct types of muscle: skeletal muscle, which is responsible for all voluntary movements; cardiac muscle, which enables the heart to contract and relax; and smooth muscle, which lines many of our tissues and organs such as the stomach, intestine, and blood vessels. For example, as you read this book, you are experiencing contraction of all three types of muscle, allowing your body to simultaneously sit in your chair, turn the pages of the book, breath, and perhaps even drink a cup of coffee or eat a snack. Nearly half of the body's protein is present in skeletal muscle, and adequate protein intake is required to form and maintain muscle mass and function throughout life. Although there are many proteins related to movement, perhaps the most important are actin and myosin, which make up much of the machinery needed to contract and relax your muscles. This is why protein deficiency can cause muscle wasting and weakness.

SOME PROTEINS PROVIDE A TRANSPORT SERVICE

You also need amino acids to make transport proteins, which are responsible for escorting substances into and around the body as well as across cell membranes. For example, absorption of many nutrients (such as calcium) requires transport proteins to help the nutrients cross the cell membranes of enterocytes. Protein deficiency can decrease the body's production of intestinal transport proteins, resulting in secondary malnutrition of a variety of nutrients. Other transport proteins that you have previously learned about are the glucose transporters, which move glucose from the blood into tissues. In addition, many nutrients and other substances are circulated in the blood bound to even more transport proteins. Examples of circulating transport proteins include hemoglobin, which transports gases (oxygen and carbon dioxide), and

© David Sacks/Stone/Getty Images

Proteins are essential for movement.

catalyst A substance that increases the rate by which a chemical reaction occurs, without being consumed in the process.

a variety of "binding proteins," which carry hormones and fat-soluble vitamins in the blood. If the diet does not provide adequate amounts of the essential amino acids, the synthesis of these proteins will decrease, and health consequences (such as anemia and vitamin deficiencies) can result.

HORMONES AND CELL-SIGNALING PROTEINS ARE CRITICAL COMMUNICATORS

Tissues and organs have a variety of ways to communicate with each other, and most of these involve proteins. Although not all hormones are proteins, most are—for example, secretin, gastrin, insulin, and glucagon.[‡] There are also proteins embedded in cell membranes that communicate information about the extracellular environment to the intracellular space. Some of these proteins are involved in the cell-signaling process that initiates protein synthesis itself. Others regulate cellular metabolism. Together, hormones and cell-signaling proteins make up part of the body's critical communication network. Thus, protein deficiency can have profound effects on the body's ability to coordinate its myriad functions.

PROTEINS PROTECT THE BODY

Perhaps one of the most basic functions of the proteins in the body is protecting it from physical danger and infection. For instance, skin is made mainly of proteins that form a barrier between the outside world and the internal environment. In addition to making up the skin, proteins provide other forms of vital protection. For example, if you cut yourself, blood clots (made possible by the presence of specialized proteins) close off this possible entry point for infection. In this way, blood clotting acts as a second level of defense. However, if a bacterium or other foreign substance does enter the body, the immune system fights back by producing proteins called **antibodies** or immunoglobulins. Antibodies bind foreign substances so they can be destroyed. Protein deficiency can make it difficult for the body to prevent and fight certain diseases, because its natural defense systems become weakened.

FLUID BALANCE IS REGULATED IN PART BY PROTEINS

Another function of proteins is regulating how fluids are distributed in the body. Most of the body consists of water, which is found both inside of cells (intracellular space) and outside of cells (extracellular space). In addition, the extracellular space can be subdivided into fluid found in blood and lymph vessels (intravascular fluid) and fluid found between cells (interstitial fluid). The amount of fluid in these spaces is highly regulated in a variety of ways, some of which involve proteins. For example, **albumin,** a protein present in the blood in relatively high concentrations, plays such a role. As the heart contracts, blood is propelled through blood vessels that become increasingly narrower. As the intravascular pressure builds, the fluid portion of the blood is forced out of the tiny capillaries and into the interstitial space. Albumin, which remains in the blood vessels, becomes more concentrated as more fluid accumulates in the interstitial space sourrounding tissues. The high intravascular concentration of albumin creates a powerful force that, like a sponge, draws the fluid from the interstitial space back into the blood vessel (Figure 5.7). Severe protein deficiency can impair albumin synthesis, resulting in low levels of this important protein in the blood and, in turn, accumulation

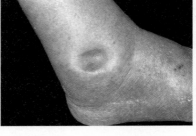

Because protein is needed for fluid balance, severe protein deficiency can cause edema. This photo depicts what happens when finger pressure is applied to an ankle with severe edema—a condition called pitting edema.

antibody A protein, produced by the immune system, that helps fight infection.

albumin A protein important in regulating fluid balance between intravascular and interstitial spaces.

[‡] The non-protein hormones, many of which are steroid hormones, include some of the reproductive hormones such as estrogen and testosterone.

FIGURE 5.7 Regulation of Fluid Balance by Albumin Albumin is a protein in the blood that helps regulate fluid balance between the intravascular space and the interstitial space.

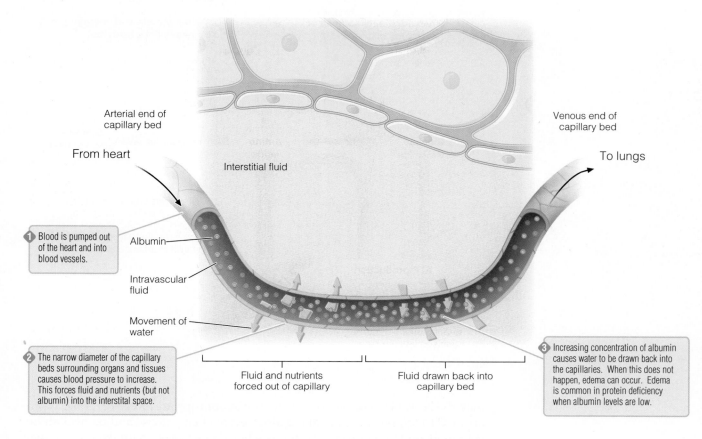

Arterial end of capillary bed

Venous end of capillary bed

From heart

To lungs

Interstitial fluid

1 Blood is pumped out of the heart and into blood vessels.

Albumin

Intravascular fluid

Movement of water

2 The narrow diameter of the capillary beds surrounding organs and tissues causes blood pressure to increase. This forces fluid and nutrients (but not albumin) into the interstitial space.

Fluid and nutrients forced out of capillary

Fluid drawn back into capillary bed

3 Increasing concentration of albumin causes water to be drawn back into the capillaries. When this does not happen, edema can occur. Edema is common in protein deficiency when albumin levels are low.

of fluid in the interstitial space. This condition is called **edema** and sometimes can be seen as swelling in the hands, feet, and abdominal cavity in severely malnourished individuals.

PROTEINS HELP REGULATE pH

Proteins also regulate how acidic or basic the body fluids are. Recall that the body must maintain a certain pH for optimal function. As you learned in Chapter 3, the pH of a fluid is determined by its hydrogen ion (H^+) concentration, and the body's fluids are kept in specific pH ranges. For example, the pH range for stomach fluid is 1.5 to 3.5, while the pH of blood is 7.3 to 7.5. One way that blood's pH is maintained is through the buffering action of proteins such as hemoglobin. Recall that the components of amino acids often have a net positive or negative charge. This is because they can readily accept and donate charged hydrogen ions. When the hydrogen ion concentration in the blood is too high (acidic), negatively charged proteins can bind excess hydrogen ions, restoring the blood to its proper pH. Conversely, proteins can release hydrogens into the blood when the hydrogen concentration is too low (basic). As such, the body can have difficulty maintaining optimal pH balance during periods of severe protein deficiency.

PROTEINS ARE SOURCES OF GLUCOSE AND ENERGY (ATP)

Aside from their role in protein structure, the body can use some amino acids for glucose synthesis and energy (ATP) production. Excess protein can also be converted to fats for more long-term energy storage. Together, these processes

edema (e – DE – ma) The buildup of fluid in the interstitial spaces.

FIGURE 5.8 Protein and Energy Metabolism The fate of dietary protein depends on the body's need for glucose and energy (ATP).

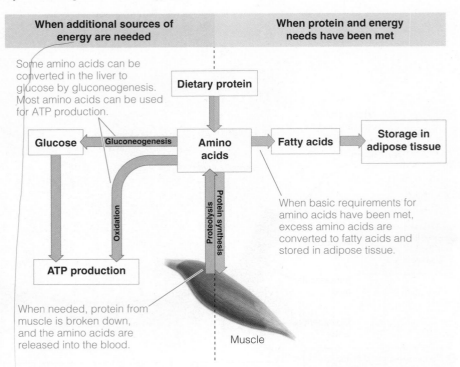

help the body (1) maintain blood glucose at appropriate levels, (2) generate energy (ATP) to power chemical reactions even when glucose and fat availability is limited, and (3) store excess energy when dietary protein intake is more than adequate (Figure 5.8).

Some Amino Acids Can Be Converted to Glucose When energy (ATP) availability is low, the body first turns to stored glycogen for a source of glucose. However, when glycogen stores are depleted, the body turns to protein (amino acids) as an alternate source of glucose. To convert amino acids to glucose, muscle tissue is broken down, and the liver takes up the amino acids. Gluconeogenesis is the process whereby glucose-yielding amino acids, called **glucogenic amino acids,** are converted to glucose. As it implies, the term **gluconeogenesis** refers to metabolic pathways that make glucose from noncarbohydrate sources. To do this, the nitrogen-containing amino group of the glucogenic amino acid is removed in the liver—a process called **deamination.** The remaining carbon skeleton is converted to glucose, which can then be metabolized to produce ATP. In addition, many cells can harvest the energy stored in amino acids by oxidizing them directly. Like carbohydrates, oxidation of 1 g of protein yields approximately 4 kcal of energy. Thus, consuming 10 g of dietary protein is equivalent to consuming 40 kcal. You will learn much more detail about these energy metabolism pathways in Chapter 7.

Excess Amino Acids Are Converted to Fat Although some people may think that eating large amounts of protein or taking amino acid supplements will increase their muscle mass, this is generally not the case. In fact, eating extra protein during times of glucose and energy sufficiency generally contributes to more fat storage, not muscle growth. This is because, during times of glucose and energy excess, the body redirects the flow of amino acids away from gluconeogenesis and ATP-producing pathways and instead converts

glucogenic amino acid An amino acid that can be converted to glucose via gluconeogenesis.

gluconeogenesis (glu – co – ne – o – GE – ne – sis) (*neo-*, new; *-genesis*, bringing forth) Synthesis of glucose from noncarbohydrate sources.

deamination The removal of an amino group from an amino acid.

them to lipids. For this to happen, the nitrogen-containing amino group of each amino acid must first be removed via deamination. However, instead of being converted to glucose as would happen in gluconeogenic pathways, the remaining carbon skeletons are diverted to lipid-producing pathways. The resulting lipids can subsequently be stored as body fat for later use.

AMINO ACIDS SERVE MANY ADDITIONAL PURPOSES

In addition to their role as protein components, amino acids themselves serve many other purposes in the body. Some regulate protein breakdown, others are involved in cell signaling, and still others are converted to neurotransmitters, which function as messengers in your nervous system. Some amino acids can also stimulate or inhibit the activity of enzymes involved in metabolism and provide nitrogen for the synthesis of many important nonprotein, nitrogen-containing compounds such as DNA and RNA. Thus, you need amino acids not only for protein synthesis and as an energy source, but for a multitude of other functions as well.

Protein Turnover, Urea Excretion, and Nitrogen Balance

Proteins are the body's "workhorses" and like any working livestock, how long they are useful is finite. Fortunately, the body can recycle and reuse most of the amino acids from these "retired" proteins to synthesize new ones. The process of continuously breaking down and resynthesizing protein is known as **protein turnover.** By regulating protein turnover, the body can adapt to periods of growth and development during childhood and maintain relatively stable amounts of protein during adulthood without requiring enormous amounts of dietary protein.

PROTEIN TURNOVER HELPS MAINTAIN AN ADEQUATE SUPPLY OF AMINO ACIDS

In addition to amino acids from foods, amino acids from recycled proteins can also be used by the body for protein synthesis. In fact, about half of the amino acids the body uses each day come from worn-out proteins that have "served their time."[21] Protein degradation (proteolysis), which is catalyzed by special protein-cleaving enzymes, releases amino acids into what is called the body's **labile amino acid pool.** Dietary amino acids also contribute to the body's labile amino acid pool. As such, new proteins are produced from a mixture of newly obtained and recycled amino acids.

Protein turnover is regulated mainly by hormones, which coordinate the appropriate balance between protein degradation and synthesis. For example, after you eat a meal, high concentrations of the hormone insulin inhibit the breakdown of your body's protein and stimulate overall protein synthesis. In this way, insulin promotes protein accumulation when energy is abundant. In contrast, when you experience various types of stress, thyroid hormone and cortisol are released, stimulating protein degradation and inhibiting protein synthesis. This results in increased protein turnover. As such, these stress-related hormones help your body maintain an immediate supply of labile, or "free" amino acids, which can be used for gluconeogenesis and ATP production.

protein turnover The cycle involving both protein synthesis and protein degradation in the body.

labile amino acid pool Amino acids that are immediately available to cells for protein synthesis and other purposes.

FIGURE 5.9 **Urea Synthesis and Excretion** Urea is synthesized in the liver and excreted by the kidneys in urine.

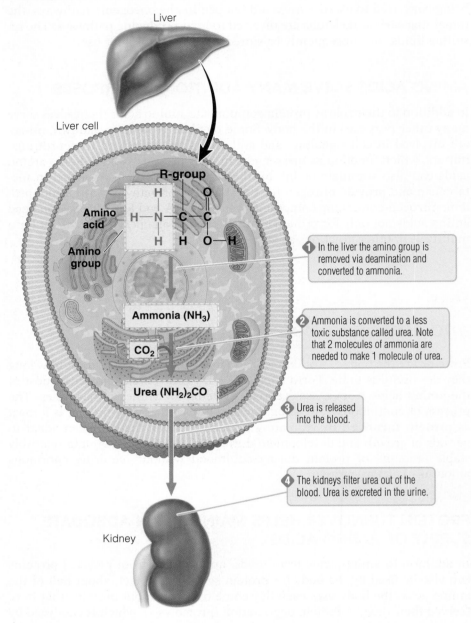

Liver

Liver cell

R-group

Amino acid

Amino group

Ammonia (NH₃)

CO₂

Urea (NH₂)₂CO

Kidney

1 In the liver the amino group is removed via deamination and converted to ammonia.

2 Ammonia is converted to a less toxic substance called urea. Note that 2 molecules of ammonia are needed to make 1 molecule of urea.

3 Urea is released into the blood.

4 The kidneys filter urea out of the blood. Urea is excreted in the urine.

NITROGEN IS EXCRETED AS UREA

As previously stated, amino acids must first be deaminated before they can be converted to glucose or used as a source of energy. This process produces ammonia (NH_3), which is toxic to cells. In response, your liver quickly converts ammonia to a less toxic substance called **urea.** As shown in Figure 5.9, urea is released into the blood, filtered out by the kidneys, and excreted in the urine. This metabolic transformation requires both energy (ATP) and carbon dioxide.

WHAT IS NITROGEN BALANCE?

Protein turnover results in a somewhat complex flux of amino acids, or "remodeling," in the body and measuring protein turnover can tell a health

urea (u – RE – a) A relatively nontoxic, nitrogen-containing compound that is produced from ammonia.

professional important information about overall protein status. More specifically, protein status can be assessed by comparing protein intake with the amount of nitrogen lost in body secretions such as urine, sweat, and feces.[22] When protein (or nitrogen) loss equals protein intake, a person is in **nitrogen balance. Negative nitrogen balance** occurs when nitrogen loss exceeds intake; this can occur during starvation, illness, or stress. When nitrogen intake exceeds the loss of nitrogen from the body, as occurs during childhood or recovery from illness, a person is in **positive nitrogen balance.** Knowing whether an individual is in positive or negative nitrogen balance can help clinicians diagnose and treat certain diseases and physiologic conditions. For example, people with kidney failure who are on dialysis often experience negative nitrogen balance (nitrogen loss > nitrogen intake) and therefore require specialized nutritional support. Conversely, growing children should be in a state of positive nitrogen balance (nitrogen intake > nitrogen loss). If this is not the case, protein intake may not be adequate to support growth.

Healthy growing children are in a state of positive nitrogen balance.

You have now learned important details concerning how the body adjusts nitrogen turnover to support nitrogen balance and optimal health. This process is largely regulated to supply constant and adequate amounts of all the amino acids needed to synthesize the many proteins you need. Dietary recommendations for proteins and amino acids therefore have been developed to take into account both nitrogen balance and the varying requirements for specific amino acids during different periods of the lifespan.

How Much Protein Do You Need?

You need to consume dietary protein for two major reasons: (1) to supply adequate amounts of the essential amino acids; and (2) for the additional nitrogen needed to make the nonessential amino acids and other nonprotein, nitrogen-containing compounds such as DNA. As such, recommendations for dietary amino acids as well as overall protein consumption reflect these needs. These recommendations are described next.

DIETARY REFERENCE INTAKES (DRIs) FOR AMINO ACIDS

To begin with, consider how much of each essential amino acid you should eat every day. Currently, the best estimates can be obtained from the Institute of Medicine's Recommended Dietary Allowances (RDAs), which are shown in Figure 5.10.[2] Note that these values have the unit of "milligrams per kilogram per day" (mg/kg/day) because they represent requirements of the essential amino acids *relative* to body size; the larger you are, the more essential amino acids you need. For example, because the RDA for the essential amino acid valine in adults is 4 mg/kg/day, a woman weighing 140 lb (~64 kg) would require 256 milligrams (4 × 64) of this amino acid in her diet each day. A person weighing 200 lb (~91 kg) would require 364 milligrams (4 × 91) of valine each day.

Because the DRI committee concluded that there is no compelling evidence that high intakes of any of the essential amino acids pose known health risks, the Institute of Medicine did not establish Tolerable Upper Intake Levels (ULs) for them.

nitrogen balance The condition in which protein (nitrogen) intake equals protein (nitrogen) loss by the body.

negative nitrogen balance The condition in which protein (nitrogen) intake is less than protein (nitrogen) loss by the body.

positive nitrogen balance The condition in which protein (nitrogen) intake is greater than protein (nitrogen) loss by the body.

FIGURE 5.10 The RDAs for the Essential Amino Acids in Adults

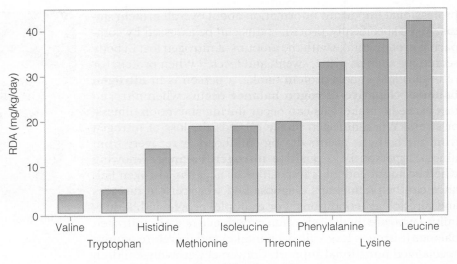

SOURCE: Institute of Medicine. Dietary reference intakes for energy, carbohydrate, fiber, fat, fatty acids, cholesterol, protein, and amino acids. Washington, DC: National Academies Press; 2005.

DIETARY REFERENCE INTAKES (DRIs) FOR PROTEINS

⟨**CONNECTIONS**⟩ Dietary Reference Intakes (DRIs) are reference values for nutrient intake. Recommended Dietary Allowances (RDAs) are set so that 97% of people have their needs met. Adequate Intake (AI) values are provided when RDAs cannot be established. Tolerable Upper Intake Levels (ULs) indicate intake levels that should not be exceeded (Chapter 2, page 38).

DRIs for *overall* protein intake have also been published. These are listed inside the cover of this book. Although not all protein sources are created equal, researchers generally agree that most diets in affluent countries such as the United States provide a balanced mix of all the essential amino acids. Therefore, the dietary recommendations for protein intake do not distinguish between people who consume high-quality proteins and those who do not.

The RDA values for protein are provided in two ways—the first is expressed as grams per day (g/day) and is considered to reflect requirements for a "typical" person in a particular life-stage group. These recommended protein intakes increase with age and are somewhat higher for males than females because, in general, males are larger than females and have more muscle mass. Using this set of values, a typical college-age male needs 56 g/day protein, whereas a comparable female needs 46 g/day.

The second type of RDA values for protein is expressed as grams per kilogram body weight per day (g/kg/day); like those for the essential amino acids, these recommendations adjust for body weight. The Institute of Medicine recommends that healthy adults consume 0.8 g/kg/day of protein. For example, adults weighing 140 lb (~64 kg) would require about 51 g of protein (0.8 × 64) each day regardless of whether they are male or female. You would easily get this much protein by eating a bowl of wheat flake cereal and low-fat milk (12 g protein) for breakfast, a hamburger (24 g protein) for lunch, and a bean burrito (15 g protein) for supper.

During infancy, the most rapid phase of growth in the life cycle, protein requirements (when adjusted for body weight) are relatively high. Protein requirements also increase during pregnancy and lactation, because additional protein is needed to support growth and milk production.[22] Note that these RDA values apply only to healthy individuals. For a variety of reasons, people recovering from trauma (such as burns) or illness may require more protein.[23] People who are healthy show little evidence of harmful effects of high protein intake, and therefore no UL values are set for this macronutrient.

EXPERTS DEBATE WHETHER ATHLETES NEED MORE PROTEIN

Although many people believe that athletes have higher protein requirements than nonathletes, this is a topic of active debate. The DRI committee that established the recommendations for amino acid and protein intake carefully considered this question. The committee concluded that physically active people likely require similar amounts of protein *on a body-weight basis*, and adult athletes can generally estimate their protein requirement the same way as other adults by using the same mathematical formula of 0.8 g/kg/day. On the other hand, in a position statement published in 2007, the International Society of Sports Nutrition concluded that protein intakes of 1.4 to 2.0 g/kg/day for physically active individuals are not only safe but may improve the training adaptations to exercise.[24] The American College of Sports Medicine likewise recommends that endurance and strength-trained athletes consume protein in the range of 1.2–1.7 g/kg/day.[25] Thus, scientists continue to grapple with this issue. You can read more about the use of protein supplements by athletes in the Focus on Sports Nutrition feature.

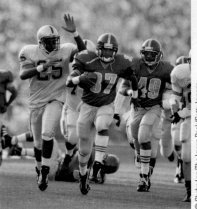

Athletes with increased muscle mass, such as football players and body builders, can probably estimate their protein requirements using the same equations used by nonathletes; they need about 0.8 grams protein for each kilogram of body weight daily. However, this is an area of active debate among scientists.

ADDITIONAL RECOMMENDATIONS FOR PROTEIN INTAKE

Aside from the RDA values, several other sets of recommendations for protein intake are also available. For example, the Institute of Medicine's Acceptable Macronutrient Distribution Ranges (AMDRs) recommend that you consume 10 to 35% of your energy as protein. Using this recommendation, consider a moderately active college student with an energy requirement of 2,000 kcal/day. How much protein should this student consume? To answer this question, you must first determine that 200 to 700 kcal (0.10 × 2,000 and 0.35 × 2,000) should come from protein. This translates to 50 to 175 g of protein (200 ÷ 4 and 700 ÷ 4).[§] Note that 1 medium hamburger patty and 1 cup of skim milk provide approximately 25 and 10 g, respectively, of protein to the diet, making the recommended amount of protein quite easy to obtain—especially at the lower end of the range.

The 2010 Dietary Guidelines for Americans and the accompanying MyPlate food guidance system provide additional recommendations concerning intake of high-protein foods such as dairy products, meat, and dried beans and peas (legumes). Aside from supporting the AMDR for protein (10 to 35% of calories from protein), the Guidelines specifically encourage a range of intakes of fat-free or low-fat milk, lean meats, seafood, eggs, nuts, seeds, and soy products to support optimal health. These food groups represent nutrient-dense, high-protein foods. More specifically, the Dietary Guidelines for Americans, Food Patterns, and MyPlate recommend 1.5 to 5 ounces of lean meat, poultry, or eggs; 0.5 to 1.5 ounces of seafood; 0.14 to 0.7 ounces of nuts, seeds, or soy products; and 2 to 3 cups of fat-free or low-fat dairy products daily, depending on caloric needs. Periodic consumption of beans and peas is also encouraged. Note that 3 ounces of lean meat is generally equivalent to a small steak, lean hamburger patty, chicken breast, or piece of fish, and that portions of meat served in restaurants and cafeterias are often much larger than this. To determine precisely how many servings are recommended for you, visit the MyPlate website (http://www.choosemyplate.gov). You can learn about including a balance of healthful, high-protein foods in your daily food choices in the Food Matters section.

[§]Remember that 1 g of protein supplies 4 kcal of energy.

FOCUS ON SPORTS NUTRITION
Do Protein and Amino Acid Supplements Enhance Athletic Performance?

Optimizing performance is the ultimate goal for most athletes. Indeed, coaches and trainers have long sought training regimens and dietary plans that increase strength, speed, and agility. Methods and products used to optimize athletic performance are called ergogenic aids. One example of an ergogenic aid that has gained significant popularity is the use of protein and amino acid supplements—especially those containing branched-chain amino acids (BCAA). Of the essential amino acids, leucine, isoleucine, and valine are classified as BCAA because they each contain a chemically "branched" R-group. BCAA are naturally present in all protein-containing foods, but the best sources are red meat and dairy products.

Because protein is the major constituent of muscle, one could argue that increased protein or amino acid intake should result in greater muscle mass and, ultimately, increased strength and performance. Furthermore, BCAA appear to be preferentially used for ATP synthesis by muscle tissue during strenuous exercise.[26] These relatively simple facts have motivated the nutritional supplement industry to produce and sell a variety of protein and amino acid powders and drinks, especially those that are high in BCAA. However, is there evidence that protein or amino acid supplementation increases protein synthesis, decreases the breakdown of muscle, or provides especially usable energy?

The answers to these questions are not entirely clear. In fact, supporting evidence—at least for the impact of protein supplementation on protein synthesis—is weak at best. Keep in mind that, before muscle actually grows, the rate of protein synthesis must increase, and the synthesis of specific proteins is highly regulated, requiring the initial step of cell signaling. In other words, something must first initiate the complex processes resulting in DNA transcription and translation before a new protein is made. Although some of the amino acids might be involved in initiating protein transcription, increased dietary protein or amino acid intake does not, *by itself*, signal this process. In fact, contrary to long-held belief, research now suggests that physical activity, and especially resistance exercise, may actually decrease a person's dietary protein requirements.[27] This is because physical activity may trigger the body to become more efficient in its use of amino acids and proteins, resulting in

decreased protein turnover and ultimately a decreased requirement for dietary protein. Clearly, muscle growth and maintenance are complex processes, and more research is needed before we fully understand how exercise affects the protein requirements of athletes.

There are, however, a handful of convincing studies that suggest that amino acid supplementation—especially BCAA—may help inhibit muscle breakdown during intensive training.[28] In response to this, the International Society of Sports Nutrition recently published a position paper on protein and exercise stating that, under certain circumstances, supplementation with BCAA may improve exercise performance and recovery from exercise.[24]

Can protein or amino acid supplements give you that winning edge? Right now, there is little evidence to support this notion, although BCAA may play a special role in preventing muscle breakdown during strenuous activities.[29] The bottom line: (1) eat a well-balanced diet providing sufficient energy with an appropriate mix of carbohydrates, fats, and protein, and (2) train long and hard.

© David Young-Wolff/PhotoEdit

It is important for you to understand basic recommendations concerning how much of each essential amino acid you should consume. In addition, you can calculate your total protein requirement and can refer to the MyPlate food guidance system for suggestions that will help you meet these goals. But what if you were to decide not to eat meat or other high-quality

Working Toward the Goal: Obtaining Sufficient Protein While Minimizing Fats

The 2010 Dietary Guidelines recommend that we meet our protein requirements by choosing and preparing a balance of lean meat and poultry, eggs, seafood, legumes, nuts, and seeds in our meals, along with low-fat or fat-free milk or milk products. This approach helps ensure adequate intake of amino acids and many other micronutrients while minimizing calories, total fat, saturated fat, and cholesterol. The following selection and preparation tips will help you meet your nutrient requirements with high-protein foods while avoiding unnecessary fat consumption.

- Try to consume significant amounts of protein from at least three different food groups daily. This is especially true for vegetarians, who need to consider protein complementation.

- When selecting high-protein dairy foods such as ice cream and yogurt, choose those that are labeled "reduced fat," "low fat," or "fat free," when possible.

- When it is time for a snack, instead of foods with lower nutrient densities such as chips, choose reasonable amounts of high-protein items such as nuts and seeds.

- When comparing similar foods, choose higher protein, lower fat options by comparing Daily Values found on Nutrition Facts labels.

- Add slivered almonds or other nuts to steamed vegetables and fresh salads.

- Experiment with ethnic cuisines—such as Indian or Mexican—that frequently utilize a variety of "pulses" (legumes). These foods are great sources of protein, while being nutrient dense and low in fat.

- To maximize the nutrient density of meat products, trim excess fat prior to cooking.

protein sources? How would this affect your health, and what might you do to make sure your diet was adequate? The answers to these questions are discussed next.

Vegetarian Diets: Healthier Than Other Dietary Patterns?

People have many different reasons for deciding which foods they will and will not eat. This seems especially true about meat and other animal products. For example, some religious groups avoid some or all meats and animal products. Economic considerations and personal preference can also determine whether people eat meat and/or which meats they choose. Because animal products provide high-quality protein in our diets as well as a multitude of other essential nutrients, it is important to consider the effect of animal food consumption (or lack thereof) on issues related to nutritional status. Some issues you might want to consider if you ever think about becoming a vegetarian are discussed next.

© Beth Segal/StockFood Creative/Getty Images

Vegetarian diets can provide all the essential nutrients, but care must be taken to make sure sufficient protein, iron, calcium, zinc, and vitamin B_{12} are consumed.

THERE ARE SEVERAL FORMS OF VEGETARIANISM

The term **vegetarian** (from the Latin *vegetus*, meaning "whole," "sound," "fresh," "lively") was first used in 1847 by the Vegetarian Society of the United Kingdom to refer to a person who does not eat any meat, poultry, or fish or their related products, such as milk and eggs. Today most "vegetarians"

vegetarian Someone who does not consume any or selected foods and beverages made from animal products.

consume dairy products and eggs and are called **lacto-ovo-vegetarians.**[28] Alternatively, **lactovegetarians** include dairy products, but not eggs, in their diets. Vegetarians who avoid all animal products are referred to as **vegans.** Thus, you might want to ask a vegetarian what type he or she is.

VEGETARIAN DIETS SOMETIMES REQUIRE THOUGHTFUL CHOICES

But do vegetarians run any special nutritional risks? The answer to this question depends on what kind of "vegetarian" a person is. In general, a well-balanced lacto-ovo- or lactovegetarian diet can easily provide adequate protein, energy, and micronutrients. Dairy products and eggs are convenient sources of high-quality protein and many vitamins and minerals. However, because meat is often the primary source of bioavailable iron, eliminating it can lead to iron deficiency. Furthermore, vegans may be at increased risk of being deficient in several micronutrients, including calcium, zinc, iron, and vitamin B_{12}.[29] This risk increases further during pregnancy, lactation, and periods of growth and development such as infancy and adolescence.[30] Thus, it is especially important that vegetarians consume sufficient amounts of plant-based foods rich in these micronutrients. You will learn more about their dietary sources as well as their functions in Chapters 10 to 13.

Special Dietary Recommendations for Vegetarians Because some types of vegetarian diets pose certain nutritional risks, it is important to follow special dietary strategies if you decide to make this dietary choice. The MyPlate food guidance system specifically recognizes protein, iron, calcium, zinc, and vitamin B_{12} as nutrients for vegetarians to focus on, making specific recommendations as to how to get adequate amounts. In addition, it is pointed out that some meat replacements, such as cheese, can be very high in calories, saturated fat, and cholesterol. Thus, lower-fat versions should be chosen, and they should be consumed in moderation. The following comments and suggestions are offered to help ensure optimal health in individuals who choose—for whatever reason—to become vegetarians. Note that these recommendations are especially pertinent to vegans.

- Select sources of protein that are naturally low in fat, such as skim milk and legumes.
- Minimize the amount of high-fat cheese used as meat replacements.
- If you do not consume dairy products, consider drinking calcium-fortified, soy-based beverages. These can provide calcium in amounts similar to milk and are usually low in fat and do not contain cholesterol.
- Add vegetarian meat substitutes (such as tofu) to soups and stews to boost protein without adding saturated fat and cholesterol.
- Recognize that most restaurants can accommodate vegetarian modifications to menu items by substituting meatless sauces, omitting meat from stir-fries, and adding vegetables or pastas in place of meat.
- Consider eating out at Asian or Indian restaurants, as they often offer a varied selection of high-protein vegetarian dishes.
- Be mindful of getting enough vitamin B_{12}; because this vitamin is naturally found only in foods that come from animals, fortified foods or dietary supplements may be necessary for vegans.

The key to a healthy vegetarian diet, like any diet, is to enjoy a wide assortment of foods and to consume them in moderation. Because no single food provides all the nutrients the body needs, eating a variety of foods can help ensure that vegetarians get the necessary nutrients and other substances that promote good health.

lacto-ovo-vegetarian A type of vegetarian who consumes dairy products and eggs in an otherwise plant-based diet.

lactovegetarian A type of vegetarian who consumes dairy products (but not eggs) in an otherwise plant-based diet.

vegan (VE – gan) A type of vegetarian who consumes no animal products.

What Are the Consequences of Protein Deficiency?

Although generally not a concern in industrialized countries (even in people consuming a vegetarian diet), protein deficiency is common in regions where the amount and variety of foods is limited. Children are especially likely to be affected because their protein requirements are higher (per unit body weight) than those of adults. Protein deficiency is also seen in adults with some debilitating conditions such as cancer. Because of the importance of proteins and amino acids for optimal health, protein deficiency can have significant health implications, some of which are described next.

PROTEIN DEFICIENCY IS MOST COMMON IN EARLY LIFE

Even in economically poor countries, protein deficiency is rare during the first months of life when infants are consuming human milk or infant formula. However, once weaned from these high-quality protein sources to foods that lack adequate protein, infants become at greater risk for protein deficiency. Because protein-deficient diets are almost always also lacking in energy, protein deficiency is often referred to as **protein-energy malnutrition (PEM).** Children with PEM are typically deficient in one or more micronutrients, as well. Thus, PEM is a condition of overall malnutrition and has many implications for child health. For example, children with PEM are at great risk for infection and illness. Recall that protein is needed to make several components of your immune system, as well as skin and membranes that keep pathogens from entering the body. The World Health Organization (WHO) estimates that PEM plays an important role in at least one-third of all child deaths each year, many of which are complicated by infection.[31]

Severe PEM actually encompasses a spectrum of malnutrition: at the extremes are two distinct types of severe PEM, and between them are conditions that combine features of both.[32] At one end of the spectrum is a condition called **marasmus,** which results from severe, chronic, overall malnutrition. In marasmus, fat and muscle tissue are depleted, and the skin hangs in loose folds, with the bones clearly visible beneath the skin. Children with marasmus tend at first to be alert and ravenously hungry, although with increasing severity they become apathetic and lose their appetites. Clinicians often say that marasmus represents the body's survival response to long-term, chronic dietary insufficiency.

The other extreme type of PEM, called **kwashiorkor,** is often distinguished from marasmus by the presence of severe edema. Note that edema sometimes is present in children with marasmus, but those with kwashiorkor usually have more extensive edema, which typically starts in the legs but often involves the entire body. When fluid accumulates in the abdominal cavity, it is referred to as **ascites.** Recall that one of the roles of protein in the body is regulation of fluid balance. Children with kwashiorkor sometimes have large, distended abdomens due to ascites. Because malnourished children often have intestinal parasites, the presence of worms sometimes contributes to this abdominal distension as well. Children with kwashiorkor often are apathetic and have cracked and peeling skin, enlarged, fatty livers, and sparse, unnaturally blond or red hair. Although many characteristics of kwashiorkor were once thought simply to be caused by protein deficiency, this does not appear to be the case.[33] Researchers now believe that many of the signs and symptoms of kwashiorkor are the result of micronutrient deficiencies, for example, vitamin A deficiency, in combination with infection or other environmental stressors.

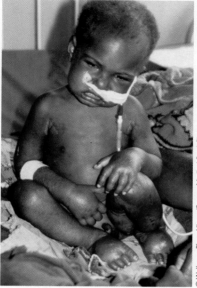

Children with kwashiorkor often have swollen abdomens (ascites), edema in their hands and feet, cracked and peeling skin, and an apathetic nature. These children are at increased risk for infection.

protein-energy malnutrition (PEM) Protein deficiency accompanied by inadequate intake of protein and often of other essential nutrients as well.

marasmus (ma – RAS – mus) (*marainein*, to waste away) A form of PEM characterized by extreme wasting of muscle and adipose tissue.

kwashiorkor (kwa – she – OR – kor) (Kwa, a language of Ghana; refers to "what happens to the first child when the next is born") A form of PEM often characterized by edema in the extremities (hands, feet).

ascites (a – SCI – tes) Abnormal accumulation of fluid in the abdominal cavity.

PROTEIN DEFICIENCY IN ADULTS

PEM can also occur in adults. Unlike children, however, adults with PEM rarely experience kwashiorkor. Instead, adult PEM generally takes the form of marasmus. There are many causes of PEM in adulthood, including inadequate dietary intake (such as occurs in alcoholics and those with eating disorders), protein malabsorption (such as occurs with some gastrointestinal disorders such as celiac disease), excessive and chronic blood loss, cancer, infection, and injury (especially burns).[34]

Adults with PEM can experience extreme muscle loss because the body's muscles are broken down to provide glucose and energy. In addition, fatty liver and edema are common. Adults with severe PEM experience decreased function of many vital physiological systems, including the cardiovascular system, renal system (kidneys), digestive system (gastrointestinal tract and accessory organs), and endocrine and immune systems. There are many causes of PEM in adults, and treatment is often long and difficult. For example, if the cause is infection, treatment may involve both dietary intervention and use of antibiotics. In contrast, if protein deficiency is a result of an eating disorder, psychological counseling becomes a key component of the health care plan. Regardless of its etiology, effective treatment of adult PEM presents a special challenge to any medical team.

Protein Excess: Is There Cause for Concern?

Protein deficiency can result in serious health concerns. But what about protein excess? Can this also be unhealthy or even life-threatening? Contrary to popular belief, high-protein diets do not cause adverse health outcomes such as osteoporosis, kidney problems, heart disease, obesity, and cancer in most people. This conclusion has been confirmed by the DRI committee, which carefully considered the peer-reviewed literature related to the potential health consequences of high-protein diets. This panel of experts concluded that, although epidemiologic studies offer limited evidence that high protein intake is associated with some adverse health outcomes, experimental data do not support this contention. In other words, the association between high protein consumption and poor health is likely not a causal relationship. In fact, the upper limit of the AMDR for protein intake (35% of energy from protein) was developed not because there was evidence that additional protein might pose a health risk, but solely to complement the recommendations for carbohydrate and fat intakes.

〈CONNECTIONS〉 Recall that causal relationships are typically determined from intervention studies, whereas epidemiologic studies are powerful in showing relationships or associations within a population (Chapter 1, page 14).

Nonetheless, high intakes of protein are often accompanied by high intakes of fat, saturated fat, and cholesterol. Because these dietary components are risk factors for heart disease, it is important to choose a variety of lean and low-fat protein foods, such as those recommended by the MyPlate food guidance system.

HIGH RED MEAT CONSUMPTION MAY BE RELATED TO INCREASED RISK FOR CANCER

Growing evidence suggests that high, chronic intake of red meat (beef, lamb, and pork) or processed meats (bacon, sausage, hot dogs, ham, and cold cuts) is associated with increased risk for colorectal cancer.[35] As a result, the World Cancer Research Fund and the American Institute for Cancer Research in 2007 recommended that we limit our intake of red meat to no more than 500 g (18 oz) each week and eat very little processed meat.[36] On average, this would be about 70 g (2.6 ounces) of meat each day—an amount less than that recommended by the 2010 Dietary Guidelines and MyPlate food guidance system. Importantly, the panel of experts who made this recommendation emphasized that they do not recommend avoiding all meat or foods of animal origin. Clearly, these foods can be a valuable source of many essential nutrients and should be considered part of a healthy diet. Like several other issues related to protein nutrition, this topic continues to be one of active debate. A more complete discussion of how nutrition—including high-protein foods—does or does not influence risk of cancer is provided in the Nutrition Matters section following Chapter 11.

Critical Thinking: Tyler's Story Consider again Tyler, the little boy who has to avoid legumes, which are generally considered excellent sources of dietary protein. Given the severity of peanut allergy for children like Tyler, do you think a food manufacturer that markets a product with peanuts without proper labeling should be held legally responsible? What about restaurants that serve a peanut-containing food without revealing this fact to the consumer? In other words, who is responsible?

Diet Analysis PLUS ✚ Activity

The current DRI recommendation for protein intake is 0.8 grams of protein per kilogram of body weight. However, Americans generally consume more protein than is recommended. Some college students may have misconceptions about protein intake and think they need to eat a diet higher in protein than is recommended. This activity provides students with an opportunity to critically appraise their protein intake and compare it to both the RDA and the AMDR.

First, they will calculate their personal RDA for protein based on their healthy body weight. Another important part of this activity demonstrates the importance of using a healthy weight range when determining an RDA for protein. Students are directed to determine the weight range for a healthy body mass index (BMI), between 18.5 and 24.9 kg/m². They will then determine the protein RDA based on that range. This may be particularly enlightening to underweight and overweight students. They will then compare their RDA for protein to the International Society of Sports Nutrition (ISSN) recommendations. They will also determine an adequate protein intake for their calorie intake using the AMDR recommendations. Then, they will compare their RDA and the ISSN recommendations to the AMDR range. Finally, they will determine where their average intake, as determined using Diet Analysis Plus (DA+), fits within that range.

1. Since calculations for protein recommendations are based on body weight, start by calculating your healthy weight range and BMI range. Healthy BMI range is 18.5–24.9 kg/m².
   - lowest healthy weight (lb) = 18.5 × your height in inches²/703

   - highest healthy weight (lb) = 24.9 × your height in inches²/703
   - Healthy weight range: _____ (low) to _____ (high) lb.

2. Determine a range of protein intake based on your healthy weight range.
   - Lowest weight (lb)/2.2 × 0.8 g = lowest recommended protein in grams for your height.
   - Highest weight (lb)/2.2 × 0.8 g = highest recommended protein in grams for your height.
   - Protein intake range: _____ g to _____ g.

3. From your Profile DRI Goals report, indicate below how many grams of protein are recommended for you. Use the value that represents the daily requirement based on grams per kilogram of body weight (g/kg/day): _____ g. This will fall in the range that you calculated above.

4. Using your current weight, and to reconfirm the DA+ program's calculation, solve the following equation by filling in the blanks.
   a. Convert your body weight in pounds to kilograms.
      _____ lb ÷ 2.2 = _____ kg.
   b. Using kilograms of body weight, calculate your daily protein requirement in grams using your RDA (g/kg/day).
      _____ kg body weight × _____ (g/kg/day) based on sex/age and life stage = _____ g/day.

5. Another way to determine an acceptable protein intake is using the AMDR (Acceptable Macronutrient Distribution Range). To do this, find your recommended calorie

intake in your DA+ program. Then multiply it by the % total energy intake recommended in the AMDRs (10–35% of total energy should come from protein).

- Recommended energy intake × 0.10 = kilocalories from protein. Kilocalories from protein/4 = grams of protein.
- Recommended energy intake × 0.35 = kilocalories from protein. Kilocalories from protein/4 = grams of protein.
- Example: 2,000 kilocalories × 0.10 = 200 kilocalories 200 kcal/4 = 50 grams of protein. 2,000 kilocalories × 0.35 = 700 kilocalories. 700 kcal/4 = 175 grams of protein.
- AMDR for protein for 2,000 kilocalories is 50 to 175 grams.

6. The International Society of Sports Nutrition (ISSN) has concluded that "protein intake of 1.4 to 2.0 g/kg/day for physically active individuals is safe and could improve the training adaptation to exercise." Assuming that you are physically active, calculate your recommended protein intake using this range of protein intakes.

- Your body weight (lb) weight/2.2 = _____ kg body weight × 1.4 g/kg/day = _____ g protein/day.
- Your body (lb) weight/2.2 = _____ kg body weight × 2.0 g/kg/day = _____ g protein/day.

Recommended range of intake for physically active individual = _____ to _____

7. How does that intake range compare with the RDA and the AMDR calculation? _____

8. How does your protein intake as determined in your DA+ activity compare with these protein intake ranges? _____

9. Using the DA+ program, find foods that contribute protein in your diet.

Name of Food	Standard Serving Size	Grams of Protein per Serving

10. Based on your findings, use DA+ to plan a diet that meets the recommended protein intake range of your choosing. Use only fruits, vegetables, dairy products, grains and meat, fish, eggs, legumes, nuts, and seeds in your meal plan. Do not include special drinks and supplements.

11. Use DA+ and protein complementation principles to plan one day's vegetarian meals that meet the recommended protein intake.

Fill in this chart describing the digestion, absorption, and circulation of dietary proteins (blue areas only).

Organs	Secretions: Enzymes, Hormones, and Other	Digestion of Protein	Absorption of Amino Acids	Circulation of Amino Acids
Mouth	Saliva	Chewing and mixing with saliva		
Stomach	HCl, pepsinogen to pepsin	Proteins are denatured by HCl, and then peptide bonds are hydrolyzed. Polypeptide chains are broken down into smaller peptide units.		
Small intestine			Transport into the small intestine cells requires both passive and active transport mechanisms.	Amino acids enter the capillaries where they travel to the liver.
Accessory organs: Pancreas, liver, and gallbladder	Pancreatic enzymes specific for protein digestion: trypsin, chymotrypsin elastase, and carboxypeptidase. Also bicarbonate, to neutralize chyme.			

Notes

1. Carpenter KJ. Short history of nutritional science: Part 1 (1785–1885). Journal of Nutrition. 2003;133:638–45.

2. Institute of Medicine. Dietary reference intakes for energy, carbohydrate, fiber, fat, fatty acids, cholesterol, protein, and amino acids. Washington, DC: National Academies Press; 2005.

3. Furst P, Stehle P. What are the essential elements needed for the determination of amino acid requirements in humans? Journal of Nutrition. 2004;134:1558S–65S.

4. Erlandsen H, Patch MG, Gamez A, Straub M, Stevens RC. Structural studies on phenylalanine hydroxylase and implications toward understanding and treating phenylketonuria. Pediatrics. 2003;112:1557–65.

5. American Academy of Pediatrics. Pediatric nutrition handbook, 6th ed. Kleinman RE, ed. Elk Grove Village, IL: American Academy of Pediatrics; 2008.

6. Reeds PJ, Garlick PJ. Protein and amino acid requirements and the composition of complementary foods. Journal of Nutrition. 2003;133:2953S–61S.

7. Schaafsma, G. The protein digestibility-corrected amino acid score. Journal of Nutrition. 2000;130:1865S–7S.

8. Santerre CR. Food biotechnology, 9th ed. Bowman BA, Russell RM, editors. Washington, DC: ILSI Press; 2006.

9. Huang S, Adams WR, Zhou Q, Malloy KP, Voyles DA, Anthony J, Kriz AL, Luethy MH. Improving nutritional quality of maize proteins by expressing sense and antisense zein genes. Journal of Agricultural and Food Chemistry. 2004;52:1958–64. Zarkadas CG, Hamilton RI, Yu ZR, Choi VK, Khanizadeh S, Rose NGW, Pattison PL. Assessment of the protein quality of 15 new northern adapted cultivars of quality protein maize using amino acid analysis. Journal of Agricultural and Food Chemistry. 2000;48:5351–61.

10. Schnog JB, Duits AJ, Muskeit FAJ, ten Cate H, Rojer RA, Brandjes DPM. Sickle cell disease: A general overview. Journal of Medicine. 2004;62:364–74.

11 National Institutes of Health. National Heart, Lung, and Blood Institute. NIH Publication No. 96-4057. 1996.

12. Bank A. On the road to gene therapy for beta-thalassemia and sickle cell anemia. Pediatric Hematology and Oncology. 2008;25:1–4. Lebensburger J, Persons DA. Progress toward safe and effective gene therapy for beta-thalassemia and sickle cell disease. Current Opinion in Drug Discovery and Development. 2008;11:225–32.

13 Taber KH, Hurley RA. Mercury exposure: effects across the lifespan. Journal of Neuropsychiatry and Clinical Neuroscience. 2008;20:iv–389.

14. U.S. Department of Health and Human Services and U.S. Environmental Protection Agency. What you need to know about mercury in fish and shellfish. 2004 EPA and FDA advice for women who might become pregnant, women who are pregnant, nursing mothers, young children. Available from: http://www.cfsan.fda.gov/~dms/admehg3.html.

15 Ballestar E. An introduction to epigenetics. Advances in Experimental Medicine and Biology. 2011;711:1–11.

16 Martin-Subero JI, Esteller M. Profiling epigenetic alterations in disease. Advances in Experimental Medicine and Biology. 2011;711:162–77. Wierda RJ, Geutskens SB, Jukema JW, Quax PH, van den Elsen PJ. Epigenetics in atherosclerosis and inflammation. Journal of Cellular and Molecular Medicine. 2010;15(6A):1225–50.

17. Burdge GC, Hanson MA, Slater-Jefferies JL, Lillycrop KA. Epigenetic regulation of transcription: A mechanism for inducing variations in phenotype (fetal programming) by differences in nutrition during early life? British Journal of Nutrition. 2007;97(6):1036–46. Mathers JC. Early nutrition: Impact on epigenetics. Forum in Nutrition. 2007;60:42–8.

18. DeBusk R. The role of nutritional genomics in developing an optimal diet for humans. Nutrition in Clinical Practice. 2010;25:627–33. Stover PJ. Nutritional genomics. Physiological Genomics. 2004;16:161–5.

19. Taylor SL, Hefle SL. Food allergy. In: Present knowledge in nutrition, 9th ed. Bowman BA, Russell RM, editors. Washington, DC: ILSI Press; 2006.

20. National Institute of Allergy and Infectious Diseases. Food allergy. Report of the NIH expert panel on food allergy research. 2006. Available at: http://www.niaid.nih.gov/topics/foodallergy/research/pages/reportfoodallergy.aspx.

21. Fuller MF, Reeds PJ. Nitrogen cycling in the gut. Annual Review of Nutrition. 1998;18:385–411. Rand WM, Pellet PL, Young VR. Meta-analysis of nitrogen balance studies for estimating protein requirements in healthy adults. American Journal of Clinical Nutrition. 2003;77:109–27.

22. Dewey KG. Energy and protein requirements during lactation. Annual Review of Nutrition. 1997;17:19–36.

23. Dickerson RN. Estimating energy and protein requirements of thermally injured patients: Art or science? Nutrition. 2002;18:439–42. Gudaviciene D, Rimdeika R, Adamonis K. Nutrition of burned patients. Medicina. 2004;40:1–8. Wilmore DW.

24. Campbell B, Kreider RB, Zeigenfuss T, La Bounty P, Roberts M, Burke D, Landis J, Lopez H, Antonio J. International Society of Sports Nutrition position stand: Protein and exercise. Journal of the International Society of Sports Nutrition. 2007;4:8.

25. American Dietetic Association; Dietitians of Canada; American College of Sports Medicine, Rodriguez NR, DiMarco NM, Langley S. American College of Sports Medicine position stand. Nutrition and athletic performance. Medicine and Science in Sports Exercise. 2009;41:709–31.

26. Mero A. Leucine supplementation and intensive training. Sports Medicine. 1999;27(6):347–58.

27. Phillips SM. Protein requirements and supplementation in strength sports. Nutrition. 2004;20:689–95. Wilson J, Wilson GJ. Contemporary issues in protein requirements and consumption for resistance trained athletes. Journal of the International Society for Sports Nutrition. 2006;5:7–27.

28. Bedford JL, Barr SI. Diets and selected lifestyle practices of self-defined adult vegetarians from a population-based sample suggest they are more 'health conscious.' International Journal of Behavior, Nutrition, and Physical Activity. 2005;2:4. Haddad EH, Tanzman JS. What do vegetarians in the United States eat? American Journal of Clinical Nutrition. 2003;78:626S–32S.

29. Antony AC. Vegetarianism and vitamin B-12 (cobalamin) deficiency. American Journal of Clinical Nutrition. 2003;78:3–6. Hunt JR. Bioavailability of iron, zinc, and other trace minerals from vegetarian diets. American Journal of Clinical Nutrition. 2003;78:633S–9S.

30. Mangels AR, Messina V. Considerations in planning vegan diets: Infants. Journal of the American Dietetic Association. 2001;101:670–7. Messina V, Mangels AR. Considerations in planning vegan diets: Children. Journal of the American Dietetic Association. 2001;101:661–9.

31. United Nations Children's Fund. Tracking progress on child and maternal nutrition. 2009. UNICEF, New York, NY.

32. Jeejeebhoy KN. Protein nutrition in clinical practice. British Medical Bulletin. 1981;37:11–17. Waterlow JC. Classification and definition of protein-calorie malnutrition. British Medical Journal. 1972;3:566–9.

33. Golden M. The development of concepts of malnutrition. Journal of Nutrition. 2002;132:2117S–22S.

34. Hansen RD, Raja C, Allen BJ. Total body protein in chronic diseases and in aging. Annals of the New York Academy of Sciences. 2000;904:345–52.

35. Chao A, Thun MJ, Connell CJ, McCullough ML, Jacobs EJ, Flanders D, Rodriguez C, Sinha R, Calle EE. Meat consumption and risk of colorectal cancer. JAMA (Journal of the American Medical Association). 2005;293:172–82.

36. World Cancer Research Fund/American Institute for Cancer Research. Food, nutrition, physical activity, and the prevention of cancer: A global perspective. Washington, DC: American Institute for Cancer Research; 2007.

Food Safety

© AP Photo/L. G. Patterson

Most of us are fortunate to have an abundant supply of healthful food. Yet there are times when the food we eat causes serious illness. Clearly, it is important to avoid eating food that is spoiled, unclean, or stored improperly. However, food can appear, smell, and taste safe to eat but still harbor dangerous disease-causing agents.

Although every effort is made to ensure that our food is safe, food safety remains an important public health concern. The U.S. Centers for Disease Control and Prevention (CDC) estimate that each year 1 out of 6 Americans (48 million people) gets sick, 128,000 are hospitalized, and 3,000 die from foodborne diseases.[1] Understanding foodborne illness requires a basic knowledge of how foods come to contain disease-causing agents and how the body reacts to them. In this Nutrition Matters, you will gain an understanding of food risks and how to avoid them so that you can prevent foodborne illness for years to come.

What Causes Foodborne Illness?

We are exposed to thousands of microscopic organisms (microbes) each day. These microbes populate the world we live in, frequently serving useful purposes. For example, helpful microbes reside in your GI tract; these assist in food digestion and prevention

of some diseases. But other microbes are pathogenic (disease-causing) and make us sick. Consuming pathogenic microbes in foods and beverages is the main cause of foodborne illness.[2] A **foodborne illness** is a disease caused by ingesting unsafe food and is sometimes referred to as "food poisoning." There are many forms and causes of foodborne illness, and you will learn about them next.

FOODBORNE ILLNESSES ARE CAUSED BY INFECTIOUS AND NONINFECTIOUS AGENTS

You can get a foodborne illness by ingesting either "infectious" or "noninfectious" substances. **Infectious agents,** or pathogens, include living microorganisms such as bacteria, viruses, molds, fungi, and parasites. Some people also consider prions, which are not living but are the cause of mad cow disease, infectious agents of foodborne illness. **Noninfectious agents** that cause foodborne illness include nonbacterial toxins; chemical residues from processing, pesticides, and antibiotics; and physical hazards such as glass and plastic.

DIFFERENT STRAINS OF A MICROORGANISM ARE CALLED SEROTYPES

Although most foodborne illnesses are caused by microbes, not all microbes are harmful. In fact, even related pathogenic microbes can cause very different signs and symptoms. To really understand how to prevent foodborne illness, you must first have a basic knowledge of microbiology and understand how various organisms can make you sick.

Each general group of closely related microorganisms can have several genetic strains or types called **serotypes.** Some serotypes are harmless,

---

foodborne illness A disease caused by ingesting unsafe food.

infectious agent of foodborne illness A pathogen in food that causes illness and can be passed or transmitted from one infected animal or person to another.

noninfectious agent of foodborne illness An inert (nonliving) substance in food that causes illness.

serotype A specific strain of a larger class of organism.

such as the ones living in our GI tracts, whereas others are pathogenic (disease-causing). For example, some of the various serotypes of the bacterium *Escherichia coli* (*E. coli*) live without any risk to us in our GI tract, while others (such as *E. coli* O157:H7) cause foodborne illness.[3] Furthermore, different serotypes of a single type of pathogenic bacterium can cause illness in different ways. For instance, some serotypes of pathogenic *E. coli* cause mild intestinal discomfort within one to three days, while others result in more severe symptoms, sometimes taking up to eight days to develop. The time elapsed between when a person eats an infected food and when he or she becomes sick is called the **incubation period.** Different pathogens have different incubation periods, and the length of the incubation period is frequently used by health care providers to help determine which pathogen is likely involved. Some of the most common infectious agents (including their incubation periods) are summarized in Table 1, and the ways in which they cause illness are described next.

incubation period The time between when infection occurs and signs or symptoms begin.

TABLE 1 Infectious Agents of Foodborne Illness, Food Sources, and Symptoms of Infection

Organism	Incubation Period	Duration of Illness	Commonly Associated Foods	Signs and Symptoms
Bacteria				
Campylobacter jejuni	2–5 days	2–10 days	Raw or undercooked poultry, untreated water, unpasteurized milk	Diarrhea (often bloody), abdominal cramping, nausea, vomiting, fever, fatigue
Clostridium botulinum	12–72 hours	From days to months	Home-canned foods with low acid content, improperly canned commercial foods, herb-infused oils	Vomiting, diarrhea, blurred vision, drooping eyelids, slurred speech, dry mouth, difficulty swallowing, weak muscles
Clostridium perfringes	8–16 hours	24–48 hours	Raw or undercooked meats, gravy, dried foods	Abdominal pain, watery diarrhea, vomiting, nausea
Escherichia coli O157:H7	1–8 days	5–10 days	Raw or undercooked meat, raw fruits and vegetables, unpasteurized milk and juice, contaminated water	Nausea, abdominal cramps, severe diarrhea (often bloody)
Escherichia coli (enterotoxigenic)	1–3 days	Variable	Water or food contaminated with human feces	Watery diarrhea, abdominal cramps, vomiting
Listeria monocytogenes	9–48 hours for GI symptoms, 2–6 weeks for invasive disease	Variable	Raw or inadequately pasteurized dairy products; ready-to-eat luncheon meats and frankfurters	Fever, muscle aches, nausea, diarrhea, premature delivery, miscarriage, or stillbirths
Salmonella	1–3 days	4–7 days	Raw poultry, eggs, and beef; fruit and alfalfa sprouts; unpasteurized milk	Diarrhea, fever, abdominal cramps, severe headaches
Shigella	24–48 hours	4–7 days	Raw or undercooked foods or water contaminated with human fecal material	Fever, fatigue, watery or bloody diarrhea, abdominal pain
Staphylococcus aureus	1–6 hours	24–48 hours	Improperly refrigerated meats, potato and egg salads, cream pastries	Severe nausea and vomiting, diarrhea, abdominal cramps
Vibrio cholerae	1–7 days	2–8 days	Contaminated water; undercooked foods; shellfish	Watery diarrhea, vomiting

TABLE 1 (continued)

Organism	Incubation Period	Duration of Illness	Commonly Associated Foods	Signs and Symptoms
Viruses				
Hepatitis A virus	15–50 days	2 weeks–3 months	Mollusks (oysters, clams, mussels, scallops, and cockles)	Jaundice, fatigue, abdominal pain, loss of appetite, nausea, diarrhea, fever
Norovirus	12–48 hours	12–60 hours	Raw or undercooked shellfish, contaminated water	Nausea, vomiting, diarrhea, abdominal pain, headache, fever
Parasites				
Trichinella (worm)	1–2 days for initial symptoms; others begin 2–8 weeks after infection	Months	Raw or undercooked pork or meats of carnivorous animals	Acute nausea, diarrhea, vomiting, fatigue, fever, and abdominal pain
Giardia intestinalis (protozoan)	1–2 weeks	Days to weeks	Contaminated water, any uncooked food	Diarrhea, flatulence, stomach cramps
Molds				
Aspergillus flavus	Days to weeks	Weeks to months	Wheat, flour, peanuts, soybeans	Liver damage

Adapted from Centers for Disease Control and Prevention. Diagnosis and management of foodborne illnesses: A primer for physicians and other health care professionals. Morbidity and Mortality Weekly Reports. 2004; 53:1–33. Available at: http://www.cdc.gov/mmwr/PDF/rr/rr5304.pdf. Murano PS. Understanding food science and technology. Thomson/Wadsworth, 2003.

SOME ORGANISMS MAKE TOXINS BEFORE WE EAT THEM

Some pathogenic organisms produce toxic substances while they are growing in foods. These toxins are called **preformed toxins,** because they have already contaminated the foods before we eat them. When we consume these foods, the toxins cause serious and rapid (one to six hours) reactions such as nausea, vomiting, diarrhea, and sometimes neurological damage.

***Staphylococcus aureus* (*S. aureus*)** One bacterium that produces preformed toxins is *Staphylococcus aureus* (*S. aureus*), which causes nearly 250,000 foodborne illnesses annually in the United States. Foods commonly infected with *S. aureus* include raw or undercooked meat and poultry, cream-filled pastries, and unpasteurized dairy products. Symptoms of *S. aureus* infection include sudden onset of severe nausea and vomiting, diarrhea, and abdominal cramps, typically occurring within one to six hours of consuming the contaminated food. Note that, because the toxin produced by *S. aureus* is quite heat-stable, it is not easily destroyed by cooking.

Methicillin-Resistant *S. aureus* (MRSA)—An Emerging Concern You may have heard about an antibiotic-resistant strain of *S. aureus* called **methicillin-resistant *Staphylococcus aureus* (MRSA),** which has received considerable attention recently by public health officials. MRSA was identified more than 40 years ago and was thought to be spread only by direct contact with an infected person—usually in a hospital setting. However, it is now recognized that MRSA can also be "community acquired," meaning the infection is not linked to contact with a health care facility.[4] Most cases of community-acquired MRSA are attributed to sharing towels and equipment in athletic and school facilities. In addition, it now appears that community-acquired infection can also occur via consumption of MRSA-infected foods.

To date, there has been only one documented outbreak of foodborne illness caused by MRSA in the United States. In this case, it is thought that an infected food preparer transmitted the bacterium to coleslaw. Three family members later consumed the coleslaw and became ill within three to four hours, having severe nausea, vomiting, and stomach cramps.[5] Because MRSA cannot be treated

preformed toxin Poisonous substance produced by microbes while they are in a food (prior to ingestion).

methicillin-resistant *Staphylococcus aureus* (MRSA) A type of *S. aureus* that is resistant to most antibiotics.

effectively by antibiotics, public health officials are carefully watching this bacterium, especially as it relates to foodborne illness. To help prevent foodborne MRSA infections, proper hand washing is essential.

Clostridium botulinum (C. botulinum) Another bacterium that produces preformed toxins, *Clostridium botulinum (C. botulinum)*, is found mainly in inadequately processed, low-acid, home-canned foods such as green beans. You can also become infected with this microbe by eating improperly canned commercial foods. Unlike *S. aureus*, however, cooking contaminated foods at sufficiently high temperatures can destroy the toxin produced by *C. botulinum*.

Infection with *C. botulinum* causes a disease called **botulism.** Mild cases result in vomiting and diarrhea, whereas symptoms of severe botulism include double vision, blurred vision, drooping eyelids, slurred speech, difficulty swallowing, dry mouth, and muscle weakness. In severe cases, botulism can cause paralysis, respiratory failure, and death.

In 2007, the CDC reported an outbreak of eight cases of *C. botulinum* infection in Indiana, Texas, and Ohio.[6] All infected persons had consumed a particular brand of hot dog chili sauce that was quickly recalled by the manufacturer. Although all infected persons recovered, they endured several days of painful GI symptoms. Because the botulism toxin is destroyed by high temperatures, some experts recommend that home-canned foods be boiled for 10 minutes before they are consumed.

Aspergillus While most molds that grow on foods such as cheese and bread are not dangerous to eat, others can produce dangerous preformed toxins. An example is **aflatoxin,** which is a toxin produced by the *Aspergillus* mold found on some agricultural crops such as peanuts. If the crop is not dried properly before storage, *Aspergillus* can continue to grow and produce toxic levels of aflatoxin. Consuming food contaminated with aflatoxin is of great concern because it can cause liver damage, cancer, and is often fatal.[7] Although many agricultural practices (such as sufficient drying) in the United States are used to help minimize contamination of our food with aflatoxin, it remains a significant public health issue in many other regions of the world.

SOME ORGANISMS MAKE ENTERIC (INTESTINAL) TOXINS AFTER WE EAT THEM

In contrast to organisms that produce preformed toxins in a food *before* it is consumed, others produce harmful toxins *after* they enter the GI tract. These toxins are called **enteric** or **intestinal toxins.** Enteric toxins draw water into the intestinal lumen, resulting in diarrhea. Although the symptoms are variable, the incubation period is generally one to five days, substantially longer than that for most preformed toxins.

Noroviruses **Noroviruses*** are examples of pathogens (viruses) that produce enteric toxins. Symptoms of norovirus infection usually include nausea, vomiting, diarrhea, and stomach cramping. Sometimes people also develop low-grade fevers, chills, headaches, muscle aches, and a general sense of tiredness. Symptoms usually begin one to two days after ingestion of the contaminated food.

An example of a norovirus outbreak occurred in 2008 on three college campuses in California, Michigan, and Wisconsin.[8] This outbreak resulted in approximately 1,000 cases of reported illness,

* Previously called Norwalk and Norwalk-like Virus.

Improper canning of low-acid foods can result in the finished product containing live *Clostridium botulinum*, which can cause severe illness.

iStockphoto.com/fcutrara

---

botulism The foodborne illness caused by *Clostridium botulinum*.

aflatoxin (a – fla – TOX – in) A toxic compound produced by certain molds, such as *Aspergillus*, that grow on peanuts, some grains, and soybeans.

enteric (intestinal) toxin A toxic agent produced by an organism after it enters the gastrointestinal tract.

norovirus A type of infectious pathogen (virus) that often causes foodborne illness.

including 10 hospitalizations, and prompted closure of one of the three campuses. Although it was never determined what caused these related outbreaks and if they were due to a common food, college campuses are at particularly high risk for norovirus outbreaks because of the extensive opportunities for transmission created by shared living and dining areas. Many evacuees from Hurricane Katrina also developed norovirus infections, and drinking contaminated water was the likely source.[9]

Unlike bacteria, viruses—such as the norovirus—cannot be treated with antibiotics. The only way to avoid norovirus infection is to follow the food safety guidelines—such as frequent hand washing and decontamination of food preparation utensils—outlined later in this section.

Some Serotypes of *E. coli* Although most forms of *E. coli* are harmless or even beneficial, some (called **enterotoxigenic** *E. coli*) produce enteric toxins, and consumption of food or water contaminated with these bacteria can cause severe GI upset. Symptoms typically occur within 6 to 48 hours after food consumption and include diarrhea, abdominal cramps, and nausea. Foods and beverages typically contaminated with these forms of *E. coli* include uncooked vegetables, fruits, raw or undercooked meats and seafood, unpasteurized dairy products, and untreated tap water. This type of *E. coli* is the primary cause of what is often referred to as "traveler's diarrhea," which is a clinical syndrome resulting from consuming microbially contaminated food or water while traveling. If you have experienced severe diarrhea and abdominal cramps during or shortly after traveling, you might have had this type of foodborne illness.

SOME ORGANISMS INVADE INTESTINAL CELLS

Some pathogens invade the cells of the intestine, seriously irritating the mucosal lining and causing severe abdominal discomfort and bloody diarrhea (dysentery); fever is also common. These types of pathogens are called **enterohemorrhagic** and include *Salmonella* and two especially dangerous serotypes of *E. coli* called *E. coli* O157:H7 and *E. coli* O104:H4. Incubation periods for enterohemorrhagic pathogens are generally several days.

Salmonella *Salmonella* is one of the most common causes of foodborne illness in the United States,

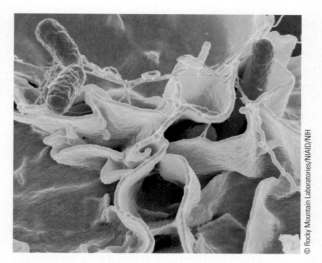

Salmonella is one of the most common causes of foodborne illness.

being typically found in raw poultry, eggs, beef, improperly washed fruit, alfalfa sprouts, and unpasteurized milk. The incubation period for this organism is one to three days, and symptoms include severe GI upset and headaches.

In 2007, *Salmonella*-infected frozen potpies caused at least 272 cases of foodborne illness in 35 states.[10] Unfortunately, the original source of the contamination was never determined. In 2010, over 1,900 *Salmonella*-related illnesses were reported in 11 states in the United States. This outbreak occurred in response to consumption of contaminated eggs, and resulted in a massive nationwide recall of eggs produced by two distributors located in Iowa.[11] Numerous *Salmonella* infections have also been reported from consuming raw alfalfa and mung bean sprouts. In fact, because they are particularly prone to a variety of bacterial infections, the FDA recommends raw sprouts not be consumed—especially by those with compromised immune systems such as children, the elderly, and persons with autoimmune conditions.[12]

***E. coli* O157:H7 and *E. coli* O104:H4** As previously mentioned, *E. coli* O157:H7 is an enterohemorrhagic serotype of *E. coli*. Since it was first identified as a human pathogen in 1982, many outbreaks of infection have been reported. Unpasteurized milk, apple juice, and apple cider can harbor this

---

enterotoxigenic Producing a toxin while in the GI tract.

enterohemorrhagic Causing bloody diarrhea.

pathogenic organism, as can improperly prepared meat (including poultry). Infection with this serotype of *E. coli* has an incubation period of one to eight days and results in nausea, abdominal cramps, and severe diarrhea that is often bloody.

In 2006, 183 persons were infected with *E. coli* O157:H7 in 26 states.[13] This outbreak rapidly gained national attention, as affected individuals had to be hospitalized, and at least three people eventually died. Fresh spinach was identified as the source of contamination, and the FDA quickly advised consumers not to eat bagged fresh spinach or fresh spinach–containing products unless they were cooked at 160°F for 15 seconds. Partly because of this outbreak, the FDA has issued new guidance to the food industry to minimize microbial contamination of fresh-cut fruits and vegetables.[14] In 2011, an outbreak of a different pathogenic serotype of *E. coli* (O104:H4) resulted in hundreds—if not thousands—of people becoming ill and many deaths throughout the European Union. Several cases (most of whom had recently traveled to Germany) were also reported in the United States. Authorities identified raw sprouts grown in Germany as the likely source of contamination.[15]

PROTOZOA AND WORMS ARE TYPES OF PARASITES

Parasites are complex one-celled or multicellular organisms that rely on other organisms to survive and are typically dangerous to the host. Not surprisingly, consumption of parasite-infested foods can cause foodborne illness. Because parasites often take up long-term residence in the body, their incubation periods are generally quite long. Several types of parasites can cause foodborne illness, the most common being protozoa and worms.

Protozoa Among the most ordinary of parasites are the **protozoa,** which are one-celled organisms that can live as parasites in the intestinal tracts of animals and humans. As part of their reproductive cycle, protozoa form **cysts** that are excreted in the feces. If cyst-containing feces come in contact with plants or animals, these food products can be contaminated as well. A foodborne illness can develop if foods or beverages contaminated with protozoan cysts are consumed.

One such parasitic protozoan, *Giardia intestinalis*, causes diarrhea, abdominal discomfort, and cramping; symptoms typically begin one to two weeks after infection. *Giardia intestinalis*

Drinking untreated water is not advised because of the possibility of it being contaminated with *Giardia intestinalis*.

can be found in chemically untreated swimming pools and hot tubs, and in rivers, ponds, and streams that have been contaminated with feces of an infected animal or person. You may already be aware that this is why chlorine (which kills this organism) is typically added to water in public swimming and bathing areas. This is also why boiling, chemically treating, or filtering water from ponds, streams, and lakes is recommended before drinking.

Worms Consuming foods that contain worms can also cause foodborne illness. Like protozoa, worms form cysts as part of their life cycles. Once ingested, cysts mature into worms that can cross the intestinal lining, travel through the blood, and eventually settle in various locations in the body, including muscles, eyes, and brain. An example of a worm is *Trichinella*, which is a roundworm that can invade a variety of animals, including pigs and some fish. Eating undercooked *Trichinella*-contaminated pork and seafood can result in the worm entering the body, causing muscle pain,

---

parasite An organism that, during part of its life cycle, must live within or on another organism without benefiting its host.

protozoa Very small (single-cell) organisms that are sometimes parasites.

cyst A stage of the life cycle of some parasites.

swollen eyelids, and fever. Although *Trichinella* infection is rare in the United States, it is still seen in areas of the world where wild game is consumed more frequently.

Anisakis simplex is another type of roundworm that causes foodborne illness. Within hours after ingestion of an *Anisakis simplex*–contaminated food, violent abdominal pain, nausea, and vomiting may occur. Some studies show that raw fish (sushi and sashimi) can contain this type of roundworm, as well as other parasites.[16] Consequently, the FDA advises that you always cook seafood thoroughly.[17] Because freezing kills many parasites, choosing to consume only sushi that has been made with previously frozen fish may be prudent.

PRIONS ARE INERT, NONLIVING PROTEINS THAT *MAY BE* INFECTIOUS

Although **prions** are not living organisms—in fact, they are not organisms at all—they may pose food safety concerns. Prions are altered proteins created when the secondary structure of a normal protein is disrupted, transforming α-helix coils into β-folded sheets. The resulting deformed protein is called a prion. Prions can cause other normal proteins to unravel, setting off a cascade of similar reactions converting hundreds of normal proteins into abnormal prions. When high levels of prions build up in a cell, the cell ruptures and releases the prions into the surrounding area, where they destroy other cells. Eventually, this kills the surrounding cells and gives the infected tissue a spongy appearance. Unfortunately, prions are extremely resilient and retain their ability to infect other cells even after exposure to extreme heat or acids. Therefore, cooking foods does not destroy prions. Furthermore, prions are not destroyed by the acidic conditions of the stomach, and if a prion-containing food is consumed, the prions can be absorbed into the bloodstream and begin the process of transforming normal proteins into prions in their new host. This is why some experts consider some prion-related diseases to be infectious.

Mad Cow Disease Several diseases in farm animals are known to be caused by prions, including scrapie in sheep and **bovine spongiform encephalopathy (BSE),** or **mad cow disease,** in bovine species.[18] Mad cow disease is characterized by loss of motor control, confusion, paralysis, wasting, and eventually death.

Creutzfelt-Jakob Disease and Variant Creutzfeldt-Jakob Disease A deadly human disease called **Creutzfeldt-Jakob disease,** caused by prions, is very rare and occurs in only one in a million people each year.[19] This disease has usually been attributed to direct infection by prions from contaminated medical equipment—for example, during surgery. However, during a BSE outbreak in the United Kingdom that occurred in the 1980s, researchers discovered a new form of Creutzfeldt-Jakob disease, which they called **variant Creutzfeldt-Jakob disease.** This disease has a relatively long incubation period (years) and is also fatal.

Considerable evidence links variant Creutzfeldt-Jakob disease to consumption of BSE-contaminated products.[20] To date, nearly 200 human cases have been reported worldwide, three of which were in the United States. No treatment exists for either the "classic" form of Creutzfeldt-Jakob disease or the variant form, and nothing can slow the progression of either disease.

Prions and Public Policy Because prions are found mainly in nerves and brains, the World Heath Organization (WHO) recommends that all governments prohibit the feeding of highly innervated tissue (such as the brain and spinal cord) from slaughtered cattle to other animals. Similarly, both Canada and the United States have banned the use of such products in human food, including dietary supplements, and in cosmetics. In 2006, Canada also banned inclusion of cattle tissues capable of transmitting BSE in all animal feeds, pet foods, and fertilizers. Partly in response to the concern about BSE, the U.S. Department of Agriculture (USDA) does not allow the slaughter of non-ambulatory ("downer") cattle for food purposes.[21] These measures, along with those established by the USDA, aim to provide a uniform national BSE policy with the goal of ensuring the safety of human food.

---

prion A misshapen protein that causes other proteins to also become distorted, damaging nervous tissue.

bovine spongiform encephalopathy (BSE; mad cow disease) A fatal disease in cattle caused by ingesting prions.

Creutzfeldt-Jakob disease A fatal disease in humans caused by a genetic mutation or surgical contamination with prions.

variant Creutzfeldt-Jakob disease A form of Creutzfeldt-Jakob disease that may be caused by consuming BSE-contaminated foods.

"Red tides" occur when marine algae produce brightly colored pigments; algae can also produce toxins that cause shellfish poisoning.

© Bill Bachman/Photo Researchers, Inc.

An example of shellfish poisoning is a phenomenon referred to as **red tide** that occurs when a particular marine alga begins to grow quickly and produce brightly colored pigments. These colorful "blooms" make the surrounding water appear red or brown. During this period of rapid growth, the algae produce potent toxins called **brevetoxins** which, when consumed by humans, causes classic shellfish poisoning. Every year, approximately 30 cases of shellfish poisoning are reported in the United States, typically in the coastal regions of the Atlantic Northeast and Pacific Northwest.[22]

How Can Noninfectious Substances Cause Foodborne Illness?

Consuming foods containing infectious pathogens poses the greatest risk of foodborne illness. However, noninfectious agents in food can also make you sick. These inert (nonliving) compounds include physical contaminants, such as glass and plastic, and other substances such as toxins, heavy metals and pesticides.

ALGAE TOXINS CAN MAKE SOME FISH AND SHELLFISH POISONOUS

One type of noninfectious foodborne illness, collectively called **shellfish poisoning,** can result from eating certain types of contaminated fish and shellfish (for example, clams and oysters). This is because these marine animals consume large amounts of algae that sometimes produce poisonous compounds called **marine toxins.** Consuming marine toxin–contaminated shellfish or fish can cause neurologic symptoms, including tingling, burning, numbness, drowsiness, and difficulty breathing. People with shellfish poisoning may also experience a strange phenomenon called hot–cold inversion, in which cold is perceived as hot and vice versa.

SOME PESTICIDES, HERBICIDES, ANTIBIOTICS, AND HORMONES ARE DANGEROUS

In addition to marine toxins, other noninfectious agents of foodborne illness can make their way into the food chain, including pesticides, herbicides, antibiotics, and hormones. Because exposure to some of these substances can make you ill, many federal and international agencies work together to ensure that their presence in foods is negligible or poses no risk to the consumer and the environment. These agencies include the Food and Agriculture Organization (FAO) of the United Nations, the U.S. Environmental Protection Agency (EPA), the FDA, and the USDA.

An example of such a compound is dichlorodiphenyl-trichloroethane (DDT), a pesticide once

---

shellfish poisoning A group of foodborne illnesses caused by consuming shellfish that contain marine toxins.

marine toxin A poison produced by ocean algae.

red tide Phenomenon in which certain ocean algae grow profusely, causing reddish discoloration of the surrounding water.

brevetoxin The toxin produced by red tide–causing algae, which, when consumed by humans, causes shellfish poisoning.

used to kill mosquitoes and increase crop yields. DDT was banned in the United States in 1972 when it was found to damage wildlife. However, the use of DDT to control malaria continues in many developing countries. More recently, **bovine somatotropin (bST),** otherwise known as bovine growth hormone, has attracted much public attention. This hormone is used in the dairy industry to increase milk production. Although some people are concerned about the safety of bST, a substantial amount of research suggests that it is safe for both cow and consumer.[23] Thus, the FDA allows its use, and there are no laws requiring milk produced by cows treated with bST be labeled as such. Even after a compound such as bST has been approved, however, the FDA continues to evaluate its safety.

FOOD ALLERGIES AND SENSITIVITIES CAN ALSO CAUSE FOODBORNE ILLNESS

Some noninfectious compounds in foods can cause illness in small segments of the population, but not because they are really toxic or poisonous in the way we have already discussed. Instead, they cause illness in a small percentage of people who are especially sensitive or even allergic to them. Because the percentage of individuals who have an adverse reaction to these compounds is small, their presence in food is not prohibited—but must be disclosed on the label. For example, monosodium glutamate (MSG), which is used as a flavor enhancer, causes severe headaches, facial flushing, and a generalized burning sensation in some people.[24] Because MSG is used extensively in Asian cuisines, this reaction is often referred to as "Chinese restaurant syndrome." The presence of MSG must be stated on the label of any food to which it has been added.

Similarly, **sulfites** are added to some foods (such as wine and dried fruits) to enhance color and prevent spoilage. Unfortunately, consuming them causes breathing difficulties in sulfite-sensitive people, especially in those with asthma. In response, the FDA requires food manufacturers to label all foods containing at least 10 parts per million (ppm) sulfites. You can check to see if a food has added sulfites by looking for the following terms: sodium bisulfite, sodium metabisulfite, sodium sulfite, potassium bisulfite, and potassium metabisulfite.

Exposure to some proteins naturally present in foods may also cause an allergic reaction in some people. For example, people who are allergic to peanuts and other legumes can experience a life-threatening allergic response when they eat these foods. You can read more about foodborne allergies (such as peanut allergy) in Chapter 5.

NEW FOOD SAFETY CONCERNS ARE ALWAYS EMERGING

Scientists and public health officials are continually trying to identify compounds in foods that may cause illness. Often these investigations receive attention in the popular press, causing public concern. Generally this is good, because public concern results in scientific scrutiny. In this section, you will learn about four foodborne substances that have garnered significant public health attention in the last few years.

Acrylamide One recently emergent food safety concern relates to a substance called **acrylamide.** Although acrylamide has probably always been in the food supply, concerns regarding its presence in foods were first reported in 2002.[25] Acrylamide is used to make polyacrylamide, found in some food-packaging materials, and small amounts of it may be present in food. It can also form in starchy foods exposed to very high temperatures, such as french fries and potato chips. Because very high levels of acrylamide cause cancer in laboratory animals, there is concern that dietary acrylamide may be harmful to humans as well.[26] However, most evidence suggests that humans would need to eat an unreasonably large amount of fried foods to consume a cancer-causing dose of acrylamide.[27] Nonetheless, both the FDA and researchers continue to study this issue.

Melamine Globalization of the world's food supply has led to increasing concerns that lower standards for food production in other countries may lead to unsafe food being imported to the United States. For example, in 2007 it was determined that pet food contaminated with **melamine** had caused

---

bovine somatotropin (bST; bovine growth hormone) A protein hormone produced by cattle and used in the dairy industry to enhance milk production.

sulfite A naturally occurring compound that is sometimes used as a food additive to prevent discoloration and bacterial growth.

acrylamide (a – CRYL – a – mide) A compound that is formed in starchy foods (such as potatoes) when heated to high temperatures.

melamine A nitrogen-containing chemical used to make lightweight plastic objects.

illness in and death of numerous dogs and cats.[28] Melamine is a nitrogen-containing chemical typically used to make lightweight plastic objects such as dishes. Investigations by the FDA determined that this compound had been added to gluten, a protein derived from wheat, which is used to produce animal foods. Because melamine is rich in nitrogen, some Chinese gluten manufacturers added it to their products to make them appear higher in protein than they actually were. The melamine-tainted gluten was then used to make pet foods in the United States.

Public attention quickly shifted from concern about pets to the possibility that melamine could enter the human food supply as well. The FDA and USDA originally concluded that consuming meat and milk from animals fed melamine-tainted feed posed very low risk to human health.[29] However, this conclusion was quickly revised when melamine-contaminated milk products caused nearly 60,000 infants and children in China to become sick in 2008 and 2009.[30] The FDA now advises consumers worldwide to avoid using any infant formula products made in China, as well as any milk products or products with milk-derived ingredients made in China. Unfortunately, this experience highlights the importance of a strong national food security monitoring system.

Bisphenol A Another somewhat recent concern receiving considerable attention relates to the chemical **bisphenol A (BPA),** which is found in some plastic food and beverage containers including baby bottles. This is because researchers have shown that BPA may influence long-term risk for cancer and reproductive abnormalities in laboratory animals.[31] Thus, scientists and public health officials are currently studying whether consuming foods and beverages from these types of containers poses health risks to humans as well.[32] In 2010, after reviewing the growing literature on BPA, the FDA announced that it did have some concern about the potential effects of BPA on the brain, behavior, and development of fetuses, infants, and young children. In response, the FDA has begun supporting efforts to decrease BPA exposure in the United States and is shifting to a more robust regulatory framework for its oversight.[32] Other countries, such as Canada, have prohibited use of some BPA-containing products such as plastic baby bottles. Individuals wishing to decrease their personal exposure to BPA can choose glass or metal containers over polycarbonate ones, not put plastics in the microwave, and wash plastic containers by hand instead of

in the dishwasher. In addition, parents of infants are advised by the U.S. Department of Health and Human Services to discard scratched plastic baby bottles and infant feeding cups, and not put boiling or very hot water, infant formula, or other liquids into bottles containing BPA while preparing them for use.[32] Without a doubt, you can be assured that public interest in compounds such as BPA will be intense for years to come.

How Do Food Manufacturers Prevent Contamination?

Many pathogens and inert compounds cause foodborne illness, and it is beyond the scope of this book to describe each one in detail. However, it is very important for you to understand how disease-causing agents are transmitted from food to food, also known as **cross-contamination,** and from person to person. It is also important for you to understand how food manufacturers strive to keep food safe.

CAREFUL FOOD-HANDLING TECHNIQUES HELP KEEP FOOD SAFE

To prevent foodborne illness, it is critical that people who handle food do so safely. This is because pathogens can be easily transmitted to almost any food by an infected person. In this way, a "perfectly clean" food can be made unsafe. Therefore, it is important that food handlers, including those who harvest, process, and prepare foods, avoid coughing or sneezing on foods. In addition, because many pathogens can pass through the GI tract and be excreted in the feces, fecal contamination of foods can cause foodborne illness. Consequently, people who process and prepare food are generally required to wear gloves while handling food and thoroughly wash their hands after using the toilet.

PROPER FOOD PRODUCTION, PRESERVATION, AND PACKAGING CAN PREVENT ILLNESS

In addition to safe food-handling techniques, the food-processing industry adheres to additional practices that help to keep your food safe. The

bisphenol A (BPA) A chemical used in the production of many plastic items including baby bottles.

cross-contamination The transfer of microorganisms from one food to another or from one surface or utensil to another.

goal is to produce a product that is as pathogen-free as possible and then to store it in a way that does not allow pathogenic growth. In 2011, the FDA Food Safety Modernization Act was signed into law to help ensure that the U.S. food supply is safe. These new regulations shifted the focus of federal regulators from *responding* to contamination to *preventing* it. For example, the FDA now has the authority to recall food products—a process that previously was only done voluntarily by food manufacturers.[33]

The USDA and the FDA have long had guidelines for food processors and handlers. These guidelines, called the **Hazard Analysis Critical Control Points (HACCP) system,** were developed to identify the critical points in food-processing where contamination is most likely. The USDA also requires that meat and poultry be inspected before sale and that guidelines for safe handling appear on packages. However, this inspection does not guarantee that the meat is pathogen free. Although we have no sure way to prevent a food from harboring dangerous pathogens or toxins, many techniques help keep food safe. These include drying, salting, smoking, fermentation, heating, freezing, and irradiation—all of which are described next.

Drying, Salting, Smoking, and Fermentation Meat has long been preserved by salting, smoking, and/or drying because these techniques—when done correctly—inhibit bacterial growth. Another technique called **fermentation** involves adding selected microorganisms to foods and is used in making sauerkraut, yogurt, pickles, and wine. Fermentation promotes the growth of nonpathogenic organisms, which minimize the growth of pathogenic organisms.

Heat Treatment: Cooking, Canning, and Pasteurization Because most foodborne pathogens prefer living in an environment between 40 to 140°F, cooling or heating foods below and above this range helps inhibit microbial growth. This is why food manufacturers often heat-treat their products and why you should also avoid this temperature range (called the **danger zone**) when storing and serving food. Table 2 provides temperature guidelines for cooking, serving, and reheating foods. As you can see, to be kept safe, different foods sometimes require different temperatures. Note that the USDA revised its recommended cooking temperature for pork, steaks, roasts, and chops to 145°F in 2011. The USDA now recommends cooking *all whole cuts of meat* (excluding poultry) to 145°F as measured

with a food thermometer placed in the thickest part of the meat, then allowing the meat to rest for three minutes before carving or consuming.

Heating food can also preserve it for later consumption. For example, during the canning process, foods are packaged (or "canned") in sanitized jars or cans and then heated at high temperatures. Pathogens are killed, and a vacuum is created within the jar. The air-free environment helps preserve the food, because most foodborne illness-causing organisms require oxygen to grow.

TABLE 2 Guidelines for Cooking, Serving, and Reheating Foods to Prevent Foodborne Illness

Note that internal temperatures should be measured with a thermometer.

Cooking	Beef and Pork
	• Cook beef roasts and steaks to a minimum of 145°F.
	• Cook ground beef and pork to at least 165°F.
	• Cook pork roasts and chops to a minimum of 145°F.
	Poultry
	• All poultry should be cooked to a minimum internal temperature of 165°F.
	Eggs
	• Cook eggs until the yolks and whites are firm.
	• Don't use recipes in which eggs remain raw or only partially cooked.
	Fish
	• Cook fish to 145°F or until the flesh is opaque and separates easily with a fork.
	• Avoid eating uncooked oysters and other shellfish. People with liver disorders or weakened immune systems are especially at risk for getting sick.
Serving	• Keep foods at 140°F or higher until served.
	• Keep foods hot with chafing dishes, slow cookers, and warming trays.
Reheating	• Reheat leftovers to at least 165°F.
	• When using a microwave oven, make sure food is evenly heated.

SOURCE: Adapted from the Partnership for Food Safety Education and FightBAC!® Available from: http://www.fightbac.org.

Hazard Analysis Critical Control Points (HACCP) system A USDA food safety protocol used to decrease contamination of foods during processing.

fermentation Metabolism, by bacteria, which occurs under relatively anaerobic conditions.

danger zone The temperature range between 40 and 140°F in which pathogenic organisms grow most readily.

The strawberries on the left were irradiated after they were picked, whereas those on the right were not.

Cordelia Molloy / Photo Researchers, Inc.

FIGURE 1 The Radura Symbol Irradiated foods must include this symbol on their food label.

Recall, however, that some organisms, such as *Clostridium botulinum,* can survive in low-acid, anaerobic conditions. Making sure that home-canned foods have the proper acidity and are heated sufficiently helps keep them safe. You can find out more about canning foods safely on the USDA website at http://www.fsis.usda.gov/Help/FAQs_Hotline_Preparation/index.asp.

Pasteurization Perhaps the most common form of heat treatment used in food preservation is **pasteurization.** This process was named for the French scientist Louis Pasteur, who discovered in the 1860s that is was possible to prevent spoilage in wine and beer by exposing them to intense heat for a short period of time. Indeed, pasteurization is still commonly used today. Foods that are typically pasteurized include dairy products, juice, and spices. Pasteurization is especially good at killing some forms of *E. coli, Salmonella, Campylobacter jejuni,* and *Listeria monocytogenes.* Several outbreaks of *Salmonella* poisoning from unpasteurized apple cider prompted laws that now require commercially available apple cider to be pasteurized or labeled as being nonpasteurized, and the FDA recommends that homemade cider be heated for 30 minutes at 155°F or 15 seconds at 180°F.

Cold Treatment: Cooling and Freezing Foods also spoil less quickly if kept cold. To slow or halt the growth of microorganisms, you should always refrigerate foods at 40°F or freeze them soon after they are prepared. This helps prevent foods from staying in the "danger zone" for an extended period of time.

Irradiation In the 1950s, the National Aeronautics and Space Administration (NASA) first used irradiation to preserve food for space travel. Shortly thereafter, the FDA approved a form of food processing called **irradiation,** sometimes called "cold pasteurization," as a form of food preservation. Today, irradiation is approved for meat, poultry, shellfish, eggs, and other foods such as fresh fruits, vegetables, and spices. During irradiation, foods are exposed to radiant energy that damages or kills bacteria. It is important to understand that irradiation neither damages nutrients nor makes the foods radioactive—just as the use of X-rays to inspect luggage does not make it radioactive. Irradiation makes foods safer to eat and can dramatically increase their shelf life. For example, compared with nonirradiated strawberries, which have a brief shelf life, irradiated strawberries last for several weeks without spoiling. Irradiated foods must be labeled with the radura symbol (Figure 1).

What Steps Can You Take to Reduce Foodborne Illness?

There are many ways you can reduce your risk of foodborne illness. First, you should familiarize yourself with recommendations by regulatory groups and keep abreast of food safety alerts and recalls. Second, it is helpful to understand the basic concepts related to safe food handling. In this section, you will learn about a variety of consumer advisory bulletins that are available from the FDA and USDA. You will also learn about a national campaign, called FightBac!®, developed by the U.S.

pasteurization (pas – ter – i – ZA – tion) A food preservation process that subjects foods to heat to kill bacteria, yeasts, and molds.

irradiation A food preservation process that applies radiant energy to foods to kill bacteria.

government to provide useful tips for consumers on how to avoid foodborne illness.

CHECK CONSUMER ADVISORY BULLETINS

Both the USDA and FDA maintain user-friendly websites containing information about current food safety recommendations. In addition, there are toll-free phone numbers that can be used to find out important food safety information. Some of these are listed here.

- **General information**: www.foodsafety.gov
- **News and safety alerts**: www.foodsafety.gov/ ~fsg/fsgnews.html
- **Recalls**: www.fsis.usda.gov/FSIS_Recalls/index. asp
- **To report a foodborne illness**: http://www.fsis. usda.gov/FSIS_Recalls/problems_with_food_ products/index.asp
- **USDA's Meat and Poultry Hotline**: 1-888-MPHotline
- **FDA's Food Safety Information Hotline**: 1-888-SAFEFOOD

THE FIGHTBAC!® CAMPAIGN PROVIDES BASIC FOOD SAFETY ADVICE

Although issues related to specific risk for various foodborne illnesses may change over time, consumers are encouraged to follow some basic rules. The USDA and the Partnership for Food Safety Education have developed a set of food safety guidelines, called **FightBAC!®**. The Fight-BAC!® campaign is a public education program focused on reducing foodborne illness. It specifically addresses avoidance of infectious agents of foodborne illness.

Clean Hands, Surfaces, and Cooking Utensils To remove and/or kill pathogens on hands, surfaces, and cooking utensils, you should frequently and adequately wash them with hot, soapy water. It is also recommended that you clean cutting boards thoroughly after each use and wash kitchen towels, cloths, and sponges frequently. Sponges can be best decontaminated by heating in a microwave for 1 minute.[34] Currently, there is no definitive answer as to whether plastic or wood cutting boards are best when it comes to food safety, although some studies suggest that wood might be preferable.[35] The USDA recommends that you periodically sanitize counters, equipment, utensils, and cutting boards with a solution of 1 tablespoon of unscented, liquid chlorine bleach in 1 gallon of water.

To prevent contamination of foods with pathogens found in fecal materials, you should always wash your hands after using the bathroom, changing diapers, or handling pets. To be effective, wash hands vigorously with soap for at least 20 seconds and rinse them thoroughly under clean, running warm water. Dry your hands using clean paper or cloth towels. Using antimicrobial gels can also help ensure that hands are pathogen free, although you should not consider this a substitute for proper hand washing.

Wash Fresh Fruits and Vegetables—Not Meat Washing fresh fruits and vegetables provides an important protection from foodborne pathogens and noninfectious agents such as pesticides. To do this effectively, you should wash produce under clean, cool, running water and, if possible, scrub with a clean brush. Fruits and vegetables should then be dried with a clean paper or cloth towel. There is some evidence that commercial fruit and vegetable cleaners may help remove *E. coli* 0157:H7 and *Salmonella* from produce,[36] but the USDA has not taken a stand on the effectiveness of these products. However, it is recommended that you *do not* wash raw meat, poultry, and fish, because doing so increases the danger of cross-contaminating otherwise noninfected surfaces and foods.

Separate Foods to Prevent Cross-Contamination To prevent transferring disease-causing pathogens from food to food (cross-contamination), you should separate raw meat and seafood from other foods in your grocery cart, in your refrigerator, and while preparing a meal. For example, put raw meats in separate, sealed plastic bags in your grocery cart and use separate cutting boards when preparing fresh produce and raw meat. Also, you should not put cooked meat back on the same plate that was used to hold the raw meat unless you have washed the plate thoroughly with hot, soapy water.

Cook Foods to the Proper Temperature As previously stated, heat kills most dangerous organisms. It can also alter the chemical composition of some preformed toxins, making them less dangerous. The temperatures and cooking times required to kill pathogens depend on the particular food and the organism or toxin. It is recommended that you measure the *internal temperature* of a cooked food and thoroughly clean meat thermometers between uses.

Also, just because a food has been previously cooked does not mean it is pathogen free. Before

---

FightBAC!® A public education program developed to reduce foodborne bacterial illness.

eating, you should reheat foods to their appropriate internal temperatures. This step is especially important for pregnant women, infants, older adults, and people with impaired immunity such as those with HIV or cancer, because these individuals are more likely than others to become seriously ill when exposed to pathogens. In fact, a *Listeria monocytogenes* outbreak in the late 1990s that resulted in at least six deaths and two miscarriages has prompted the recommendation that lunch meats and frankfurters be heated before eating.[37]

Keep Foods Cold, Chill Them Quickly, and *If in Doubt, Throw It Out* Because cold temperatures can slow the growth of microorganisms, you should always keep perishable foods refrigerated at 40°F or colder. Also, do not marinate or thaw foods at room temperature. Instead, marinate them in the refrigerator. Similarly, frozen foods should not be allowed to thaw at room temperature but instead should thaw in the refrigerator or microwave. It is also safe to thaw frozen foods in cold water in an airtight plastic wrapper or bag, changing the water every 30 minutes or so until the food is thawed. After a meal, you should refrigerate foods *as quickly as possible*, separating large amounts of foods into small, shallow containers to allow the food to cool rapidly.

Even properly chilled foods can become sources of foodborne illness, and for this reason you should consume leftovers within three to four days. Indeed, "If in doubt, throw it out!" You can see some of the recommended storage times for refrigerated foods by visiting the USDA website (http://www.fsis.usda.gov/PDF/Refrigeration_and_Food_Safety.pdf).

BE ESPECIALLY CAREFUL WHEN EATING OUT

Making sure your food is safe to eat can be especially difficult when you eat at restaurants, picnics, and buffets. However, you can decrease your risk of foodborne illness in these situations. In general, ask yourself whether the four basic concepts of the Fight Bac!® program have been followed.

- How likely was it that the people handling and preparing your foods used sanitary practices?
- What evidence is there that raw foods were kept separate from cooked foods?
- Do you think that foods were properly cooked and kept out of the thermic "danger zone"?
- Is there sufficient evidence that cold foods were kept cold?

Unfortunately, it is common at picnics and social gatherings to neglect these basic rules of food safety, resulting in foodborne illness. Therefore, use good judgment and eat only foods you know to be safe. Better safe than sorry!

What About Avoiding Foodborne Illness While Traveling or Camping?

When traveling or camping, there are added concerns related to food safety. To keep current on the latest information, visit the CDC website (http://wwwn.cdc.gov/travel/). In general, water and fresh produce pose the greatest risks to you, especially when you travel to foreign countries, as these foods often are the causes of traveler's diarrhea. You may not be able to completely prevent foodborne illness while traveling, but you can greatly lower your risk by being vigilant in choosing which foods and beverages to consume—and which to avoid. Finally, if a traveler does get traveler's diarrhea, it is important to replace lost fluids and electrolytes as soon as symptoms begin to develop. Clear liquids are routinely recommended for adults. Travelers who develop three or more loose stools in an eight-hour period may benefit from antibiotic therapy.

DRINK ONLY PURIFIED OR TREATED WATER

In general, when traveling out of the country (especially in less industrialized regions of the world) or camping in remote regions of the United States, it is advisable to drink bottled water and avoid using ice. Before drinking bottled beverages, be sure all containers have fully sealed caps. If seals are not intact, the bottles may have been refilled. If bottled water is not available, you should boil your water for one minute or, if at high altitudes (>2,000 m), three minutes. In addition, water can be chemically treated to kill pathogens that may be present, and portable water filters can also be used to remove some pathogens.

AVOID OR CAREFULLY WASH FRESH FRUIT AND VEGETABLES

Although fresh produce contaminated with bacteria may not cause illness for local people, it can cause serious illness (usually diarrhea) in visitors. Thus, when traveling to foreign locations, avoid or carefully wash fresh fruit (e.g., grapes) and

Traveling abroad requires special precautions regarding food safety.

and Bioterrorism Preparedness and Response Act" (also called the **Bioterrorism Act**) in 2002.[38] Since then, additional regulations have been approved to help ensure that foods grown and produced both in the United States and abroad cannot be tampered with. Of particular interest in this regard is the possibility that *Clostridium botulinum* could be used by terrorists as a widespread disease-causing agent. As such, a supply of antitoxin against botulism is maintained by CDC in the event such an at-

vegetables (e.g., peppers) that are eaten without peeling them. Of course, only water known to be clean should be used.

TRAVELING IN AREAS WITH VARIANT CREUTZFELT-JAKOB DISEASE

When traveling in Europe or other areas that have reported cases of BSE in cows, the CDC recommends that concerned travelers consider either avoiding beef and beef products altogether or selecting only beef products composed of solid pieces of muscle meat. These considerations, however, should be balanced with the knowledge that the risk of disease transmission is very low. Milk and milk products from cows are not believed to pose any risk for transmitting BSE to consumers—even in areas of the world with emerging variant Creutzfelt-Jakob disease incidence.

What Are Some Emerging Issues of Food Biosecurity?

Recent worldwide events have raised concerns about the safety of our food. Indeed, terrorists and other malicious people could certainly do serious and widespread harm by contaminating the nation's food supply. As a result, **food biosecurity**—the prevention of terrorist attacks on our food supply—has gained intense national attention, and Congress authorized the "Public Health Security

tack occurs. Clearly, this focus of deep individual and governmental interest will remain strong in coming years.

Changes in food production and distribution patterns both nationally and internationally may influence national food safety and risk of foodborne illness. For example, a contaminated food produced in Chile can easily find its way into the American food chain, be distributed to dozens of states, and perhaps appear on your table. In response, the origin of all foods must now be included on the food's label. As was clearly shown in the recent outbreak of melamine contamination of our pet food supply, the U.S. government has little power over whether optimal and honest food production policies are adhered to in other locations. Although the full impact of these changes in international commerce is not known, some experts warn that fewer restrictions on food importation into the United States may increase our risk for foodborne illness. For these changes to be beneficial, the health benefits imparted by the globalization of our food supply must greatly outweigh the associated risks.

food biosecurity Measures aimed at preventing the food supply from falling victim to planned contamination.

Bioterrorism Act Federal legislation aimed to ensure the continued safety of the U.S. food supply from intentional harm by terrorists.

Key Points

What Causes Foodborne Illness?

- Infectious foodborne illness can be caused by pathogenic organisms such as bacteria, viruses, parasites, molds, and fungi.

- Some organisms produce toxins while they grow in foods (preformed toxins), others produce toxins inside the intestinal tract (enteric toxins), and others invade intestinal cells.

How Can Noninfectious Substances Cause Foodborne Illness?

- Inert (noninfectious) compounds in foods can also cause foodborne illness.

- Shellfish poisoning is caused when seafood is contaminated with toxins produced by algae.

- Some people are especially sensitive or allergic to selected components of foods.

- Emerging concerns related to noninfectious agents include presence of acrylamide, melamine, and bisphenol A (BPA) in our food supply.

How Do Food Manufacturers Prevent Contamination?

- Salting, drying, and smoking remove water from foods, inhibiting pathogenic growth; fermentation involves adding nonpathogenic organisms that inhibit the growth of dangerous ones.

- Methods that slow the growth of or kill pathogens include heat and cold treatment as well as irradiation.

What Steps Can You Take to Reduce Foodborne Illness?

- You can help prevent foodborne illness by ensuring that you use proper cleaning, handling, heating, chilling, and storing techniques.

- Many consumer food safety recommendations are even more important for people who eat out often, attend picnics frequently, and travel.

What Are Some Emerging Issues of Food Biosecurity?

- Concern that terrorists can cause serious outbreaks of foodborne illness has prompted governmental programs targeted at ensuring food biosecurity in the United States.

Notes

1. Centers for Disease Control and Prevention. Estimates of foodborne illness in the United States. December 15, 2010. Available at http://www.cdc.gov/foodborneburden/index.html.

2. Centers for Disease Control and Prevention. Surveillance for foodborne disease outbreaks—1993–1997. Morbidity and Mortality Weekly Review. 2000;49:1–51. Available from: http://www.cdc.gov/mmwr/PDF/ss/ss4901.pdf.

3. Trabulsi LR, Keller R, Tardelli Gomes TA. Typical and atypical enteropathogenic Escherichia coli. Emerging Infectious Diseases. 2002;8:508–13.

4. Centers for Disease Control and Prevention. Methicillin-resistant Staphylococcus aureus (MRSA) infections. Available from: http://www.cdc.gov/mrsa/index.html.

5. Jones TF, Kellum ME, Porter SS, Bell M, Schaffner W. An outbreak of community-acquired foodborne illness caused by methicillin-resistant Staphylococcus aureus. Emerging Infectious Diseases. 2002;8:82–4.

6. Centers for Disease Control and Prevention. Botulism associated with canned chili sauce, July–August 2007. Updated August 24, 2007. Available from: http://www.cdc.gov/botulism/botulism.htm.

7. Abnet CC. Carcinogenic food contaminants. Cancer Investigation. 2007;25:189–96.

8. U.S. Centers for Disease Control and Prevention. Norovirus outbreaks on three college campuses—California, Michigan, and Wisconsin, 2008. Morbidity and Mortality Weekly Reports. 58:1095–1100, 2009. Available from: http://www.cdc.gov/mmwr/preview/mmwrhtml/mm5839a2.htm.

9. Centers for Disease Control and Prevention. Norovirus outbreak among evacuees from hurricane Katrina—Houston, Texas, September 2005. Morbidity and Mortality Weekly Report. 2005;54;1016–8. Available from: www.cdc.gov/mmwr/preview/mmwrhtml/mm5440a3.htm.

10. Centers for Disease Control and Prevention. Investigation of outbreak of human infections caused by Salmonella I 4,[5],12:i:-. Available from: http://www.cdc.gov/salmonella/4512eyeminus.html.

11. U.S. Centers for Disease Control and Prevention. Investigation update: Multistate outbreak of human Salmonella enteritidis infections associated with shell eggs. December 2, 2010. Available at http://www.cdc.gov/salmonella/enteritidis/.

12. US Food and Drug Administration. Safe handling of raw produce and fresh-squeezed fruit and vegetable juices. Available from: http://www.fda.gov/downloads/Food/ResourcesForYou/Consumers/UCM174142.pdf.

13. Centers for Disease Control and Prevention. Ongoing multistate outbreak of Escherichia coli serotype O157: H7 infections associated with consumption of fresh spinach—Unites States, September 2006. Morbidity and Mortality Weekly Reports. 2006;55:1–2. Available

from: http://www.cdc.gov/mmwr/preview/mmwrhtml/mm5538a4.htm.

14. US Food and Drug Administration. Guidance for industry: guide to minimize microbial food safety hazards of fresh-cut fruits and vegetables. February 2008. Available from: http://www.fda.gov/Food/GuidanceComplianceRegulatoryInformation/GuidanceDocuments/ProduceandPlanProducts/ucm064458.htm.

15. Federal Institute for Risk Assessment, Federal Office of Consumer Protection and Food Safety, and Robert Koch Institute. Information update on EHEC outbreak. June 10, 2011. Available from: http://www.rki.de/cln_144/nn_217400/EN/Home/PM082011.html. U.S. Centers for Disease Control and Prevention. Investigation update: outbreak of shiga toxin—producing *E. coli* O104 (STEC O104:H4) infections associated with travel to Germany. June 15, 2011. Available from: http://www.cdc.gov/print.do?url=http://www.cdc.gov/ecoli/2011/ecoliO104/.

16. Sakanari JA, McKerrow JH. Anisakiasis. Clinical Microbiology Reviews. 1989;2:278–84.

17. US Food and Drug Administration. 2005 Food code. Available from: http://www.cfsan.fda.gov/~dms/fc05-toc.html.

18. Mostl K. Bovine spongiform encephalopathy (BSE): The importance of the food and feed chain. Forum in Nutrition. 2003;56:394–6.

19. Ryou C. Prions and prion diseases: Fundamentals and mechanistic details. Journal of Microbiology and Biotechnology. 2007;17:1059–70.

20. Belay ED, Schonberger LB. The public health impact of prion diseases. Annual Review of Public Health. 2005;26:191–212. Centers for Disease Control and Prevention. vCJD (variant Creutzfeldt-Jakob disease). Available from: http://www.cdc.gov/ncidod/dvrd/vcjd/risk_travelers.htm. Roma AA, Prayson RA. Bovine spongiform encephalopathy and variant Creutzfeldt-Jakob disease: How safe is eating beef? Cleveland Clinic Journal of Medicine. 2005;72:185–94.

21. US Department of Agriculture. Food Safety and Inspection Service. Prohibition of the use of specified risk materials for human food and requirements for the disposition of non-ambulatory disabled cattle. Federal Register: January 12, 2004 (Volume 69, Number 7). Available from http://www.fsis.usda.gov/Frame/FrameRedirect.asp?main=http://www.fsis.usda.gov/OPPDE/rdad/FRPubs/03-025IF.htm.

22. National Center for Zoonotic, Vector-Borne, and Enteric Diseases. Marine toxins: general information. Last updated July 20, 2010. Available at http://www.cdc.gov/nczved/divisions/dfbmd/diseases/marine_toxins/.

23. Anonymous. Bovine somatotropin and the safety of cows' milk: National Institutes of Health technology assessment conference statement. Nutrition Reviews. 1991;49:227–32. Etherton TD, Kris-Etherton PM, Mills EW. Recombinant bovine and porcine somatotropin: Safety and benefits of these biotechnologies. Journal of the American Dietetic Association. 1993;93:177–80.

24. Walker R, Lupien JR. The safety evaluation of monosodium glutamate. Journal of Nutrition. 2000;130:1049S–52S.

25. Stadler RH, Blank E, Varga N, Robert F, Hau J, Guy PA, Robert MC, Riediker S. Food chemistry:

26. Bull RJ, Robinson M, Laurie RD, Stoner GD, Greisiger E, Meier RJ, Stober J. Carcinogenic effects of acrylamide in Sencar and A/J mice. Cancer Research. 1984;44:107–11.

27. Mucci LA, Dickman PW, Steineck G, Adami H-O, Augustsson K. Dietary acrylamide and cancer of the large bowel, kidney, and bladder: Absence of an association in a population-based study in Sweden. British Journal of Cancer. 2003;88:84–9.

28. US Food and Drug Administration. Melamine pet food recall of 2007. Available from: http://www.fda.gov/animal-veterinary/safetyhealth/recallswithdrawals/ucm129575.htm.

29. US Food and Drug Administration. FDA/USDA joint news release: Scientists conclude very low risk to humans from food containing melamine. Available from: http://www.fda.gov/NewsEvents/Newsroom/Press-Announcements/2007/default.htm.

30. US Food and Drug Administration. Melamine contamination in China. (Updated: January 5, 2009) Available from: http://www.fda.gov/NewsEvents/PublicHealthFocus/ucm179005.htm.

31. Durando M, Kass L, Piva J, Sonnenschein C, Soto AM, Luque EH, Muñoz-de-Toro M. Prenatal bisphenol A exposure induces preneoplastic lesions in the mammary gland in Wistar rats. Environmental Health Perspectives. 2007;115:80–6. Newbold RR, Jefferson WN, Padilla-Banks E. Long-term adverse effects of neonatal exposure to bisphenol A on the murine female reproductive tract. Reproductive Toxicology. 2007;24:253–8.

32. U.S. Food and Drug Administration. Bisphenol A (BPA). Update on bisphenol A(BPA) for use in food: January 2010. Available from: http://www.fda.gov/NewsEvents/PublicHealthFocus/ucm064437.htm.

33. U.S. Food and Drug Administration. FDA Food Safety Modernization Act. Public Law 111-353, Jan. 4. 2011. Available at http://www.gpo.gov/fdsys/pkg/PLAW-111publ353/pdf/PLAW-111publ353.pdf.

34. Sharma, M, Eastridge, J, Mudd, C. Effective disinfection methods of kitchen sponges [abstract]. Institute of Food Technologists. 2007;Control No. 3310.

35. Cliver DO. Cutting boards and Salmonella cross-contamination. Journal of AOAC International. 89:538–42, 2006.

36. Kenney SJ, Beuchat LR. Comparison of aqueous commercial cleaners for effectiveness in removing Escherichia coli O157:H7 and Salmonella meunchen from the surface of apples. International Journal of Food Microbiology. 2002;25:47–55.

37. Centers for Disease Control and Prevention. Update: Multistate outbreak of Listeriosis—United States, 1998–1999. Morbidity and Mortality Weekly Report. 1999;47:1117–8. Available from: http://www.cdc.gov/mmwr/preview/mmwrhtml/00056169.htm.

38. United States Congress. Public health security and bioterrorism preparedness and response act of 2002. June 12, 2002. H.R. 3448. Available from: http://www.gpo.gov/fdsys/pkg/BILLS-107hr3448enr/pdf/BILLS-107hr3448enr.pdf.

Acrylamide from Maillard reaction products. Nature. 2002;419:449–50.

Lipids

Lipids are required for hundreds, if not thousands, of physiological functions in the body. For example, body fat (which contains lipids) protects vital organs, and lipids that surround cells provide a protective yet highly functional barrier. In addition, lipids make many foods flavorful—just think of butter, cream, olive oil, and well-marbled beef. However, for many people the thought of fatty foods simply conjures up images of unhealthy living. We often shop for "fat-free" foods and try to avoid fats altogether. Food manufacturers have even developed "fat substitutes" to replace the naturally-occurring fats found in our foods. Nonetheless, it is important to recognize that, although diets high in fat can lead to health complications such as obesity and heart disease, getting enough of the right *types* of fat is just as essential for optimal health as avoiding excess fat and the wrong kinds of fat. In this chapter, you will learn about the variety of fats and oils in foods and how the body uses them to make other substances vital for health. Guidelines for lipid intake that promotes optimal health are also discussed.

Gallbladder Surgery—When Things Do Not Go Smoothly

Nancy had noticed for quite some time that eating rich, high-fat meals frequently caused her to experience cramping and diarrhea. But these unpleasant effects did not always occur and were easily avoided if she chose her foods wisely. So when Nancy learned that she had mild gallbladder disease, she was not completely surprised by this news, and it did not cause much concern. Although Nancy's doctor advised her to have her gallbladder surgically removed, she decided to wait. After all, she was a working mother of three active sons and had very little free time. Nancy did not feel as if she could afford the amount of time required to have and recover from gallbladder surgery.

At first Nancy's plan seemed to work. She watched her diet carefully and avoided fatty foods. However, 10 years after her initial diagnosis Nancy woke up one morning with severe abdominal pain. The decision to not have surgery was no longer an option. Her gallbladder was inflamed, and after consultation with her doctor Nancy decided it was time to have it removed. This normally simple outpatient procedure, however, turned out to be anything but routine.

At first the surgery seemed to have gone smoothly, and although Nancy's abdomen was tender, she was discharged from the hospital five hours later. But the next morning she was feeling even worse. Not even strong narcotic medications could relieve the relentless abdominal pain she was experiencing. After calling her doctor, Nancy returned to the emergency room. Upon examination the problem became clear; whereas most people have one duct transporting bile from the liver to the gallbladder, Nancy had multiple ducts. Because her doctor did not know this when he performed the surgery, he had inadvertently severed these "accessory" ducts without closing them off. As a result, bile leaked from Nancy's liver into her abdomen—a very serious condition. An emergency surgery was quickly performed to repair

the damage, and Nancy remained in the hospital for an additional two weeks.

The road to recovery was long, requiring a slow transition from a simple diet (e.g., rice, apples, bananas), to one containing small amounts of fats and oils. However, within two months Nancy was able to eat normally and once again enjoy an active life with her family. We asked her if she had garnered any wisdom from her experience that she would like to share, and she told us, *"I highly recommend that people with gallbladder disease do not put off getting treatment and be very careful to report abnormal signs and symptoms after their surgery if it is needed."*

Courtesy of Shelley McGuire

Critical Thinking: Nancy's Story

Why do you think that people with mild gallbladder disease, which causes indigestion and diarrhea, often mistake it for other conditions? If you found out that you had gallbladder disease that could be handled effectively with dietary changes, would you agree to have your gallbladder removed? What factors might influence your decision?

What Are Lipids?

Lipids are important macronutrients because they provide a major source of energy and are critical for optimal health in many ways. Lipids come in many varieties, and it is important to understand how they differ both chemically and physiologically. In the following section, you will learn the fundamental concepts needed to grasp these basics.

FATS AND OILS ARE TYPES OF LIPIDS

It is helpful to begin with some basic lipid-related terminology and definitions. For example, fats and oils are examples of what chemists call **lipids,** which are relatively water-insoluble, organic molecules consisting mostly of carbon, hydrogen, and oxygen atoms. In other words, lipids are hydrophobic ("water fearing"). Lipids that are liquid at room temperature are called **oils,** and those that are solid at room temperature are called **fats.** The major lipids include fatty acids, triglycerides, phospholipids, sterols, and fat-soluble vitamins. Fatty acids, triglycerides, phospholipids, and sterols are discussed in detail in this chapter, and the fat-soluble vitamins are presented in Chapter 11.

Oils and fats are both lipids. Whereas oils (like corn oil) are liquid at room temperature, fats (like margarine) are solid.

FATTY ACIDS ARE THE MOST COMMON TYPE OF LIPID

The most abundant lipids in your body and in the foods you eat are **fatty acids,** which are made entirely of carbon, hydrogen, and oxygen atoms (Figure 6.1). A chain of carbon atoms forms the backbone of each fatty acid. One end of this carbon chain, called the **alpha (α) end,** contains a carboxylic acid group (–COOH); the other end, called the **omega (ω) end,** contains a methyl group (–CH₃). Most fatty acids do not exist in their "free" (unbound) form. Instead, they are components of larger molecules, such as triglycerides and phospholipids (Figure 6.2). Fatty acids are also bound to cholesterol, forming cholesteryl esters. You will learn more about triglycerides, phospholipids, and cholesteryl esters later in this chapter.

There are hundreds of unique fatty acids, and they differ in the number of carbons they contain as well as the types and locations of the chemical bonds holding the carbon atoms together. These variations influence the physical properties of the fatty acids and the roles they play in foods and in the body.

lipid (*lipos*, fat) Organic molecule that is relatively insoluble in water and soluble in organic solvents.

oil (*oleum*, olive) A lipid that is liquid at room temperature.

fat A type of lipid that is solid at room temperature.

fatty acid A type of lipid consisting of a chain of carbons with a methyl (–CH₃) group on one end and a carboxylic acid group (–COOH) on the other.

alpha (α) end (*alpha*, the first letter in the Greek alphabet) The end of a fatty acid, which consists of a carboxylic acid (–COOH) group.

omega (ω) end (*omega*, the final letter in the Greek alphabet) The end of a fatty acid, which consists of a methyl (–CH₃) group.

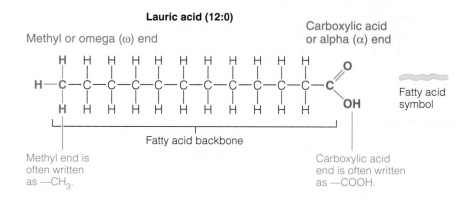

Lauric acid (12:0)

Methyl or omega (ω) end

Carboxylic acid or alpha (α) end

Fatty acid symbol

Fatty acid backbone

Methyl end is often written as —CH₃.

Carboxylic acid end is often written as —COOH.

FIGURE 6.1 Fatty Acid Structure All fatty acids have three components: a carboxylic acid or alpha (α) end (–COOH), a methyl or omega (ω) end (–CH₃), and a fatty acid backbone.

FIGURE 6.2 Fatty Acids Can Take Several Forms In the body, fatty acids are found in several forms including being "free" (unbound) molecules and components of larger molecules such as triglycerides, phospholipids, and cholesteryl esters.

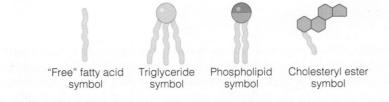

"Free" fatty acid symbol | Triglyceride symbol | Phospholipid symbol | Cholesteryl ester symbol

Number of Carbons (Chain Length) The number of carbon atoms in the backbone of a fatty acid determines its **chain length,** as shown in Figure 6.3. In fact, it is helpful to think of a fatty acid as if it were a chain, with each link representing a carbon atom. Most naturally occurring fatty acids have an even number of links or carbons—usually 12 to 22—although some may be as short as 4 or as long as 26 carbons. Fatty acids with fewer than 8 carbons are called **short-chain fatty acids;** those with 8 to 12 carbons are **medium-chain fatty acids;** those with more than 12 carbons are **long-chain fatty acids.**

The chain length of a fatty acid affects its chemical properties and physiological functions. For example, it influences the temperature at which the fatty acid melts (melting point). Fatty acids with low melting points, such as short-chain fatty acids, typically take less heat to liquify (melt) compared with fatty acids with longer chain lengths. Lipids made predominantly of short-chain fatty acids are therefore likely to be oils or even gases. Conversely, lipids containing mostly longer-chain fatty acids can have high melting points and tend to be solids (fats) at room temperature. Chain length also affects the extent to which a fatty acid is soluble in water, with short-chain fatty acids generally being more water-soluble than long-chain fatty acids. Because humans are mostly water, short-chain fatty acids are more easily absorbed and transported in the body than are long-chain fatty acids.

Number and Positions of Double Bonds Aside from the *number* of carbon atoms they contain, fatty acids also differ in the *types* of chemical bonds between their carbon atoms (Figure 6.4). These carbon–carbon bonds can be either single bonds or double bonds. If a fatty acid contains all single carbon–carbon bonds, it is a **saturated fatty acid (SFA);** those containing one or more double bonds are **unsaturated fatty acids.** Fatty acids with one double bond are **monounsaturated fatty acids (MUFAs);** those with two or more double bonds are **polyunsaturated fatty acids (PUFAs).** Note that,

chain length The number of carbons in a fatty acid's backbone.

short-chain fatty acid A fatty acid having <8 carbon atoms in its backbone.

medium-chain fatty acid A fatty acid having 8–12 carbon atoms in its backbone.

long-chain fatty acid A fatty acid having >12 carbon atoms in its backbone.

saturated fatty acid (SFA) (*saturare*, to fill or satisfy) A fatty acid that contains only carbon–carbon single bonds in its backbone.

unsaturated fatty acid A fatty acid that contains at least one carbon–carbon double bond in its backbone.

monounsaturated fatty acid (MUFA) A fatty acid that contains one carbon–carbon double bond in its backbone.

polyunsaturated fatty acid (PUFA) A fatty acid that contains more than one carbon–carbon double bond in its backbone.

FIGURE 6.3 Fatty Acids Can Have Different Chain Lengths The number of carbon atoms making up the backbone of a fatty acid determines whether it is a short-, medium-, or long-chain fatty acid.

Medium-chain fatty acid

An 8-carbon fatty acid

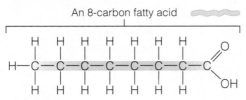

Long-chain fatty acid

A 16-carbon fatty acid

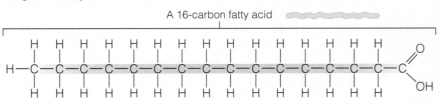

FIGURE 6.4 Saturated and Unsaturated Fatty Acids The types of carbon–carbon bonds in a fatty acid determine whether it is saturated or unsaturated.

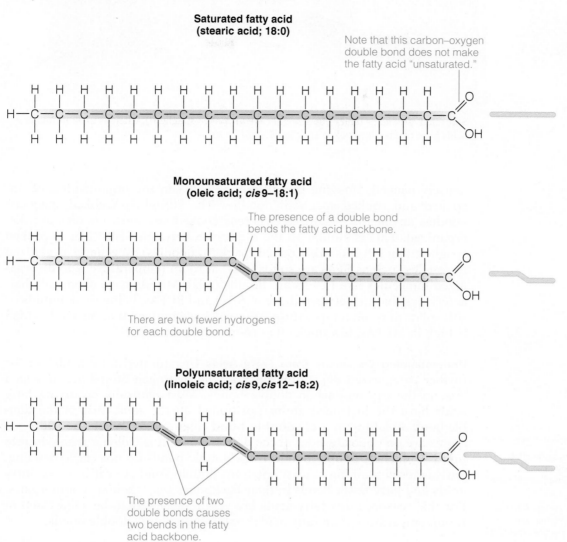

**Saturated fatty acid
(stearic acid; 18:0)**

Note that this carbon–oxygen double bond does not make the fatty acid "unsaturated."

**Monounsaturated fatty acid
(oleic acid; *cis*9–18:1)**

The presence of a double bond bends the fatty acid backbone.

There are two fewer hydrogens for each double bond.

**Polyunsaturated fatty acid
(linoleic acid; *cis*9,*cis*12–18:2)**

The presence of two double bonds causes two bends in the fatty acid backbone.

for each carbon–carbon double bond, two hydrogen atoms are lost from the fatty acid backbone. This is because carbon atoms can only have four chemical bonds. A double bond "counts" as two chemical bonds, so one hydrogen atom must be given up from each of the carbon atoms in the carbon–carbon double bond.

Like chain length, the number of double bonds can influence the physical nature of the fatty acid. As illustrated in Figure 6.4, each carbon atom in an SFA is surrounded (or "saturated") by hydrogen atoms. Being "saturated" with hydrogen atoms prevents the fatty acid from bending. Because of this rigidity, SFAs are highly organized and dense, making them solid at room temperature. Foods containing large amounts of SFAs (such as butter or coconut "oil"), therefore, tend to be solid fats at room temperature.

Compared with SFAs of similar chain length, unsaturated fatty acids (in other words, those with double bonds) have fewer hydrogen atoms and can bend. In fact, whenever there is a carbon–carbon double bond, there is a kink or a bend in the fatty acid backbone. These bends cause unsaturated fatty acids to become disorganized, preventing them from becoming

Some animal products, such as beef, contain a relatively high amount of saturated fatty acids.

Saturated fatty acids tend to be straight (like uncooked spaghetti), whereas unsaturated fatty acids have bends (like cooked spaghetti).

densely packed. Imagine the difference between the organization of uncooked and cooked spaghetti noodles. The straight uncooked spaghetti noodles are neatly organized, whereas cooked spaghetti noodles are disorganized. This is similar to the difference between SFAs (like uncooked spaghetti) and PUFAs (like cooked spaghetti noodles). In general, organized molecules such as SFAs are solids (fats) at room temperature, and disorganized molecules like PUFAs are liquids (oils). MUFAs have chemical characteristics that lie between those of SFAs and PUFAs, being thick liquids or soft solids at room temperature. You may have noticed that olive oil, which is high in MUFAs, is a thick oil at room temperature.

Understanding *Cis* versus *Trans* Fatty Acids Unsaturated fatty acids can be further categorized depending on how the hydrogen atoms are arranged around the carbon–carbon double bonds. Most naturally occurring fatty acids have the hydrogen atoms positioned on the same side of the double bond, resulting in a **cis double bond** (Figure 6.5). When the hydrogen atoms are on opposite sides of the double bond, it is called a **trans double bond.** Unlike *cis* double bonds, *trans* double bonds do not cause bending. Fatty acids containing at least one *trans* double bond are called **trans fatty acids** and have fewer bends in their backbones than their *cis* counterparts. For this reason, *trans* fatty acids are also more likely to be solid (fats) at room temperature than fatty acids containing only *cis* double bonds.

Trans Fatty Acids in Food *Trans* fatty acids are found naturally in some foods such as dairy and beef products. However, most dietary *trans* fatty acids are

cis double bond (*cis*, on this side of) A carbon–carbon double bond in which the hydrogen atoms are positioned on the same side of the double bond.

trans double bond (*trans*, across) A carbon–carbon double bond in which the hydrogen atoms are positioned on opposite sides of the double bond.

trans fatty acid A fatty acid containing at least one *trans* double bond.

FIGURE 6.5 *Cis* versus *Trans* Fatty Acids Unsaturated fatty acids can differ by whether they have *cis* or *trans* carbon–carbon double bonds. *Cis* bonds cause the fatty acid to bend, whereas *trans* bonds do not.

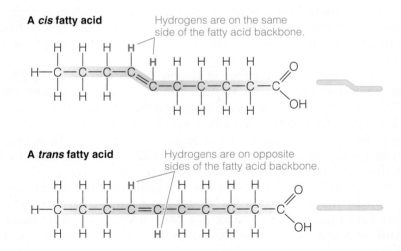

A *cis* fatty acid

Hydrogens are on the same side of the fatty acid backbone.

A *trans* fatty acid

Hydrogens are on opposite sides of the fatty acid backbone.

produced commercially via a process called **partial hydrogenation.** Partial hydrogenation converts the majority of carbon–carbon double bonds into carbon–carbon single bonds, causing oils (such as corn oil) to become fats (such as margarine or shortening). This is done by the chemical addition of hydrogen atoms. Aside from decreasing the number of double bonds and increasing the number of single bonds, the process of partial hydrogenation converts some of the *cis* double bonds to *trans* double bonds. As a result, the lipid is high in both saturated and *trans* fatty acids.

Partially hydrogenated lipids are often used in food manufacturing because they impart desirable food texture and reduce spoilage. Crackers, pastries, bakery products, shortening, and margarine are the main sources of the *trans* fatty acids in our diets.[1] However, the current focus on decreasing *trans* fatty acid intake has resulted in new food preparation and processing methods that decrease or eliminate *trans* fatty acids in many foods. For example, some fast-food chains have switched from frying their foods in high–*trans* fatty acid shortening to *trans* fatty acid–free vegetable oils.

Many public health agencies, including the Institute of Medicine and the U.S. Department of Agriculture, suggest that we limit our intake of *trans* fatty acids. These recommendations (which are described in more detail later) for lowering *trans* fatty acid intake have been made because some *trans* fatty acids may increase the risk for cardiovascular disease.[2] As of 2006, food manufacturers have been required to state the *trans* fatty acid content on their Nutrition Facts panels.* Because of the considerable concern about *trans* fatty acids in the diet, a number of cities in the United States, have become "*trans* fat–free zones." You can read more about this in the Focus on Diet and Health feature.

FATTY ACIDS ARE NAMED FOR THEIR STRUCTURES

There are several methods used to name or describe fatty acids. In general, these methods are based on the number of carbons, the number and types of double bonds, and the positions of the double bonds. Some fatty acids also have "common names."

Alpha (α) Naming System The "alpha (α)" naming system for fatty acids is based on the positions and types of double bonds relative to the carboxylic acid (α) end of the fatty acid. As an example, consider a fatty acid with the following characteristics:

- 18 carbons in length
- 2 *cis* double bonds
- first *cis* double bond between the 9th and 10th carbons from the carboxylic acid end
- second *cis* double bond between the 12th and 13th carbons from the carboxylic acid end.

The fatty acid's name is constructed beginning with an "18," signifying that there are 18 carbons. A "2" is added to form "18:2," signifying that there are two double bonds. Next, where the double bonds are located is designated as "9,12"–18:2, with the locations determined by counting *from the carboxylic acid (α) end*. Finally, because both double bonds are in the *cis* configuration,

partial hydrogenation A process whereby some carbon–carbon double bonds found in PUFAs are converted to carbon–carbon single bonds, resulting in the production of a lipid containing saturated and *trans* fatty acids.

*Because they are thought to not cause health problems, most naturally occurring *trans* fatty acids (such as those found in dairy products) are not included in this value.

Societies function best when they establish laws and regulations that limit dangerous behavior. For example, we have laws against driving under the influence of alcohol and smoking in public places. This concept sometimes carries over to nutrition, as well—such as the U.S. Food and Drug Administration's (FDA's) numerous regulations concerning accurate food labeling and safety of our food supply. The notion, however, of imposing laws concerning what kind of foods can and cannot be served in restaurants is relatively new. Such is the case for New York City, whose Board of Health voted in 2006 to ban the use of *trans* fats in the metropolis's restaurants. Specifically, restaurants must not serve foods containing *trans* fatty acids produced via partial hydrogenation. This ban, however, contains some exceptions. For instance, it still allows restaurants to serve *trans* fatty acid–containing foods that come in the manufacturer's original packaging. Thus, not all food sold must be *trans* fatty acid free. A similar rule

was passed in Seattle, and one has been considered in Chicago, although some residents of this city have ridiculed it as unnecessary government meddling.[3] This trend is not confined to just a few large U.S. cities. For example, Tiburon (a small town in California), claims to have become the first "*trans* fat–free city" in 2004, and in 2003 Denmark issued national regulations limiting the amount of *trans* fat that could be used in processed foods.

Of course, these actions come on the heels of significant scientific evidence that consumption of *trans* fatty acids found in partially hydrogenated oils can increase risk for heart disease.[2] Still, many people argue that local municipalities have no business "outlawing" foods that the FDA has deemed safe enough to allow in the public food system. Clearly, this is an area of justifiable debate, and only time will tell whether this type of local food regulation will catch on in other cities—or, more important, improve the health of the local citizens.

the name is modified to *cis*9,*cis*12–18:2. This fatty acid (*cis*9,*cis*12–18:2) is shown here.

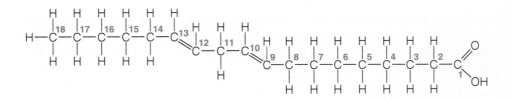

Omega (ω) Naming System An alternate system for naming a fatty acid is the omega (ω) system. In this system, the numbers of carbons and double bonds are again distinguished (for example, 18:2). However, in the omega (ω) naming system, fatty acids are described on the basis of where the first double bond is located *relative to the methyl (ω) end* of the molecule. If the first double bond is between the third and fourth carbons from the ω end, the fatty acid is an **omega-3 (ω-3) fatty acid.** If the first double bond is between the sixth and seventh carbons, it is an **omega-6 (ω-6) fatty acid.** There are also ω-7 and ω-9 fatty acids. Unlike alpha nomenclature, omega nomenclature does not usually identify whether the double bonds are in the *cis* or *trans* configuration or the location of the other double bonds in the molecule.

omega-3 (ω-3) fatty acid A fatty acid in which the first double bond is located between the third and fourth carbons from the methyl or omega (ω) end.

omega-6 (ω-6) fatty acid A fatty acid in which the first double bond is located between the sixth and seventh carbons from the methyl or omega (ω) end.

Common Names Sometimes, fatty acids are referred to by their common names, which often reflect prominent food sources where they are found. For example, *palm*itic acid (16:0) is found in *palm* oil, and *arach*idonic acid

TABLE 6.1 Names and Food Sources of Some Important Fatty Acids in the Body

Alpha (α) Nomenclature	Omega (ω) Family[a]	Common Name	Food Sources
Saturated Fatty Acids			
12:0	—	Lauric acid	Coconut and palm oils
14:0	—	Myristic acid	Coconut and palm oils; most animal and plant fats
16:0	—	Palmitic acid	Animal and plant fats
18:0	—	Stearic acid	Animal fats, some plant fats
20:0	—	Arachidic acid	Peanut oil
Unsaturated Fatty Acids			
*cis*9–16:1	ω-7	Palmitoleic acid	Marine animal oils
*cis*9–18:1	ω-9	Oleic acid	Plant and animal fats, olive oil
*cis*9,*cis*12–18:2	ω-6	Linoleic acid	Nuts, corn, safflower, soybean, cottonseed, sunflower seeds, and peanut oil
*cis*9,*cis*12,*cis*15–18:3	ω-3	Linolenic acid (α-linolenic acid)	Canola, soybean, flaxseed, and other seed oils
*cis*5,*cis*8,*cis*11,*cis*14–20:4	ω-6	Arachidonic acid	Small amounts in plant and animal oils
*cis*5,*cis*8,*cis*11,*cis*14,*cis*17–20:5	ω-3	Eicosapentaenoic acid (EPA)	Marine algae, fish oils
*cis*4,*cis*7,*cis*10,*cis*13,*cis*16,*cis*19–22:6	ω-3	Docosahexaenoic acid (DHA)	Animal fats as phospholipid component, fish oils

[a]The omega (ω) nomenclature only applies to unsaturated fatty acids, because it refers to the position of the first carbon–carbon double bond in relation to the methyl (ω) end of the fatty acid.

SOURCE: Adapted from Gropper SS, Smith JL, Groff JL. Advanced nutrition and human metabolism, 5th ed. Belmont, CA: Thomson/Wadsworth; 2007.

(*cis*5,*cis*8,*cis*11,*cis*14–20:4; from *arachis,* meaning legume or peanut) is found in peanut butter. Some common names of fatty acids are listed in Table 6.1.

Which Fatty Acids Do We Need, and Where Do They Come From?

There are literally hundreds of different fatty acids—each with its own distinct structure, chemical properties, and name. Although we need many of them for optimal health, there are only two fatty acids that adults cannot synthesize, and are therefore essential to get from food. In this section, you will learn about several fatty acids—including the essential fatty acids—as well as their sources and functions.

⟨**CONNECTIONS**⟩ Essential nutrients are needed by the body but cannot be synthesized in sufficient quantities to meet our needs (Chapter 1, page 5).

THERE ARE TWO ESSENTIAL FATTY ACIDS: LINOLEIC ACID AND LINOLENIC ACID

Although there is great diversity in the fatty acids found in foods, only two are dietary essentials: **linoleic acid** and **linolenic acid** (also called α-linolenic acid). Linoleic acid has 18 carbons, two *cis* double bonds, and is an ω-6 fatty acid. Linolenic acid has 18 carbons, three *cis* double bonds, and is an ω-3 fatty acid. Linoleic acid and linolenic acid are essential nutrients, because the body cannot create double bonds in the ω-3 and ω-6 positions.

The functions of the essential fatty acids are numerous. For example, they both serve as precursors of hormone-like compounds called eicosanoids (described below). Linolenic acid also is an important component of biological membranes, particularly in nerve tissue and the retina. In order for this to occur, these fatty acids must first be modified as described next.

linoleic acid An essential ω-6 fatty acid with 18 carbons and 2 double bonds.

linolenic acid An essential ω-3 fatty acid with 18 carbons and 3 double bonds.

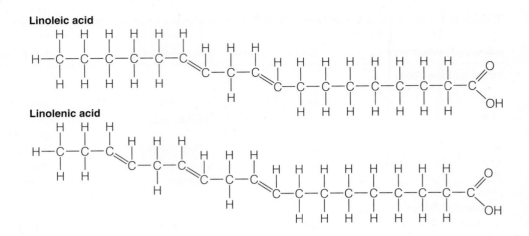

Linoleic acid

Linolenic acid

Converting Linoleic and Linolenic Acids to Longer-Chain Fatty Acids Dietary linoleic acid and linolenic acid provide the basic building blocks needed to make longer-chain ω-3 and ω-6 fatty acids. The synthesis of these newly formed (longer-chain) fatty acids is accomplished by increasing the number of carbon atoms (**elongation**) and the number of double bonds (**desaturation**). As shown in Figure 6.6, the essential fatty acid linoleic acid is used to make **arachidonic acid** (a 20-carbon, ω-6 fatty acid). This is accomplished by adding two carbon atoms via elongation and two double bonds via desaturation. Similarly, linolenic acid can be elongated and desaturated to **eicosapentaenoic acid (EPA;** a 20-carbon, ω-3 fatty acid with 5 double bonds), which can subsequently be elongated and desaturated further to **docosahexaenoic acid (DHA;** a 22-carbon, ω-3 fatty acid with 6 double bonds). These long-chain PUFAs have many functions in the body. For example, they are necessary for normal epithelial cell function and also assist in regulating gene expression. Because the conversion of linolenic acid to DHA is much lower than previously thought, dietary sources of this fatty acid may be especially important.[4]

Metabolism of the Essential Fatty Acids to Eicosanoids As you have just learned, the essential fatty acids can be elongated and desaturated to a variety of other

elongation The process whereby carbon atoms are added to a fatty acid, increasing its chain length.

desaturation The process whereby carbon–carbon single bonds are transformed into carbon–carbon double bonds in a fatty acid.

arachidonic acid (a – rach – i – DON – ic) A long-chain, polyunsaturated ω-6 fatty acid produced from linoleic acid.

eicosapentaenoic acid (EPA) (ei – co – sa – pen – ta – NO – ic) A long-chain, polyunsaturated ω-3 fatty acid produced from linolenic acid.

docosahexaenoic acid (DHA) (do – cos – a – hex – a – NO – ic) A long-chain, polyunsaturated ω-3 fatty acid produced from eicosapentaenoic acid (EPA).

FIGURE 6.6 Metabolism of the Essential Fatty Acids Linoleic Acid and Linolenic Acid Linoleic acid and linolenic acid can be elongated and desaturated to make other fatty acids. Some of these longer-chain fatty acids are used to make eicosanoids.

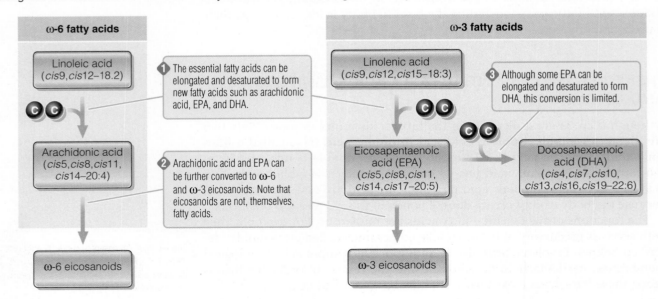

important fatty acids. Moreover, the essential fatty acids and some of their longer-chain metabolites can also be converted to other important compounds that are not, themselves, fatty acids but are lipid-like substances vital for health. One example is the **eicosanoids,** which act as chemical messengers that direct myriad physiologic functions. For example, eicosanoids play a particularly important role in regulating the immune and cardiovascular systems.[5] As shown in Figure 6.6, linoleic acid (an ω-6 fatty acid) is converted to ω-6 eicosanoids, whereas linolenic acid (an ω-3 fatty acid) is converted to ω-3 eicosanoids. An example of eicosanoids are the **prostaglandins,** which control dilation and constriction of blood vessels and therefore are important regulators of blood pressure.

The body produces both ω-3 and ω-6 eicosanoids, which have somewhat opposing actions. For instance, the ω-6 eicosanoids tend to cause inflammation and constriction of blood vessels, whereas the ω-3 eicosanoids are anti-inflammatory and stimulate dilation (or relaxation) of blood vessel walls. Also, ω-3 eicosanoids inhibit blood clotting. Both ω-3 and ω-6 eicosanoids are important for health, and the body can shift its relative production in response to its needs.

Your diet can also influence the amount and types of eicosanoids that you make. For example, Alaska natives consuming high amounts of ω-3 fatty acids from fish and marine mammals (e.g., seals) have enhanced physiologic responses stimulated by ω-3 eicosanoids. Consequently, Alaska natives consuming traditional diets tend to form blood clots more slowly than people who consume lesser amounts of ω-3 fatty acids.[6] Many studies suggest that alterations in the balance of ω-3 to ω-6 eicosanoids may influence a person's risk for conditions related to inflammation such as heart disease and cancer.[7] This is why experts recommend that we consume fish regularly, and these recommendations will be described in more detail later in this chapter.

Essential Fatty Acid Deficiency Essential fatty acid deficiencies are rare because of the almost endless supply of linoleic and linolenic acids stored in adipose tissue. Indeed, primary essential fatty acid deficiency generally occurs only in hospitalized patients receiving poor nutritional care. Secondary fatty acid deficiencies can occur with diseases that disrupt lipid absorption or utilization such as cystic fibrosis. Within two to four weeks, individuals with essential fatty acid deficiencies develop irritated and flaky skin, gastrointestinal problems, and impaired immune function. As a result, infections are common and wound healing may be slow. Children with essential fatty acid deficiencies also exhibit slow growth.

SOME FATTY ACIDS ARE CONDITIONALLY ESSENTIAL

In addition to linoleic and linolenic acids, other fatty acids may be conditionally essential during infancy. These include arachidonic acid and DHA. As described previously, in adults these fatty acids are made from linoleic acid and linolenic acid, respectively. However, babies may not be able to readily make them because they cannot produce the needed enzymes in sufficient amounts. The conditional essentiality of arachidonic acid and DHA during infancy is described in more detail in the Focus on Life Cycle Nutrition feature.

DIETARY SOURCES OF DIFFERENT TYPES OF FATTY ACIDS

Most foods contain a mixture of fatty acids. However, some foods are especially good sources of a particular fatty acid. For example, nuts (such as walnuts), seeds, and certain oils (such as those made from soybean, safflower, or corn) are generally abundant in linoleic acid. Linolenic acid is also found in

eicosanoids (ei – COS – a – noids) Biologically active compounds synthesized from arachidonic acid and eicosapentaenoic acid (EPA).

prostaglandins A group of eicosanoids involved in regulation of blood pressure; there are both ω-6 and ω-3 prostaglandins, having somewhat opposite effects.

Most young infants rely solely on either human milk or infant formula for all their nutritional needs. Although manufacturers strive to produce formulas that are similar to human milk, the lipids provided by these two infant foods are sometimes quite dissimilar. For example, human milk contains at least 47 different fatty acids, most of which are not found in infant formulas.[8] Scientists do not know precisely which of these fatty acids promote optimal growth and development during this time. However, some of the long-chain PUFAs found in human milk may be conditionally essential nutrients during infancy. These fatty acids, arachidonic acid and docosahexaenoic acid (DHA), are likely produced in sufficient amounts from linoleic and linolenic acids, respectively, in older children and adults. Therefore, they are not considered to be essential nutrients during these periods of the life cycle. However, infants (especially those born prematurely) have very low stores of arachidonic acid and DHA and may not be able to synthesize them in adequate amounts.[9] These fatty acids are thought to be important for growth and development of the eyes, nervous system, and brain.[10] Thus, many scientists believe that infants should consume adequate amounts of them during early life to achieve optimal growth and development.

Until relatively recently, only breastfed babies received these fatty acids, because infant formulas were not fortified with long-chain PUFAs. Because some research suggests that fortifying infant formula with long-chain PUFAs may be advantageous, many companies now produce infant formula fortified with arachidonic acid and DHA, and these have been marketed in the United States since 2002.[11] Because little is known about the long-term effects of these long-chain PUFA-fortified formulas on human health, the FDA has asked manufacturers to closely monitor these products in the marketplace. Somewhat surprisingly, the American Academy of Pediatrics has no official position on the use of these fortified formulas in infant feeding.

Scientists are interested in the importance of other dietary lipids during infancy as well. An example is cholesterol. Although researchers have long known that human milk contains high amounts of cholesterol, infant formula generally lacks this lipid.[12] Whether early exposure to dietary cholesterol or the multitude of "nonessential" fatty acids and other lipids in human milk is important for optimal growth and development is unknown but continues to interest the medical and scientific communities.[13]

© 2011 Blend Images/JupiterImages Corporation

some oils (such as those made from canola, soybean, or flaxseed) as well as some nuts (such as walnuts). Some foods, such as soybean oil and walnuts, are good sources of both essential fatty acids. Because many of these foods and oils are common in the American diet, getting adequate amounts of linoleic acid and linolenic acid is easy.

Many experts also believe that it is important for adults to consume adequate amounts of the longer-chain ω-3 fatty acids such as EPA and DHA because higher intakes are related to lower risk for heart disease and stroke. This may be due to the potent anti-inflammatory effects of these compounds in the body. You can read more about the relationship between longer-chain ω-3 fatty and cardiovascular disease in the Nutrition Matters at the end of this chapter. EPA and DHA are plentiful in fatty fish and other seafood; smaller amounts are found in meat and eggs. Longer-chain ω-6 fatty acids are also found in meat, poultry, and eggs. Omega-6 fatty acids are typically more plentiful in our diets than are ω-3 fatty acids.

In general, animal foods contribute the majority of dietary SFAs (saturated fatty acids), whereas plant-derived foods supply the majority of PUFAs (polyunsaturated fatty acids). MUFAs (monounsaturated fatty acids) come from both plant and animal foods. However, some tropical oils, such as coconut and palm oils, contain relatively high amounts of SFAs, and many oily fish

© Michael Mahovlich/Masterfile

Many fish are excellent sources of ω-3 fatty acids.

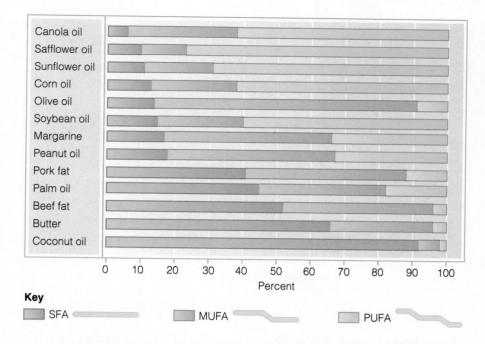

FIGURE 6.7 **Distribution of Fatty Acid Classes in Commonly Consumed Lipids** Dietary lipids contain different relative amounts of saturated fatty acids (SFAs), monounsaturated fatty acids (MUFAs), and polyunsaturated fatty acids (PUFAs).

have high levels of PUFAs. The relative amounts of SFAs, MUFAs, and PUFAs in commonly consumed fats and oils are summarized in Figure 6.7.

Mono-, Di-, and Triglycerides: What's the Difference?

As previously mentioned, most fatty acids do not exist in their free (unbound) form in foods or in the body. Instead, they are part of larger, more complex molecules such as triglycerides, diglycerides, and monoglycerides. The number of fatty acids attached to the glycerol backbone determines whether the molecule is a mono-, di-, or triglyceride.[†] A **triglyceride** has three fatty acids, a **diglyceride** consists of two fatty acids, and a **monoglyceride** contains only one fatty acid. In all cases, the fatty acid molecules are attached to the glycerol backbone via "ester" linkages (Figure 6.8). These fatty acids can be saturated, monounsaturated, polyunsaturated, or a mixture of fatty acid types. Regardless of which types of fatty acids are attached to the glycerol molecule, most mono-, di-, and triglycerides are relatively hydrophobic.

TRIGLYCERIDES PLAY MANY ROLES IN THE BODY

Triglycerides—or, more accurately, their associated fatty acids—serve many purposes in the body. Perhaps most importantly, they are sources of the essential fatty acids needed for the body to function. Triglycerides are also vitally important for meeting both immediate and long-term energy needs.

Triglycerides as an Energy Source Compared with the other energy-yielding macronutrients, triglycerides represent the body's richest source of energy. As you may recall from Chapter 1, the complete breakdown of 1 g of fatty acids yields approximately 9 kcal of energy, which is more than twice the yield from 1 g of carbohydrate or protein (4 kcal). Therefore, gram for gram, high-fat foods contain more calories than do other foods.

[†]Triglycerides, diglycerides, and monoglycerides are also called triacylglycerols, diacylglycerols, and monoacylglycerols, respectively.

triglyceride (also called triacylglycerol) (*tri*, three) A lipid composed of a glycerol molecule bonded to three fatty acids.

diglyceride (also called diacylglycerol) (*di*, two) A lipid made of a glycerol molecule bonded to two fatty acids.

monoglyceride (also called monoacylglycerol) (*monos*, single) A lipid made of a glycerol molecule bonded to a single fatty acid.

FIGURE 6.8 A **Triglyceride Molecule** A triglyceride molecule consists of a glycerol molecule bonded to three fatty acids via "ester" linkages.

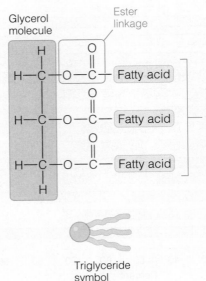

Glycerol molecule

Ester linkage

These fatty acids can be saturated (SFA), monounsaturated (MUFA), polyunsaturated (PUFA), or a combination.

Triglyceride symbol

Regardless of where they originate (from diet or adipose tissue), for triglycerides to be used as a source of energy, they must first be disassembled into glycerol and fatty acids. This process, called **lipolysis,** is catalyzed by enzymes called **lipases.** When triglycerides circulate in the blood, the form of lipase that removes the fatty acids from a triglyceride's backbone is called **lipoprotein lipase.** This is described in more detail in a later section of this chapter. When triglycerides within adipose tissue are being broken down for energy, the form of lipase that cleaves the fatty acids from their glycerol backbone is called **hormone-sensitive lipase.** To facilitate lipolysis in adipose tissue during times of starvation, low insulin-to-glucagon ratios stimulate hormone-sensitive lipase activity. Recall that insulin levels decrease and glucagon levels increase within a few hours of eating. Lipolysis via hormone-sensitive lipase is also stimulated by exercise and other forms of physiological stress.

Fatty Acids Can Also Be Converted to Energy-Yielding Ketones In addition to using fatty acids directly as an energy source, the body can convert them to other energy-yielding compounds called ketones. Recall from Chapter 4 that the production of ketones from fatty acids, a process called ketogenesis, occurs when glucose availability is low. Ketogenesis is important because some tissues such as brain, heart, skeletal muscle, and kidneys can use ketones for energy (ATP). In this way, ketones produced from fatty acids can serve as an important source of energy during times of severe glucose insufficiency (such as starvation) and can reduce the need for amino acids to be converted to glucose via the process of gluconeogenesis. This helps protect the body from having to use protein from its muscles and other tissues for energy.

Storage of Excess Energy as Triglycerides in Adipose Tissue But what happens when the energy available to the body exceeds its energy needs? During these times, excess fatty acids are stored as triglycerides in adipose tissue and, to a lesser extent, skeletal muscle. Adipose tissue consists of specialized cells called **adipocytes,** which can accumulate large amounts of lipid. Adipose tissue is found in many parts of the body, including under the skin (**subcutaneous adipose tissue**) and around the vital organs in the abdomen (**visceral adipose tissue** also called intra-abdominal fat). Many of the body's organs and tissues (such as the kidneys and breasts) have visceral adipose tissue associated with them, giving them ready access to fatty acids for their immediate energy

⟨**CONNECTIONS**⟩ Remember that gluconeogenesis is the synthesis of glucose from noncarbohydrate substances (Chapter 4, page 140).

lipolysis The breakdown of triglycerides into fatty acids and glycerol.

lipases Enzymes that cleave fatty acids from the glycerol backbones of triglycerides, phospholipids, and cholesteryl esters.

lipoprotein lipase An enzyme that hydrolyzes the ester linkage between a fatty acid and glycerol in a triglyceride, diglyceride, and monoglyceride molecule as they circulate in the bloodstream.

hormone-sensitive lipase An enzyme that catalyzes the hydrolysis of ester linkages that attach fatty acids to the glycerol molecule; mobilizes fatty acids stored in adipose tissue.

adipocyte (a – DIP – o – cyte) A specialized cell that makes up the majority of adipose tissue.

subcutaneous adipose tissue Adipose tissue found directly under the skin.

visceral adipose tissue Adipose tissue surrounding the vital organs.

needs. You will learn more about subcutaneous and visceral adipose tissue and how their distribution may influence health in Chapter 8.

Insulin Stimulates Lipogenesis and Triglyceride Storage The hormone insulin stimulates storage of triglycerides during times of energy excess—for instance, after a high-calorie meal. Insulin causes adipocytes, and to a lesser extent skeletal muscle cells, to take up glucose and fatty acids. It also stimulates the conversion of excess glucose to fatty acids. In turn, these newly formed fatty acids and excess dietary fatty acids can be incorporated into triglycerides. The synthesis of fatty acids and triglycerides is called **lipogenesis**. Because insulin inhibits the action of hormone-sensitive lipase, high levels of insulin also inhibit lipolysis, or the breakdown of lipids. Together, increased lipogenesis and decreased lipolysis after a meal help direct excess fatty acids to adipose tissue, where they are deposited for later use.

Energy Storage as Lipid Has Its Advantages Compared to glycogen, storage of excess energy as triglyceride has several advantages. First, because triglycerides are not stored with water (a bulky molecule), large amounts can fit in small spaces. Consequently, the body can store about six times more energy in one pound of adipose tissue than in one pound of liver glycogen. Indeed, the body has a seemingly infinite ability to store excess energy in adipose tissue, whereas its capacity to store glycogen is limited. In addition, gram for gram, lipids store more than twice the energy than glycogen.

Triglycerides Needed for Insulation and Protection Aside from their important role as an energy source, triglycerides stored in adipose tissue also protect your internal organs from injury. We also rely on adipose tissue for insulation, which keeps us warm. People with very little body fat can have difficulty regulating body temperature. In fact, one physiological response to being severely underweight is that very fine hair covering the body can develop. This hair, called *lanugo*, partially makes up for the absence of subcutaneous adipose tissue by providing a layer of external insulation. The presence of lanugo is common in people who have the eating disorder anorexia nervosa.[14]

What Are Phospholipids and Sterols?

In addition to triglycerides, the two other major lipid categories are phospholipids and sterols, both of which are essential components of cell membranes and are involved in the transport of lipids in the bloodstream. Because the body can synthesize all that it needs, there are no dietary requirements for either of these lipid classes. Nonetheless, they are commonly found in food.

PHOSPHOLIPIDS ARE CONSIDERED "AMPHIPATHIC"

Phospholipids are similar to triglycerides in that they contain a glycerol molecule bonded to fatty acids (Figure 6.9). However, instead of having three fatty acids, a phospholipid has only two fatty acids. Replacing the third fatty acid is a phosphate-containing **polar head group**. There are many different types of polar head groups, but the most common are choline, ethanolamine, inositol, and serine.

Phospholipids contain both polar (hydrophilic) and nonpolar (hydrophobic) regions. **Polar** substances are those with an unequal charge distribution. In other words, one portion of a polar molecule is positively charged, whereas another is negatively charged. Polar molecules, such as water and sodium chloride (NaCl; table salt), are attracted to other polar molecules. This

lipogenesis (*lipos*, fat; *genus*, birth) The metabolic processes that result in fatty acid and, ultimately, triglyceride synthesis.

phospholipid A type of lipid composed of a glycerol bonded to two fatty acids and a polar head group.

polar head group A phosphate-containing charged chemical structure that is a component of a phospholipid.

polar molecule A molecule (such as water) that has both positively and negatively charged portions.

FIGURE 6.9 A Phospholipid Molecule
A phospholipid consists of a glycerol molecule, a polar head group, and two fatty acids.

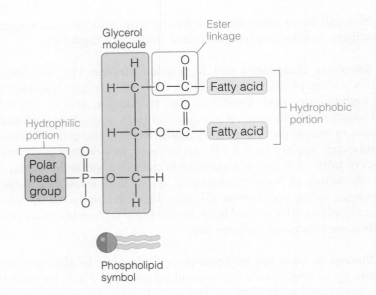

is why table salt dissolves readily in water. Because polar compounds dissolve in water, they are referred to as **hydrophilic** (or "water-loving"). Conversely, **nonpolar** compounds (such as triglycerides) have even charge distributions and do not dissolve easily in polar compounds. This is why oil and water do not mix, and why nonpolar compounds are often referred to as **hydrophobic** (or "water-fearing").

Having both nonpolar and polar portions makes *phospholipids* **amphipathic** compounds. The prefix *"amphi-"* means "both" and is the same as the prefix in the word *amphibian,* which refers to an animal capable of living in two environments, land and water. Similarly, amphipathic substances typically straddle two environments—in this case, lipids and water. In terms of a phospholipid, the two fatty acids are hydrophobic and therefore drawn to lipids, while the polar head group is hydrophilic and drawn to water. These opposing attractions allow phospholipids to act as major components of cell membranes and to play roles in the digestion, absorption, and transport of lipids in the body.

Most Foods Contain Phospholipids Because they are components of cell membranes (animals) and cell walls (plants), phospholipids are found naturally in most foods. They are also used as food additives. An example is **phosphatidylcholine** (the phospholipid with a choline polar head group), which is often added to foods such as mayonnaise to prevent them from separating. You can check whether a food has phosphatidylcholine by looking for the word "lecithin" (a common name for this compound) on the food label.

PHOSPHOLIPIDS ARE CRITICAL FOR CELL MEMBRANES AND LIPID TRANSPORT

As previously mentioned, phospholipids are the main structural component of membranes surrounding cells and organelles (Figure 6.10). For example, cell membranes consist of two layers (a bilayer) of phospholipids with the hydrophilic polar head groups pointing to the extra- and intracellular spaces, both of which are predominantly water. The hydrophobic tails face toward the interior of the membrane, forming a water-free lipid-rich zone. If the cell membrane were completely hydrophilic, it would dissolve and not create a barrier. If the cell membrane were completely hydrophobic, there would be no

hydrophilic substance (hy – dro – PHIL – ic) One that dissolves in or mixes with water.

nonpolar molecule One that does not have differently charged portions.

hydrophobic substance (hy – dro – PHO – ic) One that does not dissolve in or mix with water.

amphipathic (amphi-, on both sides) Having both nonpolar (noncharged) and polar (charged) portions.

phosphatidylcholine (also called lecithin) (PHOS – pha – tid – yl – CHO – line) A phospholipid that contains choline as its polar head group; commonly added to foods as an emulsifying agent.

FIGURE 6.10 A Cell Membrane Consists of a Phospholipid Bilayer, Proteins, and Cholesterol The amphipathic nature of phospholipids allows cell membranes to carry out their functions.

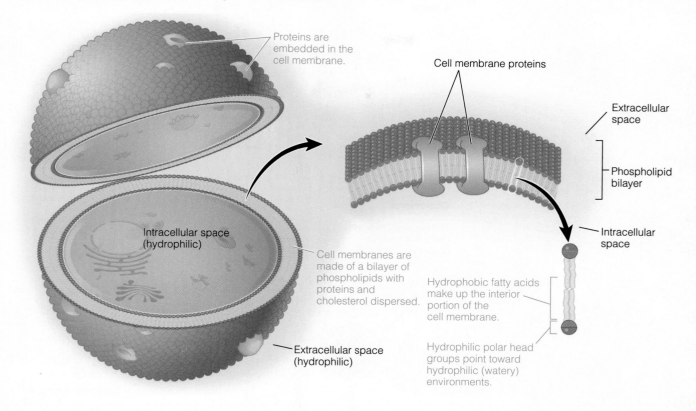

Proteins are embedded in the cell membrane.

Cell membrane proteins

Extracellular space

Phospholipid bilayer

Intracellular space

Intracellular space (hydrophilic)

Cell membranes are made of a bilayer of phospholipids with proteins and cholesterol dispersed.

Hydrophobic fatty acids make up the interior portion of the cell membrane.

Extracellular space (hydrophilic)

Hydrophilic polar head groups point toward hydrophilic (watery) environments.

chance of communication between the watery extra- and intracellular compartments. Thus, the incorporation of two layers of amphipathic phospholipids allows cell membranes to effectively carry out their functions.

Phospholipids also supply fatty acids for cellular metabolism and can act as biologically active compounds. For example, some activate enzymes important for energy metabolism, blood clotting, and cell turnover. Phospholipids also donate their fatty acids for eicosanoid production and can act as carriers of hydrophobic substances in the body. One example is the incorporation of phospholipids in the outer surface of a group of particles called lipoproteins. Lipoproteins transport lipids in the blood and are described later in this chapter.

STEROLS AND STEROL ESTERS ARE LIPIDS WITH RING STRUCTURES

Sterols are structurally different from other lipids in that they consist of multi-ring structures (Figure 6.11). A sterol can either be free (not bonded to another molecule) or attached to a fatty acid via an ester linkage. The latter is called a **sterol ester**.

Functions of Cholesterol and Cholesteryl Esters There are many types of sterols, but the most abundant and widely discussed is **cholesterol.** Some free cholesterol is found in the body, although most is bonded to a fatty acid (recall Figure 6.2). This cholesterol–fatty acid complex is called a **cholesteryl ester** and is an example of a sterol ester. Because they contain fatty acids,

sterol A type of lipid with a distinctive multi-ring structure; a common example is cholesterol.

sterol ester A chemical compound consisting of a sterol molecule bonded to a fatty acid via an ester linkage.

cholesterol (*kholikos*, bile; *stereos*, hard or solid) A sterol found in animal foods and made in the body; required for bile acid and steroid hormone synthesis.

cholesteryl ester A sterol ester made of a cholesterol molecule bonded to a fatty acid via an ester linkage.

FIGURE 6.11 Structures of a Sterol, Cholesterol, and a Cholesteryl Ester

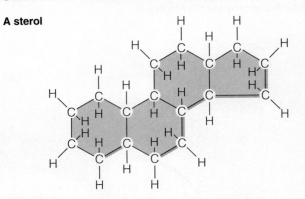

A sterol

Cholesterol

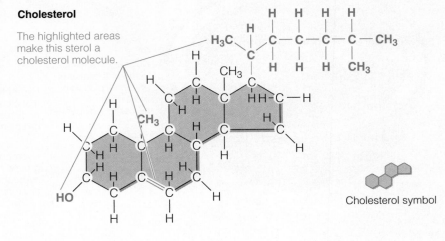

The highlighted areas make this sterol a cholesterol molecule.

Cholesterol symbol

A cholesteryl ester

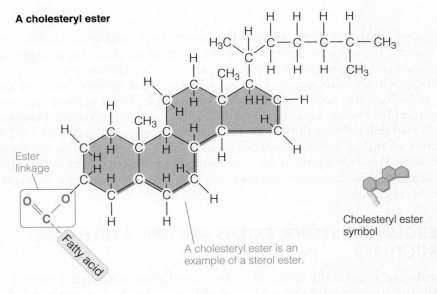

Ester linkage

Fatty acid

A cholesteryl ester is an example of a sterol ester.

Cholesteryl ester symbol

cholesteryl esters are more hydrophobic than is free cholesterol, which is weakly polar. The unhealthy relationship between cholesterol and heart disease is widely discussed, and you will learn more about this later in the chapter. You may not, however, be as familiar with the many essential roles cholesterol plays in the body.

For instance, cholesterol is needed to synthesize bile acids, which are important for the digestion and absorption of lipids. A **bile acid** consists of a

bile acid Amphipathic substance made from cholesterol in the liver; a component of bile important for lipid digestion and absorption.

cholesterol molecule attached to a very hydrophilic subunit, making it amphipathic, much like a phospholipid. Similar to phospholipids, cholesterol and cholesteryl esters are also components of membranes, such as those surrounding cells, where they help maintain fluidity. In addition, cholesterol is needed for the synthesis of the **steroid hormones,** which are important for reproduction, energy metabolism, calcium homeostasis and electrolyte balance. Cholesteryl esters also serve as crucial carrier molecules for fatty acids in the blood.

Sources of Cholesterol in the Body: Synthesis and Diet Almost every tissue in the body, especially the liver, can make cholesterol from glucose and fatty acids. Many dietary factors, however, influence how much cholesterol you make. For example, eating a low-calorie or low-carbohydrate diet can decrease cholesterol synthesis in some people.[15] This is not always the case, however, because dietary factors can interact with genetics to influence how much cholesterol a person makes. In other words, carbohydrate intake may only affect cholesterol synthesis in people with certain genetic variations. When dietary intervention is not effective in lowering blood cholesterol levels, medications are sometimes recommended. This is often the case for people at risk for heart disease. For example, the statin drugs Lipitor® (atorvastatin calcium) and Zocor® (simvastatin) decrease blood cholesterol by inhibiting one of the liver enzymes needed for its synthesis.

In addition to being made by the body, cholesterol is also obtained from animal-derived foods, such as shellfish, meat, butter, eggs, and liver (Figure 6.12). Plants do not produce substantial amounts of cholesterol, so exclusively plant-based foods are relatively low in cholesterol. However, because cholesterol is made by the body, vegans who eat no animal products are not at risk of cholesterol deficiency.

Animal-derived foods provide the vast majority of cholesterol to the diet.

steroid hormone (*stereos*, hard or solid; *hormon*, to urge on) A hormone made from cholesterol.

FIGURE 6.12 Cholesterol Content of Selected Foods

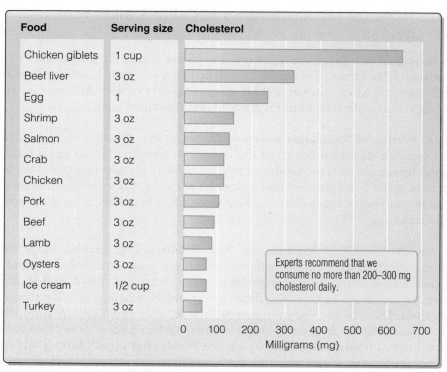

Food	Serving size	Cholesterol
Chicken giblets	1 cup	
Beef liver	3 oz	
Egg	1	
Shrimp	3 oz	
Salmon	3 oz	
Crab	3 oz	
Chicken	3 oz	
Pork	3 oz	
Beef	3 oz	
Lamb	3 oz	
Oysters	3 oz	
Ice cream	1/2 cup	
Turkey	3 oz	

Experts recommend that we consume no more than 200–300 mg cholesterol daily.

0   100   200   300   400   500   600   700
Milligrams (mg)

SOURCE: USDA National Nutrition Database for Standard Reference, Release 17, 2005.

Phytosterols and Phytostanols Are Sterol-Like Plant Compounds Although plants make very small amounts of cholesterol, some contain relatively large amounts of sterol-like compounds called **phytosterols** and **phytostanols.** One interesting group of phytosterols is found naturally in corn, wheat, rye, and other plants. Similar compounds are also produced commercially and are marketed under various names, such as Benecol®. These products are often found on the grocery shelf in butter substitutes, yogurt drinks, salad dressings, and even dietary supplements. Some studies suggest that consuming products containing phytosterols or stanols may decrease blood cholesterol, lowering the risk for cardiovascular disease.[16] As such, the FDA has approved the following health claim.

"Diets low in saturated fat and cholesterol that include two servings of foods that provide a daily total of at least 3.4 grams of plant sterols/stanols in two meals may reduce the risk of heart disease."[17]

Because a typical serving of a plant sterol–fortified table spread contains about 1.1 gram of the sterol, you would need to consume about three servings daily to reach this goal. The mechanisms by which plant sterols/stanols decrease blood cholesterol are not well understood.[16] However, because they are not readily absorbed and appear to bind cholesterol in the intestine, consumption of sterols and stanols may increase cholesterol elimination in the feces.

How Are Dietary Lipids Digested?

Once lipids have been ingested, they must be digested, absorbed, and circulated away from the small intestine. As you will learn, digestion of the various types of dietary lipids is accomplished by enzymes and other secretions produced by the gastrointestinal tract and accessory organs.

DIGESTION OF TRIGLYCERIDES REQUIRES BILE AND LIPASES

The basic goal of triglyceride digestion is to cleave two of the fatty acids from the glycerol backbone. This is accomplished in your mouth, stomach, and small intestine when you eat a meal containing fats or oils. An overview of lipid digestion is illustrated in Figure 6.13 and described next.

A Small Portion of Triglyceride Digestion Occurs in Your Mouth The first stage of triglyceride digestion begins in the mouth. As chewing breaks apart food, **lingual lipase** (an enzyme produced by your salivary glands) begins to hydrolyze fatty acids from glycerol molecules. After the food is swallowed, lingual lipase accompanies the bolus into your stomach, where this enzyme continues to function.

Triglyceride Digestion Continues in Your Stomach The second stage of triglyceride digestion begins when food enters your stomach, stimulating the release of the hormone gastrin from specialized cells found in the gastric pits. Gastrin in turn circulates in the blood, where it quickly stimulates the release of the enzyme **gastric lipase,** also produced in stomach cells. Gastric lipase is a component of the "gastric juices" and continues where lingual lipase left off, breaking the bonds that attach fatty acids to glycerol molecules.

⟨**CONNECTIONS**⟩ Remember that hydrolysis is the breaking of chemical bonds by the addition of water. As a result, larger compounds are broken down into smaller subunits (Chapter 3, page 74).

phytosterol (*phuto*, plant) Sterol made by plants.

phytostanol Sterol-like compound made by plants.

lingual lipase (*lingua*, tongue; *lipos*, fat) An enzyme, produced in the salivary glands, that hydrolyzes ester linkages between fatty acids and glycerol molecules.

gastric lipase (*gaster*, belly; *lipos*, fat) An enzyme, produced in the stomach, that hydrolyzes ester linkages between fatty acids and glycerol molecules.

FIGURE 6.13 Overview of Triglyceride Digestion

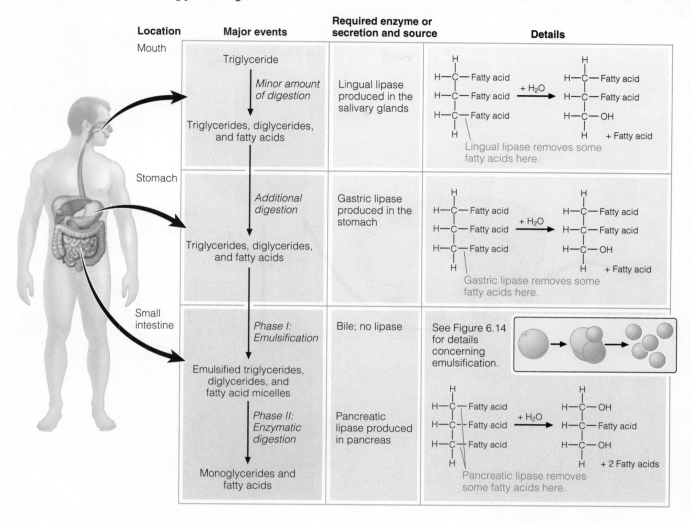

Triglyceride Digestion Is Completed in Your Small Intestine Although some fatty acids are removed from the glycerol backbones in your stomach via gastric lipase, triglyceride digestion in the stomach is incomplete. This is in part because the watery environment of your GI tract causes lipids to clump together in large lipid globules, which are difficult to digest. To overcome this problem, the final stage of triglyceride digestion in your small intestine occurs in two complementary and consecutive phases.

Phase I: Emulsification of Lipids by Bile—Micelle Formation Phase I of intestinal lipid digestion begins with the arrival of lipids into the small intestine (Figure 6.14). This stimulates the release of the enteric hormone cholecystokinin (CCK), which in turn signals the gallbladder to contract and release bile. Recall that bile consists of a mixture of bile acids, cholesterol, and phospholipids. When bile acids and phospholipids are released into the duodenum, their hydrophobic portions are drawn toward the lipid globules, while their hydrophilic portions pull in the opposite direction, toward the surrounding water. These opposing forces disperse the large lipid globules into smaller droplets, a process called **emulsification.** Emulsification makes the ester linkages more accessible to the digestive enzymes in the small intestine.

emulsification The process whereby large lipid globules are broken down and stabilized into smaller lipid droplets.

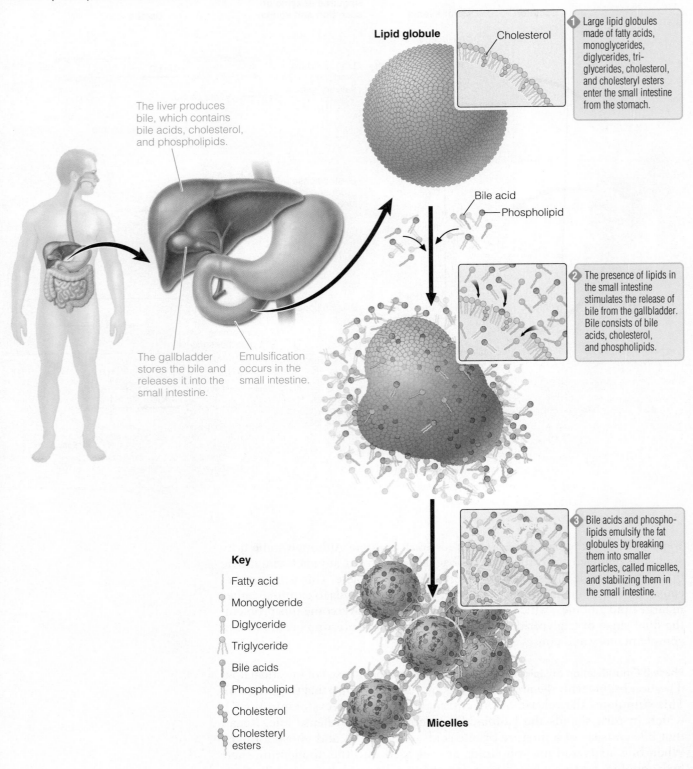

The liver produces bile, which contains bile acids, cholesterol, and phospholipids.

The gallbladder stores the bile and releases it into the small intestine.

Emulsification occurs in the small intestine.

Lipid globule

Cholesterol

1 Large lipid globules made of fatty acids, monoglycerides, diglycerides, tri-glycerides, cholesterol, and cholesteryl esters enter the small intestine from the stomach.

Bile acid

Phospholipid

2 The presence of lipids in the small intestine stimulates the release of bile from the gallbladder. Bile consists of bile acids, cholesterol, and phospholipids.

Key

	Fatty acid
	Monoglyceride
	Diglyceride
	Triglyceride
	Bile acids
	Phospholipid
	Cholesterol
	Cholesteryl esters

3 Bile acids and phospho-lipids emulsify the fat globules by breaking them into smaller particles, called micelles, and stabilizing them in the small intestine.

Micelles

Bile also disperses phospholipids and cholesteryl esters in much the same way. Bile acids and phospholipids then stay with the newly formed drop-lets, which are now referred to as **micelles.** Because they are coated with amphipathic substances, micelles do not gel together to reform larger lipid globules.

micelle A water-soluble, spherical structure formed in the small intestine via emulsification.

Gallbladder disease can result when the gallbladder becomes inflamed or when bile contains an excess amount of cholesterol in relation to its other components. The accumulation of calcium and cellular debris around the cholesterol results in the formation of gallstones, which can range in size from 5 mm to more than 25 mm in diameter. For a point of reference, 1 millimeter is roughly the thickness of a dime. Some people with gallstones have no symptoms, while others can experience tenderness or extreme pain. Gallstones can also become lodged in the cystic or common bile ducts leading from the gallbladder to the small intestine, obstructing the flow of bile and pancreatic secretions into the intestine. Surgical removal of the gallbladder is the most common treatment for persistent gallstone-related problems. Because some people may have difficulty digesting fat after the gallbladder is removed, doctors often recommend that people initially avoid high-fat meals. However, the liver continues to supply bile to the small intestine even after the gallbladder is removed, so most people do not have to adjust their diet for very long.

Gallbladder disease is more common in women than men, and its prevalence increases with age.[18] Other risk factors include obesity, rapid weight loss, and pregnancy. It remains unclear if particular foods or dietary practices influence the formation of gallstones, but some studies report the prevalence to be lower in vegetarians than nonvegetarians.[19] In addition, moderate alcohol consumption (1–2 drinks/day), exercise, and aspirin use appear to have protective roles.

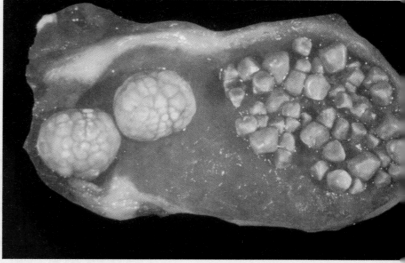

© Biophoto Associates/Photo Researchers, Inc

Gallstones are composed of calcium, cellular debris, and cholesterol.

Another way to think about lipid emulsification is to consider what happens when you shake water and cooking oil together with and without added soap. If you do not add soap, the water and oil initially appear to mix, but the oil eventually separates from the water. This is because oil is nonpolar (not charged), whereas water is polar (charged), and nonpolar and polar compounds do not willingly mix. However, if you add a few drops of dish soap and shake the mixture again, the oil becomes suspended as small droplets within the water. This is because soap is an amphipathic emulsifier, breaking up large lipid droplets into smaller ones and allowing them to disperse. Although most people never need to think about the importance of bile in lipid digestion, those with gallbladder disease fully understand. The Focus on Clinical Applications feature describes this disease in more detail.

© iStop/SuperStock

The emulsification of lipids in the small intestine by bile acids is similar to the way dish soap makes grease combine with water.

Courtesy of Shelley McGuire

Critical Thinking: Nancy's Story Recall Nancy, the woman featured at the beginning of the chapter. Now that you understand the importance of bile in the process of lipid digestion, can you explain why eating high-fat foods caused her to experience indigestion and abdominal pain? After her gallbladder surgery, why was it so important that she add back lipids *very slowly* to her diet?

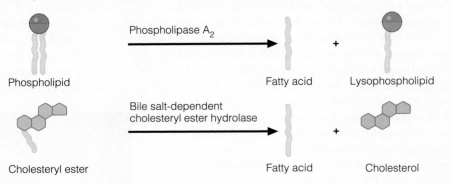

FIGURE 6.15 **Phospholipid and Cholesteryl Ester Digestion** Both phospholipids and cholesteryl esters rely on pancreatic enzymes (phospholipase A₂ and bile salt-dependent cholesteryl ester hydrolase, respectively) for their digestion in the small intestine.

Phospholipase A₂

Phospholipid

Fatty acid

Lysophospholipid

Bile salt-dependent cholesteryl ester hydrolase

Cholesteryl ester

Fatty acid

Cholesterol

Phase II: Digestion of Triglycerides by Pancreatic Lipase In response to lipid-containing chyme entering the duodenum, the small intestine releases the hormones secretin and cholecystokinin (CCK). These enteric hormones signal the pancreas to release pancreatic juices containing the enzyme **pancreatic lipase.** Pancreatic lipase completes triglyceride digestion by hydrolyzing additional fatty acids from glycerol molecules in the micelles. In general, two of the three fatty acids are removed from the triglyceride molecules, resulting in the release of a monoglyceride and two free (unbound) fatty acids.

DIGESTION OF PHOSPHOLIPIDS AND CHOLESTERYL ESTERS ALSO REQUIRES PANCREATIC ENZYMES

As is the case with triglycerides, very little digestion of the other major fatty acid-containing lipids (phospholipids and cholesteryl esters) occurs in the mouth and stomach. Instead, their digestion begins in the small intestine, where—like the triglycerides—phospholipids and cholesteryl esters contained in lipid globules are emulsified into micelles by bile. Phospholipids are then digested by the enzyme **phospholipase A₂,** produced by the pancreas and released in response to the intestinal hormone secretin. The products of digestion of a single phospholipid molecule are one fatty acid and a compound called a **lysophospholipid,** consisting of a glycerol molecule bonded to a fatty acid and a polar head group (Figure 6.15).

Although free cholesterol (not bonded to a fatty acid) does not need to be digested prior to absorption, cholesteryl esters must be broken down into cholesterol and fatty acids. This is accomplished by yet another pancreatic enzyme: **bile salt–dependent cholesteryl ester hydrolase.**

As you have learned, the pancreas is vitally important for the body to digest the many lipids found in foods. But for these lipids to be used by the body, they must be absorbed from the gastrointestinal tract and circulated to other tissues. These processes are described next.

pancreatic lipase An enzyme, produced in the pancreas, that hydrolyzes ester linkages between fatty acids and glycerol molecules.

phospholipase A₂ An enzyme, produced in the pancreas, that hydrolyzes fatty acids from phospholipids.

lysophospholipid A lipid composed of a glycerol bonded to a polar head group and a fatty acid; one of the final products of phospholipid digestion.

bile salt–dependent cholesteryl ester hydrolase An enzyme, produced in the pancreas, that cleaves fatty acids from cholesteryl esters.

How Are Dietary Lipids Absorbed and Circulated in the Body?

The products of lipid digestion are absorbed into the enterocytes and circulated away from the small intestine. This requires special handling because many of the products of digestion are hydrophobic, while both the interior of the enterocyte and the circulatory system are hydrophilic. It is the amphipathic properties of phospholipids that make both lipid absorption and circulation possible.

DIETARY LIPIDS ARE ABSORBED IN THE SMALL INTESTINE

Lipid absorption is accomplished in two ways, depending on how water-soluble (hydrophilic) the lipid is (Figure 6.16). Because they are relatively water soluble, short- and medium-chain fatty acids can be transported into the enterocytes unassisted. However, more hydrophobic compounds such as long-chain fatty acids, monoglycerides, and cholesterol must first be packaged into a form of micelle that can deliver them to the enterocytes. Once this micelle comes into contact with the brush border surface, its contents are released and transported into the enterocytes.

DIETARY LIPIDS ARE CIRCULATED AWAY FROM THE SMALL INTESTINE IN TWO WAYS

The process of lipid circulation in the body is complicated because these hydrophobic substances must somehow be transported in an overwhelmingly watery environment. Depending on how hydrophobic they are, lipids are initially circulated away from the small intestine in either the blood or the lymph.

Circulation of Relatively Hydrophilic Lipids in Your Blood The simplest case of dietary lipid circulation involves the short- and medium-chain fatty acids, which are somewhat water soluble (hydrophilic). As such, these fatty acids

FIGURE 6.16 Absorption and Circulation of Lipids in the Small Intestine How various lipids are absorbed and circulated depends on how hydrophilic they are. In general, it is easier for the body to absorb and circulate more hydrophilic lipids (such as short- and medium-chain fatty acids) than more hydrophobic lipids (such as long-chain fatty acids and monoglycerides).

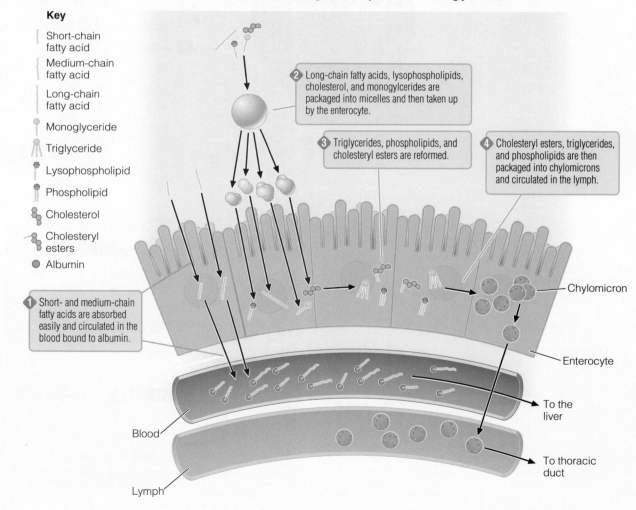

Key

- Short-chain fatty acid
- Medium-chain fatty acid
- Long-chain fatty acid
- Monoglyceride
- Triglyceride
- Lysophospholipid
- Phospholipid
- Cholesterol
- Cholesteryl esters
- Albumin

2 Long-chain fatty acids, lysophospholipids, cholesterol, and monoglycerides are packaged into micelles and then taken up by the enterocyte.

3 Triglycerides, phospholipids, and cholesteryl esters are reformed.

4 Cholesteryl esters, triglycerides, and phospholipids are then packaged into chylomicrons and circulated in the lymph.

1 Short- and medium-chain fatty acids are absorbed easily and circulated in the blood bound to albumin.

Chylomicron

Enterocyte

To the liver

To thoracic duct

Blood

Lymph

can be circulated away from your small intestine in the blood. However, they are first bound to the protein albumin. Fatty acid–albumin complexes circulate in the blood from the small intestine to the liver, where they are either metabolized or rerouted for delivery to other cells in the body.

Circulation of More Hydrophobic Lipids in Lymph via Chylomicrons The circulation of larger lipids away from the GI tract is more involved. Long-chain fatty acids, monoglycerides, and lysophospholipids (products of phospholipid digestion) first enter the enterocyte, where they are reassembled into triglycerides and phospholipids. These large lipids, along with cholesterol and cholesteryl esters, are then incorporated into particles called **chylomicrons** (also called chylomicra), which are released into the lymph for initial circulation. Chylomicrons package their hydrophobic lipids (such as the triglycerides) within a hydrophilic exterior or "shell" formed mainly by the hydrophilic head groups of the phospholipids (Figure 6.17). The lymph carrying the

⟨**CONNECTIONS**⟩ Lymph is the fluid circulating in the lymphatic system (Chapter 3, page 83).

chylomicron A lipoprotein, made in the enterocyte, that transports large lipids away from the small intestine in the lymph.

FIGURE 6.17 The Lipoproteins The ratio of lipids to proteins determines a lipoprotein's density and, for most, its name.

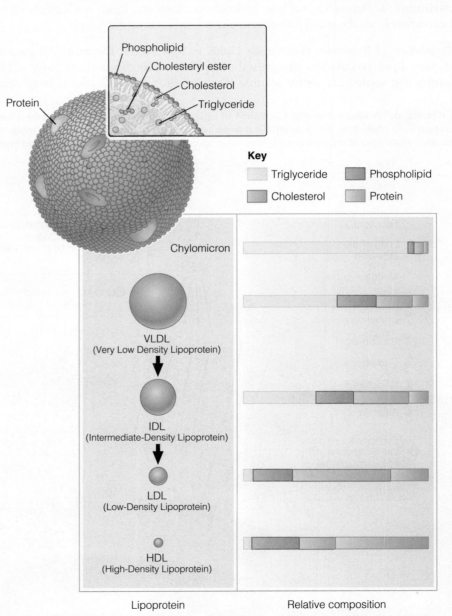

Adapted from: Skipski VP. Blood lipids and lipoproteins. Quantitation, composition, and metabolism. Nelson GJ, editor. 1972. Wiley-Interscience, New York; Christie W. American Oil Chemist Society Lipid Library. Lipoproteins. Available from: http://lipidlibrary.aocs.org/lipids/lipoprot/index.htm.

chylomicrons eventually mixes with the blood via the thoracic duct in the neck region. In this way, chylomicrons gradually enter the bloodstream, where they travel to cells that take up their contents. Note that chylomicrons are an example of a lipoprotein, which the body makes to transport dietary lipids. You will learn more about lipoproteins in the next section.

Chylomicrons in the blood deliver dietary fatty acids to cells via the enzyme lipoprotein lipase. Lipoprotein lipase is produced in many tissues, especially adipose and muscle. After it is produced in the cell's cytoplasm, this enzyme is relocated out of the cell and into the lumen (inside) of the neighboring capillary blood vessels. As the chylomicrons circulate in the blood, they are "attacked" by lipoprotein lipase. This enzymatic action releases fatty acids from the chylomicrons' triglycerides, allowing their uptake into surrounding cells. After delivering dietary fatty acids to cells, the chylomicron fragments remaining in the blood are called **chylomicron remnants.** These triglyceride-depleted particles are taken up by the liver, where they are broken down and their contents reused or recycled.

What Is the Role of Other Lipoproteins in Lipid Transport and Delivery?

As just described, small- and medium-chain dietary lipids are circulated in the blood away from the small intestine to the liver. The larger, more hydrophobic lipids are packaged into chylomicrons, circulating first in the lymph and then in the bloodstream and ultimately delivering dietary fatty acids to all of the tissues that need them. Eventually, fatty acid-depleted chylomicron remnants return to the liver, where they are broken down. As such, your liver serves as both the "central command center" and "recycling center" for lipid metabolism, receiving dietary lipids, as well as synthesizing and metabolizing other lipids as needed. To deliver these newly synthesized lipids as well as some dietary lipids to the body, the liver makes a series of lipoproteins that circulate in the blood. A summary of the origins and functions of the various lipoproteins (including the chylomicrons) is presented in Figure 6.18 and described next.

LIPOPROTEINS CONTAIN LIPIDS IN THEIR CORES

Because most lipids are very hydrophobic, their transport in the hydrophilic blood is somewhat complex. To aid in this process, the liver produces particles called **lipoproteins,** whose job is to transport lipids in the blood. Lipoproteins are complex globular structures containing varying amounts of triglycerides, phospholipids, cholesteryl esters, cholesterol, and proteins. The proteins embedded within the outer shell of the lipoproteins are called **apoproteins** or apolipoproteins. Like chylomicrons made in the small intestine, the lipoproteins made by the liver are also constructed so that their hydrophilic components (such as the apoproteins and the polar head groups of phospholipids) are situated on the outer surface, and their hydrophobic components (such as triglycerides) are facing inward (recall Figure 6.17).

You have already learned about one type of lipoprotein—the chylomicron, which is the largest and least dense member of the lipoprotein family. Whereas chylomicrons exclusively transport *dietary* lipids away from the small intestine, other lipoproteins carry lipids originating in various tissues and organs. For example, lipoproteins carry lipids that have been synthesized in the liver and lipids released from storage in adipose tissue.

chylomicron remnant The lipoprotein particle that remains after a chylomicron has lost most of its fatty acids.

lipoprotein A spherical particle made of varying amounts of triglycerides, cholesterol, cholesteryl esters, phospholipids, and proteins.

apoproteins Proteins embedded in the surface of lipoproteins.

FIGURE 6.18 The Origins and Major Functions of Lipoproteins Both the liver and the small intestine make lipoproteins that circulate lipids in the body.

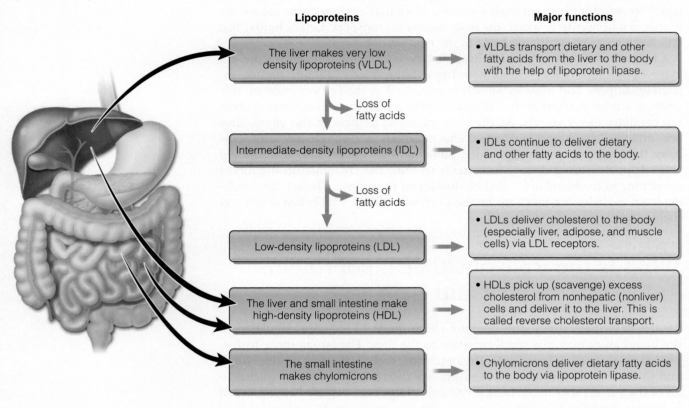

Lipoproteins / **Major functions**

The liver makes very low density lipoproteins (VLDL)
→ • VLDLs transport dietary and other fatty acids from the liver to the body with the help of lipoprotein lipase.

Loss of fatty acids ↓

Intermediate-density lipoproteins (IDL)
→ • IDLs continue to deliver dietary and other fatty acids to the body.

Loss of fatty acids ↓

Low-density lipoproteins (LDL)
→ • LDLs deliver cholesterol to the body (especially liver, adipose, and muscle cells) via LDL receptors.

The liver and small intestine make high-density lipoproteins (HDL)
→ • HDLs pick up (scavenge) excess cholesterol from nonhepatic (nonliver) cells and deliver it to the liver. This is called reverse cholesterol transport.

The small intestine makes chylomicrons
→ • Chylomicrons deliver dietary fatty acids to the body via lipoprotein lipase.

Because lipid is less dense than protein, the densities of the lipoproteins depend on their relative amounts (or percentages) of lipids and apoproteins. Lipoproteins with relatively more lipid than protein have lower densities than those with more protein and less lipid. With the exception of chylomicrons, lipoproteins are named according to their densities.

Fatty Acid Delivery by Very Low Density Lipoproteins (VLDLs) Whereas the small intestine produces chylomicrons and small amounts of other lipoproteins, the liver produces most of the lipoproteins responsible for lipid delivery in the body. One such lipoprotein is called a **very low density lipoprotein (VLDL).** VLDLs are similar to chylomicrons in that they contain triglycerides and cholesteryl esters in their cores surrounded by phospholipids, free cholesterol, and apoproteins on their surfaces. However, VLDLs have a lower lipid-to-protein ratio than do chylomicrons, making them smaller and denser. Like chylomicrons, the primary function of VLDLs is to deliver fatty acids to cells via the enzymatic action of lipoprotein lipase. However, unlike chylomicrons, which only deliver *dietary* fatty acids to the body, VLDLs also deliver fatty acids derived from liver and adipose tissue. When VLDLs in the blood circulate past cells that produce lipoprotein lipase, the enzyme cleaves fatty acids, which in turn are taken up by the surrounding cells.

Intermediate-Density Lipoproteins (IDLs) and Low-Density Lipoproteins (LDLs) As VLDLs lose fatty acids via the action of lipoprotein lipase, they become denser and are called **intermediate-density lipoproteins (IDLs).** Some IDLs

very low density lipoprotein (VLDL) A lipoprotein, made by the liver, that contains a large amount of triglyceride; its major function is to deliver fatty acids to cells.

intermediate-density lipoprotein (IDL) A lipoprotein that results from the loss of fatty acids from a VLDL; many IDLs are ultimately converted to LDLs.

are taken up by the liver, whereas others remain in the circulation, where they continue to lose additional fatty acids. Eventually, most IDLs become cholesterol-rich **low-density lipoproteins (LDLs).** Specialized proteins called **LDL receptors** on cell membranes—especially those of liver, adipose tissue, and muscle—bind to the LDLs' apoproteins allowing the LDLs to be taken up and broken down by the cell. In this way, cholesterol is delivered to many tissues that use it for structural and metabolic purposes.

LDLs can also be taken up and degraded by white blood cells (macrophages) that have been drawn to a major blood vessel due to some sort of injury or infection as part of the body's inflammatory response. Some uptake of LDLs into macrophages is necessary, because the cells use the LDL's contents to synthesize important substances such as the eicosanoids and immune factors. However, uptake of too much LDL can result in buildup of a fatty substance called **plaque** within the vessel wall, slowing or even blocking blood flow. Epidemiologic studies suggest that high levels of LDL cholesterol (LDL-C) in the blood are related to increased risk for cardiovascular disease.[20] Thus, LDL-C has been deemed "bad cholesterol." More detail concerning LDL-C, plaque formation, and cardiovascular disease is presented in the Nutrition Matters at the end of this chapter.

Effect of Diet on LDL Cholesterol Concentrations Consuming high amounts of certain SFAs or *trans* fatty acids (particularly those in partially hydrogenated oils) can increase circulating LDL-C concentration in some people.[21] The exact mechanism by which this occurs is not known, although it may be related to variations in apoprotein production.[22] In some cases, high cholesterol intake is also associated with high LDL-C concentration.[23] The relationship between cholesterol intake and LDL-C is complex, however, because some people show no effect of high cholesterol intake on LDL-C levels. This is most likely because of genetic (or epigenetic) variation related to cholesterol absorption, metabolism, and other dietary and lifestyle factors (such as smoking) that also influence circulating LDL-C levels.[24] Conversely, diets high in PUFAs, ω-3 fatty acids, or dietary fiber can lower LDL-C levels in some people.[25] Clearly, many factors influence the concentration of LDL-C in a person's blood, with different diets having different effects on various people.

Cholesterol Uptake by High-Density Lipoproteins (HDLs) The liver and, to a lesser extent, the small intestine make another series of lipoproteins called **high-density lipoproteins (HDLs).** Compared with other lipoproteins, HDLs have the lowest lipid-to-protein ratio; thus, they have the highest densities. HDLs salvage excess cholesterol from cells, transporting it back to the liver. This transfer of cholesterol from nonhepatic (nonliver) cells back to the liver is called **reverse cholesterol transport.**[26] It is well established that high levels of HDL cholesterol (HDL-C) in blood are generally associated with lower risk for cardiovascular disease.[27] This is why HDL-C is often referred to as the "good cholesterol."

There are several types of HDL, and not all forms are equally effective in removing excess cholesterol. More specifically, different HDLs have different apoproteins, resulting in somewhat different functions. The presence of particular apoproteins makes some HDLs less efficient at reverse cholesterol transport than others.

Effects of Diet on HDL Cholesterol Concentrations Although it has long been thought that diets high in carbohydrates offer protection from cardiovascular disease, a considerable amount of research now suggests that these diets

low-density lipoprotein (LDL) A lipoprotein that delivers cholesterol to cells.

LDL receptor Membrane-bound protein that binds LDLs, causing them to be taken up and dismantled.

plaque A complex of cholesterol, fatty acids, cells, cellular debris, and calcium that can form inside blood vessels and within vessel walls.

high-density lipoprotein (HDL) A lipoprotein, made primarily by the liver, that circulates in the blood to collect excess cholesterol from cells.

reverse cholesterol transport Process whereby HDLs remove cholesterol from nonhepatic (nonliver) tissues for transport to the liver.

can sometimes lower HDL-C levels.[28] In other words, very high carbohydrate diets may actually increase risk for heart disease in some people. Conversely, research suggests that high MUFA intake or moderate alcohol consumption (1–2 drinks/day) can raise HDL-C levels in some people, lowering risk for cardiovascular disease.[29] Like LDL-C, many lifestyle and genetic factors influence HDL-C concentrations, and researchers continue to study these important interactions.

What Is the Relationship between Lipid Intake and Health?

After lipids have been digested, absorbed, and circulated, they can be used by the body for myriad purposes. As described previously, fatty acids and their metabolites regulate metabolic processes within cells and orchestrate a variety of physiological responses. Phospholipids are vital components of membranes, aid in lipid digestion and absorption, and also contribute fatty acids for cellular use. Cholesterol is incorporated into cell membranes, acts as a precursor for some hormones, and is involved in lipid digestion and absorption via its role in bile.

Although many lipids are vital for good health, certain types of lipids can be associated with poor health. For example, high dietary intake of some lipids is associated with increased risks for cardiovascular disease and certain cancers. These topics are discussed briefly here, but you will also learn more about them elsewhere in this book.

EXCESS LIPID INTAKE CAN LEAD TO OBESITY

Although consumption of a high-fat diet can increase risk for obesity, there are many other factors (such as genetics and exercise) that contribute equally to this health problem.

Obesity is defined as the overabundance of body fat, and its causes and consequences are described in detail in Chapter 8. Although obesity is complex, excess energy intake is a major factor. Because fatty acids provide more than twice as many calories per gram as carbohydrates and proteins, high fat intake is likely an important piece of the obesity puzzle. Whatever its causes, obesity is a major public health concern worldwide and is associated with increased risk for many diseases such as cardiovascular disease, type 2 diabetes, and some forms of cancer. In order to decrease the risk for obesity, experts recommend that we limit our energy intake. In response to consumer demand, many food manufacturers produce low-fat and fat-free items in addition to foods that contain fat substitutes. You can read more about fat substitutes in the following Focus on Food feature.

DIETARY LIPIDS MAY BE RELATED TO RISK OF CARDIOVASCULAR DISEASE

Like obesity, cardiovascular disease is caused by a complex web of factors, including genetics, physical inactivity, and poor diet. Cardiovascular disease risk is not only influenced by *total* dietary lipid intake (and obesity), but may also be affected by specific *types* of dietary lipids. As noted, high intakes of SFAs, *trans* fatty acids, and cholesterol can increase the risk for disease in some people, whereas MUFAs may have the opposite effect. However, a person's genetic makeup often interacts with his or her diet to influence health, and this is certainly true for lipids and cardiovascular disease.

Because many people want to consume "low-fat" diets, some food manufacturers have developed substances that have desired characteristics of lipids but with fewer or no calories. These "fat substitutes" are diverse in structure, some made from complex carbohydrates, some made from proteins, and some made from blends of carbohydrates and fatty acids.

Olestra is an example of a fat substitute made from sucrose (table sugar) bonded to six to eight fatty acids. Olestra cannot be digested by human lipases or colonic bacteria, and therefore provides no usable energy to the body. In fact, it passes through the small intestine relatively intact. In 1996, the FDA approved the use of olestra in savory snacks such as potato chips, cheese puffs, and crackers. Because olestra interferes with the absorption of the fat-soluble vitamins in the GI tract, the FDA requires that food manufacturers add vitamins A, D, E, and K to olestra-containing foods. There was initial concern that consuming large quantities of olestra would cause intestinal distress such as gas and diarrhea. However, review of many human studies led the FDA to conclude that, when people consume reasonable portions of olestra-containing foods, they are likely to experience only infrequent and mild gastrointestinal upset.[30] Thus, in most situations, consuming olestra poses no concern. However, as with any food, one should consume those containing olestra or other fat substitutes in moderation.

Other fat substitutes are made from carbohydrates or protein. Examples commonly used by food manufacturers include cellulose, vegetable gum fibers (for example, guar gum), Maltrin®, and Stellar®. These substances are used in many low-fat or fat-free foods such as salad dressings, table spreads, and frozen desserts. Some of these ingredients are not heat-stable, and therefore

they cannot be used in foods that must be baked or fried. Note that, although cellulose and vegetable gums are considered dietary fiber and therefore do not contribute calories to the diet, Maltrin® and Stellar® are glucose polymers, providing 4 kcal per gram. Another fat substitute is Simplesse®, which is made from milk protein. Like Maltrin® and Stellar®, this substance contains 4 kcal per gram. Thus, foods containing these products may be "fat free," but they are certainly not "calorie free."

Fat-free ice cream often contains fat substitutes such as gum fiber.

Many studies have shown that consuming diets high in cholesterol increases blood cholesterol and risk for cardiovascular disease.[21] However, some people can eat very high amounts of cholesterol-rich foods without experiencing this effect. This difference may arise because some people absorb dietary cholesterol more efficiently than others.[31] Similarly, genetic variation can influence a person's HDL and LDL levels and functionality. For example, the ability of cells to take up LDL particles (and thus form plaque) can be influenced by variations in the genes that code for LDL receptor proteins.[32] The Nobel Prize in Physiology or Medicine was awarded in 1985 to Michael Brown and Joseph Goldstein for this discovery.[33] As scientists learn more about how genetics interacts with diet to influence health (the field of nutrigenomics), health professionals may someday be able to "prescribe" the most heart-healthful diet given a person's individual genetic makeup.

THE RELATIONSHIP BETWEEN DIETARY LIPIDS AND CANCER IS UNCLEAR

Some studies show a link between high-fat diets and risk for cancer, but the data are inconclusive.[34] Obesity, however, is a risk factor for several types of cancer, including breast and colorectal cancers.[35] As obesity is often associated with consumption of high-fat diets, dietary lipids may play an indirect role. In 2007 the World Cancer Research Fund and the American Institute for Cancer Research jointly put forth their overall recommendations concerning diet and cancer prevention.[36] These experts advised that, to decrease risk of cancer, we should be within the normal range of body weight and avoid weight gain and increases in waist circumference as we age. They also recommended that we limit consumption of energy-dense foods and consume high-fat "fast foods" sparingly, if at all. You can read more about the relationship between diet and cancer in the Nutrition Matters following Chapter 11.

What Are the Dietary Recommendations for Lipids?

Although we have dietary requirements for only the essential fatty acids, we rely on additional fatty acids as important sources of energy. However, consuming too much lipid—or the wrong kind of lipid—can be associated with health problems. It is therefore important to maintain an optimal balance of lipid intake throughout life. Recommendations concerning how to best choose dietary lipids to help promote optimal health are discussed next.

CONSUME ADEQUATE AMOUNTS OF THE ESSENTIAL FATTY ACIDS

It is important to consume adequate amounts of the essential fatty acids, and the Dietary Reference Intakes (DRIs) for them are presented inside the cover of this book. Adequate Intake levels (AIs) for linoleic acid are 17 and 12 g/day for adult males and females, respectively. As a reference, this is the amount contained in approximately 1¾ tablespoons of corn oil. For linolenic acid, the AIs are 1.6 and 1.1 g/day for adult males and females, respectively. This is the amount contained in about 1½ tablespoons of soybean oil.

PAY SPECIAL ATTENTION TO THE LONG-CHAIN OMEGA-3 FATTY ACIDS

To help lower risk for cardiovascular disease, many organizations urge consumption of additional omega-3 fatty acids in our diet. For instance, the American Heart Association recommends eating fish (particularly fatty fish) at least two times a week. Each serving should be at least 3.5 ounces of cooked fish or about 3/4 cup of flaked fish. Higher amounts are recommended for individuals who already have cardiovascular disease. The Dietary Guidelines for Americans recommend weekly fish consumption, partly because of the benefits of omega-3 fatty acids. The Dietary Guidelines also recommend that women who are pregnant or breastfeeding consume at least 8 and up to 12 ounces (2–3 servings) of a variety of seafood per week to ensure that intakes of the omega-3 fatty acids are adequate for fetal growth and development, as well as

© Rita Maas/Getty Images

Even high-fat foods can be part of a healthy diet if eaten in moderation.

〈CONNECTIONS〉 Chapter 2 discussed the Dietary Reference Intakes (DRIs), which are reference values for nutrient intake that meets the needs of about 97% of the population. When adequate information is available, Recommended Dietary Allowances (RDAs) are established, but when less information is available, Adequate Intake (AI) values are provided. Tolerable Upper Intake Levels (ULs) indicate intake levels that should not be exceeded (Chapter 2, p. 38).

for development during early infancy and childhood. Special care, however, should be taken so that only fish known to be low in mercury are consumed.

DIETARY GUIDELINES AND INSTITUTE OF MEDICINE RECOMMEND LIMITING SATURATED FATTY ACIDS

Scientists have long known that SFA intake is positively related to risk for cardiovascular disease for some people. As such, many dietary guidelines are aimed at decreasing SFA intake. For example, the Institute of Medicine recommends that "intake of SFAs should be minimized while consuming a nutritionally adequate diet." In addition, the Dietary Guidelines suggest that SFA should constitute no more than 10% of total calories, and the American Heart Association recommends that SFA intake represent 7% or less of total calories.[37] Reading food labels such as those shown in Figure 6.19 can help you keep track of your SFA intake. You can read more about general suggestions for decreasing SFA intake in the Food Matters feature, and the relationship between SFA and heart disease is discussed in the Nutrition Matters section following this chapter.

FIGURE 6.19 Using Nutrition Facts Panels to Determine Fat Intake You can tell how much total fat, saturated fat, and *trans* fat is in a food by checking its Nutrition Facts panel.

Working Toward the Goal: Getting the Right Lipids in Your Diet

The 2010 Dietary Guidelines recommend that we limit our intake of saturated fatty acids, *trans* fatty acids, and cholesterol. The following selection and preparation tips will help you reach this goal.

Saturated Fats

- Limit intake of animal fats (e.g., beef and butter fat) and tropical oils (e.g., coconut and palm oils).

Trans Fats

- Decrease intake of cookies, crackers, cakes, pastries, doughnuts, and french fries made with partially hydrogenated shortening.

- Use oils instead of partially hydrogenated shortening.
- Use tub or "*trans* fat–free" margarine instead of stick margarine—or use butter.

Cholesterol

- Limit intake of high-cholesterol foods such as liver, eggs, cheesecake, and custards.
- Replace high-fat animal products with lower-fat products, such as lean cuts of meat and fat-free or low-fat milk.

TRANS FATTY ACIDS AND CHOLESTEROL SHOULD BE MINIMIZED

Although there are no absolute guidelines as to how much *trans* fatty acids we should (or should not) consume, the Institute of Medicine recommends that people "minimize their intakes of *trans* fatty acids." Similarly, the Dietary Guidelines recommend that *trans* fatty acid intake should be "as low as possible while consuming a nutritionally adequate diet." These guidelines apply to commercially produced *trans* fatty acids, not to naturally occurring ones.

No DRIs are established for cholesterol. However, the 2010 Dietary Guidelines and American Heart Association recommend that we consume no more than 200–300 mg of cholesterol daily. This is the amount found in one to two eggs or two servings of beef. The Institute of Medicine recommends that "cholesterol intake should be minimized," and the American Heart Association recommends it be limited to <300 milligrams/day.

National Cholesterol Education Program Recommends the Therapeutic Lifestyle Change (TLC) Diet The National Cholesterol Education Program (NCEP), which is part of the National Institutes of Health, promotes the **Therapeutic Lifestyle Change (TLC) diet** for individuals with increased risk for cardiovascular disease. This diet recommends that less than 7% of your day's total calories come from SFA and that you consume no more than 200 mg of cholesterol daily. Note that this advice for cholesterol intake is similar to that put forth by the Dietary Guidelines. You can find out more about the TLC diet by visiting its interactive website at http://www.nhlbi.nih.gov/cgi-bin/chd/step2intro.cgi.

Therapeutic Lifestyle Changes (TLC) diet A set of heart-healthy diet recommendations put forth by the National Cholesterol Education Program.

GUIDELINES SET FOR TOTAL LIPID CONSUMPTION

The Institute of Medicine has not established RDA or AI values for total lipid intake, except during infancy when AIs are set at approximately 30 g per day. However, the AMDRs recommend that healthy adults consume

20 to 35% of their energy from lipid. Based on a caloric requirement of 2,000 kcal per day, a person should consume between 400 to 700 kcal from dietary lipids. Considering that 1 g of lipid contains about 9 kcal of energy, the daily lipid intake for most adults should be between 44 and 78 g. This amount is easy to obtain. For example, a typical day's menu—including three servings of low-fat milk, a bagel with cream cheese, a peanut butter sandwich, a serving of spaghetti with meatballs, and a salad with ranch dressing—contains about 56 g of lipid.

Critical Thinking: Nancy's Story What would you do if you suspected that you had gallbladder disease at this stage of your life? Who would you call at your university to discuss your symptoms and get the care you would need? What dietary changes could you make to decrease the complications of gallbladder disease until you could be seen by a medical professional?

Diet Analysis PLUS ✚ Activity

How Healthy Is Your Fat Intake?

Part A: Total Fat

1. Your AMDR for total fat can be stated in many ways, which can often be confusing. The primary way that it can be represented is as a **percentage of calories.** Referring to your *Profile DRI Goals* report, you will see in italics the recommended percentage of 20–35% of total calories. To manually calculate your percentage, fill in the blanks below:

   _____ g × 9 kcal/gram = _____ kcal of total fat you consumed *(Grams of fat you consumed)*

   _____ kcals of total fat you consumed ÷ _____ kcals × 100 = _____ % *(Total kcals you consumed)*

2. Does your percentage for *total fat* fall between the AMDR for total fat of 20–35%? _____

3. Look at your individual day's *Intake Spreadsheets*. List the 10 foods having the highest content of total fat.

Part B: Types of Fat

Refer to your *Fat as Percentage of Total Calories* report to answer the first two questions.

1. Add up the percentages of all of the types of fat you consumed. What is the value? _____%

2. Does the percent you calculated in #1 equal the total fat percent on your Energy Nutrient Intake and AMDR Goals Ranges Compared report?_____

3. Using Chapter 6 in your text and for *each* TYPE of fat you consumed, list two significant food sources.

Part C: Saturated Fat

The Institute of Medicine recommends that less than 10% of your total calories come from saturated fat.

1. Referring to your *Intake and DRI Goals Compared* report, how many GRAMS of saturated fat did you consume? _____ g

2. Calculate the percent of saturated fat calories using the formula below:

   _____ g saturated fat × 9 kcal/g = _____ kcal saturated fat

   _____ kcal of saturated fat ÷ _____ × 100 = _____ % saturated fat *Total kilocalories consumed*

3. Compare your answer in #2 for saturated fat with that on *Fat as Percentage of Total Calories* report. Are they the same? _____

4. Is the percent of saturated fat in your diet less than 10% of calories? _____

5. Looking at your individual days' Intake Spreadsheets, identify the one food that contributed the most saturated fat for that day below.

Part D: Summary and Conclusions

1. What is your assessment of your fat intake? Is there anything you may need to change according to the AMDR guidelines? Please be specific in your answer.

Notes

1. Elias SL, Innis SM. Bakery foods are the major dietary source of trans-fatty acids among pregnant women with diets providing 30 percent energy from fat. Journal of the American Dietetic Association. 2002;102:46–51. Remig V, Franklin B, Margolis S, Kostas G, Nece T, Street JC. Trans fats in America: a review of their use, consumption, health implications, and regulation. Journal of the American Dietetic Association. 2010;110:585–92.

2. Bendesen NT, Christensen R, Bartels EM, Astrup A. Consumption of industrial and ruminant trans fatty acids and risk of coronary heart disease: a systematic review and meta-analysis of cohort studies. European Journal of Clinical Nutrition. 2011; March 23 (Epub ahead of print). Judd JT, Clevidence BA, Muesing RA, Wittes J, Sunkin ME, Podczasy JJ. Dietary trans fatty acids: Effects on plasma lipids and lipoproteins of healthy men and women. American Journal of Clinical Nutrition. 1994;59:861–8. Stender S, Dyerberg J. Influence of trans fatty acids on health. Annals of Nutrition and Metabolism. 2004;48:61–6.

3. Davey M. Chicago weighs new prohibition: bad-for-you fats. The New York Times. July 18, 2006.

4. Brenna JT, Salem N Jr, Sinclair AJ, Cunnane SC; International Society for the Study of Fatty Acids and Lipids, ISSFAL alpha-Linolenic acid supplementation and conversion to n-3 long-chain polyunsaturated fatty acids in humans. Prostaglandins, Leukotrienes, and Essential Fatty Acids. 2009;80:85–91. Saldanha LG, Salem N Jr, Brenna JT. Workshop on DHA as a required nutrient: overview. Prostaglandins, Leukotrienes, and Essential Fatty Acids. 2009;81:233–6.

5. Calder PC. Polyunsaturated fatty acids and inflammation. Biochemical Society Transactions. 2005;33:423–7. Hansen SN, Harris WS. New evidence for the cardiovascular benefits of long chain omega-3 fatty acids. Current Atherosclerosis Reports. 2007;9:434–40.

6. Vanschoonbeek K, de Maat MP, Heemskerk JW. Fish oil consumption and reduction of arterial disease. Journal of Nutrition. 2003;133:657–60. Wood DA, Kotseva K, Connolly S, Jennings C, Mead A, Jones J, Holden A, De Bacquer D, Collier T, De Backer G, Faergeman O, EUROACTION Study Group. Nurse-coordinated multidisciplinary, family-based cardiovascular disease prevention programme (EUROACTION) for patients with coronary heart disease and asymptomatic individuals at high risk of cardiovascular disease: A paired, cluster-randomised controlled trial. Lancet. 2008;371:1999–2012.

7. Mori TA, Beilin LJ. Omega-3 fatty acids and inflammation. Current Atherosclerosis Reports. 2004;6:461–7. Shahidi F, Miraliakbari H. Omega-3 (n-3) fatty acids in health and disease. Part 1: cardiovascular disease and cancer. Journal of Medicinal Foods. 2004; 7:387–401. Wijendran V, Hayes KC. Dietary n-6 and n-3 fatty acid balance and cardiovascular health. Annual Review of Nutrition. 2004;24:597–615. Zampelas A. Eicosapentaenoic acid (EPA) from highly concentrated n-3 fatty acid ethyl esters is incorporated into advanced atherosclerotic plaques and higher plaque EPA is associated with decreased plaque inflammation and increased stability. Atherosclerosis. 2010;212:34–5.

8. Jensen RG. Handbook of Milk Composition. New York: Academic Press; 1995.

9. Dangour AD, Uauy R. N-3 long-chain polyunsaturated fatty acids for optimal function during brain development and ageing. Asia Pacific Journal of Clinical Nutrition. 2008;17:185–8.

10. Alessandri JM, Guisnet P, Vancassel S, Astorg P, Denis I, Langelier B, Aid S, Poumes-Ballihaut C, Champeil-Potokar G, Lavialle M. Polyunsaturated fatty acids in the central nervous system: Evolution of concepts and nutritional implications throughout life. Reproduction, Nutrition, and Development. 2004;44:509–38. Heird WC. Infant feeding and vision. American Journal of Clinical Nutrition. 2008;87:1120. Heird WC, Lapillone A. The role of essential fatty acids in development. Annual Review of Nutrition. 2005;25:549–71.

11. Afleith M, Clandinin MT. Dietary PUFA for preterm and term infants: Review of clinical studies. Critical Review of Food Science and Nutrition. 2005;45:205–29. Smithers LG, Gibson RA, McPhee A, Makrides M. Effect of long-chain polyunsaturated fatty acid supplementation of preterm infants on disease risk and neurodevelopment: A systematic review of randomized controlled trials. American Journal of Clinical Nutrition. 2008;87:912–20.

12. Picciano MF, Guthrie HA, Sheehe DM. The cholesterol content of human milk. A variable constituent among women and within the same woman. Clinical Pediatrics. 1978;17:359–62. Shahin AM, McGuire MK, Anderson N, Williams J, McGuire MA. Effects of margarine and butter consumption on distribution of trans-18:1 fatty acid isomers and conjugated linoleic acid in major serum lipid classes in lactating women. Lipids. 2006;41:141–7.

13. German JB. Dietary lipids from an evolutionary perspective: sources, structures and functions. Maternal and Child Nutrition. 2011;7:2–16. Pond WG, Mersmann HJ, Su D, McGlone JJ, Wheeler MB, Smith EO. Neonatal dietary cholesterol and alleles of cholesterol 7-alpha hydroxylase affect piglet cerebrum weight, cholesterol concentration, and behavior. Journal of Nutrition. 2008;138:282–6.

14. Strumia R. Dermatologic signs in patients with eating disorders. American Journal of Clinical Dermatology. 2005;6:165–73.

15. Vidon C, Boucher P, Cachefo A, Peroni O, Diraison F, Beylot M. Effects of isoenergetic high-carbohydrate compared with high-fat diets on human cholesterol synthesis and expression of key regulatory genes of cholesterol metabolism. American Journal of Clinical Nutrition. 2001;73:878–84.

16. Gupta AK, Savopoulos CG, Ahuia J, Hatzitolios AI. Role of phytosterols in lipid-lowering: current perspectives. QMJ: monthly journal of the Association of Physicians. 2011;104:301–8.

17. U.S. Food and Drug Administration. A food labeling guide—Appendix C: health claims. October 2009. Available from http://www.fda.gov/Food/GuidanceComplianceRegulatory Information/GuidanceDocuments/FoodLabelingNutrition/ FoodLabelingGuide/ucm064919.htm.

18. National Institutes of Health. Gallstones. Medline Plus. Updated 2009. Available from: http://www.nlm.nih.gov/ medlineplus/ency/article/000273.htm.

19. Walcher T, Haenle MM, Mason RA, Koenig W, Imhof A, Kratzer W; EMIL Study Group. The effect of alcohol, tobacco and caffeine consumption and vegetarian diet on gallstone prevalence. European Journal of Gastroenterology and Hepatology. 2010;22:1345–51. Leitzmann C. Vegetarian diets: What are the advantages? Forum in Nutrition. 2005;57:147–56.

20. Adiels M, Olofsson SO, Taskinen MR, Borén J. Overproduction of very low-density lipoproteins is the hallmark of the dyslipidemia in the metabolic syndrome. Arteriosclerosis Thrombosis Vascular Biology. 2008;28:1225–36.

Holvoet P. Oxidized LDL and coronary heart disease. Acta Cardiologica. 2004;59:479–84. Knopp RH, Paramsothy P, Atkinson B, Dowdy A. Comprehensive lipid management versus aggressive low-density lipoprotein lowering to reduce cardiovascular risk. American Journal of Cardiology. 2008;101:48B–57B.

21. Mozaffarian D, Willett WC. Trans fatty acids and cardiovascular risk: A unique cardiometabolic imprint? Current Atherosclerosis Reports. 2007;9:486–93. Stender S, Dyerberg J. Influence of trans fatty acids on health. Annals of Nutrition and Metabolism. 2004;48:61–6.

22. Lands B. A critique of paradoxes in current advice on dietary lipids. Progress in Lipid Research. 2008;47:77–106. Yang Y, Ruiz-Narvaez E, Kraft P, Campos H. Effect of apolipoprotein E genotype and saturated fat intake on plasma lipids and myocardial infarction in the Central Valley of Costa Rica. Human Biology. 2007;79:637–47.

23. U.S. Department of Agriculture and U.S. department of Health and Human Services Dietary Guidelines Advisory Committee. Report of the Dietary Guidelines Advisory Committee on the Dietary Guidelines for Americans. 2010. National Technical Information Service, Springfield VA. Available at: http://www.cnpp.usda.gov/Publications/ DietaryGuidelines/2010/DGAC/Report/2010DGACReport- camera-ready-Jan11-11.pdf.

24. Kathiresan S, Musunuru K, Orho-Melander M. Defining the spectrum of alleles that contribute to blood lipid concentrations in humans. Current Opinions in Lipidology. 2008;19:122–7. Wilson PW. Assessing coronary heart disease risk with traditional and novel risk factors. Clinical Cardiology. 2004;27:7–11.

25. Anderson JW, Randles KM, Kendall CW, Jenkins DJ. Carbohydrate and fiber recommendations for individuals with diabetes: A quantitative assessment and meta-analysis of the evidence. Journal of the American College of Nutrition. 2004;23:5–17. Christensen JH, Christensen MS, Dyerberg J, Schmidt EB. Heart rate variability and fatty acid content of blood cell membranes: A dose–response study with n-3 fatty acids. American Journal of Clinical Nutrition. 1999;70:331–7. Vanschoonbeek K, de Maat MP, Heemskerk JW. Fish oil consumption and reduction of arterial disease. Journal of Nutrition. 2003;133:657–60.

26. Cavigiolio G, Shao B, Geier EG, Ren G, Heinecke JW, Oda MN. The interplay between size, morphology, stability, and functionality of high-density lipoprotein subclasses. Biochemistry. 2008;47:4770–9. Florentin M, Liberopoulos EN, Wierzbicki AS, Mikhailidis DP. Multiple actions of high-density lipoprotein. Current Opinions in Cardiology. 2008;23:370–8.

27. Chiesa G, Parolini C, Sirtori CR. Acute effects of high-density lipoproteins: Biochemical basis and clinical findings. Current Opinions in Cardiology. 2008;23:379–85. Fernandez ML, Webb D. The LDL to HDL cholesterol ratio as a valuable tool to evaluate coronary heart disease risk. Journal of the American College of Nutrition. 2008;27:1–5.

28. Nettleton JA, Volcik KA, Hoogeveen RC, Boerwinkle E. Carbohydrate intake modifies associations between ANGPTL4[E40K] genotype and HDL-cholesterol concentrations in white men from the Atherosclerosis Risk in Communities (ARIC) study. Atherosclerosis. 2009;203:214–20. Samaha FF, Foster GD, Makris AP. Low-carbohydrate diets, obesity, and metabolic risk factors for cardiovascular disease. Current Atherosclerosis Reports. 2007;9:441–7. Tay J, Brinkworth GD, Noakes M, Keogh J, Clifton PM. Metabolic effects of weight loss on a very-low-carbohydrate diet compared with an isocaloric high-carbohydrate diet

in abdominally obese subjects. Journal of the American College of Cardiology. 2008;51:59–67.

29. Fan JG, Cai XB, Li L, Li XJ, Dai F, Zhu J. Alcohol consumption and metabolic syndrome among Shanghai adults: A randomized multistage stratified cluster sampling investigation. World Journal of Gastroenterology. 2008;14:2418–24. Joosten MM, Beulens JW, Kersten S, Hendriks HF. Moderate alcohol consumption increases insulin sensitivity and ADIPOQ expression in postmenopausal women: A randomised, crossover trial. Diabetologia. 2008;51:1375–81. Paniagua JA, de la Sacristana AG, Sánchez E, Romero I, Vidal-Puig A, Berral FJ, Escribano A, Moyano MJ, Pérez-Martinez P, López-Miranda J, Pérez-Jiménez F. A MUFA-rich diet improves posprandial glucose, lipid and GLP-1 responses in insulin-resistant subjects. Journal of the American College of Nutrition. 2007;26:434–44.

30. Department of Health and Human Services. Food and Drug Administration. Food additives permitted for direct addition to food for human consumption; olestra; final rules. Federal Register 68:46363–402, 2003. Available at http://www.fda.gov/ OHRMS/DOCKETS/98fr/03-19508.pdf.

31. Lammert F, Wang DQ. New insights into the genetic regulation of intestinal cholesterol absorption. Gastroenterology. 2005;129:718–34. Yang Y, Ruiz-Narvaez E, Kraft P, Campos H. Effect of apolipoprotein E genotype and saturated fat intake on plasma lipids and myocardial infarction in the Central Valley of Costa Rica. Wu K, Bowman R, Welch AA, Luben RN, Wareham N, Khaw KT, Bingham SA. Apolipoprotein E polymorphisms, dietary fat and fibre, and serum lipids: The EPIC Norfolk study. European Heart Journal. 2007;28:2930–6.

32. Brown MS, Goldstein JL. How LDL receptors influence cholesterol and atherosclerosis. Scientific American. 1984;251:52–60. Dedoussis GV, Schmidt H, Genschel J. LDL-receptor mutations in Europe. Human Mutation. 2004;443–59.

33. The Nobel Assembly at the Karolinska Institute. The 1985 Nobel Prize in Physiology or Medicine press release. Available from: http://nobelprize.org/medicine/laureates/1985/ press.html.

34. Prentice RL. Women's health initiative studies of postmenopausal breast cancer. Advances in Experimental Medicine and Biology. 2008;617:151–60. Van Horn L, Manson JE. The Women's Health Initiative: Implications for clinicians. Cleveland Clinic Journal of Medicine. 2008;75:385–90. Wang J, John EM, Horn-Ross PL, Ingles SA. Dietary fat, cooking fat, and breast cancer risk in a multiethnic population. Nutrition and Cancer. 2008;60:492–504.

35. Al-Serag HB. Obesity and disease of the esophagus and colon. Gastroenterology Clinics of North America. 2005;34:63–82. Key TJ, Schatzkin A, Willett WC, Allen NE, Spencer EA, Travis RC. Diet, nutrition and the prevention of cancer. Public Health Nutrition. 2004;7:187–200. McTiernan A, Yang XR, Chang-Claude J, Goode EL, et al. Associations of breast cancer risk factors with tumor subtypes: a pooled analysis from the Breast Cancer Association Consortium studies. Journal of the National Cancer Institute. 2011;103:250–63. Obesity and cancer: The risks, science, and potential management strategies. Oncology. 2005;19:871–81.

36. World Cancer Research Fund/American Institute for Cancer Research. Food, nutrition, physical activity, and the prevention of cancer: A global perspective. Washington, DC: AICR; 2007.

37. Lichtenstein AH et al., Diet and lifestyle recommendations revision 2006. A scientific statement from the American Heart Association Nutrition Committee. Circulation. 2006;114:82–96.

Nutrition and Cardiovascular Health

The steadfast function of the cardiovascular system is likely one of the body's most impressive physiologic systems. Indeed, without you having to give it a single thought, your heart beats about 100,000 times a day and 40 million times a year. The resultant movement of blood keeps your body supplied with oxygen and nutrients while removing harmful waste products via 60,000 miles of blood vessels. To maintain your health and vigor, it is important to keep your cardiovascular system in top condition. However, as we enjoy longer lives, chronic degenerative diseases are becoming more common. Of these, heart disease and stroke are the first- and third-leading causes of death in the United States.[1] Heart disease and stroke are forms of **cardiovascular disease** (CVD), a term used to describe a variety of diseases of the heart and blood vessels. In this section, you will learn about the pathophysiology of CVD and the importance of nutrition in its prevention and treatment.

How Does Cardiovascular Disease Develop?

Cardiovascular disease is generally caused by a slowing or complete obstruction of blood flow to the heart or other parts of the body, including the brain (Figure 1). Remember that blood flows through blood vessels delivering oxygen and nutrients to cells. When blood flow is restricted, cells do not receive adequate oxygen and nutrients, ultimately causing cell death. Restriction of blood flow results from a condition called **atherosclerosis,** characterized by a narrowing and hardening of the blood vessels. Atherosclerosis can reduce the blood supply to the heart muscle (coronary arteries), brain (cerebral arteries), and other parts of the body. When it occurs in coronary arteries, it can cause **heart disease,** also called coronary heart disease, whereas atherosclerosis in cerebral arteries can cause a **stroke.**

Atherosclerosis can lead to other complications, which can also contribute to heart disease or stroke. For example, a blood vessel can become weak and distended, forming an **aneurysm.** Having an aneurysm is dangerous because the arterial walls become stretched and can rupture. When an aneurysm in a major blood vessel ruptures, blood pours into the body cavity, resulting in a rapid drop in blood pressure, which can deprive tissues of oxygen and nutrients. **Blood clots,** or thromboses, are another complication associated with atherosclerosis. When small pieces of clotted blood become lodged in an artery, blood flow to the target tissue can be reduced or cut off. The risks of a blood clot are even greater when the artery has been narrowed by atherosclerosis. Regardless of whether it occurs via atherosclerosis, aneurysm, blood clot, or a combination of factors, CVD can be life threatening.

ATHEROSCLEROSIS CAN LEAD TO CARDIOVASCULAR DISEASE

Atherosclerosis, a slowly developing, chronic, degenerative disease, is one of the most important risk factors associated with heart attacks and strokes. Atheroclerosis develops when fatty deposits called **plaque** accumulate within the walls of arteries. Plaques contain fatty acids, cholesterol, lipid-filled immune cells (called **foam cells**),

---

cardiovascular disease (car – di – o – VAS – cu – lar) (*kardia*, heart; *vascellum*, vessel) A disease of the heart or vascular system.

atherosclerosis (a – ther – o – scler – O – sis) (*atheroma*, cyst full of pus; *sklerosis*, hardening) The hardening and narrowing of blood vessels caused by buildup of fatty deposits (plaques) and inflammation in the vessel walls.

heart disease (also called **coronary heart disease**) A condition that occurs when the heart muscle does not receive enough blood.

stroke A condition that occurs when a portion of the brain does not receive enough blood.

aneurysm (AN – eu – rysm) (*aneurusma*, dilation) The outward bulging of a blood vessel.

blood clot (also called **thrombosis**) A small, insoluble particle made of blood cells and clotting factors.

plaque Fatty deposit that accumulates within walls of blood vessels, sometimes leading to atherosclerosis.

foam cell A type of cell—usually an immune cell—that contains large amounts of lipid.

FIGURE 1 Causes of Cardiovascular Disease Atherosclerosis, blood clots, and aneurysms can all reduce or stop blood flow, causing cardiovascular disease.

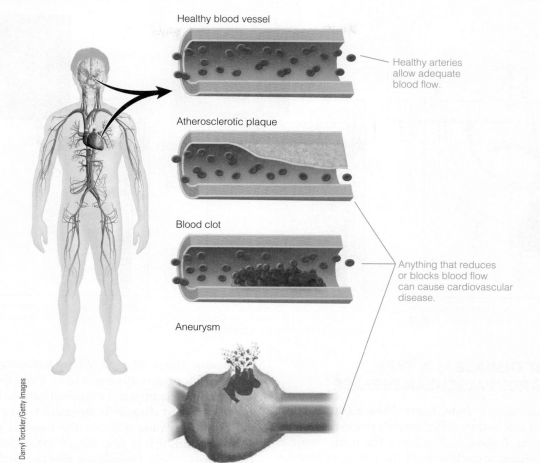

Healthy blood vessel

Healthy arteries allow adequate blood flow.

Atherosclerotic plaque

Blood clot

Anything that reduces or blocks blood flow can cause cardiovascular disease.

Aneurysm

Darryl Torckler/Getty Images

cellular waste products, calcium, and a variety of other substances. The accumulation of plaque within the arterial wall reduces blood flow. Plaque deposits can also dislodge (break off) from the arterial wall and enter the blood. When this happens, more white blood cells migrate to the injured area, in turn promoting inflammation that might be especially important in the etiology of CVD.

Chronic Inflammation Although scientists do not know what triggers the process of atherosclerosis, emerging research suggests that **chronic inflammation** may play a critical role.[2] The term *inflammation* refers to the body's response to a noxious (poisonous) stimulus, injury, or infection. Inflammation causes dilation of blood vessels, movement of white blood cells into the injured tissue, redness, and pain. Most experts now believe that many people with CVD may be in a state of chronic inflammation. This is likely one reason why taking low doses of aspirin (an anti-inflammatory drug) reduces the risk of heart disease and stroke.[3]

One potentially useful biological marker for inflammation is **C-reactive protein (CRP),** which is released into the blood by the liver, adipose tissue, and smooth muscle as part of the inflammatory response to injury or infection. High circulating concentration of CRP is related to increased risk for CVD.[4] The American Heart Association recommends the use of CRP screening only for individuals at intermediate risk for CVD.[5] This is because those at low risk are not likely to need intervention, and those at high risk should seek treatment regardless of their CRP levels. You can determine your risk for CVD by visiting their website at http://www.americanheart.org.

chronic inflammation A response to cellular injury that is characterized by chronic capillary dilation, white blood cell infiltration, release of immune factors, redness, heat, and pain.

C-reactive protein (CRP) A protein produced in the liver, adipose tissue, and smooth muscles in response to injury or infection that, when elevated, can indicate risk for cardiovascular disease.

FIGURE 2 Diagnosing Heart Disease Electrocardiograms, echocardiograms, and angiograms are all useful in diagnosing heart disease.

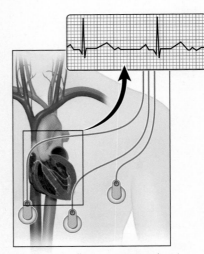

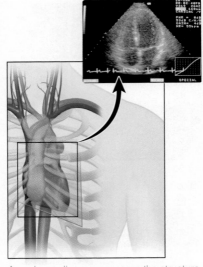

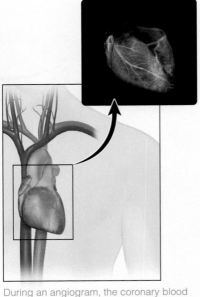

An electrocardiogram assesses heart function by recording electrical activity.

An echocardiogram assesses the structure and function of the heart using sound waves that create a moving image.

During an angiogram, the coronary blood vessels are visualized with the use of dyes.

Darryl Torckler/Getty Images; SPL/Photo Researchers, Inc.

HEART DISEASE IS A TYPE OF CARDIOVASCULAR DISEASE

In its less severe state, heart disease causes chest pain and discomfort called **angina pectoris,** or simply angina. Angina occurs when the heart tissue is still receiving some blood via the coronary arteries but not as much as it needs. Because the heart muscle does not receive adequate amounts of oxygen, the heart muscle can spasm, causing chest pain and shortness of breath. Although angina is not usually life threatening, it could indicate something more serious, such as an impending heart attack.

Heart attacks occur when the blood supply to the heart muscle is severely reduced or stopped. The medical term for a heart attack is "myocardial infarction." Because cardiac tissue can survive only for a few minutes without oxygen, heart attacks can permanently damage the heart muscle. If the damage is severe, the results can be fatal. Warning signs and symptoms of a heart attack include chest discomfort; pain radiating down one or both arms, back, neck, jaw, or stomach; shortness of breath; cold sweats; nausea; and lightheadedness. Women tend to experience somewhat different symptoms including unusual fatigue, sleep disturbances, shortness of breath, indigestion, and anxiety. It is important that anyone experiencing these symptoms contact emergency medical personnel immediately, as heart attacks are the leading cause of death in the United States.

Diagnosing Heart Disease When assessing a patient's risk for heart disease, a doctor will likely perform a complete physical examination and medical history. If heart disease is suspected, certain tests can help determine whether the heart is working normally and, if it is not, where the problem lies. These tests might include an **electrocardiogram** (EKG or ECG), which measures the electrical impulses in the heart; an **echocardiogram,** which uses sound waves to examine the heart's structure and motion; or an **angiogram,** in which the coronary arteries are visualized with the help of an injected dye. Angiograms enable physicians to determine if and where there are blockages in coronary arteries. These procedures are illustrated in Figure 2. Some doctors may also order blood tests such as that for C-reactive protein.

angina pectoris (an – GI – na pec – TOR – is) (*ankhone,* to strangle; *pectos,* breast) Pain in the region of the heart, caused by a portion of the heart muscle receiving inadequate amounts of blood.

heart attack (also called **myocardial infarction)** An often life-threatening condition in which blood flow to some or all of the heart muscle is completely blocked.

electrocardiogram A procedure during which the heart's electrical activity is recorded.

echocardiogram A visual image, produced using ultrasound waves, of the heart's structure and movement.

angiogram A procedure in which dye is injected into the blood, allowing the flow of blood through cardiac arteries to be visualized.

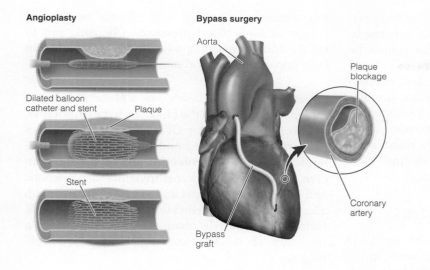

Angioplasty

Dilated balloon
catheter and stent

Plaque

Stent

Bypass surgery

Aorta

Plaque
blockage

Coronary
artery

Bypass
graft

FIGURE 3 Surgically Treating Heart Disease Surgeries such as angioplasty (the placing of a stent) and coronary bypass surgery are often life-saving for heart-disease patients.

Surgical Treatment of Heart Disease Sometimes heart disease is treated by performing surgical procedures such as placing a **stent** (wire mesh tube) in the affected artery during a procedure called **angioplasty** (or "balloon angioplasty"). During an angioplasty, the stent is initially collapsed to a small diameter and put over a ballooning catheter, which is inserted into the area of the blockage. When the balloon is inflated, the stent expands, locks in place, and forms a scaffold. The stent stays in the artery permanently, holding it open. An alternative to angioplasty is **coronary artery bypass surgery,** sometimes called CABG ("cabbage"). This type of surgery reroutes, or "bypasses," blood around clogged arteries to improve blood flow and aid in the delivery of oxygen to the heart. This is done by the removal of a segment of a blood vessel from another part of the body. This "graft" is then used to bypass blood flow around the blocked coronary arteries. These procedures are illustrated in Figure 3.

STROKE IS ANOTHER FORM OF CARDIOVASCULAR DISEASE

A stroke occurs when a portion of the brain is deprived of oxygen and critical nutrients. The extent of the resulting brain damage depends on the magnitude, duration, and area of the brain affected. Strokes often result in speech impairment or partial paralysis on one side of the body. However, if the stroke affects a large or critical part of the brain, it can be life threatening. One important warning sign for stroke is the occurrence of **transient ischemic attacks (TIAs),** sometimes referred to as "ministrokes." TIAs can occur when blood flow to the brain is temporarily disrupted.

Signs and symptoms of a TIA or stroke include sudden numbness or weakness (especially on one side of the body), confusion, slurred speech, dizziness, loss of balance, or a severe headache. When a TIA or stroke is suspected, a person should seek immediate medical assistance.

Once a stroke has been diagnosed, treatment depends on its type and location. Sometimes, medications that break up blood clots can be given. Other times, surgery is needed to place a stent in the affected artery. In all cases, time is of the essence, as quick treatment not only improves chances of survival but may also reduce the amount of resulting disability.

What Are the Risk Factors for Cardiovascular Disease?

Having any of the biological, lifestyle, and environmental risk factors associated with the development of atherosclerosis increases a person's chances of having a heart attack or stroke. The American Heart Association has categorized these factors as being either major risk factors or contributing risk

stent A device made of rigid wire mesh that is threaded into an atherosclerotic blood vessel to expand and provide support for a damaged artery.

angioplasty A procedure used to widen the heart's blood vessels by inserting a stent.

coronary bypass surgery A procedure in which a healthy blood vessel obtained from the leg, arm, chest, or abdomen is used to bypass blood from a diseased or blocked coronary artery to a healthy one.

transient ischemic attack (TIA) A "ministroke" that is caused by a temporary decrease in blood flow to the brain.

factors, depending on the availability of supporting evidence. Furthermore, they have subcategorized these risk factors into those that cannot be changed (nonmodifiable or biological risk factors) and those that can be modified, treated, or controlled by lifestyle choices or medication (modifiable risk factors).[6] The American Heart Association recommends that all adults—young and old—know their risk for CVD, and as previously mentioned, you can determine your risk by visiting their website at http://www.americanheart.org.

NONMODIFIABLE RISK FACTORS

The major nonmodifiable risk factors for CVD include age, sex, genetics (including race), prior stroke or heart attack, and having been born with low birth weight. For instance, CVD can strike at any age, but the older you get the more likely you are to develop it. In addition, men have a greater risk of CVD than women, and tend to develop it earlier in life. Further, compared with Caucasians, heart disease and stroke are more common in Mexican Americans, American Indians, native Hawaiians, and African Americans. This is thought to partly be due to higher rates of obesity and diabetes in these groups, although other genetic factors are likely important. People who were born with low birth weights also have increased risk for CVD later in life.[6] Remember that these factors cannot easily be controlled or treated. However, if you fall into any of these high-risk categories, it is even more important to take action.

MODIFIABLE RISK FACTORS

Modifiable or lifestyle risk factors include smoking, high blood pressure, elevated blood lipids, physical inactivity and obesity, diabetes, stress, and excessive alcohol consumption. Although many of these factors depend somewhat on biological influences, there are ways we can modify them. Some of these, especially those related to nutrition, are briefly described next.

Hypertension High blood pressure, also called hypertension, increases a person's risk for CVD partly because it puts additional demands on the heart muscle and damages blood vessels. High blood pressure can also cause plaque to break away from arterial walls. Dislodged plaque can both initiate inflammation and act like a blood clot, restricting blood flow even more. Biological and lifestyle factors can both contribute to high blood pressure.

For example, smoking and obesity are major causes of high blood pressure.[7] Choosing not to smoke and maintaining a healthy body weight can help prevent high blood pressure and thus CVD.

Elevated Blood Lipid Levels Are a Major Risk Factor
A high level of lipid in the blood, or **hyperlipidemia**, is also a risk factor for CVD.[8] Excess triglycerides (**hypertriglyceridemia**) and high levels of cholesterol (**hypercholesterolemia**) are of specific concern. Both forms of hyperlipidemia have genetic and lifestyle components and are often treated with a combination of medication (to address genetic factors) and lifestyle changes. Table 1 provides a summary of healthy and unhealthy blood lipid values.

hyperlipidemia (hy – per – li – pid – EM – i – a) Elevated levels of lipids in the blood.

hypertriglyceridemia (hy – per – tri – gly – cer – EM – i – a) Elevated levels of triglycerides in the blood.

hypercholesterolemia (hy – per – chol – est – er – ol – EM – i – a) Elevated levels of cholesterol in the blood.

TABLE 1 Reference Values for Blood Lipids

Blood Lipid Category and Value	Interpretation
Total Cholesterol	
≤200 mg/dL	Desirable
200–239 mg/dL	Borderline high
≥240 mg/dLs	High
Low-Density Lipoprotein (LDL)	
≤100 mg/dL	Optimal
100–129 mg/dL	Near optimal/above optimal
130–159 mg/dL	Borderline high
160–189 mg/dL	High
≥190 mg/dL	Very high
High-Density Lipoprotein (HDL)	
≤40 mg/dL	Low
≥60 mg/dL	High
Triglyceride	
≤150 mg/dL	Normal
150–199 mg/dL	Borderline high
200–499 mg/dL	High
≥500 mg/dL	Very high

SOURCE: U.S. Department of Health and Human Services. National Cholesterol Education Program. ATPIII guidelines at-a-glance quick desk reference, 2001. NIH Publication No. 01-3305. Available from: http://www.nhlbi.nih.gov/guidelines/cholesterol/atglance.pdf. Note that these reference values are endorsed by the American Heart Association.

Review of Lipid Transport in the Blood To better understand how blood lipid levels are related to CVD, it is important to briefly review how lipids (particularly cholesterol) are circulated in the blood. Recall that blood lipids are generally packaged as components of lipoproteins and that cholesterol is circulated mainly by two lipoproteins: low-density lipoproteins (LDLs) and high-density lipoproteins (HDLs). LDLs deliver cholesterol to cells, whereas HDLs pick up excess cholesterol for transport back to the liver. Many studies show that high levels of LDL cholesterol (LDL-C) are related to increased risk for plaque formation and CVD.[9] Because of this, LDL-C is called "bad cholesterol." Conversely, high levels of HDL cholesterol (HDL-C) protect people from CVD, and HDL-C is called "good cholesterol."[10]

Studies suggest that *oxidized* LDLs are much more atherogenic than *nonoxidized* LDLs.[11] Oxidized LDLs form when they are subjected to the damaging effects of free radicals. Experts believe that oxidized LDLs are especially atherogenic because foam cells (lipid-containing immune cells) are more likely to engulf them than their nonoxidized counterparts.[12] As such, it is the lipids contained in *oxidized* LDL that tend to form the core of the plaque that leads to atherosclerosis. This is one reason why dietary antioxidants, abundant in many fruits and vegetables, that decrease oxidation of LDL are likely important in preventing CVD. Because of differences in genetic make-up, some people are more susceptible than others to LDL oxidation and uptake of oxidized LDL into arterial walls.

What Is a Cholesterol Ratio? Of particular importance in terms of blood lipids is the distribution of cholesterol among the various lipoproteins—especially HDL (HDL-C) and LDL (LDL-C). As such, medical professionals sometimes use what is called a cholesterol ratio to assess a person's risk for CVD. The **cholesterol ratio** is calculated by dividing total cholesterol by HDL-C. The higher the ratio, the greater percentage of the total cholesterol is present in LDL relative to HDL. As an example, a person with a total cholesterol level of 200 mg/dL and an HDL-C of 50 mg/dL would have a cholesterol ratio of 4:1. Doctors recommend that this ratio be below 5:1 and optimally be 3.5:1 or less.

Obesity and Physical Inactivity—Related Risk Factors Of all the modifiable risk factors for CVD, likely the most important is being overweight or obese.[13] This is especially true if the excess body weight is stored in the waist area—in other words, if the person is apple-shaped. People with excess body weight are at higher risk of heart disease even if they have no other risk factors. Related to this is physical inactivity, which is an independent risk factor for CVD but also contributes to other risk factors such as obesity, hypertension, elevated blood lipids, and type 2 diabetes.

Unfortunately, there is no quick fix or magic bullet to help people shed unwanted pounds. However, being physically active and consuming a diet relatively low in fat and high in nutrient-dense foods are important in maintaining a healthy body weight. Even small reductions in weight can lower blood lipid levels and blood pressure, in turn lowering CVD risk.[14] Thus, doctors recommend that all people watch their caloric intake, maintain a healthy body weight, and engage in regular physical activity. The American Heart Association recommends at least 150 minutes per week of moderate exercise or 75 minutes per week of vigorous exercise (or a combination of moderate and vigorous activity). In other words, a goal of 30 minutes a day, five times a week is recommended.[15]

Diabetes—Another Modifiable Risk Factor Having either type 1 or type 2 diabetes is also a major risk factor for CVD. As previously discussed, diabetes is a condition in which the body can no longer effectively regulate blood glucose. Although genetics can play a critical role in determining who gets diabetes, other factors are also involved. For example, being overweight and physically inactive are risk factors for type 2 diabetes. Keeping these important health parameters in check can help prevent type 2 diabetes and, in turn, can reduce the risk for CVD. Furthermore, poorly treated diabetes can result in elevated blood glucose levels, which can damage blood vessels and lead to CVD. Controlling blood glucose levels with diet, exercise, and medication is therefore important in preventing heart disease and stroke.

cholesterol ratio The mathematical ratio of total blood cholesterol to high-density lipoprotein cholesterol (HDL-C).

How Does Nutrition Influence Cardiovascular Risk?

Consuming a varied diet in moderation can go a long way in preventing and treating CVD. This is because you need many micro- and macronutrients to maintain a healthy heart and vascular system. In this section, you will learn what scientists currently know about how diet influences several of the major risk factors for heart disease and stroke. You will also learn about how phytochemicals and functional foods are beginning to take center stage in the quest for optimal cardiovascular health.

HYPERTENSION CAN BE PARTIALLY CONTROLLED WITH DIET

Your dietary choices can make a big difference in determining whether you develop high blood pressure. For example, heavy alcohol consumption can cause blood pressure to rise, and people with hypertension are typically advised to limit their intake.[16] Further, some people's blood pressure is highly sensitive to sodium, or salt.[17] In other words, high sodium intakes can cause high blood pressure in salt-sensitive individuals. This is why most people diagnosed with high blood pressure are advised to consume relatively low-sodium diets. Consuming a low-sodium diet has become easier as food manufacturers have developed sodium-free and low-sodium food products, and recommendations concerning how to reduce dietary sodium are provided in Table 2.

The DASH Diet May Be Beneficial Many essential dietary minerals such as potassium, calcium, and magnesium are needed for maintaining healthy blood pressure levels. Because these minerals help lower blood pressure, it is important that people with increased risk for CVD consume enough foods that provide them. For example, legumes, seafood, dairy products, and many fruits and vegetables tend to be rich sources of these minerals and should be included in a heart-healthy diet.

One way to ensure sufficient intake of all these important nutrients is to follow what is called the **DASH (Dietary Approaches to Stop Hypertension) diet**, which emphasizes fruits, vegetables, and low-fat dairy products. This diet plan has been shown to lower blood pressure, especially in salt-sensitive people.[18] Although it is not clear whether a single nutrient (such as calcium or potassium) in the DASH diet is responsible for its healthy effects, the diet appears to work by increasing sodium excretion in the urine. Many organizations, such as the American Heart Association and the U.S. Department of Agriculture (USDA) recommend the DASH diet, especially for people predisposed to hypertension, and its components are summarized in Table 3.

CONTROLLING BLOOD LIPID LEVELS WITH DIET

Several dietary factors can influence circulating triglyceride and cholesterol levels, although not all people are equally affected. For example, diets high in soluble fiber (such as in oat or rice bran, oatmeal, legumes, barley, citrus fruits, and strawberries) may help lower LDL cholesterol.[19] Indeed, choosing wisely when it comes to what macronutrients we eat is important to a healthy heart.

Low-Carbohydrate Diets—Is There a Risk? Although people have long thought high-carbohydrate diets are heart-healthy, more recent studies show that very low total carbohydrate diets may be more beneficial.[20] However, because these types of diets are generally high in protein and fat, some professional organizations have cautioned against their use. The concern is that some people might restrict foods that provide essential nutrients and therefore not get the variety of foods needed to meet nutritional needs.[21] In addition, these dietary patterns tend to have excessive amounts of animal fats that might increase blood lipid levels. A recent review of the literature, however, concluded that low-carbohydrate diets can be used safely and effectively for short-term weight loss without adversely affecting cardiovascular risk factors.[22] In fact, most studies conducted in free-living populations show they are beneficial.[23] Unfortunately, little is known about the

TABLE 2 Strategies for Reducing Salt Intake

- Avoid adding salt to foods during cooking or dining.
- Choose fresh, frozen, or canned foods without salt.
- When dining out, ask for foods to be prepared without salt or with half the usual amount of salt.
- Cut down on highly salted snack foods, such as chips.
- Choose low-salt prepared foods when possible.
- Do not drink "sports drinks" unless specifically needed.

DASH (Dietary Approaches to Stop Hypertension) diet A dietary pattern emphasizing fruits, vegetables, and low-fat dairy products designed to lower blood pressure.

TABLE 3 Dietary Approaches to Stop Hypertension (DASH) Diet Basics

Food Group	Daily Servings	Examples of Foods
Grains*	6–8	1 slice bread 1 oz dry cereal† ½ cup cooked rice, pasta, or cereal
Vegetables	4–5	1 cup raw leafy vegetable ½ cup cut-up raw or cooked vegetable ½ cup vegetable juice
Fruits	4–5	1 medium fruit ¼ cup dried fruit ½ cup fresh, frozen, or canned fruit ½ cup fruit juice
Fat-free or low-fat milk and milk products	2–3	1 cup milk or yogurt 1½ oz cheese
Lean meats, poultry, and fish	6 or less	1 oz cooked meats, poultry, or fish 1 egg
Nuts, seeds, and legumes	4–5 per week	⅓ cup or 1½ oz nuts 2 Tbsp peanut butter 2 Tbsp or ½ oz seeds ½ cup cooked legumes (dry beans and peas)
Fats and oils	2–3	Use sparingly
Sweets and added sugars	5 or less per week	1 Tbsp sugar 1 Tbsp jelly or jam ½ cup sorbet, gelatin 1 cup lemonade

*Whole grains are recommended for most grain servings as a good source of fiber and nutrients.

†Serving sizes vary between ½ cup and 1¼ cups, depending on cereal type. Check the product's Nutrition Facts label.

SOURCE: Department of Health and Human Services and U.S. Department of Agriculture. Facts about the DASH eating plan. Washington, DC: U.S. Government Printing Office; 2006; available at http://www.nhlbi.nih.gov/health/public/heart/hbp/dash/new_dash.pdf.

long-term effects of these diets on the cardiovascular system, and a recently conducted clinical trial suggests that consumption of low-carbohydrate diets for one year increases total cholesterol and LDL cholesterol in persons trying to lose weight.[24] Clearly, more studies are needed.

Trans Fatty Acids, Saturated Fatty Acids, and Cholesterol Various dietary lipids can also influence blood lipid levels. For some people, high intake of dietary cholesterol can raise blood cholesterol and triglyceride levels.[25] Consuming high levels of saturated fatty acids and *trans* fatty acids can cause a rise as well.[26] This is why food labels list the content of these types of fatty acids. Although saturated fatty acids used to be considered the "worst" type of dietary lipid for maintaining a healthy cardiovascular system, mounting evidence suggests that *trans* fatty acids are equally bad if not more detrimental to health.[27] Studies show that saturated and *trans* fatty acids increase risk for heart disease, at least in part, by increasing circulating LDL concentrations.[28] *Trans* fatty acids do additional harm, however, because they not only decrease HDL but also promote inflammation.[29] However, the harmful effects of *trans* fatty acids on blood lipids may be true only for industrially produced *trans* fatty acids.[30] Those found naturally in foods—usually beef and dairy products—may actually be heart-healthy.[31]

Monounsaturated and ω-3 Fatty Acids Compared to saturated and *trans* fatty acids, which may *increase* blood lipids, some monounsaturated fatty acids (such as those in olive oil) and polyunsaturated fatty acids may help *lower* levels.[32] Studies suggest that the beneficial effects of the ω-3 fatty acids, such as those in fish and flaxseed oils, may be particularly important as they appear to decrease VLDL and triglyceride synthesis in the liver.[33] In addition, increased ω-3 fatty acid consumption may decrease blood pressure in some people,[34] reduce the risk of atherosclerotic plaque rupture,[35] and have anti-inflammatory properties.[36] However, the optimal level of ω-3 fatty acids (especially when also considering ω-6 fatty acids) is an area of active debate among scientists.[37]

Alcohol—Friend and Foe Consuming alcoholic beverages can have both detrimental and beneficial effects on blood lipid concentrations. Specifically, alcohol consumption tends to increase triglyceride levels (a negative effect), whereas moderate alcohol consumption (one to two drinks per day) can raise HDL cholesterol levels (a beneficial effect).[38] Therefore, if you drink, do so in moderation. It is never recommended, however, that nondrinkers begin to drink for health benefits.

OTHER DIETARY FACTORS AND PATTERNS ARE ASSOCIATED WITH LOWER RISK

In addition to dietary components *known* to influence cardiovascular risks such as saturated and *trans* fatty acids, other nutrients may be important as well. For example, the B vitamin folate (folic acid), vitamin B_6, and vitamin B_{12} help maintain

Garlic is a functional food because it contains sulfur-containing compounds thought to lower LDL cholesterol (LDL-C).

Consumption of a Mediterranean-type diet is associated with a lower risk for cardiovascular disease.

a healthy cardiovascular system. A deficiency in any of these vitamins can cause a compound called homocysteine to accumulate in the blood. This is important, because raised levels of homocysteine are strongly associated with increased risk for CVD.[39]* Studies designed to determine how these vitamins are related to cardiovascular risk have not been conclusive, and more research is needed.[40]

As previously mentioned, dietary antioxidants may help decrease chronic inflammation. Indeed, emerging research suggests that dietary antioxidants may help inhibit the oxidation of LDLs, thus decreasing risk for CVD.[41] Examples of antioxidant nutrients include vitamins C and E, β-carotene (a precursor of vitamin A), zinc, and selenium. However, controlled clinical studies have not shown that increasing consumption of antioxidant supplements decreases risk for this disease, although higher vitamin E intake has been associated with lower circulating concentration of C-reactive protein (CRP).[42]

Functional Foods, Phytochemicals, and Zoonutrients Are Also Important It is important to recognize that the health benefits of foods extend beyond their classic micro- and macronutrient contents. Indeed, foods have other biologically active components thought to influence health. You may recall that these foods are referred to as functional foods because they contain phytochemicals or zoonutrients, and some of these are listed in Table 4. Phytochemicals that may influence heart health include plant sterols and stanols, isoflavones found mainly in soy products, compounds found in red wine and grapes, and a variety of sulfur-containing compounds abundant in garlic,

TABLE 4 Examples of Heart-Healthy Functional Foods and Their Biologically Active Components

Functional Food	Possible Biologically Active Component
Black tea	Polyphenols
Blueberries	Anthocyanin
Cocoa	Flavanols
Dairy foods	Proteins, calcium, potassium, conjugated linoleic acid
Fish	ω-3 fatty acids
Flaxseed	ω-3 fatty acids
Garlic	Sulfur-containing compounds
Nuts	Unsaturated fatty acids, vitamin E, selenium
Olive oil	Monounsaturated fatty acids, phenolic compounds
Psyllium	Soluble fiber
Red wine and grapes	Resveratrol, quercetin
Soy	Proteins and flavonoids
Stanol-sterol-fortified foods	Plant stanols and sterols
Tomatoes	Lycopene, lutein
Whole grains	Soluble fiber, folate, antioxidants
Whole oats	β-glucan, soluble fiber

SOURCE: Adapted from Hasler CM, Bloch AS, Thomson CA, Enrione E, Manning C. Position of the American Dietetic Association: Functional foods. Journal of the American Dietetic Association. 2004;104:814–26.

onions, and leeks.[43] Animal-derived foods that may provide protection from CVD are fatty fish and milk, which appear to contain fatty acids and proteins that are especially heart-healthy.[44]

---

*You can read more about homocysteine and its relationship with the B vitamins in Chapter 10.

However, not enough data are available to determine which components of these foods are beneficial or whether higher intakes of these compounds reduce the risk of heart attacks or strokes in all people. In the meantime, consuming a variety of foods—especially fruits, vegetables, whole-grain cereals, low-fat fish, and dairy products—is recommended to ensure adequate intake of these dietary factors.

HEART-HEALTHY DIETARY *PATTERNS*

In addition to knowing which nutrients, phytochemicals, or zoonutrients are important for heart health, it is also useful to consider which overall *types* of diets are associated with decreased risk for CVD. Dietary patterns that may help lower risk for CVD include the DASH diet, vegetarian diets, and the Mediterranean diet. As previously described, the DASH diet emphasizes fruits, vegetables, and low-fat dairy products. Vegetarian diets are typically those that exclude meat products but include milk, eggs, and sometimes fish. In addition, there is much interest in whether low- or high-carbohydrate diets are best for people with elevated risk for CVD.

Mediterranean Diets **Mediterranean diets** are not "diets" *per se*, but rather eating patterns common to traditional cuisines of those living in countries surrounding the Mediterranean Sea, such as Spain, Italy, and Greece. These diets emphasize fruits, fish, vegetables, whole grains, nuts, and seeds as well as olive oil and red wine; meats are consumed in moderation. Epidemiologic studies show that people consuming a Mediterranean-type diet have lower rates of heart disease.[45] However, it is difficult to determine which dietary components are protective and whether the reduced risk for disease is due to diet alone. More likely, a variety of confounding factors are involved, such as low rates of obesity and high rates of physical activity.

What Are the General Nutrition Guidelines for Healthy Hearts?

The multifaceted and complex relationship between diet and cardiovascular risk is yet another example of how dietary variety, moderation, and balance are essential to health and well-being. Clearly, many dietary components are related to risk for CVD. In response, agencies such as the American Heart Association, the U.S. Department of Agriculture, and the National Institutes of Health have set forth guidelines regarding dietary intakes that help lower risk for these diseases. These are outlined as follows.

BE MINDFUL OF ENERGY INTAKE AND MACRONUTRIENT BALANCE

The primary goal of energy consumption recommendations is to consume only enough calories to maintain a healthy body weight. People who need to lose weight are advised to decrease their energy intake and increase their energy expenditure to meet their body weight goals. In addition, there are several general recommendations concerning intake of lipids and other nutrients, some of which are listed here.

- **Total fat**—limit to 20 to 35% of total calories.
- **Saturated fat**—intake should be ≤10% of total calories; people who already have heart disease should consume <7% of total calories as saturated fats.
- ***Trans* fat**—intake should be minimal.
- **Omega-3 (ω-3) fatty acids**—recommended that everyone consume fish at least twice weekly; people with heart disease or elevated blood triglyceride should consume more.
- **Cholesterol**—limit to 300 mg/day; people with heart disease should consume ≤200 mg/day;

Mediterranean diet A dietary pattern, originating from the region surrounding the Mediterranean Sea, which is related to lower risk for cardiovascular disease.

Mixed salads provide many nutrients needed to maintain heart health.

those with very high levels of blood cholesterol should consider reducing intake even more.

- **Complex carbohydrates**—consume ≥3 servings of whole-grain products daily; emphasize complex carbohydrates such as those found in vegetables, fruits, and whole grains rather than refined carbohydrates, such as table sugar.

VITAMINS AND MINERALS ALSO MATTER

Although less is known about the relationship between micronutrients and CVD, many of these compounds are likely important for maintaining good health. Some recommendations concerning intakes of these substances are provided below.

- **Sodium**—intake should not exceed about 2,300 mg/day—the amount equivalent to 1 teaspoon of salt.
- **Calcium**—consume at least 1,000–1,200 mg/day; low-fat dairy products should be emphasized.
- **Potassium**—choose potassium-rich foods such as legumes, potatoes, seafood, and some fruits (such as bananas) regularly.
- **B vitamins**—select foods high in folate, vitamin B_6, and vitamin B_{12}; eating a variety of both animal- and plant-based foods on a regular basis is critical to meeting this goal.

Key Points

How Does Cardiovascular Disease Develop?

- Impaired blood flow to the heart or brain can cause heart attacks or strokes, respectively.
- Atherosclerosis, aneurysms, and blood clots are the major causes of heart disease and strokes.
- Chronic inflammation appears to be an important factor in CVD.

What Are the Risk Factors for Cardiovascular Disease?

- High levels of LDL—especially oxidized LDL—and low levels of HDL can increase risk for atherosclerosis.
- When medication and changes in lifestyle choices (e.g., nutrition) are not sufficient to lower risk for CVD, there are several surgical interventions that can help.
- Nonmodifiable factors for CVD include age, sex, genetics, and low birth weight.
- Modifiable risk factors include smoking, high blood pressure, elevated blood lipid levels, diabetes, obesity, and poor diet.
- Diets high in saturated fats or *trans* fats can increase risk for CVD, whereas those high in monounsaturated fatty acids decrease risk.

- Prevention and treatment of both obesity and diabetes involve restriction of caloric intake and increased activity levels.

How Does Nutrition Influence Cardiovascular Risk?

- Some dietary patterns such as the DASH diet, vegetarian diet, and Mediterranean diet are associated with decreased risk for CVD.

What Are the General Nutrition Guidelines for Healthy Hearts?

- Experts recommend that we consume a diet low in fat (<35% of calories), saturated fatty acids (≤10% of calories), *trans* fatty acids (minimal), cholesterol (<300 mg/d), and salt (<1 teaspoon per day).
- Fish and other foods rich in omega-3 fatty acids should be consumed on a regular basis as well as whole-grain foods and those containing soluble fiber.
- Consuming adequate amounts of some vitamins and minerals, such as calcium, potassium, and certain B vitamins, can reduce risk for CVD.

Notes

1. American Heart Association. Heart disease and stroke statistics—2011 update. Circulation. 2011;123:e18-e209.
2. Steinberg D, Witztum JL. Oxidized low-density lipoprotein and atherosclerosis. Atherosclerosis and Thrombosis in Vascular Biology. 2010;30:2311–6. Stocker R, Keaney JF, Jr. Role of oxidative modifications in atherosclerosis. Physiological Reviews. 2004;84:1381–478.

3. Berger JS, Brown DL, Becker RC. Low-dose aspirin in patients with stable cardiovascular disease: A meta-analysis. American Journal of Medicine. 2008;121:43–9.

4. Ridker PM, Rifai N, Rose L, Buring JE, Cook NR. Comparison of C-reactive protein and low-density lipoprotein cholesterol levels in the prediction of first cardiovascular events. New England Journal of Medicine. 2002;347:1557–65. Sakkinen P, Abbott RD, Curb JD, Rodriguez BL, Yano K, Tracy RP. C-reactive protein and myocardial infarction. Journal of Clinical Epidemiology. 2002;55:445–51. Alizadeh Dehnavi R, de Roos A, Rabelink TJ, van Pelt J, Wensink MJ, Romijn JA, Tamsma JT. Elevated CRP levels are associated with increased carotid atherosclerosis independent of visceral obesity. Atherosclerosis. 2008;200:417–23.

5. Pearson AP, Mensah GA, Alexander RW, Anderson JL, Cannon RO, Criqui M, Fadl YY, Fortmann SP, Hong Y, Myers GL, Rifai N, Smith SC, Taubert K, Tracy RP, Vinicor. Markers of inflammation and cardiovascular disease. Application to clinical and public health practice. A statement for healthcare professionals from the Centers for Disease Control and Prevention and the American Heart Association. Circulation. 2003;107:499–511.

6. American Heart Association. Risk factors and coronary heart disease. Available from: http://www .americanheart.org/presenter.jhtml?identifier=4726.

7. Katcher HI, Gillies PJ, Kris-Etherton PM. Atherosclerotic cardiovascular disease. In: Present knowledge in nutrition, 9th ed. Bowman BA, Russell RM, editors. Washington, DC: ILSI Press; 2006.

8. Hokanson JE, Austin MA. Plasma triglyceride level is a risk factor for cardiovascular disease independent of high-density lipoprotein cholesterol level: A meta-analysis of population-based prospective studies. Journal of Cardiovascular Risk. 1996;3:213–9. Krauss RM. Triglycerides and atherogenic lipoproteins: Rationale for lipid management. American Journal of Medicine. 1998;105:S58–S62.

9. Hawkins MA. Markers of increased cardiovascular risk. Obesity Research. 2004;12:107S–14S. Holvoet P. Oxidized LDL and coronary heart disease. Acta Cardiologica. 2004;59:479–84.

10. Gordon T, Castelli WP, Hjotland MC, Kannel WB, Dawber TR. High density lipoprotein as a protective factor against coronary heart disease. The Framingham study. American Journal of Medicine. 1977;62:707–14. Watson AD, Berliner JA, Hama SY, LaDu BN, Faull KF, Fogelman AM, Navab M. Protective effect of high density lipoprotein associated paraoxonase. Inhibition of the biological activity of minimally oxidized low density lipoprotein. Journal of Clinical Investigation. 1995;96:2882–91.

11. Hofnagel O, Luechtenborg B, Weissen-Plenz G, Robenek H. Statins and foam cell formation: Impact on LDL oxidation and uptake of oxidized lipoproteins via scavenger receptors. Biochimica Biophysica Acta. 2007;1771:1117–24. Pennathur S, Heinecke JW. Oxidative stress and endothelial dysfunction in vascular disease. Current Diabetes Reports. 2007;7:257–64.

12. Pereira MA, O'Reilly E, Augustsson K, Fraser GE, Goldbourt U, Heitmann BL, Hallmans G, Knekt P, Liu S, Pietinen P, Spiegelman D, Stevens J, Virtamo J, Willett WC, Ascherio A. Dietary fiber and risk of coronary heart disease. A pooled analysis of cohort studies. Archives of Internal Medicine. 2004;164: 370–6.

13. Clarke R, Smulders Y, Fowler B, Stehouwer CD. Homocysteine, B-vitamins, and the risk of cardiovascular disease. Seminars in Vascular Medicine. 2005;5:75–6. Eckel RH. Obesity and heart disease. A statement for health care professionals from the Nutrition Committee. American Heart Association. Circulation. 1997;96:3248–50.

14. Aucott L, Gray D, Rothnie H, Thapa M, Waweru C. Effects of lifestyle interventions and long-term weight loss on lipid outcomes—a systematic review. Obesity Reviews. 2011;12:e412–25.

15. American Heart Association. Guidelines for physical activity. Available at http://www.heart.org/HEARTORG/ GettingHealthy/PhysicalActivity/GettingActive/ American-Heart-Association-Guidelines_UCM_307976_ Article.jsp. Updated January 19, 2011.

16. Lenz TL, Monaghan MS. Lifestyle modifications for patients with hypertension. Journal of the American Pharmacological Association. 2008;48:e92-9.

17. Weinberger MH. Sodium and blood pressure 2003. Current Opinions in Cardiology. 2004;19:353–6.

18. Miller ER 3rd, Erlinger TP, Appel LJ. The effects of macronutrients on blood pressure and lipids: an overview of the DASH and OmniHeart trials. Current Atherosclerosis Reports. 2006;8:460–5.

19. Mattson FH, Erickson BA, Kligman AM. Effect of dietary cholesterol on serum cholesterol in man. American Journal of Clinical Nutrition. 1972;25:589–94.

20. Mozaffarian D, Katan MB, Ascherio A, Stampfer MJ, Willett WC. Trans fatty acids and cardiovascular disease. New England Journal of Medicine. 2006;354:1601–13. Woodside JV, McKinley MC, Young IS. Saturated and trans fatty acids and coronary heart disease. Current Atherosclerosis Reports. 2008;10:460–6.

21. StJeor ST, Howard BV, Preweitt TE, Bovee V, Bazzarre T, Eckel RH. Dietary protein and weight reduction: A statement for healthcare professionals from the Nutrition Committee of the Council on Nutrition, Physical Activity, and Metabolism of the American Heart Association. Circulation. 2001;104:1869–74.

22. Katcher HI, Gillies PJ, Kris-Etherton PM. Atherosclerotic cardiovascular disease. In: Present knowledge in nutrition, 9th ed. Bowman BA, Russell RM, editors. Washington, DC: ILSI Press; 2006.

23. Kerksick CM, Wismann-Bunn J, Fogt D, et al. Changes in weight loss, body composition and cardiovascular disease risk after altering macronutrient distributions during a regular exercise program in obese women. Nutrition Journal. 2010;9:59–64. Larsen TM, Dalskov SM, van Baak M, et al. Diets with high or

low protein content and glycemic index for weight-loss maintenance. New England Journal of Medicine. 2010;363:2102–13.

24. Brinkworth GD, Noakes M, Buckley JD, Keogh JB, Clifton PM. Long-term effects of a very-low-carbo-hydrate weight loss diet compared with an isocaloric low-fat diet after 12 mo. American Journal of Clinical Nutrition. 2009; 90:23–32.

25. Katcher HI, Hill AM, Lanford JL, Yoo JS, Kris-Etherton PM. Lifestyle approaches and dietary strategies to lower LDL-cholesterol and triglycerides and raise HDL-cholesterol. Endocrinology and Metabolism Clinics North America. 2009;38:45–78.

26. Zaloga GP, Harvey KA, Stillwell W, Siddiqui R. Trans fatty acids and coronary heart disease. Nutrition in Clinical Practice. 2006;21:505–12. Katcher HI, Hill AM, Lanford JL, Yoo JS, Kris-Etherton PM. Lifestyle approaches and dietary strategies to lower LDL-cho-lesterol and triglycerides and raise HDL-cholesterol. Endocrinology and Metabolism Clinics North America. 2009;38:45–78.

27. Ascherio A, Katan MB, Zock PL, Stampfer MJ, Willett WC. Trans fatty acids and coronary heart disease. New England Journal of Medicine. 1999;340:1994–8.

28. Brouwer IA, Wanders AJ, Katan MB. Effect of animal and industrial trans fatty acids on HDL and LDL cho-lesterol levels in humans—a quantitative review. PLoS One. 2010;5:e9434. Remig V, Franklin B, Margolis S, Kostas G, Nece T, Street JC. Trans fats in America: a re-view of their use, consumption, health implications, and regulation. Journal of the American Dietetic Associa-tion. 2010;110:585–92.

29. Ascherio A. Trans fatty acids and blood lipids. Athero-sclerosis Supplements. 2006;7:25–7. Mozaffarian D. Trans fatty acids—effects on systemic inflammation and endothelial function. Atherosclerosis Supplements. 2006;7(2):29–32.

30. Gebauer SK, Psota TL, Kris-Etherton PM. The diversity of health effects of individual trans fatty acid isomers. Lipids. 2007;42:787–99. Mensink RP. Metabolic and health effects of isomeric fatty acids. Current Opinions in Lipidology. 2005;16:27–30.

31. Pfeuffer M, Schrezenmeir J. Bioactive substances in milk with properties decreasing risk of cardiovascular diseases. British Journal of Nutrition. 2000;84:S155–9.

32. Moorandian AD, Haas MJ, Wong NCW. The effect of se-lect nutrients on serum high density lipoprotein choles-terol and apolipoprotein A-I levels. Endocrine Reviews. 2006;27:2–16.

33. Riediger ND, Othman RA, Suh M, Moghadasian MH. A systemic review of the roles of n-3 fatty acids in health and disease. Journal of the American Dietetic Assoco-ciation. 2009;109:668–79.

34. Geleijnse JM, Giltay EF, Grobbee DE, Donders AR, Kok FJ. Blood pressure response to fish oil supplementation: Metaregression analysis of randomized trials. Journal of Hypertension. 2002;20:1493–9.

35. Galli C, Risé P. Fish consumption, omega 3 fatty acids and cardiovascular disease. The science and the clinical trials. Nutrition and Health. 2009;20:11–20.

36. Mori TA, Beilin LJ. Omega-3 fatty acids and inflammation. Current Atherosclerosis Reports. 2004;5:461–7. Zhao G, Etherton TD, Martin KR, et al. Anti-inflammatory effects of polyunsaturated fatty acids in THP-1 cells. Biochemical and Biophysi-cal Research Communications. 2005;336:909–17. Zampelas A. Eicosapentaenoic acid (EPA) from highly concentrated n-3 fatty acid ethyl esters is incorpo-rated into advanced atheroscleroptic plaques and higher plaque EPA is associated with decreased plaque inflammation and increased stability. Altherosclerosis. 2010;212:34–5.

37. Nettleton JA, Koletzko B, Hornstra G. ISSFAL 2010 dinner debate: Healthy fats for healthy hearts—annotated report of a scientific discussion. Annals of Nutrition and Metabolism. 2011;58:59–65.

38. Pearson TA. Alcohol and heart disease. Circulation. 1996;94:3023–25. Renaud S, de Logeril M. Wine, alco-hol, platelets and the French paradox for coronary heart disease. Lancet. 1992;339:1523–6.

39. McCully KS. Homocysteine, vitamins, and vascular disease prevention. American Journal of Clinical Nutrition. 2007;86:1563S–8S. Wang X, Qin X, Demirtas H, Li J, Mao G, Huo Y, Sun N, Liu L, Xu X. Efficacy of folic acid supplementation in stroke preven-tion: A meta-analysis. Lancet. 2007;369:1876–82.

40. Ciaccio M, Bellia C. Hyperhomocysteinemia and cardio-vascular risk: effect of vitamin supplementation in risk reduction. Current Clinical Pharmacology. 2010;5:30–6.

41. Knekt P, Ritz J, Pereira MA, O'Reilly EJ, Augustsson K, Fraser GE, Goldbourt U, Heitmann BL, Hallmans G, Liu, S, Pietinen P, Spiegelman D, Stevens J, Virtamo J, Willett WC, Rimm EB, Ascherio A. Antioxidant vita-mins and coronary heart disease risk: A pooled analysis of 9 cohorts. American Journal of Clinical Nutrition. 2004;80:1508–20. Wenger NK. Do diet, folic acid, and vitamins matter? What did we learn from the Women's Health Initiative, the Women's Health Study, the Women's Antioxidant and Folic Acid Cardiovascular Study, and other clinical trials? Cardiology Reviews. 2007;15:288–90.

42. Liepa GU, Basu H. C-reactive proteins and chronic dis-ease: What role does nutrition play? Nutrition in Clini-cal Practice. 2003;18:227–33. Singh U, Devaraj S, Jialal I. Vitamin E, oxidative stress, and inflammation. Annual Review of Nutrition. 2005;25:151–74.

43. Badimon L, Vilahur G, Padro T. Nutraceuticals and atherosclerosis: human trails. Cardiovascular Therapies. 2010;28:202–15. Castro IA, Barroso LP, Sinnecker P. Functional foods for coronary heart disease risk reduc-tion: A meta-analysis using a multivariate approach. American Journal of Clinical Nutrition. 2005;82:32–40. Gylling H, Miettinen TA. The effect of plant stanol- and sterol-enriched foods on lipid metabolism, serum lipids and coronary heart disease. Annals of Clinical Biochem-istry. 2005;42:254–63.

44. American Heart Association Nutrition Committee. Diet and lifestyle recommendations revision 2006: A scientific statement from the American Heart Association Nutrition Committee. Circulation.

2006;114:82–96. Wang L, Manson JE, Buring JE, Lee IM, Sesso HD. Dietary intake of dairy products, calcium, and vitamin D and the risk of hypertension in middle-aged and older women. Hypertension. 2008;51:1073–9.

45. Bendinelli B, Masala G, Saieva C, Salvini S, Calonico C, Sacerdote C, Agnoli C, Grioni S, Frasca G, Mattiello A, Chiodini P, Tumino R, Vineis P, Palli D, Panico S. Fruit, vegetables, and olive oil and risk of coronary heart disease in Italian women: the EPICOR Study. American Journal of Clinical Nutrition. 2011;93:275–83. Bhupathiraju SN, Tucker KL. Coronary heart disease prevention: Nutrients, foods, and dietary patterns. Clinica Chimica Acta. 2011;412:1493-514. Oliveira A, Rodríguez-Artalejo F, Gaio R, Santos AC, Ramos E, Lopes C. Major habitual dietary patterns are associated with acute myocardial infarction and cardiovascular risk markers in a southern European population. Journal of the American Dietetic Association. 2011;111:241–50.

Energy Metabolism

At any moment in time, thousands of chemical reactions are taking place around us and within us. These reactions all have a common purpose—to sustain life. To survive, all living organisms must obtain energy from their surroundings. In humans, that energy comes from the foods we eat. In previous chapters, you learned about the energy-yielding macronutrients—carbohydrates, lipids, and proteins. In this chapter, you will learn how cells actually use these nutrients as energy sources. It is important to understand the basics of how the body extracts energy from glucose, amino acids, and fatty acids. To appreciate how metabolic pathways are interrelated and how various tissues use different nutrients in response to the body's need for energy, it is helpful to examine energy metabolism using an integrated approach. You will also learn how the body regulates metabolism and how different macronutrients are used in various states of energy availability. Although you have encountered a variety of metabolic reactions throughout this book, the focus of this chapter is on energy metabolism, the chemical reactions that enable cells to obtain and use energy from nutrients.

The Importance of Newborn Screening

Tracy was thrilled to land her first job so soon after graduating from college. In addition, it was her dream to live overseas. So, when the offer came to teach English in a small, rural school in Mexico she was ecstatic. There were many work-related challenges, but nothing she could not handle. After a year, Tracy's soon-to-be husband, Michael, was also offered a teaching job; so together they made a commitment to continue living and working in Mexico. In fact, their first child, Anna, was born there. Unlike the United States, however, most hospitals in Mexico do not screen newborns for inherited metabolic disorders. And had Anna been tested, they would have discovered very early that she had a metabolic disorder called phenlyketonuria—PKU for short. Unbeknownst to Tracy and Michael, both were carriers of the PKU gene, increasing the chances that their children will inherit the disorder. Sadly, this was the case with Anna.

Tracy had no reason to suspect something was wrong, at least not at first. However, she did notice that Anna had pale skin, blonde hair, and light eye color; this seemed odd because both Tracy and Michael both had brown hair and brown eyes. In addition to Anna's paleness, Tracy noticed that she was growing slowly. Then, around one year of age, Anna started having seizures. Tracy now knew something was very wrong. Upon returning to the United States, they had their daughter examined by a pediatrician. Blood tests confirmed that Anna had PKU.

Had this been detected at birth, Anna would have been put immediately on a special diet to restrict her intake of the essential amino acid phenylalanine. This is because PKU is a metabolic disorder that both impairs the ability to metabolize phenylalanine and prevents the body from making another amino acid called tyrosine. Because tyrosine is needed to form the pigment that provides color to skin, hair, and eyes, the lack of tyrosine explained why Anna was so pale. But having a fair-skinned daughter was the least of Tracy's concerns. Anna's inability to metabolize phenylalanine had caused brain damage and the seizures she had been prone to in Mexico. Tracy's world stood still.

Anna is now six years old and has started school. Although the brain damage cannot be reversed, Anna is making amazing progress—mostly because of critical dietary changes. Because phenylalanine is found in meat, eggs, dairy products, and legumes, Anna must adhere to a special diet low in these foods for the rest of her life. Tracy also buys special phenylalanine-free protein mixes so that Anna receives proper protein nourishment. Luckily, being on this phenylalanine-restricted diet has stopped the seizures, and although she is still small for her age, Anna's growth has started to catch up to that of other children. While this situation has been challenging for Tracy and Michael, they are determined to keep Anna healthy by helping her follow her special diet.

Aletia/Shutterstock.com

Critical Thinking: Tracy's Story
Anna has an inherited disease that impacts the ability to metabolize an essential amino acid. What early indications were there that Anna might have a metabolic disease? Even though this condition is somewhat rare (1 in 10,000), explain why it is important to test all newborn infants at birth.

What Is Energy Metabolism?

Metabolism is defined as the sum of all chemical reactions that take place in the body. The term **energy metabolism** refers to hundreds of chemical reactions involved in the breakdown, synthesis, and transformation of the energy-yielding nutrients—glucose, amino acids, and fatty acids—that enable the body to store and use energy. In addition to these nutrients, cells can also metabolize alcohol for a source of energy, as you will learn in the Nutrition Matters following this chapter. Metabolic pathways work together in a complex way to maintain a steady supply of energy in the form of adenosine triphosphate (ATP), the high-energy molecule used to fuel cellular activities. These pathways occur simultaneously and are highly coordinated, allowing the body to use different combinations of nutrients in response to various physiological states. This versatility helps ensure that raw materials and energy are available to cells at all times.

⟨**CONNECTIONS**⟩ Recall that ATP is a molecule uniquely suited to transfer the energy contained in its chemical bonds to chemical reactions, that, to occur, require energy. (Chapter 1, page 10)

METABOLIC PATHWAYS CONSIST OF LINKED CHEMICAL REACTIONS

A **metabolic pathway** consists of a series of interrelated, enzyme-catalyzed chemical reactions. Some metabolic pathways are simple, consisting of 3 to 4 chemical reactions, whereas others are more complex, with 10 to 15 chemical reactions. Chemical reactions transform molecules, sometimes breaking them down and at other times forming new ones. A molecule that enters a chemical reaction is a **substrate,** also called a reactant, and the resulting molecule is a **product.** A chemical reaction can be expressed as an equation with the substrate(s) written on the left of an arrow that points to the product(s): $A + B \rightarrow C + D$. In this example, A and B are substrates, and C and D are products. The arrow means "yields." As shown in Figure 7.1, the product of one chemical reaction in a pathway becomes the substrate in the reaction that follows. Products formed before a metabolic pathway reaches completion are called **intermediate products,** whereas the final product(s) in a pathway is/are the **end product(s).** It is common for intermediate products and end product(s) of one metabolic pathway to enter other metabolic pathways.

metabolism (me – TAB – o – lism) Chemical reactions that take place in the body.

energy metabolism Chemical reactions that enable cells to store and use energy from nutrients.

metabolic pathway A series of interrelated enzyme-catalyzed chemical reactions that take place in cells.

substrate (reactant) (SUB – strate) A molecule that enters a chemical reaction.

product A molecule produced in a chemical reaction.

intermediate product A product formed before a metabolic pathway completion, often serving as a substrate in the next chemical reaction.

end product The final product in a metabolic pathway.

FIGURE 7.1 Metabolic Pathways A metabolic pathway consists of a series of interrelated, enzyme-catalyzed chemical reactions.

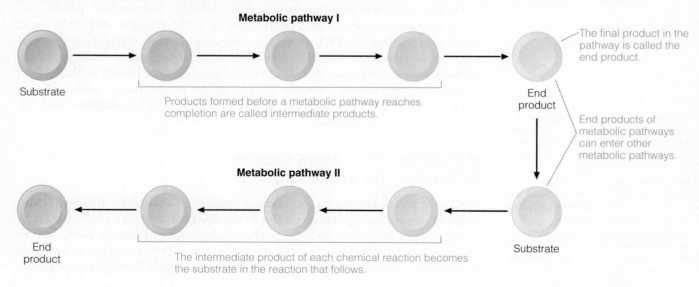

Metabolic pathway I

Substrate

Products formed before a metabolic pathway reaches completion are called intermediate products.

The final product in the pathway is called the end product.

End product

End products of metabolic pathways can enter other metabolic pathways.

Metabolic pathway II

End product

The intermediate product of each chemical reaction becomes the substrate in the reaction that follows.

Substrate

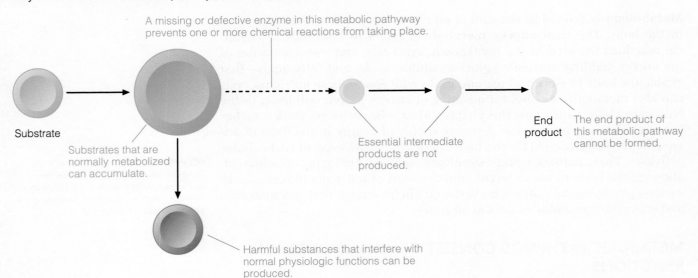

A missing or defective enzyme in this metabolic pathyway prevents one or more chemical reactions from taking place.

Substrate

Substrates that are normally metabolized can accumulate.

Essential intermediate products are not produced.

End product

The end product of this metabolic pathway cannot be formed.

Harmful substances that interfere with normal physiologic functions can be produced.

Inherited Metabolic Diseases Metabolic disorders that result from genetic defects are called **inherited metabolic diseases** (also called inborn errors of metabolism). These genetic abnormalities are caused by a deficiency in or absence of one or more enzymes needed for a metabolic pathway to function properly (Figure 7.2). As a result, the altered metabolic pathway can produce harmful substances that interfere with normal physiologic functions. Problems can also arise when the metabolic pathway is unable to produce essential compounds and when substrates that are normally broken down accumulate in the blood.

The number of inherited metabolic diseases is enormous, each with a wide range of effects. Because some inherited metabolic diseases can result in serious health problems and even death, a variety of newborn screening tests are routinely administered in hospitals across the United States. Early detection can mean early treatment, which often can prevent serious complications from occurring. In fact, some inherited metabolic diseases can be effectively managed by the elimination or restriction of certain dietary components. This is certainly the case for a condition called phenylketonuria (PKU).

People with PKU lack the enzyme phenylalanine hydroxylase, which converts the essential amino acid phenylalanine to the nonessential amino acid tyrosine. If untreated, PKU results in dangerously high circulating levels of phenylalanine as well as insufficient levels of tyrosine. Because this can cause serious health problems, including brain damage, most babies in the United States are tested for PKU shortly after birth. An infant found to have PKU is fed a special formula containing minimal amounts of phenylalanine and adequate amounts of tyrosine.

METABOLIC PATHWAYS CAN BE CATABOLIC OR ANABOLIC

Metabolic pathways can be catabolic or anabolic. **Catabolic pathways** (catabolism) release energy through the breakdown of complex molecules into simpler ones, whereas **anabolic pathways** (anabolism) require energy (ATP) to construct complex molecules from simpler ones. The energy made available by catabolic reactions often drives anabolic reactions. Likewise, the products

inherited metabolic disease (also called inborn error of metabolism) Genetic condition caused by a deficiency or absence of one or more enzymes needed for a metabolic pathway to function properly.

catabolic pathway (ca – ta – BOL – ic) A series of metabolic reactions that break down a complex molecule into simpler ones, often releasing energy in the process.

anabolic pathway (an – a – BOL – ic) A series of metabolic reactions that require energy to make a complex molecule from simpler ones, often requiring energy in the process.

TABLE 7.1 Summary of Major Energy Metabolism Pathways

Pathway	Description	Primary Tissue(s)	Anabolic, Catabolic, or Amphibolic
β-Oxidation	Breakdown of fatty acids to acetyl-CoA	Liver, muscle	Catabolic
Citric acid cycle	A central metabolic pathway that oxidizes acetyl-CoA to yield carbon dioxide, NADH + H$^+$, FADH$_2$, and GTP	All tissues except red blood cells	Amphibolic
Gluconeogenesis	Glucose synthesis from noncarbohydrate sources	Liver, muscles	Anabolic
Glycogenesis	Formation of glycogen	Liver, muscle	Anabolic
Glycogenolysis	Breakdown of glycogen for glucose production	Liver, muscle	Catabolic
Glycolysis	Anaerobic breakdown of glucose, forming two molecules of pyruvate	All	Catabolic*
Ketogenesis	Formation of ketones, an alternative energy source, from acetyl-CoA	Liver	Anabolic
Lipogenesis	Synthesis of fatty acids and triglycerides	Liver, adipose	Anabolic
Lipolysis	Breakdown of triglycerides to fatty acids and glycerol	Adipose, muscle	Catabolic
Oxidative phosphorylation	A coupled process whereby reduced coenzymes NADH + H$^+$ and FADH$_2$ are oxidized to NAD$^+$ and FAD, and ADP is phosphorylated to ATP	All tissues except red blood cells	—
Proteolysis	Breakdown of protein to amino acids	Muscle	Catabolic

*Primarily catabolic but also considered to be amphibolic.

of catabolic pathways provide many of the building blocks needed for anabolic pathways. Metabolic pathways that can be used for both catabolism and anabolism are called **amphibolic pathways.**

The availability of metabolic fuels (glucose, fatty acids, and amino acids) fluctuates throughout the day. After a meal, these energy sources are readily available and tend to exceed the body's immediate energy needs. During this time, anabolic pathways favor the storage of excess glucose, amino acids, and fatty acids as glycogen, protein, and triglycerides, respectively. Conversely, catabolic pathways increase fuel availability by breaking down the body's stored energy reserves—glycogen, protein, and triglycerides—into glucose, amino acids, and fatty acids, respectively. In this way, the body shifts between anabolic to catabolic pathways in response to energy availability and need. The major anabolic and catabolic pathways involved in energy metabolism are reviewed in Table 7.1.

CHEMICAL REACTIONS REQUIRE ENZYMES

For a chemical reaction to occur, reacting molecules (substrates) must make contact with each other and then be chemically transformed into one or more products. Because these reactions are not likely to occur on their own, the body relies on catalysts to speed them up. In cells, biological catalysts are proteins called **enzymes.** If it were not for enzymes, metabolic reactions would occur very slowly, if at all. Although enzymes increase the rate at which chemical reactions occur, they themselves do not undergo change.

The names of most enzymes end in –*ase*, such as the names of the digestive enzymes sucrase, maltase, and lactase, but there are exceptions. When

amphibolic pathway Metabolic pathway that generates intermediate products that can be used for both catabolism and anabolism.

enzymes Biological catalysts that facilitate chemical reactions.

⟨**CONNECTIONS**⟩ A hydrolysis reaction breaks chemical bonds by the addition of water (Chapter 3, page 74).

active site An area on an enzyme that binds substrates in a chemical reaction.

enzyme–substrate complex A substrate attached to an enzyme's active site.

cofactor A nonprotein component of an enzyme, often a mineral, needed for its activity.

coenzyme Organic molecule, often derived from vitamins, needed for some enzymes to function.

nicotinamide adenine dinucleotide (NAD⁺) (nic – o – TIN – a – mide AD – e – nine di – NU – cle – o – tide) The oxidized form of the coenzyme that is able to accept two electrons and two hydrogen ions, forming NADH + H⁺.

flavin adenine dinucleotide (FAD) The oxidized form of the coenzyme that is able to accept two electrons and two hydrogen ions, forming FADH₂.

nicotinamide adenine dinucleotide phosphate (NADP⁺) The oxidized form of the coenzyme that is able to accept two electrons and two hydrogen ions, forming NADPH + H⁺.

scientists first discovered that enzymes were proteins, enzymes were often given names ending in –in. The names of the enzymes pepsin, trypsin, and chymotrypsin—needed for protein digestion—reflect this older practice. Enzymes can be classified according to the type of chemical reaction they catalyze. For example, enzymes that catalyze hydrolysis reactions are called hydrolases. Similarly, enzymes that transfer a chemical group from one molecule to another are categorized as transferases.

Substrates attach to a special surface on the enzyme called the **active site,** forming an **enzyme–substrate complex.** Researchers previously thought that the shape of the active site was complementary to the shape of the substrate—like the pieces of a jigsaw puzzle. It was later realized that this was not the case, and that the shape of the active site changes to fit the shape of the substrate. The active site wraps around the substrate, altering its chemical structure and transforming it into the product. The product is released from the active site, and the enzyme is then free to bind yet another substrate (Figure 7.3). Enzymes have specificity, meaning that each interacts with only certain substrates. Because of this specificity, the body makes thousands of different enzymes to catalyze thousands of different reactions.

Cofactors and Coenzymes Assist Enzymes Some enzymes require assistance to carry out their catalytic functions. These enzyme "helpers" are called cofactors and coenzymes. **Cofactors** are inorganic substances such as zinc, potassium, iron, and magnesium. In order to function, some enzymes require cofactors to be attached to their active site. **Coenzymes** are organic molecules, some of which are derived from vitamins such as niacin and riboflavin. Examples of coenzymes include **nicotinamide adenine dinucleotide (NAD⁺), flavin adenine dinucleotide (FAD),** and **nicotinamide adenine dinucleotide phosphate (NADP⁺).** Unlike cofactors, coenzymes are not actually a part of the enzyme structure. Rather, they assist enzymes by accepting and donating

FIGURE 7.3 Enzymes Are Biological Catalysts Enzymes increase the rate at which chemical reactions occur.

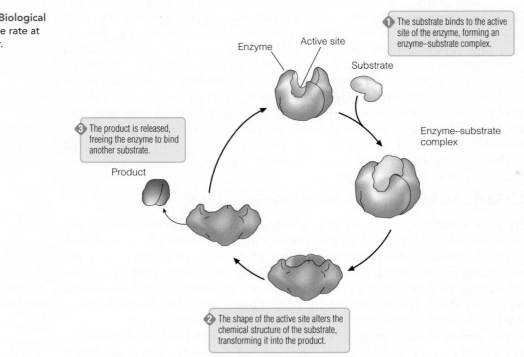

1 The substrate binds to the active site of the enzyme, forming an enzyme–substrate complex.

2 The shape of the active site alters the chemical structure of the substrate, transforming it into the product.

3 The product is released, freeing the enzyme to bind another substrate.

hydrogen ions (H^+), electrons (e^-), and other molecules during chemical reactions. This is why, while vitamins and minerals themselves do not provide cells with a source of energy (ATP), they play a critical role in energy metabolism by acting as cofactors and coenzymes.

Coenzymes and Energy Transfer Reactions Recall from Chapter 3 that reduction–oxidation (redox) reactions involve the gain and loss of electrons. Specifically, oxidation is the loss of electrons, whereas reduction is the gain of electrons. These reactions often occur simultaneously and are referred to as **coupled reactions.** In other words, when one molecule is oxidized (loss of electrons and hydrogen ions), another is reduced (gain of electrons and hydrogen ions). Coupled redox reactions allow energy to be transferred from one molecule to another.

Many coenzymes exist in two forms—oxidized (NAD^+, FAD, and $NADP^+$) and reduced (NADH + H^+, $FADH_2$, and NADPH + H^+). When energy-rich molecules are oxidized, their electrons and hydrogen ions are transferred to NAD^+ and FAD. The coenzyme NAD^+ can accept two electrons (2 e^-) and two hydrogen ions (2 H^+), forming NADH + H^+.* Similarly, $FADH_2$ is formed when two electrons (2 e^-) and two hydrogen ions (2 H^+) are transferred to FAD. The energy carried by these reduced coenzymes (NADH + H^+ and $FADH_2$) is used to produce the body's most important energy source—adenosine triphosphate (ATP). The oxidation and reduction of coenzymes are illustrated in Figure 7.4.

The reduced form of the coenzyme NADPH + H^+ also plays an important role in energy-requiring anabolic pathways. Specifically, NADPH + H^+ is needed for the synthesis of new compounds in the body such as fatty acids, cholesterol, DNA, and RNA. NADPH + H^+ is transformed to $NADP^+$ when it releases two electrons (2 e^-) and two hydrogen ions (2 H^+).

coupled reactions Chemical reactions that take place simultaneously often involving the oxidation of one molecule and the reduction of another.

*Although two hydrogen ions are transferred to NAD^+, only one actually attaches. The second hydrogen ion remains in solution. For this reason, the reduced form of NAD^+ is written as NADH + H^+. Both hydrogen ions attach to FAD, forming $FADH_2$.

FIGURE 7.4 The Role of Coenzymes in Energy Metabolism Energy is transferred to and from energy-yielding nutrients with the assistance of coenzymes.

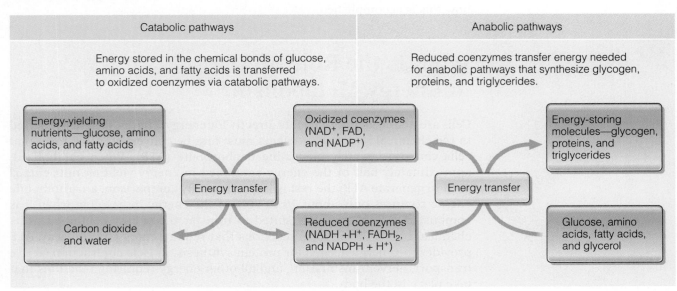

ENERGY METABOLISM IS REGULATED BY CHANGES IN ATP LEVELS

Energy metabolism can be thought of as an elaborate molecular highway system made up of anabolic and catabolic pathways. Whereas the anabolic pathways are involved in the synthesis of molecules that store energy, the catabolic pathways are involved in the breakdown of those molecules to release energy. The pathway or combination of pathways used by the body at any given time depends primarily on the cellular need for energy. When ATP is readily available, the activity of energy-yielding catabolic pathways decreases, and the activity of energy-storing anabolic pathways increases. Conversely, low ATP availability increases energy-yielding catabolic activity and decreases energy-storing anabolic activity. In simple terms, anabolic (synthesis) and catabolic (breakdown) pathways work hand in hand to ensure that the energy (ATP) needs of cells are met.

Hormones Also Regulate Energy Metabolism Hormones are important regulators of energy metabolism, helping the body shift between anabolic and catabolic pathways. Endocrine tissues can detect changes in substrate availability and respond by secreting appropriate hormones. These hormones then suppress or activate key enzymes in metabolic pathways. As a result, metabolic pathways can be "switched" on or off.

The primary hormones involved in the regulation of catabolic and anabolic pathways are insulin, glucagon, cortisol, and epinephrine. As you learned in Chapter 4, insulin is an anabolic hormone that promotes energy storage in the forms of glycogen, triglyceride, and protein. To do this, insulin increases the activity of anabolic pathways and decreases the activity of catabolic pathways. When energy availability is limited, the pancreatic hormone glucagon promotes catabolic pathways and inhibits anabolic pathways. In this way, glucagon increases energy availability by mobilizing energy-yielding molecules that have been "stored for a rainy day."

During times of stress and starvation, cortisol and epinephrine, hormones released from the adrenal glands, also play important roles in directing energy metabolism. These hormones stimulate catabolic pathways that help increase fuel availability. For example, cortisol and epinephrine stimulate the breakdown of glycogen stored in muscles to increase glucose availability.

The energy in the food we eat must ultimately be converted to ATP before cells can use it to perform the functions that sustain life. Next, you will learn how this is accomplished.

What Is the Role of ATP in Energy Metabolism?

Cells are unable to use nutrients directly for energy. That is, the energy stored in the chemical bonds of nutrients must first be converted into a form that cells can use—namely adenosine triphosphate (ATP). When catabolized, approximately half of the energy contained in energy-yielding nutrients is used to generate ATP; the rest is lost as heat. By comparison, an automobile engine captures only about 10 to 20% of the energy in gasoline when it is combusted. ATP is uniquely suited to transfer the energy contained in its chemical bonds to chemical reactions that require energy. In this way, ATP provides the energy needed for protein synthesis, muscle contraction, active transport, nerve transmission, and all other energy-requiring reactions that take place in the body.

FIGURE 7.5 Adenosine Triphosphate (ATP) Adenosine triphosphate (ATP) is a high-energy molecule that transfers its energy to chemical reactions.

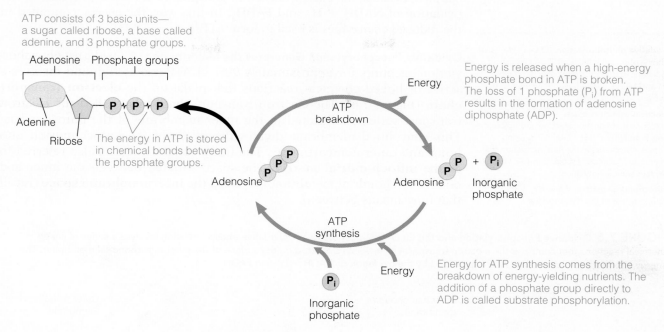

ATP consists of 3 basic units—a sugar called ribose, a base called adenine, and 3 phosphate groups.

Adenosine   Phosphate groups

Adenine

Ribose

The energy in ATP is stored in chemical bonds between the phosphate groups.

Adenosine

ATP breakdown

Energy

Energy is released when a high-energy phosphate bond in ATP is broken. The loss of 1 phosphate (P_i) from ATP results in the formation of adenosine diphosphate (ADP).

Adenosine   +   Inorganic phosphate

ATP synthesis

Energy

Inorganic phosphate

Energy for ATP synthesis comes from the breakdown of energy-yielding nutrients. The addition of a phosphate group directly to ADP is called substrate phosphorylation.

HIGH-ENERGY BONDS ENABLE ATP TO STORE AND RELEASE ENERGY

As illustrated in Figure 7.5, a molecule of ATP consists of three basic units: the sugar ribose, a base called adenine, and three phosphate groups (hence, ATP is a triphosphate). Together, ribose and adenine are referred to as adenosine. Of particular importance is the energy contained in the chemical bonds holding the phosphate groups together. These high-energy bonds enable ATP to both store and release energy. When cells need energy, a phosphate group is broken off of ATP, releasing energy and inorganic phosphate (symbolized as P_i). This results in the formation of adenosine diphosphate (ADP). The energy released when a phosphate bond is split from ATP is used to drive metabolic reactions. ATP is then regenerated by the addition of a phosphate group to ADP.

ATP IS SYNTHESIZED BY SUBSTRATE PHOSPHORYLATION AND BY OXIDATIVE PHOSPHORYLATION

Because ATP is not stored to any extent in the body, it is important for cells to be able to make ATP as it is needed. In fact, cells use ATP almost as quickly as it is made. ATP can be synthesized in two ways—substrate phosphorylation and oxidative phosphorylation. Because **substrate phosphorylation** does not require oxygen, this process is particularly important when tissues have little oxygen available to them. Although all cells have the capability to carry out substrate phosphorylation, the process produces relatively little ATP. Instead, most ATP is synthesized via oxidative phosphorylation.

Because the energy contained within energy-yielding nutrients cannot be transferred directly to ADP, electrons (e^-) and hydrogen ions (H^+) are transferred to the oxidized coenzymes NAD^+ and FAD. Once reduced, these coenzymes ($NADH + H^+$ and $FADH_2$) can be reoxidized back to NAD^+ and FAD. This process releases energy and is accomplished by a series of chemical reactions that link the oxidation of $NADH + H^+$ and $FADH_2$ to the synthesis of ATP. The process whereby $NADH + H^+$ and $FADH_2$ are oxidized and ADP

substrate phosphorylation (SUB – strate phos – pho – ryl – A – tion) The transfer of an inorganic phosphate (P_i) group to ADP to form ATP.

is phosphorylated is called **oxidative phosphorylation.** These reactions are coupled, because the energy needed to phosphorylate ADP is provided by the oxidation of NADH + H$^+$ and FADH$_2$. In this way, the energy contained in the reduced coenzymes is used to form ATP.

Oxidative Phosphorylation Generates the Majority of ATP Oxidative phosphorylation accounts for approximately 90% of ATP production and involves a series of linked chemical reactions that make up the **electron transport chain** (Figure 7.6). The electron transport chain (also called the electron transport system) is located in the inner membrane of the mitochondria. This convoluted membrane divides the inside of the mitochondrion into inner and outer compartments. The inner compartment is also referred to as the **mitochondrial matrix.** The space situated between the inner and outer mitochondrial membranes is called the **intermembrane space** (recall that *inter* means between).

oxidative phosphorylation (OX – i – da – tive phos – pho – ryl – A – tion) The chemical reactions that link the oxidation of NADH + H$^+$ and FADH$_2$ to the phosphorylation of ADP to form ATP and water.

electron transport chain A series of chemical reactions that transfers electron and hydrogen ions from NADH + H$^+$ and FADH$_2$ along protein complexes in the inner mitochondrial membrane, ultimately producing ATP via oxidative phosphorylation.

mitochondrial matrix The inner compartment of the mitochondrion.

FIGURE 7.6 Oxidative Phosphorylation and the Electron Transport Chain Oxidative phosphorylation involves a series of linked chemical reactions that comprise the electron transport chain, which is located on the surface of the inner mitochondrial membrane. The energy needed to phosphorylate ADP is provided by the oxidation of NADH + H$^+$ and FADH$_2$.

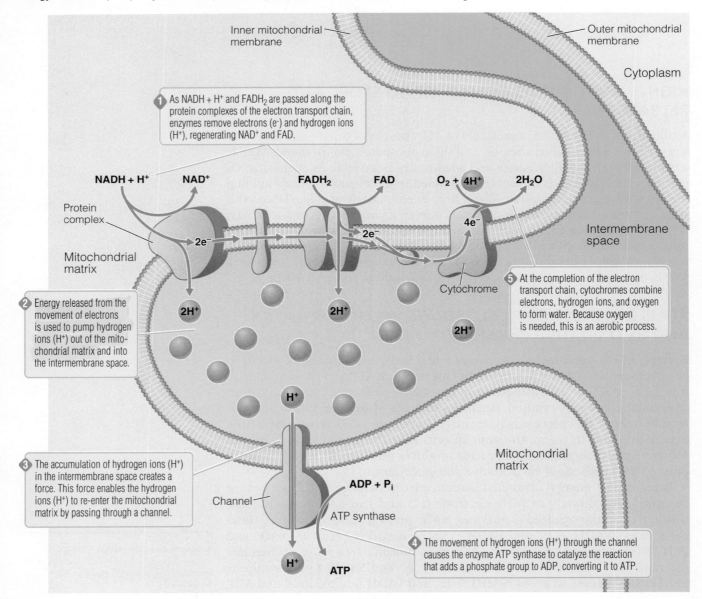

The electron transport chain consists of a series of protein complexes that are embedded in the inner mitochondrial membrane. When NADH + H$^+$ and FADH$_2$ enter the electron transport chain, enzymes remove their electrons (e$^-$) and hydrogen ions (H$^+$), regenerating NAD$^+$ and FAD (Figure 7.6, step 1). The released electrons (e$^-$) and hydrogen ions (H$^+$) take separate routes. The electrons (e$^-$) pass along the protein complexes, much like water in a bucket brigade. This movement releases energy that is used to pump the hydrogen ions (H$^+$) out of the mitochondrial matrix and into the intermembrane space (Figure 7.6, step 2).

The accumulation of hydrogen ions in the intermembrane space creates a powerful force. When this force gains sufficient strength, the hydrogen ions re-enter the mitochondrial matrix by passing through narrow channels—somewhat like tunnels (Figure 7.6, step 3). The movement of hydrogen ions through these channels releases energy that is used by the enzyme **ATP synthase** to attach a phosphate group to ADP. Thus, ATP is produced (Figure 7.6, step 4). Because NADH + H$^+$ and FADH$_2$ enter the electron transport chain at different locations along the protein complexes, the amount of ATP generated by these coenzymes differs. Nonetheless, the ATP yield is approximately three ATPs and two ATPs for each NADH + H$^+$ and FADH$_2$, respectively.[†]

At the completion of the electron transport chain, a group of iron-containing protein complexes called **cytochromes** reunites the electrons (e$^-$) and hydrogen ions (H$^+$) to form hydrogen (Figure 7.6, step 5). The hydrogen molecules then combine with oxygen (O$_2$) to form water (H$_2$O). This is why the electron transport chain and its accompanying oxidative phosphorylation reaction are considered aerobic (oxygen-requiring) metabolic processes.

In summary, cellular metabolic pathways convert the energy contained in NADH + H$^+$ and FADH$_2$ to usable energy in the form of ATP. But where do the NADH + H$^+$ and FADH$_2$ come from? As you will learn next, the answer lies in the breakdown of macronutrients.

How Do Catabolic Pathways Release Stored Energy?

Energy-yielding nutrients such as glucose, fatty acids, and amino acids store energy in their chemical bonds. For cells to produce ATP, these nutrients must undergo a series of chemical reactions that make up catabolic pathways. These pathways are interrelated in such a way that intermediate products or end products of one pathway often become substrates for other pathways. At first glance, the many catabolic pathways can appear overwhelming. However, they can be simplified by grouping them into four stages, as illustrated in Figure 7.7.

- Stage 1: The first stage of catabolism breaks down complex molecules into their fundamental building blocks. That is, protein to amino acids, glycogen to glucose, and triglycerides to fatty acids and glycerol molecules. These metabolic pathways are proteo*lysis*, glycogeno*lysis*, and lipo*lysis*, respectively. Notice that each of these metabolic processes (proteolysis, glycogenolysis, and lipolysis) all have the same ending—lysis. Lysis means separation or breaking apart, which is exactly what is happening here.
- Stage 2: During the second stage, the basic building blocks (amino acids, glucose, fatty acids, and glycerol) enter specific pathways whereby each is converted into an intermediate product that can enter a common pathway called the citric acid cycle—the third stage of catabolism.

intermembrane space The space between the inner and outer mitochondrial membranes.

ATP synthase (SYNTH – ase) A mitochondrial enzyme that adds a phosphate group (P$_i$) to ADP to form ATP during the process of oxidative phosphorylation.

cytochromes (CY – to – chromes) Iron-containing protein complexes that, as part of the electron transport chain, combine electrons, hydrogen ions, and oxygen to form water.

[†]By some estimates, the ATP yield from NADH + H$^+$ and FADH$_2$ are 2.5 and 1.5, respectively. For convenience, ATP yields are often estimated on the basis of 3 ATP per NADH + H$^+$ and 2 ATPs per FADH$_2$.

FIGURE 7.7 Four Stages of Energy Catabolism

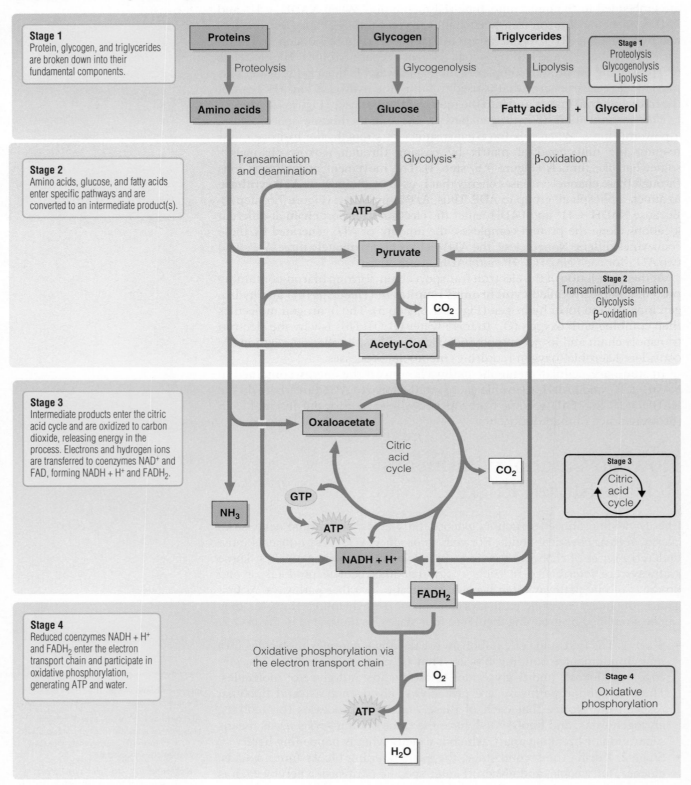

Stage 1
Protein, glycogen, and triglycerides are broken down into their fundamental components.

Stage 2
Amino acids, glucose, and fatty acids enter specific pathways and are converted to an intermediate product(s).

Stage 3
Intermediate products enter the citric acid cycle and are oxidized to carbon dioxide, releasing energy in the process. Electrons and hydrogen ions are transferred to coenzymes NAD^+ and FAD, forming NADH + H^+ and $FADH_2$.

Stage 4
Reduced coenzymes NADH + H^+ and $FADH_2$ enter the electron transport chain and participate in oxidative phosphorylation, generating ATP and water.

Proteins — Proteolysis → Amino acids
Glycogen — Glycogenolysis → Glucose
Triglycerides — Lipolysis → Fatty acids + Glycerol

Stage 1
Proteolysis
Glycogenolysis
Lipolysis

Transamination and deamination
Glycolysis*
β-oxidation

ATP

Pyruvate

CO_2

Acetyl-CoA

Stage 2
Transamination/deamination
Glycolysis
β-oxidation

Oxaloacetate

Citric acid cycle

CO_2

GTP

ATP

NH_3

NADH + H^+

$FADH_2$

Stage 3
Citric acid cycle

Oxidative phosphorylation via the electron transport chain

O_2

ATP

H_2O

Stage 4
Oxidative phosphorylation

*Note that the catabolism of glucose via glycolysis actually yields 2 pyruvate molecules and 2 ATPs.

- Stage 3: The third stage of catabolism begins when intermediate products formed during Stage 2 enter the citric acid cycle and are broken down further to form carbon dioxide, releasing energy in the process. Much of the energy released during Stage 3 is transferred to the coenzymes NAD^+ and FAD, forming $NADH + H^+$ and $FADH_2$, respectively. In addition, small amounts of ATP are formed via substrate phosphorylation.
- Stage 4: The fourth and final stage of catabolism begins when $NADH + H^+$ and $FADH_2$ enter the electron transport chain. It is here where most ATP production occurs via oxidative phosphorylation.

CATABOLIC PATHWAYS METABOLIZE GLUCOSE FOR ENERGY

Glucose is a rich source of energy that can be used by all cells in the body to produce ATP by means of catabolic energy pathways. Most glucose in the body comes from carbohydrate-rich foods. However, when additional glucose is needed, the hormone glucagon stimulates the breakdown of glycogen in the liver and the release of glucose into the blood. This metabolic process, called glycogenolysis, results in the breakdown of glycogen to glucose and takes place in Stage 1 of carbohydrate catabolism. The hormones epinephrine and cortisol also stimulate glycogenolysis in skeletal muscle.

⟨CONNECTIONS⟩ Recall that glycogenolysis is the breakdown of liver and muscle glycogen into glucose (Chapter 4, page 140).

Glycolysis Splits Glucose into Subunits The word glycolysis literally means the "splitting of sugar," which is what happens in Stage 2 of carbohydrate metabolism. **Glycolysis** is a chemical pathway that ultimately splits the six-carbon glucose molecule into two three-carbon subunits called **pyruvate.** These reactions occur in the cytoplasm of cells. Because oxygen is not required, glycolysis is an anaerobic metabolic pathway.

As you can see from Figure 7.8, glycolysis yields small amounts of energy—two ATPs and two molecules of $NADH + H^+$ per glucose. However, this represents only a small amount of the total energy available in each glucose

glycolysis The metabolic pathway that splits glucose into two three-carbon molecules called pyruvate.

pyruvate (py – RU – vate) An intermediate product formed during metabolism of carbohydrates and some amino acids.

FIGURE 7.8 Overview of Glycolysis Glycolysis is an anaerobic metabolic pathway that consists of a series of chemical reactions that split a six-carbon glucose molecule into two three-carbon molecules called pyruvate. The fate of pyruvate depends on oxygen availability in the cell.

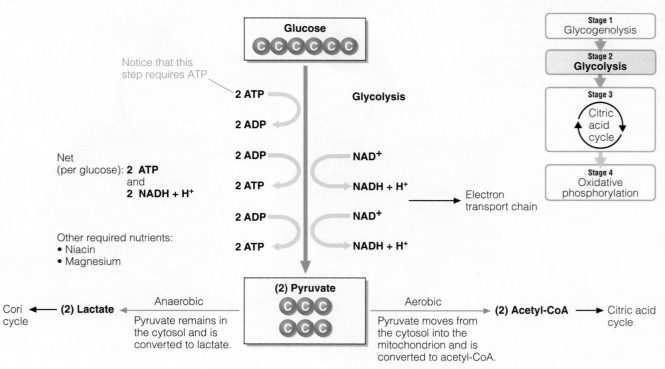

molecule. This is because the chemical bonds in pyruvate have not yet been broken. What happens next is a "major intersection" in glucose metabolism. That is, the subsequent direction taken by the pyruvate molecules produced in Stage 2 of carbohydrate catabolism is determined by oxygen availability, specifically whether pyruvate encounters anaerobic (oxygen-poor) or aerobic (oxygen-rich) conditions in the cell.

Oxygen Availability and Pyruvate The metabolic direction of pyruvate is largely determined by oxygen availability and energy requirements in the cell. For example, during high-intensity exercise, muscles have increased energy (ATP) requirements. To support high rates of ATP production via oxidative phosphorylation, large amounts of oxygen are needed. However, as exercise intensity increases, it becomes increasingly more difficult for the lungs and cardiovascular system to deliver adequate amounts of oxygen to active muscles. As a result, muscle cells can experience limited oxygen availability.

Under relatively anaerobic conditions, pyruvate remains in the cytoplasm and is converted to lactate (also called lactic acid). Lactate is then released into the blood and taken up by the liver, where it is converted to glucose via gluconeogenesis. The glucose that is formed from pyruvate can then undergo glycolysis (Figure 7.9). This sequence of chemical reactions, called the **Cori cycle,** provides a means by which small amounts of ATP can be produced in the relative absence of oxygen. Because very little ATP is actually produced, the Cori cycle cannot sustain vigorous physical activity for very long.

When oxygen is readily available (aerobic conditions), the two molecules of pyruvate (the end products of glycolysis) move from the cytoplasm into the mitochondria. Here pyruvate is chemically transformed into an intermediate product called acetyl-CoA, a two-carbon unit attached to a compound called coenzyme A (CoA). This irreversible reaction requires several enzymes and vitamin B–derived coenzymes. As shown in Figure 7.10, the formation

⟨**CONNECTIONS**⟩ Recall from Chapter 4 that gluconeogenesis means the synthesis of glucose from noncarbohydrate sources (Chapter 4, page 140).

Cori cycle The metabolic pathway that regenerates glucose by circulating lactate from muscle to the liver, where the lactate undergoes gluconeogenesis.

FIGURE 7.9 The Cori Cycle The Cori cycle is a metabolic pathway that involves both glycolysis and gluconeogenesis. Under conditions of limited oxygen availability, muscles break down glucose to lactate via glycolysis. Lactate is released into the blood and is converted to glucose by the liver via gluconeogenesis.

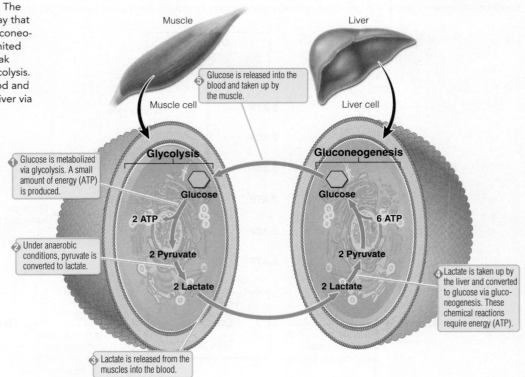

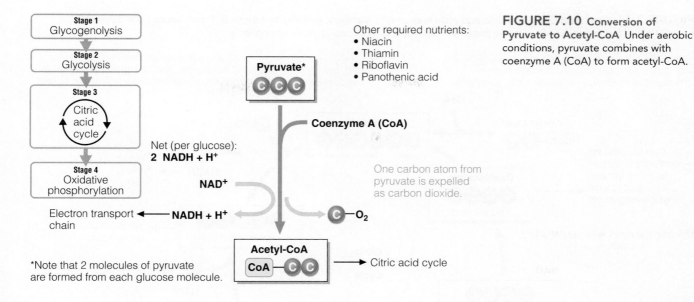

Stage 1
Glycogenolysis

Stage 2
Glycolysis

Stage 3
Citric acid cycle

Stage 4
Oxidative phosphorylation

Net (per glucose):
2 NADH + H⁺

NAD⁺

Electron transport ◄── **NADH + H⁺**
chain

*Note that 2 molecules of pyruvate are formed from each glucose molecule.

Other required nutrients:
• Niacin
• Thiamin
• Riboflavin
• Panothenic acid

Pyruvate*
C C C

Coenzyme A (CoA)

One carbon atom from pyruvate is expelled as carbon dioxide.

C—O₂

Acetyl-CoA
CoA—C C ──► Citric acid cycle

FIGURE 7.10 Conversion of Pyruvate to Acetyl-CoA Under aerobic conditions, pyruvate combines with coenzyme A (CoA) to form acetyl-CoA.

of acetyl-CoA from pyruvate also results in production of carbon dioxide and NADH + H⁺. Under aerobic conditions, acetyl-CoA is now ready to enter Stage 3 of carbohydrate catabolism—the **citric acid cycle.**

The Citric Acid Cycle and Oxidative Phosphorylation The citric acid cycle (also called the tricarboxylic acid [TCA] cycle or the Krebs cycle) is a major pathway used during aerobic conditions. It consists of a series of chemical reactions that take place within mitochondria. These enzyme-catalyzed reactions are often depicted as a circle, because the product of the last reaction of the pathway **(oxaloacetate)** becomes the substrate for **citrate** (the first reaction)—like a chemical carousel. Although the citric acid cycle is primarily a catabolic pathway, it serves other purposes. For example, intermediate products of the citric acid cycle can "leave" and enter anabolic pathways. Thus, the citric acid cycle is both catabolic and anabolic and is considered an amphibolic pathway. Amphibolic pathways provide important intersections between catabolism and anabolism on the metabolic "superhighway." The citric acid cycle is illustrated in Figure 7.11.

The citric acid cycle begins when acetyl-CoA combines with oxaloacetate to form citrate (also known as citric acid). In the process, coenzyme A (CoA) is released. The formation of citrate is followed by a series of chemical reactions that transfer the chemical energy contained in acetyl-CoA to NAD⁺ and FAD. Carbon atoms are released at several points along the citric acid cycle, combining with oxygen to form carbon dioxide. In the end, the citric acid cycle generates NADH + H⁺ and FADH₂, and these highly energized compounds enter the electron transport chain for ATP production. Note that, in the citric acid cycle, ATP is not formed directly. Rather, it is formed from another high-energy compound called **guanosine triphosphate (GTP)** via substrate phosphorylation. In all, the oxidation of 2 molecules of acetyl-CoA (formed from 1 molecule of glucose) produces 6 NADH + H⁺, 2 FADH₂, and 2 ATPs.

Although small amounts of ATP are produced via substrate phosphorylation, most (90%) of the ATP generated from glucose results from oxidative phosphorylation via the electron transport chain (Stage 4). In total, the complete oxidation of 1 molecule of glucose (via glycolysis, the citric acid cycle, and oxidative phosphorylation) generates up to 38 ATPs, depending on the source of glucose—10 NADH + H⁺ (30 ATPs via oxidative phosphorylation), 2 FADH₂ (4 ATPs via oxidative phosphorylation), and 4 ATPs formed via substrate phosphorylation (Figure 7.12).

citric acid cycle An amphibolic pathway that oxidizes acetyl-CoA to yield carbon dioxide, NADH + H⁺, FADH₂, and ATP via substrate phosphorylation.

oxaloacetate The final product of the citric acid cycle, which becomes the substrate for the first reaction in this pathway.

citrate (CIT – rate) The first intermediate product in the citric acid cycle formed when acetyl-CoA joins with the end product, oxaloacetate.

guanosine triphosphate (GTP) A high-energy compound similar to ATP.

FIGURE 7.11 Citric Acid Cycle The citric acid cycle is an aerobic metabolic pathway that forms GTP and reduced coenzymes NADH + H⁺ and FADH₂. It is an amphibolic pathway, because it has both catabolic and anabolic components.

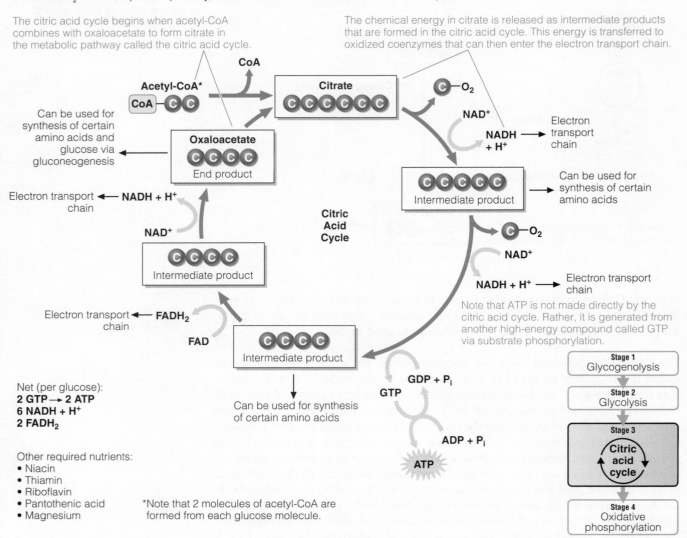

The citric acid cycle begins when acetyl-CoA combines with oxaloacetate to form citrate in the metabolic pathway called the citric acid cycle.

The chemical energy in citrate is released as intermediate products that are formed in the citric acid cycle. This energy is transferred to oxidized coenzymes that can then enter the electron transport chain.

Acetyl-CoA*

CoA

Can be used for synthesis of certain amino acids and glucose via gluconeogenesis

Oxaloacetate
End product

Citrate

O_2

NAD^+

NADH + H⁺ → Electron transport chain

Electron transport chain ← NADH + H⁺

NAD^+

Intermediate product

Can be used for synthesis of certain amino acids

Citric Acid Cycle

Intermediate product

O_2

NAD^+

NADH + H⁺ → Electron transport chain

Electron transport chain ← FADH₂

FAD

Intermediate product

Note that ATP is not made directly by the citric acid cycle. Rather, it is generated from another high-energy compound called GTP via substrate phosphorylation.

GDP + Pᵢ

GTP

Can be used for synthesis of certain amino acids

ADP + Pᵢ

ATP

Net (per glucose):
2 GTP → 2 ATP
6 NADH + H⁺
2 FADH₂

Other required nutrients:
• Niacin
• Thiamin
• Riboflavin
• Pantothenic acid
• Magnesium

*Note that 2 molecules of acetyl-CoA are formed from each glucose molecule.

Stage 1
Glycogenolysis

Stage 2
Glycolysis

Stage 3
Citric acid cycle

Stage 4
Oxidative phosphorylation

Glucose derived from carbohydrate is the body's main source of energy for immediate use. However, when glucose availability is limited, the body can break down protein and metabolize amino acids for energy, which we will discuss next.

CATABOLIC PATHWAYS CAN METABOLIZE PROTEIN FOR ENERGY

The major role of amino acids is to serve as the building blocks for proteins and other nitrogen-containing compounds. For this reason, amino acids are not considered a major source of energy, supplying less than 10% of our daily ATP needs. Nonetheless, at times the body must depend on them for energy, particularly during starvation. In addition to dietary amino acids, the body can break down its own stored protein (primarily skeletal muscle) for a source of amino acids. Stage 1 of protein catabolism is the process of **proteolysis,** which breaks down protein into amino acids. The liver then takes up amino acids, where they can enter the next stage of protein catabolism.

Transamination and Deamination Stage 2 of protein catabolism involves removing the nitrogen-containing amino group and using the remainder of the

⟨**CONNECTIONS**⟩ Transamination is the transfer of an amino group from an amino acid to an α-keto acid, usually α-ketoglutarate. Deamination is the removal of an amino group from an amino acid (Chapter 5, page 164).

proteolysis The breakdown of protein into amino acids.

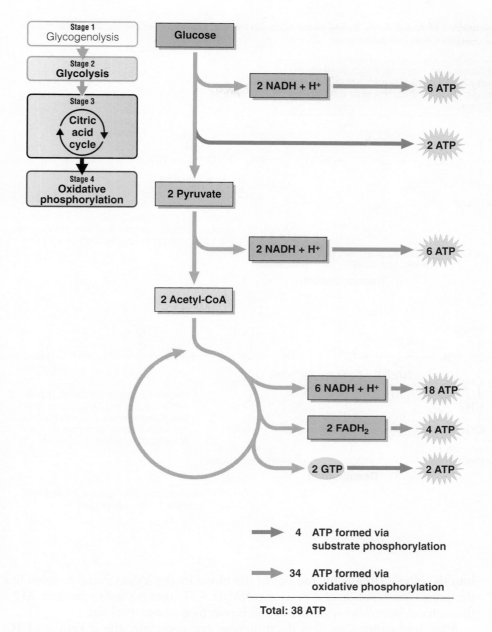

FIGURE 7.12 ATP Formation from Glucose Catabolism Although some ATP is formed directly via substrate phosphorylation, most is formed via oxidative phosphorylation. The complete oxidation of glucose yields up to 38 ATPs.

Stage 1
Glycogenolysis

Stage 2
Glycolysis

Stage 3
Citric acid cycle

Stage 4
Oxidative phosphorylation

Glucose

2 NADH + H⁺ → 6 ATP

2 ATP

2 Pyruvate

2 NADH + H⁺ → 6 ATP

2 Acetyl-CoA

6 NADH + H⁺ → 18 ATP

2 FADH₂ → 4 ATP

2 GTP → 2 ATP

→ 4 ATP formed via substrate phosphorylation

→ 34 ATP formed via oxidative phosphorylation

Total: 38 ATP

amino acid to generate substances that can enter the citric acid cycle. Recall from Chapter 6 that this is accomplished in two steps. First, as illustrated in Figure 7.13, the amino group from an amino acid is transferred to a compound called an **α-keto acid,** forming a new amino acid. This reaction is called transamination. The structure of the α-keto acid determines which amino acid is subsequently formed. Typically, the α-keto acid used is α-ketoglutarate, forming the amino acid glutamate. However, other α-keto acids may be used as well. The carbon skeleton remaining from the original amino acid, which is now itself an α-keto acid, is then used to generate ATP by entering the citric acid cycle.

The second step of amino acid catabolism, called deamination, occurs primarily in the liver. The process of **deamination** results in the removal of the amino group from the newly formed amino acid, leaving another α-keto acid. In this case, α-ketoglutarate is formed when the amino group is removed from glutamate. Thus α-ketoglutarate is regenerated and is available to again take part in the first step of amino acid catabolism—transamination. The amino group removed from the amino acid is converted to ammonia. Because ammonia is toxic to cells, it is quickly converted to a substance called urea, which is released

α-keto acid The structure remaining after the amino group has been removed from an amino acid.

deamination The removal of the amino group from an amino acid that results in the formation of an α-keto acid.

FIGURE 7.13 Transamination and Deamination of Amino Acids Before amino acids can be used as a source of energy, the amino groups must be removed. This occurs in two steps: transamination and deamination.

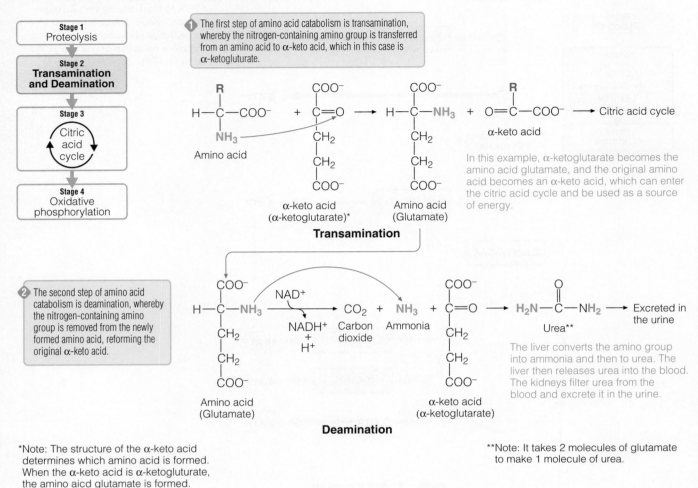

Stage 1
Proteolysis

Stage 2
Transamination and Deamination

Stage 3
Citric acid cycle

Stage 4
Oxidative phosphorylation

1. The first step of amino acid catabolism is transamination, whereby the nitrogen-containing amino group is transferred from an amino acid to α-keto acid, which in this case is α-ketogluturate.

In this example, α-ketoglutarate becomes the amino acid glutamate, and the original amino acid becomes an α-keto acid, which can enter the citric acid cycle and be used as a source of energy.

Transamination

2. The second step of amino acid catabolism is deamination, whereby the nitrogen-containing amino group is removed from the newly formed amino acid, reforming the original α-keto acid.

The liver converts the amino group into ammonia and then to urea. The liver then releases urea into the blood. The kidneys filter urea from the blood and excrete it in the urine.

Deamination

*Note: The structure of the α-keto acid determines which amino acid is formed. When the α-keto acid is α-ketoglutarate, the amino aicd glutamate is formed.

**Note: It takes 2 molecules of glutamate to make 1 molecule of urea.

into the blood. Urea is filtered out of the blood by the kidneys and excreted in the urine. Deamination also produces NADH + H$^+$ that is used to produce ATP through oxidative phosphorylation via the electron transport chain.

After transamination and deamination are complete, the α-keto acid is ready to enter Stage 3 of protein catabolism—the citric acid cycle. Unlike glucose which always enters the citric acid cycle as acetyl-CoA, α-keto acids can enter at various points. As such, the total number of ATP molecules produced from different amino acids varies. Depending on the needs of cells, the α-keto acid can be metabolized for energy or used to synthesize glucose or fatty acids. When used as an energy source, α-keto acids are oxidized to produce the reduced coenzymes NADH + H$^+$ and FADH$_2$. These, in turn, enter Stage 4 of protein catabolism—the electron transport chain.

TRIGLYCERIDES ARE AN IMPORTANT SOURCE OF ENERGY

Fatty acids also provide an important source of energy. When readily available energy sources are low, the body can break down triglyceride molecules into glycerol and fatty acids, as shown in Figure 7.14. Recall from Chapter 7 that the first stage of lipid catabolism, called lipolysis, is catalyzed by the enzyme hormone-sensitive lipase. Low glucose availability causes glucagon levels to rise, which in turn stimulates the activity of hormone-sensitive lipase. Hormone-sensitive

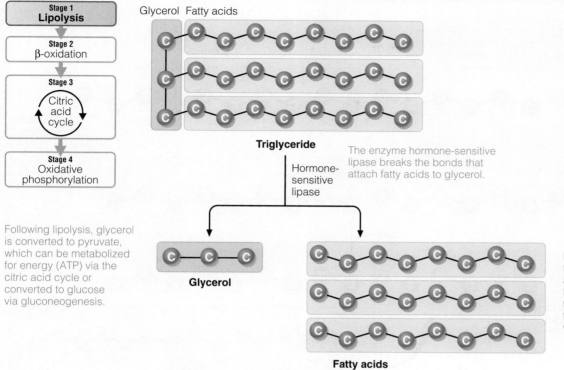

The enzyme hormone-sensitive lipase breaks the bonds that attach fatty acids to glycerol.

Following lipolysis, glycerol is converted to pyruvate, which can be metabolized for energy (ATP) via the citric acid cycle or converted to glucose via gluconeogenesis.

Fatty acids cleaved from the triglyceride are released into the blood, where they are taken up by some cells for further breakdown via beta-oxidation.

lipase is also stimulated by low levels of insulin, by exercise, and during times of stress by the adrenal hormones epinephrine and cortisol. The glycerol released during lipolysis can be converted to glucose and therefore used for energy, as well. Once fatty acids have been cleaved from the triglyceride, they are released into the blood. When taken up by cells, the fatty acids can be broken down further for energy via beta-oxidation as described next.

Beta (β)-oxidation Breaks Down Fatty Acids Stage 2 of lipid catabolism involves the arduous process of disassembling fatty acids into two-carbon subunits. This takes place in mitochondria by a series of reactions collectively called **beta (β)-oxidation.** However, before β-oxidation can take place, the fatty acid must be activated in the cytoplasm by the addition of coenzyme A (CoA) to its carboxylic acid end (–COOH). After this activation step, the fatty acid can be transported across the outer mitochondrial membrane by a molecule called **carnitine.** Fueled by attention in the media, athletes have long been hopeful that carnitine supplements might enhance athletic performance by improving the efficiency with which muscle tissue uses fatty acids for energy. However, studies have neither shown conclusively that carnitine supplements enhance athletic performance, nor that they have any negative effects.[1]

β-oxidation takes place in all cells except red blood cells, which lack the organelles (mitochondria) required for this metabolic process. β-oxidation begins when a two-carbon subunit is removed from the fatty acid chain, resulting in the formation of acetyl-CoA. The remaining (shorter) fatty acid is reactivated by the addition of another CoA and then cleaved again. This process repeats itself until the entire fatty acid has been broken down, two carbon atoms at a time. Each cleavage releases electrons (e^-) and hydrogen ions (H^+), which are used to form the reduced coenzymes NADH + H^+ and $FADH_2$. Thus, β-oxidation of an 18-carbon fatty acid requires 8 cleavages, producing 9 molecules of acetyl-CoA, 8 NADH + H^+, and 8 $FADH_2$. The

〈**CONNECTIONS**〉 Lipolysis is the breakdown of triglycerides into glycerol and fatty acids catalyzed by the enzyme hormone-sensitive lipase (Chapter 6, page 230).

β-oxidation The series of chemical reactions that break down fatty acids to molecules of acetyl-CoA.

carnitine A molecule that transports fatty acids across the mitochondrial membrane.

FIGURE 7.15 **Beta (β)-oxidation of Fatty Acids** β-oxidation is an aerobic metabolic process that consists of a series of enzyme-catalyzed chemical reactions that cleave off 2-carbon subunits from a fatty acid.

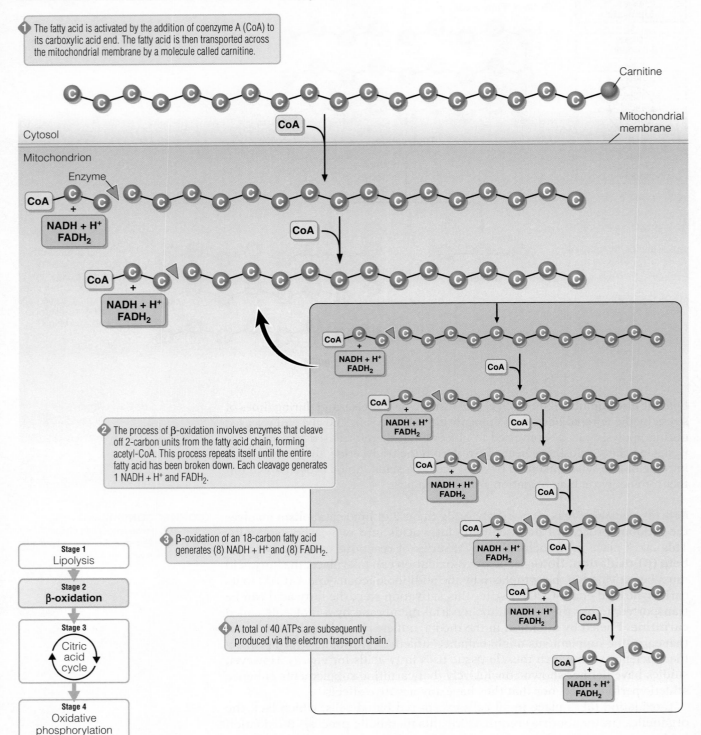

1 The fatty acid is activated by the addition of coenzyme A (CoA) to its carboxylic acid end. The fatty acid is then transported across the mitochondrial membrane by a molecule called carnitine.

Carnitine

Mitochondrial membrane

Cytosol

Mitochondrion

Enzyme

CoA

NADH + H⁺
FADH₂

CoA

NADH + H⁺
FADH₂

CoA

NADH + H⁺
FADH₂

CoA

NADH + H⁺
FADH₂

CoA

NADH + H⁺
FADH₂

CoA

NADH + H⁺
FADH₂

CoA

NADH + H⁺
FADH₂

CoA

NADH + H⁺
FADH₂

2 The process of β-oxidation involves enzymes that cleave off 2-carbon units from the fatty acid chain, forming acetyl-CoA. This process repeats itself until the entire fatty acid has been broken down. Each cleavage generates 1 NADH + H⁺ and FADH₂.

Stage 1
Lipolysis

Stage 2
β-oxidation

3 β-oxidation of an 18-carbon fatty acid generates (8) NADH + H⁺ and (8) FADH₂.

Stage 3
Citric acid cycle

4 A total of 40 ATPs are subsequently produced via the electron transport chain.

Stage 4
Oxidative phosphorylation

reduced coenzymes resulting from β-oxidation are used to generate ATP via oxidative phosphorylation. The process of β-oxidation is shown in Figure 7.15.

Stage 3 of lipid catabolism begins when molecules of acetyl-CoA resulting from β-oxidation enter the citric acid cycle. A total of 3 NADH + H⁺, one FADH₂, and 1 ATP (produced from GTP by substrate phosphorylation) are generated per acetyl-CoA. Recall that a fatty acid with 18 carbon atoms produces 9 molecules of acetyl-CoA. During Stage 4 of lipid catabolism, the

TABLE 7.2 Overview of ATP Production from an 18-Carbon Fatty Acid

Pathway	ATP Yield[a]
β-oxidation	
By oxidative phosphorylation of 8 NADH + H$^+$	24
By oxidative phosphorylation of 8 FADH$_2$	16
	40
Citric acid cycle (9 molecules of acetyl-CoA)	
By substrate phosphorylation via 9 GTP	9
By oxidative phosphorylation of 27 NADH + H$^+$	81
By oxidative phosphorylation of 9 FADH$_2$	18
	108

The total yield from the complete catabolism of an 18-carbon fatty acid (40 + 108) = 148 ATPs.

[a] NOTE: The ATP yield for fatty acid oxidation is based on integral values of 3 ATP per NADH + H$^+$ and 2 ATP per FADH$_2$. Based on nonintegral values of 2.5 ATP per NADH + H$^+$ and 1.5 ATP per FADH$_2$, the ATP yield for an 18-carbon fatty acid is 122 ATP.

reduced coenzymes produced via β-oxidation and the citric acid cycle enter the electron transport chain, producing large amounts of ATP. Thus, an 18-carbon fatty acid generates considerably more ATPs than 3 molecules‡ of glucose (148 versus 114 ATPs, respectively). The ATP yield from an 18-carbon fatty acid is summarized in Table 7.2.

It is important to understand how the body breaks down energy-yielding nutrients to generate ATP. Next we look at the body's other use for nutrients: as building blocks for constructing other important molecules.

Critical Thinking: Tracy's Story Recall Tracy's story about her daughter Anna, who has an inherited metabolic disease that affects her ability to metabolize the amino acid phenylalanine. Now that you have seen how metabolic pathways are interconnected, you should also see that a defect in a single enzyme can alter an entire metabolic pathway. As we just discussed, fatty acid metabolism involves various steps including fatty acid activation, carnitine transfer, and the action of various enzymes required for fatty acid oxidation. How would an inherited defect in an enzyme involved in any of these three steps impact fatty acid metabolism? What might be the clinical manifestations of such a disorder?

How Do Anabolic Pathways Contribute to Energy Metabolism?

It is important to recognize that anabolic pathways are not simply catabolic pathways in reverse. Catabolic pathways that break down compounds are distinctly different from anabolic pathways that synthesize compounds. In addition, anabolism tends to take place in the cytoplasm, whereas catabolism takes place primarily in mitochondria. This separation or compartmentalization is important because it enables both anabolic and catabolic pathways to function simultaneously.

In the previous section, you learned that when the body needs energy, cells break down energy-yielding macronutrients to synthesize ATP. When energy intake exceeds requirements, however, anabolic pathways predominate over catabolic pathways. During times of energy abundance, anabolic pathways

‡Note that three molecules of glucose have the equivalent number of carbon atoms as an 18-carbon fatty acid.

convert energy-yielding nutrients into forms that can be stored—glycogen and triglycerides. However, anabolic pathways serve additional functions. When the diet does not supply the proper balance or amounts of nutrients needed for ATP generation, anabolic pathways provide alternate routes for their production. Of particular importance is the ability of some cells to synthesize glucose from noncarbohydrate sources when needed via gluconeogenesis. The role of anabolic pathways in energy metabolism is discussed next.

GLYCOGENESIS GENERATES GLYCOGEN FROM GLUCOSE

The anabolic process whereby glycogen is formed from glucose in liver and muscle tissue is called glycogenesis, and is stimulated by the hormone insulin. An average person has approximately 70 g of liver glycogen and 200 g of muscle glycogen—enough glucose to provide energy for 8 to 12 hours. Although glycogen in liver and muscle play different roles, both tissues break down glycogen when there is a need for glucose.

〈CONNECTIONS〉 Recall that glycogen is formed during the process of glycogenesis (Chapter 4, page 124).

LIPOGENESIS FORMS FATTY ACIDS AND TRIGLYCERIDES

Because there is a limit to how much glycogen the body can store, insulin also promotes the uptake of excess glucose by adipose tissue where the glucose can be turned into fatty acids and ultimately triglycerides. This process, called lipogenesis, takes place mainly in liver and adipose tissue. Note that, unlike the reversible conversion of glucose to glycogen, the conversion of glucose to a fatty acid is irreversible. That is, although glucose can be transformed into a fatty acid, the fatty acid cannot be converted back to glucose.

〈CONNECTIONS〉 Lipogenesis is the synthesis of lipids (Chapter 6, page 231).

Both glucose and amino acids can enter into the process of lipogenesis by two intersecting pathways (Figure 7.16). In "pathway 1," glucose is broken down via glycolysis into two molecules of pyruvate, which in turn are converted to two molecules of acetyl-CoA. Next, molecules of acetyl-CoA are joined together to make a fatty acid—a process called fatty acid synthesis. After that, three fatty acids are attached to a molecule of glycerol to make a triglyceride. This entire process is costly in terms of energy, requiring approximately 20 to 25% of the energy originally in the glucose molecule. By comparison, the energy required to transform dietary fatty acids into stored triglyceride is only about 5%. It is easy to see that excess calories from fatty foods are more efficiently stored in adipose tissue than are those from foods containing mostly carbohydrates.

FIGURE 7.16 Lipogenesis Excess glucose and amino acids can be used to synthesize fatty acids, which subsequently join with a molecule of glycerol to a make a triglyceride.

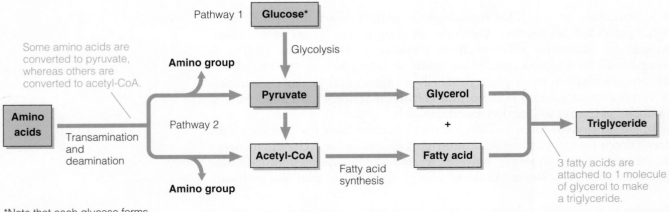

*Note that each glucose forms 2 molecules of pyruvate.

In "pathway 2," amino acids can be taken up by the liver and converted into pyruvate and acetyl-CoA. These molecules are then used to form triglyceride molecules. The energy required to convert amino acids to triglycerides is greater than that required to convert glucose to triglycerides. This is because the nitrogen-containing amino group must be removed before the remaining structure can be converted to acetyl-CoA.

GLUCONEOGENESIS FORMS GLUCOSE FROM NONCARBOHYDRATE SOURCES

Although most cells can use glucose and fatty acids for energy, the brain and central nervous system use glucose preferentially, and red blood cells use glucose exclusively. To ensure that these and other tissues have a continual supply of glucose, small amounts of glucose are stored as glycogen in the liver and skeletal muscle. However, when glycogen stores are depleted, noncarbohydrate molecules are transformed into glucose by various anabolic pathways. This process, called gluconeogenesis, occurs primarily within liver cells and, to a lesser extent, kidney cells. During periods of starvation, gluconeogenesis provides an important source of glucose to cells that depend on it as their major or sole source of energy.

Gluconeogenesis, which is stimulated by the hormones glucagon and cortisol and inhibited by insulin, uses noncarbohydrate sources—most amino acids, lactate, and glycerol—to make glucose (Figure 7.17). These compounds are first converted to oxaloacetate (the final product of the citric acid cycle) and then to glucose. The energy needed to fuel these energetically expensive reactions comes from ATP and the oxidation of NADH + H$^+$ to NAD$^+$.

Gluconeogenesis from Amino Acids Amino acids used to generate glucose via gluconeogenesis are referred to as **glucogenic amino acids.** All but two amino acids (leucine and lysine) can be used for this purpose. There are several routes by which glucogenic amino acids can be converted to glucose;

glucogenic amino acids Amino acids that can be used to make glucose.

FIGURE 7.17 **Gluconeogenesis** Noncarbohydrate sources used for gluconeogenesis include some amino acids, lactate, and glycerol.

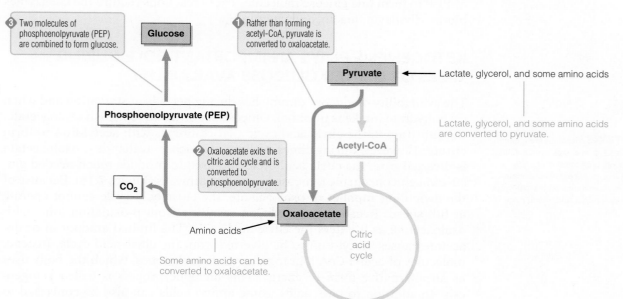

FIGURE 7.18 Gluconeogenesis and Ketogenesis Gluconeogenesis and ketogenesis typically occur simultaneously. When oxaloacetate is used for gluconeogenesis, acetyl-CoA resulting from catabolism of fatty acids and amino acids is converted to ketones via ketogenesis. Ketones serve as an important source of energy for some tissues during times of glucose insufficiency.

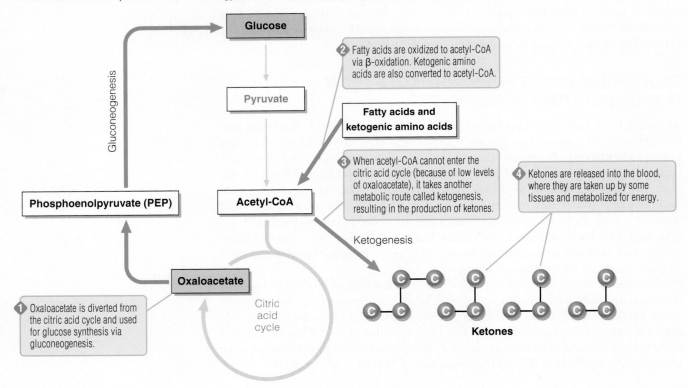

all involve removing the amino group and converting the remaining carbon skeleton to oxaloacetate. (The carbon skeletons of some amino acids are first converted to pyruvate and then to oxaloacetate.) Next, oxaloacetate is converted to the compound phosphoenolpyruvate (PEP). It takes two molecules of PEP to form one glucose molecule. These reactions require the coenzymes biotin, riboflavin, niacin, and vitamin B_6.

KETOGENESIS PLAYS AN IMPORTANT ROLE DURING TIMES OF LIMITED GLUCOSE AVAILABILITY

The availability of glucose diminishes during periods of starvation and when carbohydrate intake is minimal. Glucose is needed to replenish oxaloacetate, the substrate in the citric acid cycle that combines with acetyl-CoA to form citrate. However, during times of limited glucose availability, oxaloacetate is diverted from the citric acid cycle and used to produce much-needed glucose via gluconeogenic (glucose-forming) pathways (Figure 7.18). Because of the dwindling supplies of oxaloacetate, the citric acid cycle cannot operate at full speed. Recall that acetyl-CoA resulting from β-oxidation joins with oxaloacetate as it enters the citric acid cycle. The limited amount of oxaloacetate causes acetyl-CoA to be diverted from the citric acid cycle. Instead, molecules of acetyl-CoA join together to form ketones, which the body uses as an alternative form of energy. This anabolic process is called ketogenesis. In addition to fatty acids, some amino acids can also be converted to acetyl-CoA. These amino acids are referred to as **ketogenic amino acids** because they can be used to generate ketones during times of limited glucose availability.

⟨CONNECTIONS⟩ Ketones are organic compounds used as an energy source during starvation, fasting, consumption of a low-carbohydrate diet, or uncontrolled diabetes. Ketogenesis refers to the metabolic pathways used to produce ketones (Chapter 4, page 141).

ketogenic amino acids Amino acids that can be used to make ketones.

Approximately 1% of children in the United States have epilepsy, a condition characterized by repeated seizures. Although in most cases seizures can be controlled with medication, some children are unresponsive to medical intervention of this sort. With few other treatment options available, some neurologists are turning to an old medical practice—dietary manipulation. While more commonly associated with a popular approach to weight loss, therapeutic ketogenic diets (those stimulating ketone production in the body) have also long been recognized as a method for controlling seizures in children with difficult-to-control epilepsy. First developed in the early 1900s, ketogenic diets were the standard treatment modality until antiseizure medications became widely available.

Ketogenic diets are intended to mimic starvation through dietary manipulation so that brain cells rely heavily on ketones instead of other energy-yielding substrates, such as glucose, for a source of energy. To deplete glycogen stores in the body, the child is first put on a 24-hour medically supervised fast, typically done in the hospital. Without glucose to supply energy, the body turns to stored fat. Next, the child is fed a high-fat diet, providing approximately 80% of total calories from fat.[2] In addition, carbohydrate-rich foods such as grains, fruits, and vegetables must be eliminated in order for the child to maintain a state of ketosis. Understandably,

some parents have concerns about feeding their children a difficult and demanding diet consisting largely of high-fat foods such as eggs, cream, bacon, coconut oil, nuts, and cheese.

Doctors are not clear as to why a diet that mimics starvation helps prevent seizures in some children, although some theorize that forcing the brain to use ketones as an energy source makes areas of the brain more seizure-resistant.[3] Nonetheless, the ketogenic diet has proved successful in some cases. Because it is such a dramatic change, however, this dietary plan must be implemented gradually. It also requires that parents work closely with a medical team that often includes a neurologist and a dietitian. Vitamin and mineral supplements are critical to ensure the child is adequately nourished. Because there is a potential that the diet can impair normal growth, the child's height and weight are carefully monitored.

Clearly, maintaining a child on a ketogenic diet requires medical supervision. However, for the diet to be successful, it also takes parental involvement. Parents who use this approach to control seizures in their children recognize that it takes serious commitment and understanding to implement. It is not an easy diet plan to follow, but many parents feel that the benefits far outweigh the effort it takes to adhere to such a restrictive dietary plan.

Ketogenesis, which occurs mostly in the liver, provides the body with an important source of energy during conditions of glucose insufficiency (such as starvation and diabetes). Only muscle, brain, and kidney tissue have the enzymes needed to metabolize ketones for energy. Nonetheless, ketones can spare the body from having to use amino acids to synthesize large amounts of glucose via gluconeogenesis.

Although the use of ketones by the body has important survival implications, at times ketone production exceeds ketone use. This can lead to high levels of ketones in the blood, a condition called **ketosis.** Sometimes you can detect when a person has become ketonic by the characteristic "fruity" odor of their breath. Ketosis can occur during prolonged fasting, consumption of a very low–carbohydrate diet, strenuous exercise, and some diseases such as type 1 diabetes. Severe ketosis, called **ketoacidosis,** can cause a variety of complications, including lowered blood pH, nausea, coma, and, in extreme circumstances, death. Although ketoacidosis can be a life-threatening condition, therapeutic ketogenic diets are sometimes used to control seizures in some children with epilepsy.[4] Researchers do not understand why a diet that mimics starvation can, in some cases, control seizures, but most experts say the diet is worth trying when medications have failed. You can read more about ketogenic diets and epilepsy in the Focus on Diet and Health.

ketosis An accumulation of ketones in body tissues and fluids.

ketoacidosis A rise in ketone levels in the blood, characterized by a decrease in the pH of the blood.

The body must continuously balance anabolism and catabolism to meet its energy needs. Next, you will learn how it adapts its energy metabolism in times of feeding and fasting.

How Is Energy Metabolism Influenced by Feeding and Fasting?

The availability of metabolic fuels fluctuates throughout the day. After a meal, energy sources are readily available and tend to exceed the immediate needs of cells. During this time, anabolic pathways favor the synthesis and storage of glycogen, protein, and triglycerides. However, after a period of time without eating, the availability of circulating fuels diminishes. At this point, the body is primarily in a catabolic state, relying on ATP generated from the breakdown of the body's stored energy reserves. Thus, energy availability, in part, determines the balance of anabolic and catabolic activity (Figure 7.19). In extreme catabolic states such as prolonged fasting and starvation, the body makes additional metabolic adjustments to maintain physiological functions and prevent further deterioration.

Whether the body uses glycogen, protein, or triglycerides for energy depends on a person's nutritional and physiological state. In fact, energy availability may be the single most influential factor that governs energy metabolism. To describe energy availability, scientists often define physiological states based on the time since the previous meal as—the fed state, the postabsorptive state, the fasting state, and starvation. As you can see in Figure 7.20, relative fuel availability and hormone levels change in response to each of these states. In the next section, you will learn how energy metabolism changes in response to the first two physiological states (the fed and postabsorptive states), which are ones that most of us experience daily. You will also learn about adaptive metabolic responses to the states of fasting and prolonged starvation.

THE FED STATE FAVORS ENERGY STORAGE

The first four hours after a meal are referred to as the **fed state,** or the postprandial period. During this time, absorbed nutrients enter the blood, stimulating the release of insulin and decreasing the release of glucagon. In turn, energy metabolism in the liver, adipose tissue, and skeletal muscle is affected. Many of the major anabolic pathways operate during the fed state, including those involved in the synthesis of proteins, fatty acids, triglycerides, and glycogen. Glucose is the body's major source of energy during this time. As long as there is sufficient oxygen, most of the ATP produced during the fed state is provided by glycolysis, the citric acid cycle, and oxidative phosphorylation.

Throughout the fed state, amino acids are used primarily for protein synthesis, and little protein degradation (proteolysis) takes place. Once the body's need for protein is met, excess amino acids are converted to fatty acids and ultimately to triglycerides for storage. In addition, insulin stimulates the uptake of glucose and fatty acids into adipose tissue, where they are converted to triglycerides for storage.

CELLS RELY ON STORED ENERGY DURING THE POSTABSORPTIVE STATE

fed state (postprandial period) The first four hours after a meal.

postabsorptive state The period of time (4 to 24 hours after a meal) when dietary nutrients are not being absorbed.

Most people typically consume at least three distinct meals each day. The period of time between meals, when nutrients are not being absorbed, is referred to as the **postabsorptive state.** The postabsorptive state begins about three

FIGURE 7.19 Integration of Anabolic and Catabolic Energy Metabolism The relative activities of anabolic and catabolic pathways are determined by the availability of substrates, intermediate compounds, and energy needs.

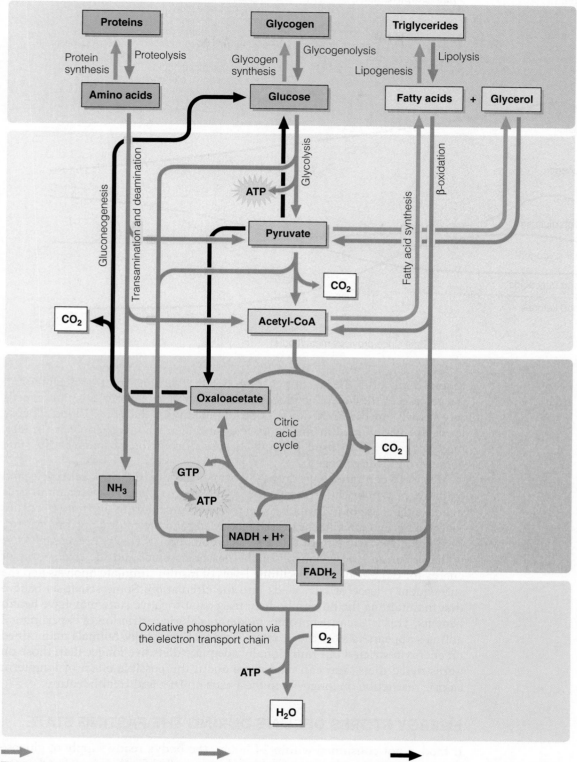

These pathways predominate during times of energy sufficiency and excess. Amino acids, glucose, and fatty acids are converted to protein, glycogen, and triglycerides, respectively.

These pathways predominate during times of energy need, breaking down stored proteins, glycogen, and triglycerides to amino acids, glucose, and fatty acids, respectively.

These pathways predominate when there is a lack of available glucose. Oxaloacetate can leave the citric acid cycle and be used to synthesize glucose via gluconeogenesis.

FIGURE 7.20 Shifts in Fuel Availability and Hormone Levels after Eating

Fed state:
• The relative concentration of insulin is higher than that of glucagon, favoring energy storage.
• Blood glucose is elevated.
• Glycogen storage increases.

Postabsorptive state:
• Insulin levels decrease, and glucagon increases.
• Blood glucose decreases.
• Use of liver glycogen to provide glucose increases.
• Use of fatty acids increases.

Fasting state:
• Relative concentration of glucagon is higher than insulin.
• Liver glycogen stores are depleted.
• Glucose is supplied mainly by gluconeogenesis.
• Stored triglycerides are broken down to allow increased use of fatty acids.
• Ketone formation (ketogenesis) increases.

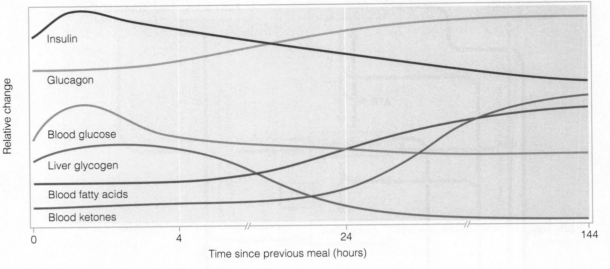

to four hours after the last intake of food, when insulin secretion declines and the release of the hormone glucagon increases. Throughout this phase, cells rely heavily on the breakdown of stored energy in the body. Blood glucose levels are maintained primarily by liver glycogen via glycogenolysis. In fact, most glucose (75%) used by body tissues during the postabsorptive state comes from liver glycogen.

Most cells continue to produce ATP from glycolysis, the citric acid cycle, and oxidative phosphorylation during the postabsorptive state. However, many cells increase their use of fatty acids as an energy source during this time to ensure that there is enough glucose for red blood cells and the central nervous system. Remember that while most cells use a combination of energy sources, glucose is the preferred energy source for the nervous system and the sole source of energy for red blood cells. Declining levels of insulin stimulate lipolysis and the subsequent release of fatty acids into the circulation. Some scientists believe that maintaining the human body in the postabsorptive state may have health benefits. This is because chronic, moderate caloric restriction of experimental animals appears to slow the process of aging. Specifically, animals maintained on energy-restricted but nutritionally adequate diets live longer than those on unrestricted diets. You can read more about the possible effect of long-term caloric restriction on longevity in the Focus on Diet and Health feature.

ENERGY STORES DECLINE DURING THE FASTING STATE

If food is not consumed within 24 hours, the body's ready supply of glucose begins to dwindle. The next nutritional state, called **fasting,** is defined as the first five days beginning 24 hours after a person's last meal. Within the first 24 to 36 hours of fasting, glycogen stores are depleted, and the insulin-to-glucagon ratio declines even further. The body's priority is to supply the nervous system and red blood cells with sufficient glucose. Because glycogen stores have been depleted, glucose must now be synthesized from noncarbohydrate

fasting state The first five days of fasting or minimal food intake, beginning 24 hours after the last meal.

Can eating less extend the human lifespan? Given the results of animal studies, some researchers believe so. There are many theories as to what causes aging, and scientists have long been interested in understanding factors that may slow this process and extend life. Growing evidence suggests that the lifelong or chronic practice of limiting energy intake while consuming adequate amounts of nutrients required by the body may improve health and longevity.[5] Studies suggest that caloric restriction reduces overall metabolic activity, which in turn reduces cell damage.[6] So far, the impact of caloric restriction on health and disease has been studied in a variety of animals, including monkeys and rats. These experiments show that animals on nutrient-dense, energy-restricted diets live longer, have a more youthful appearance, and have fewer age-related health problems compared with their well-fed counterparts. In fact, caloric restriction has been shown to be one of the few effective ways of extending the longevity of experimental animals.[7]

The benefits of reduced energy intakes on longevity may be the result of an overall decline in metabolic activity, which in turn may cause less oxidative damage to cells and tissues. Specifically, oxidative metabolism can cause the formation of unstable molecules called free radicals. Free radicals are destructive and damage cell membranes, DNA, and cell organelles such as mitochondria. Damage caused by free radicals is thought to be one of the leading causes of cellular aging and cancer. In addition to reducing free radical damage, restricting calories may also prevent damage caused by hyperglycemia. Like free radicals, an overabundance of glucose molecules can damage DNA and proteins, impairing cell function. Caloric restriction is thought to reduce cell damage by preventing elevated levels of glucose in the blood.

It is difficult to assess the impact of reduced energy intakes on longevity in humans. This is because humans have a longer life expectancy than most experimental animals and live in a largely uncontrolled environment. Furthermore, to really examine the effect of caloric restriction on longevity, human intervention trials would need to begin early in life, and participants would need to adhere to caloric restriction for many years. Thus, it may not be possible to conduct appropriate clinical trials to assess the effect of chronic caloric restriction on humans. Although living longer through caloric restriction may not appeal to some, there is reason to believe that eating fewer calories and more nutrient-dense foods likely has much to offer in terms of health and longevity.

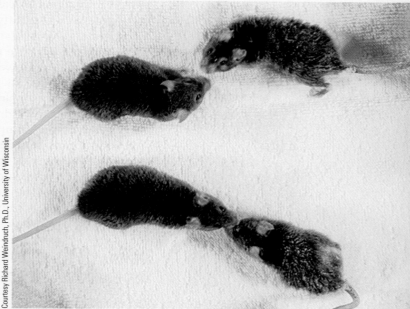

Courtesy Richard Weindruch, Ph.D., University of Wisconsin

These mice are the same age, 40 months. The two mice on the left, which were fed a low-calorie diet since 12 months of age (early middle age), look younger and are healthier than the normally fed animals on the right.

substances via gluconeogenesis. Muscles contribute to gluconeogenesis by supplying amino acids, lactate, and glycerol, all of which the liver uses to generate small amounts of glucose.

During the fasting state, the body relies more and more on mobilized fatty acids for a source of energy. However, when glucose is severely limited and oxaloacetate is being diverted away from the citric acid cycle to support gluconeogenesis, cells cannot completely oxidize acetyl-CoA. As a result, ketogenic pathways in the liver are stimulated. The ability of some tissues, especially the brain, to use ketones for ATP production has important survival implications. For example, this spares glucose for other tissues, and helps preserve lean body tissue, because the body does not have to rely as extensively on gluconeogenesis to support energy requirements.

THE BODY CONSERVES ENERGY STORES DURING THE STATE OF PROLONGED STARVATION

The loss of lean tissue during early stages of acute starvation reduces the body's total energy requirements. The body goes into "survival mode" by dramatically reducing its metabolic activity (basal metabolism) associated with lean tissue. This defensive action is important because it helps prolong survival if starvation continues. In this way, less energy is needed to maintain lean body mass, thus decreasing the body's overall energy needs. Similarly, self-imposed energy restriction, as occurs with some eating disorders and very low–calorie diets, also sacrifices lean tissue to support gluconeogenesis. This is one reason why successful weight-loss diets advocate slow and gradual weight loss to minimize the loss of lean tissue.

If food deprivation lasts longer than one week, the body enters a state of prolonged **starvation.** Maintaining adequate glucose availability to cells and the preservation of lean body mass now take on an even higher priority for the body. During this time, only red blood cells continue to rely solely on glucose for ATP production. Recall that this is because they do not have mitochondria, the organelles where fatty acids and ketones are metabolized for energy.

Under conditions of starvation, the reliance on ketones as a major source of energy is an adaptive response that helps extend life. Thus, ketogenesis is further stimulated. However, when stored fat becomes extremely limited, the body has no choice but to break down muscle tissue as its only remaining energy source. This includes muscle associated with vital body organs such as the heart and the kidneys. Eventually the ability of these organs to carry out their functions is impaired. The body continues to catabolize muscle and vital body organs until death ensues. This may take months, depending on the amount of fat reserves that a person has. Much of what is known about the physical and psychological effects of starvation comes from a classic study conducted in 1944 by Ancel Keys. In the Focus on the Process of Science feature, you can read more about this study and what it taught us about the physical and psychological effects of prolonged energy restriction.

VERSATILE SOLUTIONS HELP THE BODY MEET ITS ENERGY NEEDS

The use of various energy metabolism pathways by the body is affected by what, how much, and how frequently we consume food. The versatility of these pathways enables our bodies to respond to a variety of different circumstances that ensure energy (ATP) needs are always met. Understanding the integrated and coordinated nature of energy metabolism provides important insights as to how nutrient consumption affects life at the most basic cellular level. However, these countless metabolic reactions can be easily disrupted by imbalances in nutrient and caloric intakes. In the case of potentially life-threatening conditions such as prolonged fasting and starvation, energy metabolism pathways must meet the challenge of securing adequate glucose, while minimizing the loss of muscle tissue. The body's ability to cope with extreme situations such as these demonstrates the extent to which energy metabolism pathways can respond to preserve life. In fact, some researchers believe that crucial metabolic adaptations enabled our prehistoric ancestors to survive during periods of food scarcity.[8]

Starvation Response—A Survival Advantage Our prehistoric ancestors lived in a time of feast or famine, during which food availability was in a constant state of flux. The uncertainty of whether food would be plentiful or scarce meant that people historically had to eat today in order to survive tomorrow. Those most likely to survive the hardships of famine were those best able to adjust physiologically. For example, the ability to store excess energy during times of food abundance meant

starvation state Food deprivation for an extended period of time, typically lasting longer than one week.

During World War II, the U.S. military wanted to know more about how to feed and care for malnourished soldiers returning from war. To investigate the physical and psychological affects of starvation, the military commissioned a study called the Keys Starvation Experiment. Thirty-six healthy men agreed to participate.[9] Throughout the first three months of the study, measurements were taken to assess the subjects' physical health. In addition, nutrient and caloric requirements were estimated, and tests were given to evaluate psychological variables. For the next six months, the men were put on a semistarvation diet, consisting of 1,600 kcal per day. Living together in a dormitory, the men were regularly monitored for physical and psychological changes.

By the end of the six-month semistarvation diet, participants had lost an average of 24% of their body weight, and total energy requirements had decreased substantially. Weight loss was associated with decreases in both adipose and lean tissues. The men became lethargic, lacked endurance, and experienced diminished physical strength. Also striking to the researchers was the extent to which participants' mental health deteriorated. Many of the men developed depression, emotional instability, and what was described as "neurotic tendencies." Participants became apathetic, irritable, moody, and easily distracted. Furthermore, researchers noted unkempt personal appearance and lack of grooming. Also of interest was the fact that participants talked frequently about food. Food cravings increased, food dislikes disappeared, and participants became possessive about food rations. It became apparent that food restriction had devastating effects, both physically and psychologically.

The next three months of the study focused on refeeding. Participants were free to eat what and how much they wanted. Many men found it difficult to stop eating, even after they reported feeling full—eating 50 to 200% more calories than before the onset of the study. Most participants gained weight quickly, with the added weight mostly deposited as fat, not muscle. As weight was regained, mental state also improved. Although this study was conducted more than 50 years ago, the description of the physical and psychological changes that accompany energy restriction and weight regain is no different from what most dieters experience today.

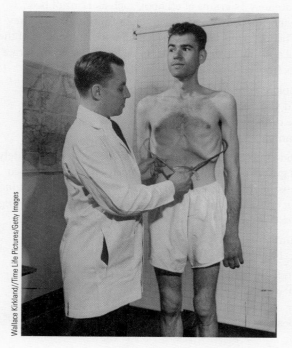

Wallace Kirkland//Time Life Pictures/Getty Images

To better understand the effects of semi-starvation, subjects' physical and psychological health was assessed throughout the study.

having an energy reserve to fall back on when needed. Conversely, during times of famine, the ability to conserve body mass by using energy sparingly also helped to ensure survival. These adaptive mechanisms, or what is called the **starvation response,** served early humans well. Today, however, we live in a dramatically different world where food is readily available for most people. Whereas the starvation response was advantageous during times of food scarcity, it is now proving to be detrimental during modern times of food abundance. It is easy to understand how the adaptive starvation response of increased appetite and energy conservation in an environment of plentiful food can lead to obesity.

starvation response Adaptive physiological mechanisms that promote the storage of excess energy during times of food abundance and conserve energy during times of limited food availability.

Aleria/Shutterstock.com

Critical Thinking: Tracy's Story The metabolism of one macronutrient can affect the metabolism of others. This is why it is important that our diets provide not only the right amount of calories, but also the recommended *balance* of calories. This is clearly reflected in the Dietary Guidelines for Americans. But what happens when someone like Anna must limit her intake of one or more of these important energy sources? How would the restriction of carbohydrates, fat, or protein impact overall energy metabolism?

Diet Analysis PLUS ✚ Activity

Although alcohol is not a nutrient, it is metabolized by many of the same metabolic pathways. If you drink alcohol, does it have an acceptable place in your diet? The following problems will help you answer this question. First of all, alcohol is considered a drug because it modifies body functions. It is NOT a food because it does not provide nutrients. Alcohol does, however, provide 7 kcal of energy per gram—but only for the pure ethanol part of the beverage. One "drink" is defined as a beverage providing ½ oz or approximately 14 grams of ethanol, such as 12 oz of beer, 5 oz of wine, and 1.5 oz of distilled liquor (gin, vodka, whiskey, etc.). The total calories in the beverage is a combination of the ethanol and any other components, such as the carbohydrate calories from the grapes in wine or the grains in beer.

1. Look at your *Intake and DRI goals Compared* report. Did you consume any alcohol (ethanol) during the days you kept a food record? If so how many grams? _____ g

   Knowing that a standard drink contains approximately 14 grams of ethanol, how many actual drinks did you consume? *(If you did **not** consume any alcohol, assume you consumed two glasses of wine during your reporting period for the purpose of practicing the calculations.)*

   Number of drinks = g of ethanol consumed ÷ 14 g/drink = _____ number of drinks.

2. Moderate drinking is defined as up to 1 drink per day for women (on average) and up to two drinks per day for men (on average). How many grams of ethanol does this allow per day for women and for men?

Women: One drink per day = _____ g of ethanol per day

Men: Two drinks per day = _____ g of ethanol per day

Knowing that each gram of ethanol has 7 kcal/gram, how many kilocalories does this component of the drink(s) provide?

Women: _____ g ethanol per day × 7 kcal/gram = _____ kcal

Men: _____ g ethanol per day × 7 kcal/day = _____ kcal

3. Knowing that the total calories in 12 oz of regular beer is approximately 150 kcal, calculate the number of kilocalories provided by sources other than the ethanol.

   150 kcal/12 oz beer = _____ kcal from ethanol = _____ kcal from other sources.

4. If an average person can metabolize about 0.5 oz of ethanol per hour, how long will it take an average-sized man or woman who drinks moderately to metabolize this amount of alcohol?

   Women: 1 drink per day × 1 hour/0.5 oz of ethanol = _____ number of hours

   Men: 2 drinks per day × 1 hour/0.5 oz of ethanol = _____ number of hours

5. After reading this section and doing these exercises, do you feel that moderate alcohol consumption is worth its calorie cost in your diet? Why or why not?

Notes

1. Karlic H, Lohninger A. Supplementation of L-carnitine in athletes: Does it make sense? Nutrition. 2004;20:709–15. Brass EP. Carnitine and sports medicine: Use or abuse? Annals of the New York Academy of Sciences. 2004;1033:67–78.

2. Neal EG, Chaffe H, Schwartz RH, Lawson MS, Edwards N, Fitzsimmons G, Whitney A, Cross JH. The ketogenic diet for the treatment of childhood epilepsy: A randomised controlled trial. The Lancet Neurology. 2008;7:500–6.

3. Cross JH, McLellan A, Neal EG, Philip S, Williams E, Williams RE. The ketogenic diet in childhood epilepsy: where are we now? Archives of Disease in Childhood. 2010; 95:550–3. Epilepsy Foundation. Treatment Options: Ketogenic Diets. Available from http://www.epilepsyfoundation.org/about/treatment/ketogenicdiet/.

4. Lefevre F, Aronson N. Ketogenic diet for the treatment of refractory epilepsy in children: A systematic review of efficacy. Pediatrics. 2000; 105:1–7. Available from http://pediatrics.aappublications.org/cgi/reprint/105/4/e46.

5. Gredilla R, Barja G. Minireview. The role of oxidative stress in relation to caloric restriction and longevity. Endocrinology. 2005;146:3713–7. Piper MD, Mair W, Partridge LJ. Counting the calories: The role of specific nutrients in extension of life

span by food restriction. Journals of Gerontology Series A: Biological Sciences and Medical Sciences. 2005;60:549–55. Sinclair DA. Toward a unified theory of caloric restriction and longevity regulation. Mechanisms of Ageing and Development. 2005;126:987–1002.

6. Redman LM, Ravussin E. Caloric restriction in humans: impact on physiological, psychological, and behavioral outcomes. Antioxidants and Redox Signaling. 2011;14:275–87.

7. Huffman DM. Exercise as a calorie restriction mimetic: implications for improving healthy aging and longevity. Interdisciplinary Topics in Gerontology. 2010;37:157–74.

8. Chakravarthy MV, Booth FW. Eating, exercise, and "thrifty" genotypes: connecting the dots toward an evolutionary understanding of modern chronic diseases. Journal of Applied Physiology. 2004;96:3–10.

9. Kalm LM, Semba RD. They starved so that others would be better fed: Remembering Ancel Keys and the Minnesota Experiment. Journal of Nutrition. 2005;135:1347–52. Keys A, Brozek J, Henschel A, Mickelsen O, Taylor HL. The biology of human starvation, Vols I–II. Minneapolis: University of Minnesota Press; 1950.

Alcohol, Health, and Disease

The ritual toast is one of the oldest celebratory traditions.

The raising of a celebratory toast is an ancient tradition. Yet alcoholic beverages, with their rich history of pageantry and ritual, have also brought misery and suffering. Although most people who drink do so without harm to themselves or others, longtime abusers know all too well that alcohol can lead to psychological and physical dependency. Today, millions of people worldwide seek help in their effort to abstain from alcohol. There is no easy explanation as to why some people can control their drinking whereas others cannot. Although genetics may predispose some people to alcoholism, cultural factors play a major role as well.

Scientists have long debated whether alcohol is beneficial or detrimental for health. Studies show that moderate alcohol consumption can reduce the risk of heart disease in middle-aged and older adults, and it may even provide some protection against type 2 diabetes and gallstones.[1] However, when it comes to alcohol, more is clearly not better. In excess, alcohol alters judgment; can lead to dependency; and can damage the liver, pancreas, heart, and brain. Heavy drinking also increases the risk for accidents and some types of cancer, and maternal alcohol consumption can seriously harm an unborn child.

Consumed responsibly, alcohol poses little threat physically, socially, or psychologically, and may even be beneficial. However, not all drinkers are responsible, and estimated annual costs associated with alcohol abuse in the United States are more than $180 billion.[2] For this reason, the Dietary Guidelines for Americans clearly state that people who choose to drink alcoholic beverages should do so sensibly and in moderation.[3] In this Nutrition Matters section, you will learn how alcohol is produced, absorbed, and circulated. You will also learn about the many effects that alcohol has on health and disease.

What Is Alcohol and How Is It Absorbed?

If you have ever been around someone who has had too much to drink, you know alcohol is a drug with mind-altering effects. Indeed, most people act and behave differently when under the influence of alcohol. Alcohol is a rather simple molecule—but has profound effects on the body.

ALCOHOL IS PRODUCED BY FERMENTATION

From a chemical perspective, **alcohol** is a broad term used to describe a class of organic compounds that have common properties. For example, all variations of alcohol have a distinctive chemical group (an –OH group bonded to a carbon atom); are quite volatile; and tend to be soluble in water. There are many types of alcohol, and most are not safe to drink. For instance, methanol, which is used to make antifreeze, can be lethal if ingested. The form of alcohol found in alcoholic beverages is called **ethanol,** which has a chemical formula of C_2H_5OH. Although alcohol is not considered a nutrient it is a source of energy—7 kcal per gram. Table 1 shows the caloric contents of selected alcoholic beverages.

alcohol An organic compound containing one or more hydroxyl (–OH) groups attached to carbon atoms.

ethanol (ETH – a – nol) An alcohol produced by the chemical breakdown of sugar by yeast.

TABLE 1 Serving Sizes, Energy Contents, and Alcohol Contents of Selected Alcoholic Beverages

Beverages	Serving Size (fl oz)	Energy (kcal/serving)[a]	Alcohol (g/serving)
Light beer	12	103	11
Beer	12	139	13
White wine	5	98	14
Red wine	5	98	14
Distilled beverages (e.g., gin, rum, vodka, whiskey)			
80 proof	1.5	97	14
86 proof	1.5	105	15
90 proof	1.5	110	16
94 proof	1.5	116	17
100 proof	1.5	124	18
Crème de menthe	1.5	186	15
Daiquiri	4	224	14
Whiskey sour	4.5	226	19
Piña colada	4.5	245	14

[a] Note that some alcoholic beverages, such as beer, contain energy-yielding nutrients. Therefore, caloric content cannot always be calculated simply by multiplying alcohol content by 7.
SOURCE: USDA National Nutrient Database for Standard Reference. Available from: http://www.nal.usda.gov/fnic/foodcomp/search/.

Alcohol is produced by a process called **fermentation.** Discovered thousands of years ago, fermentation occurs when single-cell microorganisms called yeast metabolize sugar found in fruits and grains. This involves the metabolic pathway glycolysis. Under anaerobic conditions, yeast converts glucose to pyruvate, which in turn is converted to ethanol. In the process, carbon dioxide is released (Figure 1). Whereas the coenzyme NAD^+ is converted to $NADH + H^+$ during glycolysis, fermentation converts $NADH + H^+$ to NAD^+. Once the alcohol content reaches 11 to 14%, fermentation stops naturally.

Distillation Increases the Alcohol Content A process called **distillation** can be used to increase the alcohol content of some beverages. During distillation, fermented beverages are heated, causing the alcohol to become a vaporous gas. The alcohol vapors are collected and cooled until they are liquid again. This pure alcohol concentrate is used to produce distilled

fermentation The process whereby yeast chemically breaks down sugar to produce ethanol and carbon dioxide.

distillation A process used to make a concentrated alcoholic beverage by condensing and collecting alcohol vapors.

FIGURE 1 Fermentation During fermentation, yeast metabolizes sugar to produce alcohol.

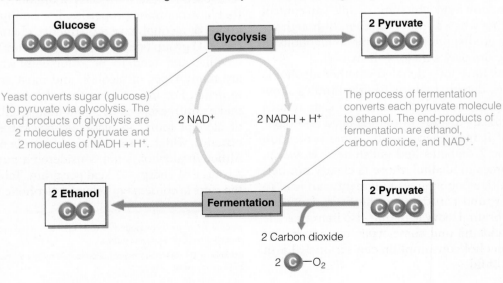

alcoholic beverages such as gin, vodka, and whiskey. Distilled alcoholic beverages are also called "hard liquors," and their alcohol content can be determined from their **proof.** A beverage's proof is twice its alcohol content. For example, distilled liquors labeled as 80 proof are 40% alcohol.

In recent years, caffeinated alcoholic beverages have become very popular. According to many experts, this combination of alcohol and caffeine is troubling because the stimulating effects of caffeine can mask intoxication. As a result, individuals may drink more alcohol than intended, leading to hazardous and life-threatening behaviors. In response to these safety and health concerns, the U.S. Food and Drug Administration (FDA) launched an investigation, and concluded that these products cannot stay on the market in their current form.

SEVERAL FACTORS INFLUENCE THE RATE OF ALCOHOL ABSORPTION

When alcohol is consumed, it requires no digestion and is readily absorbed into the blood. Although some alcohol is absorbed from the stomach, most (80%) is absorbed from the small intestine by simple diffusion. The rate of alcohol absorption is influenced by several factors. For example, it is absorbed more quickly when a person drinks on an empty stomach. This is because the presence of food in the stomach dilutes the alcohol and delays gastric emptying. As a result, alcohol cannot reach the small intestine as quickly. The type of food consumed does not have a measurable effect on the rate of alcohol absorption, but the concentration of alcohol in the beverage does. In general, the rate of alcohol absorption increases as the alcohol concentration of the beverage increases. However, beverages with high alcohol concentrations (>60 proof) can irritate the stomach lining, causing increased mucus production. This can delay gastric emptying (the passage of alcohol from the stomach into the small intestine), which slows absorption.[4]

ALCOHOL CIRCULATES AND ACCUMULATES IN THE BLOOD

Once absorbed, alcohol enters the bloodstream and circulates throughout the body. The term **blood alcohol concentration (BAC)** refers to the amount of alcohol in the blood. For example, a person with a BAC of 0.10 has one-tenth of a gram of alcohol per deciliter of blood. Within 20 minutes of consuming one standard drink (12 ounces

of beer, 5 ounces of wine, or 1.5 ounces of 80-proof distilled liquor), BAC begins to rise, and it peaks within 30 to 45 minutes. As more alcohol is consumed, it accumulates in the blood because the rate of alcohol metabolism is slower than the rate of its absorption. For this and many other reasons, people who drink should do so slowly and in moderation. The accumulation of alcohol in the blood leads to intoxication, and in most states a BAC of 0.08 g/dL is the legal limit for driving. Figure 2 provides an estimate of blood alcohol concentration based on the number of drinks consumed per hour. Although the exact amount varies depending on a person's sex and weight, it does not take many drinks for someone to exceed the legal driving limit. Also, continued consumption of alcohol at these rates (drinks/hour) may result in even higher blood alcohol concentrations.

Factors Affecting Blood Alcohol Concentration Alcohol readily disperses in the water-filled environments inside and outside of cells. Because the tissues in the body vary greatly in their water content, some take up alcohol more quickly than others. For example, muscle tissue contains more water associated with it than adipose tissue, and therefore more readily takes up alcohol. As a result, a person's body composition can influence his or her BAC. In other words, if two people of similar body weight ingest the same amount of alcohol, the leaner person will likely have a lower BAC. This occurs because alcohol diffuses from the blood into lean tissue until equilibrium is reached. In contrast, adipose tissue takes up very little alcohol, so more remains in the blood. Body size can also influence BAC. An individual with a large body tends to have more blood and body fluids than someone with a smaller body, diluting the concentration of alcohol. Therefore, the BAC of a large person is often lower than that of a small person after drinking the same amount of alcohol.

ALCOHOL AFFECTS THE CENTRAL NERVOUS SYSTEM

Alcohol is classified as a central nervous system depressant, because it sedates brain activities. This is surprising to many people, because consuming small amounts of alcohol is often associated with feelings of euphoria. These pleasant feelings arise when alcohol selectively depresses certain parts of

---

proof A measure of the alcohol content of distilled liquor.

blood alcohol concentration (BAC) A unit of measurement that reflects the level of alcohol in the blood.

FIGURE 2 **Blood Alcohol Concentration** Estimated blood alcohol concentrations based on alcohol consumption, body weight, and sex.

Males

Weight (lbs)	1	2	3	4	5	6	7	8	9
100	.04	.08	.11	.15	.19	.23	.26	.30	.34
120	.03	.06	.09	.12	.16	.19	.22	.25	.28
140	.03	.05	.08	.11	.13	.16	.19	.21	.24
160	.02	.05	.07	.09	.12	.14	.16	.09	.21
180	.02	.04	.06	.08	.11	.13	.15	.17	.19
200	.02	.04	.06	.08	.09	.11	.13	.15	.17
220	.02	.03	.05	.07	.09	.10	.12	.14	.15
240	.02	.03	.05	.06	.08	.09	.11	.13	.14

Number of drinks consumed per hour

Females

Weight (lbs)	1	2	3	4	5	6	7	8	9
100	.05	.09	.14	.18	.23	.27	.32	.36	.41
120	.04	.08	.11	.15	.19	.23	.27	.30	.34
140	.03	.07	.10	.13	.16	.19	.23	.26	.29
160	.03	.06	.09	.11	.14	.17	.20	.23	.26
180	.03	.05	.08	.10	.13	.15	.18	.20	.23
200	.02	.05	.07	.09	.11	.14	.16	.18	.20
220	.02	.04	.06	.08	.10	.12	.14	.17	.19
240	.02	.04	.06	.08	.09	.11	.13	.15	.17

Number of drinks consumed per hour

Key

▢ Driving skills impaired   ▢ Legally intoxicated   1 drink = 1.5 ounces of 80-proof hard liquor; 5 ounces of wine; 12 ounces of beer.

SOURCE: Adapted from The Pennsylvania Liquor Control Board. Available from: http://www.lcbapps.lcb.state.pa.us/webapp/edu/0.8PercentUPDATE.ASP.

the brain that normally "censor" thoughts and behaviors. The inhibitory effect of alcohol on these portions of the brain is called **disinhibition.** In other words, alcohol can cause a temporary loss of inhibition, making people feel relaxed and more outgoing. However, disinhibition also impairs judgment and reasoning. As BAC increases, areas of the brain that control speech, vision, and voluntary muscular movement become depressed as well. If drinking continues, the person may lose consciousness, and the alcohol concentration may potentially reach lethal levels. Table 2 summarizes a person's response to different levels of alcohol in the blood.

How Is Alcohol Metabolized?

Once a person consumes alcohol, the body must metabolize it into components that can be safely eliminated from the body. Contrary to popular belief, there is nothing a person can eat or drink to accelerate this process. Although small amounts of unmetabolized alcohol are eliminated from the body by the lungs (expired air), skin (sweat), and kidneys (urine), alcohol metabolism is accomplished primarily by the liver. However, there is a limit to how much alcohol the liver can metabolize in any given time. The average person can metabolize 0.5 oz of pure alcohol each hour. This is equivalent to drinking about 12 oz of beer per hour. Two major pathways are used to metabolize alcohol: the alcohol dehydrogenase (ADH) pathway and the microsomal ethanol-oxidizing system (MEOS), and these are described next.

THE ALCOHOL DEHYDROGENASE (ADH) PATHWAY METABOLIZES THE MAJORITY OF ALCOHOL

During light to moderate drinking, most alcohol is metabolized by a two-step enzyme system called the **alcohol dehydrogenase pathway.** The first step in this metabolic pathway requires the enzyme **alcohol dehydrogenase (ADH)** found primarily in liver cells. However, ADH is also

---

disinhibition A loss of inhibition.

alcohol dehydrogenase pathway (de – hy – DRO – gen – ase) The primary metabolic pathway that chemically breaks down alcohol in the liver.

alcohol dehydrogenase (ADH) An enzyme, found mostly in the liver, that metabolizes ethanol to acetaldehyde.

TABLE 2 Stages of Alcohol Intoxication

BAC (g/dL)	Stage of Intoxication	Effects
0.01–0.05	Subclinical	• Behavior nearly normal by ordinary observation
0.03–0.12	Euphoria	• Sociable; talkative; increased self-confidence • Decreased inhibitions; diminution of attention; altered judgment • Beginning of sensory-motor impairment; loss of efficiency in finer performance tests
0.09–0.25	Excitement	• Emotional instability; loss of critical judgment • Impairment of perception, memory, and comprehension • Decreased sensory response; increased reaction time • Reduced visual acuity, peripheral vision, and recovery from flashes of bright light • Impaired motor coordination and balance • Drowsiness
0.18–0.30	Confusion	• Disorientation; mental confusion and dizziness • Exaggerated emotional states • Disturbances of vision and of perception of color, form, motion, and dimensions • Increased pain threshold • Decreased muscular coordination; staggering gait; slurred speech • Apathy, lethargy
0.25–0.40	Stupor	• Lethargy; approaching loss of motor functions • Markedly decreased response to stimuli • Lack of muscular coordination; inability to stand or walk • Vomiting; incontinence • Impaired consciousness; sleep or stupor
0.35–0.50	Coma	• Complete unconsciousness • Depressed or abolished reflexes • Low body temperature • Impairment of circulation and respiration • Possible death
0.45+	Death	• Death from respiratory arrest

SOURCE: Dubowski KM. Stages of acute alcoholic influence/intoxication. 2006. Available from: http://www.borkensteincourse.org/faculty%20documents/dub_stages.pdf.

produced by gastric cells, and for this reason a small amount of alcohol is actually metabolized in the stomach before it ever reaches the blood. As shown in Figure 3, ADH converts alcohol to acetaldehyde, generating NADH + H⁺. Because acetaldehyde is a toxic molecule, it must be quickly metabolized to something less harmful. If acetaldehyde accumulates, some will pass from the liver into the blood, causing unpleasant side effects (headache, nausea, and vomiting) often associated with heavy drinking—what is commonly referred to as a hangover.

The next step in the ADH pathway is catalyzed by the enzyme **acetaldehyde dehydrogenase (ALDH),** which converts acetaldehyde to acetic acid (also called acetate). This reaction requires NAD⁺ for the transfer of hydrogen and forms NADH + H⁺. Acetic acid combines with a molecule of coenzyme A (CoA) to form acetyl-CoA. This step requires the enzyme acetyl-CoA synthase. Acetyl-CoA enters the citric acid cycle and is metabolized further. The metabolism of one molecule of alcohol to acetic acid yields

approximately six molecules of ATP. The metabolism of acetyl-CoA via the citric acid cycle yields additional energy (ATP).

Genetics Can Influence Alcohol Metabolism Studies have found that genetic alterations (polymorphisms) appear to affect the enzymatic activities of ADH and ALDH. Consequently, some people have difficulty metabolizing alcohol.[5] For example, a high percentage of Asians have a less functional form of ALDH. When alcohol is consumed, acetaldehyde levels increase quickly, causing rapid dilation of blood vessels, headaches, and facial flushing. Differences in alcohol-metabolizing enzymes have also been found between men and women.[6] Because women tend to have lower ADH activity in gastric cells than men, they may be less tolerant. This may also be why women are more likely than men to develop alcohol-related health problems.[7]

---

acetaldehyde dehydrogenase (ALDH) (ac – et – AL – de – hyde) An enzyme that converts acetaldehyde to acetic acid.

FIGURE 3 Alcohol Metabolism The alcohol dehydrogenase (ADH) pathway and the microsomal ethanol-oxidizing system (MEOS) are used to metabolize alcohol.

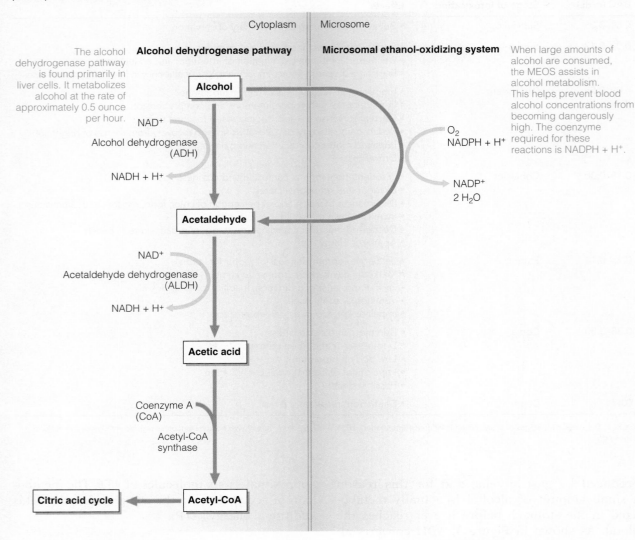

THE MICROSOMAL ETHANOL-OXIDIZING SYSTEM (MEOS) CAN ALSO METABOLIZE ALCOHOL

In addition to the alcohol dehydrogenase pathway, an alternate pathway can also be used to metabolize excess alcohol. This pathway, called the **microsomal ethanol-oxidizing system (MEOS),** is used when large amounts of alcohol are consumed. The primary purpose of this pathway is to prevent alcohol from reaching dangerously high levels in the blood. The components of this accessory pathway are up-regulated in response to frequent intoxication, which is why some heavy drinkers develop a **tolerance** to alcohol. People with high alcohol tolerances metabolize alcohol rapidly, allowing them to drink large amounts before becoming intoxicated. MEOS

is also used to metabolize other drugs besides alcohol. For this reason, drinkers often develop a **cross-tolerance** to drugs metabolized by MEOS enzymes. In other words, when tolerance to alcohol develops, a person can become more tolerant (less responsive) to some other drugs as well. In addition, the need to metabolize alcohol can out-compete the need to metabolize other drugs in people who drink heavily, resulting in drug concentrations reaching dangerously high levels. This is why taking certain drugs

microsomal ethanol-oxidizing system (MEOS) A pathway used to metabolize alcohol when it is present in high amounts.

tolerance Response to high and repeated alcohol exposure that results in reduced alcohol-related effects.

cross-tolerance Tolerance to one substance that causes tolerance to other similar substances.

FIGURE 4 Chronic Alcohol Intake Interferes with Conversion of Pyruvate to Acetyl-CoA

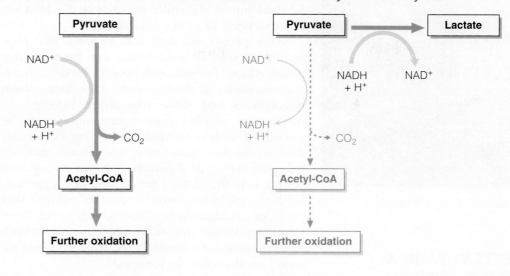

Normally, pyruvate is converted to acetyl-CoA with the aid of NAD⁺.

Heavy alcohol consumption depletes NAD⁺ levels and inhibits the conversion of pyruvate to acetyl-CoA in the liver. The liver converts pyruvate to lactate (lactic acid) rather than to acetyl-CoA, and NAD⁺ is regenerated.

in combination with alcohol is not only dangerous, but also deadly.

Whereas the ADH pathway takes place in the cytosol of the cell, the MEOS takes place inside **microsomes,** specialized organelles within the cell. The MEOS uses oxygen and the coenzyme NADPH + H⁺ to convert alcohol to acetaldehyde. Interestingly, the ADH pathway yields NADH + H⁺, whereas the MEOS actually uses energy. In addition, much of the energy released from alcohol when metabolized by the MEOS is lost as heat rather than used to supply the body with energy. This may, in part, explain why heavy drinkers often experience weight loss despite a high intake of energy.

ALCOHOL METABOLISM CAN DAMAGE THE LIVER

Over time, heavy drinking can lead to liver damage, which impairs normal liver function. This is partly caused by the accumulation of acetaldehyde and NADH + H⁺, the two main products of alcohol metabolism. Alcohol metabolism raises NADH + H⁺ levels and lowers NAD⁺ levels in the liver. As a result, the activities of numerous NAD-linked enzymes become limited. You may recall that the end product of glycolysis is pyruvate and that NAD⁺ is needed to convert pyruvate to acetyl-CoA. With limited NAD⁺, pyruvate is instead converted to lactate (also called lactic acid), as shown in Figure 4. High levels of lactate in the blood can interfere with the ability to excrete uric acid in the urine. Uric acid is

a nitrogen-containing waste product formed in some metabolic pathways. Rising levels of uric acid in the blood can exacerbate a condition called **gout,** which is caused by the excess production and deposit of uric acid crystals in joints. For this reason, people with gout are advised not to drink alcoholic beverages.

Alcohol Metabolism Can Cause Lipid Accumulation in the Liver Alcohol metabolism also causes lipids to accumulate in the liver.[8] This lipid accumulation occurs for several reasons, as listed here.

- Decreased fatty acid breakdown (β-oxidation)
- Increased uptake of fatty acids from the blood
- Increased fatty acid synthesis and triglyceride formation (lipogenesis)
- Decreased transport of triglycerides from the liver into the blood

High concentration of NADH + H⁺ is an important cause of lipid accumulation in the liver. When NADH + H⁺ concentrations are high, the amount of acetyl-CoA entering the citric acid cycle decreases. Instead, acetyl-CoA is used to synthesize fatty acids, which then form triglycerides. The lipogenic effects of chronic and heavy alcohol consumption cause triglycerides to accumulate in the liver, a condition referred to as **fatty liver.**

---

microsomes Cell organelles associated with endoplasmic reticula.

gout A condition caused by the accumulation of uric acid crystals in joints.

fatty liver A condition caused by excess alcohol consumption; characterized by the accumulation of triglycerides in the liver.

Normal, fatty, and cirrhotic livers (from left to right).

ALCOHOL AFFECTS VITAMIN A METABOLISM IN THE LIVER

Another consequence of heavy drinking involves the effects of alcohol on vitamin A metabolism in the liver.[9] Vitamin A plays an important role in vision, especially at night. One form of vitamin A is retinol. Like ethanol, retinol is an alcohol, and ADH is involved in its metabolism. Specifically, ADH converts retinol (the alcohol form of vitamin A) into retinal (the aldehyde form of vitamin A), which is needed for night vision. When a person drinks heavily, alcohol is given preferential access to ADH. As a result, less retinal is made by the body, leading to vision problems, especially at night.

What Are the Health Benefits Associated with Moderate Alcohol Consumption?

Although the detrimental health effects of heavy alcohol consumption have long been known, research now suggests that alcohol consumption in "moderation" may have health benefits. However, what some people consider to be moderate intakes of alcohol might be considered excessive by others. Therefore, it is important to define what is meant by "moderate" alcohol consumption. The 2010 Dietary Guidelines for Americans state that if alcohol is consumed, it should be consumed in moderation—up to one drink per day for women and two drinks per day for men—and only by non-pregnant adults of legal drinking age. These recommendations, however, do not apply to individuals who have difficulty restricting their alcohol intake to moderate levels.[3] In the United States, a standard drink is equivalent to 12 fluid ounces of beer, 5 fluid ounces of wine, or 1.5 fluid ounces of distilled liquor, each of which has approximately 12 to 14 g of alcohol.

Many people are surprised to learn that consuming alcohol in moderation can have positive health effects. For example, adults who drink moderate amounts of alcohol tend to live longer than nondrinkers and those who drink heavily.[10] In moderation, alcohol consumption appears to be associated with decreased risk for cardiovascular disease, strokes, gallstones, age-related memory loss, and even type 2 diabetes.[11] Some researchers believe it is the alcohol *per se*, and not a particular beverage (wine, beer, or distilled spirits) that provides cardioprotective effects.[12] However, there is some evidence that other phytochemicals such as the antioxidants found in red wine may provide added cardioprotective benefits.[13]

It is possible that people who drink in moderation may differ from nondrinkers and heavy drinkers in ways that independently influence health. For example, moderate drinkers are likely to be better educated, wealthier, and more physically active than nondrinkers and heavy drinkers, and also tend to have more regular sleep habits and healthier weights.[14] Nonetheless, even when these confounding variables are accounted for, the evidence supporting the relationship between moderate alcohol intake and health remains compelling. Only a clinical intervention trial could determine directly the health risks and benefits associated with moderate alcohol intakes. However, this type of study would require randomly assigning hundreds of healthy adults to either a treatment group (alcohol consumption) or control group (no alcohol consumption). Such a study would be difficult to carry out and unlikely to be conducted. Clearly, health professionals will need to rely on less controlled studies to assess the effects of alcohol on health.

MODERATE ALCOHOL CONSUMPTION IS RELATED TO LOWER RISK OF CARDIOVASCULAR DISEASE

The relationship between moderate alcohol consumption and reduced risk of cardiovascular disease (CVD) has been confirmed by hundreds of epidemiological studies.[15] More specifically, adults who consume an average of one to two alcoholic drinks daily have a 30 to 35% lower risk of CVD than adults who do not consume alcohol (Figure 5).[16] However, these benefits are lost when intakes of alcohol exceed these amounts. Although studies consistently show cardioprotective benefits of light to moderate alcohol consumption, the American Heart

FIGURE 5 Relationship between Average Alcohol Consumption and Coronary Heart Disease At low levels (1–2 drinks/day), alcohol lowers a person's risk of coronary heart disease. At high levels (7–8 drinks/day), the risk of coronary heart disease increases.

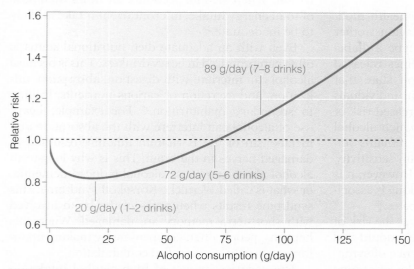

SOURCE: Corrao G, Rubbiati L, Bagnardi V, Zambon A, Poikolainen K. Alcohol and coronary heart disease: A meta–analysis. Addiction. 2000;95:1505–23.

Association discourages individuals from beginning or continuing to drink alcohol for this purpose.[17]

How Moderate Alcohol May Benefit the Heart Scientists have long speculated as to how alcohol might lower a person's risk of CVD. First, recall that CVD results from the formation and accumulation of plaque in the lining of arteries.[18] Furthermore, high-density lipoproteins (HDLs) help protect against heart disease. Studies show that light to moderate daily intake of alcohol is associated with increased HDL-cholesterol levels and therefore may offer some protection from CVD.[19] The rise in HDL-cholesterol is dose dependent.[20] For example, one drink a day is associated with a 5% increase in HDL and two to three drinks per day with a 10% increase. Light to moderate alcohol consumption is related to lower triglycerides in a similar dose-dependent fashion.

There is also evidence that alcohol may decrease levels of a protein (fibrinogen) that promotes blood clot formation and increase levels of an enzyme that dissolves blood clots.[21] Blood clots can block the flow of blood in arteries and are common causes of heart attacks and strokes. Last, because chronic inflammation is also associated with CVD, alcohol's anti-inflammatory effect may offer moderate drinkers protection. Both experimental and observational studies show a reduction in an important inflammatory marker called C-reactive protein in response to light to moderate intakes of alcohol.[22]

Moderate alcohol intake provides few, if any, health benefits among young adults. Their risk of CVD is lower than that of middle-aged adults, so the protective effects of moderate alcohol consumption are of little benefit. In fact, alcohol consumption in the young adult group is associated with increased risk of alcohol-related injury and death. Studies show that many adults with alcohol-related problems started drinking at a young age.[23] The frequency of risky drinking behavior as young adults is associated with the frequency of hazardous drinking patterns later in life.

Added Health Benefits Associated with Red Wine In addition to health benefits associated with alcohol, biologically active compounds found in grapes and red wine may provide additional protection from CVD. This has been largely attributed to the antioxidant **resveratrol,** which is found in the skin of red grapes.[24] There is considerable evidence that resveratrol helps to reduce inflammation and atherosclerosis, thus providing CVD protection.[25] In fact, there is some speculation that resveratrol might explain what scientists refer to as the "French Paradox"—a term used to describe the observation that French people have a relatively low occurrence of cardiovascular disease despite other unhealthy behaviors such as smoking and high intake of saturated fats.[26] However, there is no direct evidence that drinking red wine is actually more beneficial to health than drinking other alcoholic beverages. Studies comparing CVD metabolic risk factors of red wine drinkers to those consuming alcoholic beverages other than wine are inconsistent.[27] While some studies show red wine drinkers to have fewer heart attacks than those consuming alcoholic beverages other than wine, it is important to recognize that these benefits could reflect socioeconomic and lifestyle differences. When compared with people who prefer alcoholic beverages other than wine, wine drinkers tend to have higher incomes, consume less saturated fat, eat more fruits and vegetables, and be nonsmokers.[28]

resveratrol An antioxidant found in the skin of red grapes and abundant in red wine and purple grape juice.

MODERATE ALCOHOL CONSUMPTION MAY PROVIDE OTHER HEALTH BENEFITS

In addition to protecting cardiovascular health, moderate alcohol intake may help guard against other age-related chronic diseases such as type 2 diabetes, gallstones, and dementia. Guidelines released by the American Diabetes Association state that light to moderate intakes of alcohol in individuals with diabetes is associated with a decreased risk of CVD.[29] Although the mechanism by which alcohol modulates CVD in these individuals remains unclear, it appears that it increases insulin sensitivity, subsequently lowering blood glucose. However, it is important to recognize that heavy drinking is associated with increased risk for type 2 diabetes.[30]

Alcohol consumption may also lower the risk of gallstone formation. Regardless of the amount and frequency of consumption, the incidence of symptomatic gallstone disease tends to be lower among drinkers than nondrinkers.[31] Moderate drinking may also reduce the risk of age-related memory loss, or what is called dementia.[32] It is important to remember that these possible benefits of moderate alcohol consumption require further study.

Why Does Heavy Alcohol Consumption Pose Serious Health Risks?

In the case of alcohol, more is not better. While moderate alcohol intake may provide health benefits for middle-aged adults, when consumed in excess, alcohol is clearly hazardous to health. Over time, heavy drinking can lead to impaired nutritional status, liver damage, gout, certain cancers, heart problems, and pancreatitis (Figure 6). High-risk drinking also takes a heavy toll on families, friends, and society.

EXCESSIVE ALCOHOL INTAKE CAN IMPAIR NUTRITIONAL STATUS

Many studies show that heavy drinking leads to decreased nutrient availability and impaired nutritional status via both primary and secondary malnutrition.[33] Alcoholic beverages have calories but very few essential nutrients. For many adults, alcohol contributes substantially to total daily energy intakes (108 kcal/day), which is why even light to moderate drinking can result in weight gain.[3] Furthermore, heavy drinkers often show many other signs of malnutrition. People who abuse alcohol also tend to eat very poor diets, which can lead to essential nutrient deficiencies, or primary malnutrition. When alcohol accounts for more than 30% of total energy intake, micronutrient intake is likely to be inadequate.[34]

Even with an adequate diet, nutritional status is often compromised in heavy drinkers. This is because alcohol can interfere with digestion, absorption, utilization, and excretion of various nutrients, leading to secondary malnutrition.[35] For example, heavy use of alcohol can interfere with the absorption and metabolism of the B vitamin thiamin, resulting in damaged nerves in the brain. This is why long-term alcohol abuse can cause alcohol-induced dementia, or what is called Wernicke-Korsakoff syndrome. This syndrome results when areas of the brain involved with short-term memory are damaged. When this happens, people often compensate by making up information, a behavior called confabulation.

The negative impact of high alcohol intake on nutritional status appears more common among people of low socioeconomic status than among those of high socioeconomic status. This is likely because overall nutritional quality of diets eaten by less economically advantaged people is generally lower than that of diets eaten by those of greater means. The impact of heavy alcohol consumption on nutritional status for selected nutrients is summarized in Table 3.

LONG-TERM ALCOHOL ABUSE CAN LEAD TO LIVER DISEASE

Chronic alcohol abuse can interfere with normal liver function.[36] For example, approximately 20% of heavy drinkers develop fatty livers, which can be treated if the person stops drinking. In most cases, no clinical signs are associated with having a fatty liver, but over time this can lead to more serious conditions such as alcoholic hepatitis and cirrhosis. A person with **alcoholic hepatitis** typically has an enlarged, inflamed liver. Complications from alcoholic hepatitis can range from mild to severe, and are due, in part, to reduced blood flow through the liver. When this occurs, liver cells are deprived of oxygen and nutrients, causing them to die.[37] In addition, about 10% of heavy drinkers develop a condition called **cirrhosis,** characterized by the

alcoholic hepatitis (*hepatic*, pertaining to the liver; *-itis*, inflammation) Inflammation of the liver caused by chronic alcohol abuse.

cirrhosis (cir – RHO – sis) The formation of scar tissue in the liver caused by chronic alcohol abuse.

FIGURE 6 Effects of Habitual Alcohol Abuse on the Body Alcohol interferes with digestion, absorption, use, and excretion of nutrients. Chronic alcohol intake can lead to malnutrition, cancer, gout, and heart disease.

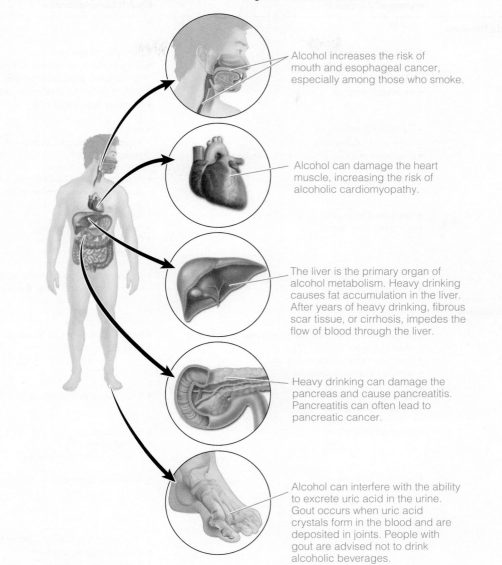

Alcohol increases the risk of mouth and esophageal cancer, especially among those who smoke.

Alcohol can damage the heart muscle, increasing the risk of alcoholic cardiomyopathy.

The liver is the primary organ of alcohol metabolism. Heavy drinking causes fat accumulation in the liver. After years of heavy drinking, fibrous scar tissue, or cirrhosis, impedes the flow of blood through the liver.

Heavy drinking can damage the pancreas and cause pancreatitis. Pancreatitis can often lead to pancreatic cancer.

Alcohol can interfere with the ability to excrete uric acid in the urine. Gout occurs when uric acid crystals form in the blood and are deposited in joints. People with gout are advised not to drink alcoholic beverages.

presence of scar tissue in the liver. This can eventually lead to liver failure and even death. However, alcoholic hepatitis does not progress to cirrhosis in all heavy drinkers, and it is not clear why it develops in some people but not in others. Researchers have estimated that 10 to 15% of heavy drinkers develop cirrhosis of the liver.[38]

LONG-TERM ALCOHOL ABUSE INCREASES CANCER RISK

The association between alcohol consumption and cancer has been extensively studied. There is conclusive evidence that heavy drinking (>50 g/day or approximately four drinks) increases a person's risk of developing certain types of cancer, especially those of the mouth, esophagus, colon, liver, and breast.[39] Even low to moderate intakes of alcohol may increase a woman's risk for developing cancer, particularly breast cancer.[40]

Almost 50% of cancers of the mouth and esophagus are associated with heavy drinking.[41] It is not clear how alcohol increases the risk of cancer, although it may act as both a carcinogen and as a cocarcinogen. Carcinogens are compounds that initiate the formation of cancer, whereas co-carcinogens are substances that enhance the carcinogenicity of other cancer-causing chemicals. For example, it is well known that heavy drinkers who smoke are at particularly high risk for cancers of the mouth, esophagus,

TABLE 3 Impact of Heavy Alcohol Consumption on Nutritional Status

Nutrient	Impact of Alcohol	Health Consequences
Water-Soluble Vitamins		
Thiamin	• Impaired absorption • Increased urinary loss • Altered metabolism • Reduced storage	• Paralysis of eye muscles • Degeneration of nerves with loss of sensation in lower extremities • Loss of balance; abnormal gait • Memory loss; psychosis
Vitamin B$_6$	• Decreased activation • Increased urinary loss • Displacement of vitamin from its binding protein	• Anemia • Impaired metabolic reactions involving amino acids
Vitamin B$_{12}$	• Impaired absorption	• Because of large body stores, vitamin B$_{12}$ deficiency is not common in alcoholics
Folate	• Decreased absorption • Increased urinary loss	• Anemia • Nutrient malabsorption • Diarrhea • Malabsorption of nutrients
Riboflavin	• Impaired absorption • Decreased activation	• Deficiency not typical in isolation but occurs in conjunction with other B vitamins
Fat-Soluble Vitamins		
Vitamin A	• Decreased synthesis of protein needed to transport vitamin A in the blood • Decreased conversion to retinal • Reduced vitamin A absorption	• Impaired vision; night blindness • Liver disease
Vitamin D	• Decreased activation • Impaired absorption	• Bone fractures and osteoporosis
Vitamin E	• Impaired absorption	• Nerve damage • Tunnel vision • Fragility of cell membranes
Vitamin K	• Impaired absorption	• Bruising and prolonged bleeding
Minerals		
Magnesium	• Increased urinary loss	• Muscle rigidity, cramps, and twitching • Irregular cardiac function • Possible hallucinations
Iron	• Increased storage • Impaired absorption • Increased loss in feces	• Both iron deficiency and overload are possible
Zinc	• Decreased absorption • Increased urinary loss • Impaired utilization	• Altered taste; loss of appetite • Impaired wound healing • Night blindness

and trachea. It is possible that alcohol interacts with cigarette smoke to make it more dangerous.

Scientists also hypothesize that poor nutritional status associated with heavy drinking may increase cancer risk. For example, because alcohol interferes with folate availability, cell division and DNA repair may be compromised.[42] Similarly, reduced levels of iron, zinc, vitamin E, certain B vitamins, and vitamin A have been experimentally linked to certain types of cancer. Because adequate intake of many of these nutrients may offer protection against cancer, overall dietary inadequacy from chronic alcohol intake may weaken a person's natural defense mechanisms.

ALCOHOL ABUSE CAN HARM THE CARDIOVASCULAR SYSTEM

Although the cardiovascular system appears to benefit from *moderate* alcohol intake, heavy alcohol consumption has harmful effects. Consequences associated with long-term, heavy drinking include hypertension, stroke, irregular heart rhythms, cardiomyopathy, and increased risk of fatal heart attacks. These detrimental effects can begin to develop in as little as five years of heavy, chronic intake (seven to eight drinks a day) of alcohol.[43]

One of the most serious cardiovascular consequences associated with heavy drinking is **alcoholic cardiomyopathy.** This is caused by the direct toxic effect of alcohol on heart muscle cells.[44] Over time, chronic exposure to high alcohol levels causes the heart muscle to weaken, making it unable to contract forcibly. Blood flow to vital organs such as the lungs, liver, kidneys, and brain is thereby reduced. If drinking continues, alcoholic cardiomyopathy can ultimately result in heart failure. For reasons that are not clear, women are more susceptible to alcohol-induced cardiomyopathy than are males.

Hypertension and Cardiac Arrhythmia The relationship between heavy drinking and hypertension is well established. That is, as alcohol intake increases, so does blood pressure.[45] Alterations in liver function may, in part, contribute to a rise in blood pressure. Fortunately, when heavy drinkers refrain from alcohol, blood pressure often returns to normal.

Alcohol can also trigger disturbances in heart rhythms, a condition called **cardiac arrhythmia.** Cardiac arrhythmia is sometimes referred to as "holiday heart syndrome," because it is associated with binge drinking, which often occurs on weekends and around holidays.[46] **Binge drinking** is defined as the consumption of four or more drinks for women and five or more drinks for men within two hours.[3] Although the cause is not clear, cardiac arrhythmia may be a response to rising levels of acetaldehyde or disturbances in electrolyte balance. Alcohol-induced cardiac arrhythmia can lead to sudden heart attacks.

ALCOHOL ABUSE CAN IMPAIR PANCREATIC FUNCTION

Excessive drinking is one of the leading causes of acute and chronic pancreatitis. In fact, 70 to 90% of all cases of chronic pancreatitis are associated with heavy drinking. **Pancreatitis** is a painful condition characterized by inflammation of the pancreas. Chronically high consumption of alcohol (seven drinks a day for more than five years) increases the risk of developing this condition. Although it is uncertain why some people develop pancreatitis while others do not, some believe that excessive alcohol consumption causes the formation of "protein plugs" that can block pancreatic ducts.[47] Pancreatitis makes it difficult for the pancreas to release pancreatic juice, which supplies the enzymes needed for digestion. The inability to completely digest food impairs nutrient absorption.

How Does Alcohol Abuse Contribute to Individual and Societal Problems?

Although the majority of people who drink alcohol do so responsibly and without negative consequences, the line between alcohol use and abuse is often blurry. When alcohol is the cause of significant problems in a person's life, drinking is problematic. **Alcohol use disorders** (AUDs), a term that encompasses a wide spectrum of problematic drinking, arise when drinking leads to negative physical, legal, or social consequences.[48]

Frequently misdiagnosed, the occurrence of AUDs in the United States is high, with annual costs exceeding $185 billion in terms of hospitalizations, premature deaths, and alcohol-related illness and injury.[49] Habitual alcohol abuse affects virtually every aspect of a person's life, including family relationships and performance at work and at school. Arrests related to drinking while under the influence of alcohol can also lead to legal problems. Although alcohol use can lead to both physical and emotional dependence, many individuals with AUDs are not able to stop drinking.

Popular culture has created a stereotypic view of "alcoholics," yet in reality alcoholism affects men and women, young and old, professionals and nonprofessionals, rich and poor. In fact, most health

---

alcoholic cardiomyopathy (car – di – o – my – O – path – y) Condition that results when the heart muscle weakens in response to heavy alcohol consumption.

cardiac arrhythmia (ar – RHYTH – mi – a) (*cardiac*, pertaining to the heart) Irregular heartbeat caused by high intake of alcohol.

binge drinking Consumption of five or more drinks in males and four or more drinks in females in a period of *two* hours.

pancreatitis (pan – cre – a – TI – tis) Inflammation of the pancreas.

alcohol use disorders A term that encompasses habitual, excessive intakes of alcohol that lead to negative physical, legal, or social consequences.

Alcoholics Anonymous is a fellowship of men and women who share their experience, strength, and hope with each other so that they may solve their common problem and help others to recover from alcoholism. The only requirement for membership is a desire to stop drinking.

professionals recognize alcoholism as a widespread, complex disease that can be treated but not cured. One of the oldest and most reputable organizations for the treatment of alcoholism is **Alcoholics Anonymous® (AA).** This organization promotes a 12-step program that provides fellowship and support for individuals who wish to stop drinking. More than 2 million people consider themselves lifelong members of AA, whose purpose is to help alcoholics achieve sobriety and stay sober. Other organizations such as **Al-Anon®** and **Alateen®** provide support to family members of alcoholics.

ALCOHOL USE ON COLLEGE CAMPUSES

Numerous studies have documented the extent to which alcohol use affects college students and campuses nationwide.[50] Alcohol abuse on many college campuses is rampant and associated with a wide range of negative consequences such as vandalism, violence, acquaintance rape, unprotected sex, and death.[51] To give perspective to the magnitude of alcohol-related problems, data from the U.S. Centers for Disease Control and Prevention (CDC) indicate that alcohol-related unintentional injuries among college students increased from nearly 1,440 to more than 1,825 per year between 1998 and 2005.[52] Individuals between 21 and 29 years of age have the highest percentages of alcohol-related fatal traffic accidents.[53] If you find yourself wondering if you or someone you know has a problem with alcohol,

you can use the self-assessment tool shown in Table 4.

Although one out of five college students reports that he or she abstains from drinking alcohol, one out of four describes himself or herself as a binge drinker.[54] According to a recent study of more than 100 college campuses, binge drinkers are more likely than non–binge drinkers to miss class, have lower academic rankings, be in trouble with campus law enforcement, and drive while under the influence of alcohol.[55] College students are more likely to binge drink compared with peers not attending college. In fact, most college students who binge drink report that these behaviors started after entering college.[56] Clearly, more research is needed to better understand how the collegiate environment inadvertently encourages risky drinking behaviors and what actions can help prevent alcohol-related problems on university campuses.

College Administrators Debate the Legal Drinking Age The issue of underage drinking on college campuses continues to perplex college administrators, with little agreement regarding how to effectively address this problem. While there is convincing evidence that lowering the minimum legal drinking age is not helpful, some college administrators and health professionals believe that keeping the legal age of drinking at 21 years is ineffective and does not curtail problem drinking. Recently, a coalition of more than 100 college presidents from across the country signed a petition asking lawmakers to examine whether the minimum legal drinking age should be changed. The statement says: *"Our experience as college and university presidents convinces us that 21 is not working. A culture of dangerous, clandestine 'binge-drinking'—often conducted off-campus—has developed."*[57] Most public health agencies, including the American Medical Association, the Centers for Disease Control and Prevention, and the National

---

Alcoholics Anonymous (AA) An organization dedicated to helping people achieve and maintain sobriety.

Al-Anon An organization dedicated to helping people cope with alcoholic family members and friends.

Alateen An organization dedicated to helping children cope with an alcoholic parent.

TABLE 4 The Johns Hopkins University Hospital Alcohol Screening Quiz

1. Do you lose time from work due to drinking?
2. Is drinking making your home life unhappy?
3. Do you drink because you are shy with other people?
4. Is drinking affecting your reputation?
5. Have you ever felt remorse after drinking?
6. Have you had financial difficulties as a result of drinking?
7. Does your drinking make you careless of your family's welfare?
8. Do you turn to inferior companions and environments when drinking?
9. Has your ambition decreased since drinking?
10. Do you crave a drink at a definite time daily?
11. Do you want a drink the next morning?
12. Does drinking cause you to have difficulty in sleeping?
13. Has your efficiency decreased since drinking?
14. Is drinking jeopardizing your job or business?
15. Do you drink to escape from worries or trouble?
16. Do you drink alone?
17. Have you ever had a loss of memory as a result of drinking?
18. Has your physician ever treated you for drinking?
19. Do you drink to build up your self-confidence?
20. Have you ever been to a hospital or institution on account of drinking?

If you answered three or more of the questions with a "yes," there is a strong possibility that your drinking patterns are detrimental to your health and that you may be alcohol dependent. You should consider getting an evaluation of your drinking behavior by a healthcare professional.

SOURCE: Office of Health Care Programs, Johns Hopkins University Hospital. Available at: www.alcohol-addiction-info.com/Alcohol_Addiction_Self_Assessment_Tools.html

Binge drinking is a major problem on college campuses throughout the United States.

Highway Traffic Safety Board, support the higher drinking age.

Taking Action In an effort to reduce alcohol abuse, many colleges are developing and implementing new comprehensive programs.[58] Clearly, traditional educational approaches that distribute pamphlets and conduct one-day alcohol awareness programs are not enough. Colleges and universities must work together with communities to create a culture that discourages high-risk drinking.[59] For example, increasing campus recreational activities that do not involve drinking offers students alternatives to bars and clubs. There are also initiatives to curb practices that encourage drinking among college students.[60] These include such things as discouraging price discounts on alcohol in the form of two-for-one drink specials, inexpensive beer pitcher sales, and other types of "happy hour" promotions. Alcohol abuse among college students is a problem that affects the entire campus community.

RECOMMENDATIONS FOR RESPONSIBLE ALCOHOL USE

Both health benefits and risks are associated with alcohol consumption by adults. Although most Americans who drink alcohol do so safely and responsibly, this is certainly not the case for all those who drink, especially young people. For this reason, one of the goals identified in Healthy People 2020 is to reduce alcohol abuse to protect the health, safety, and quality of life for all, especially children.[61] Healthy People 2020 includes a number of alcohol-related objectives such as a reduction in cirrhosis deaths, in binge drinking, and decreases in average annual alcohol consumption and alcohol-impaired driving.

The Dietary Guidelines for Americans also address the harmful effects of alcohol when consumed in excess.[3] For example, the guidelines clearly state that adults who consume alcohol should do so in moderation. Again, moderation is defined as up to *one* drink per day for women and up to *two* drinks per day for men. Importantly, some people should avoid alcohol completely, including individuals younger than the legal drinking age, those who have difficulty restricting their alcohol intake, those taking medications that can interact with alcohol, and individuals with specific medical conditions. It is also important for people to abstain from drinking when driving, operating machinery, or taking part in activities that require attention, skill, or coordination. Although pregnant women are advised to abstain from alcohol throughout pregnancy, it is especially important during the first trimester. Furthermore, women who are breastfeeding should refrain from drinking until their infant is at least three months of age; after this time women are advised to wait at least 4 hours after consuming a single alcoholic drink to breastfeed.[3] In Chapter 14, you can read more about the negative consequences of alcohol consumption during pregnancy and lactation.

Key Points

What Is Alcohol and How Is It Absorbed?

- The type of alcohol in beer, wine, and distilled liquor is called ethanol, and it is produced by a process called fermentation.

- During fermentation, yeast produces ethanol by metabolizing sugar. To make a more concentrated alcohol, fermented beverages can be distilled.

- Alcohol requires no digestion and is absorbed into the blood mainly from the small intestine.

- Once absorbed, alcohol circulates in the blood, and its concentration is expressed as blood alcohol concentration (BAC).

- Alcohol is a central nervous system depressant and acts as a sedative in the brain.

How Is Alcohol Metabolized?

- The liver metabolizes most of the alcohol that is consumed, and the two pathways used to metabolize alcohol are the alcohol dehydrogenase pathway and the microsomal ethanol-oxidizing system (MEOS).

- The alcohol deyhydrogenase pathway uses the enzymes alcohol dehydrogenase (ADH) and acetaldehyde dehydrogenase (ALDH). ADH converts alcohol to acetaldehyde, generating $NADH + H^+$. Next, ALDH converts acetaldehyde to acetic acid, also forming $NADH + H^+$.

- The accumulation of acetaldehyde and $NADH + H^+$ can create an imbalance between NAD^+ and $NADH + H^+$, which affects other NAD-linked enzymes. Heavy drinking can also increase the synthesis of lactate, interfere with vitamin A metabolism, and cause triglycerides to accumulate in the liver.

What Are the Health Benefits Associated with Moderate Alcohol Consumption?

- When consumed in moderation, alcohol is associated with decreased risk for cardiovascular disease, gallstones, age-related memory loss, and even type 2 diabetes.

- Women and men who consume an average of one or two alcoholic drinks daily, respectively, have lower risk of cardiovascular disease.

Why Does Heavy Alcohol Consumption Pose Serious Health Risks?

- Aside from increasing energy intake, which leads to weight gain, habitual drinking can adversely affect nutritional status by causing both primary and secondary malnutrition.

- Long-term alcohol abuse can alter liver function, leading to alcoholic hepatitis.

- When liver cells are deprived of oxygen and nutrients, the formation of fibrous scar tissue can lead to a condition called cirrhosis.

- Alcohol consumption can also increase the risk of developing cancers of the mouth, esophagus, colon, liver, and breast.

- Heavy alcohol consumption can damage heart tissue, causing a condition known as alcoholic cardiomyopathy.

How Does Alcohol Abuse Contribute to Individual and Societal Problems?

- The majority of people who drink alcohol do so responsibly and without negative consequences.

- The organization Alcoholics Anonymous (AA) helps people achieve sobriety and stay sober. Other organizations such as Al-Anon and Alateen provide support to family members and friends of alcoholics.

- Alcohol abuse on many college campuses is associated with a wide range of negative repercussions, and many college campuses are developing programs to help curb high-risk drinking behaviors among students.

Notes

1. French MT, Zavala SK. The health benefits of moderate drinking revisited: Alcohol use and self-reported health status. American Journal of Health Promotion. 2007;21:484–91.

2. Centers for Disease Control and Prevention. Excessive Alcohol Use. Addressing a Leading Risk for Death, Chronic Disease, and Injury at a Glance. 2011 Available from: http://www.cdc.gov/chronicdisease/resources/publications/aag/alcohol.htm.

3. U.S. Department of Agriculture and U.S. Department of Health and Human Services. Dietary guidelines for Americans 2010 (7th ed.). Washington, DC: U.S. Government Printing Office; 2010.

4. Roberts C, Robinson SP. Alcohol concentration and carbonation of drinks: The effect on blood alcohol levels. Journal of Forensic and Legal Medicine. 2007;14:398–405.

5. Zintzaras E, Stefanidis I, Santos M, Vidal F. Do alcohol-metabolizing enzyme gene polymorphisms increase the risk of alcoholism and alcoholic liver disease? Hepatology. 2006;43:352–61.

6. Sumida KD, Hill JM, Matveyenko AV. Sex differences in hepatic gluconeogenic capacity after chronic alcohol consumption. Clinical Medicine and Research. 2007;5:193–202.

7. Nolen-Hoeksema S. Gender differences in risk factors and consequences for alcohol use and problems. Clinical Psychology Review. 2004;24:981–1010.

8. Forgione A, Miele L, Cefalo C, Gasbarrini G, Grieco A. Alcoholic and nonalcoholic forms of fatty liver disease. Minerva Gastroenterologica Dietologica. 2007;53:83–100.

9. Molotkov A, Duester G. Retinol/ethanol drug interaction during acute alcohol intoxication in mice involves inhibition of retinol metabolism to retinoic acid by alcohol dehydrogenase. Journal of Biological Chemistry. 2002;277:553–7.

10. Ellison RC. Balancing the risks and benefits of moderate drinking. Annals of the New York Academy of Sciences. 2002;957:1–6.

11. Djoussé L, Gaziano JM. Alcohol consumption and risk of heart failure in the Physicians' Health Study I. Circulation. 2007;115:34–9. Elkind MS, Sciacca R, Boden-Albala B, Rundek T, Paik MC, Sacco RL. Moderate alcohol consumption reduces risk of ischemic stroke: The Northern Manhattan Study. Stroke. 2006;37(1):13–9.

12. Pai JK, Hankinson SE, Thadhani R, Rifai N, Pischon T, Rimm EB. Moderate alcohol consumption and lower levels of inflammatory markers in U.S. men and women. Atherosclerosis. 2006;186:113–20.

13. Wollin SD, Jones PJH. Alcohol, red wine and cardiovascular disease. Journal of Nutrition. 2001;131:1401–4.

14. Hill EM, Chow K. Life-history theory and risky drinking. Addiction. 2002;97:401–13.

15. Mukamal KJ, Chiuve SE, Rimm EB. Alcohol consumption and risk for coronary heart disease in men with healthy lifestyles. Archives of Internal Medicine. 2006:166:2145–50.

16. Kabagambe EK, Baylin A, Ruiz-Narvaez E, Rimm EB, Campos H. Alcohol intake, drinking patterns, and risk of nonfatal acute myocardial infarction in Costa Rica. American Journal of Clinical Nutrition. 2005;82:1336–45.

17. Goldberg IJ, Mosca L, Piano MR, Fisher EA. AHA Science Advisory. Wine and your heart: A science advisory for healthcare professionals from the Nutrition Committee, Council on Epidemiology and Prevention, and Council on Cardiovascular Nursing of the American Heart Association. Stroke. 2001;32:591–4.

18. Mukamal KJ, Mackey RH, Kuller LH. Alcohol consumption and lipoprotein subclasses in older adults. Journal of Clinical Endocrinology and Metabolism. 2007;92:2559–62.

19. Kurth T, Everett BM, Buring JE, Kase CS, Ridker PM, Gaziano JM. Lipid levels and the risk of ischemic stroke in women. Neurology. 2007;68:556–62.

20. Gaziano JM, Gaziano TA, Glynn RJ, Sesso HD, Ajani UA, Stampfer MJ, Manson JE, Hennekens CH, Buring JE. Light-to-moderate alcohol consumption and mortality in the Physicians' Health Study enrollment cohort. Journal of American College of Cardiology. 2000;35:96–105.

21. Agarwal DP. Cardioprotective effects of light–moderate consumption of alcohol: A review of putative mechanisms. Alcohol and Alcoholism. 2002;37:409–15.

22. Albert MA, Glynn RJ, Ridker PM. Alcohol consumption and plasma concentration of C-reactive protein. Circulation. 2003;107:443–7.

23. Schmid B, Hohm E, Blomeyer D, Zimmermann US, Schmidt MH, Esser G, Laucht M. Concurrent alcohol and tobacco use during early adolescence characterizes a group at risk. Alcohol and Alcoholism. 2007;42:219–25.

24. Gresele P, Pignatelli P, Guglielmini G, Carnevale R, Mezzasoma AM, Ghiselli A, Momi S, Violi F. Resveratrol, at concentrations attainable with moderate wine consumption, stimulates human platelet nitric oxide production. Journal of Nutrition. 2008;138:1602–08.

25. German JB, Walzem RL. The health benefits of wine. Annual Review of Nutrition. 2000;20:561–93.

26. Goldfinger TM. Beyond the French paradox: the impact of moderate beverage alcohol and wine consumption in the prevention of cardiovascular disease. Cardiology Clinics. 2003;21:449–57.

27. Renaud SC, Gueguen R, Siest G, Salamon R. Wine, beer, and mortality in middle-aged men from eastern France. Archives of Internal Medicine. 1999;159:1865–70.

28. Mortensen EL, Jensen HH, Sanders SA, Reinisch JM. Better psychological functioning and higher social status may largely explain the apparent health benefits of wine: A study of wine and beer drinking in young Danish adults. Archives of Internal Medicine. 2001;161:1844–48.

29. Howard AA, Arnsten JH, Gourevitch MN. Effect of alcohol consumption on diabetes mellitus: A systematic review. Annals of Internal Medicine. 2004;140:211–19. American Diabetes Association. Nutrition Recommendations and

Interventions for Diabetes-2006. A position statement of the American Diabetes Association. Diabetes Care. 2006;29:2140–57.

30. Carlsson S, Hammar N, Grill V. Alcohol consumption and type 2 diabetes. Meta-analysis of epidemiological studies indicates a U-shaped relationship. Diabetologia. 2005;48:1051–4.

31. Leitzmann MF, Giovannucci EL, Stampfer MJ, Spiegelman D, Colditz GA, Willett WC, Rimm EB. Prospective study of alcohol consumption patterns in relation to symptomatic gallstone disease in men. Alcoholism, Clinical and Experimental Research. 1999;23:835–41.

32. Espeland MA, Gu L, Masaki KH, Langer RD, Coker LH, Stefanick ML, Ockene J, Rapp SR. Association between reported alcohol intake and cognition: Results from the Women's Health Initiative Memory Study. American Journal of Epidemiology. 2005;161:228–38.

33. Manari AP, Preedy VR, Peters TJ. Nutritional intake of hazardous drinkers and dependent alcoholics in the UK. Addiction Biology. 2003;8:201–10. Salaspuro M. Nutrient intake and nutritional status in alcoholics. Alcohol and Alcoholism. 1993;28:85–8.

34. Kesse E, Clavel-Chapelon F, Slimani N, van Liere M; E3N Group. Do eating habits differ according to alcohol consumption? Results of a study of the French cohort of the European Prospective Investigation into Cancer and Nutrition (E3N-EPIC). American Journal of Clinical Nutrition. 2001;74:322–7.

35. González-Reimers E, García-Valdecasas-Campelo E, Santolaria-Fernández F, Milena-Abril A, Rodríguez-Rodríguez E, Martínez-Riera A, Pérez-Ramírez A, Alemán-Valls MR. Rib fractures in chronic alcoholic men: Relationship with feeding habits, social problems, malnutrition, bone alterations, and liver dysfunction. Alcohol. 2005;37:113–7.

36. Donohue TM Jr. Alcohol-induced steatosis in liver cells. World Journal of Gastroenterology. 2007;13:4974–8.

37. French SW. The role of hypoxia in the pathogenesis of alcoholic liver disease. Hepatology Research. 2004;29:69–7.

38. Mann RE, Smart RG, Govoni R. The epidemiology of alcoholic liver disease. Alcohol Research and Health. 2003;27:209–19.

39. Baan R, Straif K, Grosse Y, Secretan B, El Ghissassi F, Bouvard V, Altieri A, Cogliano V. WHO International Agency for Research on Cancer Monograph Working Group. Carcinogenicity of alcoholic beverages. Lancet Oncology. 2007;8:292–3.

40. Allen NE, Beral V, Casabonne D, Kan SW, Reeves GK, Brown A, Green J. Moderate alcohol intake and cancer incidence in women. Million Women Study Collaborators. Journal of the National Cancer Institute. 2009;101:296–305.

41. Blot WJ. Alcohol and cancer. Cancer Research. 1992;52:2119S–23S. Seitz HK, Meier P. The role of acetaldehyde in upper digestive tract cancer in alcoholics. Translational Research. 2007;149:293–7.

42. Hamid A, Kaur J. Long-term alcohol ingestion alters the folate-binding kinetics in intestinal brush border membrane in experimental alcoholism. Alcohol. 2007;41:441–6.

43. Piano MR. Alcoholic cardiomyopathy: Incidence, clinical characteristics, and pathophysiology. Chest. 2002;121:1638–50.

44. Iacovoni A, De Maria R, Gavazzi A. Alcoholic cardiomyopathy. Journal of Cardiovascular Medicine. 2010;11:884–92.

45. Sesso HD, Cook NR, Buring JE, Manson JE, Gaziano JM. Alcohol consumption and the risk of hypertension in women and men. Hypertension. 2008;168:884–90.

46. Djoussé L, Levy D, Benjamin EJ, Blease SJ, Russ A, Larson MG, Massaro JM, D'Agostino RB, Wolf PA, Ellison RC. Long-term alcohol consumption and the risk of atrial fibrillation in the Framingham study. American Journal of Cardiology. 2004;93:710–3.

47. Hanck C, Whitcomb DC. Alcoholic pancreatitis. Gastroenterology Clinics of North America. 2004;33:751–65.

48. American Psychiatric Association. Diagnostic and statistical manual of mental disorders, 4th ed., DSM-IV-TR (Text Revision). Washington, DC: American Psychiatric Publishing; 2000.

49. National Institute on Alcohol Abuse and Alcoholism (NIAA). Updating estimates of the economic costs of alcohol abuse in the United States: Estimates, update methods, and data. Report prepared by the Lewin Group for the National Institute on Alcohol Abuse and Alcoholism. Rockville, Md: NIAAA, National Institutes of Health, Department of Health and Human Services; 2000.

50. Hingson R, Heeren T, Winter M, Wechsler H. Magnitude of alcohol-related mortality and morbidity among U.S. college students ages 18–24: Changes from 1998 to 2001. Annual Review of Public Health. 2005;26:259–79.

51. Hingson RW, Wenxing Z, Weitzman ER. Magnitude of and Trends in Alcohol-Related Mortality and Morbidity Among U.S. College Students Ages 18–24, 1998–2005. Journal of Studies on Alcohol and Drugs. 2009;16:12–20.

52. Yi H, Williams GD, Dufour MC. Surveillance Report #65: Trends in alcohol-related traffic crashes, United States, 1977–2001. Bethesda, MD: NIAAA, Division of Biometry and Epidemiology, Alcohol Epidemiologic Data System, August 2003.

53. Turrisi R, Mallett KA, Mastroleo NR, Larimer ME. Heavy drinking in college students: Who is at risk and what is being done about it? Journal of General Psychology. 2006;133:401–20.

54. U.S. Department of Health and Human Services. National Institutes of Health. National Institute on Alcohol Abuse and Alcoholism. What colleges need to know now: An update on college drinking research. NIH Publication No. 07-5010, November 2007.

55. Lange JE, Clapp JD, Turrisi R, Reavy R, Jaccard J, Johnson MB, Voas RB, Larimer M. College binge drinking: What is it? Who does it? Alcoholism, Clinical and Experimental Research. 2002;26:723–30.

56. Timberlake DS, Hopfer CJ, Rhee SH, Friedman NP, Haberstick BC, Lessem JM, Hewitt JK. College attendance and its effect on drinking behaviors in a

longitudinal study of adolescents. Alcoholism, Clinical and Experimental Research. 2007;31:1020–30.

57. DeJong W, Larimer ME, Wood MD, Hartman R. NIAAA's Rapid Response to College Drinking problems initiative: Reinforcing the use of evidence-based approaches in college alcohol prevention. Journal of Studies on Alcohol and Drugs. 2009;16S:5–11.

58. Beck KH, Arria AM, Caldeira KM, Vincent KB, O'Grady KE, Wish ED. Social context of drinking and alcohol problems among college students. American Journal of Health Behavior. 2008;32:420–30.

59. Dejong W. Finding common ground for effective campus-based prevention. Psychology of Addictive Behaviors. 2001;15:292–6.

60. Kuo M, Wechsler H, Greenberg P, Lee H. The marketing of alcohol to college students: The role of low prices and special promotions. American Journal of Preventive Medicine. 2003;25:204–11.

61. Office of Disease Prevention and Health Promotion, U.S. Department of Health and Human Services. Healthy People 2020. Available from: http://www.healthypeople.gov/2020/default.aspx.

Energy Balance and Body Weight Regulation

8
CHAPTER

Some count them, others curse them—but what role do calories actually play in body weight regulation? Moreover, while most people struggle to lose weight, some actually find it difficult to gain weight. If you have ever pondered these issues, you are not alone. In this chapter, you will learn about the relationships among energy intake, energy expenditure, and body weight. You will also understand how the foods you eat and the activities you engage in influence body weight. Factors related to body weight regulation, the growing prevalence of obesity in the United States, and weight loss and weight maintenance are addressed as well.

Meike Bergmann/Getty Images

The Decision to Have Gastric Bypass Surgery

As far back as she can remember, August always had a weight problem. She certainly was larger and heavier than most of her peers. Although she tried to control her eating, it seemed the more August restricted her food intake, the hungrier she became. When the hunger became overwhelming, August felt out of control, and the cycle of food restriction followed by overeating would start again. Looking back, she recognizes that she never felt full. Hunger was a continuous force that drove August to eat.

As an adult, August felt tremendous guilt about her inability to lose weight. The drive to eat was overpowering. By the time she was 30, she weighed 275 pounds, making her feel self-conscious and sad. She found it difficult to look for a job, knowing that she might be judged on the basis of her weight, rather than her skill and ability.

In spite of these problems, August and her husband started a family, and had two beautiful children. August enjoyed staying home and raising them, but her weight continued to make life challenging. The simple act of sitting on the floor with her children was becoming increasingly difficult. Even walking them to school made her knees ache and her heart pound. Her health was deteriorating because of the excess weight. Diet after diet, nothing seemed to work.

Feeling hopeless, August finally found the help she needed. After a routine health exam, she and her physician discussed her long struggle with weight, and he suggested something she never considered before—gastric bypass surgery. He cautioned her that this was not the solution for everyone and gave her educational material to read. She soon learned that gastric bypass surgery would help her to feel full and eat less by reducing the size of her stomach. Food would also be rerouted through her digestive tract so that it bypassed a portion of her small intestine. Although there were clear benefits to gastric bypass surgery, like any surgical procedure, it was not without risks.

The following year, August was approved for gastric bypass surgery. She was fully aware that this radical procedure would change her life forever. It would mean eating differently and having to be careful that her nutrient needs were met. She was also told that she might experience recurring bouts of diarrhea and vomiting. Nonetheless, August decided this was her best hope.

It took several weeks for August to recover from the surgery and even longer to adjust to the physiological changes to her digestive system. Most importantly, however, August experienced something she had never experienced before—she did not feel the persistent sensation of hunger. In fact, she felt satisfied after each meal. Soon, the weight started coming off—50 pounds, 75 pounds, and eventually 125 pounds. Her physical and mental health improved. Today, August is back in school and studying to be a nurse. We asked her if she had any thoughts about obesity she wanted to share with others. In August's words:

"Gastric bypass surgery is a last resort, but for some people it is the lifeline we need. My quality of life has improved dramatically since having this procedure. Also, I want you to know that obesity is very complicated, and it is not something I would wish on anyone. Please don't judge people or think that because someone is obese they simply lack willpower. If I had one wish in life, it would be to not battle my body anymore."

© Ryan McVay/Lifesize/Getty Images

Critical Thinking: August's Story

What insights did you gain from reading August's story about what it is like to be obese? If you heard that someone you know was having gastric bypass surgery, what would you think?

What Is Energy Balance?

What determines whether the energy in the foods you eat is used to fuel your body or stored for later use? It all comes down to balance—the balance between energy intake and energy expenditure. In other words, the amount of energy you consume should be in balance with the amount of energy your body requires. Energy intake and energy expenditure are two important components of energy balance, with the third being energy stored (or mobilized). This relationship is often expressed as Energy stored = Energy consumed − Energy expended. Note that "Energy stored" can be either positive or negative depending on whether the amount of energy consumed is equal to, less than, or greater than that expended. When energy intake equals energy expenditure, a person is in a state of **energy balance** (energy intake = energy expenditure). When a person is in energy balance, body weight tends to be relatively stable. However, an **energy imbalance** can arise when the amount of energy consumed does not equal the amount of energy the body uses. Consuming more energy than the body needs puts you in a state of **positive energy balance,** which results in weight gain. Conversely, someone is in a state of **negative energy balance** when energy intake is less than energy expenditure, which results in weight loss. A change in body weight is a useful indicator of whether you are in positive or negative energy balance.

ENERGY BALANCE AFFECTS BODY WEIGHT

Body weight increases during periods of positive energy balance and decreases in response to negative energy balance. This relationship is illustrated in Figure 8.1. When a person is in positive energy balance, the amount of muscle or adipose tissue—or both—increases because excess energy is stored as protein (lean tissue) and/or fat (adipose tissue). This is desirable during periods of growth such as infancy, childhood, adolescence, and pregnancy. However, when increased body weight is primarily associated with increased body fat, positive energy balance often unhealthy.

Negative energy balance is a result of insufficient energy intake, excessive energy expenditure, or both. Under this condition, stored energy reserves are broken down, resulting in decreased body weight. Adipose tissue is the body's primary energy reserve, and is broken down during negative energy balance. One pound of adipose tissue, including supporting lean tissue, is equivalent to approximately 3,500 kcal. However, it is important to recognize that weight loss associated with negative energy balance is not always due entirely to loss of body fat. Water and muscle loss can also contribute to weight loss.

A Closer Look at Adipose Tissue There is a tendency to think of adipose tissue as a passive reservoir that buffers imbalances between energy intake and expenditure. However, adipose tissue is anything but passive. The source of several hormones, it plays an active role in regulating energy balance. The discovery that adipocytes are a source of hormones and hormone-like substances collectively called **adipokines,** has led many researchers to consider adipose tissue to be the largest endocrine organ in the body.[1] Adipose-derived hormones and adipokines provide an important communication link between adipocytes and other tissues and organs. In addition, many adipokines regulate immune and inflammatory processes, specifically those thought to be involved in weight-related diseases, such as type 2 diabetes.[2]

Adipose tissue is comprised largely of specialized cells called **adipocytes,** which contain a lipid-filled core that consists primarily of triglycerides. The number and size of adipocytes determine the amount of adipose tissue in

Zia Soleil/Iconica/Getty Images

Weighing yourself over time is a good way to determine if you are in positive or negative energy balance.

---

energy balance A state in which energy intake equals energy expenditure.

energy imbalance A state in which the amount of energy consumed does not equal the amount of energy used by the body.

positive energy balance A state in which energy intake is greater than energy expenditure.

negative energy balance A state in which energy intake is less than energy expenditure.

adipokines Hormone-like substances produced and released by adipocytes.

adipocytes Cells found in adipose tissue and used mainly for fat storage.

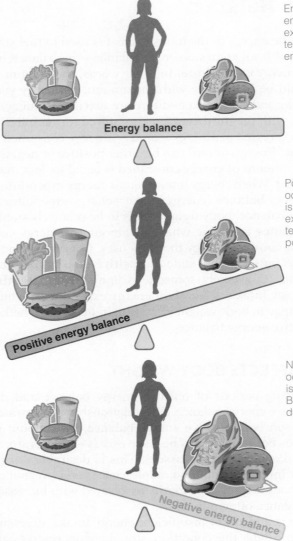

FIGURE 8.1 The Three States of Energy Balance A change in body weight can indicate if a person is in positive or negative energy balance.

Energy balance

Energy balance occurs when energy intake equals energy expenditure. Body weight tends to be stable during energy balance.

Positive energy balance

Positive energy balance occurs when energy intake is greater than energy expenditure. Body weight tends to increase during positive energy balance.

Negative energy balance

Negative energy balance occurs when energy intake is less than energy expenditure. Body weight tends to decrease during negative energy balance.

© Susumu Nishinaga/Photo Researchers, Inc.

Adipose tissue consists of cells called adipocytes.

hypertrophic growth Growth associated with an increase in cell size.

hyperplastic growth Growth associated with an increase in cell number.

visceral adipose tissue (VAT) Adipose tissue deposited between the internal organs in the abdominal area.

subcutaneous adipose tissue (SCAT) Adipose tissue found directly beneath the skin.

the body. As an adipocyte fills with triglycerides, its size increases, a process called **hypertrophic growth.** To accommodate large amounts of lipid, the diameter of an adipocyte can increase 20-fold. When existing cells are full, new adipocytes are formed, a process called **hyperplastic growth.** When a person loses body fat, enlarged adipocytes return to normal size; however, the number of adipocytes remains constant.[3]

Some scientists believe that the number and size of adipocytes may influence our ability to maintain a healthy weight. For example, people with fewer, larger adipocytes may have less difficulty maintaining weight loss than those with a greater number of smaller adipocytes.[4] Figure 8.2 shows how adipocyte number and size change in response to weight gain and loss.

Visceral Adipose Tissue vs. Subcutaneous Adipose Tissue Although adipose tissue is found throughout the body, **visceral adipose tissue (VAT),** also called intra-abdominal fat, refers to adipose tissue surrounding the internal organs in the torso (Figure 8.3). However, much of our body fat reserve, which is called **subcutaneous adipose tissue (SCAT),** is found directly beneath the skin. SCAT is found throughout the body but is most predominant in the thighs, hips, and buttocks.

FIGURE 8.2 Hypertrophic and Hyperplastic Growth of Adipose Tissue The amount of adipose tissue a person has depends on the number and size of his or her adipose cells (adipocytes).

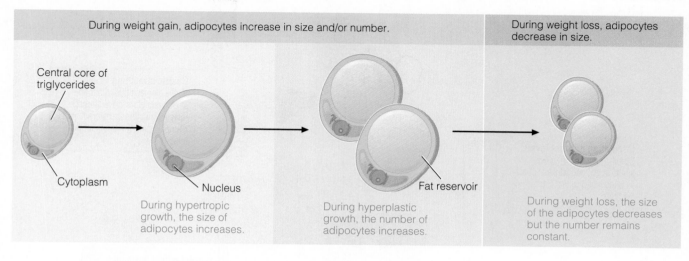

During weight gain, adipocytes increase in size and/or number.

During weight loss, adipocytes decrease in size.

Central core of triglycerides

Cytoplasm

Nucleus

Fat reservoir

During hypertropic growth, the size of adipocytes increases.

During hyperplastic growth, the number of adipocytes increases.

During weight loss, the size of the adipocytes decreases but the number remains constant.

What Determines Energy Intake?

Although it seems simple, understanding energy balance is more complicated than one might think. Your body gets energy from the foods you eat, but understanding what makes you eat and why you make certain choices is complex. Your food choices are related to many facets of your life, and the choices you make serve numerous purposes beyond providing your body with nourishment. It is also important to recognize that foods we enjoy as adults may be the very foods we avoided as children. Clearly, eating behavior is complicated, and it is important to consider the psychological, physical, social, and cultural forces that influence it.

HUNGER AND SATIETY ARE PHYSIOLOGICAL INFLUENCES ON ENERGY INTAKE

Hunger and satiety are complex physiological states that greatly influence energy intake. **Hunger** is defined as the basic physiological drive to consume food, whereas **satiety** is the physiological response to having eaten enough, resulting in food intake cessation. Humans tend to be periodic eaters, meaning that they usually eat meals at predictable and discrete times throughout the day. Most of the time, people eat to the point of comfort. In other words, they eat until they are satiated. Still, people sometimes continue to eat after they are full, even to the point of discomfort. Regardless of one's level of fullness, however, a person is usually ready to eat again within three to four hours of when they last ate.

We tend to think that factors associated within our stomachs are the main determinants of hunger and satiety. However, years ago researchers discovered that hunger and satiety could be controlled in mice by stimulating certain regions within their brains. These areas soon became known as the hunger and satiety centers. Scientists now recognize that energy intake is regulated, in part, by specific neural connections (neurons) to and within the brain, rather than by distinct hunger and satiety "centers."[5]

FIGURE 8.3 Visceral Adipose Tissue (VAT) and Subcutaneous Adipose Tissue (SCAT) Visceral adipose tissue (VAT) refers to adipose tissue deposited between the internal organs in the abdomen. Subcutaneous adipose tissue (SCAT) is found directly beneath the skin.

Front of the body

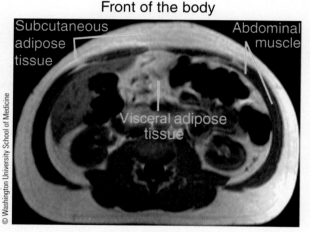

Subcutaneous adipose tissue

Abdominal muscle

Visceral adipose tissue

© Washington University School of Medicine

Back of the body

hunger The physiological drive to consume food.

satiety The state in which hunger is satisfied and a person feels he or she has had enough to eat.

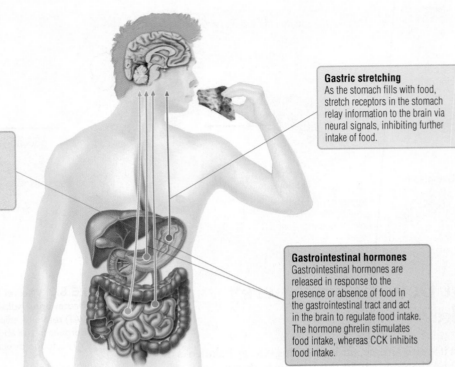

Circulating nutrient levels
Specific nutrients (glucose, amino acids, and fatty acids) act as hunger and satiety signals in the brain.

Gastric stretching
As the stomach fills with food, stretch receptors in the stomach relay information to the brain via neural signals, inhibiting further intake of food.

Gastrointestinal hormones
Gastrointestinal hormones are released in response to the presence or absence of food in the gastrointestinal tract and act in the brain to regulate food intake. The hormone ghrelin stimulates food intake, whereas CCK inhibits food intake.

One part of the brain that is critical to regulating hunger and satiety is the **hypothalamus.** The hypothalamus receives signals that influence hunger and satiety from different parts of the body, such as the gastrointestinal tract. In response, the brain releases neurotransmitters, which in turn influence hunger and satiety. **Neurotransmitters** are hormone-like chemical messengers, produced by neurons, which affect the central nervous system. Some neurons in the brain release **catabolic neurotransmitters** that promote satiety and meal termination, whereas other neurons release **anabolic neurotransmitters** that promote hunger and initiate eating. In addition to hunger and satiety, neurotransmitters also influence energy expenditure and play important roles in long-term regulation of body weight.

Both hunger and satiety are influenced by a variety of factors, including gastric distention (stretching), circulating nutrient levels, and GI hormones. As shown in Figure 8.4, these sensations exert their influence by sending signals to the hypothalamus, which, in turn, controls the body's hunger and satiety responses. However, hunger and satiety are also influenced by other factors such as stress and emotions.

Gastric Stretching—A Satiety Signal The distention of the stomach wall in response to the presence of food, called gastric stretching, provides a powerful satiety signal. The GI tract has specialized receptors (mechanoreceptors) that are responsive to this type of stretching. When you consume small amounts of food, gastric stretching is minimal. However, as the volume of food increases, stretch receptors are stimulated. Stretching of the stomach wall initiates the release of neural signals, which convey information to the brain. The brain responds by releasing neurotransmitters that elicit the sensation of satiety.[6]

hypothalamus An area of the brain that controls many involuntary functions by the release of hormones and neurotransmitters.

neurotransmitter A hormone-like chemical messenger released by nerve cells.

catabolic neurotransmitter A substance released by nerve cells that inhibits hunger and/or stimulates energy expenditure.

anabolic neurotransmitter A substance released by nerve cells that stimulates hunger and/or decreases energy expenditure.

Bariatrics is the branch of medicine concerned with treating obesity and obesity-related conditions. The term *bariatric surgery* refers to a surgical procedure that promotes weight loss. There are different types of bariatric surgery, each having potential side effects and risks. However, the health benefits associated with weight loss often outweigh the risks associated with surgery. The Food and Drug Administration (FDA) recently released new guidelines that lowered the eligibility criteria for the gastric band procedure. Patients with a BMI 30 kg/m² or greater can now be considered for gastric banding if weight-loss diets or drugs have not worked and if they have obesity-related health problems.[7]

In general, bariatric surgery alters the digestive tract so that only small amounts of food can be consumed and/or digested. For example, a minimally invasive procedure called gastric banding creates a small stomach pouch with a food-holding capacity of about 2 to 3 tablespoons (about the size of an egg). The pouch is formed by surgical implantation of a band-like device that fits around the upper portion of the stomach. The size of the opening between the upper and lower portions of the stomach can be increased or decreased by adjusting the diameter of the band. The movement of food is slower when the diameter of the band is made smaller, helping the individual feel full.

Remember that the stomach normally holds 4 to 6 cups of food. In individuals with gastric banding, it only takes a small amount of food to signal the feeling of satiety. It is important with this type of surgery that the person is careful not to overeat, to eat foods that are soft and moist, and to thoroughly chew food before swallowing. Common side effects include heartburn, abdominal pain, and vomiting. Because food intake is severely restricted, it may be necessary to take nutrient supplements to ensure that nutritional needs are met. The gastric band can be removed at any time, and the original form and function of the stomach is restored.

Another type of bariatric surgery is called gastric bypass. In addition to reducing the size of the stomach, this procedure also reroutes the GI tract to bypass a segment of the small intestine. To do this, the lower portion of the reduced stomach pouch is connected to the middle portion of the small intestine, bypassing much of the duodenum and jejunum. Digestive secretions from the remaining portion of the stomach and those from the gallbladder and pancreas are rerouted such that they are released into the lower portion of the small intestine. This procedure not only helps reduce food intake but also limits nutrient digestion and absorption. The risk of developing nutritional deficiencies is much greater after this procedure than after gastric banding. The rapid movement of food through the small intestine causes many adverse side effects, including **dumping syndrome.** Dumping syndrome causes cramps, nausea, diarrhea, and dizziness. Surgical reversal of gastric bypass is difficult and can pose certain risks to the individual.

About 20% of people who have bariatric surgery experience complications.[8] Furthermore, depending on an individual's age, sex, and general health, the death rate within one year of surgery is 1 to 5%.[9] A recent study followed more than 4,000 obese subjects, half of whom underwent bariatric surgery while the other half received conventional treatment, over an 11-year period.[10] The study found that bariatric surgery significantly improved survival in obese individuals. Other studies suggest that gastric bypass surgery may also help lower death rates associated with type 2 diabetes, heart disease, and cancer.[11] Still, one must give serious consideration to the risks before undergoing either of these procedures.

The number of weight-loss surgeries performed yearly in the United States is on the rise. Most people lose weight immediately following the surgery, and within five years have lost 60% of their excess weight. Although patients with gastric banding lose weight more slowly than those who have gastric bypass, the banding procedure has a lower mortality rate. Long-term effectiveness depends on a person's willingness to eat a healthy, well-balanced diet and exercise regularly. For many severely obese individuals, weight-loss surgery can improve their quality of life, both physically and emotionally.[12] It is important to remember that any type of bariatric surgery is a life-altering procedure that requires a commitment to adhere to strict dietary changes.

Obese people tend to have larger stomachs than do lean people and therefore can accommodate larger volumes of food before gastric stretching triggers satiety. This may be why some severely obese people do not readily feel full (satiated) after a meal. Some severely obese people opt to have weight-loss surgery, otherwise known as **bariatric surgery.** As shown in Figure 8.5, **gastric banding** reduces the size of the stomach, whereas **gastric bypass** also reroutes the flow of food through the GI tract. Although both types of procedures limit how much food can be consumed, the latter also reduces nutrient digestion and absorption. Because the size of the stomach has been drastically

bariatrics The branch of medicine concerned with the treatment of obesity.

dumping syndrome A condition whereby food moves too rapidly from the stomach into the small intestine.

bariatric surgery Surgical procedure performed to treat obesity.

gastric banding A type of bariatric surgery in which an adjustable, fluid-filled band is wrapped around the upper portion of the stomach, dividing it into a small upper pouch and a larger lower pouch.

FIGURE 8.5 Types of Weight-Loss Surgery
The term bariatric surgery refers to medical procedures that promote weight loss. To be considered for weight-loss surgery, a person must meet strict eligibility guidelines such as being unable to achieve or maintain a healthy weight through diet and exercise and also have weight-related health problems.

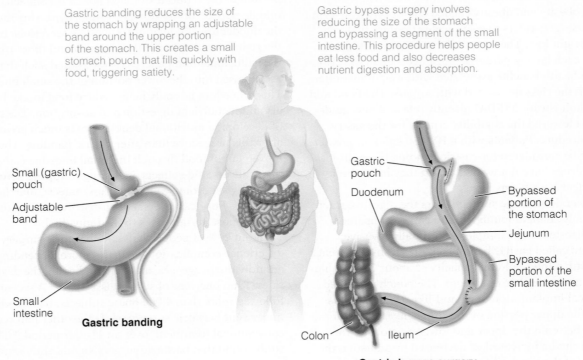

Gastric banding reduces the size of the stomach by wrapping an adjustable band around the upper portion of the stomach. This creates a small stomach pouch that fills quickly with food, triggering satiety.

Gastric bypass surgery involves reducing the size of the stomach and bypassing a segment of the small intestine. This procedure helps people eat less food and also decreases nutrient digestion and absorption.

Small (gastric) pouch

Adjustable band

Small intestine

Gastric banding

Gastric pouch

Duodenum

Bypassed portion of the stomach

Jejunum

Bypassed portion of the small intestine

Colon   Ileum

Gastric bypass surgery

High-volume, low-energy-dense foods such as fruits and vegetables have high water and fiber contents and are low in fat. These foods may trigger satiety more effectively than other foods.

gastric bypass A surgical procedure that reduces the size of the stomach and bypasses a segment of the small intestine so that fewer nutrients are absorbed.

reduced, the stomach quickly fills with food, triggering satiety. You can read more about bariatric surgery in the Focus on Clinical Applications feature.

High-Volume Foods and Satiety Recent studies show that eating high-volume foods can help people feel full longer than when they eat low-volume foods.[13] High-volume foods are those with high water and/or fiber content, such as fruits and vegetables. These types of foods increase gastric stretching, which can help people feel full and satisfied, even when they have not consumed a significant amount of calories. For example, two cups of grapes have the same amount of energy (~100 kcal) as ¼ cup of raisins. However, because the volume of the grapes is greater than raisins, a person is more likely to feel satisfied after eating the grapes. Some studies suggest that long-term consumption of a low-energy-dense diet may help people lose weight.

Circulating Nutrient Concentrations The concentration of certain nutrients in the blood can also influence hunger and satiety. For example, the brain is very sensitive to changes in blood glucose levels. When blood glucose increases following a meal, the brain responds by releasing neurotransmitters that stimulate satiety, providing a signal to stop eating. Conversely, when the brain detects a decrease in blood glucose, it releases neurotransmitters that stimulate hunger.

Amino acids can also play a role in short-term regulation of food intake. In general, elevated levels of circulating amino acids promote satiety by signaling neurons in the brain to release catabolic neurotransmitters. Of particular interest is the effect of the amino acid tryptophan on food intake. The brain uses tryptophan to synthesize the neurotransmitter serotonin, which conveys the sensations of satiety and relaxation. Some people find that eating tryptophan-rich foods, such as turkey and dairy products, promotes sleepiness in addition to satiety.[14] There is also evidence that disturbances in serotonin production can disrupt appetite regulation and may lead to obesity.[15]

TABLE 8.1 Gastrointestinal Hormones and Effect on Food Intake

Gastrointestinal Hormone	Stimulus for Release	Site of Production	Effect on Food Intake
Cholecystokinin (CCK)	Protein and fatty acids	Small intestine	↓
Glucagon-like peptide 1 (GLP-1)	Nutritional signals and neural/hormonal signals from the gastrointestinal tract	Small and large intestine	↓
Ghrelin	Empty stomach	Stomach	↑
Enterostatin	Fatty acids	Stomach and small intestine	↓
Peptide YY (PYY)	Food in the GI tract	Small and large intestine	↓

Elevated levels of circulating lipids can also promote satiety. However, this effect appears to be weak and is easily overcome by pleasant sensations associated with fatty foods such as enhanced flavor and texture. As a result, consuming high-fat foods tends to lead to excess calorie intake and weight gain, not satiety.

Gastrointestinal Hormones The presence of food in the stomach and small intestine can trigger the release of several GI hormones, the majority of which promote satiety (Table 8.1).[16] Examples include cholecystokinin (CCK) and peptide YY (PYY). Of these, CCK is the best understood. CCK is released from intestinal cells (enterocytes), particularly in response to high-fat and protein foods, signaling the brain to decrease food intake. Although the mechanisms remain unclear, CCK appears to play a role in short-term satiety. The hormone PYY, also released from the intestinal lining following a meal, plays a similar role.

Not all GI hormones signal the sensation of satiety. For example, researchers have recently become interested in the role of **ghrelin,** a newly discovered hormone dubbed the "hunger hormone." This potent hunger-stimulating hormone is secreted by cells in the stomach lining and circulates in the blood. High levels of ghrelin in the blood serve as a pre-meal signal that stimulates the sensation of hunger. Ghrelin levels are highest prior to meals when the stomach is empty, but after food is consumed, ghrelin levels decrease. Recent evidence shows that some obese people may overproduce ghrelin, which could explain why they do not always experience a feeling of satiety following a meal.[17] There is also evidence that gastric bypass surgery may lower ghrelin levels, which in turn causes hunger to diminish.[18]

APPETITE ALSO AFFECTS ENERGY INTAKE

Two important responses that help maintain a healthy body weight are eating when you hungry and stopping when you are full. However, sometimes we eat for reasons other than hunger, and at times we eat past satiety. The stimuli that override hunger and satiety tend to be more psychological than physiological. Whereas **appetite** is the psychological longing or desire for food, a **food aversion** is a strong psychological dislike of a particular food or foods. Sometimes, even the thought or smell of certain foods can trigger an adverse physical response. Our appetite can be an important determinant of what and when we eat. When the desire for a specific food is especially compelling, it is often referred to as a **food craving.** You can read more about food cravings in the Focus on Food feature.

ghrelin A hormone, secreted by cells in the stomach, that stimulates food intake.

appetite A psychological desire for food.

food aversion A strong psychological dislike of a particular food.

food craving A strong psychological desire for a particular food.

FOCUS ON FOOD
Food Cravings and Food Aversions

Are food aversions the flip side of food cravings? Even the experts are unsure. The reality is that there are no definitive answers when it comes to explaining food cravings and aversions. Whereas a food craving is a powerful, irresistible, intense desire for a particular food, food aversions develop when certain foods are viewed as repugnant. Although certain emotional states can provoke food aversions, most are conditioned responses resulting from a paired association between physical discomfort and a particular food. Food aversions tend to be persistent and long lasting, whereas food cravings can come and go.

Almost everyone experiences food cravings, because food is rewarding and pleasurable. However, some people may be more sensitive than others to the pleasurable effects triggered by certain foods. The most commonly craved foods tend to be calorie-rich ones, such as cookies, cakes, chips, and chocolate.[19]

Women tend to experience food cravings more frequently than do men.[20] This may, in part, be due to hormonal fluctuations during the menstrual cycle. In fact, many women crave certain foods around the time of their menstrual flow. Hormonal changes associated with pregnancy may also cause some women to crave particular foods. In general, food cravings tend to occur at specific times during the day (late afternoon or early evening) and in response to stressful situations, such as when a person feels anxious.[21]

People once thought food cravings were caused by a lack of specific nutrients in the diet. However, this is not always the case. For example, a craving for potato chips is not necessarily caused by a lack of salt, nor do we crave steak because we simply need more protein. Food cravings can be strong, and scientists do not have a clear explanation for their occurrence. Nonetheless, new information on the role of neurochemicals is unfolding.

During pregnancy, many women experience cravings for certain foods and/or aversions to others. Although some experts believe that food cravings help pregnant women satisfy their needs for nutrients and that food aversions may provide protection from harmful substances, there is very little scientific data to support either of these claims. In other words, a craving for ice cream does not necessarily mean that a woman is calcium deficient.

For some people, food cravings can be overwhelming, and learning how to tame them is important. Most health experts agree that chronic overconsumption of food is more likely than food cravings, *per se*, to cause weight gain. In fact, overly restrictive food regimens may actually cause food cravings. Experts believe it is healthier to indulge our cravings within reason, rather than making certain foods "off limits" or becoming preoccupied with lingering thoughts and desires for particular foods. Furthermore, getting adequate sleep, participating in regular exercise, practicing relaxation techniques to reduce stress, and eating a healthy diet may help food cravings become less persistent.

© Mark Andersen/Rubberball Productions/Getty Images

Scientists do not fully understand what causes a person to have food cravings. Calorie-rich foods such as cookies, ice cream, cakes, and chocolate are the foods people most often crave.

Appetite can easily be aroused by sensory factors such as the appearance, taste, or smell of food. For example, the pleasing smell of baked bread makes many people want to eat, regardless of whether they are hungry. Likewise, unpleasant odors can "spoil" our appetite, even when we are hungry. Emotional states can also dramatically affect appetite. Whereas some people respond to emotions such as fear, depression, disappointment, excitement, and stress by eating, others respond by not eating at all. Clearly our emotional states can have a profound influence on appetite, overriding the normal physiological cues of hunger and satiety. However, when psychological factors are the primary determinant of hunger and satiety, eating disorders such as anorexia nervosa and bulimia nervosa can occur. You can read more about eating disorders in the Nutrition Matters following this chapter.

Now that you understand how your body regulates energy intake, we turn to the other component of energy balance: energy expenditure.

What Determines Energy Expenditure?

As you have learned, a complex interplay of factors influences when and how much you eat. However, energy intake is only one component of energy balance; another is energy expenditure. The body expends energy to maintain physiological functions, support physical activity, and process food, all of which collectively make up **total energy expenditure (TEE)**. As illustrated in Figure 8.6, TEE has three main components: (1) basal metabolism, (2) physical activity, and (3) thermic effect of food. TEE also includes **adaptive thermogenesis** and **nonexercise activity thermogenesis (NEAT)**. Adaptive thermogenesis is a temporary change in energy expenditure that enables the body to adapt to such things as changes in the environment or to physiological conditions such as trauma, starvation, or stress. Shivering in response to cold is an example of adaptive thermogenesis. NEAT is energy expended for spontaneous movements such as fidgeting and maintaining posture. The contributions of adaptive thermogenesis and NEAT to TEE have yet to be determined, but they are likely small. The three main components of TEE are discussed next.

BASAL METABOLISM ACCOUNTS FOR MOST OF TEE

Basal metabolism is the energy (kcal) expended to sustain basic, involuntary life functions such as respiration, beating of the heart, nerve function, and muscle tone. **Basal metabolic rate (BMR)**, defined as the amount of

total energy expenditure (TEE) Total energy expended or used by the body.

adaptive thermogenesis Energy expended in response to changes in the environment or to physiological conditions.

nonexercise activity thermogenesis (NEAT) Energy expended for spontaneous movement such as fidgeting and maintaining posture.

basal metabolism Energy expended to sustain metabolic activities related to basic vital body functions such as respiration, muscle tone, and nerve function.

basal metabolic rate (BMR) Energy expended to support basal metabolism (expressed as kcal/hour).

FIGURE 8.6 Major Components of Total Energy Expenditure (TEE)

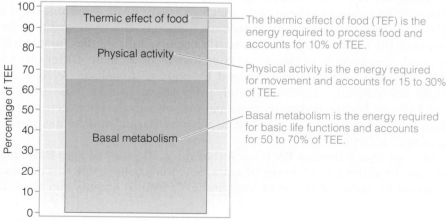

The thermic effect of food (TEF) is the energy required to process food and accounts for 10% of TEE.

Physical activity is the energy required for movement and accounts for 15 to 30% of TEE.

Basal metabolism is the energy required for basic life functions and accounts for 50 to 70% of TEE.

SOURCE: Adapted from: Lowell BB, Spiegelman. Towards a molecular understanding of adaptive thermogenesis. Nature. 2000;404:652–60.

energy expended per hour (kcal/hour) for these functions, accounts for most of TEE—approximately 50 to 70%. **Basal energy expenditure (BEE)** is basal metabolism expressed over a 24-hour period (kcal/day).

BMR is measured in such a way that energy expenditure associated with the processing of food and physical activity is eliminated. To accomplish this, BMR is typically measured in the morning, after eight hours of sleep, in a temperature-controlled room, and in a fasting state. These stringent conditions make measuring BMR difficult, which is why clinicians often measure **resting metabolic rate (RMR)** instead. Measuring RMR is not as difficult as measuring BMR, requiring only a brief resting period and no fasting. Because of this, RMR is approximately 10% higher than BMR. When resting metabolism is expressed over a 24-hour period, the term **resting energy expenditure (REE)** is used.

Using the Harris-Benedict Equation to Estimate Resting Energy Expenditure (REE) The Harris-Benedict equation is a mathematical formula developed almost 100 years ago and is still used by clinicians to estimate a person's REE. This equation is based on sex, age, height, and weight. It is calculated as follows.

Males: $REE = 66.5 + [13.8 \times weight\ (kg)] + [5 \times height\ (cm)] - [6.8 \times age\ (y)]$

Females: $REE = 655.1 + [9.6 \times weight\ (kg)] + [1.8 \times height\ (cm)] - [4.7 \times age\ (y)]$

The following sample calculation provides an example of how the Harris-Benedict equation is used to estimate REE. Consider a 200 lb (91 kg),* 20-year-old male, who is 6 feet, 3 inches (75 inches, 190.5 cm)[†] tall. Using the Harris-Benedict equation for males, he would require about 2,140 kcal/day to maintain basic body functions. Note that this value is just an estimate and does not take into account effects of physical fitness or body composition on energy expenditure.

Example of Resting Energy Expenditure (REE) Calculation

$REE = 66.5 + [13.8 \times weight\ (kg)] + [5 \times height\ (cm)] - [6.8 \times age\ (y)]$

$REE = 66.5 + 13.8(91) + 5(190.5) - 6.8(20)$

$REE = 66.5 + 1255.8 + 952.5 - 136$

$REE = 2,138.8\ kcal/day$

Factors Influencing Basal Metabolic Rate (BMR) The major factors influencing BMR include body composition, age, sex, nutritional status, and genetics.[22] The impact of these and other factors on BMR is summarized in Table 8.2. Tall, thin people tend to have higher BMRs than short, stocky people of the same weight. This is partly because tall, slender people have more surface area, resulting in greater loss of body heat. Differences in BMR can also be partially explained by body composition. Weight and height being equal, people with high proportions of lean mass (muscle) tend to have higher BMRs than do people with more fat mass (adipose tissue). This is because, compared to adipose tissue, muscle has greater metabolic activity.

Age can also influence BMR such that after 30 years of age, BMR may decrease by about 2 to 5% every 10 years. Scientists believe that this decrease is caused by age-related loss of lean tissue.[23] Physical activity can slow the

basal energy expenditure (BEE) Energy expended for basal metabolism over a 24-hour period.

resting metabolic rate (RMR) A measure of energy expenditure assessed under less stringent conditions than is BMR.

resting energy expenditure (REE) Energy expended for resting metabolism over a 24-hour period.

* To convert weight in pounds (lb) to kilograms (kg), divide weight (lb) by 2.2.
† To convert height in inches (in) to centimeters (cm), multiple the height in inches by 2.54.

TABLE 8.2 Factors Affecting Basal Metabolic Rate (BMR)

Factor	Effect on BMR
Age	After physical maturity, BMR decreases with age.
Sex	Males have higher BMR than do females of equal size and weight.
Growth	After adjusting for body size, BMR is higher during periods of growth.
Body weight	BMR increases with increasing body weight.
Body shape	Tall, thin people have higher BMRs than do short, stocky people of equal weight.
Body composition	Because muscle requires more energy to maintain than does adipose tissue, people with more lean tissue have higher BMRs than do people of equal weight with more adipose tissue.
Body temperature	Increased body temperature causes a transient increase in BMR.
Stress	Stress increases BMR.
Thyroid function	Elevated levels of thyroid hormones increase BMR, whereas low levels decrease BMR.
Energy restriction	Loss of body tissue associated with fasting and starvation decreases BMR.
Pregnancy	BMR increases during pregnancy.
Lactation	Milk synthesis increases BMR.

rate of muscle loss associated with aging and thus minimize the decline in BMR. Sex-related differences in BMR are also attributed to body composition. Women tend to have lower BMRs than men partly because they usually are smaller in size and have lower amounts of lean mass. However, when these differences are taken into account, average BMR in males still remains slightly higher than in females. This is likely due to hormonal differences between men and women.

Other factors affect BMR as well. For example, pound for pound, infants have higher BMRs than do adults. This is because infants are growing. BMR also increases during pregnancy and lactation. Fever and stress can cause a transient increase in BMR. Thyroid function also affects BMR such that thyroid overactivity—called hyperthyroidism—causes BMR to increase. Likewise, thyroid underactivity—hypothyroidism—causes BMR to decrease.

Perhaps the most striking factor that influences BMR is food restriction or dieting.[24] Severe energy restriction over time can decrease BMR because of the loss of lean body tissue. This energy-sparing response to negative energy balance protects the body's energy reserve and is an important survival mechanism when food is scarce. As caloric restriction continues, some dieters find further weight loss difficult, and in some cases experience weight gain. Weight regain that follows weight loss is called **rebound weight gain.**

PHYSICAL ACTIVITY IS THE SECOND-LARGEST COMPONENT OF TEE

After BMR, energy expended to support physical activity is the second-largest component of TEE. The amount of energy required for physical activity is quite variable, accounting for 15 to 30% of TEE.[25] Sedentary people are at the lower end of this estimate, whereas physically active people are at the upper end. Some elite athletes may require as much as 2,000 to 3,000 extra kilocalories each day to support the demands of physical activity.

Many factors affect the amount of energy expended for physical activity. Certainly, rigorous activities such as biking, swimming, and running have higher energy costs than less demanding activities. Body size also affects

rebound weight gain Weight regain that often follows successful weight loss.

The amount of energy required for physical activity is quite variable. Some elite athletes may require as many as 2,000 to 3,000 extra kilocalories each day to support the demands of physical activity.

energy expended for physical activity. Larger people have more body mass to move than smaller people and therefore expend more energy to accomplish the same activity.

THERMIC EFFECT OF FOOD (TEF) IS A MINOR COMPONENT OF TEE

Another component of TEE is the **thermic effect of food (TEF)**: the energy expended to digest, absorb, transport, metabolize, and store nutrients following a meal. In other words, it is the metabolic cost associated with processing food for utilization or storage.

The amount of energy associated with TEF depends on the amount of food consumed and the types of nutrients present in food. Because more energy is needed to process large amounts of food, TEF increases as food consumption increases. Some nutrients require more energy to process than others. High-protein foods have the highest TEF, and high-fat foods have the lowest. These differences reflect the metabolic cost associated with processing different nutrients. Because meals generally supply a mixture of nutrients, TEF is estimated to be about 5 to 10% of total energy intake.[26] For example, after consuming a 500-kcal meal, a person typically expends 25 to 50 kcal as TEF. In some ways, TEF is like having a "caloric sales tax" on energy intake.

METHODS OF ASSESSING TOTAL ENERGY EXPENDITURE (TEE)

An accurate assessment of TEE can be difficult because it requires expensive equipment and a high degree of expertise. However, simpler methods yield a good approximation of TEE. The techniques for assessing TEE include direct and indirect calorimetry, use of stable isotopes, and mathematical equations.

DIRECT AND INDIRECT CALORIMETRY USED TO ESTIMATE TEE

Total energy expenditure (TEE) can be estimated by assessing heat loss from the body. This is because much of the energy (calories) the body uses is eventually lost to our environment as heat. This method, called **direct calorimetry,** requires specialized and very expensive equipment, so it is not often used. An alternative to direct calorimetry is **indirect calorimetry.** Rather than measuring heat loss, indirect calorimetry measures the exchange of respiratory gases—oxygen intake and carbon dioxide output. The use of indirect calorimetry to estimate TEE is based on the assumption that the body uses 1 liter of oxygen to metabolize 4.8 kcal of energy-containing compounds (that is, glucose, amino acids, and fatty acids). Thus, measuring oxygen consumption and carbon dioxide production permits an estimate of TEE. Indirect calorimetry is often used in clinical settings to determine caloric requirements for patients.

thermic effect of food (TEF) Energy expended for the digestion, absorption, and metabolism of nutrients.

direct calorimetry A measurement of energy expenditure obtained by assessing heat loss.

indirect calorimetry A measurement of energy expenditure obtained by assessing oxygen consumption and carbon dioxide production.

stable isotope A form of an element that contains additional neutrons.

Indirect calorimetry measures the exchange of respiratory gases—oxygen consumption and carbon dioxide production—using portable equipment. These measures are used to estimate total energy expenditure (TEE).

STABLE ISOTOPES CAN BE USED TO ESTIMATE TEE

Another way to estimate TEE involves the use of **stable isotopes**—nonradioactive forms of certain elements. Isotopes have extra neutrons in their nuclei, making them heavier than the more common forms. Because stable isotopes of hydrogen (^{2}H) and oxygen (^{18}O) are chemically distinct from normal hydrogen (^{1}H) and oxygen (^{16}O), they can be measured in body fluids and expired air.

One technique that uses stable isotopes to estimate TEE is the **doubly labeled water** method. This method requires a person to drink two forms of water that have been labeled with stable isotopes—^{2}H$_2$O and H$_2$^{18}O. TEE can be estimated by measuring the elimination of oxygen and hydrogen isotopes from the body as water and carbon dioxide. This technique is considered the "gold standard" for estimating TEE.

TEE CAN BE ESTIMATED USING MATHEMATICAL FORMULAS

Because the equipment and expertise needed to use indirect calorimetry and stable isotopes are not readily available, clinicians commonly use relatively simple mathematical formulas to estimate TEE. Mathematical formulas developed by the Institute of Medicine as part of the Dietary Reference Intakes (DRIs)[27] are used to calculate what is called Estimated Energy Requirements (EERs), and take into account lifestages that include periods of growth. Formulas have been developed to calculate EERs for infants, children, adolescents, pregnant women, lactating women, and adults. Note that the EERs are intended to help adults maintain a healthy body weight. Once they are determined, adjustments in energy intake and physical activity can be made to support weight loss or gain. The following formulas are used to calculate EER for adult males and females. In this equation, PA refers to physical activity, which is categorized as sedentary, low, active, or very active. You can review this concept in Chapter 2.

$$\text{Males: EER} = 662 - [9.53 \times \text{age (y)}] + \text{PA} \times [15.91 \times \text{weight (kg)} + 539.6 \times \text{height (m)}]$$

$$\text{Females: EER} = 354 - [6.91 \times \text{age (y)}] + \text{PA} \times [9.36 \times \text{weight (kg)} + 726 \times \text{height (m)}]$$

You now understand how the balance of energy intake and energy expenditure affects body weight. Next we look at how body weight and body composition (the balance of fat and lean tissue) affect health.

How Are Body Weight and Body Composition Assessed?

Trying to keep energy intake and energy expenditure equal is key to maintaining a stable body weight. For many people, this can be difficult, resulting in unwanted weight gain. But at what point does added weight gain become unhealthy? And where do we draw the line between a few extra pounds and a serious health concern? After all, what appears to be a healthy weight for one person may cause health problems for another. To answer these questions, it is important to first understand how body weight and composition are defined, measured, and assessed.

BEING OVERWEIGHT MEANS HAVING EXCESS WEIGHT; BEING OBESE MEANS HAVING EXCESS FAT

Although the terms overweight and obese are often used interchangeably, they have very different meanings. Being **overweight** refers to having excess weight for a given height, whereas being **obese** refers to having an abundance

doubly labeled water Water that contains stable isotopes of hydrogen and oxygen atoms.

overweight Having excess weight for a given height.

obese Having excess body fat.

of body fat. Because simply knowing a person's weight does not provide information about the different components of the body, it is possible for muscular people, such as athletes, to be considered overweight but not obese. Conversely, some inactive people may not be considered overweight yet may still be obese. Most people who are overweight are obese as well, because weight gain in adults is generally caused by an increase in adipose tissue, rather than muscle. For this reason, body weight is often used as an indirect indicator of obesity.

TABLES ARE A GUIDE TO ASSESSING BODY WEIGHT

Because there is substantial variation in body weight for any given height, defining an "ideal body weight" may not be possible. Consequently, recommended body weights are simply reference values and not necessarily ideal for all people. The reference standards most commonly used to assess body weight are height–weight tables and body mass index (BMI).

Height–Weight Tables To determine the relationship between body weight and life expectancy, life insurance companies analyzed data from thousands of people. Weights associated with the longest life expectancies were deemed "desirable" and the companies published tables that listed "ideal weight ranges" for adult males and females. Because height–weight tables reflect a rather narrow segment of the population, they do not provide a good indication of whether a person's weight is within a healthy range.

Body Mass Index Even though height–weight tables are often used to assess body weight, body mass index (BMI) is a more widely used measure. BMI is based on the ratio of weight to height and is calculated using either of the following formulas.

$$BMI = [\text{weight (kg)}] / [\text{height (m)}]^2$$

$$BMI = [\text{weight (lb)}] / [\text{height (in)}]^2 \times 703.1$$

Based on these formulas, BMI for a person who weighs 150 lb (68.2 kg) and is 5 feet, 5 inches (65 inches, 1.65 m) tall is 25 kg/m². You can use the chart on the inside back cover of your book to determine your BMI. People with low BMIs typically have low amounts of body fat, whereas those with high BMIs tend to have higher amounts of body fat. Because each BMI unit represents 6 to 8 pounds for a given height, an increase in just 2 BMI units represents a 12- to 16-pound increase in body weight.

⟨CONNECTIONS⟩ Recall that *mortality* refers to death and *morbidity* refers to illness (Chapter 1, pages 22, 23).

Cutoff values for BMI classifications (that is, underweight, healthy weight, overweight, and obese) are based on the association between BMI and weight-related mortality and morbidity. Figure 8.7 shows that higher BMI values are strongly associated with increased risk of death attributed to weight-related health problems.[28] It is estimated that more than 80% of weight-related deaths in the United States occur among individuals with a BMI greater than 30 kg/m². Most medical organizations, including the US Centers for Disease Control and Prevention (CDC), use the following criteria to assess body weight in adults based on BMI.

- Underweight: <18.5 kg/m²
- Healthy weight: 18.5–24.9 kg/m²
- Overweight: 25.0–29.9 kg/m²
- Obese: ≥30 kg/m²

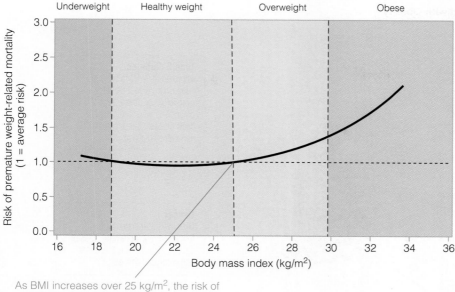

As BMI increases over 25 kg/m², the risk of weight-related mortality increases.

FIGURE 8.7 Body Mass Index (BMI) and Weight-Related Mortality The relationship between BMI and health is "J-shaped," meaning that risk of poor health outcomes increases at both low and high BMI values.

SOURCE: Adapted from: Kivimäki M, Ferrie JE, Batty GD, Smith ME, Marmot MF, Shipley MJ. Optimal form of operationalizing BMI in relation to all-cause and cause-specific mortalilty: The original Whitehall study. Obesity. 2008:16:1926–32.

CLINICIANS USE SEVERAL TECHNIQUES TO ASSESS BODY COMPOSITION

Using BMI to interpret a person's weight for height can be informative. However, health professionals sometimes want to know a person's actual body composition values. The body can be thought of as having two main compartments: (1) the fat compartment and (2) the fat-free (lean) compartment. The fat compartment (adipose tissue) consists mostly of stored triglyceride and supporting structures, whereas the fat-free compartment is mostly muscle, water, and bone. How much fat and fat-free mass a person has is determined by many factors, including sex, genetics, physical activity, hormones, and diet.

The amount of fat stored in the body changes throughout the life cycle. For example, nearly 30% of total body weight in a healthy six-month-old infant is fat, whereas this percentage may be cause for concern in adults. According to some experts, it is recommended that body fat levels be around 12 to 20% and 20 to 30% of total body weight in adult males and females, respectively (Table 8.3). Body fat over 25% in males and over 33% in females indicates obesity.[29] As women approach menopause, many experience an increase in body fat. This may, in part, be related to hormonal changes, but it could also reflect a decline in physical activity and energy expenditure.

TABLE 8.3 Percent Body Fat Classifications

Classification	Body Fat (% total body weight)	
	Males	Females
Normal	12–20	20–30
Borderline obese	21–25	31–33
Obese	>25	>33

SOURCE: Bray G. What is the ideal body weight? Journal of Nutritional Biochemistry. 1998;9:489–92.

FIGURE 8.8 Health Problems Associated with Obesity

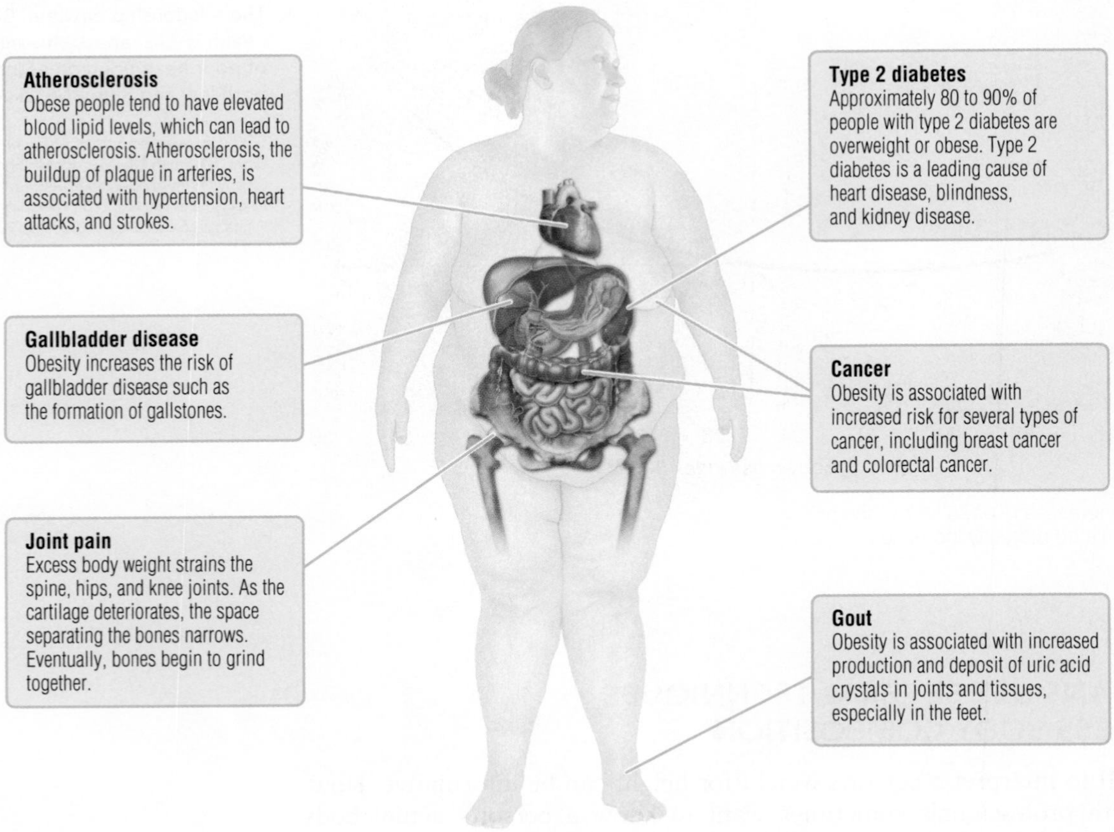

Atherosclerosis
Obese people tend to have elevated blood lipid levels, which can lead to atherosclerosis. Atherosclerosis, the buildup of plaque in arteries, is associated with hypertension, heart attacks, and strokes.

Gallbladder disease
Obesity increases the risk of gallbladder disease such as the formation of gallstones.

Joint pain
Excess body weight strains the spine, hips, and knee joints. As the cartilage deteriorates, the space separating the bones narrows. Eventually, bones begin to grind together.

Type 2 diabetes
Approximately 80 to 90% of people with type 2 diabetes are overweight or obese. Type 2 diabetes is a leading cause of heart disease, blindness, and kidney disease.

Cancer
Obesity is associated with increased risk for several types of cancer, including breast cancer and colorectal cancer.

Gout
Obesity is associated with increased production and deposit of uric acid crystals in joints and tissues, especially in the feet.

Just as too much body fat can cause health problems, too little body fat can also be detrimental. Body fat below 5% for males and 12% for females is considered too low. In women, low body fat can lead to a wide range of health problems such as bone loss and impaired fertility. Low body fat levels in women can be a consequence of disordered eating patterns. In some female athletes, long hours of training and extensive workouts can also cause body fat levels to become dangerously low.

Obesity increases risk for a variety of health problems, including type 2 diabetes, sleep irregularities, joint pain, stroke, heart disease, gallstones, and certain cancers (Figure 8.8).[30] Because of this, clinicians use a variety of methods to estimate body fat. Whereas some techniques are expensive and quite complex, others are more readily available and easy to use. Methods used to estimate body composition are summarized in Figure 8.9 and discussed next.

Densitometry Underwater weighing, or **hydrostatic weighing**, is a form of densitometry, which simply means measuring the density of something. Hydrostatic weighing involves measuring a person's weight both in and out of water. The more fat a person has, the less dense he or she is, and the less he or she weighs underwater. Hydrostatic weighing is not routinely used to estimate body composition. Not only does it require special equipment, but it is not very practical or convenient.

Dual-Energy X-Ray Absorptiometry **Dual-energy X-ray absorptiometry (DEXA)** is used to estimate total body fat, distribution of fat among different areas of

hydrostatic weighing (underwater weighing) Method for estimating body composition that compares weight on land to weight underwater.

dual-energy X-ray absorptiometry (DEXA) A method used to assess body composition by passing X-ray beams through the body.

FIGURE 8.9 Methods Used to Estimate Body Composition

Hydrostatic weighing

Hydrostatic weighing requires the subject to exhale air from the lungs and then be submerged in water. It is important to remain motionless while weight underwater is measured.

Dual-energy X-ray absorptiometry (DEXA)

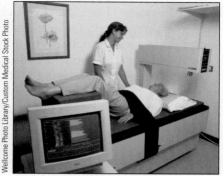

While the person lies on a table, a scanning device passes over the body. The X-ray beams emitted differentiate between fat mass, lean mass, and skeletal mass. A two-dimensional image of the body is displayed on a computer screen.

Bioelectrical impedance

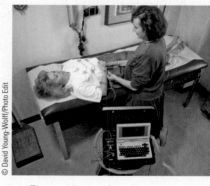

Electrodes are placed on a person's hand and foot, and a weak electrical current is passed through the body. The conductivity of the current is measured, which provides an estimate of the fat mass and fat-free mass.

Skinfold thickness

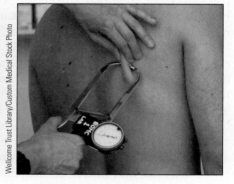

Using a measuring device called a caliper, the heath care worker measures the thickness of a fold of skin with its underlying layer of fat at precise locations on the body.

the body, lean body mass, and bone mass. During a DEXA measurement, a person must lie still while a scanning device passes over his or her body. The X-ray beams emitted differentiate between fat and fat-free mass. DEXA is considered the "gold standard" of body composition analysis and is often used by researchers and clinicians.

Bioelectrical Impedance **Bioelectrical impedance** is a measure of electrical conductivity—in other words, how easily a weak electric current travels through the body. Electrodes are placed on a person's hand and foot, and a weak electric current is emitted. Lean tissue, which contains a great deal of water, conducts electric currents better than adipose tissue, which has little water associated with it. The accuracy of bioelectrical impedance can be affected by hydration status—estimates of body fat will be incorrectly high if a person is dehydrated. This technique is relatively accurate, simple to use, and is considered an acceptable method for estimating body composition in clinical settings. Bathroom scales with built-in bioelectrical impedance

bioelectrical impedance A method used to assess body composition based on measuring the body's electrical conductivity.

systems are now available for home use, although little is known about their accuracy.

Skinfold Thickness The skinfold thickness method has been used for many years to estimate body fat. A measuring device called a **skinfold caliper** is used to measure the thickness of the subcutaneous fat. The examiner gently pinches the skin at various anatomical sites on the body. The double layer of skin includes the underlying fat tissue, but not the muscle. Body fat is estimated using mathematical formulas based on the skinfold thickness measures. Although the accuracy of this method depends on a person's ability to use a caliper, it is a fairly accurate and inexpensive way to assess body fat.

BODY FAT DISTRIBUTION AFFECTS HEALTH

⟨CONNECTIONS⟩ Recall that lipolysis is the breakdown of triglyceride molecules into glycerol and fatty acids (Chapter 6, page 230).

The distribution of body fat holds important information about health. The term **central obesity** (or central adiposity) describes the accumulation of adipose tissue predominantly in the abdominal region, or what is referred to as visceral adipose tissue (VAT). Because central obesity typically results in an increase in waist circumference, people with this body fat distribution pattern are commonly described as being "apple-shaped." In contrast, people with a predominance of subcutaneous adipose tissue (SCAT) in lower parts of the body are typically referred to as "pear-shaped." In general, males tend to be more "apple-shaped" and women tend to be more "pear-shaped."

Although the reasons are unclear, people with central obesity are at increased risk for developing weight-related health problems (such as type 2 diabetes, hypertension, and atherosclerosis).[31] However, there is some evidence that intra-abdominal fat is more likely to undergo lipolysis, compared with adipose tissue stored elsewhere in the body.[32] This can result in elevated blood levels of LDL cholesterol, triglycerides, glucose, and insulin and lower HDL cholesterol levels. Collectively, these signs and symptoms are associated with a condition called **metabolic syndrome,** which increases risk for developing type 2 diabetes and cardiovascular disease.

Reproductive hormones such as estrogen and testosterone influence where body fat is stored. Testosterone (a male hormone) encourages VAT, whereas estrogen (a female hormone) favors SCAT. As women age, declining estrogen levels can cause body fat to shift toward an accumulation of VAT.[33] Approximately 20% of total body fat in men is VAT, whereas it accounts for 5 to 8% of total body fat in women.

Waist Circumference Reflects Central Adiposity Evaluating your body fat distribution pattern is easier than you might think. A large **waist circumference** is a good indicator of central adiposity. This measurement is easily obtained by placing a tape measure around the narrowest area of the waist. However, a health professional is likely to use more precise anatomical landmarks as shown in Figure 8.10. A waist circumference of 40 inches or greater in males and 35 inches or greater in females indicates central adiposity.[34] Although age, ethnicity, and BMI influence the relationship between waist circumference and health outcomes, studies suggest that cardiovascular risk may be higher in those with the highest waist circumference.[35] Some clinicians prefer to use the ratio of waist circumference to hip circumference (waist-to-hip ratio) to determine body fat distribution. However, studies show that waist circumference alone provides a simple yet effective measure of central adiposity.[36]

You are now familiar with how weight and obesity can be assessed. But to truly understand what causes obesity, it is important to consider factors that influence eating habits.

skinfold caliper An instrument used to measure the thickness of subcutaneous fat.

central obesity Accumulation of body fat within the abdominal cavity.

metabolic syndrome Condition characterized by an unhealthy metabolic profile, abdominal body fat, and insulin resistance, that increases risk for developing type 2 diabetes and cardiovascular disease.

waist circumference A measure used as an indicator of central adiposity.

FIGURE 8.10 Waist Circumference Is Used to Assess Central Adiposity Central adiposity increases a person's risk for weight-related health problems. A large waist circumference is an important indicator of central adiposity.

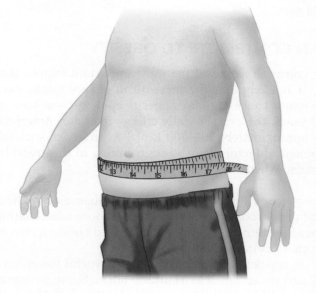

- Person should stand with their feet 6 to 7 inches apart, with their weight evenly distributed across both legs.

- Person should be relaxed, and the measurement should be taken while person is breathing out.

- The tape should be positioned midway between the top of the hip bone and the bottom of the rib cage.

- The tape should be loose enough so that there is one finger's width between the tape and the body.

- A waist circumference of 40 inches or more in males and 35 inches or more in females indicates central adiposity.

How Does Lifestyle Contribute to Obesity?

When it comes to our bodies, we tend to be our own worst critics. While we might wish that our bodies looked or were shaped differently, it is important to recognize that the issue of obesity is of concern not because of societal norms but rather because excess weight can be injurious to our health. In the late 19th century, only 3% of adults in the United States were overweight and very few were obese. Today, nearly 35% of adults are classified as obese (Table 8.4). Recent estimates by the Centers for Disease Control and Prevention estimate that the prevalence of obese adults exceeds 25% in nearly one-third of the U.S.[37] Similar trends are evident in children and adolescents. In the past 20 years, the number of overweight American children has doubled, and the number of

TABLE 8.4 Estimated Prevalence of Obesity during the Past 40 Years in the United States

| Age (years) | (% of U.S. population) | |
	1971–1974	2005–2008
2–5	...*	11
6–11	4	17
12–19	6	18
20–74	15	35

*Reliable national data not available.

SOURCE: National Center for Health Statistics. Health, United States, 2010: With special feature on death and dying. Hyattsville, MD. 2011.

⟨**CONNECTIONS**⟩ The National Health and Nutrition Examination Survey (NHANES) is an ongoing study to monitor nutrition and health in the U.S. population (Chapter 1, page 16).

overweight adolescents has tripled. An overview of obesity trends in the United States is presented in Table 8.4. These alarmingly high rates of obesity are cause for concern. Obesity is not only on the rise in the United States but around the world as well.[38] What is causing this global epidemic? To answer this question, it is important to consider lifestyle-related factors that cause us to consume more calories and/or to expend less energy, behaviors that together shift energy balance in favor of weight gain.

EATING HABITS CAN CONTRIBUTE TO OBESITY

The amount of food we consume is one of the most important factors that influences body weight. It is likely that the increase in energy intake over the last few decades is a major cause of today's obesity epidemic—at least in adults. Over the past 40 years, the average daily energy intake of American adults has increased by approximately 200–300 kcal.[39] A variety of societal, sociocultural, and psychological factors may be contributing to this trend by influencing what, how much, when, and where we eat.

Societal Influences on Eating Habits Food-related societal influences can affect what and how much food people consume. These include more adults eating away from home, increased portion sizes of foods, increased consumption of energy-dense foods, and changes in snacking habits.

In the United States, energy-dense, inexpensive, and flavorful foods have become readily available and are accepted as a cultural norm. Some estimates indicate that 37% of adults and 42% of children eat fast food daily.[40] Although consuming fast food is not the sole cause of obesity, it may certainly contribute. Many fast-food meals are high in fat, refined starchy carbohydrates, and calories. As such, a super-sized "value" meal can provide more than half the calories required in a day. In response to significant public pressure, many fast-food restaurants also offer healthy food choices such as salads and sandwiches made with lean meats and whole-grain breads. Selecting these menu options is one way for individuals and families to eat more healthily.

Studies also show that serving size influences food consumption. That is, when larger food portions are served, people tend to eat more.[41] For example, participants in a study ate 39% more M&Ms® when they were given a two-pound bag than when they received a one-pound bag.[42] It appears some people depend more on visual cues to judge how much to eat than on physiological cues such as hunger and satiety. This is particularly important in light of the recent trend toward "super-sized" food portions.

Factors related to food itself can also influence what and how much food a person consumes. People tend to consume larger quantities of tastier foods than bland-tasting foods.[43] In other words, adding extra items such as butter and sour cream to baked potatoes is more likely to lead to overconsumption than eating plain baked potatoes. Similarly, food variety has also been shown to influence food intake. When presented with a variety of food choices, people tend to eat more than when they are offered single foods.[44] Scientists have also demonstrated that when people consume or perceive they have consumed fewer calories at one meal, they tend to "reward" themselves by eating more

When people are presented with a variety of food choices, they tend to eat more than when they are offered fewer choices. Thus, buffet-style eating may be more conducive to overeating than eating foods chosen from a more limited menu.

A "value meal" consisting of a double cheeseburger (760 kcal), medium french fries (360 kcal), and a medium soft drink (230 kcal) provides more than half of a person's daily energy requirements.

A regular meal consisting of a hamburger (200 kcal), small french fries (230 kcal), and a small soft drink (140 kcal) has less than half the calories of a "value meal."

at subsequent meals.[45] This may, in part, explain why increased consumption of reduced-calorie foods and beverages does not necessarily lead to a reduction in total calorie intake.

Sociocultural and Psychological Factors Affect Obesity Sociocultural factors such as cultural norms, economic status, marital status, and education can also influence our risk for obesity. Although these factors do not directly cause obesity, some may indirectly contribute to the problem. For example, obesity in the United States is most prevalent in Southern states and among certain racial and ethnic groups. For example, the obesity rates in adults between the ages of 20 and 74 years are 52% for black women, 44% for Mexican-American women, and 34% for non-Hispanic white women.[46]

Beyond sociodemographic factors, some psychological disorders are also related to risk for obesity. It is not clear if obesity predisposes individuals to these disorders or if some psychological profiles lead to obesity. All scientists can say for certain is that some personality factors appear to be associated with the risk for obesity. For example, obese individuals are more likely to experience clinical depression and panic attacks than nonobese people.[47] This may stem from the fact that obesity can lower a person's self-esteem and confidence. In addition, individuals who are impulsive and have difficulty coping with stress are likely to turn to food for emotional comfort and gratification, making them more susceptible to obesity. It also appears that weight gain can be an adaptive response to emotional trauma, providing emotional isolation and psychological protection from hurtful circumstances.[48]

New research suggests that a person's social network may also relate to his or her risk for obesity.[49] People with friends who have experienced weight gain are more likely to gain weight than those with weight-stable friends. It is not clear how social networks are linked to weight gain, but researchers believe that weight gain among close friends and family members might serve as a permissive cue for others to gain weight as well.

Critical Thinking: August's Story Think back to August's story about her experience being obese. Many factors influence a person's risk of becoming obese. How does August fit the profile of risk factors associated with obesity?

SEDENTARY LIFESTYLES CONTRIBUTE TO WEIGHT GAIN

Choosing what and how much to eat is not the only lifestyle choice related to obesity. At the same time we are eating more, Americans have also become less physically active. Fewer jobs that require physical work and an increase in labor-saving devices make daily life less physically demanding. To compound matters, almost 60% of adults fail to engage in any kind of leisure-time physical activity lasting 10 minutes or more in a given week.[50] Together, these changes over time have likely had a significant negative impact on the nation's health.

It is important to realize that physical activity is not limited to formal exercise. **Physical activity** is defined as any bodily movement produced by skeletal muscles that results in a substantial increase over resting expenditure of energy, whereas **exercise** comprises planned, structured, and repetitive activities that are done to improve or maintain physical fitness.[51] Physical activities such as such as walking up the stairs or mowing the lawn can be just as beneficial as formal exercise. Not surprisingly, the vast majority of studies show that lack of physical activity increases the risk of being overweight or obese.[52] For most people, walking an extra mile each day would increase energy expenditure by 100 kcal/day.[53] This is equivalent to taking about 2,000 to 2,500 extra steps each day. If all other factors remained constant, a person could lose 1 pound of body weight per month by making this simple change. In addition to helping us maintain a healthy body weight, physical activity and exercise also help us stay healthy and physically fit.

Physical Inactivity: A Growing Problem Physical inactivity has contributed to the growing rates of obesity, especially among children and adolescents. The long-term effect of childhood obesity on health is a growing concern. There is a direct correlation between regular physical activity and health among children and adolescents. The National Association for Sports and Physical Education recommends that schools provide 150 minutes per week of instructional physical education for elementary school children, and 225 minutes per week for middle and high school students.[54] Although the majority of states mandate physical education, few states meet these minimum standards.

Even though over 90% of high schools nationwide require school-based physical education classes, the majority of American youth still spend more time watching television, playing video games, or surfing the Internet than engaging in physical activities and sports. Studies show that overweight children spend more time doing screen-related activities than children who are not overweight.[55] It is likely that the more time children spend watching television and playing video/computer games, the less likely they are to be physically active. In addition, weight gain may also be due to greater exposure to food advertisements on children's television programming, which in turn may alter food choices.[56] The average child views about 40,000 commercials a year on television alone, exposing them to extensive advertising and marketing promotions for candy, sugar-sweetened cereal, soda, and fast food.[57]

Physical Activity Recommendations The American College of Sports Medicine (ACSM) recently released new physical activity guidelines for all healthy adults (18 to 64 years old).[58] Depending on physical fitness, most adults should engage in moderate- to high-intensity aerobic exercise three to five days a week and resistance training two to three days a week. By some measures, moderate-intensity physical activity is defined as physical activity that is done at 3.0 to 5.9 times the intensity of rest.[51] It is also recognized that physical activity above this minimum recommendation can provide

physical activity Bodily movement that uses skeletal muscles that results in a substantial increase in energy expenditure over resting energy expenditure (REE).

exercise Planned, structured activities done to improve or maintain physical fitness.

even greater health benefits. These physical activity guidelines for Americans are summarized below.[58]

- Children and adolescents (6–17 years): Engage in 60 minutes (1 hour) or more of physical activity daily. Most of this activity should be either moderate- or vigorous-intensity aerobic physical activity, and should include vigorous-intensity physical activity at least three days a week.
- Adults (18–64 years): Engage in at least 150 minutes (2 hours and 30 minutes) a week of moderate-intensity, or 75 minutes (1 hour and 15 minutes) a week of vigorous-intensity aerobic physical activity, or an equivalent combination of moderate- and vigorous-intensity aerobic activity. In addition to aerobic activities, moderate- or high-intensity muscle-strengthening activities that involve all major muscle groups should be performed two or more days per week.
- Older adults (65 years and older) should follow the adult guidelines. When older adults cannot meet the adult guidelines, they should be as physically active as their abilities and condition will allow.

Can Genetics Influence Body Weight?

Although there is general agreement that lifestyle factors—and not genetics—are the driving force behind the obesity epidemic, both likely influence body weight. In an environment where energy-dense foods are abundant and physical activity is low, certain genetic factors can make some people susceptible to weight gain. Even individuals not genetically predisposed to obesity are likely to gain weight in this type of environment. People have long suspected that genetics plays a role in influencing body weight. The good news is that, although genetics influences our susceptibility to obesity, lifestyle choices go a long way toward overriding these potentially negative influences.

IDENTICAL TWIN STUDIES HELP SCIENTISTS UNDERSTAND ROLE OF GENETICS

Adoption studies have helped researchers to distinguish between genetic and lifestyle influences in the development of obesity. Studies of genetically identical twins separated at birth suggest that genetic makeup directly influences body weight.[59] When body weights of children adopted at birth are compared with those of their adoptive parents, there is little similarity.[60] However, when compared to their biological parents, the similarity in weight is most striking. Scientists estimate that at least 50% of our risk for becoming overweight or obese is determined by genetics or epigenetics. Recall that epigenetic alterations in DNA can regulate gene expression—that is, which genes are switched on or off. Researchers believe that conditions in the womb can prompt epigenetic changes that later influence a person's body weight as well as that of future generations.[61]

⟨CONNECTIONS⟩ Recall that a gene is a section of DNA that contains hereditary information needed for cells to make a protein (Chapter 5, page 168).

DISCOVERY OF THE "OBESITY GENES" PROVIDES FIRST GENETIC MODEL OF OBESITY

Although scientists long believed that genetic makeup can influence body weight, direct evidence was lacking until the discovery of an obese mouse, named the *ob/ob* mouse. This mouse appeared to have a genetic mutation causing it to consume large amounts of food (in other words, become hyperphagic), be inactive, and therefore gain weight easily. Soon after the *ob/ob* mouse was discovered, another chance occurrence took place—the discovery

ob/ob mouse Obese mouse with mutations in genes that code for the hormone leptin.

A genetic mutation made the mouse on the left obese.

⟨CONNECTIONS⟩ Recall that a mutation is an alteration in a gene resulting in the synthesis of an altered protein (Chapter 5, page 174).

of a mouse that was both obese and diabetic. Using breeding records, researchers found that this second mouse, called the **db/db mouse,** was not genetically related to the *ob/ob* mouse.[62]

To better understand what might be causing these mice to become obese, researchers performed some classic scientific experiments (Figure 8.11).[63] Using a technique called parabiosis, the *ob/ob* mouse and the *db/db* mouse were surgically joined together so that blood circulated between them. Curiously, the *ob/ob* mouse lost weight, while the *db/db* mouse showed no weight change. When an *ob/ob* mouse was joined to a normal mouse, the *ob/ob* mouse lost weight, but the normal mouse experienced no weight change. Last, they joined a *db/db* mouse to a normal mouse. Although there was no weight change in the *db/db* mouse, the normal mouse lost weight. This experiment was the first step in a series of investigations that revealed an internal system of body weight regulation. However, how such a system worked remained elusive until 1994 when scientists discovered the first "obesity genes." These genes were called the *ob* gene and the *db* gene, after the mice from which they were discovered.

Obesity Genes Lead to the Discovery of Leptin Scientists soon learned the **ob gene** codes for a hormone called leptin.[64] **Leptin** is a potent satiety signal produced by adipose tissue, and the lack of this hormone in *ob/ob* mice causes them to eat uncontrollably. Soon thereafter, researchers discovered that the **db gene** codes for the leptin receptor, found primarily in the hypothalamus. Without a functioning leptin receptor, leptin cannot exert its effect. To determine if leptin played a role in regulating body weight, *ob/ob* and *db/db* mice were injected with leptin. When the leptin-deficient *ob/ob* mice received leptin, there was significant weight loss. However, there was no change in *db/db* mice in response to leptin injections. This is because *db/db* mice have defective leptin receptors, making them unresponsive to this hormone.

People hoped leptin—touted as the antiobesity hormone—would become the miracle cure for obesity. Yet the vast majority of obese people produce appropriate or even elevated amounts of leptin. Although rare, leptin deficiency has been reported in humans. However, these individuals have provided scientists with a unique opportunity to study the effects of leptin.[65] When the severely obese, leptin-deficient individuals received leptin injections, all experienced dramatic weight loss. Because leptin deficiency is an uncommon occurrence in humans, scientists have turned their investigation toward trying to better understand why some people appear unresponsive to this hormone. Although leptin has not proved to be effective for treating human obesity in most cases, its discovery has led to important insights into body weight regulation.

How Does the Body Regulate Energy Balance and Body Weight?

Although the discovery of leptin has not solved the obesity dilemma, it has profoundly deepened our understanding of body weight regulation. While much remains to be discovered about body weight regulation and possible defects in these key energy-regulating activities, it is clear that the body plays an active role in influencing how much energy is consumed, how much energy is expended, and how much energy the body stores. Physiological systems require homeostatic regulatory mechanisms that maintain "checks and balances" via hormonal signaling pathways. Whether body weight is regulated by such a system has long been the subject of much debate. However, the discovery of the obesity genes and the hormone leptin has helped scientists gain insights into how the body monitors and maintains energy stores.

db/db mouse Obese mouse with mutations in genes that code for the leptin receptor.

ob gene The gene that codes for the protein leptin.

leptin A hormone, produced mainly by adipose tissue, that helps regulate body weight.

db gene The gene that codes for the leptin receptor.

FIGURE 8.11 Parabiosis Experiments Using *ob/ob* and *db/db* Mice *ob/ob* mice lack the satiety signal leptin, whereas *db/db* mice produce large amounts of leptin but have defective leptin receptors.

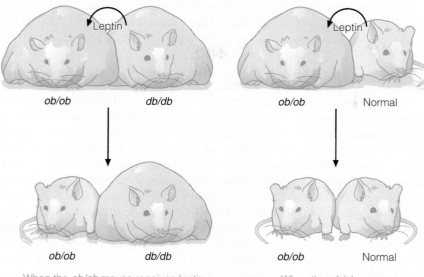

When the *ob/ob* mouse receives leptin from the *db/db* mouse, the ob/ob mouse loses weight. There is no weight change in the *db/db* mouse, because it is unable to respond to leptin.

When the *ob/ob* mouse receives leptin from the normal mouse, the *ob/ob* mouse loses weight. There is no weight change in the normal mouse.

When the normal mouse receives large amounts of leptin from the *db/db* mouse, the normal mouse loses weight. There is no weight change in the *db/db* mouse, because it is unable to respond to leptin.

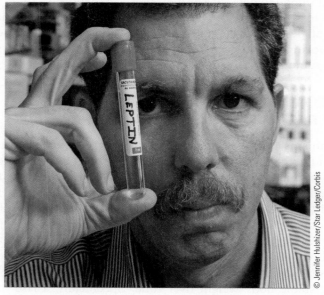

Jeffrey Friedman, shown here holding a test tube of leptin, discovered the "obesity hormone" in his laboratory at The Rockefeller University in the mid-1990s.

ADJUSTING ENERGY INTAKE AND ENERGY EXPENDITURE MAINTAINS ENERGY BALANCE

The ability of the body to adjust energy intake and energy expenditure on a long-term basis serves an important purpose—body weight regulation. If the body had no means of regulating energy balance, the consequences would be serious. Even a slight imbalance in daily energy intake and energy expenditure would result in a substantial weight gain or loss over time. To prevent such imbalances, long-term energy balance regulatory signals communicate the body's energy reserves to the brain, which in turn releases neurotransmitters that influence energy intake and/or energy expenditure.[66] If this long-term system functions effectively, body weight remains somewhat stable. This system, sometimes called the set point theory, is perhaps the most widely accepted theory of body weight regulation.

Set Point Theory of Body Weight Regulation Body weight is a function of the balance between energy intake and energy expenditure. However, increases or decreases in food intake do not always produce the expected change in body weight. Sometimes, a person can increase energy intake without gaining weight or reduce consumption and still not lose weight. For many years, scientists have suspected that a complex signaling system regulates body weight by making adjustments in energy intake and energy expenditure.[67] To test this theory, researchers observed weight-gain and weight-loss cycles in food-restricted mice. When food-restricted, the mice lost weight. Not surprisingly, when taken off food restriction, the mice increased their food consumption and soon returned to their original weight. However, what was surprising to scientists was the fact that after returning to their original weight, the mice maintained this weight by spontaneously decreasing their intake of food. This phenomenon was called the set point theory of body weight regulation.

Proponents of the **set point theory** of body weight regulation believe that hormones circulating in the blood (such as leptin) regulate body weight by communicating the amount of adipose tissue in the body to the brain. When the amount of adipose tissue increases beyond a "set point," a signal causes food intake to decrease and/or energy expenditure to increase, favoring weight loss. Conversely, when the amount of adipose tissue decreases below a "set point," food intake increases and/or energy expenditure decreases, favoring weight gain. In this way, body weight is restored to its "set point" and remains relatively stable on a long-term basis.

LEPTIN COMMUNICATES THE BODY'S ENERGY RESERVE TO THE BRAIN

The mechanisms that regulate long-term energy balance are complex and not well understood. However, the discovery of the *ob* gene and leptin provided the first real evidence that hormones produced in adipocytes can communicate adiposity to the brain, which in turn influences hunger and satiety (Figure 8.12). When body fat increases, the concentration of circulating leptin (produced primarily by adipose tissue) increases as well. Conversely, when body fat decreases, leptin production decreases. Thus, fluctuations in leptin levels reflect changes in the body's primary energy reserve.

Leptin is thought to be part of a communication loop that helps maintain a relatively stable body weight over time. Specifically, a rise in leptin concentrations in response to increased adiposity signals the brain to increase its release of catabolic neurotransmitters. Catabolic neurotransmitters help the body resist further weight gain by decreasing food intake and increasing energy expenditure. Conversely, decreased adiposity causes blood concentrations of leptin to decrease. When this occurs, the brain increases its release of anabolic neurotransmitters. The release of these neurotransmitters stimulates food intake and decreases energy expenditure, thus protecting the body against further weight loss. In this way, leptin is thought to be part of a long-term homeostatic system that helps prevent large shifts in body weight.

Defects in Leptin Signaling May Lead to Obesity Some researchers believe that defects in this leptin signaling system may lead to impaired body weight

set point theory A theory suggesting that hormones regulate body weight by making adjustments in energy intake and energy expenditure.

FIGURE 8.12 Leptin and Body Weight Regulation

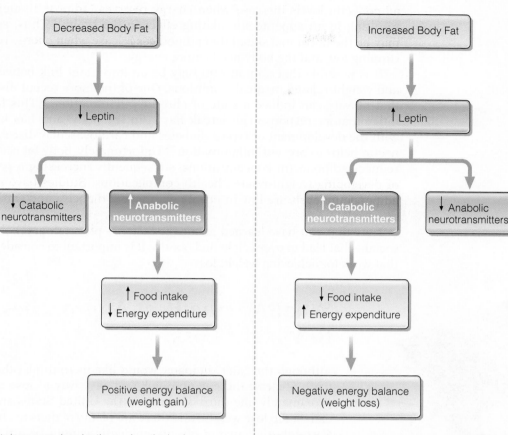

When body fat decreases, less leptin reaches the brain. Anabolic neurotransmitters are released, and catabolic neurotransmitters are suppressed. This condition favors increased food intake, decreased energy expenditure, and weight gain.

When body fat increases, more leptin reaches the brain. Anabolic neurotransmitters are suppressed, and catabolic neurotransmitters are released. This condition favors decreased food intake, increased energy expenditure, and weight loss.

regulation in some people.[68] In other words, the brain may not be able to recognize and respond to the appetite suppression signal that comes from leptin. In times when food is scarce, a decline in leptin may help the body conserve energy, increasing a person's chance of survival.[69] However, you may be wondering, "If leptin curbs food intake, why are so many people obese? If obese people have elevated leptin levels, why do they continue to gain weight?" These are very good questions. A number of researchers believe that some people may have disconnected leptin-signaling pathways.[70] In other words, the brains of some obese people may not be responsive to leptin's appetite suppression signal, regardless of how much leptin is being produced. Clearly, there is more to learn about this regulatory system. As scientists continue to study the role of leptin and other adipokines in long-term body weight homeostasis, we will gain a better understanding of the underlying mechanisms associated with body weight regulation.

ADIPONECTIN MAY PROVIDE A LINK BETWEEN OBESITY AND WEIGHT-RELATED DISEASES

Currently, leptin is the only hormone in humans that has been shown to be a long-term mediator of body weight regulation.[71] Other "obesity hormones" are being discovered on a regular basis. One such hormone is **adiponectin,**

adiponectin A hormone secreted by adipose tissue that appears to be involved in energy homeostasis; also appears to promote insulin sensitivity and suppress inflammation.

also produced by adipose tissue. Although its physiological role is not fully understood, adiponectin appears to have the opposite effect of leptin. In mice, adiponectin levels increase when energy reserves (adipose tissue) decrease, resulting in an appetite-stimulating effect in the brain.[72] Thus, adiponectin may be the hormonal signal that adipocytes release when energy reserves are running low and the body needs more energy.

It is possible that adiponectin may be an important link between obesity and weight-related medical conditions. One of the more recent discoveries is that obesity can induce a state of chronic inflammation.[73] This fat-induced inflammatory response can wreak havoc in the body, and has been linked with the development of type 2 diabetes and cardiovascular disease.[74] Adiponectin helps to prevent inflammation.[75] Unfortunately, body fat accumulation reduces adiponectin concentrations, subsequently increasing a person's risk of developing weight-related health complications. Studies have shown that adiponectin levels are low in people with type 2 diabetes and cardiovascular disease.[76]

Now that you have learned about the factors—physical, psychological, and social—that lead to overweight and obesity, it is important to consider strategies that work for achieving weight loss.

What Are the Best Approaches to Weight Loss?

Although the "diet" industry would like us to think otherwise, the truth is clear: there is no quick and easy way to lose weight. Approximately one-third of adults in the United States are on a special diet to lose weight, and there are plenty of diets to choose from. Certainly the current obesity epidemic cannot be attributed to a lack of weight-loss advice. In addition, a variety of nonprescription weight-loss products are aggressively marketed. Some of them can be hazardous to your health.

Most popular weight-loss plans have specific rules for people to follow. Some restrict the types of food that can be eaten, others recommend a strict exercise regimen, and still others acclaim periods of fasting or cleansing. The real issue that many weight loss plans fail to address is not what it takes to lose weight but what it takes to keep it off. Rather than succumbing to a fad diet or quick fix weight loss approach, consider what actual weight loss experts have to say.

HEALTHY FOOD CHOICES PROMOTE OVERALL HEALTH

Although there are many reasons why people want to lose weight, the most important is to improve health. Achieving and maintaining weight loss requires making lasting lifestyle changes, including changes to what we choose to eat and how much physical activity we engage in. Most health experts suggest that people focus less on weight loss and more on healthy eating and overall fitness. Misguided efforts toward weight loss at any cost present food as the enemy, rather than as a means to good health. This is unfortunate, because most people who successfully lose weight and keep it off do so by eating a balanced diet of nutrient-dense foods and by maintaining a moderately high level of physical activity.[77] Key recommendations for weight

The current obesity epidemic cannot be attributed to a lack of weight-loss advice. Amazon.com lists over 8,000 books related to weight-loss diets.

Peter Tobia/Philadelphia Inquirer/KRT/Newscom

⟨CONNECTIONS⟩ Recall that nutrient-dense foods are high in nutrients relative to the amount of calories (Chapter 2, page 52).

management based on the 2010 Dietary Guidelines for Americans are provided in the Food Matters that follows.

A healthy weight-loss and weight-maintenance program consists of three components: (1) setting reasonable goals, (2) choosing nutritious foods in moderation, and (3) increasing energy expenditure by daily physical activity.

Setting Reasonable Goals Setting reasonable and attainable goals is an important component of any successful weight-loss program. A realistic weight-loss goal is to reduce current weight by 5 to 10%. For someone weighing 180 lbs, this would amount to an initial weight loss of approximately 9 to 18 pounds. Studies show that even this modest reduction in weight can improve overall health.[78] When it comes to weight loss, slow and steady is advised and weight loss should not exceed one to two pounds a week. This can be achieved by decreasing energy intake by 100 to 200 kcal each day or by walking one to two miles each day. Over time, these small changes can result in significant reductions in body weight. Rather than making dramatic dietary changes, making small changes such as reducing portion sizes or cutting back on energy-dense snack foods can make a big difference in overall energy intake.

Once body weight stabilizes and the new lower weight is maintained for a few months, individuals can decide whether additional weight loss is needed. Some people benefit from joining weight management programs such as Weight Watchers® or Take Off Pounds Sensibly®. These types of programs provide long-term support and motivate people to maintain a healthy diet and lifestyle.

Choosing Nutritious Foods Weight-loss plans that drastically reduce calories and offer limited food choices leave people feeling hungry and dissatisfied. Instead, those that encourage people to eat foods that are healthy and appealing tend to be more successful. Contrary to popular belief, it is not necessary to avoid foods that contain fat in order to lose weight. The 2010 Dietary Guidelines for Americans recommend that we choose our fats as carefully as we choose our carbohydrates. In general, it is best to limit intake of foods containing *trans* fatty acids and saturated fatty acids. Instead, we should emphasize foods containing relatively more polyunsaturated and monounsaturated fatty acids.

A common misconception is that dairy products and meat are high-fat foods and therefore should be avoided when trying to lose weight. Again, the key to good nutrition is moderation and choosing wisely. For example, switching from whole to reduced-fat milk is one way to lower caloric intake without losing out on the many vitamins and minerals in dairy products. Likewise, the types of meat you choose and the methods by which the meat is prepared can greatly affect how many calories you consume. Lean meats prepared by broiling or grilling are both nutritious and satisfying.

Reducing energy intake is best achieved by eliminating or cutting back on energy-dense foods that have little nutritional value such as potato chips, cookies, and cakes. Aside from their lower energy densities, nutrient-dense foods such as whole grains, legumes, fruits, and vegetables offer many beneficial substances such as micronutrients and fiber. Furthermore, because these foods tend to have greater volume compared to more energy-dense foods, they help people feel more satisfied after they eat. In the Focus on Food feature, you can read about how eating high-volume, low-energy-dense foods can help people feel full and satisfied.

Healthy eating also requires people to pay attention to hunger and satiety cues. However, people often let the amount of food served or packaged,

Working Toward the Goal: Maintaining a Healthy Body Weight by Balancing Caloric Intake with Energy Expenditure

The 2010 Dietary Guidelines for Americans recommend that we maintain a healthy body weight by balancing caloric intake with energy expenditure. The following tips will help you maintain a healthy body weight and prevent gradual weight gain over time.

- Consume foods that are nutrient dense while limiting foods that are high in saturated fats, *trans* fats, refined carbohydrates, and added sugars.
- Select foods high in dietary fiber, such as fruit, vegetables, and whole grains.
- Follow an eating plan that emphasizes variety, moderation, and balance.
- Select low-fat dairy products and lean meats whenever possible.
- Pay attention to internal cues of hunger and satiety.
- Make food portions smaller than normal. If still hungry, eat a second portion.
- Avoid distractions while eating, such as watching television, which can lead to overconsumption of food.
- Alcohol is a source of calories (7 kcal/g), and it is therefore important to monitor your intake of alcoholic beverages.

- Eat regularly and avoid skipping meals, which can lead to extreme hunger.
- Avoid eating food directly from containers or packages. Pay attention to food portions by serving food on a plate or in a bowl.
- Many restaurants serve generous food portions—enough for two people. Try sharing a meal or take the uneaten portion of the meal home.
- Be aware of beverages that contain high-calorie ingredients such as sugar, high fructose corn syrup, cream, and syrups.
- Increase your physical activity by using the stairs instead of elevators, walking instead of driving, and parking so as to walk farther. If you have children, include them in fun family activities such as hiking, riding bicycles, and swimming.
- Join a fitness center or take a fitness class.
- Take up a hobby, such as gardening, that is both rewarding and physically challenging.
- Make exercise interesting by changing your routine or by doing it with a friend.

rather than hunger and satiety, determine how much they eat. That is, visual cues, rather than internal cues, have a greater influence on the quantity of food we consume. For instance, some commercially made muffins are extremely large, containing as many calories as eight slices of bread. Learning to recognize and choose reasonable portions of food is a critical component to successful weight management. In fact, reducing portion sizes by as little as 10 to 15% can reduce daily energy intake by as much as 300 kcal. One way to limit serving size is to consider sharing a large meal the next time you eat at a restaurant.

Physical Activity In addition to eating less, people can also help tip energy balance toward negative energy balance by engaging in physical activity. Walking one mile a day, which takes most people about 15 to 20 minutes, uses about 100 kcal. This adds up to 700 kcal per week. Being physically active throughout the day can make a big difference in promoting weight loss and weight maintenance. When possible, take the stairs rather than the elevator, walk or bike rather than drive, and incorporate chores into your routine daily activities. Even without weight loss, individuals who are physically active show improved physical fitness. Regardless of one's weight, a lack of exercise may prove the greatest health hazard of all.

It is important to realize that overweight people can still be physically fit, and healthy-weight people can be physically unfit. In fact, studies show that obese individuals who are physically fit have fewer health problems than do healthy-weight individuals who are unfit.[79] Normal blood pressure, normal blood glucose regulation, and healthy levels of blood lipids are important indicators of physical fitness.[80] Exercise also promotes a positive self-image and helps people take charge of their lives. Although people often resist the idea of starting an exercise program, they rarely regret it once they begin. Physical activity is an effective strategy for preventing unhealthy weight gain in normal, overweight, and obese individuals.

An expert panel assembled by the International Association for the Study of Obesity concluded that daily exercise can help prevent healthy-weight individuals from becoming overweight, overweight people from becoming obese, and obesity from worsening in already obese individuals.[81] For every pound of body weight lost, a person requires 8 fewer kcal/day. Thus, when a person loses 10 lbs, energy requirements decrease by approximately 80 kcal/day. Thus, to maintain a 10-lb weight loss, a person must further reduce energy intake or increase energy expenditure.

Not all obese people are physically unfit and/or unhealthy. Although physical activity does not totally eliminate health risks associated with obesity, fitness level is strongly associated with health.

CHARACTERISTICS OF PEOPLE WHO SUCCESSFULLY LOSE WEIGHT

One of the best ways to learn what works in terms of weight loss is to study people who have been successful. The National Weight Control Registry (NWCR) is a large, prospective study of individuals who have lost significant amounts of weight and have been successful at keeping it off.[82] To be eligible to participate in the study, individuals must have maintained a weight loss of 30 pounds or more for one year or longer. Researchers interviewed this unique group of "successful losers" to learn about their weight loss and weight-loss management practices. However, identifying common characteristics among NWCR participants proved difficult. Whereas some reported counting calories or grams of fat, others used prepared, prepackaged foods. Some participants preferred losing weight on their own, while others sought assistance from weight-loss programs. Researchers identified one striking common thread: almost 90% of the NWCR participants reported that both diet and physical activity were part of their weight-loss plan. Only 10% reported using diet alone, and only 1% used exercise alone. On average, study participants engaged in 60 to 90 minutes of physical activity daily, which was equivalent to 2,500 kcal/week for women and 3,300 kcal/week for men. In addition to a high level of physical activity, participants tended to follow a low-fat, high-carbohydrate diet, ate breakfast regularly, and monitored their weight frequently. The low-fat, high-carbohydrate diet that prevailed in this study has wide support, but experts continue to debate the value of diets with different macronutrient distributions.

Does Macronutrient Distribution Matter?

With all the weight-loss advice, it can be difficult to sort truth from fiction. Even experts have differing opinions about the best balance of macronutrients—proteins, fats, and carbohydrates—for achieving weight loss. Today, one of the biggest controversies is the role of dietary carbohydrate versus dietary fat in promoting weight loss and weight gain.[83] Weight-loss diets that are low in fat and high in carbohydrates have long been considered the most effective in terms of weight loss and weight maintenance. In fact, the 2010 Dietary Guidelines for Americans advocate low-fat food choices with an emphasis on whole grains, fruits, and vegetables. Similarly, the Acceptable Macronutrient Distribution Range (AMDR) suggests that we consume 45 to 65% of energy

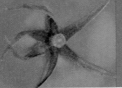
People trying to lose weight tend to have the most success when they select foods that are both nutritious and satisfying. Too often, weight-loss plans emphasize food restriction, causing a state of perpetual hunger. As a result, people typically revert to former eating practices. However, new studies show that eating more can sometimes help people weigh less.[84] This is because eating high-volume, low-energy-dense foods can help people feel full longer than when they eat low-volume, high-energy-dense foods.[85] In other words, it is possible for people to consume larger volumes of foods while eating fewer calories.

The energy density of food reflects the relationship between energy (kcal) and weight (grams). High-energy-dense foods are those that provide 4 to 9 kcal per gram, whereas low-energy-dense foods have less than 1.5 kcal per gram.[86] High-energy-dense foods typically have low moisture or high fat content. These include cookies, cakes, crackers, cheesecake, and butter. Examples of foods with medium energy density (1.5 to 4 kcal/gram) include eggs, dried fruits, breads, and cheese. Low-energy-dense foods have a high water and/or fiber content and are low in fat, such as most fruits and vegetables, broth-based soups, low-fat cheese, and certain types of meat such as roasted turkey.

Studies show that people feel more satisfied and tend to consume fewer calories when they eat low-energy-dense foods.[87] For example, people who eat salads or broth-based soups prior to a high-energy-dense main course consume fewer calories overall. Similarly, the effect of volume on satiety was demonstrated by feeding people milkshakes with the same number of calories but with different volumes (300 mL, 450 mL, and 600 mL). Participants consumed 12% fewer calories when fed 600-mL (approximately 20 ounces) milkshakes than when they consumed those with less volume. These and other studies show that meal volume provides a stronger satiety signal than the total calories in the food.[88] Most important, people consuming low-energy-dense foods lose more weight over time than those following a fat-reduction weight-loss plan.

Here are some suggestions to increase your intake of low-energy-dense foods.

- Eat a low-energy-dense salad before a main course. Decrease the energy density of your salad by adding plenty of greens, vegetables, and low-fat salad dressing.

- As an alternative to cream-based soup, choose broth-based soup containing vegetables, high-fiber grains (such as barley), and legumes (such as split peas or lentils).

- Reduce energy density by cutting back on the amount of high-fat meat in recipes and adding more vegetables.

- Avoid beverages that have high fat content. Instead, add a small amount of fruit juice to carbonated water.

- Read the Nutrition Facts panel on food labels, and compare the number of calories in relation to the number of grams. In general, foods with low-energy density are those with less than 1.5 kcal per gram.

Scott Bauer/ARS/USDA

Foods that are not energy dense, such as these grapes, have a high water and/or fiber content and are low in fat.

from carbohydrate, 10 to 35% from protein, and 20 to 35% from fat. However, Robert Atkins, one of the first pioneers of the low-carbohydrate diet, turned the nutritional world upside down in 1972 when he proposed that eating too much carbohydrate, rather than too much fat, may actually cause people to gain weight. The caloric distribution advocated by the AMDRs is compared to those of typical high- and low-carbohydrate diets in Table 8.5. We next

TABLE 8.5 Caloric Distribution of High- and Low-Carbohydrate Diets and the Acceptable Macronutrient Distribution Ranges (AMDR)

Nutrient	AMDR[a]	% of Total Calories High-Carbohydrate Diet[b]	Low-Carbohydrate Diet[c]
Fat	20–35	10–15	55–65
Carbohydrate	45–65	65–75	5–20
Protein	10–35	10–25	20–40

[a]Source: Institute of Medicine. Dietary Reference Intakes for energy, carbohydrate, fiber, fat, fatty acids, cholesterol, protein, and amino acids. Washington, DC: National Academies Press; 2005.
[b]Source: Ornish D. Eat more weigh less: Dr. Dean Ornish's life choice diet for losing weight safely while eating abundant. New York: Harper Collins; 1993.
[c]Source: Atkins RC. Dr. Atkins' new diet revolution, revised. National Book Network; 2003.

examine the scientific evidence concerning the effects of macronutrient distribution on weight loss and overall health.

HIGH-CARBOHYDRATE, LOW-FAT WEIGHT-LOSS DIETS

Many researchers believe diets high in carbohydrates and low in fat promote weight loss and have an overall beneficial effect on health. To maintain a relatively low intake of fat (10 to 15% of total calories), high-carbohydrate weight-loss diet advocates advise dieters to avoid meat, dairy, oils, and olives; low-fat meat and dairy products can be eaten in moderation. With an emphasis on fruit, vegetables, and whole grains, this weight-reduction and maintenance plan provides about 65 to 75% of total calories from carbohydrates, with protein and fat making up the remainder. Examples of popular high-carbohydrate, low-fat weight-loss diets include Dr. Dean Ornish's Program for Reversing Heart Disease[89] and The New Pritikin Program for Diet and Exercise.[90]

There are several reasons why supporters of low-fat diets believe they help prevent obesity. First, gram for gram, fat has more than twice as many calories as carbohydrate and protein. Therefore, it is reasonable to assume that consuming less fat may lead to lower energy intake, which in turn results in weight loss. Fat can also make food more flavorful, contributing to overconsumption. Last, excess calories from fat are more efficiently stored by the body than those from carbohydrate or protein. Converting excess glucose and amino acids into fatty acids for storage takes energy and "subtracts" some of the energy originally in the foods. Some experts also claim that low-fat diets benefit overall health as well, by lowering total and LDL cholesterol concentrations, increasing HDL cholesterol concentrations, and improving blood glucose regulation.[91] However, these benefits may be due to weight loss in general rather than to reduced dietary fat.

Do High-Carbohydrate, Low-Fat Weight-Loss Diets Work? Long-standing dietary advice aimed at helping people lose weight has consistently focused on reducing dietary fat. Although total energy intake has increased, the percentage of total calories from fat has declined from 45% in the 1960s to approximately 33% today, which is equivalent to an average of 10 to 20 fewer grams of fat per day.[92] Yet decreased fat intake has not resulted in a decreased prevalence of obesity. In fact, obesity rates have increased under the low-fat regime.[93] It is not clear which dietary factors contributed to this trend; however, some researchers believe failure to make healthy carbohydrate food choices may be contributing to weight gain.

Although clinicians hoped Americans would replace fatty foods in their diet with more nutritious items such as whole grains, fruits, and vegetables,

High-carbohydrate/low-fat weight-loss diets (left) emphasize foods such as whole grains, lean meats, pasta, fruits, vegetables, and low-fat dairy products. Low-carbohydrate weight-loss diets (right) emphasize lean meats, fish, nonstarchy fruits and vegetables, eggs, low-fat dairy, and unsaturated fats.

they have not done so. The availability of low-fat snack products has made it possible for people to eat snack foods minus the fat. Typically, the ingredient that replaces fat in these products is refined carbohydrate (such as white flour and sugar), and people mistakenly think that low-fat snack foods are healthy snack alternatives. Overconsuming fat-free snack foods may, in part, contribute to weight gain, because many have the same amount of calories as the original product. Eating foods high in refined carbohydrate, especially those low in fat, may make us hungrier and therefore can make us heavier.[94] The theory that a low-fat diet is our best defense against weight gain is not without debate, and many health experts believe there is now enough solid evidence to lift the "ban" on dietary fat.[95]

LOW-CARBOHYDRATE WEIGHT-LOSS DIETS

On the opposite end of the weight-loss diet spectrum are low-carbohydrate diets. Health claims made by advocates of these weight-loss plans include weight loss without hunger and improved cardiovascular health.[96] Some experts believe that people are more likely to gain weight from excess carbohydrates as opposed to excess fats or proteins because high-carbohydrate foods cause insulin levels to rise, which favors fat storage (lipogenesis). Thus, limiting starch and refined sugars in the diet should theoretically help people lose weight.

There are many low-carbohydrate weight-loss diets, each differing in terms of the types of foods allowed. Although some exclude nearly all carbohydrates, others take a more moderate approach by allowing healthy, carbohydrate-rich foods such as fruit, vegetables, and whole grains. Indeed, low-carbohydrate diets continue to be an extremely popular approach to weight loss.[97] Low-carbohydrate diets appear safe in the short term, although some experts have raised doubts about their long-term effectiveness.[98] Examples of popular low-carbohydrate, weight-loss diets include Dr. Atkins' Diet Revolution[99], The Carbohydrate Addict's Diet[100], Life Without Bread[101], and Protein Power[102].

Do Low-Carbohydrate Diets Work? The lure of a diet plan that does not require counting calories or limiting portion sizes is appealing to many people. However, do low-carbohydrate diets really help one lose weight? When a person restricts his or her intake of carbohydrates, the body breaks down glycogen to provide glucose. Approximately 500 g (about 1 lb) of glycogen is stored in the liver and skeletal muscle combined. Glycogen is bound to molecules of water, approximately 3 g of water per gram of glycogen. Thus, the combined weight of stored glycogen and associated water is about 4 lb. Consequently, when glycogen is broken down, its associated water is eliminated from the body. Because of this, the initial weight loss associated with low-carbohydrate diets is largely attributed to water loss.

Once glycogen stores are depleted, the body begins to rely heavily on triglycerides for energy. However, recall from Chapter 7 that when glucose is limited, fatty acids can only be only partially broken down. As a result, ketone formation increases, a sign that the body is using fat as major sources of energy. Limiting carbohydrate intake causes ketosis, which is why these diets are referred to as **ketogenic diets.** Ketosis often results in a loss of appetite, further promoting weight loss.

Studies comparing weight loss associated with low-carbohydrate diets to that associated with low-fat diets show that, at six months, greater weight loss is achieved on low-carbohydrate diets. One of the largest and longest studies comparing three different weight-loss plans—low-carbohydrate, high-carbohydrate, and the macronutrient distribution ranges recommended in the Dietary Reference Intakes—found that the low-carbohydrate plan resulted in the greatest amount of weight loss.[103] By the end of one year, those on the low-carbohydrate weight-loss plan had lost an average of 10.4 lb, compared with those on the high-carbohydrate (4.8 lb) and standard (5.7 lb) diets. Those in the low-carbohydrate study group showed no change in blood lipids, despite high intakes of dietary fat.

A more recent study that evaluated the effectiveness and safety of low-carbohydrate weight-loss diets reported that they were effective alternatives to low-fat diets and a Mediterranean diet.[104] Even though weight loss was achieved with all three approaches, those following a low-carbohydrate approach lost on average 10 lb after two years. Even more surprising was that the low-carbohydrate diet had more favorable effects on blood lipids and glucose, compared with the low-fat and Mediterranean weight-loss regimens.

Still, weight loss associated with low-carbohydrate diets may not necessarily be caused by alterations in the macronutrient composition of the diet but rather by a reduction in caloric intake.[105] That is, people may eat less on the diet because of limited food choices and decreased appetite associated with ketosis. There is no consistent evidence that carbohydrate restriction causes the body to burn energy more efficiently. These findings cast doubt on the claim that a person can maintain a high intake of calories and still lose weight while on a low-carbohydrate diet.

Nutritional Adequacy of Low-Carbohydrate Diets The amount of carbohydrates allowed on most low-carbohydrate diet plans varies from 5 to 20% of total energy. By comparison, current recommendations suggest that 45 to 65% of total calories come from carbohydrates. Low-carbohydrate diets have been criticized as containing too much total fat, saturated fat, and cholesterol. Current recommendations regarding optimal levels of fat in the diet suggest that 20 to 35% of total calories should come from fat. By comparison, most low-carbohydrate diets provide 55 to 65% of calories from fat. Many low-carbohydrate diets also fail to distinguish between "healthy" and "unhealthy" fats. Finally, the amount of protein in some low-carbohydrate diets exceeds recommended amounts. It is currently recommended that we consume 10 to 35% of calories from protein. Many low-carbohydrate diets provide 20 to 40% of total calories from protein, so some dieters may be exceeding the upper limit of the AMDR for protein (10–35% of total calories for adults >18 years of age).

Perhaps one of the biggest concerns regarding low-carbohydrate diets is the restriction of healthy, high-carbohydrate foods such as fruit, vegetables,

⟨CONNECTIONS⟩ Recall that ketosis is a condition resulting from the accumulation of ketones in the blood (Chapter 4, page 141).

⟨CONNECTIONS⟩ Recall that the Mediterranean diet is an eating pattern rich in monounsaturated fatty acids, a balanced ratio of omega-6:omega-3 essential fatty acids, fiber, and antioxidants (Chapter 6, page 263).

ketogenic diets Diets that stimulate ketone production.

and whole grains—a valid criticism.[106] Low-carbohydrate diets may lack essential micronutrients, dietary fiber, and beneficial phytochemicals. Although most low-carbohydrate diet plans recommend that people take dietary supplements, these cannot replace the many other substances (such as fiber and phytochemicals) in these restricted foods. People on low-carbohydrate weight-loss eating plans should eat fresh fruits and nonstarchy vegetables to help ensure that their diets are nutritious and balanced.

Some Health Concerns Regarding Low-Carbohydrate Diets Are Unfounded
Because some cells (such as red blood cells) require glucose, consuming a very-low–carbohydrate diet causes the body to break down protein (muscle), so that the resulting amino acids can be used for glucose synthesis (gluconeogenesis). Therefore, some experts have raised concerns that low-carbohydrate diets may cause loss of lean tissue. However, studies suggest that weight loss associated with low-carbohydrate diets is largely attributed to a loss of body fat, with little loss of lean tissue.[107] It appears that protein intakes associated with some low-carbohydrate diets are high enough to prevent the loss of muscle.

Some health professionals have also expressed concern about the effect of low-carbohydrate diets on bone health and kidney function. This is because low-carbohydrate diets can also be high in protein. However, limited research shows no harmful effects of low-carbohydrate/high-protein diets on bone health.[108] Some professionals worry that low-carbohydrate diets may impair kidney function, because high protein intakes and ketones are thought to overburden the kidneys.[109] Although this has not been adequately studied, low-carbohydrate diets are not advised for people with impaired kidney function.

Low-carbohydrate diets also do not appear to have an adverse effect on cardiovascular health, at least in the short term.[110] Rather, several studies suggest that low-carbohydrate diets result in favorable changes in blood lipid levels, glycemic control, and blood pressure, despite increased intakes of fat.[111] These improvements may be due to a reduction in body weight rather than to a direct effect of the macronutrient composition of the diet.

Most popular diets look at energy balance in a simplistic fashion, as though there were one magic key that opens the door to easy weight control. On the contrary, maintaining energy balance is quite complex. The reality is that nobody knows for certain the "ideal" distribution of macronutrients for weight loss. The AMDR provides an array of relative proportions of macronutrients associated with a healthy diet. Therefore, it is up to individuals to determine the eating pattern that best suits their needs within parameters of the AMDR. What the research strongly suggests, however, is that reducing calorie intake is the critical component for successful weight loss.

In this chapter, you have learned that, energy balance is quite complex and that our eating behaviors are shaped by many factors, including genetics, the physiological states of hunger and satiety, and the psychological and social determinants of appetite. You have also learned how a combination of healthy eating and exercise can lead to successful long-term weight control. However, this simple formula can be far from easy to follow, because of influences ranging from social pressures to genetics to the foods available in a given society.

© Ryan McVay/Lifesize/Getty Images

Critical Thinking: August's Story Now that you understand that both lifestyle and genetics can influence body weight, think back to August's story about her difficulty losing weight. Many people assume that someone like August could and should lose weight by diet and exercise alone. Do you think these views are correct?

Diet Analysis **PLUS** ✚ Activity

Part A: Estimated Energy Requirements (EER), Total Energy Expenditure (TEE), and Resting Energy Expenditure (REE)

1. The EER represents the average dietary energy intake that will maintain energy balance in healthy persons of a given sex, age, weight, height, and physical activity level (PAL). Use the following formulas to estimate your energy expenditure. Information to assist you in making these calculations can be found on page 45 in your text. Use Table 2.2 on page 45 to determine your physical activity level (PA).

   Males EER (kcal/day) = 662 − [9.53 × age (y)] + PA × [15.91 × weight (kg)] + [539.6 × height (m)]

   Females EER (kcal/day) = 354 − [(6.91) × age (y)] + PA × [9.36 × weight (kg) + [726 × height (m)]

2. Now look at your Profile DRI Goals, and compare your calculated value with the printed "Energy" value. Are they the same?

3. Total energy expenditure (TEE) is based on the energy needed to maintain involuntary, physiological functions, support physical activity, and process food. Knowing that basal metabolism makes up 50 to 70% of the TEE, use your calculated EER value to estimate how many kilocalories are used to support vital body functions (basal energy expenditure; BEE). Calculate your upper level first by multiplying by 0.70 and then your lower level by multiplying by 0.50.

   _____ (EER) kcal/day × 0.70 = _____ kcal/day. Upper limit of kilocalories needed for involuntary, vital body functions

   _____ (EER) kcal/day × 0.50 = _____ kcal/day. Lower limit of kilocalories needed for involuntary, vital body functions

4. Recall that resting metabolism is similar to basal metabolism but is estimated under less stringent conditions. Thus, it is likely to be higher than basal metabolism. The Harris-Benedict equation can be used to predict your REE. Compare your calculated REE to your calculated BEE. Remember the REE estimates energy needed for your basic functions in addition to the energy needed to digest your last meal.

   Male: REE = 66.5 + [13.8 × weight (kg)] + [5 × height (cm)] − [6.8 × age (y)] = _____ kcal/day

   Female: REE = 655.1 + [9.6 × weight (kg)] + [1.8 × height (cm)] − [4.7 × age (y)] = _____ kcal/day

Part B: Physical Activity

Total energy expenditure (TEE) also includes energy expenditure associated with physical activity. This factor can vary greatly from person to person and may account for 15 to 30% of your TEE.

1. Calculate your EER using the next-highest physical activity(PA) value. How much would your EER increase if you were to increase your physical activity?

Part C: Thermic Effect of Food (TEF)

1. The third significant component of total energy expenditure is the thermic effect of food (TEF), or the energy needed to digest, absorb, transport, metabolize, and store nutrients after meals. Using the 5 to 10% range mentioned in your text, calculate the range representing your own TEF from the EER determined in Part A.

   EER kcal/day × 0.05 = _____ kcal/day

   EER kcal/day × 0.10 = _____ kcal/day

Notes

1. Prins JB. Adipose tissue as an endocrine organ. Best Practice and Research Clinical Endocrinology and Metabolism. 2002;16:639–51.

2. Lau DCW, Dhillon B, Yan H, Szmitko PE, Verma S. Adipokines: molecular links between obesity and atheroslcerosis. American Journal of Physiology. Heart and Circulatory Physiology. 2005;288:H2031–41.

3. Spalding KL, Arner E, Westermark PO, Buchholz BA, Bergmann O, Blomqvist L, Hoffstedt J, Näslund E, Britton T, Concha H, Hassan M, Rydén M, Frisén J, Arner P. Dynamics of fat cell turnover in humans. Nature. 2008;453:783–7.

4. Arner P, Spalding KL. Fat cell turnover in humans. Biochemical and Biophysical Research Communications. 2010;396:101–4.

5. Pénicaud L. Relationships between adipose tissues and brain: what do we learn from animal studies? Diabetes & Metabolism. 2010;36:S39–44.

6. de Graaf C, Blom WA, Smeets PA, Stafleu A, Hendriks HF. Biomarkers of satiation and satiety. American Journal of Clinical Nutrition. 2004;79:946–61.

7. U.S. Food and Drug Administration. FDA expands use of banding system for weight loss. 2011. Available from: http://www.fda.gov/NewsEvents/Newsroom/PressAnnouncements/ucm245617.htm

8. Ali MR, Fuller WD, Choi MP, Wolfe BM. Bariatric surgical outcomes. Surgical Clinics of North America. 2005;85:835–52.

9. Flum DR, Salem L, Broeckel-Elrod JA, Patchen-Dellinger E, Cheadle A, Chan L. Early mortality among Medicare beneficiaries undergoing bariatric surgical procedures. Journal of the American Medical Association. 2005;294:1903–8.

10. Sjöström L, Narbro K, Sjöström CD, Karason K, Larsson B, Wedel H, Lystig T, Sullivan M, Bouchard C, Carlsson B, Bengtsson C, Dahlgren S, Gummesson A, Jacobson P, Karlsson J, Lindroos AK, Lönroth H, Näslund I, Olbers T, Stenlöf K, Torgerson J, Agren G, Carlsson LM. Effects of bariatric surgery on mortality in Swedish obese subjects. New England Journal of Medicine. 2007;357:741–52.

11. Moo TA, Rubino F. Gastrointestinal surgery as treatment for type 2 diabetes. Current Opinion in Endocrinology, Diabetes and Obesity. 2008;15:153–8.

12. Coelho JC, Campos AC. Surgical treatment of morbid obesity. Current Opinion in Clinical Nutrition and Metabolic Care. 2001;4:201–6. Pope GD, Finlayson SR, Kemp JA, Birkmeyer JD. Life expectancy benefits of gastric bypass surgery. Surgical Innovations. 2006;13:265–73.

13. Ello-Martin JA, Ledikwe JH, Rolls BJ. The influence of food portion size and energy density on energy intake: Implications for weight management. American Journal of Clinical Nutrition. 2005;82:236S–41S.

14. Wurtman RJ, Wurtman JJ. Brain serotonin, carbohydrate-craving, obesity and depression. Obesity Research. 1995;3:477S–80S.

15. Hainer V, Kabrnova K, Aldhoon B, Kunesova M, Wagenknecht M. Serotonin and norepinephrine reuptake inhibition and eating behavior. Annals of the New York Academy of Science. 2006;1083:252–69.

16. Orr J, Davy B. Dietary influences on peripheral hormones regulating energy intake: Potential applications for weight management. Journal of the American Dietetic Association. 2005;105:1115–24.

17. Inui A, Asakawa A, Bowers CY, Mantovani G, Laviano A, Meguid MM, Fujimiya M. Ghrelin, appetite, and gastric motility: The emerging role of the stomach as an endocrine organ. Federation of American Societies for Experimental Biology Journal. 2004;18:439–56.

18. Cummings DE. Ghrelin and the short- and long-term regulation of appetite and body weight. Physiological Behavior. 2006;89:71–84.

19. Yanovski S. Sugar and fat: Cravings and aversions. Journal of Nutrition. 2003;133:835S–7S.

20. Lafay L, Thomas F, Mennen L, Charles MA, Eschwege E, Borys JM, Basdevant A. Gender differences in the relation between food cravings and mood in an adult community: Results from the Fleurbaix Laventie Ville Sante study. International Journal of Eating Disorders. 2001;29:195–204.

21. Dye L, Warner P, Bancroft JJ. Food craving during the menstrual cycle and its relationship to stress, happiness of relationship and depression; a preliminary enquiry. Journal of Affective Disorders. 1995;34:157–64.

22. Hulbert AJ, Else PL. Basal metabolic rate: History, composition, regulation, and usefulness. Physiological and Biochemical Zoology. 2004;77:869–76.

23. Henry CJ. Mechanisms of changes in basal metabolism during ageing. European Journal of Clinical Nutrition. 2000;54:S77–91.

24. Luke A, Schoeller DA. Basal metabolic rate, fat-free mass, and body cell mass during energy restriction. Metabolism. 1992;41:450–6.

25. Brooks GA, Butte NF, Rand WM, Flatt JP, Caballero B. Chronicle of the Institute of Medicine physical activity recommendation: How a physical activity recommendation came to be among dietary recommendations. American Journal of Clinical Nutrition. 2004;79:921S–30S.

26. Nair KS, Halliday D, Garrow JS. Thermic response to isoenergetic protein, carbohydrate or fat meals in lean and obese subjects. Clinical Science. 1983;65:307–12.

27. Institute of Medicine. Dietary reference intakes for energy, carbohydrate, fiber, fat, fatty acids, cholesterol, protein, and amino acids. Washington, DC: National Academies Press; 2005.

28. Aronne LJ. Classification of obesity and assessment of obesity-related health risks. Obesity Research. 2002;10:105S–15S.

29. Friedl KE. Can you be large and not obese? The distinction between body weight, body fat, and abdominal fat in occupational standards. Diabetes Technology and Therapeutics. 2004;6:732–49.

30. Katzmarzyk PT, Janssen I, Ardern CI. Physical inactivity, excess adiposity and premature mortality. Obesity Reviews. 2003;4:257–901.

31. Lafontan M, Berlan M. Do regional differences in adipocyte biology provide new pathophysiological insights? Trends in Pharmacological Sciences. 2003;24:276–83.

32. Ibrahim MM. Subcutaneous and visceral adipose tissue: structural and functional differences. Obesity Reviews. 2010;11:11–8.

33. Toth MJ, Tchernof A, Sites CK, Poehlman ET. Menopause-related changes in body fat distribution. Annals of the New York Academy of Sciences. 2000;904:502–6.

34. National Institutes of Health. Clinical guidelines on the identification, evaluation, and treatment of overweight and obesity in adults. National Institutes of Health, National Heart, Lung, and Blood Institute, Obesity Education Initiative. Available from http://www.nhlbi.nih.gov/guidelines/obesity/practgde.htm.

35. See R, Abdullah SM, McGuire DK, Khera A, Patel MJ, Lindsey JB, Grundy SM, de Lemos JA. The association of differing measures of overweight and obesity with prevalent atherosclerosis: The Dallas Heart Study. American College of Cardiology. 2007;50:752–9.

36. Heinrich KM, Jitnarin N, Suminski RR, Berkel L, Hunter CM, Alvarez L, Brundige AR, Peterson AL, Foreyt JP, Haddock CK, Poston WS. Obesity classification in military personnel: A comparison of body fat, waist circumference, and body mass index measurements. Military Medicine. 2008;17:67–73.

37. National Center for Health Statistics. Health, United States, 2010: With Special Feature on Death and Dying. Hyattsville, MD. 2011. Available from: http://www.cdc.gov/nchs/data/hus/hus10.pdf.

38. Popkin BM, Gordon-Larsen P. The nutrition transition: Worldwide obesity dynamics and their determinants. International Journal of Obesity and Related Metabolic Disorders. 2004;3:S2–9.

39. Austin GL, Ogden LG, and O Hill J. Trends in carbohydrate, fat, and protein intakes and association with energy intake in normal-weight, overweight, and obese individuals: 1971–2006. American Journal of Clinical Nutrition. 2011;93:836–43.

40. Briefel RR, Johnson CL. Secular trends in dietary intake in the United States. Annual Review of Nutrition. 2004;24:401–31.

41. Ello-Martin JA, Ledikwe JH, Rolls BJ. The influence of food portion size and energy density on energy intake: Implications for weight management. American Journal of Clinical Nutrition. 2005;82:236 S–41S. Wansink B, Kim J. Bad popcorn in big buckets: Portion size can influence intake as much as taste. Journal of Nutrition Education and Behavior. 2005;37:242–5.

42. Rolls BJ, Roe LS, Kral TVE, Meengs JS, Wall DE. Increasing the portion size of a packaged snack increases energy intake in men and women. Appetite. 2004;42:63–9.

43. Mourao DM, Bressan J, Campbell WW, Mattes RD. Effects of food form on appetite and energy intake in lean and obese young adults. International Journal of Obesity. 2007;31:1688–95.

44. Hetherington MM, Foster R, Newman T, Anderson AS, Norton G. Understanding variety: Tasting different foods delays satiation. Physiology and Behavior. 2006;87:263–71.

45. Chandon P and Wansink B. The biasing health halos of fast-food restaurant health claims: Lower calorie estimates and higher side-dish consumption intentions. Journal of Consumer Research. 2007;34:301–14.

46. National Center for Health Statistics (NCHS). Health, United States, 2010. Available from: http://www.cdc.gov/nchs/data/hus/hus10.pdf.

47. Kim JY, Oh DJ, Yoon TY, Choi JM, Choe BK. The impacts of obesity on psychological well-being: A cross-sectional study about depressive mood and quality of life. Preventive Medicine and Public Health. 2007;40:191–5.

48. Mamun AA, Lawlor DA, O'Callaghan MJ, Bor W, Williams GM, Najman JM. Does childhood sexual abuse predict young adult's BMI? A birth cohort study. Obesity. 2007;15:2103–10. Noll JG, Zeller MH, Trickett PK, Putnam FW. Obesity risk for female victims of childhood sexual abuse: A prospective study. Pediatrics. 2007;120:61–7. bes Gustafson TB, Sarwer DB. Childhood sexual abuse and obesity. Obesity Reviews. 2004;5:129–35.

49. Christakis NA and Fowler JH. The spread of obesity in a large social network over 32 years. New England Journal of Medicine. 2007;57:370–9.

50. Brownson RC, Boehmer TK, Luke DA. Declining rates of physical activity in the United States: What are the contributors? Annual Review of Public Health. 2005;26:421–43.

51. American College of Sports Medicine (ACSM). ACSM's Guidelines for Exercise Testing and Prescription, 8th ed. Philadelphia: Lippincott Williams & Wilkins; 2010.

52. Wareham NJ, van Sluijs EM, Ekelund U. Physical activity and obesity prevention: A review of the current evidence. Proceedings of the Nutrition Society. 2005;64:229–47.

53. Hill JO, Wyatt HR, Reed GW, Peters JC. Obesity and the environment: Where do we go from here? Science. 2003;299:853–5.

54. National Association for Sport and Physical Education and American Heart Association. 2010 Shape of the

Nation Report. Available at: http://www.aahperd.org/naspe/publications/upload/Shape-of-the-Nation-2010-Final.pdf

55. Caroli M, Argentieri L, Cardone M, Masi A. Role of television in childhood obesity prevention. International Journal of Obesity and Related Metabolic Disorders. 2004;28:S4–108.

56. Kaiser Family Foundation. The role of media in childhood obesity; 2004. Publication number 7030. Available from http://www.kff.org/entmedia/7030.cfm.

57. Halford JC, Boyland EJ, Hughes GM, Stacey L, McKean S, Dovey TM. Beyond-brand effect of television food advertisements on food choice in children: The effects of weight status. Public Health Nutrition. 2007;1–8. Powell LM, Szczypka G, Chaloupka FJ. Exposure to food advertising on television among US children. Archives of Pediatric and Adolescent Medicine. 2007;16:553–60.

58. U.S. Department of Health and Human Services. 2008 Physical Activity Guidelines for Americans. Washington (DC): U.S. Department of Health and Human Services; 2008. ODPHP Publication No. U0036. Available from: http://www.health.gov/paguidelines.

59. Stunkard AJ, Harris JR, Pedersen NL, McClearn GE. The body-mass index of twins who have been reared apart. New England Journal of Medicine. 1990;322:1483–7. Wardle J, Carnell S, Haworth CM, Plomin R. Evidence for a strong genetic influence on childhood adiposity despite the force of the obesogenic environment. American Journal of Clinical Nutrition. 2008;87:398–404. Lee K, Song YM, Sung J. Which obesity indicators are better predictors of metabolic risk? Healthy twin study. Obesity. 2008;16:834–40.

60. Sorensen TI, Holst C, Stunkard AJ. Adoption study of environmental modifications of the genetic influences on obesity. International Journal of Obesity and Related Metabolic Disorders. 1998;22:73–81.

61. Gillman MW, Barker D, Bier D, Cagampang F, Challis J, Fall C, Godfrey K, Gluckman P, Hanson M, Kuh D, Nathanielsz P, Nestel P, Thornburg KL. Meeting report on the 3rd International Congress on Developmental Origins of Health and Disease (DOHaD). Pediatric Research. 2007;61:625–9.

62. Ingalls A, Dickie M, Snell GD. Obese, a new mutation in the house mouse. Journal of Heredity. 1950;41:317–8.

63. Coleman D, Hummel KP. Effects of parabiosis of normal with genetically diabetic mice. American Journal of Physiology. 1969;217:1298–304. Coleman DL. Effects of parabiosis of obese with diabetes and normal mice. Diabetologia. 1973;9:294–8.

64. Zhang Y, Proenca R, Maffei M, Leopold L, Friedman JM. Positional cloning of the mouse obese gene and its human homologue. Nature. 1994;372:125–32.

65. Farooqi IS, O'Rahilly S. Monogenic obesity in humans. Annual Review of Medicine. 2005;56:443–58.

66. Marx J. Cellular warriors at the battle of the bulge. Science. 2003;299:846–9.

67. Kennedy AG. The role of the fat depot in the hypothalamic control of food intake in the rat. Proceedings of the Royal Society of London. 1953;140:578–92.

68. Popovic V, Duntas LH. Brain somatic cross-talk: Ghrelin, leptin and ultimate challengers of obesity. Nutrition and Neuroscience. 2005;8:1–5. Scarpace PJ, Zhang Y. Elevated leptin: Consequence or cause of obesity? Frontiers in Bioscience. 2007;12:3531–44.

69. Jequier E. Leptin signaling, adiposity, and energy balance. Annals of the New York Academy of Sciences. 2002;967:379–88.

70. Couce ME, Green D, Brunetto A, Achim C, Lloyd RV, Burguera B. Pituitary. Limited brain access for leptin in obesity. 2001;4:101–10.

71. Klok MD, Jakobsdottir S, Drent ML. The role of leptin and ghrelin in the regulation of food intake and body weight in humans: A review. Obesity Reviews. 2007;8:21–34. Farooqi S, O'Rahilly S. Genetics of obesity in humans. Endocrine Reviews. 2006;27:710–8.

72. Nishida M, Funahashi T, Shimomura I. Pathophysiological significance of adiponectin. Medical Molecular Morphology. 2007;40:55–67.

73. Torres-Leal FL, Fonseca-Alaniz MH, Rogero MM, Tirapegui. The role of inflamed adipose tissue in the insulin resistance. Cell Biochemistry and Function. 2010;28:623–31.

74. Matsuzawa Y. Adiponectin: a key player in obesity related disorders. Current Pharmaceutical Design. 2010;16:1896–901.

75. Oh DK, Ciaraldi T, Henry RR. Adiponectin in health and disease. Diabetes, Obesity and Metabolism. 2007;9:282–9.

76. Mehta S, Farmer JA. Obesity and inflammation: A new look at an old problem. Current Atherosclerosis Report. 2007;9:134–8.

77. Wing RR, Phelan S. Long-term weight loss maintenance. American Journal of Clinical Nutrition. 2005;82:222S–5S.

78. National Institutes of Health, and National Heart, Lung and Blood Institute. Clinical guidelines on the identification, evaluation and treatment of overweight and obesity in adults—the evidence report. National Institutes of Health Publication Number 00–4084. Bethesda, MD: National Institutes of Health; October 2000.

79. Farrell SW, Braun L, Barlow CE, Cheng YJ, Blair SN. The relation of body sass index, cardiorespiratory fitness, and all-cause mortality in women. Obesity Research. 2002;10:417–23.

80. Barlow CE, Kohl HW III, Gibbons LW, Blair SN. Physical fitness, mortality and obesity. International Journal of Obesity and Related Metabolic Disorders. 1995;19:S41–4.

81. International Association for the Study of Obesity. Diet, nutrition and the prevention of chronic diseases. Report of the joint WHO/FAO expert consultation. WHO Technical Report Series, No. 916 (TRS 916); 2003. Available from: http://biotech.law.lsu.edu/obesity/who/trs_916.pdf.

82. Catenacci VA, Ogden LG, Stuht J, Phelan S, Wing RR, Hill JO, Wyatt HR. Physical activity patterns in the National Weight Control Registry. Obesity. 2008;16:153–61. Butryn ML, Phelan S, Hill JO, Wing RR. Consistent self-monitoring of weight: A key component of successful weight loss maintenance. Obesity. 2007;15:3091–6. Phelan S, Wyatt H, Nassery S, Dibello J, Fava JL, Hill JO, Wing RR. Three-year weight change in successful weight losers who lost weight on a low-carbohydrate diet. Obesity. 2007;15:2470–7.

83. Westman EC, Yancy WS Jr, Vernon MC. Is a low-carb, low-fat diet optimal? Archives of Internal Medicine. 2005;165:1071–2.

84. Bell EA, Castellanos VH, Pelkman CL, Thorwart ML, Rolls BJ. Energy density of foods affects energy intake in normal-weight women. American Journal of Clinical Nutrition. 1998;67:412–20. Yao M, Roberts SB. Dietary energy density and weight regulation. Nutrition Reviews. 2001;59:247–58.

85. Ello-Martin JA, Ledikwe JH, Rolls BJ. The influence of food portion size and energy density on energy intake: Implications for weight management. American Journal of Clinical Nutrition. 2005;82:236S–41S.

86. National Center for Chronic Disease Prevention and Health Promotion Division of Nutrition and Physical Activity, Department of Health and Human Services, Centers for Disease Control and Prevention. Can eating fruits and vegetables help people to manage their weight? Research to Practice Series, No.1. Available from http://www.cdc.gov/nccdphp/dnpa/nutrition/pdf/rtp_practitioner_10_07.pdf.

87. Rolls BJ, Bell EA, Waugh BA. Increasing the volume of a food by incorporating air affects satiety in men. American Journal of Clinical Nutrition. 2000;72:361–8.

88. Norton GN, Anderson AS, Hetherington MM. Volume and variety: relative effects on food intake. Physiology and Behavior. 2006;87(4):714–22.

89. Ornish, Dean. Dr. Dean Ornish's Program for Reversing Heart Disease. New York, New York: Random House Publishing Group, 1996.

90. Pritikin, Nathan and McGrady Patrick. The Pritikin Program for Diet and Exercise. New York, New York: Random House Publishing Group, 1984.

91. Lovejoy JC, Bray GA, LeFevre M, Smith SR, Most MM, Denkins YM, Volaufova J, Rood JC, Eldrige AL, Peters JC. Consumption of a controlled low-fat diet containing olestra for 9 months improves health risk factors in conjunction with weight loss in obese men: The Olé Study. International Journal of Obesity and Related Metabolic Disorders. 2003;27:1242–9.

92. Center for Nutrition Policy and Promotion and the U.S. Department of Agriculture. Nutrition Insights. Is fat consumption really decreasing? Insight 5 April 1998. Available from www.cnpp.usda.gov/Publications/NutritionInsights/insight5.pdf.

93. Willett WC. Dietary fat and body fat: Is there a relationship? Journal of Nutritional Biochemistry. 1998;9:522–4. Willett WC. Is dietary fat a major determinant of body fat? American Journal of Clinical Nutrition. 1998;67:556S–625S.

94. Busetto L, Marangon M, De Stefano F. High-protein low-carbohydrate diets: what is the rationale? Diabetes Metabolism Research and Reviews. 2011;27:230–2.

95. Taubes G. The soft science of dietary fat. Science. 2001;291:2536–45.

96. Foster GD, Wyatt HR, Hill JO, McGuckin BG, Brill C, Selma B, Szapary PO, Rader DJ, Edman JS, Klein S. A randomized trial of a low-carbohydrate diet for obesity. New England Journal of Medicine. 2003;248:2082–90.

97. Crowe TC. Safety of low-carbohydrate diets. Obesity Reviews. 2005;6:235–45.

98. Bravata DM, Sanders L, Huang J, Krumholz HM, Olkin I, Gardner CD, Bravata DM. Efficacy and safety of low-carbohydrate diets: A systematic review. Journal of the American Medical Association. 2003;289:1838–49. Foster GD, Wyatt HR, Hill JO, McGuckin BG, Brill C, Mohammed BS, Szapary PO, Rader DJ, Edman JS, Klein S. A randomized trial of a low-carbohydrate diet for obesity. New England Journal of Medicine. 2003;348:2082–90.

99. Atkins, Robert. Dr. Atkins' New diet Revolution. New York, New York: HarperCollins Publishers, 2002.

100. Heller Rachael and Heller Richard. The Carbohydrate Addict's Diet. Penguin Group, 1993.

101. Allen, Christopher. Life without Bread: How a Low-Carbohydrate Diet Can Save Your Life. New York, New York: McGraw-Hill Companies, 2000.

102. Eades Michael R. Protein Power. New York, New York: Bantam Book, 1999.

103. Gardner CD, Kiazand A, Alhassan S, Kim S, Stafford RS, Balise RR, Kraemer CK, King AC. Comparison of the Atkins, Zone, Ornish, and LEARN diets for change in weight and related risk factors among overweight premenopausal women: The A to Z weight loss study: A randomized trial. Journal of the American Medical Association. 2007;297:969–77.

104. Shai I, Schwarzfuchs D, Henkin Y, Shahar DR, Witkow S, Greenberg I, Golan R, Fraser D, Bolotin A, Vardi H, Tangi-Rozental O, Zuk-Ramot R, Sarusi B, Brickner D, Schwartz Z, Sheiner E, Marko R, Katorza E, Thiery J, Fiedler GM, Blüher M, Stumvoll M, Stampfer MJ. Weight loss with a low-carbohydrate, Mediterranean, or low-fat diet; Dietary Intervention Randomized Controlled Trial (DIRECT) group. New England Journal of Medicine. 2008;359:229–41.

105. Bravata DM, Sanders L, Huang J, Krumholz HM, Olkin I, Gardner CD, Bravata DM. Efficacy and safety of low-carbohydrate diets: A systematic review. Journal of the American Medical Association. 2003;289:1838–49.

106. Schwenke DC. Insulin resistance, low-fat diets, and low-carbohydrate diets: Time to test new menus. Current Opinion in Lipidology. 2005;16:55–60.

107. Astrup A, Meinert Larsen T, Harper A. Atkins and other low-carbohydrate diets: Hoax or an effective tool for weight loss? Lancet. 2004;364:897–9.

108. Farnsworth E, Luscombe ND, Noakes M, Wittert G, Argyiou E, Clifton PM. Effect of a high-protein, energy-restricted diet on body composition, glycemic control, and lipid concentrations in overweight and hyperinsulinemic men and women. American Journal of Clinical Nutrition. 2003;78:31–9.

109. Martin WF, Armstrong LE, Rodriguez NR. Dietary protein intake and renal function. Nutrition and Metabolism. 2005;2:25.

110. Acheson KJ. Carbohydrate and weight control: Where do we stand? Current Opinion in Clinical Nutrition and Metabolic Care. 2004;7:485–92.

111. Aude YW, Agatston AS, Lopez-Jimenez F, Lieberman EH, Almon M, Hansen M, Rojas G, Lamas GA, Hennekens CH. The national cholesterol education program diet vs. a diet lower in carbohydrates and higher in protein and monounsaturated fat: A randomized trial. Archives of Internal Medicine. 2004;164:2141–6.

Disordered Eating

Our bodies have natural boundaries that tell us when to eat, what to eat, and how much to eat. Paying attention to and trusting these internal cues help us have a healthy relationship with food. Healthy eating means eating without fear, guilt, or shame. However, for some people, preoccupation with food and weight loss can reach obsessive proportions. These troublesome disturbances in eating behaviors can be signs of disordered eating. If disordered eating patterns continue, they can eventually progress into an eating disorder.

In the past 25 years, the number of people diagnosed with eating disorders has increased. Although eating disorders in the general population are somewhat rare, they are relatively common among adolescent girls and young women. The American Psychological Association estimates that approximately 8 million females (8%) in the United States battle with some form of eating disorder. Although males can also develop eating disorders, the occurrence is more difficult to estimate because they may be more reluctant to seek help for what is commonly perceived to be a "female" condition. Nonetheless, the prevalence of eating disorders among males has been estimated to be as high as 2%.[1]

Not surprisingly, clinicians are very interested in learning more about eating disorders and how they can be treated effectively. In this Nutrition Matters, you will learn about disordered eating behaviors and the different types, causes, and complexities of eating disorders.

© Plush Studios/Getty Images

People with a distorted body image perceive themselves to be fat even if they are very thin.

How Do Eating Disorders Differ from Disordered Eating?

Disordered eating behaviors include a wide variety of unhealthy eating patterns such as irregular eating, consistent undereating, and consistent overeating.[2] These behaviors are common and often occur in response to stress, illness, or dissatisfaction with personal appearance. Although disordered eating patterns can be disturbing to others, they typically do not persist long enough to cause serious physical harm. However, in some people, disordered eating can progress into a full-blown eating disorder such as anorexia nervosa, an extreme pursuit of thinness; bulimia nervosa, a pattern of bingeing and purging; or binge-eating disorder. **Eating disorders** are characterized by extreme disturbances in eating behaviors that can be both physically and psychologically harmful.

---

disordered eating Unhealthy eating patterns such as eating irregularly, consistent undereating, and/or consistent overeating.

eating disorder Extreme disturbance in eating behaviors that can result in serious medical conditions, psychological consequences, and dangerous weight loss.

People with eating disorders often feel isolated, and their relationships with family and friends become strained.

To understand what causes eating disorders, it is important to consider a variety of factors that include but go beyond food intake. That is, eating disorders are complex behaviors that arise from a combination of physical, psychological, and social issues.

As shown in Table 1, the American Psychiatric Association classifies eating disorders into three distinct categories: anorexia nervosa (AN), bulimia nervosa (BN), and eating disorders not otherwise specified (EDNOS).[3] There are also subcategories that depend on the presence or absence of specific behaviors. Because behaviors associated with the different eating disorders often overlap, it can be difficult to identify which type of disorder a person has (Table 2).[4]

PEOPLE WITH ANOREXIA NERVOSA PURSUE EXCESSIVE THINNESS

Anorexia nervosa (AN) is characterized by an irrational fear of gaining weight or becoming obese. As a consequence, individuals with AN are unwilling to maintain a minimally normal body weight. In addition, there is often a disconnect between actual and perceived body weight and shape, such that people with AN believe they are "fat" even though they may be dangerously thin. When people with AN look in the mirror, they tend to be very critical of their body shape and size. Because of this distorted self-perception, weight loss becomes an obsession. Individuals with AN often spend hours scrutinizing

---

anorexia nervosa (AN) An eating disorder characterized by an irrational fear of gaining weight or becoming obese.

TABLE 1 American Psychiatric Association Classification and Diagnostic Criteria for Eating Disorders

Anorexia Nervosa (AN)

A. Refusal to maintain body weight at or above a minimally healthy weight for age and height.

B. Intense fear of gaining weight or becoming fat, even though underweight.

C. Disturbance in body weight or shape is experienced, or denial of the seriousness of the current low body weight.

D. Absence of at least three consecutive menstrual cycles (postmenarchal, premenopausal females).

Two types:
- *Restricting type:* Does not regularly engage in binge eating or purging behavior.
- *Binge-eating/purging type:* Regularly engages in binge eating or purging behavior.

Bulimia Nervosa (BN)

A. Recurrent episodes of binge eating. An episode of binge eating is characterized by both of the following: (1) eating, in a discrete period of time, an amount of food that is definitely larger than what most people would eat during a similar period of time and under similar circumstances, and (2) a sense of lack of control over eating during the episode.

B. Recurrent compensatory behavior to prevent weight gain, such as self-induced vomiting; misuse of laxatives, diuretics, enemas, or other medications; fasting; or excessive exercise.

C. Binge eating and inappropriate compensatory behaviors occurring at least twice a week for three months.

D. Self-evaluation is unduly influenced by body shape and weight.

E. Disturbance does not occur exclusively during episodes of anorexia nervosa.

Two types:
- *Purging type:* Regularly engages in self-induced vomiting or the misuse of laxatives, diuretics, or enemas.
- *Nonpurging type:* Regularly engages in inappropriate compensatory behaviors, such as fasting or excessive exercise; but not in self-induced vomiting or the misuse of laxatives, diuretics, or enemas.

Eating Disorders Not Otherwise Specified (EDNOS) is for disorders of eating that do not meet the criteria for any specific eating disorder. Examples include:

A. For females, criteria for anorexia nervosa are met, except the individual has regular menses.

B. Criteria for anorexia nervosa are met except that, despite significant weight loss, current weight is in the healthy range.

C. Criteria for bulimia nervosa are met except that the binge eating and inappropriate compensatory mechanisms occur at a frequency of less than twice a week or a duration of less than three months.

D. Regular use of inappropriate compensatory behavior by an individual of healthy body weight after eating small amounts of food.

E. Repeatedly chewing and spitting out, but not swallowing, large amounts of food.

F. Binge-eating disorder: recurrent episodes of binge eating in the absence of the regular use of inappropriate compensatory behaviors characteristic of bulimia nervosa.

SOURCE: Adapted from Diagnostic and Statistical Manual of Mental Disorders, 4th ed. (DSM-IV). Washington, DC: American Psychiatric Association, 2004. Used by permission of American Psychiatric Association.

TABLE 2 Behaviors Associated with Different Eating Disorders and Disordered Eating

Type of Eating Disorder and Disordered Eating	Food Restriction	Compensatory Mechanism	Bingeing
Anorexia nervosa, restricting type		Excessive exercise	
Anorexia nervosa, binge-eating/purging type		Excessive exercise	
Bulimia nervosa, purging type		Self-induced vomiting and/or the misuse of laxatives, diuretics, excessive exercise, or enemas	
Bulimia nervosa, nonpurging type		Excessive exercise	
Binge-eating disorder			
Restrained eating			

NOTE: Dark blue indicates a primary behavior associated with specific eating disorders and disordered eating, whereas light blue indicates that a behavior occurs to a lesser extent.

their bodies, paying excessive attention to their appearance.

People with AN tend to view their self-worth in terms of weight and body shape. Because they perceive starvation as an accomplishment rather than a problem, they have little motivation to change. Although the symptoms of AN center on food, the causes are much more complex. Like other eating disorders, AN stems mainly from psychological issues. In the case of AN, the denial of food coupled with the relentless pursuit of thinness become ways of coping with emotions, conflict, and stress.[5]

The American Psychiatric Association recognizes two types of AN: **restricting type** and **binge-eating/purging type** (Figure 1). People with restricting type AN maintain a low body weight by food restriction and/or excessive exercise, whereas people with binge-eating/purging type AN engage in periods of bingeing and purging (that is, self-induced vomiting and/or abuse of laxatives, diuretics, exercise, or enemas) in addition to severe food restriction.

Rituals Associated with Anorexia Nervosa Typically, people with AN

limit their food intake as well as the variety of foods consumed. For example, they often categorize foods as either "safe" or "unsafe" depending on whether or not consuming the food is thought to cause weight gain. Foods often perceived as unsafe, and typically the first to be eliminated, are meat, high-sugar foods, and fatty foods. Eventually the diet becomes so restricted that nutritional needs are no longer met.

anorexia nervosa, restricting type An eating disorder characterized by food restriction.

anorexia nervosa, binge-eating/purging type An eating disorder characterized by food restriction as well as bingeing and purging.

FIGURE 1 Types of Anorexia Nervosa There are two forms of anorexia nervosa, an eating disorder characterized by self-starvation and a relentless pursuit of weight loss.

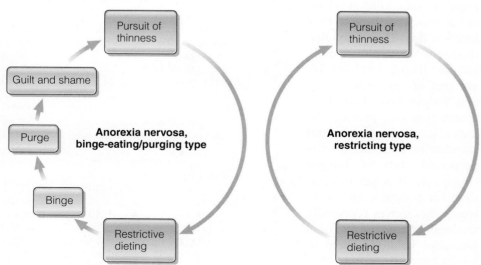

367

TABLE 3 Physical Signs and Symptoms, Behavioral Characteristics, and Health Consequences Associated with Anorexia Nervosa

Physical Signs and Symptoms	Behavioral and Emotional Signs and Symptoms	Health Consequences
• At or below 15% of ideal body weight • Thin appearance • Fainting, dizziness, fatigue, overall weakness • Intolerance to cold • Significant loss of fat and lean body mass • Brittle nails • Dry skin, dry hair, thinning hair, hair loss • Low blood pressure • Formation of fine hair on the body (lanugo) • Irregular or absence of menstruation	• Performance of food rituals such as excessive chewing or cutting food into small pieces • Restriction of amount and types of foods consumed • Rigid schedule and routine • Refusal to eat • Denial of hunger • Excessive exercise • Depression or lack of emotion • Food preoccupation • Tendency toward perfectionism • Rigidity • Frequent monitoring of body weight	• Reproductive problems • Loss of bone mass • Electrolyte imbalance • Irregular heartbeat • Bruises • Injuries such as stress fractures • Impaired iron status • Impaired immune status • Slow heart rate and low blood pressure

In addition to limiting food intake, individuals with AN may also have food rituals such as chewing food but not swallowing, overchewing, obsessive cooking for others, and cutting food into unusually small pieces. These behaviors provide a display of self-restraint and control over the urge to eat. These and other traits often associated with AN are summarized in Table 3.

In addition to dietary restrictions, other compulsive behaviors are commonly associated with AN.[6] For example, many individuals with AN maintain a rigid exercise schedule, working out as much as three to four hours a day. They may also maintain meticulous daily records listing food intake, amount of time exercising, and weight loss. This daily monitoring can provide the anorexic person the mental stamina to continue in this perpetual state of hunger (Figure 2). It is not uncommon for people with AN to weigh themselves repeatedly throughout the day. These obsessive behaviors help to lessen concerns of not being sufficiently lean or obsessive thoughts of being "too fat." A page taken from a journal of a person with AN reflects the inability to recognize these problematic behaviors (Figure 3).

FIGURE 2 Food, Weight, and Exercise Record of a Person with Anorexia Nervosa
Some people with eating disorders keep meticulous records regarding exercise and diet.

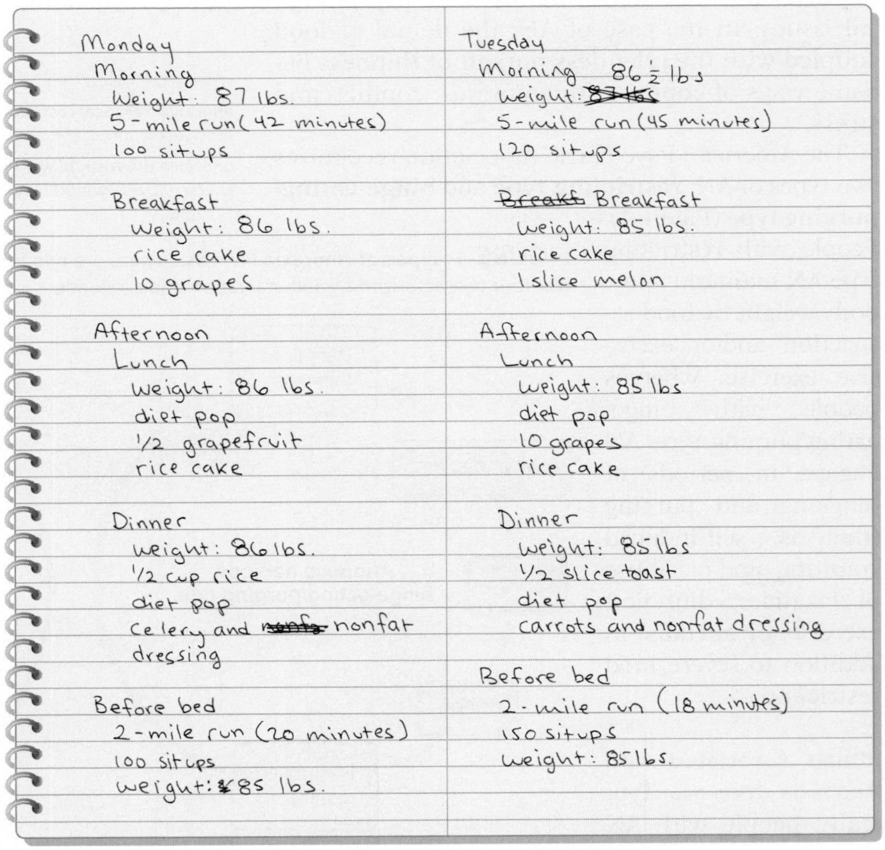

Monday
Morning
Weight: 87 lbs.
5-mile run (42 minutes)
100 situps

Breakfast
Weight: 86 lbs.
rice cake
10 grapes

Afternoon
Lunch
Weight: 86 lbs.
diet pop
1/2 grapefruit
rice cake

Dinner
Weight: 86 lbs.
1/2 cup rice
diet pop
Celery and ~~nonfat~~ nonfat dressing

Before bed
2-mile run (20 minutes)
100 situps
Weight: ~~85~~ lbs.

Tuesday
Morning 86½ lbs
Weight: ~~85 lbs~~
5-mile run (45 minutes)
120 situps

~~Breakf~~ Breakfast
Weight: 85 lbs.
rice cake
1 slice melon

Afternoon
Lunch
Weight: 85 lbs
diet pop
10 grapes
rice cake

Dinner
Weight: 85 lbs
1/2 slice toast
diet pop
carrots and nonfat dressing

Before bed
2-mile run (18 minutes)
150 situps
Weight: 85 lbs.

FIGURE 3 Thoughts and Behaviors Associated with Anorexia Nervosa

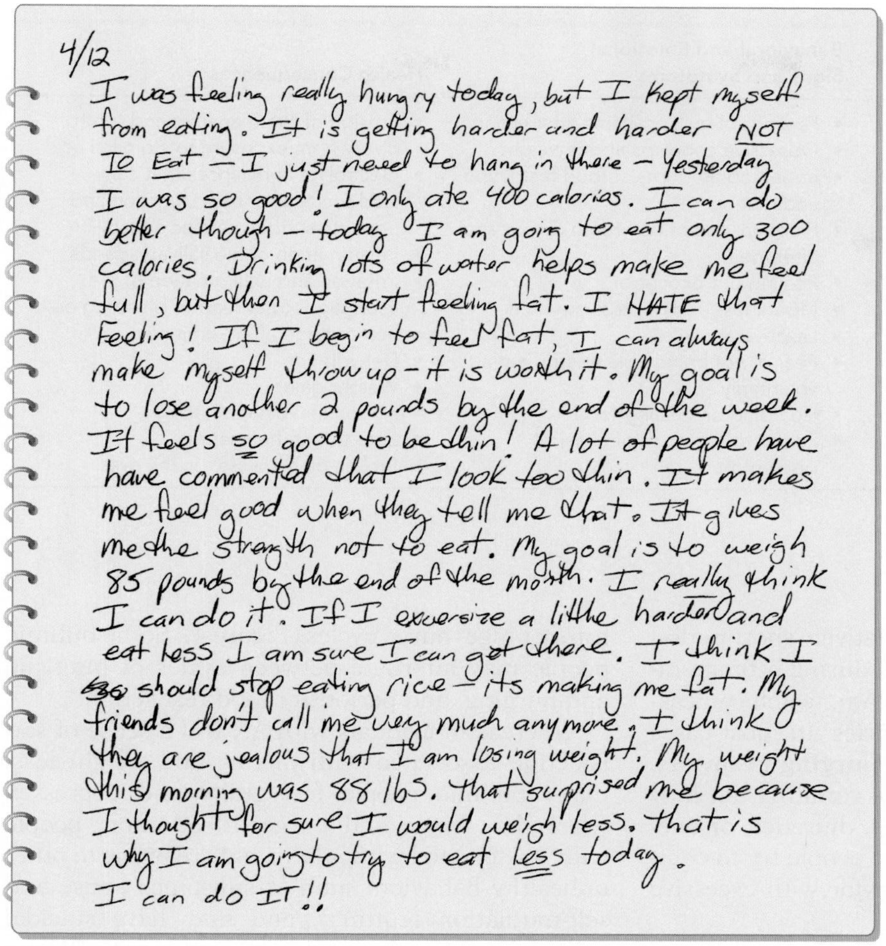

4/12

I was feeling really hungry today, but I kept myself from eating. It is getting harder and harder NOT To Eat — I just need to hang in there — Yesterday I was so good! I only ate 400 calories. I can do better though — today I am going to eat only 300 calories. Drinking lots of water helps make me feel full, but then I start feeling fat. I HATE that feeling! If I begin to feel fat, I can always make myself throw up — it is worth it. My goal is to lose another 2 pounds by the end of the week. It feels so good to be thin! A lot of people have have commented that I look too thin. It makes me feel good when they tell me that. It gives me the strength not to eat. My goal is to weigh 85 pounds by the end of the month. I really think I can do it. If I exercise a little harder and eat less I am sure I can get there. I think I should stop eating rice — its making me fat. My friends don't call me very much anymore. I think they are jealous that I am losing weight. My weight this morning was 88 lbs. That surprised me because I thought for sure I would weigh less. that is why I am going to try to eat less today. I can do IT!!

Health Concerns Associated with Anorexia Nervosa

Although people with AN eventually become underweight and appear gaunt, health problems can begin even before this happens. For instance, nutrient deficiencies can develop as well as electrolyte imbalances and hair loss. If AN is left untreated, serious health problems such as dehydration, cardiac abnormalities, appearance of unusual body hair (lanugo), muscle wasting, and bone loss can also occur. These problems are largely the result of chronic malnutrition. Once healthy eating habits and body weight are restored, most of these medical concerns can be reversed.

In females, the loss of body fat leads to a decline in the reproductive hormone estrogen, which can lead to disrupted menstruation and poor reproductive function.[7] These physiological responses undoubtedly evolved as a means to protect women from pregnancy during periods of low food availability. Estrogen also plays an important role in bone health by preventing bone loss. Therefore, the longer a woman goes without menstruating—a condition called amenorrhea—the greater the loss of bone. Although estrogen levels return to normal when body weight is restored, bones probably do not completely recover. For this reason, women with eating disorders who have experienced amenorrhea remain at increased risk for bone disease even after they have regained healthy eating patterns.[8]

AN is a serious condition that affects a person's physical and mental well-being and requires immediate medical, psychological, and nutritional intervention. Although researchers have reported that the death rate for AN is high, this is true only for the most severe cases that require hospitalization.[9] Nonetheless, the death rate associated with AN is higher than that of other eating disorders. It is estimated that between 5 and 20% of those who develop AN die from complications within 10 years of the initial diagnosis.[10] Causes of death are primarily attributed to cardiac arrest or suicide.[11]

PEOPLE WITH BULIMIA NERVOSA BINGE AND PURGE

Like those with AN, people with **bulimia nervosa (BN)** also use food as a coping mechanism. However, those with BN turn to, rather than away from, food during times of stress and emotional conflict. In fact, the origin of the word *bulimia* comes from the Greek word *boulimia*, meaning "the hunger of an ox." People with bulimia nervosa outnumber those with AN by about two to one.[12]

The most common type of BN is characterized by repeated cycles of "bingeing and purging."[13] **Bingeing** is defined as the compulsive consumption

---

bulimia nervosa (BN) An eating disorder characterized by repeated cycles of bingeing and purging.

bingeing Uncontrolled consumption of large quantities of food in a relatively short period of time.

TABLE 4 Physical Signs and Symptoms, Behavioral Characteristics, and Health Consequences Associated with Bulimia Nervosa

Physical Signs and Symptoms	Behavioral and Emotional Signs and Symptoms	Health Consequences
• Fluctuations in body weight • Swollen or puffy face • Odor of vomit on breath or in bathroom • Sores around mouth • Irregular bowel function • Inducement of vomiting after eating	• Feelings of guilt or shame after eating • Obsessive concerns about weight • Repeated attempts at food restriction and dieting • Frequent use of bathroom during and after meals • Feeling out of control • Moodiness and depression • Laxative abuse • Fear of not being able to stop eating voluntarily • Hoarding or stealing food • Eating to the point of physical discomfort	• Erosion of tooth enamel and tooth decay from exposure to stomach acid • Electrolyte imbalances that can lead to irregular heart function and possible sudden cardiac arrest • Inflammation of the salivary glands • Irritation and inflammation of the esophagus (may lead to bleeding or hemorrhage after vomiting) • Dehydration • Weight gain • Abdominal pain, bloating • Sore throat, hoarseness • Broken blood vessels in the eyes

of large amounts of food in a relatively short period of time. Foods often consumed during a binge include cakes, cookies, and ice cream, which can easily add up to thousands of calories. In most cases of BN, bingeing is followed by **purging** behaviors, which typically take the form of vomiting but may also include the use of laxatives, diuretics, or enemas. Alternatively, some bulimic people try to compensate for their bingeing behavior with excessive exercise.

Not all people with BN purge, and for this reason the American Psychiatric Association recognizes nonpurging as a specific type of BN.[14] People with nonpurging BN exercise or fast after a binge, rather than vomit. These and other characteristics commonly observed in people with BN are presented in Table 4.

Rituals Associated with Bulimia Nervosa Because bingeing and purging are often carried out secretly, family and friends may be unaware that someone has BN.[15] In fact, binges often occur when nobody is around. Although bulimics may consume several thousand calories during a binge, purging behaviors help to maintain energy balance. However, even purging immediately after bingeing cannot totally prevent nutrient absorption, and some weight gain is likely. People with BN tend to be within or slightly above their recommended weight range, and there may be no outward change in physical appearance.

Individuals with BN often experience regret and feelings of loss of control after bingeing. This can lead to depression and increased likelihood of future binge–purge cycles (Figure 4). Some bulimic people may alternate between cycles of bingeing and purging, and periods of food restriction.

Whereas individuals with AN feel a sense of satisfaction associated with dieting and weight loss, many bulimic people feel guilty and depressed when they binge and purge. In addition, people with BN tend to be impulsive and are prone to other unhealthy behaviors such as substance abuse and self-mutilation (cutting), and may have suicidal tendencies.[16] For these reasons, people with BN are more likely to seek treatment than are those with AN. Help for people with BN often comes after getting caught in the act of bingeing and purging.

Health Concerns Associated with Bulimia Nervosa Over time, repeated cycles of bingeing and vomiting can damage the body. Much of this damage is caused when the delicate lining of the esophagus and mouth becomes irritated by frequent exposure to stomach acid. The acidity of gastric juice can also damage dental enamel, causing tooth decay. Dentists and dental hygienists are often the first to notice these signs. Some people with bulimia induce vomiting by inserting their fingers deep into the mouth. As a result, hands can become scraped from striking the teeth—another characteristic sign of BN. Frequent vomiting and/or overuse of laxatives and diuretics can cause dehydration and electrolyte

purging Self-induced vomiting and/or misuse of laxatives, diuretics, and/or enemas.

FIGURE 4 Thoughts and Behaviors Associated with Bulimia Nervosa

> ~ 5/9 ~
>
> \# I feel so fat and ugly. Yesterday I saw this guy that I like walking with another girl — it made me feel sad. When I got to class, the teacher passed back our exam. I didn't get a very good grade. I WISH I WAS PRETTY AND SMART!!!
>
> Yesterday I felt so bad that I binged and purged most of the afternoon. I hate when I do that but I can't help myself. I didn't intend for it to happen. I bought all this food thinking that it would last. It lasted for about an hour. Before I knew it the cookies and ice cream were gone. Then I went to the dining hall and ate dinner. !! As soon as I got back to my room I made myself vomit again. I was glad that my roommate wasn't there. She would think I am so GROSS. That is because I AM GROSS!!! I wish I could stop this. Tomorrow I am going to diet. I am going to get thin. Yeah, right ....

imbalance. This can lead to irregular heart function and can even result in sudden cardiac arrest. Fortunately, many people with BN seek help and medical assistance, and most improve with treatment.

MOST EATING DISORDERS ARE CLASSIFIED AS "NOT OTHERWISE SPECIFIED"

Certainly, AN and BN are the most familiar eating disorders. However, other habitual, unhealthy eating behaviors that do not meet precise diagnostic criteria associated with AN or BN can also be problematic. **Eating disorders not otherwise specified (EDNOS)** is a general category of eating disorders that are not as well-defined or thoroughly researched as AN and BN. Individuals with EDNOS may display a spectrum of behaviors and traits associated with AN or BN but do not meet all of the diagnostic criteria. For example, a person may exhibit behaviors associated with AN but have normal weight and (if a woman) still menstruate.

Binge-Eating Disorder Only recently has the American Psychiatric Association recognized **binge-eating disorder (BED)** as a disordered eating pattern distinct from bulimia nervosa. There is some uncertainty as to how BED should be classified. Currently, BED falls under the broad category of EDNOS. Although many practitioners and researchers believe that BED should be officially recognized with its own diagnostic category, others feel doing so is premature and that the condition needs further study.[17]

The number of people who fall into this category is much greater than the number of people with AN and BN combined.[18] Although people with BED can be of normal weight, most are overweight. In fact, some weight-loss treatment programs estimate that between 20% and 40% of obese patients experience BED.[19] The prevalence of BED in the general population is approximately 1 to 5%.[20]

Binge-eating disorder is characterized by recurring episodes of consuming large amounts of food within a short period of time. The American

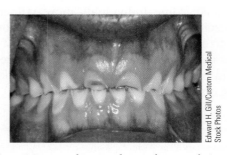

Edward H. Gill/Custom Medical Stock Photos

Repeated vomiting can damage the tooth enamel. As a result, teeth appear mottled. This can occur in people with bulimia nervosa and in individuals with other disordered eating behaviors.

eating disorders not otherwise specified (EDNOS) A category of eating disorders that includes some, but not all, of the diagnostic criteria for anorexia nervosa and/or bulimia nervosa.

binge-eating disorder (BED) A subcategory of EDNOS characterized by recurring consumption of large amounts of food within a short period of time, but not followed by purging.

Psychiatric Association's provisional diagnostic criteria for BED requires that a person eat excessively large amounts of food in a two-hour period at least twice per week for at least six months; feels a lack of control over the episodes; and experiences feelings of disgust, depression, or guilt in response to overeating.[21] Other proposed diagnostic criteria include eating much more rapidly than normal, eating until uncomfortably full, eating large amounts of food (even when not physically hungry), and eating alone out of embarrassment at the quantity of food being eaten.[22]

Although most people overeat from time to time, binges associated with BED are distinctly different. For many people with BED, binges provide an escape from stress and emotional pain. In other words, food has a psychological numbing effect and induces a state of emotional well-being. Some studies indicate that people with BED are likely to have been raised in families affected by alcohol abuse.[23] In the case of BED, however, food rather than alcohol becomes their "drug of choice." Anger, sadness, anxiety, and other types of emotional distress often trigger binges. Many struggle with clinical depression, although it is not clear if depression triggers BED or if BED causes depression.[24]

As in BN, binges in BED typically take place in private and are often accompanied by feelings of shame. In general, individuals with BED do not purge after bingeing, and therefore tend to be in positive energy balance. As a result, BED can lead to unwanted weight gain, increasing a person's risk for weight-related health problems such as type 2 diabetes, gallstones, and cardiovascular disease.

Some People with Binge-Eating Disorder Are Restrained Eaters Individuals who suppress their desire for food and avoid eating for long periods of time between binges are called **restrained eaters.**[25] Restrained eaters limit their food intake to lose weight. However, after an extended period of food restriction, restrained eaters find themselves feeling out of control and respond by bingeing. This cycle of fasting and bingeing can be difficult to stop. Many restrained eaters perceive themselves as overweight, and consuming large amounts of food generates further feelings of inadequacy. Feelings of self-contempt can cause such people to turn back to food for emotional comfort. Similar to other disordered eating patterns, restrained eaters find themselves in a vicious cycle resulting in poor physical and psychological health.

Are There Other Disordered Eating Behaviors?

In addition to the eating disorders formally recognized by the American Psychiatric Association, there are many other troublesome disordered eating behaviors. These include not eating in public, situational purging, chewing food but spitting it out before swallowing, and obsessive dieting. Although there is insufficient information for these to be classified as "true" eating disorders, these food-related disturbances can be debilitating and disruptive. If these behaviors continue, they could progress into a full-blown eating disorder. As with any newly recognized pattern of dysfunctional behavior, there is a need for more research and a better understanding of these conditions. Some of the more recognized food-related disturbances include excessive night-time eating, avoidance of new foods, and obsessions concerning muscularity, which are described next.

SOME FOOD-RELATED DISTURBANCES INVOLVE NOCTURNAL EATING

Nocturnal sleep-related eating disorder (SRED) is characterized by eating while asleep.[26] Reportedly, individuals leave their bed and walk to the kitchen to prepare and eat food. This entire episode takes place without any recollection of having done so. Individuals may begin to suspect that they have SRED when they notice unexplained missing food and see evidence of these late-night activities left behind in the kitchen. Foods most typically consumed are those high in fat and calories and may involve unusual food combinations. It is common for people with SRED to experience weight gain. Although both men and women can experience SRED, it is far more common in women. Some researchers believe that stress, dieting, and depression can trigger episodes of SRED and that food restriction during the day may make some individuals more vulnerable to unconscious binge eating at night.[27] Although difficult to estimate, it has been reported that 1 to 3% of the general population may experience SRED, with rates as high as 10 to 15% among those with eating disorders.[28]

---

restrained eaters People who experience cycles of fasting followed by bingeing.

nocturnal sleep-related eating disorder (SRED) A disordered eating pattern characterized by eating while asleep without any recollection of having done so.

Night eating syndrome (NES) is a disorder closely related to SRED, except individuals are fully aware of their eating. It is best characterized by a cycle of daytime food restriction, excessive food intake in the evening, and nighttime insomnia.[29] In fact, people with NES typically consume more than half of their daily calories after dinner. Most people report that night eating causes them to feel depressed, anxious, and guilty. In adults, the prevalence of NES is between 1% and 2%, and it is more common among females than males. Reportedly, 9 to 15% of individuals participating in weight-loss treatment programs experience episodes of NES.[30] Signs and symptoms associated with NES include the following.

- Not feeling hungry for the first several hours after waking
- Overeating in the evening, with more than one-half of daily food intake consumed after dinner
- Difficulty falling asleep, accompanied by the urge to eat
- Waking during the night and finding it necessary to eat before falling back asleep
- Feeling guilty, ashamed, moody, tense, and agitated, especially at night

FOOD NEOPHOBIA: AVOIDANCE OF TRYING NEW FOODS

It is not unusual for parents of young children to complain about having a picky eater. In fact, some children are reluctant to try new foods, and even slight changes in food routines can be upsetting. Most children outgrow these behaviors along with their avoidance of new foods. However, some people are not able to rid themselves of these irrational fears and may be at increased risk of developing food neophobia later in life.

Food neophobia is an eating disturbance defined as an irrational fear, or avoidance, of trying new foods. Individuals with food neophobia typically eat a very limited range of foods and often have well-defined food rituals and practices.[31] For example, some may refuse to eat foods made from two or more food items. In other words, foods when eaten separately may be enjoyed but when joined together are considered disgusting. In adults, these types of behaviors are socially restricting and embarrassing. In extreme cases, they can lead to nutritional inadequacies, although some individuals may take supplements to compensate for their poor eating habits. People with food neophobia tend to be slightly overweight because they often limit themselves to comfort foods such as hamburgers, french fries, and macaroni and cheese,

Night eating syndrome (NES) is best characterized by a cycle of daytime food restriction, excessive food intake in the evening, and nighttime insomnia.

which are high in calories.[32] Yet, most adults with food neophobia are not interested in therapy or getting treatment. Some researchers believe that food neophobia is a type of obsessive–compulsive disorder that may require extensive psychological therapy.[33]

People with food neophobia have an irrational fear of trying new foods and engage in unusual food rituals and food-related practices.

---

night eating syndrome (NES) A disordered eating pattern characterized by a cycle of daytime food restriction, excessive intake in the evening, and nighttime insomnia.

food neophobia An eating disturbance characterized as an irrational fear, or avoidance, of trying new foods.

MUSCLE DYSMORPHIA: PREOCCUPATION WITH MUSCULARITY

Although societal pressure to be thin tends to be more intense for females than males, it is not uncommon for males to also have issues related to body image. However, rather than wanting to become thin, males more often want to be strong and muscular. This desire may be reinforced by the media, which portrays an "ideal" male body as having strong, well-defined muscles. Once referred to as "reverse anorexia," **muscle dysmorphia** is a disorder seen primarily in men with an intense fear of being too small, weak, or skinny.[34] The term *dysmorphia* comes from two Greek words: *dys*, which means "bad" or "abnormal," and *morphos*, which means "shape" or "form." Thus, people with muscle dysmorphia are preoccupied with perceived defects with their bodies.

Activities and tendencies associated with muscle dysmorphia include working out for hours each day, allowing exercise to interfere with family and social life, paying excessive attention to diet, and having unusual food rituals and eating practices. Reportedly, people with muscle dysmorphia may also engage in health-threatening practices, such as use of anabolic steroids, to gain sufficient muscle mass.[35]

Individuals affected by muscle dysmorphia often have personality traits similar to those with recognized eating disorders.[36] One study reported that one-third of men with muscle dysmorphia had a history of eating disorders.[37] Studies also show that men with muscle dysmorphia tend to have low self-esteem and concerns regarding their masculinity.[38] To compensate for these insecurities, individuals are driven to achieve the "perfect male body." Haunted by obsessive thoughts of being "puny" and "weak," individuals

John Lamb/Getty Images

Muscle dysmorphia has been described as the opposite of anorexia nervosa and is characterized by an obsessive preoccupation with increasing muscularity.

with muscle dysmorphia can have difficulties maintaining personal and social relationships with others.

What Causes Eating Disorders?

Most females with eating disorders are adolescents or young adults. Nonetheless, a pattern of eating disorders among middle-aged or older adult women has also begun to emerge.[39] This trend may be a response to a youth-oriented society that intensifies insecurities associated with aging. Because the occurrence of age-related health problems increases by middle-age, developing an eating disorder during this stage of life may be particularly concerning. For example, bone loss associated with menopause could be made worse by food restriction.

The prevalence of eating disorders in females is well-documented, but far less is known about males. Researchers estimate that males account for approximately 5 to 10% of people with AN and 10 to 15% of people with BN.[40] Because the small number of males with eating disorders makes this group difficult to study, it is not clear what factors are related to eating disorders in males. However, they may include history of obesity, participation in a sport that emphasizes thinness, and heightened emphasis on physical appearance.

Scientists have many theories as to what causes eating disorders, but there are no simple answers. People develop eating disorders for a variety of reasons, and why some are more vulnerable than others is not clear. However, several factors are likely contributors, including sociocultural characteristics, family dynamics, personality traits, and biological (genetic) factors.[41]

SOCIOCULTURAL FACTORS

Eating disorders are more prevalent in some cultures than others. Cultures where food is abundant and slimness is valued are the ones most likely to foster eating disorders. This is one reason why eating disorders are more widespread in Western industrialized countries than in regions of Africa, China, and many Arab nations.[42]

Differences in sociocultural environments within American society may also influence the

---

muscle dysmorphia Pathological preoccupation with increasing muscularity.

AP Photo/Mark Lennihan

© Beepstock/Alamy

Many popular dolls have body proportions that are unrealistic, emphasizing extreme muscularity or thinness.

development of eating disorders. However, beliefs that may have one time protected certain ethnic groups against eating disorders may be eroding as multicultural youth acculturate to mainstream American values. It appears that the desire to fit into white, middle-class society has led to a rise in the occurrence of eating disorders across racial and ethnic groups. Data show that the prevalence of eating disorders among Latina- and African-American women is catching up to that of their Caucasian counterparts.[43]

The Role of the Media The media, which often portray unrealistic physiques and glamorize unnaturally thin bodies, are criticized for invoking a sense of inadequacy in young, impressionable people. Even the body shapes of popular dolls are not realistic. All of these factors, and more, likely play an important role in predisposing susceptible individuals to eating disorders. For example, celebrity role models shown on television and in the movies can give teenagers an unrealistic standard of thinness, causing some to engage in unhealthy eating practices to achieve this thin and glamorous appearance.[44]

Based on media images, the perfect body is tall, is lean, and has well-defined muscles. Whereas the average woman is 5 feet 4 inches tall and 140 lbs, the average model is 5 feet 11 inches tall and 117 lbs.[45] In fact, female fashion models, beauty pageant contestants, and actresses have become increasingly thinner over the years. Even the body weights of Miss America beauty contestants have decreased. When the pageant first began in the 1920s, BMI averaged 20 to 25 kg/m². Today, nearly all the participants have BMIs below what is considered healthy. Furthermore, nearly half of the contestants have BMIs consistent with one of the diagnostic criteria for AN ($<18.5 \, \text{kg/m}^2$).[46] One study reports that women who participated in beauty pageants as children were more likely to experience body dissatisfaction, interpersonal distrust, and impulse dysregulation than nonparticipants.[47]

Because dissatisfaction with body weight and shape is thought to be an essential precursor to the development of eating disorders, it is important for children to understand that healthy bodies come in many shapes and sizes. Just as important, older children must be prepared for the physical and emotional changes associated with puberty. Changes in body dimensions and weight gain during adolescence can make females feel embarrassed and uncomfortable with their maturing bodies.

Although eating disorders frequently begin with a desire to lose weight, relatively few people on weight-loss diets develop them. Furthermore, not everyone living in an affluent society that stigmatizes obesity and advocates extreme slenderness develops an eating disorder. So these factors alone are not sufficient to cause eating disorders. Rather, the social and cultural environment in industrialized countries appears to foster the development

The BMIs of Miss America beauty contestants have decreased over the last 50 years.

of eating disorders in individuals who are already vulnerable in some other way.

Social Networks A person's peers may play a contributing role in the development of eating disorders. This is because attitudes and behaviors about slimness and appearance are often learned from those we associate with. Body dissatisfaction and dieting in adolescents often stem from the desire to gain acceptance among their friends. Some teenagers, especially girls, may feel that—to be liked and to belong—they must be thin.

FAMILY DYNAMICS

Parents influence many aspects of their children's lives. Thus, it is not surprising that family dynamics can play a role in the development and perpetuation of eating disorders. Although no single family type necessarily leads to the development of eating disorders, researchers have found certain distinguishing behaviors and characteristics that increase the likelihood of this occurring.[48] These include overprotectiveness, rigidity, conflict avoidance, abusiveness, chaotic family dynamics,

and the presence of a mother with an eating disorder.[49]

Enmeshed Families **Enmeshment** is a term used to describe family members who are overly involved with one another and have little autonomy.[50] An enmeshed family has no clear boundaries among its members. This environment can make it difficult for children to develop independence and individualism. Children raised in such families often feel tremendous pressure to please their parents and meet expectations. Rather than doing things for themselves, they strive to please others. Enmeshed family dynamics promote dependency, which may lay the foundation for the emergence of eating disorders. Under these circumstances, food may become the only component in the child's life over which he or she can exert control.

Chaotic Families In contrast to enmeshed families in which family connections are exceedingly tight,

enmeshment Families whose interaction is overly involved with one another and have little autonomy.

Children who participate in beauty pageants may be more likely to experience body dissatisfaction later in life.

chaotic families have exceedingly loose family structures. **Chaotic families,** also called disengaged families, lack cohesiveness, and there is little parental involvement.[51] The roles of family members are loosely defined; children often have a sense of abandonment; and parents may be depressed, alcoholic, or emotionally absent. A child growing up in this type of home may later develop eating disorders as a way to fill an emotional emptiness, gain attention, or suppress emotional conflict.

Mothers with Eating Disorders In addition to overall family dynamics, it appears that the presence of a mother with an eating disorder or body dissatisfaction can negatively influence eating behaviors in her children.[52] The inability of a mother to demonstrate a healthy relationship with food and model healthy eating to her children is a serious concern. Furthermore, mothers with eating disorders are more likely to criticize their daughters' appearance and encourage them to lose weight. As a result, children of women with eating disorders are at increased risk for developing eating disorders themselves.

PERSONALITY TRAITS AND EMOTIONAL FACTORS CAN TRIGGER EATING DISORDERS

Scientists have long thought that certain personality traits make some people more prone than others to eating disorders. Some of these characteristics include low self-esteem; lack of self-confidence; and feelings of helplessness, anxiety, and depression.[53] Individuals with eating disorders are often described as being **food preoccupied.** That is, they spend an inordinate amount of time thinking about food. Another personality trait commonly associated with eating disorders is perfectionism.[54] Such people have difficulty dealing with shortcomings in themselves. Thus, an imperfect body is not easily tolerated.

BIOLOGICAL AND GENETIC FACTORS MAY ALSO PLAY A ROLE IN EATING DISORDERS

Seeking to better understand eating disorders, researchers have investigated biological influences that may play a role in their development. Because certain personality traits and eating behaviors are, in part, determined by the nervous and endocrine systems, it makes sense that brain chemicals may play a role in

the development of eating disorders. However, it is also possible that disordered eating may disrupt neuroendocrine regulation. For example, studies show that individuals with eating disorders are often clinically depressed.[55] It is therefore difficult to determine if clinical depression leads to eating disorders or vice versa. In any case, because medication used to treat clinical depression is often effective in the treatment of certain eating disorders, depression may be a contributing factor.

Studies of identical and fraternal twins provide evidence that eating disorders may, in part, be inherited, and scientists have become interested in identifying genes that might influence susceptibility.[56] Although these types of studies cannot completely differentiate the contribution of genes versus environment, some research suggests that the contribution of genetics may actually be greater than that of the environment. How this occurs, however, is unclear.[57]

Are Athletes at Increased Risk for Eating Disorders?

There are more competitive female athletes today than ever before, and perhaps not coincidentally, the number of female athletes with eating disorders

---

chaotic families Families whose interaction is characterized by a lack of cohesiveness and little parental involvement.

food preoccupation Spending an inordinate amount of time thinking about food.

John Kelly/Getty Images

Sports that demand a thin physical appearance are likely to have more athletes with eating disorders than activities where large size is thought to be an advantage.

has increased as well. Some studies indicate that the prevalence of eating disorders among female student-athletes and nonathletes do not differ,[58] while other studies suggest otherwise.[59] At any rate, losing weight can have serious health consequences for an athlete, above and beyond affecting athletic performance. It is, therefore, important for coaches and trainers to recognize early warning signs and symptoms associated with eating disorders in athletes.

ATHLETICS MAY FOSTER EATING DISORDERS IN SOME PEOPLE

The prevalence of disordered eating and eating disorders among collegiate athletes is estimated to be somewhere between 15 and 60%.[60] Disagreement and inconsistent estimates may be due, in part, to the reluctance of athletes to admit that they have such a problem. In addition, some athletes may exhibit disordered eating behaviors yet do not satisfy all the criteria needed for diagnosis. Regardless of the exact number, athletes (especially females) are considered by many experts to be a group at risk for developing eating disorders.

For some athletes, physical performance is not only determined by speed, strength, and coordination but also by body weight. Athletes such as ski jumpers, cyclists, rock climbers, and long-distance runners may deliberately try to achieve a low body weight to gain a competitive advantage. In addition, sports such as gymnastics that demand a thin physical appearance are likely to have more athletes with eating disorders than are activities where greater size may be beneficial.[61] This may, in part, be due to the fact that judges often consider size and appearance when rating performance. Athletes who participate in dancing, figure skating, synchronized swimming, gymnastics, and diving also perceive that body size can affect how judges rate their performance.

Because athletes are invariably competitive people and may equate their self-worth with athletic success, they may be especially willing to engage in risky weight-loss practices. In addition, coaches and trainers often believe that excess weight can hinder performance. According to the National Collegiate Athletic Association (NCAA), sports with the highest number of female athletes with eating disorders are cross-country, gymnastics, swimming, and track and field.[62] Sports with the highest number of male athletes with eating disorders are wrestling and cross-country.

THE FEMALE ATHLETE TRIAD

Athletes with eating disorders are at extremely high risk for developing medical complications. This is largely because the rigor of athletic training alone is very stressful on the body, and adequate nourishment is required to meet these physical demands. Serious health problems can arise when an athlete is restricting food intake and bingeing and/or purging. For example, female athletes are at increased risk for developing a syndrome known as the **female athlete triad.** The female athlete triad is a combination of interrelated conditions including disordered eating (or eating disorders), amenorrhea, and osteopenia—a condition characterized by a loss of calcium from bones (Figure 5).[63]

How the three components of the female athlete triad are related is complex. Disordered eating can lead to very low levels of body fat, which can cause estrogen levels to decrease. Without adequate

---

female athlete triad A combination of interrelated conditions: disordered eating/eating disorder, menstrual dysfunction, and osteopenia.

FIGURE 5 Female Athlete Triad The female athlete triad is a combination of interrelated conditions: disordered eating or eating disorder, menstrual dysfunction, and osteopenia.

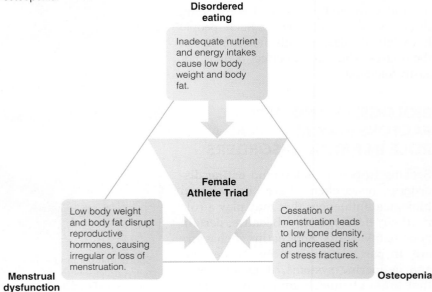

Disordered eating

Inadequate nutrient and energy intakes cause low body weight and body fat.

Female Athlete Triad

Low body weight and body fat disrupt reproductive hormones, causing irregular or loss of menstruation.

Cessation of menstruation leads to low bone density, and increased risk of stress fractures.

Menstrual dysfunction

Osteopenia

FIGURE 6 Stress Fractures and Menstrual History in Collegiate Female Distance Runners

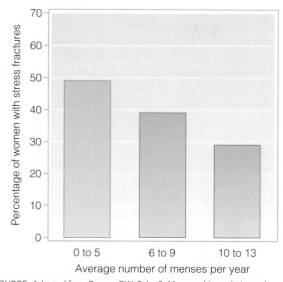

SOURCE: Adapted from Barrow GW, Saha S. Menstrual irregularity and stress fractures in collegiate female distance runners. American Journal of Sports Medicine. 1988;3:209–16.

estrogen levels, menstrual cycles can become irregular and, in some cases, stop completely. A lack of estrogen can cause bones to become dangerously weak, leading to osteopenia. In more severe cases, the entire matrix of the bone can begin to deteriorate, a condition called osteoporosis. As shown in Figure 6, the risk of stress fractures in collegiate female runners increased as menstrual cycles became more irregular.[64]

The risk of stress fractures in female runners increases when menstrual cycles become irregular or stop altogether.

Determining the prevalence of women with the female athlete triad can be difficult, because some athletes may feel unburdened when menstruation stops and therefore are unlikely to report it. However, these athletes often begin to experience repeated injuries such as stress fractures, which can draw attention to the fact that they may have a problem. It is important for parents, coaches, and health care providers to be aware of the spectrum of disordered eating and eating disorders among athletes, so that assistance can be provided.

How Can Eating Disorders Be Prevented and Treated?

Those with eating disorders are at risk for serious medical and/or emotional problems and need treatment from qualified health professionals. Typically, this team includes mental health specialists who can help address and treat underlying psychological issues, medical doctors who can treat physiological complications, and a dietitian who helps the person make better food choices. An important treatment goal for individuals with eating disorders is to learn how to enjoy food without fear and guilt and to rely on the physiological cues of hunger and satiety to regulate food intake. For this to become possible, specialists help people with eating disorders recognize and appreciate their self-worth. Many treatment options are available for people who have eating disorders. However, the first step is to recognize that there is a problem and to seek help.

PREVENTION PROGRAMS MUST PROMOTE A HEALTHY BODY IMAGE

Programs and educational curricula designed to increase awareness and prevent eating disorders in young girls produce varied results. Although it is important to reach school-age children before eating disorders begin, too many programs simply focus on deterring dangerous eating-disorder behaviors rather than on encouraging healthy attitudes toward food, dieting, and body image. To prevent eating disorders from developing, educational strategies must focus on issues related to overall health and self-esteem. Suggestions for promoting a healthy body image among children and adolescents are presented in Table 5.

TABLE 5 Promoting a Healthy Body Image among Children and Adolescents

- Encourage children to focus on positive body features.
- Help children understand that everyone has a unique body size and shape.
- Be a good role model for children by demonstrating healthy eating behaviors.
- Resist making negative comments about your own weight or body shape.
- Focus on positive *non*physical traits such as generosity, kindness, and a friendly laugh.
- Do not criticize a child's appearance.
- Never associate self-worth with physical attributes.
- Prepare a child in advance for puberty by discussing physical and emotional changes.
- Enjoy meals together as a family.
- Discuss how the media can negatively affect body image.
- Avoid using food as a reward or punishment.

SOURCE: Adapted from Story, M., Holt, K., Sofka, D., "Bright Futures in Practice: Nutrition" (2nd ed.). Arlington, VA: National Center for Education in Maternal and Child Health, 2002. Reprinted by permission of the National Center for Education in Maternal and Child Health.

TREATMENT STRATEGIES MUST FOCUS ON PSYCHOLOGICAL ISSUES

People with eating disorders may not recognize or admit they have a problem. Concerns expressed by friends and family members are often repeatedly ignored or dismissed, making them feel confused and frustrated by their inability to help—a completely normal reaction to a very difficult situation. It is important to remember that, even though people with eating disorders may resist getting help, family and friends play important supportive roles. Experts recommend that they not focus on the eating disorder *per se*, because this may make the person feel more defensive. Instead, expressing concerns regarding the person's unhappiness and encouraging him or her to seek help may be more effective.

It is important that the medical team or program chosen to treat a person with an eating disorder has specific training and expertise in this area. Treatment goals for people with AN include restoration of lost weight, resolution of psychological issues such as low self-esteem and distorted body image, and long-term recovery of healthy eating patterns. Because some people with AN are severely malnourished, they may benefit from intensive care provided by inpatient facilities.[65] These types of facilities are staffed by professionals who can provide care in all aspects of recovery, including medical needs, nutritional problems, and psychological issues. This is best achieved by a team of medical professionals, including physicians, nurses, social workers, mental health therapists, and dietitians.

People with BN, like those with AN, often have nutritional, medical, and psychological issues that must be addressed during recovery. Treatment goals for people with BN also include the reduction and eventual elimination of bingeing and purging. For example, establishing a healthy relationship with food, developing strategies to help resist the urge to binge and purge, and maintaining a healthy body weight without bingeing and purging are all important.

The sooner a person with an eating disorder gets help, the better chance of full recovery. However, recovery can be a long and slow process. Well-meaning advice such as "just eat" only makes matters worse. It is important for families and friends to know that most people with eating disorders do recover. Treatment often involves counseling for the entire family, which helps everyone to heal and to move forward in life.

Key Points

How Do Eating Disorders Differ from Disordered Eating?

- The three main types of eating disorders are anorexia nervosa (AN), bulimia nervosa (BN), and eating disorders not otherwise specified (EDNOS).
- AN is characterized by a fear of weight gain, a distorted body image, and food restriction.
- BN involves cycles of bingeing and purging.
- Binge-eating disorder (BED) has been provisionally classified as an EDNOS and is characterized

by having at least two binges per week for at least six months.

Are There Other Disordered Eating Behaviors?

- Nocturnal sleep-related eating disorder (SRED) is characterized by eating while asleep, without recollection of having done so.
- Night eating syndrome (NES) is characterized by a cycle of daytime food-restriction, excessive food intake in the evening, and nighttime insomnia.

- Food neophobia is an eating disturbance characterized by an irrational fear, or avoidance of trying new foods and by unusual food rituals and practices.

- Muscle dysmorphia occurs primarily in men who see themselves as weak and/or small, even though they have lean, well-defined physiques.

What Causes Eating Disorders?

- Eating disorders are more prevalent in cultures where food is abundant and slimness is valued.

- Dysfunctional family dynamics are associated with eating disorders.

- Personality traits associated with eating disorders include low self-esteem, lack of self-confidence, obsessiveness, and feelings of helplessness, anxiety, and depression.

Are Athletes at Increased Risk for Eating Disorders?

- Sports that value and reward a thin physical appearance are likely to have more athletes with eating disorders than activities for which size is not as important.

- Female athletes are at increased risk for developing the "female athlete triad": disordered eating/eating disorder, menstrual dysfunction, and osteopenia.

How Can Eating Disorders Be Prevented and Treated?

- Eating disorder prevention involves educational strategies that focus on self-esteem and encourage healthy behaviors, rather than just the dangers of eating disorders.

- Once an eating disorder has developed, it is important to seek treatment from qualified professionals.

- Treatment goals for people with AN include restoration of lost weight and resolution of psychological issues such as low self-esteem.

- Treatment goals for people with BN include the reduction and eventual elimination of bingeing and purging.

Notes

1. Woodside BD, Garfinkel PE, Lin E, Goering P, Kaplan AS, Goldbloom DS, Kennedy SH. Comparisons of men with full or partial eating disorders, men without eating disorders, and women with eating disorders in the community. American Journal of Psychiatry. 2001;158:570–4.

2. Paxton SJ. Body dissatisfaction and disordered eating. Journal of Psychosomatic Research. 2002;53:961–2.

3. American Psychiatric Association. Diagnostic and statistical manual of mental disorders, 4th ed. (DSM-IV). Washington, DC: American Psychiatric Association; 2004.

4. Eddy KT, Dorer DJ, Franko DL, Tahilani K, Thompson-Brenner H, Herzog DB. Diagnostic crossover in anorexia nervosa and bulimia nervosa: Implications for DSM-V. American Journal of Psychiatry. 2008;165:245–50.

5. Keel PK, Klump KL, Miller KB, McGue M, Iacono WG. Shared transmission of eating disorders and anxiety disorders. International Journal of Eating Disorders. 2005;38:99–105.

6. Halmi KA, Tozzi F, Thornton LM, Crow S, Fichter MM, Kaplan AS, Keel P, Klump KL, Lilenfeld LR, Mitchell JE, Plotnicov KH, Pollice C, Rotondo A, Strober M, Woodside DB, Berrettini WH, Kaye WH, Bulik CM. The relation among perfectionism, obsessive-compulsive personality disorder and obsessive-compulsive disorder in individuals with eating disorders. International Journal of Eating Disorders. 2005;38:371–4.

7. Wolfe BE. Reproductive health in women with eating disorders. Journal of Obstetrics and Gynecology in Neonatal Nursing. 2005;34:255–63.

8. Tudor-Locke C, McColl RS. Factors related to variation in premenopausal bone mineral status: A health promotion approach. Osteoporosis International. 2000;11:1–24.

9. Birmingham CL, Su J, Hlynsky JA, Goldner EM, Gao M. The mortality rate from anorexia nervosa. International Journal of Eating Disorders. 2005;38:143–6.

10. Patrick L. Eating disorders: A review of the literature with emphasis on medical complications and clinical nutrition. Alternative Medical Review. 2002;7:184–202.

11. Holm-Denoma JM, Witte TK, Gordon KH, Herzog DB, Franko DL, Fichter M, Quadflieg N, Joiner TE Jr. Deaths by suicide among individuals with anorexia as arbiters between competing explanations of the anorexia-suicide link. Journal of Affective Disorders. 2008;107:231–6.

12. Williams PM, Goodie J, Motsinger CD. Treating eating disorders in primary care. American Family Physician. 2008;77:187–95.

13. Hay P, Bacaltchuk J. Bulimia nervosa. Clinical Evidence. 2004;12:1326–47.

14. Cooper Z, Fairburn CG. Refining the definition of binge eating disorder and nonpurging bulimia nervosa. International Journal of Eating Disorders. 2003;34:S89–95.

15. Kruger D. Bulimia nervosa: Easy to hide but essential to recognize. Journal of the American Academy of Physician Assistants. 2008;21:48–52.

16. Thompson-Brenner H, Eddy KT, Franko DL, Dorer D, Vashchenko M, Herzog DB. Personality pathology and substance abuse in eating disorders: A longitudinal study. International Journal of Eating Disorders. 2008;41:203–8.

17. Bulik CM, Brownley KA, Shapiro JR. Diagnosis and management of binge eating disorder. World Psychiatry. 2007;6:142–8.

18. Striegel-Moore RH, Franko DL. Epidemiology of binge eating disorder. International Journal of Eating Disorders. 2003;34:S19–S29.

19. Niego SH, Kofman MD, Weiss JJ, Geliebter A. Binge eating in the bariatric surgery population: A review of the literature. International Journal of Eating Disorders. 2007;40:349–59.

20. Pagoto S, Bodenlos JS, Kantor L, Gitkind M, Curtin C, Ma Y. Association of major depression and binge eating disorder with weight loss in a clinical setting. Obesity. 2007;15:2557–9.

21. American Psychiatric Association. Diagnostic and Statistical Manual of Mental Disorders. 4th ed. Text Revision. Washington, DC: American Psychiatric Association; 2004.

22. Latner JD, Clyne C. The diagnostic validity of the criteria for binge eating disorder. International Journal of Eating Disorders. 2008;41:1–14.

23. Dansky BS, Brewerton TD, Kilpatrick DG. Comorbidity of bulimia nervosa and alcohol use disorders: Results from the National Women's Study. International Journal of Eating Disorders. 2000;27:180–90.

24. Vanderlinden J, Dalle Grave R, Fernandez F, Vandereycken W, Pieters G, Noorduin C. Which factors do provoke binge eating? An exploratory study in eating disorder patients. Eating and Weight Disorders. 2004;9:300–5.

25. Masheb RM, Grilo CM. On the relation of attempting to lose weight, restraint, and binge eating in outpatients with binge eating disorder. Obesity Research. 2000;8:638–45.

26. Schenck CH, Mahowald MW. Review of nocturnal sleep-related eating disorders. International Journal of Eating Disorders. 1994;15:343–56.

27. Winkelman JW. Sleep-related eating disorder and night eating syndrome: Sleep disorders, eating disorders, or both? Sleep. 2006;29:949–54.

28. The Cleveland Clinic. Sleep-related eating disorders. Available from http://my.clevelandclinic.org/disorders/sleep_disorders/hic_sleep-related_eating_disorders.aspx.

29. O'Reardon JP, Peshek A, Allison KC. Night eating syndrome: diagnosis, epidemiology and management. CNS Drugs. 2005;19:997–1008.

30. Tanofsky-Kraff M, Yanovski SZ. Eating disorder or disordered eating? Non-normative eating patterns in obese individuals.Obesity Research. 2004;12:1361–66.

31. Marcontell DK, Laster AE, Johnson J. Cognitive-behavioral treatment of food neophobia in adults. Journal of Anxiety Disorders. 2003;17:243–51.

32. Nicklaus S, Boggio V, Chabanet C, Issanchou S. A prospective study of food variety seeking in childhood, adolescence and early adult life. Appetite. 2005;44:289–97.

33. Pelchat ML. Of human bondage: Food craving, obsession, compulsion, and addiction. Physiology and Behavior. 2002;76:347–52.

34. Pope CG, Pope HG, Menard W, Fay C, Olivardia R, Phillips KA. Clinical features of muscle dysmorphia among males with body dysmorphic disorder. Body Image. 2005;2:395–400.

35. Wroblewska AM. Androgenic-anabolic steroids and body dysmorphia in young men. Journal of Psychosomatic Research. 1997;42:225–34.

36. Grieve FG. A conceptual model of factors contributing to the development of muscle dysmorphia. Eating Disorders. 2007;15:63–80.

37. Olivardia R, Pope HG Jr, Hudson JI. Muscle dysmorphia in male weightlifters: A case-control study. American Journal of Psychiatry. 2000;157:1291–6.

38. Grieve FG. A conceptual model of factors contributing to the development of muscle dysmorphia. Eating Disorders. 2007;15:63–80.

39. Clarke LH. Older women's perceptions of ideal body weights: The tensions between health and appearance motivations for weight loss. Ageing and Society. 2002;22:751–3.

40. Sharp CW, Clark SA, Dunan JR, Blackwood DH, Shapiro CM. Clinical presentation of anorexia nervosa in males: 24 new cases. International Journal of Eating Disorders. 1994;15:125–34.

41. Becker AE, Keel P, Anderson-Fye EP, Thomas JJ. Genes and/or jeans? Genetic and socio-cultural contributions to risk for eating disorders. Journal of Addictive Disorders. 2004;23:81–103.

42. Eddy KT, Hennessey M, Thompson-Brenner H. Eating pathology in East African women: The role of media exposure and globalization. Journal of Nervous and Mental Disorders. 2007;195:196–202.

43. George JB, Franko DL. Cultural issues in eating pathology and body image among children and adolescents. Journal of Pediatric Psychology. 2010;35:231–42.

44. Brown JD, Witherspoon EM. The mass media and American adolescents' health. Journal of Adolescent Health. 2002;31:153–70.

45. National Eating Disorders Association. Statistics: Eating disorders and their precursors. Available from http://www.nationaleatingdisorders.org/uploads/statistics_tmp.pdf.

46. Rubinstein S, Caballero B. Is Miss America an undernourished role model? Journal of the American Medical Association. 2000;283:1569.

47. Wonderlich AL, Ackard DM, Henderson JB. Childhood beauty pageant contestants: Associations with adult disordered eating and mental health. Eating Disorders. 2005;13:291–301.

48. Fernández-Aranda F, Krug I, Granero R, Ramón JM, Badia A, Giménez L, Solano R, Collier D, Karwautz A, Treasure J. Individual and family eating patterns during childhood and early adolescence: An analysis of associated eating disorder factors. Appetite. 2007;49:476–85.

49. Coulthard H, Blissett J, Harris G. The relationship between parental eating problems and children's feeding behavior: A selective review of the literature. Eating Behavior. 2004;5:103–15.

50. Humphries LL, Wrobel S, Wiegert HT. Anorexia nervosa. American Family Physician. 1982;26:199–204.

51. Kluck AS. Family factors in the development of disordered eating: Integrating dynamic and behavioral explanations. Eating Behavior. 2008;9:471–83.

52. Mazzeo SE, Zucker NL, Gerke CK, Mitchell KS, Bulik CM. Parenting concerns of women with histories

of eating disorders. International Journal of Eating Disorders. 2005;37:S77–S9.

53. Peterson CB, Thuras P, Ackard DM, Mitchell JE, Berg K, Sandager N, Wonderlich SA, Pederson MW, Crow SJ. Comprehensive Psychiatry. Personality dimensions in bulimia nervosa, binge eating disorder, and obesity. 2010;51:31–6.

54. Cassin SE, von Ranson KM. Personality and eating disorders: A decade in review. Clinical Psychology Review. 2005;25:895–916.

55. Mischoulon D, Eddy KT, Keshaviah A, Dinescu D, Ross SL, Kass AE, Franko DL, Herzog DB. Depression and eating disorders: treatment and course. Journal of Affective Disorders. 2011;130:470–7.

56. Kaye WH, Bulik CM, Plotnicov K, Thornton L, Devlin B, Fichter MM, Treasure J, Kaplan A, Woodside DB, Johnson CL, Halmi K, Brandt HA, Crawford S, Mitchell JE, Strober M, Berrettini W, Jones I. The genetics of anorexia nervosa collaborative study: Methods and sample description. International Journal of Eating Disorders. 2008;41:289–300. Bulik CM, Slof-Op't Landt MC, van Furth EF, Sullivan PF. The genetics of anorexia nervosa. Annual Review of Nutrition. 2007;27:263–75.

57. Bulik CM, Reba L, Siega-Riz AM, Reichborn-Kjennerud T. Anorexia nervosa: Definition, epidemiology, and cycle of risk. International Journal of Eating Disorders. 2005;37:S2–S9.

58. Cox LM, Lantz CD, Mayhew JL. The role of social physique anxiety and other variables in predicting eating behaviors in college students. International Journal of Sport Nutrition. 1997;7:310–7.

59. Reinking MF, Alexander LE. Prevalence of disordered-eating behaviors in undergraduate female collegiate athletes and nonathletes. Journal of Athletic Training. 2005;40:47–51.

60. Sudi K, Ottl K, Payerl D, Baumgartl P, Tauschmann K, Muller W. Anorexia athletica. Nutrition. 2004;20:657–61.

61. Salbach H, Klinkowski N, Pfeiffer E, Lehmkuhl U, Korte A. Body image and attitudinal aspects of eating disorders in rhythmic gymnasts. Psychopathology. 2007;40:388–93.

62. Johnson C, Powers PS, Dick R. Athletes and eating disorders: The National Collegiate Athletic Association study. International Journal of Eating Disorders. 1999;26:179–88.

63. Beals KA, Hill AK. The prevalence of disordered eating, menstrual dysfunction, and low bone mineral density among US collegiate athletes. International Journal of Sport Nutrition and Exercise Metabolism. 2006;16:1–23.

64. Barrow GW, Saha S. Menstrual irregularity and stress fractures in collegiate female distance runners. American Journal of Sports Medicine. 1988;3:209–16.

65. Lock J, Agras WS, Bryson S, Kraemer HC. A comparison of short- and long-term family therapy for adolescent anorexia nervosa. Journal of the American Academy of Child and Adolescent Psychiatry. 2005;44:632–9.

Physical Activity and Health

People exercise for different reasons: some to stay in shape or to improve overall physical fitness, others because they enjoy the challenge of training and participating in athletic competitions. Regardless of the reason, exercise offers many health benefits. There is no doubt that routine physical activity is vital to staying healthy and physically fit. Even moderate amounts of physical activity can impart tremendous benefits in terms of health-related quality of life. Yet, despite common knowledge that physical activity is important to our health, nearly half of American adults and youths lead a sedentary lifestyle.[1] Furthermore, many people are unclear about the types and amounts of physical activity needed to stay healthy.

For most physically active people, dietary recommendations do not differ much from those suggested for the general population. However, for competitive athletes who sometimes push themselves to the extreme, dietary requirements may be greater. Indeed, what a competitive athlete eats and drinks before, during, and after exercise can impact athletic performance.

In the previous chapter, you learned that obesity and a sedentary lifestyle are risk factors for a multitude of health concerns. In this chapter, you will learn about the many health benefits associated with physical activity. We will also explore how the body responds to the physical demands of exercise and the current public health recommendations related to physical activity and health promotion. In addition, the overall importance of nutrition during athletic training and competition will be discussed.

Turning a Spark into an Inferno

What can turn a sedentary person into a committed athlete? For Tom, that turning point came when he saw the excitement and joy on the faces of triathletes as they crossed the finish line. As a middle-aged man, Tom had not exercised since high school. In fact, the reality of his sedentary lifestyle was evident. Tipping the scale at 320 pounds and habitually eating a diet mainly of meat, potatoes and fast food, Tom admitted his health was poor. According to Tom, "I ate what I liked, and I liked what I ate. Nutrition meant nothing to me. I had no awareness of what was a healthy diet." But after his wife died, Tom found himself sinking into a depression. His health was suffering, and he knew something had to be done. So he joined a gym.

Volunteering for another local triathlon, Tom was once again struck by the charged atmosphere as athletes streaked past him on the racecourse. As he cheered them on, Tom marveled at their fitness, determination, and athleticism. This was Tom's defining moment, as he silently set a goal for himself—to compete in Ironman® Coeur D'Alene, one of the most grueling endurance athletic competitions there is. Consisting of three parts—a 2.4-mile swim, a 112-mile bike ride, and a 26.2-mile run—this "extreme" triathlon was going to challenge Tom beyond anything he had ever experienced. But where does a 62-year-old man who is 100 pounds overweight and who has never biked, swum, or run even begin?

Tom's big break came when he asked Shawn, a personal trainer, for advice. Shawn quickly recognized that Tom was highly motivated and that he needed someone who was willing to guide him through this arduous journey. Many physical trainers might have discouraged Tom from taking on such a lofty goal. However, Shawn recognized that Tom was serious and motivated for all the right reasons: "I saw a spark of life in Tom. My job was to turn it into an inferno, to blaze a path for others to follow."

And so the training began. Tom recalls his first open-water swim, when his wet suit fell apart because his large frame caused the seams to give way. His new (smaller) suit is comfortable and fits him like a glove. Over the past year, Tom's determination, discipline, and commitment have transformed both his health and outlook on life. A daily regimen that begins with a healthy breakfast and workout routine has resulted in Tom losing nearly 100 pounds and decreasing his body fat from 40 to 29%. Not surprisingly, his blood glucose and lipid levels have decreased as well. This journey "although demanding" has enriched Tom's life in ways he never could have imagined. There is no going back for Tom, and he wishes that everyone who thinks about that first life-changing step would take it. According to Tom, anything is possible. Thinking about what Tom has accomplished with serious commitment and hard work can be summarized in a single word. Inspiring!

Courtesy of U Aim High Fitness

Critical Thinking: Tom's Story
Although many people have given up on exercise, Tom's story is an inspiration to all—both young and old. What would you say to someone who is convinced that he or she does not have time for exercise? Do you have athletic goals? If so, what would it take for you to accomplish them? If not, why not?

What Are the Health Benefits of Physical Activity?

Numerous studies show that as little as 30 minutes of sustained physical activity on most days can substantially improve health and quality of life.[2] More specifically, physical activity helps reduce a person's risks for obesity and certain chronic diseases such as stroke, cardiovascular disease, type 2 diabetes, osteoporosis, and some forms of cancer. Unfortunately, these benefits are quickly lost once a person again becomes inactive.[3] This is why it is important to make exercise, along with good nutrition, a permanent part of your lifestyle. The many far-reaching benefits of being physically active and the components of a well-rounded physical fitness program are discussed next.

PHYSICAL ACTIVITY IMPROVES HEALTH AND PHYSICAL FITNESS

It does not matter if a person is young and healthy or old and disabled, everyone can benefit from physical activity. But what does it mean to be physically active? Recall from Chapter 8 that physical activity is defined as bodily movement that is produced by the contraction of skeletal muscles that increases energy expenditure above a basal level.[4] Thus, physical activity includes day-to-day and leisure-time activities such as gardening, household chores, and walking. **Exercise** is a subcategory of physical activity and is defined as planned, structured, and repetitive bodily movement done to improve or maintain physical fitness.[4]

Physical fitness is a relative term, and individuals must determine their own goals for the level of fitness they wish to achieve. Clearly, a competitive athlete will have different fitness goals than someone just starting a physical activity program. The U.S. Department of Health and Human Services (DHHS), in its 2008 Physical Activity Guidelines for Americans, defines **physical fitness** as the ability to carry out daily tasks with vigor and alertness, and to perform leisure-time pursuits without undue fatigue.[5] Regardless of where we are on the physical fitness continuum, regular physical activity may be one of the most important things we can do to maintain health and overall well-being.

There is overwhelming evidence that physical activity reduces the risk of many debilitating diseases and health conditions.[6] The strength of the scientific evidence supporting health benefits associated with physical activity is summarized in Table 9.1. Overall benefits of exercise on health and fitness include the following.

- *Weight Management.* Regular physical activity helps to increase muscle mass and decrease body fat, in large part because it increases energy expenditure and improves appetite regulation. When combined with a healthy diet, physical activity can promote weight loss and unwanted weight gain.
- *Prevention of High Blood Pressure.* Physical activity can improve cardiovascular function, which in turn can help control and, in some cases, reduce blood pressure in hypertensive individuals. In addition, weight loss associated with physical activity can lower body weight, which also helps to lower blood pressure.
- *Decreased Risk of Cardiovascular Disease.* There are several ways that physical activity can lower a person's risk of cardiovascular disease. In addition to strengthening the heart muscle, physical activity can lower resting heart rate and improve blood lipids by decreasing triglyceride and LDL cholesterol and increasing HDL cholesterol. These beneficial changes help slow the progression of atherosclerosis, maintain healthy blood pressure, and assist the heart to function more efficiently.

exercise Planned, structured, and repetitive bodily movement done to improve or maintain physical fitness.

physical fitness The ability to carry out daily tasks with vigor and alertness, and to perform leisure-time pursuits without undue fatigue.

TABLE 9.1 Strength of the Scientific Evidence for Health Benefits of Physical Activity

---

Children and Adolescents

Strong Evidence
- Improved cardiorespiratory endurance and muscular fitness
- Favorable body composition
- Improved bone health
- Improved cardiovascular and metabolic health biomarkers

Moderate Evidence
- Reduced symptoms of anxiety and depression

Adults and Older Adults

Strong Evidence
- Lower risk of:
  - Early death
  - Heart disease
  - Stroke
  - Type 2 diabetes
  - High blood pressure
  - Adverse blood lipid profile
  - Metabolic syndrome
  - Colon and breast cancers
- Prevention of weight gain
- Weight loss when combined with diet
- Improved cardiorespiratory and muscular fitness
- Prevention of falls
- Reduced depression
- Better cognitive function (older adults)

Moderate to Strong Evidence
- Better functional health (older adults)
- Reduced abdominal obesity

Moderate Evidence
- Weight maintenance after weight loss
- Lower risk of hip fracture
- Increased bone density
- Improved sleep quality
- Lower risk of lung and endometrial cancers

SOURCE: U.S. Department of Health and Human Services. 2008 Physical activity guidelines for Americans. Available from http://www.health.gov/paguidelines/default.aspx.

- *Reduced Risk of Cancer.* Increased physical activity decreases a person's risk of developing colon, lung, endometrial, and breast (postmenopausal) cancers.
- *Decreased Back Pain.* Exercise can increase muscle strength, endurance, flexibility, and overall posture, all of which can help prevent injuries and alleviate back pain associated with muscle strain.
- *Lower Risk of Type 2 Diabetes.* Regular physical activity can help decrease blood glucose, insulin resistance, and long-term health complications associated with type 2 diabetes. In addition, physical activity can help prevent and control this type of diabetes by helping maintain a healthy body weight.
- *Optimized Bone Health.* Weight-bearing physical activity can stimulate bone formation and slow the progression of age-related bone loss. Because physical activity can improve muscle strength and flexibility, staying

physically active can reduce the likelihood of injuries when falls occur.

- *Enhanced Self-Esteem, Stress Management, and Quality of Sleep.* Being physically active provides a sense of achievement and empowerment. This can help reduce depression and anxiety, which in turn can foster improved self-esteem. Regular physical activity provides a positive and effective means to manage stress, and it can even improve mood and promote restful sleep.

Physical Activity Recommendations What is "regular physical activity," and how much physical activity do we need for optimal health? What may be the right amount of physical activity for one person may not be enough for another. Because physical *inactivity* is a major public health concern, several health organizations have released specific recommendations regarding how much activity we need to stay healthy and prevent health problems associated with a sedentary lifestyle.[7] However, the many different (and sometimes contradictory) guidelines can seem confusing. For example, the 2008 Physical Activity Guidelines for Americans stress the importance of regular physical activity and recommend that adults get at least 150 minutes a week of moderate-intensity or 75 minutes a week of vigorous-intensity aerobic physical activity.[5] Yet physical activity guidelines issued by the American College of Sports Medicine in 2010 recommend that apparently healthy, active adults need moderate- to vigorous-intensity physical activity for at least 200–300 minutes a week.[4] In addition, they recommend that every adult engage in resistance training two to three days each week. Table 9.2 provides a summary of exercise recommendations by the American College of Sports Medicine and the DHHS.

Although there are some inconsistencies among physical activity recommendations, there are many points of similarity. Most important is the general consensus that at least 30 minutes of moderate-intensity physical activity on most days of the week provides substantial health benefits for most adults. However, depending on energy intake, this amount of physical activity may or may not be adequate to maintain or obtain a healthy body weight.[8] To determine how much physical activity is right for you, it is important to consider your current body weight, fitness levels, and personal goals.

AP Photo/Elaine Thompson

Sister Madonna Buder, from Spokane, Washington, has competed in hundreds of triathlons, more than 30 Ironman® races, and two Boston Marathons®. In 2005, she became the first female over the age of 75 years to complete an Ironman. At 81 years of age, she is still competing.

FIVE COMPONENTS OF PHYSICAL FITNESS

Regardless of whether a person exercises to stay physically fit, play recreational sports, or participate in competitive athletic events, the health-enhancing effect of exercise is the same—improved physical fitness. There are many types of physical activities, most of which are beneficial to health. Some activities such as weight lifting help strengthen muscles, whereas others such as jogging improve cardiovascular fitness. Even bending and stretching exercises such as yoga help keep muscles strong and joints flexible. When starting a workout program, it is important to begin slowly with activities that you enjoy, and remember that physical activity does not have to be strenuous to improve overall fitness. A physical activity program that includes a variety of activities can improve your capacity in all five components of physical fitness—cardiovascular fitness, muscle strength,

TABLE 9.2 Summary of Selected Recommendations for Physical Activity for Adults

Year	Organization/Agency	Health Focus	Activity/Exercise Recommendation
2010	American College of Sports Medicine[a]	Guidelines for exercise testing and recommendations	• Depending on physical fitness level, 60–200 min/week of light-moderate intensity (sedentary and low-active individuals), or 200–300 min/week of moderate-hard intensity aerobic activity (active to very active). • In addition, resistance training of each major muscle group 2–3 days/week with at least 48 hours separating the exercise training sessions for the same muscle group. • Overweight/obese individuals should participate in aerobic activities at least 5 days/week, with a goal of eventually being able to exercise for 300 min/week at moderate intensity or 150 min/week at vigorous intensity (or an equivalent combination). Resistance-training exercise should also be included.
2008	U.S. Department of Health and Human Services[b]	Health promotion, disease prevention, and weight control recommendations for children, adolescents, adults, pregnant/lactating women, older adults, and those with physical disabilities	• 150 min/week of moderate-intensity, or 75 min/week of vigorous-intensity aerobic physical activity, or an equivalent combination of moderate- and vigorous-intensity aerobic physical activity. • In addition, muscle-strengthening activities that are moderate or high intensity and involve all major muscle groups on 2 or more days each week. • Adults who are at a healthy body weight but slowly gaining weight or wishing to lose weight should gradually increase the level of physical activity toward the equivalent of 300 min/week, or reduce caloric intake, or both until their weight is stable or they reach their desired body weight.

[a] Adapted from: American College of Sports Medicine. ACSM's guidelines for exercise testing and prescription. 8th ed. Wolters Kluewer, Lippincott Williams and Wilkins, New York, 2010.
[b] Adapted from: U.S. Department of Health and Human Services. 2008 Physical activity guidelines for Americans. Available from http://www.health.gov/paguidelines/default.aspx.

endurance, flexibility, and body composition. All of these components contribute to your overall health.

- **Cardiovascular fitness** (also called cardiorespiratory endurance) refers to the ability of the circulatory and respiratory systems to supply oxygen and nutrients to working muscles during sustained physical activity. Activities that safely elevate your heart rate can help improve cardiovascular fitness. These include activities such as walking at a brisk pace, swimming, jogging, and bicycling.
- **Muscle strength** refers to the maximal force exerted by muscles during an activity. Activities such as lifting weights or walking up stairs force muscles to work against resistance and, therefore, can help improve muscle strength. The purpose of strength-building activities, some of which are referred to as **resistance training**, is to gradually challenge specific groups of muscles to make them work harder. These types of activities help to keep muscles and bones strong.
- **Endurance** refers to the ability to exercise for an extended period of time without becoming fatigued. Activities such as walking at a brisk pace, swimming, jogging, and bicycling can improve endurance.
- **Flexibility** refers to the range of motion around a joint. Activities such as yoga, stretching, and swimming can help lengthen muscles and improve flexibility.
- **Body composition** refers to the relative amounts of muscle, fat tissue, and bone in the body. Body fat is best reduced through cardiovascular exercise, whereas muscle mass can be increased through strength-building activities (such as lifting weights). Weight-bearing activities such as walking and jogging can help increase bone mass.

cardiovascular fitness The ability of the circulatory and respiratory systems to supply oxygen and nutrients to working muscles during sustained physical activity.

muscular strength The maximal force exerted by muscles during an activity.

resistance training Physical activities that, using weights or other methods of physical resistance, overload specific groups of muscles to make them work harder.

muscle endurance Ability to exercise for an extended period of time without becoming fatigued.

flexibility Range of motion around a joint.

body composition The proportion of fat tissue and fat-free tissue in the body.

GETTING "FITT" INVOLVES FREQUENCY, INTENSITY, TYPE, AND TIME

To get the most out of a fitness program, it is important to consider several established guidelines collectively known as the **FITT principle**—frequency, intensity, type, and time. According to the American College of Sports Medicine, considering all four components of the FITT principle forms the foundation of any sound physical fitness plan.[9] An example of how the four FITT principle components can be used to plan a physical activity program is shown in Table 9.3

Frequency of Physical Activity In terms of the FITT principle, **frequency** refers to how often you exercise. To improve cardiovascular fitness, three to five workout sessions per week are optimal.[9] Sedentary people should start gradually, perhaps with no more than two workouts per week. To achieve muscular strength, only two to three workouts per week of resistance activities are necessary.

Intensity of Physical Activity The **intensity** of a physical activity refers to the amount of physical exertion expended during the activity. The intensity of physical activities is typically classified as low, moderate, or vigorous, depending on how hard your body is working. Increases in breathing, sweating, and heart rate are often used to judge the intensity of physical activity. It is

FITT principle A method of planning a physical fitness program that considers frequency, intensity, type, and time spent exercising each week.

frequency Component of the FITT principle that addresses how often a person exercises.

intensity Component of the FITT principle that refers to the amount of physical exertion expended during physical activity.

TABLE 9.3 FITT Guidelines for Planning a Physical Activity Program in Healthy Adults

	Type of Physical Activity		
	Cardiovascular	**Strength**	**Flexibility**
Frequency	5 or more days per week	2 or more days per week	2 or more days per week
Intensity	Moderate (level of effort is a 5 or 6 on a scale of 0 to 10 where 0 is the level of effort of sitting and 10 is maximal effort)	Involves all major muscle groups; enough to enhance muscle strength and improve body composition	Enough to develop and maintain a full range of motion
Time	Minimum of 150 minutes a week; as a person moves from 150 minutes a week toward 300 minutes (5 hours) a week, he or she gains additional health benefits	Muscle-strengthening exercises should be performed to the point at which it would be difficult to do another repetition without help; to enhance muscle strength, one set of 8 to 12 repetitions of each exercise is effective, although two or three sets may be more effective	2 to 4 repetitions, holding each stretch for 10–30 seconds, accumulating 60 seconds per stretch
Type	Running, bicycling, dancing, swimming, tennis, volleyball, soccer, skiing, jumping rope, cross-country skiing	Pull-ups, weight lifting, resistance bands	Yoga, pilates, stretching

SOURCE: Garber EC, Blissmer B, Deschenes MR, Franklin BA, Lamonte, MJ, Lee IM, Nieman DC, Swain DP. Quantity and quality of exercise for developing and maintaining cardiorespiratory, musculoskeletal, and neuromotor fitness in apparently health adults: guidance for prescribing exercise. Medicine and Science in Sports and Exercise. 2011;43:1334-59.

important to exercise at an intensity level that is right for your level of fitness. This will ensure that you are training at an intensity level that benefits health while maintaining safety. There are several ways to judge the intensity level of your workouts, and these are discussed next.

Determining Exercise Intensity According to the U.S. Centers for Disease Control and Prevention (CDC), one of the simplest methods to assess the intensity level of your workout is the **talk test**.[10] This is based on a person's ability to converse while exercising. Being too winded to carry on a conversation while exercising is an indication that the intensity level of your workout is vigorous.

You can also gauge the intensity of physical activity by monitoring your pulse or heart rate. To do this, you must first estimate your **maximum heart rate** by subtracting your age from 220. For example, the maximum heart rate for a 20-year-old would be 200 beats/min (220 – 20). For moderate-intensity exercise, the CDC recommends that people stay between 50 and 70% of their maximum heart rate. In our example, a heart rate between 100 and 140 beats/min (200 × 0.50 = 100 beats/min; 200 × 0.70 = 140 beats/min) would be optimal. As such, this would be considered this person's **target heart rate** range for moderate-intensity exercise. However, to exercise vigorously, the CDC recommends maintaining a heart rate of 70 to 85% of maximum. As such, this individual's target heart rate range would be 140 to 170 beats/min to achieve a vigorous activity intensity. You can use the heart rate training chart shown in Table 9.4 to estimate your personal optimal target heart rate ranges for moderate and vigorous activities.

Perceived level of exertion can also be used to judge the intensity level of physical activity. This method uses the **Borg Scale of Perceived Exertion** and is based on monitoring your body's response to physical activity. This includes changes in heart rate, breathing rate, sweating, and muscle fatigue. For example, people who exercise at low intensity would experience a slight increase in their heart and breathing rates (compared with resting rates), and their skin might feel slightly damp from sweating. Depending on your fitness goal, workouts can be adjusted to achieve the desired level of intensity. The

talk test A method used to judge the intensity level of workouts by assessing a person's ability to converse while exercising.

maximum heart rate Estimated by subtracting age from 220 (220 – age).

target heart rate That needed to achieve a desired activity intensity; 50–70% and 70–85% of maximum heart rate are often used as targets for moderate- and vigorous-level intensities, respectively.

Borg Scale of Perceived Exertion (rating of perceived exertion) Self-monitoring of the body's response to physical activity based on changes in heart rate, breathing rate, sweating, and muscle fatigue.

TABLE 9.4 Target Heart Rate Ranges for Moderate- and Vigorous-Intensity Activity Moderate-level physical activity is that which increases heart rate to 50–70% of maximum, whereas vigorous-level activity increases heart rate to 70–85% of maximum.

Age (years)	Maximum Heart Rate (beats/min)*	Target Heart Rate Ranges for Moderate-Intensity Activity (beats/min)**	Target Heart Rate Ranges for Vigorous-Intensity Activity (beats/min)***
20	200	100–140	140–170
25	195	98–137	137–166
30	190	95–133	133–162
35	185	93–130	130–157
40	180	90–126	126–153
45	175	88–123	123–149
50	170	85–119	119–145
55	165	83–116	116–140
60	160	80–112	112–136
65	155	78–109	109–132
70	150	75–105	105–128

SOURCE: Centers for Disease Control and Prevention. Target heart rate and estimated maximum heart rate. Available at: http://www.cdc.gov/physicalactivity/everyone/measuring/heartrate.html.
*Maximum heart rate is 220 minus age (y).
**CDC recommends 50–70% maximum heart rate for moderate-intensity activity.
***CDC recommends 70–85% maximum heart rate for vigorous-intensity activity.

TABLE 9.5 Borg Scale of Perceived Exertion The Borg rating scale can be used to estimate exertion intensity based on perceived changes in heart rate, breathing, sweat production, and muscle fatigue.

Perceived Exertion	Physical Signs/ Symptoms	Workout Intensity
No exertion to low exertion	None	Sedentary
Very light to light	Feeling of motion	Low intensity
Somewhat hard	Warmth on cold day, slight sweat on warm day	Moderate intensity
Hard to very hard	Sweating but can still talk without difficulty	High intensity
Extremely hard	Heavy sweating, difficulty talking	
Maximal exertion	Feeling of near exhaustion	

SOURCE: The Borg RPE Scale®. Copyright 1985, 1994 by Gunnar Borg. The scale with correct instructions can be obtained from Borg Perception; see the home page: www.borgperception.se/index.html.

next time you engage in physical activity, you can use this rating scale to judge the intensity of your workout (Table 9.5).

While heart rates and perceptions can be used to estimate energy exertion levels, researchers and clinicians often need a more accurate measure. This can be done based on the fundamental principle of indirect calorimetry, which is the energy expenditure during exercise that is proportional to oxygen consumption and carbon dioxide production. As exercise intensity increases, there is an increase in oxygen consumption until a steady state is reached. By measuring the exchange of oxygen and carbon dioxide while exercising, an estimate of energy expenditure relative to energy exertion can be obtained.

Type of Activity The FITT principle also includes consideration of the *type* of exercise. There are many ways to incorporate physical activity into your daily routine. Regardless of the type of activity, however, it is important to involve as many muscle groups as possible, particularly the larger muscles such as those in the legs and arms. To improve cardiovascular fitness, the movement should be rhythmic and continuous, while gradually increasing the intensity. Examples of these types of activities include walking, cycling, swimming, hiking, cross-country skiing, and dancing. In addition to activities that strengthen cardiovascular fitness, a sound physical fitness program should also include strength-building activities, such as weight lifting and climbing. In addition, physical activities such as static stretching and yoga can help to improve range of motion and flexibility.

Time of Activity Finally, *time* is important to consider when aiming to address all aspects of the FITT principle. Time of activity refers to how long the physical activity lasts, and is only applicable to aerobic activities. For cardiovascular fitness, the American College of Sports Medicine recommends 30 minutes of moderate-intensity physical activity on at least 5 days/week to a total of at least 150 minutes each week. Alternatively, individuals can strive for 20–25 minutes of vigorous-intensity physical activity on at least 3 days/week to a total of at least 75 minutes each week. A combination of intensity levels and times can also be used to reach the same workout goal.[4] Individuals trying to lose weight or prevent weight regain may need longer workouts

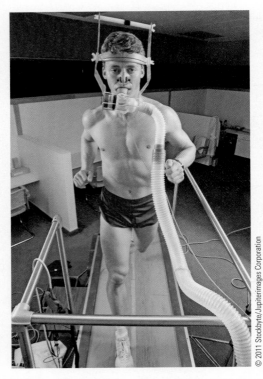

Indirect calorimetry is used to estimate energy expenditure during exercise by measuring oxygen consumption and carbon dioxide production.

(300 minutes/week of moderate-intensity or 150 minutes/week of vigorous-intensity activities), and those with lower fitness levels should exercise at a comfortable pace and within their target heart rate zone. As fitness levels increase, it becomes easier to exercise longer and at a greater intensity level.

It is important to remember that exercising too frequently or too intensely can lead to injuries. Exercise can be physically demanding, and the body needs time to recover. To help prevent injuries, experts recommend warm-up and cool-down sessions such as easy, light stretches before and, especially, after exercise.

You now have an appreciation for the importance of physical activity in your daily life, and the types and amount of physical activity needed to gain or maintain a healthy level of physical fitness. Next, we will examine how the body uses energy-yielding nutrients to fuel physical activity and exercise.

How Does Physical Activity Impact Energy Metabolism?

Throughout the day, we are in a state of perpetual motion. Whether we are standing, sitting, walking, or running—energy is required. Thus, physical activity greatly influences energy expenditure. On average, a person expends approximately 15 to 30% of his or her total daily energy intake on physical activity. This percentage is even higher for people who exercise. At rest, the body expends approximately 1 to 1.5 kcal/minute. During physical exertion, however, energy expenditure can increase to as much as 15 to 36 kcal/minute. As illustrated in Table 9.6, physical activity level can contribute substantially to total energy requirements.

TABLE 9.6 Effect of Physical Activity Level on Total Energy Requirements in Healthy-Weight Adults[a] Total energy requirements (and therefore expenditure) increase as physical activity level increases in healthy-weight adults who are maintaining energy balance.

Sex and Age (years)	Physical Activity Level (PAL)			
	Sedentary	Low Active	Active	Very Active
	(kcal/day)			
Females[b]				
20	1,879	2,085	2,342	2,651
30	1,784	1,989	2,247	2,555
40	1,688	1,894	2,151	2,460
50	1,593	1,799	2,056	2,365
60	1,498	1,703	1,961	2,270
Males				
20	2,542	2,770	3,060	3,536
30	2,447	2,675	2,964	3,441
40	2,352	2,579	2,869	3,345
50	2,256	2,484	2,774	3,250
60	2,161	2,389	2,679	3,155

[a]Calculated using Estimated Energy Requirement (EER) equations; reference man of 5 feet 10 inches and 154 lb and a reference woman of 5 feet 4 inches and 126 lb. EER equations and physical activity level (PAL) designations are from the Institute of Medicine. Dietary reference intakes for energy, carbohydrate, fiber, fat, fatty acids, cholesterol, protein, and amino acids. Washington (DC): The National Academies Press; 2005.
[b]Estimates for females do not include women who are pregnant or breastfeeding.

ATP CAN BE GENERATED BY AEROBIC AND ANAEROBIC PATHWAYS

But why does physical activity have such a profound effect on energy expenditure? First, remember that it requires motor units within skeletal muscle fibers to contract and relax in response to neural signals from the brain. To perform this complex action, individual muscle fibers convert the chemical energy in adenosine triphosphate (ATP) to mechanical energy. In this way, ATP fuels muscle contraction. The speed of muscle movement is determined primarily by the availability of fuel sources and the metabolic pathways used by muscle fibers to generate ATP.

During exercise, ATP is generated from the metabolic breakdown of glucose, fatty acids, and, to a lesser extent, amino acids. The availability of these nutrients is determined by the foods we eat and by those stored in the body. Because the amount of available ATP in muscles is generally low, active muscle cells must continuously generate ATP. To fuel physical activity, multiple metabolic pathways work together to provide ATP to muscle fibers. Some of these pathways operate in aerobic (oxygen-requiring) conditions, whereas others are used during anaerobic (oxygen-poor) conditions. Both aerobic and anaerobic energy pathways offer certain advantages.

Think about how long it would take your muscles to fatigue while taking a vigorous walk. In contrast, how long would it take your muscles to fatigue while running at your fastest pace? Certainly the harder you exert yourself, the more quickly your muscles become tired. There are several reasons why muscles can sustain high-intensity workouts for only a short period of time, but limited oxygen availability is one of the most important. Muscles use both anaerobic and aerobic metabolic pathways to fuel physical activity. However, the intensity and duration of the activity determines the relative contribution of each. For example, short, high-intensity exercise, such as a 100-meter sprint, relies heavily on anaerobic pathways that quickly deliver ATP to muscles. By contrast, endurance sports, such as a marathon, rely more on aerobic pathways, which deliver a more sustained supply of ATP.

Aerobic metabolic pathways include the citric acid cycle, β-oxidation, and oxidative phosphorylation, which were discussed in detail in Chapter 7. Because these pathways are complex and consist of multiple metabolic steps, they generate ATP relatively slowly. However, aerobic metabolic pathways can generate a tremendous amount of ATP over an extended period of time. During low-intensity activities such as leisurely walking, cells rely mainly on these pathways because oxygen availability is usually adequate.

As exercise intensity increases, the rate of ATP production by aerobic means cannot keep pace with energy needs of muscles. Therefore, during brief, high-intensity activities such as running a 100-meter sprint, muscles rely heavily on anaerobic metabolism for ATP. Anaerobic pathways that generate ATP without relying on oxygen include the ATP–creatine phosphate (ATP–CP) pathway and glycolysis. Because these anaerobic metabolic pathways are relatively simple, ATP production is very fast. However, anaerobic pathways cannot maintain a high rate of ATP production for very long. This is why the body can rely on anaerobic metabolism only for brief periods of time.

At the onset of exercise, both aerobic and anaerobic pathways are activated to some extent. The relative contribution of each energy system during exercise is shown in Figure 9.1. When vigorous exercise begins, muscles primarily use anaerobic pathways to generate ATP. As such, anaerobic pathways (ATP–CP and glycolysis) can be thought of as *immediate* and *short-term energy systems*, respectively. However, these pathways cannot sustain physical activity for

⟨**CONNECTIONS**⟩ Recall that the citric acid cycle is a central metabolic pathway that oxidizes acetyl-CoA to yield carbon dioxide, NADH + H$^+$, FADH$_2$, and GTP (Chapter 7, page 283).

⟨**CONNECTIONS**⟩ β-oxidation metabolizes fatty acids, forming acetyl-CoA (Chapter 7, page 287).

⟨**CONNECTIONS**⟩ Oxidative phosphorylation consists of a series of reactions that use electrons and hydrogen ions from NADH + H$^+$ and FADH$_2$ to generate water, carbon dioxide, and ATP (Chapter 7, page 277).

⟨**CONNECTIONS**⟩ Recall that glycolysis is a series of chemical reactions that split the six-carbon glucose molecule into two three-carbon molecules called pyruvate (Chapter 7, page 281).

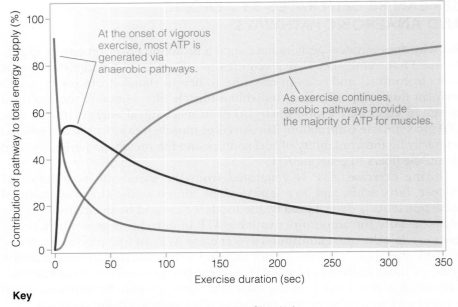

Key

— ATP–creatine phosphate pathway   — Glycolysis

— Aerobic pathways (includes citric acid cycle, β-oxidation, and oxidative phosphorylation)

SOURCE: Adapted from Gastin PB. Energy system interaction and relative contribution during maximal exercise. Sports Medicine. 2001;31:725-41.

long. With the continuation of exercise, the energy demands of active muscles quickly exceed the ATP supply available from anaerobic metabolism. If physical activity is to continue, muscles must utilize oxygen-requiring (aerobic) pathways (such as the citric acid cycle and oxidative phosphorylation) for ATP production. These pathways can be thought of as *long-term energy systems*.

During exercise, each of these energy systems contributes to ATP production. However, the relative contribution of each depends on a variety of factors such as exercise intensity, the duration of exercise, and oxygen availability in muscles. Anaerobic and aerobic metabolic pathways used by muscles to fuel physical activity are examined next.

ATP–Creatine Phosphate (ATP–CP): A Pathway for Immediate Energy Only a small amount of ATP is available in resting muscle, and this is depleted within the first few seconds of vigorous exercise. Therefore, muscles must quickly replenish this energy source. The anaerobic **ATP–creatine phosphate (ATP–CP) pathway** is the simplest and most rapid means by which active muscles can generate ATP. This pathway can be thought of as an *immediate energy system* because it acts quickly but cannot sustain physical activity for long.

Creatine phosphate (CP) is a high-energy, phosphate-containing compound found in muscle tissue. Creatine is synthesized in the liver, kidneys, and pancreas from amino acids, and it can also be obtained by eating meat and fish. Creatine phosphate is formed when creatine combines with inorganic phosphate (P_i). When ATP supplies diminish, the enzyme **creatine phosphokinase** quickly breaks CP down, forming inorganic phosphate (P_i) and creatine. The released P_i is attached to ADP to form ATP, which can then fuel muscle activity. This metabolic pathway is illustrated in Figure 9.2.

The ATP–CP pathway is ideally suited to meet the immediate energy demands of intense activities such as power lifting, speed skating, and track or field events that last for only short periods of time. This energy system can sustain physical exertion for only 3 to 15 seconds. High-intensity activities rapidly deplete CP, and resting muscles take several minutes to replenish the

ATP–creatine phosphate (ATP–CP) pathway An anaerobic metabolic pathway that uses ADP and creatine phosphate (CP) to generate ATP.

creatine phosphate (CP) (CRE– a – tine) A high-energy compound, consisting of creatine and phosphate, used to generate ATP.

creatine phosphokinase An enzyme that splits creatine phosphate to generate creatine and inorganic phosphate (P_i).

FIGURE 9.2 ATP–creatine phosphate (ATP–CP) pathway The ATP–creatine phosphate (ATP–CP) pathway enables active muscles to generate ATP rapidly in the relative absence of oxygen.

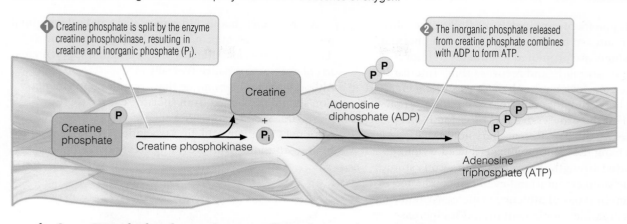

1 Creatine phosphate is split by the enzyme creatine phosphokinase, resulting in creatine and inorganic phosphate (P_i).

2 The inorganic phosphate released from creatine phosphate combines with ADP to form ATP.

Creatine phosphate → Creatine phosphokinase → Creatine + P_i → Adenosine diphosphate (ADP) → Adenosine triphosphate (ATP)

supply. Once CP is depleted, muscles must then rely on other metabolic pathways for ATP formation. For this reason, some athletes take creatine supplements as ergogenic aids to enhance athletic endurance. Whether this practice is beneficial is described in greater detail in the Focus on Sports Nutrition.

Glycolysis: A Short-Term Energy System The continuation of high-intensity exercise activities increases the demand for ATP. Until sufficient oxygen is made available to fuel aerobic energy metabolism, active muscles must turn to another anaerobic pathway called glycolysis that can be thought of as a *short-term energy system*. Recall that glycolysis is the metabolic pathway that breaks down glucose into pyruvate molecules. Glycolysis can briefly sustain activity beyond the immediate burst of energy provided by the ATP–CP pathway. At the onset of exercise, oxygen availability in muscles is somewhat limited, even if the activity is of low intensity. Thus, it is difficult for aerobic metabolic pathways to initially meet the energy demands of exercising muscle. At this stage, the ATP–CP pathway and glycolysis predominate. However, it is incorrect to think that anaerobic metabolism occurs only under conditions of low oxygen availability. Muscles also use anaerobic pathways when there is a need to generate ATP rapidly. This is certainly the case during high-intensity, short, explosive bursts of activity such as a 100-meter swimming event. To generate ATP quickly, muscles rely on the rapid breakdown of glucose via glycolysis (Figure 9.3).

Courtesy of Kathy Beerman

When there is a need to generate ATP rapidly, muscles rely primarily on anaerobic metabolic pathways.

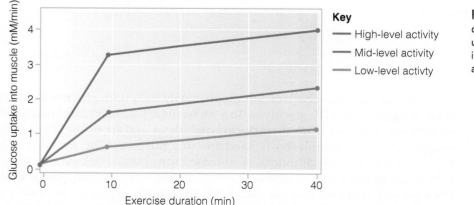

Key
— High-level activity
— Mid-level activity
— Low-level activty

FIGURE 9.3 Effects of exercise duration and intensity on glucose uptake As the intensity of exercise increases, muscles rely more on glucose as an energy source.

SOURCE: Adapted from Rose, AJ and Richter EA. Skeletal muscle glucose uptake during exercise: How is it regulated? Physiology. 2005;20:260–70.

FOCUS ON SPORTS NUTRITION
Do Creatine Supplements Enhance Athletic Performance?

Creatine supplements are widely available and sold over-the-counter in most grocery stores and pharmacies. These products are promoted to athletes to improve muscle mass, strength, and recovery time. Interest in creatine as a performance enhancer was first generated by researchers studying the effect of creatine supplementation on the progressive loss of vision in people with a rare eye disease. Researchers noted that participants receiving the treatment experienced weight gain and increased strength. Since then, numerous studies on creatine supplementation have been conducted in many different types of athletes.[11] One consistent finding is that creatine supplementation does increase creatine levels in muscle tissue, especially in people who had low creatine levels to begin with. Whether higher muscle creatine translates into improved performance, however, is less clear.

While large-scale, well-controlled clinical trials are lacking, the majority of studies assessing the effect of creatine on exercise performance show that creatine supplementation may enhance high-intensity, short-duration athletic performance such as sprinting or power lifting.[12] There is little evidence of a benefit to athletes participating in low-intensity, endurance sports such as long-distance running. In fact, creatine supplementation may hinder performance because it appears to cause weight gain that is mostly attributed to fluid retention. Although some athletes believe that creatine supplements stimulate muscle strength, there is insufficient evidence to support this claim.

The safety of creatine supplementation has been widely debated. While some claim that it can cause dehydration, muscle cramps, and possibly impair kidney function, the majority of studies suggest that creatine supplementation is safe and poses very little risk, at least for short-term use.[13] However, it has not been extensively studied in younger athletes—those under 18 years of age. Because data are largely lacking, creatine supplementation is not advised for this age group. Furthermore, little is known about the effects of long-term creatine supplementation.

A typical diet that includes meat, fish, and other animal products provides approximately 1 to 2 g of creatine a day. By comparison, athletes taking creatine supplements to enhance performance typically ingest 20 to 30 g/day. Although some might argue that creatine supplementation should be banned from sporting events, there is no reliable method to test for its use. Nonetheless, the National Collegiate Athletic Association (NCAA) rules prohibit institutions from providing creatine to athletes.[14]

Although a typical diet provides adequate amounts of creatine, creatine supplements remain popular among athletes.

Unlike aerobic metabolism, which primarily uses both glucose and fatty acids to generate ATP, glycolysis (an anaerobic pathway) depends solely on glucose. Glucose from dietary carbohydrates and from the breakdown of glycogen (glycogenolysis) is used for this purpose. The liver and kidneys also make small amounts of glucose from noncarbohydrate precursors (such as some amino acids, glycerol, and lactate) via gluconeogenesis. Muscles rely more on glucose derived from muscle glycogen than on glucose circulating in the blood. Thus, muscle glycogen stores can become depleted quickly during intense exercise. By some estimates, the rate of

The relative contribution of aerobic and anaerobic pathways during exercise depends on the intensity of the exercise. Muscles utilize anaerobic pathways when there is a need to generate ATP rapidly, such as during high-intensity, short, explosive bursts of activity. Activities such as long-distance running rely more on aerobic pathways.

muscle glycogen depletion is 30 to 40 times higher during high-intensity activities such as sprinting, compared with low-intensity activities such as a leisurely walk.[15] Depletion of muscle glycogen during anaerobic metabolism can contribute to fatigue because the rate at which skeletal muscles take up glucose from the blood is not rapid enough to sustain a high rate of physical exertion.

Lactate Production Recall from Chapter 7 that glycolysis is an anaerobic pathway whereby glucose is broken down to produce two molecules of pyruvate, forming two molecules of ATP in the process. Oxygen availability then determines the fate of pyruvate. If oxygen is available, pyruvate can enter aerobic metabolic pathways (such as the citric acid cycle and oxidative phosphorylation) to generate more ATP. If oxygen is limited, pyruvate is instead converted to lactate (also called lactic acid), as is the case during high-intensity workouts. Consequently, as high-intensity exercise continues, so does lactate formation.

Because very little ATP is actually produced during glycolysis, anaerobic metabolism can fuel a high-intensity workout only for a short period of time. Moreover, the accumulation of lactate and other intermediate metabolic products creates an acidic environment (a lower pH), which in turn inhibits the activity of enzymes needed for glycolysis. Without glycolysis, the muscle can no longer make ATP under anaerobic conditions, contributing to muscle fatigue. Furthermore, the drop in pH causes muscle contractions to weaken, contributing even more to fatigue. When this occurs, athletes often experience a burning sensation in their muscles.

Until recently, the accumulation of lactic acid in muscles was thought to be the major cause of muscle soreness. Although lactic acid buildup may be associated with acute muscle pain, it does not adequately explain muscle soreness that can linger for days after a vigorous workout. Scientists now believe that delayed-onset muscle soreness, which occurs up to 36 hours after exercise, may be caused by microscopic tears in muscles in response to new activities or working out with greater intensity.[16] Tiny tears cause muscle inflammation, which in turn causes localized soreness. This type of muscle soreness typically occurs after beginning a new exercise program or after increasing duration or intensity of an existing exercise routine. The good news is that muscle fibers that sustain microtears and then repair themselves become stronger and better able to resist future tearing and, therefore, better equipped at handling the same type of exercise in the future.

⟨CONNECTIONS⟩ Recall that the Cori cycle is a metabolic pathway that regenerates glucose by circulating lactate from muscle to the liver, where it undergoes gluconeogenesis (Chapter 7, page 282).

The Cori Cycle There is a tendency to view lactate as a metabolic waste product. However, lactate can be "recycled" into glucose and then "reused" as an energy source. For this to occur, muscles release lactate into the blood where it circulates to the liver and is converted to glucose via gluconeogenesis. Once released into the blood, the glucose is then taken up by muscles and other tissues. The release of lactate from the muscles into the blood, and its subsequent conversion to glucose by the liver, is called the Cori cycle. During periods of intense exercise, the Cori cycle helps increase glucose availability.

Aerobic Pathways: Long-Term Energy Systems After several minutes of low-intensity activity (such as leisurely walking) to moderate-intensity activity (such as briskly walking and doubles tennis), breathing becomes faster and harder. The heart begins to beat more frequently and forcefully. These changes in pulmonary and cardiovascular function help deliver needed oxygen to muscles. With sufficient oxygen now available, muscle cells are better able to use aerobic metabolic pathways to produce ATP. Aerobic metabolic pathways are important in that substantial amounts of ATP can be generated for an extended period of time—a critical advantage over anaerobic pathways.

Aerobic pathways use glucose, fatty acids, and—to a lesser extent—amino acids to generate ATP.[17] These energy sources are catabolized to produce NADH + H$^+$ and FADH$_2$, which in turn are used to generate ATP via oxidative phosphorylation. Although these pathways are relatively slow in terms of the rate of ATP production, the energy yield is rich. Fatty acids used for energy during physical activity are derived primarily from triglycerides stored in adipose tissue. However, intramuscular triglycerides also help fuel activity. Aerobic pathways also use stored muscle glycogen and glucose circulating in the blood.

Energy Metabolism during Prolonged Activity During prolonged activity such as running a marathon, the combined efforts of gluconeogenesis and glycogenolysis help provide sustained glucose to active muscles. However, once glycogen stores are exhausted, glycogenolysis can no longer contribute to the glucose pool. During this time, gluconeogenesis alone may not be able to provide glucose at the rate needed by skeletal muscles, either. When this happens, some athletes lose their stamina or "hit the wall," a term used to describe the feeling of profound fatigue that can occur during an athletic event.

Consuming sports drinks, energy bars, or energy gels during exercise can help provide an additional source of glucose. These products provide carbohydrates in a form that is easily digested and rapidly absorbed. Sports drinks are particularly helpful for some people because they not only provide glucose but also help keep athletes hydrated and replace electrolytes. However, too much fluid can cause cramping, so it is important for athletes to find out what products work best for them.

You now understand that movement requires energy and that, depending on the intensity and duration of the activity, muscles use different energy sources and metabolic pathways to generate ATP. However, the ability to sustain these metabolic tasks also depends on long-term physiologic adaptations that occur in response to repeated and sustained exercise. The next section provides an overview of how physiologic systems respond and adapt to the demands of exercise.

What Physiologic Adaptations Occur in Response to Athletic Training?

The body undergoes physiologic adaptations in response to the physical demands of frequent exercise. These changes, collectively called the **adaptation response,** enable the body to adjust to the rigors of exercise more efficiently and forcefully. Of course, the magnitude of these changes depends on a person's level of fitness and the intensity and duration of training. These training-induced adaptations cause muscles to become larger and stronger, a response referred to as muscle **hypertrophy.** Other adaptive responses involve the cardiovascular system. This is why it is important for athletes to train in order to optimize athletic performance. When a person stops training, these gains quickly diminish. For instance, muscles that had previously experienced hypertrophy but are subsequently not physically challenged can undergo **atrophy,** meaning that they become smaller again.

BOTH STRENGTH AND ENDURANCE TRAINING IMPROVE ATHLETIC PERFORMANCE

Different types of exercise place different physical demands on the body. For example, **strength training** (such as weight lifting) promotes muscle growth by challenging muscles with exercise that is difficult but brief in duration. These types of activities help increase muscle size and strength (in other words, hypertrophy). Conversely, **endurance training** (such as running and swimming) involves steady, low- to moderate-intensity activities that are more sustained in duration. Endurance training is especially beneficial to pulmonary and cardiovascular function and can help improve **aerobic capacity** (the maximum amount of oxygen the body can use during a specified period). It also increases the number and size of mitochondria in cells, improving the ability to produce ATP.

Some athletic trainers recommend **interval training,** which involves alternating short, fast bursts of high-intensity exercise with slower, less demanding activity. This type of training helps improve both aerobic and anaerobic capacities. **Anaerobic capacity** is defined as the ability to perform repetitive, high-intensity activity with little or no rest.

The anatomical and physiological changes that result from training are very important and help to improve athletic conditioning and performance. For example, the expansion of blood vessels in response to exercise increases your ability to take in and deliver oxygen to muscles. In addition, training causes muscles to develop a higher tolerance for lactate, allowing them to exercise longer before fatigue sets in.

Each individual is different, however, and depending on the type and intensity of training, the adaptation response will differ. Most athletic trainers recommend following a varied training regimen, alternating between workouts that increase strength and those that enhance stamina. However, it is important to avoid overtraining because there are potential risks associated with doing too much too soon. Specifically, training beyond the body's ability to recover can result in injury, muscle soreness, and joint pain.

Training Helps Muscles Use Glucose More Sparingly Training increases the ability of muscles to metabolize fatty acids for energy. Increased capacity to use fatty acids as an energy source helps spare glucose, in turn, delaying the onset of muscle fatigue. Recall that only small amounts of glycogen are stored in the liver and skeletal muscles. The ability of the body to "conserve glucose" is particularly important for endurance athletes such as marathon runners,

adaptation response The body's physiologic response to being challenged by frequent physical exertion.

hypertrophy An increase in size of a tissue or organ due to an increase in cell size and/or number.

atrophy A decrease in size of a tissue or organ due to a decrease in cell size and/or number.

strength training Athletic workout that enhances muscle strength.

endurance training Athletic workout that improves pulmonary and cardiovascular function.

aerobic capacity The maximum amount of oxygen the body can use during maximal physical exertion.

interval training Athletic workout that alternates between fast bursts of high-intensity exercise and slower, less demanding activity.

anaerobic capacity The ability to perform repetitive, high-intensity activity with little or no rest.

long-distance cyclists, and distance swimmers, who depend on stored glycogen over an extended period of time. Training boosts the capacity for fatty acid metabolism by increasing the number and size of mitochondria, the location of aerobic metabolism in muscle.[18] The use of fatty acids for energy involves β-oxidation and the citric acid cycle—both of which occur in mitochondria.

Training Improves Cardiovascular Fitness As previously described, increased intensity and duration of workouts necessitate that muscles utilize more oxygen for ATP production via aerobic pathways. In response, the cardiovascular system must work even harder to increase oxygen delivery to muscle cells. During a hard physical workout, for example, increased heart rate (beats per minute), **ventilation rate** (breaths per minute), and **stroke volume** (the amount of blood pumped per heartbeat) are required to deliver adequate oxygen and nutrients to active muscles. Physical training improves cardiac function and blood circulation to muscle tissues, making oxygen delivery more efficient. Equally important, increased blood flow facilitates the removal of metabolic waste products, such as lactic acid. Clearly, trained athletes with peak cardiovascular function have a competitive edge over untrained athletes.

Because of these adaptation responses, training increases what is called **maximal oxygen consumption,** also called VO_2 max, which is the maximum ability of the cardiovascular system to deliver oxygen to muscles and the ability of cells to use oxygen to generate ATP. A trained athlete has a higher VO_2 max than does an untrained individual, and this can both increase strength and help delay the onset of fatigue.

There are many reasons why training improves VO_2 max. For example, training strengthens the cardiac (heart) muscle, increasing its size and ability to contract forcefully. Increased stroke volume enables the heart of trained athletes to pump blood more efficiently. Thus, athletes often have lower resting heart rates than nonathletes—a sign of physical fitness. In addition, training causes an expansion of capillary vessels supplying blood to muscles—another example of the adaptation response. This aids blood flow and subsequent delivery of nutrients and oxygen to muscle cells. The production of red blood cells also increases in response to training, leading to greater oxygen-carrying capacity.

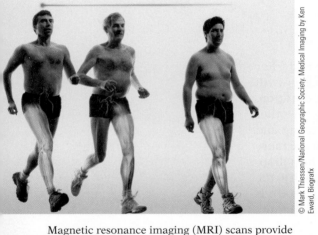

Magnetic resonance imaging (MRI) scans provide images of muscle shape, bones, and blood vessels. The two men on the left are physically fit (ages 37 and 60 years, respectively) and exercise regularly. The man on the right is 35 years of age and leads a sedentary lifestyle. After an hour on the treadmill, color differences in the leg muscles, indicate that the sedentary person worked harder.

© Mark Thiessen/National Geographic Society. Medical Imaging by Ken Eward, Biografx

Courtesy of U Aim High Fitness

Critical Thinking: Tom's Story Think back to Tom's story about his transition from being sedentary to training for an Ironman competition. What physiological adaptations have likely taken place in Tom's body as a result of his training? How will these adaptations help him endure even longer and harder workouts in the future?

ventilation rate Rate of breathing (breaths/minute).

stroke volume The amount of blood pumped by the heart with each contraction.

maximal oxygen consumption (VO_2 max) Maximum volume of oxygen used by tissues per minute.

ergogenic aid Substance taken to enhance physical performance beyond the gains obtained from physical training alone.

SOME ATHLETES USE "PERFORMANCE-ENHANCING" AIDS

Some athletes believe that taking certain substances, collectively referred to as **ergogenic aids,** can enhance physical performance beyond the gains obtained from physical training alone. The hope is that the combination of rigorous workouts and ergogenic substances will provide a competitive edge. Ergogenic aids encompass a wide range of substances including vitamins, minerals, herbal products, botanical agents, and extracts from organs or glands.[19] Although these products are largely untested, they remain popular among athletes and are widely available.[20] Whereas substances such as vitamins, minerals, bee pollen, creatine, and protein supplements are likely safe if taken as recommended, others may pose serious adverse side-effects.[21] For example, deaths associated with products containing ephedrine, which was marketed to boost metabolism, resulted in the Food and Drug Administration (FDA) banning its sale. It is important for athletes and sports enthusiasts to be aware that, in most cases, little effort is made to study the safety and efficacy of these products. In fact, the FDA does not require them to be tested prior to being placed in the marketplace. For these reasons, several athletic-governing organizations have banned the use of many "performance-enhancing" products by collegiate and professional athletes. A summary of commonly used "performance-enhancing" dietary substances is presented in Table 9.7.

The body's response to the physical demands of exercise is quite remarkable. Next, it is important to consider the dietary needs of athletes and how the physical demands of exercise can affect nutrient and energy requirements.

TABLE 9.7 Dietary Substances Used for Performance Enhancement

Substance	Physiologic Role in Body	Desired Effects	Research-Based Evidence	Adverse Effects
Amino acids/protein[a]	Required nutrients used for growth, maintenance, and repair of body structures including muscle	Enhance endurance; increase or maintain muscle mass to improve strength and size	No beneficial effects found	Possible increased stress placed on the kidneys; increased nitrogen excretion can cause fluid imbalance leading to dehydration; ingestion of large amounts of some amino acids can interfere with the absorption of other amino acids leading to nutrient imbalances; branched-chain amino acids may stimulate release of insulin, which can decrease blood glucose
Beta-hydroxy-beta-methylbutyrate (HMB)[b]	Metabolite of the amino acid leucine and plays a role in protein synthesis	Increase strength and lean body mass; suppress breakdown of muscle	Supplementation studies show small gains in strength and lean body mass in untrained athletes	No adverse effects reported with short-term use
Caffeine[c]	Naturally occurring compound, found in a variety of foods and beverages, that stimulates the central nervous system.	Increase mental alertness; improve athletic performance; enhance fatty acid metabolism; delay fatigue	Consumption (3 to 9 mg/kg body weight) 1 hour prior to exercise can enhance endurance and performance	Relatively safe with no negative performance effects; does not cause significant dehydration or electrolyte imbalance; may cause tolerance or dependence; abrupt discontinuation produces irritability, mood shifts, headache, drowsiness, or fatigue
Carnitine[d,e]	Amino acid found in nearly all cells of the body; plays an important role in energy metabolism	Increase long-chain fatty acid metabolism in skeletal muscle during exercise	Does not improve fatty acid metabolism during exercise; no increase in carnitine levels in muscle	Dosages exceeding 5 g/day, may cause nausea, vomiting, abdominal cramps, and "fishy" body odor
Chromium[e]	Trace mineral thought to enhance the action of insulin; involved in carbohydrate, fat, and protein metabolism	Promote weight loss; increase lean body mass; enhance performance	No benefits on body mass or composition	No adverse effects associated with high intakes from food or supplements

(continued)

TABLE 9.7 (continued)

Substance	Physiologic Role in Body	Desired Effects	Research-Based Evidence	Adverse Effects
Coenzyme Q$_{10}$[a,f]	Vitamin-like substance that functions as a coenzyme or as a precursor to coenzymes; synthesized from the amino acid tyrosine; used by mitochondrial enzymes and is important for energy metabolism and the production of ATP	Reduce oxidative damage in tissues; delay fatigue; improve cardiac function	May lower blood pressure, slow progression of dementia; may lower blood glucose; variable results on athletic performance enhancement	Few serious reported side effects but may include nausea, vomiting, stomach upset, heartburn, diarrhea, loss of appetite, skin itching, rash, insomnia, headache, dizziness, irritability, increased light sensitivity of the eyes, fatigue, or flu-like symptoms
Creatine[a]	A nitrogenous organic acid involved in anaerobic ATP production.	Supply ATP to muscle and nerve cells	Increases creatine levels in muscle tissue	Studies suggest that creatine supplementation is safe and poses very little risk, although may cause fluid retention
Dyhydroepiandrosterone (DHEA)[a]	Hormone used by the body to produce estrogen and testosterone	Improve energy, strength, immunity; increase muscle and decrease fat; slow aging process	No evidence to support effectiveness as a "rejuvenating" hormone	May cause high levels of estrogen or testosterone; no conclusive evidence on safety or efficacy; potential role in promotion of prostate and breast cancers
Ginseng[e]	Root that contains active chemical components thought to stimulate the immune system and to increase energy, longevity, and vitality.	Increase endurance and muscle strength; hasten recovery from exercise; improve energy metabolism during exercise; speed reaction time; improve mental concentration; improve coordination	May help lower blood glucose; possible effects on immune function	Usually well-tolerated; possible nervousness, insomnia, high blood pressure, GI discomfort, headaches, and allergic reactions
Glucosamine/chondroitin[f,g,h]	Glucosamine is an amino-sugar compound that the body uses to produce cartilage and other connective tissue; chondroitin sulfate is a complex carbohydrate that helps cartilage retain water	Repair damaged joints; treat osteoarthritis	Clinical trials provide evidence both for and against efficacy in treatment of joints and osteoarthritis	None reported
Glutamine[a]	Essential amino acid needed for synthesis of numerous vital proteins in the body	Lymphocyte (white blood cell) proliferation; support for immune system; increase glycogen synthesis; anticatabolic effect	Little support from well-controlled human studies for any ergogenic effects	No harm or ill effects reported

[a] Mayo Clinic Drugs and Supplements. Available from: http://www.mayoclinic.com/health/drug-information/DrugHerbIndex
[b] Palisin T, Stacy JJ. Beta-hydroxy-beta-methylbutyrate and its use in athletics. Current Sports Medicine Reports. 2005;4:220–3.
[c] Paluska SA. Caffeine and exercise. Current Sports Medicine Reports. 2003;2:213–9.
[d] University of Maryland Medical Center. Medical Alternative Medicine Index. Available from: http://www.umm.edu/altmed/.
[e] National Institutes of Health, Office of Dietary Supplements. Available from: http://ods.od.nih.gov.
[f] Reginster JY, Deroisy R, Rovati LC, Lee RL, Lejeune E, Bruyere O, Giacovelli G, Henrotin Y, Dacre JE, Gossett C. Long-term effects of glucosamine sulphate on osteoarthritis progression: A randomized, placebo-controlled clinical trial. Lancet. 2001;357:251–6.
[g] Pavelka K, Gatterova J, Olejarova M, Machacek S, Giacovelli G, Rovati LC. Glucosamine sulfate use and delay of progression of knee osteoarthritis: A 3-year, randomized, placebo-controlled, double-blind study. Archives of Internal Medicine. 2002;162:2113–23.

How Does Physical Activity Influence Dietary Requirements?

The unrelenting stress of training and competition can deplete the body of important substances and energy needed for cellular processes. It is important, therefore, for competitive athletes to have the right balance of nutrients, energy, and fluids during training, competition, and recovery. In fact, many collegiate and professional sport teams employ nutrition specialists such as dietitians to help in this regard.

Although no single dietary plan is right for everyone, it is helpful for athletes to be familiar with general guidelines of meal planning and food selection. In addition, physically active individuals should take special care to consume enough water to replace that lost during training and competition, and enough energy to meet the demands of physical activity and prevent weight loss. Although exercise may slightly increase micronutrient requirements,[22] these needs are readily met by an adequate and balanced diet. Athletes can optimize athletic performance by understanding the basics of energy metabolism and applying this information to their dietary and training strategies.

ENERGY REQUIREMENTS TO SUPPORT PHYSICAL ACTIVITY

It is important for people engaged in high levels of physical activity to have adequate energy intakes. For example, you can see from Table 9.6 that a sedentary 30-year-old man weighing 148 pounds needs approximately 2,447 kcal/day to simply maintain his body weight. However, if this same individual were to become very active, he would need approximately 1,000 kcal more each day. These values are based on the Institute of Medicine's Estimated Energy Requirement (EER) calculations, which were discussed in Chapter 2 and are available in Appendix B.[24] Without enough energy to meet the added demands of physical activity, body fat and protein are mobilized to meet energy needs. When muscle tissue is broken down to supply energy, strength and endurance can be severely compromised. In addition, athletes who consume low-energy diets often lack important micronutrients needed for energy metabolism, hemoglobin synthesis, bone health, protection from oxidative stress, and immune function.[23]

People who are physically active must consume enough energy to support normal daily activity as well as that required for exercise. To satisfy energy requirements, it is important to consider the type, frequency, duration, and intensity of exercise. Estimated energy expenditure associated with various types of activities is listed in Table 9.8. For example, if the 30-year-old, 148-pound male discussed previously decided to run for one hour at a 10-minute per mile pace, he would have expended about 710 kilocalories in addition to the amount of energy required to support normal daily activities over the course of the day. As a result, he would need to consume approximately 3,157 kcal (2,447 + 710) to maintain body weight on that day. Athletes who are training on a regular basis can have very high total energy requirements, especially if they are exercising for long periods of time. In fact, some athletes may require 4,000 to 5,000 kcal/day.

Competitive athletes and other highly active individuals are encouraged to eat a variety of nutrient-dense foods such as whole-grain cereal products, legumes, fruits, vegetables, lean meat, fish, poultry, and low-fat dairy products. In fact, there are essentially no differences in diets recommended for athletes and nonathletes—athletes simply require *more* of the same types of foods. To plan a diet that provides adequate energy and nutrients, athletes should visit the MyPlate food guidance system website (http://www.choosemyplate.gov).

TABLE 9.8 Average Energy Expenditure for Selected Physical Activities in Relation to Body Weight and Sex

	Body Weight, kg (lb)						
	50 (110)	57 (125)	64 (140)	70 (155)	77 (170)	84 (185)	91 (200)
Activity and Sex	(kcal/hour)[a]						
Aerobic Dance							
Male	480	488	506	531	556	582	607
Female	394	413	433	453	472	492	511
Biking (12 mile/hour)							
Male	380	401	422	443	464	486	507
Female	329	345	361	379	394	410	427
Gardening							
Male	303	320	337	354	371	388	405
Female	263	276	289	302	315	328	341
Golf							
Male	425	448	472	496	519	543	567
Female	368	386	404	422	441	459	477
Running (10 min/mile)							
Male	619	653	688	722	757	791	826
Female	536	562	589	615	642	669	695
Swimming (slow crawl)							
Male	364	384	405	425	445	465	486
Female	315	331	346	362	378	393	409
Walking (15 min/mile)							
Male	257	271	285	300	314	328	342
Female	222	233	244	255	266	277	288
Weight Lifting							
Male	340	359	378	397	415	434	453
Female	294	309	323	338	352	367	382

[a]Food Processor SQL exercise calculations expended during the exercise period are based on resting energy expenditure (REE) formulas (Available from: National Academy of Science, Food and Nutrition Research Council. Recommended Dietary Allowances, 10th edition. Washington, DC: National Academy Press, 1989) and metabolic equivalent (Available from: Ainsworth BE, Haskell WL, Whitt MC, Irwin ML, Swartz AM, Strath SJ, O'Brien WL, Bassett DR, Schmitz KH, Emplaincourt PO, Jacobs DR, Leon AS. Compendium of physical activities: an update of activity codes and MET intensities. Medicine and Science in Sports and Exercise. 2000;32:S498–S516.).
SOURCE: Food Processor Nutritional Analysis and Fitness Software, ESHA Research, Salem, Oregon. version 10.3.0, 2008.

RECOMMENDATIONS FOR MACRONUTRIENT INTAKE ARE SIMILAR FOR PHYSICALLY ACTIVE AND SEDENTARY INDIVIDUALS

Physically active people must consider not only their total energy requirements but also the source of these calories. In recent years, considerable attention has been given to the "ideal" distribution of total calories in the diet of athletes. Recall that the Institute of Medicine's Acceptable Macronutrient Distribution Ranges (AMDRs) suggest that 45 to 65% of total calories come from carbohydrates, 20 to 35% from fat, and 10 to 35% from protein. As you will learn, these recommendations are likely the same for all people—regardless of their activity level.

Carbohydrates It is important for physically active individuals to consume enough carbohydrates to prevent fatigue and maintain liver and muscle glycogen stores. Some athletes try to increase the amount of glycogen stored in muscles, using a technique called carbohydrate loading, which is described in

Glycogen is an important energy reserve that can help sustain athletes participating in distance or endurance sports. Even the demands of routine training can cause the body's glycogen stores to become quickly depleted. This is why it is important that physically active individuals consume adequate amounts of carbohydrates between workouts to replenish and maintain glycogen stores. In addition, some athletes try to increase the glycogen-storing capacity of their muscles through a technique called **carbohydrate loading.** Studies suggest that carbohydrate loading can increase the amount of glycogen stored in muscles, delaying the onset of fatigue during strenuous physical activities and those that last longer than 60 minutes.[25]

Carbohydrate loading strategists propose that athletes can stimulate their muscles to store more glycogen by combining the right workout intensity with the right level and timing of carbohydrate intake. The current recommended carbohydrate loading regime involves two stages.[26] First, athletes are encouraged to consume enough carbohydrate to replenish and maintain glycogen stores while training. This amounts to approximately 6 to 8 g carbohydrate/kg body weight/day. Approximately 36 to 48 hours prior to competition, athletes should begin to taper their workout intensity, while increasing their carbohydrate intake to between 10 and 12 g/kg body weight/day. For example, a man weighing 200 lb (91 kg) would increase his carbohydrate intake from between 546 and 728 g/day of carbohydrate to approximately 910 to 1,092 g/day, which is equivalent to about 40 slices of bread. This sequence largely replaces the original carbohydrate-loading regime developed in the 1960s, which included an initial phase of high-intensity workouts while simultaneously reducing carbohydrate intake.

For most recreational athletes, extensive carbohydrate loading is not necessary. Nor is it beneficial for athletic competitions that are short in duration (less than 60 minutes). A healthy diet that satisfies both nutrient and energy requirements is more than adequate. However, serious athletes participating in high-intensity or endurance competitions may find that carbohydrate loading gives them a competitive edge. Still, glycogen stores alone do not make an athlete. True athletic success requires a combination of natural talent, long hours of training, and a healthful balanced diet.

---

greater detail in the Focus on Sports Nutrition. The Institute of Medicine does not provide any special recommendations concerning carbohydrate intake for athletes as part of the DRIs. However, the American Dietetic Association recommends that athletes in training consume 5 to 10 g of carbohydrates/kg of body weight/day.[27] Thus, recommended daily carbohydrate intake for a 200-lb (91-kg) athlete is approximately 455 to 910 g, an amount equal to that found in 30 to 60 slices of bread. Clearly, carbohydrate requirements will depend somewhat on daily energy expenditure, sex, and the type of sport.

Protein Although considerable amounts of protein are needed to maintain, build, and repair muscle, there is an ongoing debate as to whether especially active people (and even competitive athletes) require additional protein.[28] The Institute of Medicine has concluded that the calculation used to estimate protein requirements for healthy adults (0.8 g of protein/kg/day) is the same as that which should be used to estimate protein requirements of physically active individuals as well.[24] This conclusion is in part based on several studies showing that athletes in training use amino acids more efficiently than nonathletes. Whereas the International Society of Sports Nutrition recommends protein intakes of 1.4 to 2 g/kg/day, others claim that endurance athletes should consume 1.2 to 1.4 g protein/kg/day, and that athletes engaged in strength training may require as much as 1.6 to 1.7 g protein/kg/day.[27] Until more information suggests otherwise, most experts agree that as long as athletes consume adequate amounts of energy and high-quality protein sources—such as meat, dairy, and eggs—the amount of protein will likely be adequate to maintain, build, and repair tissue.[30]

Lipids In addition to carbohydrate and protein, physically active people must also consume adequate amounts of fat. Fat is an important source of energy,

carbohydrate loading A technique used by some athletes to increase glycogen stores in muscles by combining a certain workout intensity with a level of carbohydrate intake.

provides essential fatty acids, and is necessary for the absorption of fat-soluble substances (e.g., vitamin A) in the body. Recommendations regarding the amount and types of dietary fats are the same for everyone—regardless of activity level. That is, foods rich in monounsaturated fatty acids (such as olive and canola oils) and polyunsaturated fatty acids (such as vegetable and fish oils) should be emphasized, whereas saturated and *trans* fatty acids should be limited. Furthermore, some experts believe that long-chain omega-3 fatty acids may be especially important for athletes.[30] This is because exercise can lead to inflammation that can possibly be mitigated by a diet high in eicosapentaenoic acid (EPA) and docosahexaenoic acid (DHA).

AN ADEQUATE DIET IS LIKELY TO SATISFY MICRONUTRIENT REQUIREMENTS

Vitamins and minerals are needed for many functions, including energy metabolism, repair and maintenance of body structures, protection from oxidative damage, and immunity. As long as an individual consumes a varied diet that supplies adequate energy, vitamin and mineral needs will likely be met as well. Supplemental nutrient intake to enhance athletic performance is controversial, and there is little evidence regarding the efficacy of this practice.[31]

However, people who consume foods with low nutrient densities, as well as athletes who restrict or avoid animal products such as meat, dairy, and eggs, are at increased risk for low intakes of several micronutrients—especially iron, calcium, and zinc. As these nutrients are especially important in terms of supporting physical activity, deficiencies may have more severe complications for some athletes. It is important for coaches and trainers to be aware of nutritional issues that affect athletes who follow strict vegetarian diets.[32]

Iron Iron is a component of hemoglobin and myoglobin, the iron-containing portions of red blood cells and muscles, respectively. Hemoglobin transports oxygen to cells and picks up carbon dioxide, a waste product of energy metabolism. Similarly, myoglobin is involved in the delivery of oxygen to muscles. Iron deficiency can therefore compromise oxygen delivery to and carbon dioxide removal from tissues. Because oxygen is required for aerobic energy metabolism, poor iron status in athletes can seriously affect training and performance. Although iron is available in cereals, grains, and some vegetables, it is not as readily absorbed as that in meat. All individuals, including athletes, who do not eat meat are at increased risk for impaired iron status and should have their blood periodically tested.

Endurance athletes may have higher iron requirements than nonathletes because of increased loss of blood in the feces, iron loss associated with excessive sweating, and rupturing of red blood cells as a result of the repeated striking of the feet against hard surfaces.[33] Because of blood loss associated with menstruation, female athletes are more likely than males to experience impaired iron status. Unfortunately, very few collegiate athletic programs routinely screen for iron deficiency in female athletes.[34]

In addition to primary iron deficiency caused by a lack of iron in the diet, some athletes experience a temporary form of anemia referred to as **sports anemia.** Sports anemia often occurs at the onset of training and is caused by a disproportionate increase in plasma volume compared with the synthesis of new red blood cells. This **hemodilution** may make the person appear iron deficient when he or she is not. Because sports anemia is due to a healthy adaptive response to training rather than inadequate nutrient intake, there is no need for additional iron in the diet. Unlike true iron deficiency anemia caused by dietary factors (such as insufficient iron or folate), sports anemia does not appear to hinder athletic performance.[35]

⟨**CONNECTIONS**⟩ Recall that plasma is the liquid portion of blood (Chapter 3, page 79).

sports anemia A physiological (adaptive) response to training caused by a disproportionate increase in plasma volume relative to the increase in number of red blood cells.

hemodilution A decrease in the number of red blood cells per volume of plasma caused by plasma volume expansion.

Calcium Although recommended calcium intakes for athletes and nonathletes are the same, adequate calcium intake is especially important for optimal athletic performance. As calcium is needed for building and repairing bone tissue, diets low in calcium can increase an athlete's risk for low bone density, which in turn increases the likelihood of stress fractures. Stress fractures are small cracks that occur in bones, especially of the feet, in response to repeated jarring.

Female athletes who have menstrual irregularities are at particularly high risk for bone loss and fractures. In fact, female athletes who exercise excessively while restricting their intake of calories are at risk for a syndrome referred to as the female athlete triad, which was addressed in the Nutrition Matters following Chapter 8. The female athlete triad is characterized by bone loss, disordered eating practices, and menstrual dysfunction (irregular or cessation of menses).

Athletes who restrict their intake of dairy products may benefit from foods that are fortified with calcium and vitamin D, which facilitates calcium absorption. These include fortified orange juice, soy products, and breakfast cereals. If their diet is lacking in calcium-rich foods, athletes should consider taking calcium supplements.[36]

Zinc Most (70%) of the zinc in the American diet comes from animal products, primarily meat. This is why athletes who restrict their intake of meat, dairy, or eggs are at increased risk for not getting enough zinc. Zinc plays an important role in energy metabolism and is needed for the maintenance, repair, and growth of muscles, making it an especially important mineral for athletes. Because zinc plays a role in more than 200 enzyme-catalyzed reactions, impaired zinc status has wide-reaching effects. Plant-based sources of zinc include legumes, whole-grain products, wheat germ, some fortified cereals, nuts, and soy products. Since supplemental zinc can affect the bioavailability of other nutrients such as iron and copper, athletes should be careful not to exceed the UL for zinc (40 mg/day).

© BananaStock/SuperStock

Low body fat can increase bone loss, which in turn can lead to stress fractures. It is important for all female athletes, especially those who have menstrual irregularities, to make sure they are meeting their recommended intake for calcium.

EXERCISE INCREASES THE NEED FOR FLUID AND ELECTROLYTES

Exercise increases body temperature, and the body dissipates this heat by sweating. Because sweat evaporation is the body's primary mode of heat loss during vigorous exercise, water loss can be substantial. If this water is not replaced, dehydration can result. Dehydration is caused by a lack of fluid intake relative to fluid loss from the body. To meet the body's need for water, most adult men and women need approximately 3.7 and 2.7 liters of water per day, respectively. However, athletes can require substantially more water than this.[37]

Although sweat consists mainly of water, it also contains electrolytes (important ions) such as sodium, chloride, and potassium, and, to a lesser extent, minerals such as iron and calcium. For this reason, extensive sweating can disrupt the balance of electrolytes in the body as well as fluids.

Dehydration Can Cause Serious Heat-Related Illness Decreased blood volume caused by excessive sweating reduces blood flow to the skin, which in turn affects the body's ability to dissipate heat. As a result, core body temperature can rise, increasing the risk of heat exhaustion and heat stroke. **Heat exhaustion** can occur when a person loses 5% of his or her body weight due to sweating.[38] When this happens, an athlete often feels ill and experiences muscle spasms, a rapid and weak pulse, low blood pressure, disorientation, and profuse sweating (Table 9.9). Heat exhaustion can quickly progress to heat stroke

heat exhaustion Moderate rise in body temperature that occurs when the body has difficulty dissipating heat.

TABLE 9.9 Signs and Symptoms of Mild to Severe Dehydration

Early or Mild Dehydration	Moderate to Severe Dehydration
• Flushed face • Extreme thirst • Dry, warm skin • Impaired ability to urinate; production of reduced amounts of dark, yellow urine • Dizziness made worse when standing • Weakness • Cramping in arms and legs • Crying with few or no tears • Difficulty concentrating • Sleepiness • Irritability • Headache • Dry mouth and tongue, with thick saliva	• Low blood pressure • Fainting • Disorientation, confusion, slurred speech • Severe muscle contractions in the arms, legs, stomach, and back • Convulsions • Bloated stomach • Heart failure • Sunken, dry eyes • Skin that has lost its firmness and elasticity • Rapid and shallow breathing • Fast, weak pulse

AP Photo/Jeff Roberson

It is important for athletes to prevent dehydration by consuming adequate amounts of fluids before, during, and after exercise.

if fluid loss continues. **Heat stroke** occurs when 7 to 10% of body weight is lost due to sweating. This very serious condition is characterized by dry skin, confusion, and loss of consciousness. A person experiencing heat stroke requires immediate medical assistance.

Dehydration not only impairs body temperature regulation, but can also cause a number of other problems. For example, blood sodium concentration can decrease when an athlete experiences excessive sweating or when large amounts of fluids are consumed without adequate sodium intake. This condition, called **hyponatremia,** is more common in endurance events such as marathons and triathlons or other exercise lasting several hours. Hyponatremia can cause fluids to leave the blood and accumulate in surrounding tissues. The accumulation of fluids in lung tissue can impair gas exchange, raising carbon dioxide levels in the blood.

Individuals must be particularly careful when exercising in hot or humid weather, as these conditions can increase sweating and subsequent water and electrolyte loss from the body. A study of marathon runners found that 13% of the nearly 500 runners had hyponatremia by the time the race was completed.[39] Because drinking large amounts of plain water may actually increase risk for hyponatremia, many experts recommend that endurance athletes replenish body fluids by consuming drinks that contain both carbohydrates and electrolytes, such as sports drinks.[40]

Preventing Dehydration and Electrolyte Imbalance To prevent dehydration, athletes must consume adequate amounts of fluids before, during, and after exercise. The National Athletic Trainers' Association recommends that athletes consume two to three cups of fluid two to three hours before exercise.[41] This ensures that the athlete is fully hydrated while allowing time for excess fluids to be excreted from the body. Similarly, the American College of Sports Medicine stresses the importance of prehydration several hours prior to competition.[42] As previously noted, athletes who exercise in hot and humid conditions may require more fluid.

heat stroke Severe rise in body temperature due to excess loss of body fluids (dehydration) associated with sweating.

hyponatremia Low level of sodium in the blood is often associated with deyhydration.

Because dehydration can compromise performance, it may be necessary for athletes to consume fluids during exercise. The purpose of fluid intake during athletic events is not to rehydrate but rather to stay hydrated by replacing fluid loss associated with sweating. This is why race organizers set up water stations along the course, allowing participants to replace fluids lost during competition. However, during training, it may be necessary for athletes to carry water with them. There are many hydration devices to choose from, and it is important to find one that is comfortable and can hold adequate amounts of fluids.

For exercise that lasts less than one hour, plain water is adequate to replenish body fluids. When workouts last more than one hour, drinking fluids (such as sports drinks) that contain small amounts of carbohydrates and electrolytes is beneficial.[43] Sports drinks are readily available, well tolerated during exercise, and provide roughly 10 to 20 g of carbohydrates (40 to 80 kcal) per 8 fluid ounces.

It is important to be aware that some sports beverages, dubbed as "energy" drinks, contain caffeine. Caffeine is a stimulant and, in moderation, generally has no adverse health effects. Contrary to popular belief, consuming caffeinated beverages and foods in relatively small doses (<180 mg/day or about 2 cups of coffee a day) is not likely to cause dehydration.[44] Nonetheless, caffeine affects people differently; whereas some people enjoy the mild stimulatory effect of caffeine, others find that it causes stomach upset, nervousness, and jitters. The International Olympic Committee does not classify caffeine as a prohibited substance.

NUTRITION PLAYS AN IMPORTANT ROLE IN POST-EXERCISE RECOVERY

Nutrition is certainly an important aspect of post-exercise recovery. Hard training is physically demanding, and athletes need to ingest the right amounts and types of food following exercise. A well-executed dietary recovery plan is important to help the body repair muscle tissue, replace lost nutrients, and fully restore its energy reserves such as glycogen.

Carbohydrates Help Replenish Glycogen Stores Long, intense athletic workouts can deplete liver and muscle glycogen stores, and it takes approximately 20 hours to fully replenish these. Although the rate of glycogen synthesis during the post-exercise period is variable, studies show that, for optimal muscle glycogen resynthesis, there are several recommendations. First, consumption of carbohydrate-rich foods is advised as soon as possible after exercise.[45] This is because the rate at which muscles synthesize glycogen is highest within the first two hours following exercise. In addition, it is also recommended that small, frequent amounts of carbohydrates be consumed, rather than a single large carbohydrate-rich meal.[46] This helps to sustain a more even rise in blood glucose, rather than a rapid rise followed by a quick drop.

It is also important to consume high glycemic-index foods after exercise, because they are digested and absorbed more rapidly than low-glycemic index foods. Studies have found that this also helps to promote more rapid glycogen resynthesis.[47] Furthermore, some studies show that, when carbohydrates and proteins are consumed together during the post-exercise period, the rate of muscle glycogen synthesis may be higher, compared with carbohydrate consumption alone.[48] While the ideal ratio of carbohydrate-to-protein has not been firmly established, some athletic trainers recommend a ratio of 4:1.[49] This mix of carbohydrate and protein may promote greater muscle glycogen, but there is no evidence that it improves overall athletic performance.

Because glycogen is synthesized from glucose molecules, athletes should select carbohydrate-rich foods.[50] For example, some sports drinks are

Long, intense workouts can deplete glycogen stores. For glycogen repletion, athletes should consume small, frequent amounts of carbohydrates within the first two hours following exercise. Energy bars provide athletes with needed calories, carbohydrates, and protein.

sweetened with fructose, rather than glucose or sucrose. Studies comparing the effect of glucose or fructose consumption on glycogen resynthesis show that the rate of glycogen synthesis is higher when glucose-containing beverages are consumed.[51]

Physically active people must also consider the *amount* of carbohydrate they consume during the post-workout recovery period when the rate of muscle glycogen synthesis is highest. Consuming a large amount of carbohydrates (50 to 70 g) within the first hour of post-recovery can "jump start" the metabolic process of glycogen synthesis.[52] Although protein is essential for muscle recovery, consuming protein-rich foods alone does not appear to enhance glycogen repletion.[27]

Protein and Muscle Recovery Because prolonged, intense exercise increases protein catabolism, high-quality protein sources should be consumed during the post-exercise recovery phase. In fact, the first few hours following exercise are often referred to as the "anabolic window," an important time to reverse the catabolic state associated with physical activity and to increase the rate of protein synthesis. By consuming protein-rich foods, the concentration of amino acids in the blood increases, in turn providing the necessary building blocks for muscle maintenance and recovery.

As previously discussed, experts disagree as to whether athletes require more protein (relative to body weight) than nonathletes. There is some evidence, however, that certain types of protein may be more advantageous in post-workout muscle recovery than other sources. For example, whey protein, which comes from milk, has a high biological value, and its popularity among athletes has grown in recent years. Some athletes believe that whey protein is particularly important for building and retaining muscle because it provides the right balance of amino acids. Studies comparing the efficacy of whey protein with other sources of dietary protein suggest that it can speed tissue repair, which in turn accelerates recovery from exercise-induced muscle damage.[53]

Water Loss Replacement Exercise can cause substantial water loss from the body. Even if athletes consume fluids while exercising, they can still lose up to 2 to 6% of their body weight during strenuous activities. Therefore, fluid replacement is critical during the post-workout recovery phase. This is best done by consuming 2 to 3 cups of fluids for each pound of weight lost during exercise. One simple way to monitor hydration status is to observe the color of the urine the morning after a hard workout. Light-colored urine is a general indicator that a person is well-hydrated, whereas dark-colored urine may indicate that additional fluids are needed. Although sweat contains significant amounts of sodium, enough sodium is usually present in foods consumed after exercise to restore electrolyte balance. Therefore, it is neither necessary nor recommended that physically active individuals take sodium (salt) supplements.[54]

Critical Thinking: Tom's Story Diet plays an important role in athletic training and performance. What dietary recommendations would you give someone like Tom that could help sustain and replenish his body before, during, and after his rigorous workouts?

Diet Analysis PLUS ✚ Activity

Physical Activity and Health

The following activity assumes you have kept a log of your food/beverage intakes for at least 3 days and you have input this information into the Diet Analysis Plus software program. You may or may not have kept account of your physical activity. The purpose of this exercise is for you to evaluate whether the 2008 Physical Activity Guidelines for Americans established by the U.S. Department of Health and Human Services (DHHS) (see Table 9.2) would be effective for achieving their stated goals: *Adults (aged 18–64 years) 150 min/week of moderate intensity physical activity or 75 min/week of vigorous-intensity aerobic physical activity or an equivalent combination of moderate-and vigorous-intensity aerobic physical activity.*

The following guidelines are intended for sedentary individuals and describe "moderate" intensity activity (described indirectly as the "ability to converse while exercising" on page 392 of this chapter) done daily for increasing amounts of time.

1. Adjust your activity level in your profile to *"sedentary,"* while leaving your input of food/beverage intakes unchanged.

2. For a baseline value of your Energy Balance, using the **Track Activity** feature, do not put any form of exercise down. The program will assume that any time you do not account for in a 24-hour period will be counted as "resting."

3. Print out the **Energy Balance** report making sure to enter the *starting date* and *ending date* of your record-keeping. Also, make sure to check all the boxes in the **Choose Meals** box.

4. Next, click on the **Track Activity** button near the top of the screen and input 30 minutes of any *moderate form* of exercise (look for the word "moderate" in the description of the activity to make sure). Make sure to add this exercise session to each of your days in the reporting period.

5. Click on the **Reports** button, and print out another Energy Balance report, again making sure your start and end dates are different and encompass your entire reporting period.

6. Return again to the **Track Activity** section, and change the duration of the same moderate form of exercise to 60 minutes for each day. Print out the Energy Balance report. Finally, repeat steps 4–5 changing the duration of the same activity to 90 minutes per day.

7. You should have *four* energy balance reports at this point, one representing no exercise, one representing 30 minutes of moderate exercise per day, another representing 60 minutes per day, and the final representing 90 minutes per day.

Analysis of Results

1. Examining each of the Energy Balance reports, list the average energy expenditure for each physical activity level. Describe what occurs when exercise increases from zero minutes/day to 30 minutes/day. Which incremental increase in duration resulted in the greatest difference in average energy expenditure?

2. Next, examine each report's Average Net Gain/Loss and list below. Does the value for the 30 minutes/day result in any net calorie deficit? If not, do you feel that beneficial health claims reported in the 2008 physical activity guidelines for Americans are valid? Why or why not?

3. What is the difference in Average Net Gain/Loss values between the 30 min/day and 60 min/day report? In your opinion, does the change in net gain/loss values support the goal of "prevention of weight gain" for everyone the majority of the time?

4. Finally, examine the Average Net Gain/Loss value for the 90 minutes per day duration of the exercise. Why do you think the highest duration of exercise is necessary to prevent weight regain?

Notes

1. National Center for Health Statistics (NCHS). Health, United States, 2010. Available from: http://www.cdc.gov/nchs/data/hus/hus10.pdf.

2. Warburton DE, Nicol CW, Bredin SS. Health benefits of physical activity: The evidence. Canadian Medical Association Journal. 2006;174:801–9.

3. Vuori IM. Health benefits of physical activity with special reference to interaction with diet. Public Health Nutrition. 2001;4:517–28.

4. American College of Sports Medicine (ACSM). ACSM's guidelines for exercise testing and prescription. 8th ed. Wolters Kluewer, Lippincott Williams and Wilkins, New York, 2010.

5. U.S. Department of Health and Human Services. 2008 Physical activity guidelines for Americans. Available from: http://www.health.gov/PAguidelines/guidelines/chapter4.aspx.

6. Williams MA, Haskell WL, Ades PA, Amsterdam EA, Bittner V, Franklin BA, Gulanick M, Laing ST, Stewart KJ; American Heart Association Council on Clinical Cardiology; American Heart Association Council on Nutrition, Physical Activity, and Metabolism. Resistance exercise in individuals with and without cardiovascular disease: 2007 update: A scientific statement from the American Heart Association Council on Clinical Cardiology and Council on Nutrition, Physical Activity, and Metabolism. Circulation. 2007;116:572–84.

7. Blair SN, LaMonte MJ, Nichaman MZ. The evolution of physical activity recommendations: How much is enough? American Journal of Clinical Nutrition. 2004;79:913S–20S.

8. Donnelly JE, Blair SN, Jakicic MJ, Manore MM, Rankin JW, Smith BK. Appropriate physical activity intervention strategies for weight loss and prevention of weight regain for adults. American College of Sports Medicine Position Stand. Medicine and Science in Sports and Exercise. 2009;41:459–71.

9. American College of Sports Medicine (ACSM). ASCM's Resource Manual for Guidelines for Exercise Testing and Prescription. 6th ed. Philadelphia: Lippincott Williams and Wilkins; 2010.

10. Centers for Disease Control and Prevention. Physical activity for everyone. Available from: http://www.cdc.gov/physicalactivity/everyone/measuring/index.html. Persinger R, Foster C, Gibson M, Fater DCW, Porcari JP. Consistency of the talk test for exercise prescription. Medicine and Science in Sports and Exercise. 2004;36:1632–36.

11. Volek JS, Rawson ES. Scientific basis and practical aspects of creatine supplementation for athletes. Nutrition. 2004;20:609–14.

12. Kreider RB. Effects of creatine supplementation on performance and training adaptations. Molecular and Cellular Biochemistry. 2003;244:89–94.

13. Bemben MG, Lamont HS. Creatine supplementation and exercise performance: Recent findings. Sports Medicine. 2005;35:107–25.

14. LaBotz M, Smith BW. Creatine supplement use in an NCAA Division I athletic program. Clinical Journal of Sports Medicine. 1999;9:167–9.

15. Antonutto G, Di Prampero PE. The concept of lactate threshold. A short review. Journal of Sports Medicine and Physical Fitness. 1995;35:6–12.

16. Clarkson PM, Hubal MJ. Exercise-induced muscle damage in humans. American Journal of Physical Medicine and Rehabilitation. 2002;81:S52–69.

17. Hagerman FC. Energy metabolism and fuel utilization. Medicine and Science in Sports and Exercise. 1992;24:S309–14.

18. Jones TE, Baar K, Ojuka E, Chen M, Holloszy JO. Exercise induces an increase in muscle UCP3 as a component of the increase in mitochondrial biogenesis. American Journal of Physiology, Endocrinology, and Metabolism. 2003;284:E96–101.

19. U.S. Food and Drug Administration. Overview of dietary supplements. Available from: http://www.cfsan.fda.gov/-dms/ds-oview.html.

20. DeAngelis CD, Fontanarosa PB. Drugs alias dietary supplements. Journal of the American Medical Association. 2003;290:1519–20.

21. Estes M, Kloner R, Olshansky B, Virmani R. Task force 9: Drugs and performance-enhancing substances. Journal of the American College of Cardiology. 2005;45:1368–9.

22. Volpe SL. Micronutrient requirements for athletes. Clinics in Sports Medicine. 2007;26:119–30.

23. Loucks AB. Low energy availability in the marathon and other endurance sports. Sports Medicine. 2007;37:348–52.

24. Institute of Medicine. Dietary reference intakes for energy, carbohydrate, fiber, fat, fatty acids, cholesterol, protein, and amino acids. Washington, DC: National Academies Press; 2005.

25. Pizza FX, Flynn MG, Duscha BD, Holden J, Kubitz ER. A carbohydrate loading regimen improves high intensity, short duration exercise performance. International Journal of Sport Nutrition. 1995;5:110–6.

26. Burke LM. Nutrition strategies for the marathon: Fuel for training and racing. Sports Medicine. 2007;37:344–7.

27. Campell B, Kreider RB, Ziegenfuss T, La Bounty P, Roberts M, Burke D, Landix J, Lopez H, Antonio J. International Society of Sports Nutrition position stand: protein and exercise. Journal of the International Society of Sports Nutrition. 2007;4:1–7. American Dietetic Association; Dietitians of Canada; American College of Sports Medicine. Rodriguez NR, Di Marco NM, Langley S. American College of Sports Medicine position stand. Nutrition and athletic performance. Medicine and Science in Sports and Exercise. 2009;41:709–31.

28. Tipton KD, Witard OC. Protein requirements and recommendations for athletes: Relevance of ivory tower arguments for practical recommendations. Clinics in Sports Medicine. 2007;26:17–36.

29. Tipton KD, Wolfe RR. Protein and amino acids for athletes. Journal of Sports Science. 2004;22:65–79.

30. Simopoulos AP. Omega-3 fatty acids and athletics. Current Sports Medicine Reports. 2007;6:230–6.

31. Maughan RJ, King DS, Lea T. Dietary supplements. Journal of Sports Science. 2004;22:95–113.

32. Barr SI, Rideout CA. Nutritional considerations for vegetarian athletes. Nutrition. 2004;20:696–703.

33. Suedekum NA, Dimeff RJ. Iron and the athlete. Current Sports Medicine Reports. 2005;4:199–202.

34. Cowell BS, Rosenbloom CA, Skinner R, Summers SH. Policies on screening female athletes for iron deficiency in NCAA division I-A institutions. International Journal of Sport Nutrition and Exercise Metabolism. 2003;13:277–85.

35. Smith JA. Exercise, training and red blood cell turnover. Sports Medicine. 1995;19:9–31.

36. Kunstel K. Calcium requirements for the athlete. Current Sports Medicine Reports. 2005;4:203–6.

37. Sawka MN, Cheuvront SN, Carter R 3rd. Human water needs. Nutrition Reviews. 2005;63:S30–9

38. Shirreffs SM. The importance of good hydration for work and exercise performance. Nutrition Reviews. 2005;63:S14–S21.

39. Almond CS, Shin AY, Fortescue EB, Mannix RC, Wypij D, Binstadt BA, Duncan CN, Olson DP, Salerno AE, Newburger JW, Greenes DS. Hyponatremia among runners in the Boston Marathon. New England Journal of Medicine. 2005;352:1550–6.

40. Hsieh M. Recommendations for treatment of hyponatraemia at endurance events. Sports Medicine. 2004;34:231–8.

41. Casa DJ, Armstrong LE, Hillman SK, Montain SJ, Reiff RV, Rich BS, Roberts WO, Stone JA. National Athletic Trainers' Association position statement: Fluid replacement for athletes. Journal of Athletic Training. 2000;35:212–24.

42. Sawka MN, Burke LM, Eichner ER, Maughan RJ, Montain SJ, Stachenfeld NS. American College of Sports Medicine position stand. Exercise and fluid replacement. American College of Sports Medicine. Medicine and Science in Sports and Exercise. 2007;39:377–90.

43. American College of Sports Medicine. Position stand: Exercise and fluid replacement. Medicine and Science in Sports and Exercise. 2007;39:377–90.

44. Armstrong LE, Pumerantz AC, Roti MW, Judelson DA, Watson G, Dias JC, Sokmen B, Casa DJ, Maresh CM, Lieberman H, Kellogg M. Fluid, electrolyte, and renal indices of hydration during 11 days of controlled caffeine consumption. International Journal Sport Nutrition and Exercise Metabolism. 2005;15:252–65.

45. Kerksick C, Harvey T, Stout J, Campbell B, Wilborn C, Kreider R, Kalman D, Ziegenfuss T, Lopez H, Landis J, Ivy JL, Antonio J. International Society of Sports Nutrition position stand: Nutrient timing. Journal of the International Society of Sports Nutrition. 2008;5:17.

46. Burke LM, Kiens B, Ivy JL. Carbohydrates and fat for training and recovery. Journal of Sports Science. 2004;22:15–30.

47. Walton P, Rhodes EC. Glycaemic index and optimal performance. Sports Medicine. 1997;23:164–72.

48. Baty JJ, Hwang H, Ding Z, Bernard JR, Wang B, Kwon B, Ivy JL. The effect of a carbohydrate and protein supplement on resistance exercise performance, hormonal response, and muscle damage. Journal of Strength and Conditioning Research. 2007;21:321–9.

49. Howarth KR, Moreau NA, Phillips SM, Gibala MJ. Co-ingestion of protein with carbohydrate during recovery from endurance exercise stimulates skeletal muscle protein synthesis in humans. Journal of Applied Physiology. 2009;106:1394–402.

50. Campbell C, Prince D, Braun M, Applegate E, Casazza GA. Carbohydrate-supplement form and exercise performance. International Journal of Sport Nutrition and Exercise Metabolism. 2008;18:179–90.

51. Blom PC, Høstmark AT, Vaage O, Kardel KR, Maehlum S. Effect of different post-exercise sugar diets on the rate of muscle glycogen synthesis. Medicine and Science in Sports and Exercise. 1987;19:491–6.

52. Sherman WM, Doyle JA, Lamb DR, Strauss RH. Dietary carbohydrate, muscle glycogen, and exercise performance during 7 d of training. American Journal of Clinical Nutrition. 1993;57:27–31.

53. Tang JE, Manolakos JJ, Kujbida GW, Lysecki PJ, Moore DR, Phillips SM. Minimal whey protein with carbohydrate stimulates muscle protein synthesis following resistance exercise in trained young men. Applied Physiology, Nutrition, and Metabolism. 2007;32:1132–8.

54. Speedy DB, Thompson JM, Rodgers I, Collins M, Sharwood K, Noakes TD. Oral salt supplementation during ultradistance exercise. Clinical Journal of Sport Medicine. 2002;12:279–84.

Water-Soluble Vitamins

So far, you have learned how the body uses energy-yielding macronutrients for a multitude of functions. However, carbohydrates, proteins, and lipids represent only some of the nutrients needed for life. It is also important to learn about vitamins and minerals, as these important micronutrients are essential components of a healthy and balanced diet as well.

We will begin the study of micronutrients with vitamins. Recall that vitamins are organic molecules found in both plant- and animal-derived foods, required in very small amounts to maintain the body's basic functions. There are 14 known vitamins, and they are classified as water soluble or fat soluble, depending on their chemical solubility. Vitamins participate in a variety of functions, from regulating energy metabolism to repairing cell damage.

In this chapter, you will learn about the water-soluble vitamins, including their structures, dietary sources, regulation, major functions, and dietary recommendations. How to recognize the signs and symptoms of each vitamin's deficiency and toxicity states will also be presented. Ways to prepare and store fresh and cooked foods to optimize their vitamin contents and when it is appropriate—or inappropriate—to get vitamins from supplements will be discussed as well.

"I'm Sorry, But Your Baby Has a Neural Tube Defect"

Beth and her husband, Jeff, first met when they were undergraduates. Jeff was studying math, while Beth majored in chemistry. Although Beth had enjoyed her coursework, after she graduated she decided on a new career path—dietetics. Graduate school was not easy, but Beth dealt with all of its challenges and successfully completed her master's degree. Luckily, Beth quickly found a job working as a dietitian at a health clinic. However, throughout the years of classes and late-night studying, Beth had often thought about the day when she and Jeff could start a family. Now that day had finally come.

The prenatal checkups went well, and Beth enjoyed being pregnant. But halfway through her pregnancy, Beth's world suddenly turned upside down. She will always remember the words her doctor whispered when he delivered the news—"I am sorry to have to tell you this, but your baby has a neural tube defect." Feeling breathless, she struggled to understand what her doctor was telling her.

As a dietitian, Beth was well aware of the importance of good nutrition during pregnancy. For example, she knew that alcohol consumption could cause birth defects, and low levels of the B vitamin folate were linked to increased risk of neural tube defects. As such, she abstained from drinking alcohol and made sure she was eating healthy foods. But still, Beth kept wondering if she had done something to cause her baby to have a neural tube defect. Once Beth started doing some research, she was surprised to learn that not all neural tube defects are caused simply by a lack of dietary folate. In fact, many cases are thought to be completely unrelated to diet. Knowing this helped ease the guilt she was feeling.

As Beth's due date approached, she and Jeff were well prepared for the birth of their baby boy. Delivered by caesarean-section, Daniel was welcomed into the world by a team of doctors and nurses. The extent of his neural tube defect was not as severe as originally predicted, and after surgery to close the opening in Daniel's spine, Jeff and Beth joined their son in the neonatal intensive care unit. It was hard seeing their baby connected to tubes and monitors, but the worst was now behind them.

Beth and Jeff understand that many challenges lie ahead, but they are reassured by knowing there are many resources available to assist them. We asked Beth if there was anything she wanted to share about her experience, and this is what she said: "Please know that sometimes neural tube defects can be prevented by consuming adequate amounts of folate. However, not *all* cases involving a neural tube defect are preventable."

© Sara Press/Getty Images

Critical Thinking: Beth's Story

Do you know someone with a neural tube defect or other birth defect? If so, are there any outward signs of his or her condition? How might having a birth defect, especially one that affects the spinal cord, affect someone's ability to go to college or participate in extracurricular activities? Does your university or college provide assistance to students with special needs such as this?

The Water-Soluble Vitamins: A Primer

There are nine essential water-soluble vitamins: eight B vitamins and vitamin C. Because some people consider choline and carnitine to be conditionally essential, water-soluble, "vitamin-like" compounds, we have included discussions of these substances in this chapter as well. As you previously learned, each macronutrient class has some distinguishing characteristics that make it chemically unique. For instance, proteins are made from nitrogen-containing amino acids, whereas carbohydrates consist of sugar units. The situation for the vitamins, however, is somewhat different. This is because there is no single distinguishing characteristic that sets them apart from the rest of the nutrient classes. Instead, each vitamin is chemically distinct from the others except for the fact that they are all structurally "organic." There is, however, a basic set of biological properties that define and characterize these compounds, and some of these are described next.

⟨CONNECTIONS⟩ Essential nutrients are substances needed for health but either not synthesized by the body or synthesized in inadequate amounts. Consequently, essential nutrients must be obtained from the diet (Chapter 1, page 5).

WATER-SOLUBLE VITAMINS TEND TO HAVE SIMILAR PROPERTIES

One attribute that all water-soluble vitamins share is that they dissolve in water (as opposed to lipids). Second, although there are exceptions, the body generally digests, absorbs, and transports the water-soluble vitamins in similar ways. For example, they are absorbed mostly in the small intestine and, to a lesser extent, the stomach. Further, many water-soluble vitamins found in foods are bound to proteins that enzymes must remove before the vitamins can be absorbed. In addition, their bioavailability (i.e., absorption) can be influenced by many factors, including nutritional status, other nutrients and substances in foods, medications, age, and illness. And, except for choline which is circulated away from the GI tract in the lymph, all of the water-soluble compounds are circulated directly to the liver in the blood. Finally, because the body does not actively store water-soluble vitamins, they generally do not have toxic effects when consumed in large amounts. Thus, although there is no common *chemical* structure for these compounds, they share myriad *biological* commonalities.

WATER-SOLUBLE VITAMINS FUNCTION IN DIVERSE WAYS

Table 10.1 lists the names, basic functions, and some dietary sources of each of the water-soluble vitamins. As you can tell from Table 10.1, in contrast to the many shared physiological properties, the *functions* of the water-soluble vitamins are diverse and often unique. Nonetheless, most of the water-soluble vitamins or compounds made from them serve as coenzymes in energy metabolism pathways. Others—like folate and vitamin C—are involved in a wide variety of metabolic activities such as serving as carriers of single-carbon (methyl) units and protecting cells from free radical damage, respectively.

SOME VITAMINS HAVE SEVERAL NAMES

If you have ever shopped for vitamins, you may have noticed that there are hundreds of products with various names—some of them are simple (like vitamin C), and others have more technical names (like pantothenic acid).

TABLE 10.1 Major Functions, Deficiency Diseases, Toxicity Symptoms, and Food Sources of the Essential Water-Soluble Vitamins, Choline, and Carnitine

Vitamin	Major Functions	Deficiency	Toxicity	Food Sources
Thiamin (B$_1$)	• Coenzyme (TPP)‡ • Energy metabolism • Synthesis of DNA, RNA, and NADPH + H$^+$ • Nerve function	Beriberi	—	Pork Whole grains Legumes Tuna Soy milk
Riboflavin (B$_2$)	• Coenzyme (FAD and FMN)§ • Energy metabolism (redox reactions) • Metabolism of folate, vitamin A, niacin, vitamin B$_6$, and vitamin K • Neurotransmitter metabolism	Ariboflavinosis	—	Liver Mushrooms Dairy products Tomatoes Spinach
Niacin (B$_3$)	• Coenzyme (NAD$^+$ and NADP$^+$)** • Energy metabolism (redox reactions) • Protein synthesis • Glucose homeostasis • Cholesterol metabolism • DNA repair	Pellagra	Skin inflammation and flushing	Liver Fish Meat (including poultry) Tomatoes Mushrooms
Pantothenic acid (B$_5$)	• Coenzyme (CoA) • Energy metabolism • Heme synthesis • Cholesterol, fatty acid, steroid, and phospholipid synthesis	Burning feet syndrome (possibly)	—	Liver Mushrooms Sunflower seeds Yogurt Turkey
Vitamin B$_6$	• Coenzyme (PLP)†† • Amino acid metabolism (transamination) • Neurotransmitter and hemoglobin synthesis • Glycogenesis • Regulation of steroid hormone function	Microcytic hypo-chromic anemia	—	Fish Chickpeas Liver Potatoes Bananas
Biotin	• Coenzyme • Energy metabolism (carboxylation) • Regulation of gene expression	Depression, loss of muscle control, and skin irritations	—	Peanuts Almonds Mushrooms Egg yolk Tomatoes
Folate	• Coenzyme (THF)‡‡ • Single-carbon transfers • Amino acid metabolism • DNA and RNA synthesis	Macrocytic anemia, increased cardiovascular disease risk, increased risk for fetal malformations	Neurological problems	Organ meats Legumes Okra Leafy vegetables Orange juice
Vitamin B$_{12}$	• Coenzyme • Homocysteine metabolism • Energy metabolism	Macrocytic anemia	—	Mollusks Liver Salmon Meat Cottage cheese

TABLE 10.1 *(continued)*

Vitamin	Major Functions	Deficiency	Toxicity	Food Sources
Vitamin C	• Antioxidant • "Recharging" enzymes • Collagen synthesis • Tyrosine, neurotransmitter, and hormone synthesis • Protection from free radicals	Scurvy	Gastrointestinal problems	Peppers Papayas Citrus fruits Broccoli Strawberries
Choline	• Phospholipid synthesis • Neurotransmitter synthesis	Liver damage	Fishy body odor	Eggs Liver Legumes Pork Fish
Carnitine	• Phospholipid synthesis • Neurotransmitter synthesis	Muscle weakness, hypoglycemia, heart irregularities	—	Meat Eggs Dairy foods Avocados

† TPP, thiamin pyrophosphate
§ FAD⁺, flavin adenine dinucleotide; FMN, flavin mononucleotide
** NAD⁺, nicotinamide adenine dinucleotide; NADP⁺, nicotinamide adenine dinucleotide phosphate
†† PLP, pyridoxal phosphate
‡‡ THF, tetrahydrofolate

SOURCE: Institute of Medicine. Dietary reference intakes for thiamin, riboflavin, niacin, vitamin B₆, folate, vitamin B₁₂, pantothenic acid, biotin, and choline. National Academy Press, Washington, DC: 1998.

Have you ever wondered why some of these compounds have more than one name? One of the reasons is that scientists initially thought there were only two vitamins: vitamin A (fat-soluble) and vitamin B (water-soluble).[1] Later, researchers discovered that there were several vitamins in each of these groups, and some of the vitamins were given different letter designations (such as vitamins C, D, and E). To complicate matters, what researchers initially thought was a single B vitamin turned out to be several vitamins. Each of these vitamins consequently retained its "vitamin B" classification but was also given a number to differentiate it (such as vitamin B₁). Note that B vitamins are often collectively referred to as the **B-complex vitamins.** Many vitamins also have "common names," such as *thiamin* and *riboflavin,* and most have chemical names; for example, vitamin C is ascorbic acid. Thus, it is possible for a single vitamin to go by numerous names.

For some students, memorizing details can make a topic difficult to master. For example, remembering which common names go with which B vitamins can be challenging. However, there are some "tricks of the trade" for learning basic facts such as these. One "trick" you may already use is to come up with a phrase or sentence that helps you remember a certain sequence of words. Table 10.2 not only lists the B vitamins and their common names, but also provides an example of such a phrase that will help you remember the B vitamins.

© 2011 Hill Street Studios/Jupiterimages Corporation

Purchasing vitamin supplements is sometimes challenging because many vitamins have more than one name.

B-complex vitamins A term used to describe all the B vitamins.

TABLE 10.2 Naming B Vitamins

Note that folate does not typically have a corresponding number.

Vitamin	Common Name	Mnemonic
B_1	Thiamin	*The*
B_2	Riboflavin	*Romans*
B_3	Niacin	*never*
B_5	Pantothenic acid	*painted*
B_6	Pyridoxine	*pyramids*
B_7	Biotin	*before*
B_{12}	Cobalamin	*college*

SOME FOODS ARE "ENRICHED" OR "FORTIFIED" WITH MICRONUTRIENTS

It is important to understand what it means when a food is labeled "enriched" or "fortified." As you know, foods contain a mixture of nutrients. While some foods contain nutrients that are naturally occurring, others contain nutrients added during processing. Foods that have nutrients added during processing are said to be **fortified.** For example, the B vitamin thiamin is often added to milled rice and other processed cereal products such as flour. This is important because the outer coating and the inner portion of the rice grain contain most of the thiamin, and these parts are removed during processing of the grain into refined products such as "white (polished) rice." Adding back vitamins via fortification helps make processed grain products, such as pasta, bread, and flour, more nutritious.

A special form of fortification is **enrichment.** Only food products fortified with a specified set of nutrients to certain levels suggested by the U.S. Food and Drug Administration (FDA) can be labeled as being "enriched."[2] Enrichment often makes the nutritional content of the food similar to what it was before processing. Other times, an enriched food is actually more nutrient-dense than its unprocessed counterpart. The nutrients that must be added to a food for it to be labeled "enriched" are four B vitamins (thiamin, niacin, riboflavin, and folate) and the mineral iron. Only select foods can be labeled as enriched. These are rice, flour, breads and rolls, farina (such as cream of wheat), pasta, cornmeal, and corn grits. Looking at food packaging and Nutrition Facts panels (Figure 10.1) makes it relatively easy to determine whether a food is enriched, contains enriched cereal products, or has been fortified in other ways.

Unless it has been fortified, processed white rice is not a good source of micronutrients such as thiamin.

WATER-SOLUBLE VITAMINS CAN BE DESTROYED BY COOKING AND IMPROPER STORAGE

A diet that provides the right balance and variety of foods and beverages can provide all the nutrients needed to maintain health. Unfortunately, no matter how carefully foods are prepared and stored, some nutrients can be destroyed—especially water-soluble vitamins. For example, many can be lost by exposure to water, air, heat, or light, whereas others are affected by acidity (pH). Proper food preparation and storage methods can help prevent excessive

fortified food A food to which nutrients have been added.

enrichment The fortification of a select group of foods (rice, flour, bread or rolls, farina, pasta, cornmeal, and corn grits) with FDA-specified levels of thiamin, niacin, riboflavin, folate, and iron.

FIGURE 10.1 Using Food Labels to Determine If a Product Is "Enriched" or "Fortified" Although only certain foods such as pasta and cornmeal can be labeled as being "enriched," other foods can contain enriched ingredients or be fortified to other levels or with other nutrients.

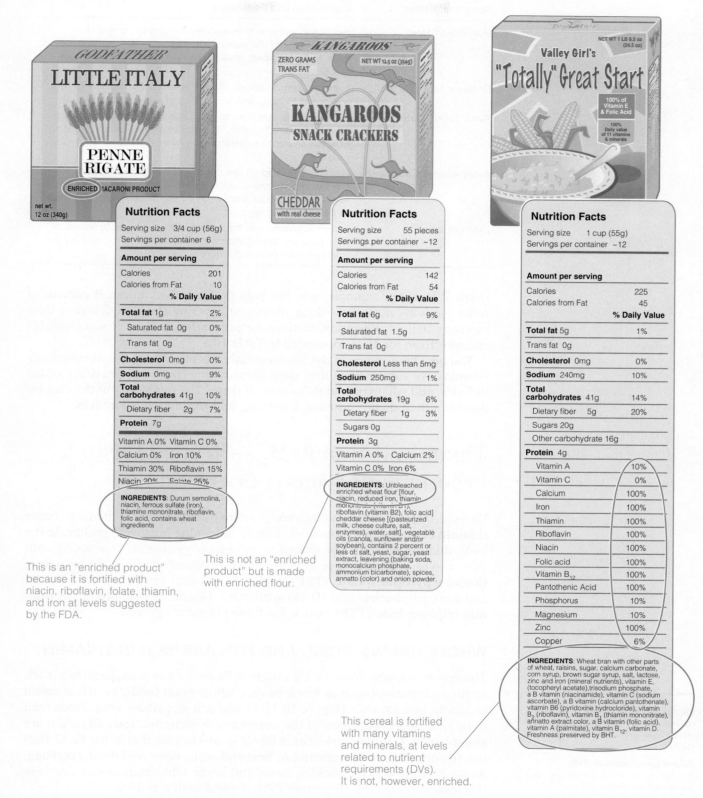

GODFATHER

LITTLE ITALY

PENNE RIGATE

ENRICHED MACARONI PRODUCT

net wt. 12 oz (340g)

Nutrition Facts

Serving size 3/4 cup (56g)
Servings per container 6

Amount per serving

Calories	201
Calories from Fat	10

% Daily Value

Total fat 1g		2%
Saturated fat 0g		0%
Trans fat 0g		
Cholesterol 0mg		0%
Sodium 0mg		9%
Total carbohydrates 41g		10%
Dietary fiber 2g		7%
Protein 7g		

Vitamin A 0%	Vitamin C 0%
Calcium 0%	Iron 10%
Thiamin 30%	Riboflavin 15%
Niacin 20%	Folate 25%

INGREDIENTS: Durum semolina, niacin, ferrous sulfate (iron), thiamine mononitrate, riboflavin, folic acid, contains wheat ingredients

This is an "enriched product" because it is fortified with niacin, riboflavin, folate, thiamin, and iron at levels suggested by the FDA.

KANGAROOS

ZERO GRAMS TRANS FAT   NET WT 12.5 oz (354g)

KANGAROOS
SNACK CRACKERS

CHEDDAR with real cheese

Nutrition Facts

Serving size 55 pieces
Servings per container ~12

Amount per serving

Calories	142
Calories from Fat	54

% Daily Value

Total fat 6g		9%
Saturated fat 1.5g		
Trans fat 0g		
Cholesterol Less than 5mg		
Sodium 250mg		1%
Total carbohydrates 19g		6%
Dietary fiber 1g		3%
Sugars 0g		
Protein 3g		

Vitamin A 0%	Calcium 2%
Vitamin C 0%	Iron 6%

INGREDIENTS: Unbleached enriched wheat flour [flour, niacin, reduced iron, thiamin mononitrate (vitamin B1), riboflavin (vitamin B2), folic acid] cheddar cheese [(pasteurized milk, cheese culture, salt, enzymes), water, salt], vegetable oils (canola, sunflower and/or soybean), contains 2 percent or less of: salt, yeast, sugar, yeast extract, leavening (baking soda, monocalcium phosphate, ammonium bicarbonate), spices, annatto (color) and onion powder.

This is not an "enriched product" but is made with enriched flour.

NET WT 1 LB 8.5 oz (24.5 oz)

Valley Girl's
"Totally" Great Start

100% of Vitamin E & Folic Acid

100% Daily value of 11 vitamins & minerals

Nutrition Facts

Serving size 1 cup (55g)
Servings per container ~12

Amount per serving

Calories	225
Calories from Fat	45

% Daily Value

Total fat 5g		1%
Trans fat 0g		
Cholesterol 0mg		0%
Sodium 240mg		10%
Total carbohydrates 41g		14%
Dietary fiber 5g		20%
Sugars 20g		
Other carbohydrate 16g		
Protein 4g		

Vitamin A	10%
Vitamin C	0%
Calcium	100%
Iron	100%
Thiamin	100%
Riboflavin	100%
Niacin	100%
Folic acid	100%
Vitamin B12	100%
Pantothenic Acid	100%
Phosphorus	10%
Magnesium	10%
Zinc	100%
Copper	6%

INGREDIENTS: Wheat bran with other parts of wheat, raisins, sugar, calcium carbonate, corn syrup, brown sugar syrup, salt, lactose, zinc and iron (mineral nutrients), vitamin E, (tocopheryl acetate),trisodium phosphate, a B vitamin (niacinamide), vitamin C (sodium ascorbate), a B vitamin (calcium pantothenate), vitamin B6 (pyridoxine hydrochloride), vitamin B2 (riboflavin), vitamin B1 (thiamin mononitrate), annatto extract color, a B vitamin (folic acid), vitamin A (palmitate), vitamin B12, vitamin D. Freshness preserved by BHT.

This cereal is fortified with many vitamins and minerals, at levels related to nutrient requirements (DVs). It is not, however, enriched.

TABLE 10.3 Recommendations to Help Prevent Water-Soluble Vitamin Loss from Foods

Recommendation	Explanation and Application
Use Minimal Heat	Heating at extreme temperatures for long periods of time destroys most of the water-soluble vitamins. Do not overcook fruits and vegetables. Instead, cook or steam them until just tender.
Decrease Exposure to Light	Light can destroy several of the B vitamins. For example, if milk is exposed to light, destruction of riboflavin can occur.
Avoid Excess Water	Water-soluble vitamins can move out of foods when they are placed in water. You should therefore avoid excessive soaking of foods and cook vegetables in pieces that are as large as possible.
Avoid Alkaline Conditions	High pH (low acidity) can destroy some vitamins such as thiamin and vitamin C. As such, avoid adding baking soda to vegetables to retain their color during cooking or to legumes to decrease cooking time.
Limit Oxidation	Vitamin C and many of the B vitamins are destroyed by exposure to air (oxygen). Storing vitamin-rich foods in air-tight containers and wraps can help prevent this from happening.

nutrient loss. For example, you can help preserve the vitamin B content of some foods by not overcooking them and covering them to decrease their exposure to light. Recommendations for preventing the loss of water-soluble vitamins from foods are presented in Table 10.3.

You now know some of the fundamental facts about how the water-soluble vitamins are similar, how they were named, why they are sometimes added to foods, and what you can do to protect them. The remainder of the chapter describes their dietary sources, functions, and recommended intakes.

Thiamin (Vitamin B₁)—Needed for Production of Acetyl Coenzyme A

Vitamin B₁ (one of the B-complex vitamins) was given the chemical name **thiamin** because it contains thiol (sulfur) and amine (nitrogen) groups. In the body, thiamin must first be altered to serve its physiological process. Specifically, two phosphate groups must be added to thiamin to form the coenzyme **thiamin pyrophosphate (TPP),** which is essential for energy metabolism and deoxyribonucleic acid (DNA) synthesis.* Thiamin is also present as **thiamin triphosphate (TTP),** which has three phosphate groups.

WHOLE GRAINS, PORK, AND FISH ARE RICH IN THIAMIN

Thiamin is available from a wide variety of foods such as pork, peas, fish (such as tuna), legumes (such as black beans), whole-grain foods, enriched cereal products, and soy milk (Figure 10.2). As you will see, whole-grain foods tend to be good sources of many of the water-soluble vitamins. You can read more about how to include nutrient-rich whole grains in your diet in the Food Matters feature. Thiamin is sensitive to heat and easily destroyed during cooking. As such, using shorter cooking times and lower temperatures can decrease this loss, although it is important to cook meat until it is done.

*Thiamin pyrophosphate (TPP) is also called thiamin diphosphate (TDP).

thiamin (vitamin B₁) An essential water-soluble vitamin involved in energy metabolism; synthesis of DNA, RNA, and NADPH + H⁺; and nerve function.

thiamin pyrophosphate (TPP) The coenzyme form of thiamin that has two phosphate groups.

thiamin triphosphate (TTP) A form of thiamin with three phosphate groups.

Working Toward the Goal: Consuming Whole-Grain Foods to Optimize Your Intake of Water-Soluble Vitamins

The 2010 Dietary Guidelines for Americans recommend that most adults consume three or more servings of whole-grain products per day, with the rest of the grains coming from enriched products. In all, at least half of the cereal grain products consumed should be of the whole-grain variety. This is very important in terms of obtaining adequate amounts of micronutrients, including the water-soluble vitamins. The following selection and preparation tips will help you meet your goal.[3]

- Choose foods that list one of the following whole-grain ingredients *first* on their ingredient lists: brown rice, bulgur, graham flour, oatmeal, whole-grain corn, whole oats, whole rye, whole wheat, or wild rice.

- Remember that color is not an indication of whether a food is whole grain. Bread can be brown because of molasses or other added ingredients.

- Substitute a whole-grain product for a refined product—such as whole-wheat bread for white bread or whole-wheat pasta for plain pasta.

- Substitute whole-wheat or oat flour for up to half of the flour in pancake, waffle, muffin, or other flour-based recipes.

- Choose whole-grain chips and snacks, such as baked tortilla chips or popcorn, instead of potato chips or cheese puffs.

- Remember that many foods labeled as "multigrain," "stone-ground," "100% wheat," "cracked wheat," "seven-grain," or "bran" are not whole-grain products. Foods are usually labeled as "whole grain" if they are.

FIGURE 10.2 Food Sources and Structure of Thiamin

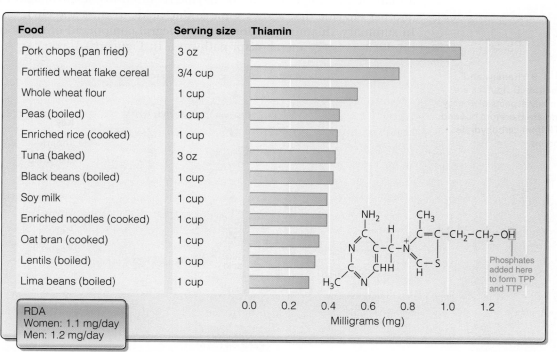

Food	Serving size	Thiamin
Pork chops (pan fried)	3 oz	
Fortified wheat flake cereal	3/4 cup	
Whole wheat flour	1 cup	
Peas (boiled)	1 cup	
Enriched rice (cooked)	1 cup	
Tuna (baked)	3 oz	
Black beans (boiled)	1 cup	
Soy milk	1 cup	
Enriched noodles (cooked)	1 cup	
Oat bran (cooked)	1 cup	
Lentils (boiled)	1 cup	
Lima beans (boiled)	1 cup	

RDA
Women: 1.1 mg/day
Men: 1.2 mg/day

Milligrams (mg)

Phosphates added here to form TPP and TTP

SOURCE: USDA Nutrient Database for Standard Reference, Release 16-1.

"ANTITHIAMIN FACTORS" DECREASE THIAMIN BIOAVAILABILITY

There are several factors that influence thiamin's bioavailability. For example, thiamin absorption increases when thiamin intake is chronically low. To accomplish this, thiamin is actively transported from the intestinal lumen into the absorptive epithelial cells of the small intestine. When thiamin intake is chronically high, its absorption proceeds more slowly by simple diffusion. In addition, consuming certain foods can influence thiamin's bioavailability. For instance, compounds found in raw fish, coffee, tea, berries, Brussels sprouts, and cabbage can destroy thiamin or interfere with its absorption. These "antithiamin factors" work in several ways, one of which is by inactivating thiamin via oxidation. Consuming vitamin C–rich foods can help prevent thiamin oxidation because vitamin C acts as an antioxidant. Alcohol consumption can also decrease the absorption of thiamin by interfering with its active transport across the basolateral membrane of enterocytes.

Once absorbed from the small intestine, thiamin circulates in the blood to the liver, where the vitamin is metabolized, used, or recirculated in the blood to other cells in the body. Excess thiamin is not stored. Rather, it is further metabolized and excreted in the urine.

THIAMIN IS CRITICAL FOR ATP PRODUCTION

Thiamin has many functions within cells, including both coenzyme and noncoenzyme roles. For example, although thiamin, *per se*, is not an "energy-yielding nutrient[†]," its coenzyme form TPP is required to make ATP. Specifically, thiamin-containing TPP, which is made in the liver, is needed for the activation of an enzyme that converts pyruvate to acetyl coenzyme A (acetyl-CoA). As discussed in Chapter 7, acetyl-CoA is an important compound situated at the crossroads of ATP production. Thus, consuming enough thiamin is critical in order to meet your energy needs to support all your daily activities. TPP is also a mandatory component of an enzyme required in the citric acid cycle and another enzyme that allows some amino acids to enter the citric acid cycle. In summary, thiamin functions as an integral component of the body's ATP-producing energy metabolism pathways. In Figure 10.3, you can see where

⟨CONNECTIONS⟩ Recall that simple diffusion does not require energy (ATP) and moves a substance from a region of higher concentration to one of lower concentration. Active transport requires energy (ATP) and moves a substance from a low to a high concentration (Chapter 3, pages 76–77).

⟨CONNECTIONS⟩ Coenzymes are organic compounds that are required for the actions of some enzymes (Chapter 7, page 274).

⟨CONNECTIONS⟩ The citric acid cycle is an energy metabolism pathway that breaks down intermediate metabolites, generating ATP, NADH + H⁺, FADH₂, and carbon dioxide (Chapter 7, page 283).

FIGURE 10.3 B Vitamins and Energy Metabolism B vitamins are important for many aspects of energy metabolism and are therefore needed to produce ATP from carbohydrates, lipids, and proteins.

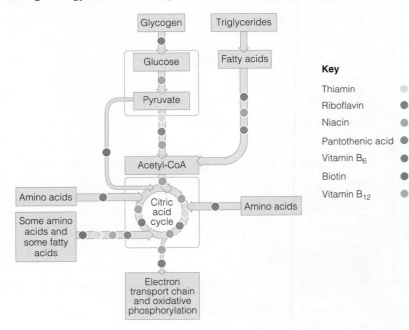

[†]"Energy-yielding nutrients" are carbohydrates, proteins, and lipids.

thiamin and the other B vitamins are instrumental in energy metabolism pathways.

TPP also plays a role in the synthesis of deoxyribonucleic acid (DNA) and ribonucleic acid (RNA). Remember that DNA is the genetic material that, together with RNA, directs protein synthesis. Without TPP acting as a coenzyme, the enzymes that make DNA and RNA cannot function properly, and protein synthesis is halted. TPP is also required for the synthesis of nicotinamide adenine dinucleotide phosphate (NADPH + H$^+$), which is needed for triglyceride synthesis.

Thiamin also has noncoenzyme functions; for example, its triphosphate form (TTP) is needed for nerve function. Although the mechanisms are not well understood, TTP appears to be involved in either neurotransmitter production or the generation of nerve impulses.

One of the roles thiamin plays in the body is as a coenzyme in energy metabolism pathways.

THIAMIN DEFICIENCY RESULTS IN BERIBERI

As early as 2600 B.C., Chinese physicians described the life-threatening condition of thiamin deficiency called **beriberi**. Although this disease is now uncommon in the United States, beriberi is still prevalent in regions of the world that rely heavily on unfortified, milled rice as a major source of energy. The term *beriberi* is thought by some to have been derived from the Indonesian word meaning "*I can't; I can't*" (what someone might have said in response to weakness and impaired functions brought on by thiamin deficiency).

There are four forms of beriberi. **Dry beriberi** is found mostly in adults and is characterized by severe muscle wasting, leg cramps, tenderness, and decreased feeling in the feet and toes. **Wet beriberi** involves severe edema (swelling) in the arms and legs as well as enlargement of the heart and respiratory problems, which can result in heart failure. It is not clear why some people develop dry beriberi and others develop wet beriberi. **Infantile beriberi** occurs in babies breastfed by thiamin-deficient mothers, and can cause cardiac arrest. **Cerebral beriberi** (also called Wernicke-Korsakoff syndrome) is typically associated with alcoholism, because alcohol decreases absorption of thiamin. Furthermore, thiamin intake by alcoholics can be low, and alcoholism-induced liver damage decreases the liver's ability to convert thiamin to TPP. There also appear to be genetic polymorphisms predisposing some people to cerebral beriberi. Recall that genetic polymorphisms are relatively common variations in genes that occur in a population. In the case of cerebral beriberi, polymorphisms influencing thiamin transport and metabolism appear to be involved.[4] Cerebral beriberi is characterized by paralysis of the eye muscles, involuntary eye movement, poor muscle coordination, confusion, and short-term memory loss.

RECOMMENDED INTAKES OF THIAMIN

The Recommended Dietary Allowances (RDAs) for thiamin are 1.2 and 1.1 milligrams (mg) per day for adult males and females, respectively. Because thiamin toxicity is almost unheard of, there is no Tolerable Upper Intake Level (UL) for this vitamin. A complete list of the Dietary Reference Intakes (DRIs) for all vitamins is provided inside the front cover of this book.

beriberi (BER – i – BER – i) A disease that results from thiamin deficiency.

dry beriberi A form of thiamin deficiency characterized by muscle loss and leg cramps.

wet beriberi A form of thiamin deficiency characterized by severe edema.

infantile beriberi A form of thiamin deficiency that occurs in infants breastfed by thiamin-deficient mothers.

cerebral beriberi (Wernicke-Korsakoff syndrome) A form of thiamin deficiency characterized by poor muscle control and paralysis of the eye muscles.

Riboflavin (Vitamin B₂)—Coenzyme Required for Reduction–Oxidation Reactions

Riboflavin (vitamin B₂) consists of a multi-ring structure attached to a sugar, ribose. This vitamin got its name because of its ribose component and its yellow color (*flavus* means "yellow" in Latin). The body uses riboflavin to produce two coenzymes—**flavin mononucleotide (FMN)** and **flavin adenine dinucleotide (FAD)**—needed for energy metabolism.

⟨CONNECTIONS⟩ Recall that ribose is a five-carbon monosaccharide that is also a constituent of RNA (Chapter 4, page 135).

MEATS AND DAIRY PRODUCTS ARE RICH SOURCES OF RIBOFLAVIN

Riboflavin is found in a variety of foods including liver, meat, dairy products, enriched cereals, and other fortified foods (Figure 10.4). Most fruits and vegetables provide only marginal amounts, although whole-grain foods contribute some riboflavin to the diet. Although riboflavin is relatively stable during cooking, it is easily destroyed when exposed to excessive light. This is why milk is generally packaged in cardboard or opaque (cloudy) plastic containers. Storing other foods in a similar manner can also help prevent the loss of riboflavin.

In food, riboflavin can be in its free form, bound to protein, or as a coenzyme (FMN or FAD). However, because only free (unbound) riboflavin can be absorbed, stomach acid and intestinal enzymes must first convert all bound riboflavin to its free form prior to absorption. As with thiamin, absorption of riboflavin occurs in the small intestine via simple diffusion when intake is chronically high and via active transport when intake is chronically low. In this way, bioavailability increases at low intakes. Riboflavin in animal

riboflavin (vitamin B₂) An essential water-soluble vitamin involved in energy metabolism, the synthesis of a variety of vitamins, nerve function, and protection of biological membranes.

flavin mononucleotide (FMN) (FLA – vin) A coenzyme form of riboflavin.

flavin adenine dinucleotide (FAD) A coenzyme form of riboflavin.

FIGURE 10.4 Food Sources and Structure of Riboflavin

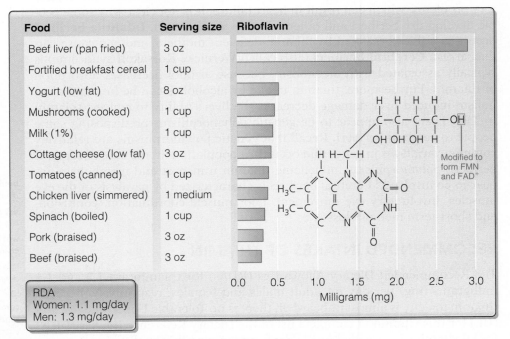

SOURCE: USDA Nutrient Database for Standard Reference, Release 16-1.

foods is somewhat more bioavailable than that in plant sources. Alcohol can inhibit its absorption. Upon absorption, riboflavin circulates in the blood to the liver, where the vitamin is used or transported in the blood to other tissues. Riboflavin is not readily stored in the body, and excess is excreted in the urine.

RIBOFLAVIN ASSISTS IN REDUCTION–OXIDATION (REDOX) REACTIONS

Once it is taken up by cells, riboflavin is converted to a series of coenzymes needed for several chemical reactions involved in energy metabolism, many of which are reduction–oxidation (redox) reactions. For example, riboflavin-derived FAD participates in the citric acid cycle, being reduced to $FADH_2$ (recall Figure 10.3). $FADH_2$ is reoxidized to FAD in the electron transport chain, ultimately resulting in the formation of ATP, water, and carbon dioxide. $FADH_2$ is also needed for the breakdown, or oxidation, of fatty acids into acetyl-CoA molecules—a process called β-oxidation—allowing the body to use fatty acids for energy. Thus, like niacin, riboflavin is critical for ATP production.

The B vitamin riboflavin is abundant in meat and dairy products.

〈**CONNECTIONS**〉 Oxidation is the loss of electrons, whereas reduction is the gain of electrons (Chapter 3, pages 72–73).

FAD is also required for the synthesis of other compounds. For example, it is needed to convert vitamin A and folate (a B vitamin) to their active forms; convert tryptophan (an amino acid) to niacin (a B vitamin); and form vitamins B_6 and K. FAD is also needed for the metabolism of some critical neurotransmitters (such as dopamine) and is involved in several important reactions that protect biological membranes from oxidative damage. Riboflavin can also be converted to flavin mononucleotide (FMN), a coenzyme needed for activating vitamin B_6.

In summary, riboflavin and its associated coenzymes (FAD and FMN) are important in energy metabolism; the biosynthesis or activation of several compounds, including some vitamins; and protection of biological membranes from oxidative damage. Thus, its deficiency can result in numerous and diverse complications.

RIBOFLAVIN DEFICIENCY CAUSES ARIBOFLAVINOSIS

Riboflavin deficiency, called **ariboflavinosis,** is typically associated with broader nutrient deficiencies and overall malnutrition. In other words, people with ariboflavinosis generally have other nutrient deficiencies as well. The signs and symptoms of ariboflavinosis are multifaceted, including muscle weakness, sores around the mouth and lips (**cheilosis**), inflammation of the mouth (**stomatitis**), enlarged and inflamed tongue (**glossitis**), and confusion. Riboflavin deficiency is rare in the United States but can occur in alcoholics consuming poor diets and in people with diseases that interfere with riboflavin utilization, such as thyroid disease.

RECOMMENDED INTAKES OF RIBOFLAVIN

The RDAs for riboflavin for adult males and females are 1.3 and 1.1 mg/day, respectively. Because there are no known toxic effects associated with high intakes of riboflavin, no ULs have been established. However, riboflavin supplementation can cause the urine to become bright yellow. This is due to the intense yellow color of riboflavin and has no apparent health consequences.[5]

ariboflavinosis (a – ribo – flav – i – NO – sis) A disease caused by riboflavin deficiency.

cheilosis (chei – LO – sis) Sores occurring on the outsides and corners of the lips.

stomatitis (stom – a – TI – tis) Inflammation of the mucous membranes of the mouth.

glossitis (gloss – I – tis) Inflamed tongue.

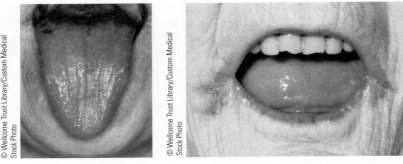

Glossitis

Cheilosis and stomatitis

Symptoms of ariboflavinosis (riboflavin deficiency) include glossitis, cheilosis, and stomatitis.

Niacin (Vitamin B₃)—Required for Energy Metabolism

Meat provides both niacin and the amino acid tryptophan, which is used by the body to make niacin. This makes meat a particularly good source of niacin.

You have now learned that two of the B vitamins—thiamin and riboflavin—are required to synthesize important coenzymes critical for energy metabolism. Similarly, **niacin** (vitamin B₃) is also needed to make ATP. Specifically, the niacin-derived coenzymes nicotinamide adenine dinucleotide (NAD⁺) and nicotinamide adenine dinucleotide phosphate (NADP⁺) are involved in numerous reactions, many of which are required for energy metabolism.

MEAT AND MUSHROOMS ARE GOOD SOURCES OF NIACIN

In foods, niacin takes two forms—nicotinic acid and nicotinamide. The body can also make niacin from the dietary essential amino acid tryptophan. About 1 mg of niacin can be made from 60 mg of tryptophan. As such, both niacin and tryptophan are considered "dietary sources" of niacin, and a unit of measure called the **niacin equivalent (NE)** refers to the combined amounts of niacin and tryptophan in foods. The niacin contents—or NE—of some foods are listed in Figure 10.5. Niacin or tryptophan is found in a variety of foods such as liver, poultry, fish, tomatoes, beef, and mushrooms. Whole-grain foods, enriched cereal products, and other fortified foods are also important sources of this vitamin. Niacin is quite stable, and is not easily destroyed by heat or exposure to light.

Compared to that in grain products, the niacin in animal products is more bioavailable. This is because the niacin in grains is bound to proteins, making the vitamin unavailable for absorption. Treating grain products with alkaline (basic) substances such as lime water or baking soda can cleave the protein from the niacin, increasing its bioavailability. The traditional technique of soaking corn and cornmeal in lime water—a historically common practice in preparing tortillas in Mexico—is an example of this. Note that lime *water* derived from limestone is not the same as lime *juice* derived from the citrus fruit. As shown in Figure 10.6, lime is still added to some forms of cornmeal. Using lime-treated cornmeal to prepare foods such as tamales makes them good sources of niacin. However, because alkaline conditions can destroy other vitamins in foods, it is not recommended that we routinely increase the pH of our foods to enhance niacin absorption.

Small amounts of niacin can be absorbed in the stomach, but most absorption occurs in the small intestine. Like thiamin and riboflavin, niacin is absorbed by simple diffusion when intake is high and by active transport

niacin (vitamin B₃) (NI – a – cin) An essential water-soluble vitamin involved in energy metabolism, synthesis of fatty acids and proteins, metabolism of vitamin C and folate, glucose homeostasis, and cholesterol metabolism.

niacin equivalent (NE) A unit of measure that describes the niacin and/or tryptophan content in food.

FIGURE 10.5 Food Sources and Structure of Niacin

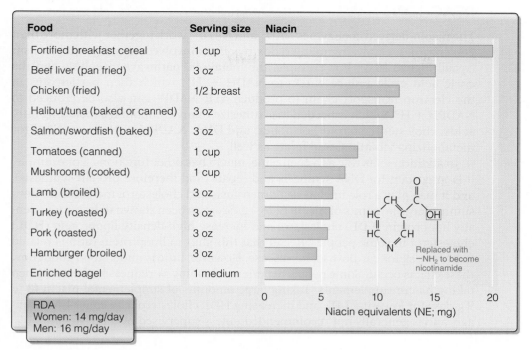

Food	Serving size	Niacin
Fortified breakfast cereal	1 cup	
Beef liver (pan fried)	3 oz	
Chicken (fried)	1/2 breast	
Halibut/tuna (baked or canned)	3 oz	
Salmon/swordfish (baked)	3 oz	
Tomatoes (canned)	1 cup	
Mushrooms (cooked)	1 cup	
Lamb (broiled)	3 oz	
Turkey (roasted)	3 oz	
Pork (roasted)	3 oz	
Hamburger (broiled)	3 oz	
Enriched bagel	1 medium	

Niacin equivalents (NE; mg)

Replaced with $-NH_2$ to become nicotinamide

RDA
Women: 14 mg/day
Men: 16 mg/day

SOURCE: USDA Nutrient Database for Standard Reference, Release 16-1.

when intake is low. Niacin then circulates in the blood to the liver, where most of the vitamin is attached to transport proteins. The liver then releases niacin into the blood, where the vitamin travels to cells that need it. Alternatively, the liver can convert dietary niacin to NAD^+ or $NADP^+$, its coenzyme forms, which are then put into the circulation. When more niacin is needed, the liver converts the amino acid tryptophan to niacin as well. This reaction requires the assistance of riboflavin and vitamin B_6.

Lime—from lime *water*, not lime *juice*—is added to increase the bioavailability of niacin found naturally in corn.

Ingredients: Specially ground and dehydrated whole kernel corn and lime.
No preservatives added.

FIGURE 10.6 Increasing Niacin Bioavailability with "Lime Water" Ground corn is often treated with lime (from limestone; not lime juice) to increase the bioavailability of niacin. This practice makes foods made with cornmeal good sources of this vitamin.

Tamales

Corn tortillas

Atole

NIACIN IS INVOLVED IN REDUCTION–OXIDATION (REDOX) REACTIONS

The niacin-derived coenzymes NAD^+ and $NADP^+$ work together with more than 200 enzymes, catalyzing many redox reactions involved in energy metabolism (recall Figure 10.3). For example, the metabolic pathways glycolysis and the citric acid cycle reduce NAD^+ to $NADH + H^+$, which then participates in the electron transport chain to produce ATP. $NADP^+$ can also be reduced to $NADPH + H^+$, which is needed to synthesize many compounds, including fatty acids, cholesterol, steroid hormones, and DNA. $NADPH + H^+$ is required for metabolizing vitamin C and folate as well.

In addition to its role as a coenzyme, niacin has other functions. For example, it is important for DNA synthesis and repair (and therefore protein synthesis), and it may play a role in glucose homeostasis and cholesterol metabolism. Consuming large amounts of niacin (2 to 4 g/day) has been shown to lower low-density lipoprotein (LDL) cholesterol and increase high-density lipoprotein (HDL) cholesterol in some people.[6] Recall that this shift in lipoprotein ratio is related to lower risk for cardiovascular disease. However, little is understood about how these effects occur. Some people experience flushing or redness in the face when taking niacin supplements. Taking large amounts of supplemental niacin (3 to 9 g/day) for lowering LDL and increasing HDL cholesterol can cause liver damage, and is generally not recommended unless supervised by a physician.

⟨CONNECTIONS⟩ Remember that low levels of LDL and high levels of HDL are associated with lower risk of cardiovascular disease (Chapter 6, page 245)

NIACIN DEFICIENCY RESULTS IN PELLAGRA

Niacin deficiency was originally given the name *mal del sol,* which in Italian means *"illness of the sun."* This is because niacin deficiency causes a variety of skin problems, including severe skin irritation, which is made worse by exposure to sunlight. Later, niacin deficiency was named **pellagra,** Italian for *"rough skin."* The symptoms of pellagra are often referred to as the "four D's": dermatitis, dementia, diarrhea, and death. Indeed, pellagra results in rough, red skin that eventually thickens and turns dark. It also causes neurological problems (that is, dementia) including depression, anxiety, irritability, and inability to concentrate. The associated gastrointestinal disturbances cause loss of appetite, diarrhea, and a characteristically red and swollen tongue. If not treated, severe pellagra can eventually cause death.

Pellagra was once endemic in some portions of the southern United States where corn (which is low in both niacin and its precursor tryptophan) provided the primary source of protein to the diet. In fact, pellagra was so common that public health officials believed it was caused by an infectious agent, such as a bacterium. Pellagra is now typically seen only in conjunction with poverty, general malnutrition, or chronic alcoholism. Diets providing limited amounts of both niacin and tryptophan can put a person at especially high risk for pellagra. Genetic makeup and some medications can also cause secondary niacin deficiency by inhibiting the synthesis of niacin from tryptophan. For example, people with Hartnup disorder have a genetic abnormality that impairs the absorption of several amino acids including tryptophan. As a result, people with Hartnup disorder cannot rely on tryptophan as a source of niacin.[7]

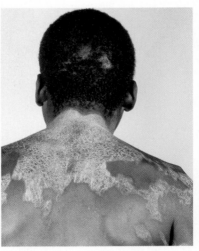

As shown here, pellagra (niacin deficiency) can cause serious dermatitis.

RECOMMENDED INTAKES OF NIACIN

The RDAs for men and women are 16 and 14 mg, respectively, of niacin (or niacin equivalents, NE) each day. Because large doses of nicotinic acid can have toxic effects, such as increasing plasma glucose and damaging the liver, a UL for niacin has also been established. Experts recommend that we consume no more than 35 mg/day of niacin from supplements and fortified foods.

pellagra (pell – A – gra) A disease caused by niacin and/or tryptophan deficiency.

Pantothenic Acid (Vitamin B₅)—A Component of Coenzyme A

The next B vitamin to learn about is **pantothenic acid** (vitamin B₅), which is a nitrogen-containing compound named for the Greek word *pantos,* meaning "everywhere." This is because pantothenic acid is found in almost every plant and animal tissue. Pantothenic acid functions as a component of coenzyme A (CoA) in a variety of metabolic reactions—especially those involved in energy metabolism. Recall from Chapter 7 that one of the pivotal steps in energy metabolism involves converting pyruvate to acetyl-CoA.

PANTOTHENIC ACID IS FOUND IN MOST PLANT AND ANIMAL FOODS

Pantothenic acid is found in many foods, including mushrooms, organ meats (such as liver), and sunflower seeds (Figure 10.7). Dairy products, turkey, fish, and coffee are also good sources. High temperatures can destroy pantothenic acid in foods, so cooking them at more moderate temperatures is advisable. Bioavailability of pantothenic acid increases via active transport when pantothenic acid intake is low. Once absorbed, pantothenic acid is circulated in the blood to the liver where the vitamin is metabolized and, in turn, circulated to the rest of the body. Although pantothenic acid itself is not stored in the body, concentrations of its coenzyme form (CoA) are especially high in the liver, kidneys, heart, adrenal glands, and brain.

Pantothenic acid is found in a diverse group of foods, including mushrooms.

pantothenic acid (vitamin B₅) (pan – to – THE – nic) A water-soluble vitamin involved in energy metabolism, hemoglobin synthesis, and phospholipid synthesis.

FIGURE 10.7 Food Sources and Structure of Pantothenic Acid

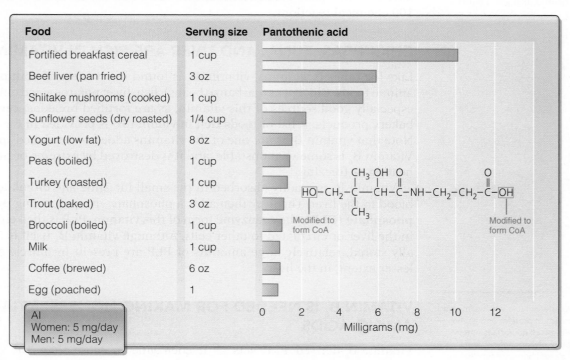

Food	Serving size	Pantothenic acid
Fortified breakfast cereal	1 cup	
Beef liver (pan fried)	3 oz	
Shiitake mushrooms (cooked)	1 cup	
Sunflower seeds (dry roasted)	1/4 cup	
Yogurt (low fat)	8 oz	
Peas (boiled)	1 cup	
Turkey (roasted)	1 cup	
Trout (baked)	3 oz	
Broccoli (boiled)	1 cup	
Milk	1 cup	
Coffee (brewed)	6 oz	
Egg (poached)	1	

AI
Women: 5 mg/day
Men: 5 mg/day

Milligrams (mg)

SOURCE: USDA Nutrient Database for Standard Reference, Release 16-1.

PANTOTHENIC ACID IS NEEDED FOR GLYCOLYSIS AND THE CITRIC ACID CYCLE

Like some of the other B-complex vitamins, pantothenic acid is critical for ATP production. Specifically, it is needed to synthesize CoA, a coenzyme required to metabolize glucose, amino acids, and fatty acids via glycolysis and the citric acid cycle (recall Figure 10.3). You also require this vitamin to synthesize many other critical compounds in your body, including heme (the iron-containing portion of hemoglobin), cholesterol, bile salts, phospholipids, fatty acids, and steroid hormones.

RECOMMENDED INTAKES FOR PANTOTHENIC ACID

Because pantothenic acid is found in almost all foods, deficiencies are rare. Nonetheless, a condition called "burning feet syndrome" has historically been thought to be due to severe pantothenic acid deficiency. As its name implies, burning feet syndrome causes a tingling in the feet and legs as well as fatigue, weakness, and nausea. It is noteworthy, however, that there are likely many causes of "burning feet" aside from pantothenic acid deficiency. Pantothenic acid toxicity has not been reported, but very high intakes of this vitamin have been associated with nausea and diarrhea. There is not enough information to establish RDAs for pantothenic acid, but an AI of 5 mg/day has been set for adults. Because no evidence exists of toxicity from this nutrient, there is no UL for pantothenic acid.

Vitamin B$_6$—Critical for Metabolism of Amino Acids

There are three forms of **vitamin B$_6$**—pyridoxine, pyridoxal, and pyridoxamine—all made of a modified, nitrogen-containing ring structure. All three forms have similar biological activities in the body, and are involved in more than 100 chemical reactions.

CHICKPEAS, TUNA, AND LIVER ARE RICH IN VITAMIN B$_6$

Like the other B vitamins, vitamin B$_6$ is found in a variety of both plant and animal foods. Chickpeas (garbanzo beans), fish, liver, potatoes, and chicken are especially good sources of this vitamin, as are fortified breakfast cereals and bakery products. A list of foods rich in vitamin B$_6$ is provided in Figure 10.8. Note that vitamin B$_6$ is *not* one of the vitamins added to "enriched" products. Vitamin B$_6$ is somewhat unstable, and it is destroyed by extreme or prolonged heating or freezing.

Vitamin B$_6$ is readily absorbed in the small intestine and circulated in the blood to the liver. The liver then adds a phosphate group, forming **pyridoxal phosphate (PLP),** the coenzyme form of this vitamin. PLP is then either used in the liver or circulated to other cells. Although vitamin B$_6$ itself is not actually stored, relatively large amounts of PLP are present in muscle and, to a lesser extent, in the liver.

VITAMIN B$_6$ IS NEEDED FOR MAKING NONESSENTIAL AMINO ACIDS

Vitamin B$_6$-derived PLP acts as a coenzyme in more than 100 chemical reactions related to the metabolism of proteins and amino acids via transamination. These reactions are required for synthesizing nonessential amino acids

A traditional Middle Eastern dip called hummus, made from chickpeas (garbanzo beans), is a good source of vitamin B$_6$.

vitamin B$_6$ A water-soluble vitamin involved in the metabolism of proteins and amino acids, the synthesis of neurotransmitters and hemoglobin, glycogenolysis, and regulation of steroid hormone function.

pyridoxal phosphate (PLP) (pyr – i – DOX – al) The coenzyme form of vitamin B$_6$.

FIGURE 10.8 Food Sources and Structure of Vitamin B6

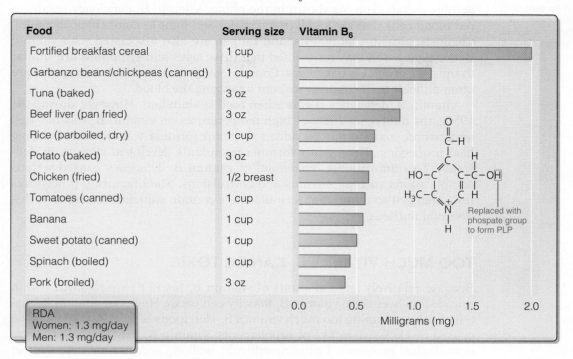

Food	Serving size	Vitamin B6
Fortified breakfast cereal	1 cup	
Garbanzo beans/chickpeas (canned)	1 cup	
Tuna (baked)	3 oz	
Beef liver (pan fried)	3 oz	
Rice (parboiled, dry)	1 cup	
Potato (baked)	3 oz	
Chicken (fried)	1/2 breast	
Tomatoes (canned)	1 cup	
Banana	1 cup	
Sweet potato (canned)	1 cup	
Spinach (boiled)	1 cup	
Pork (broiled)	3 oz	

RDA
Women: 1.3 mg/day
Men: 1.3 mg/day

(x-axis: Milligrams (mg), 0.0 to 2.0)

Replaced with phospate group to form PLP

SOURCE: USDA Nutrient Database for Standard Reference, Release 16-1.

from essential amino acids. Recall that only 9 amino acids—the "essential" ones—must be obtained from foods. The others can typically be made by the body. However, without vitamin B_6, all 20 amino acids would be essential. Vitamin B_6 is also involved in producing nonprotein substances, such as the neurotransmitters serotonin and dopamine as well as heme. For example, you need vitamin B_6 to convert tryptophan to niacin, break down glycogen to its glucose subunits, and regulate some of the steroid hormones. It is also needed to prepare some amino acids to enter into the citric acid cycle.

⟨CONNECTIONS⟩ Recall that transamination involves the transfer of an amino group from one amino acid to an α-keto acid, forming a different amino acid (Chapter 5, page 164).

VITAMIN B6 DEFICIENCY CAUSES MICROCYTIC HYPOCHROMIC ANEMIA

Vitamin B_6 deficiency results in inadequate heme production and, thus, lowers concentrations of hemoglobin in red blood cells. This condition, called **microcytic hypochromic anemia,** interferes with the ability of red blood

microcytic hypochromic anemia (mic – ro – CYT – ic hy – po – CHRO – mic a – NE – mi – a) A condition in which red blood cells are small and light in color due to inadequate hemoglobin synthesis; can be due to vitamin B_6 deficiency.

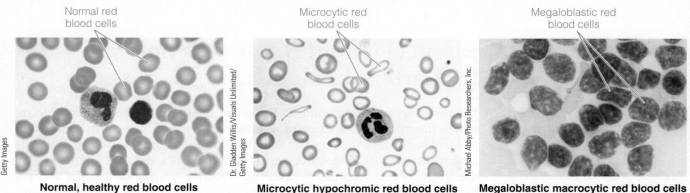

Normal red blood cells

Microcytic red blood cells

Megaloblastic red blood cells

Normal, healthy red blood cells

Microcytic hypochromic red blood cells

Megaloblastic macrocytic red blood cells

Red blood cells from healthy people can be visually distinguished from those of people with microcytic hypochromic anemia or megaloblastic macrocytic anemia.

〈CONNECTIONS〉 Because oxygen is the final electron acceptor in the electron transport chain, aerobic ATP production is dependent on oxygen availability and, thus, on hemoglobin (Chapter 7, pages 278–279).

cells to deliver oxygen to tissues and impairs their ability to produce ATP via aerobic metabolism. As shown in the photo, Vitamin B_6 deficiency results in red blood cells that are small (microcytic) and light in color (hypochromic). Cheilosis, glossitis, stomatitis, and fatigue are also associated with vitamin B_6 deficiency. You may have noted that these signs and symptoms are similar to those of riboflavin deficiency. Consequently, these nutrient deficiencies are often difficult to distinguish without analyzing the blood.

Vitamin B_6 deficiency is rare when food is abundant. However, during the 1950s, the damaging effect of high temperatures on vitamin B_6 was not well understood, and vitamin B_6 added to infant formula was destroyed during heat processing. Thus, many formula-fed infants developed vitamin B_6 deficiency. This unfortunate "epidemic" of vitamin B_6 deficiency caused serious complications such as seizures and convulsions. Manufacturing procedures have changed so that infant formulas now contain sufficient amounts of this essential nutrient.

TOO MUCH VITAMIN B_6 CAN BE TOXIC

Because relatively large amounts of vitamin B_6 (as PLP) are retained within muscle and liver cells, vitamin B_6 toxicity can occur. However, vitamin B_6 toxicity from consuming too much vitamin B_6-rich foods is rare. Rather, it can be caused by excessive intake of supplements. Vitamin B_6 toxicity causes severe neurological problems, including difficulty walking and numbness in the feet and hands. These effects are due largely to excess PLP reacting with amine groups on proteins, altering their function, as well as to nerve damage.

Limited studies suggest that very high dosages of vitamin B_6 (1,000 mg/day) may relieve the symptoms of premenstrual syndrome (PMS) and carpal tunnel syndrome.[8] Other data suggest that relatively large doses of vitamin B_6 may decrease the risk of heart disease.[9] However, most studies do not support a helpful effect of vitamin B_6 supplements on these conditions.[10] Because toxicity is likely at these high doses, treating these conditions with vitamin B_6 is generally not recommended.

RECOMMENDED INTAKES OF VITAMIN B_6

RDAs for vitamin B_6 in adults vary from 1.3 to 1.7 mg/day. To prevent the neurological problems associated with vitamin B_6 toxicity, a UL has been established at 100 mg/day. Note that supplements containing 500 mg of vitamin B_6 are widely available, making it relatively easy to consume amounts above that which is recommended.

Biotin (Vitamin B_7)—Coenzyme for Carboxylation Reactions

The discovery of **biotin** (vitamin B_7) in the 1930s occurred in response to research investigating the cause of what was then called "egg white injury," and you will soon learn the reason for this. Biotin is a sulfur-containing molecule with two connected ring structures. This nutrient is somewhat unique among the water-soluble vitamins because we obtain it from both the diet and via biotin-producing bacteria in our large intestines. Biotin is required for gluconeogenesis and the citric acid cycle, which in turn are important for ATP production.

biotin (vitamin B_7) (BI – o – tin) A water-soluble vitamin involved in energy metabolism.

NUTS, MUSHROOMS, AND EGGS ARE RICH IN BIOTIN

As shown in Figure 10.9, sources of biotin include peanuts, tree nuts (such as almonds and cashews), mushrooms, eggs, and tomatoes. Like some of the other B vitamins, biotin in foods is frequently bound to proteins. Typically, enzymes in the small intestine cleave the protein–biotin complexes, releasing free (unbound) biotin, which is then absorbed. However, sometimes biotin binds exceptionally tightly to a food protein, making the biotin difficult to absorb. As previously mentioned, the discovery of this vitamin was linked to a condition called "egg white injury." This is because biotin's bioavailability can be greatly reduced when it is consumed with foods containing the protein **avidin,** present in large quantities in egg whites. Avidin attaches to biotin in the intestinal tract, making the vitamin difficult to absorb. However, this is not the case when eggs are cooked, because heat destroys (denatures) avidin. Thus, eating raw eggs can decrease biotin bioavailability, whereas eating cooked eggs does not. It is for this reason that experimental diets high in raw egg whites are sometimes used to induce biotin deficiency in human research subjects as well as laboratory animals.[11] Consuming alcohol can also decrease biotin absorption, and high temperatures can destroy biotin in foods.

As previously noted, biotin is also produced by bacteria in the GI tract. However, intestinal bacteria do not make enough biotin to meet our needs, thus making biotin an essential nutrient. Regardless of whether it is obtained from the diet or intestinal bacteria, biotin circulates in the blood away from the intestine to the liver where the vitamin is then distributed to the rest of the body. Biotin is not readily stored, but small amounts are present in muscle, liver, and brain tissue.

Avidin, a protein found in raw egg whites, binds tightly to biotin, making it difficult for the body to absorb.

avidin (AV – i – din) A protein present in egg whites that binds biotin, making it unavailable for absorption.

FIGURE 10.9 Food Sources and Structure of Biotin

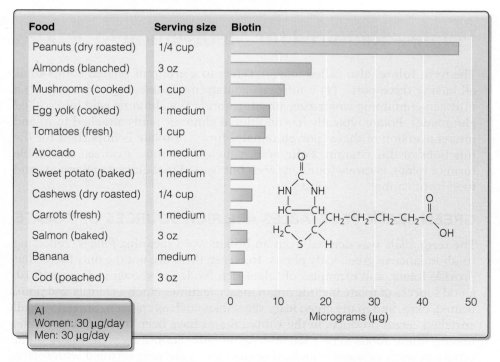

Food	Serving size	Biotin
Peanuts (dry roasted)	1/4 cup	
Almonds (blanched)	3 oz	
Mushrooms (cooked)	1 cup	
Egg yolk (cooked)	1 medium	
Tomatoes (fresh)	1 cup	
Avocado	1 medium	
Sweet potato (baked)	1 medium	
Cashews (dry roasted)	1/4 cup	
Carrots (fresh)	1 medium	
Salmon (baked)	3 oz	
Banana	medium	
Cod (poached)	3 oz	

AI
Women: 30 µg/day
Men: 30 µg/day

Microgram axis: 0 10 20 30 40 50
Micrograms (µg)

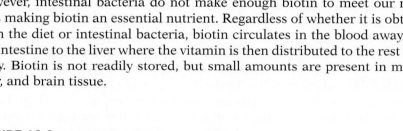

SOURCE: Hands E. Food Finder Vitamin and Mineral Source Guide, 3rd ed. Salem, OR: ESHA Research, 1995.

BIOTIN ADDS BICARBONATE (HCO$_3^-$) SUBUNITS IN CARBOXYLATION REACTIONS

The body uses biotin to make enzymes that catalyze **carboxylation reactions.** As such, biotin acts as a coenzyme in these reactions. Specifically, each biotin-requiring enzyme causes a bicarbonate subunit (HCO$_3^-$) to be added to another molecule, and these enzymes are generally involved in energy metabolism pathways (recall Figure 10.3). For example, a biotin-requiring enzyme converts pyruvate to oxaloacetate, a key step in gluconeogenesis. Consequently, inadequate biotin intake limits your ability to produce glucose when carbohydrate intake is low. Biotin also activates enzymes that allow some amino acids and fatty acids to enter the citric acid cycle for subsequent ATP production. In addition to biotin's role in activating these enzymes, the vitamin has noncoenzyme functions related to gene expression, especially influencing cell growth and development.

RECOMMENDED INTAKES FOR BIOTIN

Although biotin deficiency is uncommon, it can occur—especially in people who routinely consume large quantities of raw egg whites (containing avidin). In theory, however, it would take daily consumption of at least 12 raw egg whites for a prolonged period of time to cause biotin deficiency. Biotin deficiency can also be caused by medical conditions that impair nutrient absorption (such as inflammatory bowel disease) and some genetic disorders. Signs and symptoms of biotin deficiency are poorly understood but include depression, hallucinations, skin irritations, infections, hair loss, poor muscle control, seizures, and developmental delays in infants. Because there is insufficient information for the development of RDAs for biotin, an AI level for adults of 30 micrograms (μg) per day has been set. Very high biotin intake has no known detrimental effects, so a UL has not been established for this vitamin.

Folate—Required for Methylation Reactions

The term **folate,** also called folacin, refers to a group of related compounds all having three parts: (1) a nitrogen-containing double-ring structure, (2) a nitrogen-containing single-ring structure, and (3) a glutamic acid (also called glutamate). Folate typically has additional glutamic acids attached to it, and interconversion of these "polyglutamate" forms of folate is important for the functions of this vitamin. **Folic acid,** which is the most oxidized and stable form of folate, is rarely found in foods but is used in dietary supplements and food fortification.

GREEN LEAFY VEGETABLES ARE RICH SOURCES OF FOLATE

The term *folate* was derived from an Italian word meaning *foliage,* reflecting its abundance in green leafy plants. However, these are not the only foods that provide folate, and examples of folate-rich foods are shown in Figure 10.10. Good sources of folate include organ meats, legumes (such as lentils and pinto beans), okra, and many green leafy vegetables such as spinach. Since 1998, all enriched cereal products in the United States have been fortified with folate, making these foods very good sources of this vitamin as well. Many other products, such as some brands of orange juice, are now fortified with folate. Because this vitamin can be destroyed by excessive heat, light, and oxygen, cooked foods often have less folate than raw foods.

carboxylation reaction A metabolic reaction in which a bicarbonate subunit (HCO$_3$) is added to a molecule.

folate (also called folacin) (FO – late) A water-soluble vitamin involved in single-carbon transfer reactions; needed for amino acid metabolism and DNA synthesis.

folic acid The form of folate commonly used in vitamin supplements and food fortification.

FIGURE 10.10 Food Sources and Structure of Folate

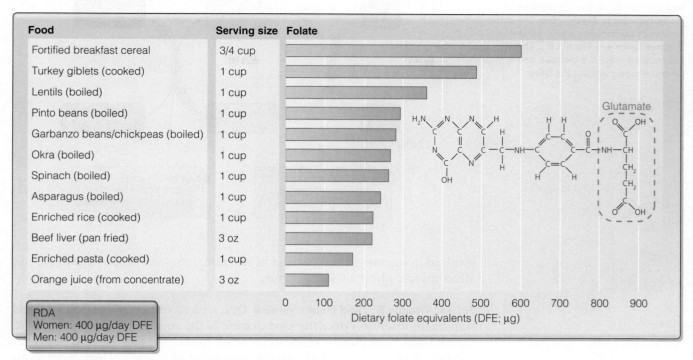

Food	Serving size	Folate
Fortified breakfast cereal	3/4 cup	
Turkey giblets (cooked)	1 cup	
Lentils (boiled)	1 cup	
Pinto beans (boiled)	1 cup	
Garbanzo beans/chickpeas (boiled)	1 cup	
Okra (boiled)	1 cup	
Spinach (boiled)	1 cup	
Asparagus (boiled)	1 cup	
Enriched rice (cooked)	1 cup	
Beef liver (pan fried)	3 oz	
Enriched pasta (cooked)	1 cup	
Orange juice (from concentrate)	3 oz	

Dietary folate equivalents (DFE; µg)

RDA
Women: 400 µg/day DFE
Men: 400 µg/day DFE

SOURCE: USDA Nutrient Database for Standard Reference, Release 16-1.

The bioavailability of folate varies, depending on the form of folate present in the food, genetic factors, and use of certain medications. In general, however, folic acid from supplements and fortified foods is more readily absorbed than folate naturally occurring in foods. This is because glutamate units typically must be cleaved from food folate prior to its absorption. In addition, some foods (such as cabbage) contain compounds that inhibit this process, thus decreasing folate bioavailability. Because folate absorption is variable, amounts present in foods are expressed as **dietary folate equivalents (DFEs),** an estimate of the amount actually absorbed by the body.

Once taken up by the intestinal cells, folate is converted to **tetrahydrofolate (THF)** by the addition of four hydrogen atoms. Finally, a methyl group (–CH₃) is added, producing an inactive form of folate, **5-methyltetrahydrofolate (5-methyl THF),** which is released into the blood and circulated to the liver. The liver then distributes 5-methyl THF to the rest of the body, where the vitamin is reactivated to its functional THF form. A limited amount of folate is generally present in the liver, although it is not actively stored there.

Folate (from the Italian word meaning *foliage*) is found in many plant foods, including green leafy vegetables.

dietary folate equivalent (DFE) A unit of measure used to describe the amount of bioavailable folate in a food or supplement.

tetrahydrofolate (THF) The active form of folate.

5-methyltetrahydrofolate (5-methyl THF) An inactive form of folate.

FOLATE FACILITATES SINGLE-CARBON TRANSFERS

Folate, in its active form of THF, acts as a coenzyme for many reactions, all involving the transfer of single-carbon groups (–CH₃) called "methyl groups." These reactions shift carbon atoms from one molecule to another to form the many organic substances required for life. Because single-carbon transfers are

The conversion of homocysteine to methionine requires folate and vitamin B₁₂. These are coupled reactions because one cannot happen without the other.

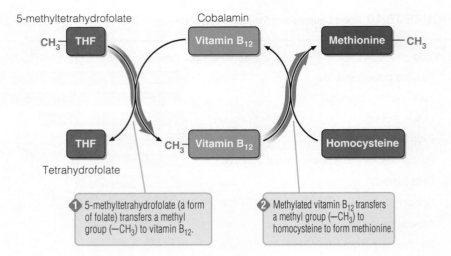

1. 5-methyltetrahydrofolate (a form of folate) transfers a methyl group (–CH₃) to vitamin B₁₂.

2. Methylated vitamin B₁₂ transfers a methyl group (–CH₃) to homocysteine to form methionine.

involved in so many different types of reactions, consuming too little folate can affect myriad physiological systems.

Folate, Vitamin B₁₂, and Homocysteine One area of research that has received significant attention over the past decade is the role of folate in amino acid metabolism. Specifically, folate is involved in the metabolism and interconversion of amino acids that require movement of single-carbon groups. An example is the conversion of the amino acid **homocysteine** to the amino acid methionine (Figure 10.11). This reaction transfers a methyl group (–CH₃) from 5-methyl THF to homocysteine, providing the body with both THF (active folate) and the essential amino acid methionine. The reaction also requires vitamin B₁₂, which accepts the methyl group (–CH₃) from 5-methyl THF and transfers it to homocysteine. Because the production of methionine from homocysteine requires both folate and vitamin B₁₂, a deficiency in either of these B vitamins results in the accumulation of homocysteine. High levels of homocysteine are associated with increased risk for heart disease and may have serious consequences.[12] You can read more about this in the Nutrition Matters section concerning cardiovascular disease following Chapter 6.

Folate and Synthesis of DNA and RNA Folate, in its THF form, is also involved in single-carbon transfer reactions required to make **purines** and **pyrimidines**—the molecules that make up DNA and RNA. Consequently, because DNA must be synthesized each time a new cell is made, folate is essential for the growth, maintenance, and repair of all tissues in the body. This is especially critical during periods of rapid growth, like fetal growth, and in cells with short life spans, such as those lining the GI tract.

Folate, Neural Tube Defects, and Spina Bifida Folate is required for the normal growth and development of nerve tissue in the fetus, and increased maternal intake of folate during the reproductive period has been shown to decrease the risk of **neural tube defects** in some newborns. Folate is involved in the development of the neural tube, a structure that eventually forms the brain and spinal cord. In early fetal development, the neural tube is actually not a tube but a flat sheet of nerve tissue. Consider what would happen if you cut a garden hose lengthwise and pressed it flat in your hand. This "open" form would represent the early stage of neural tube development. During fetal growth, the flat sheet of neural tissue "closes," forming a tube, much like what would happen if you let go of the cut garden hose.

homocysteine (ho – mo – CYS – teine) A compound that is converted to methionine in a coupled reaction that requires both folate and vitamin B₁₂.

purine (PUR – ine) A constituent of DNA and RNA.

pyrimidine (pyr - i – MID – ine) A constituent of DNA and RNA.

neural tube defect A malformation in which the neural tissue does not form properly during fetal development.

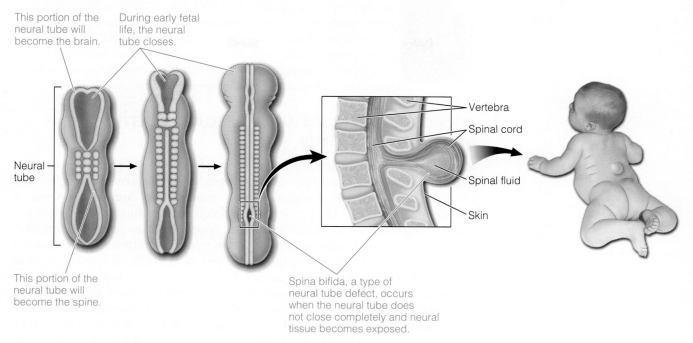

FIGURE 10.12 Neural Tube Defects and Spina Bifida Spina bifida is a form of neural tube defect that occurs very early in pregnancy when the neural tube does not form properly.

This portion of the neural tube will become the brain.

During early fetal life, the neural tube closes.

Neural tube

This portion of the neural tube will become the spine.

Vertebra

Spinal cord

Spinal fluid

Skin

Spina bifida, a type of neural tube defect, occurs when the neural tube does not close completely and neural tissue becomes exposed.

Neural tube defects result when closure of the neural tissue or "tube" is incomplete. When this happens, the spinal cord or brain sometimes protrudes outside of the body. There are various types of neural tube defects, affecting different locations. One type of neural tube defect affecting the brain is called **anencephaly,** a condition that results in an infant being born without a major portion of the brain. Brain development is severely impaired in babies born with this type of neural tube defect, and they do not survive. However, the most common form of neural tube defect, called **spina bifida** (Latin for *split spine*), occurs when the neural tube fails to close properly (Figure 10.12).[13] Approximately 5 to 6 out of every 10,000 babies born in the United States each year have a neural tube defect, with around 63% of them having spina bifida.[14] Although some forms of spina bifida do not cause problems, others are more severe. For example, they can result in mental impairment, partial paralysis, and poor bowel and bladder control.

Human dietary intervention studies show that maternal folate supplementation or increased consumption of folate-rich foods decreases the risk of neural tube defects in some women.[15] Why folate is required for normal neural tube development probably involves many folate-containing enzymes, including those required for DNA synthesis. Certain genetic variations (polymorphisms) in the mother or child also appear to predispose some fetuses to neural tube defects.[16] For example, some studies suggest that transfer of folate from the mother to the fetus is impaired in pregnant women with certain gene polymorphisms.[17] Clearly, however, neural tube defects can occur for reasons completely unrelated to folate intake or folate metabolism. Future research on nutrigenomics will undoubtedly shed more light on the complex interaction between folate intake and genetics in fetal development. Until then, it is important that all women capable of becoming pregnant consume plenty of folate-rich foods. The Institute of Medicine also recommends that sexually active women take folic acid supplements or consume folic acid–fortified foods in addition to a folate-rich, varied diet.

anencephaly A type of neural tube defect in which an infant is born without a major portion of the brain.

spina bifida A form of neural tube defect in which the spine does not properly form.

Critical Thinking: Beth's Story Recall Beth and her baby, Daniel, whom you met earlier in this chapter's Everybody Has a Story piece. You now know more about folate and how it is associated with risk of neural tube defect. Explain, at a very basic physiologic level, how inadequate folate is thought to result in this condition. In Beth's case, she was likely consuming what is thought to be an adequate amount of folate. What other factors do you think might have contributed to Daniel developing a neural tube defect?

FOLATE DEFICIENCY CAUSES MEGALOBLASTIC MACROCYTIC ANEMIA

Mild folate deficiency results in a variety of symptoms, including fatigue, weakness, irregular heart function, and headaches. Severe folate deficiency causes a condition called **megaloblastic macrocytic anemia.** Because of folate's importance in DNA synthesis and cell maturation, severe folate deficiency causes many cells, including red blood cells, to remain immature. Immature red blood cells—called *megaloblasts*—tend to be large and contain organelles not typically found in mature red blood cells. For example, mature red blood cells do not have nuclei, whereas immature red blood cells do. Because they are large and contain nuclei, the abnormal red blood cells are easily recognized under a microscope. However, folate deficiency is only one cause of this form of anemia, which can also result from deficiencies of other nutrients, such as vitamin B_{12}. Consequently, it is sometimes difficult to determine whether megaloblastic macrocytic anemia is due to folate or vitamin B_{12} deficiency. Examples of red blood cells from healthy people and those with various forms of anemia are shown in the photos on page 435.

Folate deficiency was once relatively common in the United States. This is no longer the case, partly because enriched cereal products are now fortified with this vitamin.[18] Instead, folate deficiency more commonly occurs in alcoholics, people with intestinal diseases, and people taking certain medications. Folate deficiency is also common in the elderly, who frequently take medications that inhibit folate absorption and use.

megaloblastic macrocytic anemia A condition in which red blood cells are large and immature, due to folate deficiency or vitamin B_{12} deficiency.

vitamin B_{12} (cobalamin) A water-soluble vitamin involved in energy metabolism and the conversion of homocysteine to methionine.

RECOMMENDED INTAKES OF FOLATE

It is recommended that adults consume 400 µg DFE of folate daily. This increases to 600 µg/day DFE for pregnant women. Women capable of or planning to become pregnant are encouraged to consume 400 µg/day DFE of folate (folic acid) from supplements, fortified foods, or both, *in addition to* consuming foods naturally rich in folate. This recommendation has been made in response to the previously discussed findings that folate supplementation decreases risk for neural tube defects in some women. Because excessive intakes of folate may make it difficult to detect vitamin B_{12} deficiency, a UL for folate has been set at 1,000 µg/day DFE from fortified foods or supplements. There is no evidence that high intake of naturally occurring folate poses any risk.

Vitamin B_{12} (Cobalamin)—Vitamin Made Only by Microorganisms

Vitamin B_{12}, also called cobalamin, was the last of the B vitamins to be discovered. This vitamin is a complex molecule and gets its name from the fact that it contains the trace element cobalt

Shellfish are excellent sources of vitamin B_{12}.

(Co) and several nitrogen (N) atoms. Like several of the other B vitamins, vitamin B_{12} is needed for ATP production and, as you just learned, functions with folate in amino acid metabolism.

MICROORGANISMS PRODUCE VITAMIN B_{12}

Vitamin B_{12} is a unique vitamin because it cannot be made by plants or higher animals (such as mammals and birds) but only by microorganisms (such as bacteria and fungi). Therefore, the presence of vitamin B_{12} in food is actually the result of its being made by microorganisms living in the food source's environment (generally the case for plants) or GI tract (more typical with animals). Some of these foods are listed in Figure 10.13; they include shellfish (such as clams and crabs), meat (including poultry and organ meats), fish, and dairy products. In addition, many ready-to-eat breakfast cereals are fortified with vitamin B_{12}.

Most vitamin B_{12} in foods is bound to proteins that must be cleaved before the vitamin can be absorbed. These proteins are removed in the stomach via acids and the enzyme pepsin. Free vitamin B_{12} is then bound to two proteins, **R protein** and **intrinsic factor,** that are also made in the stomach. R protein is thought to protect vitamin B_{12} from destruction in the stomach. In the intestine, R protein is released while intrinsic factor remains bound to vitamin B_{12}. This vitamin B_{12}–intrinsic factor complex is then transported into the absorptive cell, and the inability to produce intrinsic factor can result in severe vitamin B_{12} deficiency.

Once absorbed, vitamin B_{12} is released from intrinsic factor and bound to another protein called **transcobalamin,** which transports the vitamin in the blood to the liver. As with the other vitamins, the liver then uses vitamin B_{12} or circulates it to other tissues. Because it sequesters relatively large amounts of vitamin B_{12}–containing proteins and enzymes, the liver typically contains several years' supply of this vitamin.

R protein A protein, produced in the stomach, that binds to vitamin B_{12}.

intrinsic factor A protein, produced by the stomach, needed for vitamin B_{12} absorption.

transcobalamin The protein that transports vitamin B_{12} in the blood.

FIGURE 10.13 Food Sources and Structure of Vitamin B_{12}

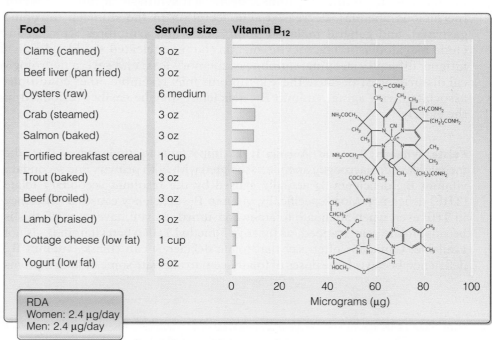

Food	Serving size	Vitamin B_{12}
Clams (canned)	3 oz	
Beef liver (pan fried)	3 oz	
Oysters (raw)	6 medium	
Crab (steamed)	3 oz	
Salmon (baked)	3 oz	
Fortified breakfast cereal	1 cup	
Trout (baked)	3 oz	
Beef (broiled)	3 oz	
Lamb (braised)	3 oz	
Cottage cheese (low fat)	1 cup	
Yogurt (low fat)	8 oz	

0   20   40   60   80   100
Micrograms (µg)

RDA
Women: 2.4 µg/day
Men: 2.4 µg/day

SOURCE: USDA Nutrient Database for Standard Reference, Release 16-1.

VITAMIN B$_{12}$ IS INVOLVED IN ATP AND METHIONINE PRODUCTION

Vitamin B$_{12}$ participates as a coenzyme in only two reactions. One reaction catalyzes the production of succinyl-CoA, an intermediate in the citric acid cycle. This ultimately allows the body to use some amino acids and fatty acids for energy (ATP) production. The other reaction catalyzes the conversion of homocysteine to the amino acid methionine. This was described previously in the section on folate and was shown in Figure 10.11. Recall that, during the conversion of homocysteine to methionine, the inactive form of folate (5-methyl THF) is converted to its active form (THF). Without adequate vitamin B$_{12}$, homocysteine levels build up in the blood, "trapping" folate in its inactive 5-methyl THF form. In this way, vitamin B$_{12}$ deficiency can cause secondary folate deficiency.

VITAMIN B$_{12}$ DEFICIENCY CAUSES PERNICIOUS ANEMIA

Vitamin B$_{12}$ deficiency can result from either inadequate dietary intake or poor absorption. Primary vitamin B$_{12}$ deficiency is sometimes seen in vegans or infants breastfed by vitamin B$_{12}$–deficient mothers. Secondary vitamin B$_{12}$ deficiency can occur when gastric cells stop producing intrinsic factor or hydrochloric acid (HCl); this often happens as we age and results in poor absorption of vitamin B$_{12}$. An example of secondary vitamin B$_{12}$ deficiency is **pernicious anemia,** an autoimmune disease in which antibodies destroy the stomach cells that produce intrinsic factor.[19]* Because intrinsic factor is essential for vitamin B$_{12}$ absorption, pernicious anemia can occur even when large amounts of vitamin B$_{12}$ are consumed. This is why people with pernicious anemia cannot be treated with oral vitamin B$_{12}$ supplements. Instead, they must receive vitamin B$_{12}$ by injection. Secondary vitamin B$_{12}$ deficiency can also be caused by other conditions such as genetic defects, GI infections, surgeries, and some medications.

Vitamin B$_{12}$ deficiency is especially common in the elderly population, affecting as many as 15 to 40%.[20] This is due to a variety of factors, including inadequate vitamin B$_{12}$ intake, decreased synthesis of intrinsic factor (pernicious anemia), decreased acid secretion in the stomach (atrophic gastritis), and general malabsorption. Signs and symptoms typically include megaloblastic macrocytic anemia (large, nucleated red blood cells); fatigue; difficulty sleeping; numbness; memory loss; and severe neurological disturbances. Because these symptoms may resemble other conditions associated with aging, vitamin B$_{12}$ deficiency may be easily overlooked in the elderly.

Folate, Vitamin B$_{12}$, and Anemia It is important to understand that the megaloblastic, macrocytic anemia associated with both primary and secondary vitamin B$_{12}$ deficiency is actually caused by the resultant secondary folate (THF) deficiency. More specifically, vitamin B$_{12}$ deficiency causes a deficiency in THF even when adequate folate is consumed. As you have learned, this is because folate gets "trapped" as inactive 5-methyl THF when vitamin B$_{12}$ is not available. Clinicians often say that folate deficiency can "mask" vitamin B$_{12}$ deficiency, because large doses of folate seem to alleviate some of its symptoms

---

pernicious anemia An autoimmune disease that causes vitamin B$_{12}$ deficiency due to lack of intrinsic factor.

*Note that the term *pernicious anemia* is often used synonymously with vitamin B$_{12}$ deficiency regardless of its etiology. However, pernicious anemia specifically refers to anemia resulting from vitamin B$_{12}$ deficiency caused by an autoimmune disease.

(such as the anemia). However, other complications of vitamin B_{12} deficiency are not alleviated with high doses of folate. Because the other symptoms involve severe neurologic damage that can be fatal, misdiagnosing vitamin B_{12} deficiency as folate deficiency is dangerous. Some researchers are worried that the enrichment of cereal grains with folic acid might lead to increased prevalence of undiagnosed vitamin B_{12} deficiency, especially in the elderly.[21] Time will tell if this concern is legitimate.

RECOMMENDED INTAKES FOR VITAMIN B_{12}

The RDA for vitamin B_{12} is 2.4 μg/day for adults. Although this recommendation does not change for older people, it is important for those older than 50 years of age to choose vitamin B_{12}–fortified foods when possible or consume vitamin B_{12} supplements if needed. This is especially true for older individuals with atrophic gastritis. Vegans who do not eat any animal products should consider taking a supplement or eating foods that have been fortified with vitamin B_{12}. No ULs are established for this vitamin.

Vitamin C (Ascorbic Acid)—Critical Antioxidant

Of all the water-soluble vitamins, it is likely that you have heard the most about **vitamin C,** also called **ascorbic acid.** Indeed, there is much public interest in vitamin C, and scientific evidence suggests that it likely plays a role in almost every physiological system. For example, it is important for the immune, cardiovascular, neurological, and endocrine systems. This relatively simple compound can be made from glucose by all plants and most animals, but not humans. In fact, primates (including humans), fruit bats, and guinea pigs are some of the few animals for which vitamin C is an essential dietary nutrient.

MANY FRUITS AND VEGETABLES ARE RICH SOURCES OF VITAMIN C

As shown in Figure 10.14, vitamin C is found in many fruits and vegetables, such as citrus fruits, peppers, papayas, broccoli, strawberries, and peas. The bioavailability of vitamin C is generally high, although it can be easily destroyed in food by heat, exposure to oxygen, and acidity. This is why freshly peeled or prepared fruits and vegetables tend to provide more vitamin C than cooked, processed, or stored ones.

Absorption of vitamin C occurs mainly in the small intestine via active transport using glucose transport proteins. This is because vitamin C is structurally similar to glucose. However, at very high intakes, vitamin C is also absorbed by simple diffusion in both the stomach and small intestine. Vitamin C then circulates in the blood to the liver, which subsequently distributes the vitamin to the rest of the body. As in the intestine, the uptake of vitamin C into the body's cells relies, in part, on glucose transporters. Excess vitamin C is rapidly metabolized and excreted in the urine.

VITAMIN C IS A POTENT ANTIOXIDANT

Unlike the B vitamins, vitamin C is not a coenzyme. Instead, it acts as a nonenzymatic **antioxidant.** Recall from Chapter 3 that atoms contain negatively charged electrons, and an abundance of electrons in an atom or molecule can

〈**CONNECTIONS**〉 Remember from Chapter 4 that glucose transporters are proteins that escort glucose molecules across cell membranes (Chapter 4, page 133).

vitamin C (ascorbic acid) A water-soluble vitamin that has antioxidant functions in the body.

antioxidant A compound that readily gives up electrons (and hydrogen ions) to other substances.

FIGURE 10.14 Food Sources and Structure of Vitamin C

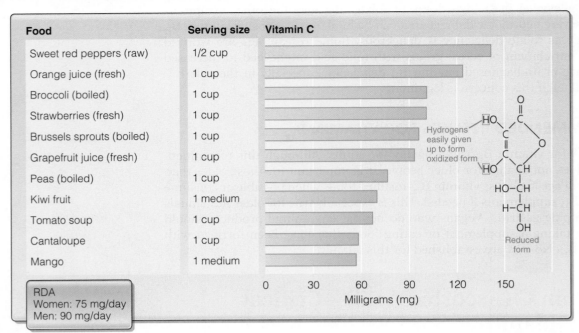

Food	Serving size	Vitamin C
Sweet red peppers (raw)	1/2 cup	
Orange juice (fresh)	1 cup	
Broccoli (boiled)	1 cup	
Strawberries (fresh)	1 cup	
Brussels sprouts (boiled)	1 cup	
Grapefruit juice (fresh)	1 cup	
Peas (boiled)	1 cup	
Kiwi fruit	1 medium	
Tomato soup	1 cup	
Cantaloupe	1 cup	
Mango	1 medium	

RDA
Women: 75 mg/day
Men: 90 mg/day

Milligrams (mg)

Hydrogens easily given up to form oxidized form

Reduced form

SOURCE: USDA Nutrient Database for Standard Reference, Release 16-1.

result in an overall net negative charge. Such an atom or molecule is therefore said to be "reduced." When electrons are removed, the atom or molecule has a more positive charge and is said to be "oxidized." Because vitamin C can easily accept and donate electrons, it is involved in a variety of reduction–oxidation (redox) reactions.

One way in which vitamin C functions as an antioxidant is in "recharging" enzymes. Recall that enzymes are not used up or destroyed while catalyzing reactions. However, sometimes a portion of the enzyme may need to be restored or regenerated. This is especially true when the enzyme contains an element, such as copper or iron, that gets oxidized during the reaction. These types of enzymes must be restored or "re-reduced" to catalyze subsequent reactions. Enzymes that need restoration are like flashlights that require batteries. A flashlight may work for years, as long as its batteries are replaced or recharged periodically. Using this analogy, you can think of an enzyme as a flashlight, copper or iron atoms as batteries, and vitamin C as a battery recharger. During the chemical reaction, copper or iron atoms are oxidized (made more positively charged) and then must be reduced (or "recharged") for the enzyme to catalyze another reaction. This recharging via reduction is the function of vitamin C and is illustrated in Figure 10.15.

Collagen Production The process of collagen production provides a real-life example of how vitamin C "recharges" an enzyme. **Collagen** is an important protein found in connective tissues. For collagen to function properly, three strands of protein must twist together in just the right way—a process requiring a copper-containing enzyme to be oxidized. Vitamin C is needed to reduce this copper-containing enzyme between reactions so that it can continue to form collagen. This is why vitamin C is required for the health of all the collagen-containing connective tissues such as those that make up the skin, muscles, cartilage, tendons, and gums.

⟨**CONNECTIONS**⟩ The quaternary structure of a protein forms when several smaller polypeptides come together to form a larger protein (Chapter 5, page 172).

collagen A structural protein found in connective tissue, including skin, bones, teeth, cartilage, and tendons.

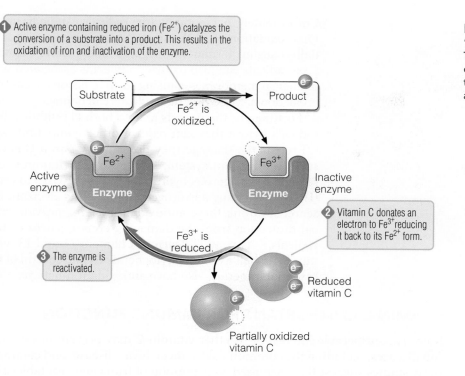

1 Active enzyme containing reduced iron (Fe^{2+}) catalyzes the conversion of a substrate into a product. This results in the oxidation of iron and inactivation of the enzyme.

Substrate

Product

e⁻

Fe^{2+} is oxidized.

Active enzyme

e⁻ Fe^{2+}

Enzyme

Fe^{3+}

Enzyme

Inactive enzyme

2 Vitamin C donates an electron to Fe^{3+} reducing it back to its Fe^{2+} form.

Fe^{3+} is reduced.

3 The enzyme is reactivated.

e⁻

e⁻ e⁻ Reduced vitamin C

Partially oxidized vitamin C

FIGURE 10.15 Vitamin C "Recharges" Enzymes by Reducing Them Vitamin C reduces iron- and copper-containing enzymes back to their original forms, so they can catalyze additional reactions.

Carnitine, Tyrosine, Neurotransmitter, and Hormone Synthesis Vitamin C also donates electrons to two iron-containing enzymes involved in the synthesis of carnitine. Recall from Chapter 6 that carnitine allows the body to use fatty acids for energy metabolism. Thus, vitamin C is indirectly required for ATP synthesis. Vitamin C also "recharges" iron-containing enzymes involved in making tyrosine (a nonessential amino acid) from phenylalanine (an essential amino acid), and it "recharges" enzymes involved in the synthesis of the neurotransmitters norepinephrine and serotonin. The reducing action of vitamin C is also needed to activate enzymes involved in synthesizing the hormones cholecystokinin (CCK) and gastrin.

Vitamin C Enhances Iron, Copper, and Chromium Bioavailability The reducing power of vitamin C also plays an important role in the GI tract. Minerals such as iron, copper, and chromium are better absorbed in their reduced states. Therefore, consuming vitamin C along with these essential minerals can increase their bioavailability by chemically reducing them in the GI tract. For example, drinking orange juice along with iron-fortified cereal can increase the amount of iron absorbed from the cereal. This is because vitamin C in the orange juice reduces the cereal's ferric iron (Fe^{3+}) to ferrous iron (Fe^{2+}), the form that is more readily absorbed.

Vitamin C Protects You from Free Radical Damage Vitamin C also protects the body from **free radicals,** which are highly reactive molecules with an unequal balance of charges. In most cases, these free radicals are produced during normal cellular metabolism. Exposure to toxic substances such as smog, cigarette smoke, and ozone, as well as some drugs and intense sunlight, can also increase free radical production. Free radicals have unpaired electrons, making them unstable and reactive. Like most substances, free radicals strive to have an even number of electrons and will take electrons from other molecules to do so. In other words, free radicals readily oxidize other molecules to stabilize themselves.

free radical A reactive molecule with one or more unpaired electrons. Free radicals are destructive to biological membranes, DNA, and proteins.

Vitamin C, found in citrus fruits and many other fruits and vegetables, protects the body from the damaging effects of free radicals.

Unfortunately, oxidation by free radicals can be harmful. First, oxidation of DNA can break and damage it, potentially causing mutations in genes and possibly cancer. Free radicals can also oxidize fatty acids associated with biological membranes, causing them to become weak and break down. Moreover, free radicals can damage proteins.

Fortunately, the body is able to both (1) stabilize free radicals before they can oxidize other compounds and (2) repair the damage they do cause. Vitamin C is one such "antioxidant system." For example, vitamin C can reduce the dangerous hydroxyl free radical (OH⁻) to water (H_2O) by donating a hydrogen atom and its electron. Researchers think these antioxidant systems provide critical protection from the damaging consequences of free radicals.[22] As you will learn in Chapters 11 and 13, vitamin E, as well as some phytochemicals and several of the essential minerals, also have antioxidant functions.

VITAMIN C IS IMPORTANT FOR IMMUNE FUNCTION

There is considerable interest in whether vitamin C may prevent or cure certain diseases, including the common cold, cancer, heart disease, and cataracts. Many studies suggest that increased consumption of fruits and vegetables rich in vitamin C is associated with decreased risk for these diseases. Yet controlled clinical intervention studies often have not demonstrated a protective effect of vitamin C.[23] Growing evidence, however, shows that large doses of vitamin C can benefit the immune system.[24] Because the immune system is directly involved in defending against many of these diseases, it is possible that vitamin C indirectly affects disease by strengthening immune function. Vitamin C most likely works with other antioxidant nutrients such as vitamin E.

VITAMIN C DEFICIENCY CAUSES SCURVY

Vitamin C deficiency can cause a sometimes-deadly condition called **scurvy.** In the 18th century, a British medical doctor named James Lind conducted what was perhaps the first controlled nutrition intervention experiment when he determined that consuming citrus fruits would prevent this disease. Lind studied 12 sailors suffering from scurvy by treating them with lemons, limes, cider, nutmeg, seawater, or vinegar. He found that consuming lemons or limes, but not the other treatments, prevented and cured scurvy.

One of the signs of scurvy (vitamin C deficiency) is the presence of very small red spots on the skin caused by internal bleeding.

Signs and symptoms associated with scurvy include bleeding gums, skin irritations, bruising, and poor wound healing, many of which are due to inadequate collagen production. Although scurvy was once common, increased availability of fruits and vegetables has made it rare. However, scurvy is still seen in developing countries, alcoholics, and in children and elderly individuals on very restricted diets.[25] Although most people can consume very large doses (2 to 4 g/day) of supplemental vitamin C without experiencing harmful effects, it can cause nausea, diarrhea, cramping, and kidney stones in other people. Caution, therefore, is advised when consuming vitamin C supplements.

scurvy A condition caused by vitamin C deficiency; symptoms include bleeding gums, bruising, poor wound healing, and skin irritations.

RECOMMENDED INTAKES FOR VITAMIN C

The RDAs for vitamin C in adult men and women are 90 and 75 mg/day, respectively. Because cigarette smoke increases exposure to free radicals and consequently the body's need for antioxidants, smokers are advised to increase their vitamin C intake by an additional 35 mg/day. Whether this recommendation also applies to people exposed to secondhand smoke is not clear. To avoid possible GI distress, the UL for vitamin C intake from supplements has been set at 2 g/day (2,000 mg/day).

Is Choline a "New" Conditionally Essential Nutrient?

Choline is a water-soluble, nitrogen-containing compound found in many foods. Because we can typically make choline in sufficient amounts, scientists assumed it was a nonessential nutrient. However, there is now evidence that some people might not be able to make sufficient choline if they were to eat a choline-free diet.[26] In addition, recent studies suggest that postmenopausal women may require dietary choline for optimal health.[27] The Institute of Medicine has, therefore, designated choline as an essential nutrient—at least in some situations—and choline is considered by many experts to be a conditionally essential nutrient.[28] Other scientists consider it a "vitamin-like substance." Clearly, we still have much to learn about the way choline functions.

⟨**CONNECTIONS**⟩ Phospholipids are amphipathic compounds, are important components of cell membranes and lipoproteins, and help emulsify lipids in the small intestine (Chapter 6, page 231).

EGGS ARE RICH SOURCES OF DIETARY CHOLINE

As shown in Figure 10.16, choline is abundant in many plant and animal foods, particularly eggs, liver, poultry, and fish. In addition, because the choline-containing compound called lecithin (also called phosphatidylcholine) is often added to foods as an emulsifier, you consume relatively large amounts of choline if you eat products such as mayonnaise and salad dressings.

choline (CHO – line) A water-soluble compound used by the body to synthesize acetylcholine (a neurotransmitter) and a variety of phospholipids needed for cell membrane structure; considered a conditionally essential nutrient.

FIGURE 10.16 Food Sources and Structure of Choline

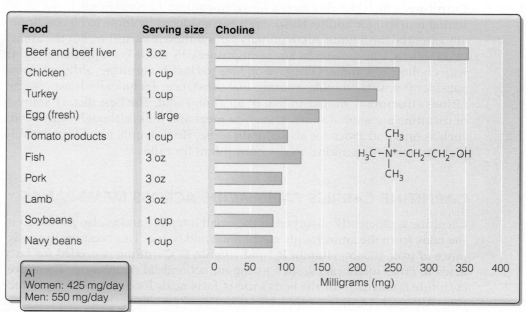

SOURCE: USDA Nutrient Database for Standard Reference, Release 20.

Remember that lecithin is a phospholipid containing a glycerol backbone, two fatty acids, and a choline-containing polar group.

Free (unbound) choline found in food is absorbed in the small intestine and circulates to the liver via the blood. However, the choline component of lecithin must be cleaved from the glycerol backbone by pancreatic enzymes before it can be absorbed. Choline is then taken up by intestinal cells and reconstituted (reformed) into lecithin molecules, which are released into the lymph. A large portion of the lecithin that is ultimately delivered to the liver becomes a component of lipoproteins (such as low-density lipoprotein, or LDL). In most healthy adults, choline is readily made in the body from the essential amino acid methionine, with the assistance of vitamin B_{12} and folate.

Choline is needed for synthesizing a variety of phospholipids (such as lecithin), as well as the neurotransmitter acetylcholine, and is an important component of biological membranes and lipoproteins. Choline is also needed for muscle control. Other compounds synthesized from choline play structural roles in the body and act as single-carbon donors in metabolic reactions.

CHOLINE DEFICIENCY IS RARE

Because choline deficiency is rare in humans, little is known about deficiency signs and symptoms. However, there is limited evidence that choline deficiency resulting from choline-free diets can cause liver damage in adult men.[26] Without adequate choline, the body may be unable to produce very low density lipoproteins (VLDLs), resulting in triglyceride accumulation in the liver.

Very high choline intake from foods can cause a fishy body odor, excess perspiration, salivation, low blood glucose, and liver damage. Although these symptoms may not be desirable, they are generally not fatal. Currently, there is not enough information to establish RDAs for choline. However, AIs for men and women are 550 and 425 mg/day, respectively. The UL for choline is 3.5 g/day (3,500 mg/day).

Carnitine—Needed for Fatty Acid Transport

Carnitine, which like choline is a nitrogen-containing compound, is not an essential nutrient for adults. However, it is considered by some to be conditionally essential for infants. All mammals can synthesize carnitine, but the ability to do this may be lacking in newborns, especially those born prematurely. As with choline, researchers disagree on how to classify carnitine, although some consider it a water-soluble, vitamin-like substance. Technically, however, carnitine's structure is similar to that of an amino acid. The best dietary sources of carnitine are animal sources, such as meat and milk, although other foods such as nuts and avocados also contain some. Human milk provides relatively high amounts of carnitine, as do many infant formulas.[29]

CARNITINE CARRIES FATTY ACIDS ACROSS MEMBRANES

Carnitine is efficiently absorbed in the small intestine and is also produced in the body from the amino acids methionine and lysine. This requires the assistance of iron, niacin, vitamin B_6, and vitamin C. Carnitine is needed for fatty acids to cross membranes, such as the mitochondrial membrane. Therefore, carnitine is essential for the body's use of fatty acids for energy (ATP) production. Although carnitine supplements are sold as aids to help build muscle and increase athletic performance, there is little or no scientific evidence to support this claim.

carnitine A nitrogen-containing compound needed for fatty acid transport across membranes; may be conditionally essential in infancy.

Carnitine deficiency is rare, but it can occur in some genetic conditions, resulting in muscle weakness, hypoglycemia, and heart irregularities. Carnitine deficiency is rare even in vegans who consume very little dietary carnitine.[30] This supports the concept that carnitine is not an essential nutrient for adults. No toxic effects of high doses of carnitine are known. Because researchers do not consider carnitine an essential nutrient and high intakes of carnitine do not seem detrimental, no DRIs are established for this substance.

Summary of the Water-Soluble Vitamins and Use of Supplements

You now know that the water-soluble vitamins serve numerous purposes. Although it is not possible to list all the functions of each vitamin, it is useful to group some of these roles into functional categories. For example, most water-soluble vitamins participate as coenzymes in energy metabolism pathways, several are involved in DNA synthesis, and others serve antioxidant roles. A general classification system of the major roles of the water-soluble vitamins is summarized in Table 10.4.

Because water-soluble vitamins have so many functions, consuming them is vitally important for health. Consequently, several federal agencies, such as the U.S. Department of Agriculture (USDA) and the FDA, have developed guidelines and policies that help consumers choose foods wisely. For example, the MyPlate food guidance system helps people determine how many servings of each food group are needed to obtain these essential nutrients. To customize these recommendations for yourself, visit the MyPlate website (http://www.choosemyplate.gov). Remember that, although whole-grain products, fruits, and vegetables tend to be excellent sources of water-soluble vitamins, these compounds are found in all the food groups. Clearly, dietary balance remains a key component of good health—even when it comes to the water-soluble vitamins.

You have now learned the fundamentals related to each of the water-soluble vitamins. As such, you should be able to choose foods that specifically contain each of these important dietary components. But you may be wondering about vitamin supplements and when it might be helpful to consider taking them. You may also be wondering, since vitamins are found naturally

TABLE 10.4 General Functions of the Water-Soluble Vitamins, Choline, and Carnitine

Functions	Thiamin	Riboflavin	Niacin	Pantothenic Acid	Vitamin B_6	Biotin	Folate	Vitamin B_{12}	Vitamin C	Choline[b]	Carnitine
Coenzyme	■	■	■	■	■	■	■	■			
Energy metabolism[a]	■	■	■	■	■	■					■
Antioxidant function		■							■		
Interconversion or activation of nutrients		■			■		■	■			
Blood health		■			■		■	■	■		
DNA or RNA synthesis	■	■					■	■			
Nerve/muscle function	■	■			■			■		■	

[a]Note that these vitamins are not, themselves, energy-yielding nutrients but are involved in energy metabolism via their coenzyme roles.
[b]The roles that choline plays in the body are still being investigated.

in foods, why so many people take supplements. Although a thorough review of the dietary supplement industry and potential effects of supplement use is beyond the scope of this textbook, there are some basic concepts related to supplementation that can help you make wise choices about what to take—and what not to take. These are briefly described next.

DIETARY SUPPLEMENTS CAN CONTAIN MANY SUBSTANCES

At times, we may not get enough of an essential vitamin from foods, and taking a vitamin supplement could be beneficial. But how do you know when, and how can you determine which type of supplement to consume? Although answering these questions may seem overwhelming, knowing some basic information about dietary supplements can help you make educated decisions about whether supplements are right for you.

First, it is important to realize that the term **dietary supplement** describes a broad group of products. As defined by the FDA, dietary supplements can contain one or more of the following ingredients: vitamins, minerals, amino acids, herbs, or other plant-derived substances (botanicals), and a multitude of other compounds derived from animals and foods.[31] Thus, only some dietary supplements contain vitamins.

Unlike drugs, dietary supplements do not have to be "preapproved" by the FDA for safety and effectiveness, although their labels can display approved health claims. It is the responsibility of dietary supplement manufacturers and distributors to ensure that their products are safe and that their label claims are accurate and truthful. However, once a product enters the marketplace, the FDA has the authority to take action against any dietary supplement it believes presents a risk of illness or injury.

Sorting Fact from Fiction Can Be Challenging It is important to make informed decisions about which supplements to take and which to avoid. As described in Chapter 1, proving that a nutrient influences health requires proper use of the scientific method. Because it is sometimes difficult to determine whether claims regarding vitamin supplements and health are accurate, you should consult a reliable source for expert advice. An excellent source of up-to-date information is the Office of Dietary Supplements, which is part of the National Institutes of Health. This office was established in 1994 to strengthen knowledge and understanding of dietary supplements by evaluating scientific information and educating the public. The Office of Dietary Supplements maintains an excellent, user-friendly website (http://ods.od.nih.gov) that contains information concerning most dietary supplements. It also provides the following tips for buying and using any dietary supplement.

- *Consider safety first.* Some supplement ingredients can be toxic—especially in high doses. Do not hesitate to check with a health professional before taking any dietary supplement.
- *Think twice about chasing the latest headline.* Sound health advice is generally based on research over time, not a single study touted by the media. Be wary of results claiming a "quick fix." And remember, if something sounds too good to be true, it probably is.
- *More may not be better.* Some products can be harmful when consumed in high amounts for a long time. Do not assume that more is better—it might be toxic.
- *"Natural" does not always mean safe.* The term "natural" simply means something is not synthetic or human-made. Do not assume that this term ensures wholesomeness or safety.

dietary supplement Product intended to supplement the diet and that contains vitamins, minerals, amino acids, herbs or other plant-derived substances, or a multitude of other compounds.

When Should You Consider Taking a Supplement? There are no hard-and-fast rules about when you should take a dietary supplement. However, if your diet lacks variety and balance, taking a dietary supplement may help you consume appropriate amounts of the essential nutrients. This is likely in several situations, as listed here.

- When your food availability or variety is limited by time limitations or cooking constraints (such as college life)
- If you decide, for one reason or another, *not* to consume certain foods (e.g., red meat or dairy products)
- During periods of rapid growth and development such as infancy, childhood, adolescence, and pregnancy
- When the consequences of normal aging (e.g., loss of calcium from bone or atrophic gastritis) make it difficult to consume adequate amounts of a nutrient from foods (e.g., calcium and vitamin B_{12})
- When you consume a low-calorie diet for weight loss
- If you suffer from certain health conditions that increase your nutrient requirements

Certainly, taking a supplement containing reasonable amounts of the essential vitamins and minerals can cause no harm and may be advantageous in some situations. However, supplements should never replace prescribed medications or the variety of foods important to a healthful diet. Also, care should be taken such that daily intakes of vitamins and minerals do not exceed UL values. Clinicians advise that we keep a record of all dietary supplements that we take and periodically share this information with our health care providers. This is especially important when medications are taken on a regular basis, as is often the case in the elderly.

After deciding that you would like to take a dietary supplement, you then need to consider which type of supplement is best. Determining this is sometimes difficult, because there are often many choices. Again, one of your best resources in this regard is the Office of Dietary Supplements' website, which contains dozens of "Supplement Fact" sheets. Reading these carefully and consulting with your health care provider should provide valuable information to help you determine which type of supplement best fits your needs.

Critical Thinking: Beth's Story You now should have much more knowledge about why consuming adequate amounts of the water-soluble vitamins is important during the lifespan—especially during critical periods of growth and development such as pregnancy. Recall Beth from the Everybody Has a Story feature at the beginning of this chapter. Can you list at least two reasons why each of the water-soluble vitamins is needed to support a healthy pregnancy? Why do AI/RDA levels for these substances generally increase during this time?

Diet Analysis PLUS ✚ Activity

This chapter introduced you to the nine essential water-soluble vitamins including the eight B vitamins and vitamin C (along with their alternate names), and two of the conditionally essential nutrients, choline and carnitine. Using your *Intake and DRI Goals Compared* report representing the average of your intakes, fill in the following table to assess your intakes of some of these nutrients. *(Remember that your intakes may fluctuate dramatically from day to day, so make sure to use the report that averages these fluctuations over the time period.)*

You will notice that a few of the following vitamins and compounds (biotin, pantothenic acid, and choline) are NOT reported on your *Intake vs. Goals Compared* report. Instead, use the Recommended Dietary Allowance and Adequate Intake tables in your text to fill in at least your RDA/AI for these three nutrients.

1. Use the following table to compare your dietary intake with your RDA/AI values.

Vitamin/Choline	Alternate name *(if any)*	Your RDA/AI *(include units)*	Your estimated intake *(include units)*	The % of RDA/AI you consumed	Did you consume too little (−), too much (+), or equal (=) of your RDA/AI?
Thiamin					
Riboflavin					
Niacin					
Pantothenic acid					
Vitamin B$_6$					
Biotin					
Folate					
Vitamin B$_{12}$					
Vitamin C					
Choline					

2. You will notice that some of these nutrients have a Recommended Dietary Allowance (RDA) while others have an Adequate Intake Level (AI); can you explain why this is?

3. For the following, list the foods from your individual days' *Intake Spreadsheets* that provided the largest amounts of the following water-soluble vitamins.

   Thiamin

   Name of food consumed   **Amount of nutrient in food consumed**
                                                              (include units)

   _____   _____
   _____   _____
   _____   _____

   Riboflavin

   Name of food consumed   **Amount of nutrient in food consumed**
                                                              (include units)

   _____   _____
   _____   _____
   _____   _____

   Niacin

   Name of food consumed   **Amount of nutrient in food consumed**
                                                              (include units)

   _____   _____
   _____   _____
   _____   _____

Vitamin B$_6$

Name of food consumed   **Amount of nutrient in food consumed**
                                              (include units)

_____

_____

_____

Folate

Name of food consumed   **Amount of nutrient in food consumed**
                                              (include units)

_____

_____

_____

Vitamin B$_{12}$

Name of food consumed   **Amount of nutrient in food consumed**
                                              (include units)

_____

_____

_____

Vitamin C

Name of food consumed   **Amount of nutrient in food consumed**
                                              (include units)

_____

_____

_____

4. Using the information you found from answering question 3, what do you notice about the food sources of these water-soluble nutrients, especially the B vitamins? Are there foods (or food groups) that are especially good sources of several of these substances?

5. Based on the assessments of your water-soluble vitamin intake, do you need to change your intakes of these nutrients? If so, please list at least three ways by which this can be accomplished.

Notes

1. Carpenter KJ. A short history of nutritional science: Part 1 (1785–1885). Journal of Nutrition. 2003;133:638–45. Carpenter KJ. A short history of nutritional science: Part 2 (1885–1912). Journal of Nutrition. 2003;133:975–84. Carpenter KJ. A short history of nutritional science: Part 3 (1912–1944). Journal of Nutrition. 2003;133:3023–32. Carpenter KJ. A short history of nutritional science: Part 4 (1945–1985). Journal of Nutrition. 2003;133:3331–42.

2. Institute of Medicine. Dietary Reference Intakes: Guiding principles for nutrition labeling and fortification. Washington, DC: National Academy Press; 1998. Park YK, McDowell MA, Hanson EA, Yetley E. History of cereal-grain product fortification in the United States. Nutrition Today. 2001;36:124–37.

3. Adapted from U.S. Department of Agriculture. Building blocks for fun and healthy meals. 2000. Available from: http://www.fns.usda.gov/tn/Resources/buildingblocks.html.

4. Thomson AD, Cook CC, Guerrini I, Sheedy D, Harper C, Marshall EJ. Wernicke's encephalopathy revisited. Translation of the case history section of the original manuscript by Carl Wernicke 'Lehrbuch der Gehirnkrankheiten fur Aerzte and Studirende' (1881) with a commentary. Alcohol. 2008;43:174–9.

5. West DW, Owen EC. The urinary excretion of metabolites of riboflavin in man. British Journal of Nutrition. 1963;23:889–98.

6. Asztalos BF. High-density lipoprotein particles, coronary heart disease, and niacin. Journal of Clinical Lipidology. 2010;4:405–10. Krauss RM. Lipids and lipoproteins in patients with type 2 diabetes. Diabetes Care. 2004;27:1496–504. Keener A, Sanossian N. Niacin for stroke prevention: Evidence and rationale. Central Nervous System Neuroscience and Therapy. 2008;14:287–94.

7. Azmanov DN, Kowalczuk S, Rodgers H, Auray-Blais C, Giguère R, Rasko JE, Bröer S, Cavanaugh JA. Further evidence for allelic heterogeneity in Hartnup disorder. Human Mutations. 2008;29:1217–21. Orbak Z, Ertekin V, Selimoglu A, Yimaz N, Tan H, Konak M. Hartnup disease masked by swashiorkor. Journal of Health, Population, and Nutrition. 2010;28:413–5.

8. Bendich A. The potential for dietary supplements to reduce premenstrual syndrome (PMS) symptoms. Journal of the American College of Nutrition. 2000;19:3–12. Chocano-Bedova PO, Manson JE, Hankinson SE, Willett WC, Johnson SR, Chasan-Taber, L. Ronnenberg AG, Bigelow C, Bertone-Johnson ER. Dietary B vitamin intake and incident premenstrual syndrome. American Journal of Clinical Nutrition. 2011;93:1080–6.

9. Schnyder G, Roffi M, Pin R, Flammer Y, Lange H, Eberli FR, Meier B, Turi ZG, Hess OM. Decreased rate of coronary restenosis after lowering of plasma homocysteine levels. New England Journal of Medicine. 2001;345:1593–60. Schnyder G, Roffi M, Flammer Y, Pin R, Hess OM. Effect of homocysteine-lowering therapy with folic acid, vitamin B$_{12}$, and vitamin B6 on clinical outcome after percutaneous coronary intervention: The Swiss Heart study: A randomized controlled trial. Journal of the American Medical Association. 2002;28:973–9.

10. Bleie Ø, Semb AG, Grundt H, Nordrehaug JE, Vollset SE, Ueland PM, Nilsen DW, Bakken AM, Refsum H, Nygård OK. Homocysteine-lowering therapy does not affect inflammatory markers of atherosclerosis in patients with stable coronary artery disease. Journal of Internal Medicine.

2007;262:244–53. Jamison RL, Hartigan P, Kaufman JS, DS, Goldfarb SR, Warren SR, Gaziano PD. Veterans Affairs Site Investigators. Effect of homocysteine lowering on mortality and vascular disease in advance chronic kidney disease and end-stage renal disease: A randomized controlled trial. Journal of the American Medical Association. 2007;289:1163–70. Ray JG, Kearon C, Yi Q, Sheridan P, Lonn E. Heart Outcomes Prevention Evaluation 2 (HOPE-2) Investigators. Homocysteine-lowering therapy and risk for venous thromboembolism: A randomized trial. Annals of Internal Medicine. 2007;146:761–7. Spence JD. Homocysteine-lowering therapy: A role in stroke prevention? Lancet Neurology. 2007;6:830–8.

11. Mock DM, Mock NI, Stewart CW, LaBorde JB, Hansen DK. Marginal biotin deficiency is teratogenic in ICR mice. Journal of Nutrition. 2003;133:2519–25.

12. Agoston-Coldea L, Mocan T, Gatfosse M, Lupu S, Dumitrascu DL. Plasma homocysteine and the severity of heart failure in patients with previous myocardial infarction. Cardiology Journal. 2011;18:55–62. Scott JM. Homocysteine and cardiovascular risk. American Journal of Clinical Nutrition. 2000;72:333–4.

13. Mitchell LE, Adzick NS, Melchionne J, Pasquariello PS, Sutton LN, Whitehead AS. Spina bifida. Lancet. 2004;364:1885–95. Stoll C, Dott B, Alembik Y, Roth MP. Associated malformations among infants with neural tube defects. American Journal of Medical Genetics. Part A. 2011; Feb 18. [Epub ahead of print]

14. Parker SE, Mai CT, Canfield MA, Rickard R, Wang Y, Meyer RE, Anderson P, Mason CA, Collins JS, Kirby RS, Correa A; for the National Birth Defects Prevention Network. Updated national birth prevalence estimates for selected birth defects in the United States, 2004–2006. Birth Defects Research. Part A, Clinical and Molecular Teratology. 2010;88:1008–16.

15. Delany C, McDonnell R, Robson M, Corcoran S, Fitzpatrick C, De La Harpe D. Folic acid supplement use in the prevention of neural tube defects. Irish Medical Journal. 2011;104:12–15. Moyers S, Bailey LB. Fetal malformation and folate metabolism: Review of recent evidence. Nutrition Reviews. 2001;7:215–24.

16. Esfahani S, Cogger EA, Caudill MA. Heterogeneity in the prevalence of methylenetetrahydrofolate reductase gene polymorphisms in women of different ethnic groups. Journal of the American Dietetic Association. 2003;103:200–1. O'Byrne MR, Au KS, Morrison AC, Lin JI, Fletcher JM, Ostermaier KK, Tyerman GH, Doebel S, Northrup H. Association of folate receptor (FOLR1, FOLR2, FOLR3) and reduced folate carrier (SLC19A1) genes with meningomyelocele. Birth Defects Research. Part A. Clinical and Molecular Teratology. 2010;88:689–94. Thomas P, Fenech M. Methylenetetrahydrofolate reductase, common polymorphisms, and relation to disease. Vitamins and Hormones. 2008;79:375–92.

17. Polymorphisms in genes related to folate and cobalamin metabolism and the associations with complex birth defects. Prenatal Diagnosis. 2008;28:485–93.

18. Pfeiffer CM, Caudill SP, Gunter EW, Osterloh J, Sampson EJ. Biochemical indicators of B vitamin status in the US population after folic acid fortification: Results from the National Health and Nutrition Examination Survey 1999–2000. American Journal of Clinical Nutrition. 2005;82:442–50. Pfeiffer CM, Johnson CL, Jain RB, Yetley EA, Picciano MF,

Rader JI, Fisher KD, Mulinare J, Osterloh JD. Trends in blood folate and vitamin B-12 concentrations in the United States, 1988–2004. American Journal of Clinical Nutrition. 2007;86:718–27. Yang Q, Cogswell ME, Hamner HC, Carriquiry A, Bailey LB, Pfeiffer CM, Berry RJ. Folic acid source, usual intake, and folate and vitamin B-12 status in US adults: National Health and Nutrition Examination Survey (NHANES) 2003–2006. American Journal of Clinical Nutrition. 2010;91:64–72.

19. Cattan D. Pernicious anemia: What are the actual diagnosis criteria? World Journal of Gastroenterology. 2011;17:543–4. Toh BH, Alderuccio F. Pernicious anaemia. Autoimmunity. 2004;37:357–61.

20. Baik HW, Russell RM. Vitamin B_{12} deficiency in the elderly. Annual Review of Nutrition. 1999;19:357–77.

21. Morris MS, Jacques PF, Rosenberg IH, Selhub J. Circulating unmetabolized folic acid and 5-methyltetrahydrofolate in relation to anemia, macrocytosis, and cognitive test performance in American seniors. American Journal of Clinical Nutrition. 2010;91:1733–44; Yang Q, Cogswell ME, Hamner HC, Carriquiry A, Bailey LB, Pfeiffer CM, Berry RJ. Folic acid source, usual intake, and folate and vitamin B-12 status in US adults: National Health and Nutrition Examination Survey (NHANES) 2003–2006. American Journal of Clinical Nutrition. 2010;91:64–72.

22. Bowen DJ, Beresford SAA. Dietary interventions to prevent disease. Annual Review of Public Health. 2002;23:255–86. Kris-Etherton PM, Hecker KD, Bonanome A, Coval SM, Binkowski AM, Hilpert KF, Griel AE, Etherton TD. Bioactive compounds in foods: Their role in the prevention of cardiovascular disease and cancer. American Journal of Medicine. 2002;113:71S–88S.

23. Douglas RM, Hemilä H, Chalker E, Treacy B. Vitamin C for preventing and treating the common cold. Cochrane Database Systematic Reviews. 2007;Jul 18:CD000980. Hemilä H. Randomised trials on vitamin C. British Journal of Nutrition. 2011;105:485–7. Jacob RA, Aiello GM, Stephensen CB, Blumberg JB, Milbury PE, Wallock LM, Ames BN. Moderate antioxidant supplementation has no effect on biomarkers of oxidant damage in healthy men with low fruit and vegetable intakes. Journal of Nutrition. 2003;133:740–3. Padayatty SJ, Katz A, Wang Y, Eck P, Kwon O, Lee J-H, Chen S, Corpe C, Dutta A, Dutta SK, Levine M. Vitamin C as an antioxidant: evaluation of its role in disease prevention. Journal of the American College of Nutrition. 2003;22:18–35.

24. Bhaskaram P. Micronutrient malnutrition, infection, and immunity: An overview. Nutrition Reviews. 2002;60:S60–45.

25. Stephen R, Utecht T. Scurvy identified in the emergency department: a case report. J Emergency Medicine. 2001;21:235–37. Weinstein M, Babyn P, Zlotkin S. An orange a day keeps the doctor away: scurvy in the year 2000. Pediatrics. 2001;108:E55.

26. Cole LK, Dolinsky VW, Dyck JR, Vance DE. Impaired Phosphatidylcholine Biosynthesis Reduces Atherosclerosis and Prevents Lipotoxic Cardiac Dysfunction in ApoE-/- Mice. Circulation Research. 2011;108:686–94. Zeisel SH, daCosta KD, Franklin PD, Alexander EA, Lamont JT, Sheard NF, Beiser A. Choline, an essential nutrient for humans. FASEB Journal. 1991;5:2093–98. Buchman AL, Dubin M, Jenden D, Moukarzel A, Roch MH, Rice K, Gornbein J, Ament ME, Eckhert CD. Lecithin increases plasma free choline and decreases hepatic steatosis in long-term total parenteral nutrition patients. Gastroenterology. 1992;102:1363–70.

27. Fischer LM, daCosta KA, Kwock L, Stewart PW, Lu TS, Stabler SP, Allen RH, Zeisel SH. Sex and menopausal status influence human dietary requirements for the nutrient choline. American Journal of Clinical Nutrition. 2007;85:1275–85.

28. Institute of Medicine. Dietary Reference Intakes for thiamin, riboflavin, niacin, vitamin B_6, folate, vitamin B_{12}, pantothenic acid, biotin, and choline. Washington, DC: National Academy Press; 1998.

29. Ferreira I. Quantification of non-protein nitrogen components of infant formulae and follow-up milks: Comparison with cows' and human milk. British Journal of Nutrition. 2003;90:127–33.

30. Lombard KA, Olson AL, Nelson SE, Rebouche CJ. Carnitine status of lactoovovegetarians and strict vegetarian adults and children. American Journal of Clinical Nutrition. 1989;50:301–6.

31. U.S. Food and Drug Administration and Center for Food Safety and Applied Nutrition. Dietary supplement health and education act of 1994. Available from: http://www.cfsan.fda.gov/~dms/ dietsupp.html.

Fat-Soluble Vitamins

What would you think of an advertisement promoting a product that could boost your immune system, fight cancer and heart disease, improve your vision, and keep your bones strong? This may sound too good to be true, but all these health benefits can be attributed to the fat-soluble vitamins. In contrast to the water-soluble vitamins you learned about in Chapter 10, fat-soluble vitamins are substances that *do not* dissolve in water. But like the water-soluble vitamins, the fat-soluble vitamins are found in a wide variety of foods and are vital for health. Because the body needs fat-soluble vitamins in very small quantities, they—like the water-soluble vitamins—are considered micronutrients.

There are four fat-soluble vitamins: vitamins A, D, E, and K. Whereas water-soluble vitamins function primarily as coenzymes in energy metabolism, the fat-soluble vitamins are needed for cell growth, reproduction, and gene regulation. In this chapter, you will learn about the chemical structures, food sources, and functions of the fat-soluble vitamins in the body. Information concerning deficiencies, toxicities, and dietary recommendations will also be discussed.

Living Successfully with Factor V Leiden Thrombophilia

Muayyad, a native of Jordan, had always been interested in eating healthy. However, it took a life-altering event for him to rethink his original career path in business administration and instead pursue a new profession in health and nutrition.

While on a hike, Muayyad lost his footing and fell. Although he was able to finish the hike, he knew that he had injured his leg. The bone did not feel broken, but the injury was painful enough that he thought it best to see his doctor. Muayyad's doctor prescribed anti-inflammatory medications and told him to rest his leg. Not thinking much more about what seemed a minor injury, Muayyad followed through with his plans to visit his family in Jordan—an 18-hour flight. By the time Muayyad's plane landed, his left leg was severely swollen and painful. Again, doctors suggested he stay off his leg and let it heal. When he returned to the United States, however, the situation was even worse. This time, his doctor determined that there were blood clots not only in his leg but also in his lungs—a very unusual condition for an otherwise healthy, young person. Clearly, something was seriously wrong.

After Muayyad underwent numerous blood tests and spent a week in the hospital, doctors determined that he had a rare inherited disorder called Factor V Leiden thrombophilia. Factor V is a protein that helps your blood clot after you have been injured. When blood clots form, your body releases a molecule called "activated protein C" (APC) that prevents the blood clots from growing too large. APC works by inactivating Factor V, resulting in an effective negative feedback loop. In people with the Factor V "Leiden mutation," APC is unable to inactivate Factor V, increasing the development of large blood clots. Genetic tests showed that Muayyad was homozygous for this mutation (meaning that both genes that code for APC were affected), which put him at especially high risk for blood clot formation. After finding out that Muayyad had this condition, his three brothers also underwent genetic testing. The results indicated that his younger brother was also homozygous; another brother was heterozygous (meaning that only one gene was affected); and his oldest brother did not have the condition at all.

Muayyad was prescribed blood-thinning medications to prevent further blood clots and was instructed not to take drugs like aspirin that deter blood clot formation. In addition, because vitamin K is critical in the blood-clotting process, Muayyad was counseled to avoid large amounts of vitamin K–rich foods. This is because vitamin K supports blood clot formation. Since his diagnosis, Muayyad has become vigilant about eating only moderate amounts of foods like broccoli, cucumber, kale, and spinach, which are high in vitamin K. Perhaps most important, he has become keenly aware of how his diet is critical to his everyday health. Muayyad now hopes to combine his previous business-related education with his studies in nutrition to someday have a dietetic practice developing products for people with nutrition-related medical conditions such as his.

© Shelley McGuire

Critical Thinking: Muayyad's Story

Having Factor V Leiden thrombophilia can be somewhat scary. However, learning more about why some foods—such as those containing high levels of vitamin K—must be eaten only in moderation can help an affected person live comfortably and safely. What do you think it would be like to have to take a blood-thinning medication and be careful about your vitamin K intake? What information would you need to be able to do this effectively?

What Makes the Fat-Soluble Vitamins Unique?

Although each of the fat-soluble vitamins is chemically unique, this group of nutrients shares certain general characteristics in terms of how they are absorbed and circulated (Table 11.1). For example, like the water-soluble vitamins, fat-soluble vitamins are absorbed mostly in the small intestine. Unlike their water-soluble counterparts, however, absorption requires the presence of dietary lipids as well as the action of bile. In other words, if you were to eat a very low-fat diet, you might have difficulty absorbing the fat-soluble vitamins. Unlike water-soluble vitamins, which are circulated in the blood, fat-soluble vitamins are circulated away from the small intestine in the lymph via chylomicrons before eventually entering the blood either as components of lipoproteins or bound to transport proteins.

Like water-soluble vitamins, each fat-soluble vitamin has several forms, some of which are more biologically active than others. Because the body can store most of the fat-soluble vitamins, consuming large amounts of them (especially in supplement form) can result in toxicities, sometimes with serious consequences. Once absorbed and circulated to cells in the body, the fat-soluble vitamins are involved in many processes such as vision, blood coagulation, regulation of gene expression, cell maturation, and stabilization of free radicals. Before delving into each of the fat-soluble vitamins in detail, you may find it helpful to review the information in Table 11.2, which provides an overview of the functions and sources of these compounds.

⟨**CONNECTIONS**⟩ Recall that bile is a fluid produced in the liver and stored in the gallbladder and aids in lipid digestion (Chapter 3, page 82).

⟨**CONNECTIONS**⟩ Remember that chylomicrons are lipoproteins made in the intestinal epithelial cell (enterocyte) (Chapter 6, page 242).

EACH FAT-SOLUBLE VITAMIN HAS SEVERAL NAMES

You may recall that researchers initially thought there were only two vitamins, and they were first referred to as *vitamin A* (fat-soluble) and *vitamin B* (water-soluble). However, as with the water-soluble vitamins (which were eventually distinguished as vitamin C, niacin, riboflavin, etc.), scientists eventually discovered that there were several fat-soluble vitamins. To distinguish each of the fat-soluble vitamins from the others, scientists assigned each one a different letter (vitamin A, vitamin D, and so forth).

Because each fat-soluble vitamin has several forms, numbers are sometimes used to distinguish them from each other (such as vitamin K_1, vitamin K_2). Most of these compounds, like the water-soluble vitamins, also have names that

TABLE 11.1 General Characteristics of the Fat-Soluble Vitamins

	General Characteristics
Food sources	• Typically found in fatty portions of foods. • Some are easily destroyed by heat and light.
Digestion	• Very little needed.
Absorption	• Occurs mostly in small intestine. • Requires incorporation into micelles and the actions of bile. • Once transported into the intestinal cell, vitamins are packaged with other lipids into chylomicrons.
Circulation away from gastrointestinal tract	• Initially circulated in lymph, then in the blood.
Functions	• Gene regulation and various other functions such as being antioxidants and participating in the blood-clotting cascade.
Toxicity	• Except for vitamin K, toxicities are dangerous and sometimes fatal.

TABLE 11.2 Functions, Sources, Deficiency and Toxicity Diseases, and Food Sources of the Essential Fat-Soluble Vitamins

Vitamin	Major Functions	Deficiency	Toxicity	Selected Food Sources
Vitamin A	• Growth • Reproduction • Vision • Cell differentiation • Immune function • Bone health	• Vitamin A deficiency disorder (VADD) • Night blindness • Xerophthalmia • Hyperkeratosis	• Hypervitaminos A • Blurred vision • Birth defects • Liver damage • Osteoporosis	• Liver • Pumpkin • Sweet potato • Carrot
Vitamin D	• Calcium homeostasis • Bone health • Cell differentiation	• Rickets • Osteomalacia • Osteoporosis	• Hypercalcemia • Calcification of soft tissues	• Fish • Shiitake mushrooms • Fortified milk • Fortified cereals
Vitamin E	• Antioxidant properties • Cell membrane stability • Eye health • Heart health	• Neuromuscular problems • Hemolytic anemia	• Hemorrhage	• Tomatoes • Nuts, seeds • Spinach • Fortified cereals
Vitamin K	• Coenzyme • Blood clotting • Bone health • Tooth health	• Vitamin K deficiency bleeding (VKDB)	• No known effects	• Kale • Spinach • Broccoli • Brussels sprouts

reflect their chemical composition or function. For instance, the active form of vitamin D is called calcitriol, partly because of its role in calcium homeostasis.

Vitamin A and the Carotenoids—Needed for Eyesight and Much More

Vitamin A was first identified in 1916 as an important substance for the health of experimental animals.* The term *vitamin A* is now used to refer to a series of compounds called **retinoids,** which include retinol, retinoic acid, and retinal. Retinoids, shown in Figure 11.1, are also referred to as **preformed vitamin A.** Although all three forms are physiologically important, retinol is the most biologically active form of vitamin A, and the body can synthesize it from retinal. Retinal can also be converted to retinoic acid, but retinoic acid cannot be converted to any other retinoid.

In addition to preformed vitamin A, the vitamin A "family" also contains compounds called **carotenoids,** which have structures somewhat similar to the retinoids. The carotenoids include two types—those that can be converted to vitamin A and those that cannot. Carotenoids that can be converted to vitamin A are called **provitamin A carotenoids. Beta-carotene (β-carotene)** is one of the most common provitamin A carotenoids in foods, and the body can split it to form two molecules of retinal. Carotenoids that cannot be converted to vitamin A are called **nonprovitamin A carotenoids.** These include lycopene, astaxanthin, zeaxanthin, and lutein. Note that the nonprovitamin A carotenoids are not technically vitamins but instead are considered phytochemicals or zoonutrients depending on whether they come from plant- or animal-based foods, respectively.

retinoid (preformed vitamin A) (RE – ti – noid) A term used to describe all forms of preformed vitamin A.

carotenoids (car – O – te – noids) Dietary compounds with structures similar to that of vitamin A; some, but not all, can be converted to vitamin A.

provitamin A carotenoid A type of carotenoid that can be converted to vitamin A.

β-carotene A provitamin A carotenoid commonly found in yellow and orange foods.

nonprovitamin A carotenoid A type of carotenoid that cannot be converted to vitamin A.

* The discovery of vitamin A is attributed to Elmer V. McCullum and Marguerite Davis at the University of Wisconsin and Thomas Osborne and Lafayette Mendel at Yale University.

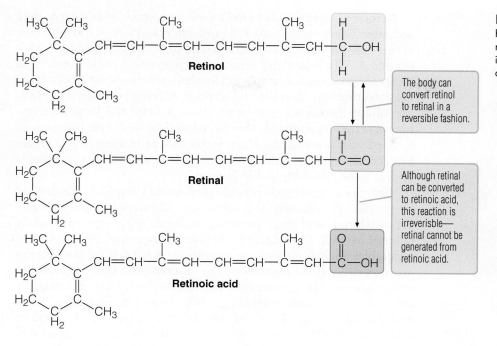

FIGURE 11.1 Vitamin A Vitamin A has three forms: retinol, retinal, and retinoic acid. Retinol and retinal can be interconverted; retinoic acid cannot be converted to either retinol or retinal.

The body can convert retinol to retinal in a reversible fashion.

Although retinal can be converted to retinoic acid, this reaction is irreverisble—retinal cannot be generated from retinoic acid.

VITAMIN A AND PROVITAMIN A CAROTENOIDS ARE FOUND IN DIVERSE PLANT AND ANIMAL FOODS

There are several forms of vitamin A, each with its own biological potency. Therefore, scientists use a unit of measure called the **retinol activity equivalent (RAE)** to describe the overall vitamin A content of foods. The RAE also makes it possible to compare the nutritional content of foods containing preformed vitamin A to foods containing provitamin A carotenoids, such as β-carotene. Approximately 12 µg of β-carotene or 1 µg of retinol equals 1 µg RAE. Foods containing vitamin A or provitamin A carotenoids are listed in Figure 11.2.

⟨CONNECTIONS⟩ Recall that tryptophan can be converted to niacin in the body (Chapter 10, page 429).

retinol activity equivalent (RAE) A unit of measure used to describe the combined amount of preformed vitamin A and provitamin A carotenoids in foods.

FIGURE 11.2 Food Sources of Preformed Vitamin A or Provitamin A Carotenoids

Food	Serving size	Vitamin A
Beef liver (pan fried)	3 oz	
Pumpkin (canned)	1 cup	
Sweet potato (boiled)	8 oz	
Carrots (boiled)	1 cup	
Spinach (boiled)	1 cup	
Kale (boiled)	3 oz	
Chicken liver	1 cup	
Squash (baked)	1 cup	
Red peppers (boiled)	1 cup	
Fortified breakfast cereal	1 cup	
Herring	3 oz	
Milk, fortified	1 cup	

RDA
Women: 700 µg RAE/day
Men: 900 µg RAE/day

Retinol activity equivalents (µg RAE)
0   1,000   2,000   3,000   4,000   5,000   6,000   7,000

SOURCE: USDA Nutrient Database for Standard Reference, Release 16-1.

barbaradudzinska/Shutterstock.com

Peppers are excellent sources of provitamin A carotenoids.

Preformed vitamin A is found mostly in animal foods such as liver and other organ meats, fatty fish, and dairy products. A single serving of beef liver provides more than seven times your daily requirement for vitamin A. Whole-fat dairy products such as whole milk, cheese, and butter are good sources of vitamin A, whereas reduced-fat products are not good sources unless they are fortified. Although vitamin A fortification of any food is not mandatory, the U.S. Food and Drug Administration (FDA) recommends that all reduced-fat fluid milk products be fortified with this vitamin. Many foods such as margarine and breakfast cereals are also fortified with vitamin A. Vitamin A compounds appear to be relatively stable in food.

Whereas animal foods provide preformed vitamin A, plants tend to provide provitamin A carotenoids. Carotenoids are typically yellowish-red, and their presence in foods makes them brightly colored as well. Consequently, yellow, orange, and red fruits and vegetables such as cantaloupe, carrots, and red *and* yellow peppers are especially good sources of the carotenoids. Leafy greens are also good sources. Carotenoids can also be found in some brightly colored animal foods, such as egg yolks. In addition lobsters, crabs, and shrimp have pink- or red-colored carotenoids in their bodies and shells. Carotenoids are also frequently added to foods as coloring agents. An example is cheese, which is sometimes orange because a colorant called annatto is added. Annatto is produced from the carotenoid-rich bright red berries of the achiote plant. It is also used to color foods in Mexican, Puerto Rican, and Indian cooking. Interestingly, processing and heating may actually increase the bioavailability of some carotenoids (such as β-carotene).

ABSORPTION OF VITAMIN A AND THE CAROTENOIDS REQUIRES ADEQUATE LIPIDS

Vitamin A and the carotenoids are absorbed in the small intestine, and this requires the presence of dietary lipid and bile. Because of this, a low-fat diet can decrease vitamin A and carotenoid absorption (and thus their bioavailability). Upon entering the enterocyte, vitamin A and the carotenoids are incorporated into chylomicrons, which then enter the lymph and subsequently the blood.

Once in the blood, circulating vitamin A and the carotenoids (as part of the chylomicrons) can be taken up by many cell types including those in adipose, muscle, and eye tissue. What is not taken up in this initial transport is delivered to the liver as part of the chylomicron remnant. In the liver, the carotenoids and vitamin A undergo additional metabolism and/or packaging. For example, the carotenoids can be packaged into other lipoproteins (such as VLDL) for recirculation or converted to retinol (in the case of provitamin A carotenoids). In the liver, vitamin A is attached to the transport proteins **retinol-binding protein** and **transthyretin,** and then released into the blood. Vitamin A and carotenoids not taken up by tissues at this stage are stored mainly in the liver and adipose tissue.

© JPTenor / Alamy

Adalberto Rios Lanz/Photo Library

Achiote paste (sometimes called annatto; shown on the right), which gets its vibrant color from carotenoids in the achiote plant (shown on the left), is often used in Puerto Rican, Mexican, and Indian cuisines to add color to food. It is also used to make some cheeses orange.

retinol-binding protein and **transthyretin**
Proteins that carry retinoids in the blood.

VITAMIN A IS CRITICAL FOR VISION, GROWTH, AND REPRODUCTION

Vitamin A has many functions and plays important roles in vision, growth, and reproduction. In addition, it is needed for maintaining a healthy immune system and building strong bones. The role of vitamin A in these important physiological functions is discussed next.

Vitamin A and Vision How many times have you heard "*eat your carrots because they're good for your eyes*"? Historical evidence suggests that, even thousands of years ago, Egyptian physicians prescribed vitamin A–rich liver to treat poor vision. However, not until the 20th century did scientists begin to understand the mechanisms by which consuming vitamin A–rich foods improved

FIGURE 11.3 Vitamin A and Vision Vitamin A (retinal) is essential for vision, especially in darkness, when the function of rod cells is most important.

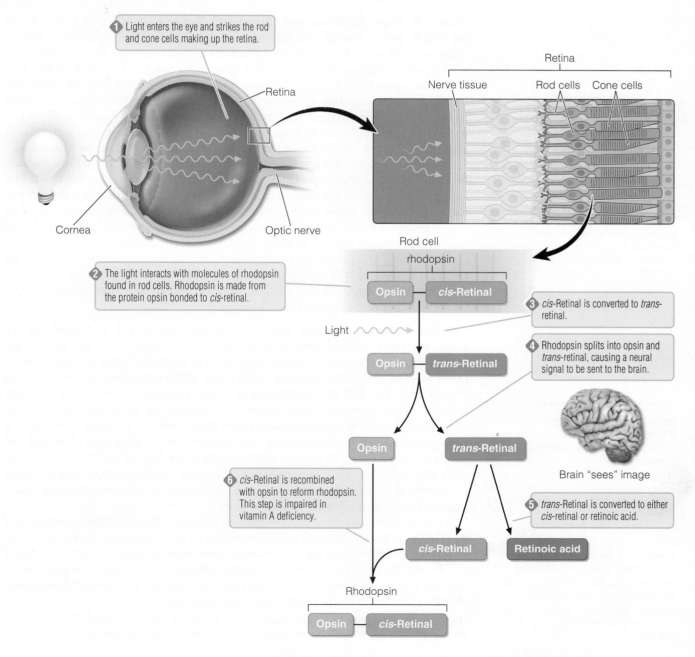

1. Light enters the eye and strikes the rod and cone cells making up the retina.

Retina

Cornea

Optic nerve

Retina

Nerve tissue   Rod cells   Cone cells

Rod cell

rhodopsin

2. The light interacts with molecules of rhodopsin found in rod cells. Rhodopsin is made from the protein opsin bonded to *cis*-retinal.

Opsin   *cis*-Retinal

3. *cis*-Retinal is converted to *trans*-retinal.

Light

Opsin   *trans*-Retinal

4. Rhodopsin splits into opsin and *trans*-retinal, causing a neural signal to be sent to the brain.

Opsin   *trans*-Retinal

Brain "sees" image

6. *cis*-Retinal is recombined with opsin to reform rhodopsin. This step is impaired in vitamin A deficiency.

5. *trans*-Retinal is converted to either *cis*-retinal or retinoic acid.

cis-Retinal   Retinoic acid

Rhodopsin

Opsin   *cis*-Retinal

eyesight.[†] As illustrated in Figure 11.3, when light enters your eye, the light immediately encounters the inner back lining, called the **retina.** The retina consists of a layer of nerve tissue as well as millions of cells called **cones** and **rods.** The cones help you see color, whereas the rods are needed to see black and white, making them particularly important for night vision. Both cones and rods require vitamin A to function effectively, although scientists have a better understanding of the role of vitamin A in the rod cells.

To function, the rods must contain thousands of molecules of a substance called **rhodopsin.** Specifically, rhodopsin is made of *cis*-retinal (a form of vitamin A) bonded to the protein **opsin.** When light strikes rhodopsin, *cis*-retinal is converted to *trans*-retinal, subsequently separating from the opsin. This metabolic conversion, in turn, causes a neural signal to be sent to the brain. In this way, the brain "sees" light and then transforms it into an image that you recognize. Thus, vitamin A plays a critical role in allowing you to process light into sight. Vitamin A is also important for maintaining the health of the outermost tissue layer of the eye, called the **cornea.**

Vitamin A, Night Blindness, and "Bleaching" Not only is vitamin A important for vision, but it also allows the eyes to adapt to low-light situations. This is because *trans*-retinal must be reconverted to *cis*-retinal and then recombined with opsin to reform rhodopsin each time light reaches the back of the eye. However, *trans*-retinal is not recycled to *cis*-retinal with 100% efficiency. Instead, some *trans*-retinal is converted to retinoic acid, which cannot be used to form rhodopsin. Extra retinal must therefore always be available for vision to remain optimal. Without enough retinal to reform rhodopsin, night vision becomes especially difficult—resulting in a condition called **night blindness.**

Adequate vitamin A is also needed to prevent prolonged "bleaching," which is a phenomenon that sometimes occurs after exposure to intense light. The rapid dissociation of rhodopsin in response to a flash of bright light temporarily disrupts the visual cycle. In fact, many people see spots in their field of vision until their eyes are able to reform rhodopsin. Although a brief delay in vision is normal, a lack of vitamin A can impair the ability to reform rhodopsin, extending the period of time needed for the eyes to adjust. This is why people with vitamin A deficiency have an especially difficult time driving in the dark. When exposed to the bright flash of light from an oncoming vehicle, the person becomes temporarily "blinded."

Cell Differentiation, Growth, and Reproduction Vitamin A is important for some nonvision-related processes as well. For example, retinoic acid helps some immature, undifferentiated cells become epithelial cells instead of some other type of cell.[1] This process, called **cell differentiation,** is illustrated in Figure 11.4. For cell differentiation to occur, vitamin A is taken up into the nucleus of the immature cell, where the vitamin regulates gene expression. Specifically, vitamin A upregulates genes that code for proteins that cause a newly formed cell to differentiate and mature into a functional *epithelial* cell (such as an enterocyte), as opposed to some other type of cell (such as a liver cell). Because older epithelial cells are constantly sloughed off, a steady supply of vitamin A is needed to make new epithelial cells—a process critical to maintaining the gastrointestinal tract and skin.

Vitamin A is also important for growth and reproduction. The mechanisms by which vitamin A influences these processes are numerous, including regulation of cell differentiation. For example, during embryonic growth, vitamin A directs the differentiation and maturation of various cell types that give rise to specific tissues and organs. For these reasons, vitamin A is needed throughout the life cycle for growth, repair, and reproduction.

[†]In 1967, Drs. George Wald, Haldan Hartline, and Ragnar Granit were awarded the Nobel Prize in Physiology or Medicine for their work on vitamin A and night blindness.

⟨**CONNECTIONS**⟩ Remember that epithelial cells make up our skin and form the lining of our organs and blood vessels (Chapter 3, page 79).

⟨**CONNECTIONS**⟩ Recall that *cis* refers to a carbon–carbon double bond with hydrogens positioned on the same side, and *trans* refers to a carbon–carbon double bond with the hydrogens on opposite sides of the double bond (Chapter 6, page 222).

retina (RE – ti – na) The inner lining of the back of the eye.

cones and **rods** Cells, in the retina, that are needed for vision.

rhodopsin (rhod – OPS – in) A compound in the retina that consists of the protein opsin and the vitamin A derivative *cis*-retinal; needed for vision.

opsin (OPS – in) The protein component of rhodopsin.

cornea (COR – ne – a) The outermost layer of tissue covering the front of the eye.

night blindness A condition characterized by impaired ability to see in the dark.

cell differentiation The process by which an immature cell becomes a specific type of mature cell.

FIGURE 11.4 Vitamin A Is Important for Cell Differentiation and Maturation Vitamin A is involved in cell signaling, stimulating synthesis of proteins needed for differentiation and maturation of epithelial cells, such as the enterocyte shown here.

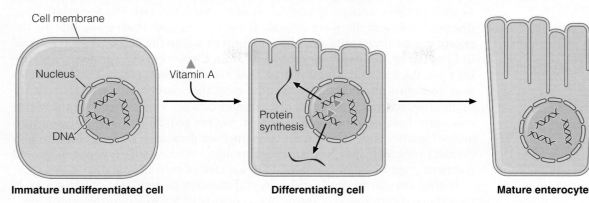

Immature undifferentiated cell　　　**Differentiating cell**　　　**Mature enterocyte**

Optimal Vitamin A Status and Immune Function Vitamin A also has a variety of immunological functions, including the maintenance of protective barriers such as the skin and lining of the intestinal and respiratory tracts, most of which are epithelial tissue.[2] When vitamin A is not available, these important tissues break down, allowing bacteria and viruses to enter the body and cause infection. The body also requires vitamin A to produce functional immune cells, called lymphocytes, as well as antibodies. Without lymphocytes and antibodies, the body cannot mount an effective immune response when exposed to pathogens. Because vitamin A is important for the maintenance of the intestinal tract lining, production of immune cells, and synthesis of antibodies, children with vitamin A deficiency commonly experience health problems such as chronic diarrhea and respiratory disease.[3]

Vitamin A, the Carotenoids, and Cancer Because the immune system works to rid the body of abnormal cells that can lead to uncontrolled cell growth, scientists are interested in whether optimal vitamin A and carotenoid status may help reduce the risk for certain cancers. Epidemiologic studies have shown that consuming a diet rich in vitamin A or provitamin A carotenoids is associated with lower risk of some types of cancer.[4] However, long-term, high-dose supplementation with β-carotene by itself or in combination with other dietary factors has been shown to increase the incidence of lung and colorectal cancers in some people—especially in those who smoke.[5] The Institute of Medicine, therefore, cautions people *against* taking vitamin A or β-carotene supplements to reduce the risk for cancer.[6]

Vitamin A Needed for Strong Bones—But More Is Not Always Better Vitamin A is also required for healthy bones. Bone tissue is continually being broken down by cells called osteoclasts and reformed by cells called osteoblasts. Adequate levels of vitamin A appear necessary for a healthy balance of these processes. However, very high vitamin A intake may lead to greater risk for bone fractures in older people.[7] This effect seems to be true for preformed vitamin A but not for the provitamin A carotenoids. This may be because high intake of vitamin A interferes with calcium absorption.[8] Clearly, more research is needed to understand how vitamin A functions in the body to influence bone health—and, perhaps, bone disease.

NONPROVITAMIN A CAROTENOIDS ARE POTENT ANTIOXIDANTS

Scientists once thought the health-promoting effects of the carotenoids were due solely to their conversion to vitamin A, although this does not appear to

〈**CONNECTIONS**〉 Antibodies are proteins produced by the immune cells to help fight infection (Chapter 5, page 132).

〈**CONNECTIONS**〉 Protein synthesis is initiated when specific genes are "turned on." This step is followed by transcription and then translation (Chapter 5, page 167).

〈CONNECTIONS〉 Remember that free radicals are reactive molecules with one or more unpaired electrons (Chapter 10, page 447).

be the case. Whereas several nonprovitamin A carotenoids (such as lycopene) are themselves associated with decreased risk for cancer—at least in animal models—others appear beneficial in decreasing the risk for age-related eye disease.[9,10] Researchers now think these effects and others are due to the potent antioxidant functions of some carotenoids in the body.[11] As you learned in Chapter 10, antioxidants such as vitamin C protect DNA, proteins, and lipids from the harmful effects of free radicals. Consuming fruits and vegetables that contain antioxidant carotenoids may therefore decrease your risk for many diseases, such as heart disease and cancer. This is because these diseases are thought to be caused, in part, by free radical damage. For example, animal studies show that lutein consumption decreases growth of mammary (breast) tumors,[12] and human epidemiologic studies suggest that increased lycopene consumption is related to lower risk of prostate cancer.[13]

Higher circulating levels of lutein and zeaxanthin may also decrease risk of age-related deterioration of the retina—a disease called **macular degeneration.**[14] Macular degeneration affects the **macula,** a highly sensitive portion of the retina needed for vision. Studies suggest that carotenoids exert their effects on the macula by positively influencing the immune system.[15] There are likely additional antioxidant effects involved as well.

Although research shows that consuming high amounts of carotenoid-containing fruits and vegetables is associated with reduced risk of developing certain chronic diseases, this does not seem to be the case for carotenoid supplements. In fact, as previously mentioned, dietary intervention trials often suggest a *negative* effect of carotenoid supplements on health.[16] Thus, scientists believe that the health-promoting effects of fruits and vegetables may be due to carotenoids working in synergy with each other or with other nutrients found in foods.[17]

VITAMIN A DEFICIENCY CAUSES VITAMIN A DEFICIENCY DISORDER (VADD)

Primary vitamin A deficiency is uncommon in industrialized countries such as the United States. Nonetheless, even in these countries, secondary vitamin A deficiency can occur in people with diseases affecting the pancreas, liver, or GI tract. An example is cystic fibrosis, which can impair both lipid digestion and absorption. Vitamin A deficiency is also prevalent in alcoholics who have poor diets and suffer from liver damage. Excessive alcohol consumption also depletes the body's stores of vitamin A by mechanisms that are not well understood.

Primary vitamin A deficiency tends to be more pervasive in developing societies, and because it has many health consequences, it is sometimes referred to as **vitamin A deficiency disorder (VADD).** VADD has important implications for worldwide health, especially among children. In its milder form, VADD causes night blindness. More severe VADD damages the cornea and other portions of the eye, leading to dry eyes, scarring, and even blindness. This complex disease, called **xerophthalmia,** is often accompanied by the presence of **Bitot's spots,** which are caused by accumulations of dead cells and secretions on the surface of the eye (Figure 11.5). Vitamin A deficiency can also cause a condition called **hyperkeratosis,** in which skin and nail cells produce too much of the protein keratin, a process that causes them to become bumpy, rough, and irritated. This is because immature skin and nail cells, which are epithelial cells, produce more keratin than do differentiated mature cells.

Because of the importance of vitamin A in supporting a healthy immune system, people (especially children) with VADD have increased risk for infection. Furthermore, protein energy malnutrition (PEM), which is also endemic

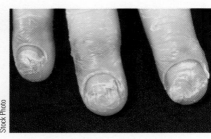

Vitamin A deficiency can cause hyperkeratosis, a condition in which skin and nails become rough and scaly. Hyperkeratosis occurs in response to the inability of epithelial cells constituting the skin and nails to mature.

macular degeneration (MA – cu – lar) A chronic disease that results from deterioration of the retina.

macula A portion of the retina important for sight.

vitamin A deficiency disorder (VADD) A multifaceted disease resulting from vitamin A deficiency.

xerophthalmia (xer – o – PHTHAL – mi – a) A condition caused by vitamin A deficiency and characterized by serious damage to the cornea. Xeropthalmia can lead to blindness.

Bitot's spots A result of vitamin A deficiency characterized by white spots on the eye; caused by buildup of dead cells and secretions.

hyperkeratosis (hy – per – ker – a – TO – sis) A complication of vitamin A deficiency in which immature skin and nail cells overproduce the protein keratin, causing them to be rough and scaly.

FIGURE 11.5 Vitamin A Deficiency Vitamin A deficiency can cause xerophthalmia, a major cause of blindness in many regions of the world.

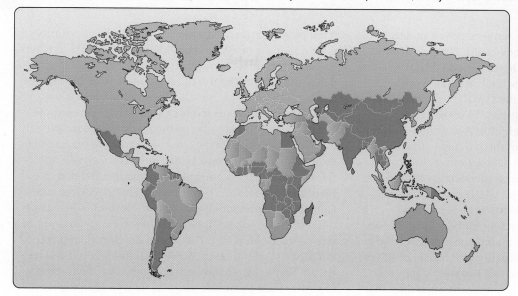

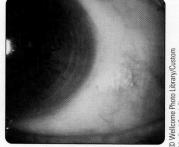

Xerophthalmia

© Wellcome Photo Library/Custom Medical Stock Photo

Areas of the world where vitamin A deficiency is common, especially in children.

SOURCE: Adapted from World Health Organization. Global prevalence of vitamin A deficiency in populations at risk 1995–2005: WHO global database on vitamin A deficiency. World Health Organization. Geneva, World Health Organization. 2009.

in vitamin A–deficient parts of the world, can make vitamin A deficiency even worse. This is partly because the body cannot produce the proteins required to transport vitamin A in the blood (such as retinol-binding protein). International programs aimed at vitamin A supplementation and food fortification have helped decrease the prevalence of VADD in certain parts of the world, as you will see in the Focus on Diet and Health feature.

VITAMIN A TOXICITY CAUSES HYPERVITAMINOSIS A

Chronic consumption of large doses of preformed vitamin A (three to four times the RDA) can lead to vitamin A toxicity, called **hypervitaminosis A.** Hypervitaminosis A can cause serious complications including blurred vision, liver abnormalities, and reduced bone strength. In addition, consuming very high doses of either naturally occurring or synthetic vitamin A during pregnancy can lead to birth defects such as neurological damage and physical deformities, and high-dose carotenoid supplementation increases risk for lung cancer in smokers.

Consuming high amounts of carotenoids from foods and supplements can cause the skin to become yellow-orange—a condition called **carotenodermia.** This is mainly because these brightly colored compounds are deposited in the skin. Although carotenodermia can cause the skin to have an orange-yellow appearance, it is not considered dangerous and can be reversed with a low-carotenoid diet.

RECOMMENDED INTAKES FOR VITAMIN A AND THE CAROTENOIDS

The Recommended Dietary Allowances (RDAs) for vitamin A intake in adult men and women are 900 and 700 µg RAE/day, respectively. This amount can be consumed as either preformed vitamin A or provitamin A carotenoids. Because preformed vitamin A is found mostly in animal products, vitamin A supplementation may be necessary for vegans not consuming sufficient amounts of provitamin A carotenoids or vitamin A–fortified foods. No DRIs have been established for any of the nonprovitamin A carotenoids, because there is insufficient evidence to support their essentiality.

⟨CONNECTIONS⟩ Dietary Reference Intakes (DRIs) are reference values for nutrient intake. When adequate information is available, Recommended Dietary Allowances (RDAs) are established, but when less information is available, Adequate Intake (AI) values are provided. Tolerable Upper Intake Levels (ULs) indicate intake levels that should not be exceeded (Chapter 2, page 33).

⟨CONNECTIONS⟩ A conditionally essential nutrient is one that is nonessential in most situations but becomes essential in others (Chapter 1, page 5).

hypervitaminosis A (hy – per – vit – a – min – O – sis) A condition in which elevated circulating vitamin A levels result in blurred vision, liver damage, and reduced bone strength.

carotenodermia A condition in which carotenoids accumulate in the skin, causing it to become yellow-orange.

Although vitamin A deficiency is not common in developed countries, it poses a serious health concern around the world. The World Health Organization (WHO) estimates that 250 million preschool children worldwide have vitamin A deficiency disorder (VADD)—the majority of these are in developing countries.[18] Furthermore, an estimated 250,000 to 500,000 vitamin A–deficient children become blind every year, half of them dying within 12 months of losing their sight. Of particular concern is the effect of VADD on morbidity and mortality rates from severe illnesses, such as malaria. VADD is common in regions of the world where malaria is endemic, and vitamin A supplementation decreases both its incidence and severity.[19] Although researchers do not know precisely how vitamin A affects the risk and severity of malaria, its effect on the immune system likely plays a key role. Vitamin A is important for many aspects of immunity, including the maintenance of epithelial cells such as those lining the lungs and GI tract as well as the production of antibodies.

Current strategies to combat VADD worldwide include supplementation, food fortification, and education regarding the importance of vitamin A–rich foods and health. In addition, the WHO recommends vitamin A supplementation at the earliest possible opportunity after 6 months of age in many regions of the world where VADD is common. Another strategy to reduce the incidence of VADD is the use of biotechnology or bioengineering to produce crops (genetically modified organisms, or GMOs) with higher levels of vitamin A. For example, VADD is common in areas of the world where people rely on rice, which is a poor source of vitamin A, for most of their calories. In the late 1990s, however, German and Swiss researchers produced a genetically modified rice plant that synthesizes β-carotene.[20] This rice, later named Golden Rice, provides β-carotene to the diet, whereas ordinary processed rice does not. More recently, a strain of rice called Golden Rice 2 has been developed, and it contains 20 times more β-carotene than the original Golden Rice.[21] Some people have ethical concerns about the use of GMOs such as Golden Rice. In fact, Golden Rice and other such GMOs have been banned in the European Union. Nonetheless, many scientists and public health experts hope solutions can be found to these issues and concerns so that these nutrient-dense foods can be used to prevent and treat a multitude of nutritional deficiencies worldwide—such as VADD.[22]

International efforts to decrease vitamin A deficiency have improved the health of children worldwide.

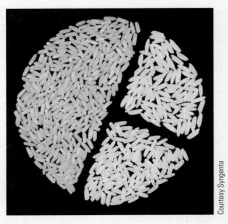

Golden Rice (upper right) and Golden Rice 2 (left) contain β-carotene, whereas plain milled rice (lower right) does not. Golden Rice and Golden Rice 2 are examples of genetically modified organisms (GMOs).

To prevent the known toxic effects of excessive consumption of preformed vitamin A, a Tolerable Upper Intake Level (UL) has been set at 3,000 µg RAE/day for adults. As previously noted, the Institute of Medicine advises *against* supplementation of the diet with carotenoids for most people, although there are no UL values for these compounds. A complete list of the Dietary Reference Intake (DRI) values for all vitamins is provided inside the front cover of this book.

Vitamin D—The "Sunshine Vitamin"

Vitamin D is an interesting and unique vitamin because it is considered by many to be both a nutrient and a *prohormone*, a compound that the body can convert to an active hormone. This is because, although vitamin D is found in food (and is therefore a nutrient), significant amounts can also be produced by the body, which uses the compound to produce a hormone. Recall from Chapter 3 that a hormone is a molecule produced by one cell type that is released into the blood, where it communicates with other types of cells to stimulate an appropriate physiological response. In the case of vitamin D, the prohormone form is produced in the skin and activated in the liver and kidneys; then the active hormone communicates with a variety of other tissues such as those of the small intestine. Because many people produce sufficient amounts of vitamin D in their bodies, some scientists consider vitamin D to be a nonessential nutrient. Other people are not able to produce adequate amounts of vitamin D; therefore, under some circumstances, vitamin D must be obtained in sufficient amounts from the diet. There are two forms of vitamin D in foods. **Ergocalciferol (vitamin D$_2$)** is found in plant sources, whereas **cholecalciferol (vitamin D$_3$)** is found in animal foods. Cholecalciferol is also the form of vitamin D made by the body. Some dietary supplements and fortified foods contain vitamin D$_2$, whereas others contain vitamin D$_3$.

ergocalciferol (vitamin D$_2$) (er – go – cal – Cl – fer – ol) The form of vitamin D found in plant foods, vitamin D–fortified foods, and supplements.

cholecalciferol (vitamin D$_3$) (cho – le – cal – Cl – fer – ol) The form of vitamin D in animal-derived foods, fortified foods, and supplements; also made by the human body.

VITAMIN D IS FOUND NATURALLY IN ONLY A FEW FOODS

Egg yolks, butter, whole milk, fatty fish, and mushrooms are some of the few foods that naturally contain vitamin D (Figure 11.6). In addition, many dairy

FIGURE 11.6 Food Sources and Structure of Vitamin D Vitamin D is found in two forms: ergocalciferol (vitamin D$_2$) and cholecalciferol (vitamin D$_3$).

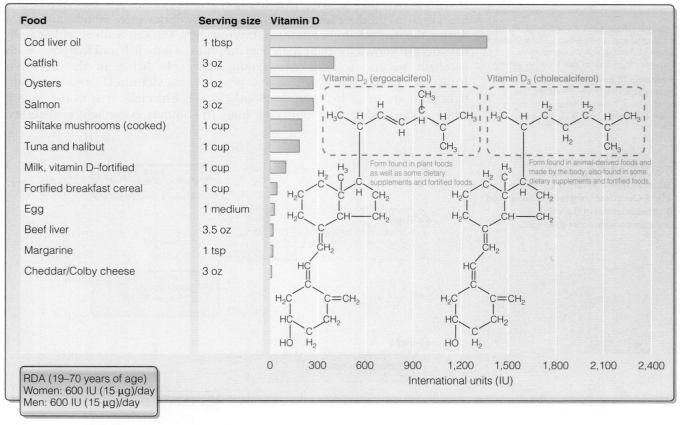

SOURCE: Hands E. Food Finder Vitamin and Mineral Source Guide, 3rd ed. Salem, OR: ESHA Research.

Sunlight exposure is needed for endogenous synthesis of vitamin D.

products as well as breakfast cereals are fortified with vitamin D, and most dietary vitamin D comes from these foods. Vitamin D is needed for calcium absorption, and it is largely removed from reduced-fat milk (an excellent source of calcium) during processing. For these reasons, the FDA encourages manufacturers to add vitamin D to reduced-fat fluid milk. However, the addition of vitamin D to milk is optional unless it is labeled as "fortified milk." Fortification of other products with vitamin D is also voluntary. You can determine whether a food is fortified with this vitamin by reading its label. Vitamin D is quite stable and is not destroyed during food preparation, processing, or storage.

VITAMIN D IS ALSO MADE IN THE SKIN

As well as coming from the diet, vitamin D is also produced by the body. As illustrated in Figure 11.7, vitamin D synthesis is accomplished in two steps. First, the cholesterol metabolite **7-dehydrocholesterol** is converted by ultraviolet light to **previtamin D₃** (also called **precalciferol**) in the skin. This is why vitamin D is sometimes called the "sunshine vitamin." Next, previtamin D₃ is converted to vitamin D₃ (cholecalciferol), which then diffuses into the blood and circulates to the liver. Approximately 5–30 minutes of midday sun exposure at least twice a week to the face, arms, legs, or back without sunscreen provides sufficient amounts of cholecalciferol for most people.[23] Note that cholecalciferol produced in your skin is chemically the same as cholecalciferol found in animal-derived foods.

Many environmental, genetic, and lifestyle factors can influence how much vitamin D can be made endogenously. For example, people who live in areas with persistent smog, overcast skies, or limited amounts of sunlight may require additional sun exposure to synthesize adequate amounts of vitamin D. This is why people living in extreme northern regions (such as Alaska and Scandinavia) produce less vitamin D during the dark winter months and more during the sunny summer months. Interestingly, Alaskan Natives and Scandinavians are known for their high consumption of fatty fish and fish oils—both excellent natural sources of vitamin D. People living in the northern United States and Canada are also exposed to less vitamin D–producing sunlight than are those living in the South. This is especially true in the winter, when days are short and nights are long. In addition to sunlight availability,

7-dehydrocholesterol A metabolite of cholesterol that is converted to cholecalciferol (vitamin D₃) in the skin via exposure to ultraviolet light such as that found in sunshine.

previtamin D₃ (precalciferol) An intermediate product made during the conversion of 7-dehydrocholesterol to cholecalciferol (vitamin D₃) in the skin.

FIGURE 11.7 Synthesis of Cholecalciferol (Vitamin D₃) in the Skin 7-Dehydrocholesterol is converted to cholecalciferol in the skin via exposure to ultraviolet (UV) light.

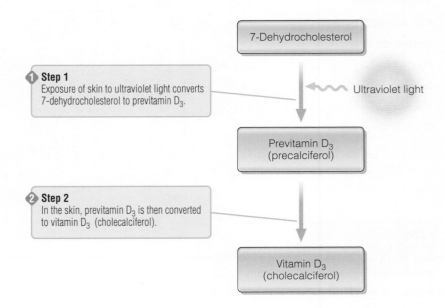

people with darker skin may need up to three times more sun exposure to produce enough vitamin D. This is because dark skin contains more melanin, a pigment that blocks synthesis of precalciferol.[24] Sunscreen can also block the ultraviolet rays needed for vitamin D formation, and age-related changes decrease vitamin D production in the elderly.

You may be wondering if tanning beds result in vitamin D synthesis in the skin. Some, but not all, tanning machines emit the right kind of light needed for vitamin D production. However, because of the damaging effects on the skin and the increased risk for skin cancer, relying on tanning sessions for vitamin D is not recommended. Importantly, the American Academy of Dermatology and the World Health Organization both advise against the use of tanning beds.[25]

DIETARY VITAMIN D ABSORPTION AND VITAMIN D ACTIVATION

In the small intestine, dietary vitamin D is packaged into micelles with the help of bile. It is then taken up into the enterocytes, where it is incorporated into chylomicrons and circulated away from the intestine—first in the lymph and then in the blood. Chylomicrons deliver vitamin D to cells, and what is not taken up is delivered to the liver in chylomicron remnants. But before vitamin D can be utilized by cells, it must be further metabolized. This two-step activation process, illustrated in Figure 11.8, is discussed next.

Activation of Vitamin D Occurs in the Liver and Kidneys The activation of vitamin D occurs in liver and kidney cells. First, cholecalciferol (vitamin D_3) is converted to **25-hydroxyvitamin D (25-[OH]D_3)** in the liver. Then 25-(OH)D_3 is circulated in the blood to the kidneys, where it is converted to **1,25-dihydroxyvitamin D (1,25-[OH]$_2$D$_3$).** It is the 1,25-(OH)$_2$D$_3$ form, also called **calcitriol,** that is active in the body. Calcitriol is important for calcium absorption, and conversion of 25-(OH)D_3 to calcitriol (1,25-[OH]$_2$D$_3$) increases when calcium concentration in the blood is low. This increased synthesis is stimulated by the actions of **parathyroid hormone (PTH)** whose production in the parathyroid glands increases when more calcium is needed.

VITAMIN D REGULATES CALCIUM HOMEOSTASIS, GENE EXPRESSION, AND CELL DIFFERENTIATION

Vitamin D plays an important role in regulating blood calcium concentrations. This involves several organs, including the small intestine, kidneys, and bones. Vitamin D is also involved in a wide variety of other functions, such as regulation of gene expression and cell differentiation.

Calcium Homeostasis Involves the Intestine, Kidneys, and Bones Calcitriol, the active form of vitamin D, helps maintain healthy levels of calcium in the blood, ensuring that calcium is always available to the body's tissues. To make this happen, calcitriol simultaneously increases calcium absorption in the small intestine, decreases calcium excretion in the urine, and facilitates the release of calcium from bones (Figure 11.9). An overview of these processes is provided below.

- *Small intestine.* Calcitriol upregulates several genes that code for calcium transport proteins in the small intestine. Without vitamin D, these proteins are not made, and calcium absorption is limited.

25-hydroxyvitamin D (25-[OH]D_3) An inactive form of vitamin D; made from cholecalciferol in the liver.

1,25-dihydroxyvitamin D (1,25-[OH]$_2$D$_3$) (calcitriol) The active form of vitamin D in the body; produced in the kidneys from 25-(OH)D_3.

parathyroid hormone (PTH) A hormone, produced in the parathyroid glands, that stimulates the conversion of 25-(OH)D_3 to 1,25-(OH)$_2$D$_3$ in the kidneys.

FIGURE 11.8 Activation of Vitamin D The conversion of cholecalciferol to calcitriol takes place sequentially in the liver and kidneys, the latter being stimulated by parathyroid hormone (PTH) during periods of low blood calcium.

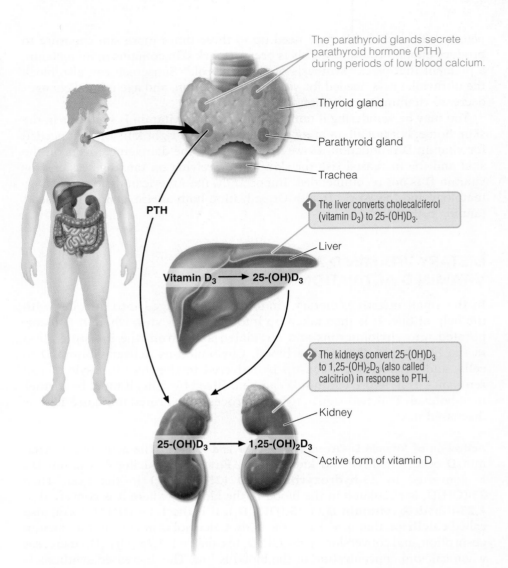

The parathyroid glands secrete parathyroid hormone (PTH) during periods of low blood calcium.

Thyroid gland

Parathyroid gland

Trachea

PTH

1 The liver converts cholecalciferol (vitamin D_3) to 25-(OH)D_3.

Liver

Vitamin D_3 ⟶ 25-(OH)D_3

2 The kidneys convert 25-(OH)D_3 to 1,25-(OH)$_2D_3$ (also called calcitriol) in response to PTH.

Kidney

25-(OH)D_3 ⟶ 1,25-(OH)$_2D_3$

Active form of vitamin D

- *Kidney.* Calcitriol and PTH cause the kidneys to reduce their excretion of calcium into the urine. As a result, more calcium remains in the blood.
- *Bone.* Calcitriol and PTH together stimulate osteoclasts to break down bone, resulting in the release of calcium from the skeleton into the blood.

It is important to note that vitamin D's role in maintaining adequate levels of calcium in the blood goes beyond maintaining healthy bones. For example, blood calcium is delivered to muscle cells where it is needed for muscle contractions. Calcium is also vital in blood pressure regulation and the conduction of neural impulses. Without vitamin D to help maintain adequate levels of calcium in the blood, these essential functions would be impaired.

Vitamin D and Cell Differentiation Vitamin D also plays a critical role in stimulating immature cells to become mature, functioning cells—in other words, cell differentiation. You might recall that vitamin A is also important in orchestrating this process, especially in epithelial cells. Like vitamin A, vitamin D moves into the nucleus of a cell and upregulates selected genes coding for specific proteins involved in cell differentiation. For example, vitamin D causes immature (undifferentiated) bone cells to become mature bone marrow cells and causes certain intestinal epithelial cells to differentiate into mature enterocytes. In this way, vitamin D plays additional roles in bone health

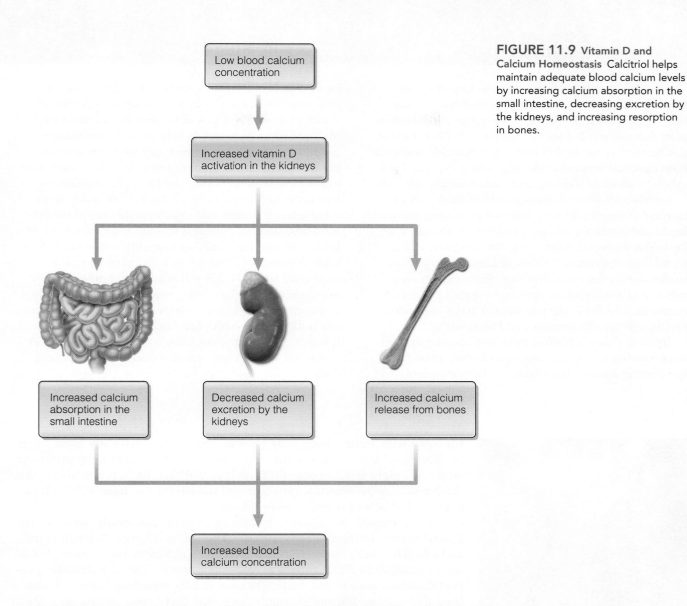

Low blood calcium
concentration

↓

Increased vitamin D
activation in the kidneys

Increased calcium
absorption in the
small intestine

Decreased calcium
excretion by the
kidneys

Increased calcium
release from bones

Increased blood
calcium concentration

FIGURE 11.9 Vitamin D and
Calcium Homeostasis Calcitriol helps
maintain adequate blood calcium levels
by increasing calcium absorption in the
small intestine, decreasing excretion by
the kidneys, and increasing resorption
in bones.

and GI function. Some evidence also suggests that vitamin D may help prevent certain types of cancers such as those of the colon, breast, skin, and prostate.[26] Although this connection warrants further investigation, vitamin D's role in cell differentiation may be involved. This is described more fully in the Focus on Diet and Health feature.

VITAMIN D DEFICIENCY MAY BE RELATIVELY COMMON

Vitamin D deficiency can cause inadequate bone mineralization during early life, and it is also associated with increased demineralization of bone later in the life cycle. In addition, vitamin D deficiency may affect the risk for a variety of chronic degenerative diseases such as Alzheimer's disease. These myriad and lifelong effects have gained significant recent public health attention because scientists have become aware that vitamin D deficiency may be more common in the United States than previously believed. Additional research is needed to understand better the health impacts of vitamin D deficiency throughout the entire life cycle.

Vitamin D Deficiency Can Cause Rickets in Children Vitamin D deficiency in infants and children can result in improper bone mineralization—a disease called **rickets**—and is a significant public health concern worldwide.[27]

rickets A condition caused by vitamin D deficiency in young children and characterized by deformed bones, especially in the legs.

Vitamin D has long been known to be important for forming healthy bones in childhood and maintaining them in adulthood. However, more recent evidence suggests that vitamin D does this—and much more. Indeed, this nutrient is involved in many important physiological functions, and optimal levels may reduce the risks for a variety of diseases such as cancer and multiple sclerosis.

For instance, there is considerable evidence that an increase in vitamin D intake, optimal circulating vitamin D levels, and increased sunlight exposure are all related to lower risk for breast cancer.[28] Similar results have been found for colon cancer.[29] Although clinical intervention studies are needed to confirm these findings, there is a substantial amount of research that suggests that this association may be due to the role vitamin D has in regulating healthy cellular differentiation and maturation.

Optimal vitamin D levels have also been implicated in preventing multiple sclerosis, a chronic, potentially debilitating disease that affects the fatty sheath of nerve cells in the brain, spinal cord, and optic nerve. Symptoms can range from mild to severe, but typically the disease results in loss of movement and mobility. Multiple sclerosis is widely believed to be an autoimmune disease, and it is exceedingly more common in populations living in more extreme latitudes where sun exposure is limited.[30] These are the same populations that are at greatest risk for vitamin D deficiency. Both observational human and experimental animal studies have provided evidence that vitamin D deficiency might lead to greater susceptibility to multiple sclerosis.[31] As with its potential relationship with cancer, however, controlled clinical trials will need to be conducted to ascertain whether increasing vitamin D intake or endogenous vitamin D synthesis in the body via sunlight can affect the risk for multiple sclerosis.[32] In the meantime, it remains prudent to consume a diet rich in vitamin D and to get 10 to 15 minutes of moderate sun exposure daily.

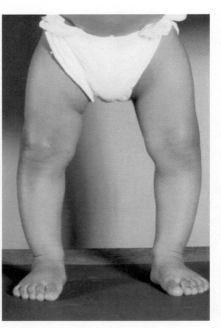

Vitamin D deficiency can cause rickets in infants and children.

© Wellcome Photo Library /Custom Medical Stock Photo

Children with rickets have slow growth and characteristically bowed legs or "knocked" knees. This occurs when the long bones cannot support the stress of weight-bearing activities, such as walking. Rickets can also cause the breastbone to protrude outward and the rib cage to narrow, which can lead to cardiac and respiratory problems.

Rickets caused by vitamin D deficiency was commonly seen in the United States during the early 1900s.[33] The use of vitamin D–fortified milk and infant formulas, however, nearly eradicated rickets in this country; that is, until recently.[34] In 2000, a medical research group published a disturbing report documenting a rise in the incidence of rickets, especially in dark-skinned, breastfed babies. Further research suggested that some infants nourished solely on human milk are at increased risk of inadequate vitamin D intake. This may be due to a lack of sun exposure in the mothers, resulting in low amounts of vitamin D in their milk. In addition, insufficient sun exposure in infants may contribute to the rise in rickets, especially for those with dark skin.

Because of concerns that increased sunlight exposure may lead to skin cancer, clinicians are hesitant to recommend that parents expose their infants and children to additional sunlight. Instead, the American Academy of Pediatrics recommends that all breastfed infants be supplemented with 400 IU/day (10 µg/day) vitamin D beginning within the first few days of life.[35] Formula-fed infants consuming <1 L/day vitamin D–fortified formula also need an alternative way to get 400 IU/day vitamin D, such as through vitamin supplements. It is recommended that this level of supplementation be continued through childhood and adolescence.

Vitamin D Deficiency Can Cause Osteomalacia and Osteoporosis in Adults In adults, vitamin D deficiency results in both poor bone mineralization and increased bone demineralization. The risk of vitamin D deficiency in adults is highest in people with dark skin, those who live in areas with little sunlight, those who habitually cover themselves in clothing (such as some Muslim women), and those who do not consume vitamin D–fortified milk products. Inadequate bone mineralization resulting from vitamin D deficiency causes bones to become soft and weak—a condition called **osteomalacia.** Symptoms of osteomalacia include diffuse bone pain and muscle weakness. People with osteomalacia are also at increased risk for bone fractures. Vitamin D deficiency can also result in demineralization of previously healthy bone, ultimately leading to a disease called **osteoporosis.** Osteoporosis is a serious chronic disease—especially in the elderly—affecting more than 28 million Americans (1 in 10). You can read more about both healthy bone growth and osteoporosis in the Nutrition Matters following Chapter 12. To help prevent both osteomalacia and osteoporosis, people over 50 years of age are advised to get at least 15 minutes of sun exposure each day and to increase their vitamin D intake. In some cases, vitamin D supplements may be necessary.

Is Vitamin D Deficiency Linked to Alzheimer's Disease? Some evidence suggests that vitamin D deficiency may be related to increased risk for Alzheimer's disease, a debilitating condition that causes personality changes and cognitive decline in older adults.[36] Research suggests that this may be related to gene polymorphisms that result in secondary vitamin D deficiency.[37] It is also possible that Alzheimer's disease causes vitamin D deficiency, not the other way around. Clearly, additional research is required to more fully understand this relationship.

RECOMMENDED INTAKES AND VITAMIN D TOXICITY

The RDA for vitamin D in adults is 15 µg (600 IU)/day. This is the amount in about 1 liter (4.7 cups) of vitamin D–fortified milk. To help decrease bone loss in the elderly, the RDA increases to 20 µg (800 IU)/day at 71 years of age.

Vitamin D toxicity from food sources is uncommon. However, supplementation with high doses of vitamin D can cause calcium levels in the blood and urine to rise, resulting in hypercalcemia and hypercalciuria, respectively. This is due to vitamin D's multiple roles in calcium absorption, excretion, and mobilization. Hypercalcemia can result in the deposit of calcium in organs such as the heart and lungs, and it can affect the function of the central nervous system. Hypercalciuria can also cause kidney stone formation. Furthermore, vitamin D toxicity promotes bone loss and, when severe, can be fatal. Because of these dangers, the Institute of Medicine has established a UL of 50 µg (2,000 IU) for vitamin D in supplemental form.

Vitamin E—Antioxidant That Protects Biological Membranes

The term *vitamin E* refers collectively to eight different compounds that all have somewhat similar chemical structures. Of these, **α-tocopherol** is the most biologically active. Scientists first recognized that vitamin E was essential in 1922 when they discovered that vitamin E–deficient animals could not reproduce. The name *tocopherol* was consequently derived from the Greek *tokos* (childbirth) and *phero* (to bear). However, vitamin E is now known to have many functions not related to reproduction, such as providing important antioxidant protection and maintaining cardiovascular health.

VITAMIN E IS ABUNDANT IN OILS, NUTS, AND SEEDS

Although vitamin E is found in both plant and animal foods, it is especially abundant in vegetable oils, nuts, and seeds. Some dark green vegetables such as broccoli and spinach contain vitamin E as well. Figure 11.10 lists food sources

osteomalacia (os – te – o – ma – LA – ci – a) Softening of the bones in adults that can be due to vitamin D deficiency.

osteoporosis (os – te – o – por – O – sis) A serious bone disease resulting in weak, porous bones.

α-tocopherol (to – CO – pher – ol) The most biologically active form of vitamin E.

FIGURE 11.10 Food Sources
of Vitamin E

Food	Serving size	Vitamin E
Fortified breakfast cereal	1 cup	
Tomatoes	1 cup	
Almonds	1 oz	
Spinach (boiled)	1 cup	
Sunflower seeds	1 oz	
Vegetable oil	1 tbsp	
Hazelnuts	1 oz	
Pine nuts	1 oz	
Pumpkin (canned)	1 cup	
Sweet potato (canned)	1 cup	
Broccoli (boiled)	3 oz	
Red peppers (raw)	1 cup	

RDA
Women: 15 mg/day
Men: 15 mg/day

0   3   6   9   12   15
Milligrams (mg)

SOURCE: USDA Nutrient Database for Standard Reference, Release 16-1.

of vitamin E. Vitamin E is easily destroyed during food preparation, processing, and storage. Vitamin E in foods is best preserved by exposing them to as little heat as possible during cooking. Storage of foods in airtight containers is also important.

Like absorption of vitamins A and D, absorption of vitamin E occurs in the small intestine and requires the presence of lipids, bile and the formation of micelles. Vitamin E is then circulated in chylomicrons via the lymph, eventually reaching the liver via the blood. Dietary vitamin E not taken up from chylomicrons by cells is subsequently repackaged into VLDLs in the liver for further delivery in the body. Excess vitamin E is deposited mainly in adipose tissue.

VITAMIN E IS A POTENT ANTIOXIDANT

Like the carotenoids, vitamin E acts as an antioxidant that protects biological membranes from the destructive effects of free radicals. Vitamin E may also protect the cell's genetic material (DNA) from oxidative harm.

Protecting Biological Membranes As illustrated in Figure 11.11, vitamin E is often associated with biological membranes. Recall that many of these membranes (such as cell membranes) consist of a phospholipid bilayer. Cell organelles, such as mitochondria where ATP is produced, are also encased in phospholipid bilayers. Protecting biological membranes is vital to the stability and function of cells and their organelles. Vitamin E plays a major role in this by preventing free radical–induced oxidative damage. By donating electrons, vitamin E stabilizes free radicals and prevents them from damaging fatty acids embedded in membranes. This protection is especially important in cells that are exposed to high levels of oxygen, such

Pakhnyushcha /Shutterstock.com

Nuts and seeds are good sources of vitamin E.

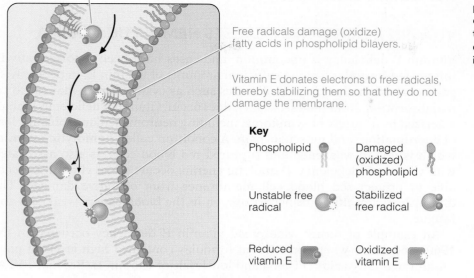

Free radicals are unstable because they have unpaired electrons in their outer shells.

Free radicals damage (oxidize) fatty acids in phospholipid bilayers.

Vitamin E donates electrons to free radicals, thereby stabilizing them so that they do not damage the membrane.

Key

Phospholipid

Damaged (oxidized) phospholipid

Unstable free radical

Stabilized free radical

Reduced vitamin E

Oxidized vitamin E

FIGURE 11.11 Vitamin E Protects Biological Membranes Vitamin E protects biological membranes (such as cell membranes) by donating electrons, thereby neutralizing free radicals that could otherwise damage the fatty acids in phospholipid bilayers.

as those in the lungs and red blood cells. The ability of vitamin E to act as an antioxidant is enhanced in the presence of other antioxidant micronutrients, such as vitamin C and selenium. This is one of many reasons why it is important to consume a variety of foods that provide all these needed nutrients.

Vitamin E and Cancer Because antioxidant nutrients protect DNA from cancer-causing free radical damage, health professionals are interested in the possibility that vitamin E might prevent or cure cancer. Diets high in vitamin E are associated with decreased cancer risk, although there is little experimental evidence that vitamin E by itself is responsible.[38] In other words, consuming foods rich in vitamin E is likely better than taking vitamin E supplements in this regard.

Vitamin E May Help Prevent Cataracts Vitamin E is also important for eye health, specifically for the prevention of cataracts. **Cataracts** are opaque growths that develop on the lens of the eye, causing vision to become clouded. Cataracts tend to develop as people age, often leading to progressively impaired vision. If left untreated, cataracts can cause blindness. Because of increased free radical damage, people who smoke or are exposed to excessive amounts of sunlight are at increased risk for developing cataracts. Dietary antioxidants may prevent or delay the growth of cataracts, and epidemiologic studies have shown that older people who consume a vitamin E–rich diet or take vitamin E supplements are at lower risk for developing this type of eye disease.[39] The physiological mechanisms underlying these relationships are not well understood, however, and researchers continue to study how vitamin E and other antioxidants influence eye health.[40]

Vitamin E Supplementation Is Not Recommended for Cardiovascular Health As you learned in Chapter 6, atherosclerosis can lead to heart disease or stroke via the accumulation in the arteries of fatty material called *plaque*. Many animal studies have shown that vitamin E slows the rate of plaque formation. Because of vitamin E's antioxidant function, increased dietary intakes of it may help slow the progression of atherosclerosis.[41] However, human intervention studies suggest that vitamin E supplementation (400 mg/day) can increase overall risk of mortality in people with chronic disease.[42] For this reason, although dietary vitamin E may be important in maintaining a

⟨**CONNECTIONS**⟩ Remember that phospholipids are amphipathic lipids containing both polar (phosphate group) and nonpolar (fatty acids) components (Chapter 6, page 231).

⟨**CONNECTIONS**⟩ Recall that polyunsaturated fatty acids (PUFAs) are those that have more than one double bond between the carbons making up the fatty acid backbone (Chapter 6, page 245).

cataract (CAT – a – ract) Age-related cloudiness that develops on the lens of the eye, causing impaired vision.

healthy heart, taking vitamin E supplements to decrease risk for cardiovascular disease is discouraged.

VITAMIN E DEFICIENCY CAUSES HEMOLYTIC ANEMIA

Vitamin E deficiency is uncommon, and cases have been reported only in infants fed formulas lacking sufficient amounts of vitamin E, in people with genetic abnormalities, and in diseases such as cystic fibrosis that cause fat malabsorption. Nonetheless, when it does occur, vitamin E deficiency is characterized by a variety of symptoms, including neuromuscular problems, loss of coordination, and muscular pain. A condition called **hemolytic anemia,** which is due to weakened and ruptured red blood cells, is also associated with vitamin E deficiency. Hemolytic anemia occurs when vitamin E is not able to protect red blood cell membranes from oxidative damage. The diminished capacity to transport oxygen in the blood causes weakness and fatigue.

An example of rather widespread vitamin E deficiency occurred in the 1960s and 1970s when some infant formulas contained high levels of polyunsaturated fatty acids (PUFAs) and low levels of vitamin E. Because PUFAs are easily damaged (oxidized) by free radicals, consumption of these formulas caused infants to have increased need for antioxidant nutrients such as vitamin E. As a result, some babies developed hemolytic anemia, especially premature infants who were born with very low stores of vitamin E.[43] This unfortunate event led to vitamin E fortification of infant formulas.

RECOMMENDED INTAKES FOR VITAMIN E

The RDA for vitamin E is 15 mg/day, an amount easily obtained from the diet. Unlike with the other fat-soluble vitamins, vitamin E toxicity is rare even with high intakes of vitamin E supplements. However, in some people, very high doses of vitamin E supplements can cause dangerous bleeding or hemorrhage. Why some people respond this way and others do not is not understood but probably involves genetic differences. And, as mentioned previously, some evidence suggests a potentially harmful effect of vitamin E supplementation on cardiovascular health. As such, it is best to be cautious when taking vitamin E supplements, and the UL for vitamin E from all sources is set at 1,000 mg/day.

Vitamin K—Critical for Coagulation

Named for its role in coagulation (*koagulation* in Danish) by Henrik Dam, a Danish physiologist who found that vitamin K deficiency in chickens caused excessive bleeding, *vitamin K* refers to three compounds that have similar structures and functions.[§] Vitamin K found naturally in plant foods is called **phylloquinone** (vitamin K_1); this form is also found in some vitamin K supplements and is often given to infants to prevent bleeding. The second form is **menaquinone** (vitamin K_2), which is produced by bacteria present in the large intestine. Because we cannot get enough vitamin K from this bacterial production, vitamin K is an essential nutrient. A third form of vitamin K called **menadione** (vitamin K_3) is neither found naturally in food nor made by intestinal bacteria but is produced commercially.

hemolytic anemia Decreased ability of the blood to carry oxygen and carbon dioxide due to rupturing of red blood cells; caused by vitamin E deficiency.

phylloquinone (vitamin K_1) (phyll – o – quin – ONE) A form of vitamin K found in foods and dietary supplements.

menaquinone (vitamin K_2) (men – a – quin – ONE) A form of vitamin K produced by bacteria.

menadione (vitamin K_3) (men – a – DI – one) A form of vitamin K produced commercially.

[§] Dam received the Nobel Prize for Physiology or Medicine in 1943 for this discovery.

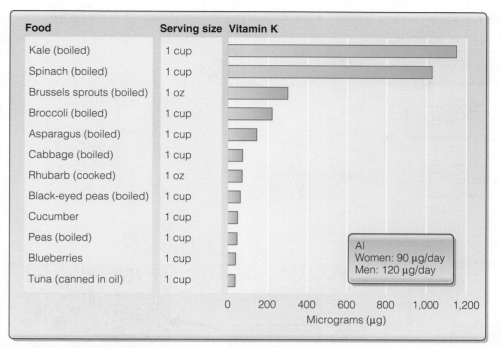

Food	Serving size	Vitamin K
Kale (boiled)	1 cup	
Spinach (boiled)	1 cup	
Brussels sprouts (boiled)	1 oz	
Broccoli (boiled)	1 cup	
Asparagus (boiled)	1 cup	
Cabbage (boiled)	1 cup	
Rhubarb (cooked)	1 oz	
Black-eyed peas (boiled)	1 cup	
Cucumber	1 cup	
Peas (boiled)	1 cup	
Blueberries	1 cup	
Tuna (canned in oil)	1 cup	

AI
Women: 90 µg/day
Men: 120 µg/day

Micrograms (µg)

FIGURE 11.12 Food Sources of Vitamin K

SOURCE: USDA Nutrient Database for Standard Reference, Release 16-1.

VITAMIN K IS FOUND NATURALLY IN DARK GREENS, FISH, AND LEGUMES

As listed in Figure 11.12, food sources of vitamin K include green vegetables such as kale, spinach, broccoli, and Brussels sprouts. You can also obtain this vitamin from fish and legumes. Like many of the vitamins, vitamin K can be destroyed by exposing food to excessive light or heat.

Dietary vitamin K is absorbed along with other fat-soluble vitamins in the small intestine. It is then incorporated into chylomicrons, which are released into the lymphatic system, eventually entering the blood. Alternatively, vitamin K produced by bacteria in the large intestine is transported into intestinal epithelial cells by simple diffusion and then circulated to the liver in the blood. The liver then packages both dietary and bacterially produced forms of vitamin K into lipoproteins for additional delivery to the rest of the body.

Many dark green vegetables, such as Brussels sprouts, provide vitamin E to the diet.

VITAMIN K INVOLVED IN BLOOD CLOTTING CASCADE

Vitamin K functions as a coenzyme in a variety of enzymatic reactions that ultimately result in the addition of calcium to molecules. These vitamin K–dependent reactions are needed for the life-and-death process of blood clotting (Figure 11.13). When you injure a blood vessel, your body rapidly forms a blood clot to stop the bleeding. Without this process, called **coagulation,** you might bleed to death even after a minor scrape.

For your blood to coagulate and form a clot, a cascade of chemical reactions must take place. This cascade can be thought of as a line of dominoes strategically positioned to—one by one—fall over until the entire line has collapsed. In other words, the last domino will not fall unless all the previous dominoes have already fallen. In the case of vitamin K and coagulation, one set of chemical reactions must happen before the next set can occur. Specifically, vitamin K

coagulation (co –ag – u – LA – tion) The process by which blood clots are formed.

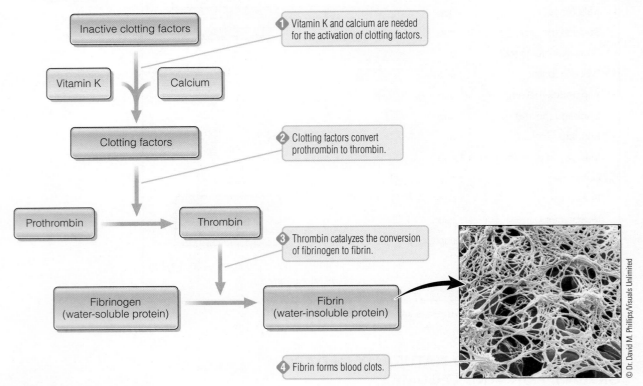

FIGURE 11.13 Vitamin K and Blood Clotting Vitamin K acts as a coenzyme in enzymatic reactions involved in the formation of blood clots.

Inactive clotting factors

1 Vitamin K and calcium are needed for the activation of clotting factors.

Vitamin K

Calcium

Clotting factors

2 Clotting factors convert prothrombin to thrombin.

Prothrombin → Thrombin

3 Thrombin catalyzes the conversion of fibrinogen to fibrin.

Fibrinogen (water-soluble protein) → Fibrin (water-insoluble protein)

4 Fibrin forms blood clots.

© Dr. David M. Phillips/Visuals Unlimited

must first activate clotting factors, which subsequently bind calcium. Binding to calcium activates the clotting factors and allows the next reaction in the cascade to take place. These reactions ultimately convert the protein **prothrombin** to **thrombin,** which then catalyzes the conversion of **fibrinogen** to **fibrin.** Fibrin is a water-insoluble protein that forms a web-like clot to stop bleeding. Without adequate vitamin K, this cascade of events shuts down, and clots cannot form. A drug called Coumadin® delays blood clot formation by decreasing the activity of vitamin K, as you will see in the Focus on Clinical Applications feature.

Vitamin K is also necessary for enzymatic reactions that enable skeletal proteins to bind calcium. Some studies show that consuming foods high in vitamin K is associated with decreased risk for hip fracture.[44] However, further studies are needed to determine whether increased vitamin K intake results in increased bone strength.[45]

© Shelley McGuire

Critical Thinking: Muayyad's Story Now that you understand the role of vitamin K in blood clotting, can you explain why Muayyad had to limit his intake of vitamin K–rich foods? Given that vitamin K has multiple physiological functions, what impacts could a vitamin K–restricted diet have on health, other than to increase the possibility of bleeding?

prothrombin A clotting factor (protein) that is converted to the enzyme thrombin.

thrombin The enzyme that catalyzes the conversion of fibrinogen to fibrin.

fibrinogen (fi – BRIN – o – gen) A water-soluble protein that is converted to the water-insoluble protein fibrin.

fibrin An insoluble protein that forms blood clots.

VITAMIN K DEFICIENCY CAN CAUSE SEVERE BLEEDING

Although rare in healthy adults, vitamin K deficiency occurs in some infants. This is because even healthy infants are born with very low stores of vitamin K. Moreover, at birth the newborn's large intestine completely lacks vitamin K–producing bacteria, and human milk can contain very

In many situations, the use of a medication can influence the metabolism or function of a nutrient. This is especially the case for vitamin K. For example, long-term use of antibiotics can kill bacteria in your large intestine that normally produce vitamin K, increasing risk for vitamin K deficiency. Another example of a nutrient–drug interaction occurs with Coumadin (the generic name is warfarin), a drug often prescribed to prevent blood clots in people with cardiovascular disease. Coumadin works by decreasing the activity of vitamin K, especially in its role in blood clot formation. This is why Coumadin is often said to "thin the blood." Because blood clots can block the flow of blood to the heart or brain, the use of Coumadin can be lifesaving. However, people using this drug must take special care not to injure themselves because they risk serious hemorrhaging and blood loss even with a seemingly minor cut or bruise.

Because a sudden increase or decrease in vitamin K intake can interfere with the effectiveness of Coumadin, it is important to keep vitamin K intake constant. In general, clinicians advise people taking this medication (such as Muayyad, whom you read about earlier in this chapter) to eat a balanced diet and to limit consumption of foods containing high amounts of vitamin K to one serving per day. These include many dark green vegetables such as kale, spinach, and broccoli. You can learn more about this and other drug–nutrient interactions by visiting the NIH's Clinical Center website (http://www.cc.nih.gov/ccc/patient_education/important_drug_food_info.html). You can also get accurate information about the vitamin K content of hundreds of foods from the USDA's food composition database (http://www.ars.usda.gov/Services/docs.htm?docid=17477). This information might be especially useful if you or someone you know is taking a blood-thinning drug like Coumadin.

---

low levels of this vitamin. Thus, babies (especially those who are breastfed) have minimal amounts of vitamin K during the first few weeks of life. Although this does not present a problem to most infants, some develop severe vitamin K deficiency—a condition called **vitamin K deficiency bleeding**.** Because of vitamin K's vital role in blood clotting, babies with vitamin K deficiency bleeding can experience uncontrolled internal bleeding (hemorrhage). In response to this, the American Academy of Pediatrics recommends that, as a precautionary measure, all babies be given vitamin K injections at birth.[46] Vitamin K deficiency also occurs in children and adults with diseases that cause lipid malabsorption. In addition, prolonged use of antibiotics can kill bacteria residing in the large intestine, resulting in vitamin K deficiency.[47]

Although RDAs have not been set for vitamin K, AIs have been established. The AI values for men and women are 120 and 90 µg/day, respectively. Vitamin K is rarely toxic even in very high amounts, so a UL has not been established.

Fat-Soluble Vitamins: Summary and Overall Recommendations

In summary, the fat-soluble vitamins are involved in a wide variety of processes encompassing all the physiological systems in the body. In addition, the carotenoids may impart additional health benefits both by being converted to vitamin A (provitamin A carotenoids) or by exerting important antioxidant properties.

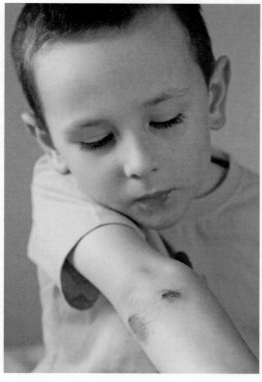

Vitamin K is essential for blood coagulation, helping the body to heal after being cut. People taking drugs to decrease blood clotting (like Coumadin) need to be mindful about their vitamin K intake.

vitamin K deficiency bleeding A disease that occurs in newborn infants; characterized by uncontrollable internal bleeding (hemorrhage) from inadequate vitamin K.

** Vitamin K deficiency bleeding was previously called "hemorrhagic disease of the newborn."

Working Toward the Goal: Increasing Fat-Soluble Vitamin Intake via Fruits and Vegetables

The 2010 Dietary Guidelines recommend that you choose a variety of fruits and vegetables each day, selecting from all five vegetable subgroups (dark green, red/orange, legumes, starchy vegetables, and other vegetables). This is, in part, to ensure adequate intake of fat-soluble vitamins and carotenoids. The following selection and preparation tips will help you meet this goal.

- When grocery shopping, buy one fruit and one vegetable that you do not normally eat. For example, when was the last time you ate a pomegranate or prepared Brussels sprouts? This way, you will be more likely to get all the important vitamins, minerals, and phytochemicals needed by your body.

- Instead of high-fat snacks, choose dried fruits, banana chips, or vegetable chips. These snacks are nutrient dense, providing many vitamins and phytochemicals. Choosing brightly colored snacks of this sort will also help increase intake of carotenoids.

- Try to eat at least one salad each day, and vary what you put into it. For example, alternate between lettuce salads and spinach salads, and try not to use the same salad toppings each time. Add mushrooms and cheese today for vitamin D, and tomorrow sprinkle on a spoonful of sunflower seeds for a vitamin E boost.

- Think about color when you plan a meal. Because many of the fat-soluble vitamins as well as the carotenoids tend to be found in brightly colored fruits and vegetables, expanding the color palette on your plate will help you get a variety of nutrients and phytochemicals.

However, it is important to remember that these dietary components do not work independently of each other. In this way, they are similar to many of the B vitamins, which coordinately regulate the body's energy metabolism pathways. For example, vitamins A, D, and E, as well as the carotenoids, work together in orchestrating gene transcription. Vitamin E and the carotenoids act as antioxidants, while vitamins A, D, and K contribute to bone health. You can review, compare, and contrast the general functions of all the fat-soluble vitamins and carotenoids in Table 11.3. You might want to compare this with the overlapping functions of the water-soluble vitamins you learned about in Chapter 10. Like their water-soluble counterparts, the fat-soluble vitamins are found in a wide variety of foods—including those from both plant- and animal-derived sources. Because of the wide distribution of these nutrients in foods, it is important to consume a balance of food from

TABLE 11.3 General Functions of the Fat-Soluble Vitamins and Carotenoids

	Vitamin or Phytochemical				
Functions	Vitamin A	Carotenoids[a]	Vitamin D	Vitamin E	Vitamin K
Regulation of gene transcription	X	X	X		
Immune function	X	X			
Antioxidant function		X		X	
Bone/tooth health	X		X		X
Blood health	X		X	X	X
Eye health	X	X		X	
Cardiovascular health		X		X	

[a]Note that these are proposed functions of the carotenoids.

all the food groups and a variety of foods from within each group. You can also help ensure adequate intakes of these vitamins by following the 2010 Dietary Guidelines and MyPlate food guidance system, and you can find tips in the Food Matters feature on how you can increase your intake of fat-soluble vitamins by consuming more fruits and vegetables. As with all the essential nutrients, choosing a diet high in nutrient-dense foods and emphasizing moderation, variety, and balance can go a long way to help maintain optimal health in terms of the fat-soluble vitamins.

© Shelley McGuire

Critical Thinking: Muayyad's Story Do you think it would be difficult for a student with Factor V Leiden thrombophilia, or another condition that requires blood-thinning medications, to eat nutritiously and safely in college? In what ways might university foodservice staff assist such an individual to avoid large portions of vitamin K–rich foods? Should these measures be voluntary or mandatory in the college setting?

Diet Analysis PLUS ✚ Activity

By now, you have gained a better understanding of the importance of the fat-soluble vitamins A, D, E, and K in your own diet. Using the *Intake and DRI Goals Compared* report, representing the average of your intakes over the reporting period, fill in the following table to assess your intakes of the fat-soluble vitamins. *(Remember that your intakes may fluctuate dramatically from day to day, so make sure to use the report that averages these fluctuations over the time period.)*

1. Evaluate your intake of these fat-soluble vitamins.

Vitamin	Your RDA/AI	Your estimated intake	The % RDA/AI you consumed	Did you consume too little, too much, or just about your RDA/AI?
A				
D				
E				
K				

2. For the following, list the foods from your individual days' *Intake Spreadsheets* that provided the greatest amounts of the fat-soluble vitamins.

Vitamin A

Name of food consumed   **Amount of nutrient in food consumed (include units)**

_____

_____

_____

Vitamin D

Name of food consumed   **Amount of nutrient in food consumed (include units)**

_____

_____

_____

_____

Vitamin E

Name of food consumed   **Amount of nutrient in food consumed (include units)**

_____

_____

_____

_____

Vitamin K

Name of food consumed   **Amount of nutrient in food consumed (include units)**

_____

_____

_____

_____

3. Based on the assessments of your intake of the fat-soluble vitamins included in your Diet Analysis Plus report, do you need to change your diet to get either more or less of these nutrients? If the answer is *yes*, how would you go about doing this?

Notes

1. Gudas LJ, Wagner JA. Retinoids regulate stem cell differentiation. Journal of Cellular Physiology. 2011;226:322–30. Ziegler TR, Evans ME, Fernandez-Estivariz C, Jones DP. Trophic and cytoprotective nutrition for intestinal adaptation, mucosal repair, and barrier function. Annual Review of Nutrition. 2003;23:229–61.

2. Duriancik DM, Lackey DE, Hoag KA. Vitamin A as a regulator of antigen presenting cells. Journal of Nutrition. 2010;140:1395–9. Montrone M, Martorelli D, Rosato A, Dolcetti R. Retinoids as critical modulators of immune functions: New therapeutic perspectives for old compounds. Endocrine Metababolic Immune Disorders Drug Targets. 2009;9:113–31; Villamor E, Fawzi WW. Effects of vitamin A supplementation on immune responses and correlation with clinical outcomes. Clinical Microbiology Reviews. 2005;18:446–64.

3. Sommer A. Vitamin A deficiency and clinical disease: an historical overview. Journal of Nutrition. 2008;138:1835–9.

4. von Lintig J. Colors with functions: Elucidating the biochemical and molecular basis of carotenoid metabolism. Annual Review of Nutrition. 2010;30:35–56.

5. Baron JA, Cole BF, Mott L, Haile R, Grau M, Church TR, Beck GJ, Greenberg ER. Neoplastic and antineoplastic effects of β-carotene on colorectal adenoma recurrence: Results of a randomized trial. Journal of the National Cancer Institute. 2003;95:717–22. Omenn GS, Goodman GE, Thornquist MD, Balmes J, Cullen MR, Glass A, Keogh JP, Meyskens FL, Valanis B, Williams JH, Barnhart S, Hammar S. Effects of a combination of beta carotene and vitamin A on lung cancer and cardiovascular disease. New England Journal of Medicine. 1996;334:1150–55.

6. Institute of Medicine. Dietary reference intakes for vitamin C, vitamin E, selenium, and carotenoids. National Academy Press, Washington, DC, 2000.

7. Genaro Pde S, Martini LA. Vitamin A supplementation and risk of skeletal fracture. Nutrition Reviews. 2004;62:65–7.

8. Johansson S, Melhus H. Vitamin A antagonizes calcium response to vitamin D in man. Journal of Bone Mineral Research. 2001;16:1899–905.

9. Trottier G, Boström PJ, Lawrentschuk N, Fleshner NE. Nutraceuticals and prostate cancer prevention: A current review. Nature Reviews, Urology. 2010;7:21–30.

10. Beatty S, Nolan J, Kavanagh H, O'Donovan O. Macular pigment optical density and its relationship with serum and dietary levels of lutein and zeaxanthin. Archives of Biochemistry and Biophysics. 2004;430:70–6. Stringham JM, Hammond BR. Dietary lutein and zeaxanthin: Possible effects on visual function. Nutrition Reviews. 2005;63:59–64.

11. Namitha KK, Negi PS. Chemistry and biotechnology of carotenoids. Critical Reviews in Food Science and Nutrition. 2010;50:728–60.

12. Sahin K, Sahin N, Kucuk O. Lycopene and chemotherapy toxicity. Nutrition and Cancer. 2010;62:988–95. Terry P, Jain M, Miller A, Howe GR, Rohan TE. Biological activity of lycopene metabolites: Implications for cancer prevention. Mein JR, Lian F, Wang XD. Nutrition Reviews. 2008;66:667–83.

13. Wertz K, Siler U, Goralczyk R. Lycopene: Modes of action to promote prostate health. Archives of Biochemistry and Biophysics. 2004;430:127–34. Trottier G, Bostrom PJ, Lawrentschuk N, Fleshner NE. Nutraceuticals and prostate cancer prevention: a current review. Nature Reviews. Urology. 2010;7:21–30.

14. Stringham JM, Hammond BR. Dietary lutein and zeaxanthin: Possible effects on visual function. Nutrition Reviews. 2005;63:59–64.

15. Chew BP, Park JS. Carotenoid action on the immune response. Journal of Nutrition. 2004;134:257S–61S. Palozza P, Parrone N, Catalano A, Simone R. Tomato lycopene and inflammatory cascade: Basic interactions and clinical implications. Current Medicinal Chemistry. 2010;17:2547–63. Sharoni Y, Danilenko M, Dubi N, Ben-Dor A, Levy J. Carotenoids and transcription. Archives of Biochemistry and Biophysics. 2004;430:89–96.

16. The Alpha-Tocopherol, Beta Carotene Cancer Prevention Study Group. The effect of vitamin E and beta carotene on the incidence of lung cancer and other cancers in male smokers. New England Journal of Medicine. 1994;330:1029–35. Baron JA, Cole BF, Mott L, Haile R, Grau M, Church TR, Beck GJ, Greenberg ER. Neoplastic and antineoplastic effects of β-carotene on colorectal adenoma recurrence: Results of a randomized trial. Journal of the National Cancer Institute. 2003;95:717–22. Omenn GS, Goodman GE, Thornquist MD, Balmes J, Cullen MR, Glass A, Keogh JP, Meyskens FL, Valanis B, Williams JH, Barnhart S, Hammar S. Effects of a combination of beta carotene and vitamin A on lung cancer and cardiovascular disease. New England Journal of Medicine. 1996;334:1150–55.

17. Liu RH. Potential synergy of phytochemicals in cancer prevention: Mechanism of action. Journal of Nutrition. 2004;134:3479S–85S.

18. World Health Organization. Micronutrient deficiencies. Combating vitamin A deficiency. Available from: http://www.who.int/nutrition/topics/vad/en/.

19. Caulfield LE, Richard SA, Black RE. Undernutrition as an underlying cause of malaria morbidity and mortality in children less than five years old. American Journal of Tropical Medicine and Hygiene. 2004;71:55–63. Underwood B. Vitamin A deficiency disorders: International efforts. Sommer A. Vitamin a deficiency and clinical disease: An historical overview. Journal of Nutrition. 2008;138:1835–9.

20. Beyer P. Golden Rice and 'Golden' crops for human nutrition. New Biotechnology. 2010;27:478–81. Beyer P, Al-Babili S, Ye X, Lucca P, Schaub P, Welsch R, Potrykus I. Golden rice: Introducing the β-carotene biosynthesis pathway into rice endosperm by genetic engineering to defeat vitamin A deficiency. Journal of Nutrition. 2002;132:506S–10S. Ye X, Al-Babili A, Kloti A, Zhang J, Lucca P, Beyer P, Potrykus I. Engineering the provitamin A (β-carotene) biosynthetic pathway into (carotenoid-free) rice endosperm. Science. 2000;287:303–5.

21. Paine JA, Shipton CA, Chaggar S, Howells RM, Kennedy MJ, Vernon G, Wright SY, Hinchliffe E, Adams JL, Silverstone AL, Drake R. Improving the nutritional value of Golden Rice through increased pro-vitamin A content. Nature Biotechnology. 2005;23:482–7.

22. Qaim M. Benefits of genetically modified crops for the poor: Household income, nutrition, and health. New Biotechnology. 2010;27:552–7. Welch RM, Graham RD. Breeding for

micronutrients in staple food crops from a human nutrition perspective. Journal of Experimental Botany. 2004;55:353–64.

23. Holick MF. Vitamin D deficiency. New England Journal of Medicine (JAMA). 2007;357:266–81.

24. Parra EJ. Human pigmentation variation: Evolution, genetic basis, and implications for public health. American Journal Physical Anthropology. 2007;28:85–105.

25. American Academy of Dermatology. American Academy of Dermatology Association reconfirms need to boost vitamin D intake through diet and nutritional supplements rather than ultraviolet radiation. 2011. Available from http://www.aad .org/. World Health Organization. Sunbeds, tanning and UV exposure. 2010. Available from http://www.who.int/ mediacentre/factsheets/fs287/en/print.html.

26. Shahriari M, Kerr PE, Slade K, Grant-Kels JE. Vitamin D and the skin. Clinics in Dermatology. 2010;28:663-8. Zhang R, Naughton DP. Vitamin D in health and disease: Current perspectives. Nutrition Journal. 2010;9:65–71.

27. Calvo MS, Whiting SJ, Barton CN. Vitamin D intake: A global perspective of current status. Journal of Nutrition. 2005;135:310–6.

28. Cui Y, Rohan TE. Vitamin D, calcium, and breast cancer risk: A review. Cancer Epidemiology, Biomarkers, and Prevention. 2006;15:1427–37.

29. Giovannucci E. The epidemiology of vitamin D and colorectal cancer: Recent findings. Current Opinion in Gastroenterology. 2006;22:24–9. Ingraham BA, Bragdon B, Nohe A. Molecular basis of the potential of vitamin D to prevent cancer. Current Medical Research Opinions. 2008;24:139–49. Kim YS, Milner JA. Dietary modulation of colon cancer risk. Journal of Nutrition. 2007;137:2576S–9S.

30. Pierrot-Deseilligny C, Souberbielle JC. Is hypovitaminosis D one of the environmental risk factors for multiple sclerosis? Brain. 2010;133:1869–88.

31. Munger KL, Zhang SM, O'Reilly E, Hernan MA, Olek MJ, Willett WC et al. Vitamin D intake and incidence of multiple sclerosis. Neurology. 2004;62:60–5. Yang S, Smith C, Prahl JM, Luo X, Deluca HF. Vitamin D deficiency suppresses cell-mediated immunity in vivo. Archives of Biochemistry and Biophysics. 1993;303:98–106.

32. Broadley SA. Could vitamin D be the answer to multiple sclerosis? Multiple Sclerosis. 2007;13:825–6. Solomon AJ, Whitham RH. Multiple sclerosis and vitamin D: A review and recommendations. Current Neurology and Neuroscience Reports. 2010;10:389–96.

33. Rajakumar K. Vitamin D, cod-liver oil, sunlight, and rickets: A historical perspective. Pediatrics. 2003;112:132–5.

34. Kreiter SR, Schwartz RP, Kirkman HN, Charlton PA, Calikoglu AS, Davenport ML. Nutrition rickets in African American breast-fed infants. Journal of Pediatrics. 2000;137:153–7. Pugliese MF, Blumberg DL, Hludzinski J, Kay S. Nutritional rickets in suburbia. Journal of the American College of Nutrition. 1998;17:637–41. Prentice A. Vitamin D deficiency: A global perspective. Nutrition Reviews, 2008;66:S153–64.

35. American Academy of Pediatrics. Pediatric nutrition handbook, 6th ed., Elk Grove Village, IL; 2008. Wagner CL, Greer FR. American Academy of Pediatrics Section on Breastfeeding. American Academy of Pediatrics Committee on Nutrition. Prevention of rickets and vitamin D deficiency in infants, children, and adolescents. Pediatrics. 2008;122:1142–52.

36. Sato Y, Kanoko T, Satoh K, Iwamoto J. The prevention of hip fracture with risedronate and ergocalciferol plus calcium supplementation in elderly women with Alzheimer disease: A randomized controlled trial. Archives of Internal Medicine. 2005;165:1737–42. Sato Y, Iwamoto J, Kanoko T, Satoh K. Amelioration of osteoporosis and hypovitaminosis D by sunlight exposure in hospitalized, elderly women with Alzheimer's disease: A randomized controlled trial. Journal of Bone Mineral Research. 2005;20:1327–33.

37. Gezen-Ak D, Dursun E, Ertan T, Hanagai H, Gurvit H, Emre M, Eker E, Ozturk M, Engin F, Yilmazer S. Association between vitamin D receptor gene polymorphism and Alzheimer's disease. Tohoku Journal of Experimental Medicine. 2007;212:275–82.

38. Bostick RM, Potter JD, McKenzie DR, Sellers TA, Kushi LH, Steinmetz KA, Folsom AR. Reduced risk of colon cancer with high intakes of vitamin E: The Iowa Women's Health Study. Cancer Research. 1992;15:4230–7. Kirsh VA, Hayes RB, Mayne ST, Chatterjee N, Subar AF, Dixon LB, Albanes D, Andriole GL, Urban DA, Peters U. PLCO Trial. Supplementation and dietary vitamin E, beta-carotene, and vitamin C intakes and prostate cancer risk. Journal of the National Cancer Institute. 2006;98:245–54. Kline K, Yu W, Sanders BG. Vitamin E and breast cancer. Journal of Nutrition. 2004;134:3458S–62S. Peters U, Littman AJ, Kristal AR, Patterson RE, Potter JD, White E. Vitamin E and selenium supplementation and risk of prostate cancer in the vitamins and lifestyle (VITAL) study cohort. Cancer Causes and Control. 2008;19:75–87. Slatore CG, Littman AJ, Au DH, Satia JA, White E. Long-term use of supplemental multivitamins, vitamin C, vitamin E, and folate does not reduce the risk of lung cancer. American Journal of Respiratory Critical Care Medicine. 2008;177:524–30.

39. Jacques PF, Taylor A, Moeller S, Hankinson SE, Rogers G, Tung W, Ludovico J, Willett WC, Chylack LT. Long-term nutrient intake and 5-year change in nuclear lens opacities. Archives of Ophthalmology. 2005;123:517–26. Teikari JM, Virtamo J, Rautalahti M, Palmgren J, Liesto K, Heinonen OP. Long-term supplementation with alpha-tocopherol and beta-carotene and age-related cataract. Acta Ophthalmologica Scandinavica. 1997;75:634–40.

40. Ferrigno L, Aldigeri R, Rosmini F, Sperduto RD, Maraini G; The Italian-American Cataract Study Group. Associations between plasma levels of vitamins and cataract in the Italian-American clinical trial of nutritional supplements and age-related cataract (CTNS): CTNS report #2. Ophthalmic Epidemiology. 2005;12:71–80.

41. Jialal I, Devaraj S. Scientific evidence to support a vitamin E and heart disease health claim: Research needs. Journal of Nutrition. 2005;135:348–53. Saremi A, Arora R. Vitamin E and cardiovascular disease. American Journal of Therapeutics. 2010;17:e56–e65. Stampfer MJ, Hennekens CH, Manson JE, Colditz GA, Rosner B, Willett WC. Vitamin E consumption and the risk of coronary disease in women. New England Journal of Medicine. 1993;328:1444–9.

42. Miller ER, Pastor-Barriuso R, Dalal D, Riemersma RA, Appel LJ, Guallar E. Meta-analysis: High-dosage vitamin E supplementation may increase all-cause mortality. Annals of Internal Medicine. 2005;142:37–46.

43. Oski FA. Nutritional anemias. Seminars in Perinatology. 1979;3:381–95.

44. Radecki TE. Calcium and vitamin D in preventing fractures: Vitamin K supplementation has powerful effect. British Medical Journal. 2005;331:108. Sasaki N, Kusano E, Takahashi H, Ando Y, Yano K, Tsuda E, Asano Y. Vitamin K2 inhibits glucocorticoid-induced bone loss partly by preventing the

reduction of osteoprotegerin (OPG). Journal of Bone and Mineral Metabolism. 2005;23:41–47.

45. Francucci CM, Rilli S, Fiscaletti P, Boscaro M. Role of vitamin D on biochemical markers, bone mineral density, and fracture risk. Journal of Endocrinological Investigation. 2007;30:24–8.

46. Committee on Fetus and Newborn (American Academy of Pediatrics). Controversies concerning vitamin K and the newborn. Pediatrics. 2003;112:191–2.

47. Bhat RV, Deshmukh CT. A study of vitamin K status in children on prolonged antibiotic therapy. Indian Pediatrics. 2003;40:36–40.

Nutrition and Cancer

It is not surprising that the thought of cancer can be frightening. Cancer is, after all, the second leading cause of death in the United States, exceeded only by heart disease. Over 11 million Americans have either had or currently have cancer, causing one in every four deaths annually.[1] However, our understanding of ways to prevent and cure cancer—for example, by consuming an optimal diet—is greater now than ever before. To put this into perspective, scientists estimate that about one-third of all cancers are diet related, and the American Cancer Society estimates that a healthy diet and regular exercise might prevent almost 200,000 cancer deaths in the United States each year.[2] In this Nutrition Matters, you will learn about the causes of cancer and the importance of nutrition and other lifestyle choices in its prevention.

How Does Cancer Develop?

Life is created when an egg and a sperm join together to form a new cell. This single new cell contains the basic genetic material (deoxyribonucleic acid, or DNA) that will be passed on to the millions of cells that ultimately make up the human body. For this to occur, the single cell must replicate its DNA and divide into two cells. In turn, those two cells replicate their DNA and divide into four cells. This cycle of DNA replication and cell division, called the **cell cycle,** occurs continually throughout a person's lifetime as new cells are needed for growth, development, and repair of the body's tissues. Another component of the cell cycle is **apoptosis,** which refers to the process in which old, damaged, or poorly functioning cells undergo programmed cell death. Each step of the cell cycle is illustrated in Figure 1. The appropriate balance of cell replication and cell death ensures that your cells are healthy and functional.

Cancer can occur when this delicate balance between cell growth and cell death is disrupted. In other words, cancer results from unregulated cell growth and division and/or an inability to initiate apoptosis. When this happens, a cancerous mass can rapidly grow and spread to other parts of the body. For example, liver cancer initially disrupts

© Stede Preis/Jupiterimages Corporation

liver function. As the cancer grows and spreads, however, the disease worsens because it impairs function of multiple organs in the body. Ultimately, some types of cancer can seriously damage critical processes and become life threatening.

Technically, cancer is caused by an alteration in various regions of the DNA that code for specific proteins regulating the cell cycle. Although experts believe that about 5% of cancer-causing DNA alterations are inherited from our parents at conception, most are thought to occur during our lifetime. "External" factors that can initiate or promote cancer include tobacco use, poor diet, exposure to harmful chemicals, radiation, and infectious organisms.

cell cycle The process by which cells grow, mature, replicate their DNA, and divide.

apoptosis (a – po – TO – sis) The normal process by which a cell leaves the cell cycle and dies; programmed cell death.

cancer A condition characterized by unregulated cell division.

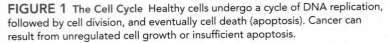

FIGURE 1 **The Cell Cycle** Healthy cells undergo a cycle of DNA replication, followed by cell division, and eventually cell death (apoptosis). Cancer can result from unregulated cell growth or insufficient apoptosis.

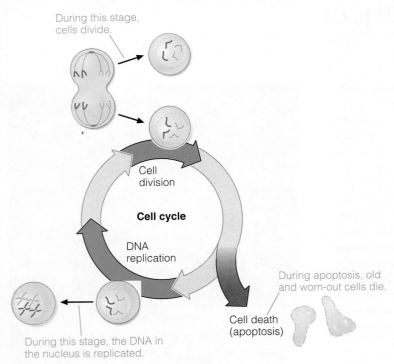

During this stage, cells divide.

Cell division

Cell cycle

DNA replication

During apoptosis, old and worn-out cells die.

Cell death (apoptosis)

During this stage, the DNA in the nucleus is replicated.

There are also "internal" factors such as hormones, immune conditions, and mutations that occur spontaneously and initiate cancer. Cancer can occur in almost any tissue type. However, it is most common in the prostate gland (males only), lungs, colon or rectum, and mammary (breast) tissue. Of these types of cancers, the mortality rate from lung cancer is highest.[1]

Carcinogens are agents that can cause cancer. There are many known carcinogens, such as X-rays, pollutants, and some of the chemicals in tobacco. Exposure to known chemical carcinogens tends to be highly regulated by public health agencies. However, exposure to other types of carcinogens, such as excessive sunlight, certain bacteria and viruses, and unknown chemical carcinogens, is more difficult to control.

CANCER DEVELOPS IN A MULTISTEP MANNER

You may be wondering how cancer occurs and progresses. Like type 2 diabetes and cardiovascular disease, cancer is generally a chronic, degenerative disease that can take years to develop. Many scientists believe that some cancers develop early in life but may not become evident until much later. For example, repeated sunburns during childhood might result in skin cancer during adulthood. The process by which normal, healthy cells develop into a cancerous growth occurs in steps, as illustrated in Figure 2 and described next.

Step 1: Initiation In the first step, called **initiation,** normal cells become "initiated" cells via a mutation in their DNA. An altered gene that is capable of transforming a normal cell into a cancer cell is called an **oncogene.** Although the formation of oncogenes is thought to be caused primarily by carcinogens, oncogenes can also be genetically inherited from our parents. The body has many defense mechanisms to help guard against the formation of oncogenes and ensuring that the altered cells that contain this mutated DNA do not continue to divide and grow. For example, antioxidant vitamins (such as vitamin C) and minerals (such as selenium) can help prevent and repair some forms of DNA damage.[3] The immune system is also involved in ridding the body of defective cells. In this way, antioxidants and the immune system can help halt cancer growth at the initiation stage, preventing mutated cells from passing on their defective genes.

Step 2: Promotion When damaged DNA is not repaired or destroyed, a cell can enter the second step of cancer, called **promotion.** During promotion, cancer cells replicate their mutated DNA and divide over and over, ultimately forming a growth, or **tumor.** Cancer cells that continue to grow are referred to as **malignant** and can spread to other parts of the body. However, not all tumors are dangerous; those that tend to be less life threatening are said to be **benign.** For example, breast tissue can

carcinogen (car – CIN – o – gen) Compound or condition that causes cancer.

initiation The first stage of cancer, in which a normal gene is transformed into a cancer-forming gene (oncogene).

oncogene (ON – co – gene) An abnormal gene that transforms normal cells to cancer cells.

promotion The second stage of cancer, in which the initiated cell begins to replicate itself, forming a tumor.

tumor A growth; sometimes caused by cancer.

malignant tumor A cancerous growth.

benign tumor A growth that may or may not have minimal physiological consequences but does not invade surrounding tissue or metastasize.

FIGURE 2 **The Development of Cancer** The stages of cancer involve initiation, promotion, progression, and metastasis.

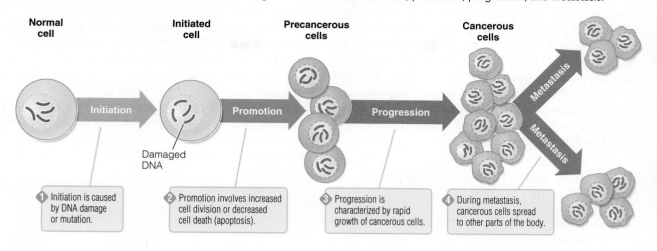

Normal cell → Initiation → Initiated cell (Damaged DNA) → Promotion → Precancerous cells → Progression → Cancerous cells → Metastasis / Metastasis

1. Initiation is caused by DNA damage or mutation.
2. Promotion involves increased cell division or decreased cell death (apoptosis).
3. Progression is characterized by rapid growth of cancerous cells.
4. During metastasis, cancerous cells spread to other parts of the body.

develop benign noncancerous growths (consisting of fibrous tissue). Although benign tumors do not spread to other tissues, some can grow quite large and can seriously impact a person's health. Other benign tumors can produce hormones or other damaging substances, or can become transformed into cancerous cells. Thus, it is important to seek medical advice to determine if a tumor is benign or malignant and whether or not it is dangerous.

Step 3: Progression and Metastasis Malignant tumors can enter a third step of cancer called **progression.** During progression, the tumor continues to grow rapidly and invades the surrounding tissue. At this stage, tumors also produce substances that encourage the growth of blood vessels, ensuring that the rapidly growing cancerous cells have a constant supply of energy, nutrients, and oxygen. In addition, cancerous cells can break away and travel in the blood or lymph to other tissues, forming new tumors. The movement of cancerous cells from their original site to other locations in the body is called **metastasis.** For example, cells associated with prostate cancer can break away, or metastasize, from the prostate gland to the other organs such as the bones and lungs.

The progression of a healthy cell to a cancerous one, and eventually metastasis, involves many steps, and anything that interrupts these steps can halt or slow the growth of cancer. Millions of dollars are spent annually to discover new ways to prevent and treat cancer. Scientists now know that lifestyle alterations such as good nutrition, surgery, radiation, and chemotherapy all can influence the various stages of cancer growth. These preventive lifestyle choices and cancer treatments are described in the following sections.

How Is Cancer Detected and Treated?

Although a thorough understanding of medical screening and cancer treatment is beyond the scope of this book, it is important for you to understand basic information about how cancer is detected and treated. In addition, it is helpful to know why some forms of cancer as well as their treatment options can greatly impact a person's appetite and nutritional status. This is why cancer patients are typically treated by a team of medical professionals that includes a dietitian who specializes in the various aspects of cancer and nutrition-related issues. The branch of medicine that pertains to the treatment and management of cancer is called **oncology.** Some of the fundamentals related to the detection, treatment, and nutritional implications of cancer are discussed next.

ROUTINE CANCER SCREENING IS RECOMMENDED

The old adage "An ounce of prevention is worth a pound of cure" is particularly true when it comes to cancer. Fortunately, there are many lifestyle choices you can make to decrease your cancer risk. Perhaps the most important aspect of cancer

progression The third stage of cancer, in which tumor cells rapidly divide and invade surrounding tissues.

metastasis The spreading of cancer cells to other parts of the body via the blood or lymph.

oncology The branch of medicine related to cancer and its treatment.

prevention is getting regular medical checkups to screen for cancer, because early detection often leads to a more positive prognosis. Cancer screening is especially important for people with several risk factors for this disease, and recommended guidelines can be found on the American Cancer Society's website (http://www.cancer.org/Healthy/FindCancerEarly/index). Depending on the cancer, clinicians recommend that baseline screening begin as early as 20 years of age.

THERE ARE MANY TREATMENT OPTIONS FOR CANCER

When cancer is detected, there are many treatment options including surgery, radiation therapy, and chemotherapy. However, which treatment or combination of treatments is used depends on many factors, such as the form, stage, and location of cancer and the health and age of the patient. Regardless of the therapeutic approach, the goal is always the same—to physically remove the tumor or interrupt the uncontrolled growth of the cancerous cells.

Surgical treatment involves removing the cancerous growth with the hope that it has not spread to other parts of the body. If the cancer has metastasized (spread), surgery is often followed by radiation and/or chemotherapy. Radiotherapy involves exposing any residual or remaining cancer cells to high-energy radiation to stop them from growing.* **Chemotherapy** involves the use of drugs (or *chemi*cals, hence *chem*otherapy) that halt the cycle of cell division. Some chemotherapeutic agents work by interfering with the function of nutrients, such as folate. Recall that folate is essential for cell division. By targeting the enzymes needed for folate to function, the growth of certain cancer cells is halted. This is one reason why some cancer treatments can impact a person's nutritional status. Unfortunately, chemotherapy agents are largely nonspecific, meaning that they kill noncancerous cells as well. Consequently, people receiving chemotherapy often experience many side effects such as loss of appetite, poor nutrient absorption, weight loss, and hair loss.

CANCER AND ITS TREATMENTS CAN INFLUENCE NUTRITIONAL STATUS

There are many nutritional complications associated with cancer and its treatments. The severity of these complications depends on the kind of cancer and type of treatment. For example, cancer often results in a type of anorexia: loss of appetite and an aversion to selected foods. Cancer can also cause severe wasting of lean (muscle) tissue, a condition called **cachexia.** The causes of cancer-associated anorexia and cachexia are unclear. However, there is growing evidence that a variety of substances produced by the tumors themselves may be responsible.[4] Together, anorexia and cachexia can cause severe undernutrition that must be addressed by the medical team. Fortunately, there are many ways that cancer-related malnutrition can be managed.

Although it may be difficult, it is important for people with cancer to maintain a healthy, balanced diet. This goal is often met by consuming small meals and sufficient amounts of fluids throughout the day. For example, a cancer patient might consider eating six small meals rather than three large meals in a day. Because people respond to cancer and its treatments differently, each person needs to develop his or her own strategies to ensure adequate nutrition. Nonetheless, several guidelines are often recommended. These include eating dry toast or crackers before getting out of bed in the morning, avoiding food prior to a treatment, and sipping liquids throughout the day using a straw. Because metal silverware can leave a bad taste in their mouths, some cancer patients prefer eating with plastic utensils.[5] Many cancer patients also have difficulty digesting the milk sugar lactose, and therefore find it necessary to avoid dairy products. Because chemotherapy can cause diarrhea and loss of electrolytes, it is important for cancer patients to consume foods and liquids containing sodium and potassium such as potatoes, bananas, apricot nectar, and bouillon or broth.

What Other Factors Are Related to Risk of Cancer?

Many factors influence a person's likelihood of developing cancer, some of which are summarized in Table 1 and Figure 3. Some factors, such as genetic predisposition, sex (gender), and age, cannot be changed. However, lifestyle-related risk factors, such as poor nutrition, tobacco use, and physical

---

chemotherapy The use of drugs to stop the growth of cancer.

cachexia (ca – CEX – i – a) A condition in which a person loses lean body mass (muscle tissue).

---

*While certain types of radiation initiate cancer, other types are used to treat cancer.

FIGURE 3 Genetic, Environmental, and Lifestyle Factors Interact to Influence Cancer Risk When genetic susceptibility is combined with environmental factors and high-risk behaviors, the overall risk for cancer increases.

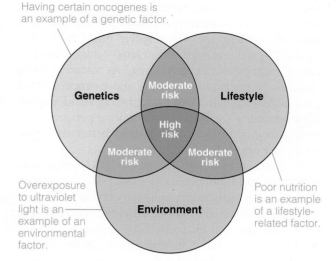

Having certain oncogenes is an example of a genetic factor.

Overexposure to ultraviolet light is an example of an environmental factor.

Poor nutrition is an example of a lifestyle-related factor.

In this section, you will learn about biological, lifestyle, and environmental risk factors associated with cancer that are *not* diet-related. Later, we will focus on what is known about *dietary* patterns related to this disease. Keep in mind, however, that many of these risk factors interact to influence your cancer risk. For example, your genetic predisposition may impact whether you can or cannot alter your cancer risk by eating (or not eating) certain foods.

BIOLOGICAL FACTORS ARE RELATED TO CANCER RISK

Biological (nonmodifiable) factors related to cancer risk include sex, ethnicity, race, and a variety of additional genetic variables. In addition, although cancer can develop in children, it is typically diagnosed in adulthood—making age an important biological risk factor as well. Researchers estimate that 25% of all cancer cases are caused by biological (unmodifiable) factors.

Sex (Gender) Whether you are male or female can be a risk factor for some types of cancer. For example, prostate cancer is found only in men, whereas breast cancer is far more common in women. Basic anatomical and physiological differences between males and females account for these differences. Other cancers, such as lung cancer, are slightly more common in males than females. In this case, the difference is more likely due to lifestyle choices, such as tobacco use, and less to biological (genetic)

inactivity, are well within our control. Environmental factors like excessive exposure to sunlight and pollutants can also affect your risk for cancer. The interaction of these factors and their collective influence on cancer risk can be considerable. Remember that just because you may have a risk factor for a disease does not mean that you will develop it. In fact, most people never develop cancer, yet have several known risk factors. Nonetheless, being aware of risk factors and making healthy lifestyle choices can help lower your risk.

Ethnicity and/or race can influence risk of some cancers.

differences. The risk of developing most cancers, such as colorectal cancer, is similar for both sexes.

Ethnicity and Race Although scientists do not know why, ethnicity and/or race can also influence risk for some cancers. Note that although these terms are often used interchangeably, they actually have different meanings. "Race" refers to a human population considered distinct based on physical characteristics. For example, being Caucasian is a racial characteristic. As such, a person's race is dependent solely on his or her genetic makeup. On the other hand, "ethnicity" refers to social groups with a shared history, sense of identity, geography, and cultural roots that may occur despite racial difference. For instance, black men of African descent living in the United States are at the greatest risk of developing prostate cancer, yet this form of cancer is somewhat rare among black men living in Africa.[1] Whether this ethnic difference is due to genetic, lifestyle, or environmental differences is difficult to determine, although it is most likely an interaction among these factors.

Other Genetic and Epigenetic Factors In addition to age, sex, race, and ethnicity, other genetic factors can also predispose a person to various cancers. For example, the presence of certain gene polymorphisms can increase or decrease overall risk of developing cancer. You may recall that gene polymorphisms are variations of a particular gene that are somewhat common in the population. Polymorphisms are not considered mutations, because polymorphisms have a frequency of at least 1%. An example of a polymorphism that can increase risk of cancer is a genetic variation called BRCA1.[6] People with this polymorphism are more likely than those with other gene variants to develop breast cancer. Another example of a polymorphism is GSTP1, which increases breast cancer risk in premenopausal women. Interestingly, there is some epidemiological evidence that consumption of cruciferous vegetables (like broccoli) may help ameliorate the negative effect of having the GSTP1 gene variant, illustrating how nutritional choices and genetics can interact.[7] As scientists learn more about the human genome, more cancer-related gene polymorphisms will undoubtedly be identified. Understanding the interactions among genetics, lifestyle, and environment may lead to better strategies for cancer prevention and treatment.

In addition to alterations in the DNA, *epi*genetic variations can influence cancer risk.[8] You may recall that the term *epigenetics* refers to changes in gene function that occur without actually changing the DNA sequence. For example, if a gene is flanked with methyl groups ($-CH_3$), its expression can be altered. Scientists are now learning that epigenetic patterns—including some related to cancer risk—can be passed from one generation to the next and can also be altered by nutritional status. Clearly, this is an area of important current scientific research.

LIFESTYLE AND ENVIRONMENTAL FACTORS IMPACT CANCER RISK

Although you have no control over the biological factors that predispose you to or protect you from cancer, the good news is that leading a healthy lifestyle and avoiding environmental carcinogens can help prevent some types of cancer. The major lifestyle and environmental risk factors for cancer, listed in Table 1,

TABLE 1 Major Biological, Lifestyle, and Environmental Factors Related to Cancer Risk

Risk Factors	Comments
Biological (Nonmodifiable) Risk Factors	
Age	Most cancers are detected in people age 50 years and older.
Sex	Some cancers (e.g., prostate) occur only in men, whereas others (e.g., cervical) occur only in women; some cancers occur in both sexes but are more prevalent in one.
Ethnicity and/or race	Some cancers are more common in people of a certain race and/or ethnicity.
Genetic predisposition	Many genetic factors can put a person at greater risk for various types of cancers.
Age at first menstruation	Breast cancer is more common in women with early onset of menses; note that diet may play a role in this relationship.
Lifestyle and Environmental (Modifiable) Risk Factors	
Body weight	Being overweight or obese increases risk for many cancers (e.g., breast, colorectal, and ovarian).
Physical activity	Being physically inactive increases risk for many cancers (e.g., breast and colorectal).
Tobacco use	Cancers of the lung, colon and rectum, mouth, larynx, bladder, kidney, cervix, esophagus, and pancreas are related to tobacco use, including cigarettes, cigars, chewing tobacco, and snuff; secondhand smoke can also increase risk for cancer.

TABLE 1 (*Continued*)

Risk Factors	Comments
Lactation	Having never breastfed or having only breastfed for a short period of time increases risk for breast cancer.
Reproductive history	Breast and uterine cancers are more common in women who never have children or begin having children after the age of 30 years.
Alcohol consumption	Alcohol consumption over 1 to 2 drinks per day increases risk of some cancers (e.g., breast, colorectal, and mouth).
Dietary fat intake	Diets high in fat are related to increased risk for several cancers (e.g., breast, colorectal, and pancreatic).
Fruit and vegetable	Diets high in fruits and vegetables appear to be protective against some cancers (e.g., colorectal and bladder).
Red meat or processed meat	High intakes of red or processed meat are related to increased risk for colorectal cancer.
Exposure to strong sunlight or tanning beds	Skin cancer is related to unprotected exposure to strong sunlight or tanning beds.
Infection with certain bacteria or viruses	Some bacteria and viruses (e.g., hepatitis B and C) can cause cancer.
Radiation exposure	Radiation from occupational, medical, and environmental sources increases the risk for lung cancer.
Exposure to environmental carcinogens	Some compounds, such as benzene, radon, asbestos, and lead, can increase the risk for cancer.

SOURCE: Adapted from American Cancer Society. Cancer facts and figures 2010. Atlanta, GA: American Cancer Society; 2010.

include tobacco use (especially smoking), chemical carcinogens, radiation (from excessive sun exposure or X-rays), infectious disease, certain reproductive patterns, and physical inactivity. Of course, many dietary variables are also related to risk of cancer, and these will be discussed later in this Nutrition Matters.

Smoking and Tobacco Use One of the most dangerous lifestyle choices a person can make in terms of cancer risk is smoking or chewing tobacco. Indeed, smoking accounts for at least 30% of all cancer deaths and 87% of lung cancer deaths. Further, the risk of developing lung cancer is about 23 times higher in male smokers and 13 times higher in female smokers, compared to lifelong nonsmokers.[1] It is also important to recognize that, even if you do not smoke, living in close proximity to a smoker can increase your risk of developing lung cancer. It is estimated that each year, about 3,400 nonsmoking adults die of lung cancer as a result of breathing secondhand smoke.[9] Experts agree that the best way to prevent lung cancer is to not smoke and to avoid exposure to secondhand smoke. Using smokeless tobacco (such as chewing tobacco) also increases the risk of developing certain types of cancer such as those of the mouth, throat, and esophagus.[10] This risk is magnified further in smokeless-tobacco users who drink large amounts of alcohol.

Other Environmental and Lifestyle Factors In addition to tobacco use, there are many other environmental carcinogens, such as excessive exposure to X-rays, strong ultraviolet light (from sunlight and tanning beds), and some pesticides. Also, some cancers are caused by viral or bacterial infection. For example, the hepatitis viruses can promote liver cancer.[11] Similarly, cervical cancer can be caused by the human papilloma virus,[12] and stomach cancer can be due to the bacterium *Helicobacter pylori*.[13] Because of the compelling evidence of a causal relationship between the human papilloma virus and cervical cancer, the U.S. Centers for Disease Control and Prevention in 2007 recommended the routine vaccination of females.

Many modifiable lifestyle choices can also influence cancer risk. This is especially true for breast cancer, which is related to many factors including a woman's reproductive patterns. For example, the use of oral contraceptives, not having children or having your first child after 30 years of age, and never having breastfed all increase a woman's risk.[14] Researchers think these factors contribute to breast cancer risk in part by increasing lifetime exposure to the cancer-promoting hormone estrogen. However, research continues to enable scientists to more fully understand the complex relationship between reproductive patterns and cancer risk.

Can Optimal Nutrition Help Prevent Cancer?

The National Cancer Institute and the American Cancer Society estimate that approximately one-third of all cancer deaths that occur in the United States each year are due to poor nutrition and physical inactivity. This estimate is both staggering and hopeful, because it means that many cancers can be prevented by dietary change—something most people can control.

Scientists are still unraveling the myriad complementary and overlapping mechanisms by which nutrition and energy balance influence cancer. Following the dietary recommendations developed by the American Institute of Cancer Research and the World Cancer Research Fund can help decrease your risk of cancer.[2] As outlined in Table 2, there are eight general recommendations and two special recommendations jointly put forth by these groups. Once again, the multifaceted and complex relationship between diet and risk of cancer demonstrates the importance of dietary variety and moderation in relation to health and well-being.

RECOMMENDATION #1: MAINTAIN A HEALTHY BODY WEIGHT

Perhaps the most important way to decrease your risk for cancer is to maintain a healthy body weight. Being overweight or obese increases the risk for several types of cancer including those of the esophagus, pancreas, colon, breast (postmenopausal), uterus, and kidney.[2] There is also some evidence that greater body fat contributes to liver and gallbladder cancers. Because there is a tendency for overweight

TABLE 2 The World Cancer Research Fund and American Institute for Cancer Research Recommendations for Preventing Cancer through Food, Nutrition, and Physical Activity

General Recommendations	Personal Recommendations
Be as lean as possible within the normal range of body weight.	• Ensure that body weight through childhood and adolescent growth projects toward the lower end of the normal body mass index (BMI) range at 21 years of age. • Maintain body weight within the normal range after 21 years of age. • Avoid age-related weight gain and increases in waist circumference throughout adulthood.
Make physical activity a part of everyday life.	• Be moderately physically active, equivalent to brisk walking, for at least 30 minutes every day. • As fitness improves, aim for 60 minutes or more of moderate, or for 30 minutes or more of vigorous physical activity every day. • Limit sedentary habits such as watching television.
Limit consumption of energy-dense or sugary foods.	• Consume energy-dense foods sparingly. • Avoid sugary drinks. • Consume "fast foods" sparingly, if at all.
Eat mostly foods of plant origin.	• Eat at least five portions/servings (at least 400 g or 14 oz) of a variety of nonstarchy vegetables and of fruit every day. • Eat relatively unprocessed cereals (grains) and/or pulses (legumes) with every meal. • Limit refined starchy foods.
Limit intake of red meat and avoid processed meat.	• If you eat red meat, consume less than 500 g (18 oz) each week, and eat little or no processed meat.
Limit alcoholic drinks.	• If you drink, limit your alcohol intake to no more than 2 drinks a day for men and 1 drink daily for women.
Limit consumption of salt and avoid moldy cereals (grains) or legumes.	• Consume salt at levels within the ranges suggested by the Dietary Reference Intakes, and do not consume any cereal grain or legume that might have been improperly dried.
Aim to meet nutritional needs through diet alone.	• Consume a well-balanced diet providing all needed nutrients instead of relying on dietary supplements.
Mothers should breastfeed, and children should be breastfed.	• Most mothers should exclusively breastfeed their infants for 6 months and continue to breastfeed with complementary feeding thereafter.
Cancer survivors should follow the dietary recommendations put forth for cancer prevention.	• Cancer survivors should receive nutritional care from an appropriately trained professional. • If possible, cancer survivors should follow the other recommendations listed here for diet, healthy weight, and physical activity.

SOURCE: Adapted from World Cancer Research Fund/American Institute for Cancer Research. Food, nutrition, physical activity, and the prevention of cancer: A global perspective. Washington, DC: American Institute of Cancer Research. Washington, DC, 2007.

children to become overweight adults, it is especially important for children and adolescents to maintain a healthy body weight. The expert panel suggests that you follow these personal recommendations.

- Maintain a healthy weight through childhood and adolescence so that, at 21 years of age, body mass index (BMI) is between 18.5 and 24.9 kg/m².
- Maintain a body weight within the normal range after 21 years of age.*
- Avoid age-related weight gain and increases in waist circumference throughout adulthood.

The physiological mechanisms by which obesity is related to cancer risk are not well understood. However, it is possible that the same genetic factors that increase our risk for obesity may also influence our risk for cancer.[15] It is likely that excessive energy intake makes it easier for cancer cells to grow and metastasize. As scientists learn more about how being overweight or obese increases cancer risk, it is likely that they will also know more about how to prevent and treat weight-related cancers.

RECOMMENDATION #2: BE PHYSICALLY ACTIVE

Being sedentary can also contribute to certain cancers. For example, weight gain resulting from physical inactivity can contribute to the development of some cancers such as breast cancer. Moreover, there is convincing evidence that physical activity decreases the risk for colorectal cancer.[2] As with the physiological mechanisms by which obesity is related to cancer risk, researchers continue to study why increased physical activity seems to decrease risk for some forms of cancer. However, recent studies have suggested

that the effect of exercise on lowering chronic inflammation might be involved.[16] You can read more about recommendations for physical activity in Table 2.

RECOMMENDATION #3: LIMIT CONSUMPTION OF ENERGY-DENSE FOODS AND SWEETENED BEVERAGES

Overconsumption of energy-dense foods and sweetened beverages is likely contributing to the global increase in obesity. Recall that *energy density* is the amount of energy (in kcal) per weight (usually 100 grams) of food. Foods made primarily of highly processed ingredients often contain substantial amounts of fat or sugar and tend to be energy dense. Energy-dense foods (including sweetened beverages) not only contribute large amounts of calories but tend to not be very nutrient dense, therefore contributing little to overall nutrient requirements. Taken together, the evidence shows that the specific dietary constituents in energy-dense foods and sweetened beverages are not as problematic as the calories they contain. As such, the committee recommends that you consume energy-dense foods and sweetened beverages sparingly.

RECOMMENDATION #4: EAT MOSTLY FOODS OF PLANT ORIGIN

Upon reviewing hundreds of studies, the National Cancer Institute and the American Cancer Society concluded that overall dietary patterns often associated with decreased cancer risk—such as Mediterranean and vegetarian diets—consist mainly of foods of plant origin. In addition, increasing consumption of several specific types of plant foods may be particularly protective against some forms of cancer. For example, increasing consumption of fruit and nonstarchy vegetables may help decrease the risk for cancers of the mouth, esophagus, lungs, and stomach. Nonstarchy vegetables include green, leafy vegetables (like spinach), carrots, turnips, broccoli, okra, eggplant, and bok choy, whereas starchy vegetables include potatoes, yams, and cassava. In addition, greater consumption of high-fiber foods may lower risk of colorectal cancer.

Being physically active is a powerful way to help prevent cancer.

* "Normal range" refers to appropriate ranges issued by national governments or the World Health Organization.

Micronutrients in Plants Are Important The mechanisms by which increased consumption of plant-based foods help decrease cancer risk are likely many and complex. First, minimally processed plant-based foods such as fruits, vegetables, and whole grains tend to be excellent sources of many essential vitamins and minerals needed to maintain healthy cells. In other words, they are *nutrient dense*. As you learned in previous chapters, these foods tend to be excellent sources of the antioxidant vitamins and minerals (such as vitamins C and E and selenium), as well as folate, vitamin A, and vitamin D. There are many ways that these nutrients might inhibit cancer. For example, the antioxidant function of vitamin C, vitamin E, and selenium may inhibit free-radical formation or repair damaged DNA, thus inhibiting the initiation phase of cancer. Similarly, vitamin A, vitamin D, and folate are all required for cell differentiation and growth, and many different vitamins are needed for maintaining a healthy immune system. In addition to the "traditional" nutrients, some plant foods also are rich sources of phytochemicals that might influence cancer. We will discuss these next.

Phytochemicals—Potential Cancer Fighters There are literally hundreds, if not thousands, of biologically active compounds found in plant foods. As you have learned, plant-based compounds called phytochemicals, although not considered traditional nutrients, may nonetheless influence health. Some of the phytochemicals thought to be important in decreasing risk for cancer are listed in Table 3. Examples of possible anticarcinogenic phytochemicals include β-carotene,[†] lycopene, and isoflavones. Many foods are known to contain phytochemicals, and such "functional foods" include tomatoes, green tea, garlic, soy products, certain spices, and cruciferous vegetables. For example, the bright red compound lycopene (a nonprovitamin A carotenoid) found in tomatoes is a potent antioxidant, and some studies show that increased consumption of tomato products decreases risk for prostate cancer.[17] Other compounds, called isothiocyanates (found in cruciferous vegetables such as broccoli, cauliflower, cabbage, and Brussels sprouts), are also potent antioxidants that may help keep cells healthy.[18]

⟨**CONNECTIONS**⟩ Free radicals are atoms or molecules with unpaired electrons, making them reactive and harmful to DNA, lipids, and proteins (Chapter 10, page 447).

[†]Note that, although it can be converted in the body to vitamin A, β-carotene is also considered a phytochemical because it has activities in the body that are independent of vitamin A.

TABLE 3 Some Potentially Anticarcinogenic Phytochemicals

Phytochemical	Selected Food Sources	Proposed Actions[a]
β-carotene	Carrots, pumpkin, sweet potatoes	Antioxidant; cell cycle; antibacterial
Caffeic acid phenethyl ester	Honey	Immune modulation; antioxidant
Capsaicin	Chili peppers	Immune modulation; apoptosis
Curcumin	Turmeric	Immune modulation
Diallyl sulfide	Garlic, onions	Modification of enzymes; antibacterial
Gingerol	Ginger	Antioxidant; immune modulation
Isoflavones	Soybeans	Anti-estrogenic activity
Isothiocyanates	Broccoli, Brussels sprouts, cauliflower	Modification of enzymes; antioxidant
Lignins	Flaxseed	Estrogen antagonist
Limonoids	Oranges, lemons, limes, grapefruit	Apoptosis
Lutein	Tomatoes, spinach	Antioxidant
Lycopene	Tomatoes, pink grapefruit	Antioxidant
Phenylpropanoids	Cinnamon, cloves, vanilla	Antioxidant; antibacterial
Polyphenols	Green tea	Immune modulation; apoptosis
Resveratrol	Grapes	Antioxidant; immune modulation

[a]Note that these should be viewed as simply proposed actions of these compounds; in some cases, data are inconsistent and/or weak.

SOURCES: Lampe JW. Spicing up a vegetarian diet: Chemopreventive effects of phytochemicals. American Journal of Clinical Nutrition. 2003;78:579S–83S. Nishino H, Murakoshi M, Mou XY, Wada S, Masuda M, Ohsaka Y, Satomi Y, Jinno K. Cancer prevention by phytochemicals. Oncology. 2005;69(Suppl 1):38–40. Rafter JJ. Scientific basis of biomarkers and benefits of functional foods for reduction of disease risk: cancer. British Journal of Nutrition. 2002;88:S219–24. Surh Y-J. Cancer chemoprevention with dietary phytochemicals. Nature Reviews. 2003;3:768–80. Talalay P, Fahey JW. Phytochemicals from cruciferous plants protect against cancer by modulating carcinogen metabolism. Journal of Nutrition. 2001;131:3027S–33S.

Fruits and vegetables contain many nutrients and phytochemicals that are likely important in preventing cancer.

Researchers continue to actively study the roles of phytochemicals and functional foods in preventing and treating cancer.

RECOMMENDATION #5: LIMIT INTAKE OF RED MEAT AND PROCESSED MEAT

Although the American Institute for Cancer Research and the World Cancer Research Fund recognize that many foods of animal origin are nourishing and health-promoting if consumed in modest amounts, they recommend that the consumption of red meat be limited to no more than 500 g (18 oz) each week.[2] In this context, "red meat" refers to beef, pork, lamb, and goat. They also recommend that we avoid processed meat if possible. In this instance, "processed meat" refers to meat preserved by smoking, curing or salting, or addition of chemical preservatives, and this type of meat includes bacon, sausage, hot dogs, ham, and cold cuts.

Nitrites, Nitrosamines, Heterocyclic Amines, and Polycyclic Aromatic Hydrocarbons Researchers do not know why excessive consumption of red meat is related to increased cancer risk, but there are several possible explanations. For example, the **nitrites** sometimes added to processed meats to preserve color and enhance flavor may be carcinogenic. This is because they are converted to compounds called **nitrosamines** that are known to cause cancer.[19] Charring meat at high temperatures may also lead to the formation of **heterocyclic amines,** shown to be carcinogenic in laboratory animals.[20] Small amounts of other carcinogens called **polycyclic aromatic hydrocarbons** are produced when meat is grilled.[21] There is also evidence that high intake of iron from red meat may be the factor related to increased cancer risk.[2] However, further research is needed to understand the physiological relationships among red meat intake, consumption of nitrites, heterocyclic amines, polycyclic aromatic hydrocarbons, iron, and cancer risk.

RECOMMENDATION #6: LIMIT ALCOHOL

Studies show that even small amounts of alcohol consumption can increase risk for oral, esophageal, colorectal, and breast cancers.[22] Thus, based solely on the cancer evidence, even small amounts of alcoholic drinks should be avoided. However, because moderate alcohol consumption may lower the risk for cardiovascular disease, the expert panel suggests that, if you drink, you should limit consumption to no more than two drinks per day if you are male and one drink per day if you are female. A drink is defined as 12 ounces of beer, 5 ounces of wine, or 1.5 ounces of distilled spirits such as gin or vodka. This recommendation is consistent with that of the 2010 Dietary Guidelines for Americans.

RECOMMENDATION #7: LIMIT SALT AND AVOID MOLDY GRAINS AND LEGUMES

Improper food preservation methods, such as oversalting and inadequate drying, may also increase cancer risk. For example, the expert committee concluded that excessive consumption of salt and salt-preserved foods is a probable cause of stomach cancers. Similarly, inadequate drying of foods—in particular grains and some legumes such as

---

nitrites Nitrogen-containing compounds that are often added to processed meats to enhance color and flavor.

nitrosamines Nitrogen-containing chemical carcinogens, produced from nitrites, that have been shown to cause cancer.

heterocyclic amines Cancer-causing compounds that can be formed when meat is cooked at high temperatures.

polycyclic aromatic hydrocarbons Cancer-causing compounds that can be formed when meat is grilled.

Breastfeeding helps protect women from breast cancer and decreases risk of obesity in infants. For this and many other reasons, it is recommended that all women breastfeed their infants.

peanuts—can lead to growth of a fungus (mold) that produces dangerous toxins such as aflatoxin. Aflatoxin has been shown to cause liver cancer, and there is evidence that it may also lead to cancers of the mouth, esophagus, and stomach.[23] You can read more about aflatoxin in the Nutrition Matters following Chapter 5. In summary, it is recommended that we limit salt consumption and avoid moldy grains and legumes.

RECOMMENDATION #8: CHOOSE FOODS OVER SUPPLEMENTS

Research shows that certain high-dose nutrient supplements can be protective against some forms of cancer in especially vulnerable people.[24] For example, there is considerable evidence that consumption of calcium supplements can decrease risk for colorectal cancer.[25] Unfortunately, use of other high-dose dietary supplements might also cause cancer.[26] For instance, studies show that smokers who take supplements of β-carotene have increased risk for lung cancer. Thus, the expert group concluded that consumption of nutrients through usual diet—as compared to supplements—is preferred. As such, dietary supplements are not recommended for cancer prevention.

SPECIAL RECOMMENDATION #1: WOMEN SHOULD BREASTFEED THEIR INFANTS

This joint report is the first to specifically make cancer-related recommendations concerning breastfeeding. Because breastfeeding helps both protect a woman from breast cancer (premenopausal) and prevent the later development of obesity in the infant that may lead to cancer in later life, it is recommended that all women aim to exclusively breastfeed their infants for up to six months.

SPECIAL RECOMMENDATION #2: CANCER SURVIVORS SHOULD FOLLOW THE SAME RECOMMENDATIONS

Except in special situations, the panel recommends that cancer survivors follow the same dietary recommendations as those suggested for the general population. This includes all recommendations for consuming a healthy diet, maintaining healthy body weight, and pursuing sufficient physical activity levels. In addition, this group of experts recommends that all cancer survivors receive nutritional care from an appropriately trained nutrition professional such as a dietitian.

Diet and Cancer: What Is in the Future?

There is still much to learn about how dietary choices influence risk of cancer. This is partially due to the inherent difficulties in studying the relationship between nutrition and the development of this disease. In addition, many genetic factors can modulate the relationship between dietary intake and cancer risk. In other words, genetic factors interact with diet to influence the risk of cancer. The medical community hopes that as we learn more about genes, lifestyle, and other factors that influence cancer risk, it will be possible to offer specialized dietary advice to people who are at greatest risk for developing certain types of cancers. Perhaps more than any other chronic disease, cancer holds the greatest hope in the area of nutritional prevention and treatment.

Key Points

How Does Cancer Develop?

- Cancer results from the uncontrolled growth or inadequate death (apoptosis) of cells caused by an alteration or mutation in the DNA.

- During cancer initiation, DNA is altered due to a carcinogen or simply due to a mistake in DNA replication during the cell cycle.

- If the mutation in the DNA is not repaired or if the cell does not die via apoptosis, it can enter the second stage of cancer, called promotion.

- In the third stage, called progression, cancerous cells break away and lodge elsewhere in the body. This is called metastasis.

How Is Cancer Detected and Treated?

- It is not always possible to prevent cancer, but routine screening helps detect it earlier, when it is most treatable.

- When cancer is detected, there are many options for treatment, such as surgery, radiation, and chemotherapy.

- Both cancer and cancer treatment have many implications for nutritional status.

What Other Factors Are Related to Risk of Cancer?

- Cancer can be caused by biological, lifestyle, and environmental variables.

- Biological factors include genetic predisposition, age, race, and sex.

- Environmental carcinogens include chemicals, cigarette smoke, radiation, and pathogens (bacteria and viruses).

- Nondietary lifestyle factors that influence cancer risk include a variety of reproductive patterns in women that increase the body's exposure to the cancer-promoting hormone estrogen.

Can Optimal Nutrition Help Prevent Cancer?

- To help prevent cancer, maintain a healthy body weight, exercise, emphasize plant-based foods, limit intake of red or processed meats, if you drink do so in moderation, avoid moldy legumes, and get the majority of your nutrients from foods. It is also recommended that women breastfeed their infants.

Notes

1. American Cancer Society. Cancer Facts and Figures 2010. Atlanta: American Cancer Society, 2010.

2. World Cancer Research Fund/American Institute for Cancer Research. Food, nutrition, physical activity, and the prevention of cancer: A global perspective. Washington, DC: American Institute for Cancer Research; 2007.

3. Bardia A, Tleyjeh IM, Cerhan JR, Sood AK, Limburg PJ, Erwin PJ, Montori VM. Efficacy of antioxidant supplementation in reducing primary cancer incidence and mortality: Systematic review and meta-analysis. Mayo Clinic Proceedings. 2008;83:23–34. Ni J, Yeh S. The roles of alpha-vitamin E and its analogues in prostate cancer. Vitamins and Hormones. 2007;76:493–518.

4. Lelbach A, Muzes G, Feher J. Current perspectives of catabolic mediators of cancer cachexia. Medical Science Monitor. 2007;13:168–173. Tisdale MJ. Tumor-host interactions. Journal of Cellular Biochemistry. 2004;93:871–7. Ströhle A, Zänker K, Hahn A. Nutrition in oncology: The case of micronutrients (review). Oncology Reports. 2010;24:815–28. Tisdale MJ. Cancer cachexia. Langenbecks Archives of Surgery. 2004; 389:299–305.

5. American Institute for Cancer Research. Nutrition of the cancer patient. Washington, DC: American Institute for Cancer Research; 2007. Available from: http://www.aicr.org.

6. Nelson HD, Huffman LH, Fu R, Harris EL. Genetic risk assessment and BRCA mutation testing for breast and ovarian cancer susceptibility: Systematic evidence review for the U.S. preventive services task force. Annals of Internal Medicine. 2005;143:362–79. Pasche B. Recent advances in breast cancer genetics. Cancer Treatment Research. 2008;141:1–10. Paradiso A, Formenti S. Hereditary breast cancer: Clinical features and risk reduction strategies. Annals of Oncology. 2011;22:i31–6. U.S. Preventive Services Task Force. Genetic risk assessment and BRCA mutation testing for breast and ovarian cancer susceptibility: Recommendation statement. Annals of Internal Medicine. 2005;143:355–61.

7. Lee SA, Fowke JH, Lu W, Ye C, Zheng Y, Cai Q, Gu K, Gao YT, Shu XO, Zheng W. Cruciferous vegetables, the GSTP1 Ile105Val genetic polymorphism, and breast cancer risk. American Journal of Clinical Nutrition. 2008;87:753–60.

8. Prins GS, Tang WY, Belmonte J, Ho SM. Developmental exposure to bisphenol A increases prostate cancer susceptibility in adult rats: Epigenetic mode of action is implicated. Fertility and Sterility. 2008; 89(2 Suppl):e41. Vucic EA, Brown CJ, Lam WL. Epigenetics of cancer progression. Pharmacogenomics. 2008;9:215–34.

9. Centers for Disease Control and Prevention. Smoking-attributable mortality, years of potential life lost, and productivity losses—United States, 2000–2004. Morbidity and Mortality Weekly Reports. 2008;57:1226–8. U.S. Department of Health and Human Services. The health consequences of involuntary exposure to tobacco smoke: A report of the Surgeon General. Rockville, MD: U.S. Department of Health and Human Services, Public Health Service, Centers for Disease Control and Prevention. National Center for Chronic Disease Prevention and Health Promotion, Office on Smoking and Health; 2006.

10. U.S. Department of Health and Human Services. The health consequences of using smokeless tobacco: A report of the advisory committee to the surgeon general. Atlanta, GA: U.S. Department of Health and Human Services, National Institutes of Health, National Cancer Institute; 1986.

11. He B, Zhang H, Shi T. A comprehensive analysis of the dynamic biological networks in HCV induced hepatocarinogenesis. PLoS One. 2011;6:e18516.

12. Hung CF, Ma B, Monie A, Tsen SW, Wu TC. Therapeutic human papilloma virus vaccines: Current clinical trials and future directions. Expert Opinion on Biological Therapy. 2008;8:421–39. Paavonen J. Human papilloma virus infection and the development of cervical cancer and related genital neoplasias. International Journal of Infectious Disease. 2007;11(Suppl.2):S3–9.

13. Plummer M, Franceschi S, Munoz N. Epidemiology of gastric cancer. International Agency for Research on Cancer Scientific Publications. 2004;157:311–26.

14. Jernstrom H, Lubinski J, Lynch HT, Ghadirian P, Neuhausen S, Isaacs C, Weber BL, Horsman D, Rosen B, Foulkes WD, Friedman E, Gershon-Baruch R, Ainsworth P, Daly M, Garber J, Olsson H, Sun P, Narod SA. Breast-feeding and the risk of breast cancer in BRCA1 and BRCA2 mutation carriers. Journal of the National Cancer Institute. 2004;96:1094–8.

15. Hursting SD, Lashinger LM, Wheatley KW, Rogers CJ, Colbert LH, Nunez NP, Perkins SN. Reducing the weight of cancer: Mechanistic targets for breaking the obesity-carcinogenesis link. Best Practical Research in Clinical Endocrinology and Metababolism. 2008;22:659–69. Wei EK, Wolin KY, Colditz GA. Time course of risk factors in cancer etiology and progression. Journal of Clinical Oncology. 2010;28:4052–7.

16. Fair AM, Montgomery K. Energy balance, physical activity, and cancer risk. Methods in Molecular Biology. 2009;472:57–88. Basen-Engquist K, Chang M. Obesity and cancer risk: Recent review and evidence. Current Oncology Reports. 2011;13:71–5. Friedenreich CM. Physical activity and breast cancer: Review of the epidemiologic evidence and biologic mechanisms. Recent Results in Cancer Research. 2011;188:125–39. Walsh NP, Gleeson M, Shephard RJ, Gleeson M, Woods JA, Bishop NC, Fleshner M, Green C, Pedersen BK, Hoffman-Goetz L, Rogers CJ, Northoff H, Abbasi A, Simon P. Position statement. Part one: Immune function and exercise. Exercise Immunology Review. 2011;17:6–63.

17. Campbell JK, Canene-Adams K, Lindshield BL, Boileau TWM, Clinton SK, Erdman JW. Tomato phytochemicals and prostate cancer risk. Journal of Nutrition. 2004;134:3486S–92S. Seren S, Lieberman R, Bayraktar UD, Heath E, Sahin K, Andic F, Kucuk O. Lycopene in cancer prevention and treatment. American Journal of Therapeutics. 2008;15:66–81.

18. Juge N, Mithen RF, Traka M. Molecular basis for chemoprevention by sulforaphane: a comprehensive review. Cellular and Molecular Life Sciences. 2007; 64:1105–27.

19. Ferguson LR. Meat and cancer. Meat Science. 2010;84:308–13. Jakszyn P, Gonzalez CA. Nitrosamine and related food intake and gastric and oesophageal cancer risk: A systematic review of the epidemiological evidence. World Journal of Gastroenterology. 2006; 21;12:4296–303. Bingham SA, Hughes R, Cross AJ. Effect of white versus red meat on endogenous N-nitrosation in the human colon and further evidence of a dose response. Journal of Nutrition. 2002;132:3522–5.

20. Shut HA, Snyderwise EG. DNA adducts of heterocyclic amine food mutagens: Implications for mutagenesis and carcinogenesis. Carcinogenesis. 1999;20:353–68.

21. Stacewicz-Sapuntzakis M, Borthakur G, Burns JL, Bowen PE. Correlations of dietary patterns with prostate health. Mol Nutr Food Res. 2008;52:114-30. Van Maanen JM, Moonen EJ, Maas LM, Kleinjans JC, van Schooten FJ. Formation of aromatic DNA adducts in white blood cells in relation to urinary excretion of 1-hydroxypyrene during consumption of grilled meat. Carcinogenesis. 1994;15:2263–8.

22. Seitz HK, Maurer B. The relationship between alcohol metabolism, estrogen levels, and breast cancer risk. Alcohol Research and Health. 2007;30:42–3. Moderate alcohol intake and cancer incidence in women. Allen NE, Beral V, Casabonne D, Kan SW, Reeves GK, Brown A, Green J; Million Women Study Collaborators. Journal of the National Cancer Institute. 2009;101:296–305.

23. Kew MC. Prevention of hepatocellular carcinoma. Annals of Hepatology. 2010;9:120–32.

24. Grau MV, Baron JA, Sandler RS, Haile RW, Beach ML, Church TR, Heber D. Vitamin D, calcium supplementation, and colorectal adenomas: Results of a randomized trial. Journal of the National Cancer Institute. 2003;95:1765–71. Wactawski-Wende J, Kotchen JM, Anderson GL. Calcium plus vitamin D supplementation and the risk of colorectal cancer. New England Journal of Medicine. 2006;354:684–96.

25. Carroll C, Cooper K, Papaioannou D, Hind D, Pilgrim H, Tappenden P. Supplemental calcium in the chemoprevention of colorectal cancer: A systematic review and

meta-analysis. Clinical Therapeutics. 2010;32:789-301. Weingarten MA, Zalmanovici A, Yaphe J. Dietary calcium supplementation for preventing colorectal cancer and adenomatous polyps. Cochrane Database Systematic Reviews. 2008;23:CD003548.

26. Bardia A, Tleyjeh IM, Cerhan JR, Sood AK, Limburg PJ, Erwin PJ, Montori VM. Efficacy of antioxidant supplementation in reducing primary cancer incidence and mortality: Systematic review and meta-analysis. Mayo Clinic Proceedings. 2008;83:23–34. Lawson KA, Wright ME, Subar A, Mouw T, Hollenbeck A, Schatzkin A, Leitzmann MF. Multivitamin use and risk of prostate cancer in the National Institutes of Health-AARP Diet and Health Study. Journal of the National Cancer Institute. 2007;99:754–64.

The Major Minerals and Water

CHAPTER **12**

You have now learned about the carbon-containing (organic) nutrients—proteins, carbohydrates, lipids, and vitamins—that are needed for many functions in the body. It is also essential that you consume *inorganic* substances such as minerals and water. Minerals, which like vitamins are considered micronutrients, make up the structure of bones and teeth, and participate in hundreds of chemical reactions that take place in the body. However, unlike vitamins that are categorized by their solubility (water-soluble and fat-soluble), minerals are classified by how much the body requires. Minerals required in relatively large quantities include calcium, phosphorus, magnesium, potassium, sodium, and chloride. These are called the major minerals and are covered in this chapter. Other minerals, which are needed in smaller quantities—the trace minerals—are presented in Chapter 13.

The major minerals play both structural and functional roles in the body. They act as cofactors, regulate gene expression, are involved in the formation of bones and teeth, and influence every physiological system in the body. In addition, several of the major minerals are important for water balance. Over the past decade, scientists have discovered that many, if not all, of the major minerals are involved in reducing the risk for certain chronic diseases. In this chapter, you will learn about the fundamentals related to each major mineral, including dietary sources, regulation of absorption and excretion, function, deficiency, toxicity, and requirements. The importance and regulation of water are also discussed.

JoAnn's Challenge with Bone Health

JoAnn lives an incredible life, which has been and continues to be focused on family and public service. For example, although she now calls Idaho home, JoAnn has lived around the world in places such as Sweden, Malawi, Ecuador, Chile, and Colombia, where she was involved in international development and education. Now in her 70s, she continues to be an active community volunteer. For instance, she is a key member of a grassroots organization whose goal is to restore and utilize the local vintage movie theater. She also travels frequently to Boise to accompany her husband, an elected member of the Idaho legislature. By all accounts, JoAnn is a woman committed to living purposefully and well.

However, JoAnn lives every day with a significant health challenge. That is, she has somewhat severe curvature and weakness in her spine. Although JoAnn has always been physically active and has—throughout her entire life—eaten nutritiously, several years ago she was diagnosed with scoliosis and osteoporosis, both serious bone diseases. These painful conditions have had a major impact on JoAnn's lifestyle, requiring constant and time-consuming physical therapy, medication, and continued attention to her diet and exercise routine.

As might be expected, JoAnn faces this challenge with determination and grace. For example, since her diagnosis, she continues to strengthen her bones and slow her bone loss by eating well and exercising daily. She attends aquatic exercise classes three days a week and takes yoga classes four mornings a week. JoAnn also hikes and walks whenever possible, and she eats a balanced diet built around dairy foods, legumes, salads, and fish. These foods are excellent sources of protein, calcium, phosphorus, and magnesium—nutrients needed to keep her bones healthy. Clearly, JoAnn understands the importance of keeping her bones as strong as possible so that she can continue to travel and do the many activities she enjoys.

We talked with JoAnn about living successfully with an aging skeleton, and she told us that one of the most difficult things for her is not being able to exercise while she is traveling. To her, exercise is mandatory when it comes to her bone health, and JoAnn is convinced she would be in a wheelchair if she were to give up any of her activities. But finding the time and place to exercise while traveling is often a challenge. JoAnn also emphasized her belief that mindful food choices, such as eating a variety of fruits and vegetables daily and making sure her calcium intake is adequate, have helped keep her bones as strong and healthy as possible. Not surprisingly, JoAnn gave us the following advice about living a long, healthy, and productive life: *"Be passionate about what you do, and make the best better. Set standards, and live by them."* She should know.

© Shelley McGuire

Critical Thinking: JoAnn's Story

Does anyone in your family or someone you know have bone disease, such as osteoporosis? If so, how has this affected his or her quality of life? In what way has having bone disease influenced that person's lifestyle choices, such as activity level or diet?

What Are Minerals?

In nutrition, the term **mineral** is used to describe inorganic atoms or molecules, other than water. Because the body requires them in very small amounts, dietary minerals are considered micronutrients. All minerals needed for health are essential nutrients because the body cannot make them from other compounds. Thus, we rely on the foods we eat to provide us with these important substances. Minerals can be neither created nor destroyed; even if you completely combust (burn) a food, the minerals will remain in the ash. This is true of the minerals in our bodies as well.

The periodic table depicts all the elements that make up our world. Although many elements are listed in the periodic table, only some of them are minerals, and only a few of these are necessary for life (Figure 12.1). Essential minerals are classified as major minerals or trace minerals, depending on how much the body needs. **Major minerals** (sometimes called macrominerals) are those required in amounts greater than 100 mg/day (equivalent to the weight of 4–5 grains of rice). The body requires six major minerals: calcium, phosphorus, magnesium, sodium, chloride, and potassium.

Some minerals, like the sodium chloride shown here, are inorganic substances that are essential for health.

COMMON CHARACTERISTICS OF MAJOR MINERALS

Minerals have diverse structural and functional roles. In fact, many serve so many roles that they influence virtually every physiological system in the body. It is also important to recognize that most minerals work together to carry out

mineral Inorganic substance, other than water, that is required by the body for basic functions or structure.

major mineral An essential mineral that is required in amounts greater than 100 mg daily.

FIGURE 12.1 The Periodic Table and the Required Trace and Major Minerals Six major minerals and at least eight trace minerals are vital for health and are therefore essential nutrients.

> The essential major minerals are calcium (Ca), phosphorus (P), magnesium (Mg), sodium (Na), chloride (Cl), and potassium (K).

> The essential trace minerals are iron (Fe), copper (Cu), iodine (I), selenium (Se), chromium (Cr), manganese (Mn), molybdenum (Mo), and zinc (Zn).

1 H Hydrogen																	2 Helium He
3 Lithium Li	4 Beryllium Be											5 Boron B	6 Carbon C	7 Nitrogen N	8 Oxygen O	9 Fluorine F	10 Neon Ne
11 Sodium Na	12 Magnesium Mg											13 Aluminum Al	14 Silicon Si	15 Phosphorus P	16 Sulfur S	17 Chlorine Cl	18 Argon Ar
19 Potassium K	20 Calcium Ca	21 Scandium Sc	22 Titanium Ti	23 Vanadium V	24 Chromium Cr	25 Manganese Mn	26 Iron Fe	27 Cobalt Co	28 Nickel Ni	29 Copper Cu	30 Zinc Zn	31 Gallium Ga	32 Germanium Ge	33 Arsenic As	34 Selenium Se	35 Bromine Br	36 Krypton Kr
37 Rubidium Rb	38 Strontium Sr	39 Yttrium Y	40 Zirconium Zr	41 Niobium Nb	42 Molybdenum Mo	43 Technetium Tc	44 Ruthenium Ru	45 Rhodium Rh	46 Palladium Pd	47 Silver Ag	48 Cadmium Cd	49 Indium In	50 Tin Sn	51 Antimony Sb	52 Tellurium Te	53 Iodine I	54 Xenon Xe
55 Cesium Cs	56 Barium Ba	71 Lutetium Lu	72 Hafnium Hf	73 Tantalum Ta	74 Tungsten W	75 Rhenium Re	76 Osmium Os	77 Iridium Ir	78 Platinum Pt	79 Gold Au	80 Mercury Hg	81 Thallium Tl	82 Lead Pb	83 Bismuth Bi	84 Polonium Po	85 Astatine At	86 Radon Rn
87 Francium Fr	88 Radium Ra	103 Lawrencium Lr	104 Rutherfordium Rf	105 Dubnium Db	106 Seaborgium Sg	107 Bohrium Bh	108 Hassium Hs	109 Meitnerium Mt									

57 Lanthanum La	58 Cerium Ce	59 Praseodymium Pr	60 Neodymium Nd	61 Promethium Pm	62 Samarium Sm	63 Europium Eu	64 Gadolinium Gd	65 Terbium Tb	66 Dysprosium Dy	67 Holmium Ho	68 Erbium Er	69 Thulium Tm	70 Ytterbium Yb
89 Actinium Ac	90 Thorium Th	91 Protactinium Pa	92 Uranium U	93 Neptunium Np	94 Plutonium Pu	95 Americium Am	96 Curium Cm	97 Berkelium Bk	98 Californium Cf	99 Einsteinium Es	100 Fermium Fm	101 Mendelevium Md	102 Nobelium No

interrelated tasks. For example, calcium, magnesium, and phosphorus function together to form and strengthen the structure of the skeleton. Some minerals are involved in the metabolic breakdown of energy-yielding nutrients, while others are essential to nerve function, muscle contraction, and maintenance of bodily fluids. Minerals also participate in chemical reactions by serving as cofactors. In addition, the electrolytes (sodium, chloride, and potassium) are essential for maintaining a healthy balance of fluids in the body.

Although minerals serve diverse physiological roles in the body, they share some similarities. These include their dietary sources and the body's ways of regulating them, as summarized in Table 12.1 and described next.

Regulation of Minerals in the Body As it does with most nutrients, the body maintains optimal levels of minerals, in large part by regulating absorption (in the small intestine) and excretion (in the kidneys). Changes in these regulatory processes ensure mineral availability and prevent toxicity. The body can also store small amounts of certain minerals in the liver, bones, and other tissues. Because these three processes (absorption, excretion, and storage) are tightly regulated, toxicity is rare for most minerals. Still, genetic disorders and overconsumption of mineral-containing supplements or medications can sometimes have serious health consequences.

Dietary Sources of the Major Minerals Minerals are abundant in both plant- and animal-based foods, but in general animal-based products have higher mineral content than do plant-based foods. For example, a 1-cup serving of milk contains approximately 300 mg calcium, while a 1-cup serving of broccoli contains approximately 43 mg calcium. The location in which a plant is grown or where an animal grazes can also influence mineral contents of the foods made from them. For instance, a plant grown in selenium-deficient soil will have lower selenium content than one grown in selenium-rich soil. Not surprisingly, selenium deficiency is more common in people living in geographic regions with selenium-poor soil. The extent to which a food is processed can also influence its mineral content. Cereal grains generally contain minerals such as copper, selenium, and zinc, but these important nutrients are lost during milling. Food manufacturers sometimes fortify their products with minerals lost during processing, but most health care professionals recommend choosing foods made with whole grains whenever possible.

TABLE 12.1 General Characteristics of the Major Minerals

Food sources	• Seafood, meat, and dairy products tend to be the best sources • Vegetables and legumes are sometimes good sources • Whole-grain products are better sources than milled products
Digestion	• Very little needed
Absorption	• Occurs mostly in small intestine but sometimes also in large intestine • Bioavailability sometimes influenced by nutritional status and interactions with other dietary components
Circulation	• Circulated from the small intestine to the liver in the blood, and then to the rest of the body
Functions	• Many are cofactors for enzymes, some of which are involved in energy metabolism • Some have major structural roles such as maintaining bone and tooth health • Involved in nerve and muscle function • Electrolytes involved in fluid balance
Toxicity	• Rare and usually associated with excess supplemental or medicinal intake

Calcium—The Body's Most Abundant Mineral

Calcium (Ca) is the most abundant mineral in the body, making up about 1 kg (2.2 pounds) of an average adult's body weight. Although more than 99% of the body's calcium is in the skeleton, this mineral is also present in blood and other tissues, where it participates in muscle contraction, neural signaling, blood clot formation, and blood pressure regulation.

DIETARY AND SUPPLEMENTAL SOURCES OF CALCIUM

Calcium is abundant in a variety of plant- and animal-derived foods, although the best and most common sources tend to be dairy products (Figure 12.2). Other good sources of calcium include dark green leafy vegetables such as collard greens and spinach, salmon and sardines (with bones), and some legumes. Calcium-fortified foods such as breakfast cereals, orange juice, and soy products can also provide considerable amounts of calcium. It is relatively easy to assess the calcium content of most packaged foods, because the Nutrition Facts panels on food labels are required to list calcium content, and foods fortified with calcium are frequently labeled as such. You can read more tips on how to increase your calcium intake, by consuming low-fat dairy products, in the accompanying Food Matters feature.

⟨**CONNECTIONS**⟩ Recall that the term *fortified* refers to any food to which a nutrient or combination of nutrients has been added (Chapter 10, page 422).

Calcium Supplements In addition to calcium-containing food, there are many forms of calcium supplements available.[1] At one time, there was concern regarding some "natural" forms of calcium such as oyster shell and bonemeal, because they could potentially contain high levels of lead. Changes in manufacturing processes now ensure that these products are safe to consume.[2] In general,

calcium (Ca) A major mineral found in bones, teeth, and blood; needed for skeletal structure, blood clotting, muscle and nerve function, and energy metabolism.

FIGURE 12.2 Food Sources of Calcium

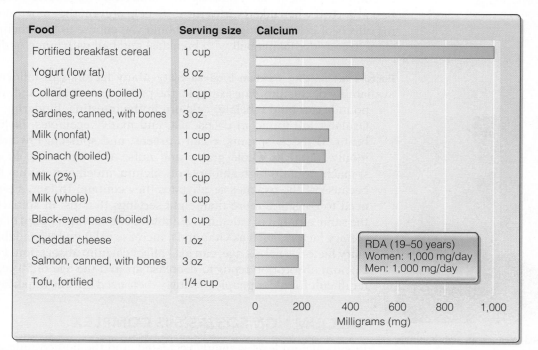

SOURCE: USDA Nutrient Database for Standard Reference, Release 17.

Working Toward the Goal: Increasing Calcium and Potassium Intakes by Consuming Low-Fat Dairy Foods

The 2010 Dietary Guidelines for Americans recommend increasing calcium and potassium intakes, largely by consuming sufficient amounts of dairy foods. This can be accomplished by making sure you get 3 cups per day of fat-free or low-fat milk or equivalent dairy products, for example 1½ oz hard cheese or 1½ cups low-fat ice cream. These products are calcium-rich yet provide fewer calories than their full-fat alternatives. The following selection and preparation tips will help you meet this goal.

- When grocery shopping, visit the cheese section and buy a variety that you do not usually purchase. When was the last time you enjoyed chévre (goat cheese)?

- Carry a single-serving container of plain or flavored milk or a calcium-fortified milk alternative. They make quick and appealing snacks and are more nutrient dense than soft drinks.

- Instead of jelly, consider topping your toast or bagel with a slice of cheese. Many Europeans

routinely include cheese in their morning meals, providing yet one more opportunity to consume this nutrient-dense food.

- Use fat-free or low-fat milk instead of water when making instant oatmeal or other hot cereals.

- Rather than having a candy bar or bag of chips, snack on fat-free or low-fat yogurt or cheese.

- Make a dip for fruits or vegetables from yogurt, cottage cheese, or low-fat cream cheese. Not only is this delicious, but it also adds calcium to your diet.

- Experiment with making calcium-rich smoothies by blending various fruits with yogurt or reduced-fat ice cream.

- Top cut-up fruit with yogurt for a quick and nutritious dessert.

- If you drink cappuccinos or lattes, ask that they be made with fat-free milk instead of more calorie-dense alternatives.

⟨CONNECTIONS⟩ Recall that *bioavailability* refers to the extent to which a nutrient is absorbed in the gastrointestinal tract (Chapter 3, page 101).

calcium supplements are relatively well utilized by the body, although they are best absorbed when taken with a meal. To help prevent the sometimes constipating effects of calcium supplements, should you choose to take them, you might want to consider taking half with breakfast and half with lunch or dinner.

Factors Influencing Calcium Bioavailability Many factors can influence the bioavailability of calcium. For example, the presence of oxalate and phytate compounds can bind (chelate) calcium in the intestine, hindering absorption. Oxalates are found in cocoa, tea, and many vegetables, including green beans, Brussels sprouts, collard greens, and spinach. Phytates are commonly found in whole grains and nuts. Although some foods, such as spinach, contain high amounts of calcium, much of it cannot be absorbed because of the oxalates or phytates they contain. In fact, a person would need to consume more than eight servings (8 cups) of spinach to absorb the same amount of calcium available in one serving (1 cup) of milk.[3] Other dietary factors, such as vitamin D, increase calcium bioavailability. Nondietary factors, such as age, can also affect calcium absorption. For example, calcium absorption tends to decrease around the age of 50 years, making it difficult for older people to satisfy their need for this important mineral.[4]

© Stockbyte Platinum/Getty Images

You would need to eat about eight servings of spinach to obtain the amount of calcium available from a single serving of milk.

CALCIUM HOMEOSTASIS IS COMPLEX

Blood calcium level is tightly regulated, and the homeostatic mechanisms are complex.[5] Calcium homeostasis involves a well-orchestrated system that

FIGURE 12.3 **Regulation of Blood Calcium** Blood calcium concentration is regulated by three hormones: parathyroid hormone (PTH), calcitriol (vitamin D), and calcitonin.

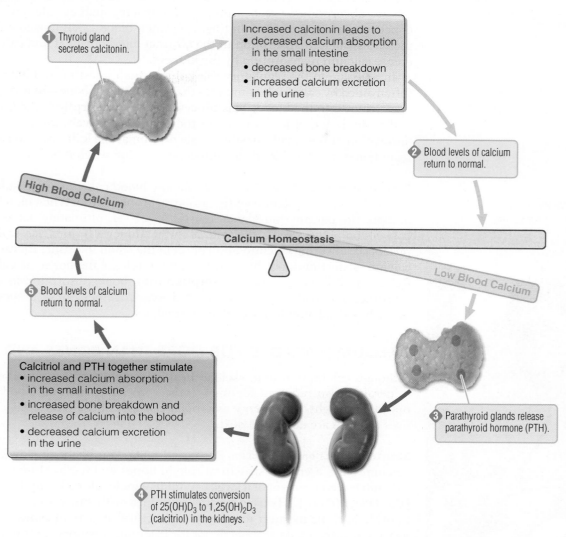

1 Thyroid gland secretes calcitonin.

Increased calcitonin leads to
• decreased calcium absorption in the small intestine
• decreased bone breakdown
• increased calcium excretion in the urine

2 Blood levels of calcium return to normal.

High Blood Calcium

Calcium Homeostasis

Low Blood Calcium

5 Blood levels of calcium return to normal.

Calcitriol and PTH together stimulate
• increased calcium absorption in the small intestine
• increased bone breakdown and release of calcium into the blood
• decreased calcium excretion in the urine

3 Parathyroid glands release parathyroid hormone (PTH).

4 PTH stimulates conversion of 25(OH)D$_3$ to 1,25(OH)$_2$D$_3$ (calcitriol) in the kidneys.

includes three hormones: calcitriol (the active form of vitamin D), parathyroid hormone (PTH), and calcitonin. As shown in Figure 12.3, these hormones work together to maintain blood calcium levels within a healthy range.

Responses to Low Blood Calcium When blood calcium level is low, the parathyroid glands (located within the thyroid gland) release a hormone called **parathyroid hormone (PTH).** PTH circulates in the blood to the kidneys, where it stimulates the conversion of the inactive form of vitamin D to its active form, calcitriol [25(OH)D$_3$ → 1,25(OH)$_2$D$_3$]. Calcium absorption in the small intestine, in part, requires the assistance of several calcium transport proteins, one of which is **calbindin,** whose synthesis is upregulated by vitamin D. As such, calcium absorption is increased in response to PTH release, vitamin D activation, and calbindin synthesis.

To further increase circulating levels of calcium, both PTH and calcitriol decrease calcium loss in the urine by a process called reabsorption. It is important to differentiate between absorption and reabsorption. Whereas *absorption* refers to the movement of substances from the GI tract into the

calbindin A transport protein, made in enterocytes, that assists in calcium absorption; synthesis is stimulated by calcitriol.

parathyroid hormone (PTH) A hormone, produced in the parathyroid glands, that is released in response to low blood calcium concentration; stimulates the conversion of 25(OH)D$_3$ to 1,25(OH)$_2$D$_3$, the active form of vitamin D.

⟨CONNECTIONS⟩ In Chapter 11, you learned that vitamin D is both a nutrient and a prohormone, being converted to calcitriol (Chapter 11, page 471).

⟨CONNECTIONS⟩ Remember that cell differentiation is the process by which immature cells become mature, functioning cells (Chapter 11, page 466).

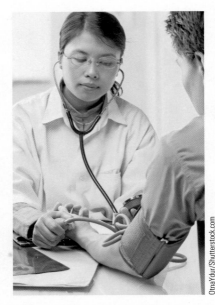

Adequate calcium consumption appears to be important in maintaining healthy blood pressure, thus lowering risk for cardiovascular disease.

resorption The breakdown and assimilation of a substance in the body.

calcitonin (cal – ci – TO – nin) A hormone produced in the thyroid gland in response to high blood calcium levels.

hydroxyapatite [$Ca_{10}(PO_4)_6(OH)_2$] The mineral matrix of bones and teeth.

osteoblast A bone cell that promotes bone formation.

osteoclast A bone cell that promotes bone breakdown.

blood, *reabsorption* refers to the movement of previously absorbed substances from tissues (such as kidneys) back into the blood. Thus, anything that increases absorption or reabsorption of a nutrient, such as calcium, raises its concentration in the blood. In this case, when blood calcium is low, calcitriol increases calcium absorption in the GI tract and both PTH and calcitriol increase calcium reabsorption by the kidneys.

In addition to the regulatory roles of the small intestine and kidneys, PTH (often aided by calcitriol) stimulates the breakdown of bone—in turn releasing its calcium into the blood. This process is called **resorption,** the breakdown and assimilation of a substance in the body. Collectively, increased calcium absorption at the small intestine, reabsorption at the kidneys, and resorption from bones increase blood calcium concentration when it is low.

Responses to High Blood Calcium While low blood calcium can result in poor health, high blood calcium levels can cause problems too. When blood calcium is high, the parathyroid glands release less PTH, ultimately decreasing the conversion of vitamin D to calcitriol in the kidneys. In turn, less calcitriol is produced, lowering calcium absorption in the small intestine. Elevated blood calcium also stimulates the thyroid gland to release the hormone **calcitonin.** Calcitonin decreases calcium resorption from bone, absorption in the small intestine, and reabsorption by the kidneys. Together, these processes help lower blood calcium levels back to normal.

CALCIUM IS NEEDED FOR MORE THAN BONES AND TEETH

Although calcium's role in skeletal health is likely the best understood, this mineral is also required for many other processes such as blood clotting, optimal eyesight, and energy metabolism. Some of the many ways the body uses calcium are described next.

Structural Functions of Calcium: Bones and Teeth You likely already know that calcium plays a critical structural role in bones and teeth. More specifically, calcium is a component of a large crystal-like molecule called **hydroxyapatite** [$Ca_{10}(PO_4)_6(OH)_2$]**.** Hydroxyapatite combines with other minerals such as fluoride and magnesium to form the structural matrix of bones and teeth, as well as providing a storage depot for calcium. It is important to recognize that bone is a complex, living tissue composed of two different kinds of cells: osteoblasts and osteoclasts. **Osteoblasts** promote bone formation, whereas **osteoclasts** facilitate the breakdown of older bone (resorption). Together, these cells help keep bones healthy and strong—largely by synthesizing or breaking down calcium-containing hydroxyapatite as needed. You will learn more about how this occurs later in this chapter and in the Nutrition Matters following this chapter.

Regulatory Functions of Calcium In addition to maintaining the structure of the skeleton, calcium regulates dozens of other activities in the body. For example, it works with vitamin K to stimulate blood clot formation, and facilitates both muscle contractions and the transmission of nerve impulses.[6] Calcium is also needed for healthy vision, regulation of blood glucose, and cell differentiation.[7] And although calcium is not used directly for energy, it is a cofactor for several enzymes needed for energy metabolism. In addition to all these functions, scientists are just beginning to understand additional roles that calcium might play in the body. For example, there is growing (albeit controversial) evidence that adequate calcium consumption may help reduce risk for cardiovascular disease and some forms of cancer.[8] You can read more about this in the Focus on Food feature.

The relationship between dairy products and bone health has been part of standard nutritional advice for decades. Recent research, however, shows that consuming high amounts of low-fat dairy products may also be associated with decreased risk for many other chronic degenerative diseases, including heart disease, cancer, type 2 diabetes, and obesity.[9]

How dairy foods exert these effects is not completely understood, and there remains significant scientific debate in this area. As for osteoporosis, although the protective effect of adequate calcium consumption from dairy products on bone disease (especially in childhood) is well accepted by medical experts, it is likely that this mineral—by itself—is not solely responsible for all of milk's positive health effects. Indeed, a recently published study of more than 200,000 participants found no direct relationship between calcium intake and risk of hip fractures.[10] In fact, the study actually suggested an increased risk of hip fracture with calcium supplementation. Thus, it is likely that the unique *combination* of micro- and macronutrients in dairy products is important in promoting optimal bone health.[11] Regarding cardiovascular health, studies suggest again that a combination of milk's nutrients may be important for optimal health. In particular, calcium and milk protein have both been shown to lower blood pressure.[12,13] Similarly, there is some evidence that intake of milk and milk products is associated with lower risk for type 2 diabetes, an important risk factor for cardiovascular disease.

Increased consumption of dairy products has also been linked with decreased risk for breast, prostate, and colon cancer.[14] One study suggested that women who drink at least three glasses of milk each day have a lower occurrence of breast cancer than women who do not drink milk.[15] Similarly, consuming even one additional serving of low-fat dairy foods is associated with a 40 to 50% reduction of colon cancer risk.[16] Although it is not known which factors in milk provide these health benefits,

calcium, vitamin D, certain proteins, and conjugated linoleic acid (CLA—a fatty acid) are likely candidates.[17]

Some research also suggests that consuming low-fat dairy products can help prevent obesity, support weight loss in overweight adults, and promote fat loss and muscle gain in exercisers. For example, people consuming three to four daily servings of milk, yogurt, or cheese were found to lose over 50% more body fat than those who consumed one or fewer servings of dairy products.[18] Others studies have shown that, compared with drinking a soy beverage or a carbohydrate-rich beverage after a workout, consumption of 2 cups of skim milk resulted in the loss of nearly twice as much body fat.[19] Because of research like this, some athletes are now being encouraged to drink milk during and after workouts and games. The effects of dairy foods on weight maintenance, weight loss, and muscle gain are probably linked to a combination of the 9 essential nutrients and unique zoonutrients and proteins found in dairy products.[20] However, not all studies have supported the beneficial effect of dairy consumption on weight loss and muscle gain, and research continues in this area.[21] Nonetheless, and for myriad reasons, milk is a healthy food choice.

Calcium-rich dairy foods are considered by some experts to be *functional foods*, because consuming them is linked with health benefits that go beyond those provided by just the essential nutrients they contain.

CALCIUM AND BONE HEALTH

Adequate calcium intake is important throughout the life cycle. In children, calcium deficiency can result in rickets, a disease that is more commonly caused by vitamin D deficiency (see Chapter 11). Children with rickets have poor bone mineralization and characteristically "bowed" bones, especially in the legs. In adults, a calcium-poor diet is associated with **osteopenia,** the moderate loss of bone mass. **Bone mass** is a measure of the amount of minerals (mostly calcium and phosphorus) contained in bone. When bone loss is severe, it can result in a condition called **osteoporosis.** As illustrated in Figure 12.4, osteoporosis is characterized by progressive bone loss that, over time, causes bones to become weak and fragile. Having osteoporosis increases the risk of bone fractures, which can be serious in older people. The National Osteoporosis Foundation estimates that this condition is a major health concern for 44 million Americans, or 55% of people 50 years of age and older.[22]

osteopenia (os – te – o – PE – ni – a) A condition whereby bone mineral density is lower than normal.

bone mass A measure reflecting the amount of minerals contained in bone.

osteoporosis A bone disease characterized by reduced bone mineral density and disruption of the microarchitecture of the bone.

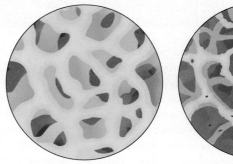

FIGURE 12.4 Normal and Osteoporotic Bone Calcium deficiency in adults can cause bones to weaken and ultimately become osteoporotic.

Normal bone   Osteoporotic bone

Although adequate calcium intake is certainly needed to maintain healthy bones, results from the Women's Health Initiative suggest that consumption of calcium and vitamin D supplements results in only small improvements in bone health.[23] Another study suggests that calcium supplementation might actually increase risk for hip fractures in some people.[10] Clearly, how calcium and other factors interact to influence bone health is complex, and you can read more about bone growth and its relationship with diet and osteoporosis in the Nutrition Matters following this chapter.

CALCIUM DEFICIENCY ALSO AFFECTS NERVE AND MUSCLE FUNCTION

In addition to bones, calcium deficiency affects several other tissues. For instance, calcium plays an important role in muscle contraction and nerve function. For this reason, low blood calcium can cause muscle pain, muscle spasms, and a tingling sensation in the hands and feet. More serious calcium deficiency causes muscles to tighten and become unable to relax, a condition called **tetany.** However, tetany is typically not caused by dietary calcium insufficiency alone but by a combination of factors such as disease combined with improper use of medication.

CALCIUM TOXICITY CAN CAUSE KIDNEY STONES AND CALCIFICATION OF SOFT TISSUES

While too little dietary calcium can cause health problems, too much calcium can be harmful as well. Overconsumption can cause calcium to be deposited in soft tissues such as muscle and kidney. High calcium intakes are also associated with impaired kidney function, formation of certain types of kidney stones, and can interfere with the bioavailability of other nutrients, such as iron and zinc.[24] It is unusual for excessive consumption of calcium-containing foods to cause high levels of calcium in the blood. Instead, this is more commonly associated with consumption of calcium supplements or certain diseases. Excessive consumption of vitamin D supplements can also result in secondary calcium toxicity. This is mainly due to the stimulatory effect of vitamin D on bone resorption and subsequent release of calcium into the blood.

RECOMMENDED INTAKES FOR CALCIUM

The Recommended Dietary Allowance (RDA) for calcium has been set at 1,000 mg/day for adults. This is approximately the amount contained in 3 to 4 cups of milk. Because calcium absorption tends to decrease with age, and women are especially prone to bone loss, the RDA increases to 1,200 mg/day at 51 years of age for women and at 71 years of age for men. To decrease the risk of calcium

⟨**CONNECTIONS**⟩ The Dietary Reference Intakes (DRIs) are reference values for nutrient intake. When adequate information is available, Recommended Dietary Allowances (RDAs) are established to reflect intakes that meet the needs of about 97% of the population. When less information is available, Adequate Intake (AI) values are provided. Tolerable Upper Intake Levels (ULs) indicate intakes that should not be exceeded (Chapter 2, pages 38–44).

tetany A condition in which muscles tighten and are unable to relax.

toxicity, a Tolerable Upper Intake Level (UL) has been set at 2,500 mg/day until age 51 years of age, when it decreases to 2,000 mg/day. A complete list of the Dietary Reference Intake (DRI) values for all major minerals is provided inside the front cover of this book.

Phosphorus—A Component of Biological Membranes

Phosphorus (P) is a component of phospholipids and therefore is associated with virtually all body structures, including biological membranes, the skeleton, and muscle. It also serves numerous functional roles and is critical to the transport of lipids, energy metabolism, and protein synthesis.

PHOSPHORUS IS ABUNDANT IN PROTEIN-RICH FOODS

Phosphorus is naturally abundant in most foods, especially those rich in protein. As illustrated in Figure 12.5, good sources include dairy products, meat (including poultry), seafood, nuts, and seeds. Bioavailability of phosphorus is high in most foods, although phosphorus in seeds and grains is more difficult to absorb than that in animal-derived foods. This is because seeds and grains contain phytates, which bind phosphorus in the GI tract.

In addition to the phosphorus found *naturally* in foods, dietary phosphorus can also be obtained from processed foods and some soft drinks. This is because phosphorus is sometimes added to foods to promote moisture retention, smoothness, and taste. For instance, a typical cola drink (12 ounces) contains about 50 mg of phosphorus. You can find out if a soft drink contains added phosphorus by looking for the ingredient "phosphoric acid" on the label. Reliance on soft drinks to supply dietary phosphorus is not recommended because these beverages have low nutrient densities and therefore contribute very little to nutrient requirements. There is also some epidemiological

⟨**CONNECTIONS**⟩ Recall that the term *nutrient density* refers to the relative ratio of nutrients to energy in a food (Chapter 2, page 52).

phosphorus (P) A major mineral needed for cell membranes, bone and tooth structure, DNA, RNA, ATP, lipid transport, and a variety of metabolic reactions requiring phosphorylation.

FIGURE 12.5 Food Sources of Phosphorus

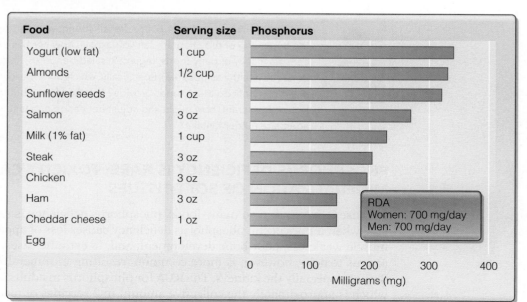

Food	Serving size	Phosphorus
Yogurt (low fat)	1 cup	
Almonds	1/2 cup	
Sunflower seeds	1 oz	
Salmon	3 oz	
Milk (1% fat)	1 cup	
Steak	3 oz	
Chicken	3 oz	
Ham	3 oz	
Cheddar cheese	1 oz	
Egg	1	

RDA
Women: 700 mg/day
Men: 700 mg/day

0   100   200   300   400
Milligrams (mg)

SOURCE: USDA Nutrient Database for Standard Reference, Release 17.

evidence that consumption of cola beverages is associated with demineralization of bones, especially during childhood and adolescence.[25]

REGULATION OF BLOOD PHOSPHORUS LEVELS IS SIMILAR TO THAT OF CALCIUM

⟨**CONNECTIONS**⟩ Active transport is an energy (ATP)-requiring process. Simple diffusion does not require energy and transports substances from a region of higher concentration to that of a lower concentration (Chapter 3, page 77).

Phosphorus is readily absorbed in the small intestine by both vitamin D–dependent active transport and by simple diffusion mechanisms.[26] As with calcium, the concentration of blood phosphorus is regulated mainly by calcitriol (vitamin D), PTH, and calcitonin. When blood phosphorus levels are low, calcitriol together with PTH increase phosphorus absorption in the small intestine and phosphorus resorption from bones. These actions help return blood phosphorus to normal.[27] During periods of high blood phosphorus, calcitonin stimulates the bone-building activity of osteoblasts. The osteoblasts in turn take up phosphorus from the blood, thus helping lower blood phosphorus back to healthy levels.

ADDITIONAL ROLES FOR PHOSPHORUS

⟨**CONNECTIONS**⟩ Phospholipids are made of a glycerol backbone bonded to two fatty acids and a polar head group via a phosphate linkage (Chapter 6, page 231).

Because phosphorus is an important component of phospholipids, it serves numerous structural and functional roles. For example, it is needed for the structure of biological membranes (such as those surrounding cells) and is critical for the transport of lipids in the body via lipoproteins. In addition, phosphorus is a component of each cell's genetic material (DNA) and adenosine triphosphate (ATP), making this mineral essential for protein synthesis and energy metabolism. Phosphorus-containing compounds also help maintain the blood's pH (acid–base balance) by acting as buffers that accept and donate hydrogen ions (H^+). Phosphorus is also involved in hundreds of metabolic reactions in the body. In these reactions, phosphate groups are transferred from one molecule to another, producing "phosphorylated" molecules. Some molecules are activated when they are phosphorylated, whereas others are inactivated. Finally, phosphorus, along with calcium, is required to form hydroxyapatite [$Ca_{10}(PO_4)_6(OH)_2$]—the mineral matrix making up bones and teeth.

Critical Thinking: JoAnn's Story Recall JoAnn, who was featured in the beginning of this chapter. Can you explain how her basic diet, which consists mainly of dairy foods, legumes, salads, and fish, helps keep her bones as healthy as possible? Specifically, which bone-building micro- and macronutrients do these foods provide? If JoAnn found out that she was lactose intolerant, how might she adjust her eating habits to maintain adequate nutrient intake?

PHOSPHORUS DEFICIENCY IS RARE; TOXICITY CAUSES MINERALIZATION OF SOFT TISSUES

Because it is found in so many foods, phosphorus deficiency is rare. Nonetheless, when it does occur, phosphorus deficiency causes loss of appetite, anemia, muscle weakness, poor bone development, and, in extreme cases, death. Phosphorus toxicity, however, is more common, resulting in mineralization of soft tissues, especially the kidneys. The RDA for phosphorus in adults is 700 mg/day, which is approximately the collective amount in 2 servings of meat and 2 cups of milk. Because of the potentially toxic effects of high doses of phosphorus, a UL of 4,000 mg/day has been established.

Magnesium—Needed for Building Bones and Stabilizing Enzymes

Magnesium (Mg), is important for many physiological processes, including energy metabolism and enzyme function. More recently, magnesium has also gained attention because of the possible associations between magnesium deficiency and several chronic diseases, such as heart disease and type 2 diabetes.[28]

BEANS, NUTS, AND SEEDS ARE EXCELLENT SOURCES OF MAGNESIUM

Magnesium is found in a variety of foods including whole grains, green leafy vegetables, seafood, legumes, nuts, and chocolate (Figure 12.6). In general, foods made with whole grains contain more magnesium than do refined-grain products. This is because the mineral-rich bran portion of grains is removed during the processing of many refined flours. Bioavailability of magnesium is variable, and it is likely that several dietary factors influence its absorption. For example, some studies suggest that diets high in calcium or phosphorus decrease the bioavailability of magnesium.[29] In addition, protein deficiency may decrease magnesium absorption,[30] and high levels of dietary fiber can decrease its bioavailability.[31]

BLOOD LEVELS OF MAGNESIUM ARE REGULATED BY THE SMALL INTESTINE AND KIDNEYS

The amount of magnesium in the blood is regulated by both the intestine and kidneys, although the latter appears to be the main site. As with most minerals, magnesium absorption in the small intestine and reabsorption by the kidneys increase when blood levels are low. The opposite occurs when circulating magnesium levels are high.

© Image Source/Getty Images

Whole-grain products, like brown rice, contain more magnesium than do refined cereal grains, like white rice.

magnesium (Mg) A major mineral needed for stabilizing enzymes and ATP and as a cofactor for many enzymes.

FIGURE 12.6 Food Sources of Magnesium

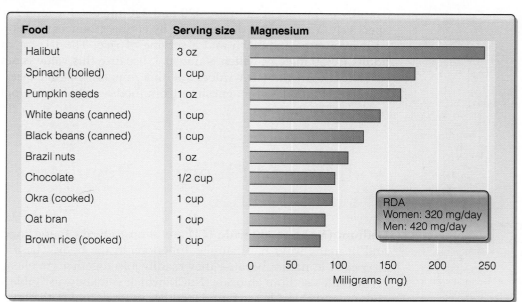

Food	Serving size	Magnesium
Halibut	3 oz	
Spinach (boiled)	1 cup	
Pumpkin seeds	1 oz	
White beans (canned)	1 cup	
Black beans (canned)	1 cup	
Brazil nuts	1 oz	
Chocolate	1/2 cup	
Okra (cooked)	1 cup	
Oat bran	1 cup	
Brown rice (cooked)	1 cup	

RDA
Women: 320 mg/day
Men: 420 mg/day

Milligrams (mg)

SOURCE: USDA Nutrient Database for Standard Reference, Release 17.

MAGNESIUM IS IMPORTANT FOR BONES AND TO STABILIZE ANIONS

The majority of magnesium in the body is associated with the bones, where the mineral helps provide structure. Magnesium typically exists in the body as a positively charged ion (cation)—Mg^{2+}. Because of this charge, magnesium can be used to stabilize enzymes and neutralize negatively charged ions (that is, anions). Among its substrates are ATP and ADP, which typically exist as the negatively charged ATP^{2-} and ADP^{3-}. Because magnesium helps stabilize these high-energy compounds, it is vital for energy metabolism. In all, magnesium participates in more than 300 chemical reactions—most notably those involved in DNA and RNA metabolism—and also influences nerve and muscle function, especially in heart tissue.[32]

MAGNESIUM DEFICIENCY AND TOXICITY ARE RARE

Magnesium deficiency is rare but sometimes is seen in alcoholics, who tend to eat nutrient-poor diets (primary malnutrition) and have impaired nutrient absorption (secondary malnutrition). Other individuals at risk for magnesium deficiency include those with Crohn's disease, celiac disease, and the elderly. Severe magnesium deficiency causes abnormal nerve and muscle function, especially in cardiac tissue. Because of this, there is concern that mild magnesium deficiency may increase risk for cardiovascular disease, type 2 diabetes, and even migraine headaches.[33,34] However, convincing evidence that higher intakes of magnesium decrease the incidence of these conditions is not available.

Magnesium toxicity only occurs in people taking large doses of supplemental magnesium or medications containing magnesium, such as "milk of magnesia," which is used to treat heartburn, indigestion, and constipation. Symptoms of magnesium toxicity include diarrhea, intestinal cramping, and nausea. Because severe magnesium toxicity can cause heart failure, magnesium supplementation should be carefully monitored.

RECOMMENDED INTAKES FOR MAGNESIUM

The RDAs for magnesium are 420 and 320 mg/day for men and women, respectively. You can get this amount by eating 3 ounces of halibut accompanied by a cup of spinach and a serving of rice. The UL for magnesium in adults is 350 mg/day, which is unusual, because this value is actually similar to or less than the RDA values for this mineral. In this case, the UL does not apply to intake of magnesium from foods—only from supplemental and medicinal forms.

Sodium and Chloride—Regulators of Fluid Balance

sodium (Na) A major mineral important for regulating fluid balance, nerve function, and muscle contraction.

chloride (Cl) A major mineral important for regulating fluid balance, protein digestion in the stomach (via HCl), and carbon dioxide removal by the lungs.

Sodium (Na) and **chloride (Cl)** are almost always found together in foods and in many ways have similar functions in the body. These minerals tend to coexist in nature because they readily join together via chemical bonds to form a salt—sodium chloride (NaCl), which you know as "table salt." Because sodium and chloride are found together in foods and both play major roles in fluid balance, they are discussed together here.

SODIUM AND CHLORIDE CONTENTS OF FOODS ARE READILY AVAILABLE

Foods that contribute sodium to the diet are also good sources of chloride. You can easily determine the sodium content of packaged foods by reading Nutrition Facts panels. Because sodium consumption can lead to hypertension in some people, many health care professionals recommend that we limit the amount of sodium in our diet. As such, the following health claim can be found on the label of some low-sodium foods: "Diets low in sodium may reduce the risk of high blood pressure, a disease associated with many factors." The following terms are also used on food packaging to describe salt content:

- Salt free—less than 5 mg sodium per serving
- Very low salt—less than 35 mg sodium per serving
- Low salt—less than 140 mg sodium per serving

The sodium contents of some commonly consumed foods are listed in Figure 12.7. Because salt is added to many processed foods, they provide the majority of both sodium and chloride to the American diet. One teaspoon of salt contains more than 2,000 mg of each of these minerals. Other food additives, such as monosodium glutamate (MSG), which is commonly used in Asian cuisine, contain sodium as well (gram for gram, about one-third of the sodium of table salt). Condiments, such as soy sauce and ketchup, also tend to be high in sodium. In general, unprocessed foods such as fresh

© studiomode / Alamy

Many condiments, like some of those shown here, contain sodium.

FIGURE 12.7 Food Sources of Sodium

Food	Serving size	Sodium
Table salt	1 tsp	
Sauerkraut	1 cup	
Ham	3 oz	
Macaroni and cheese	1 cup	
Cheeseburger with bun	1 medium	
Chili con carne (canned)	1 cup	
Cottage cheese	1 cup	
Crab	3 oz	
Soy sauce	1 tbsp	
Dill pickle	1 medium	
Fast food taco	1 medium	
French fries	10 pieces	
Potato chips	1 oz	
Hamburger patty	3 oz	

AI
Women: 1,500 mg/day
Men: 1,500 mg/day

Milligrams (mg): 0 – 500 – 1,000 – 1,500 – 2,000 – 2,500

SOURCE: USDA Nutrient Database for Standard Reference, Release 17 and Release 23.

fruits and vegetables contain small amounts of sodium and chloride, whereas manufactured and highly processed foods such as fast foods, frozen entrees, and savory snacks often contain substantial amounts. Some meats (including poultry), dairy products, and seafood naturally contain moderate amounts of both sodium and chloride.

SODIUM ABSORPTION REQUIRES GLUCOSE

The bioavailability of sodium and chloride is very high, and almost all that you consume is absorbed in the small intestine. However, diseases and GI infections that affect the lining of the small intestine can disrupt sodium absorption. In these conditions, large amounts of sodium and chloride can be lost in the stool.

There are several intestinal transport mechanisms, but in general sodium is absorbed first, with chloride following close behind.[35] One of the most important ways that sodium is absorbed requires glucose to be cotransported with it. In other words, glucose must be present for sodium to be absorbed. This is one of the reasons why sports drinks designed to replace salt lost during physical activity were originally formulated to contain both glucose and salt. Sodium is also actively absorbed in the colon, resulting in water reabsorption. Without colonic absorption, large amounts of water would be lost in the feces, causing potentially dangerous diarrhea and dehydration.

BLOOD SODIUM LEVELS ARE CAREFULLY REGULATED

The sodium and chloride concentrations in the blood are carefully regulated, as shown in Figure 12.8. When blood sodium concentration decreases, a condition called **hyponatremia** develops. Hyponatremia can be caused by inadequate salt intake or loss of sodium chloride from the body via extreme sweating. It can also be caused by excessive water consumption not accompanied by adequate sodium intake. Hyponatremia stimulates the adrenal glands to secrete the hormone **aldosterone,** in turn causing the kidneys to retain (reabsorb) sodium. In other words, aldosterone decreases sodium excretion in the urine. As a result, blood sodium levels are restored to normal. Low blood pressure also increases aldosterone release from the adrenal glands by stimulating the kidneys to produce an enzyme called **renin.** In the blood, renin converts the liver-derived protein **angiotensinogen** into **angiotensin I.** Angiotensin I is then converted to **angiotensin II** in the lungs. Angiotensin II stimulates the adrenal glands to release additional aldosterone; blood sodium concentration then increases, drawing more water into the blood and helping raise blood pressure back to normal.

When blood sodium concentration increases above normal levels (**hypernatremia**), aldosterone and renin secretion decreases, in turn diminishing salt reabsorption by the kidneys. In this way, renin, angiotensin II, and aldosterone work hand-in-hand to maintain healthy levels of blood sodium. This is often referred to as the renin–angiotensin–aldosterone system.

SODIUM AND CHLORIDE ARE ELECTROLYTES

Sodium and chloride are the body's principal electrolytes.* When the NaCl molecule dissociates in water, sodium is released as a cation (Na^+), and chloride is released as an anion (Cl^-). Both minerals play major roles in fluid balance. Because water naturally moves to areas that have high Na^+ and/or Cl^- concentrations, the body can maintain fluid balance by selectively moving these electrolytes where more water is needed. This process is described in

*Note that sodium, chloride, and potassium are technically ions, not electrolytes. However, because most nutrition literature refers to them as electrolytes, we will refer to them in this way as well.

⟨CONNECTIONS⟩ Remember that electrolytes are substances that produce charged particles, or ions, when dissolved in fluids. The terms *electrolyte* and *ion* are often used interchangeably (Chapter 3, page 77).

hyponatremia Low blood sodium concentration.

aldosterone (al – DO – ster – one) A hormone produced by the adrenal glands in response to low blood sodium concentration (hyponatremia) and angiotensin II.

renin (RE – nin) An enzyme produced in the kidneys in response to low blood pressure; converts angiotensinogen to angiotensin I.

angiotensinogen (an – gi – o – ten – SIN – o – gen) An inactive protein, made by the liver, that is converted by renin into angiotensin I.

angiotensin I The precursor of angiotensin II.

angiotensin II A protein derived from angiotensin I in the lungs; stimulates aldosterone release.

hypernatremia High blood sodium concentration.

FIGURE 12.8 Regulation of Blood Sodium Blood sodium concentration is regulated by the kidneys via aldosterone, angiotensin II, and renin.

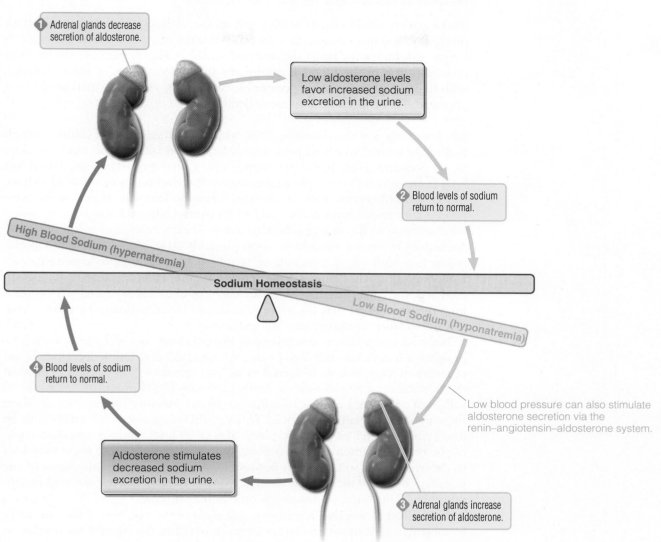

1. Adrenal glands decrease secretion of aldosterone.

Low aldosterone levels favor increased sodium excretion in the urine.

2. Blood levels of sodium return to normal.

High Blood Sodium (hypernatremia)

Sodium Homeostasis

Low Blood Sodium (hyponatremia)

4. Blood levels of sodium return to normal.

Aldosterone stimulates decreased sodium excretion in the urine.

Low blood pressure can also stimulate aldosterone secretion via the renin–angiotensin–aldosterone system.

3. Adrenal glands increase secretion of aldosterone.

more detail later in this chapter. Sodium is also important for nerve function and muscle contraction, both of which also involve potassium (K^+). In addition, chloride is needed for production of hydrochloric acid (HCl) in the stomach, for elimination of carbon dioxide (CO_2) by the lungs, and for optimal immune function.

SODIUM AND CHLORIDE DEFICIENCIES CAN OCCUR DURING ILLNESS AND PHYSICAL EXERTION

Although sodium and chloride deficiencies are rare, they can occur in response to severe diarrhea and vomiting. These conditions result in loss of sodium and chloride through the GI tract (diarrhea) or before they even enter the lower intestine (vomiting). Because of the rapid loss of both electrolytes and water, diarrhea and vomiting can be life threatening, especially in infants and children. Less severe sodium and chloride deficiencies can occur during endurance sports such as marathon running. Symptoms of electrolyte deficiency include nausea, dizziness, muscle cramps, and, in severe cases, coma.

OVERCONSUMPTION OF SODIUM CHLORIDE INCREASES BLOOD PRESSURE IN SOME PEOPLE

There are no "toxic" effects of high salt intake, although in some people, high intakes of sodium chloride are associated with increased blood pressure, a major risk factor for heart disease and stroke.[36] This is because the sodium concentration of the blood is one of the major regulators of blood volume, with higher sodium causing greater blood volume—and thus increased blood pressure. This is presented in more detail next.

Salt Sensitivity and Hypertension Sodium helps maintain blood volume, which is directly related to blood pressure—when blood volume increases, so does blood pressure. High blood pressure is also called **hypertension,** and it has long been known that people who consume large amounts of salt *tend* to have higher blood pressure than those who consume less salt. However, because people with hypertension also *tend* to be overweight and smoke, it is somewhat difficult to discern which factors are causally related to blood pressure. According to recent estimates, approximately 31% of the U.S. adult population has high blood pressure or is taking medication for hypertension.[37] Understanding what causes hypertension is important because of its associated risk for cardiovascular disease. Thus, public health organizations have long recommended that we strive to decrease blood pressure by weight loss, refraining from smoking, and salt restriction.

Because many factors contribute to hypertension, not all hypertensive people benefit from a low-salt diet. In fact, whereas salt restriction decreases blood pressure in some people (referred to as "salt sensitive"), in a small portion of the population a low-salt diet actually increases blood pressure.[38] For others, salt intake does not appear to influence blood pressure one way or another; these people are "salt insensitive." This variable response to salt intake may be due to factors such as genetics, physical activity level, and the responsiveness of the renin–angiotensin–aldosterone system. Although there is no easy way to determine salt sensitivity, it is more common in certain subgroups of the population. These include the elderly, women, African Americans, and people with hypertension, diabetes, or chronic kidney disease.[39]

Regardless of whether a person is salt sensitive or not, one of the best ways to control hypertension via dietary means is to follow the Dietary Approaches to Stop Hypertension (DASH) eating plan. The DASH diet emphasizes reasonable amounts of whole grains, fruits and vegetables, low-fat dairy products, and lean meats. You can read more about this and other ways to decrease blood pressure in the Nutrition Matters concerning heart health following Chapter 6.

RECOMMENDED INTAKES FOR SODIUM AND CHLORIDE

Despite the long-standing interest in the health effects of salt, data are insufficient to establish RDAs for sodium and chloride. Thus, AIs have been set at 1,500 and 2,300 mg/day for sodium and chloride, respectively. Salt-sensitive individuals are advised to consume considerably less than this amount.[40] These AIs are not meant to apply to highly active individuals, such as competitive endurance athletes. In addition, people living in very warm climates may require additional salt in their diets. Recognizing that typical salt intake is significantly higher than the AI values, the 2010 Dietary Guidelines for Americans recommend that we reduce daily sodium intake to less than 2,300 milligrams. People who are 51 years and older and those of any age who are African

hypertension High blood pressure.

American or have hypertension, diabetes, or chronic kidney disease are advised to further reduce intake to 1,500 mg/day.

Because excessive intake of salt is associated with increased risk for high blood pressure in some people, ULs of 2,300 and 3,600 mg/day for sodium and chloride, respectively, have been established for these minerals. No ULs are set for infants, but it is recommended that salt not be added to foods during the first year of life.

Potassium—An Important Intracellular Cation

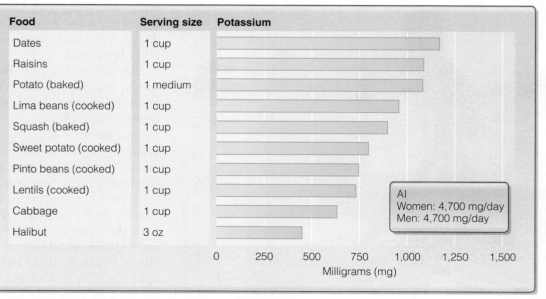

Potatoes are an excellent source of potassium.

Whereas sodium is the most abundant cation in *extra*cellular fluids, **potassium (K)** is the most abundant cation in *intra*cellular fluids. Unlike sodium, which can increase blood pressure in some people, potassium helps lower blood pressure. Potassium is a critical electrolyte, being important for fluid balance and many other physiological functions.

POTASSIUM IS FOUND IN A VARIETY OF FRUITS, VEGETABLES, DAIRY PRODUCTS, AND MEAT

Dietary sources of potassium include legumes, potatoes, seafood, dairy products, meat, and a variety of fruits and vegetables such as sweet potatoes and bananas. The bioavailability of potassium from these foods is high and is not thought to be influenced by other factors. The potassium content of various foods is illustrated in Figure 12.9.

potassium (K) A major mineral important in fluid balance, muscle and nerve function, and energy metabolism.

FIGURE 12.9 Food Sources of Potassium

Food	Serving size	Potassium
Dates	1 cup	
Raisins	1 cup	
Potato (baked)	1 medium	
Lima beans (cooked)	1 cup	
Squash (baked)	1 cup	
Sweet potato (cooked)	1 cup	
Pinto beans (cooked)	1 cup	
Lentils (cooked)	1 cup	
Cabbage	1 cup	
Halibut	3 oz	

AI
Women: 4,700 mg/day
Men: 4,700 mg/day

Milligrams (mg): 0, 250, 500, 750, 1,000, 1,250, 1,500

SOURCE: USDA Nutrient Database for Standard Reference, Release 17.

POTASSIUM IS THE BODY'S MAJOR INTRACELLULAR CATION

As with sodium and chloride, potassium absorption occurs with great efficiency in the small intestine and to a lesser extent in the colon. Potassium balance is achieved mainly by the kidneys, and is regulated primarily by the hormone aldosterone. When potassium level of the blood decreases, aldosterone is released from the adrenal glands, causing increased potassium excretion in the urine. This is in contrast to aldosterone's effects on sodium and chloride. Consequently, aldosterone causes blood levels of sodium and chloride to *increase* while simultaneously causing blood level of potassium to *decrease*.

The potassium cation (K^+) is an important electrolyte, working with sodium and chloride to maintain proper fluid balance in the body. In addition, potassium is critical for muscle function (especially in heart tissue), nerve function, and energy metabolism. Whereas increased sodium intake causes blood pressure to rise in salt-sensitive individuals, consuming high amounts of potassium can decrease blood pressure in some people.

POTASSIUM DEFICIENCY AND TOXICITY

Although potassium deficiency is rarely seen, it can occur with prolonged diarrhea and/or vomiting. Heavy use of certain diuretics can also cause excessive potassium loss in the urine. **Diuretics** are drugs that help the body eliminate water by increasing urine formation, and are often taken to lower blood pressure. However, when the body excretes excessive amounts of water, it also loses electrolytes. This can lead to low blood potassium, a condition called **hypokalemia.** People with eating disorders involving vomiting or abuse of diuretics or laxatives, such as those with bulimia nervosa, are at increased risk for hypokalemia. Potassium deficiency causes muscle weakness, constipation, irritability, and confusion, and some studies suggest it may cause insulin resistance.[41] In severe cases, potassium deficiency can cause irregular heart rhythms, muscular weakness, decreased blood pressure, and difficulty breathing.

Potassium toxicity is rare from foods, but can occur with intravenous potassium injections or very high-dosage potassium supplementation. Because high levels of potassium in the blood can cause cardiac arrest, potassium toxicity can be life threatening.[42] Less severe symptoms include tingling in the feet and hands, and muscular weakness. As with the other electrolytes, no RDAs have been established for potassium, although an AI has been set at 4,700 mg/day for adults. This is the amount of potassium in about four potatoes. No ULs have been set for this mineral.

diuretic (di – ur – E – tic) A substance or drug that causes water loss (via urine) from the body.

hypokalemia Low blood potassium concentration.

© Matthew Salacuse/The Image Bank/Getty Images

The majority of your body is made of water.

Water—The Essence of Life

Because water (H_2O) is the most abundant molecule in the body and is involved in myriad physiological functions, it is truly the essence of life. Classified as a macronutrient, water acts as a biological solvent, serves as a chemical reactant in biochemical reactions, and helps regulate body temperature.

DISTRIBUTION OF WATER IN THE BODY

In adults, approximately 50 to 70% of body weight is composed of water. And, because muscle has more water associated with it than does fat, the total volume of body water increases as the proportion of lean tissue in the body increases. For this reason, males generally have greater total water volume than females, and physically fit individuals tend to have more water than their sedentary counterparts.

Water is a principal component of biological fluids, and its distribution inside and outside of cells is shown in Figure 12.10. Recall from Chapter 3 that fluids can be categorized as intracellular fluid (inside cells) or extracellular fluid (outside cells). Extracellular fluid includes interstitial or intercellular fluid (between cells) and intravascular fluid (blood and lymph). Plasma, which is the fluid component of the blood, is an example of an intravascular fluid. A typical 70-kg man has about 42 liters of water in his body—28 liters inside of cells (intracellular fluid) and 14 liters outside of cells (extracellular fluid). Most of the extracellular fluid is in interstitial space (11 liters), with the remainder as intravascular fluid (3 liters).

〈**CONNECTIONS**〉 Extracellular fluids are those outside of cells, whereas intracellular fluids are those within cells (Chapter 3, page 76).

FIGURE 12.10 **Fluid Compartments** Water in the body can be characterized as being intracellular, extracellular, interstitial, and intravascular.

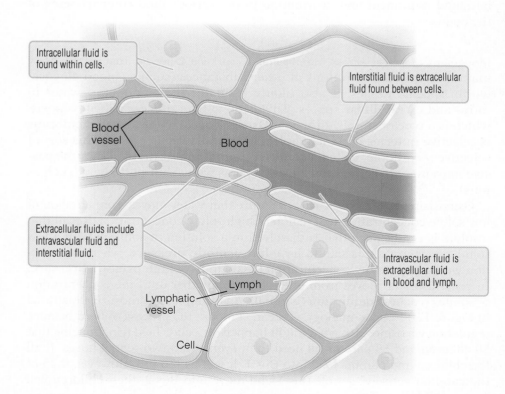

Intracellular fluid is found within cells.

Interstitial fluid is extracellular fluid found between cells.

Blood vessel

Blood

Extracellular fluids include intravascular fluid and interstitial fluid.

Intravascular fluid is extracellular fluid in blood and lymph.

Lymph

Lymphatic vessel

Cell

FLUID (WATER) BALANCE IS SHIFTED BY MOVEMENT OF ELECTROLYTES

The amount of water in the various spaces of the body is highly regulated and, as previously mentioned, is dependent on several essential minerals—namely sodium, chloride, and potassium. Plasma proteins such as albumin also help maintain a healthy fluid distribution between intravascular and interstitial compartments—as described in Chapter 5.

Importance of Electrolytes to Fluid Balance As previously described, electrolytes (such as NaCl) separate into individual, charged particles called ions when added to water. For example, sodium chloride dissociates into sodium (Na^+) and chloride (Cl^-). As such, electrolytes almost exclusively exist as dissolved particles (ions) in the various intracellular and extracellular fluid compartments of the body. Whereas sodium and chloride ions predominate in extracellular fluid, potassium (K^+) and phosphate (PO_4^-) ions are concentrated in intracellular fluid. The concentrations of these electrolytes must be maintained within certain ranges for cells to function properly. To maintain this balance, cells must make frequent adjustments. This is done by moving electrolytes and water into or out of the cells.

Recall that cell membranes control the movement of most substances into and out of cells. For example, Na^+ and Cl^- cannot cross cell membranes passively but instead need help from membrane-bound transport proteins (or "pumps") and the input of energy (ATP). Thus, movement of electrolytes into and out of cells is an *active* transport process. Conversely, water passes freely across cell membranes and between cells without requiring energy, making this a *passive* transport process. The body couples the active pumping of ions across cell membranes with the passive movement of water to regulate fluid movement and distribution in the various fluid compartments of the body.

Movement of Water across Membranes Occurs via Osmosis Just as the body maintains the concentrations of other substances (such as blood glucose) within specific ranges, it also tightly regulates the amounts of water in intra- and extracellular spaces. This is done by a process called osmosis, which is a type of simple diffusion. Osmosis occurs when a cell membrane is selective in terms of which molecules are allowed across it. This sort of selective membrane, called a **semipermeable membrane,** allows the passive movement of some substances (such as water) but not others (such as ions).

Osmosis relies on a basic rule of nature that water moves from a region of low solute concentration to a region of high solute concentration. Recall that a solute is a substance that is dissolved in a solution. A solution with a low solute concentration is said to have **low osmotic pressure,** whereas a solution with a high solute concentration is said to have **high osmotic pressure.** In general, water moves from one region to another until the concentration of solutes reaches equilibrium (is equal). An example of osmosis is illustrated in Figure 12.11. As an analogy, you can compare the process of fluid balance regulation via osmosis to what might happen on a class field trip. Imagine that 50 children and 50 parents all need to fit evenly into two school buses. If all the children and parents load into Bus 1, it will be too full. However, if 25 of the parents "actively" get off Bus 1 and move into Bus 2, their children will "passively" follow. This is somewhat similar to how the body passively moves water (like the children) via the directed, active movement of electrolytes (like the parents). Where one goes, the other will follow.

⟨**CONNECTIONS**⟩ Recall that proteins, such as albumin, in the blood help move water from the interstitial space back into the intravascular space in capillary beds (Chapter 5, page 183).

⟨**CONNECTIONS**⟩ Simple diffusion is a form of passive transport whereby substances cross cell membranes without the assistance of a transport protein or energy (Chapter 3, page 76).

⟨**CONNECTIONS**⟩ A solute is a substance dissolved in a liquid (Chapter 3, page 77).

semipermeable membrane A barrier that allows passage of some, but not all, molecules across it.

low osmotic pressure Occurs when a solution has a small amount of solutes dissolved in it.

high osmotic pressure Occurs when a solution has a large amount of solutes dissolved in it.

FIGURE 12.11 Osmosis Osmosis occurs when water moves from a region of low osmotic pressure (low solute concentration) to a region of high osmotic pressure (high solute concentration).

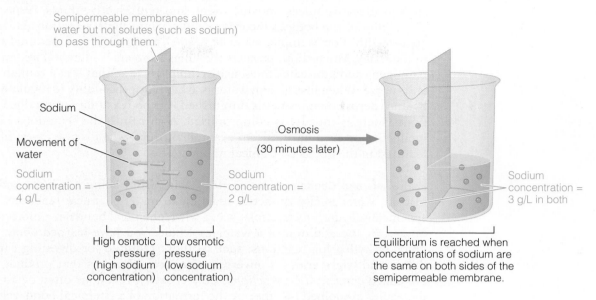

Semipermeable membranes allow water but not solutes (such as sodium) to pass through them.

Sodium

Movement of water

Sodium concentration = 4 g/L

Sodium concentration = 2 g/L

Osmosis (30 minutes later)

Sodium concentration = 3 g/L in both

High osmotic pressure (high sodium concentration)

Low osmotic pressure (low sodium concentration)

Equilibrium is reached when concentrations of sodium are the same on both sides of the semipermeable membrane.

An example of how the body uses osmosis to selectively move water is the active transport of sodium into the cells that line the colon. The increased concentration of sodium within intestinal cells creates an osmotic pressure that draws water from the lumen of the colon into the cells that line it. Without this osmosis-driven absorption, excessive amounts of water would be lost in the feces. However, osmosis may have negative health consequences such as in the regulation of blood volume and blood pressure. Recall that high salt (sodium chloride) intake can lead to increased blood volume and blood pressure in some people. This is because salt-sensitive people reabsorb less sodium in their kidneys, resulting in high levels of sodium in the blood. This causes water to move into the intravascular space, increasing blood volume—and, in turn, increasing blood pressure. Remember, where electrolytes go, water will follow.

WATER IS ESSENTIAL FOR LIFE

Water plays a critical role in many processes, including hundreds of chemical reactions. This versatile molecule also helps keep your internal body temperature constant even when your surrounding environment is very cold or hot. Moreover, water provides protection and serves as an important solvent and lubricant.

Water as a Solvent, Transport Medium, and Lubricant Because so many substances can dissolve in water, it is sometimes referred to as the universal solvent. Water is the primary solvent in all bodily fluids, such as blood, saliva, and gastrointestinal secretions. For example, blood is a solution consisting of water, cells, and a variety of dissolved solutes including nutrients and metabolic waste products. Substances (such as nutrients) dissolved in blood can move from inside the blood vessel out into the watery environment within and around tissues and cells. Conversely, waste products (such as urea) produced in cells can be dissolved in water, released into the blood, and subsequently eliminated from the body. Without water to dissolve these substances, these physiological processes would not be possible.

Water is also a lubricant. This is especially true in the GI tract, respiratory tract, skin, and reproductive system, which produce important secretions such as digestive juices, mucus, sweat, and reproductive fluids, respectively. The ability of the body to incorporate water into these secretions is vital for good health. For example, water is needed for producing functional mucus in the lungs. Mucus both protects the lungs from environmental irritants and pathogens and lubricates the sensitive lung tissue so that it can remain moist and supple. When insufficient water is available or the ability to regulate water balance across membranes is diminished, the body's ability to produce secretions (such as mucus) is compromised. As a result, the secretions become ineffective or even harmful. An example of this is the disease cystic fibrosis, featured in the Focus on Clinical Applications.

Hydrolysis and Condensation Reactions Aside from being a solvent and a lubricant, water is also an active reactant in many chemical reactions. More specifically, *hydrolysis reactions* occur when chemical bonds in a molecule are broken by the addition of a water molecule. You have learned about many kinds of hydrolytic reactions, such as those required for digesting carbohydrates and triglycerides. Conversely, chemical reactions that produce water are called *condensation reactions*. Condensation reactions often occur when molecules are joined together by the formation of a chemical bond, releasing water in the process. The typical adult produces about one and a half cups of water each day from condensation reactions.

Water Regulates Body Temperature Water is also essential for regulating body temperature. When energy-yielding nutrients such as amino acids, glucose, and fatty acids are metabolized, some of the energy is released as heat. This heat helps maintain your internal body temperature at a comfortable 98.6°F. However, excess heat generated by metabolism must be eliminated so that the temperature of the body does not rise. Hot environments can also raise the body's internal temperature, and water helps prevent this from happening as well.

There are several ways water assists with body temperature regulation. First, a given amount of water can "hold" or retain a surprisingly large amount of heat. The term **specific heat** refers to the amount of heat (energy) it takes to increase the temperature of 1 gram of a substance 1 degree Celsius. Water has a high specific heat, meaning that changing its temperature requires a large amount of heat. You have probably experienced this unique property of water if you have visited a swimming pool in the heat of the day or the cool of the evening. Although the air temperature may fluctuate around the clock, the temperature of the water in the pool remains somewhat constant. In contrast, metals tend to have low specific heats, and it takes very little heat to increase their temperature. This is why metal pots and pans heat up quickly when placed on a hot burner. Because water can accumulate a great deal of energy without heating up, the body can maintain a relatively stable internal temperature even when its metabolic rate is high or the environment is hot. In addition, heat energy contained in sweat is eliminated by a process referred to as evaporative cooling. When an individual perspires, evaporative cooling helps the body stay within its healthy temperature range during hot weather and periods of intense physical activity. For all of these reasons, it is especially important to stay fully hydrated when engaged in hard physical work—including athletics.

WATER INSUFFICIENCY CAUSES DEHYDRATION

Lack of sufficient water is called **dehydration** and can lead to serious consequences. Compared with insufficiencies of the other macronutrients, water loss is much more difficult for the body to tolerate. For example,

specific heat The heat (energy) required to raise the temperature of a substance.

dehydration A condition in which the body has an insufficient amount of water.

Secretions are produced by a variety of tissues, including the skin (sweat), salivary glands (saliva), GI tract (gastric juices), and lungs (mucus), and the inability to produce these secretions can have serious health consequences. An example is **cystic fibrosis,** which affects approximately 3,200 babies in the United States annually, is most common in people of northern European descent, and is quite rare in African and Asian Americans. Researchers estimated that, in 2007, about 30,000 U.S. adults and children (70,000 worldwide) had cystic fibrosis.[43]

Cystic fibrosis is caused by a genetic defect resulting in the faulty production of the protein that transports chloride ions (Cl^-) across cell membranes.[44] In cells that produce secretions, such as those of the lungs and pancreas, the ability to actively pump Cl^- from the intracellular space across the cell membrane into the extracellular space is crucial for incorporating water into the secretions produced there. This is because water moves from the region of low Cl^- concentration (low osmotic pressure) inside the cell to the region of high Cl^- concentration (high osmotic pressure) outside the cell. Thus, if a cell cannot actively pump Cl^- from the intracellular compartment to the extracellular compartment, water will not be incorporated into its secretion.

In the case of cystic fibrosis, this transport system is disrupted because of the production of a faulty Cl^- transport protein, resulting in the formation and accumulation of thick, sticky secretions. This is particularly problematic in the lungs and the pancreas. As a result, people with cystic fibrosis have significant—often life-threatening—respiratory difficulties and may have profound digestive complications. Cystic fibrosis also causes the sweat to be concentrated and very salty. In fact, cystic fibrosis is often first suspected when parents kiss their baby's skin and find that it tastes salty. Other complications of cystic fibrosis can include blockage of the bowel, delayed growth, coughing, frequent respiratory infection, and liver damage.

There is currently no way to prevent or cure cystic fibrosis. However, scientists now know precisely what genes contain the mutations responsible for this disease and are working to develop ways to insert copies of the normal gene into the cells of cystic fibrosis patients. In the meantime, clinicians recommend that people with cystic fibrosis maintain a healthy diet, engage in regular exercise, consume adequate fluids, use medications as appropriate, and take nutritional supplements (especially the fat-soluble vitamins and calcium) to help ensure optimal health.[45]

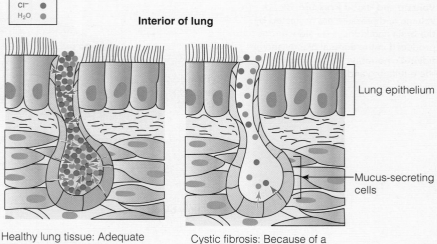

Interior of lung

Lung epithelium

Mucus-secreting cells

Healthy lung tissue: Adequate amount of healthy mucus is produced due to transport of chloride out of cells and subsequent osmosis.

Cystic fibrosis: Because of a faulty chloride transporter, only small amounts of thick, sticky mucus are released into the lungs.

as little as a 2% loss of a person's body weight in water can lead to many complications, listed here.

- Mental confusion
- Decreased motor control and ability to engage in exercise
- Difficulty with short- and long-term memory
- Decreased attention span
- Inability to maintain healthy body temperature, especially when exercising
- Increased risk for urinary tract infections
- Fatigue

These complications pose especially serious health concerns for children and the elderly, who have little tolerance for the negative consequences of dehydration.[46] Importantly, signs of dehydration can easily be mistaken for those of dementia in the elderly. And, as previously discussed, athletes who lose large amounts of water in sweat are at increased risk for dehydration.

cystic fibrosis A genetic (inherited) disease in which the production of a defective chloride transporter results in the inability of the body to transport chloride out of cells.

As a consequence, ensuring adequate water intake in infants, children, the elderly, and athletes is especially important.

The Body's Response to Dehydration When water intake is inadequate to replace water loss, the body tries to retain as much fluid as possible (Figure 12.12). The pituitary gland, located at the base of the brain, responds to dehydration (in other words, low blood volume) by releasing **antidiuretic hormone (ADH).** ADH, also called vasopressin, circulates in the blood to the kidneys, where it stimulates water conservation by decreasing urine production and output. ADH also causes blood vessels to constrict, raising blood pressure.

At the same time the pituitary gland is releasing ADH in response to low blood *volume*, low blood *pressure* caused by dehydration stimulates the kidneys to produce the enzyme renin. Renin causes the adrenal glands to release aldosterone, in turn stimulating the kidneys to retain sodium and chloride. Increased sodium and chloride reabsorption by the kidneys helps conserve

antidiuretic hormone (ADH) (also called vasopressin) A hormone produced in the pituitary gland and released during periods of low blood volume; stimulates the kidneys to decrease urine production, thus conserving water.

FIGURE 12.12 Regulation of Blood Volume and Blood Pressure Blood volume and pressure are regulated by the brain and kidneys via aldosterone produced in the adrenal glands and antidiuretic hormone (ADH) produced in the pituitary gland.

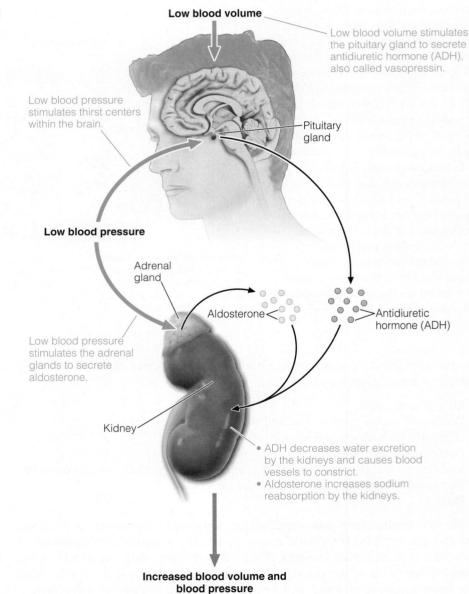

Low blood volume

Low blood volume stimulates the pituitary gland to secrete antidiuretic hormone (ADH), also called vasopressin.

Low blood pressure stimulates thirst centers within the brain.

Pituitary gland

Low blood pressure

Adrenal gland

Aldosterone

Antidiuretic hormone (ADH)

Low blood pressure stimulates the adrenal glands to secrete aldosterone.

Kidney

• ADH decreases water excretion by the kidneys and causes blood vessels to constrict.
• Aldosterone increases sodium reabsorption by the kidneys.

Increased blood volume and blood pressure

water by drawing it back into the blood. Angiotensin II (produced via renin) works at the level of the brain to cause ADH release as well as the sensation of thirst—increasing blood volume even more. Recall that the initial stimulus in this scenario was a decrease in blood volume. As such, the combined actions of ADH and the renin–angiotensin–aldosterone system represent an integrated physiological system with a negative feedback loop—the end result (increased blood volume) being the opposite of the initial stimulus (decreased blood volume).

RECOMMENDATIONS FOR WATER INTAKE

In 2004, the Institute of Medicine released its first recommendations for water intake. These guidelines suggest that women should get about 90 ounces (2.7 liters or 11 cups) of water daily in the foods and drinks they consume. The recommended water intake for males is 125 ounces (3.7 liters or 16 cups) daily. These amounts may sound high, but they include water from all beverages and foods consumed. In general, about 80% of our total water intake comes from beverages, with the remaining 20% coming from foods. A small amount of water (300–400 mL/day) is also generated by metabolism. Note that, upon careful consideration, the DRI committee concluded that caffeinated beverages such as coffee and cola appear to contribute to total water intake to the same degree as noncaffeinated fluids. This is in contrast to previous suggestions that these beverages could not be relied upon for supplying water to the body. You can read about the potential relationship between coffee and health in the Focus on Diet and Health feature. You can also learn more about the different types of bottled water available on the market, and whether bottled water is actually preferable to tap water, on the next page.

Special Recommendations for Active Populations Physical activity, especially in warm environments, increases your need for water. Studies show that very active people living or working in warm environments may have daily water requirements of up to about 240 ounces (7 liters, or 30 cups).[47] No specific guidelines are established for these individuals, but the Institute of Medicine recommends that very active people take special care to consume enough fluids every day.

Sports Drinks, Hydration, and Active Children As previously discussed, consuming sports drinks during endurance events, such as marathons, can help replace the water and electrolytes that are lost. In addition, many sports drinks also contain glucose, which not only facilitates electrolyte absorption but also provides a small amount of energy. However, for less active people, water is the drink of choice for simple hydration and rehydration. In addition, there is no evidence that consuming sports drinks provides a competitive edge to most athletes in training or competition—unless the activity is of high intensity or in a hot environment.[48]

Although most active adults are able to easily satisfy their fluid requirements, children may be at increased risk for dehydration during exercise. For instance, because they may not automatically feel thirsty during prolonged exercise, children may need to be encouraged to take a water break.[49] To prevent dehydration in children, the American Academy of Pediatrics recommends that special care be taken to restrict the amount of time young athletes train and compete, especially in warm environments, and that coaches and parents make sure that clothing and athletic gear is not overly hot.[50] They also recommend that children be well hydrated prior

Some research suggests that coffee, when consumed in moderation, may be beneficial to health.

Use of sports drinks may be especially important for athletic youth.

to physical activity and that 9- to 12-year-old children consume 3 to 8 ounces (0.4 to 1 cup) of fluid every 20 minutes; older children may need up to 16 ounces (2 cups) every 20 minutes. Because children tend to prefer sports drinks over water, and to help replace lost electrolytes, providing them with this choice may help ensure healthy hydration, especially in hot weather.

Is Bottled Water Better? You may have wondered if drinking bottled water is preferable to tap water. To understand this issue better, it is important to understand the differences among the various types of bottled water you can buy. To help consumers know what they are purchasing, the U.S. Food and Drug Administration (FDA) regulates the use of various terms that can be used on the labels of bottled water. These are provided next.

- *Artesian water*—Water from a well tapping a confined aquifer in which the water level stands at a specified height above the top of the aquifer.
- *Mineral water*—Water that contains at least 250 parts per million (ppm) total dissolved solids and that originates from a geologically and physically protected underground water source.
- *Purified water*—Water that is produced by distillation, deionization, reverse osmosis, or other suitable processes. When appropriate, this type of bottled water may also be called "demineralized water," "deionized water," "distilled water," or "reverse osmosis water."
- *Sparkling water*—Water that, after treatment and possible replacement of carbon dioxide, contains the same amount of carbon dioxide that it had at emergence from the source.
- *Spring water*—Water derived from an underground formation from which water flows naturally to the surface of the earth at an identified location.

There are many kinds of bottled water available for purchase. In general, bottled water provides no nutritional benefit over tap water, unless the local water supply is not safe to drink.

The FDA requires that bottled water contain no added ingredients, except for optional antimicrobial agents or fluoride. If additional substances are added, then the product label must include this information, such as "bottled water with minerals added" or "bottled water with raspberry flavor."

But, is there any reason to believe that bottled water is better than tap water when it comes to health? The answer most commonly given by health professionals to this question is *no*. In fact, because most municipalities add fluoride to their water, drinking bottled water instead of tap water may increase risk for tooth decay. Furthermore, many argue that the environmental costs associated with packaging and the waste related to bottled water outweigh any nutritional benefits. Thus, except in situations in which the local water might be contaminated, most agree that tap water is as wholesome as bottled varieties and certainly healthier for the environment.

How Are the Functions and Food Sources of the Major Minerals Related?

To summarize, the body uses the major minerals for many purposes, including those that play both functional and structural roles. For example, some major minerals participate as cofactors for enzymes, regulate energy

Millions of people enjoy starting their day with a steaming cup of coffee. Indeed, coffee is among the most widely consumed beverages in the world. But is there more to coffee than just the pick-me-up that it provides? And are there health concerns or even health benefits to be had by drinking this beverage?

It is important to realize that coffee is a complex liquid, containing a variety of chemicals.[51] Perhaps the best understood is caffeine, which acts as both a stimulant and a diuretic.[52] Caffeine is rapidly absorbed from the stomach and small intestine and distributed to all tissues, including the brain. The caffeine content of 1 cup brewed coffee varies greatly, being anywhere from 72 to 130 milligrams; the content in espresso ranges from 58 to 76 milligrams in each shot.[53] Other compounds such as cafestol and kahweol are also present in coffee.[54] These substances appear to be highest in boiled coffee, Turkish coffee, and French press coffee, while filtered coffee, percolated coffee, and instant coffee contain low levels.[55] Still other chemicals, such as chlorogenic acid and caffeic acid, appear to have potent antioxidant activity.[56] In general, very little is known about the potential health benefits or risks of these compounds. Coffee is also a relatively good source of some micronutrients, such as magnesium, potassium, and niacin. In addition, coffee is the primary source of antioxidants for some people.[57]

Although older studies suggest that coffee consumption may pose a plethora of health risks, more recent work in general suggests just the opposite. For example, there is increasing evidence that drinking coffee may reduce the risk of developing type 2 diabetes.[58] Results from one long-term U.S. study suggest that, compared with non-coffee drinkers, men who drink more than 6 cups of caffeinated coffee per day have half the risk of developing type 2 diabetes, and women reduce their risk by nearly 30%.[59] Studies also suggest that coffee and caffeine consumption are inversely related to the risk of developing Parkinson's disease.[60] For example, a study of 8,004 Japanese American men living in Hawaii reported an inverse relationship between the incidence of Parkinson's disease and coffee consumption—those who drank more than 4 cups each day were five times less likely to develop the disease than those who drank no coffee.[61] Several studies have even provided evidence that coffee consumption is related to a lower risk of colorectal cancer. According to one study, the risk for this type of cancer was reduced as coffee drinking rose to 5 cups a day.[62] There is also evidence that higher coffee consumption is linked to lower risks for ovarian and prostate cancer.[63]

Nonetheless, some data suggest that some compounds in coffee may pose several health risks—at least when

Branislav Senic /Shutterstock.com

Researchers are discovering that coffee likely contains many health-promoting substances.

consumed in excessive amounts by some segments of the population. For example, studies provide some evidence that coffee consumption may increase circulating low-density lipoprotein (LDL) cholesterol levels.[64] This relationship appears to be dependent on the type of coffee (unfiltered being worse than filtered) and amount consumed. However, several epidemiologic studies have shown no relationship between coffee consumption and cardiovascular risk, suggesting that the relationship between coffee and LDL concentrations is complex.[65] Because coffee contains polyphenols and phytates—compounds that bind nonheme iron and zinc in the small intestine—it is theoretically possible that high intake of coffee may increase risks for iron and zinc deficiencies. Excessive coffee consumption has also been related to unwarranted sleepiness in adolescents and young adults, may cause headaches in some people, and appears to be associated with increased birth defects and perhaps miscarriages in women.[66] The safety of caffeine intake by children has also long been a public health concern, although there is little or no evidence that reasonable caffeine consumption is related to poor growth or any adverse effects in this population.[67] Of course, these potential effects are serious and should be given considerable thought when choosing whether and how much coffee one should consume.

So, the good news is that there is growing evidence that moderate coffee consumption may play a role in optimizing health. However, as with all foods and beverages, moderation is essential—especially during periods of the life cycle characterized by increased growth and development.

TABLE 12.2 General Functions of the Essential Major Minerals

Functions	Major Minerals				
	Calcium	Phosphorus	Magnesium	Sodium/Chloride	Potassium
Cofactor	X				
Immune function				X	
Bone and tooth health	X	X	X		
Blood health	X	X			
Energy metabolism[a]	X	X	X		X
Fluid balance				X	X
Nerve and muscle function	X	X	X	X	X

[a] Note that these minerals are not, themselves, energy-yielding nutrients but rather are involved in energy metabolism via other roles, such as activation of enzymes or as a cofactor.

metabolism, are important for a healthy immune system, or are involved in glucose homeostasis. In addition, several play important parts in regulating fluid balance in the body as well as in nerve and muscle function. The major minerals also make up most of the bones and teeth and therefore are important for their strength and functions. As such, these important nutrients often work together to keep us healthy, and their overlapping roles are summarized in Table 12.2. It is important that you consume sufficient amounts of each of these essential minerals from the diet, and remember that, as with the other micronutrients, no single food group can supply all of them. Consequently, consuming a balanced diet featuring a wide variety of foods is required.

© Shelley McGuire

Critical Thinking: JoAnn's Story Now that you have learned about several of the major minerals important to bone health, you might want to again consider JoAnn. How might a college student on a limited budget ensure that he or she is consuming adequate nutrients (particularly major minerals) to optimize bone strength? What bone-building nutrients would be the most difficult to obtain? How could college food-service facilities label foods that are especially helpful in this regard?

Diet Analysis PLUS ✚ Activity

Like the vitamins, major minerals are considered micronutrients. Unlike vitamins (which contain carbon atoms), however, minerals are inorganic. The major minerals differ from the trace minerals by being essential in relatively larger quantities.

Using your *Intake and DRI Goals Compared* report generated from your diet analysis project and representing the average of your intakes, fill in the following table to assess your intakes of the major minerals that are included in your Diet Analysis Plus printouts. *(Remember that your intakes may fluctuate dramatically from day to day, so make sure to use the report that averages these fluctuations over the time period.)*

1.

Mineral	Chemical symbol	Your RDA/AI *(include units)*	Your estimated intake *(include units)*	The % of RDA/AI you consumed	Did you consume too little (−), too much (+), or an amount equal (=) to your RDA/AI?
Calcium					
Magnesium					
Potassium					
Sodium					

2. As with the vitamins discussed in Chapters 10 and 11, the major minerals have either a Recommended Dietary Allowance (RDA) or an Adequate Intake (AI) level. Can you hypothesize about and come to any conclusion as to why certain minerals have an RDA while others have an AI?

3. For the following, list the foods from your individual days' *Intake Spreadsheets* that provided the greatest amounts of the following major minerals.

Calcium
Name of food consumed   **Amount of nutrient in food consumed**
                                                                                (include units)
_____
_____
_____

Magnesium
Name of food consumed   **Amount of nutrient in food consumed**
                                                                                (include units)
_____
_____
_____

Potassium
Name of food consumed   **Amount of nutrient in food consumed**
                                                                                (include units)
_____
_____
_____

Sodium
Name of food consumed   **Amount of nutrient in food consumed**
                                                                                (include units)
_____
_____
_____

4. Based on the assessment of your intake of selected major minerals, do you need to change your intakes of these nutrients? Why or why not? Given the types of foods that you like and that are available to you, how might you best increase/decrease the amounts of major minerals that you need?

Notes

1. Anderson JJB. Nutritional biochemistry of calcium and phosphorus. Journal of Nutritional Biochemistry. 1991;2:300–7. Sheikh MS, Santa Ana CA, Nicar MJ, Schiller LR, Fordtran JS. Gastrointestinal absorption of calcium from milk and calcium salts. New England Journal of Medicine. 1987;317:532–6.

2. Scelfo GM, Flegal R. Lead in calcium supplements. Environmental Health Perspectives. 2004;108:309–13. Whiting S. Safety of some calcium supplements questioned. Nutrition Reviews. 1994;52:95–7.

3. Fairbanks BW, Mitchell HH. The availability of calcium in spinach, in skim milk powder, and in calcium oxalate. Journal of Nutrition. 1938;16:79–89.

4. Heaney RP, Recker RR, Stegman MR, Moy AJ. Calcium absorption in women: Relationships to calcium intake, estrogen status, and age. Journal of Bone Mineral Research. 1989;469–75.

5. Hoenderop JG, Nilius B, Bindels RJ. Calcium absorption across epithelia. Physiological Reviews. 2005;85:373–422.

6. Chin ER. Role of Ca21/calmodulin-dependent kinases in skeletal muscle plasticity. Journal of Applied Physiology. 2005;99:414–23. Mihalyi E. Review of some unusual effects of calcium binding to fibrinogen. Biophysical Chemistry. 2004;112:31–40. Thorneloe KS, Nelson MT. Ion channels in smooth muscle: Regulators of intracellular calcium and contractility. Canadian Journal of Physiology and Pharmacology. 2005;83:215–42.

7. Barnes S, Kelly ME. Calcium channels at the photoreceptor synapse. Advances in Experimental Medicine and Biology. 2004;514:465–76. French RJ, Zamponi GW. Voltage-gated sodium and calcium channels in nerve, muscle, and heart. IEEE Transactions in Nanobioscience. 2005;4:58–69. Senin II, Koch KW, Akhtar M, Philippov, PP. Ca21-dependent control of rhodopsin phosphorylation: Recoverin and rhodopsin kinase. Advances in Experimental Medicine and Biology. 2002;514:69–99.

8. Alonso A, Zozaya C, Vazquez Z, Alfredo Martinez J, Martinez-Gonzalez MA. The effect of low-fat versus whole-fat dairy product intake on blood pressure and weight in young normotensive adults. Journal of Human Nutrition and Dietetics. 2009; 22:336–42. Manios Y, Moschonis G, Koutsikas K, Papoutsou S, Petraki I, Bellou E, Naoumi A, Kostea S, Tanagra S. Changes in body composition following a dietary and lifestyle intervention trial: The postmenopausal health study. Maturitas. 2009;62: 58–65. Wilkinson SB, Tarnopolsky MA, MacDonald MJ, MacDonald JR, Armstrong D, Phillip SM. Consumption of fluid skim milk promotes greater muscle protein accretion after resistance exercise than does consumption of an isonitrogenous and isoenergetic soy-protein beverage. American Journal of Clinical Nutrition. 2007;85:1031–40. Zemel MB, Teegarden D, Van Loan M, Schoeller DA, Matkovic V, Lyle RM, Craig BA. Dairy-rich diets augment fat loss on an energy-restricted diet: a multi-center trial. Nutrients. 2009;1:83–100.

9. Alvarez-Leon EE, Roman-Vinas B, Serra-Majem L. Dairy products and health: A review of the epidemiological evidence. British Journal of Nutrition. 2006;96:S94–9. Haug A, Hestmark AT, Harstad OM. Bovine milk in human health—a review. Lipids in Health and Disease. 2007;6:25. Huth PJ, DiRienzo DB, Miller GD. Major scientific advances with dairy foods in nutrition and health. Journal of Dairy Science. 2006;89:1207–21. Gibson RA. Milk fat and health consequences. Nestle Nutrition Workshop Series Pediatric Program. 2011;67:197–207.

10. Bischoff-Ferrari HA, Dawson-Hughes B, Baron JA, et al. Calcium intake and hip fracture risk in men and women: A meta-analysis of prospective cohort studies and randomized controlled trials. American Journal of Clinical Nutrition. 2007;86:1780–90.

11. Bucher HC, Cook RJ, Guyatt GH, Lang JD, Cook DJ, Hatala R, Hunt DL. Effects of dietary calcium supplementation on blood pressure. Journal of the American Medical Association. 1996;275:1016–22. Lanham-New SA. The balance of bone health: Tipping the scales in favor of potassium-rich, bicarbonate-rich foods. Journal of Nutrition. 2008;138:172S–7S.

12. Hilpert K, West SG. Bagsjaw DM, Fishell V, Barnhart L, Lefevre M, Most MM, Zemel MB, Chow M, Hinderliter AL, Kris-Etherton P. Effects of dairy products on intracellular caclium and blood pressure in adults with essential hypertension. Journal of the American College of Nutrition. 2009;28:103S–19S. Pfeuffer M, Schrezenmeir J. Milk and the metabolic syndrome. Obesity Reviews. 2007;8:109–18. Saito T. Antihypertensive peptides derived from bovine casein and whey proteins. Advances in Experimental Biology and Medicine. 2008;606:295–317.

13. Fumeron F, Lamri A, Abi Khalil C, Jaziri R, Porchay-Baldérelli I, Lantieri O, Vol S, Balkau B, Marre M. Dairy consumption and the incidence of hyperglycemia and the metabolic syndrome: results from a French prospective study. Data from the Epidemiological Study on the Insulin Resistance Syndrome (DESIR). Diabetes Care. 2011;34: 813–17. Warensjo E, Nolan D, Tapsell L. Dairy food consumption and obesity-related chronic disease. Advances in Food and Nutrition Research. 2010;59:1–41.

14. Newmark HL, Heaney RP. Dairy products and prostate cancer risk. Nutrition and Cancer. 2010;62:297–9. Rock CL. Milk and the risk and progression of cancer. Nestle Nutrition Workshop Series Pediatric Program. 2011;67:173–85. Cho E, Smith-Warner SA, Spiegelman D, Beeson WL, van den Brandt PA, Colditz GA, Folsom AR, Fraser GE, Freudenheim JL, Giovannucci E, Goldbohm RA, Graham S, Miller AB, Pietinen P, Potter JD, Rohan TE, Terry P, Toniolo P, Virtanen MJ, Willett WC, Wolk A, Wu K, Yaun SS, Zeleniuch-Jacquotte A, Hunter DJ. Dairy foods, calcium, and colorectal cancer: A pooled analysis of 10 cohort studies. Journal of the National Cancer Institute. 2004;96:1015–22. Hjartaker A, Laake P, Lund E. Childhood and adult milk consumption and risk of premenopausal

breast cancer in a cohort of 48,844 women. International Journal of Cancer. 2001;93:888–93. Weaver CM. Should dairy be recommended as part of a healthy vegetarian diet? Point. American Journal of Clinical Nutrition. 2009;89:1634S–7S.

15. Knekt P, Jarvinen R, Seppanen R, Pukkala E, Aromaa A. Intake of dairy products and the risk of breast cancer. British Journal of Cancer. 1998;73:687–91.

16. Wu K, Willett WC, Fuchs CS, Colditz GA, Giovannucci EL. Calcium intake and risk of colon cancer in women and men. Journal of the National Cancer Institute. 2002;94:437–46.

17. Huncharek M, Muscat J, Kupelnick B. Colorectal cancer risk and dietary intake of calcium, vitamin D, and dairy products: A meta-analysis of 26,335 cases from 60 observational studies. Nutrition and Cancer. 2009;61:47–69. Pufulete M. Intake of dairy products and risk of colorectal neoplasia. Nutrition Research Reviews. 2008;21:56–67.

18. Zemel MB. Regulation of adiposity and obesity risk by dietary calcium: mechanisms and implications. Journal of the American College of Nutrition. 2002;21:1465–1515.

19. Wilkinson SB, Tarnopolsky MA, MacDonald MJ, MacDonald JR, Armstrong D, Phillips SM. Consumption of fluid skim milk promotes greater muscle protein accretion after resistance exercise than does consumption of an isonitrogenous and isoenergetic soy-protein beverage. American Journal of Clinical Nutrition. 2007;85:1031–40.

20. Brooks BM, Rajeshwari R, Nicklas TA, Yang SJ, Berenson GS. Association of calcium intake, dairy product consumption with overweight status in young adults (1995–1996): The Bogalusa Heart Study. Journal of the American College of Nutrition. 2006;25:523–32. Silveira MB, Carraro R, Monereo S, Tébar J. Conjugated linoleic acid (CLA) and obesity. Public Health Nutrition. 2007;10:1181–6. Zemel MB. Calcium and dairy modulation of obesity risk. Obesity Research. 2005;13:192–3.

21. Parikh SJ, Yanovski JA. Calcium intake and adiposity. American Journal of Clinical Nutrition. 2003;77:281–7. Snijder MB, van der Heijden AA, van Dam RM, Stehouwer CD, Hiddink GJ, Nijpels G, Heine RJ, Bouter LM, Dekker JM. Is higher dairy consumption associated with lower body weight and fewer metabolic disturbances? The Hoorn Study. American Journal of Clinical Nutrition. 2007;85:989–95. Bortolotti M, Rudelle S, Schneiter P, Vidal H, Loizon E, Tappy L, Acheson KJ. Dairy calcium supplementation in overweight or obese persons: Its effect on markers of fat metabolism. American Journal of Clinical Nutrition. 2008;88:877–85.

22. National Osteoporosis Foundation. Fast facts. Available from: http://www.nof.org/node/40.

23. Jackson RD, LaCroix AZ, et al. Calcium plus vitamin D supplementation and the risk of fractures. New England Journal of Medicine. 2006;354:669–83.

24. Hallberg L, Rossander-Hulten L, Brune M, Gleerup A. Calcium and iron absorption: Mechanism of action and nutritional importance. European Journal of Clinical Nutrition. 1992;46:317–27. Institute of Medicine. Dietary reference intakes for calcium and vitamin D. Washington, DC: National Academies Press; 2011.

25. Tucker KL, Morita K, Qiao N, Hannan MT, Cupples LA, Kiel DP. Colas, but not other carbonated beverages, are associated with low bone mineral density in older women: The Framingham osteoporosis study. American Journal of Clinical Nutrition. 2006;84:936–42.

26. Tenenhouse HS. Regulation of phosphorus homeostasis by the type IIA Na/phosphate cotransporter. Annual Review of Nutrition. 2005;25:197–214.

27. Takeda E, Taketani Y, Sawada N, Sato T, Yamamoto H. The regulation and function of phosphate in the human body. Biofactors. 2004;21:345–55.

28. Davì G, Santilli F, Patrono C. Nutraceuticals in diabetes and metabolic syndrome. Cardiovascular Therapeutics. 2010;28:216–26. Mooren FC, Krüger K, Völker K, Golf SW, Wadepuhl M, Kraus A. Oral magnesium supplementation reduces insulin resistance in non-diabetic subjects—a double-blind, placebo-controlled, randomized trial. Diabetes, Obesity, and Metabolism. 2011;13:281–4.

29. Reinhold JG, Fardadji B, Abadi P, Ismail-Beigi F. Decreased absorption of calcium, magnesium, zinc and phosphorus by humans due to increased fiber and phosphorus consumption as wheat bread. American Journal of Clinical Nutrition. 1991;49:204–6.

30. Hunt MS, Schofield FA. Magnesium balance and protein intake level in adult human females. American Journal of Clinical Nutrition. 1969;22:367–73. Schwartz R, Walker G, Linz MD, MacKellar I. Metabolic responses of adolescent boys to two levels of dietary magnesium and protein. I. Magnesium and nitrogen retention. American Journal of Clinical Nutrition. 1973;26:510–8.

31. Seiner R, Hesse A. Influence of a mixed and a vegetarian diet on urinary magnesium excretion and concentration. British Journal of Nutrition. 1995;73:783–90. Wisker E, Nagel R, Tanudjaja TK, Feldheim W. Calcium, magnesium, zinc, and iron balances in young women: Effects of a low-phytate barley-fiber concentration. American Journal of Clinical Nutrition. 1991;54:553–9.

32. Bo S, Pisu E. Role of dietary magnesium in cardiovascular disease prevention, insulin sensitivity and diabetes. Current Opinions in Lipidology. 2008;19:50–6. Gums JG. Magnesium in cardiovascular and other disorders. American Journal of Health-System Pharmacy. 2004;61:1569–76.

33. Alghamdi AA, Al-Radi OO, Latter DA. Intravenous magnesium for prevention of atrial fibrillation after coronary artery bypass surgery: A systematic review and meta-analysis. Journal of Cardiac Surgery. 2005;20:293–9. Belin RJ, He K. Magnesium physiology and pathogenic mechanisms that contribute to the development of the metabolic syndrome. Magnesium Research. 2007;20:107–29;

34. Durlach J, Pages N, Bac P, Bara M, Guiet-Bara A. Headache due to photosensitive magnesium depletion. Magnesium Research. 2005;18:109–22. Sun-Edelstein C, Mauskop A. Foods and supplements in the management of migraine headaches. Clinical Journal of Pain. 2009;25:446–52.

35. Gamba G. Molecular physiology and pathophysiology of electroneutral cation-chloride cotransporters. Physiological Reviews. 2005;85:423–93.

36. Conlin PR. Interactions of high salt intake and the response of the cardiovascular system to aldosterone. Cardiology in Review. 2005;13:118–24. Meneton P, Jeunemaitre X, de Wardener HE, MacGregor GA. Links between dietary salt intake, renal salt handling, blood pressure, and cardiovascular diseases. Physiological Reviews. 2005;85:679–715. Jaitovich A, Bertorello AM. Salt, Na^+,K^+-ATPase and hypertension. Life Sciences. 2010;86:73–8; Sanada H, Jones JE, Jose PA. Genetics of salt-sensitive hypertension. Current Hypertension Reports. 2011;13:55–66.

37. U.S. Department of Health and Human Services, Centers for Disease Control and Prevention, National Center for Health

Statistics. Health, United States, 2010. Available from: http://www.cdc.gov/nchs/data/hus/hus10.pdf.

38. Titze J, Ritz E. Salt—its effect on blood pressure and target organ damage: New pieces in an old puzzle. Journal of Nephrology. 2009;22:177–89.

39. Slimko ML, Mensah GA. The role of diets, food, and nutrients in the prevention and control of hypertension and prehypertension. Cardiology Clinics. 2010;28:665–74.

40. Institute of Medicine. Dietary Reference Intakes for water, potassium, sodium, chloride, and sulfate. Washington, DC: National Academies Press; 2005.

41. Colussi G, Catena C, Lapenna R, Nadalini E, Chiuch A, Sechi LA. Insulin resistance and hyperinsulinemia are related to plasma aldosterone levels in hypertensive patients. Diabetes Care. 2007;30:2349–54. McCarty MF, Stumvoll M, Goldstein BJ, van Haeften TW. Type 2 diabetes: Principles of pathogenesis and therapy. Lancet. 2005;365:1333–46.

42. Kallen RJ, Rieger CHL, Cohen HS, Suter MA, Ong RT. Near-fatal hyperkalemia due to ingestion of salt substitute by an infant. Journal of the American Medical Association. 1976;235:2125–6.

43. Cystic Fibrosis Foundation. About cystic fibrosis. Available from: http://www.cff.org/AboutCF/.

44. Davies JC, Alton EW. Gene therapy for cystic fibrosis. Proceedings of the American Thoracic Society. 2010;7:408–14.

45. Stallings VA, Start LF, Robinson KA, Feranchak AP, Quinton H. Evidence-based practice recommendations for nutrition-related management of children and adults with cystic fibrosis and pancreatic insufficiency. Results of a systematic review. Journal of the American Dietetic Association. 2008;108:832–9.

46. Kent AJ, Banks MR. Pharmacological management of diarrhea. Gastroenterology Clinics of North America. 2010;39:495–507.

47. Ruby BC, Shriver TC, Zderic TW, Sharkey BJ, Burks C, Tysk S. Total energy expenditure during arduous wildfire suppression. Medicine and Science in Sports and Exercise. 2002;34:1048–54.

48. Jeukendrup AE, Jentjens RL, Moseley L. Nutritional considerations in triathlon. Sports Medicine. 2005;35:163–81. von Duvillard SP, Braun WA, Markofski M, Beneke R, Leithauser R. Fluids and hydration in prolonged endurance performance. Nutrition. 2004;20:651–6.

49. Casa DJ, Yeargin SW. Avoiding dehydration among young athletes. American Academy of Sports Medicine Health and Fitness Journal. 2005;8:20–23.

50. American Academy of Pediatrics, Council on Sports Medicine and Fitness and Council on School Health. Policy statement—Climatic heat stress and exercising children and adolescents. Pediatrics. 2011:128:e741–e747.

51. Higdon JV, Frei B. Coffee and health: A review of recent human research. Critical Reviews in Food Science and Nutrition. 2006;46:2:101–23. van Dam RM. Coffee consumption and risk of type 2 diabetes, cardiovascular diseases, and cancer. Applied Physiology, Nutrition, and Metabolism. 2008;33:1269–83.

52. Carrillo JA, Benitez J. Clinically significant pharmacokinetic interactions between dietary caffeine and medications. Clinical Pharmacokinetics. 2000;39:127–53.

53. McCusker RR, Goldberger BA, Cone EJ. Caffeine content of specialty coffees. Journal of Analytical Toxicology. 2003;27:520–2.

54. Gross G, Jaccaud E, Huggett AC. Analysis of the content of diterpenes cafestol and kahweol in coffee brews. Food Chemistry and Toxicology. 1997;35:547–54.

55. Urgert R, van der Weg G, Kosmeijer-Schuil TG, van de Bovenkamp P, Hovenier R, Katan MB. Levels of the cholesterol-elevating diterpenes cafestol and kahweol in various coffee brews. Journal of Agriculture and Food Chemistry. 1995;43:2167–72.

56. Clifford MN. Chlorogenic acids and other cinnamates—natural occurrence and dietary burden. Journal of the Science of Food and Agriculture. 1999;79:362–72. Olthof MR, Hollman PC, Katan MB. Chlorogenic acid and caffeic acid are absorbed in humans. Journal of Nutrition. 2001;131:66–71.

57. Pellegrini N, Salvatore S, Valtueña S, Bedogni G, Porrini M, Pala V, Del Rio D, Sieri S, Miglio C, Krogh V, Zavaroni I, Brighenti F. Development and validation of a food frequency questionnaire for the assessment of dietary total antioxidant capacity. Journal of Nutrition. 2007;137:93–8.

58. Campos H, Baylin A. Coffee consumption and risk of type 2 diabetes and heart disease. Nutrition Reviews. 2007;65:173–9. Greenberg JA, Boozer CN, Geliebter A. Coffee, diabetes, and weight control. American Journal of Clinical Nutrition. 2006;84:682–93. Van Dam RM, Feskens EJM. Coffee consumption and risk of type 2 diabetes mellitus. Lancet. 2002;360:1477–8.

59. Psaltopoulou T, Ilias I, Alevizaki M. The role of diet and lifestyle in primary, secondary, and tertiary diabetes prevention: a review of meta-analyses. Review of Diabetic Studies. 2010;7:26–35. Salazar-Martinez E, Willett WC, Ascherio A, Leitzmann M, Manson JE, Stampfer MJ, Hu FB. Coffee consumption and risk of type 2 diabetes in men and women. Annals of Internal Medicine. 2004;140:1–8.

60. Ascherio A, Chen H, Schwarzschild MA, Zhang SM, Colditz GA, Speizer FE. Caffeine, postmenopausal estrogen, and risk of Parkinson's disease. Neurology. 2003;60:790–5. Ascherio A, Zhang SM, Hernán MA, Kawachi I, Colditz GA, Speizer FE, Willett WC. Prospective study of caffeine consumption and risk of Parkinson's disease in men and women. Annals of Neurology. 2001;50:56–63. Gao X, Chen H, Schwarzschild MA, Logroscino G, Ascherio A. Perceived imbalance and risk of Parkinson's disease. Movement Disorders. 2008; 23:613–6.

61. Ross GW, Abbot RD, Petrovitch H, et al. Association of coffee and caffeine intake with the risk of Parkinson's disease. Journal of the American Medical Association. 2002;283:2674–9.

62. Arab L. Epidemiologic evidence on coffee and cancer. Nutrition and Cancer. 2010;62:271–83. La Vecchia C, Tavani A. Coffee and cancer risk: An update. European Journal of Cancer Prevention. 2007;16:385–9.

63. Tworoger SS, Gertig DM, Gates MA, Hecht JL, Hankinson SE. Caffeine, alcohol, smoking, and the risk of incident epithelial ovarian cancer. Cancer. 2008 112:1169–77. Wilson KM, Kasperzyk JL, Rider JR, Kenfield S, van Dam RM, Stampfer MJ, Giovannucci E, Mucci LA. Coffee consumption and prostate cancer risk and progression in the Health Professionals Follow-up Study. Journal of the National Cancer Institute. 2011 May 17. [Epub ahead of print]

64. Boekschoten MV, Engberink MF, Katan MB, Schouten EG. Reproducibility of the serum lipid response to coffee oil in healthy volunteers. Nutrition Journal. 2003;Oct 4;2:8. Jee SH, He J, Appel LJ, Whelton PK, Suh I, Klag MJ. Coffee consumption and serum lipids: A meta-analysis of randomized controlled clinical trials. American Journal of Epidemiology. 2001;153:353–62.

65. Cheung RJ, Gupta EK, Ito MK. Acute coffee ingestion does not affect LDL cholesterol level. Annals of Pharmacotherapy. 2005;39:1209–13. Du Y, Melchert HU, Knopf H, Braemer-Hauth M, Gerding B, Pabel E. Association of serum caffeine concentrations with blood lipids in caffeine-drug users and nonusers—results of German National Health Surveys from 1984 to 1999. European Journal of Epidemiology. 2005;20:311–6. Lopez-Garcia E, van Dam RM, Willett WC, Rimm EB, Manson JE, Stampfer MJ, Rexrode KM, Hu FB. Coffee consumption and coronary heart disease in men and women: A prospective cohort study. Circulation. 2006;113:2045–53.

66. Brown ML. Maternal exposure to caffeine and risk of congenital anomalies: A systematic review. Epidemiology. 2006;17:324–31. Lim SS, Noakes M, Norma RJ. Dietary effects on fertility treatment and pregnancy outcome. Current Opinions in Endocrinology, Diabetes, and Obesity. 2007;14:465–9. Savitz DA, Chan RL, Herring AH, Howards PP, Hartmann KE. Caffeine and miscarriage risk. Epidemiology. 2008;19:55–62. Roehrs T, Roth T. Caffeine: Sleep and daytime sleepiness. Sleep Medicine Reviews. 2008;12:153–62.

67. Dewey KG, Romero-Abal ME, Quan de Serrano J, Bulux J, Peerson JM, Eagle P, Solomons NW. Effects of discontinuing coffee intake on iron status of iron-deficient Guatemalan toddlers: A randomized intervention study. American Journal of Clinical Nutrition. 1997;66:168–76. Stein MA, Krasowski M, Leventhal BL, Phillips W, Bender BG. Behavioral and cognitive effects of methylxanthines. A meta-analysis of theophylline and caffeine. Archives of Pediatric and Adolescent Medicine. 1996;150:284–8.

NUTRITION MATTERS

Nutrition and Bone Health

The skeletal system makes up the basic architecture of the body, and although you might think of the skeleton as being inert, bones are made of living tissue. Bones not only provide structure but are also the site of red and white blood cell production. They produce platelets important for the process of blood clotting and synthesize several growth factors that assist in the healing process. The skeleton begins to form early in fetal development, and continues to grow and develop throughout much of your life. As you have learned previously, a multitude of macro- and micronutrients (such as protein, calcium, and phosphorus) are needed for bone growth, remodeling, and maintenance. Thus, nutrition plays a key role in keeping bones strong and healthy.

It is important to develop and maintain healthy bones in childhood and even early adulthood, especially today as we enjoy longer and more active lives. This is because having weak bones in our 20s and 30s can lead to many negative health consequences in later life. For this reason, Healthy People 2020 has identified the prevention of bone disease, or osteoporosis, as one of its major goals.[1]

Do Bones Continue to Develop and Grow Throughout Life?

The skeletal system begins to develop early in life—in fact, in about the fifth week of gestation. However, during early fetal life, bones are soft and rubbery because they are made of tissue called **cartilage.** At around the seventh week of development, this cartilage is slowly replaced by hard, mineralized bone. This process of converting cartilage to bone, called **ossification,** continues through childhood. In fact, ossification is not complete until early adult life.

BONE TISSUE IS COMPLEX AND LIVING

Bones consist of several components, which are concentrically arranged somewhat like the layers of an onion. On their outer surface, they are covered by a thin membrane called the **periosteum,** made of connective tissue and hundreds of small blood vessels and nerves. Beneath the periosteum is a layer of hard, dense bone called **cortical bone** (also called compact

© creativ collection/Age Fotostock

bone), which surrounds another layer called **trabecular bone.** Trabecular bone is not as dense as cortical bone, instead consisting of a web-like matrix. The major function of cortical bone is to provide strength, whereas trabecular bone contains the **bone marrow**

---

cartilage (CAR – ti – lage) The soft, nonmineralized precursor of bone.

ossification The process by which minerals are added to cartilage, ultimately resulting in bone formation.

periosteum (per – i – O – ste – um) The outer covering of bone, consisting of blood vessels, nerves, and connective tissue.

cortical bone (compact bone) The dense, hard layer of bone found directly beneath the periosteum.

trabecular bone (tra – BE – cu – lar) Inner, less dense layer of bone; contains the bone marrow.

bone marrow Soft, spongy, inner part of bone; makes red blood cells, white blood cells, and platelets.

that produces red blood cells, white blood cells, and platelets. All bones in the body contain both cortical and trabecular bone, although the ratios differ. For example, the long bones of the forearm have less trabecular bone than do those of the vertebral column (backbone) and pelvis. Trabecular bone is also abundant in the ends of long bones. Positioned in the very core of each bone are large blood vessels, lymphatic vessels, and nerves that allow transport of hormones, nutrients, and neural impulses to and from bone tissue.

BONES ARE MADE OF OSTEOBLASTS AND OSTEOCLASTS

There are two basic types of bone cells: osteoblasts and osteoclasts. Recall from earlier in Chapter 12 that **osteoblasts** are bone-forming cells that deposit minerals and proteins needed for the formation, maintenance, growth, and repair of bones. **Osteoclasts** work in an opposite fashion, breaking down old bone to make way for new bone. In addition, osteoclasts function somewhat like calcium-storage depots, helping maintain blood calcium concentrations at optimal levels. For instance, when blood calcium levels are low, the body stimulates osteoclasts and, thus, bone breakdown. This, in turn, releases calcium into the blood and increases blood calcium levels back to normal. The actions of osteoblasts and osteoclasts are highly coordinated so that rates of bone formation and breakdown change as needed. This cycle of bone breakdown and rebuilding is called **bone remodeling** (or bone turnover) and can be compared to the remodeling of a house, during which the demolition team must remove existing walls, electrical wiring, insulation, and other materials before the builder can replace them with new ones. In the case of bones, osteoclasts (like the demolition team) break down and remove existing bone, and osteoblasts (like the builder) build new bone. As such, osteoclasts and osteoblasts work together to remodel bones and keep them strong. This remodeling is orchestrated by a variety of hormones and nutrients, including vitamin D and calcium, as described in Chapters 11 and 12.

Bone Mass and Peak Bone Mass During periods of growth, such as infancy, childhood, and puberty, the actions of osteoblasts exceed those of osteoclasts—lengthening and reshaping bones. In other words, bone growth is greater than bone loss. Most bone growth is complete before age 20, although a small amount (about 10%) of bone is laid down between 20 and 30 years of age. While bones are in their growth phase, the total amount of minerals contained in the skeleton, called **bone mass,** increases. Bone mass is typically at its highest level at about 30 years of age (Figure 1). **Peak bone mass** is the greatest total amount of bone that a person will have in his or her lifetime. People with larger frames tend to have higher peak bone mass than do people with smaller frames, and women

---

osteoblast A bone cell that promotes bone formation.

osteoclast A bone cell that promotes bone breakdown.

bone remodeling (bone turnover) The process by which older and damaged bone is removed and replaced by new bone.

bone mass The total amount of bone mineral in the body.

peak bone mass The greatest amount of bone mineral that a person has during his or her life.

FIGURE 1 Changes in Bone Mass during the Life Cycle Women are at greater risk of osteoporosis than men because women are more likely to have low bone mass in later life.

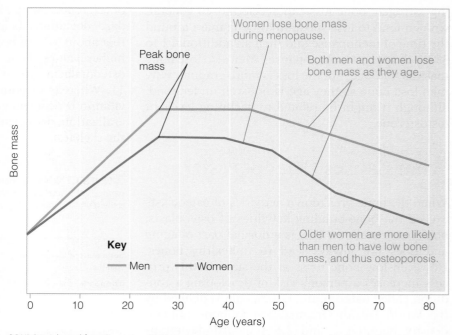

SOURCE: Adapted from Compston JE. Osteoporosis, corticosteroids and inflammatory bowel disease. Alimentary Pharmacology and Therapeutics. 1995;9:237–250.

generally have somewhat lower peak bone mass than do men.

Another measure of bone health is **bone mineral density,** which is the amount of mineral *in a given area or volume of bone.* The greater the bone mineral density, the stronger the bone. This concept is similar to the various densities found in wood. For example, oak is very dense and therefore quite strong, whereas cedar is less dense and considered a "softer" type of wood. Similarly, a bone with greater density is stronger (and therefore less likely to break) than a bone with lower density. The risk for developing bone disease later in life depends, in part, on peak bone mass and bone mineral density achieved in early adulthood. As previously stated, bone mass and bone mineral density peak around 30 years of age, after which they begin the process of gradual decline. There are many reasons for this. For example, calcium absorption decreases with age, due in part to age-related decreases in the production, absorption, and metabolism of vitamin D.[2] When calcium absorption in the small intestine declines, calcium resorption from bones increases to maintain blood calcium concentrations.

Hormone Changes and Bone Mass Another factor related to bone loss is the normal decline in reproductive hormone concentrations that occurs with age. For example, estrogen (a female reproductive hormone produced in the ovaries) is important for maintaining bone strength.[3] When a woman reaches menopause and her ovaries stop producing estrogen, bone loss can accelerate. In general, women tend to lose substantial bone mass around the time of menopause and then an additional 1 to 3% each year after menopause.[4] Men also lose bone mass as they age, but the loss is more gradual. Why men lose bone as they age is not well understood, although it might be related to declining levels of testosterone.[5]

What Causes Osteoporosis?

When the bone breakdown activities of osteoclasts exceed the bone-building activities of osteoblasts, bone mass declines. This is a normal part of aging and does not always lead to unhealthy bones. However, low bone mass at the start of this process increases a person's risk of developing a condition called **osteopenia.** Osteopenia is defined by the World Health Organization as bone mineral density between 1 and 2.5 standard deviations below that of a healthy young adult.[6,*] As a point

Because bone mass reaches its peak during early adulthood, eating right and getting plenty of physical activity during this time of life are critically important for long-term bone health.

of reference, having a bone mineral density 2 standard deviations below the mean would indicate that about 97% of healthy young adults would have higher values. Note that osteopenia differs from osteomalacia, which you learned about in Chapter 11. Whereas osteomalacia is generally caused by vitamin D deficiency and results in poor bone mineralization, osteopenia is due to *demineralization* of the skeleton.

⟨**CONNECTIONS**⟩ Remember that vitamin D stimulates the synthesis of several calcium transport proteins in the small intestine and is therefore important for calcium absorption (Chapter 11, page 473).

bone mineral density Amount of minerals contained in a given area or volume of bone.

osteopenia (os – te – o – PE – ni – a) A condition characterized by low bone mineral density

* "Standard deviation" is a statistical term that relates to variability or "spread" around a mean value.

FIGURE 2 Healthy and Osteoporotic Bone

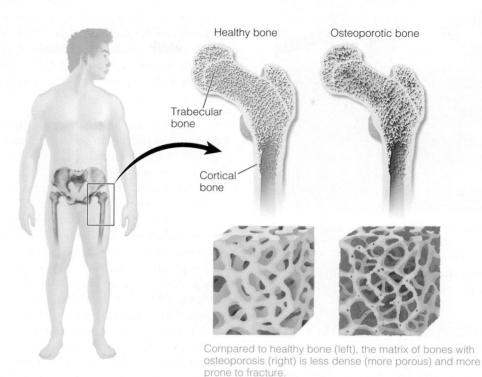

Healthy bone   Osteoporotic bone

Trabecular bone

Cortical bone

Compared to healthy bone (left), the matrix of bones with osteoporosis (right) is less dense (more porous) and more prone to fracture.

As more bone is lost, the entire matrix of the bone tissue begins to weaken. This can lead to **osteoporosis,** a type of bone disease characterized by reduced bone mineral density and disruption of the microarchitecture of the bone. Bones of a person with osteoporosis are, as the name would imply, porous. As you can see from Figure 2, the matrix of an osteoporotic bone has a less dense lattice system than does that of a healthy bone. Osteoporotic bones are weak and break easily. Both cortical and trabecular bone can become osteoporotic, although it is more common in trabecular bone. The spongy trabecular-rich bones of the pelvis, vertebrae, and portions of long bones are at greatest risk for fracture in people with osteoporosis. Osteoporosis is diagnosed when bone mineral density is greater than 2.5 standard deviations below that of a healthy young adult.[6]

THERE ARE TWO TYPES OF OSTEOPOROSIS

There are two types of osteoporosis. **Type 1 osteoporosis,** also called postmenopausal osteoporosis, is more common in females than males, and typically occurs in women between 50 and 60 years of age. This type of osteoporosis is directly linked to declining estrogen concentrations that occur at menopause. Type 1 osteoporosis most dramatically affects trabecular bone. **Type 2 osteoporosis** occurs in both men and women, tends to be diagnosed later in life (70 to 75 years of age), and is a result of breakdown of both cortical and trabecular bone. This type of osteoporosis is due to a combination of dietary and age-related factors. People with either form of osteoporosis can lose significant height during old age and experience severe pain, especially in the spine. In addition, some people develop a curvature in the upper spine called **kyphosis,** or dowager's hump. Kyphosis can be problematic, because the bending of the backbone can compress the chest cavity resulting in

osteoporosis A bone disease characterized by reduced bone mineral density and disruption of the microarchitecture of the bone.

type 1 osteoporosis The form of osteoporosis that occurs in women, caused by hormone-related bone loss.

type 2 osteoporosis The form of osteoporosis that occurs in men and women; caused by age-related and lifestyle factors.

kyphosis (dowager's hump) A curvature of the upper spine, caused by osteoporosis.

Osteoporosis can lead to kyphosis (dowager's hump).

difficulty breathing, abdominal pain, decreased appetite, and premature satiety.

OSTEOPOROSIS CAN SERIOUSLY AFFECT HEALTH AND WELL-BEING

Osteoporosis is a serious chronic degenerative disease with numerous personal, societal, and economic implications. According to some researchers, approximately 8 million women and 2 million men in the United States have osteoporosis, and 34 million Americans have osteopenia.[7] Because bone fractures often result in hospitalization and loss of independence, preventing osteoporosis is important to our health and well-being. The U.S. Department of Health and Human Services estimates that health care costs associated with osteoporosis are currently over $18 billion each year, and the Surgeon General of the United States predicts that these costs may double or triple by the year 2040.[8]

BIOLOGICAL AND LIFESTYLE FACTORS INCREASE RISK FOR OSTEOPOROSIS

Many factors are associated with risk for osteoporosis including those related to genetics as well as lifestyle choices. Biological risk factors—such as whether you are male or female—cannot be altered. However, lifestyle factors—such as exercise, diet, and smoking—can be modified to decrease risk for bone disease.

Biological Factors Related to Osteoporosis Biological (nonmodifiable) factors associated with increased risk for osteoporosis are related to your genetic makeup.[9] Taken together, these nonmodifiable factors are thought to contribute about half of a person's risk for osteoporosis.[10] Although biological risk factors cannot be changed, it is important to be aware of which ones apply to you. Some of the more well-documented risk factors are listed here.

- *Sex.* Women are at greater risk for osteoporosis than men. This has to do with many factors such as loss of estrogen production after menopause and overall smaller body size.
- *Age.* In adults, the rate of bone loss increases with age, especially in women. The influence of age on bone health is complex, including age-related changes that cannot be avoided, as well as some that can such as declining physical activity levels and poor dietary habits.
- *Body size.* In general, people with body mass indexes (BMIs) of less than 19 kg/m^2 are at increased risk for osteoporosis. Of course, body size is influenced by both biological and lifestyle factors (such as nutrition).
- *Ethnicity/race.* People of northern European or Asian descent are at greatest risk; those of African or Hispanic descent are at lowest risk.
- *Other genetic factors.* Although genetics does not completely determine who develops or does not develop osteoporosis, those with a family history are at especially high risk.

Because biological risk factors cannot be changed, it is important to do all you can to maintain bone health by making wise lifestyle choices.

Lifestyle Factors Related to Osteoporosis While you may not be able to alter your biological risk for osteoporosis, you are in control of many other factors that increase your risk for bone disease. These include smoking, taking certain medications, being physically inactive, and eating a diet that lacks important nutrients needed for bone health. The major lifestyle factors related to osteoporosis are described next.

Smoking Smoking can increase your risk for osteoporosis, and this relationship appears to be dose-dependent (i.e., the more a person smokes, the weaker his or her bones).[11] Although the reasons for this are not completely understood, smoking may weaken bones by decreasing estrogen production. However, because smoking is also related to bone loss in men, there are probably other mechanisms by which it influences bone health. Fortunately, the

Smoking can greatly increase risk for osteoporosis.

negative impact of smoking seems to be, at least partly, reversible. Thus, it is never too late to "kick the habit."

Some Medications Chronic use of some medications (such as thyroid hormones) can weaken bones and increase bone loss. Similarly, some diuretics (such as Lasix®) increase calcium loss in the urine, leading to weak bones.[12] Other medications, such as corticosteroid therapy, can also decrease bone strength and thus increase risk for osteoporosis.[13] It is important to ask your health care provider what effects, if any, your medications might have on your bone health. This is especially important as you age.

Low Estrogen Concentrations When levels of estrogen are low, the risk for osteoporosis increases. In women, most of the estrogen is produced by the ovaries. The surgical removal of the ovaries (called an ovariectomy) before the age of 45 years without **hormone replacement therapy (HRT)** increases risk for bone disease. Although HRT therapy after menopause helps maintain bone in later life, it is important to consider other health problems—such as increased risk for certain cancers—that may be associated with these medications.[14] Before beginning HRT treatment, women should discuss the health risks and benefits of these medications with their health care provider. When HRT use is considered solely for prevention of osteoporosis, the U.S. Food and Drug Administration (FDA) recommends that approved non-estrogen treatments should first be carefully considered and that, if these medications are taken, that they be used for the shortest time needed and at the lowest effective dose.[15]

History of Eating Disorders Individuals with a history of either anorexia nervosa or bulimia are at increased risk for lower bone density, especially in the lower back and hips where bones are primarily trabecular in nature.[16] This is partly because some types of eating disorders such as anorexia nervosa can lead to very low body fat, which in turn can disrupt normal reproductive function, lowering estrogen levels in women.[17] In addition, because individuals with anorexia nervosa tend to eat insufficient amounts of food and sometimes engage in purging behaviors, poor nutritional status is common, and therefore can also contribute to bone loss.

Alcohol Consumption Chronic alcoholism is associated with an increased risk for osteoporosis.[18] Although alcohol probably affects the cycle of bone remodeling, researchers think that much of the negative influence of alcoholism on bone health is due to consumption of an inadequate diet, especially one lacking sufficient amounts of calcium, vitamin D, and protein.

Physical Inactivity Regular physical activity, especially weight-bearing exercise, increases bone density and decreases risk for osteoporosis.[19] This is because weight-bearing exercise increases the rate of bone remodeling—leading to increased bone density. Weight training, walking, running, dancing,

hormone replacement therapy (HRT) Medication typically containing estrogen and progesterone, sometimes taken by women after having their ovaries removed or after menopause.

Weight-bearing exercise can strengthen bones.

and tennis are examples of bone-strengthening exercises. As little as one hour of these types of activities each week can make a big difference in bone strength. Importantly, engaging in weight-bearing exercises results in healthier bones in both children and the elderly, reinforcing the idea that we are never too young or too old to benefit from physical activity.

Nutritional Factors Because many nutrients are needed for healthy bones, consuming a well-balanced diet may be one of the most important things you can do to prevent osteoporosis.[20] At the forefront of these nutrients is calcium. Chronically low calcium intake can decrease peak bone mass in young adulthood and increase bone loss in later life. This is because calcium is the most prominent component of bones, and without it they become weak. Phosphorus and magnesium are also important throughout the lifespan for building and maintaining healthy bones. Consequently, a diet providing inadequate amounts of these minerals can predispose a person to osteoporosis.

Protein and vitamins C, D, and K are also required for the bone-building and remodeling processes.[21] Protein is needed not only for structural purposes within the bone matrix but also for other processes related to calcium homeostasis. Vitamin D is needed for calcium absorption in the small intestine and bone cell (osteocyte) differentiation, and vitamins C and K are involved in healthy bone-building activities. Recall that vitamin C assists the enzyme needed to produce collagen, which is critical for bone structure. Vitamin K enables bone-building proteins to bind calcium. In addition, fluoride can increase bone strength because it is incorporated into bone's basic mineral matrix.[22] Fluoride also appears to stimulate maturation of osteoblasts, thus enhancing the production of new bone. Because foods provide little fluoride, many communities add it to their drinking water. You can find out if your town has fluoridated water by contacting a local dentist or your city's water department.

Are You at Risk?

It is essential that we do all that we can to prevent osteoporosis. As such, all people and especially those at highest risk for bone disease, should be screened regularly. If bone loss is detected early, some medications and lifestyle changes can help slow its

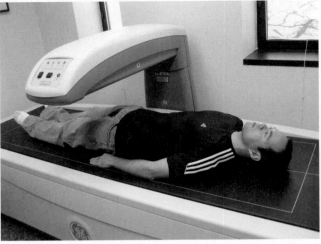

FIGURE 3 Dual-Energy X-Ray Absorptiometry (DEXA) Scan DEXA scans are used to monitor and assess bone health—especially as we age.

Fitness & Wellness, Inc.

progression and thus prevent osteoporosis. Some of these are described next.

FOR MOST PEOPLE, SCREENING SHOULD BEGIN AT AGE 65–70 YEARS

In addition to knowing your risk factors for osteoporosis, it is important to consider having your bone mineral density determined. Clinicians can measure both bone mineral density and mass using a variety of instruments. One such instrument is called a dual-energy X-ray absorptiometer (DEXA), which emits very low amounts of X-rays to produce a detailed picture of the body. A photograph of an individual undergoing a DEXA scan is shown in Figure 3. The National Osteoporosis Foundation recommends DEXA testing for the people listed in Table 1.[23]

OSTEOPOROSIS CAN BE TREATED WITH MEDICATIONS

Currently, the most effective way to treat osteoporosis in women is the use of HRT, which is typically a combination of the hormones estrogen and progesterone.[24] However, as previously described, HRT may increase the risk of other health problems. Thus, women considering HRT should consult

⟨**CONNECTIONS**⟩ Recall that dual-energy X-ray absorptiometry (DEXA) is also used to measure a person's body fat and lean mass (Chapter 8, page 338).

TABLE 1 Recommendations for Bone Mineral Density Testing

- Women age 65 years and older and men age 70 years and older, regardless of clinical risk factors
- Younger postmenopausal women and men age 50 to 69 years about whom you have concern based on their clinical risk factor profile
- Women in the menopausal transition if there is a specific risk factor associated with increased fracture risk such as low body weight, prior low-trauma fracture, or high-risk medication
- Adults who have a fracture after 50 years of age
- Adults with a condition (e.g., rheumatoid arthritis) or taking a medication (e.g., glucocorticoids in a daily dose ≥ 5 mg prednisone or equivalent for ≥ three months) associated with low bone mass or bone loss
- Anyone being considered for pharmacologic therapy for osteoporosis
- Anyone being treated for osteoporosis, to monitor treatment effect
- Anyone not receiving therapy in whom evidence of bone loss would lead to treatment
- Postmenopausal women discontinuing estrogen should be considered for bone density testing

National Osteoporosis Foundation, *Clinician's Guide to Prevention and Treatment of Osteoporosis*, pp. 13–14. Copyright © 2010 National Osteoporosis Foundation (NOF). All rights reserved. Reprinted by permission.

their health care provider to evaluate whether it is right for them. Some studies suggest that consuming **phytoestrogens,** which are estrogen-like phytochemicals found primarily in soy products, may also help maintain bone health after menopause, although not all studies have reported a beneficial effect.[25]

Medications, such as **selective estrogen receptor modulators (SERMs)** and **bisphosphonates,** can also help slow bone loss. SERMs mimic the beneficial effects of estrogen on bone density without some of the risks associated with HRT.[26] Bisphosphonates work by inhibiting bone breakdown, especially in the spine and hip regions. An example of a bisphosphonate is Fosamax®, which is commonly prescribed to treat osteopenia and osteoporosis.[27]

OPTIMAL NUTRITION IS CRITICAL FOR BONE HEALTH

The Dietary Reference Intakes (DRIs) for many nutrients required for bone health have been set at levels that both promote healthy bone growth during earlier life and minimize bone loss in later life. These nutrients include calcium, phosphorus, magnesium, vitamin D, vitamin C, vitamin K, and protein. Obtaining recommended levels of these nutrients is important throughout life, especially in individuals with increased risk for osteoporosis. Conversely, some foods contain compounds that, if consumed in large amounts, might contribute to weaker bones. These are described next.

Foods to Consume Only in Moderation Some food components can decrease absorption of calcium, including fiber, oxalates, and phytates. Phytates are commonly found in whole grains and nuts, and oxalates are present in many vegetables, especially green leafy varieties such as spinach and collard greens. People with osteopenia or osteoporosis should not consume large amounts of these compounds on a daily basis. However, because these foods are nutrient rich, it is prudent to continue to eat them in moderate amounts. Because excessive alcohol consumption is associated with increased risk of osteoporosis, clinicians recommend that alcohol consumption be kept moderate. This is generally defined as less than one drink a day for women and less than two drinks a day for men. Studies also show a weak relationship between caffeine intake and risk for bone disease.[28] However, typical coffee or tea consumption of 2 to 3 cups each day appears to pose no risk to bone health. There has also been some concern that consuming diets high in phosphorus may have a detrimental effect on bone. This is because studies using animal models suggest that high phosphorus intake, especially coupled with low calcium intake, can increase bone loss. However, after careful review of the literature, the Institute of Medicine concluded there is little or no evidence that consuming an otherwise adequate diet that is high in phosphorus influences bone health in humans.[29]

Foods to Encourage Obtaining adequate amounts of the bone-building nutrients requires that we consume a variety of foods. For instance, to help reach your goal for calcium, vitamin D, and protein, the Dietary Guidelines for Americans and accompanying MyPlate food guidance system recommend

phytoestrogen An estrogen-like phytochemical.

selective estrogen receptor modulator (SERM) A drug that does not contain the hormone estrogen but causes estrogen-like effects in the body.

bisphosphonates A class of drugs sometimes taken to help reduce bone loss.

Milk provides high-quality protein, calcium, phosphorus, and vitamin D—all important nutrients for bone health throughout the lifespan.

that you consume 2 to 3 cups of milk (or equivalent dairy product) each day. You can find out what your personal recommendation for dairy intake is by visiting the MyPlate website (www.choosemyplate .gov). If lactose intolerance is a problem for you, lactose-free milk or yogurt is suggested as a good alternative.

Comparing the calcium and protein contents of foods using Nutrition Facts panels is also a relatively easy way to choose foods that supply these nutrients. This is because the calcium and protein contents of foods are required components of food labeling. In addition, the FDA has approved the following health claim that can be included on packaging when appropriate:

"Regular exercise and a healthy diet with enough calcium help teens and young women maintain good bone health and may reduce their high risk of osteoporosis later in life."

Calcium supplements are often recommended to prevent osteoporosis. However, data from the Women's Health Initiative suggest that long-term (seven-year) daily supplementation with 1,000 mg of calcium and 400 IU of vitamin D has only small effects on bone density and risk of bone fracture.[30] Other studies show beneficial effects while supplements are taken, but this effect disappears when supplements are no longer consumed.[31] Hence, the benefits of calcium supplements are somewhat controversial.

Vitamin D is found in oily fish and, on exposure to sunlight, is made from cholesterol in the skin. To help prevent bone loss in older individuals, people over the age of 70 are advised to increase their vitamin D consumption and get at least 15 minutes of sun exposure each day, when possible.[32] Vitamin D supplements may also be required in the elderly because production of this vitamin declines as we age, dietary intake is often limited, and some older individuals find it difficult to get outside on a regular basis.

Consuming sufficient amounts of micronutrient-dense foods is also important for maintenance of strong bones throughout the lifespan. To obtain enough vitamin C, experts recommend that we consume 1 to 4 cups of vegetables and 1 to 2½ cups of fruit each day. Good sources of vitamin C are red peppers, citrus fruits, broccoli, strawberries, and Brussels sprouts. Fish and dark green leafy vegetables such as kale, spinach, broccoli, and Brussels sprouts provide the highest amounts of vitamin K in the diet. Legumes are also good sources of this nutrient. About 60% of the magnesium in your body is associated with bone, where it mainly provides structure to the crystal lattice. Foods rich in magnesium include seafood, legumes, nuts, and unprocessed (brown) rice, whereas phosphorus is found in almost all foods—especially those high in protein.

Many fruits and vegetables, like Brussels sprouts, can contribute to healthy bones. This is one reason why it is important to eat a variety of these foods on a daily basis.

Key Points

Do Bones Continue to Develop and Grow Throughout Life?

- Bones consist of an outer layer called the periosteum, and inner layers referred to as cortical and trabecular bone.

- Whereas cortical bone is dense, trabecular bone is more lattice-like in structure.

- Bones are made of osteoblasts that form bone and osteoclasts that break down old and damaged bone.

- Bone remodeling occurs throughout the entire life cycle, with bone growth being complete by about 30 years of age, when a person has the most bone mass he or she will ever have (peak bone mass).

What Causes Osteoporosis?

- Both genetic and lifestyle factors influence risk for osteoporosis.

- Biological risk factors include genetics, sex, and age, whereas lifestyle factors include smoking, medications, estrogen levels, eating disorders, alcohol consumption, physical inactivity, and inadequate nutrition.

Are You at Risk?

- Dual-energy X-ray absorptiometry (DEXA) can be used to measure bone mineral density.

- Treatment options for osteoporosis include a variety of medications.

- Consuming a diet high in many of the essential minerals (including calcium), protein, and several of the vitamins is important for lifelong bone health.

Notes

1. U.S. Department of Health and Human Services. Healthy people 2020. ODPHP Publication No. B0132. Washington, DC: U.S. Government Printing Office; 2010. Available online at http://www.healthypeople.gov/2020/.

2. Bullamore JR, Wilkinson R, Gallagher JC, Nordin BE, Marshall DH. Effect of age on calcium absorption. Lancet. 1970;2:535–7. Russell RM. Changes in gastrointestinal function attributed to aging. American Journal of Clinical Nutrition. 1992;55:1203S–1207S.

3. Syed F, Khosla S. Mechanisms of sex steroid effects on bone. Biochemical and Biophysical Research Communications. 2005;328:688–96. Tuck SP, Datta HK. Osteoporosis in the aging male: Treatment options. Clinical Interventions in Aging. 2007;2:521–36.

4. Institute of Medicine. Dietary reference intakes for calcium and vitamin D. Washington, DC: National Academies Press; 2011.

5. Ebeling PR. Clinical practice. Osteoporosis in men. New England Journal of Medicine. 2008;358:1474–82.

6. World Health Organization. Assessment of fracture risk and its application to screening for postmenopausal osteoporosis. Report of a WHO study group. WHO Press. Geneva, Switzerland, 2007 (WHO Technical Report Series, No. 843). World Health Organization. WHO scientific group on the assessment of osteoporosis at primary health care level. WHO Press, Geneva, Switzerland, 2007.

7. Sweet MG, Sweet JM, Feremiah MP, Galazka SS. Diagnosis and treatment of osteoporosis. American Family Physician. 2009;79:193–200.

8. US Department of Health and Human Services. Bone health and osteoporosis: A report of the Surgeon General. Rockville, MD: US Department of Health and Human Services, Office of the Surgeon General, 2004.

9. Dontas IA, Yiannakopoulos CK. Risk factors and prevention of osteoporosis-related fractures. Journal of Musculoskeletal and Neuronal Interactions. 2007;7:268–72. Guthrie JR, Dennerstein L, Wark JD. Risk factors for osteoporosis: A review. Medscape Women's Health. 2000;5:E1. National Osteoporosis Foundation. Physician's guide to prevention and treatment of osteoporosis. 2010. Available online at http://www.nof.org/professionals/Clinicians_Guide.htm. U.S. Preventive Services Task Force. Screening for osteoporosis in postmenopausal women: recommendations and rationale. Rockville, Md.: Agency for Healthcare Research and Quality; September 2002. Available online at http://www.ahrq.gov/clinic/3rduspstf/osteoporosis/osteorr.htm.

10. Weaver CM. Calcium. In: Present knowledge in nutrition, 9th ed. Bowman BA, Russell RM, editors. Washington, DC: ILSI Press; 2006.

11. Wong PK, Christie JJ, Wark JD. The effects of smoking on bone health. Clinical Sciences (London). 2007;113:233–41.

12. Pack AM, Gidal B, Vazquez B. Bone disease associated with antiepileptic drugs. Cleveland Clinic Journal of Medicine. 2004;71Suppl2:S42–8. Roberts CG, Ladenson PW. Hypothyroidism. Lancet. 2004;363:793–803. Wexler JA, Sharretts J. Thyroid and bone. Endocrinology and Metabolism Clinics North America. 2007;36:673–705.

13. Cranney A, Adachi JD. Corticosteroid-induced osteoporosis: A guide to optimum management. Treatments in Endocrinology. 2002;1:271–9. Hadji P. Reducing the risk of bone loss associated with breast cancer treatment. Breast. 2007;16:S10–5. Popp AW, Isenegger J, Buergi EM,

Buergi U, Lippuner K. Glucocorticosteroid-induced spinal osteoporosis: Scientific update on pathophysiology and treatment. European Spine Journal. 2006;15:1035–49.

14. Stevenson JC. Hormone replacement therapy: Review, update, and remaining questions after the Women's Health Initiative Study. Current Osteoporosis Reports. 2004;2:12–6. Toles CA. The incidence of cardiovascular disease in menopausal women on hormone replacement therapy: A clinical evidence-based medicine review. Journal of the National Black Nurses Association. 2007;18:75–80.

15. National Institute on Aging. Hormones and menopause. NIH Publication No. 09-7411. 2009. Available online at http://www.nia.nih.gov/HealthInformation/Publications/hormones.htm. U.S. Food and Drug Administration. Menopause and hormones, 2009. Available at http://www.fda.gov/ForConsumers/ByAudience/ForWomen/ucm118624.htm.

16. Abrams SA, Silber TJ, Esteban NV, Vieira NE, Stuff JE, Meyers R, Majd M, Yergey AL. Mineral balance and bone turnover in adolescents with anorexia nervosa. Journal of Pediatrics. 1993;123:326–31.

17. Jayasinghe Y, Grover SR, Zacharin M. Current concepts in bone and reproductive health in adolescents with anorexia nervosa. British Journal of Obstetrics and Gynecology. 2008;115:304–15.

18. Chakkalakal DA. Alcohol-induced bone loss and deficient bone repair. Alcoholism: Clinical and Experimental Research. 2005;29:2077–90. Laitinen K, Valimaki M. Alcohol and bone. Calcified Tissue International. 1991;49:S70–3.

19. Tremblay MS, Colley RC, Saunders TJ, Healy GN, Owen N. Physiological and health implications of a sedentary lifestyle. Applied Physiology and Nutrient Metabolism. 2010;35:725-40. Iwamoto J, Takeda T, Sato Y. Effect of treadmill exercise on bone mass in female rats. Experimental Animals. 2005;65:1–6. Jee WS, Tian XY. The benefit of combining non-mechanical agents with mechanical loading: A perspective based on the Utah paradigm of skeletal physiology. Journal of Musculoskeletal and Neuronal Interactions. 2005;5:110–18.

20. Everitt AV, Hilmer SN, Brand-Miller JC, Jamieson HA, Truswell AS, Sharma AP, Mason RS, Morris BJ, Le Couteur DG. Dietary approaches that delay age-related diseases. Clinical Interventions in Aging. 2006;1:11–31. Prentice A. Diet, nutrition and the prevention of osteoporosis. Public Health Nutrition. 2004;7:227–43.

21. Bügel S. Vitamin K and bone health in adult humans. Vitamins and Hormones. 2008;78:393–416. Holick MF, Chen TC. Vitamin D deficiency: A worldwide problem with health consequences. American Journal of Clinical Nutrition. 2008;87:1080S–6S. Francucci CM, Rilli S, Fiscaletti P, Boscaro M. Role of vitamin K on biochemical markers, bone mineral density, and fracture risk. Journal of Endocrinological Investigation. 2007;30:24–8. Isaia G, D'Amelio P, Di Bella S, Tamone C. Protein intake: The impact on calcium and bone homeostasis. Journal of Endocrinological Investigation. 2007;30 (6 Suppl):48–53. Montero-Odasso M, Duque G. Vitamin D in the aging musculoskeletal system: An authentic strength preserving hormone. Molecular Aspects of Medicine. 2005;26:203–19. Palacios C. The role of

nutrients in bone health, from A to Z. Critical Reviews in Food Science and Nutrition. 2006;46:621–8.

22. Duque G. Anabolic agents to treat osteoporosis in older people: Is there still place for fluoride? Fluoride for treating postmenopausal osteoporosis. Journal of the American Geriatric Society. 2001;49:1387–9.

23. National Osteoporosis Foundation. Physician's guide to prevention and treatment of osteoporosis. 2010. Available online at http://www.nof.org/professionals/Clinicians_Guide.htm.

24. Stevenson JC. Hormone replacement therapy: Review, update, and remaining questions after the Women's Health Initiative Study. Current Osteoporosis Reports. 2004;2:12–6.

25. Weaver CM, Cheong JM. Soy isoflavones and bone health: The relationship is still unclear. Journal of Nutrition. 2005;135:1243–7. Usui T. Pharmaceutical prospects of phytoestrogens. Endocrine Journal. 2006;53:7–20. Vatanparast H, Chilibeck PD. Does the effect of soy phytoestrogens on bone in postmenopausal women depend on the equol-producing phenotype? Nutrition Reviews. 2007;65:294–9.

26. Gennari L, Merlotti D, Valleggi F, Martini G, Nuti R. Selective estrogen receptor modulators for postmenopausal osteoporosis: Current state of development. Drugs and Aging. 2007;24:361–79. Le Goff B, Guillot P, Glémarec J, Berthelot JM, Maugars Y. A comparison between bisphosphonates and other treatments for osteoporosis. Current Pharmaceutical Design. 2010;16:3037–44. Migliaccio S, Brama M, Spera G. The differential effects of bisphosphonates, SERMS (selective estrogen receptor modulators), and parathyroid hormone on bone remodeling in osteoporosis. Clinical Interventions in Aging. 2007;2:55–64. Pinkerton JV, Dalkin AC. Combination therapy for treatment of osteoporosis: A review. American Journal of Obstetrics and Gynecology. 2007;197(6):559–65. Reginster JY. Antifracture efficacy of currently available therapies for postmenopausal osteoporosis. Drugs. 2011;71:65–78.

27. Valverde P. Pharmacotherapies to manage bone loss–associated diseases: A quest for the perfect benefit-to-risk ratio. Current Medicinal Chemistry. 2008;15:284–304. Wells GA, Cranney A, Peterson J, Boucher M, Shea B, Robinson V, Coyle D, Tugwell P. Etidronate for the primary and secondary prevention of osteoporotic fractures in postmenopausal women. Cochrane Database Systemic Reviews. 2008;1:CD003376.

28. Barger-Lux MJ, Heaney RP. Caffeine and the calcium economy revisited. Osteoporosis International. 1995;5:97–102. Barger-Lux MJ, Heaney RP, Stegman MR. Effects of moderate caffeine intake on the calcium economy of premenopausal women. American Journal of Clinical Nutrition. 1990;52:722–5. Higdon JV, Frei B. Coffee and health: A review of recent human research. Critical Reviews in Food Science and Nutrition. 2006;46:101–23.

29. Institute of Medicine. Dietary reference intakes for calcium and vitamin D. Washington, DC: National Academies Press; 2011.

30. Jackson RD, LaCroix AZ, et al. Calcium plus vitamin D supplementation and the risk of fractures. New England Journal of Medicine. 2006;354:669–83.

31. Björkman M, Sorva A, Risteli J, Tilvis R. Vitamin D supplementation has minor effects on parathyroid

hormone and bone turnover markers in vitamin D–deficient bedridden older patients. Age and Ageing. 2008;37:25–31. Lambert HL, Eastell R, Karnik K, Russell JM, Barker ME. Calcium supplementation and bone mineral accretion in adolescent girls: An 18-mo randomized controlled trial with 2-y follow-up. American Journal of Clinical Nutrition. 2008;87:455–62. Ward KA, Roberts SA, Adams JE, Lanham-New S, Mughal MZ. Calcium supplementation and weight bearing physical activity—do they have a combined effect on the bone density of prepubertal children? Bone. 2007;41:496–504. Zhu K, Devine A, Dick IM, Wilson SG, Prince RL. Effects of calcium and vitamin D supplementation on hip bone mineral density and calcium-related analytes in elderly ambulatory Australian women: A five-year randomized controlled trial. Journal of Clinical Endocrinology and Metabolism. 2008;93:743–9.

32. Dawson-Hughes B. Racial/ethnic considerations in making recommendations for vitamin D for adults and elderly men and women. American Journal of Clinical Nutrition. 2004;80:1763–6. Holick MF. Sunlight and vitamin D for bone health and prevention of autoimmune diseases, cancers, and cardiovascular disease. American Journal of Clinical Nutrition. 2004;80:1678S–8S. Institute of Medicine. Dietary reference intakes for calcium and vitamin D. Washington, DC: National Academies Press; 2011.

The Trace Minerals

In the previous chapter, you learned about several minerals, such as calcium and magnesium, that the body needs in relatively large amounts—that is, at least 100 mg each day. In this chapter, you will learn about other essential minerals that are required in very small quantities. These minerals, such as selenium, iron, and iodine, are called the trace minerals or "microminerals." Although you need only minute amounts of them, getting enough of each of the trace minerals is vital to your health. For example, selenium plays an important role in protecting cells from oxidative damage caused by free radicals; iron is essential for energy metabolism; and iodine is needed for the production of thyroid hormones. In this chapter, you will learn about the fundamentals related to each trace mineral, including dietary sources, regulation, functions, deficiency, toxicity, and dietary requirements.

Shelley McGuire

Living Life as an "Iron Man"

A few years ago, David (now a 24-year-old graduate student) and his wife were thinking about starting a family and decided to apply for life insurance. To receive a discount for having good health, David needed to have a routine physical, which included having blood work done. But there was nothing routine about the results that came back: David's liver enzymes were profoundly elevated. His doctor first asked him how much alcohol he drank, because excessive alcohol consumption is the most common reason for such high levels. But David drank alcohol only infrequently. Diagnosing his condition was difficult because David had no other signs or symptoms of disease. Two months later, however, David's doctors determined, using genetic testing, that he had a condition called hereditary hemochromatosis. People with hereditary hemochromatosis absorb excessive amounts of iron from the foods they eat. Over time the excess iron can accumulate in organs and tissues such as the heart, liver, pancreas, joints, and pituitary gland. When this happens, these tissues can become diseased.

Immediately after his diagnosis, David's life changed. He stopped taking iron supplements and cut back on the amount of iron-rich foods he consumed, like beef. He also began having therapeutic blood removal (phlebotomy) weekly to rid his body of excess iron. Not only was this inconvenient and tiring—interfering with his school and work schedules—but it was also expensive ($100 each week). To make matters worse, his student health insurance did not cover the costs, and David and his wife had to borrow money to pay for the procedures. Nonetheless, David continued with his weekly phlebotomies for six months. After this intensive treatment, David was exhausted, but his blood iron levels were back to normal. Now, he receives therapeutic phlebotomies at the local blood bank every eight weeks. His liver enzymes are normal, and David and his wife have qualified for life insurance. In fact, they are excitedly awaiting the birth of their first child, and are pondering when and if they should have him or her genetically tested for hemochromatosis.

As is often the case when people are diagnosed with a disease, David has taken a renewed interest in his health. For example, he and his wife now exercise three times each week, and David has lost 20 pounds. David has also learned to enjoy running and has recently completed

his first half marathon. In fact, his friends jokingly call him "Iron Man." David is relieved that he learned of his condition relatively early in his life and is able to manage it with phlebotomies and a balanced diet. Indeed, he has taken this life change in stride. We asked him what has been the most frustrating part of having hemochromatosis, and he told us that it was being a frequent "guinea pig" for phlebotomists-in-training. David is clearly an excellent example of how one can live well with a nutrition-related condition. It takes determination, but taking control of hemochromatosis has made all the difference in the world to David and his wife.

© Shelley McGuire

Critical Thinking: David's Story

How would having hemochromatosis or another condition that requires you to be careful about how much iron you eat change your daily life? If you found out that you had hemochromatosis or a similar disease, how would you cope with the costs and time commitment needed for treatment?

What Do the Trace Minerals Have in Common?

The body requires at least eight **trace minerals:** iron, copper, iodine, selenium, chromium, manganese, molybdenum, and zinc. Other trace minerals, such as fluoride, are not technically essential nutrients but may influence health nonetheless. Although each of the trace minerals has its own unique set of functions in the body, there are some characteristics that they share. For example, they are all absorbed primarily in the small intestine and circulated in the blood. An overview of general characteristics of the trace minerals is presented in Table 13.1 and described next.

REGULATION OF TRACE MINERALS IN THE BODY

Trace minerals are obtained from a wide variety of plant- and animal-based foods, and their bioavailability can be influenced by many factors, such as genetics, life stage, nutritional status, and interactions with other food components. Trace minerals are absorbed primarily in the small intestine and subsequently circulated to the liver in the blood.

The amounts of most trace minerals in the body are regulated, although the site of regulation varies. For example, iron is regulated through alterations in absorption in the small intestine, whereas other minerals (such as iodine and selenium) are regulated primarily by the kidneys. The liver assists in the regulation of some minerals (such as copper and manganese) by incorporating excess amounts of them into bile, which is subsequently eliminated in the feces. Although there are clear exceptions (such as iron and iodine), the occurrences of trace mineral deficiencies and toxicities are rare except in genetic disorders, excessive supplement consumption, and environmental overexposure.

⟨CONNECTIONS⟩ Recall that a free radical is a reactive molecule with one or more unpaired electrons. Free radicals are destructive to cell membranes, DNA, and proteins (Chapter 10, page 447).

TRACE MINERALS ACT AS COFACTORS AND PROSTHETIC GROUPS AND PROVIDE STRUCTURE

Some trace minerals are important components of enzymes. When a mineral is part of an enzyme, it is called a **cofactor,** and the enzyme is called a **metalloenzyme.** The function of a mineral cofactor is similar to that of a

trace mineral An essential mineral that is required in amounts less than 100 mg daily.

cofactor A mineral that combines with an enzyme, thereby activating it.

metalloenzyme An enzyme that contains a mineral cofactor.

TABLE 13.1 General Characteristics of the Trace Minerals

	General Characteristics
Food sources	• Amount often depends on mineral content of soil • Found in all food groups • Whole-grain products tend to contain more than refined cereal products • Amount not influenced by cooking
Digestion	• Very little needed
Absorption	• Occurs mostly in small intestine but also in stomach • Bioavailability often influenced by form, nutritional status, and interactions with other dietary components
Circulation	• Via blood from the GI tract to the liver and throughout the body
Functions	• Cofactors for enzymes, some of which are involved in redox reactions • Components of nonenzymatic proteins • Structural roles (such as mineralization)
Toxicity	• Rare; generally associated with genetic disorders, excessive supplement intake, or environmental exposure

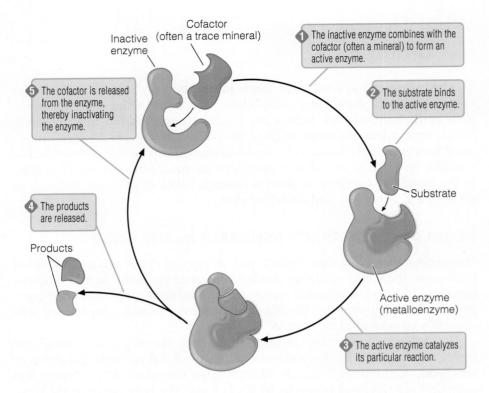

Cofactor (often a trace mineral)

Inactive enzyme

1 The inactive enzyme combines with the cofactor (often a mineral) to form an active enzyme.

2 The substrate binds to the active enzyme.

5 The cofactor is released from the enzyme, thereby inactivating the enzyme.

Substrate

4 The products are released.

Products

Active enzyme (metalloenzyme)

3 The active enzyme catalyzes its particular reaction.

vitamin coenzyme. Some trace minerals act as enzyme activators; this process allows the enzyme to bind to the substrate and carry out its function (Figure 13.1). Other times, trace minerals are components of larger nonenzymatic molecules, such as hemoglobin. You may recall from Chapter 5 that nonprotein components of proteins are called "prosthetic groups." Some trace minerals, such as fluoride and calcium, provide structure and strength to mineralized tissues (bones and teeth). In summary, the trace minerals generally act as cofactors and prosthetic groups and provide scaffolding for many of the tissues in the body.

Iron—Transporter of Oxygen

It may surprise you to learn that, although **iron (Fe)** is likely the most studied trace mineral, iron deficiency remains the most common micronutrient deficiency worldwide.[1] Indeed, this is a stark example of how science and public policy do not always move forward simultaneously. Scientists once thought that iron's only function was to transport oxygen and carbon dioxide in the blood, but it is now clear that iron is very important in other ways as well.

IRON IS IN BOTH PLANT- AND ANIMAL-DERIVED FOODS

iron (Fe) A trace mineral needed for oxygen and carbon dioxide transport, energy metabolism, stabilization of free radicals, and synthesis of DNA.

heme iron Iron that is a component of a heme group; heme iron includes hemoglobin in blood, myoglobin in muscles, and cytochromes in mitochondria.

nonheme iron Iron that is not a component of a heme group.

Iron is found in a variety of foods, and some examples are listed in Figure 13.2. There are two forms of iron in food: heme iron and nonheme iron. Iron that is bound to a heme group is called **heme iron.** Recall that heme is an iron-containing prosthetic group that is a component of several complex proteins such as hemoglobin. Heme is also a component of myoglobin and cytochromes, found in muscles and mitochondria, respectively. Excellent sources of heme iron include shellfish, beef, poultry, and organ meats such as liver. Iron that is not a component of a heme molecule is called **nonheme iron.**

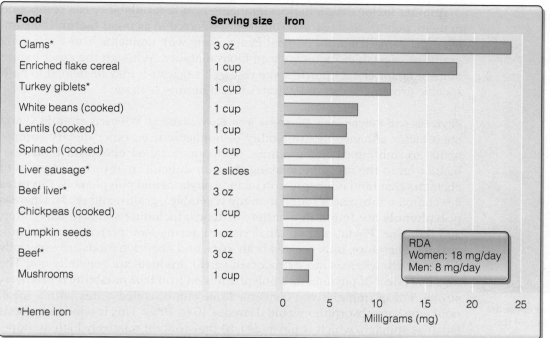

Food	Serving size	Iron
Clams*	3 oz	
Enriched flake cereal	1 cup	
Turkey giblets*	1 cup	
White beans (cooked)	1 cup	
Lentils (cooked)	1 cup	
Spinach (cooked)	1 cup	
Liver sausage*	2 slices	
Beef liver*	3 oz	
Chickpeas (cooked)	1 cup	
Pumpkin seeds	1 oz	
Beef*	3 oz	
Mushrooms	1 cup	

*Heme iron

RDA
Women: 18 mg/day
Men: 8 mg/day

Milligrams (mg)

FIGURE 13.2 Food Sources of Iron

SOURCE: USDA Nutrient Database for Standard Reference, Release 17.

Nonheme iron is found in green leafy vegetables, mushrooms, and legumes and typically accounts for more than 85% of the iron you consume. Remember that certain foods such as rice, cornmeal, and breads fortified with iron can be labeled as being "enriched." Enriched foods, as well as other iron-fortified foods, are also good sources of nonheme iron. Whether iron is in its heme or nonheme form greatly affects its bioavailability.

MANY FACTORS INFLUENCE IRON BIOAVAILABILITY

The bioavailability of iron is complex and influenced by many factors, including its form (heme vs. nonheme) in food, iron status of the person, and the presence or absence of other dietary components in the meal. Perhaps the most important factor is its form, with heme iron being two to three times more bioavailable than nonheme iron. Absorption of heme iron is generally high, being most affected by iron status—in other words, whether the body needs more or less iron—whereas many factors can greatly influence absorption of nonheme iron. One of the most important factors is its ionic state. Nonheme iron is found in two ionic forms in foods: the more oxidized **ferric iron** (Fe^{3+}) and the more reduced **ferrous iron** (Fe^{2+}), the second being more bioavailable.

Vitamin C and "Meat Factor" Increase Nonheme Iron Absorption One of the best described enhancers of nonheme iron absorption is vitamin C (ascorbic acid), which converts ferric iron (Fe^{3+}) to ferrous iron (Fe^{2+}) in the intestinal lumen. For this to occur, vitamin C donates a negatively charged electron to ferric iron (Fe^{3+}), reducing it to ferrous iron (Fe^{2+}). Consequently, consuming vitamin C and nonheme iron together in a meal enhances the absorption of the nonheme iron. Stomach acid also helps reduce ferric iron (Fe^{3+}) to ferrous iron (Fe^{2+}), and some studies suggest that the chronic use of antacids to neutralize stomach acidity can decrease nonheme iron absorption.[2]

Shellfish and beef are excellent sources of heme iron.

⟨**CONNECTIONS**⟩ Cytochromes are involved in the electron transport chain and therefore in ATP production (Chapter 7, page 278).

⟨**CONNECTIONS**⟩ Remember that the nutrients that must be added to "enriched foods" are iron, thiamin (vitamin B_1), riboflavin (vitamin B_2), niacin (vitamin B_3), and folate (Chapter 10, page 422).

ferric iron (Fe^{3+}) The oxidized form of iron.
ferrous iron (Fe^{2+}) The reduced form of iron.

Iron—Transporter of Oxygen   **557**

Consuming vitamin C with nonheme iron can increase absorption of the iron.

Another factor that enhances nonheme iron bioavailability is a compound in meat, poultry, and seafood sometimes referred to as **meat factor.** Consuming even a small amount of meat factor along with nonheme iron–containing grains or vegetables increases iron bioavailability. Although the exact nature of meat factor is not known, some research suggests it may be related to both specific proteins and carbohydrates found in muscle tissue.[3]

Phytates and Polyphenols Decrease Iron Bioavailability Whereas vitamin C and meat factor enhance bioavailability of nonheme iron, other dietary components are inhibitory. For example, compounds called **chelators** bind nonheme iron in the intestine, making the iron difficult to absorb. Examples of chelators that bind nonheme iron include phytates and polyphenols. **Phytates** are complex substances found in many vegetables, grains, and seeds, whereas **polyphenols** are found in a variety of foods, including spinach, tea, coffee, and red wine. Phytates are often destroyed during food preparation and processing. Therefore, unprocessed bran, oats, and fiber-rich foods are especially phytate-rich, whereas more processed cereal products are generally not. The negative effect of phytates and polyphenols on iron bioavailability is relatively strong. For example, if you consumed one cup of coffee or tea with a meal, nonheme iron absorption would decrease 40 to 70%.[4] This is why some foods (such as spinach, which is phytate-rich) that contain relatively high amounts of iron are not actually good sources of iron for the body. However, if you were to consume vitamin C or meat factor along with these foods, you could partly counteract the inhibitory effects of the iron chelators.

IRON ABSORPTION IS TIGHTLY REGULATED

Both iron deficiency and toxicity can cause serious problems. As such, we have complex homeostatic mechanisms that regulate how much iron is absorbed and stored in the body. These mechanisms ensure that neither too little nor too much iron is absorbed. Except for bleeding associated with menstruation, injury, or childbirth, the body loses very little iron once the iron is absorbed. Consequently, the amount of iron in the body is regulated via increases and decreases in the amount absorbed (Figure 13.3).

Iron Absorption As with other nutrients, iron absorption involves transport across two membranes in the intestinal enterocyte: the brush border membrane (facing the intestinal lumen) and the basolateral membrane (facing the submucosa). Heme iron can be transported across these membranes without modification, while nonheme iron must first be reduced to its ferrous form (Fe^{2+}) prior to transport.[5] This is why reducing agents such as vitamin C increase the bioavailability of nonheme iron.

As illustrated in Figure 13.3A, several membrane-bound iron transport proteins move iron from the intestinal lumen into the enterocyte. The synthesis of these transport proteins is regulated by the liver-derived hormone **hepcidin.** Specifically, high levels of hepcidin lead to decreased production of iron-transporting proteins, whereas low levels increase their production.[6] Once inside the enterocyte, ferrous iron (Fe^{2+}) is bound to the protein **ferritin.**[7] Ferritin temporarily stores iron in the enterocyte until the cell receives a signal that the body needs more iron. When this happens, both ferrous iron (Fe^{2+}) and heme iron are transported across the basolateral membrane into the submucosa where they enter the blood. Disruptions in the iron transport process across the enterocytes can result in either iron deficiency or iron "overload." An example of a condition that causes iron overload is hereditary hemochromatosis, which was introduced in the Everybody Has a Story feature at the beginning of the chapter. This disease is described in more detail next.

meat factor An unidentified compound, found in meat, fish, and poultry, that increases the absorption of nonheme iron.

chelator (CHE – la – tor) A substance that binds compounds (such as iron) in the gastrointestinal tract, making them unavailable for absorption.

phytates (PHY – tates) Phosphorus-containing compounds often found in the outer coating of kernels of grain as well as in vegetables and legumes; can bind dietary minerals.

polyphenols (pol – y – PHEN – ols) Organic compounds found in some plant-based foods; can bind dietary minerals.

hepcidin A hormone, produced by the liver in response to high levels of iron in the body, that helps regulate iron status by decreasing iron uptake by the intestine.

ferritin (FER – ri – tin) A protein important for iron absorption and storage in the body.

FIGURE 13.3 Effect of Iron Status on Iron Absorption The body increases iron absorption when it needs iron and decreases iron absorption when it has excess amounts.

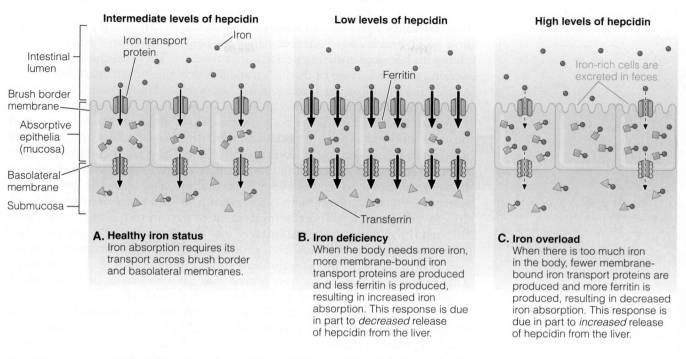

Intermediate levels of hepcidin

Iron transport protein

Iron

Intestinal lumen

Brush border membrane

Absorptive epithelia (mucosa)

Basolateral membrane

Submucosa

A. Healthy iron status
Iron absorption requires its transport across brush border and basolateral membranes.

Low levels of hepcidin

Ferritin

Transferrin

B. Iron deficiency
When the body needs more iron, more membrane-bound iron transport proteins are produced and less ferritin is produced, resulting in increased iron absorption. This response is due in part to *decreased* release of hepcidin from the liver.

High levels of hepcidin

Iron-rich cells are excreted in feces.

C. Iron overload
When there is too much iron in the body, fewer membrane-bound iron transport proteins are produced and more ferritin is produced, resulting in decreased iron absorption. This response is due in part to *increased* release of hepcidin from the liver.

Hereditary Hemochromatosis In addition to being caused by consumption of excess iron from medicine or supplements, iron toxicity can be caused by a genetic condition called **hereditary hemochromatosis.** Hemochromatosis has long been recognized by the medical community as a serious health concern, although its true causes have only recently been discovered.[8] Scientists now know that people with this disease have a defect in one of a number of genes that code for either an intestinal iron transport protein or hepcidin, the major hormone regulating iron homeostasis. As a result, large amounts of iron are absorbed into the blood. Alterations in the various genes related to hereditary hemochromatosis are commonly found in people of northern European ancestry. Researchers think that between 4 and 5 out of every 1,000 Americans may have this gene, although not all of them actually develop the disease.[9] As iron accumulates in the body, it can eventually cause joint pain, bronze skin color, arthritis, chronic fatigue, cessation of menstruation in women, impaired liver function, and organ damage. In severe cases, hereditary hemochromatosis can lead to liver failure, liver cancer, and increased risks for diabetes and heart disease.

Until recently, the only way to test for hereditary hemochromatosis was by measuring levels of iron in the liver; this process required extracting a small sample of liver tissue. However, it can now be diagnosed by genetic testing of the DNA in blood or cheek cells. Although hereditary hemochromatosis cannot currently be prevented or cured, it can be treated by frequent removal of blood. It is also highly recommended that people with this condition not

hereditary hemochromatosis (he – mo – chrom – a – TO – sis) A genetic abnormality resulting in increased absorption of iron in the intestine.

© Shelley McGuire

Critical Thinking: David's Story You now know more about what causes hemochromatosis. Recall David, the graduate student you read about in this chapter's Everybody Has a Story. Can you now explain in some detail how hemochromatosis alters normal iron regulation in his body? Why would he and his wife be contemplating having their baby tested for this disease? If you had this condition, would you have your children tested?

Iron—Transporter of Oxygen **559**

consume iron or vitamin C supplements (especially with food). In addition, they are advised to consume only small portions of iron-rich foods.

Iron Deficiency Increases Iron Absorption Iron status can also greatly influence how much iron is absorbed; specifically, absorption increases during iron deficiency and decreases during periods of iron excess. These mechanisms involve altering the production of the iron transport proteins in the brush border cell membrane as well as the amount of ferritin produced in the enterocyte. Recall that this is illustrated in Figure 13.3.

During iron deficiency, the body increases production of the iron transport proteins while decreasing production of ferritin (Figure 13.3B). As previously described, this is accomplished primarily via reduced production of hepcidin in the liver. Recall that hepcidin is a hormone that decreases synthesis of iron transport proteins in the intestine. Consequently, iron transport into the enterocyte is increased, and there is less ferritin to bind the iron within the cell. Together, these actions increase movement of iron from the intestinal lumen through the enterocyte and, ultimately, into the circulation. Conversely, if the body has adequate or excess iron, production of hepcidin increases, production of iron transport proteins decreases, and production of ferritin increases. This causes less iron to be transported into the enterocyte, and that which is transported is more likely to be retained in the cell (Figure 13.3C). These iron-rich enterocytes are eventually sloughed off into the intestinal lumen and eliminated in the feces. The presence of iron in the feces can cause stools to be black in color—a classic sign of iron overload. In this way, enterocytes lining the small intestine serve as important regulators of iron status, helping prevent both iron deficiency and toxicity.

Ferritin and Hemosiderin Store Excess Iron Once iron is released from the enterocyte, it enters the blood and binds to **transferrin,** a protein produced in the liver (Figure 13.3). Transferrin delivers the iron to cells that have **transferrin receptors,** special proteins on their cell membranes. The number of transferrin receptors on a cell's membrane can be regulated by the cell's need for iron. For example, if the cell has enough iron, the number of transferrin receptors can be decreased, inhibiting uptake of additional iron. Conversely, cells needing additional iron can produce more receptors, so they can bind or "capture" more iron for uptake.

Excess iron can be stored in the liver, bone marrow, and spleen. There are two iron storage compounds in these tissues—ferritin and hemosiderin. These large proteins contain about one-third of a healthy person's total iron. As previously mentioned, ferritin is also produced in enterocytes, where it helps regulate iron absorption. In addition, small amounts of the storage form of ferritin are present in liver, bone, and spleen. This ferritin is continually released into the blood, where its concentration is generally proportional to the amount of iron stored in the body. Because of this, blood ferritin is often measured to assess iron status. Whereas ferritin is considered the main storage form of iron, **hemosiderin** is needed for more long-term storage. Hemosiderin, thought to be a degradation product of ferritin, protects the body from iron toxicity when intake is chronically high. This occurs because hemosiderin binds excess iron, preventing it from entering the blood. Hemosiderin is found in many cell types, but is especially high in the liver.

IRON IS A COMPONENT OF HEME AND NONHEME PROTEINS

The body's iron-containing compounds include the iron-storage proteins (ferritin and hemosiderin), along with other compounds that require iron to function. The latter compounds, together making up what is commonly

transferrin (trans – FER – rin) A protein, produced in the liver, important for iron transport in the blood.

transferrin receptor Protein found on cell membranes that binds transferrin, allowing the cell to take up iron.

hemosiderin (he – mo – SID – er – in) An iron-storing protein found in many tissues, especially the liver.

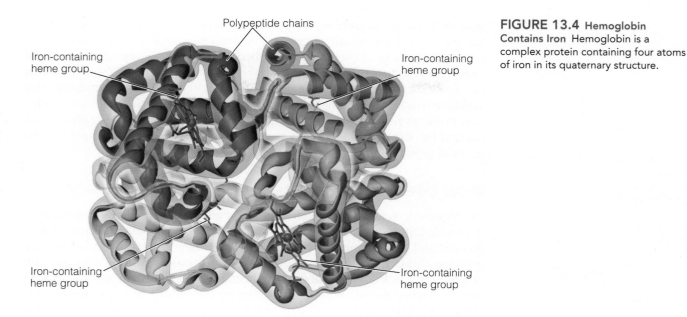

Polypeptide chains

Iron-containing heme group

Iron-containing heme group

Iron-containing heme group

Iron-containing heme group

FIGURE 13.4 Hemoglobin Contains Iron Hemoglobin is a complex protein containing four atoms of iron in its quaternary structure.

called "functional iron," include a variety of proteins that you will learn about next. Heme-containing proteins include hemoglobin, myoglobin, and the cytochromes. Nonheme proteins include those enzymes for which iron is a cofactor.

Oxygen and Carbon Dioxide Transport: Hemoglobin The most abundant protein in red blood cells (erythrocytes) is **hemoglobin,** made up of four protein subunits (called *globins*) and four iron-containing heme groups (Figure 13.4). Red blood cells typically contain about two-thirds of the body's total iron. As they circulate through the blood vessels of the lungs, red blood cells come in contact with oxygen, which then attaches to iron atoms in hemoglobin. Oxygenated blood then circulates from the lungs to the heart, which in turn pumps it to other tissues in the body. Ultimately, hemoglobin delivers the oxygen to cells that need it for aerobic energy metabolism. Once the oxygen is released, hemoglobin can bind carbon dioxide, a waste product of this metabolic pathway. Deoxygenated, carbon dioxide–rich blood is circulated back to the lungs, where the carbon dioxide is eliminated from the body upon exhalation. Without sufficient hemoglobin, oxygen availability to tissues decreases, in turn decreasing the rate of aerobic energy metabolism. This can result in severe fatigue. Because the body has such a high demand for iron, when red blood cells are degraded, most of their iron is recycled or stored for later use.

Oxygen Reservoir: Myoglobin **Myoglobin** is another iron-containing heme protein, but unlike hemoglobin found in red blood cells, myoglobin is found in muscle cells. It consists of a single heme group and a single protein subunit (another form of *globin*). Myoglobin acts as a reservoir of oxygen, releasing it to muscle cells when they need to produce ATP to fuel physical activity.

Cellular Energy Metabolism In addition to delivering oxygen to cells, iron plays a key role in other aspects of energy metabolism. For example, it is a basic component of the cytochromes, heme-containing protein complexes that function in the electron transport chain. Cytochromes serve as electron carriers. As such, iron is indirectly involved in the conversion of adenosine diphosphate (ADP) to adenosine triphosphate (ATP) and the production of water. Iron also serves as a cofactor for a variety of nonheme-containing enzymes involved in the electron transport chain, citric acid cycle, and gluconeogenesis.

〈CONNECTIONS〉 Recall that a prosthetic group such as heme is a nonprotein component important for the quaternary structure and function of a protein (Chapter 5, page 172).

〈CONNECTIONS〉 Recall that oxygen is the final electron acceptor in the electron transport chain and is thus essential for ATP production (Chapter 7, page 278).

hemoglobin (HE – mo – glob – in) A complex protein, composed of four iron-containing heme groups and four protein subunits, found in the blood and needed for oxygen and carbon dioxide transport in the body.

myoglobin (MY – o – glob – in) A heme protein found in muscle.

Iron—A Cofactor for Detoxifying and Antioxidant Enzymes Iron is also a cofactor for several important enzymes that help metabolize drugs and remove toxins. Activity of these enzymes, collectively called the **cytochrome P450** enzymes, is thought to influence risk for many chronic diseases such as cancer and cardiovascular disease.[10] As a cofactor for antioxidant enzymes that stabilize free radicals, iron is important for protecting DNA, cell membranes, and proteins from oxidative damage. Iron is also a cofactor for an enzyme needed for DNA synthesis. This is one reason why iron is so important for optimal growth and development.

IRON DEFICIENCY CAUSES ANEMIA AND MUCH MORE

Inadequate iron status represents the most common nutritional deficiency in the United States and the world.[11] Because iron requirements increase during growth and development, iron deficiency is more typically seen in infants, growing children and teenagers, and pregnant women. Women are especially at increased risk for iron deficiency because iron is lost in the blood each month during menstruation. Interestingly, the eating of nonnutritive substances such as ice, dirt, and clay, referred to as **pica,** is thought by some to be associated with iron deficiency.[12] Pica is mostly seen in pregnant women, and it is reported that between 8 and 65% of pregnant women engage in this behavior.[13] Whereas some researchers believe pica *results* in iron deficiency, others believe iron deficiency may somehow *cause* pica or be completely unrelated to pica.

Iron deficiency was once thought to cause only anemia. Recall that the term *anemia* refers to a decreased ability of the blood to carry oxygen, or what is called hypoxia. Scientists now know, however, that iron deficiency can influence many aspects of health, some of which are described next.

Even mild iron deficiency can cause fatigue.

Mild Iron Deficiency Leads to Fatigue and Other Problems Impaired iron status is associated with fatigue and impaired physical work performance. In addition, it can affect behavior and intellectual abilities in children.[14] Unfortunately, some of these effects, which often begin prior to the onset of noticeable iron deficiency anemia, are irreversible, even after iron supplementation. Mild iron deficiency also impairs body temperature regulation,[15] especially in cold conditions, and may negatively influence the immune system.[16] Studies also suggest that mild iron deficiency during pregnancy increases the risk of premature delivery, low birth weight, and maternal mortality.[17]

Detecting Mild Iron Deficiency As previously mentioned, serum ferritin levels are correlated with the amount of iron stored in the body. Because serum ferritin levels decrease long before anemia develops, measuring blood ferritin concentrations allows early detection of iron deficiency. A serum ferritin level lower than 12 μg/L is often used to indicate depleted iron stores.[18]

Another measure of mild iron deficiency is **total iron-binding capacity (TIBC).** Iron is transported in the blood attached to the protein transferrin, and each transferrin molecule can hold two iron atoms. Total iron-binding capacity is a laboratory measurement related to the total number of free (unbound) iron-binding sites on transferrin. When iron stores are depleted, TIBC increases because the transferrin molecules are not saturated with iron and have the "capacity" to bind more. A TIBC greater than 400 μg/dL is indicative of iron depletion.[18]

cytochrome P450 An iron-containing enzyme that helps metabolize toxins.

pica (PI – ca) An abnormal eating behavior that involves consuming nonfood substances such as dirt or clay.

total iron-binding capacity (TIBC) A measure of iron status that relates to the total number of free (unbound) iron-binding sites on transferrin.

© Jupiter Images/Photos.com/Alamy

A related biochemical marker of mild to moderate iron deficiency is **serum transferrin saturation.** As just described, most of the transferrin in the blood of healthy people contains (or is "saturated" with) iron. However, during early stages of iron deficiency, some transferrin does not contain iron. In this way, serum transferrin saturation reflects the percentage of transferrin that contains iron, and low levels reflect mild iron deficiency. Serum transferrin saturation is related mathematically to TIBC and is calculated as follows: serum iron ÷ TIBC × 100. As with TIBC, low levels can be detected before iron deficiency anemia develops, and a value of less than 16% indicates mild iron deficiency.[18]

Severe Iron Deficiency Causes Microcytic Hypochromic Anemia Severe iron deficiency causes microcytic hypochromic anemia, a condition characterized by small, pale red blood cells. Microcytic hypochromic anemia due to iron deficiency occurs when the body is unable to produce enough heme and, thus, hemoglobin. As with other forms of anemia, signs and symptoms include fatigue, difficulties in mental concentration, and compromised immune function.

Detecting Severe Iron Deficiency As iron deficiency progresses, adequate amounts of hemoglobin can no longer be made, and its concentration in red blood cells decreases. However, because other nutrient deficiencies (such as protein or vitamin B_6) can also inhibit hemoglobin synthesis, low hemoglobin concentration cannot by itself be used to diagnose iron deficiency per se. Instead, hemoglobin concentration is usually assessed because it is an inexpensive and easy way to test for *any* type of anemia. A value of less than 13 or 12 g/dL suggests the possibility of iron deficiency for men and women, respectively.[18]

Another indicator of more severe iron deficiency is **hematocrit,** which represents the percentage of total blood volume that is made of red blood cells (Figure 13.5). Hematocrit values decrease during iron deficiency because microcytic anemia is, by definition, associated with the presence of small red blood cells. As with hemoglobin concentration, however, there are many other factors that can influence hematocrit. For example, to help compensate for low oxygen availability, living in high altitudes increases both hemoglobin and hematocrit.[19] In addition, smoking can cause elevated hemoglobin and hematocrit levels.[20] Recent evidence has also shown that infection and inflammation are related to decreased hemocrit and hemoglobin levels, even when adequate iron is consumed.[21] Nonetheless, hematocrit is often used as an initial screen of iron status. A hematocrit of less than 39% for men or 36% for women generally indicates anemia, although this finding should be followed up with other tests (such as TIBC) to determine if iron deficiency is actually the cause.[18]

⟨**CONNECTIONS**⟩ Recall that microcytic hypochromic anemia can also be caused by vitamin B_6 deficiency (Chapter 10, page 435).

serum transferrin saturation A measure of iron status that reflects the percentage of transferrin that contains iron.

hematocrit The percentage of whole blood that comprises red blood cells (erythrocytes).

FIGURE 13.5 What Is a Hematocrit? Hematocrit is a measure of the percentage of blood consisting of red blood cells. Iron deficiency anemia lowers hematocrit.

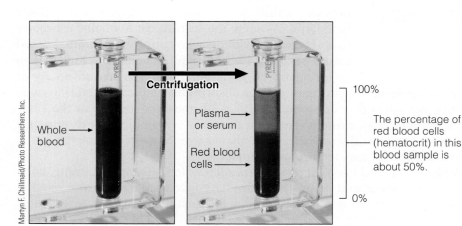

Martyn F. Chillmaid/Photo Researchers, Inc.

Whole blood

Centrifugation

Plasma or serum

Red blood cells

100%

The percentage of red blood cells (hematocrit) in this blood sample is about 50%.

0%

Basics of Iron Supplementation When diet alone cannot maintain iron status, iron supplementation may be needed. Supplemental iron is available in two general forms: ferrous iron (Fe^{2+}) and ferric iron (Fe^{3+}). Of these, the ferrous forms are best absorbed.[22] You can find these types of supplements by looking for the terms "ferrous fumarate," "ferrous sulfate," and "ferrous gluconate" on product labels. The U.S. Centers for Disease Control and Prevention (CDC) recommends taking 50 to 60 mg of iron (the approximate amount in one 300-mg tablet of ferrous sulfate) twice daily for three months for treatment of iron deficiency anemia.[23] Starting with half this dose and gradually increasing to the full dose will help minimize potential side effects such as GI upset and dark stools. Taking half the dose in the morning and the other half in the evening with food may also help alleviate these complications.

IRON TOXICITY CAN BE FATAL

Iron toxicity (also called iron overload) from foods is rare, although it can occur as a result of medicinal or supplemental iron overdoses—especially in children. Symptoms include vomiting, diarrhea, constipation, and black stools. In severe cases, iron toxicity can cause death. This is because excess iron is deposited in soft tissues such as the liver, heart, and muscles, impairing their function. Some researchers have also suggested that excessive iron supplementation can lead to cardiovascular disease and some forms of cancer.[24] However, more research is needed before conclusions can be made. Iron toxicity can also be caused by genetic abnormalities, as is the case with hereditary hemochromatosis.

RECOMMENDED INTAKES FOR IRON

Recommended Dietary Allowances (RDAs) for men and women are 8 and 18 mg/day, respectively, increasing to 27 mg/day during pregnancy. A serving (3 oz) of beef contains only about 3 mg of iron, so it can be difficult for some women to meet the RDA, especially while pregnant. For this reason, iron supplements are often recommended. To prevent GI distress and other complications from iron excess, a Tolerable Upper Intake Level (UL) has been set at 45 mg/day. A complete list of the Dietary Reference Intake (DRI) values for all trace minerals is provided inside the front cover of this book.

Iron requirements may increase by as much as 70% in endurance athletes.

Special Recommendations for Vegetarians and Endurance Athletes Vegetarians may have difficulty consuming enough iron because only meat (which they do not eat) provides substantial amounts of highly bioavailable heme iron. As such, the Institute of Medicine estimates that vegans (individuals who consume no animal-based foods) have dietary iron requirements 80% higher than requirements for nonvegetarians.[18] To take advantage of the positive effect of meat factor on iron absorption, lacto-ovovegetarians who eat fish are advised to consume it with nonheme iron sources in the same meal. For example, eating even small amounts of fish along with enriched pasta can increase the amount of nonheme iron absorbed from the pasta. Some vegetarians may require iron supplements.

Athletes engaged in endurance sports such as long-distance running, competitive swimming, and cycling may also have increased dietary iron requirements. There are many possible reasons for this, including increased blood loss in feces and urine as well as chronic rupture of red blood cells due to the impact of running on hard surfaces. The Institute of Medicine suggests that iron requirements for endurance athletes may be increased by as much as 70%, although specific DRI values have not been set for this group.[18]

Copper—Cofactor in Redox Reactions

Copper (Cu) is present in two forms: its oxidized **cupric** form (Cu^{2+}) and its reduced **cuprous** form (Cu^{1+}). As with iron, the ending "-ous" represents the more reduced form (recall ferr*ous* iron), whereas "-ic" represents the more oxidized form (recall ferr*ic* iron). Copper is a cofactor for enzymes involved in a wide variety of processes, such as ATP production and protection from free radical–induced oxidative damage. Note that copper and iron share many similarities in terms of food sources, absorption, and functions in the body.[25]

ORGAN MEATS ARE EXCELLENT SOURCES OF COPPER

Although organ meats, such as liver, are likely the best sources of copper, it is also found in shellfish, whole-grain products, mushrooms, nuts, and legumes (Figure 13.6). As you will see throughout this chapter, nuts tend to be good sources of the trace minerals, and you can learn more about including nuts in your diet by reading the Food Matters feature. Copper bioavailability can decrease in response to heavy use of antacids, which causes it to form insoluble complexes that are not readily absorbed. Excessive intake of dietary iron can also decrease copper absorption because these two minerals compete for the same transport proteins in enterocyte cell membranes.

EXCESS COPPER IS ELIMINATED IN BILE

Copper is absorbed in the small intestine and, to a lesser extent, the stomach, and absorption is influenced by copper status. As with iron, copper absorption

Nuts and seeds tend to be especially good sources of the trace minerals.

copper (Cu) An essential trace mineral that acts as a cofactor for nine enzymes involved in reduction–oxidation (redox) reactions.

cupric ion (Cu^{2+}) The more oxidized form of copper.

cuprous ion (Cu^{1+}) The more reduced form of copper.

FIGURE 13.6 Food Sources of Copper

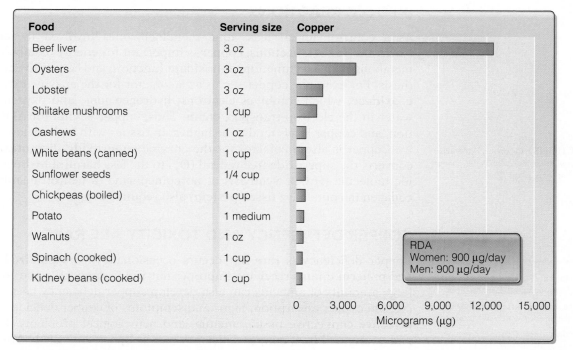

Food	Serving size	Copper
Beef liver	3 oz	
Oysters	3 oz	
Lobster	3 oz	
Shiitake mushrooms	1 cup	
Cashews	1 oz	
White beans (canned)	1 cup	
Sunflower seeds	1/4 cup	
Chickpeas (boiled)	1 cup	
Potato	1 medium	
Walnuts	1 oz	
Spinach (cooked)	1 cup	
Kidney beans (cooked)	1 cup	

RDA
Women: 900 µg/day
Men: 900 µg/day

0   3,000   6,000   9,000   12,000   15,000
Micrograms (µg)

SOURCE: USDA Nutrient Database for Standard Reference, Release 17.

Working Toward the Goal: Increasing Trace
Mineral Intake via Nuts and Seeds

The 2010 Dietary Guidelines for Americans and the
MyPlate food guidance system recommend that we
consume a variety of nuts and seeds in moderation as
part of a nutritious diet. This is partly because they
tend to be excellent sources of trace minerals. The
following selection and preparation tips can help you
meet this goal.

- Use basil–and–pine nut pesto or garlic stem–and–
  walnut pesto on pasta or as a nutritious and tasty
  spread on bread and crackers.

- Consider stirring slivered almonds or sliced Brazil
  nuts into steamed vegetables for a hearty and
  mineral-packed side dish.

- Add toasted peanuts or cashews to a vegetable
  stir-fry.

- Sprinkle chopped nuts on top of low-fat ice cream
  or frozen yogurt.

- Add toasted walnuts or macadamia nuts to a green
  salad.

- Pack a snack bag of sunflower or pumpkin seeds
  in your backpack. They make a great substitute for
  nutrient-poor chips.

- Try toasted hazelnuts or pecans on your oatmeal
  for a crunchy and delicious breakfast.

- Choose a slice of pecan pie instead of a less
  nutrient-dense variety.

To help minimize calories and sodium intake, select
nuts that are dry roasted and avoid those with excessive amounts of added sugar and salt.

increases when stores are low and decreases when copper status is adequate.
The mechanisms by which this occurs are poorly understood. Once absorbed,
copper circulates in the blood to the liver, where it is bound to its transport protein **ceruloplasmin.** Excess copper is not readily stored in the body. Instead,
the liver incorporates it into bile and eliminates it in the feces.

COPPER IS INVOLVED IN REDUCTION–OXIDATION (REDOX) REACTIONS

As a cofactor for at least nine metalloenzymes involved in reduction–
oxidation (redox) reactions, copper is important for energy metabolism, iron
metabolism, neural function, antioxidant function, and connective tissue synthesis. For example, copper serves as a cofactor for the enzyme **cytochrome
c oxidase,** which combines electrons, hydrogen ions, and oxygen to form
water in the electron transport chain. Thus, copper is vital for ATP production, and copper levels tend to be highest in tissues with high metabolic activity. Copper is also a cofactor for the enzyme **superoxide dismutase,** which
converts the superoxide free radical (O_2^-) to the less harmful hydrogen peroxide molecule (H_2O_2). Synthesis of norepinephrine (a neurotransmitter) and
collagen (a connective tissue protein) also requires copper.

COPPER DEFICIENCY AND TOXICITY ARE RARE

Copper deficiency is rare but occurs occasionally in hospitalized patients
and preterm infants receiving improper nutrition support. People consuming
large amounts of antacids can also develop copper deficiency because of decreased copper absorption. Signs and symptoms of copper deficiency include
defective connective tissue, anemia, and neurological problems. In infants
and young children, copper deficiency causes bones to weaken. Mild copper
deficiency has also been associated with impaired blood glucose regulation[26]

⟨CONNECTIONS⟩ Oxidation is the loss
of electrons, whereas reduction is the gain of
electrons (Chapter 3, page 72).

ceruloplasmin (ce – RU – lo – plas – min) The
protein that transports copper in the blood.

cytochrome c oxidase A copper-containing
enzyme needed in the electron transport chain.

superoxide dismutase A copper-containing
enzyme that reduces the superoxide free
radical to form hydrogen peroxide.

and depressed immune function,[27] but further studies are needed to better understand these relationships.

Because copper toxicity is rare in humans, very little is known about its consequences. However, there have been situations in which people have ingested large quantities of copper in drinking water or contaminated soft drinks.[28] In these cases, copper toxicity caused cramping, nausea, and diarrhea. In addition, copper toxicity may cause liver damage. An RDA for copper has been established at 900 µg/day. To prevent possible liver damage due to copper toxicity, a UL of 10 mg (10,000 µg)/day has been set.

Iodine (Iodide)—An Essential Component of the Thyroid Hormones

The body needs **iodine (I)** for only one reason: It is an essential component of the hormones produced by the thyroid gland. Thyroid hormones regulate growth, reproduction, and energy metabolism. In addition, they influence the immune system and neural development. Technically, most iodine in the body is in the anion form of **iodide (I⁻).** However, consistent with much of the nutrition literature, we refer to this mineral as "iodine."

Marine foods such as salmon and seaweed (nori) are naturally iodine-rich.

MARINE FOODS SUPPLY IODINE NATURALLY

The iodine content of foods frequently depends on the iodine content of the soil and water in which they are grown (Figure 13.7). Ocean fish and mollusks tend to have particularly high amounts because they concentrate the iodine from seawater into their tissues. Seaweed (such as nori), which is used in many Asian cuisines, also contains iodine. Dairy products are excellent sources because iodine is used in milk processing. However, most of the iodine we consume comes from iodized table salt. There are two general kinds

iodine (I) An essential trace mineral that is a component of the thyroid hormones.

iodide (I⁻) The most abundant form of iodine in the body.

FIGURE 13.7 Food Sources of Iodine

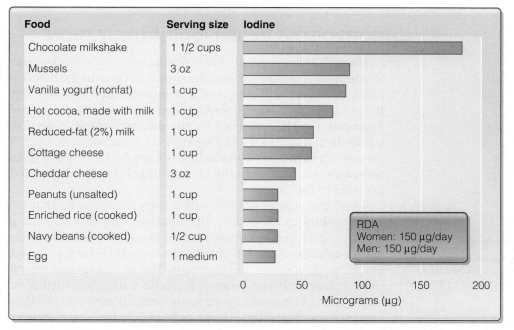

SOURCE: Hands E. *Food finder. Vitamin and mineral source guide*, 3rd ed. Salem, OR: ESHA Research 1995.

Although iodine deficiency is now relatively rare in the United States, this was not always the case. Prior to the fortification of salt with iodine in the 1920s, iodine deficiency (which causes goiter formation) was a major public health problem in some regions of the country. In fact, the Great Lakes and Rocky Mountains were once referred to as the "Goiter Belt." A survey done in Michigan prior to iodination of salt found that about 39% of schoolchildren had visible goiters, indicating relatively severe iodine deficiency.[29] When scientists discovered that iodine deficiency was the cause of goiter, statewide campaigns for goiter prevention and treatment were launched promoting iodized salt as the hoped-for solution. Within four years, there was a 75% reduction in childhood goiter,[30] and by the 1950s, fewer than 1% of children were found to have goiter.[31] Clearly, consuming iodized salt has drastically reduced the occurrence of goiters in the United States. Today, we have the choice to purchase noniodized or iodized salt, with the difference in cost typically being less than 10¢.

This success story is a prime example of how scientists can team up with public health officials and food manufacturers to improve the well-being of millions of people via better nutrition. It is the hope of many charitable organizations (such as Kiwanis International) that are working with international health organizations (such as the World Health Organization) to eradicate iodine deficiency around the world.

Courtesy The Illustrated Gallery

of table salt available in the United States—noniodized and iodized. Iodized salt has been fortified with iodine, and about half the salt that Americans put on their food is iodized. Note that kosher salt and sea salt are not iodine fortified. You can read more about the iodine fortification of salt in the Focus on the Process of Science feature.

GOITROGENS INHIBIT IODINE UTILIZATION

In general, iodine is highly bioavailable, being mostly absorbed in the small intestine and, to a lesser extent, the stomach. Once in the blood, iodine is rapidly taken up by the thyroid gland and incorporated into the thyroid hormones. Iodine uptake by the thyroid gland is regulated by **thyroid-stimulating hormone (TSH)**, produced by the pituitary gland located directly beneath the brain (Figure 13.8). During periods of iodine deficiency, TSH release increases, in turn increasing iodine uptake by the thyroid gland. TSH release decreases when excess iodine is consumed. Iodine not needed by the body is excreted in the urine.

There are several examples of dietary components that can influence the body's ability to use iodine. Specifically, compounds called **goitrogens** can inhibit iodine uptake by the thyroid gland. Goitrogens are found in cassava (a root eaten worldwide); cruciferous vegetables such as cabbage, cauliflower, and Brussels sprouts; and soybeans. The term *goitrogen* refers to the fact that these compounds can potentially cause a disease called *goiter*, described in detail below. Consuming goitrogens typically poses no problem except in situations of very low iodine intake or in people who have thyroid dysfunction.

thyroid-stimulating hormone (TSH) A hormone, produced in the pituitary gland, that stimulates uptake of iodine by the thyroid gland.

goitrogens (GOIT – ro – gens) Compounds found in some vegetables that decrease iodine utilization by the thyroid gland.

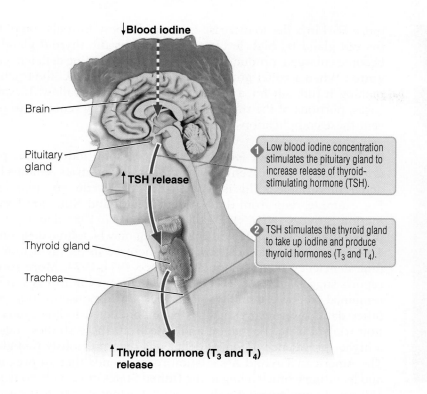

↓**Blood iodine**

Brain

Pituitary gland

↑**TSH release**

Thyroid gland

Trachea

1 Low blood iodine concentration stimulates the pituitary gland to increase release of thyroid-stimulating hormone (TSH).

2 TSH stimulates the thyroid gland to take up iodine and produce thyroid hormones (T_3 and T_4).

↑**Thyroid hormone (T_3 and T_4) release**

FIGURE 13.8 Regulation of Iodine Uptake by the Thyroid Gland Thyroid-stimulating hormone (TSH), made in the pituitary gland, stimulates iodine uptake by the thyroid gland and production of thyroid hormones (T_3 and T_4). TSH release increases during iodine deficiency.

IODINE IS A COMPONENT OF THYROID HORMONES

Iodine is an essential component of the thyroid hormones **thyroxine (T_4)** and **triiodothyronine (T_3)**. Thyroxine, which contains four iodine atoms, is produced in the thyroid gland and is converted to the more active form, triiodothyronine, which contains three iodine atoms. T_3 and T_4 help regulate energy metabolism, growth, and development. These hormones are also essential for proper development of the brain, spinal cord, and skeleton during fetal growth. Because thyroid hormones are involved in regulating energy metabolism, low levels cause severe fatigue. However, the mechanisms by which T_3 and T_4 influence such a vast number of physiological systems are not completely understood.

IODINE DEFICIENCY CAUSES CRETINISM AND GOITER

Iodine deficiency is a significant public health problem worldwide. It is most prevalent in countries without iodized salt and those that are not near an ocean or sea.[32] Manifestations of iodine deficiency take many forms, collectively called **iodine deficiency disorders (IDDs).** Which type of IDD a person has depends on various factors, including genetics, severity of the deficiency, and its timing in the life cycle. The two most studied forms of IDD are cretinism and goiter.

Fetal Iodine Deficiency: Cretinism **Cretinism** occurs when a baby is born to an iodine-deficient mother. During fetal growth, especially in the first few months, the baby relies on the mother's thyroid hormones for its own growth and development. If the mother is iodine deficient, she does not produce sufficient amounts of the thyroid hormones, and the baby's growth and development are affected. Cretinism causes severe mental retardation, poor growth, infertility, and increased risk for mortality.

Childhood and Adult Iodine Deficiency: Goiter When iodine deficiency occurs in children and adults, low levels of T_3 and T_4 signal the pituitary gland to release

Wellcome Trust Library/Custom Medical Stock Photo

Cretinism occurs when an iodine-deficient mother becomes pregnant, causing her child (as shown here) to experience iodine deficiency *in utero.*

thyroxine (T_4) The less-active form of thyroid hormone; contains four atoms of iodine.

triiodothyronine (T_3) The more-active form of thyroid hormone; contains three atoms of iodine.

iodine deficiency disorders (IDDs) A broad spectrum of conditions caused by inadequate iodine.

cretinism A form of iodine deficiency disorder (IDD) that affects babies born to iodine-deficient mothers.

Goiter is a result of compensatory growth of the thyroid gland in response to iodine deficiency.

more TSH in order to increase their production. Stimulation of the thyroid gland by high levels of TSH causes the thyroid gland to become enlarged, producing the classic sign of iodine deficiency—a **goiter.** When a goiter grows very large, it can obstruct the trachea, making it difficult for a person to breathe and swallow. In some cases, portions of the thyroid gland are surgically removed to prevent these complications.

Recent Concern about Iodine Deficiency Despite iodinization programs, iodine intake even in some developed countries remains low and quite variable.[32] This is of particular concern during pregnancy. For example, data from the National Health and Nutrition Examination Surveys (NHANES) collected between 1988 and 1994 reported that 6.7% of pregnant U.S. women may be iodine deficient.[33] This represents an almost sevenfold increase in prevalence of iodine deficiency in this population since 1971–1974. More recent reports suggest that the level of iodine deficiency in this group has remained somewhat stable since 2003.[34] In response to these and other data, several investigators have conducted iodine intervention trials during pregnancy, and data from these studies suggest a higher prevalence of iodine deficiency than previously thought.[35] The American Thyroid Association recommends that all pregnant and lactating women living in the United States or Canada consume 150 μg of supplemental iodine daily.[36] Further research is needed to determine if this recommendation is applicable to all pregnant women or just those at greatest risk for iodine deficiency.

RECOMMENDED INTAKES FOR IODINE

Iodine toxicity resulting from high intakes of iodine-rich food, water, and supplements can take several forms, including both underactive thyroid activity (hypothyroidism) and overactive thyroid activity (hyperthyroidism). In fact, iodine toxicity can actually cause goiters to form in some people, although the mechanisms are poorly understood. It should be noted, however, that both hypo- and hyperthyroidism can be caused by factors other than iodine toxicity. For example, an autoimmune disease (Hashimoto's thyroiditis) can cause hypothyroidism. This occurs when antibodies destroy the thyroid hormone–producing cells in the thyroid gland. The RDA for iodine intake is 150 μg/day. To prevent iodine toxicity, a UL of 1,100 μg/day has been established.

Selenium—A Mineral with Antioxidant Functions

The essentiality of **selenium (Se)** was quite recently discovered—only in the 1950s. Since that time, much has been learned about how the body uses this trace mineral. Selenium is thought to be involved in decreasing risk for cancer, protecting the body from toxins and free radicals, activating vitamin C, slowing the aging process, and enhancing immunity.

NUTS, SEAFOOD, AND MEATS ARE RICH IN SELENIUM

Although foods can contain several forms of selenium, it is typically associated with the amino acid methionine, which usually contains sulfur. When methionine contains selenium, it is called **selenomethionine.** The best sources

goiter (GOIT – er) A form of iodine deficiency disorder (IDD) that affects children and adults; characterized by an enlarged thyroid gland.

selenium (Se) An essential trace mineral that is important for reduction–oxidation (redox) reactions, thyroid function, and activation of vitamin C.

selenomethionine The amino acid methionine that has been altered to contain selenium instead of sulfur.

FIGURE 13.9 Food Sources of Selenium

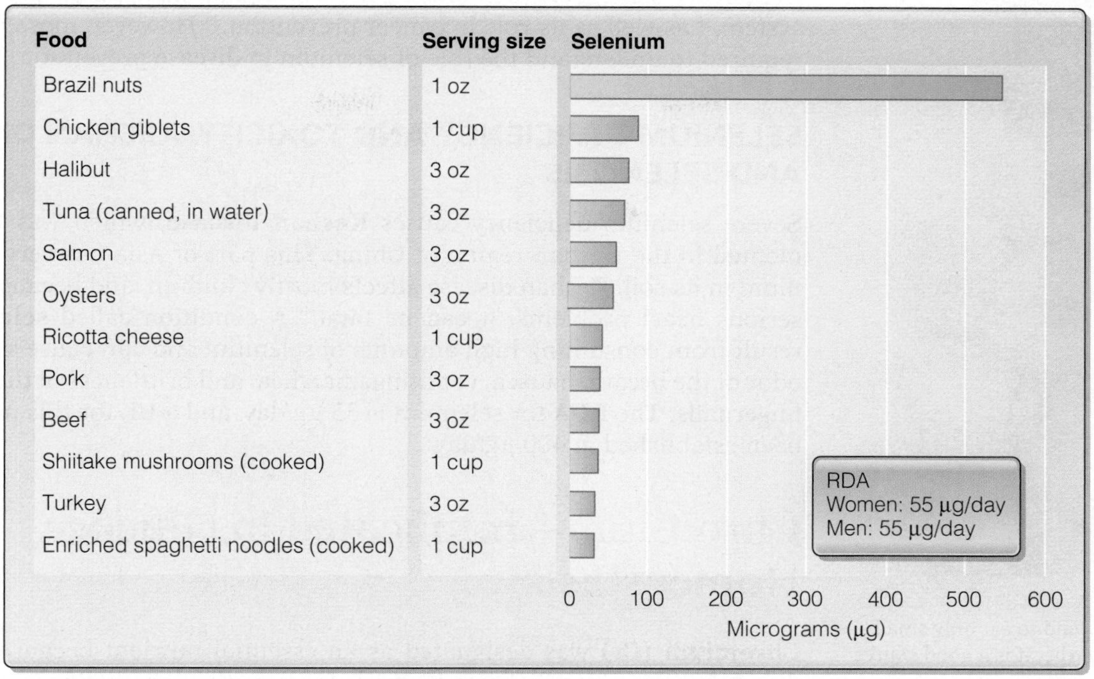

Food	Serving size	Selenium
Brazil nuts	1 oz	
Chicken giblets	1 cup	
Halibut	3 oz	
Tuna (canned, in water)	3 oz	
Salmon	3 oz	
Oysters	3 oz	
Ricotta cheese	1 cup	
Pork	3 oz	
Beef	3 oz	
Shiitake mushrooms (cooked)	1 cup	
Turkey	3 oz	
Enriched spaghetti noodles (cooked)	1 cup	

RDA
Women: 55 µg/day
Men: 55 µg/day

0   100   200   300   400   500   600
Micrograms (µg)

SOURCE: USDA Nutrient Database for Standard Reference, Release 17.

of selenium are nuts, seafood, and meats (Figure 13.9). Garlic is also a good source of selenium, although it does not provide a large amount to the diet. Water can also be a source of this mineral if the ground from which the water is drawn is selenium-rich. Grains can contain selenium, but their selenium content is highly dependent on the soil in which they are grown. For example, in some areas of China where the soil selenium content is very low, grains contain negligible amounts. Conversely, in some portions of the western United States where the selenium content in soil is very high, grains can contain toxic levels. Grazing animals have been known to develop selenium toxicity from eating grass grown in soils with high levels of selenium.

SELENIUM IS INCORPORATED INTO SELENOPROTEINS

The bioavailability of selenium in foods is high, and absorption of this mineral in the intestine is not regulated. Therefore, almost all selenium consumed enters the blood. Once absorbed and circulated, selenium is taken up by cells, incorporated into selenomethionine, and subsequently used to make proteins. Proteins that contain selenomethionine are called **selenoproteins.** Metabolism and breakdown of selenoproteins releases selenium that can then be inserted into other proteins. In this way, the body maintains a relatively large amount of selenium in muscles as selenomethionine and selenoproteins. Blood concentrations of selenium are maintained by increasing or decreasing the amount excreted in the urine. When consumption is high, selenium can also be expelled in the breath, causing a garlicky odor.

SELENIUM IS AN IMPORTANT ANTIOXIDANT

There are at least 14 selenoproteins in the body. One group of enzymatic selenoproteins called **glutathione peroxidases** has redox functions, protecting against oxidative damage.[37] Other selenoproteins are needed for thyroid

selenoprotein A protein that contains selenomethionine instead of sulfur-containing methionine.

glutathione peroxidases A group of selenoprotein enzymes that have reduction–oxidation (redox) functions in the body.

ultimathule /Shutterstock.com

Although we tend to eat only small amounts of garlic, it is a good source of selenium.

function and for vitamin C metabolism. Because selenium has many protective effects, scientists are interested in its potential importance to the immune system,[38] as well as its role in cancer prevention.[39] However, more research is required to understand the role of selenium in disease prevention.

SELENIUM DEFICIENCY AND TOXICITY: KESHAN DISEASE AND SELENOSIS

Severe selenium deficiency causes **Keshan disease,** which was first documented in the Keshan region of China. This part of Asia has very little selenium in its soil. Keshan disease affects mostly children, and because it causes serious heart problems, it can be fatal.[40] A condition called **selenosis** can result from consuming high amounts of selenium and can cause a garlic-like odor of the breath, nausea, vomiting, diarrhea, and brittleness of the teeth and fingernails. The RDA for selenium is 55 µg/day, and a UL for this mineral has been established at 400 µg/day.

Chromium—Implicated in Glucose Homeostasis

Chromium (Cr) was designated as an essential nutrient because scientists discovered that its deficiency in animals caused a diabetic-like state. Later, it was determined that chromium may be involved in glucose homeostasis and insulin function in humans as well. There is, however, considerable debate as to whether chromium is a required nutrient for humans.[41]

CHROMIUM CONTENT OF FOODS DEPENDS ON CHROMIUM CONTENT OF SOIL

Chromium is found in a variety of foods. However, it is difficult to list those that are especially good sources, partly because the chromium content of a food can vary greatly depending on the chromium content of the soil where the food was grown. In addition, chromium is found in such small quantities that it can be difficult to accurately measure. Nonetheless, whole-grain products, fruits, and vegetables tend to be good sources of chromium, whereas refined cereals and dairy foods contain very little. Some processed meats have relatively high amounts of chromium, as do some beers and wines.

Chromium bioavailability in the small intestine is increased by vitamin C and acidic medications such as aspirin (acetylsalicylic acid) and decreased by antacids. Regardless of chromium intake, very little (<2%) is absorbed; most is excreted in the feces. Once absorbed, chromium is transported in the blood to the liver, and excess chromium is excreted in the urine. When the body is unable to excrete excess chromium in the urine, it is deposited mainly in the liver. Although scientists do not know why, consuming large amounts of simple sugars can increase urinary chromium excretion.[42]

CHROMIUM MAY BE INVOLVED IN GLUCOSE HOMEOSTASIS

Studies suggest that chromium is needed for the hormone insulin to function properly in the body, especially in people with type 2 diabetes.[43] Chromium may also be required for normal growth and development in children, and it appears to increase lean mass and decrease fat mass—at least in laboratory animals.[44] Because of this, chromium in a form called **chromium picolinate**

Keshan disease A disease that affects mostly children, resulting from selenium deficiency.

selenosis Selenium toxicity.

chromium (Cr) A trace mineral that may be needed for proper insulin function in some people.

chromium picolinate A form of chromium taken as an ergogenic aid by some athletes.

has been widely marketed as an ergogenic aid for athletes.[45] Other types of chromium supplements are promoted as products that help regulate blood glucose. In fact, estimated sales of chromium-containing supplements are approximately $85 million, representing 5.6% of the total mineral-supplement market.[46] However, most controlled studies investigating the effect of chromium-containing supplements on athletic performance or blood glucose regulation have shown no beneficial outcomes.[47]

Whole-grain foods can provide chromium to the diet.

CHROMIUM DEFICIENCY AND TOXICITY ARE RARE

Chromium deficiency is seen only in hospitalized patients receiving inadequate nutrition support, and it results in elevated blood glucose levels, decreased sensitivity to insulin, and weight loss.[18] Chromium toxicity is also rare, even when supplemental chromium is consumed. This is probably due to low chromium absorption. However, toxic levels are sometimes seen when people are exposed to high levels of industrially released chromium. For example, when stainless steel is heated to a very high temperature during welding, chromium can be released into the air. Environmental exposure of this kind causes skin irritations and may increase risk for lung cancer.[48] These effects are never seen when chromium is obtained from the diet.

RECOMMENDED INTAKES FOR CHROMIUM

There is insufficient information for establishing RDAs for chromium. However, AIs have been set. These recommendations are quite variable (20 to 35 μg/day), and are higher for men than for women. Because there are no known adverse effects of high intake of chromium from diet or supplements, the Institute of Medicine has not established ULs for this mineral.

Manganese—Important for Gluconeogenesis and Bone Formation

Like most other trace minerals, **manganese (Mn)** is a necessary cofactor for many enzymes in the body. As with chromium, manganese deficiency is rare in humans, and toxicity is typically found only after excessive environmental exposure.

PLANT FOODS ARE THE BEST SOURCES OF MANGANESE

Good sources of manganese include whole-grain products, pineapples, nuts, and legumes. Dark green leafy vegetables such as spinach are high in manganese, and water can also contain significant amounts. Regardless of manganese intake, very little (<10%) is absorbed in the small intestine. Excess manganese is delivered to the liver, where it is incorporated into bile and excreted in the feces.

Manganese is a cofactor for numerous metalloenzymes. For example, it is needed for enzymes involved in glucose production (gluconeogenesis) and bone formation. Manganese also binds ADP and ATP in a variety of reactions, and is therefore important for energy metabolism. Like copper, manganese is a cofactor for superoxide dismutase, an enzyme that protects cells from free radicals.

Pineapples are excellent sources of the trace mineral manganese.

manganese (Mn) An essential trace mineral that is a cofactor for enzymes needed for bone formation, glucose production, and energy metabolism.

SEVERE MANGANESE DEFICIENCY CAUSES WEAK BONES AND SLOW GROWTH

Because manganese intake is almost always sufficient, manganese deficiency is rare. Nonetheless, when it does occur, it results in scaly skin, poor bone formation during early life, and growth faltering. Manganese toxicity from foods and supplements is uncommon, but exposure to high levels of environmental manganese (as can occur in mining) can cause serious neurological problems. Manganese toxicity can also occur in people with liver disease and in those consuming water with high manganese levels. AIs for manganese have been set at 2.3 and 1.8 mg/day for men and women, respectively. To prevent nerve damage caused by manganese toxicity, a UL has been established at 11 mg/day. This UL includes manganese intake from food, water, and supplements.

Molybdenum—Required in Very Small Quantities

Like many of the trace minerals, the body needs **molybdenum (Mo)** because it is a cofactor for several important metalloenzymes. However, this trace mineral is required in especially minute amounts. Molybdenum deficiency is almost unheard of, and there is little concern among health professionals about the dietary inadequacy of this mineral.

LEGUMES, NUTS, AND GRAINS PROVIDE MOLYBDENUM

Although the molybdenum content of foods depends on its content in soils, legumes (peas and lentils) as well as grains and nuts tend to be good sources of this mineral. Molybdenum appears to be almost completely absorbed in the intestine, where it is circulated to the liver in the blood. Molybdenum functions in redox reactions and is a cofactor for several enzymes in the body. These enzymes are involved in the metabolism of sulfur-containing amino acids (such as methionine and cysteine) as well as purines that make up DNA and RNA. Molybdenum-requiring enzymes are also involved in detoxifying drugs in the liver.

MOLYBDENUM DEFICIENCY IS EXCEEDINGLY RARE

Only one case of primary molybdenum deficiency has been documented, and this was in a hospitalized patient receiving intravenous nutrition support that was devoid of molybdenum. In this case, molybdenum deficiency was accompanied by abnormal heart rhythms, headache, and visual problems.[49] There are no known adverse effects of high intakes of molybdenum in humans. However, in experimental animals, toxicity can seriously harm reproduction. The RDA for molybdenum is 45 µg/day, and to prevent potential reproductive problems, a UL of 2,000 µg/day has been established.

Zinc—Involved in RNA Synthesis and Gene Expression

Zinc (Zn) is a cofactor for more than 300 enzymes in the body, and the mechanisms by which it influences human health are still being explored. However, researchers know that it is needed for growth, reproduction, immunity, gene expression, and protein synthesis.

molybdenum (Mo) (mo – LYB – de – num) An essential trace mineral that is a cofactor for enzymes needed for amino acid and purine metabolism.

zinc (Zn) An essential trace mineral involved in gene expression, immune function, and cell growth.

FIGURE 13.10 Food Sources of Zinc

Food	Serving size	Zinc
Oysters	3 oz	
Enriched breakfast cereal	1 cup	
Beef	3 oz	
Crab	3 oz	
Lamb	3 oz	
Chicken giblets	1 cup	
Beef liver	3 oz	
Pork	3 oz	
Turkey	3 oz	
Ricotta cheese	1 cup	
White beans (canned)	1 cup	
Chocolate	1 oz	

RDA
Women: 8 mg/day
Men: 11 mg/day

Milligrams (mg) — 0 10 20 30 40 50 60 70 80

SOURCE: USDA Nutrient Database for Standard Reference, Release 17.

ZINC IS FOUND IN SHELLFISH, ORGAN MEATS, AND DAIRY FOODS

Zinc is found in high concentrations in shellfish, meat, organ foods (such as liver), dairy products, legumes, and chocolate (Figure 13.10). It is also frequently added to fortified cereal-grain products. Zinc bioavailability is influenced by a variety of dietary factors, many of which are similar to those that influence iron absorption.[50] Bioavailability of zinc from animal sources is greater than from plant sources, and acidic substances (such as vitamin C) increase its absorption. Because zinc absorption relies on the same transport protein as several other divalent cations (e.g., Fe^{2+}, Cu^{2+}, and Ca^{2+}), excessive intakes of these minerals can decrease zinc bioavailability.

ZINC ABSORPTION IS REGULATED SIMILARLY TO IRON

Similar to that of iron, zinc bioavailability is highly regulated, and its absorption in the small intestine requires at least two proteins. The first protein transports zinc across the brush border into the enterocyte, where the second (metallothionine) then binds it. Recall that ferritin in the enterocyte binds excess iron, blocking its absorption during periods of high iron status. Similarly, **metallothionine** binds zinc, making it unavailable for transport across the basolateral membrane. To facilitate zinc absorption, synthesis of metallothionine decreases during periods of low zinc availability. Conversely, its synthesis increases during periods of zinc excess. Because intestinal cells are continually sloughed off, excess zinc bound to metallothionine is excreted in the feces. Zinc absorption is also influenced by genetic factors and other dietary constituents, such as phytates, which decrease bioavailability.

Acrodermatitis Enteropathica A genetic abnormality called **acrodermatitis enteropathica** is caused by a defect in the protein that transports zinc into

Chocolate is a good source of zinc. Remember that all foods can be part of a healthy diet.

metallothionine (me – tall – o – THI – o – nine) A protein in the enterocyte that regulates zinc absorption and elimination.

acrodermatitis enteropathica A genetic abnormality resulting in decreased absorption of dietary zinc.

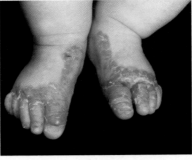

Acrodermatitis enteropathica is a genetic disorder that causes decreased zinc absorption and, therefore, secondary zinc deficiency.

⟨CONNECTIONS⟩ Recall that, in the second step of protein synthesis, called gene transcription, the information contained in genes (DNA) is transferred to mRNA molecules in the nucleus (Chapter 5, page 168).

FIGURE 13.11 Zinc Finger Special zinc-containing portions of proteins, called zinc fingers, modulate gene expression and, ultimately, protein synthesis.

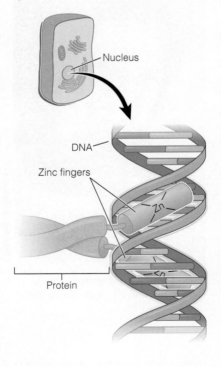

zinc finger Zinc-containing three-dimensional structure of some proteins that allows them to regulate gene expression.

fluoride (F⁻) A nonessential trace mineral that strengthens bones and teeth.

enterocytes. Because of this defective protein, less zinc is transported into enterocytes, causing it to be lost in the feces and resulting in secondary zinc deficiency. Acrodermatitis enteropathica is characterized by severe zinc deficiency even when zinc is seemingly adequate in the diet, and it is usually seen in the first few months of life. Babies with acrodermatitis enteropathica fail to grow properly and have severely red and scaly skin, especially around the scalp, eyes, and feet.[51] They also suffer from diarrhea. Untreated acrodermatitis enteropathica is fatal, and treatment requires lifelong zinc supplementation. When excessive levels of zinc are consumed, adequate amounts of zinc can be absorbed, even in the presence of the faulty transport protein. Researchers estimate that 1 in 500,000 babies is born with this disease.

ZINC IS INVOLVED IN RNA SYNTHESIS AND GENE EXPRESSION

Zinc is involved in hundreds of biological activities, many of which require zinc as a cofactor. For example, it is needed for enzymes involved in RNA synthesis and alcohol metabolism. Zinc also stabilizes the structures of many proteins that regulate gene expression. These proteins contain three-dimensional components called **zinc fingers** that turn on and off specific genes, thus regulating transcription (Figure 13.11). Without zinc, these proteins cannot function. In this way, zinc fingers are important in initiating protein synthesis critical to cell maturation, growth, and proper immune function.[52] Zinc also acts as a potent antioxidant,[53] and it appears important in stabilizing cell membranes. Note that zinc supplements are often touted as helping "cure" the common cold. Although some studies do not support this claim, a recent systematic review of the published literature concluded that zinc was effective in this regard, especially in children.[54]

VEGETARIANS MAY NEED ADDITIONAL ZINC

Zinc deficiency in humans was first documented in the 1960s in regions of Egypt and Iran, where people tend to eat plant-based diets high in phytates that inhibit zinc absorption.[55] Mild zinc deficiency appears to decrease appetite, thus increasing illness and decreasing growth—especially in children. Severe zinc deficiency causes skin irritation, diarrhea, and delayed sexual maturation.

Although dietary zinc toxicity is uncommon, high levels can be obtained from supplements. Symptoms include poor immune function, depressed levels of high-density lipoprotein (HDL) cholesterol, and impaired copper status. Nausea, vomiting, and loss of appetite have also been reported. RDAs for zinc are set at 11 and 8 mg/day for men and women, respectively. The Institute of Medicine recommends that vegetarians, particularly vegans, consume up to 50% more zinc than nonvegetarians. This is mostly because zinc bioavailability from plant-based diets is lower than that from mixed diets. A UL for zinc has been set at 40 mg/day.

Fluoride—Nonessential Mineral That Strengthens Bones and Teeth

Fluoride (F⁻) is not actually *required* for any basic body functions. However, dietary fluoride strengthens bones and teeth. Therefore, knowing about sources of fluoride and how it works in the body is important. Note that the nonionic form of fluoride is *fluorine*, which is a poisonous gas.

MANY COMMUNITIES FLUORIDATE THEIR WATER

Very few foods are good sources of fluoride, although potatoes, tea, and legumes contain some fluoride, as do fish with intact bones (e.g., some types of canned salmon and anchovies). In some areas, water naturally contains high amounts of fluoride and is therefore a good source. In addition, swallowing fluoridated toothpaste can contribute fluoride to the body. Because foods provide little fluoride, many communities add fluoride to their drinking water. The American Dental Association recommends that water be fluoridated at 0.7 to 1.2 parts fluoride per million parts water (0.7 to 1.2 ppm).[56] Note that the U.S. Department of Health and Human Services and the U.S. Environmental Protection Agency have recently proposed lowering this recommended level to 0.7 ppm.[57] This proposal is in response to increased levels of fluoride available through nonwater sources and concerns about excessive fluoride exposure. You can find out whether your town has fluoridated water by contacting a local dentist or the city water department, or by visiting the website of the CDC at http://apps.nccd.cdc.gov/MWF/Index.asp. Bottled water typically contains little fluoride and is therefore a poor source of this mineral. Because the GI tract does not tightly regulate fluoride's absorption, almost all fluoride consumed is absorbed in the small intestine. It is then circulated in the blood to the liver, and ultimately is taken up by bones and teeth. Excess fluoride is excreted in the urine.

FLUORIDE STRENGTHENS BONES AND TEETH

Fluoride affects the health of bones and teeth in several ways. First, it strengthens them by being incorporated into their basic mineral matrices. Bones and teeth that contain fluoride are stronger than those without it. Teeth that contain extra fluoride are especially resistant to bacterial breakdown and cavities (dental caries). Consuming fluoride also appears to stimulate maturation of osteoblasts, the cells that build new bone. Second, *topical* (not dietary) application of fluoride-containing toothpaste and fluoride treatments works directly on the cavity-causing bacteria in the mouth to decrease acid production. The damaging effects of acids produced by these bacteria are largely responsible for tooth decay. Thus, lower acid production means fewer cavities.

FLUORIDE TOXICITY RESULTS IN FLUOROSIS

Aside from the beneficial effects of fluoride on tooth and bone strength, there are no known biologically important effects of this mineral. However, fluoride toxicity is well documented. Signs and symptoms include GI upset, excessive production of saliva, watering eyes, heart problems, and in severe cases, coma. In addition, very high fluoride intake causes pitting and mottling (discoloration) of teeth, called **dental fluorosis,** and a weakening of the skeleton called **skeletal fluorosis.** Fluoride toxicity is a special concern in small children, who sometimes swallow large amounts of toothpaste on a daily basis. Thus, parents should carefully monitor tooth-brushing routines. Because fluoride is not an essential nutrient, no RDAs are established. However, AIs have been set at 4 and 3 mg/day for men and women, respectively. The UL is 10 mg/day.

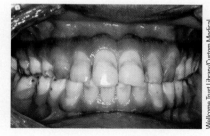

Fluoride toxicity can cause discoloration of the teeth—a condition called dental fluorosis.

Are There Other Important Trace Minerals?

In addition to the trace minerals already covered in this chapter, others may influence your health. These include nickel, aluminum, silicon, vanadium, arsenic, and boron. Whether these minerals are required by the body or influence

dental fluorosis Discoloration and pitting of teeth caused by excessive fluoride intake.

skeletal fluorosis Weakening of the bones caused by excessive fluoride intake.

health and well-being has been neither confirmed nor denied. In fact, most of what is known about them is based on animal studies, not human studies. Therefore, we will simply mention each of these minerals and comment on possible biological functions as well as food sources. Like fluoride, there are no recommended intakes for these minerals, although some of them have ULs. These are listed inside the front cover of this book.

- Nickel (Ni) may be involved in protein and lipid metabolism, redox reactions, and gene regulation. It may also influence calcium metabolism, especially in terms of bone formation. Nickel is found in legumes, nuts, and chocolate.
- Aluminum (Al) may be important for reproduction, bone formation, DNA synthesis, and behavior. It is found in many baked goods, grains, some vegetables, tea, and many antacids.
- Silicon (Si) is important for bone growth in laboratory animals. Dietary sources include whole-grain products. It is also used as a food additive.
- Vanadium (V) acts much like insulin in some laboratory animals and cell culture systems. It also influences cell growth and differentiation and may affect thyroid hormone metabolism. Sources include mushrooms, shellfish, and black pepper.
- Arsenic (As) may be a cofactor for a variety of enzymes and may be important for DNA synthesis. Seafood and whole grains are good sources of arsenic.
- Boron (B) is especially important for reproduction in laboratory animals, and deficiency causes severe fetal malformations. Sources of boron include fruits, leafy vegetables, legumes, and nuts.

Integration of Functions and Food Sources

The functions of the trace minerals are diverse. Aside from participating as cofactors for hundreds of metalloenzymes, trace minerals also regulate gene expression and energy metabolism, protect against free radicals via redox systems, and are important for blood health, immune function, and bone health (Table 13.2). Thus, it is important to consume sufficient—albeit small—amounts of each of these trace minerals on a regular basis. As with the vitamins, no single food group can supply all the trace minerals to your diet. Therefore, consumption of a balanced diet featuring a variety of foods is required to ensure adequate intake of each mineral.

TABLE 13.2 General Functions of the Required Trace Minerals

Functions	Trace Minerals							
	Iron	Copper	Iodine	Selenium	Chromium	Manganese	Molybdenum	Zinc
Cofactor (metalloenzyme)	X	X		X		X	X	X
Regulation of gene transcription								X
Immune function				X				X
Antioxidant (redox) function	X	X		X		X	X	X
Bone/tooth health						X		
Blood health	X							
Energy metabolism[a]	X	X	X	X	X	X		

[a] Note that these minerals are not themselves energy-yielding nutrients but are involved in energy metabolism via other mechanisms.

Critical Thinking: David's Story You have now learned why consumption of all the essential trace minerals is critical for health. You have also learned about some conditions, such as hemochromatosis, that require individuals (like David) to make sure they do not consume *too much* of particular nutrients (like iron). Now that you know why David must limit his intake of iron, can you explain how his dietary choices could impact his status for other nutrients in his body?

Diet Analysis PLUS ✚ Activity

Trace minerals are needed in very small quantities, and yet getting them from foods is vital to your health. Using your *Intake and DRI Goals Compared* report generated from your diet analysis project, please fill in the following table to assess your intakes of the trace minerals. *(Remember that your intakes may fluctuate dramatically from day to day. Therefore, make sure to use the report that averages these fluctuations over the time period.)*

1.

Mineral	Symbol	Your RDA/AI *(include units)*	Your estimated intake *(include units)*	The % of RDA/AI you consumed	Did you consume too little, too much, or an amount close to your RDA/AI?
Iron					
Zinc					

2. For the following, list the foods from your individual days' *Intake Spreadsheets* that provided the largest amounts of the following trace minerals.

Iron
Name of food consumed   **Amount of nutrient in food consumed (include units)**

_____

_____

_____

Zinc
Name of food consumed   **Amount of nutrient in food consumed (include units)**

_____

_____

_____

3. Based on the assessment of your intake of the trace minerals iron and zinc, do you need to change your consumption of these nutrients? Why or why not?

4. Look over your foods that you ate and listed in the exercises for Chapters 10–12 as well as from question 2 above. List *three* foods having high micronutrient content for the greatest number of the vitamins and minerals you tracked. In other words, list the three foods that you most often listed in these exercises. Besides being nutrient dense, are there any other characteristics they share?

Notes

1. World Health Organization. Iron supplementation in young children in regions where malaria transmission is intense and infectious disease highly prevalent; WHO Statement. 2006. Available from: http://www.who.int/child_adolescent_health/documents/pdfs/who_statement_iron.pdf.

2. Pruchnicki MC, Coyle JD, Hoshaw-Woodard S, Bay WH. Effect of phosphate binders on supplemental iron absorption in healthy subjects. Journal of Clinical Pharmacology. 2002;42:1171–6.

3. Chul Huh E, Hotchkiss A, Broulette J, Glahn RP. Carbohydrate fractions from cooked fish promote iron uptake by Caco-2 cells. Journal of Nutrition. 2004;134:1681–9. Hurrell RE, Reddy, MB, Juillerat M, Cook JD. Meat protein fractions enhance nonheme iron absorption in humans. Journal of Nutrition. 2006;136:2808–12.

4. Lopez MAA, Martos FC. Iron availability: An updated review. International Journal of Food Sciences and Nutrition. 2004;55:597–606. Theil EC. Iron homeostasis

and nutritional iron deficiency. Journal of Nutrition. 2011;141:724S–28S.

5. Miret S, Simpson RJ, McKie AT. Physiology and molecular biology of dietary iron absorption. Annual Review of Nutrition. 2003;23:283–301.

6. Franchini M, Montagnana M, Lippi G. Hepcidin and iron metabolism: From laboratory to clinical implications. Clinica Chimica Acta. 2010;411:1565–9. Nemeth E, Ganz T. The role of hepcidin in iron metabolism. Acta Haematologica 2009;122:78–86. Wang J, Pantopoulos K. Regulation of cellular iron metabolism. Biochemical Journal. 2011;434:365–81.

7. Theil EC. Iron, ferritin, and nutrition. Annual Review of Nutrition. 2004;24:327–43.

8. Pietrangelo A. Hereditary hemochromatosis: Pathogenesis, diagnosis, and treatment. Gastroenterology. 2010;139:393–408. Singh B, Arora S, Agrawal P, Gupta SK. Hepcidin: A novel peptide hormone regulating iron metabolism. Clinica Chimica Acta. 2011;412:823–30.

9. Phatak PD, Sham RL, Raubertas RF, Dunnigan K, O'Leary MT, Braggins C, Cappuccio JD. Prevalence of hereditary hemochromatosis in 16 031 primary care patients. Annals of Internal Medicine. 1998;129:954–61.

10. Coon MJ. Cytochrome P450: Nature's most versatile biological catalyst. Annual Review of Pharmacology and Toxicology. 2005;45:1–25. Masson LF, Sharp L, Cotton SC, Little J. Cyrochrome P-450 1A1 gene polymorphisms and risk of breast cancer: A HuGE review. American Journal of Epidemiology. 2005;161:901–15.

11. Centers for Disease Control and Prevention. Recommendations to prevent and control iron deficiency in the United States. Mortality and Morbidity Weekly Report. 1998;47:1–29. Available from: http://www.cdc.gov/mmwr/pdf/rr/rr4703.pdf. Stoltzfus RJ. Defining iron-deficiency anemia in public health terms: Reexamining the nature and magnitude of the public health problem. Journal of Nutrition. 2001;131:565S–7S.

12. Young SL. Pica in pregnancy: New ideas about an old condition. Annual Review of Nutrition. 2010;30:403–22.

13. Young S. Craving earth: Understanding pica—the urge to eat clay, starch, ice, and chalk. New York: Columbia University Press; 2011.

14. Best C, Neufingerl N, Del Rosso JM, Transler C, van den Briel T, Osendarp S. Can multi-micronutrient food fortification improve the micronutrient status, growth, health, and cognition of schoolchildren? A systematic review. Nutrition Reviews. 2011;69:186–204.

15. Rosenzweig PH, Volpe SL. Iron, thermoregulation, and metabolic rate. Critical Reviews in Food Science and Nutrition. 1999;39:131–48.

16. Cunningham-Rundles S, McNeeley DF, Moon A. Mechanisms of nutrient modulation of the immune response. Journal of Allergy and Clinical Immunology. 2005;115:1119–28. Wang W, Knovich MA, Coffman LG, Torti FM, Torti SV. Serum ferritin: Past, present and future. Biochimica Biophysica Acta. 2010; 1800:760–9.

17. Gambling L, Danzeisen R, Fosset C, Andersen HS, Dunford S, Srai SKS, McArdle HJ. Iron and copper interactions in development and the effect on pregnancy outcome. Journal of Nutrition. 2003;133:1554S–6S.

18. Institute of Medicine. Dietary Reference Intakes for vitamin A, vitamin K, arsenic, boron, chromium, copper, iodine, iron, manganese, molybdenum, nickel, silicon, vanadium, and zinc. Washington, DC: National Academy Press; 2001.

19. Siques P, Brito J, Leon-Verlarde F, Barrios L, De La Cruz JJ, Lopez V, Hurruzo R. Hematological and lipid profile changes in sea-level natives after exposure to 3550-m altitude for 8 months. High Altitude Medicine and Biology. 2007;8:286–95.

20. Kurata C. Medical check-up findings characteristic of smokers: Aimed at improving smoking cessation interventions by physicians. Internal Medicine. 2006;45:1027–32.

21. Ganz T, Nemeth E. Iron sequestration and anemia of inflammation. Seminars in Hematology. 2009;46:387–93.

22. Hoffman R, Benz E, Shattil S, Furie B, Cohen H, Silberstein L, McGlave P. Hematology: Basic principles and practice, 3rd ed. New York: Churchill Livingstone, Harcourt Brace; 2000.

23. U.S. Centers for Disease Control and Prevention. Recommendations to prevent and control iron deficiency in the United States. Morbidity and Mortality Weekly Report. 1998;47:1–29. Available online at http://www.cdc.gov/mmwr/preview/mmwrhtml/00051880.htm.

24. Ganz T, Nemeth E. Hepcidin and disorders of iron metabolism. Annual Review of Medicine. 2011;62:347–60.

25. Puig S, Thiele D. Molecular mechanisms of copper uptake and distribution. Current Opinion in Chemical Biology. 2002;6:171–80.

26. Cooper GJS, Chan Y-K, Dissanayake AM, Leahy FE, Koegh GF, Frampton CM, Bamble GD, Brunton DH, Baker JR, Poppitt SD. Demonstration of a hyperglycemia-driven pathogenic abnormality of copper homeostasis in diabetes and its reversibility by selective chelation. Diabetes. 2005;54:1468–76. Zheng Y, Li XK, Wang Y, Cai L. The role of zinc, copper and iron in the pathogenesis of diabetes and diabetic complications: Therapeutic effects by chelators. Hemoglobin. 2008;32:135–45.

27. Failla ML. Trace elements and host defense: Recent advances and continuing challenges. Journal of Nutrition. 2003;133:1443S–7S. Turnlund JR, Jacob RA, Keen CL, Strain JJ, Kelley DS, Domek JM, Keyes WR, Ensunsa JL, Lykkesfeldt J, Coulter J. Long-term high copper intake: Effects on indexes of copper status, antioxidant status, and immune function in young men. American Journal of Clinical Nutrition. 2004;79:1037–44.

28. Chuttani H, Gupta P, Gulati S, Gupta D. Acute copper sulfate poisoning. American Journal of Medicine. 1965;39:849–54. Bremner I. Manifestations of copper excess. American Journal of Clinical Nutrition. 1998;67:1069S–73S.

29. Olin RM. Iodine deficiency and prevalence of simple goiter in Michigan. Journal of the American Medical Association. 1924;82:1328–32.

30. Kimball OP. The efficiency and safety of the prevention of goiter. Journal of the American Medical Association. 1928; 91:454–60. Kimball OP. Prevention of goiter in Michigan and Ohio. Journal of the American Medical Association. 1937;108:860–4.

31. Altland JK, Brush BE. Goiter prevention in Michigan, results of thirty years' voluntary use of iodized salt. Journal of the Michigan Medical Society. 1952;51:985–9.

32. Zimmermann MB. Assessing iodine status and monitoring progress of iodized salt programs. Journal of Nutrition. 2004;134:1673–7.

33. Hollowell JG, Staehling NW, Hannon WG, Flanders DW, Gunter EW, Maberly GF, Braverman LE, Pino S, Miller DT, Garbe PL, DeLozier DM, Jackson RJ. Iodine nutrition in the United States. Trends and public health implications: Iodine excretion data from National Health and Nutrition Examination Surveys I and III (1971–1974 and 1988–1994). Journal of Clinical Endocrinology and Metabolism. 1998;83:3401–8.

34. Caldwell KL, Jones R, Hollowell JG. Urinary iodine concentration: United States National Health and Nutrition Examination Survey 2001–2002. Thyroid. 2005;15:692–9; Caldwell KL, Miller GA, Wang RY, Jain RB, Jones RL. Iodine status of

the U.S. population, National Health and Nutrition Examination Survey 2003–2004. Thyroid. 2008;18:1207–14.

35. Berbel P, Obregon MJ, Bernal J, Escobar del Rey F, Morreale de Escobar G. Iodine supplementation during pregnancy: A public health challenge. Trends in Endocrinology and Metabolism. 2007;18:338–43. National trends in iodine nutrition: Is everyone getting enough? Thyroid. 2007;17:823–7.

36. Becker DV, Braverman LE, Delange F, Dunn JT, Franklyn JA, Hollowell JG, Lamm SH, Mitchell ML, Pearce E, Robbins J, Rovet JF. The Public Health Committee of the American Thyroid Association. Iodine supplementation for pregnancy and lactation—United States and Canada: Recommendations of the American Thyroid Association. Thyroid. 2006;16:949–51.

37. Valko M, Izakovic M, Mazur M, Rhodes CJ, Telser J. Role of oxygen radicals in DNA damage and cancer incidence. Molecular and Cellular Biochemistry. 2004;266:37–56.

38. Augustyniak A, Bartosz G, Cipak A, Duburs G, Horáková L, Luczaj W, Majekova M, Odysseos AD, Rackova L, Skrzydlewska E, Stefek M, Strosová M, Tirzitis G, Venskutonis PR, Viskupicova J, Vraka PS, Zarković N. Natural and synthetic antioxidants: An updated overview. Free Radical Research. 2010;44:1216–62. Selmi C, Tsuneyama K. Nutrition, geoepidemiology, and autoimmunity. Autoimmunity Reviews. 2010;9:A267–70.

39. Ramoutar RR, Brumaghim JL. Antioxidant and anticancer properties and mechanisms of inorganic selenium, oxosulfur, and oxo-selenium compounds. Cell Biochemistry and Biophysics. 2010;58:1–23.

40. Thomson CD. Assessment of requirements for selenium and adequacy of selenium status: A review. European Journal of Clinical Nutrition. 2004;58:391–402.

41. Balk EM, Tatsioni A, Lichtenstein AH, Lau J, Pittas AG. Effect of chromium supplementation on glucose metabolism and lipids: A systematic review of randomized controlled trials. Diabetes Care. 2007;30:2154–63. Wang ZQ, Cefalu WT. Current concepts about chromium supplementation in type 2 diabetes and insulin resistance. Current Diabetes Reports. 2010;10:145–51.

42. Kozlovsky AS, Moser PB, Reiser S, Anderson RA. Effects of diets high in simple sugars on urinary chromium losses. Metabolism. 1986;35:515–8.

43. Cefalu WT, Hu FB. Role of chromium in human health and in diabetes. Diabetes Care. 2004;27:2741–51. Hopkins LL, Jr. Ransome-Kuti O, Majaj AS. Improvement of impaired carbohydrate metabolism by chromium (III) in malnourished infants. American Journal of Clinical Nutrition. 1968;21:203–11. Mertz W. Interaction of chromium with insulin: A progress report. Nutrition Reviews. 1998;56:174–7.

44. McNamara JP, Valdez F. Adipose tissue metabolism and production responses to calcium proprionate and chromium proprionate. Journal of Dairy Science. 2005;88:2498–507. Page TG, Southern LL, Ward TL, Thompson DLJ. Effect of chromium picolinate on growth and serum and carcass traits of growing-finishing pigs. Journal of Animal Science. 1993;71:656–62.

45. Volpe SL. Minerals as ergogenic aids. Current Sports Medicine Reports. 2008;7:224–9.

46. Nutrition Business Journal. NJ's Supplement Business Report 2006. San Diego, CA: Penton Media Inc; 2006.

47. Pittler MH, Stevinson C, Ernst E. Chromium picolinate for reducing body weight: Meta-analysis of randomized trials. International Journal of Obesity and Related Metabolic Disorders. 2003;27:522–9. Vincent JB. The potential value and toxicity of chromium picolinate as a nutritional supplement, weight loss agent and muscle development agent. Sports Medicine. 2003;33:213–30.

48. Costa M, Klein CB. Toxicity and carcinogenicity of chromium compounds in humans. Critical Reviews in Toxicology. 2006;36:155–63. Coyle YM, Minahjuddin AT, Hynan LS, Minna JD. An ecological study of the association of metal air pollutants with lung cancer incidence in Texas. Journal of Thoracic Oncology. 2006;1:654–61. Michaels D, Lurie P, Monforton C. Lung cancer mortality in the German chromate industry, 1958 to 1998. Journal of Occupational and Environmental Medicine. 2006;48:995–7.

49. Abumrad NN, Schneider AJ, Steel D, Rogers LS. Amino acid intolerance during total parenteral nutrition reversed by molybdate therapy. American Journal of Clinical Nutrition. 1981;34:2551–9.

50. Ford D. Intestinal and placental zinc transport pathways. Proceedings of the Nutrition Society. 2004;63:21–29. Lonnerdal B. Dietary factors influencing zinc absorption. Journal of Nutrition. 2000;130:1378S–83S.

51. Kury S, Dreno B, Bezieau S, Giraudet S, Kharfi M, Kamoun R, Moisan JP. Identification of SLC39A4, a gene involved in acrodermatitis enteropathica. Nature Genetics. 2002;31:239–40. Wang K, Zhou B, Kuo YM, Zemansky J, Gitschier J. A novel member of a zinc transporter family is defective in acrodermatitis enteropathica. American Journal of Human Genetics. 2002;71:66–73.

52. Prasad AS. Impact of the discovery of human zinc deficiency on health. Journal of the American College of Nutrition. 2009;28:257–65.

53. Eibl JK, Abdallah Z, Ross GM. Zinc-metallothionein: A potential mediator of antioxidant defence mechanisms in response to dopamine-induced stress. Canadian Journal of Physiology and Pharmacology. 2010;88:305–12. Powell SR. The antioxidant properties of zinc. Journal of Nutrition. 2000;130:1447S–54S.

54. Caruso TJ, Prober CG, Gwaltney JM Jr. Treatment of naturally acquired common colds with zinc: A structured review. Clinical Infectious Diseases. 2007;45:569–574. Kurugöl Z, Bayram N, Atik T. Effect of zinc sulfate on common cold in children: Randomized, double blind study. Pediatrics International. 2007;49:842–7. Kurugöl Z, Akilli M, Bayram N, Koturoglu G. The prophylactic and therapeutic effectiveness of zinc sulphate on common cold in children. Acta Paediatrica. 2006;95:1175–81. Singh M, Das RR. Zinc for the common cold. Cochrane Database Systematic Reviews. 2011;2:CD001364.

55. Hambidge M. Human zinc deficiency. Journal of Nutrition. 2000;130:1344S–9S.

56. American Dental Association. American Dental Association supports fluoridation. 2002. Available from: http://www.ada.org/2092.aspx. American Dental Association. Statement on water fluoridation efficacy and safety. 2002. Available from: http://www.ada.org/219.aspx.

57. U.S. Department of Health and Human Services and U.S. Environmental Protection Agency. HHS and EPA announce new scientific assessments and actions on fluoride. January, 2011. Available at http://www.hhs.gov/news/press/2011pres/01/20110107a.html. U.S. Environmental Protection Agency. New fluoride risk assessment and relative source contribution documents. EPA-822-F-11-001. January, 2011. Available at http://water.epa.gov/action/advisories/drinking/upload/fluoridefactsheet.pdf.

Life Cycle Nutrition

From beginning to end, the human life cycle is a process of continuous change. Birth, growth, maturation, aging, and death are all part of the natural progression of life. With each stage, our bodies change in size, proportion, and composition. Because of this, nutritional requirements vary enormously. For example, energy needs during periods of growth differ vastly from those associated with the later stages of life, when the body is in a state of physical maintenance or decline.

Regardless of where you are in the life cycle, an appropriate diet is essential to good health. In fact, nutritional status early in life can influence health at later stages. Therefore, it is important to remember that the food choices you make today may have far greater consequences on your long-term health than you might think. This chapter surveys the influence of growth, development, and aging on nutritional requirements across the human life cycle. The continuum of life encompasses infancy, childhood, adolescence, adulthood, and, for women, the special life stages of pregnancy and lactation.

Backyard Harvest

Each of us should ponder the personal steps that we can take toward making our community a better place. Whether it is by contributing to that community locally or at the global level, it is important for every person to be involved. You may be wondering how the actions of one person can make a difference. Does it sound impossible? Is your life already too busy? Don't tell that to Amy Grey. Amy's motto is "one garden at a time," which is how Backyard Harvest® got its start. The idea for Backyard Harvest came when Amy accidentally grew more than 200 heads of lettuce in her first attempt at vegetable gardening. Not wanting the lettuce to go to waste, she brought some of her bounty to the local food bank. Amy was struck by the fact that her lettuce was the only fresh produce available and that only canned and processed goods filled the shelves. It was this experience that spun the idea for Backyard Harvest—what if other gardeners shared her leafy predicament? Amy approached the local environmental group with her idea of connecting local gardeners with food banks that serve families, senior meal programs, and other community members. Soon, Backyard Harvest became a reality.

Amy's mission was to develop a program that prevented excess produce from going to waste, instead making it available to people with limited access to affordable fresh fruit and vegetables. Word spread throughout the community, and volunteers were soon busy harvesting and collecting excess produce grown by local gardeners. Gleaning events were scheduled to pick fruit from trees, bushes, and vines. After a successful pilot season, the scope of the project grew. What started as a simple goal of collecting and distributing 1,000 pounds of food quickly exceeded expectations. Backyard Harvest has distributed literally tons of locally grown fresh produce to community assistance programs, food banks, and meal sites, which in turn serve thousands of low-income families and seniors. In addition to fresh produce, recipients also receive comprehensive information on how to prepare fruits and vegetables and preserve them for later use. As one woman said, knowing that the fruit and vegetables were coming from the Backyard Harvest project makes her "feel loved." Amy feels the same way.

This simple act of sharing food can happen in your community, too. All it takes is one person with a commitment to make a difference. Watching Amy take action in this small community has motivated others to become involved. For example, a joint initiative between Backyard Harvest, the University of Idaho, and the Moscow (Idaho) Food Co-op now provides education and outreach opportunities for community members regardless of income level. This includes gardening and food preservation workshops, a growers' market, and gardening and nutrition field trips for local elementary schools. For more information about Amy Grey and Backyard Harvest, go to http://www.backyardharvest.org.

© Courtesy of Amy Grey, Backyard Harvest

Critical Thinking: Amy's Story

What type of organization would you start or join if you wanted to improve the nutritional health of people in your community? In which phase or phases of the life cycle do you think people would most benefit from an organization such as Backyard Harvest?

What Physiological Changes Take Place during the Human Life Cycle?

Cells form, mature, carry out specific functions, die, and are replaced by new cells. In many ways, the life cycle of cells mirrors our own lives. That is, after a new human is conceived and born, the next 70 to 90 years are characterized by periods of growth and development, maintenance, reproduction, physical decline, and eventually death. Our ability to reproduce enables us to pass our genetic material (DNA) on to the next generation.

GROWTH AND DEVELOPMENT TAKE PLACE AT VARIOUS TIMES DURING THE LIFE CYCLE

Growth and development generally take place in a predictable and orderly manner. These important physiological events take place throughout the life cycle. Whereas **growth** refers to physical changes that result from either an increase in cell size or number, **development** is the attainment or progression of a skill or capacity to function. Knowing more about growth and developmental milestones will help you better understand why nutritional needs change throughout life.

Growth Patterns Growth may involve an increase in either the number of cells, called **hyperplasia,** or the size of cells, called **hypertrophy** (Figure 14.1). The highest rates of growth occur during infancy, childhood, adolescence, and pregnancy. The most useful and common way to assess growth is by measuring a person's height and weight. The World Health Organization and the U.S. National Center for Health Statistics (NCHS), which is part of the Centers for Disease Control and Prevention (CDC), have compiled height and weight reference standards into growth charts.[1] These charts indicate expected growth for well-nourished infants, children, and adolescents. Growth charts include percentile curves that represent

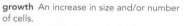

growth An increase in size and/or number of cells.

development Attainment or progression of a skill or capacity to function.

hyperplasia An increase in the number of cells.

hypertrophy An increase in the size of cells.

FIGURE 14.1 Types of Growth
Body size increases when the number and/or size of cells increase.

One type of growth occurs when cells increase in size. This is called hypertrophy.

Another type of growth occurs when cells increase in number. This is called hyperplasia.

Sometimes growth occurs by both hypertrophy and hyperplasia.

growth patterns from birth through 20 years of age. To evaluate adequacy of growth, a child's measurement (such as weight) is assessed in relation to age and sex. For example, if a 5-year-old girl's weight falls at the 60th percentile, 40% of healthy children of similar age and sex weigh more, and 60% weigh less. In this way, growth can be monitored over time and used as a general indicator of health throughout these important phases of the life cycle.

Development Follows a Predictable Pattern The assessment of a child's development is just as important as the assessment of his or her growth. Some of the major developmental accomplishments during the first year of life include such things as vocalization, facial expressions, and motor control over various regions of the body. One of the most significant developmental milestones occurs when an infant transitions from crawling to walking. Although there is a great deal of variability in terms of when these developments take place, the pattern is usually predictable. For example, infants generally crawl before they walk. However, some infants walk as early as 10 months, whereas others may not take their first step until several months later. Failure to reach major developmental milestones by certain ages is cause for concern and may indicate a problem such as illness or poor nutrition.

Growth and development continue steadily throughout infancy and childhood, increasing markedly during adolescence. Adolescence begins when hormonal changes trigger the physical transformation of a child into an adult. Most dramatic is the maturation of reproductive organs and the subsequent ability to reproduce.

Physical Maturity and Senescence As a person reaches physical maturity, the rates of growth and development begin to slow. **Cell turnover,** the cycle by which cells form and break down, reaches equilibrium at this time. During this phase of the life cycle, which is typically the longest, growth ceases and the body enters a phase of maintenance. As a person continues to age, the rate at which new cells form decreases, resulting in a loss of some body tissue. Remaining cells become less effective at carrying out their functions. Gradually, the physical changes characteristically associated with aging, called **senescence,** become apparent. Senescence brings about a slow decline in physical function and health, which eventually can influence a person's nutrient and caloric requirements.

NUTRIENT REQUIREMENTS CAN CHANGE FOR EACH STAGE OF THE LIFE CYCLE

Aging is an inevitable process, and although individuals grow and develop in different ways, physical changes tend to coincide with various stages of the human life cycle. Age-related physical changes affect body size and composition, which in turn influence nutrient and energy requirements. For this reason, the Dietary Reference Intakes (DRIs) recommend specific nutrient and energy intakes for each life-stage group, including infants (0 to 6 months and 7 to 12 months), toddlers (1 to 3 years), early childhood (4 to 8 years), and adolescence (9 to 13 years and 14 to 18 years). Four life-stage groups are used to distinguish nutrient and energy requirements for adults. These include young adulthood (19 to 30 years), middle age (31 to 50 years), adulthood (51 to 70 years), and older adults (over 70 years). Along with these stages, the DRI recommendations consider the special conditions of pregnancy and lactation. Figure 14.2 depicts how the DRI life-stage groups are divided.

cell turnover The cycle of cell formation and cell breakdown.

senescence The phase of aging during which function diminishes.

FIGURE 14.2 Stages in the Human Life Cycle and DRI Life-Stage Groups
Growth and development at different stages of the life cycle affect body size and composition, which in turn influences nutrient and energy requirements.

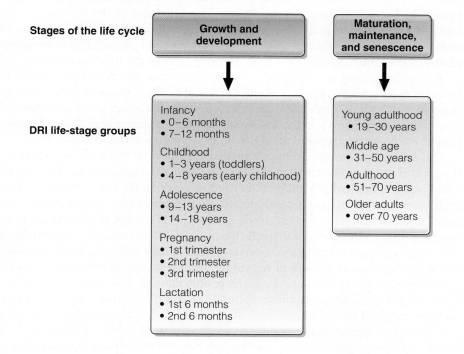

Stages of the life cycle

Growth and development

Maturation, maintenance, and senescence

DRI life-stage groups

Infancy
• 0–6 months
• 7–12 months

Childhood
• 1–3 years (toddlers)
• 4–8 years (early childhood)

Adolescence
• 9–13 years
• 14–18 years

Pregnancy
• 1st trimester
• 2nd trimester
• 3rd trimester

Lactation
• 1st 6 months
• 2nd 6 months

Young adulthood
• 19–30 years

Middle age
• 31–50 years

Adulthood
• 51–70 years

Older adults
• over 70 years

What Are the Major Stages of Prenatal Development?

Although it is important for all women to meet their nutritional needs, nutrition is particularly important during pregnancy. Poor nutritional status before and during pregnancy can have serious and long-term effects on the unborn child. For example, a pregnant woman with poor nutritional status is at increased risk for having a baby born too early or too small. Because a mother's nutritional choices can have profound effects on the life of her unborn child, early prenatal care and ongoing assessment are important.

PRENATAL DEVELOPMENT IS DIVIDED INTO EMBRYONIC AND FETAL PERIODS

There may be no other time in a woman's life when her body experiences such extensive changes as during pregnancy. These physiological transformations are needed to support the new, emerging life within her. While every woman is different, these pregnancy-related events tend to occur in a predictable and organized way. The time shortly before and after conception is referred to as the **periconceptional period.** Conception takes place when the two gametes, an ovum and a sperm, unite. The product of conception is referred to as the **zygote.** Once conception has occurred, prenatal development takes place in two periods—the embryonic period and the fetal period. The formation of the zygote signifies the beginning of the **embryonic period.** This stage of prenatal development spans the first 8 weeks of pregnancy, and is subdivided into pre-embryonic and embryonic phases. The embryonic period is followed by the **fetal period,** which starts at the beginning of the ninth week postconception and concludes at birth.

periconceptional period Time shortly before and after conception.

zygote (ZY – gote) An ovum that has been fertilized by a sperm.

embryonic period The period of prenatal development from conception through the eighth week of gestation.

fetal period Period of prenatal development, which starts at the beginning of the ninth week of gestation and continues until birth.

Embryonic Period Cell division begins immediately after the formation of the zygote, which eventually forms into a dense cellular sphere called a **blastocyst.** Around 2 weeks after conception, the blastocyst implants itself into the endometrium, the innermost lining of the uterus. The cell mass of the blastocyst then begins to differentiate, giving rise to specific tissues and organs. This stage of the embryonic period is referred to as the **pre-embryonic phase.** The **embryonic phase** of prenatal development spans from the start of the third week to the end of the eighth week after fertilization. During this phase, the developing child, referred to as an **embryo,** grows to about the size of a kidney bean. Cell division continues throughout the embryonic period, forming rudimentary structures that will eventually develop into specific tissues and organs. By the end of the embryonic period, the basic structures of all major body organs are formed.

The developing embryo follows a precise timetable in terms of development. If a critical nutrient is lacking at this time, a tissue or organ may not form properly. The term **critical period** is often used to describe the time when an organ undergoes rapid growth and development. Because organ formation occurs very rapidly during this time, the embryo is extremely vulnerable to adverse environmental influences that could disrupt this process. Unfortunately, abnormalities that occur during critical periods are irreversible. For example, maternal drug abuse during critical periods of neural development can have harmful, lasting effects on fetal brain function. This irreversible damage can result in severe behavioral and learning problems as the child grows older. In some cases, abnormalities can lead to embryonic/fetal demise, or what is commonly known as a miscarriage. A **miscarriage** is defined as the death of a fetus during the first 20 weeks of pregnancy. Although critical periods are most likely to occur during the early stages of pregnancy, they can also occur at later stages.

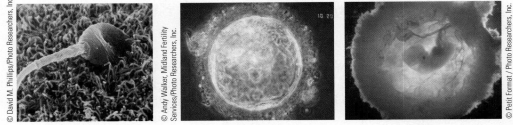

Conception   Blastocyst   Embryo

blastocyst (BLAS – to – cyst) Early period of gestational development that lasts approximately 8 to 13 days after conception.

pre-embryonic phase The early phase of the embryonic period that begins with fertilization and continues through implantation.

embryonic phase The latter phase of the embryonic period during which time organs and organ systems first begin to form.

embryo The developing human from two through eight weeks after fertilization.

critical period Period in development when cells and tissue rapidly grow and differentiate to form body structures. Alteration of growth or development during this period is irreversible.

miscarriage The death of a fetus during the first 20 weeks of pregnancy.

teratogen (te – RAT – o – gen) Environmental agent that can alter normal cell growth and development, causing a birth defect.

fetal alcohol spectrum disorder (FASD) A range of alcohol-related problems that can result from prenatal alcohol exposure.

Teratogens Although most pregnancies result in the birth of a healthy baby, approximately 3 to 4% of babies (about 150,000) in the United States are born with a birth defect each year.[2] A birth defect is an abnormality related to a structure or function that may result in mental or physical impairment. Some birth defects are mild, but others can be severe and life threatening. Birth defects occur for many reasons, such as genetics, environment, lifestyle, or a combination of these factors. Approximately 60% of birth defects have no known cause.

Some birth defects are caused by **teratogens,** a term used to describe a broad group of environmental agents that negatively affect the normal course of cell growth and development in the unborn child. Teratogens, which include chemicals, drugs, infections, and radiation, are responsible for about 4 to 5% of all birth defects.[3] Even excessive intakes of certain nutrients have been found to be teratogenic. For example, high intakes of preformed vitamin A during pregnancy can cause fetal malformations.

The harmful effects of teratogens are usually apparent at birth, although some may not be detected until much later. There are many teratogens, but perhaps the most familiar is alcohol. The term **fetal alcohol spectrum disorder (FASD)** is used to describe the range of alcohol-related problems that can result from prenatal alcohol exposure. Fetal alcohol syndrome is perhaps the most familiar form of FASD, and results in severe lifelong consequences. Babies born with fetal alcohol syndrome have distinctive characteristics such as a small head circumference, unusual facial features, and other physical deformities.[4] In addition, many of these infants are developmentally delayed. Despite widespread awareness of the deleterious effects of alcohol on an unborn child, approximately 1 to 2 babies per 1,000 live births in the United States are born with fetal alcohol syndrome. A less severe form of FASD, called **fetal alcohol effect** is associated with learning and behavior problems, which are often not apparent until later in life. Fortunately, FASD is preventable. That is, if a woman does not consume alcohol while pregnant, there is no risk of having a baby with FASD. Because no amount of alcohol is considered safe during pregnancy, all women should abstain from drinking alcohol during this time.

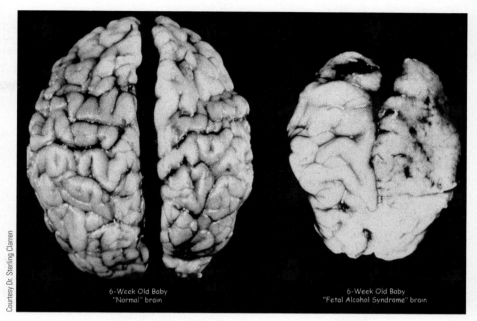

Courtesy Dr. Sterling Clarren

6-Week Old Baby "Normal" brain

6-Week Old Baby "Fetal Alcohol Syndrome" brain

Alcohol exposure during development can induce significant structural changes to the developing brain. The brain on the left is that of a normal six-week-old infant, whereas the brain on the right is that of an infant born with fetal alcohol syndrome.

Fetal Period By the end of the embryonic period, the basic structures of all major body structures and organs are formed, and the embryo is now referred to as a **fetus.** However, for the fetus to survive outside the womb, much additional growth and development is needed. Fetal weight increases by a factor of almost 500 during this period. At term, the fetus weighs approximately 7 to 8 pounds (3.2 to 3.6 kg) and is roughly 20 inches (51 cm) long. Inadequate weight gain and poor nutritional health during this period of pregnancy can dramatically affect fetal growth. Figure 14.3 shows the progressive stages of prenatal development and critical periods for selected organs and structures.

The Formation of the Placenta Within two weeks following conception, the blastocyst implants itself in the lining of the uterus, called the endometrium. Shortly thereafter, embryonic and maternal tissues begin to form the **placenta.** Although the placenta develops early in pregnancy, it takes several weeks before it is fully functional. Weighing between 1 and 2 pounds at term, this highly vascularized structure (see Figure 14.4 on page 591) has important functions.

- The placenta transfers nutrients, hormones, oxygen, and other substances from the maternal blood to the fetus.
- Metabolic waste products formed by the fetus pass through the placenta into the mother's blood; they are then excreted by the mother's kidneys and lungs.
- The placenta is a source of several hormones that serve a variety of functions during pregnancy.

Although placental membranes prevent the fetal and maternal blood from physically mixing, the exchange of gases, nutrients, and waste products is quite efficient. Unfortunately, many potentially harmful substances can also cross from the mother's blood into the fetal circulation via the placenta. For this reason, pregnant women must be particularly careful about using medications or other substances that could harm the unborn child.

fetal alcohol effect A form of fetal alcohol spectrum disorder resulting in physical and cognitive outcomes that are less severe than those of fetal alcohol syndrome.

fetus A term used, beginning at the ninth week of pregnancy through birth, to describe a developing human.

placenta An organ, consisting of fetal and maternal tissues, that supplies nutrients and oxygen to the fetus, and aids in the removal of metabolic waste products from the fetal circulation.

FIGURE 14.3 Stages of Prenatal Growth and Development and Critical Periods of Organ Formation Prenatal growth and development is divided into two periods—the embryonic period (conception to the end of the 8th week) and the fetal period (9th week until birth).

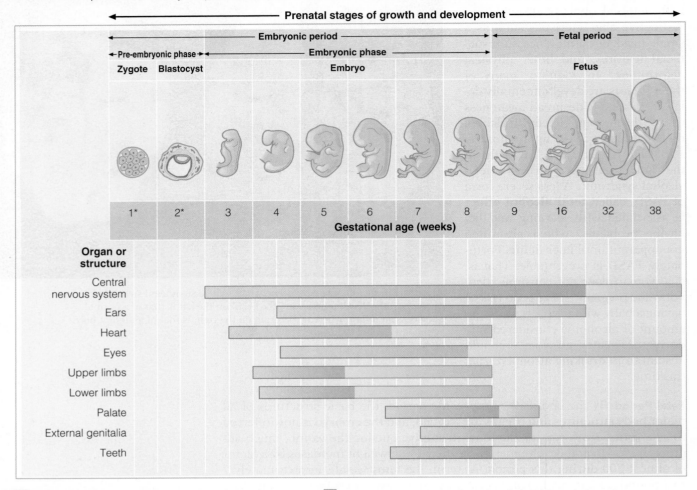

Highlighted legend:
- Highly sensitive period—disruption during this period results in major physical and functional malformations.
- Less-sensitive period—disruption during this period results in functional defects and minor malformations.

*Note that abnormal development during the pre-embryonic phase often results in death.
SOURCE: Adapted from Moore KL, Persaud TVN. The developing human: Clinically oriented embryology. 8th ed. Philadelphia: W.B. Saunders, 2008.

GESTATIONAL AGE IS IMPORTANT TO ASSESS

How long can a woman expect to be pregnant once she conceives? There are several ways this question can be answered. Whereas the terms *embryonic period* and *fetal period* refer to stages of prenatal development, pregnancy is more commonly described in terms of trimesters. The first trimester is the time from conception to the end of week 13 and includes the entire embryonic period as well as part of the fetal period. The second trimester is from week 14 to the end of week 26, and the third trimester is from week 27 to the end of pregnancy.

The duration of pregnancy, or what is called **gestation length,** is the period of time between conception and birth. Average gestation length is 38 weeks. Because most women do not know exactly when conception takes place, calculating gestation length can be difficult. A method more commonly used to assess the length of pregnancy is **gestational age,** defined as the time from the first day of the woman's last menstrual cycle to the current date.[5] Based on gestational age, the average length of pregnancy is about 40 weeks. Babies born with gestational ages between 37 and 42 weeks are considered **full-term infants,** whereas those born with gestational ages less than 37 weeks are called **preterm infants** (or premature infants). Babies born after 42 weeks of gestational age are called **post-term infants.**

gestation length The period of time from conception to birth.

gestational age Common measure used to assess length of pregnancy, determined by counting the number of weeks between the first day of a woman's last normal menstrual period and birth.

full-term infant Baby born with gestational age between 37 and 42 weeks.

preterm infant (premature infant) Baby born with a gestational age less than 37 weeks.

post-term infant Baby born with a gestational age greater than 42 weeks.

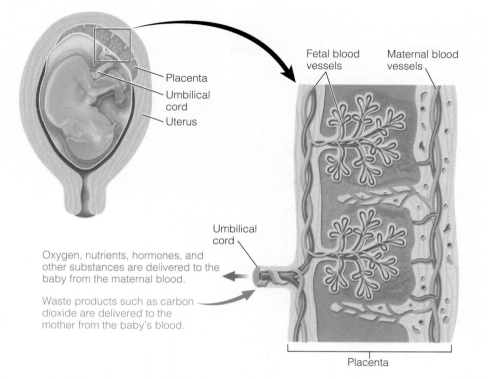

FIGURE 14.4 Structure and Functions of the Placenta The placenta forms early in pregnancy and is made of both fetal and maternal tissues.

Placenta
Umbilical cord
Uterus

Fetal blood vessels

Maternal blood vessels

Umbilical cord

Oxygen, nutrients, hormones, and other substances are delivered to the baby from the maternal blood.

Waste products such as carbon dioxide are delivered to the mother from the baby's blood.

Placenta

Gestational Age and Birth Weight Not only is it important that babies are born full term, but it is also important that they are born at a healthy weight. Thus, gestational age and birth weight are both important predictors of infant health. The earlier a baby is born, the greater the risk for complications that can affect the child's survival and long-term health. This is largely because organs may not be fully developed and may therefore be unable to sustain life.

As illustrated in Figure 14.5, growth charts are used to classify infants according to birth weight and gestational age. Babies weighing less than 5 pounds, 8 ounces (2,500 g) at birth are called **low-birth-weight (LBW) infants.** Low-birth-weight infants are small because they are either preterm or have experienced slow growth *in utero*, known as **intrauterine growth retardation (IUGR).** Babies who experience IUGR are often referred to as **small for gestational age (SGA) infants,** defined as having birth weights below the 10th percentile for gestational age. Babies born with birth weights between the 10th and 90th percentiles for gestational age are classified as **appropriate for gestational age (AGA) infants,** whereas those with birth weights above the 90th percentile are classified as **large for gestational age (LGA) infants.** There are many precautions a pregnant woman can take to help ensure that her baby is born at a healthy weight.

Low-birth-weight babies are 40 times more likely to die before 1 year of age compared with healthy-weight infants.[6] In fact, together, premature births and LBW are the leading causes of infant mortality. In 2008, approximately 12% of babies born in the United States were premature, and 8% were LBW.[7] Compared to that of other developed countries, the prevalence of babies born LBW remains somewhat high in the United States (e.g., 8% of births in the United States compared to 4% of births in Finland and Iceland). Not only does being LBW put a baby at risk in early life, but it may also have profound long-term effects. Evidence suggests that less-than-optimal conditions *in utero* may cause permanent changes in the structure or function of organs and tissues, predisposing individuals to certain chronic diseases later in life.[8] You can read

⟨**CONNECTIONS**⟩ Recall that infant mortality rate is the number of infant deaths during the first year of life per 1,000 live births (Chapter 1, page 22).

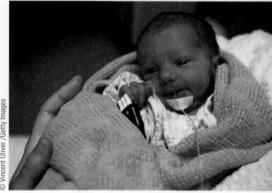

© Vincent Oliver/Getty Images

Low-birth-weight infants often require care in neonatal intensive care units.

low-birth-weight (LBW) infant A baby that weighs less than 2,500 g (5 lb 8 oz) at birth.

intrauterine growth retardation (IUGR) Slow or delayed growth *in utero*.

small for gestational age (SGA) infant A baby that weighs less than the 10th percentile for weight for gestational age.

appropriate for gestational age (AGA) infant A baby that has a weight between the 10th and the 90th percentiles for weight for gestational age.

large for gestational age (LGA) infant A baby with a weight at or above the 90th percentile for weight for gestational age.

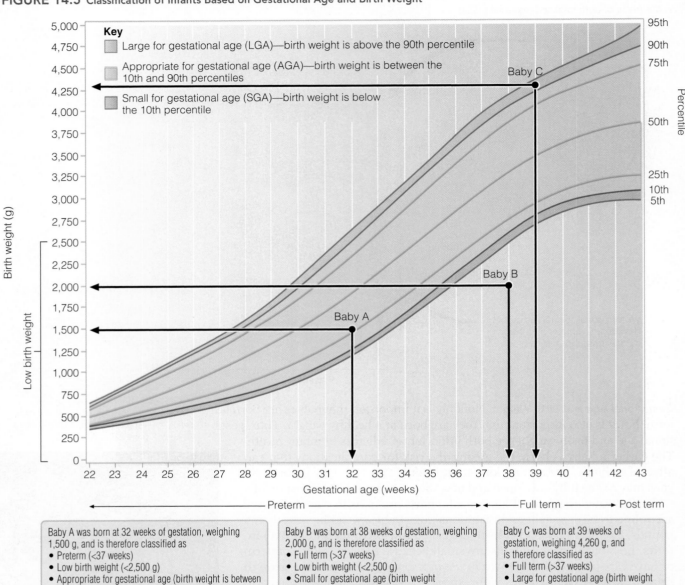

Baby A was born at 32 weeks of gestation, weighing 1,500 g, and is therefore classified as
• Preterm (<37 weeks)
• Low birth weight (<2,500 g)
• Appropriate for gestational age (birth weight is between 10th and 90th percentile).

Baby B was born at 38 weeks of gestation, weighing 2,000 g, and is therefore classified as
• Full term (>37 weeks)
• Low birth weight (<2,500 g)
• Small for gestational age (birth weight is below 10th percentile).

Baby C was born at 39 weeks of gestation, weighing 4,260 g, and is therefore classified as
• Full term (>37 weeks)
• Large for gestational age (birth weight is above the 90th percentile).

more about this phenomenon, called the **developmental origins of health and disease** (formerly called the fetal origins hypothesis) in the Focus on the Process of Science feature.

What Are the Recommendations for a Healthy Pregnancy?

Although unavoidable situations can and do affect pregnancies, there are many precautions women can take before and during pregnancy to help ensure babies are born healthy. Many of these recommendations can decrease risk of having a preterm or LBW baby. The three most important factors that are largely within a woman's control are adequate weight gain, a healthy diet, and not smoking.

developmental origins of health and disease A concept suggesting that conditions during gestation and infancy can alter risk for chronic diseases later in life.

It may be hard to imagine that experiences *in utero* can actually influence a person's risk for chronic disease as an adult. However, an overwhelming amount of evidence indicates that these early life experiences can do just that. Scientists now believe that the origins of certain diseases—diabetes, cardiovascular disease, asthma, certain cancers, and some psychiatric disorders—may be established before a person is even born and during infancy.[9] Both epidemiological and experimental data show that adult health is greatly affected by conditions before birth, soon after birth, and during childhood. This concept, called the developmental origins of health and disease, resulted from some very interesting investigative work by researchers David Barker and Clive Osmond of Southampton University in England and was based on events that took place almost 100 years ago.

In the early 1900s, lowering the infant mortality rates among England's poorest citizens was a goal of health officials. To help accomplish this, a midwife named Ethel Margaret Burnside was assigned to assist pregnant women in these communities. Nurses like Burnside made regular home visits during which they weighed and measured each baby at birth, and then again at 1 year of age. Throughout the years, information on thousands of infants was carefully recorded. Years later, the data recorded by Burnside was discovered and became of great interest to epidemiologists Barker and Osmond.

The researchers observed that regions with the highest neonatal deaths in the early 1900s were the same regions that later had the highest rates of heart disease. Curious to see if these events were related, Barker and Osmond analyzed the birth records of 13,249 men born between 1911 and 1945 and found that LBW was related to increased risk for heart disease later in life. Barker and Osmond proposed that prenatal conditions had somehow influenced adult health years later.

This concept, referred to as fetal programming, suggests that less-than-optimal prenatal conditions may alter fetal development, resulting in increased risk of certain chronic diseases later in life. Barker and Osmond's idea, originally called the fetal origins hypothesis, has now been demonstrated in many populations and animal models.[10] Furthermore, scientists now know that environmental factors (such as poor nutrition) experienced during infancy can also have long-lasting effects—many

via epigenetic alterations. Consequently, the concept of "fetal origins" was expanded to "developmental origins" to encompass both prenatal and postnatal effects on lifelong health. Considerable evidence links poor prenatal and postnatal nutrient availability to adult diseases such as cardiovascular disease, stroke, hypertension, type 2 diabetes, and obesity.[11] While the mechanisms by which very early nutrition can influence long-term health are not well understood, scientists think that, when nutrients are limited, fetal blood may be shunted to the brain to protect it, thus reducing blood flow to other organs. As a result, the course of normal fetal growth and development may be irreversibly altered during critical periods. Although this adaptive response may be beneficial to the fetus in the short term, it seems to have negative consequences during later stages of life. The long-term negative effects of fetal programming appear to be most detrimental when LBW infants subsequently experience rapid weight gain during infancy and childhood. This may be partly because rapid postnatal growth may exceed the functional capacity of underdeveloped physiological systems of LBW infants.

As scientists continue to study this phenomenon by which very early (even prenatal) nutrition influences long-term health, a better understanding of the physiologic mechanisms by which this occurs will become clearer. Undoubtedly, the emerging field of epigenetics may help elucidate the mechanisms by which prenatal and postnatal conditions influence gene expression, and how this can influence chronic disease susceptibility later in life.[12]

Ethel Margaret Burnside provided care to pregnant women in England's poorest communities. The developmental origins of health and disease concept was later developed using data she collected during the early 1900s. An excerpt from a ledger kept by Miss Burnside's nurses is shown here.

TABLE 14.1 Recommended Ranges for Total Weight Gain and Rate of Weight Gain during Pregnancy

Pre-Pregnancy Weight Classification	Body Mass Index (kg/m²)	Recommended Total Weight Gain Range (lbs)*	Recommended Rate of Weight Gain** Second and Third Trimester (lbs/wk)
Underweight	<18.5	28–40	1 (1–1.3)
Normal weight	18.5–24.9	25–35	1 (0.8–1)
Overweight	25.0–29.9	15–25	0.6 (0.5–0.7)
Obese	≥30.0	11–20	0.5 (0.4–0.6)

SOURCE: Institute of Medicine, National Research Council. Weight gain during pregnancy: Reexamining the guidelines. National Academies Press. Washington, DC. 2009.
Available online at: http://www.iom.edu/Reports/2009/Weight-Gain-During-Pregnancy-Reexamining-the-Guidelines.aspx.
* Weight-gain range for singleton pregnancies.
** Calculations assume a 0.5–2 kg (1.1–4.4 lbs) weight gain in the first trimester.

RECOMMENDED WEIGHT GAIN DEPENDS ON BMI

Because adequate weight gain during pregnancy is an important determinant of fetal growth and development, health care practitioners monitor weight gain very carefully throughout pregnancy. The current weight-gain guidelines for pregnant women, developed by the Institute of Medicine (IOM),[13] are based on maternal pre-pregnancy body mass index (BMI). These recommendations are summarized in Table 14.1. Women who gain the recommended amount of weight for their BMI range are likely to deliver full-term babies with a healthy birth weight. For example, healthy-weight women (pre-pregnancy BMI between 18.5 and 24.9 kg/m²) are advised to gain 25 to 35 pounds, whereas overweight women (pre-pregnancy BMI 25.0 to 29.9 kg/m²) are encouraged to gain less—between 15 and 25 pounds. Obese women (pre-pregnancy BMI ≥30.0 kg/m²) should gain between 11 and 20 pounds, whereas underweight women (pre-pregnancy BMI <18.5 kg/m²) are encouraged to gain between 28 and 40 pounds.

In addition to total weight gain, it is also important to monitor the rate (lbs/week) of weight gain. Whereas little weight gain is necessary during early stages of pregnancy, a steady gain of 3 to 4 pounds each month is recommended for healthy-weight women throughout the second and third trimesters. Rates of weight gain for each pre-pregnancy BMI category are listed in Table 14.1. While it is fully recognized that other factors such as age and ethnicity can impact pregnancy outcomes, these weight gain recommendations are meant to apply to pregnant teenagers as well as adults and to all racial and ethnic groups. Provisional weight-gain guidelines for women who are pregnant with twins have also been established. Healthy-weight women pregnant with twins should gain 37 to 54 pounds; overweight women, 31 to 50 pounds; and obese women, 25 to 42 pounds.

Components of Weight Gain Some women may have concerns regarding these recommended gains out of fear that the weight will be difficult to lose after the baby is born. However, this is typically not the case when women adhere to recommended weight gain allowances. Furthermore, women need to recognize that weight gain is associated with many components of pregnancy such as the fetus, breasts, uterus, and placenta. Thus, weight gain during pregnancy is not solely due to an increase in maternal fat stores. Table 14.2 lists the components of typical weight gain associated with a healthy pregnancy.

MATERNAL NUTRIENT AND ENERGY REQUIREMENTS CHANGE DURING PREGNANCY

Pregnancy results in dramatic changes in the mother's body that enable her to support and nurture her growing offspring. Many of these changes, which are

⟨**CONNECTIONS**⟩ Recall that body mass index (kg/m²) is a measure of adiposity (Chapter 8, page 336).

TABLE 14.2 Components of Typical Weight Gain during Pregnancy

Component	Approximate Weight Gain (lb) at 38 Weeks of Gestation
Fetus	7 to 8
Placenta	1½ to 2
Uterus and supporting structures	2½ to 3
Maternal adipose stores	7 to 8
Mammary tissue	1 to 2
Maternal extracellular fluids (blood and amniotic fluid)	6 to 7
Total weight gain	25 to 30

SOURCE: Institute of Medicine, National Research Council. Weight gain during pregnancy: Reexamining the guidelines. National Academies Press. Washington, DC. 2009. Available at: http://www.iom.edu/Reports/2009/Weight-Gain-During-Pregnancy-Reexamining-the-Guidelines.aspx.

listed in Table 14.3, affect nutrient requirements. Dietary recommendations for pregnant women are based on extensive research and are intended to promote optimal health in the mother and that of her unborn child. In addition to seeking regular prenatal care and guidance from health care providers, pregnant women may find the newly developed MyPlate food guidance system website (http://www.choosemyplate.gov) helpful in dietary planning. Based on a woman's age, stage of pregnancy, exercise habits, and pre-pregnancy weight, suggested dietary plans have been developed to ensure that pregnant women get the right types and amounts of all the food they need. The 2010 Dietary Guidelines for Americans also provide key recommendations that specifically address issues relevant to pregnant women, and these are highlighted in the Food Matters feature.

Recommended Energy Intake Adequate weight gain during pregnancy requires adequate energy intake. During pregnancy, additional energy is needed to support the growth of the fetus and placenta, as well as maternal tissues. Resting energy expenditure increases during pregnancy because of added

⟨**CONNECTIONS**⟩ Resting energy expenditure (REE) is the energy expended for resting metabolic activity over a 24-hour period (Chapter 8, page 332).

TABLE 14.3 Physiological Changes during Pregnancy

Cardiovascular System
- Heart enlarges slightly.
- Heart rate and cardiac output increase.
- Blood pressure decreases during the first half of pregnancy and returns to nonpregnant values during the second half of pregnancy.
- Plasma volume and red blood cell volume increase.
- Respiratory rate and oxygen consumption increase.

Gastrointestinal Tract and Food Intake
- Appetite increases.
- Senses of taste and smell are altered.
- Thirst increases.
- Gastrointestinal motility decreases.
- Efficiency of nutrient absorption increases.
- Gastroesophageal reflux becomes more common.

Renal System
- Kidney filtration rate increases.
- Sodium retention increases.
- Total body water increases.

Energy Metabolism and Energy Balance
- Basal metabolic rate (BMR) increases.
- Body temperature increases.
- Fat mass, lean mass, and body weight increase.

physiological demands on the mother. For example, the heart and lungs must work harder to deliver nutrients and oxygen to the fetus.

The energy demands of pregnancy are quite high—about 60,000 kcal over the course of the pregnancy. Although very little extra energy is needed during the first trimester, women are generally advised to increase their energy intake above pre-pregnancy Estimated Energy Requirements (EERs) by approximately 350 and 450 kcal/day during the second (14 to 26 weeks) and third (week 27 to the end of pregnancy) trimesters of pregnancy, respectively. For example, a woman with an EER of 2,000 kcal/day when she is not pregnant would require 2,350 kcal/day during her second trimester of pregnancy. During her third trimester of pregnancy, she would require approximately 2,450 kcal/day. Young or underweight women may need to increase their energy intake even more.

Recommended Macronutrient Intakes It is important for pregnant women to consume enough carbohydrates, protein, and fat to ensure that their energy needs are met. If the pregnancy is progressing normally, carbohydrates should remain the primary energy source (45 to 65% of total calories). The Recommended Dietary Allowance (RDA) for carbohydrate during pregnancy is 175 g/day, which provides adequate amounts of glucose for both the mother and the fetus. For most women, this increase represents approximately 45 g/day of additional carbohydrates, which is easily satisfied by eating 2–3 servings of carbohydrate-rich foods such as whole-grain breads or cereals.

During pregnancy, additional protein is needed for forming fetal and maternal tissues. The recommendation (RDA) is for pregnant women to increase protein intake by about 25 g/day, so that they consume approximately 70 g of total protein daily. This amount is easily obtained by eating a variety of high-quality protein sources such as meat, dairy products, and eggs. For example, 3 ounces of meat or 2 cups of yogurt provide approximately 25 g of protein. Consistent with the Acceptable Macronutrient Distribution Ranges (AMDRs), protein should continue to provide 10 to 35% of total calories.

Pregnant women who follow a vegan diet must plan their meals carefully to ensure adequate intake of essential amino acids. Plant foods that provide relatively high amounts of protein include tofu and other soy-based products and legumes such as dried beans and lentils. To make sure that all essential amino acids are consumed in adequate amounts, it is important to eat a variety of these and other protein-containing foods.

Dietary fat, also an important source of energy during pregnancy, should contribute approximately 20 to 35% of total calories during pregnancy. It is important to remember that the essential fatty acids serve other vital roles beyond the provision of energy. Linoleic acid and linolenic acid are important parent compounds used by the body to form other fatty acids and biologically active compounds. For example, linoleic acid is converted to arachidonic acid, whereas linolenic acid is converted to eicosapentaenoic acid (EPA) and docosahexaenoic acid (DHA). While all these fatty acids are critical for fetal growth and development, DHA is particularly important for brain development and formation of the retina. Although there are no DRI values for total fat during pregnancy, AIs have been established for the essential fatty acids. During pregnancy, the AIs for linoleic and linolenic acids are 13 and 1.4 g/day, respectively.

Dietary Sources of Essential Fatty Acids To ensure adequate intake of essential fatty acids during pregnancy, women should eat fish several times a week and/or use omega-3-rich oils (such as canola or flaxseed oil). Because some types of fish contain high levels of mercury, the Food and Drug Administration (FDA) and the Environmental Protection Agency (EPA) advise pregnant women to limit their consumption of fish that might contain low levels of mercury (salmon, tuna, sardines, and mackerel) and to avoid eating certain types of fish (shark, swordfish,

⟨CONNECTIONS⟩ Recall that the body is unable to make the essential fatty acids linoleic acid and linolenic acid, and therefore these fatty acids must be supplied by the diet. Docosahexaenoic acid (DHA) is derived from linolenic acid (Chapter 6, page 225).

king mackerel, and tilefish) thought to have high levels of mercury.[14] However, the importance of an adequate intake of omega-3 fatty acids during pregnancy cannot be overstated. Although pregnant women should heed the advisory issued by FDA and EPA to limit or avoid certain types of fish, it is important for them to consume other foods that are rich in omega-3 fatty acids to ensure optimal fetal and infant development. The 2010 Dietary Guidelines for Americans recommend that pregnant women consume 8 to 12 ounces of seafood per week from a variety of seafood types, while limiting white (albacore) tuna to 6 ounces per week and avoiding tilefish, shark, swordfish, and king mackerel.

Recommended Micronutrient Intakes In addition to meeting increased energy and macronutrient requirements, it is important for women to have adequate intakes of vitamins and minerals. Vitamins and minerals are needed for the formation of maternal and fetal tissues and for many reactions involved in energy metabolism. With few exceptions (calcium, phosphorus, fluoride, and vitamins D, E, and K), the requirements for most micronutrients increase during pregnancy; recommended intakes for pregnant women are listed inside the front cover of this book.

Most dietary recommendations during pregnancy are intended for women who generally consume a variety of foods from the different food groups. However, dietary restrictions can mean that some pregnant women may need to plan their diets more carefully so that all of their nutritional needs are being met. This is particularly true for pregnant women who are vegan vegetarians. Because vegans consume no foods of animal origin, they must make a special effort to get adequate amounts of vitamin B_{12}, vitamin B_6, iron, calcium, and zinc. Additional servings of whole grains, legumes, nuts, and calcium-fortified foods such as tofu and soy milk can help to satisfy these requirements. Foods that provide vitamin B_{12}, such as nutritional yeast and vitamin B_{12}-fortified cereals, are also vital.

⟨**CONNECTIONS**⟩ Recall that vegan vegetarians do not eat animal-derived foods, including meat, dairy products, and eggs (Chapter 5, page 192).

Vitamin A Although requirements for most vitamins and minerals increase during pregnancy, vitamin A is an exception. Because excessively high intakes of preformed vitamin A can be teratogenic, pregnant women should not exceed the Upper Intake Level (UL; 3,000 µg/day) from either foods or supplements. Foods that contain large amounts of preformed vitamin A include beef liver and chicken liver. However, it is not necessary for pregnant women to limit their intake of beta-carotene or other provitamin A carotenoids found mostly in plant foods.

Calcium Surprisingly, the recommended intake of calcium does not increase during pregnancy. Although extra calcium is needed for the fetus to grow and develop properly, changes in maternal physiology are able to accommodate these needs without increasing dietary intake. For example, calcium absorption increases and urinary calcium loss decreases. Therefore, the RDA for calcium during pregnancy (1,000 mg/day) is the same as that for nonpregnant women under 50 years of age.

Iron Because iron is needed for the formation of hemoglobin and the growth and development of the fetus and the placenta, the RDA for iron increases substantially during pregnancy, from 18 to 27 mg/day. Most well-planned diets provide women with approximately 15 to 18 mg of iron daily. For example, 6 ounces of beef provides 2 to 3 mg of heme iron, whereas 1 cup of iron-fortified breakfast cereal typically has 5 to 8 mg of nonheme iron. However, the bioavailability of iron in meat (heme) is higher than that in plants (nonheme). Many pregnant women have difficulty meeting the recommended intake for iron by diet alone. In addition, about 12% of women enter pregnancy with impaired iron status. For these reasons, iron supplementation is often encouraged during the second and third trimesters of pregnancy, when iron requirements are the greatest.[15] Still, recommendations for or against routine iron supplementation for non-anemic women during pregnancy remain unclear, as there is little evidence that it improves clinical outcomes for the mother or newborn.[16]

Folate Adequate folate intake is especially important during pregnancy. Recall that folate is critical for cell division, and therefore is needed for the development of all tissues including those of the nervous system. Women with poor folate status before or in early pregnancy are at increased risk of having a baby with a neural tube defect (NTD). Remember that a neural tube defect is a specific type of birth defect that affects the spinal cord or brain. As the critical period for the formation of the neural tube occurs early in pregnancy—21 to 28 days after conception—the neural tube may already be formed before a woman realizes she is pregnant. Because of the importance of folate in neural tube development, an RDA of 600 µg DFE (dietary folate equivalents)/day has been set for pregnancy. Women capable of becoming pregnant are advised to consume 400 µg DFE/day of folic acid as a supplement or in fortified foods in addition to consuming folate naturally occurring in foods. Examples of folate-rich foods include dark green leafy vegetables, lentils, orange juice, and enriched cereal grain products.

Recognizing the importance of folate in preventing NTDs, the FDA began to require folate fortification of all "enriched" cereal grain products in 1996. In addition, food manufacturers are allowed to make health claims on appropriate food labels stating that adequate intake of dietary folate or folic acid supplements may reduce the risk of NTDs. Since this nationwide effort, folate status in the United States has improved, and the incidence of NTDs has decreased.[17] While folate fortification efforts may have been successful at decreasing the incidence of NTDs, not all NTDs can be prevented by increased folate intake. Rather, some NTDs are multifactorial disorders, meaning that they are caused by a combination of environmental and genetic factors.

MATERNAL SMOKING IS HARMFUL TO THE FETUS

Smoking during pregnancy poses many risks to the unborn. There is overwhelming evidence that it increases the risk of having a preterm or LBW baby.[18] Not only is it important for pregnant women to not smoke, some studies show that secondhand smoke can also harm the fetus.[19]

In addition to the chemical compounds found in tobacco that can harm the fetus and placenta, smoking also causes maternal blood vessels to constrict, reducing blood flow from the uterus to the placenta. As a result, nutrient and oxygen availability to the fetus decreases. Smoking also increases the risk of premature detachment of the placenta, which can result in a miscarriage. If a woman smokes, she is advised to quit as early in the pregnancy as possible.

STAYING HEALTHY DURING PREGNANCY

Every pregnancy is unique. While some women do not experience much unease at all, others experience a wide variety of pregnancy-related discomforts. Pregnancy entails both physical and emotional changes, and implementation of a few simple dietary and lifestyle changes can minimize many pregnancy-related discomforts.

Pregnancy-Related Physical Complaints Hormonal changes associated with pregnancy are believed to be the underlying cause of several common physical complaints such as morning sickness, fatigue, heartburn, constipation, and food cravings or aversions. Many of these discomforts occur during the early stages of pregnancy, whereas others may persist throughout. In most cases, these discomforts are not serious and can typically be managed with simple diet-related

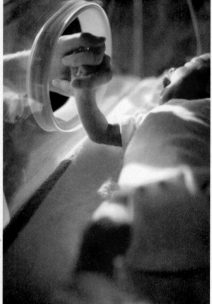

Women who smoke during pregnancy increase their risk of having a baby born prematurely. Premature birth is a leading cause of death among newborns and a major cause of long-term disability.

strategies. For example, some women who experience morning sickness, a condition characterized by queasiness, nausea, and vomiting, can find relief by avoiding foods with offensive odors. Other strategies include eating dry toast or crackers, eating small, frequent meals, and eating before getting out of bed in the morning. Heartburn, a common complaint during pregnancy, can often be managed by avoiding spicy or greasy foods, and waiting at least two to three hours after eating before lying down. To alleviate constipation, another common pregnancy-related complaint, women are advised to consume adequate amounts of fruits, vegetables, whole grains, and fluids.

Food Cravings and Aversions Food cravings and food aversions are also common during pregnancy. Powerful urges to consume or avoid certain foods may be caused by hormone-induced heightened senses of taste and smell. While most food cravings and aversions rarely pose serious problems during pregnancy, some expectant mothers develop powerful desires to consume nonfood items such as laundry starch, clay, soil, and burnt matches. The urge to consume nonfood items, called **pica,** has no known cause and can be potentially harmful to the mother and baby. Recall from chapter 13 that some researchers believe that pica may be related to iron deficiency.

PREGNANCY-RELATED HEALTH CONCERNS

While most minor physical discomforts associated with pregnancy are considered normal, it is important for all expectant women to be aware of changes that could indicate a more serious problem. Two common health concerns that can develop during the later stages of pregnancy are gestational diabetes and pregnancy-induced hypertension.

Gestational Diabetes Some women develop a form of diabetes called gestational diabetes during pregnancy—usually around 28 weeks or later. **Gestational diabetes** occurs when pregnancy-related hormonal changes cause maternal cells to become less responsive to insulin, triggering blood glucose levels to rise. To test for gestational diabetes, most pregnant women are given a routine blood test during the third trimester of pregnancy. Once diagnosed, a healthy diet and exercise regimen can help keep blood glucose levels under control. Some pregnant women who have difficulty controlling blood glucose may require insulin injections, however. Although gestational diabetes disappears within six weeks after delivery, approximately half of all women with gestational diabetes develop type 2 diabetes later in life.[20] To minimize this risk, it is especially important for a woman with a history of gestational diabetes to maintain a healthy weight, make sound food choices, and be physically active.

Pregnancy-Induced Hypertension Pregnancy-induced hypertension (also called pre-eclampsia or toxemia of pregnancy) is another serious complication that can develop during the later stages of pregnancy. **Pregnancy-induced hypertension,** which affects about 3% of pregnancies, is characterized by high blood pressure, sudden swelling and weight gain due to fluid retention, and protein in the urine.[21] If pregnancy-induced hypertension is suspected, women are typically advised to rest and limit their daily activities. Although most women who develop pregnancy-induced hypertension deliver healthy babies, some are not as fortunate. In some cases, a woman's blood pressure can increase to dangerously high levels—a condition referred to as eclampsia. When this occurs, the only effective treatment is delivery of the baby.

You now understand the important role that nutrition plays before and during pregnancy. However, a healthy diet is equally important during lactation. Successful lactation requires careful dietary planning to provide the energy and nutrient intakes needed to satisfy the nutritional requirements of both the mother and infant. The physiological changes associated with milk production and how

pica A desire to consume nonfood items such as laundry starch, clay, soil, and burned matches.

gestational diabetes A form of diabetes resulting from hormone-related changes during pregnancy.

pregnancy-induced hypertension (also called pre-eclampsia or toxemia of pregnancy) A form of pregnancy-related hypertension characterized by high blood pressure, sudden swelling and weight gain due to fluid retention, and protein in the urine.

FIGURE 14.6 Comparison of Recommended Energy and Nutrient Intakes for Nonpregnant, Pregnant, and Lactating Women

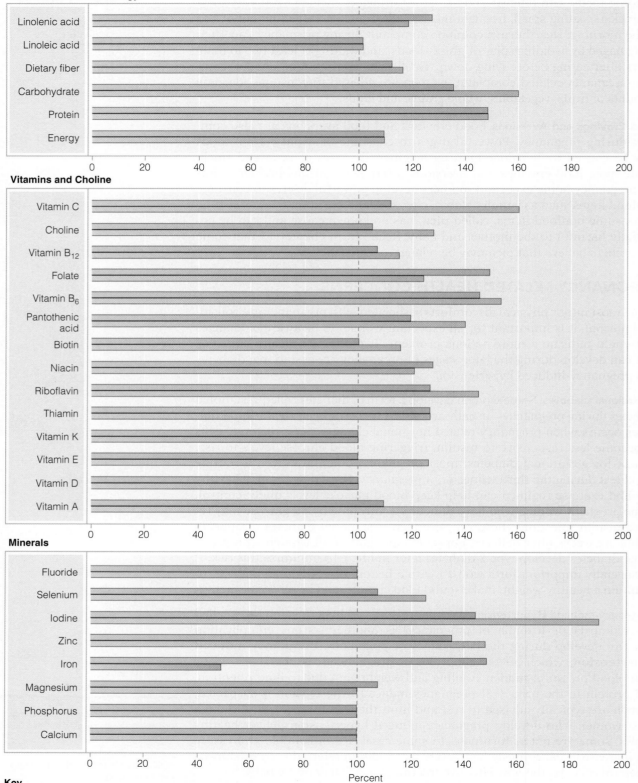

Macronutrients and Energy
- Linolenic acid
- Linoleic acid
- Dietary fiber
- Carbohydrate
- Protein
- Energy

Vitamins and Choline
- Vitamin C
- Choline
- Vitamin B$_{12}$
- Folate
- Vitamin B$_6$
- Pantothenic acid
- Biotin
- Niacin
- Riboflavin
- Thiamin
- Vitamin K
- Vitamin E
- Vitamin D
- Vitamin A

Minerals
- Fluoride
- Selenium
- Iodine
- Zinc
- Iron
- Magnesium
- Phosphorus
- Calcium

Percent

Key
■ Recommended intake values for the second trimester of pregnancy for women 19–30 years of age
■ Recommended intake values for the first 6 months of lactation for women 19–30 years of age
-- Recommended intake values for nonpregnant/nonlactating women 19–30 years of age

SOURCE: Adapted from the Dietary Reference Intakes series, National Academies Press. Copyright 1997, 1998, 2000, 2001, 2002, 2004, 2005, 2011 by the National Academy of Sciences.

this process, called lactation, impacts the nutrient requirements of women are discussed next. Recommended nutrient intakes for non-pregnant, pregnant, and lactating women are compared in Figure 14.6.

Why Is Breastfeeding Recommended during Infancy?

Women often experience noticeable changes in the size and shape of their breasts while pregnant. These changes are necessary to prepare the breasts (also called mammary glands) for milk production after the baby is born. Human milk is the ideal food for nourishing babies. Not only does human milk support optimal growth and development during infancy and early childhood, but evidence also shows that the benefits associated with breastfeeding may extend to later stages of life. In addition, human milk provides immunologic protection against pathogenic viruses and bacteria. It is not surprising that breastfed babies tend to be sick less often than babies fed formula. Moreover, breastfeeding is beneficial to the mother, decreasing the risk of certain diseases and helping women return to their prepregnant weight more easily. As with pregnancy, adequate nutrition is important during lactation, when the woman is nourishing herself and producing milk to feed her baby.

iStockphoto.com/endopack

Most women find it convenient to breastfeed their babies. It is recommended that mothers breastfeed on demand (unrestricted breastfeeding).

LACTATION IS THE PROCESS OF MILK PRODUCTION

During pregnancy, many hormones prepare the mammary glands for milk production. In particular, the hormones estrogen and progesterone stimulate an increase in the number of milk-producing cells and an expansion of ducts that transport milk out of the breast. As illustrated in Figure 14.7, milk production takes place in specialized structures of the breast called **alveoli.** Each **alveolus** (the singular of alveoli) is made up of milk-producing secretory cells. Milk is formed when the secretory cells release the various milk components into the hollow center (lumen) of the alveolus. From there, the milk is released into a

alveolus (plural, *alveoli*) A cluster of milk-producing cells that make up the mammary glands.

FIGURE 14.7 Anatomy and Physiology of the Human Breast
The human breast, or mammary gland, is a complex organ composed of many different types of tissues that produce and secrete milk.

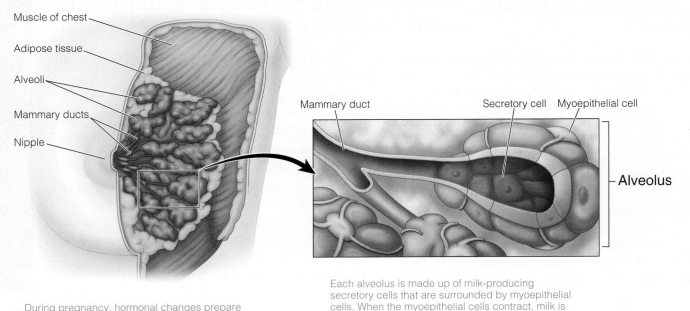

Muscle of chest

Adipose tissue

Alveoli

Mammary ducts

Nipple

Mammary duct

Secretory cell

Myoepithelial cell

Alveolus

During pregnancy, hormonal changes prepare the breast for the production of milk.

Each alveolus is made up of milk-producing secretory cells that are surrounded by myoepithelial cells. When the myoepithelial cells contract, milk is forced out of the alveoli and into the mammary ducts.

network of **mammary ducts** that eventually lead to the nipple. After the birth of the child, the mammary glands begin to produce milk via a process called **lactation.** Women are encouraged to nurse their babies soon after delivery because suckling initiates the process of **lactogenesis,** the onset of milk production.

Prolactin and Oxytocin Regulate Milk Production While the hormones estrogen and progesterone prepare the mammary glands for milk production, it is the hormones **prolactin** and **oxytocin** that actually regulate milk production and the release of milk from alveoli into the mammary ducts. Suckling stimulates nerves in the nipple that signal the hypothalamus. In turn, the hypothalamus signals the pituitary gland to release both of these hormones (Figure 14.8). Prolactin stimulates the secretory cells in the mammary gland to synthesize milk, whereas oxytocin causes small muscles (made of specialized cells called **myoepithelial cells**) that surround the alveoli to contract. These muscular contractions force the milk out of the alveoli and into the mammary ducts. This active release of milk is called **milk let-down.** As the baby suckles, the milk moves through the mammary ducts, toward the nipple, and into the baby's mouth. Anxiety, stress, and fatigue can interfere with the milk let-down reflex, sometimes making breastfeeding challenging. For this reason, it is important for all women to seek help if they experience problems with breastfeeding.

MILK PRODUCTION IS A MATTER OF SUPPLY AND DEMAND

Whereas milk production is regulated by many physiological factors, the amount of milk produced is determined largely by how much milk the infant consumes. Women who exclusively breastfeed, meaning that human milk is the sole source of infant feeding, produce more milk than those who supplement breastfeeding with infant formula or other foods. On average, women produce 26 ounces (3⅓ cups) of milk per day during the first 6 months postpartum, and 20 ounces (2½ cups) per day in the second 6 months. The reason women produce less milk during the second six months is that most infants are also fed supplemental foods at this age. Because newborns have small

mammary duct Structure that transports milk from the alveolar secretory cells toward the nipple.

lactation The production and release of milk.

lactogenesis (lac – to – GEN – e – sis) The onset of milk production.

prolactin (pro – LAC – tin) A hormone, produced in the pituitary gland, which stimulates the production of milk in alveoli.

oxytocin (ox – y – TO – sin) A hormone, produced by the hypothalamus and stored in the pituitary gland, which stimulates the movement of milk into the mammary ducts.

myoepithelial cells Muscle cells that surround alveoli and contract, forcing milk into the mammary ducts.

milk let-down The movement of milk through the mammary ducts toward the nipple.

FIGURE 14.8 Neural and Hormonal Regulation of Lactation Milk production is regulated by "supply and demand."

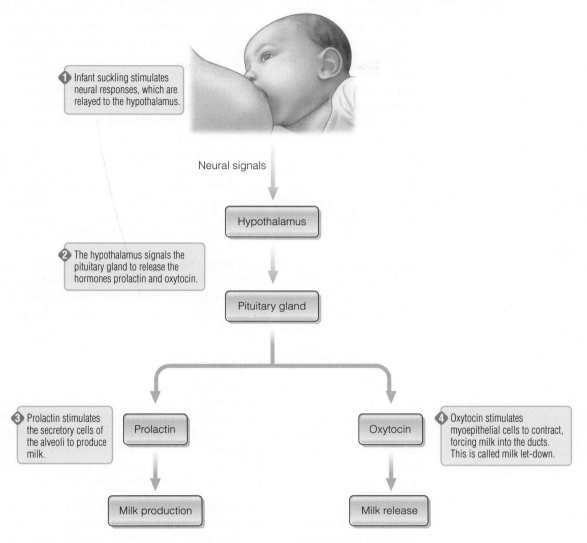

① Infant suckling stimulates neural responses, which are relayed to the hypothalamus.

Neural signals

Hypothalamus

② The hypothalamus signals the pituitary gland to release the hormones prolactin and oxytocin.

Pituitary gland

③ Prolactin stimulates the secretory cells of the alveoli to produce milk.

Prolactin

④ Oxytocin stimulates myoepithelial cells to contract, forcing milk into the ducts. This is called milk let-down.

Oxytocin

Milk production

Milk release

stomachs and can only consume small amounts of milk at each feeding, many mothers breastfeed as frequently as every two to three hours. However, as the baby grows, more milk can be consumed at each feeding, reducing the need to breastfeed as frequently. The American Academy of Pediatrics (AAP) recommends that women nurse their newborns at least 8 to 12 times each day and feed "on demand" rather than scheduling their baby's feedings.[22]

Breastfeeding requires proper positioning of the baby at the breast and the ability of the baby to latch onto the nipple. Once this occurs, the baby rhythmically coordinates sucking and swallowing. Sometimes mothers are concerned about whether their infants are receiving enough milk. The best indicators of adequate milk production are healthy infant weight gain and appropriate frequency of wet diapers. As a guideline, parents should expect at least four to five wet diapers per day.[22] Under most conditions, a baby needs no additional source of nourishment other than human milk for the first 4 to 6 months of life.[22] Most problems associated with breastfeeding can be resolved with the assistance of a medical professional such as a pediatrician or lactation specialist.

HUMAN MILK IS BENEFICIAL FOR BABIES

Experts agree that human milk is the best possible food for almost all babies, and these benefits are immediate and long lasting. Shortly after the mother gives birth, her breasts produce a special substance called **colostrum,** which nourishes the newborn and helps protect the baby from disease. This is, in part, because colostrum contains an abundance of immunoglobulins, which are proteins that fight infections. Over the next several days, there is a gradual transition from the production of colostrum to that of mature milk. Like colostrum, mature human milk has many important immunological benefits. For example, some immunological components help inhibit the growth of pathogens, whereas others promote the development of the infant's immature GI tract. This may help reduce the infant's risk of developing food allergies later in life. Not only does breastfeeding lessen the incidence of infectious diseases, but some studies also suggest that infants fed human milk are less likely to develop type 1 and type 2 diabetes, certain types of cancer, asthma, and obesity in later life.[23] Although it is clear that breastfed babies gain weight more slowly during the first year of life than formula-fed infants, it is less certain that these differences carry over to adolescence and adulthood.[24] However, patterns of rapid weight gain early in life may be related to increased risk for being overweight or obese later. [25]

Nutrient Composition of Human Milk The nutrient composition of human milk perfectly matches the nutritional needs of the infant. In addition to nutrients, human milk also contains enzymes and other compounds that make certain nutrients easier to digest and absorb. For example, the presence of the enzyme lipase in the milk helps the infant digest triglycerides. In addition, human milk contains lactoferrin, which enhances iron absorption and has antimicrobial activity that protects the infant from certain infectious diseases.

Protein Human milk has a relatively low protein content compared with that of cow milk, and is considered ideal for infant growth and development. The high protein content of cow milk makes it unsuitable for infants. Although cow milk is used to make many types of infant formulas, the protein content has been adjusted. The protein in human milk is not only present in the right amount, but is also easily digested, and its amino acid profile is uniquely suited to support optimal growth and development during infancy.

Lipids More than half the calories in human milk come from lipids. Some lipids are synthesized by the mammary gland, and others come from the maternal diet. Although there are dozens of fatty acids in human milk, scientists are particularly interested in DHA because of its important role in brain and eye development during infancy. Some studies have shown improved cognitive performances in breastfed infants compared with formula-fed infants.[26] Only in the last few years have some manufacturers started adding DHA to infant formula. Human milk also contains high amounts of cholesterol, which is an important component of cell membranes. Currently, cholesterol is not added to infant formula, but researchers are interested in understanding whether it should be included.

Carbohydrates The primary carbohydrates in human milk are lactose and oligosaccharides. Not only is lactose an important source of energy, but it also facilitates the absorption of other nutrients such as calcium, phosphorus, and magnesium. Although they cannot be digested by the baby and therefore do not provide a source of energy, oligosaccharides present in human milk may have functional roles such as inhibiting the growth of harmful bacteria in the infant's GI tract.

LACTATION INFLUENCES MATERNAL ENERGY AND NUTRIENT REQUIREMENTS

Milk production requires energy, and for this reason maternal energy requirements increase during lactation. The amount of additional energy needed by the mother during lactation depends on whether she is exclusively breastfeeding or

⟨**CONNECTIONS**⟩ Recall that oligosaccharides are carbohydrates that consist of 3 to 10 monosaccharides (Chapter 4, page 121).

colostrum (co – LOS – trum) The first secretion from the breasts after birth that provides nourishment and immunological protection to newborns.

feeding a combination of human milk and infant formula. Because women tend to produce more milk during the first 6 months of lactation compared with the second 6 months, additional energy required for milk production is approximately 500 and 400 kcal/day, respectively, during these periods. However, some of the energy needed for milk production during the first 6 months should come from the mobilization of maternal body fat that was stored during pregnancy. In other words, it is expected that breastfeeding women will be in negative energy balance during the first 6 months postpartum. This is why dietary energy recommendations during the first 6 months of lactation are lower than during the second 6 months. As the following formulas show, energy intake recommendations for lactation are based on three factors: energy required for nonpregnant, nonlactating state, energy required for milk production, and energy mobilized from body fat stores.

Estimated Energy Requirement (EER) Calculations for Lactation

- 0 to 6 months postpartum = Adult EER + energy required for milk production − energy mobilized from body fat

  = Adult EER + 500 kcal/day − 170 kcal/day

  = Adult EER + 330 kcal/day

- 6 to 12 months postpartum = Adult EER + energy required for milk production

  = Adult EER + 400 kcal/day

Thus, the Institute of Medicine (IOM) recommends an additional 330 kcal/day above nonpregnant requirements during the first 6 months of lactation and an additional 400 kcal/day in the second 6 months. Equations for calculating adult energy requirements (Estimated Energy Requirements; EERs) are provided in Appendix B.

Recommendations for micronutrient intakes during lactation are generally similar to those during pregnancy, although some (such as vitamin A) are somewhat higher and others (such as folate and iron) are lower. These are listed inside the front cover of this book. Of special interest is the recommendation for vitamin C, which increases substantially in lactation. Largely because a relatively high amount of vitamin C is secreted in milk, the RDA increases from 85 mg/day during pregnancy to 120 mg/day during lactation. In addition, it is important that lactating women take in enough fluids. Although increased fluid intake beyond basic needs does not increase milk production, a lack of fluid can decrease milk volume. The AI for total fluids for lactating women is 3.8 liters (13 cups) per day, which includes fluids in both foods and beverages.

BREASTFEEDING IS BENEFICIAL FOR MOTHERS

Most people are aware that breastfeeding is beneficial for infants, but fewer know that it is also beneficial for mothers. For example, breastfeeding shortly after giving birth stimulates the uterus to contract, helping minimize blood loss and shrink the uterus to its pre-pregnancy size. Some women also find that breastfeeding helps them return to their pre-pregnancy weight more easily.[27]

Breastfeeding is also associated with several long-term maternal health benefits. For example, it can delay the return of menstrual cycles. The span of time between the birth of the baby and the first menses, called **postpartum amenorrhea,** allows iron stores to recover, and reduces the likelihood that the mother will become pregnant again too soon. Although the duration of postpartum amenorrhea varies greatly, menstruation tends to resume around

postpartum amenorrhea (a – men – or – RHE – a) The span of time between the birth of the baby and the return of menses.

6 months postpartum in breastfeeding women and around 6 weeks postpartum in nonbreastfeeding women. Because it is possible to ovulate without menstruating, women are encouraged to use contraception until they wish to become pregnant again. In addition to prolonging postpartum amenorrhea, breastfeeding also reduces a woman's risk of developing breast cancer, ovarian cancer, and possibly osteoporosis later in life.[28]

What Are the Nutritional Needs of Infants?

The rate of growth and development during the first year of life is astonishing. At no other time in the human lifespan, aside from the prenatal period, do these processes occur so rapidly. The transition from breastfeeding to eating baby food can be challenging, and it is important for parents to be aware of signs that indicate readiness for this next phase. During infancy, providing a diet that contains all the essential nutrients and an environment that is safe, secure, and engaging helps build a solid foundation for the remainder of life.

INFANT GROWTH IS ASSESSED USING GROWTH CHARTS

During the first year of life, a healthy baby's weight almost triples, and length increases by up to 50%. At the same time, major developmental milestones such as walking and self-feeding take place. Infant growth during the first year of life follows a fairly predictable pattern. For example, within a few days after birth, most healthy, full-term infants tend to lose up to 5 to 6% of their body weight, although they usually regain this weight within two weeks. This early weight loss is attributed mostly to loss of fluids. By 4 to 6 months, infant weight doubles and length increases by 20 to 25%. Growth rates then decrease slightly from 6 to 12 months. Equations used to estimate energy requirements during the first year of life are based in part on body weight and can be found in Appendix B.

Growth Charts Infant growth and development are carefully monitored throughout the first year of life. Important measures such as weight, length, and head circumference are routinely recorded on growth charts. Until recently, growth charts developed by the National Center for Health Statistics in collaboration with the National Center for Chronic Disease Prevention and Health Promotion had been used to assess infant growth. However, the Centers for Disease Prevention and Health Promotion now recommend that clinicians use newly developed international growth charts compiled by the World Health Organization (WHO) for infants and children aged under 24 months.[29] These new standards are based on a large, healthy, globally diverse population of breastfed infants (Figure 14.9). Data collected for these growth charts show that feeding practices and environmental influences are more likely to influence a child's growth than genetics and ethnicity. The WHO believes that use of these charts will enable practitioners to better identify growth-related conditions such as under- and overnutrition worldwide.[1]

Growth charts are divided into grids that represent the distribution of weight, length, and head circumference of a representative sample of healthy infants. For instance, an infant in the 60th percentile for weight at a given age is heavier than 60% of the reference population of infants. Growth charts enable practitioners to monitor and assess infant growth over time. Because infants tend to follow a consistent growth pattern, a dramatic change in percentile could indicate a problem. For example, an infant who drops from the

FIGURE 14.9 Growth during the First Two Years of Life Weight, length, and head circumference can be monitored during infancy and childhood using growth charts. Healthy growth trajectories for weight-for-length-for-age and head circumference-for-age during the first two years are shown here.

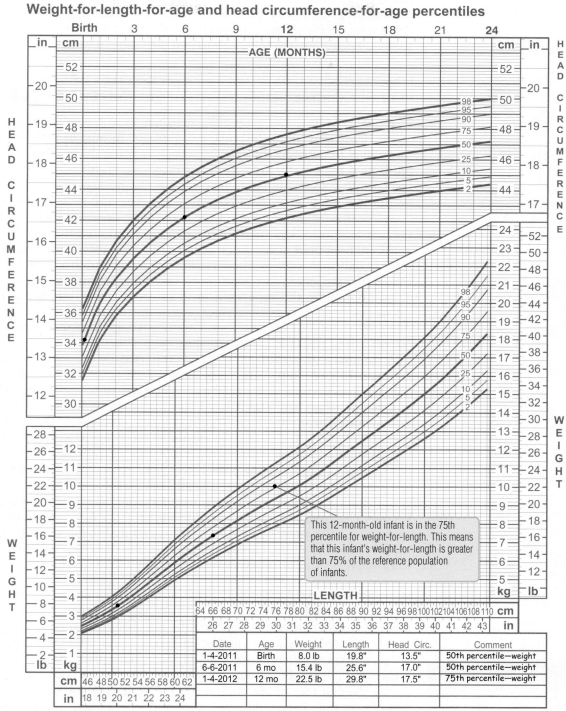

Birth to 24 months: Girls
Weight-for-length-for-age and head circumference-for-age percentiles

Date	Age	Weight	Length	Head Circ.	Comment
1-4-2011	Birth	8.0 lb	19.8"	13.5"	50th percentile—weight
6-6-2011	6 mo	15.4 lb	25.6"	17.0"	50th percentile—weight
1-4-2012	12 mo	22.5 lb	29.8"	17.5"	75th percentile—weight

This 12-month-old infant is in the 75th percentile for weight-for-length. This means that this infant's weight-for-length is greater than 75% of the reference population of infants.

SOURCE: WHO Child Growth Standards (http://www.who.int/childgrowth/en)

50th percentile for weight at birth to the 10th percentile for weight at 6 months is growing more slowly than expected. This can indicate that something is wrong and that the child should be evaluated. Poor breastfeeding technique, for example, can result in an infant not receiving adequate amounts of energy and nutrients, thus slowing or delaying growth.

Infant Development In addition to growth, many developmental changes occur during infancy. For example, newborns have little control over their bodies. However, in the first few months of life, babies begin to vocalize and are even able to return a friendly smile. Improved muscle control allows them to hold their heads steady, and by 6 months most infants can sit upright with support. These and other developmental milestones affect how and what babies should be fed. Within this first year of life, infants progress from a diet that consists solely of human milk and/or infant formula to being able to feed themselves a variety of foods.

DEVELOPMENTAL STAGES PROVIDE THE BASIS FOR RECOMMENDED INFANT FEEDING PRACTICES

In the late 1950s, fewer than 20% of babies were breastfed, partly because of the belief that infant formula was superior to human milk. Fortunately, this trend has now reversed. Over the past 30 years, the number of women in the United States who breastfeed has increased steadily.[30] This is, in part, a response to compelling scientific evidence that human milk is ideally suited for optimal infant growth and development. The AAP recommends exclusive breastfeeding for a minimum of the first 4 months of life, but preferably for the first 6 months, and the continuation of mixed breastfeeding for the second 6 months until at least one year.[31] Endorsing this recommendation is a recent "Call to Action to Support Breastfeeding" published in 2011 by the U.S. Surgeon General. In addition to endorsing exclusive breastfeeding for at least 6 months, this report put forth 20 action items designed to help U.S. women exclusively breastfeed their infants for this length of time. It is hoped that implementation of these action items will help the United States meet the breastfeeding goals outlined in Healthy People 2020.[32]

A summary of breastfeeding trends reported by the CDC indicates that, although 75% of women currently initiate breastfeeding, only 43% and 22% are still breastfeeding at 6 and 12 months postpartum, respectively. Only 13% of U.S. infants are still exclusively breastfed at 6 months.[33] Common reasons why women discontinue breastfeeding are the "perception" that milk production is inadequate and difficulties related to returning to work. However, with support and education, many women are able to overcome these obstacles. For example, understanding proper storage of expressed milk enables women to be away from their babies while still maintaining their milk supply. One organization that provides women with lactation assistance is the La Leche League. This international nonprofit group provides information and encouragement to all women who want to breastfeed their babies. While the majority of women are able to breastfeed their babies, it is important not to make someone feel guilty or embarrassed if she is not successful.

Infant Formula—An Alternative to Human Milk Although breastfeeding is the preferred method of infant feeding, there are times when it is not recommended, such as when the mother is taking chemotherapeutic drugs to treat cancer, infected with human immunodeficiency virus (HIV), using illicit drugs, or has untreated tuberculosis. The only acceptable alternative to human milk is commercial infant formula. Infants should not be fed cow milk at any time during the first year of life. As you can see from Table 14.4, the nutrient content of cow milk is very different from human milk and infant formula. The most commonly used infant formulas are derived from cow milk or soybeans. Both of these types of formula provide infants with the essential nutrients and energy needed to support growth and development. To reduce the risk of iron deficiency anemia, pediatricians recommend that parents feed their infants iron-fortified infant formula.[34] It is also important for parents to consider infant formulas that has been fortified with DHA and arachidonic acid. However, there is some debate whether adding these fatty acids to infant formula

There are many different types of infant formula to choose from.

TABLE 14.4 Nutrient/Energy Composition of Human Milk, Infant Formula, and Cow Milk (per 5 oz)

Component	Human Milk[a]	Cow Milk–Based Formula (Similac™)[b]	Soy-Based Formula (Prosobee™)[b]	Whole Cow Milk[c]
Energy (kcal)	105	100	96	93
Protein (g)	2	2	3	5
Fat (g)	6	6	5	5
Carbohydrate (g)	9	10	9	7
Cholesterol (mg)	21	3	0	15
Iron (g)	0.1	1.8	1.8	0
Calcium (mg)	45	81	90	172
Vitamin A (IU)	331	299	375	247
Vitamin D (IU)	3	59	65	78
Vitamin C (mg)	6	9	8	0
Folate (µg)	7	15	14	8

[a] Picciano MF. Representative values for constituents of human milk. Pediatric Clinics of North America. 2001;48; 1–3.
[b] Infant formula, with iron, ready to serve. Data from U.S. Department of Agriculture, Agricultural Research Service. 2010. USDA National Nutrient Database for Standard Reference, Release 23. Nutrient Data Laboratory home page. Available from http://www.ars.usda.gov/ba/bhnrc/ndl.
[c] Milk, whole, 3.25% milkfat. Data from U.S. Department of Agriculture, Agricultural Research Service. 2010. USDA National Nutrient Database for Standard Reference, Release 23. Nutrient Data Laboratory home page. Available from http://www.ars.usda.gov/ba/bhnrc/ndl.

actually has beneficial effects on the mental and psychomotor development in infants.[35] Currently, it is not mandatory for manufacturers to add DHA and arachidonic acid to infant formula. Infant formula is typically available as a powder and as a liquid concentrate. It is imperative that formula be prepared in a safe manner and that bottles and nipples be clean.

NUTRIENT SUPPLEMENTATION RECOMMENDATIONS ARE BASED ON WHETHER THE INFANT IS BREASTFED OR FORMULA FED

Are human milk or infant formula and complementary "baby foods" adequate to be the sole sources of nutrition during the first years of life, or should babies also be given nutrient supplements? The answer to this question depends on whether the infant is breastfed or formula fed, and the infant's age. Supplemental vitamin D, fluoride, iron, and fluids are sometimes recommended for infants, but the decision to use these supplements should be discussed with a health practitioner. Recommendations regarding nutrient supplementation are summarized in Table 14.5, and discussed next.

TABLE 14.5 Recommended Nutrient Supplementation during Infancy (0–12 months)

	Infant-Feeding Method	
Nutrient	**Human Milk (Exclusive)**	**Infant Formula**
Vitamin D	• Vitamin D supplements (400 IU/day) recommended beginning in the first few days of life and continuing until the infant is consuming at least 16 oz of infant formula per day	• Not needed if infant is consuming at least 16 oz formula per day, because all infant formulas manufactured in the United States meet standards for vitamin D
Fluoride	• Fluoride supplements (0.25 mg/day) recommended starting at age 6 months if local water has a fluoride concentration less than 0.3 ppm	• Not needed if formula is prepared with local water that has a fluoride concentration of at least 0.3 ppm. Fluoride supplements (0.25 mg/day) recommended starting at age 6 months if local water has a fluoride concentration less than 0.3 ppm
Iron	• Iron supplements (1 mg/day per kilogram of body weight) recommended for infants exclusively breastfed during the second 6 months of life	• Iron supplements (1 mg/day per kilogram of body weight) recommended for infants not fed iron-fortified infant formula
Fluids	• Additional fluids not needed unless infant has excessive fluid loss due to vomiting and/or diarrhea	• Additional fluids not needed unless infant has excessive fluid loss due to vomiting and/or diarrhea

SOURCE: American Academy of Pediatrics. Pediatric Nutrition Handbook, 6th ed., Elk Grove Village, IL; 2008.

⟨CONNECTIONS⟩ Rickets is a disease in children that causes bones to become soft and bend (Chapter 11, page 475).

Vitamin D Although scientists have long assumed that breastfed babies receive enough vitamin D from human milk and exposure to sunlight, apparently this is not always so. In recent years, several cases of rickets in breastfed infants have been reported nationwide, raising concerns about the adequacy of vitamin D content in human milk. However, if human milk is the "perfect food" for infants, then why would it be lacking this important nutrient?

Factors such as dark skin color, cloud cover, smog, and use of sunscreen can interfere with vitamin D synthesis. For example, sunscreens with a sun protection factor (SPF) of 8 or greater block the ultraviolet rays needed for vitamin D synthesis in skin. For these reasons, women with limited amounts of sun exposure may have low concentrations of vitamin D in their milk and may benefit from taking vitamin D supplements.[36] In fact, studies show that vitamin D levels in milk increase in response to maternal vitamin D supplementation.[37]

Because it is important to use sunscreen to help prevent skin cancer, the National Academy of Sciences and the AAP recommend that infants consuming less than 16 oz of infant formula each day receive vitamin D supplements (400 IU/day) beginning in the first few days of life. Supplementation should continue until they are consuming sufficient vitamin D from their diet.[38] Because all infant formulas manufactured in the United States meet standards for vitamin D, there is little risk of vitamin D deficiency in formula-fed infants consuming more than 16 oz of formula daily.

Fluoride Fluoride is important for forming teeth and can help prevent dental caries later in life. The AAP recommends fluoride supplements for breastfed infants starting at 6 months of age if the local water source has a fluoride concentration <0.3 parts per million (ppm).[22] Parents should check with the local water department to find out the fluoride content of the drinking water in their community. If purified bottled water is used to prepare infant formula, fluoride supplements are recommended. The recommended daily dosage of fluoride supplements is 0.25 mg/day for children between 6 months and 3 years of age.

Iron Another nutrient that is sometimes given to infants in the form of supplements is iron. A full-term infant is born with a substantial iron reserve that helps meet his or her need for iron during the first 6 months of life. After this time, infants consuming iron-fortified formulas that contain at least 1 mg iron per 100 kcal are likely to maintain adequate iron status. Although human milk contains less iron than infant formula, it is more readily absorbed. The AAP and CDC recommend iron supplementation for infants who continue to be exclusively breastfed during the second 6 months of life.[39] It is suggested that these infants be given 1 mg of iron per kilogram of body weight daily until complementary iron-rich foods are introduced. For breastfed infants who were preterm or LBW, a supplement of 2 to 4 mg of iron per kilogram of body weight daily is recommended, starting at 1 month after birth and continuing through the first year of life.

Water During the first 6 months of life water requirements are likely met if adequate amounts of human milk and/or formula are provided, so additional fluids are not needed. However, when vomiting and/or diarrhea cause excess fluid loss, water replacement is necessary to prevent dehydration. The AAP recommends plain water rather than juice when it is necessary to replace fluid loss.[40] Although over-the-counter fluid replacement products with added sugars and/or electrolytes can be given, they are not usually

COMPLEMENTARY FOODS CAN BE INTRODUCED BETWEEN 4 AND 6 MONTHS OF AGE

Many important developmental milestones take place between 4 and 6 months of age, and some of these can help parents determine whether their infant is ready for complementary feeding.[41] Signs that an infant is ready for complementary foods include sitting up with support and good head and neck control. The AAP recommends introducing nonmilk complementary foods when an infant is between 4 and 6 months of age. Until this time, younger infants are not physiologically or physically ready for foods other than human milk and infant formula.

Because iron status begins to decline at 4 to 6 months of age, pediatricians typically recommend that an infant's first complementary foods be iron-rich ones, such as iron-fortified cereal or pureed meat. Rice cereal, or other single-grain cereals, can be mixed with human milk or infant formula to give it a smooth, soft consistency. In addition to consuming sufficient amounts of human milk or iron-fortified infant formula, infants require approximately 1 oz of iron-fortified cereal (or equivalent) daily after 6 months of age to meet their iron requirements. Some infants find it difficult to consume food from a spoon, but in time most infants become quite skilled at this. After spoon-feeding is well established, the consistency of the food can be made thicker and more challenging. For example, other foods such as pureed vegetables and fruits can be introduced. During this period, complementary foods are considered *extra*, because infants still need regular feedings of human milk or formula.

Although infants should not be given complementary foods before 4 to 6 months of age, there is no evidence that delaying their introduction beyond this point is beneficial. It is important for parents to introduce new foods into the infant's diet gradually. After a new food is given, parents should wait three to five days to make sure the food is tolerated and there are no adverse reactions.[42] Signs and symptoms associated with allergic reactions or food sensitivities include a rash, diarrhea, runny nose, and in more severe cases, difficulty breathing. Although the most common food allergies are caused by cow milk, eggs, soy, nuts, wheat, fish and shellfish, there is little scientific evidence that delaying the introduction of these foods can prevent allergies later in life.[43]

Iron-fortified cereal is often recommended as a "first food."

Complementary Foods Should Not Displace Human Milk or Infant Formula
Infants should continue to be fed human milk and/or iron-fortified formula throughout the first year of life. The fact that cow milk, goat milk, and soy milk are low in iron provides further justification for delaying the introduction of these types of milk until after a baby's first birthday. Consuming large amounts of milk can displace iron-containing foods, which increases an infant's risk of developing iron deficiency anemia. Similarly, many pediatricians caution parents about giving infants too much fruit juice. Once recommended because it provided infants with a good source of vitamin C and additional water, the AAP now discourages parents from giving fruit juice to infants before 6 months of age. Like juice, milk can displace human milk and iron-fortified formula, which are much more nutrient dense.[40] It is also recommended that older infants not be given more than 4 to 6 oz of fruit juice a day. Overconsumption of fruit juice can lead to excess energy intake, dental caries, diarrhea, flatulence, and abdominal cramps. The AAP also recommends that older infants be given 100% pure fruit juice, rather than blends or fruit drinks.

TABLE 14.6 Developmental Skills and Feeding Practices during Infancy

Age	Developmental Skills	Appropriate Foods and Feeding Methods
0–1 months	Exhibits startle reflex; is able to suck and swallow	• Human milk and/or iron-fortified formula
2–3 months	Is able to support head; is interactive; reaches toward objects	• Human milk and/or iron-fortified formula
4–6 months	Is able to roll over to back; can sit with support; vocalizes; props self using forearms; grasps objects; moves tongue from side to side	• Human milk and/or iron-fortified formula • Slow introduction to iron-fortified infant cereal (or other iron-rich food source such as pureed meat) begins sometime between ages of 4 and 6 months
7–9 months	Begins to sit without support; holds objects; transfers objects between hands; pulls to standing; has improved mouth control and can begin to drink from a cup with assistance; can pick up small pieces of food and place them in mouth	• Human milk and/or iron-fortified formula; iron-fortified infant cereal; dry cereal; 100% fruit juice; pureed or mashed vegetables and fruits; soft pieces/mashed food such as meat
10–12 months	Begins to walk with assistance; crawls; refines grasp; exhibits mature chewing and improved ability to drink from a cup; develops skills needed for self-feeding	• Human milk and/or iron-fortified formula; iron-fortified cereal; soft pieces of fruits and vegetables; bread, crackers, and dry cereal; soft pieces of meat; 100% fruit juice

SOURCE: American Academy of Pediatrics. Pediatric nutrition handbook, 6th ed. Elk Grove Village, IL; 2008.

Baby bottle tooth decay can result when infants are put to sleep with bottles containing carbohydrate-rich liquids, including juice and milk.

Tooth decay can begin early in life, and it is therefore important for parents to establish good dental hygiene practices. Infants allowed to fall asleep with bottles filled with milk, formula, juice, or any other carbohydrate-containing beverage are at risk for developing **baby bottle tooth decay.** This is caused by the pooling of carbohydrate-containing beverages in the infant's mouth while asleep. Because the sugars can damage the newly formed teeth, parents are advised not to put infants to bed with bottles that contain anything but water.

Feeding Older Infants As older infants (9 to 12 months of age) become more adept at chewing, swallowing, and manipulating food in their hands, they are ready to move on to the next stages of feeding. Table 14.6 provides a list of appropriate foods that can be given to infants at this age. Because certain foods pose a risk for choking, it is important that parents not give these to their children. These include such foods as popcorn, nuts, whole grapes, raisins, raw carrots, pieces of hot dogs, and hard candy. In addition, the CDC recommends not feeding honey to children under 1 year of age, because it can contain spores that cause botulism. Even very low exposure to these spores can make young children sick. Approximately 94 cases of infant botulism occur in the United States each year.[44] Other sources that can harbor botulism spores are soil, corn syrup, and improperly canned foods.

What Are the Nutritional Needs of Toddlers and Young Children?

The life-stage groups referred to as "toddlers" and "early childhood" span the ages of 1 through 8 years. During this time, many physical, cognitive, psychological, and developmental changes take place. This is a time of growing independence as children gain the ability and confidence to

baby bottle tooth decay Dental caries that occur in infants and children who are given bottles containing carbohydrates (such as milk or juice) at bedtime.

function on their own. At this age, children become more opinionated, often expressing their likes and dislikes. Indeed, feeding toddlers and young children can be challenging. Childhood is also a time when attitudes about food are being formed, and parents play an important role in helping toddlers and young children develop a healthy relationship with food. How parents deal with the challenges of feeding children is very important.

GROWTH AND DEVELOPMENT INFLUENCE NUTRITIONAL NEEDS OF TODDLERS AND YOUNG CHILDREN

In terms of defining nutrient requirements, childhood is divided into stages: toddlers (ages 1 to 3 years) and early childhood (ages 4 to 8 years). During these periods, children grow at a steady rate but one that is considerably slower than in infancy. It is not unusual for children's eating habits to change or appetite to decrease. Regardless of a child's eating habits and preferences, it is important for parents to provide enough nutritious food for optimal growth and development.

As in infancy, sex-specific growth charts are used to monitor weight and height throughout childhood. In addition, BMI is used to assess weight for height in children over 2 years of age. Children with BMIs between the 85th and 95th percentiles are classified as overweight, whereas children with BMIs at or above the 95th percentile are classified as obese. At the other end of the spectrum, children with BMIs at or below the 5th percentile are classified as underweight, and are considered at risk for undernutrition. Note that, unlike adults for whom BMI categories are not dependent on age, whether a child is classified as having healthy weight or not is age-dependent. For instance, a 3-year-old boy with a BMI of 17.5 kg/m² would be considered overweight, whereas this same BMI would indicate a healthy weight in an 8-year-old boy. This is because healthy ranges of body fat change with age (Figure 14.10).

Both the CDC and the AAP recommend that health professionals use BMI to assess weight in children beginning at 2 years of age. However, it can be difficult to determine if a child is actually overweight or simply in a transitional part of the growth cycle. Nonetheless, if excess weight for height persists throughout childhood, there may be cause for concern. Childhood obesity increases the risk of obesity later in life, regardless of whether the parents are obese.[45] The longer obesity persists during childhood, the more likely that weight-related health problems will develop.[46] According to recent estimates, approximately 17% of children 6 to 11 years of age are at or above the 95th percentile for BMI.[47] You can read more about the increasing prevalence and health concerns of overweight children in the Focus on Diet and Health feature.

FEEDING BEHAVIORS IN CHILDREN

"Please, just one bite," begs an anxious parent. This all-too-familiar plea exemplifies the fact that feeding children is not always easy. Some parents are quite surprised at how quickly mild-mannered infants become willful and opinionated toddlers. How parents respond to feeding challenges can determine whether these behaviors persist or fade. Although forcing children to eat

© 2008 Stephanie Rausser/Getty Images

Feeding young children can be challenging for parents. Toddlers often have strong opinions about what and when they want to eat.

FIGURE 14.10 **Interpretation of Body Mass Index Values in Children** Unlike adults, how one interprets body mass index (BMI) for children and teens is dependent on sex and age. The BMI chart shown here is specific for boys.

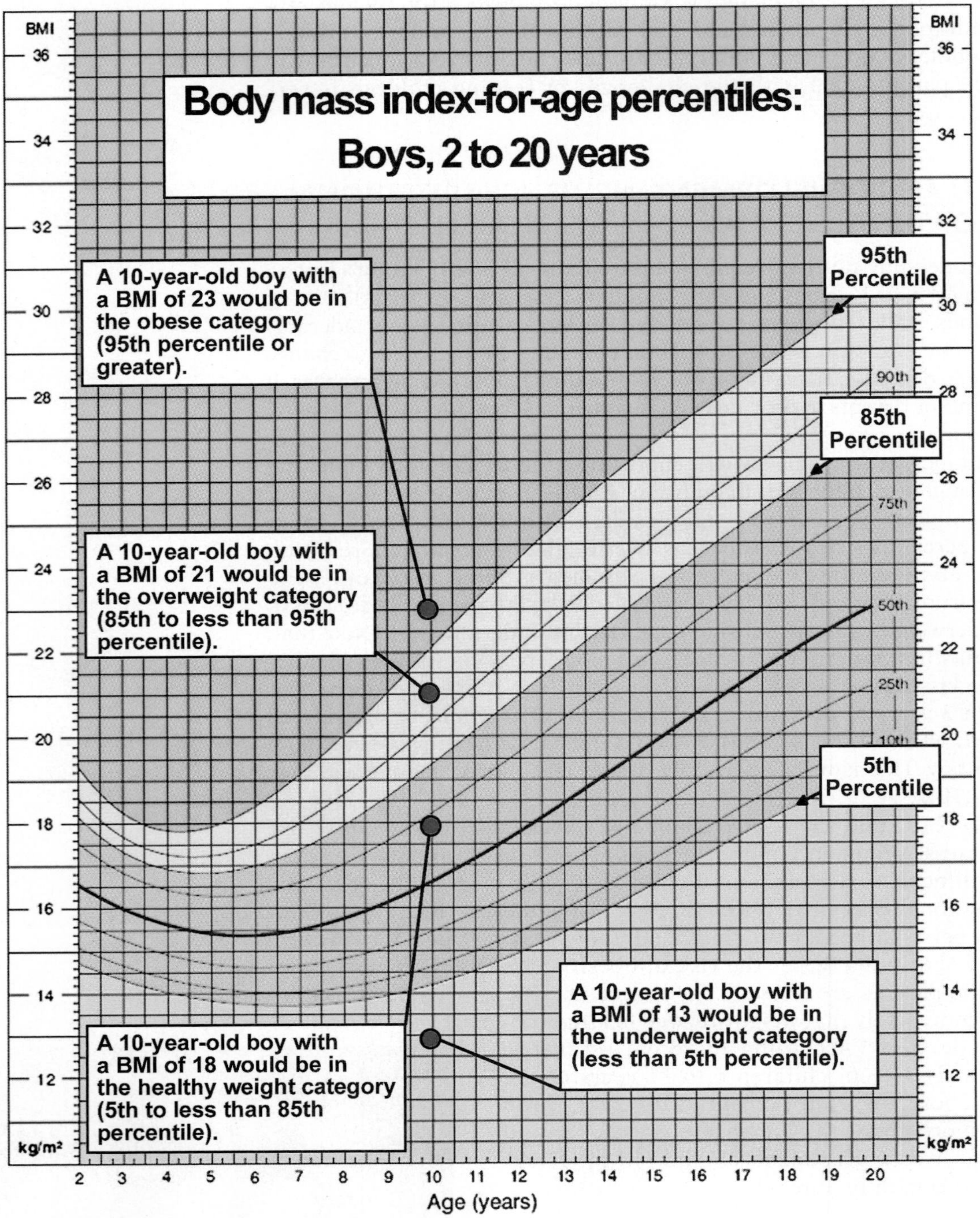

Body mass index-for-age percentiles:
Boys, 2 to 20 years
BMI
36
34
32
30
28
26
24
22
20
18
16
14
12
kg/m²
A 10-year-old boy with a BMI of 23 would be in the obese category (95th percentile or greater).
A 10-year-old boy with a BMI of 21 would be in the overweight category (85th to less than 95th percentile).
A 10-year-old boy with a BMI of 18 would be in the healthy weight category (5th to less than 85th percentile).
A 10-year-old boy with a BMI of 13 would be in the underweight category (less than 5th percentile).
95th Percentile
90th
85th Percentile
75th
50th
25th
10th
5th Percentile
2 3 4 5 6 7 8 9 10 11 12 13 14 15 16 17 18 19 20
Age (years)

SOURCE: Centers for Disease Control and Prevention. CDC Growth Charts: United States. Available from http://www.cdc.gov/healthyweight/assessing/bmi/childrens_bmi/about_childrens_bmi.html.

when they do not want to is never recommended, it is sometimes difficult for parents to remain calm when a child refuses to eat. Most experts agree that the child, not the parent, should be the one who determines if he or she eats, and how much.[48] Rather, it is the parent's job to make mealtime a pleasant experience and to provide nutritious, age-appropriate foods to choose from. The following six general guidelines regarding common childhood feeding challenges can encourage healthy eating.

The percentage of overweight children in the United States is on the rise, and unfortunately, health conditions once only common in adults are now becoming more common in children. For example, weight-related health conditions such as type 2 diabetes, high blood pressure, and elevated blood lipids are now becoming increasingly prevalent and even commonplace among America's youth.[49] Health experts estimate that 17% of children (6 to 11 years of age) are obese (BMI >95th percentile).[47] The short- and long-term health and social consequences of childhood obesity are of great concern to parents and health professionals. Because excess weight during childhood is likely to continue into adolescence and adulthood, it is important to understand factors that contribute to this trend.

Similar to those of adults, the behaviors most concretely linked to excessive weight gain in children are unhealthy eating patterns and physical inactivity. Although there is much to learn about specific meal patterns that promote childhood obesity, easy access to high-fat, energy-dense foods is a primary contributing factor. Changing the food environment alone is not sufficient to reverse this growing trend. Efforts to encourage children of all ages to participate in healthful physical activities are also important. For many American children, watching television and other electronic media has largely replaced physical activity. It is not surprising that there is an association between obesity and the amount of time spent watching television.[50] More than one-third of children in the United States watch three or more hours of television each day, and this does not include additional time playing computer and video games.[51] In addition, children often consume energy-dense foods while watching television, and controlled research suggests that caloric consumption following a period of video game play may be increased.[52]

Persuasive media messages directed at children may also be contributing to unhealthy food choices. Approximately 90% of food commercials that air during Saturday morning children's television shows are for products with limited nutritional value, and the average American child is exposed to more than 10,000 televised food advertisements yearly.[53] And not surprisingly, when given the choice among identical foods, children prefer those wrapped in fast-food packages to those wrapped in plain, unmarked packages.[54] The selection of labeled foods (fast-food packages vs. plain packages) was related to the number of television sets in the home and the frequency of visits to fast-food restaurants. For these reasons, several health organizations including the AAP encourage parents to limit the amount of time children watch television. It is recommended that children younger than 2 years of age watch no television at all and that parents limit older children to two hours of television per day.[53]

Restricting the number of hours that children watch television does not, by itself, prevent weight gain or facilitate weight loss. Children also need to be physically active, and parents and schools need to create environments that encourage children to exercise. Unfortunately, some states have relaxed physical education requirements in public schools. Although the CDC recommends that children participate in 30 minutes of physical education class daily, only 6 to 8% of public schools meet this guideline.[55] The 2008 Physical Activity Guidelines for Americans recommend that all youth participate in at least 60 minutes of moderate- to vigorous-intensity physical activity each day. This activity should consist mostly of aerobic exercise but also include muscle- and bone-strengthening activities.[56]

In addition to exercise, providing healthy food choices at home and school is vitally important for children. In the past, school cafeterias only served meals that met federal guidelines for nutritional standards. Today, many schools sell foods and beverages that are not part of the federal school meal program. Parents may also be surprised to discover that national fast-food chains operate in many school cafeterias around the country. Money generated from selling these products is often used to help support school-related activities and programs. Furthermore, vending machines offer a multitude of poor quality snacks and beverages to students in many schools.[57] Studies show that children consume, on average, 50 more cans of soda per year in schools where vending machines are available, compared with schools without vending machines.[58] Some schools now require vending machines to be stocked with nutritional snacks such as milk or fruit.

Preventing children from becoming overweight will take considerable effort at home and at school. Regardless of the many pressures and enticements, teaching children about the importance of physical activity and good nutrition is an important step in keeping them healthy. After all, there are as many healthy food choices available today as there are unhealthy ones.

CHRIS YOUNG/PA Photos /Landov

Many organizations and parent groups are encouraging schools to eliminate fat- and sugar-rich snacks from school vending machines.

Meals provide family members an important opportunity to interact, to share, and to be together.

- *Avoid using food to control behavior.* Most experts agree that using food for rewards or punishments is not a good idea. Although rewarding a child with treats may correct behaviors short term, it will likely create serious food-related issues down the road. For example, rewarding "good" behavior with a dessert may establish a connection between sweet foods and approval. Rather, parents should teach children that food is pleasurable and nourishing and not something to turn to for approval or emotional comfort.

- *Model good eating habits.* It is important for parents to serve as good role models. Studies show that if parents or siblings enjoy a particular food, the child will be more likely to enjoy it too.[59] Studies also show that families that eat meals together on a regular basis are more likely to have children who eat healthier diets than families that do not.[60] Mealtime provides a time for the entire family to be together and share quality time.

- *Recognize preference differences.* Food preferences of children do not always appear rational to adults. That is, children often judge foods as acceptable or unacceptable based on attributes such as color, texture, and appearance rather than taste and nutritional value. Also, children likely experience strong flavors such as onions and certain spices more intensely than adults.

- *Introduce new foods gradually.* Making new foods familiar to children is an important first step in food acceptance. Encouraging children to help select and prepare new foods not only helps to make foods more familiar, but may also stimulate interest. It is important for parents to remember that accepting new foods takes time. Pressuring children to eat or to try new foods is not recommended.

- *Encourage nutritious snacking.* Children have small stomachs and therefore should eat smaller portions than adults. They also need to eat more frequently. In fact, limiting food consumption to three meals a day is very difficult for some children. This is why experts recommend that parents and caregivers provide children with nutritious between-meal snacks. Although children who snack frequently are often not hungry at mealtime, this is not usually a problem as long as healthy, nutritious snacks are served.

- *Promote self-regulation.* It is important that children learn to self-regulate their food intake based on internal cues of hunger and satiety. For this reason, serving sizes of food need to be age appropriate, allowing children to ask for more if desired. Children often claim to be too "full" to eat certain foods, and experts do not recommend forcing them to eat all the food on their plate. Instead, parents need to set limits. For example, if a child declines to eat his or her meal and then asks for dessert, it is reasonable to say no.

Critical Thinking: Amy's Story Think back to Amy's story about Backyard Harvest. Family participation in Backyard Harvest offers many educational opportunities for children. What can children learn about food and healthy eating by harvesting and donating freshly grown produce? What critical nutrients might be provided by the fresh fruits and vegetables distributed by Backyard Harvest that might otherwise be lacking in a low-income child's diet?

RECOMMENDED ENERGY AND NUTRIENT INTAKES FOR TODDLERS AND YOUNG CHILDREN

During childhood, adequate energy is needed both to support total energy expenditure (TEE) and for the synthesis of new tissues. In other words, a child should be in positive energy balance. Thus, EERs are equal to the sum of TEE plus the energy associated with growth. As the rate of growth slows during childhood, the amount of energy (kcal/day) needed daily to support growth also decreases. For example, the energy needed to support growth in infants (0 to 3 months of age) is approximately 175 kcal/day. By comparison, the energy needed to support growth in toddlers is only 20 kcal/day. Although energy needed for growth decreases with age, EER values increase because some of the variables that affect TEE (such as height, weight, and physical activity level) increase. For example, the EER for a 1-month-old male infant weighing 9.7 lb is 467 kcal/day. By 35 months of age (weight = 31.3 lb), the EER increases to 1,186 kcal/day. EER equations for toddlers and young children are listed in Appendix B.

It is also important to consider the healthiest distribution of calories in the diet of children. The Institute of Medicine recommends, via the AMDRs, that children 1 to 3 years of age consume 30–40, 45–65, and 5–20% of their calories as fats, carbohydrates, and protein, respectively. Note that this would result in children consuming diets higher in fat than adults. However, by 4 years of age, the AMDRs for toddlers and young children are close to those for adults. Although the amount of protein needed per kilogram of body weight decreases during childhood, recommended total protein intakes increase. AMDRs and recommended protein intakes for toddlers and young children (g/day and g/kilogram/day) are listed inside the front cover of this book. The RDA for protein for toddlers and young children is 0.95 g/day per kg of body weight. It is especially important to provide high-quality protein foods during this time, such as meat, yogurt, cheese, and eggs.

Although there are no RDAs or AIs for total fat intake for toddlers and young children, adequate fat intake is important for growth and development. It is particularly important for children to consume adequate amounts of the essential fatty acids linoleic acid and linolenic acid. Foods that provide these types of fatty acids include vegetable oils, nuts, seeds, and fish. Carbohydrates are needed to provide the brain with glucose. After 1 year of age, the amount of glucose used by the brain is similar to that of adults, which is why carbohydrate recommendations for children and adults are the same. Dietary fiber is also important for children's health. The AI for dietary fiber is 19 g/day for toddlers and 25 g/day for young children. By comparison, it is recommended that adults consume 25 to 38 g/day of dietary fiber. Serving children fruits, vegetables, and whole-grain cereal products made from whole grains provides them with an excellent source of dietary fiber. These nutrient-dense foods that most children enjoy eating provide an array of micronutrients as well.

Calcium and Iron Are Nutrients of Special Concern for Toddlers and Young Children

Calcium-rich foods are particularly important to the development of strong, healthy bones throughout childhood. The RDA for calcium increases from 700 mg/day at 1 to 3 years of age to 1,000 mg/day at 4 to 8 years of age. Milk and other dairy products are good sources of calcium. In fact, it would take 7 cups of broccoli to provide the same amount of calcium (approximately 280 mg calcium) as 1 cup of milk. For this reason, children between the ages of 2 and 8 years should consume at least 2 cups of low-fat milk or equivalent milk products each day. In general, 1 cup of yogurt, 1½ ounces of natural cheese, and 2 ounces of processed cheese are all equivalent to 1 cup of milk.

⟨CONNECTIONS⟩ Recall that iron deficiency affects the ability to synthesize hemoglobin, resulting in anemia (Chapter 13, page 562).

Many children do not meet recommended intakes for milk and milk products, which is cause for concern.

Iron is another nutrient that is important for growing children. It is recommended that toddlers and young children consume 7 and 10 mg/day, respectively, of iron to meet their needs. Iron deficiency is one of the most common nutritional problems in childhood. This is because children have a rapid rate of growth, and blood volume is increasing. Children who drink large amounts of milk and eat a limited variety of other foods may be at increased risk of iron deficiency.[61] This may be related to over-consumption of cow's milk and juice that can displace iron-rich foods from the diet.

A lack of iron-rich foods can lead to the development of iron-deficiency anemia, causing children to become irritable and inattentive and to have decreased appetites. In fact, researchers have found that student achievement tests improved when children with iron-deficiency anemia received iron supplements.[62] To prevent iron deficiency, parents are encouraged to feed their children a variety of iron-rich foods such as meat, fish, poultry, eggs, legumes (peas and beans), enriched cereal products, whole-grain cereal products, and other iron-fortified foods.

DIETARY GUIDELINES FOR TODDLERS AND YOUNG CHILDREN

Because children have unique eating patterns and nutritional needs, it is important to provide a variety of nutrient-dense foods. To determine the number of servings from each food group needed to meet recommended nutrient and energy intakes for toddlers and young children, parents and childcare providers are encouraged to use the MyPlate food guidance system (http://www.choosemyplate.gov). This site also provides materials designed specifically for preschoolers (http://www.choosemyplate .gov/preschoolers/index.html) and children aged 6 to 11 (http://www .choosemyplate.gov/kids/index.html) that promote healthy eating. In addition to emphasizing a healthy diet, the Dietary Guidelines for Americans stress the importance of regular physical activity to promote physical health and psychological well-being. As previously noted, parents are encouraged to make sure that children engage in at least 60 minutes of moderate- to vigorous-intensity physical activity daily, including strength-building exercises.

How Do Nutritional Requirements Change during Adolescence?

Toward the end of early childhood, hormonal changes begin to transform a child into an adolescent. This transition marks the beginning of profound physical growth and psychological development. Hormones are responsible for triggering changes in height, weight, and body composition. In females, the onset of menstruation, known as **menarche,** also begins around this time. Menarche typically occurs in girls around 13 years of age. As with any stage of growth, nutrition plays an important role as an adolescent matures from a child into a young adult. Unfortunately, this is also a time when unhealthy eating practices often begin to develop. For example, weight dissatisfaction is common among teens and can lead to inappropriate dieting and other destructive weight-loss behaviors.

menarche (men – ARK – e) The first time a female menstruates.

GROWTH AND DEVELOPMENT DURING ADOLESCENCE

Adolescence is the physical and psychological bridge between childhood and adulthood, and this transformation is signified by the onset of **puberty.** Defined as the maturation of the reproductive system and the capacity to reproduce, puberty is initiated by hormonal changes that trigger the physical transformation of a child into an adult. During this time, the adolescent begins to experience many physical, psychological, and social changes. Because these changes span a relatively long period, the DRIs for this stage of the life cycle are divided into two phases: 9 to 13 years and 14 to 18 years.

The timing of puberty varies, and adolescents of the same age can differ in terms of physical maturation. For example, some adolescents are "early" developers, whereas others are "late" developers. Consequently, nutritional needs of adolescents may depend

Adolescence is the period between childhood and adulthood.

more on the stage of physical maturation than chronological age. Also, females tend to experience puberty at an earlier age than do males. Puberty begins roughly around 9 years of age in females and 11 years of age in males. However, for reasons that are not clear, more and more American girls are entering puberty at younger ages. Some researchers believe this trend may be related in part to the rise in childhood obesity rates, as overweight girls tend to develop physically and menstruate earlier than do thinner girls.[63]

Changes in Height, Weight, and Body Composition Considerable physical growth takes place during adolescence. Before the onset of puberty, males and females have attained about 84% of their adult height. During the adolescent growth spurt, females and males grow approximately 6 and 8 inches in height, respectively. Bone mass also increases, with peak bone mass nearly attained by about 20 years of age. Because bone mass increases rapidly during adolescence, it is especially important for teens to consume adequate amounts of nutrients that promote bone health, such as protein, vitamin D, calcium, magnesium, and phosphorus.

⟨**CONNECTIONS**⟩ Recall that peak bone mass is the maximum amount of minerals found in bones during the lifespan (Chapter 12, page 541).

Changes in linear growth (height) are accompanied by changes in body weight and composition during adolescence. Overall, females and males gain an average of 35 and 45 pounds, respectively. However, changes in body composition differ considerably between females and males. Whereas females experience a percentage decrease in lean mass and a relative percentage increase in fat mass, a male adolescent experiences the opposite. Although increased body fat in females is normal and healthy, it can contribute to weight dissatisfaction, which sometimes can lead to unhealthy dieting and caloric restriction. Caloric restriction during adolescence can delay growth, development, and reproductive maturation. The topic of eating disorders and related health problems was addressed in the Nutrition Matters that concluded Chapter 8. While good nutrition is important throughout the entire life cycle, the accelerated rate of growth and development during adolescence puts teens at particularly high risk for diet-related health problems.

Psychological Changes Adolescence is marked not only by rapid physical changes, but also by numerous psychological and developmental changes. For example, adolescents' newfound desire for independence often strains family relationships

puberty Maturation of the reproductive system.

and leads to rebellious behaviors. Furthermore, there is a strong need to fit in and be accepted by peers. For these and many other reasons, healthy eating can become a low priority for adolescents. In fact, peers are often more influential in determining food preferences than family members. Dieting, skipping meals, and increasing consumption of foods away from home are common occurrences among teenagers. However, with maturity comes the recognition that current health behaviors can have lasting impacts on long-term health.

NUTRITIONAL CONCERNS AND RECOMMENDATIONS DURING ADOLESCENCE

The rapid growth and development associated with adolescence increases the body's need for certain nutrients and energy. The MyPlate food guidance system can be used by teens to determine the amount of food from each food group needed to meet their recommended nutrient and energy intakes. Using this personalized guide for meal planning is particularly important because an inadequate diet during this stage of life can compromise health and have lasting effects.

Energy and Macronutrients Estimated Energy Requirement (EER) calculations during adolescence take into account energy needed to maintain health, promote optimal growth, and support a desirable level of physical activity. Again, positive energy balance is assumed during this period of time. Equations used to determine EERs in adolescents are the same as those for children except for the amount of energy needed for growth. For example, for boys and girls 9 through 18 years of age, the EER is calculated by summing total energy expenditure (TEE) and an additional 25 kcal/day for growth. Equations used to determine EERs during adolescence are provided in Appendix B. Although obesity rates have risen over the past 30 years among adolescents, average energy intakes by teens have remained remarkably stable. The only noted exception appears to be for adolescent females, who have increased their caloric intake by approximately 260 kcal/day.[64] Thus, changes in energy expenditure have likely contributed to the increased obesity rates in this age group.

Although growth is influenced by many factors, adequate intake of protein is particularly important. Younger adolescents (9 to 13 years) require more protein per kilogram of body weight than do adults—0.95 vs. 0.8 g/kg/day, respectively. For 14- to 18-year-olds, the RDA for protein decreases to 0.85 g/kg/day. Based on the RDA, recommended protein intake is 34 g/day for females 9 to 13 years of age and 46 g/day for females 14 to 18 years of age.

Similar to an adult's diet, the majority of calories in an adolescent's diet should come from nutrient-dense, carbohydrate-rich foods. These include fruits, vegetables, and whole grains. Yet, despite national public health efforts to encourage healthy eating, the number of adolescents who meet recommendations for fruit and vegetable consumption is quite low.[65] The "typical" adolescent diet is also less than optimal in terms of snack foods that are high in fat and sugar. It is not surprising that snacking frequency among teens is associated with higher intakes of total calories. In 2005–2006, adolescents consumed on average 526 kilocalories daily as snacks (23% of total energy intake).[66] Sweeteners and added sugars, many of which are associated with the consumption of carbonated beverages, contribute 16% of the total calories consumed by teens.[67] Percentages of total energy from fat and saturated fat among adolescents are approximately 34% and 12%, respectively. These amounts, which are consistent for both younger and older adolescent males and females, are well within the recommended intakes based on the AMDRs.[68]

Selected Micronutrients Too little dietary calcium at a time when skeletal growth is increasing can compromise bone health later in life. For this reason, the RDA for calcium during adolescence is set at 1,300 mg/day. Getting enough calcium is best achieved by consuming dairy products such as milk, yogurt, and cheese. The average calcium intakes for adolescent males[69] and females[70] are 1,266 and 918 mg/day, respectively (EAR = 800 mg/day). When asked, the majority of adolescents recognize the importance of calcium for proper bone health.[71]

Adolescents also require additional iron to support growth. Because of the iron loss associated with menstruation, the RDA for iron in females is higher than that of males (15 and 11 mg/day, respectively) during later adolescence. Although it appears that most adolescents are well-nourished with respect to iron, approximately 8% of teenagers in the United States have impaired iron status, and another 1 to 2% have iron-deficiency anemia.[72]

Mean total folic acid intake (diet and supplements) for adolescent females is 274 µg DFE/day (EAR = 330 µg DFE/day).[73] As you learned previously, impaired folate status around the time of conception can increase the risk of having a baby born with a neural tube defect. For this and other reasons, it is important for health care professionals to stress the importance of consuming folate-rich foods such as orange juice, green leafy vegetables, and enriched cereals.

Although the majority of teens appear to have adequate intakes of vitamin C, it is well known that smoking can increase vitamin C requirements. In spite of the fact that smoking rates among U.S. teens have been on the decline, a significant number—approximately 35% of high school–age teens—continue to smoke. Vitamin C is needed for collagen synthesis, and a lack of this important vitamin can disrupt growth and bone development.

How Do Age-Related Changes in Adults Influence Nutrient and Energy Requirements?

Adulthood spans the longest segment of the life cycle—approximately 60 to 65 years. Like individuals in other stages of life, adults experience physical, psychological, and social changes that can affect health. Managing family and career obligations can make this a particularly challenging time in a person's life. A busy schedule can be stressful, and personal health needs often go unattended. It is vital for adults to recognize the importance of taking care of their own physical and emotional well-being because a healthy lifestyle can help ensure a full and active life for many years to come.

Because many chronic, degenerative diseases can occur during this period of the life cycle, it is important for adults to stay physically and mentally fit. Although aging is inevitable, studies show that young and middle-aged adults who adopt a healthy lifestyle that includes eating nutritious foods, maintaining a healthy body weight, participating in regular physical activity, and not smoking are less likely to develop cardiovascular disease, type 2 diabetes, hypertension, and certain types of cancer, all of which can contribute to premature death.

ADULTHOOD IS CHARACTERIZED BY PHYSICAL MATURITY

Adulthood is the period in the lifespan characterized by attainment of physical maturity, which is typically achieved by 20 years of age. Some males continue to grow in their early 20s, and bone mass increases slightly until age 30 for both males and females. After achieving physical maturity in young adulthood, adults typically undergo a long period of physical stability (maintenance) before gradually transitioning to senescence. The DRIs divide adulthood into four groups—young adults (19 through 30 years), middle age (31 through 50 years), adulthood (51 through 70 years), and older adults (>70 years). There is a wide spectrum of health and independence among

Many older adults are physically fit and remain active.

⟨CONNECTIONS⟩ Recall that life expectancy is the average number of years a person can expect to live (Chapter 1, page 23).

Jeanne Louise Calment (1875–1997), photographed at the age of 122 years, was one of the oldest humans whose age was fully authenticated. She attributed her longevity to olive oil, port wine, and chocolate. Her genes probably contributed to her longevity; her father lived to the age of 94 years and her mother to 86 years.

⟨CONNECTIONS⟩ Recall that free radicals are compounds with unpaired electrons and can damage cell membranes, proteins, and DNA (Chapter 10, page 447).

most middle-aged and older adults. Whereas some people maintain active lifestyles throughout adulthood, others are frail and require a great deal of assistance. For this reason, functional status is sometimes a better indicator of nutritional needs than chronological age.

The "Graying" of America Although physical maturity is achieved early in adulthood, many productive years still remain ahead. In the United States, life expectancy continues to increase, and people over 85 years old are becoming the fastest growing segment of this age group.[74] Note that life expectancy, the expected number of years of life remaining at a given age, is different from **lifespan,** the maximum number of years an individual member of a particular species has remained alive. In humans, both life expectancy and lifespan have been rising steadily, but the graying of America (first introduced in Chapter 1) is evident now more than ever. Although there is little doubt that a person's genetics play a major role in the aging process, lifestyle choices are vitally important as well.

The graying of America reflects a shift in the age distribution of the United States (Figure 14.11). Never before have there been so many older people, and never before has life expectancy been so long. Whereas average life expectancy (at birth) was about 47 years in 1900, today a baby born in the United States will likely live to be nearly 78 years old.[47] These changes are partly due to advances in medical technology and improved health care. The graying of America is also partly due to the increased number of births in the 1940s through the early 1960s. People born at this time are often labeled the "baby boom" generation. During the baby boom years, many hospitals expanded their obstetric and gynecology units to accommodate the increase in births. Shortly thereafter, many thousands of schools were built throughout the United States to accommodate the growing number of school-age children. Today, there is an increased need for retirement communities and health care practitioners to care for the rising number of older "boomers."

THERE ARE MANY THEORIES AS TO WHY WE AGE

The branch of science and medicine dedicated to the social, behavioral, psychological, and health issues of aging is called **gerontology.** Gerontologists have many theories as to why we age. Some believe the body deteriorates because of daily wear and tear. Others believe that a combination of lifestyle, environmental, and genetic factors determines how quickly we age.

One of the greatest contributions in understanding biological aging came from experiments performed in the 1960s by the physiologist Leonard Hayflick. Using cell cultures, Hayflick demonstrated that the number of times cells can divide is limited.[75] Thus, senescence, or the process of aging, may be programmed in cells, and factors that increase cell replication may speed up this process. For example, damage caused by unstable compounds called free radicals (generated by metabolism, lifestyle practices, and environmental factors) may increase cell replication and thus promote aging. In addition, free radicals extensively damage proteins, cell membranes, and DNA. Over time, damaged cells become unable to function fully, triggering a cascade of other age-related changes. Free radical damage has been implicated in several degenerative diseases such as cancer, atherosclerosis, and cataracts.[76]

The evidence that aging and many age-related chronic diseases may, in part, be caused by free radical damage raised hopes that antioxidant nutrients could slow the aging process by protecting the body from oxidative damage. Although animal studies provide some supportive evidence, human data remain controversial.[77] Clinical trials are needed to clarify whether consumption

FIGURE 14.11 The "Baby Boom Generation" and Aging of the U.S. Population An increased number of births between 1946 and 1964 has resulted in a shift in the age structure of the U.S. population over the past century.

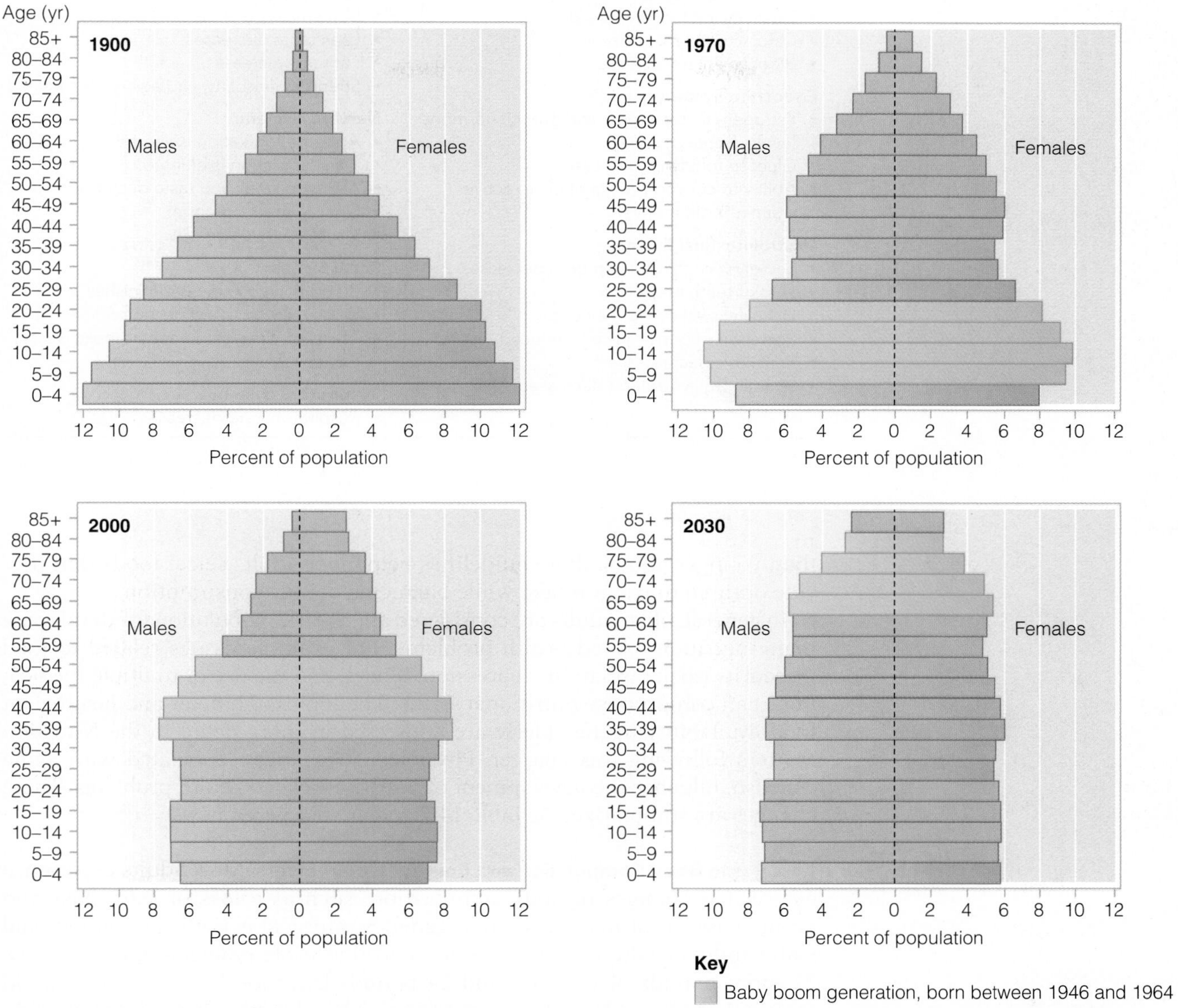

Key
Baby boom generation, born between 1946 and 1964

SOURCE: Hobbs R and Stoops N. U.S. Census Bureau. Demographic trends in the 20th century. U.S. Government Printing Office, Washington, DC. 2002. Available from: http://www.census.gov/prod/2002pubs/censr-4.pdf

of foods rich in antioxidants or antioxidant supplements can help slow the progression of age-related changes and help people live longer and healthier lives.

NUTRITIONAL ISSUES OF ADULTS

Although we cannot "turn back the hands of time," there is much that we can do to keep our bodies strong and healthy. During adulthood, nutrient intake recommendations are intended to reduce the risk of chronic disease while providing adequate amounts of essential nutrients. The MyPlate food guidance system can assist older adults in planning meals and determining the amount of food from each food group needed to meet their recommended nutrient and energy intakes. Because older adults generally require less food

⟨CONNECTIONS⟩ Recall that antioxidant nutrients (such as vitamins C and E) prevent unstable free radicals from damaging cells by donating electrons to them (Chapter 10, page 445).

lifespan Maximum number of years an individual in a particular species has remained alive.

gerontology The branch of science and medicine that focuses on health issues related to aging.

TABLE 14.7 Physiological Changes Typically Associated with Aging

Cardiovascular System
- Elasticity in blood vessels decreases.
- Cardiac output decreases.
- Blood pressure increases.

Endocrine System
- Estrogen, testosterone, and growth hormone levels decrease.
- Glucose tolerance decreases.
- Ability to convert provitamin D to active vitamin D diminishes.

Gastrointestinal System
- Secretion of saliva and mucus decreases.
- Loss of teeth may occur.
- Difficulty swallowing may occur.
- Secretion of gastric juice is reduced.
- Peristalsis decreases.
- Vitamin B_{12} absorption decreases.

Musculoskeletal System
- Bone mass decreases.
- Lean mass decreases.
- Fat mass increases.
- Strength, flexibility, and agility are reduced.

Nervous System
- Appetite regulation is altered.
- Thirst sensation is blunted.
- Ability to smell and taste decreases.
- Sleep patterns change.
- Visual acuity decreases.

Renal System
- Blood flow to kidneys diminishes.
- Kidney filtration rate decreases.
- Ability to clear blood of metabolic wastes decreases.

Respiratory System
- Respiratory rate decreases.

than younger adults, these guidelines help older adults select foods that provide optimal nutrient intake while balancing energy consumption.

In general, older adults are considered an "at-risk" population for developing many nutrition-related health problems. For example, issues related to food insecurity, social isolation, depression, illness, and the use of multiple medications can compromise nutritional status. The topics of hunger and insufficient food availability in the elderly are addressed in more detail in the Nutrition Matters following this chapter. Physiological changes associated with aging can also influence the development of nutrition-related health problems. These changes are summarized in Table 14.7 and are discussed next.

Changes in Body Composition and Energy Requirements Most adults experience age-related changes in body composition, such as a loss of lean mass and an increase in fat mass. Although genetics, physical activity, and nutritional status influence these shifts, in general, a 70-year-old male has approximately 22 more pounds of body fat and 24 pounds less muscle than a 20-year-old male.[78] Age-related loss of lean mass causes basal metabolism to decrease. As a result, energy requirements tend to decrease with age. This explains, in part, why many middle-aged and older adults experience age-related weight gain. A key recommendation included in the Dietary Guidelines for Americans is for adults to prevent age-related weight gain by decreasing energy intake and maintaining adequate physical activity.

Equations used to estimate energy requirements are the same for all adults. However, for each year of age above 19, total energy expenditure decreases 7 kcal/day for women and 10 kcal/day for men. Consequently, compared to 30-year-old women and men with similar weights, heights, and activity levels, 70-year-old women and men require 280 (40×7) and 400 (40×10) fewer kilocalories each day, respectively. Of course, energy requirements depend on many factors such as physical activity, weight, and changes in the relative amounts of muscle and body fat. Both over-consumption and under-consumption of energy can lead to weight-related health problems in older adults.

Along with weight gain, the loss of muscle mass can make older people less steady, increasing their risk for injury. According to the CDC, more than one-third of adults 65 years of age and over experience fall-related injuries annually, a common cause of most head traumas and hip fractures.[79] Exercise not only decreases fat mass and slows age-related bone loss, but it also helps strengthen muscles and improve coordination.

Adequate protein intake is important for older adults to help prevent age-related loss of skeletal muscle. Currently, the RDA for protein in older adults is 0.8 g/kg/day. However, some researchers believe that current requirements are too low and that protein requirements should be increased to 1.0–1.25 g/kg/day.[80] For an older adult weighing 150 lb, this would increase their recommended protein intake from 54 to 77 g/day.

Preventing Bone Disease Age-related bone loss can lead to osteoporosis, which is characterized by weak, fragile bones. While men can develop osteoporosis, the condition is far more common in women. Although many factors influence bone density, it is especially important for older adults to have adequate intakes of nutrients (protein, calcium, vitamin D, phosphorus, and magnesium) that support bone health. To promote lasting bone health, the RDA for calcium increases from 1,000 to 1,200 mg/day at age 51 in women and at age 71 in men. However, older adults often find it difficult to meet this recommended intake. The occurrence of lactose intolerance increases with age, inhibiting the ability of many to consume calcium-rich dairy products. To counter this problem, health care providers often recommend lactose-reduced dairy products, calcium-fortified foods such as soy milk, and/or supplements.

Considerable evidence shows that older adults, especially those who live in northern regions of the United States, are at increased risk for vitamin D deficiency. This is likely caused by limited exposure to sunlight. In addition, the ability to synthesize vitamin D from cholesterol via sunlight decreases with age. For these reasons, the RDA for vitamin D increases from 15 to 20 μg/day at age 71. Although magnesium and phosphorus are also important for bone health, aging does not seem to affect dietary requirements for these nutrients. You can read more about nutrition and bone health in the Nutrition Matters following Chapter 12.

Hormonal Changes in Women As women age, they experience a natural and progressive decline in estrogen levels. This time in a woman's life, referred to as **perimenopause,** can lead to a variety of other physical changes, including irregular menstrual cycles. Unfortunately, the decline in estrogen during this perimenopausal stage of life increases the rate of bone loss, making some women's bones weak and fragile. By the time a woman reaches **menopause,** around 50 to 60 years of age, her ovaries are producing very little estrogen and menstrual cycles completely stop. Women who have dense bones before menopause have fewer problems.

Menopause is a normal part of aging and affects each woman differently. Some women experience profound emotional and physiological changes, whereas others experience no discomfort at all. To get relief from menopausal symptoms, some women are prescribed hormone replacement therapy (HRT). Although HRT can protect against bone loss, it is associated with increased risk for cancers of the breast and uterus.[81] It is important for women to discuss the benefits and risks associated with HRT with their health care providers. Instead of HRT some women seek more natural options to ease them through the transitional period of menopause. These "natural" alternatives include certain foods such as soy, bioidentical hormone therapy, and herbal products such as black cohosh. You can read more about these alternatives in the Focus on Clinical Applications.

Because monthly blood loss associated with menstruation has stopped, menopause can improve iron status. In fact, the RDA for iron decreases for women during this life stage, from 18 to 8 mg/day—the same as recommended

⟨CONNECTIONS⟩ Recall that osteoporosis is caused by a reduction in bone mass and can lead to fractures and loss of stature in older adults (Chapter 12, page 542).

⟨CONNECTIONS⟩ Hormone replacement therapy is taken by some women to restore estrogen levels that decline as a result of menopause or the surgical removal of the ovaries (Chapter 12, page 545).

perimenopause Literally, the time "around" the time of menopause.

menopause The time in a woman's life when menstruation ceases, usually during the sixth decade of life.

Most women recognize that menopause is a natural physiological event and not a disease. Nonetheless, many struggle with the physical discomforts caused by declining estrogen levels. While some women experience no discomfort at all, others are debilitated by recurring hot flashes, night sweats, mood changes, and insomnia. Until recently, many women have chosen to ease menopausal-related discomforts with hormone replacement therapy (HRT). HRT restores hormonal balance by providing a combination of estrogen and progesterone. HRT was also thought to protect women from bone loss, Alzheimer's disease, heart attacks, and certain types of cancer. However, a landmark research study, the Women's Health Initiative, contradicted many of these health claims.[82] Rather than providing protection, HRT increased women's risks of breast cancer, heart attacks, strokes, and blood clots. These risks clearly outweighed the major benefits associated with HRT, specifically decreased risk of hip fractures and colorectal cancer. Since the release of these findings, there has been a steady decline in HRT use among women of all ages.[83] Many perimenopausal women are now seeking what they perceive to be safer alternatives to HRT such as diet, bioidentical hormone therapy, and herbal remedies.

Some of the most commonly recommended foods to be eaten during perimenopause and menopause are those made from soybeans. This is because soybeans contain phytochemicals called isoflavones that are structurally similar to estrogen. However, isoflavones do not provide relief from menopausal symptoms in all women. The North American Menopause Society recognizes that while some studies suggest a modest benefit from isoflavones in relieving menopausal symptoms, other studies fail to show such an effect.[84]

As an alternative to synthetic hormones, an increasing number of women are turning to "bioidentical" hormones—custom-formulated compounds that match the structure and function of hormones produced naturally in the body. Saliva or blood tests are used to determine the types and amounts of hormones needed by a particular woman. Although some practitioners claim that bioidentical hormones may be safer and have fewer side effects than traditional HRT, more research is needed to better understand their safety and efficacy.[85] A position paper issued by the American College of Obstetricians and Gynocologists cautions women about taking some bioidentical hormones because these compounds have not undergone rigorous testing.[86] Issues regarding purity, potency, safety, and efficacy of bioidentical hormones are also of concern. In fact, the U.S. Food and Drug Administration advises women who decide to use these hormones to use them at the lowest dose that helps and for the shortest time needed.[87]

In addition to dietary changes and hormonal preparations, some women turn to herbal products to ease menopausal symptoms. The most commonly used products are black cohosh (*Cimicifuga racemosa*), St. John's wort (*Hypericum perforatum*), and valerian (*Valeriana officinalis*). Although not conclusive, several studies have confirmed that bioactive components in black cohosh may relieve hot flashes in some women.[88] Some menopausal women also use St. John's wort to relieve mild to moderate depression. In fact, several studies show that women taking St. John's wort and black cohosh together had the most improvement in mood and anxiety during menopause.[89] Although some women report that valerian is an effective sleep aid, few studies show its effectiveness.[90] In addition, valerian does not appear to be effective in reducing the occurrence of depression or anxiety.

It is not surprising that many women are turning to alternative therapies to manage perimenopausal and menopausal symptoms. However, women must be cautioned that nonconventional treatments have not been extensively studied and little is known about their long-term effects or safety. In addition to these alternative approaches, it is equally important for women to recognize that an active and healthy lifestyle can also help ease menopausal discomforts. For example, exercise, stress management, and relaxation techniques can improve sleeplessness, mood swings, anxiety, and depression. Exercise also helps prevent the unwanted weight gain that many women experience during this phase of life.

for adult men. Regardless, adequate iron intakes remain a concern, especially for older adults, who often limit their intake of meat, poultry, and fish.

Changes in the Gastrointestinal Tract Age-related changes in the GI tract affect nutritional status in adults, especially older adults. For example, aging muscles can become less responsive to neural signals, slowing the movement of food through the GI tract. Decreased GI motility is one reason why constipation is more common in older adults than in younger ones. When fecal material remains in the colon for prolonged periods, too much water reabsorption can occur, resulting in hard, compacted stool. Drinking adequate amounts of fluids and eating fiber-rich foods can improve GI function and help prevent this problem. Although the recommended fiber intake for older adults is 20 to 35 g/day, older men and women consume on average only 18 and 14 g/day, respectively.[91] There are many reasons as to why older adults may be reluctant to consume foods rich in dietary fiber, but it is important that they find ways to incorporate fiber into their diets.

Neuromuscular changes in the GI tract can also make swallowing more difficult. Difficulty swallowing, a condition called *dysphagia*, is a common cause of choking. This frightening experience can cause older adults to avoid eating certain foods. Preparing foods so that they are moist and have a soft texture is sometimes helpful. People who have difficulty swallowing should seek medical assistance.

With increased age, there is also inflammation of the stomach accompanied by a decline in the number of cells that produce gastric secretions such as hydrochloric acid and intrinsic factor. This condition, called **atrophic gastritis,** can decrease the bioavailability of nutrients such as calcium, iron, biotin, folate, vitamin B_{12}, and zinc. Without intrinsic factor, vitamin B_{12} cannot be absorbed, which can lead to vitamin B_{12} deficiency. Recall from Chapter 10 that pernicious anemia, which is different from the type of vitamin B_{12} deficiency associated with aging, can also occur in the elderly. Symptoms associated with vitamin B_{12} deficiency include dementia, memory loss, irritability, delusions, and personality changes, all of which can easily be overlooked in older adults. To avoid the development of vitamin B_{12} deficiency, older adults are often advised to take vitamin B_{12} supplements or consume adequate amounts of foods fortified with vitamin B_{12}.

⟨**CONNECTIONS**⟩ Recall that intrinsic factor is produced by the gastric cells and is needed for the absorption of vitamin B_{12} (Chapter 10, page 443).

Other Nutritional Issues in Older Adults Other age-related physiological changes can contribute to nutritional deficiencies in older adults. For example, problems with oral health, missing teeth, or poorly fitting dentures can make food less enjoyable, often limiting the types of foods a person eats. Nutrient-dense foods that require chewing, such as meat, fruits, and vegetables, can cause pain, embarrassment, and discomfort for older adults and therefore may be avoided. It is important for older people who are experiencing problems with oral health to get proper dental care.

Changes in Sensory Stimuli Sensory changes in taste and smell can also affect food intake in older adults. With age, the ability to smell diminishes, making food tasteless and unappealing. Certain medications can also alter taste and diminish appetite. Older adults may find that adding spices to foods makes them more appealing, flavorful, and enjoyable. With advanced age, some people also experience a pattern of weight loss caused by a general decline in appetite, or what is called **anorexia of aging.** Anorexia of aging puts individuals at especially increased risk for protein-energy malnutrition.

Inadequate Fluid Intake The sensation of thirst can also become blunted with age. Because of this, many older adults do not consume enough fluid and are at increased risk for dehydration. Dehydration can upset the balance of electrolytes in cells and tissues. A lack of fluid can also disrupt bowel function and can exacerbate constipation. Not only can certain medications increase water loss from the body, but some older adults intentionally limit fluid intake because of embarrassment over loss of bladder control. Symptoms of dehydration, which are often overlooked in the elderly, include headache, dizziness, fatigue, clumsiness, visual disturbances, and confusion. Adequate fluid consumption and early detection of dehydration in older adults is very important.

atrophic gastritis (gas – TRI – tis) Inflammation of the mucosal membrane that lines the stomach, reducing the ability of the stomach to produce gastric secretions.

anorexia of aging Loss of appetite in the elderly that leads to weight loss and overall physiological decline.

The AI for total water (drinking water, beverages, and foods) for adults is 2.7 and 3.7 liters/day for women and men, respectively. To stay fully hydrated, older adults should drink 9 to 13 cups of water or other beverages every day, plus additional fluids from foods with high water contents.

Drug–Nutrient Interactions The elderly are at particularly high risk for adverse drug–nutrient interactions. This is, in part, because older adults often take multiple medications to treat a variety of chronic diseases. The most frequently prescribed medications for people over 50 years of age are for arthritis, hypertension, type 2 diabetes, cancer, and heart disease. The more medications a person takes, the greater the risk of experiencing a drug–nutrient interaction. While some drugs can cause loss of appetite, leading to inadequate food intake, others can alter taste, making food unpleasant to eat. Even nutrient absorption can be affected by medications. It is important for elderly people to be aware of such problems and seek advice regarding the nutrient-related side effects of their medications.

ASSESSING NUTRITIONAL RISK IN OLDER ADULTS

Clearly, many factors put older adults at increased nutritional risk. For this reason, several national health organizations jointly sponsored the Nutritional Screening Initiative (NSI), a collaborative effort to improve nutritional health in older adults. The NSI helped to identify risk factors closely associated with poor nutritional status in this group. Once compiled, these risk factors were used to develop a screening tool called the NSI DETERMINE checklist, shown in Figure 14.12. As you can see, each nutritional risk factor is represented in the DETERMINE acronym. For example "D" stands for "disease," and "E" refers to "eating poorly." Individuals with a nutritional score of 6 or more on the NSI DETERMINE checklist are considered to be at high nutritional risk.

A Meals on Wheels volunteer delivers meals to a home-bound elderly person.

Many services help improve nutritional status and overall health for older adults. For example, most communities have congregate meal programs where people can enjoy nutritious, low-cost meals in the company of others. Many congregate meal programs are federally subsidized, and total cost is often based on an individual's ability to pay. Another program that provides meals to senior citizens is organized and implemented by the Meals on Wheels® Association of America. This program, which relies heavily on volunteers, delivers low-cost meals to home-bound elderly adults and others who are disabled. The mission of Meals on Wheels is to provide national leadership to end senior hunger, with a vision to end senior hunger by 2020. The federally funded Supplemental Nutrition Assistance Program (SNAP; formerly known as the Food Stamp Program) also assists with food-related expenses of low-income seniors.

Critical Thinking: Amy's Story What organizations in your community are available to assist low-income families and home-bound elderly individuals? Do any of these programs distribute fresh produce? If not, what would be the barriers to launching such an organization? What would be the benefits? What makes Backyard Harvest unique in terms of assisting community members?

FIGURE 14.12 Nutritional Screening Initiative DETERMINE Checklist The Nutrition Screening Initiative was an effort by several organizations to develop screening tools to assess risk factors associated with poor nutritional health in older adults.

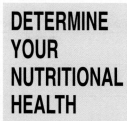

DETERMINE YOUR NUTRITIONAL HEALTH

Circle the number in the "yes" column for those that apply to you or someone you know. Total your nutritional score.

	YES
I have had an illness or condition that made me change the kind and/or amount of food I eat.	2
I eat fewer than 2 meals per day.	3
I eat few fruits or vegetables or milk products.	2
I have 3 or more drinks of beer, liquor or wine almost every day.	2
I have tooth or mouth problems that make it hard for me to eat.	2
I don't always have enough money to buy the food I need.	4
I eat alone most of the time.	1
I take 3 or more different prescribed or over-the-counter drugs a day.	1
Without wanting to, I have lost or gained 10 pounds in the last 6 months.	2
I am not always physically able to shop, cook and/or feed myself.	2
TOTAL	

Total Your Nutritional Score. If it's—

0–2	Good! Recheck your nutritional score in 6 months.
3–5	You are at moderate nutritional risk. See what can be done to improve your eating habits and lifestyle.
6 or more	You are at high nutritional risk. Bring this Checklist the next time you see your doctor, dietitian or other qualified health or social service professional.

DETERMINE: Warning signs of poor nutritional health.

DISEASE
Any disease, illness or chronic condition which causes you to change the way you eat, or makes it hard for you to eat, puts your nutritional health at risk.

EATING POORLY
Eating too little and eating too much both lead to poor health. Eating the same foods day after day or not eating fruit, vegetables, and milk products daily will also cause poor nutritional health.

TOOTH LOSS/MOUTH PAIN
A healthy mouth, teeth and gums are needed to eat. Missing, loose or rotten teeth or dentures which don't fit well, or cause mouth sores, make it hard to eat.

ECONOMIC HARDSHIP
As many as 40% of older Americans have incomes of less than $6,000 per year. Having less—or choosing to spend less—than $25–30 per week for food makes it very hard to get the foods you need to stay healthy.

REDUCING SOCIAL CONTACT
One-third of all older people live alone. Being with people daily has a positive effect on morale, well-being and eating.

MULTIPLE MEDICINES
Many older Americans must take medicines for health reasons. Almost half of older Americans take multiple medicines daily.

INVOLUNTARY WEIGHT LOSS/GAIN
Losing or gaining a lot of weight when you are not trying to do so is an important warning sign that must not be ignored. Being overweight or underweight also increases your chance of poor health.

NEEDS ASSISTANCE IN SELF CARE
Although most older people are able to eat, one of every five have trouble walking, shopping, buying food and cooking food, especially as they get older.

ELDER YEARS ABOVE AGE 80
Most older people lead full and productive lives. But as age increases, risk of frailty and health problems increase. Checking your nutritional health regularly makes good sense.

SOURCE: Adapted from Nutrition Screening Initiative. Report of Nutrition Screening I: Toward a Common View. Washington DC: Nutrition Screening Initiative; 1991.

Diet Analysis PLUS ✚ Activity

Assume that you have a sister or daughter, Sarah, who is 25 years old, weighs 130 pounds, is 5 feet 4 inches tall, is a nonsmoker, and is not a vegetarian. Sarah has recently discovered she is pregnant and needs advice as to her changing nutritional needs. She plans to breastfeed almost exclusively during the first 6 months of her baby's life. She would like to know more about her nutritional needs during that time as well. Using your Diet Analysis program, set up a profile using Sarah's information. Print out the *Profile DRI Goals* report choosing simply "Female" from the "Sex" drop-down menu to generate a report representing her nutritional needs in the nonpregnant condition. Then, go back and choose "Pregnant" from the drop-down menu, changing nothing else. Print out this second *Profile DRI Goals* report. Finally, repeat the steps choosing "Lactating" from the menu. By now you should have three *Profile DRI Goals* reports representing the three physiological conditions.

Using these print-outs and this chapter, fill in dietary recommendations for energy and essential nutrients on the following table:

Dietary Component	Not Pregnant	Pregnant	Lactating
Calories (kcal/day)			
Protein (g/day)			
Carbohydrates (g/day)			
Total fat (g/day)			
Linoleic acid (g/day)			
Linolenic acid (g/day)			
Vitamin A (µg/day)			
Folate (µg DFE/day)			
Vitamin C (mg/day)			
Calcium (mg/day)			
Iron (mg/day)			

Analysis

1. In general, what purpose(s) do the nutrient increases during the pregnant phase fulfill? Are there specific organ systems or processes (of the mother and/or the infant) that these particular nutrients support?

2. Why do you suppose that calorie needs during the lactating phase (the second 6 months) are different than those during pregnancy?

3. Why are calcium recommendations unchanged during pregnancy and throughout lactation for this 25-year-old?

Notes

1. Centers for Disease Control and Prevention. National Center for Health Statistics. Clinical growth charts. Available from: http://www.cdc.gov/nchs/about/major/nhanes/growthcharts/clinical_charts.htm. World Health Organization. The WHO child growth standards. 2006. Available from: http://www.who.int/childgrowth.

2. National Center for Health Statistics. Data on birth defects. Available from: http://www.cdc.gov/nchs/FASTATS/bdefects.htm.

3. Brent RL. Environmental causes of human congenital malformations: The pediatrician's role in dealing with these complex clinical problems caused by a multiplicity of environmental and genetic factors. Pediatrics. 2004;113:957–68.

4. Eustace LW, Kang DH, Coombs D. Fetal alcohol syndrome: A growing concern for health care professionals. Journal of Obstetrics, Gynecologic, and Neonatal Nursing. 2003;32:215–21.

5. Villar J, Merialdi M, Gulmezoglu AM, Abalos E, Carroli G, Kulier R, de Onis M. Characteristics of randomized controlled trials included in systematic reviews of nutritional interventions reporting maternal morbidity, mortality, preterm delivery, intrauterine growth restriction and small for gestational age and birth weight outcomes. Journal of Nutrition. 2003;133:1632S–9S.

6. National Center for Health Statistics. Centers for Disease Control and Prevention. Preliminary births for 2004: Infant and maternal health. Available from: http://www.cdc.gov/nchs/products/pubs/pubd/hestats/prelimbirths04/prelimbirths04health.htm.

7. Centers for Disease Control and Prevention. U.S. Department of Health and Human Services. Martin JA, Hamilton BE, Sutton PD, Ventura SJ, Mathews TH, Osterman MJK. Births: Final data for 2008. National vital statistics reports; vol 59 no 1. Hyattsville, MD: National Center for Health Statistics. 2010. Available from http://www.cdc.gov/nchs/data/nvsr/nvsr59/nvsr59_01.pdf.

8. Gillman MW. Developmental origins of health and disease. New England Journal of Medicine. 2005;353:1848–50.

9. Barker DJ. The foetal and infant origins of inequalities in health in Britain. Journal of Public Health Medicine. 1991;13:64–8.

10. Barker DJ. Fetal programming of coronary heart disease. Trends in Endocrinology and Metabolism. 2002;13:364–8.

11. Rasmussen KM. The "fetal origins" hypothesis: Challenges and opportunities for maternal and child nutrition. Annual Review of Nutrition. 2000;21:73–95.

12. Waterland RA, Michels KB. Epigenetic epidemiology of the developmental origins hypothesis. Annual Review of Nutrition. 2007;27:363–88.

13. Institute of Medicine and National Research Council. Weight gain during pregnancy: Reexamining the guidelines. Washington, DC: The National Academies Press. May 2009. Available from: http://www.iom.edu/pregnancyweightgain.

14. U.S. Department of Health and Human Services and U.S. Environmental Protection Agency. What you need to know about mercury in fish and shellfish. March 2004. Available from http://www.cfsan.fda.gov/~dms/admehg3.html.

15. Beard JL. Effectiveness and strategies of iron supplementation during pregnancy. American Journal of Clinical Nutrition. 2000;71:1288S–94S.

16. U.S. Preventive Services Task Force. Recommendation Statement. Screening for iron deficiency anemia—Including iron supplementation for children and pregnant women. Rockville, MD: Agency for Healthcare Research and Quality (AHRQ): 2006. Available from http://www.uspreventiveservicestaskforce.org/.

17. Feinleib M, Beresford SA, Bowman BA, Mills JL, Rader JI, Selhub J, Yetley EA. Folate fortification for the prevention of birth defects: Case study. American Journal of Epidemiology. 2001;154:S60–9.

18. Misra DP, Astone N, Lynch CD. Maternal smoking and birth weight: Interaction with parity and mother's own in utero exposure to smoking. Epidemiology. 2005;16:288–93.

19. Husgafvel-Pursiainen K. Genotoxicity of environmental tobacco smoke: A review. Mutation Research. 2004;567:427–45.

20. Di Cianni G, Ghio A, Resi V, Volpe L. Gestational diabetes mellitus: An opportunity to prevent type 2 diabetes and cardiovascular disease in young women. Women's Health. 2010;6:97–105.

21. Papageorghiou AT, Campbell S. First trimester screening for preeclampsia. Current Opinion in Obstetrics and Gynecology. 2006;18:594–600.

22. American Academy of Pediatrics. Pediatric nutrition handbook, 6th ed., Elk Grove Village, IL. 2008.

23. Leon DA, Ronalds G. Breast-feeding influences on later life— cardiovascular disease. Advances in Experimental and Medical Biology. 2009;639;153–66.

24. Beyerlein A von Kries R. Breastfeeding and body composition in children: will there ever be conclusive empirical evidence for a protective effect against overweight? American Journal of Clinical Nutrition. 2011.

25. Stettler N, Zemel BS, Kumanyika S, Stallings VA. Infant weight gain and childhood overweight status in a multicenter, cohort study. Pediatrics. 2002;109:194–9.

26. Agostoni C, Giovannini M. Cognitive and visual development: Influence of differences in breast and formula fed infants. Nutrition and Health. 2001;15:183–8.

27. Dewey KG. Impact of breastfeeding on maternal nutritional status. Advances in Experimental Medical Biology. 2004; 554:91–100. Winkvist A, Rasmussen KM. Impact of lactation on maternal body weight and body composition. Journal of Mammary Gland Biology and Neoplasia. 1999;4:309–18.

28. Taylor JS, Kacmar JE, Nothnagle M, Lawrence RA. A systematic review of the literature associating breastfeeding with type 2 diabetes and gestational diabetes. Journal of the American College of Nutrition. 2005;24:320–6. Labbok MH. Effects of breastfeeding on the mother. Pediatric Clinics of North America. 2001;48:143–58.

29. Centers for Disease Control and Prevention. Use of world health organization and CDC growth charts for children aged 0–59 months in the United States. Morbidity and Mortality Weekly Review. 2010;59:1-15. Available from http://www.cdc.gov/mmwr/preview/mmwrhtml/rr5909a1.htm.

30. Ryan AS, Wenjun Z, Acosta A. Breastfeeding continues to increase into the new millennium. Pediatrics. 2002;110:1103–9.

31. Grummer-Strawn LM, Scanlon KS, Fein SB. Infant feeding and feeding transitions during the first year of life. Pediatrics. 2008;122:S36–42.

32. U.S. Department of Health and Human Services. The Surgeon General's call to action to support breastfeeding. Washington, DC: U.S. Department of Health and Human Services, Office of the Surgeon General; 2011. Available from http://www.surgeongeneral.gov.

33. U.S. Centers for Disease Control and Prevention. Breastfeeding among U.S. children born 1999–2007. CDC National Immunization Survey. Available from http://www.cdc.gov/breastfeeding/data/NIS_data/.

34. Lind T, Hernell O, Lonnerdal B, Stenlund H, Domellof M, Persson LA. Dietary iron intake is positively associated with hemoglobin concentration during infancy but not during the second year of life. Journal of Nutrition. 2004;134:1064–70.

35. Beyerlein A, Hadders-Algra M, Kennedy K, Fewtrell M, Singhal A, Roesenfeld E, Lucas A, Bouwstra H, Koletzko B, von Kries R. Infant formula supplementation with long-chain polyunsaturated fatty acids has no effect on Bayley developmental scores at 18 months of age—IPD meta-analysis of 4 large clinicial trials. Journal of Pediatric Gastroenterology and Nutrition. 2010;50:79–84.

36. Camadoo L, Tibbott R, Isaza F. Maternal vitamin D deficiency associated with neonatal hypocalcaemic convulsions. Nutrition Journal. 2007;6:23.

37. Basile LA, Taylor SN, Wagner CL, Horst RL, Hollis BW. The effect of high-dose vitamin D supplementation on serum vitamin D levels and milk calcium concentration in lactating women and their infants. Breastfeeding Medicine. 2006;1:27–35.

38. Wagner CL and Greer FR. Prevention of rickets and vitamin D deficiency in infants, children, and adolescents Pediatrics. 2008;122:1142–52.

39. American Academy of Pediatrics. Breastfeeding and the use of human milk. Pediatrics. 2005;115:496–501. Brotanek JM, Gosz J, Weitzman M, Flores G. Iron deficiency in early childhood in the United States: Risk factors and racial/ethnic disparities. Pediatrics. 2007;120:568–75.

40. American Academy of Pediatrics Committee on Nutrition. The use and misuse of fruit juice in pediatrics. Pediatrics. 2001;107:1210–3.

41. Huh SY, Rifas-Shiman SL, Taveras EM, Oken EM, Gillman MW. Timing of solid food introduction and risk of obesity in preschool aged children. Pediatrics. 2011;127:544–51.

42. Kleinman RE. American Academy of Pediatrics recommendations for complementary feeding. Pediatrics. 2000;106:1274.

43. Thygarajan A, Burks AW. American Academy of Pediatrics recommendations on the effects of early nutritional interventions on the development of atopic disease. Current Opinion in Pediatrics. 2008;20:698–702. Greer FR, Sicherer SH, Burks W; American Academy of Pediatrics Committee on Nutrition; American Academy of Pediatrics Section on Allergy and Immunology. Effects of early nutritional interventions on the development of atopic disease in infants and children: The role of maternal dietary restriction, breastfeeding, timing of introduction of complementary foods, and hydrolyzed formulas. Pediatrics. 2008;121:183–91.

44. Tanzi MG, Gabay MP. Association between honey consumption and infant botulism. Pharmacotherapy. 2002;22:1479–83. Centers for Disease Control and Prevention. Botulism. Available from: http://www.cdc.gov/nczved/divisions/dfbmd/diseases/botulism/.

45. Whitaker RC, Wright JA, Pepe MS, Seidel KD, Dietz WH. Predicting obesity in young adulthood from childhood and parental obesity. New England Journal of Medicine. 1997;337:869–73.

46. Freedman DS, Mei Z, Srinivasan SR, Berenson GS, Dietz WH. Cardiovascular risk factors and excess adiposity among overweight children and adolescents: the Bogalusa Heart Study. Journal of Pediatrics. 2007;150:12–17.

47. National Center for Health Statistics. Health, United States, 2010: With Special Feature on Death and Dying. Hyattsville, MD. 2011.

48. Satter E. The feeding relationship: Problems and interventions. Journal of Pediatrics. 1990;117:181–9.

49. l'Allemand-Jander D. Clinical diagnosis of metabolic and cardiovascular risks in overweight children: early development of chronic diseases in the obese child. Internatiolnal Journal of Obesity. 2010;34:S32–6.

50. Robinson TN. Television viewing and childhood obesity. Pediatric Clinics of North America. 2001;4:1017–25.

51. Crespo CJ, Smit E, Troiano RP, Bartlett SJ, Macera CA, Andersen RE. Television watching, energy intake, and obesity in US children: Results from the third National Health and Nutrition Examination Survey, 1988–1994. Archives of Pediatric and Adolescent Medicine. 2001;155:360–5.

52. Chaput J-P, Visby T, Nyby S, Klingenberg L, Gregersen NT, Tremblay A, Astrup A, Sjödin A. Video game playing increases food intake in adolescents: a randomized crossover study. American Journal of Clinical Nutrition. 2011; 93:1196–203.

53. American Academy of Pediatrics Committee on Public Education. Children, adolescents, and television. Pediatrics. 2001;107:423–6.

54. Robinson TN, Borzekowski DL, Matheson DM, Kraemer HC. Effects of fast food branding on young children's taste preferences. Archives of Pediatric and Adolescent Medicine. 2007;161:792–7.

55. Kann L, Brener ND, Wechsler H. Overview and summary: School health policies and programs study 2006. Journal of School Health. 2007;77:385–97.

56. U.S. Department of Health and Human Services. 2008 physical activity guidelines for Americans. Washington, D.C. Available from http://www.health.gov/paguidelines/pdf/paguide.pdf.

57. American Academy of Pediatrics Committee on Public Education. Soft drinks in schools. Pediatrics. 2004;113:152–4.

58. Institute of Medicine. Schools can play a role in preventing childhood obesity. Preventing childhood obesity: Health in the balance. 2005. Available from: http://www.iom.edu.

American Academy of Pediatrics. Soft drinks in schools. Pediatrics. 2004;113:152–4.

59. Koivisto Hursti UK. Factors influencing children's food choice. Annals of Medicine. 1999;31:26–32.

60. Patrick H, Nicklas TA. A review of family and social determinants of children's eating patterns and diet quality. Journal of American College of Nutrition. 2005;2:83–92.

61. Centers for Disease Control and Prevention. Recommendations to prevent and control iron deficiency in the United States. Morbidity and Mortality Weekly Review. 1998;47(RR-3):1–29. Kazal LA. Prevention of iron deficiency in infants and toddlers. American Family Physician. 2002;66:1217–25.

62. Soemantri AG, Pollitt E, Kim I. Iron deficiency anemia and educational achievement. American Journal of Clinical Nutrition. 1985;42:1221–8. Iannotti LL, Tielsch JM, Black MM, Black RE. Iron supplementation in early childhood: Health benefits and risks. American Journal of Clinical Nutrition. 2006;84:1261–76.

63. Lee JM, Appugliese D, Kaciroti N, Corwyn RF, Bradley RH, Lumeng JC. Weight status in young girls and the onset of puberty. Pediatrics. 2007;119:e624–30.

64. Enns CW. Mickle SH, Goldman JD. Trends in food and nutrient intakes by adolescents in the United States. Family Economics and Nutrition Review. 2003;15:15–27.

65. Neumark-Sztainer D, Story M, Hannan PH, Croll J. Overweight status and eating patterns among adolescents: where do youths stand in comparison with Healthy People 2010 objectives? American Journal of Public Health. 2002;92:844–51. Larson NI, Neumark-Sztainer D, Hannan PJ, Story M. Trends in adolescent fruit and vegetable consumption, 1999–2004: Project EAT. American Journal of Preventive Medicine. 2007;32:147–50.

66. U.S. Department of Agriculture. Agricultural Research Service. Snacking patterns of U.S. adolescents. What we eat in America, NHANES 2005–2006. Food Surveys Research Group. Dietary Data Brief No. 2. October 2010. Available from http://www.ars.usda.gov/SP2UserFiles/Place/12355000/pdf/DBrief/snacking_0506.pdf.

67. Nielsen SJ, Popkin BM. Changes in beverage intake between 1977 and 2001. American Journal of Preventive Medicine. 2004;27:205–10.

68. Troiano RP, Briefel RR, Carroll MD, Bialostosky K. Energy and fat intakes of children and adolescents in the United States: Data from the National Health and Nutrition Examination Surveys 1,2,3. American Journal of Clinical Nutrition. 2000;72;1343S–53s.

69. Bailey RL, Dodd KW, Goldman JA, Gahche JJ, Dwyer JT, Moshfegh AJ, Sempos CT, Picciano MF. Estimation of total usual calcium and vitamin D intakes in the United States. Journal of Nutrition. 2010;140:817–22.

70. What We Eat In America, NHANES, 2003–2004. U.S. Department of Agriculture, Agricultural Research Service. 2007. Nutrient intakes from food: Mean amounts consumed per individual, one day, 2003–2004. Available from: http://jn.nutrition.org/cgi/content/full/133/2/609S.

71. Harel Z, Riggs S, Vaz R, White L, Menzies G. Adolescents and calcium: What they do and do not know and how much they consume. Journal of Adolescent Health. 1998;22:225–8. Martin JT, Coviak CP, Gendler P, Kim KK, Cooper K, Rodrigues-Fisher L. Female adolescents' knowledge of bone health promotion behaviors and osteoporosis risk factors. Orthopaedic Nursing. 2004;23:235–44.

72. Centers for Disease Control and Prevention. Recommendations to prevent and control iron deficiency anemia in

the United States. Morbidity Mortality Weekly Report. 2002;51:897–9.

73. Bailey RL, Dodd KW, Gahche JJ, Dwyer JT, McDowell MA, Yetley EA, Sempos CA, Burt VL, Radimer KL, Picciano MF. Total folate and folic acid intake from foods and dietary supplements in the United States: 2003–2006. American Journal of Clinical Nutrition. 2010;91:231–7.

74. Centers for Disease Control and Prevention. The state of aging and health in America. 2004. Available from: http://www.cdc.gov/aging/pdf/state_of_aging_and_health_in_america_2004.pdf.

75. Shay JW, Wright WE. Hayflick, his limit, and cellular ageing. National Review of Molecular Cellular Biology. 2000;1:72–6.

76. Willcox JK, Ash SL, Catignani GL. Antioxidants and prevention of chronic disease. Critical Reviews in Food Science and Nutrition. 2004;44:275–95.

77. Stanner SA, Hughes J, Kelly CN, Buttriss J. A review of the epidemiological evidence for the antioxidant hypothesis. Journal of Public Health Nutrition. 2004;7:407–22.

78. Chernoff R. Geriatric nutrition: The health professional's handbook, 3rd ed. Gaithersburg, MD: Aspen; 2006. Moretti C, Frajese GV, Guccione L, Wannenes F, De Martino MU, Fabbri A, Frajese G. Androgens and body composition in the aging male. Journal of Endocrinological Investigation. 2005;28:56–64.

79. Centers for Disease Control and Prevention. Falls among older adults: An overview. 2010. Available from http://www.cdc.gov/HomeandRecreationalSafety/Falls/adultfalls.html.

80. Chernoff R. Protein and older adults. Journal of the American College of Nutrition. 2004;23:627S–30S.

81. Beral V, Banks E, Reeves G. Evidence from randomised trials on the long-term effects of hormone replacement therapy. Lancet. 2002;360:942–4.

82. Nelson HD, Humphrey LL, Nygren P, Teutsch SM, Allan JD. Postmenopausal hormone replacement therapy: Scientific review. Journal of the American Medical Association. 2002;288:872–81.

83. Kim N, Gross C, Curtis J, Stettin G, Wogen S, Choe N, Krumholz HM. The impact of clinical trials on the use of hormone replacement therapy: A population-based study. Journal of General Internal Medicine. 2005;20:1026–31. Hersh AL, Stefanick ML, Stafford RS. National use of postmenopausal hormone therapy: Annual trends and response to recent evidence. Journal of the American Medical Association.2004;291:47–53.

84. The role of soy isoflavones in menopausal health: report of The North American Menopause Society. Wulf H. Utian Translational Science Symposium in Chicago, IL. 2010. Menopause. 2011.

85. Cirigliano M. Bioidentical hormone therapy: A review of the evidence. Journal of Women's Health. 2007;16:600–31. Moskowitz D. A comprehensive review of the safety and efficacy of bioidentical hormones for the management of menopause and related health risks. Alternative Medicine Review. 2006;11:208–23. Boothby LA, Doering PL, Kipersztok S. Bioidentical hormone therapy: A review. Menopause. 2004;11:356–67.

86. The American College of Obstetricians and Gynecologists. Compounded Bioidentical Hormones. Obstetrics and Gynecology. 2005;322:1139–40.

87. U.S. Food and Drug Administration. Bio-identicals: sorting myths from facts. Consumer Updates. April 8, 2008. Available from http://www.fda.gov/ForConsumers/Consumer Updates/ucm049311.htm.

88. Rotem C, Kaplan B. Phyto-Female Complex for the relief of hot flushes, night sweats and quality of sleep: Randomized, controlled, double-blind pilot study. Gynecological Endocrinology. 2007;23:117–22. Oktem M, Eroglu D, Karahan HB, Taskintuna N, Kuscu E, Zeyneloglu HB. Black cohosh and fluoxetine in the treatment of postmenopausal symptoms: A prospective, randomized trial. Advances in Therapy. 2007;24:448–61.

89. Geller SE, Studee L. Botanical and dietary supplements for mood and anxiety in menopausal women. Menopause. 2007;14:541–9. Uebelhack R, Blohmer JU, Graubaum HJ, Busch R, Gruenwald J, Wernecke KD. Black cohosh and St. John's wort for climacteric complaints: A randomized trial. Obstetrics and Gynecology. 2006;107:247–55.

90. Taibi DM, Landis CA, Petry H, Vitiello MV. A systematic review of valerian as a sleep aid: Safe but not effective. Sleep Medicine Review. 2007;11:209–30.

91. Mozaffarian D, Kumanyika SK, Lemaitre RN, Olson JL, Burke GL, Siscovick DS. Cereal, fruit, and vegetable fiber intake and the risk of cardiovascular disease in elderly individuals. Journal of the American Medical Association. 2003;289:1659–66.

Food Security, Hunger, and Malnutrition

We have all experienced hunger—the physical sensation of not having enough to eat. While most people ease their hunger by eating, for others this may not always be possible. When sufficient food is not available or accessible, hunger can lead to serious physical, social, and psychological consequences, a condition referred to as food insecurity. The prevalence of food insecurity in the world is astonishing. Although nobody knows for sure the exact number of people who experience food insecurity, the United Nations Food and Agriculture Organization

(FAO) estimates that approximately 925 million people worldwide experience persistent hunger.[1]

It is important to know that hunger and food insecurity can affect people of all ages and in every country in the world—even wealthy countries such as the United States. In this Nutrition Matters, you will learn about concerns related to the causes and complications of food insecurity, hunger, and malnutrition worldwide. You will also learn about domestic and international assistance programs designed to alleviate these devastating public health problems.

What Is Food Security?

Food security is defined as the condition whereby people are able to obtain sufficient amounts of nutritious food for an active, healthy life. Conversely, **food insecurity** exists when people do not have adequate physical, social, or economic access to food.[2] Many people are surprised to learn that there is enough food produced in the world to provide every person with at least 2,700 kcal each day.[2] In other words, food insecurity is not caused by insufficient worldwide food production. Rather, people become food insecure when they cannot obtain sufficient food to feed themselves or their families. Other constraints such as limited physical or mental functioning can also contribute to food insecurity. In reality, the principal factors related to food insecurity in the world are poverty, war, and natural disaster.

There are varying degrees of food insecurity, with chronic hunger and undernutrition representing the more severe consequences. While the word "hunger" is often used to describe the physical discomfort experienced by individuals consuming insufficient food, on a global level it is more commonly used to describe a shortage of available food. Because many people live with uncertainty as to whether they will have enough to eat, food

---

food security The condition whereby individuals are able to obtain sufficient amounts of nutritious food to live active, healthy lives.

food insecurity The condition whereby individuals or families cannot obtain sufficient food.

insecurity and its resulting hunger are major social concerns in the world today.

PEOPLE RESPOND TO FOOD INSECURITY IN DIFFERENT WAYS

People living in food-insecure households respond to the threat of hunger in different ways. Whereas some individuals take advantage of charitable organizations that assist people in need, others may resort to obtaining food by stealing, begging, or scavenging.[3] For many people, food insecurity can cause feelings of alienation, deprivation, and distress. In addition, it can adversely affect family dynamics and social interactions within the larger context of community. Of course, food insecurity that results in hunger can also lead to malnutrition and its associated health complications. Clearly, the short- and long-term consequences of food insecurity for each person in a household can be devastating.

Prevalence of Food Insecurity in the United States

How much money do you spend on food each week? On average, Americans spend $44 per person on weekly food purchases.[4] Households experiencing food insecurity spend considerably less: The U.S. Department of Agriculture (USDA) recently reported that food-insecure households spend 33% less on food than food-secure households of the same size and composition.[5] Determining the extent of food insecurity is a scientific challenge. However, only when researchers have determined the scope and nature of the problem can public health officials develop effective strategies and target them to the appropriate populations. Just as there is no single cause of food insecurity, there is no one solution to the problem.

In prosperous countries, the prevalence of food insecurity is often difficult to assess because it is usually not associated with detectable signs of malnutrition. For this reason, clinical measurements of nutritional status such as anthropometry (for example, weight and height) are not always useful indicators of food insecurity. Instead, the prevalence and incidence of food insecurity in U.S households is typically assessed using data regarding food availability and access. For example, the USDA uses a survey that asks household members to answer questions about behaviors related to food availability and access.[5] Depending on the responses, individuals or households are classified as having either low food security or very low food security. **Low food security** refers to households in which one or more members experienced disrupted eating patterns (or were worried they would do so) because money or access to food was

Food insecurity refers to the situation in which households do not have sufficient food.

insufficient. Those classified as having **very low food security** experience multiple indications of food insecurity, sometimes leading to reduced food intake. Sixty-five percent of households classified as having very low food security reported that at least one member had gone hungry at some time during the year as a result of not having enough money for food.[5]

Recent survey results indicate that 15% of Americans—more than 17 million households—struggle to provide enough food for all their family members, and more than 17 million (23%) U.S. children live in households where food is scarce at times.[5] Approximately one-third of food insecure households reported that their eating patterns were greatly affected because they lacked money or other resources to purchase food. Families with incomes below the poverty line, many of which are households headed by single mothers, were at greatest risk of hunger. With the current downward trend in the U.S. economy and the increasing cost of food, which rose nearly 3% between 2010 and 2011, the number of U.S. households experiencing food insecurity is expected to increase even more.[5]

POVERTY IS THE UNDERLYING FACTOR ASSOCIATED WITH FOOD INSECURITY

A person's risk for experiencing food insecurity in the United States is associated with income, ethnicity, family structure, and location of the home. Because many of these risk factors are intertwined,

low food security Classification of food-insecure households in which one or more members experience disrupted eating patterns (or is worried they will do so) because of insufficient resources or access to food.

very low food security Households that experience multiple indications of disrupted eating patterns due to inadequate resources or access to food, sometimes leading to reduced food intake.

Millions of Americans struggle to get enough to eat. Although food banks help provide food to those in need, many report that critical shortages of food are making it difficult to meet the increasing demands.

it can be difficult to determine how much each one independently contributes to food insecurity. Nonetheless, the link between income—specifically poverty—and food insecurity is indisputable and is often the "common thread" among these factors.

It is important to recognize that many people who live in poverty maintain steady employment. Based on data provided by the U.S. Census Bureau, there were 10 million low-income working families in 2009, and approximately 30% of working families are officially low-income.[6] In fact, experts estimate that 40% of adults requesting emergency food assistance are employed.[7] In 2009 alone, approximately 5 million households turned to private food pantries and soup kitchens to provide food for their families.[8] To make matters worse, people who at one time contributed items to food banks are now finding themselves in need.[9] It is not surprising that many food banks around the country are experiencing a sharp decline in donations as the demand for food rises.[10]

When money is limited, people are often forced to reduce food-related expenditures to pay for such things as housing, utilities, or health care.[11] Although one in three households living in poverty is food insecure, some households with incomes above the poverty line also experience food insecurity. This is because unexpected events such as a medical expense or a repair bill can cause some people, at least temporarily, to not have the financial means to purchase sufficient food. Thus, many factors in addition to income can predispose a person or family to food insecurity.

In addition to low income, food insecurity in the United States is more prevalent among certain ethnic groups. For example, black and Latino households are at higher risk for food insecurity than most other racial or ethnic groups.[12] Households headed by single women are at even higher risk. In addition, people living in urban and rural areas are more likely to experience food insecurity than those living in suburban regions. However, it is important to recognize that all three of these factors (ethnicity, head of household, and living location) are strongly associated with income status.[12]

Socioeconomically disadvantaged communities with limited access to affordable, healthy food, tend to be located in urban and rural low-income neighborhoods. These areas are referred to as **food deserts,** and residents who live in them are dependent on local food outlets that offer limited, expensive food choices with low nutritional value.[13] The lack of supermarkets within communities presents additional economic hardships to those living in food deserts. These individuals must either purchase groceries at local convenience stores or navigate public transportation to other communities that have more affordable market choices. These disparities have adverse health outcomes, and these populations tend to have high rates of obesity, type 2 diabetes, and cardiovascular disease.[14] Marketplace incentives to attract food retailers into food deserts are currently underway, with the hope that improved access to affordable healthy foods such as fruits, vegetables, whole grains, and low-fat dairy products will improve access to and affordability of quality food sources. You can learn more about where food deserts are located in the United States by visiting the USDA's Food Desert Locator at http://www.ers.usda.gov/data/fooddesert/.

What Are the Consequences of Food Insecurity?

Although food insecurity in the United States does not typically lead to hunger or nutrient deficiencies, it still represents a major public health concern. There are many consequences of food insecurity, and these have been studied most extensively in women and children. It is not uncommon for food-insecure mothers to shield their children from not having enough to eat by consuming less food themselves. Therefore, women often experience the negative consequences of food insecurity before their children do.[15] For example, some studies have found that women in food-insecure households

---

food desert Community with limited access to affordable, healthful foods; tends to be located in urban and rural low-income neighborhoods.

reduce the size of their meals in order to better provide for their children.[15] Nonetheless, it is the children raised in food-insecure households who experience the most significant and lasting, long-term effects.[16] Studies also show that these children tend to have difficulties in school, earn lower scores on standardized tests, miss more days of school, exhibit more behavioral problems and depression, and have increased risk for suicide. Of course, these factors are likely not direct consequences of food insecurity, but instead repercussions common to households experiencing this problem.[17] Although parents often try to protect their children from the realities of food insecurity, interviews with children reveal considerable circumstantial awareness that is not always apparent to other family members.[18]

It is also important to understand that the elderly are at especially high risk for experiencing food insecurity.[19] However, they often experience it differently from both children and younger adults. Elders with limited mobility and poor health, for example, may have food available to them but have difficulty or anxiety associated with meal preparation. In addition, many older people have relatively low incomes, which they are unable to supplement by additional employment. Poverty rates are highest among older women and among those who live alone.[19]

MANY ORGANIZATIONS PROVIDE FOOD-BASED ASSISTANCE IN THE UNITED STATES

Fortunately, there are many programs and services available in the United States to alleviate food insecurity. Some of these are federally funded, whereas others are community efforts staffed by volunteers. A summary of selected federally funded assistance programs is provided in Table 1 and discussed briefly here.

Supplemental Nutrition Assistance Program One example of a federal food-based assistance program is the **Supplemental Nutrition Assistance Program (SNAP),** formerly known as the Food Stamp Program. SNAP is often the first line of defense against hunger for low-income households. This program, administered by the USDA, helps provide food for more than 28 million people each month.[20] People who are eligible for SNAP are given an electronic benefit transfer (EBT) card that can be used like a debit card to make food

Supplemental Nutrition Assistance Program (SNAP) A federally funded program, formerly known as the Food Stamp Program, that helps low-income households pay for food.

TABLE 1 Examples of Food Assistance Programs in the United States

Program	Major Objective	Website
Child and Adult Care Food Program (CACFP)	• Provides families with affordable, quality day care and nutrition for children and elderly adults.	http://www.fns.usda.gov/cnd/CARE/
Expanded Food and Nutrition Education Program (EFNEP)	• Helps low-income people gain the knowledge, skills, attitudes, and behaviors necessary to maintain nutritionally balanced diets, and learn to contribute to their personal development and the improvement of the total family diet and nutritional well-being. Assists low-income people to acquire knowledge, skills, attitudes, and behaviors necessary to maintain nutritionally balanced diets and to improve family health and nutritional well-being.	http://www.csrees.usda.gov/nea/food/efnep/efnep.html
Supplemental Nutrition Assistance Program (SNAP)	• Provides benefits to low-income people so that they can buy food to improve their diets.	http://www.fns.usda.gov/snap/
Head Start	• Promotes school readiness by enhancing the social and cognitive development of children through the provision of educational, health, nutritional, social, and other services to enrolled children and families.	http://www.acf.hhs.gov/acf_services.html#hs
Meals on Wheels® Association of America	• Delivers meals to people who are elderly, home-bound, disabled, frail, or at risk of malnutrition.	http://www.mowaa.org/
National School Lunch and School Breakfast Programs	• Provides children with nutritious meals for free or at reduced cost.	http://www.fns.usda.gov/cnd/Default.htm
Special Supplemental Nutrition Program for Women, Infants, and Children (WIC)	• Assists in purchase of nutritious food and provides nutrition education to low-income women, infants, and children who are at nutritional risk.	http://www.fns.usda.gov/wic/

purchases at grocery and convenience stores and many farmers' markets. A household's monthly allotment depends on the number of people in the household and their combined income, with the average monthly allotment being approximately $126 for each household member (less than $4.00/day per person).[21] SNAP participants can only use their EBT cards to purchase food; cards cannot be used to buy tobacco, alcohol, paper products, or other non-food items. For many reasons, not all individuals eligible for SNAP actually apply to receive benefits.[22] For some people, applying for SNAP may be a daunting process, whereas others may feel stigmatized by applying for food assistance.

Special Supplemental Nutrition Program for Women, Infants, and Children (WIC) In addition to SNAP, millions of pregnant or lactating women, infants, and children in the United States benefit from the **Special Supplemental Nutrition Program for Women, Infants, and Children,** known as WIC. This federally funded program, which is also administered through the USDA, assists families in making nutritious food purchases by providing coupons that can be used to buy a variety of WIC-approved foods. These foods are generally nutrient-dense foods such as peanut butter, milk, rice, beans, cereal, fruits and vegetables, and

canned tuna. Many farmers' markets also accept WIC coupons, allowing people to purchase a variety of locally grown fresh fruits and vegetables. In addition to assisting with food purchases, WIC provides health assistance and nutrition education to eligible women and young infants and children. The program also encourages exclusive breastfeeding and other optimal infant feeding guidelines set forth by the American Academy of Pediatrics.[23]

Other Federally Funded Food-Based Assistance Programs Other federally funded food-based assistance programs available in the United States are the **National School Lunch Program** and the **School Breakfast Program,** which provide nutritionally balanced meals to school-age children, either free of charge or at a reduced cost. Administered by the USDA, these programs are available in public schools, nonprofit private schools, residential child-care institutions, and after-school enrichment programs. Because there is such a need, many schools and child-care programs also provide breakfast, lunch, and snacks to children throughout summer vacation. The National School Lunch Program provided lunch for more than 31 million children each school day in 2009.[24] Since its establishment in 1946, the National School Lunch Program has served more than 219 billion lunches.[24] In 2010, the Institute of Medicine published a comprehensive document providing guidance as to the optimal types and quantities of foods that should be served in the National School Breakfast and Lunch Programs.[25] This initiative was subsequently followed by passage of the Healthy Hunger-Free Kids Act of 2010. This legislation gives USDA, for the first time in over 30 years, the opportunity to make real reforms to the school lunch and breakfast programs by improving the critical nutrition and hunger safety net for millions of children.[26]

Privately Funded Food Assistance Programs In addition to these and other government-funded programs, the private sector also provides services to make food more available to those in need. These organizations include food recovery programs, food banks and pantries, and food kitchens—many of which are staffed by members of the community.

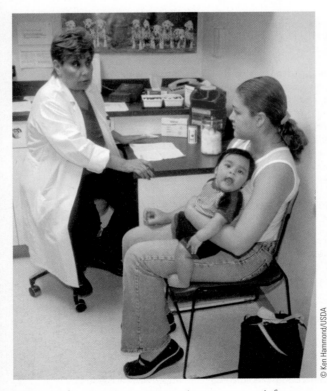

© Ken Hammond/USDA

WIC provides many important services to women, infants, and children.

Special Supplemental Nutrition Program for Women, Infants, and Children (WIC) A federally funded program that assists families in a targeted, at-risk population in making nutritious food purchases.

National School Lunch and School Breakfast Programs Federally funded programs that provide free or subsidized nutritious meals to school-age children.

Food recovery programs collect and distribute food that would otherwise go to waste.

Food recovery programs function by collecting and distributing food that otherwise would be wasted or discarded. Recovered foods are donated to food pantries, emergency kitchens, and homeless shelters, providing millions of people with food. The USDA estimates that food retailers and producers discard more than 96 billion pounds of edible food annually.[27] To minimize waste, food recovery programs such as that run by Feeding America® (formerly called America's Second Harvest) work to collect and distribute this otherwise "lost" food. There are many types of food recovery efforts, some of which are listed below.

- *Field gleaning.* These programs gather and distribute unharvested agricultural crops.
- *Food rescue.* These initiatives collect and donate unused perishable foods from grocery stores, gardens, restaurants, campus dining facilities, hotels, and caterers.
- *Nonperishable food collection.* These programs collect and distribute damaged or outdated canned or boxed foods from retail sources.

In addition to these food recovery programs, community food banks, food pantries, and food kitchens also provide food to those in need. **Food banks** and **pantries** rely on community donations to stock their shelves with nonperishable and perishable items, which are then distributed to people who need them. **Food kitchens** serve prepared meals to members of the community but mostly to those who are homeless or living in shelters.

What Causes Worldwide Hunger and Malnutrition?

Because poverty is more prevalent in developing countries than in industrialized ones, food insecurity tends to be most prevalent and severe in nations with low *per capita* incomes. The FAO estimates that 16% of people living in poor countries do not have enough to eat and experience hunger.[1] Based on the 2010 Global Hunger Index, the FAO reported that the overall number of hungry people worldwide is nearly 1 billion.[28] The Asia-Pacific region has the largest *number* (20%; 578 million) of hungry people, but sub-Saharan Africa has the highest *percentage* (30%; 239 million). Causes of global food insecurity are complex. However, its consequences are almost always hunger, malnutrition, and other adverse societal conditions.[28]

MANY FACTORS CONTRIBUTE TO GLOBAL FOOD INSECURITY

Causes of food insecurity in poor countries are often different from those seen in the United States. Most experts agree that global food insecurity is not due to lack of available food on the international level. Rather, it is caused by diminished local food supply resulting from a variety of circumstances including political instability, lack of available land for growing crops, population growth, and gender inequalities.

Political Unrest Availability of and access to food are often limited by civil strife, wars, and political unrest. This turmoil can displace millions of people from their homes, forcing them to relocate to crude facilities set up for refugees. In countries with large refugee populations, such as Sudan, Rwanda, and Pakistan, food insecurity and malnutrition are rampant. According to the United Nations (UN), those living in refugee camps have the highest rates of disease and malnutrition of any group worldwide.[29] Because of the danger associated with political unrest, it can be difficult for relief agencies to provide the aid that is needed. Despite recent progress made by repatriation movements, the number of refugees is once again on the rise. Mainly attributed to violence taking place in the Middle East, the number of forcibly displaced persons is approaching 43 million worldwide, 15 million of whom are political refugees.[30]

food recovery program Program that collects and distributes food that would otherwise be discarded.

field gleaning Gathering and distributing unharvested agricultural crops that would normally be left in the field or on the tree.

food bank Organization that collects donated foods and distributes them to local food pantries, shelters, and food kitchens.

food pantry Program that provides canned, boxed, and sometimes fresh foods directly to individuals in need.

food kitchen Program that prepares and serves meals to those in need.

Although humanitarian aid agencies try to assist those living in refugee camps, many still have insufficient food, water, and medicine.

© Reuters/STR/Landov

Impact of Urbanization on Food Availability, Diet, and Health The use of land for purposes other than feeding a region's people and supporting local economies can greatly contribute to food shortages.[31] Without land on which crops can be grown, people cannot produce adequate food for themselves and their families. As a result, many people relocate from rural to urban areas with the hope of finding employment opportunities elsewhere. This trend, called **urbanization,** is both a consequence and a cause of food shortages in many parts of the world.

Urbanization and industrialization have had profound impacts on population demographics, transforming food systems and creating new nutritional challenges.[32] For example, the expansion of large supermarket chains in urban areas has greatly impacted small food producers and retailers.[33] Rather than buying from local, small farms, supermarket chains are more likely to utilize large consolidated food distribution centers.[33] This shift in food production, procurement, and distribution systems has contributed to displacement of workers, a decline in traditional food markets, and changes in local food culture.[33]

Nutrition Transition Fueled by changes brought on by urbanization, the composition of diets among city dwellers has shifted away from traditional foods to that of processed foods. This dietary adaptation has, in part, led to what is referred to as the **nutrition transition,** a shift from undernutrition to overnutrition or unbalanced nutrition that often occurs simultaneously with the industrialization of a society. The fact that many of these regions face food shortages at the same time they are experiencing increasing prevalence of obesity, heart disease, type 2 diabetes, and other diet-related health

problems highlights the special challenges related to food insecurity worldwide. Meeting these two distinct dual nutritional challenges—food shortages and obesity—highlights the importance of addressing the needs of both the rural poor and those in urban migration.

Population Growth Population growth in some of the poorest regions of the world has also increased the challenge of providing adequate food and water worldwide. Not surprisingly, countries with the fastest growing populations tend to be those already burdened with staggering rates of hunger and malnutrition.[34] Thus, the ability to provide even the most basic of needs—food and shelter—may be compromised further.

Gender Inequality In many developing nations, a gap exists between which opportunities are made available to males and to females. Some experts believe that promoting gender equality at many levels holds the greatest promise for reversing the steady increase in persistent and widespread global hunger.[35] For instance, providing equal educational opportunities for girls in many regions of the world might increase their earning capacity and help improve maternal nutrition and health. Clearly, a lack of education sustains the vicious cycle of poverty that is passed on from one generation to the next.

GLOBAL FOOD INSECURITY RESULTS IN MALNUTRITION

Although there are many consequences of food insecurity in poor countries, perhaps the most important is malnutrition. As you have learned, malnutrition is poor nutritional status resulting from inadequate or excessive dietary intake of energy and/or nutrients. In cases of food shortages, malnutrition takes the form of undernutrition, which has negative short- and long-term effects on the health of individuals, families, and societies. In the case of the nutrition transition, malnutrition takes the form of overnutrition and unbalanced nutrition.

Forms of Global Malnutrition Whereas some malnourished people may consume enough energy but lack certain nutrients, others may lack both. Still others consume excessive calories, sometimes coupled with

urbanization A shift in a country's population from primarily rural to urban regions.

nutrition transition The shift from undernutrition to overnutrition or unbalanced nutrition that often occurs simultaneously with the industrialization of a society.

a lack of nutrients. The number of women, infants, and children in the world with micronutrient deficiencies is staggering, with iron, iodine, and vitamin A topping the list.[36] Billions of people worldwide, many of whom are infants and children, suffer from iron and iodine deficiencies, both of which can impair growth and cognitive development. In addition, the health of millions of preschool-age children in the world is further compromised by vitamin A deficiency, causing blindness and many other consequences.[37] According to the United Nations Children's Fund (UNICEF), the health and welfare of nearly one-third of the world's population is affected by vitamin and mineral deficiencies.[38] Yet the resources exist to rectify these problems. By distributing low-cost, nutrient-rich foods and/or nutrient supplements, micronutrient deficiencies throughout the world could be eradicated. Although adequate nutrition is important throughout a child's life, the window of opportunity for interventions to have the greatest impact is between conception and a child's second birthday. The effects of persistent malnutrition on a child's health and development are largely irreversible after the age of 2.[39]

Protein-energy malnutrition (PEM) can also be a severe consequence of malnutrition. Inadequate intake of food that provides protein, energy, and micronutrients seriously affects growth and development in infants and children. PEM makes infants and children less able to resist and fight off disease and infection.

Consequences of Malnutrition for Individuals Infants, children, and women are especially vulnerable to malnutrition. For example, malnutrition during pregnancy can deplete a mother's nutritional stores and increase her risk of having a low-birth-weight (LBW) baby. Poor maternal nutritional status can also increase risk of neonatal death. It has been estimated that nearly 60% of the deaths of infants and young children in the world are caused, in part, by malnutrition.[39] Because poor nutrition compromises the immune system, malnutrition can make the adverse effects of disease even worse, leading prematurely to death.[40] For example, compared to a well-nourished child, a malnourished infant or young child has a greater likelihood of death if he or she gets ill from diarrhea, malaria, or respiratory infection. When a child is slightly underweight, the risk of death increases to 2.5 times that of children with healthy weights. The risk of death increases even further when a child is severely underweight.

Malnutrition can also seriously affect a child's growth and development. As a consequence, 24% of children under 5 years of age—149 million children—are estimated to have **stunted growth.**[40] These children are in the lower percentiles for height-for-age on growth reference curves. Compared to having low weight-for-height, which can be a consequence of acute malnutrition, stunting is undoubtedly a cumulative response to living in a chronically poor environment. Africa has the highest percentage (35%) of infants and young children with stunted growth, followed by Asia, Latin America, and the Caribbean.[40] Children with stunted growth are at greater risk than those of healthy height to have suboptimal health and productivity throughout their lives.

Ready-to-Use Therapeutic Foods Save Lives People who need food the most are often those who live in remote areas that are difficult to reach. Also complicating matters are a lack of refrigeration, poor access to clean water, and limited cooking amenities in these regions. For example, both contaminated water and drought render powdered milk

Ready-to-use therapeutic foods, which generally consist of peanut butter, vegetable oil, milk powder, sugar, vitamins, and minerals, provide a nutrient-dense, high-quality protein in a product that has a long shelf life. Provision of ready-to-use therapeutic foods has been credited with saving the lives of millions of children around the world.

stunted growth Diminished height resulting from chronic undernutrition.

useless, even when it is available. Perishable foods will spoil if not stored properly, and grains and cereals are not high-quality protein sources. Consequently, it difficult for health care providers to treat severely malnourished individuals living in these conditions. The recent development of **ready-to-use therapeutic foods (RUTF),** however, promises to largely overcome many of these challenges. RUTF products are prepackaged and require no preparation. Their formulations are based mainly on peanut butter, vegetable oil, milk powder, sugar, vitamins, and minerals. To be most effective (especially in infants and young children), international aid organizations often recommend that RUTFs provide 500 kcal/serving, be nutrient dense, and provide high-quality protein. Because RUTFs require no refrigeration, are easily distributed, and generally have long shelf lives, they are becoming the standard of care when it comes to refeeding malnourished children worldwide.

Consequences of Food Insecurity for Societies Aside from affecting an individual's health and well-being, malnutrition can harm whole societies. For example, extensive food insecurity and malnutrition can result in an entire nation of adults with reduced capacity for physical work and lower work productivity. These consequences can have profound long-term adverse effects on the country's economic growth and standard of living. In this way, not only is poverty a cause of hunger, but hunger is a cause of poverty. Therefore, addressing food insecurity and malnutrition is one important way to encourage economic progress in poor countries.

INTERNATIONAL ORGANIZATIONS PROVIDE FOOD-BASED ASSISTANCE TO THE GLOBAL COMMUNITY

Unlike wealthy countries, impoverished countries often lack stable governments and have few programs in place to assist those in need. These countries typically depend instead on relief efforts provided by international organizations such as the World Health Organization (WHO), the United Nations, U.S. Peace Corps, and Heifer International®. Organizations such as these try to assist countries in making lasting changes that will ultimately improve health and food security. Interventions with the greatest impact in poverty-stricken nations include efforts that improve maternal nutrition during pregnancy and lactation, reduce societal inequities, provide access to health services, promote self-sufficiency, and reduce illiteracy.[41]

You may be wondering how one person can make a difference in the world, especially in terms of helping to alleviate world hunger. An example of how people become inspired to make a difference can be gleaned from former President John F. Kennedy, who first challenged students in 1961 to serve their country by working to improve the quality of life for others in developing countries. This challenge was transformed into a federally funded program called the **U.S. Peace Corps.** The mission of the Peace Corps is to promote world peace and friendship by the following means:

- Assisting interested countries in meeting their need for trained men and women.
- Bringing a better understanding of Americans to people in other countries.
- Helping promote a better understanding of other people on the part of Americans.

Since its inception, more than 200,000 people have served in the Peace Corps in more than

AP Photo/Mahesh Kumar A.

Children with stunted growth are proportionally small in height-for-age.

ready-to-use therapeutic food (RUTF) Packaged, convenient, nutrient-dense food products that require no preparation or refrigeration and have long shelf lives.

U.S. Peace Corps A federally funded program that sends American volunteers to live and work with people in underdeveloped countries.

139 countries. These volunteers work to help others have a better life by assisting farmers to grow crops, teaching mothers to better care for their children, and educating entire communities about health and disease prevention. Thus, the Peace Corps offers the opportunity to make a difference in the lives of others by helping address the problems of food insecurity and malnutrition in many parts of the world.

Heifer International® Heifer International is a humanitarian-focused organization with a global commitment to foster environmentally sound farming methods aimed at combating both hunger and environmental concerns. This organization recognizes that impoverished people often make decisions based on short-term needs, rather than cultivating long-term solutions. Heifer International strives to teach families how to restore and manage land in ways that can provide food and income for generations to come. This problem-solving approach leads to novel, pragmatic solutions that empower impoverished communities to provide for themselves. The focus on long-term development rather than temporary relief efforts helps to restore hope, health, and dignity. Most remarkable is the underlying philosophy of "living loans" that ensures project sustainability. For example, communities receive the gift of livestock, which bring such benefits

Peace Corps volunteers find their work both challenging and rewarding.

as food, wool, and nonmechanized power. To repay this loan, the offspring of the livestock are "gifted" to another farmer or community. The "gifting" of farm animals repays the debt, while bringing the hope of prosperity to others. This simple idea of "passing on the gift" is the foundation on which Heifer International has created a living cycle of sustainability for over 65 years.

What Are Potential Solutions for Global Food Insecurity and Malnutrition?

Experts generally agree that there is enough food in the world for everyone to have enough to eat.[5] So why are food insecurity and malnutrition so widespread—especially in the poorest countries? As you have learned, the causes of food insecurity and malnutrition vary by geographic zone, political stability, national and local economic policy, and population growth. For this reason, it is important for nutritionists and policymakers to appreciate that the causes of food insecurity are varied and intertwined. Only by viewing the complexity of this issue can the relative importance of each contributing factor be addressed and effective solutions sought.

ALLEVIATING FOOD INSECURITY AND MALNUTRITION

World health experts concur that improving food availability and access must be a global priority. Although malnutrition is a direct consequence of insufficient dietary intake, its ultimate cause often has more to do with the economic and social circumstances of the poor. For example, the high

Heifer International is a nonprofit organization whose goal is to help end world hunger and poverty by fostering self-reliance and sustainability.

prevalence of certain diseases (such as AIDS), violence, illiteracy, and political corruption are also important factors. Thus, to effect a genuine remedy for food insecurity and its related malnutrition, the underlying contributing problems must be addressed.

The United Nations Many international organizations are committed to alleviating world hunger. One such relief organization is the United Nations (UN), a multinational organization first established in 1945 to promote peace through international cooperation and collective security. Today, 192 "member states" comprise the UN, and combating international hunger is one of its most important efforts. For example, UNICEF (a component of the UN) has presented a conceptual framework that provides incentives to work toward the common goal of improving the quality of life (including relieving hunger) for people in the world's poorest countries. These incentives include providing financial reimbursement to families who make the effort to have their children attend school or who start their own small businesses.[42] This type of initiative is called a **conditional cash transfer program.**

Oportunidades, a conditional cash transfer program launched by the Mexican government, is designed to simultaneously break the poverty cycle and improve nutrition. Serving more than 25 million people, this program gives cash payments to parents or caregivers responsible for family health decisions. The money received can then be used for expenses associated with health care, nutrition supplementation, and education. Studies of children associated with Oportunidades show improved outcomes in overall health, growth, and development.[43] In fact, Mexico's Oportunidades program has been such a remarkable success that similar programs have now been implemented in more than 20 countries. It has also served as a model for our nation's first conditional cash transfer program. This initiative, called Opportunity New York City, targeted the cycle of generational poverty by making incentive payments to families for complying with sound educational choices and health care.

Another UN plan, the Millennium Development Project, has pledged to halve the proportion of people who suffer from hunger in the world by the year 2015.[44] Endorsed by the majority of the countries in the world, this unprecedented effort addresses the needs of the world's poorest countries. To achieve the goals of the Millennium Development Project, efforts are currently under way to improve education, promote gender equality, reduce infant mortality, improve maternal health, combat AIDS and other infectious diseases, promote sustainable agricultural practices, and develop global economic partnerships. Although on track to meet several of these goals, this endeavor is an ongoing challenge that requires commitment and dedication from the worldwide community. The potential for Millennium Development Project to ease the burden of food insecurity, hunger, and malnutrition in the world is substantial.

TAKING ACTION AGAINST HUNGER CAN MAKE A DIFFERENCE

Although the problem may seem staggering, it is important to remember that there is much we can do to take action against hunger. Your actions alone can make a difference in the lives of others. Working collectively toward the common goal of eliminating hunger and malnutrition is an important personal and professional priority. The American Dietetic Association (ADA) is one of several professional organizations that work to alleviate world hunger by challenging its members to take action. For instance, the ADA recently created a Dietetic Practice Group to encourage dietitians to work with other health professionals to help reduce poverty and hunger in their communities. The American Society for Nutrition (ASN) is also committed to developing international strategies and policies that can help to alleviate hunger and poverty around the world.

---

conditional cash transfer program Initiative directed at reducing poverty by making the transfer of money contingent upon the receivers' actions.

Key Points

What Is Food Security?

- Food insecurity exists when people do not have adequate physical, social, or economic access to food.

- Being black or Latino, living in a household headed by a single woman, being elderly, and being of low income are associated with increased risk of food insecurity.

What Are the Consequences of Food Insecurity?

- Food assistance programs that help at-risk individuals obtain food include the Supplemental Nutrition Assistance Program (SNAP); the Special Supplemental Nutrition Program for Women, Infants, and Children (WIC); the School Breakfast and School Lunch Programs; food recovery programs; and community food banks and kitchens.

What Causes Worldwide Hunger and Malnutrition?

- Food insecurity is more common and severe in nonindustrialized countries than in more wealthy nations.

- There is sufficient food in the world to provide nutrition to all its inhabitants. Worldwide food insecurity is usually due to a combination of poverty, inadequate food distribution, political instability, urbanization, gender inequities, and other factors.

- Consequences of food insecurity in poorer regions of the world include greater incidence of low birth weight, neonatal death, growth stunting, vitamin A deficiency, iron deficiency, and iodine deficiency.

What Are Potential Solutions for Global Food Insecurity and Malnutrition?

- The U.S. Peace Corps offers Americans opportunities to make a difference in the lives of others by helping address the problems of food insecurity and malnutrition in many parts of the world.

- Although malnutrition is a consequence of insufficient dietary intake, its ultimate cause has more to do with the economic and social circumstances of the poor.

- UNICEF helps combat international hunger by working toward the common goal of improving the quality of life for people in the world's poorest countries.

- Conditional cash transfer programs may be one way to encourage sound nutritional and educational behaviors and simultaneously alleviate poverty.

Notes

1. Food and Agriculture Organization of the United Nations. Global hunger declining, but still unacceptably high. 2010. Available from http://www.fao.org/docrep/012/al390e/al390e00.pdf.

2. Food and Agriculture Organization of the United Nations. The state of food insecurity in the world 2010. Available from http://www.fao.org/docrep/013/i1683e/i1683e.pdf.

3. Kempson KM, Palmer Keenan D, Sadani PS, Ridlen S, Scotto Rosato N. Food management practices used by people with limited resources to maintain food sufficiency as reported by nutrition educators. Journal of the American Dietetic Association. 2002;102:1795–9.

4. U.S. Department of Agriculture. Economic Research Service. Household food security in the United States, 2008. Available from http://www.ers.usda.gov/Publications/ERR83/ERR83c.pdf.

5. Nord M, Coleman-Jensen A, Andrews M, Carlson S. Household food security in the United States, 2009. ERR-108, U.S. Department of Agriculture, Economics Research Service. 2010.Available from http://www.ers.usda.gov/Publications/ERR108/ERR108.pdf. U.S. Department of Agriculture. Economic Research Service. Food CPI and expenditures: analysis and forecasts of the CPI for food. 2011. Available from: http://www.ers.usda.gov/briefing/cpifoodandexpenditures/consumerpriceindex.htm

6. DeNavas-Walt C, Proctor BD, Smith JC. U.S. Census Bureau, Current Population Reports, P60-238, Income, Poverty, and Health Insurance Coverage in the United States: 2009. U.S. Government Printing Office, Washington, DC, 2010. Available from http://www.census.gov/prod/2010pubs/p60-238.pdf.

7. U.S. Census Bureau. Income, poverty, and health insurance coverage in the United States: 2009. Current Population Reports, P60-238, U.S. Government Printing Office, Washington, D.C., 2010. Available from http://www.census.gov/prod/2010pubs/p60-238.pdf.

8. America's Second Harvest. The almanac of hunger and poverty in America 2007. Chicago, IL: 2007. Available from http://feedingamerica.org/our-network/the-studies/~/media/Files/research/almanac/section1.ashx.

9. Kim M, Ohls J, Cohen R. Hunger in America, 2001. National report prepared for America's Second Harvest. Princeton, NJ: Mathematica Policy Research Inc. 2001. Available from http://www.mathematica-mpr.com/pdfs/hunger2001.pdf.

10. The United States Conference of Mayors—Sodexho. Hunger and homelessness survey. Available from http://www.usmayors.org/HHSurvey2007/hhsurvey07.pdf.

11. Olson CM. Nutrition and health outcomes associated with food insecurity and hunger. Journal of Nutrition. 1999;129:521S–4S.

12. Furness BW, Simon PA, Wold CM, Asarian-Anderson J. Prevalence and predictors of food insecurity among low-income households in Los Angeles County. Public Health Nutrition. 2004;7:791–4.

13. Beaulac J, Kristjansson E, Cummins S. A systematic review of food deserts, 1966–2007. Preventing Chronic Disease. 2009;6:A105.

14. Whitacre P, Tsai P, Mulligan J. The public health effects of food deserts: workshop summary. The National Academies Press. Washington, D.C. 2009. Available from http://books.nap.edu/catalog.php?record_id=12623.

15. Kendall A, Olson CM, Frongillo EA Jr. Relationship of hunger and food insecurity to food availability and consumption. Journal of the American Dietetic Association. 1996;96:1019–24.

16. Rose-Jacobs R, Black MM, Casey PH, Cook JT, Cutts DB, Chilton M, Heeren T, Levenson SM, Meyers AF, Frank DA. Household food insecurity: Associations with at-risk infant and toddler development. Pediatrics. 2008;121:65–72.

17. Alaimo K, Olson CM, Frongillo EA. Family food insufficiency, but not low family income, is positively associated with dysthymia and suicide symptoms in adolescents. Journal of Nutrition. 2002;132:719–25.

18. Fram MS, Frongillo EA, Jones SJ, Williams RC, Burke MP, Deloach KP, Blake CE. Children are aware of food insecurity and take responsibility for managing food resources. Journal of Nutrition. 2011;141:1114–9.

19. Lee JS, Frongillo EA Jr. Nutritional and health consequences are associated with food insecurity among U.S. elderly persons. Journal of Nutrition. 2001;131:1503–9.

20. U.S. Department of Agriculture. Food and Nutrition Service. Supplemental Nutrition Assistance Program. Available from http://www.fns.usda.gov/snap/Default.htm.

21. U.S. Department of Agriculture. Economic Research Service. The food assistance landscape. 2011. Available from http://www.ers.usda.gov/Publications/EIB6-8/EIB6-8.pdf.

22. Zedlewski SR. Leaving welfare often severs families' connections to the Food Stamp Program. Journal of the American Medical Women's Association. 2002;57:23–6.

23. American Academy of Pediatrics. WIC Program. Provisional Section on Breastfeeding. Position Statement. Pediatrics. 2001;108:1216–7.

24. U.S. Department of Agriculture Food and Nutrition Service. National school lunch program. Available from http://www.fns.usda.gov/cnd/lunch/aboutlunch/NSLP-FactSheet.pdf.

25. Institute of Medicine. School meals: Building blocks for healthy children. 2010. Washington, DC: The National Academies Press.

26. U.S. Public Law 111-296. Congress. Healthy hunger-free kids act of 2010. Available from http://www.gpo.gov/fdsys/pkg/PLAW-111publ296/pdf/PLAW-111publ296.pdf.

27. Kantor LS, Lipton K, Manchester, A, Oliveira V. Estimating and addressing America's food losses. U.S. Department of Agriculture. Economic Research Service. FoodReview, 1996. Available from http://www.ers.usda.gov/Publications/FoodReview/Jan1997/Jan97a.pdf.

28. International Food Policy Research Institute, Concern Worldwide, and Welthungerhilfe. 2010 Global hunger index. Available from http://www.ifpri.org/sites/default/files/publications/ghi10.pdf. Food and Agriculture Organization of the United Nations. 2011 World Hunger and Poverty Facts and Statistics. Available from: http://www.worldhunger.org/articles/Learn/world%20hunger%20facts%202002.htm. United Nations Economic and Social Commission for Asia and the Pacific (UNESCAP). Eradicate extreme poverty and hunger. 2006. Available from http://www.mdgasiapacific.org/files/shared_folder/documents/fs_sa_mdg_goal1.pdf.

29. Office of the United Nations High Commissioner for Refugees. Global Report 2010. Available from http://www.unhcr.org/gr10/index.html.

30. United Nations High Commissioner for Refugees. 2009 global trends: refugees, asylum-seekers, returnees, internally displaced and stateless persons. 2010. Available from http://www.unhcr.org/4c11f0be9.html.

31. Kennedy G, Nantel G, Shetty P. Globalization of food systems in developing countries: impact on food security and nutrition. Food and Agriculture Organization of the United Nations. FAO Food and Nutrition Paper. 2004;83:1–300.

32. Reardon T, Timmer P. Barrett C, and Berdegué J. The rise of supermarkets in Africa, Asia and Latin America. American Journal of Agricultural Economics. 2003;85:1140–6.

33. The impact of global change and urbanization on household food security, nutrition, and food safety. Food and Agriculture Organization of the United Nations. Available from http://www.fao.org/ag/agn/nutrition/national_urbanization_en.stm.

34. El-Ghannam AR. The global problems of child malnutrition and mortality in different world regions. Journal of Health and Social Policy. 2003;16:1–26. Horton KD. Bringing attention to global hunger. Journal of the American Dietetic Association. 2008;108:435.

35. Task Force on Education and Gender Equality. United Nations Millennium Project 2005. Taking action: achieving gender equality and empowering women. Available from http://www.unmillenniumproject.org/documents/Gender-complete.pdf.

36. Kennedy E, Meyers L. Dietary Reference Intakes: Development and uses for assessment of micronutrient status of women—a global perspective. American Journal of Clinical Nutrition. 2005;81:1194S-7S.

37. Humphrey JH, West KP Jr., Sommer A. Vitamin A deficiency and attributable mortality among under-5-year-olds. Bulletin of the World Health Organization. 1992:70:225–32.

38. United Nations Children's Fund. The state of the world's children 2005. http://www.unicef.org/publications/files/SOWC_2005_(English).pdf

39. Population Reference Bureau. 2007 World population data sheet: Malnutrition is a major contributor to child deaths. Available from http://www.prb.org/Journalists/PressReleases/2007/2007WPDSBriefing.aspx.

40. Milman A, Frongillo EA, de Onis M, Hwang JY. Differential improvement among countries in child stunting is associated with long-term development and specific interventions. Journal of Nutrition. 2005:135:1415–22.

41. McCall E. Communication for development strengthening the effectiveness of the United Nations. United Nations Development Programme. 2011.

Available from http://www.unicef.org/cbsc/files/Inter-agency_C4D_Book_2011.pdf

42. Behrman JR, Parker SW, Todd PE. Schooling impacts of conditional cash transfers on young children: Evidence from Mexico. Economic Development and Cultural Change. 2009;57:439–77.

43. Fernald LC, Gertler PJ, Neufeld LM. Role of cash in conditional cash transfer programmes for child health, growth, and development: An analysis of Mexico's Oportunidades. Lancet. 2008; 371:828–37.

44. United Nations Millennium Development Project. United Nations Millennium development goals. Millennium Development Project report 2010. Available from http://www.un.org/millenniumgoals/pdf/MDG%20Report%202010%20En%20r15%20-low%20res%2020100615%20-.pdf.

APPENDIXES

Aids to Calculation

The study of nutrition often requires solving mathematical problems. The three most common types of calculations are related to conversions, percentages, and ratios.

Conversions

It is important to know how to convert from one unit of measure to another. For example, sometimes it is necessary to convert pounds to kilograms, inches to centimeters, ounces to grams, etc. To convert one unit of measure to another, you need a conversion factor. For example, if a book measures 1 foot in length and another measures 12 inches in length, you cannot calculate the combined length of the two books by simply adding 1 to 12. Rather, the two values must have the same units before they can be summed. In this case, you need to know the conversion factor for changing inches to feet (12 inches = 1 foot) or feet to inches (1 foot = 12 inches). Some examples of common conversion factors are listed below.

2.2 lb = 1 kg
1 oz = 28 g
1 in = 2.54 cm
1 m = 3.3 ft

In addition, it is often necessary to determine how many calories (kcal) are in a given amount of food. This calculation is similar to that done in a conversion.

1 g protein = 4 kcal protein
1 g fat = 9 kcal fat
1 g carbohydrate = 4 kcal carbohydrate
1 g alcohol = 7 kcal alcohol

Sample Conversions

Example 1. Converting weight in pounds to weight in kilograms and weight in kilograms to weight in pounds

To convert 150 lb to kg, divide by 2.2, like this:

$$150 \text{ lb} \div 2.2 \text{ lb/kg} = 68.2 \text{ kg}$$

To convert 68.2 kg to lb, multiply by 2.2, like this:

$$68.2 \text{ kg} \times 2.2 \text{ lb/kg} = 150 \text{ lb}$$

Example 2. Converting weight in ounces to weight in grams and weight in grams to weight in ounces

To convert 4 oz to g, multiply by 28, like this:

$$4 \text{ oz} \times 28 \text{ g/oz} = 112 \text{ g}$$

To convert 112 g to oz, divide 112 by 28, like this:

$$112 \text{ g} \div 28 \text{ g/oz} = 4 \text{ oz}$$

Example 3. Converting height in inches to height in centimeters and height in feet to height in meters

To convert 58 in to cm, multiply 58 by 2.54, like this:

$$58 \text{ in} \times 2.54 \text{ cm/in} = 147.3 \text{ cm}$$

To convert 147.3 cm to in, divide 147.3 cm by 2.54, like this:

$$147.3 \text{ cm} \div 2.54 \text{ cm/in} = 58 \text{ in}$$

To convert 5.3 ft to m, divide 5.3 ft by 3.3, like this:

$$5.3 \text{ ft} \div 3.3 \text{ ft/m} = 1.6 \text{ m}$$

Example 4. Calculating kilocalories in food from weight in grams and weight in grams from kilocalories in food

To calculate how many kcal are in 50 g of protein, multiply by 4, like this:

$$50 \text{ g} \times 4 \text{ kcal/g} = 200 \text{ kcal}$$

To calculate how many g of protein it would take to get 200 kcal, divide by 4, like this:

$$200 \text{ kcal} \div 4 \text{ kcal/g} = 50 \text{ g}$$

Percentages

A percentage expresses the contribution of a part to the total. The total is always 100. To calculate a percentage, you must first determine the relation of the part to the total, which is then expressed as a percentage by multiplying by 100. For example, what percentage of total kilocalories in a 400-kcal meal containing 225 kcal from lipids is from fat?

To solve this problem, divide the part (in this case, kilocalories from fat) by the total (in this case, total kilocalories in the meal), then multiply by 100, like this:

$$(225 \text{ kcal} \div 400 \text{ kcal}) \times 100 = 56\%$$

It may also be necessary to calculate percentages of total kilocalories from fat, carbohydrate, and protein. Let's say that this meal also provides 87 kcal from carbohydrate and 87 kcal from protein. Calculate the percentages of total kilocalories in this meal from fat, carbohydrate, and protein like this:

Fat: $(225 \text{ kcal} \div 400 \text{ total kcal}) \times 100 = 56\%$
Carbohydrate: $(87 \text{ kcal} \div 400 \text{ total kcal}) \times 100 = 22\%$
Protein: $(87 \text{ kcal} \div 400 \text{ total kcal}) \times 100 = 22\%$

Notice that when the percentages are added together $(56\% + 22\% + 22\%)$, they total 100%.

This type of calculation is very common. However, sometimes it is first necessary to determine how many kilocalories are in a specified weight of food. For example, nutrient content is often provided in grams (50 g fat, 50 g carbohydrate, and 35 g protein). To calculate the percentages of total kilocalories from fat, carbohydrate, and protein, you must first calculate how many calories there are in these macronutrients.

This problem requires you to:

- Calculate the kilocalories provided by each macronutrient class.
- Calculate total kilocalories in the food.
- Calculate percentages of total kilocalories provided by fat, carbohydrate, and protein.

Step 1. Calculate the kilocalories provided by each macronutrient class.

50 g fat × 9 kcal/g = 450 kcal from fat
50 g carbohydrate × 4 kcal/g = 200 kcal from carbohydrate
35 g protein × 4 kcal/g = 140 kcal from protein

Step 2. Calculate total kilocalories in the food.

450 kcal from fat + 200 kcal from carbohydrate + 140 kcal from protein = 790 kcal total

Step 3. Calculate percentages of total kilocalories provided by fat, carbohydrate, and protein.

(450 kcal from fat ÷ 790 total kcal) × 100 = 57% of kcal from fat
(200 kcal from carbohydrate ÷ 790 total kcal) × 100 = 25% of kcal from carbohydrate
(140 kcal from protein ÷ 790 total kcal) × 100 = 18% of kcal from protein

Notice that these percentages (in this case, 57% + 25% + 18%) must add up to 100%.

Ratios

Ratios reflect relative amounts of two or more entities. For example, if there is twice as much of Substance A as there is of Substance B, the ratio of Substance A to Substance B is 2:1. Because the units being compared are always the same, a ratio is not expressed in terms of units. Another example would be a diet that provides 50 g of carbohydrate and 100 g of protein. The ratio of carbohydrate to protein is 1:2; that is, for every 1 g of carbohydrate there are 2 g of protein.

Here is another example. A diet provides 3,000 mg of sodium and 2,000 mg of potassium. The ratio here (calculated by dividing milligrams of sodium by milligrams of potassium) is 1.5:1. This means there are 1.5 mg of sodium in this diet for every 1 mg of potassium.

Estimated Energy Requirement (EER) Calculations and Physical Activity Values (PA)

Estimated Energy Requirement Calculations

Sex and Life-Stage Group	Equations for Estimated Energy Requirement (EER; kcal/day)[a]
0–3 months	$[89 \times weight (kg) - 100] + 175$
4–6 months	$[89 \times weight (kg) - 100] + 56$
7–12 months	$[89 \times weight (kg) - 100] + 22$
13–36 months	$[89 \times weight (kg) - 100] + 20$
3–8 years (male)	$88.5 - [61.9 \times age (y)] + PA \times [26.7 \times weight (kg) + 903 \times height (m)] + 20$
3–8 years (female)	$135.3 - [30.8 \times age (y)] + PA \times [10.0 \times weight (kg) + 934 \times height (m)] + 20$
9–18 years (male)	$88.5 - [61.9 \times age (y)] + PA \times [26.7 \times weight (kg) + 903 \times height (m)] + 25$
9–18 years (female)	$135.3 - [30.8 \times age (y)] + PA \times [10.0 \times weight (kg) + 934 \times height (m)] + 25$
19+ years (male)	$662 - [9.53 \times age (y)] + PA \times [15.91 \times weight (kg) + 539.6 \times height (m)]$
19+ years (female)	$354 - [6.91 \times age (y)] + PA \times [9.36 \times weight (kg) + 726 \times height (m)]$
Pregnancy	
14–18 years	
1st trimester	Adolescent EER + 0
2nd trimester	Adolescent EER + 340
3rd trimester	Adolescent EER + 452
19–50 years	
1st trimester	Adult EER + 0
2nd trimester	Adult EER + 340
3rd trimester	Adult EER + 452
Lactation	
14–18 years	
1st six months postpartum	Adolescent EER + 330
2nd six months postpartum	Adolescent EER + 400
19–50 years	
1st six months postpartum	Adult EER + 330
2nd six months postpartum	Adult EER + 400
Overweight or Obese[b]	
3–18 years (male)	$114 - [50.9 \times age (y)] + PA \times [19.5 \times weight (kg) + 1161.4 \times height (m)]$
3–18 years (female)	$389 - [41.2 \times age (y)] + PA \times [15.0 \times weight (kg) + 701.6 \times height (m)]$
19+ years (male)	$1086 - [10.1 \times age (y)] + PA \times [13.7 \times weight (kg) + 416 \times height (m)]$
19+ years (female)	$448 - [7.95 \times age (y)] + PA \times [11.4 \times weight (kg) + 619 \times height (m)]$

[a] "PA" stands for the physical activity value appropriate for the age and physiological state. These values can be found in the next table.
[b] Body mass index (BMI) ≥25 kg/m²; values represent estimated total energy expenditure (TEE; kcal/day) for weight maintenance; weight loss can be achieved by a reduction in energy intake and/or an increase in energy expenditure.

Physical Activity Values (PA)

Age Group (sex)	Physical Activity Level (PAL)[a]	Physical Activity Value (PA)
3–8 years (male)	Sedentary	1.00
	Low active	1.13
	Active	1.26
	Very active	1.42
3–8 years (female)	Sedentary	1.00
	Low active	1.16
	Active	1.31
	Very active	1.56
3–18 years (overweight male)[b]	Sedentary	1.00
	Low active	1.12
	Active	1.24
	Very active	1.45
3–18 years (overweight female)[b]	Sedentary	1.00
	Low active	1.18
	Active	1.35
	Very active	1.60
9–18 years (male)	Sedentary	1.00
	Low active	1.13
	Active	1.26
	Very active	1.42
9–18 years (female)	Sedentary	1.00
	Low active	1.16
	Active	1.31
	Very active	1.56
19+ years (male)	Sedentary	1.00
	Low active	1.11
	Active	1.25
	Very active	1.48
19+ years (female)	Sedentary	1.00
	Low active	1.12
	Active	1.27
	Very active	1.45
19+ years (overweight/obese male)[b]	Sedentary	1.00
	Low active	1.12
	Active	1.29
	Very active	1.59
19+ years (overweight/obese female)[b]	Sedentary	1.00
	Low active	1.16
	Active	1.27
	Very active	1.44

[a] *Sedentary* activity level is characterized by no physical activity aside from that needed for independent living. *Low active* level is characterized by walking 1.5 to 3 miles/day at 2–4 mph (or equivalent) in addition to the light activity associated with typical day-to-day life. People who are *active* walk 3 to 10 miles/day at 2–4 mph (or equivalent) in addition to the light activity associated with typical day-to-day life. *Very active* individuals walk 10 or more miles/day at 2–4 mph (or equivalent) in addition to the light activity associated with typical day-to-day life.
[b] Body mass index (BMI) ≥25 kg/m².

SOURCE: Institute of Medicine. *Dietary reference intakes for energy, carbohydrate, fiber, fat, fatty acids, cholesterol, protein, and amino acids (macronutrients).* Washington, DC: National Academies Press; 2005.

Summary of the 2010 Dietary Guidelines for Americans

Major Concepts	Key Recommendations
Balance Calories to Manage Weight	• Prevent and/or reduce overweight and obesity through improved eating and physical activity behaviors. • Control total calorie intake to manage body weight. For people who are overweight or obese, this will mean consuming fewer calories from foods and beverages. • Increase physical activity and reduce time spent in sedentary behaviors. • Maintain appropriate calorie balance during each stage of life.
Reduce Certain Foods and Food Components	• Reduce daily sodium intake to less than 2,300 mg/day and further reduce intake to 1,500 mg/day among individuals with increased risk for hypertension. • Consume less than 10% of calories from saturated fats by replacing them with monounsaturated and polyunsaturated fats. • Consume less than 300 mg/day of dietary cholesterol. • Keep *trans* fatty acid consumption as low as possible. • Reduce intake of calories from solid fats (saturated and *trans* fats) and added sugars. • Limit consumption of foods that contain refined grains. • If you consume alcohol, consume it in moderation—up to one drink per day for women and two drinks per day for men; only adults of legal drinking age should drink alcoholic beverages.
Increase Certain Foods and Nutrients[1]	• Increase vegetable and fruit intake. • Eat a variety of vegetables, especially peas and beans and vegetables that are dark green, red, or orange. • Consume at least half of all grains as whole grains. • Increase intake of fat-free or low-fat milk and milk products or fortified soy beverages. • Choose a variety of protein foods, including seafood, lean meat and poultry, eggs, beans and peas, soy products, and unsalted nuts and seeds. • Replace protein foods high in solid fats with choices that are lower in solid fats and calories. • Choose foods that provide more potassium, dietary fiber, calcium, and vitamin D, such as vegetables, fruits, whole grains, and dairy products.
Build Healthy Eating Patterns	• Select an eating pattern that meets nutrient needs over time at an appropriate calorie level. • Account for all foods and beverages consumed and assess how they fit within a total healthy eating pattern. • Follow food safety recommendations when preparing and eating foods to reduce the risk of foodborne illness.

[1] Additional recommendations are also made for specific population groups. For example, women capable of becoming pregnant are advised to choose foods that supply heme iron, additional iron sources, and enhancers of iron absorption (e.g., vitamin C). It is also recommended that they consume synthetic folic acid (from foods or supplements) in addition to foods high in folate. Women who are pregnant or breastfeeding are advised to consume 8 to 12 oz seafood per week, being careful to avoid those types known to be high in mercury. Individuals who are 50 years or older should consume foods fortified with vitamin B_{12} or a dietary supplement.

SOURCE: U.S. Department of Agriculture and U.S. Department of Health and Human Services. Dietary guidelines for Americans, 2010. 7th edition, Washington, DC: U.S. Government Printing Office; December 2010.

The Exchange System

Chapter 4 briefly explains the exchange system, and this appendix provides details from the American Dietetic Association's *2008 Choose Your Foods: Exchange Lists for Diabetes*. Exchange lists can help people with diabetes manage their blood glucose levels by controlling the amount and kinds of carbohydrates they consume. These lists can also help when planning diets for weight management by controlling calorie and fat intake.

The Exchange System

The exchange system sorts foods into groups by their proportions of carbohydrate, fat, and protein (Table D-1). These groups may be organized into several exchange lists of foods (Tables D-2 through D-12). For example, the carbohydrate group includes these exchange lists:

- Starch
- Fruits
- Milk (fat free, reduced fat, and whole)
- Sweets, Desserts, and Other Carbohydrates
- Nonstarchy Vegetables

Then any food on a list can be "exchanged" for any other on that same list. Another group, for alcohol, has been included as a reminder that these beverages often deliver substantial carbohydrate and calories, and therefore warrant their own list.

Serving Sizes

The serving sizes have been carefully adjusted and defined so that a serving of any food on a given list provides roughly the same amount of carbohydrate, fat, and protein, and, therefore, total energy. Any food on a list can thus be exchanged, or traded, for any other food on the same list without significantly affecting the diet's energy–nutrient balance or total calories. For example, a person may select 17 small grapes or ½ large grapefruit as one fruit exchange, and either choice would provide roughly 15 grams of carbohydrate and 60 kilocalories. A whole grapefruit, however, would count as two fruit exchanges.

To apply the food exchange system successfully, users must become familiar with the specified serving sizes. A convenient way to remember the serving sizes and energy values is to keep in mind a typical item from each list (review Table D-1).

The Foods on the Lists

Foods do not always appear on the exchange list where you might first expect to find them. They are grouped according to their energy–nutrient content rather than by their source (such as milks), their outward appearance, or their vitamin and mineral contents. For example, cheeses are grouped with meats (not milk) because, like meats, cheeses contribute energy from protein and fat but provide negligible carbohydrate.

For similar reasons, starchy vegetables such as corn, green (fresh) peas, and potatoes are found on the Starch list with breads and cereals, not with the vegetables. Likewise, bacon is grouped with the fats and oils, not with the meats. Effective diet planners learn to view mixtures of foods, such as casseroles and soups, as combinations of foods from different exchange lists. They also learn to interpret food labels with the exchange system in mind.

Controlling Energy, Fat, and Sodium

The exchange lists help people control their energy intakes by paying close attention to serving sizes. People wanting to lose weight can limit foods from the Sweets, Desserts, and Other Carbohydrates and Fats lists, and they might choose to avoid the Alcohol list altogether. The Free Foods list provides low-calorie choices.

By assigning items like bacon to the Fats list, the exchange lists alert people to foods that are unexpectedly high in fat. Even the Starch list specifies which grain products contain added fat (such as biscuits, corn bread, and waffles) by marking them with a symbol to indicate added fat (the symbols are explained in the tables' keys). In addition, the exchange lists encourage consumers to think of fat-free milk as milk and of whole milk as milk with added fat, and to think of lean meats as meats and of medium-fat and high-fat meats as meats with added fat. To that end, foods on the Milk and Meat lists are separated into categories based on their fat contents (review Table D-1). The Milk list is subdivided for fat free, reduced fat, and whole; the Meat list is subdivided for lean, medium fat, and high fat. The Meat list also includes plant-based proteins, which tend to be rich in fiber. Notice that many of these foods bear the symbol for "high fiber."

People wanting to control their sodium intake can begin by eliminating any foods bearing the "high sodium" symbol. In most cases, the symbol identifies foods that, in one serving, provide 480 milligrams or more of sodium. Foods on the Combination Foods or Fast Foods lists that bear the symbol provide more than 600 milligrams of sodium. Other foods may also contribute substantially to sodium (consult Chapter 12 for details).

Planning a Healthy Diet

To obtain a daily variety of foods that provide health-promoting amounts of carbohydrates, protein, and fats, as well as vitamins, minerals, and fiber, the meal plan for adults and teenagers should include at least:

- two to three servings of nonstarchy vegetables;
- two servings of fruits;
- six servings of grains (including at least three of whole grains), dried beans, and starchy vegetables;
- two servings of low-fat or fat-free milk;
- about 6 ounces of meat or meat substitutes;
- and only *small* amounts of fat and sugar.

The actual amounts are determined by age, sex, activity level, and other factors that influence energy needs.

Table D-1   The Food Exchange Lists

Lists	Typical item/portion size	Carbohydrate (g)	Protein (g)	Fat (g)	Energy[a] (kcal)
Carbohydrates					
Starch[b]	1 slice bread	15	0–3	0–1	80
Fruits	1 small apple	15	–	–	60
Milk					
Fat free, low fat, 1%	1 cup fat-free milk	12	8	0–3	100
Reduced fat, 2%	1 cup reduced-fat milk	12	8	5	120
Whole	1 cup whole milk	12	8	8	160
Sweets, desserts, and other carbohydrates[c]	2 small cookies	15	varies	varies	varies
Nonstarchy vegetables	½ cup cooked carrots	5	2	–	25
Meat and meat substitutes					
Lean	1 oz chicken (no skin)	–	7	0–3	45
Medium fat	1 oz ground beef	–	7	4–7	75
High fat	1 oz pork sausage	–	7	8+	100
Plant-based proteins	½ cup tofu	varies	7	varies	varies
Fats	1 tsp butter	–	–	5	45
Alcohol	12 oz beer	varies	–	–	100

[a] The energy value for each exchange list represents an approximate average for the group and does not reflect the precise number of grams of carbohydrate, protein, and fat. For example, a slice of bread contains 15 grams of carbohydrate (60 kilocalories), 3 grams of protein (12 kilocalories), and a little fat—rounded to 80 kilocalories for ease in calculating. A half cup of vegetables (not including starchy vegetables) contains 5 grams carbohydrate (20 kilocalories) and 2 grams protein (8 kilocalories), which has been rounded down to 25 kilocalories.
[b] The Starch list includes cereals, grains, breads, crackers, snacks, starchy vegetables (such as corn, peas, and potatoes), and legumes (dried beans, peas, and lentils).
[c] The Sweets, Desserts, and Other Carbohydrates list includes foods that contain added sugars and fats such as sodas, candy, cakes, cookies, doughnuts, ice cream, pudding, syrup, and frozen yogurt.

Table D-2   Starch

The Starch list includes bread, cereals and grains, starchy vegetables, crackers and snacks, and legumes (dried beans, peas, and lentils). 1 starch choice = 15 grams carbohydrate, 0–3 grams protein, 0–1 grams fat, and 80 kilocalories.

Note: In general, one starch exchange is ½ cup cooked cereal, grain, or starchy vegetable; ⅓ cup cooked rice or pasta; 1 ounce of bread product; or ¾ ounce to 1 ounce of most snack foods.

Bread		Bread	
Food	**Serving size**	**Food**	**Serving size**
Bagel, large (about 4 oz)	¼ (1 oz)	Pancake, 4 inches across, ¼ inch thick	1
▼ Biscuit, 2½ inches across	1	Pita, 6 inches across	½
Bread		Roll, plain, small	1 (1 oz)
☺ reduced calorie	2 slices (1½ oz)	▼ Stuffing, bread	⅓ cup
white, whole grain, pumpernickel, rye, unfrosted raisin	1 slice (1 oz)	▼ Taco shell, 5 inches across	2
Chapatti, small, 6 inches across	1	Tortilla, corn, 6 inches across	1
▼ Corn bread, 1¾ inch cube	1 (1½ oz)	Tortilla, flour, 6 inches across	1
English muffin	½	Tortilla, flour, 10 inches across	⅓
Hot dog bun or hamburger bun	½ (1 oz)	▼ Waffle, 4-inch square or 4 inches across	1
Naan, 8 inches by 2 inches	¼		

(Continued)

Cereals and grains

Food	Serving size
Barley, cooked	1/3 cup
Bran, dry	
☺ oat	1/4 cup
☺ wheat	1/2 cup
☺ Bulgur (cooked)	1/2 cup
Cereals	1/2 cup
☺ bran	
cooked (oats, oatmeal)	1/2 cup
puffed	1½ cups
shredded wheat, plain	1/2 cup
sugar-coated	1/2 cup
unsweetened, ready-to-eat	3/4 cup
Couscous	1/3 cup
Granola	
low fat	1/4 cup
▼ regular	1/4 cup
Grits, cooked	1/2 cup
Kasha	1/2 cup
Millet, cooked	1/3 cup
Muesli	1/4 cup
Pasta, cooked	1/3 cup
Polenta, cooked	1/3 cup
Quinoa, cooked	1/3 cup
Rice, white or brown, cooked	1/3 cup
Tabbouleh (tabouli), prepared	1/2 cup
Wheat germ, dry	3 tbsp
Wild rice, cooked	1/2 cup

Starchy vegetables

Food	Serving size
Cassava	1/3 cup
Corn	1/2 cup
on cob, large	1/2 cob (5 oz)
☺ Hominy, canned	3/4 cup
☺ Mixed vegetables with corn, peas, or pasta	1 cup
☺ Parsnips	1/2 cup
☺ Peas, green	1/2 cup
Plantain, ripe	1/3 cup
Potato	
baked with skin	1/4 large (3 oz)
boiled, all kinds	1/2 cup or 1/2 medium (3 oz)
▼ mashed, with milk and fat	1/2 cup
french fried (oven-baked)[a]	1 cup (2 oz)

Starchy vegetables

Food	Serving size
☺ Pumpkin, canned, no sugar added	1 cup
Spaghetti/pasta sauce	1/2 cup
☺ Squash, winter (acorn, butternut)	1 cup
☺ Succotash	1/2 cup
Yam, sweet potato, plain	1/2 cup

Crackers and snacks[b]

Food	Serving size
Animal crackers	8
Crackers	
▼ round butter type	6
saltine type	6
▼ sandwich style, cheese or peanut butter filling	3
▼ whole-wheat regular	2–5 (3/4 oz)
☺ whole-wheat lower fat or crispbread	2–5 (3/4 oz)
Graham cracker, 2½-inch square	3
Matzoh	3/4 oz
Melba toast, about 2-inch by 4-inch piece	4
Oyster crackers	20
Popcorn	3 cups
▼ ☺ with butter	3 cups
☺ no fat added	3 cups
☺ lower fat	3 cups
Pretzels	3/4 oz
Rice cakes, 4 inches across	2
Snack chips	
fat free or baked (tortilla, potato), baked pita chips	15–20 (3/4 oz)
▼ regular (tortilla, potato)	9–13 (3/4 oz)

Beans, peas, and lentils[c]

The choices on this list count as 1 starch + 1 lean meat.

Food	Serving size
☺ Baked beans	1/3 cup
☺ Beans, cooked (black, garbanzo, kidney, lima, navy, pinto, white)	1/2 cup
☺ Lentils, cooked (brown, green, yellow)	1/2 cup
☺ Peas, cooked (black-eyed, split)	1/2 cup
🧂 ☺ Refried beans, canned	1/2 cup

KEY
☺ = More than 3 grams of dietary fiber per serving.
▼ = Extra fat, or prepared with added fat. (Count as 1 starch + 1 fat.)
🧂 = 480 milligrams or more of sodium per serving.
[a] Restaurant-style french fries are on the Fast Foods list.
[b] For other snacks, see the Sweets, Desserts, and Other Carbohydrates list. For a quick estimate of serving size, an open handful is equal to about 1 cup or 1 to 2 ounces of snack food.
[c] Beans, peas, and lentils are also found on the Meat and Meat Substitutes list.

Table D-3   Fruits

Fruit[a]

The Fruits list includes fresh, frozen, canned, and dried fruits and fruit juices. 1 fruit choice = 15 grams carbohydrate, 0 grams protein, 0 grams fat, and 60 kilocalories.

Note: In general, one fruit exchange is ½ cup canned or fresh fruit or unsweetened fruit juice; 1 small fresh fruit (4 ounces); or 2 tablespoons dried fruit.

Food	Serving size	Food	Serving size
Apple, unpeeled, small	1 (4 oz)	Mandarin oranges, canned	¾ cup
Apples, dried	4 rings	Mango, small	½ (5½ oz) or ½ cup
Applesauce, unsweetened	½ cup		
Apricots		Nectarine, small	1 (5 oz)
canned	½ cup	☺ Orange, small	1 (6½ oz)
dried	8 halves	Papaya	½ or 1 cup cubed (8 oz)
☺ fresh	4 whole (5½ oz)		
Banana, extra small	1 (4 oz)	Peaches	
☺ Blackberries	¾ cup	canned	½ cup
Blueberries	¾ cup	fresh, medium	1 (6 oz)
Cantaloupe, small	⅓ melon or 1 cup cubed (11 oz)	Pears	
		canned	½ cup
		fresh, large	½ (4 oz)
Cherries		Pineapple	
sweet, canned	½ cup	canned	½ cup
sweet, fresh	12 (3 oz)	fresh	¾ cup
Dates	3	Plums	
Dried fruits (blueberries, cherries, cran- berries, mixed fruit, raisins)	2 tbsp	canned	½ cup
		dried (prunes)	3
Figs		small	2 (5 oz)
dried	1½	☺ Raspberries	1 cup
☺ fresh	1½ large or 2 medium (3½ oz)	☺ Strawberries, whole	1¼ cup
		☺ Tangerines, small	2 (8 oz)
Fruit cocktail	½ cup	Watermelon	1 slice or 1¼ cups cubes (13½ oz)
Grapefruit			
large	½ (11 oz)	Apple juice/cider	½ cup
sections, canned	¾ cup	Fruit juice blend, 100% juice	⅓ cup
Grapes, small	17 (3 oz)	Grape juice	⅓ cup
Honeydew melon	1 slice or 1 cup cubed (10 oz)	Grapefruit juice	½ cup
		Orange juice	½ cup
		Pineapple juice	½ cup
☺ Kiwi	1 (3½ oz)	Prune juice	⅓ cup

KEY

☺ = More than 3 grams of dietary fiber per serving.

▼ = Extra fat, or prepared with added fat. (Count as 1 starch + 1 fat.)

🧂 = 480 milligrams or more of sodium per serving.

[a] The weight listed includes skin, seeds, and rind.

Table D-4   Milk

The Milk list groups milks and yogurts based on the amount of fat they have (fat free/low fat, reduced fat, and whole). Cheeses are found on the Meat and Meat Substitutes list, and cream and other dairy fats are found on the Fats list.

Note: In general, one milk choice is 1 cup (8 fluid ounces or ½ pint) milk or yogurt.

Milk and yogurts

Food	Serving size
Fat free or low fat (1%)	
1 fat-free/low-fat milk choice = 12 g carbohydrate, 8 g protein, 0–3 g fat, and 100 kilocalories	
Milk, buttermilk, acidophilus milk, lactase-treated milk	1 cup
Evaporated milk	½ cup
Yogurt, plain or flavored with an artificial sweetener	⅔ cup (6 oz)
Reduced fat (2%)	
1 reduced-fat milk choice = 12 g carbohydrate, 8 g protein, 5 g fat, and 120 kilocalories	
Milk, acidophilus milk, kefir, lactase-treated milk	1 cup
Yogurt, plain	⅔ cup (6 oz)
Whole	
1 whole milk choice = 12 g carbohydrate, 8 g protein, 8 g fat, and 160 kilocalories	
Milk, buttermilk, goat's milk	1 cup
Evaporated milk	½ cup
Yogurt, plain	8 oz

Dairy-containing and dairy-like foods

Food	Serving size	Count as
Chocolate milk		
fat free	1 cup	1 fat-free milk + 1 carbohydrate
whole	1 cup	1 whole milk + 1 carbohydrate
Eggnog, whole milk	½ cup	1 carbohydrate + 2 fats
Rice drink		
flavored, low fat	1 cup	2 carbohydrates
plain, fat free	1 cup	1 carbohydrate
Smoothies, flavored, regular	10 oz	1 fat-free milk + 2½ carbohydrates
Soy milk		
light	1 cup	1 carbohydrate + ½ fat
regular, plain	1 cup	1 carbohydrate + 1 fat
Yogurt		
and juice blends	1 cup	1 fat-free milk + 1 carbohydrate
low carbohydrate (less than 6 grams carbohydrate per choice)	⅔ cup (6 oz)	½ fat-free milk
with fruit, low fat	⅔ cup (6 oz)	1 fat-free milk + 1 carbohydrate

Table D-5 Sweets, Desserts, and Other Carbohydrates

1 other carbohydrate choice = 15 grams carbohydrate, variable grams protein, variable grams fat, and variable kilocalories.

Note: In general, one choice from this list can substitute for foods on the Starch, Fruits, or Milk lists.

Beverages, soda, and energy/sports drinks

Food	Serving size	Count as
Cranberry juice cocktail	½ cup	1 carbohydrate
Energy drink	1 can (8.3 oz)	2 carbohydrates
Fruit drink or lemonade	1 cup (8 oz)	2 carbohydrates
Hot chocolate		
regular	1 envelope added to 8 oz water	1 carbohydrate + 1 fat
sugar free or light	1 envelope added to 8 oz water	1 carbohydrate
Soft drink (soda), regular	1 can (12 oz)	2½ carbohydrates
Sports drink	1 cup (8 oz)	1 carbohydrate

Brownies, cake, cookies, gelatin, pie, and pudding

Food	Serving size	Count as
Brownie, small, unfrosted	1¼-inch square, $^7/_8$ inch high (about 1 oz)	1 carbohydrate + 1 fat
Cake		
angel food, unfrosted	$^1/_{12}$ of cake (about 2 oz)	2 carbohydrates
frosted	2-inch square (about 2 oz)	2 carbohydrates + 1 fat
unfrosted	2-inch square (about 2 oz)	1 carbohydrate + 1 fat
Cookies		
chocolate chip	2 cookies (2¼ inches across)	1 carbohydrate + 2 fats
gingersnap	3 cookies	1 carbohydrate
sandwich, with creme filling	2 small (about $^2/_3$ oz)	1 carbohydrate + 1 fat
sugar free	3 small or 1 large (¾–1 oz)	1 carbohydrate + 1–2 fats
vanilla wafer	5 cookies	1 carbohydrate + 1 fat
Cupcake, frosted	1 small (about 1¾ oz)	2 carbohydrates + 1–1½ fats
Fruit cobbler	½ cup (3½ oz)	3 carbohydrates + 1 fat
Gelatin, regular	½ cup	1 carbohydrate
Pie		
commercially prepared fruit, 2 crusts	$^1/_6$ of 8-inch pie	3 carbohydrates + 2 fats
pumpkin or custard	$^1/_8$ of 8-inch pie	1½ carbohydrates +1½ fats
Pudding		
regular (made with reduced-fat milk)	½ cup	2 carbohydrates
sugar free or sugar and fat free (made with fat-free milk)	½ cup	1 carbohydrate

Candy, spreads, sweets, sweeteners, syrups, and toppings

Food	Serving size	Count as
Candy bar, chocolate/peanut	2 "fun size" bars (1 oz)	1½ carbohydrates + 1½ fats
Candy, hard	3 pieces	1 carbohydrate
Chocolate "kisses"	5 pieces	1 carbohydrate + 1 fat
Coffee creamer		
dry, flavored	4 tsp	½ carbohydrate + ½ fat
liquid, flavored	2 tbsp	1 carbohydrate
Fruit snacks, chewy (pureed fruit concentrate)	1 roll (¾ oz)	1 carbohydrate
Fruit spreads, 100% fruit	1½ tbsp	1 carbohydrate

(Continued)

Table D-5 Sweets, Desserts, and Other Carbohydrates (Continued)

Candy, spreads, sweets, sweeteners, syrups, and toppings

Food	Serving size	Count as
Honey	1 tbsp	1 carbohydrate
Jam or jelly, regular	1 tbsp	1 carbohydrate
Sugar	1 tbsp	1 carbohydrate
Syrup		
chocolate	2 tbsp	2 carbohydrates
light (pancake type)	2 tbsp	1 carbohydrate
regular (pancake type)	1 tbsp	1 carbohydrate

Condiments and sauces[a]

Food	Serving size	Count as
Barbecue sauce	3 tbsp	1 carbohydrate
Cranberry sauce, jellied	¼ cup	1½ carbohydrates
▮ Gravy, canned or bottled	½ cup	½ carbohydrate + ½ fat
Salad dressing, fat free, low fat, cream based	3 tbsp	1 carbohydrate
Sweet and sour sauce	3 tbsp	1 carbohydrate

Doughnuts, muffins, pastries, and sweet breads

Food	Serving size	Count as
Banana nut bread	1-inch slice (1 oz)	2 carbohydrates + 1 fat
Doughnut		
cake, plain	1 medium (1½ oz)	1½ carbohydrates + 2 fats
yeast type, glazed	3¾ inches across (2 oz)	2 carbohydrates + 2 fats
Muffin (4 oz)	¼ muffin (1 oz)	1 carbohydrate + ½ fat
Sweet roll or Danish	1 (2½ oz)	2½ carbohydrates + 2 fats

Frozen bars, frozen desserts, frozen yogurt, and ice cream

Food	Serving size	Count as
Frozen pops	1	½ carbohydrate
Fruit juice bars, frozen, 100% juice	1 bar (3 oz)	1 carbohydrate
Ice cream		
fat free	½ cup	1½ carbohydrates
light	½ cup	1 carbohydrate + 1 fat
no sugar added	½ cup	1 carbohydrate + 1 fat
regular	½ cup	1 carbohydrate + 2 fats
Sherbet, sorbet	½ cup	2 carbohydrates
Yogurt, frozen		
fat free	⅓ cup	1 carbohydrate
regular	½ cup	1 carbohydrate + 0–1 fat

Granola bars, meal replacement bars/shakes, and trail mix

Food	Serving size	Count as
Granola or snack bar, regular or low fat	1 bar (1 oz)	1½ carbohydrates
Meal replacement bar	1 bar (1⅓ oz)	1½ carbohydrates + 0–1 fat
Meal replacement bar	1 bar (2 oz)	2 carbohydrates + 1 fat
Meal replacement shake, reduced calorie	1 can (10–11 oz)	1½ carbohydrates + 0–1 fat
Trail mix		
candy/nut based	1 oz	1 carbohydrate + 2 fats
dried fruit based	1 oz	1 carbohydrate + 1 fat

KEY

▮ = 480 milligrams or more of sodium per serving.
[a] You can also check the Fats list and Free Foods list for other condiments.

Table D-6 Nonstarchy Vegetables

The Nonstarchy Vegetables list includes vegetables that have few grams of carbohydrates or calories; starchy vegetables are found on the Starch list. 1 nonstarchy vegetable choice = 5 grams carbohydrate, 2 grams protein, 0 grams fat, and 25 kilocalories.

Note: In general, one nonstarchy vegetable choice is ½ cup cooked vegetables or vegetable juice or 1 cup raw vegetables. Count 3 cups of raw vegetables or 1½ cups of cooked vegetables as one carbohydrate choice.

Nonstarchy vegetables[a]	Nonstarchy vegetables[a]
Amaranth or Chinese spinach	Kohlrabi
Artichoke	Leeks
Artichoke hearts	Mixed vegetables (without corn, peas, or pasta)
Asparagus	Mung bean sprouts
Baby corn	Mushrooms, all kinds, fresh
Bamboo shoots	Okra
Beans (green, wax, Italian)	Onions
Bean sprouts	Oriental radish or daikon
Beets	Pea pods
🅸 Borscht	☺ Peppers (all varieties)
Broccoli	Radishes
☺ Brussels sprouts	Rutabaga
Cabbage (green, bok choy, Chinese)	Sauerkraut
☺ Carrots	Soybean sprouts
Cauliflower	Spinach
Celery	Squash (summer, crookneck, zucchini)
☺ Chayote	Sugar pea snaps
Coleslaw, packaged, no dressing	☺ Swiss chard
Cucumber	Tomato
Eggplant	Tomatoes, canned
Gourds (bitter, bottle, luffa, bitter melon)	🅸 Tomato sauce
Green onions or scallions	🅸 Tomato/vegetable juice
Greens (collard, kale, mustard, turnip)	Turnips
Hearts of palm	Water chestnuts
Jicama	Yard-long beans

KEY

☺ = More than 3 grams of dietary fiber per serving.

🅸 = 480 milligrams or more of sodium per serving.

[a] Salad greens (like chicory, endive, escarole, lettuce, romaine, spinach, arugula, radicchio, watercress) are on the Free Foods list.

Table D-7 Meat and Meat Substitutes

The Meat and Meat Substitutes list groups foods based on the amount of fat they have (lean meat, medium-fat meat, high-fat meat, and plant-based proteins).

Lean meats and meat substitutes

1 lean meat choice = 0 grams carbohydrate, 7 grams protein, 0–3 grams fat, and 45 kilocalories.

Food	Amount	Food	Amount
Beef: Select or Choice grades trimmed of fat: ground round; roast (chuck, rib, rump); round; sirloin; steak (cubed, flank, porterhouse, T-bone); tenderloin	1 oz	Fish, fresh or frozen, plain: catfish, cod, flounder, haddock, halibut, orange roughy, salmon, tilapia, trout, tuna	1 oz
		🅸 Fish, smoked: herring or salmon	1 oz
🅸 Beef jerky	1 oz	Game: buffalo, rabbit, venison	1 oz
Cheeses with 3 grams of fat or less per oz	1 oz	🅸 Hot dog with 3 grams of fat or less per oz (8 dogs per 14-oz package)	1
Cottage cheese	¼ cup	*Note: May be high in carbohydrate.*	
Egg substitutes, plain	¼ cup		
Egg whites	2	Lamb: chop, leg, or roast	1 oz

(Continued)

Table D-7 Meat and Meat Substitutes (Continued)

Food	Amount	Food	Amount
Organ meats: heart, kidney, liver *Note: May be high in cholesterol.*	1 oz	Salmon, canned Sardines, canned	1 oz 2 medium
Oysters, fresh or frozen	6 medium	Sausage with 3 grams of fat or less per oz	1 oz
Pork, lean		Shellfish: clams, crab, imitation shellfish, lobster, scallops, shrimp	1 oz
Canadian bacon	1 oz		
rib or loin chop/roast, ham, tenderloin	1 oz	Tuna, canned in water or oil, drained	1 oz
Poultry, without skin: Cornish hen, chicken, domestic duck or goose (well drained of fat), turkey	1 oz	Veal, lean chop, roast	1 oz
Processed sandwich meats with 3 grams of fat or less per oz: chipped beef, deli thin-sliced meats, turkey ham, turkey kielbasa, turkey pastrami	1 oz		

Medium-fat meat and meat substitutes

1 medium-fat meat choice = 0 grams carbohydrate, 7 grams protein, 4–7 grams fat, and 75 kilocalories.

Food	Amount
Beef: corned beef, ground beef, meat loaf, prime grades trimmed of fat (for example, prime rib), short ribs, tongue	1 oz
Cheeses with 4–7 grams of fat per oz: feta, mozzarella, pasteurized processed cheese spread, reduced-fat cheeses, string	1 oz
Egg *Note: High in cholesterol, so limit to 3 per week.*	1
Fish, any fried product	1 oz
Lamb: ground, rib roast	1 oz
Pork: cutlet, shoulder roast	1 oz
Poultry: chicken with skin; dove, pheasant, wild duck, or goose; fried chicken; ground turkey	1 oz
Ricotta cheese	2 oz or ¼ cup
Sausage with 4–7 grams of fat per oz	1 oz
Veal, cutlet (no breading)	1 oz

High-fat meat and meat substitutes

1 high-fat meat choice = 0 grams carbohydrate, 7 grams protein, 8+ grams fat, and 100 kilocalories. These foods are high in saturated fat, cholesterol, and calories and may raise blood cholesterol levels if eaten on a regular basis. Try to eat 3 or fewer servings from this group per week.

Food	Amount
Bacon	
pork	2 slices (16 slices per lb or 1 oz each, before cooking)
turkey	3 slices (½ oz each before cooking)
Cheese, regular: American, bleu, brie, cheddar, hard goat, Monterey jack, queso blanco, and Swiss	1 oz
Hot dog: beef, pork, or combination (10 per lb-sized package)	1
Hot dog: turkey or chicken (10 per lb-sized package)	1
Pork: ground, sausage, spareribs	1 oz
Processed sandwich meats with 8 grams of fat or more per oz: bologna, pastrami, hard salami	1 oz
Sausage with 8 grams fat or more per oz: bratwurst, chorizo, Italian, knockwurst, Polish, smoked, summer	1 oz

Plant-based proteins

1 plant-based protein choice = variable grams carbohydrate, 7 grams protein, variable grams fat, and variable kilocalories. Because carbohydrate content varies among plant-based proteins, you should read the food label.

Food	Serving size	Count as
"Bacon" strips, soy based	3 strips	1 medium-fat meat
Baked beans	⅓ cup	1 starch + 1 lean meat
Beans, cooked: black, garbanzo, kidney, lima, navy, pinto, white[a]	½ cup	1 starch + 1 lean meat
"Beef" or "sausage" crumbles, soy based	2 oz	½ carbohydrate + 1 lean meat
"Chicken" nuggets, soy based	2 nuggets (1½ oz)	½ carbohydrate + 1 medium-fat meat
Edamame	½ cup	½ carbohydrate + 1 lean meat
Falafel (spiced chickpea and wheat patties)	3 patties (about 2 inches across)	1 carbohydrate + 1 high-fat meat

[a] Beans, peas, and lentils are also found on the Starch list; nut butters in smaller amounts are found in the Fats list.

(Continued)

Table D-7 Meat and Meat Substitutes (Continued)

Plant-based proteins

1 plant-based protein choice = variable grams carbohydrate, 7 grams protein, variable grams fat, and variable kilocalories.
Because carbohydrate content varies among plant-based proteins, you should read the food label.

Food	Serving size	Count as
Hot dog, soy based	1 (1½ oz)	½ carbohydrate + 1 lean meat
😊 Hummus	⅓ cup	1 carbohydrate + 1 high-fat meat
😊 Lentils, brown, green, or yellow	½ cup	1 carbohydrate + 1 lean meat
😊 Meatless burger, soy based	3 oz	½ carbohydrate + 2 lean meats
😊 Meatless burger, vegetable and starch based	1 patty (about 2½ oz)	1 carbohydrate + 2 lean meats
"Nut" spreads: almond butter, cashew butter, peanut butter, soy nut butter ª	1 tbsp	1 high-fat meat
😊 Peas, cooked: black-eyed and split peas	½ cup	1 starch + 1 lean meat
🧂😊 Refried beans, canned	½ cup	1 starch + 1 lean meat
"Sausage" patties, soy based	1 (1½ oz)	1 medium-fat meat
Soy nuts, unsalted	¾ oz	½ carbohydrate + 1 medium-fat meat
Tempeh	¼ cup	1 medium-fat meat
Tofu, regular	4 oz (½ cup)	1 medium-fat meat
Tofu, light	4 oz (½ cup)	1 lean meat

KEY

😊 = More than 3 grams of dietary fiber per serving.

▼ = Extra fat, or prepared with added fat. (Count as 1 starch + 1 fat.)

🧂 = 480 milligrams or more of sodium per serving (based on the sodium content of a typical 3-oz serving of meat, unless 1 or 2 oz is the normal serving size).

ª Beans, peas, and lentils are also found on the Starch list; nut butters in smaller amounts are found in the Fats list.

Table D-8 Fats

Fats and oils have mixtures of unsaturated (polyunsaturated and monounsaturated) and saturated fats. Foods on the Fats list are grouped together based on the major type of fat they contain.

1 fat choice = 0 grams carbohydrate, 0 grams protein, 5 grams fat, and 45 kilocalories.

Note: In general, one fat exchange is 1 teaspoon of regular margarine, vegetable oil, or butter or 1 tablespoon of regular salad dressing.

When used in large amounts, bacon and peanut butter are counted as high-fat meat choices (see Meat and Meat Substitutes list). Fat-free salad dressings are found on the Sweets, Desserts, and Other Carbohydrates list. Fat-free products such as margarines, salad dressings, mayonnaise, sour cream, and cream cheese are found on the Free Foods list.

Monounsaturated fats		Monounsaturated fats	
Food	**Serving size**	**Food**	**Serving size**
Avocado, medium	2 tbsp (1 oz)	Nuts	
"Nut" butters (*trans* fat free): almond butter, cashew butter, peanut butter (smooth or crunchy)	1½ tsp	mixed (50% peanuts)	6 nuts
		peanuts	10 nuts
		pecans	4 halves
Nuts		pistachios	16 nuts
almonds	6 nuts	Oil: canola, olive, peanut	1 tsp
Brazil	2 nuts	Olives	
cashews	6 nuts	black (ripe)	8 large
filberts (hazelnuts)	5 nuts	green, pimiento stuffed	10 large
macadamia	3 nuts		

Polyunsaturated fats		Polyunsaturated fats	
Food	**Serving size**	**Food**	**Serving size**
Margarine: low-fat spread (30–50% vegetable oil, *trans* fat free)	1 tbsp	Mayonnaise	
		reduced fat	1 tbsp
Margarine: stick, tub (*trans* fat free) or squeeze (*trans* fat free)	1 tsp	regular	1 tsp

(Continued)

Table D-8 Fats (Continued)

Polyunsaturated fats

Food	Serving size
Mayonnaise-style salad dressing	
reduced fat	1 tbsp
regular	2 tsp
Nuts	
pignolia (pine nuts)	1 tbsp
walnuts, English	4 halves
Oil: corn, cottonseed, flaxseed, grape seed, safflower, soybean, sunflower	1 tsp
Oil: made from soybean and canola oil	1 tsp
Plant stanol esters	
light	1 tbsp
regular	2 tsp
Salad dressing	
🧂reduced fat	2 tbsp
Note: May be high in carbohydrate.	
🧂regular	1 tbsp
Seeds	
flaxseed, whole	1 tbsp
pumpkin, sunflower	1 tbsp
sesame seeds	1 tbsp
Tahini or sesame paste	2 tsp

Saturated fats

Food	Serving size
Bacon, cooked, regular or turkey	1 slice
Butter	
reduced fat	1 tbsp
stick	1 tsp
whipped	2 tsp
Butter blends made with oil	
reduced fat or light	1 tbsp
regular	1½ tsp
Chitterlings, boiled	2 tbsp (½ oz)
Coconut, sweetened, shredded	2 tbsp
Coconut milk	
light	⅓ cup
regular	1½ tbsp
Cream	
half and half	2 tbsp
heavy	1 tbsp
light	1½ tbsp
whipped	2 tbsp
whipped, pressurized	¼ cup
Cream cheese	
reduced fat	1½ tbsp (¾ oz)
regular	1 tbsp (½ oz)
Lard	1 tsp
Tropical oils: coconut, palm, palm kernel	1 tsp
Salt pork	¼ oz
Shortening, solid	1 tsp
Sour cream	
reduced fat or light	3 tbsp
regular	2 tbsp

KEY

🧂 = 480 milligrams or more of sodium per serving.

Table D-9 Free Foods

A "free" food is any food or drink choice that has less than 20 kilocalories and 5 grams or less of carbohydrate per serving.

- Most foods on this list should be limited to three servings (as listed here) per day. Spread out the servings throughout the day. If you eat all three servings at once, it could raise your blood glucose level.
- Food and drink choices listed here without a serving size can be eaten whenever you like, in any moderate amount.

Low-carbohydrate foods

Food	Serving size
Cabbage, raw	½ cup
Candy, hard (regular or sugar free)	1 piece
Carrots, cauliflower, or green beans, cooked	¼ cup
Cranberries, sweetened with sugar substitute	½ cup
Cucumber, sliced	½ cup
Gelatin	
dessert, sugar free	
unflavored	
Gum	
Jam or jelly, light or no sugar added	2 tsp
Rhubarb, sweetened with sugar substitute	½ cup

Low-carbohydrate foods

Food	Serving size
Salad greens	
Sugar substitutes (artificial sweeteners)	
Syrup, sugar free	2 tbsp

Modified-fat foods with carbohydrate

Food	Serving size
Cream cheese, fat free	1 tbsp (½ oz)
Creamers	
nondairy, liquid	1 tbsp
nondairy, powdered	2 tsp

(Continued)

Table D-9 Free Foods (Continued)

Modified-fat foods with carbohydrate

Food	Serving size
Margarine spread	
fat free	1 tbsp
reduced fat	1 tsp
Mayonnaise	
fat free	1 tbsp
reduced fat	1 tsp
Mayonnaise-style salad dressing	
fat free	1 tbsp
reduced fat	1 tsp
Salad dressing	
fat free or low fat	1 tbsp
fat free, Italian	2 tbsp
Sour cream, fat free or reduced fat	1 tbsp
Whipped topping	
light or fat free	2 tbsp
regular	1 tbsp

Condiments

Food	Serving size
Barbecue sauce	2 tsp
Catsup (ketchup)	1 tbsp
Honey mustard	1 tbsp
Horseradish	
Lemon juice	
Miso	1½ tsp
Mustard	
Parmesan cheese, freshly grated	1 tbsp
Pickle relish	1 tbsp
Pickles	
dill	1½ medium
sweet, bread and butter	2 slices
sweet, gherkin	¾ oz
Salsa	¼ cup

Condiments

Food	Serving size
Soy sauce, light or regular	1 tbsp
Sweet-and-sour sauce	2 tsp
Sweet chili sauce	2 tsp
Taco sauce	1 tbsp
Vinegar	
Yogurt, any type	2 tbsp

Drinks/mixes

Any food on the list—without a serving size listed—can be consumed in any moderate amount.

- Bouillon, broth, consommé
- Bouillon or broth, low sodium
- Carbonated or mineral water
- Club soda
- Cocoa powder, unsweetened (1 tbsp)
- Coffee, unsweetened or with sugar substitute
- Diet soft drinks, sugar free
- Drink mixes, sugar free
- Tea, unsweetened or with sugar substitute
- Tonic water, diet
- Water
- Water, flavored, carbohydrate free

Seasonings

Any food on this list can be consumed in any moderate amount.

- Flavoring extracts (for example, vanilla, almond, peppermint)
- Garlic
- Herbs, fresh or dried
- Nonstick cooking spray
- Pimento
- Spices
- Hot pepper sauce
- Wine, used in cooking
- Worcestershire sauce

KEY

= 480 milligrams or more of sodium per serving.

Table D-10 Combination Foods

Many foods are eaten in various combinations, such as casseroles. Because "combination" foods do not fit into any one choice list, this list of choices provides some typical combination foods.

Entrees

Food	Serving size	Count as
Casserole type (tuna noodle, lasagna, spaghetti with meatballs, chili with beans, macaroni and cheese)	1 cup (8 oz)	2 carbohydrates + 2 medium-fat meats
Stews (beef/other meats and vegetables)	1 cup (8 oz)	1 carbohydrate + 1 medium-fat meat + 0–3 fats
Tuna salad or chicken salad	½ cup (3½ oz)	½ carbohydrate + 2 lean meats + 1 fat

(Continued)

Table D-10 Combination Foods (Continued)

Frozen meals/entrees

Food	Serving size	Count as
🌢😊 Burrito (beef and bean)	1 (5 oz)	3 carbohydrates + 1 lean meat + 2 fats
🌢 Dinner-type meal	generally 14–17 oz	3 carbohydrates + 3 medium-fat meats + 3 fats
🌢 Entrée or meal with less than 340 kilocalories	about 8–11 oz	2–3 carbohydrates + 1–2 lean meats
Pizza		
🌢 cheese/vegetarian, thin crust	¼ of a 12-inch pie (4½–5 oz)	2 carbohydrates + 2 medium-fat meats
🌢 meat topping, thin crust	¼ of a 12-inch pie (5 oz)	2 carbohydrates + 2 medium-fat meats + 1½ fats
🌢 Pocket sandwich	1 (4½ oz)	3 carbohydrates + 1 lean meat + 1–2 fats
🌢 Pot pie	1 (7 oz)	2½ carbohydrates + 1 medium-fat meat + 3 fats

Salads (deli style)

Food	Serving size	Count as
Coleslaw	½ cup	1 carbohydrate + 1½ fats
Macaroni/pasta salad	½ cup	2 carbohydrates + 3 fats
🌢 Potato salad	½ cup	1½–2 carbohydrates + 1–2 fats

Soups

Food	Serving size	Count as
🌢 Bean, lentil, or split pea	1 cup	1 carbohydrate + 1 lean meat
🌢 Chowder (made with milk)	1 cup (8 oz)	1 carbohydrate + 1 lean meat + 1½ fats
🌢 Cream (made with water)	1 cup (8 oz)	1 carbohydrate + 1 fat
🌢 Instant	6 oz prepared	1 carbohydrate
🌢 with beans or lentils	8 oz prepared	2½ carbohydrates + 1 lean meat
🌢 Miso	1 cup	½ carbohydrate + 1 fat
🌢 Oriental noodle	1 cup	2 carbohydrates + 2 fats
Rice (congee)	1 cup	1 carbohydrate
🌢 Tomato (made with water)	1 cup (8 oz)	1 carbohydrate
🌢 Vegetable beef, chicken noodle, or other broth type	1 cup (8 oz)	1 carbohydrate

KEY

😊 = More than 3 grams of dietary fiber per serving.

▼ = Extra fat, or prepared with added fat.

🌢 = 600 milligrams or more of sodium per serving (for combination food main dishes/meals).

Table D-11 Fast Foods

The choices in the Fast Foods list are not specific fast-food meals or items, but are estimates based on popular foods. Ask the restaurant or check its website for nutrition information about your favorite fast foods.

Breakfast sandwiches

Food	Serving size	Count as
🌢 Egg, cheese, meat, English muffin	1 sandwich	2 carbohydrates + 2 medium-fat meats
🌢 Sausage biscuit sandwich	1 sandwich	2 carbohydrates + 2 high-fat meats + 3½ fats

(Continued)

Table D-11 Fast Foods (Continued)

Main dishes/entrees

Food	Serving size	Count as
🏷️😊 Burrito (beef and beans)	1 (about 8 oz)	3 carbohydrates + 3 medium-fat meats + 3 fats
🏷️ Chicken breast, breaded and fried	1 (about 5 oz)	1 carbohydrate + 4 medium-fat meats
Chicken drumstick, breaded and fried	1 (about 2 oz)	2 medium-fat meats
🏷️ Chicken nuggets	6 (about 3½ oz)	1 carbohydrate + 2 medium-fat meats + 1 fat
🏷️ Chicken thigh, breaded and fried	1 (about 4 oz)	½ carbohydrate + 3 medium-fat meats + 1½ fats
🏷️ Chicken wings, hot	6 (5 oz)	5 medium-fat meats + 1½ fats

Oriental

Food	Serving size	Count as
🏷️ Beef/chicken/shrimp with vegetables in sauce	1 cup (about 5 oz)	1 carbohydrate + 1 lean meat + 1 fat
🏷️ Egg roll, meat	1 (about 3 oz)	1 carbohydrate + 1 lean meat + 1 fat
Fried rice, meatless	½ cup	1½ carbohydrates + 1½ fats
🏷️ Meat and sweet sauce (orange chicken)	1 cup	3 carbohydrates + 3 medium-fat meats + 2 fats
🏷️ Noodles and vegetables in sauce (chow mein, lo mein)	1 cup	2 carbohydrates + 1 fat

Pizza

Food	Serving size	Count as
Pizza		
🏷️ cheese, pepperoni, regular crust	⅛ of a 14-inch (about 4 oz)	2½ carbohydrates + 1 medium-fat meat + 1½ fats
🏷️ cheese/vegetarian, thin crust	¼ of a 12-inch (about 6 oz)	2½ carbohydrates + 2 medium-fat meats + 1½ fats

Sandwiches

Food	Serving size	Count as
🏷️ Chicken sandwich, grilled	1	3 carbohydrates + 4 lean meats
🏷️ Chicken sandwich, crispy	1	3½ carbohydrates + 3 medium-fat meats + 1 fat
🏷️ Fish sandwich with tartar sauce	1	2½ carbohydrates + 2 medium-fat meats + 2 fats
Hamburger		
🏷️ large with cheese	1	2½ carbohydrates + 4 medium-fat meats + 1 fat
regular	1	2 carbohydrates + 1 medium-fat meat + 1 fat
🏷️ Hot dog with bun	1	1 carbohydrate + 1 high-fat meat + 1 fat
Submarine sandwich		
🏷️ less than 6 grams fat	6-inch sub	3 carbohydrates + 2 lean meats
🏷️ regular	6-inch sub	3½ carbohydrates + 2 medium-fat meats + 1 fat
Taco, hard or soft shell (meat and cheese)	1 small	1 carbohydrate + 1 medium-fat meat + 1½ fats

Salads

Food	Serving size	Count as
🏷️😊 Salad, main dish (grilled chicken type, no dressing or croutons)		1 carbohydrate + 4 lean meats
Salad, side, no dressing or cheese	small (about 5 oz)	1 vegetable

(Continued)

Table D-11 Fast Foods (Continued)

Sides/appetizers

Food	Serving size	Count as
▼ French fries, restaurant style	small	3 carbohydrates + 3 fats
	medium	4 carbohydrates + 4 fats
	large	5 carbohydrates + 6 fats
▮ Nachos with cheese	small (about 4½ oz)	2½ carbohydrates + 4 fats
▮ Onion rings	1 serving (about 3 oz)	2½ carbohydrates + 3 fats

Desserts

Food	Serving size	Count as
Milkshake, any flavor	12 oz	6 carbohydrates + 2 fats
Soft-serve ice cream cone	1 small	2½ carbohydrates + 1 fat

KEY

☺ = More than 3 grams of dietary fiber per serving.

▼ = Extra fat, or prepared with added fat.

▮ = 600 milligrams or more of sodium per serving (for fast-food main dishes/meals).

Table D-12 Alcohol

1 alcohol equivalent = variable grams carbohydrate, 0 grams protein, 0 grams fat, and 100 kilocalories.

Note: In general, one alcohol choice (½ ounce absolute alcohol) has about 100 kilocalories. For those who choose to drink alcohol, the 2010 Dietary Guidelines for Americans suggest limiting alcohol intake to one drink or less per day for women, and two drinks or less per day for men. To reduce your risk of low blood glucose (hypoglycemia), especially if you take insulin or other medications that lower blood glucose, always drink alcohol with food. Although alcohol, by itself, does not directly affect blood glucose, be aware of the carbo-hydrate (for example, in mixed drinks, beer, and wine) content, which may raise your blood glucose.

Alcoholic beverage	Serving size	Count as
Beer		
light (4.2% alcohol)	12 fl oz	1 alcohol equivalent + ½ carbohydrate
regular (4.9% alcohol)	12 fl oz	1 alcohol equivalent + 1 carbohydrate
Distilled spirits (80 or 86 proof): vodka, rum, gin, whiskey	1½ fl oz	1 alcohol equivalent
Liqueur, coffee (53 proof)	1 fl oz	1 alcohol equivalent + 1 carbohydrate
Sake	1 fl oz	½ alcohol equivalent
Wine		
sherry (14–20% alcohol)	3½ fl oz	1 alcohol equivalent + 1 carbohydrate
dry, red, or white (10% alcohol)	5 fl oz	1 alcohol equivalent

Answers to Study Card Review Questions

Answers to the multiple-choice questions appear directly on the study card for each chapter. Answers to the essay questions are given below.

Chapter 1

1. A nutrient is a substance found in foods that can be used by the body to support health, and the term "nutrition" refers to the interdisciplinary study of how foods impact health. The meaning of the term "nutrient" is evolving rapidly as scientists become more able to both measure small amounts of compounds (previously undiscovered or unquantifiable) in foods and assess the myriad and diverse health outcomes that they influence, such as regulation of gene expression and processes such as inflammation. **3.** Organic compounds are those that contain carbon, whereas organic foods are those that have been raised and/or processed in a certain way. Although there is limited evidence that organically produced foods may be more nutritious than conventional foods, this is an area of active scientific debate. **4.** The six nutrient classes are carbohydrates, proteins, lipids, water, vitamins, and minerals. The first four are macronutrients, and the last two are micronutrients. Carbohydrates, proteins, lipids, and vitamins are organic molecules, whereas water and minerals are inorganic molecules. Carbohydrates, proteins, and lipids are energy yielding because energy contained in them can be transferred to ATP and used to fuel the body's many chemical reactions. **6.** A serving of this food contains 200 kcal from carbohydrates, 16 kcal from protein, and 405 kcal from lipid; a total of 621 kcal. This translates to roughly 32% of energy from carbohydrate, 3% from protein, and 65% from lipid. **7.** In the most basic sense, the scientific method includes three steps carried out in the following order: making an observation, proposing a hypothesis (explanation for the observation), and appropriately collecting data to test the hypothesis. **8.** Epidemiologic studies are based on data collected strictly by observation and sample collection/analysis; subjects are not asked to change anything in their daily lives. Conversely, intervention studies involve treatments and carefully testing how a particular change (e.g., dietary intake) impacts the selected outcome variable (e.g., health parameter). Epidemiologic studies can only test correlations, whereas intervention studies can be used to assess cause-and-effect relationships. **10.** Cell culture studies are often done to investigate physiologic mechanisms at levels that are not possible in human or animal subjects. In addition, they can be used in situations where studying humans or animals would be unethical or impractical. **11.** Peer-reviewed journals are frequently the best sources of scientific findings because their articles have been read and reviewed by experts in the field. Other primary sources with credibility include publications and websites of government agencies (e.g., the USDA) and private entities (e.g., the Institute of Medicine). **12.** A randomized, double-blind, placebo-controlled study is a type of intervention study in which participants are randomly assigned to receive a treatment (intervention) or placebo (control). In this way, each participant has an equal chance of being in each group. The term *double-blind* means that neither the investigator nor the participant knows to which group participants have been assigned. People in the placebo group are given a treatment that is similar to the active treatment but does not contain or have the intervention of interest. This type of study is considered the "gold standard" in nutrition research, because it helps decrease the chance that researcher bias has occurred, helps account for placebo effect, and controls for confounding variables. It is also optimal if a researcher wishes to test for a cause-and-effect relationship. **16.** The *nutrition transition* refers to the shift from undernutrition to overnutrition (or unbalanced nutrition) that sometimes occurs when an agricultural or hunter–gatherer society makes the transition to a more industrialized economy. Whereas nutrient deficiencies have been historically less common in less-industrialized societies, chronic degenerative disease (such as heart attack and diabetes) have been more common in industrialized societies. Nutrition transition is resulting in obesity and chronic degenerative diseases becoming more prevalent in nonindustrialized societies as well.

Chapter 2

3. There are many factors influencing nutrient requirements, including age, sex, physical activity, genetic variability, and life stage. **4.** The four methods of assessing nutritional status are anthropometric measurements, biochemical measurements, clinical assessment, and dietary assessment. Anthropometric measurements are commonly used during infancy and childhood, whereas biochemical measurements are more frequently used during periods of suspected illness. Medical personnel use clinical assessment to ascertain signs and symptoms of malnutrition, and dietary assessment is frequently conducted by both clinicians and researchers. Although anthropometric measurements are often inexpensive to obtain, they cannot be used to determine specific nutrient deficiencies. Conversely, although biochemical and clinical assessment may be more costly, their results are more precise in that specific nutrient deficiencies can be diagnosed. Dietary assessment is useful in determining whether a diet has variety and balance, although it tends to be more time consuming than other methods. **6.** The four categories of Dietary Reference Intake (DRI) values are the Estimated Average Requirement (EAR), the Recommended Dietary Allowance (RDA), the Adequate Intake (AI) level, and the Tolerable Upper Intake Level (UL). Of these, the RDA and AI are meant to serve as dietary intake goals. **7.** Recommended Dietary Allowances (RDAs) were determined from information used to establish the EAR values. When EARs (and thus RDAs) could not be established, AIs were set. Consequently, RDA values were determined from more complete and conclusive research than were AIs. In addition, assessing one's nutritional adequacy is somewhat easier for nutrients that have RDAs than for those with AIs. **9.** Between 900 and 1,300 kcal should come from carbohydrates. One change that would decrease carbohydrate intake would be to vary the foods that are eaten at breakfast. Instead of having a bowl of cereal and a piece of toast each morning, alternate that with scrambled eggs or yogurt. However, it would be unwise if this change decreased the amount of whole grains consumed because these foods are good sources of dietary fiber and many of the micronutrients. **11.** This person's EER is 3,025 kcal/day. **12.** She is consuming 333% of her RDA for vitamin E. However, because she is still consuming less than her UL for vitamin E, she need not be concerned about this. **13.** The USDA is charged with overseeing U.S. agriculture, which of course is critically important for providing Americans with adequate food. In general, the first dietary recommendations put forth were focused on making sure Americans consumed adequate amounts of calories. As scientific knowledge improved, subsequent recommendations also provided guidance aimed at consuming sufficient amounts of micronutrients. Today, the Dietary Guidelines for Americans also address optimal health in terms of balanced nutrient intake, weight management, and food safety.   **14.** Weight-maintenance recommendations include improving eating and physical activity behaviors, controlling total calorie intake to manage weight (or lose weight, if necessary), increasing physical activity, and maintaining appropriate caloric balance during each stage of life. This is especially important as national obesity rates continue to climb.   **15.** The Dietary Guidelines for Americans recommend reducing intakes of sodium, saturated fats, cholesterol, *trans* fats, refined grains, added sugars, and alcohol. We are urged to increase our consumption of fruits and vegetables, whole grains, fat-free and low-fat dairy products, seafood, dietary fiber, calcium, and vitamin D. These recommendations are aimed at increasing nutrient density of our diet while decreasing caloric intake and disease-promoting food components.   **18.** Answers can vary depending on the food composition table used and individual life-stage group. However, using the table that accompanies this book, this meal would provide approximately 110% of a person's RDA for vitamin B_{12}. Basically, any animal-derived food (like chicken nuggets) could be added to this meal to increase its vitamin B_{12} content.

Chapter 3

9. The GI tract contains four tissue layers—the mucosa, submucosa, muscularis, and serosa. The mucosa is the innermost lining and consists of epithelial cells. These cells produce and release a variety of secretions needed for digestion. The submucosa is a layer of connective tissue that surrounds the mucosal layer. It contains blood vessels that provide nourishment to the mucosa and muscularis. It also contains lymphatic vessels and a network of nerves that control the release of secretions from the mucosa. The muscularis usually consists of two layers of smooth muscles. A network of nerves is embedded within the muscularis, and controls the contraction and relaxation of the smooth muscles. The function of the muscularis is to promote GI motility. The outermost layer is the serosa, which is also connective tissue. The serosa secretes a fluid that lubricates the digestive organs; the serosa also anchors the digestive organs within the abdominal cavity.   **12.** Organs that produce and release secretions needed for digestion include the salivary glands, stomach, pancreas, liver, gallbladder, and small intestine. The salivary glands release saliva, which contains water, salt, and digestive enzymes. In the mouth, saliva moistens and chemically breaks down food, and also plays a role in the process of taste. The stomach releases gastric juice, which contains a mixture of water, hydrochloric acid, intrinsic factor, and enzymes. When food mixes with gastric juice, chyme is formed. Both the enzymes and hydrochloric acid associated with gastric juice facilitate the chemical breakdown of food. Intrinsic factor is needed for the absorption of vitamin B_{12}. The pancreas releases pancreatic juice, which contains water, bicarbonate, and digestive enzymes. The bicarbonate neutralizes the acidic chyme as it passes from the stomach into the small intestine. The pancreatic enzymes are needed for the chemical breakdown of nutrients. The gallbladder releases bile (produced in the liver), which is an emulsifying agent. Bile enables fat to mix in the watery environment of the small intestine. Last, the small intestine releases a variety of enzymes, which are needed for the chemical breakdown of nutrients.   **17.** Digestive events taking place in the mouth include the physical and chemical breakdown of food. Food mixes with saliva, which helps moisten the food. The enzymes found in saliva facilitate the chemical breakdown of the food. After swallowing, food moves into the stomach. The presence of food in the stomach stimulates the release of the hormone gastrin, which stimulates the release of gastric juice and gastric motility. Increased gastric motility facilitates the mixing of food with gastric juice. As it becomes more liquid, it is called chyme. The enzymes present in gastric juice chemically break down nutrients. Small amounts of chyme pass into the small intestine, which triggers the release of enteric hormones—cholecystokinin (CCK) and secretin. These hormones coordinate the release of secretions from the pancreas and gallbladder, the relaxation of sphincters, and GI motility. Enterocytes, which make up the brush border of the small intestine, produce enzymes needed for the final stage of digestion. When nutrients are completely broken down, they are taken up into the enterocytes, the primary site of nutrient absorption.   **18.** The lining of the small intestine has a large surface area, making it well suited for nutrient absorption. First, the mucosa is arranged in large, pleated folds called plica circulares, which face toward the lumen. These folds are covered with tiny projections called villi. Each villus is lined with absorptive epithelial cells called enterocytes. The enterocytes are covered with even smaller projections called microvilli. The microvilli make up the absorptive surface, which is referred to as the brush border. These structures create an enormous surface area where nutrient absorption takes place. In addition, some enzymes are produced and released by the brush border, facilitating the final step in the digestive process.   **24.** The ileocecal sphincter regulates the passage of undigested food from the last portion of the small intestine (ileum) into the cecum (the first portion of the large intestine. Haustral contractions slowly propel the contents (feces) through the colon. During this time, water and electrolytes are extracted and returned to the blood for reuse by the body. Bacteria residing in the large intestine, referred to as the intestinal microbiota, break down the undigested food residue. The feces collect in the rectum, which signals the need to defecate. When the anal sphincter relaxes, feces move into the anal canal and are expelled from the body.

Chapter 4

4. Disaccharides consist of two monosaccharides bonded together. Sucrose is made of glucose and fructose; lactose is made of glucose and galactose; and maltose is made of two glucose molecules. A glycosidic bond forms between monosaccharides by a condensation reaction. This occurs when a hydrogen atom from one monosaccharide interacts with a hydroxyl group from another monosaccharide. In the process, a water molecule is formed.   **9.** Whereas humans can digest starch, they are unable to digest fiber. This is because fiber contains β-glycosidic bonds that are resistant to digestive enzymes.   **10.** The difference between dietary fiber and functional fiber is that dietary fiber occurs naturally in plants, whereas functional fiber, which is typically derived from natural fibrous plant sources, is added to food during manufacturing.   **14.** The digestion of amylose and that of amylopectin involves many of the same enzymes. The exception is α-dextrinase, which is needed to hydrolyze α-1,6 glycosidic bonds present in amylopectin. The digestion of amylose and amylopectin begins in the mouth. The salivary glands release α-salivary amylase, which hydrolyzes α-1,4 glycosidic bonds present in both amylose and amylopectin. This results in a partial-breakdown product called dextrins. There is no starch digestion in the stomach. Dextrins then pass into the small intestine. The pancreas releases α-pancreatic amylase, which continues to hydrolyze α-1,4 glycosidic bonds. Dextrins from amylose are broken down to maltose, whereas dextrins from amylopectin result in maltose and limit-dextrins. Limit-dextrins contain α-1,6 glycosidic bonds that were located at branch points in the original amylopectin molecule. Maltose is hydrolyzed to form glucose by the brush border enzyme maltase and limit-dextrins are hydrolyzed to form glucose by α-dextrinase, also a brush border enzyme.   **18.** The hormones insulin and glucagon play important roles in blood glucose regulation. Both of these hormones are released from the pancreas. When blood glucose increases, the pancreas increases the amount of insulin released into the blood. Some cells, such as those making up skeletal muscle and adipose tissue, require the presence of insulin to take up glucose. When insulin binds to receptors on the surface of these cells, insulin-responsive glucose transport molecules relocate from the cell's cytoplasm to the surface of the cell membrane. Glucose transport proteins enable glucose to cross the cell membrane. As a result, blood glucose concentrations decrease. Insulin also facilitates glycogen, protein, and fat synthesis. When blood glucose decreases, the pancreas increases the amount of glucagon released into the blood. Glucagon stimulates glycogenolysis in the liver. Glycogenolysis is the process by which glycogen is broken down to produce glucose molecules. When liver glycogen is broken down, the glucose is released into the blood. As a result, blood glucose rises.   **19.** The citric acid cycle both begins and ends with a molecule called oxaloacetate. Specifically, oxaloacetate combines with acetyl-CoA to produce citrate. Thus, adequate amounts of oxaloacetate are needed to keep the citric acid cycle fully functional. In addition, oxaloacetate is used to synthesize glucose by a process called gluconeogenesis. When glucose is limited, cells increase their use of fatty acids for energy. Fatty acid metabolism results in the production of many molecules of acetyl-CoA, which normally enter the citric acid cycle by combining with oxaloacetate. However, when oxaloacetate is limited, molecules of acetyl-CoA are not able to readily enter the citric acid cycle. As a result, they enter a different metabolic pathway that forms ketones.

21. 45% of 1,800 kcal = 810 kcal
810 kcal = 203 g of carbohydrate
This person should consume a minimum of 203 g/day carbohydrate.

Chapter 5

4. Protein complementation is the practice whereby incomplete proteins are consumed in combination so that the mix of foods provides sufficient amounts of all essential amino acids. Thus, a "complete protein" is consumed. Protein complementation is especially important during periods of rapid growth and development (such as early childhood) when amino acid requirements are relatively high and overall food intake may be limited.   **6.** Sickle cell anemia is an inherited disease caused by a mutation in the DNA (gene) that codes for the protein hemoglobin. This alteration in the genetic code causes an incorrect amino acid to be inserted into the protein during translation, producing a hemoglobin molecule that cannot bind iron and therefore cannot function properly. It is hypothesized that sickle cell anemia somehow protects from malaria; signs and symptoms of this disease include anemia, chest pain, swollen hands, and stunted growth.   **8.** There are four levels of protein structure. These are called primary, secondary, tertiary, and quaternary structures. Primary structure refers to the number, types, and order of amino acids in the polypeptide chain. Secondary structure is how the protein's amino and carboxylic acid groups cause folding—usually in the form of α-helices or β-folded sheets. Tertiary structure is caused by attractions between R-groups, and quaternary structure occurs when more than one peptide chain and/or a nonprotein subunit (a prosthetic group) are components of a single functional protein.   **9.** Both *genetics* and *epigenetics* refer to factors related to protein synthesis. The former refers to the sequence and structure of the DNA, whereas the latter refers to factors related to gene expression. Nutrition can influence which genes are transcribed. Moreover, nutritional status (especially very early in life) may influence epigenetic modifications that have life-long consequences.   **13.** Proteins serve many functions in the body. For example, they make up most of our muscle tissue and are therefore required for movement and basic physiological processes (such as cardiac function). The immune system also requires proteins for antibody production and for synthesizing and maintaining protective barriers such as the skin. Proteins also serve as important communicators in the body, making up many of the hormones that it produces. In addition, almost all enzymes are made from proteins. Thus, all the metabolism required for basic body functions is protein dependent as well.   **14.** Because they are gaining muscle mass, growing children and individuals engaged in weight training would be expected to be in positive protein balance.   **15.** Between 200 and 700 kcal should come from protein; this would require 50 to 175 grams of protein. Because most protein in the diet comes from meat and dairy products, one should substitute cereals, fruits, and vegetables for these foods when protein intake is higher than recommended. Conversely, these foods should be added to increase protein intake.   **16.** Answers can vary slightly depending on the food composition table used and the individual RDA for protein. If one uses the food composition table that accompanies this text, this meal provides 40 g of protein. For a person (for example, a 25-year-old female) who requires 46 g of protein daily, this would represent 87% of the RDA. Because animal products supply most of the protein in the diet, one might choose to consume milk instead of a soft drink to increase the protein content of this meal. **17.** Ascites is more common in kwashiorkor than in marasmus, and occurs in response to a deficit of albumin, a protein needed to draw water back into the circulation.

Chapter 6

7. A phospholipid is composed of a glycerol molecule bonded to two fatty acids and a polar (phosphate-containing) head group. Of these components, the fatty acids are the most hydrophobic, and the polar head group is the most hydrophilic. Because it contains both polar (hydrophilic) and nonpolar (hydrophobic) regions, a phospholipid is said to be *amphipathic*. In general, phospholipids are important in the structure of biological membranes and are used to transport lipids in the body.   **13.** Although overconsumption of specific lipid types is not thought to lead to obesity, general excessive lipid intake often results in weight gain, and ultimately obesity because lipids (specifically, triglycerides) are rich sources of calories.   **14.** The Acceptable Macronutrient Distribution Ranges (AMDRs) suggest that we obtain 20 to 35% of energy from lipids. The Dietary Guidelines for Americans recommend that we limit our intake of saturated fatty acids to less than 10% of calories and keep intake of *trans* fatty acids to as low as possible, while consuming greater amounts of polyunsaturated and monounsaturated fatty acids. Cholesterol should be limited to less than 300 mg/day.
14. This person should consume 550 to 963 kcal/day from lipids, which would be obtained from 61 to 107 grams of oil or fat in the diet.

Chapter 7

6. Some ATP is generated by a process called substrate phosphorylation. This means that an inorganic phosphate is attached directly to a molecule of ADP to form ATP. Most ATP is generated via oxidative phosphorylation. This takes place in mitochondria and involves structures that make up the electron transport chain. Oxidative phosphorylation occurs when reduced coenzymes (NADH + H$^+$ and one FADH$_2$) become oxidized. The oxidation of coenzymes provides the energy needed to phosphorylate ADP to generate ATP. **13.** Carbohydrate catabolism begins with glycogenolysis, the breakdown of glycogen to glucose. Glucose enters the metabolic pathway called glycolysis. Glycolysis splits a glucose molecule, forming two molecules of pyruvate. If oxygen is readily available, each pyruvate combines with a molecule called coenzyme A (CoA), forming acetyl-CoA. If oxygen is not readily available, pyruvate is converted to lactate. Some ATP and NADH + H⁺ are formed during glycolysis. Each molecule of acetyl-CoA enters the citric acid cycle by combining with oxaloacetate, forming citrate. Citrate is metabolized via the citric acid cycle, forming reduced coenzymes NADH + H⁺ and FADH₂. ATP is also formed via GTP. The reduced coenzymes enter the electron transport chain and undergo oxidative phosphorylation. Each NADH + H⁺ yields approximately 3 ATPs, whereas each FADH₂ yields approximately 2 ATPs. Protein catabolism begins with proteolysis, releasing amino acids. For amino acids to be used as an energy source, the amine group is removed by a two-step process—transamination and deamination. Transamination involves the transfer of the amine group from the amino acid to an α-keto acid. Often this α-keto acid is α-ketogluturate, which forms the amino acid glutamate. The remaining carbon skeleton forms another α-keto acid. Next, the newly formed glutamate is deaminated. This involves the removal of the amino group, which is used to synthesize ammonia. Ammonia is converted to urea, which is released into the blood. Urea is filtered out of the blood by the kidneys and excreted in the urine. The carbon skeleton remaining from the deaminated amino acid is used as an energy source. The structure of the α-keto acid determines where it enters the citric acid cycle and how much ATP is produced. Lipid catabolism begins with lipolysis, the breakdown of triglycerides to glycerol and fatty acids. Glycerol can be used as a source of energy or used to synthesize glucose. Fatty acids are metabolized into numerous molecules of acetyl-CoA via a metabolic process called β-oxidation. For each acetyl-CoA formed, one NADH + H⁺ and one FADH₂ are generated. Each molecule of acetyl-CoA enters the citric acid cycle by combining with oxaloacetate. On completion of this cycle, more NADH + H⁺ and FADH₂ are generated. These reduced coenzymes enter the electron transport chain and undergo oxidative phosphorylation, producing ATP. **19.** Certain cells depend on glucose for a source of energy. When glucose is limited, cells synthesize glucose from noncarbohydrate sources, mainly glucogenic amino acids, via a metabolic process called gluconeogenesis. To slow the rate of muscle catabolism, some cells decrease their glucose requirements by using an alternate energy source called ketones. In this way, ketones spare lean body mass. **22.** The period shortly after a meal is referred to as the absorptive state. Absorbed nutrients enter the blood, stimulating the release of insulin. Insulin increases the activity of anabolic pathways involved in synthesizing proteins, fatty acids, triglycerides, and glycogen. During this time, glucose is the body's major source of energy (ATP). After 24 hours without food, blood glucose decreases and the body's glycogen stores are depleted. Glucagon is the dominant hormone during this time. Glucose comes from noncarbohydrate sources, mainly glucogenic amino acids, via gluconeogenesis. There is an increase in the mobilization of fatty acids for a source of energy. Ketogenesis also increases, and cells begin to use ketones as an energy source.

Chapter 8

9. Three signals originating from the GI tract that regulate hunger and satiety include gastric stretching, circulating levels of nutrients, and the hormone ghrelin. As the stomach fills with food, the stomach walls stretch. Gastric stretching signals the brain, which initiates the feeling of satiety. Increased blood concentrations of glucose, amino acids, and fatty acids trigger satiety, whereas low concentrations of glucose in the blood trigger hunger. The hormone ghrelin is produced by the stomach and released in response to a lack of food in the stomach. Ghrelin concentrations decrease after food intake. **11.** Total energy expenditure (TEE) refers to energy expenditure associated with basal metabolism, physical activity, and thermic effect of food. Basal metabolism, the largest component of TEE, is energy expenditure associated with vital body functions. Energy expenditure associated with physical activity accounts for 15 to 30% of TEE. The small component of TEE (10%) is the thermic effect of food. This refers to energy expenditure associated with processing food by the body. **14.** Three methods used to assess body composition are hydrostatic weighing, DEXA, and skinfold thickness. Hydrostatic or underwater weighing requires a person to be submerged in water and compares a person's underwater weight to weight on land. Although this method provides an accurate assessment of body composition, the equipment is expensive and not easily accessible, and some people may not feel comfortable being submerged in water. DEXA uses X-ray beams to differentiate between fat mass and lean mass. DEXA provides an accurate estimate of fat mass, lean body mass, and bone mass. However, this body composition method can be expensive and may not be accessible. Skinfold measures are made using calipers, an instrument that measures the thickness of fat folds. A mathematical formula is then used to estimate body fat. Accuracy of this method depends on the ability of clinicians to correctly locate anatomical locations to take skinfold measures. One strength of this method is that calipers are relatively inexpensive and transportable. **15.** Central adiposity, adipose tissue accumulation in the abdominal region, poses a greater risk to health than body fat stored elsewhere in the body. This is because intra-abdominal fat, also called visceral adipose tissue, is more likely to undergo lipolysis. Waist circumference is a good predictor of central adiposity. A waist circumference of 40 inches or greater in males and 35 inches or greater in females indicates central adiposity. **16.** BMI = 25 kg/m²; this person is at the higher end of the healthy weight classification or at the lower end of the overweight classification. BMI does not take into account body composition. Therefore, physical activity and muscle mass must also be considered when making a final assessment. **20.** The set point theory suggests that body weight, in part, is regulated by internal mechanisms that help maintain a relatively stable body weight over an extended period of time. These internal mechanisms make adjustments in energy intake and energy expenditure to restore body weight to its desired weight or what is called the "set point." **21.** Adiposity is communicated to the brain in part by circulating concentrations of the hormone leptin. When adiposity increases, leptin concentrations increase, signaling neurons in the hypothalamus to release catabolic neuropeptides. Catabolic neuropeptides decrease energy intake and increase energy expenditure, resulting in negative energy balance and favoring weight loss. When adiposity decreases, leptin concentrations decrease, signaling neurons in the hypothalamus to release anabolic neuropeptides. These neuropeptides increase energy intake and decrease energy expenditure, resulting in positive energy balance and favoring weight gain. **23.** A realistic initial weight-loss goal is to reduce current weight by 5 to 10%. For someone weighing 200 lbs, this would amount to an initial weight loss of approximately 10–20 lbs (180–190 lbs). **25.** Advocates of low-carbohydrate weight-loss diets claim that the hormone insulin facilitates weight gain. The reduction of carbohydrate in the diet results in less insulin being released from the pancreas. With lower concentrations of insulin in the blood, body fat is not as readily stored.

Chapter 9

7. The ATP–creatine phosphate (ATP–CP) pathway is used for immediate energy (ATP) at the onset of vigorous physical activity. The enzyme creatine phosphokinase breaks down creatine phosphate, forming inorganic phosphate and creatine. The inorganic phosphate attaches to ADP to form ATP, which is used to fuel muscle activity. Glycolysis, a short-term energy system, is an anaerobic energy-yielding metabolic pathway that can sustain physical activity beyond the energy provided by the ATP–CP pathway. Glycolysis predominates under conditions of low oxygen availability. Muscles use this pathway when there is a need to generate ATP rapidly. This pathway depends solely on glucose, which is derived from muscle glycogen, and on glucose circulating in the blood. Muscles depend on this anaerobic pathway during high-intensity workouts because there is a need to generate ATP rapidly. During periods of prolonged, low-intensity physical activity, muscles depend on aerobic pathways to generate energy (ATP). With sufficient oxygen availability, muscles are able to utilize aerobic pathways to metabolize glucose and fatty acids (and to a lesser extent amino acids) to generate ATP. Aerobic metabolic pathways include the citric acid cycle, β-oxidation, and oxidative phosphorylation. **10.** Physical activity improves cardiac function and capacity to circulate blood to muscle tissues. In response to training, cardiac muscle becomes stronger, increasing its size and ability to contract forcefully. As a result, more blood can be pumped out of the heart with each beat. Because the heart becomes more efficient, athletes often have a lower resting heart rate than nonathletes. Training also causes an expansion of capillary vessels supplying blood to muscles. This aids blood flow and the delivery of nutrients and oxygen to muscle cells. Training increases the production of red blood cells, which increases the oxygen-carrying capacity of the blood. Collectively, these physiological adaptations to exercise increase maximal oxygen consumption, which is the maximum ability of the cardiovascular system to deliver oxygen to muscles. **13.** Glycogen is an important energy reserve for athletes participating in distance or endurance sports. This is because glycogen provides muscles with an important source of glucose. The depletion of glycogen stores during exercise can cause athletes to become fatigued. Some athletes try to increase glycogen stores in muscles by using a technique called carbohydrate loading, which involves combining the right workout intensity with the right level of carbohydrate intake. **14.** Dehydration, due to lack of adequate fluid intake and/or excessive sweating, can reduce blood volume, which interferes with the body's cooling system. Decreased blood volume reduces blood flow to the skin, which in turn affects the body's ability to dissipate heat. As a result, core body temperature rises, increasing the risk of heat exhaustion and heat stroke. Heat stroke is characterized by dry skin, confusion, and loss of consciousness. **15.** According to the American Dietetic Association, it is recommended that an athlete weighing 175 lb (80 kg) consume 400 to 800 g/day carbohydrate. **16.** Based on the recommendation that an athlete should consume 0.8 g protein/kg/body weight, a person weighing 150 lb would require approximately 55 g/day protein. Based on recommendations by the American College of Sports Nutrition that athletes should consume 1.2 to 1.4 g/day protein, this athlete would need to consume approximately 82 to 95 g/day protein.

Chapter 10

2. Thiamin is found in fish, legumes, and pork. Good sources of riboflavin include liver, meat, and dairy products. Niacin (or tryptophan, its precursor) is abundant in meat, tomatoes, and mushrooms. Good sources of pantothenic acid include mushrooms, organ meats, and sunflower seeds. Vitamin B_6 is plentiful in chickpeas, fish, and liver. Food sources of biotin include peanuts, tree nuts, and eggs. Folate is found in many fruits and vegetables such as spinach and oranges; it is also abundant in organ meats.

Vitamin B_{12} is found in shellfish, meat, and dairy products. Brightly colored fruits and vegetables such as citrus fruits, peppers, and papayas tend to be good sources of vitamin C, whereas choline is most abundant in eggs, liver, legumes, and pork. **10.** Answers can vary slightly depending on food composition table used and individual RDA for folate. However, if one uses the food composition table that accompanies this text, this meal provides approximately 60 micrograms (µg) DFE of folate. For a person (for example, a 25-year-old female) who requires 400 µg DFE of folate daily, this would represent 15% of the RDA. To increase folate consumption, a variety of fruits and vegetables, legumes, and enriched cereal products should be added to the diet. For example, a salad could be ordered instead of the fries in this meal. **13.** Vitamin B_{12} deficiency results in the inability of the body to convert the inactive form of folate—5-methyl tetrahydrofolate (5-methyl THF)—to its active form of THF. Thus, primary vitamin B_{12} deficiency can cause secondary folate deficiency. Because the signs and symptoms of folate deficiency can lead a clinician to conclude that the person has primary folate deficiency, the true cause of malnutrition (vitamin B_{12} deficiency) may be overlooked. Because of this, vitamin B_{12} deficiency is said to be "masked" when it is accompanied by folate deficiency. **15.** Antioxidants donate electrons and hydrogen ions to other compounds, thus reducing them, and are important for a variety of reasons in the body. For example, antioxidants (such as vitamin C) protect the body from free radical damage. Free radicals are compounds that have unpaired electrons. These compounds seek to take electrons from other substances, thus oxidizing them. Vitamin C readily donates its own electrons to free radicals (stabilizing them), thus protecting DNA, proteins, and lipids from oxidative damage. Vitamin C also donates electrons to enzymes that need them in order to function. These enzymes tend to be metalloenzymes, containing iron or copper.

Chapter 11

5. Answers can vary slightly depending on food composition table used and individual RDA for vitamin A. However, if one uses the food composition table that accompanies this text, this meal provides negligible amounts of vitamin A, representing 0% of any person's RDA. To increase vitamin A (or provitamin A carotenoid) consumption, milk or a milkshake could be substituted for the cola, and a mixed salad could be eaten as a side dish. **6.** Vitamin D is called the "sunshine vitamin" because its synthesis in the body depends on exposure to sunlight. This involves the conversion of 7-dehydrocholesterol to previtamin D_3 (precalciferol) in the skin via a reaction that requires ultraviolet light found in sunlight—hence, the name "sunshine vitamin." Next, in the skin, previtamin D_3 is converted to vitamin D_3 (cholecalciferol), which in turn is converted to 25-hydroxyvitamin D [25-(OH)D_3] in the liver and then [1,25-(OH$_2$)D_3 or calcitriol] in the kidneys. **9.** The answer to this question is variable, depending on which MyPlate is appropriate for you as well as your personal dairy preferences. See your instructor for additional assistance with this problem. **14.** Good sources of preformed vitamin A include liver and some fatty fish, whereas fatty fish and fortified milk provide vitamin D. Many vegetable oils and nuts are excellent sources of vitamin E, whereas green leafy vegetables and legumes provide vitamin K. Vitamins A and D are relatively stable. However, vitamins E and K can be destroyed by cooking and excessive processing. **15.** Vitamin C, the carotenoids, and vitamin E are all considered antioxidants because they help stabilize potentially dangerous free radicals in the body. This may be especially important in cancer prevention, as free radicals can harm DNA in such ways that cell division becomes unregulated.

Chapter 12

5. Phosphorus is needed for energy metabolism because it is a critical component of adenosine triphosphate (ATP), and it is required for fatty acid transport via lipoproteins. Phosphorus deficiency can cause fatigue and weakness for many reasons. For example, without sufficient phosphorus to generate ATP, energy-requiring reactions (such as those needed for muscle function) cannot occur. In addition, because fatty acids provide an important source of energy to cells, phosphorus deficiency can decrease the availability of these energy-yielding nutrients to the body by inhibiting lipoprotein synthesis.   **7.** The UL for magnesium applies only to magnesium consumed in supplemental form—not that which occurs naturally in foods.   **10.** Answers can vary slightly depending on food composition table used and individual AI for sodium. However, if one uses the food composition table that accompanies this text, this meal provides approximately 1,006 mg of sodium. For a person (for example, a 25-year-old male) who requires 1,500 mg (1.5 g) of sodium daily, this would represent approximately 67% of the AI and 44% of the maximum sodium intake recommended in the Dietary Guidelines for Americans (2,300 mg/day). As most of the sodium in this meal was from the fries, sodium intake could be decreased by substituting a baked potato, salad, or fruit salad for this component.   **11.** Reducing salt intake can be accomplished in a variety of ways, including choosing foods labeled as being "low-sodium" or "sodium-free," preparing foods without additional salt, adding less salt at the table, limiting the amount of sodium-containing condiments used, and replacing high-salt snacks (such as chips) with fresh fruits and vegetables.   **12.** Osmosis is the phenomenon by which water moves from a region of low solute concentration to a region of high solute concentration. Using the principle of osmosis, the body can actively transport sodium and chloride ions across cell membranes in such a way that water passively follows. For example, for water to move from an intracellular space to an extracellular compartment, chloride can be actively pumped from the cell's cytosol out to the surrounding interstitial space. In this way, fluids can become part of the body's secretions.   **14.** Sodium, chloride, and potassium levels in the blood are regulated primarily by the hormone aldosterone. However, whereas aldosterone release causes sodium retention, this hormone increases potassium excretion in the urine. Consequently, aldosterone causes blood levels of sodium and chloride to *increase* while simultaneously causing blood levels of potassium to *decrease*.   **19.** Calcium, phosphorus, and magnesium are integral components of the microarchitecture that makes up the skeletal system.

Chapter 13

2. In some cases, the concentration of certain minerals in the soil and water can greatly affect the amount of these minerals in the plants and animals raised in a region. This is because plants and animals obtain their minerals from the soil and water available to them.   **3.** Because the content and bioavailability of iron in animal products tends to be high, vegetarians are at greater risk of iron deficiency than omnivores. This is especially true for vegans who do not consume any animal products. Similarly, because bioavailability of zinc from animal sources is especially high, it is recommended that vegetarians (especially vegans) consume up to 50% more zinc than omnivores. To help maintain healthy iron status, vegetarians who do eat fish are advised to consume it along with iron-containing plant foods; iron supplements may be necessary. In addition, zinc-containing foods such as legumes and fortified cereal products should be emphasized.   **8.** Both cretinism and goiter are caused by iodine deficiency. However, cretinism occurs in infants of severely iodine-deficient mothers, whereas goiter occurs later in life. The United States has relatively low incidences of these conditions because much of the salt marketed has been fortified with iodine.   **12.** The answer to this question is variable, depending on which MyPlate is appropriate for you as well as your personal preferences. See your instructor for additional assistance with this problem.   **14.** Answers can vary slightly depending on the food composition table used and individual RDA values for zinc. However, if one uses the food composition table that accompanies this text, this meal provides approximately 5.3 mg of zinc. For a person (for example, a 25-year-old female) for whom the RDA is 8 mg/day, this would represent approximately 66% of the RDA for zinc. To increase the zinc content of this meal, a slice of cheese could be added to the hamburger, or a 3-bean salad substituted for the fries.

Chapter 14

10. Infant suckling causes neural signals to stimulate the hypothalamus. The hypothalamus signals the pituitary gland to release the hormones prolactin (made in the pituitary gland) and oxytocin (made in the hypothalamus). Prolactin stimulates the secretory cells in the mammary gland to synthesize milk constituents, whereas oxytocin causes the myoepithelial cells to contract. This forces milk out of the alveoli lumen and into the mammary ducts. The release of milk is called milk let-down.   **16.** Infancy has one of the highest rates of growth and development during the life cycle. By the end of the first year of life, an infant's weight triples and length doubles. The rate of growth decreases during childhood. During adolescence, hormonal changes cause the reproductive organs to mature. The onset of puberty varies, but females tend to experience puberty at an earlier age than males. During the adolescent growth spurt, female height increases by approximately 6 inches, whereas males gain an additional 8 inches. In addition to increased height, weight increases and body composition changes. Females experience a decrease in percent lean mass and a relative increase in percent fat mass, whereas males experience an increase in percent lean mass and a relative decrease in percent fat mass.   **18.** Three physical changes frequently associated with aging include changes in body composition (loss of lean mass, increased fat mass), a decline in the production of gastric secretions, and age-related bone loss. These changes can affect nutrient requirements. For example, the decline in lean mass can reduce caloric requirements. Decreased ability to produce gastric secretions can interfere with digestion and nutrient absorption. Requirements for calcium increase because age-related bone loss can lead to osteoporosis. In women, the decline in the hormone estrogen, which accompanies menopause, also accelerates bone loss.   **19.** There are many reasons why elderly people are at increased risk for nutrient deficiencies. Many elderly people are on fixed incomes, limiting money available for food. Some elderly people become isolated and depressed, which can lead to inadequate food intake. Also, loss of teeth can limit the variety and types of food consumed. This, too, can lead to nutritional problems.

GLOSSARY

1,25-dihydroxyvitamin D (1,25-[OH]₂D₃) $(1,25\text{-}[OH]_2D_3)$ **(calcitriol)** The active form of vitamin D in the body produced in the kidneys from 25-(OH)D₃.

24-hour recall A retrospective dietary assessment method that analyzes each food and drink consumed over the previous 24 hours.

25-hydroxyvitamin D (25-[OH]D₃) An inactive form of vitamin D that is made from cholecalciferol in the liver.

5-methyltetrahydrofolate (5-methyl THF) An inactive form of folate.

7-dehydrocholesterol A metabolite of cholesterol that is converted to cholecalciferol (vitamin D₃) in the skin via exposure to ultraviolet light such as that found in sunshine.

A

α-dextrinase Brush border enzyme that hydrolyzes α-1,6 glycosidic bonds.

α-helix A common configuration that makes up many proteins' secondary structures.

α-keto acid The structure remaining after the amino group has been removed from an amino acid.

absorption The passage of nutrients through the lining of the GI tract into the blood or lymphatic circulation.

Acceptable Macronutrient Distribution Ranges (AMDRs) Recommendations concerning the distribution or percentages of energy from each of the macronutrient groups.

acetaldehyde dehydrogenase (ALDH) An enzyme that converts acetaldehyde to acetic acid.

acidic Having a pH less than 7.

acrodermatitis enteropathica A genetic abnormality resulting in decreased absorption of dietary zinc.

acrylamide A compound that is formed in starchy foods (such as potatoes) when heated to high temperatures.

active site An area on an enzyme that binds substrates in a chemical reaction.

active transport mechanism Transport mechanism that enables substances to cross cell membranes, requiring the expenditure of energy (ATP).

active transport An energy-requiring mechanism by which some substances cross cell membranes.

adaptation response The body's physiologic response to being challenged by frequent physical exertion.

adaptive thermogenesis Energy expended in response to changes in the environment or to physiological conditions.

adenosine triphosphate (ATP) A chemical used by the body to perform work.

Adequate Intake (AI) level Nutrient intake of healthy populations that appears to support adequate nutritional status; established when RDAs cannot be determined.

adipocyte A specialized cell that makes up the majority of adipose tissue; used mainly for fat storage.

adipokines Hormone-like substances produced and released by adipocytes.

adiponectin A hormone secreted by adipose tissue that appears to be involved in energy homeostatis; also appears to promote insulin sensitivity and suppress inflammation.

aerobic capacity The maximum amount of oxygen the body can use during maximal physical exertion.

aflatoxin A toxic compound produced by certain molds, such as *Aspergillus*, that grow on peanuts, some grains, and soybeans.

Al-Anon An organization dedicated to helping people cope with alcoholic family members and friends.

Alateen An organization dedicated to helping children cope with an alcoholic parent.

albumin A protein important in regulating fluid balance between intravascular and interstitial spaces.

alcohol dehydrogenase (ADH) An enzyme found mostly in the liver that metabolizes ethanol to acetaldehyde.

alcohol dehydrogenase pathway The primary metabolic pathway that chemically breaks down alcohol in the liver.

alcohol use disorders A term that encompasses habitual, excessive intakes of alcohol that lead to negative physical, legal, or social consequences.

alcohol An organic compound containing one or more hydroxyl (–OH) groups attached to carbon atoms.

alcoholic cardiomyopathy Condition that results when the heart muscle weakens in response to heavy alcohol consumption.

alcoholic hepatitis Inflammation of the liver caused by chronic alcohol abuse.

Alcoholics Anonymous (AA) An organization dedicated to helping people achieve and maintain sobriety.

aldosterone A hormone produced by the adrenal glands in response to low blood sodium concentration (hyponatremia) and angiotensin II.

alpha (α) glycosidic bond A downward-facing type of glycosidic bond between two monosaccharides.

alpha (α) end The end of a fatty acid, which consists of a carboxylic acid (–COOH) group.

alveolus (plural, alveoli) A cluster of milk-producing cells that make up the mammary glands.

amino acid Nutrient composed of a central carbon bonded to an amino group, carboxylic acid group, and a side-chain group (R-group).

amino group (–NH₂) The nitrogen-containing component of an amino acid.

amphibolic pathway Metabolic pathway that generates intermediate products that can be used for both catabolism and anabolism.

amphipathic Having both nonpolar (noncharged) and polar (charged) portions.

amylopectin A type of starch consisting of a highly branched arrangement of glucose molecules.

amylose A type of starch consisting of a linear (nonbranching) chain of glucose molecules.

anabolic neurotransmitters A protein released by nerve cells that stimulates hunger and/or decreases energy expenditure.

anabolic pathway A series of metabolic reactions that require energy to make a complex molecule from simpler ones, often requiring energy in the process.

anaerobic capacity The ability to perform repetitive, high-intensity activity with little or no rest.

anal sphincters Internal and external sphincters that regulate the passage of feces through the anal canal.

anaphylaxis A severe and potentially life-threatening allergic reaction.

anencephaly A type of neural tube defect in which an infant is born without a major portion of the brain.

aneurysm The outward bulging of a blood vessel.

angina pectoris Pain in the region of the heart, caused by a portion of the heart muscle receiving inadequate amounts of blood.

angiogram A procedure in which dye is injected into the blood, allowing the flow of blood through cardiac arteries to be visualized.

angioplasty A procedure used to widen the heart's blood vessels by inserting a stent.

angiotensin I The precursor of angiotensin II.

angiotensin II A protein derived from angiotensin I in the lungs; stimulates aldosterone release.

angiotensinogen An inactive protein, made by the liver, that is converted by renin into angiotensin I.

animal study The use of experimental animal subjects such as mice, rats, or primates.

anion An ion with a net negative charge.

anorexia nervosa (AN) An eating disorder characterized by an irrational fear of gaining weight or becoming obese.

anorexia nervosa, binge-eating/purging type An eating disorder characterized by food restriction as well as bingeing and purging.

anorexia nervosa, restricting type An eating disorder characterized by food restriction.

anorexia of aging Loss of appetite in the elderly that leads to weight loss and overall physiological decline.

anthropometric measurements Measurements or estimates of physical aspects of the body such as height, weight, circumferences, and body composition.

antibody A protein, produced by the immune system, that helps fight infection.

antidiuretic hormone (ADH) (also called vasopressin) A hormone produced in the pituitary gland and released during periods of low blood volume; stimulates the kidneys to decrease urine production, thus conserving water.

antioxidant A compound that readily gives up electrons (and hydrogen ions) to other substances.

aorta The main artery that initially carries blood from the heart to all areas of the body except the lungs.

apoproteins Proteins embedded in the surface of lipoproteins.

apoptosis The normal process by which a cell leaves the cell cycle and dies; programmed cell death.

appetite A psychological desire for food.

appropriate for gestational age (AGA) infant A baby that has a weight between the 10th and the 90th percentiles for weight for gestational age.

arachidonic acid A long-chain, polyunsaturated ω-6 fatty acid produced from linoleic acid.

ariboflavinosis A disease caused by riboflavin deficiency.

arteriole Small blood vessel that branches off from arteries.

artery A blood vessel that carries blood away from the heart.

ascites Abnormal accumulation of fluid in the abdominal cavity.

atherosclerosis The hardening and narrowing of blood vessels caused by buildup of fatty deposits and inflammation in the vessel walls.

α-tocopherol The most biologically active form of vitamin E.

atom The smallest portion into which an element can be divided and still retain its properties.

ATP–creatine phosphate (ATP–CP) pathway An anaerobic metabolic pathway that uses ADP and creatine phosphate to generate ATP.

ATP synthase A mitochondrial enzyme that adds a phosphate (P_i) to ADP to form ATP during the process of oxidative phosphorylation.

atrophic gastritis Inflammation of the mucosal membrane that lines the stomach, reducing the number of cells that produce gastric secretions.

atrophy A decrease in size of a tissue or organ due to an decrease in cell size and/or number.

autoimmune disease A condition in which the immune system attacks an otherwise healthy part of the body.

avidin A protein present in egg whites that binds biotin, making it unavailable for absorption.

B

beta (β) carotene A provitamin A carotenoid commonly found in yellow and orange foods.

beta (β) complex vitamins A term used to describe all the B vitamins.

beta (β) folded sheet A common configuration that makes up many proteins' secondary structures.

beta (β) oxidation The series of chemical reactions that breaks down fatty acids to molecules of acetyl-CoA.

baby bottle tooth decay dental caries that occur in infants and children who are given bottles containing carbohydrates (such as milk or juice) at bedtime.

bariatric surgery Surgical procedure performed to treat obesity.

bariatrics The branch of medicine concerned with the treatment of obesity.

basal energy expenditure (BEE) Energy expended for basal metabolism over a 24-hour period.

basal metabolic rate (BMR) Energy expended for basal metabolism per hour (expressed as kcal/hour).

basal metabolism Energy expended to sustain metabolic activities related to basic vital body functions such as respiration, muscle tone, and nerve function.

basic (also called alkaline) Having a pH greater than 7.

basolateral membrane The cell membrane that faces away from the lumen of the GI tract and toward the submucosa.

benign tumor A growth that may or may not have minimal physiological consequences but does not invade surrounding tissue or metastasize.

beriberi A disease that results from thiamin deficiency.

beta (β) glycosidic bond An upward-facing type of glycosidic bond between two monosaccharides.

bile acid Amphipathic substance made from cholesterol in the liver; a component of bile important for lipid digestion and absorption.

bile salt–dependent cholesteryl ester hydrolase An enzyme, produced in the pancreas, that cleaves fatty acids from cholesteryl esters.

bile A fluid, made by the liver and stored in and released from the gallbladder, that contains bile salts, cholesterol, water, and bile pigments.

binge drinking Consumption of five or more drinks in males and four or more drinks in females in a period of two hours.

bingeing Uncontrolled consumption of large quantities of food in a relatively short period of time.

bioavailability The extent to which nutrients are absorbed into the blood or lymphatic system.

biochemical measurement Laboratory analysis of biological samples, such as blood and urine, used in nutritional assessment.

bioelectrical impedance A method used to assess body composition based on measuring the body's electrical conductivity.

biological marker (biomarker) A measurement in a biological sample, such as blood or urine, that reflects a nutrient's function.

Bioterrorism Act Federal legislation aimed to ensure the continued safety of the U.S. food supply from intentional harm by terrorists.

biotin (vitamin B_7) A water-soluble vitamin involved in energy metabolism.

bisphenol A (BPA) A chemical used in the production of many plastic items including baby bottles.

bisphosphonates A class of drugs sometimes taken to help reduce bone loss.

Bitot's spots A sign of vitamin A deficiency characterized by white spots on the eye; caused by buildup of dead cells and secretions.

blastocyst Early period of gestational development that lasts approximately 8 to 13 days after conception.

blood alcohol concentration (BAC) A unit of measurement that describes the level of alcohol in the blood.

blood clot (also called thrombosis) A small, insoluble particle made of blood cells and clotting factors.

body composition Components of the body such as fat, lean mass (muscle), water, and minerals.

bolus A soft, moist mass of chewed food.

bomb calorimeter A device used to measure the amount of energy in a food.

bone marrow Soft, spongy, inner part of bone; makes red blood cells, white blood cells, and platelets.

bone mass A measure reflecting the amount of minerals contained in bone.

bone mineral density The amount of bone mineral per unit volume.

bone remodeling (bone turnover) The process by which older and damaged bone is removed and replaced by new bone.

Borg Scale of Perceived Exertion Self-monitoring of the body's response to physical activity based on changes in heart rate, breathing rate, sweating, and muscle fatigue.

botulism The foodborne illness caused by *Clostridium botulinum*.

bovine somatotropin (bST; bovine growth hormone) A protein hormone produced by cattle and used in the dairy industry to enhance milk production.

bovine spongiform encephalopathy (BSE; mad cow disease) A fatal disease in cattle caused by ingesting prions.

bran The outer layer of a grain; contains most of the fiber.

brevetoxin The toxin produced by red tide–causing algae, which, when consumed by humans, causes shellfish poisoning.

brush border enzyme Enzyme, produced by enterocytes, which aids in the final steps of digestion.

brush border The absorptive surface of the small intestine made up of thousands of microvilli that cover the luminal surface of enterocytes.

buffer A substance that releases or binds hydrogen ions in order to resist changes in pH.

bulimia nervosa (BN) An eating disorder characterized by repeated cycles of bingeing and purging.

Buy Fresh Buy Local **campaign** A grass-roots movement that emphasizes purchasing foods grown, processed, and distributed locally.

C

cachexia A condition in which a person loses lean body mass (muscle tissue).

calbindin A transport protein, made in enterocytes, that assists in calcium absorption; synthesis is stimulated by calcitriol.

calcitonin A hormone produced in the thyroid gland in response to high blood calcium levels.

calcium (Ca) A major mineral found in bones, teeth, and blood; needed for skeletal structure, blood clotting, muscle and nerve function, and energy metabolism.

calorie A unit of measure used to express the amount of energy in a food.

cancer A condition characterized by unregulated cell division.

capillaries Blood vessels with thin walls, which allow for the exchange of materials between blood and tissues.

carbohydrate Organic compound made up of varying numbers of monosaccharides.

carbohydrate loading A technique used by some athletes to increase glycogen stores in muscles by combining a certain workout intensity with a level of carbohydrate intake.

carboxylation reaction A metabolic reaction in which a bicarbonate subunit (HCO_3^-) is added to a molecule.

carcinogen Compound or condition that causes cancer.

cardiac arrhythmia Irregular heartbeat caused by high intake of alcohol.

cardiovascular disease A disease of the heart or vascular system.

cardiovascular fitness The ability of the circulatory and respiratory systems to supply oxygen and nutrients to working muscles during sustained physical activity.

carnitine A molecule that transports fatty acids across the mitochondrial membrane.

carotene A provitamin A carotenoid commonly found in yellow and orange foods.

carotenodermia A condition in which carotenoids accumulate in the skin, causing it to become yellow-orange.

carotenoids Dietary compounds with structures similar to that of vitamin A; some, but not all, can be converted to vitamin A.

carrier-mediated active transport An energy-requiring mechanism whereby a substance moves from a region of lower concentration to a region of higher concentration, requiring the assistance of a carrier transport protein.

cartilage The soft, nonmineralized precursor of bone.

catabolic neurotransmitter A protein released by nerve cells that inhibits hunger and/or stimulates energy expenditure.

catabolic pathway A series of metabolic reactions that break down a complex molecule into simpler ones, often releasing energy in the process.

catalyst A substance that increases the rate by which a chemical reaction occurs, without being consumed in the process.

cataract Age-related cloudiness that develops on the lens of the eye, causing impaired vision.

cation An ion with a net positive charge.

cause-and-effect relationship (also called causal relationship) When an alteration in one variable causes a change in another variable.

cecum The first section of the large intestine.

celiac disease An autoimmune response to the protein gluten that damages the absorptive surface of the small intestine; also called gluten-sensitive enteropathy.

cell culture system Specific type of cells that can be grown in the laboratory and used for research purposes.

cell cycle The process by which cells grow, mature, replicate their DNA, and divide.

cell differentiation The process by which an immature cell becomes a specific type of mature cell.

cell signaling The first step in protein synthesis, in which the cell receives a signal to produce a protein. Note that this term is also used for a variety of other processes (aside from protein synthesis) within the cell.

cell turnover The cycle of cell formation and cell breakdown.

central nervous system The part of the nervous system made up of the brain and spinal cord.

central obesity Accumulation of body fat within the abdominal cavity.

cephalic phase The response of the central nervous system to sensory stimuli, such as smell, sight, and taste, that occurs before food enters the GI tract; characterized by increased GI motility and release of GI secretions.

certified organic foods Plant and animal foods that have been grown, harvested, and processed without conventional pesticides, fertilizers, growth promoters, bioengineering, or ionizing radiation.

cerebral beriberi (Wernicke-Korsakoff syndrome) A form of thiamin deficiency characterized by poor muscle control and paralysis of the eye muscles.

ceruloplasmin The protein that transports copper in the blood.

chain length The number of carbons in a fatty acid's backbone.

chaotic families Families whose interaction is characterized by a lack of cohesiveness and little parental involvement.

cheilosis Sores occurring on the outsides and corners of the lips.

chelator A substance that binds compounds (such as iron) in the gastrointestinal tract, making them unavailable for absorption.

chemical bonds The attractive force between atoms that are formed by the transfer or sharing of electrons.

chemoreceptor A sensory receptor that responds to a chemical stimulus.

chemotherapy The use of drugs to stop the growth of cancer.

chief cells Exocrine cells in the gastric mucosa that produce digestive enzymes.

chloride (Cl) A major mineral important for regulating fluid balance, protein digestion in the stomach (via HCl), and carbon dioxide removal by the lungs.

cholecalciferol (vitamin D₃) The form of vitamin D in animal-derived foods, fortified foods, and supplements; also made by the human body.

cholecystokinin (CCK) A hormone, produced by the small intestine, that stimulates the release of enzymes from the pancreas and contraction of the gallbladder.

cholesterol ratio The mathematical ratio of total blood cholesterol to high-density lipoprotein cholesterol (HDL-C).

cholesterol A sterol found in animal foods and made in the body; required for bile acid and steroid hormone synthesis.

cholesteryl ester A sterol ester made of a cholesterol molecule bonded to a fatty acid via an ester linkage.

choline A water-soluble compound used by the body to synthesize acetylcholine (a neurotransmitter) and a variety of phospholipids needed for cell membrane structure; considered a conditionally essential nutrient.

chromium (Cr) A trace mineral that may be needed for proper insulin function in some people.

chromium picolinate A form of chromium taken as an ergogenic aid by some athletes.

chromosome A strand of DNA and associated proteins in a cell's nucleus.

chronic degenerative disease A noninfectious disease that develops slowly and persists over time.

chronic inflammation A response to cellular injury that is characterized by chronic capillary dilation, white blood cell infiltration, release of immune factors, redness, heat, and pain.

chylomicron remnant The lipoprotein particle that remains after a chylomicron has lost most of its fatty acids.

chylomicron A lipoprotein, made in the enterocyte, that transports large lipids away from the small intestine in the lymph.

chyme The thick fluid resulting from the mixing of food with gastric secretions in the stomach.

cirrhosis The formation of scar tissue in the liver; caused by chronic alcohol abuse.

cis **double bond** A carbon–carbon double bond in which the hydrogen atoms are positioned on the same side of the double bond.

citrate The first intermediate product in the citric acid cycle formed when acetyl-CoA joins with the end product, oxaloacetate.

citric acid cycle An amphibolic pathway that oxidizes acetyl-CoA to yield carbon dioxide, NADH + H⁺, FADH₂, and ATP via substrate phosphorylation.

coagulation The process by which blood clots are formed.

coenzyme Organic molecule, often derived from vitamins, needed for enzymes to function.

cofactor A nonprotein component of an enzyme, often a mineral, needed for its activity.

collagen A structural protein found in connective tissue, including skin, bones, teeth, cartilage, and tendons.

colon The portion of the large intestine that carries material from the cecum to the rectum.

colostrum The first secretion from the breasts after birth; provides nourishment and immunological protection to newborns.

community-supported agriculture (CSA) A system that connects local food growers to local consumers such that food "bundles" are purchased on a weekly or monthly basis.

complete protein source A food that contains all the essential amino acids in relative amounts needed by the body.

complex carbohydrates Category of carbohydrate that includes oligosaccharides and polysaccharides.

complex relationship A relationship that involves one or more interactions.

compound A molecule made up of two or more different types of atoms.

computerized nutrient database Software that provides information concerning the nutrient and energy contents of foods.

condensation A chemical reaction that results in the formation of water.

conditional cash transfer program Initiative directed at reducing poverty by the transfer of money contingent upon the receivers' actions.

conditionally essential nutrient Normally nonessential nutrient that, under certain circumstances, becomes essential.

cones and **rods** Cells, in the retina, that are needed for vision.

confounding variable A factor, other than the one of interest, that might influence the outcome of an experiment.

connective tissue Tissue that supports, connects, and anchors body structures.

control group A group of people, animals, or cells in an intervention study that does not receive the experimental treatment.

copper (Cu) An essential trace mineral that acts as a cofactor for nine enzymes involved in reduction–oxidation (redox) reactions.

Cori cycle The metabolic pathway that regenerates glucose by circulating lactate from muscle to the liver, where it undergoes gluconeogenesis.

cornea The outermost layer of tissue covering the front of the eye.

coronary bypass surgery A procedure in which a healthy blood vessel obtained from the leg, arm, chest, or abdomen is used to bypass blood from a diseased or blocked coronary artery to a healthy one.

correlation (also called association) When a change in one variable is related to a change in another variable.

cortical bone (compact bone) The dense, hard layer of bone found directly beneath the periosteum.

cortisol Hormone secreted by the adrenal glands in response to stress; helps increase blood glucose availability via gluconeogenesis and glycogenolysis.

coupled reactions Chemical reactions that take place simultaneously, often involving the oxidation of one molecule and the reduction of another.

C-reactive protein (CRP) A protein produced in the liver, adipose tissue, and smooth muscles in response to injury or infection that, when elevated, can indicate risk for cardiovascular disease.

creatine phosphate (CP) A high-energy compound, consisting of creatine and phosphate, used to generate ATP.

creatine phosphokinase An enzyme that splits creatine phosphate to generate creatine and inorganic phosphate (P_i).

cretinism A form of iodine deficiency disorder (IDD) that affects babies born to iodine-deficient mothers.

Creutzfeldt-Jakob disease A fatal disease in humans caused by a genetic mutation or surgical contamination with prions.

critical period Period in development when cells and tissue rapidly grow and differentiate to form body structures.

Crohn's disease A chronic inflammatory condition that usually affects the ileum and first portion of the large intestine.

cross-contamination The transfer of microorganisms from one food to another or from one surface or utensil to another.

cross-tolerance Tolerance to one substance that causes tolerance to other similar substances.

cupric ion (Cu^{2+}) The more oxidized form of copper.

cuprous ion (Cu^{1+}) The more reduced form of copper.

cyst A stage of the life cycle of some parasites.

cystic fibrosis A genetic (inherited) disease in which a defective chloride transporter results in the inability of the body to transport chloride out of cells.

cytochrome c oxidase A copper-containing enzyme needed in the electron transport chain.

cytochrome P450 An iron-containing enzyme that helps metabolize toxins.

cytochromes Iron-containing protein complexes that, as part of the electron transport chain, combine electrons, hydrogen ions, and oxygen to form water.

cytoplasm (also called cytosol) The gel-like matrix inside cells.

D

Daily Value (DV) Recommended intake of a nutrient based on either a 2,000- or 2,500-kcal diet.

danger zone The temperature range between 40 and 140°F in which pathogenic organisms grow most readily.

DASH (Dietary Approaches to Stop Hypertension) diet A dietary pattern emphasizing fruits, vegetables, and low-fat dairy products designed to lower blood pressure.

db **gene** The gene that codes for the leptin receptor.

db/db **mouse** Obese mouse with a mutation in genes that code for the leptin receptor.

deamination The removal of an amino group from an amino acid that results in the formation of an α-keto acid.

defecation The expulsion of feces from the body through the rectum and anal canal.

dehydration (hypohydration) A condition in which the body has an insufficient amount of water.

denaturation The alteration of a protein's three-dimensional structure by heat, acid, chemicals, enzymes, or agitation.

dental fluorosis Discoloration and pitting of teeth caused by excessive fluoride intake.

desaturation The process whereby carbon–carbon single bonds are transformed into carbon–carbon double bonds in a fatty acid.

development Attainment or progression of a skill or capacity to function.

developmental origins of health and disease A concept suggesting that conditions during gestation or infancy can alter risk for chronic diseases later in life.

dextrin A partial breakdown product formed during starch digestion consisting of varying numbers of glucose units.

diabetes mellitus Medical condition characterized by a lack of insulin or impaired insulin utilization that results in elevated blood glucose levels.

diabetic ketoacidosis Severe metabolic condition resulting from the accumulation of ketones in the blood.

diet record A prospective dietary assessment method that requires the individual to write down detailed information about foods and drinks consumed over a specified period of time.

dietary assessment The evaluation of a person's dietary intake.

dietary fiber Fiber that naturally occurs in plants.

dietary folate equivalent (DFE) A unit of measure used to describe the amount of bioavailable folate in a food or supplement.

Dietary Guidelines for Americans Dietary recommendations put forth by the U.S. Department of Agriculture that give specific nutritional guidance to individuals as well as advice about physical activity, alcohol intake, and food safety.

Dietary Reference Intakes (DRIs) A set of four types of nutrient intake reference standards used to assess and plan dietary intake; these include the Estimated Average Requirements (EARs), Recommended Dietary Allowances (RDAs), Adequate Intake levels (AIs), and the Tolerable Upper Intake Levels (ULs).

dietary supplement Product intended to supplement the diet and that contains vitamins, minerals, amino acids, herbs or other plant-derived substances, or a multitude of other food-derived compounds.

dietitian A nutritionist who helps people make healthy dietary choices.

digestion The physical and chemical breakdown of food by the digestive system into a form that allows nutrients to be absorbed.

digestive enzymes Biological catalysts that facilitate chemical reactions that break chemical bonds by the addition of water (hydrolysis), resulting in the breakdown of large molecules into smaller components.

diglyceride (also called diacylglycerol) A lipid made of a glycerol molecule bonded to two fatty acids.

direct calorimetry A measurement of energy expenditure obtained by assessing heat loss.

disaccharidase Brush border enzyme that hydrolyzes glycosidic bonds in disaccharides.

disaccharide Carbohydrate consisting of two monosaccharides bonded together.

disease A condition that causes physiological or psychological discomfort, dysfunction, or distress.

disinhibition A loss of inhibition.

disordered eating Unhealthy eating patterns such as irregular eating, consistent undereating, and/or consistent overeating.

distillation A process used to make a concentrated alcoholic beverage by condensing and collecting alcohol vapors.

diuretic A substance or drug that causes water loss from the body.

diverticular disease, or diverticulosis Condition in the large intestine; characterized by the presence of pouches that form along the intestinal wall.

diverticulitis Inflammation of diverticula (pouches) in the lining of the large intestine.

docosahexaenoic acid (DHA) A long-chain, polyunsaturated ω-3 fatty acid produced from eicosapentaenoic acid.

double-blind study A human experiment in which neither the participants nor the scientists know to which group the participants have been assigned.

doubly labeled water Water that contains stable isotopes of hydrogen and oxygen atoms.

down-regulation In the context of protein synthesis, decreased expression of a gene.

dry beriberi A form of thiamin deficiency characterized by muscle loss and leg cramps.

dual-energy X-ray absorptiometry (DEXA) A method used to assess body composition by passing X-ray beams through the body.

dumping syndrome A condition that occurs when food moves too rapidly from the stomach into the small intestine.

duodenum The first segment of the small intestine.

dysphagia Difficulty swallowing.

E

eating disorder Extreme disturbance in eating behaviors that can result in serious medical conditions, psychological consequences, and dangerous weight loss.

eating disorders not otherwise specified (EDNOS) A category of eating disorders that includes some, but not all, of the diagnostic criteria for anorexia nervosa and/or bulimia nervosa.

echocardiogram A visual image, produced using ultrasound waves, of the heart's structure and movement.

edema The buildup of fluid in the interstitial spaces.

egestion The process whereby solid waste (feces) is expelled from the body.

eicosanoids Biologically active compounds synthesized from arachidonic acid and EPA.

eicosapentaenoic acid (EPA) A long-chain, polyunsaturated ω-3 fatty acid produced from linolenic acid.

electrocardiogram A procedure during which the heart's electrical activity is recorded.

electrolytes Substances such as salt that dissolve or dissociate into ions when put in water.

electron A subatomic particle that orbits around the nucleus of an atom and carries a negative charge.

electron transport chain A series of chemical reactions that transfer electron and hydrogen ions from $NADH + H^+$ and $FADH_2$ along protein complexes in the inner mitochondrial membrane, ultimately producing ATP.

element A pure substance made up of only one type of atom.

elongation The process whereby carbon atoms are added to a fatty acid, increasing its chain length.

embryo The developing human from two through eight weeks after fertilization.

embryonic period The stage of development that extends from conception through the eighth week of gestation.

embryonic phase The latter phase of the embryonic period during which time organs and organ systems first begin to form.

emulsification The process by which large lipid globules are broken down and stabilized into smaller lipid droplets.

end product The final product in a metabolic pathway.

endocrine cells Those that produce and release hormones into the blood.

endocytosis A form of vesicular active transport whereby the cell membrane surrounds extracellular substances and releases them to the cytoplasm.

endoscopy A procedure used to examine the lining of the GI tract.

endosperm The portion of a grain that contains mostly starch.

endurance training Athletic workouts that improve pulmonary and cardiovascular function.

energy balance A state in which energy intake equals energy expenditure.

energy imbalance A state in which the amount of energy consumed does not equal the amount of energy used by the body.

energy metabolism Chemical reactions that enable cells to store and use energy from nutrients.

energy The capacity to do work.

energy-yielding nutrient A nutrient that the body can use to produce ATP.

enmeshment Families whose members are overly involved with one another and have little autonomy.

enrichment The fortification of a select group of foods (rice, flour, bread or rolls, farina, pasta, cornmeal, and corn grits) with FDA-specified levels of thiamin, niacin, riboflavin, folate, and iron.

enteric (intestinal) toxin A toxic agent produced by an organism after it enters the gastrointestinal tract.

enteric nervous system Neurons located within the submucosa and muscularis layers of the digestive tract.

enterocytes Epithelial cells that make up the lumenal surface of each villus.

enterohemorrhagic Causing bloody diarrhea.

enterohepatic circulation Circulation between the small intestine and the liver used to recycle compounds such as bile.

enterotoxigenic Producing a toxin while in the GI tract.

environmental factor An element or variable in our surroundings over which we may or may not have control (such as pollution and temperature).

enzymes Biological catalysts that facilitate chemical reactions.

enzyme–substrate complex A substrate attached to an enzyme's active site.

epidemiologic study A study in which data are collected from a group of individuals who are not asked to change their behaviors in any way.

epigenetics Alterations in gene expression that do not involve changes in the DNA sequence.

epiglottis A cartilage flap that covers the trachea while swallowing.

epinephrine Hormone released from the adrenal glands in response to stress; helps increase blood glucose levels by promoting glycogenolysis.

epithelial tissue Tissue that forms a protective layer on bodily surfaces and lines internal organs, ducts, and cavities.

ergocalciferol (vitamin D_2) The form of vitamin D found in plant foods, vitamin D–fortified foods, and supplements.

ergogenic aid Substance taken to enhance physical performance beyond the gains obtained from physical training alone.

esophagus The passageway that begins at the pharynx and ends at the stomach.

essential nutrient A substance that must be obtained from the diet, because the body needs it and cannot make it in required amounts.

Estimated Average Requirement (EAR) The amount of a nutrient that meets the physiological requirements of half the healthy population of similar individuals.

Estimated Energy Requirement (EER) Average energy intake required to maintain energy balance in healthy individuals based on sex, age, physical activity level, weight, and height.

ethanol An alcohol produced by the chemical breakdown of sugar by yeast.

etiology The cause or origin of a disease.

exercise Planned, structured activities done to improve or maintain physical fitness.

exocrine cells Those that produce and release their secretions into ducts.

exocytosis A form of vesicular active transport whereby intracellular cell products are enclosed in a vesicle and the contents of the vesicle are released to the outside of the cell.

extracellular Situated outside of a cell.

F

facilitated diffusion A passive transport mechanism whereby substances cross cell membranes from a region of higher concentration to a region of lower concentration with the assistance of a transport protein.

fasting hypoglycemia Low blood glucose that occurs when the pancreas releases excess insulin during periods of low food intake.

fasting state The first five days of fasting or minimal food intake, beginning 24 hours after the last meal.

fat A lipid that is solid at room temperature.

fatty acid A lipid consisting of a chain of carbons with a methyl (–CH₃) group on one end and a carboxylic acid group (–COOH) on the other.

fatty liver A condition caused by excess alcohol consumption; characterized by the accumulation of triglycerides in the liver.

feces Waste matter consisting mostly of undigested food residue that is eliminated from the body through the anus.

fed state (postprandial period) The first four hours after a meal.

female athlete triad A combination of interrelated conditions: disordered eating, menstrual dysfunction, and osteopenia.

fermentation The process whereby yeast chemically breaks down sugar to produce ethanol and carbon dioxide; typically occurs under relatively anaerobic conditions.

ferric iron (Fe^{3+}) The oxidized form of iron.

ferritin A protein important for iron absorption and storage in the body.

ferrous iron (Fe^{2+}) The reduced form of iron.

fetal alcohol effect (FAE) A form of fetal alcohol spectrum disorder resulting in physical and cognitive outcomes that are less severe than those of fetal alcohol syndrome.

fetal alcohol spectrum disorder (FASD) A range of alcohol-related problems that can result from prenatal alcohol exposure.

fetal period Period of prenatal development, which starts at the beginning of the ninth week of gestation and continues until birth.

fetus A term used, beginning at the ninth week of pregnancy through birth, to describe a developing human.

fiber Polysaccharide found in plants that is not digested or absorbed in the human small intestine.

fibrin An insoluble protein that forms blood clots.

fibrinogen A water-soluble protein that is converted to the water-insoluble protein fibrin.

field gleaning Gathering and distributing unharvested agricultural crops that would normally be left in the field or on the tree.

FightBAC!® A public education program developed to reduce foodborne bacterial illness.

filtration The process of selective removal of metabolic waste products from the blood.

"FITT" principle A method of planning a physical fitness program that considers frequency, intensity, type, and time spent exercising each week.

flavin adenine dinucleotide (FAD) The oxidized form of the coenzyme that is able to accept two electrons and two hydrogen ions, forming FADH₂.

flavin mononucleotide (FMN) A coenzyme form of riboflavin.

flexibility Range of motion around a joint.

fluoride (F^-) A nonessential trace mineral that strengthens bones and teeth.

foam cell A type of cell—usually an immune cell—that contains large amounts of lipids.

folate (also called folacin) A water-soluble vitamin involved in single-carbon transfer reactions; needed for amino acid metabolism and DNA synthesis.

folic acid The form of folate commonly used in vitamin supplements and food fortification.

food allergy A condition in which the body's immune system reacts against a protein in food.

food aversion A strong psychological dislike of a particular food.

food bank Agency that collects donated foods and distributes them to local food pantries, shelters, and food kitchens.

food biosecurity Measures aimed at preventing the food supply from falling victim to planned contamination.

food composition table Tabulated information concerning the nutrient and energy contents of foods.

food craving A strong psychological desire for a particular food.

food desert Community with limited access to affordable, healthful foods; tends to be located in urban and rural low-income neighborhoods.

food frequency questionnaire A retrospective dietary assessment method that assesses food selection patterns over an extended period of time.

food insecurity The condition whereby individuals or families cannot obtain sufficient food.

food intolerance When the body reacts negatively to a food or food component but does not mount an immune response.

food kitchen Program that prepares and serves meals to those in need.

food neophobia An eating disturbance characterized as an irrational fear of trying new foods.

food pantry Program that provides canned, boxed, and sometimes fresh foods directly to individuals in need.

food preoccupation Spending an inordinate amount of time thinking about food.

food recovery program Program that collects and distributes food that would otherwise be discarded.

food security The condition whereby individuals are able to obtain sufficient amounts of nutritious food to live active, healthy lives.

Food Tracker A component of the MyPlate website that allows individuals to conduct a dietary self-assessment.

foodborne illness A disease caused by ingesting unsafe food.

fortified food A food to which nutrients have been added.

Framingham Heart Study A large epidemiologic study begun in the 1940s designed to assess the relationship between lifestyle factors and risk for heart disease.

free radical A reactive molecule with one or more unpaired electrons. Free radicals are destructive to biological membranes, DNA, and proteins.

frequency Component of the FITT principle that addresses how often a person exercises.

fructose A six-carbon monosaccharide found in fruits and vegetables; also called levulose.

full-term infant Baby born with gestational age between 37 and 42 weeks.

functional fiber Fiber that is added to food to provide beneficial physiological effects.

functional food A food that contains an essential nutrient, phytochemical, or zoonutrient and that is thought to benefit human health.

G

galactose A six-carbon monosaccharide found mainly bonded with glucose to form the milk sugar lactose.

gastric banding A type of bariatric surgery in which an adjustable, fluid-filled band is wrapped around the upper portion of the stomach, dividing it into a small upper pouch and a larger lower pouch.

gastric bypass A surgical procedure that reduces the size of the stomach and bypasses a segment of the small intestine so that fewer nutrients are absorbed.

gastric emptying The process by which food leaves the stomach and enters the small intestine.

gastric juice Digestive secretions produced by exocrine cells that make up gastric pits.

gastric lipase An enzyme, produced in the stomach, that hydrolyzes ester linkages between fatty acids and glycerol molecules.

gastric mucosal barrier A thick layer of mucus that protects mucosal lining of the stomach from the acidic gastric juice.

gastric phase The phase of digestion stimulated by the arrival of food into the stomach characterized by increased GI motility and release of GI secretions.

gastric pits Invaginations of the mucosal lining of the stomach that contain specialized endocrine and exocrine cells.

gastric ulcer A sore in the lining of the stomach.

gastrin A hormone, secreted by endocrine cells in the stomach, which stimulates the production and release of gastric juice.

gastroesophageal reflux disease (GERD) A condition caused by the weakening of the gastroesophageal sphincter, which enables gastric juices to reflux into the esophagus, causing irritation to the mucosal lining.

gastroesophageal sphincter A circular muscle that regulates the flow of food between the esophagus and the stomach; also called lower esophageal sphincter or cardiac sphincter.

gastrointestinal (GI) tract A tubular passage that runs from the mouth to the anus that includes several organs that participate in the process of digestion; also called the digestive tract.

gastrointestinal (GI) hormones Hormones secreted by the mucosal lining of the GI tract that regulate GI motility and secretion.

gastrointestinal (GI) motility Mixing and propulsive movements of the gastrointestinal tract caused by contraction and relaxation of the muscularis.

gastrointestinal (GI) secretions Substances released by organs that make up the digestive system that facilitate the process of digestion; also called digestive juices.

gene therapy The use of altered genes to enhance health.

gene A portion of a chromosome that codes for the primary structure of a protein.

genetic factor An inherited element or variable in our lives that cannot be altered.

genetic makeup (genotype) The particular DNA contained in a person's cells.

genetically modified organism (GMO) An organism (plant or animal) made by genetic engineering.

germ The portion of a grain that contains most of its vitamins and minerals.

gerontology The branch of science and medicine that focuses on health issues related to aging.

gestation length The period of time from conception to birth.

gestational age Length of pregnancy, determined by counting the number of weeks between the first day of a woman's last normal menstrual period and birth.

gestational diabetes A form of diabetes, characterized by insulin resistance, that develops in response to hormone-related changes during pregnancy.

ghrelin A hormone, secreted by cells in the stomach lining, that stimulates food intake.

glossitis Inflamed tongue.

glucagon Hormone secreted by the pancreatic α-cells in response to decreased blood glucose.

glucogenic amino acid Amino acids that can be used to make glucose.

glucometer A medical device used to measure the concentration of glucose in the blood.

gluconeogenesis Synthesis of glucose from noncarbohydrate sources.

glucose transporters Proteins that assist in the transport of glucose molecules across cell membranes.

glucose A six-carbon monosaccharide produced by photosynthesis in plants.

glutathione peroxidases A group of selenoprotein enzymes that have reduction–oxidation (redox) functions in the body.

gluten A protein found in cereal grains such as wheat, rye, barley, and possibly oats.

glycemic index (GI) A rating system used to categorize foods according to the relative glycemic responses they elicit.

glycemic load (GL) A rating system used to categorize foods that takes into account the glycemic index as well as the amount of carbohydrate typically found in a single serving of the food.

glycemic response The change in blood glucose following the ingestion of a specific food.

glycogen Polysaccharide consisting of a highly branched arrangement of glucose molecules; found primarily in liver and skeletal muscle.

glycogenesis Formation of glycogen.

glycogenolysis The breakdown of liver and muscle glycogen into glucose.

glycolysis The metabolic pathway that splits glucose into two three-carbon molecules called pyruvate.

glycosidic bond A type of chemical bond that forms between two monosaccharides.

goiter A form of iodine deficiency disorder (IDD) that affects children and adults; characterized by an enlarged thyroid gland.

goitrogens Compounds found in some vegetables that decrease iodine utilization by the thyroid gland.

gout A condition caused by the accumulation of uric acid in the joints.

graying of America The phenomenon occurring in the United States in which the proportion of elderly individuals in the population is increasing with time.

growth An increase in size and/or number of cells.

guanosine triphosphate (GTP) A high-energy compound similar to ATP.

H

haustral contractions Slow muscular movements that move the colonic contents back and forth, helping to compact the feces.

Hawthorne effect Phenomenon in which study results are influenced by an unintentional alteration of a behavior by the study participants.

Hazard Analysis Critical Control Points (HACCP) system A food safety protocol used to decrease contamination of foods during processing.

health claim FDA-approved claim that can be included on a food's packaging to describe a specific, scientifically supported health benefit.

Healthy People 2020 A federally developed document that provides overall health objectives for the nation.

heart attack (also called myocardial infarction) An often life-threatening condition in which blood flow to some or all of the heart muscle is completely blocked.

heart disease (also called coronary heart disease) A condition that occurs when the heart muscle does not receive enough blood.

heat exhaustion Rise in body temperature that occurs when the body has difficulty dissipating heat.

heat stroke Rise in body temperature due to excess loss of body fluids associated with sweating.

Helicobacter pylori (H. pylori) A bacterium residing in the GI tract that causes peptic ulcers.

hematocrit The percentage of whole blood that is red blood cells (erythrocytes).

heme iron Iron that is a component of a heme group; heme iron includes hemoglobin in blood, myoglobin in muscles, and cytochromes in mitochondria.

hemodialysis A medical procedure that uses a machine to filter waste products from the blood and to restore proper fluid balance.

hemodilution A decrease in the number of red blood cells per volume of plasma caused by plasma volume expansion.

hemoglobin A complex protein, composed of four iron-containing heme groups and four protein subunits, needed for oxygen and carbon dioxide transport in the body.

hemolytic anemia Decreased ability of the blood to carry oxygen and carbon dioxide due to rupturing of red blood cells; caused by vitamin E deficiency.

hemosiderin An iron-storing protein found in many tissues, especially the liver.

hepatic portal vein A blood vessel that circulates blood to the liver from the GI tract.

hepcidin A hormone, produced by the liver in response to high levels of iron in the body, that helps regulate iron status by decreasing iron uptake by the intestine.

hereditary hemochromatosis A genetic abnormality resulting in increased absorption of iron in the intestine.

heterocyclic amines Cancer-causing compounds that can be formed when meat is cooked at high temperatures.

hexoses Monosaccharide made of six carbon atoms.

high-density lipoprotein (HDL) A lipoprotein made primarily by the liver that circulates in the blood to collect excess cholesterol from cells.

high osmotic pressure Occurs when a solution has a large amount of solutes dissolved in it.

high-fructose corn syrup (HFCS) A substance derived from corn that is used to sweeten foods and beverages.

high-quality protein source A complete protein source with high amino acid bioavailability.

homeostasis A state of balance or equilibrium.

homocysteine A compound that is converted to methionine in a folate and vitamin B_{12}-requiring, coupled reaction.

hormone replacement therapy (HRT) Medication typically containing estrogen and progesterone, sometimes taken by women after having their ovaries removed or after menopause.

hormones Substances released from glands or cells in response to various stimuli that exert their effect by binding to receptors on specific tissues.

hormone-sensitive lipase An enzyme that catalyzes the hydrolysis of ester linkages that attach fatty acids to the glycerol molecule; mobilizes fatty acids stored in adipose tissue.

hunger The physiological drive to consume food.

hydrolysis A chemical reaction whereby compounds react with water and are split apart.

hydrophilic substance One that dissolves or mixes with water.

hydrophobic substance One that does not dissolve or mix with water.

hydrostatic weighing (underwater weighing) Method for estimating body composition that compares weight on land to weight underwater.

hydroxyapatite [$Ca_{10}(PO_4)_6(OH)_2$] The mineral matrix of bones and teeth.

hypercarotenemia A condition in which carotenoids accumulate in the skin, causing the skin to become yellow-orange.

hypercholesterolemia Elevated levels of cholesterol in the blood.

hyperglycemia Abnormally high level of glucose in the blood.

hyperkeratosis A symptom of vitamin A deficiency in which immature skin cells overproduce the protein keratin, causing rough and scaly skin.

hyperlipidemia Elevated levels of lipids in the blood.

hypernatremia High blood sodium concentration.

hyperplasia An increase in the number of cells.

hyperplastic growth Growth associated with an increase in cell number.

hypertriglyceridemia Elevated levels of triglycerides in the blood.

hypertrophic growth Growth associated with an increase in cell size.

hypertrophy An increase in size of a tissue or organ due to an increase in cell size or number.

hypervitaminosis A A condition in which elevated circulating vitamin A levels result in blurred vision, liver damage, and reduced bone strength.

hypoglycemia Abnormally low level of glucose in the blood.

hypokalemia Low blood potassium concentration.

hyponatremia Low blood sodium concentration.

hypothalamus An area of the brain that controls many involuntary functions by the release of hormones and neuropeptides.

hypothesis A prediction about the relationship between variables.

I

ileocecal sphincter The sphincter that separates the ileum from the cecum and regulates the flow of material between the small and large intestines.

ileum The last segment of the small intestine that comes after the jejunum.

impaired glucose regulation Condition characterized by elevated levels of glucose in the blood.

in vitro Involving the use of cells or environments that are not part of a living organism.

in vivo Involving the study of natural phenomena in a living organism.

incidence The number of people who are newly diagnosed with a condition in a given period of time.

incomplete protein source A food that lacks or contains very low amounts of one or more essential amino acids.

incubation period The time between when infection occurs and signs or symptoms begin.

indirect calorimetry A measurement of energy expenditure obtained by assessing oxygen consumption and carbon dioxide production.

infant mortality rate The number of infant deaths (<1 year of age) per 1,000 live births in a given year.

infantile beriberi A form of thiamin deficiency that occurs in infants breastfed by thiamin-deficient mothers.

infectious agent of foodborne illness A pathogen in food that causes illness and can be passed or transmitted from one infected animal or person to another.

infectious disease A contagious illness caused by a pathogen such as a bacteria, virus, or parasite.

inflammatory bowel disease (IBD) Chronic conditions such as ulcerative colitis and Crohn's disease that cause inflammation of the lower GI tract.

inherited metabolic disease (also called inborn error of metabolism) Genetic condition caused by a deficiency or absence of one or more enzymes needed for a metabolic pathway to function properly.

initiation The first stage of cancer, in which a normal gene is transformed into a cancer-forming gene (oncogene).

inorganic compound A substance that does not contain carbon–carbon bonds or carbon–hydrogen bonds.

insoluble fiber Dietary fiber that is incapable of being dissolved in water.

insulin receptors Proteins, found on the surface of certain cell membranes, that bind insulin.

insulin resistance Condition characterized by the inability of insulin receptors to respond to the hormone insulin.

insulin Hormone secreted by the pancreatic β-cells in response to increased blood glucose.

insulin-responsive glucose transporters Glucose transporters that require insulin to function.

intensity Component of the FITT principle that refers to the amount of physical exertion extended during physical activity.

interaction When the relationship between two factors is influenced or modified by another factor.

intermediate product A product formed before a metabolic pathway reaches completion, often serving as a substrate in the next chemical reaction.

intermediate-density lipoprotein (IDL) A lipoprotein that results from the loss of fatty acids from a VLDL; IDLs are ultimately converted to LDLs.

intermembrane space The space between the inner and outer mitochondrial membranes.

interstitial fluid Fluid that surrounds cells.

interval training Athletic workout that alternates between fast bursts of intensive exercise and slower, less demanding activity.

intervention study An experiment in which something is altered or changed to determine its effect on something else.

intestinal microbiota Bacteria that reside in the large intestine (also called intestinal microbiome).

intestinal phase The phase of digestion in which chyme enters the small intestine; characterized by both a decrease in gastric motility and secretion of gastric juice.

intracellular Situated within a cell.

intrauterine growth retardation (IUGR) Slow or delayed growth *in utero*.

intrinsic factor A protein, produced by the stomach, needed for vitamin B_{12} absorption.

iodide (I⁻) The most abundant form of iodine in the body.

iodine (I) An essential trace mineral that is a component of the thyroid hormones.

iodine deficiency disorders (IDDs) A broad spectrum of conditions caused by inadequate iodine.

ion An atom that has acquired an electrical charge by gaining or losing one or more electrons.

iron (Fe) A trace mineral needed for oxygen and carbon dioxide transport, energy metabolism, removal of free radicals, and synthesis of DNA.

irradiation A food preservation process that applies radiant energy to foods to kill bacteria.

irritable bowel syndrome (IBS) A condition that typically affects the lower GI tract, causing abdominal pain, muscle spasms, diarrhea, and constipation.

jejunum The midsection of the small intestine, located between the duodenum and the ileum.

K

Keshan disease A disease resulting from selenium deficiency.

ketoacidosis A rise in ketone levels in the blood, characterized by a decrease in the pH of the blood.

ketogenesis Metabolic pathway that leads to the production of ketones.

ketogenic amino acids Amino acids that can be used to make ketones.

ketogenic diets Diets that stimulate ketone production.

ketone Organic compound used as an energy source during starvation, fasting, low-carbohydrate diets, or uncontrolled diabetes.

ketosis Condition resulting from excessive ketones in the blood.

kilocalorie (kcal or Calorie) 1,000 calories.

kosher food A type of food that has been prepared and served in ways that are observant of Jewish law.

kwashiorkor (Kwa, a language of Ghana; referring to "what happens to the first child when the next is born") A form of PEM often characterized by edema in the extremities (hands, feet).

kyphosis (dowager's hump) A curvature of the upper spine, caused by osteoporosis.

L

labile amino acid pool In the body, amino acids that are immediately available to cells for protein synthesis and other purposes.

lactase Brush border enzyme that hydrolyzes lactose into glucose and galactose.

lactation The production and release of milk.

lacteal A lymphatic vessel found in an intestinal villus.

lactogenesis The onset of milk production.

lacto-ovo-vegetarian A type of vegetarian who consumes dairy products and eggs in an otherwise plant-based diet.

lactose intolerance Inability to digest the milk sugar lactose; caused by a lack of the enzyme lactase.

lactose Disaccharide consisting of glucose and galactose; produced by mammary glands.

lactovegetarian A type of vegetarian who consumes dairy products (but not eggs) in an otherwise plant-based diet.

large for gestational age (LGA) infant A baby with a weight at or above the 90th percentile for weight for gestational age.

LDL receptor Membrane-bound protein that binds LDLs, causing them to be taken up and dismantled.

leptin A hormone, produced mainly by adipose tissue, that helps regulate body weight.

let-down The movement of milk through the mammary ducts toward the nipple.

life expectancy A statistical prediction of the average number of years of life remaining to a person at a specific age.

lifespan Maximum number of years an individual in a particular species has remained alive.

lifestyle factor Behavioral component of our lives over which we may or may not have control (such as diet and tobacco use).

limit dextrin A partial breakdown product formed during amylopectin digestion that contains three to four glucose molecules and an α-1,6 glycosidic bond.

limiting amino acid An essential amino acid in the lowest concentration in an incomplete protein source.

lingual lipase An enzyme, produced in the salivary glands, that hydrolyzes ester linkages between fatty acids and glycerol molecules.

linoleic acid An essential ω-6 fatty acid with 18 carbons and 2 double bonds.

linolenic acid An essential ω-3 fatty acid with 18 carbons and 3 double bonds.

lipases Enzymes that cleave fatty acids from the glycerol backbones of triglycerides, phospholipids, and cholesteryl esters.

lipid Organic substance that is relatively insoluble in water and soluble in organic solvents.

lipogenesis The metabolic processes that result in fatty acid and, ultimately, triglyceride synthesis.

lipolysis The breakdown of triglycerides into fatty acids and glycerol.

lipoprotein lipase An enzyme that hydrolyzes the ester linkage between a fatty acid and glycerol in a triglyceride, diglyceride, and monoglyceride molecule as they circulate in the bloodstream.

lipoprotein A spherical particle made of varying amounts of triglycerides, cholesterol, cholesteryl esters, phospholipids, and proteins.

long-chain fatty acid A fatty acid having >12 carbon atoms in its backbone.

low food security Classification of food-insecure households in which one or more members experience disrupted eating patterns (or is worried they will do so) because of insufficient resources or access to food.

low osmotic pressure When a solution has a small amount of solutes dissolved in it.

low-birth-weight (LBW) infant A baby that weighs less than 2,500 g (5 lb 8 oz) at birth.

low-density lipoprotein (LDL) A lipoprotein that delivers cholesterol to cells.

low-quality protein source A food that is either an incomplete protein source or one that has low amino acid bioavailability.

lumen The cavity inside a tubular structure in the body.

lymph A fluid found in lymphatic vessels.

lymphatic system A component of the circulatory system made up of lymphatic vessels and lymph that flows from organs and tissues, drains excess fluid from spaces that surround cells, and picks up dietary fats from the digestive tract.

lysophospholipid A lipid composed of a glycerol bonded to a polar head group and a fatty acid; final product of phospholipid digestion.

M

macronutrients The class of nutrients that we need to consume in relatively large quantities.

macula A portion of the retina important for sight.

macular degeneration A chronic disease that results from deterioration of the retina.

magnesium (Mg) A major mineral needed for stabilizing enzymes and ATP and as a cofactor for many enzymes.

major mineral An essential mineral that is required in amounts greater than 100 mg daily.

malignant tumor A cancerous growth.

malnutrition Poor nutritional status caused by either undernutrition or overnutrition.

maltase Brush border enzyme that hydrolyzes maltose into two glucose molecules.

maltose Disaccharide consisting of two glucose molecules bonded together; formed during the chemical breakdown of starch.

mammary duct Structure that transports milk from the alveolar secretory cells toward the nipple.

manganese (Mn) An essential trace mineral that is a cofactor for enzymes needed for bone formation, glucose production, and energy metabolism.

marasmus A form of PEM characterized by extreme wasting of muscle and adipose tissue.

marine toxin A poison produced by ocean algae.

mastication Chewing and grinding of food by the teeth to prepare for swallowing.

maximal oxygen consumption (VO₂ max) Maximum volume of oxygen that can be used by tissues per minute.

maximum heart rate Estimated by subtracting age from 220 (220 – age).

meat factor An unidentified compound, found in meat, that increases the absorption of nonheme iron.

mechanoreceptor A sensory receptor that responds to pressure, stretching, or mechanical stimulus.

medical history Questions asked to assess overall health.

Mediterranean diet A dietary pattern, originating from the region surrounding the Mediterranean Sea, which is related to lower risk for cardiovascular disease.

medium-chain fatty acid A fatty acid having 8–12 carbon atoms in its backbone.

megaloblastic macrocytic anemia A condition in which red blood cells are large and immature, due to folate deficiency or vitamin B_{12} deficiency.

melamine A nitrogen-containing chemical used to make lightweight plastic objects.

menadione (vitamin K₃) A form of vitamin K produced commercially.

menaquinone (vitamin K₂) A form of vitamin K produced by bacteria.

menarche The first time a female menstruates.

menopause The time in a woman's life when menstruation ceases, usually during the sixth decade of life.

messenger ribonucleic acid (mRNA) A form of RNA involved in gene transcription.

metabolic pathway A series of interrelated enzyme-catalyzed chemical reactions that take place in cells.

metabolic syndrome Condition characterized by an abnormal metabolic profile, abdominal body fat, and insulin resistance that increases risk for developing type 2 diabetes.

metabolism Chemical reactions that take place in the body.

metalloenzyme An enzyme that contains a mineral cofactor.

metallothionine A protein in the enterocyte that regulates zinc absorption and elimination.

metastasis The spreading of cancer cells to other parts of the body via the blood or lymph.

methicillin-resistant *Staphylococcus aureus* (MRSA) A type of S. *aureus* that is resistant to most antibiotics.

micelle A water-soluble, spherical structure formed in the small intestine via emulsification.

microcytic hypochromic anemia A condition in which red blood cells are small and light in color due to inadequate hemoglobin synthesis; can be due to vitamin B_6 deficiency.

micronutrients The class of nutrients that we need to consume in relatively small quantities, (<1 gram/day).

microsomal ethanol-oxidizing system (MEOS) A pathway used to metabolize alcohol when it is present in high amounts.

microsomes Cell organelles associated with endoplasmic reticula.

microvilli Hairlike projections on the luminal surface of enterocytes.

milk let-down The movement of milk through the mammary ducts toward the nipple.

mineral Inorganic substance, other than water, that is required by the body for basic functions or structure.

miscarriage The death of a fetus during the first 20 weeks of pregnancy.

mitochondria Cellular organelles involved in generating energy (ATP).

mitochondrial matrix The inner compartment of the mitochondrion.

molecular formula Indicates the number and types of atoms in a molecule.

molecule A substance held together by chemical bonds.

molybdenum (Mo) An essential trace mineral that is a cofactor for enzymes needed for amino acid and purine metabolism.

monoglyceride (also called monoacylglycerol) A lipid made of a glycerol bonded to a single fatty acid.

monosaccharide Carbohydrate consisting of a single sugar.

monounsaturated fatty acid (MUFA) A fatty acid that contains one carbon–carbon double bond in its backbone.

morbidity rate The number of illnesses in a given period of time.

mortality rate The number of deaths in a given period of time.

mucosa The lining of the gastrointestinal tract that is made up of epithelial cells; also called mucosal lining.

mucus A substance that coats and protects mucous membranes.

muscle dysmorphia (MD) Pathological preoccupation with increasing muscularity.

muscle endurance Ability to exercise for an extended period of time without becoming fatigued.

muscle tissue Tissue that specializes in movement.

muscular strength The maximal force exerted by muscles during an activity.

muscularis The layer of tissue in the gastrointestinal tract that consists of at least two layers of smooth muscle.

mutation The alteration of a gene.

myoepithelial cells Muscle cells that surround alveoli and contract, forcing milk into the mammary ducts.

myoglobin A heme protein found in muscle.

MyPlate Graphic representation of the major concepts of the Dietary Guidelines for Americans; interactive website available at www.choosemyplate.gov.

N

National Center for Health Statistics (NCHS) A component of the U.S. Public Health Service whose mission is to compile statistical information to be used in improving the health of Americans.

National Health and Nutrition Examination Survey (NHANES) A federally funded epidemiologic study begun in the 1970s to assess trends in diet and health in the U.S. population.

National School Lunch and School Breakfast Programs Federally funded programs that provide free or subsidized nutritious meals to school-age children.

negative correlation (also called inverse correlation) An association between factors in which a change in one is related to change in the other in the opposite direction.

negative energy balance A state in which energy intake is less than energy expenditure.

negative feedback systems Corrective responses that oppose change and restore homeostasis.

negative nitrogen balance The condition in which protein (nitrogen) intake is less than protein (nitrogen) loss by the body.

nephron Tubule in the kidneys that filters waste materials from the blood that are later excreted in the urine.

neural tissue Tissue that specializes in communication via nerves.

neurotransmitters Chemical messengers released from nerve cells that transmit information.

neural tube defect A malformation in which the neural tissue does not form properly during fetal development.

neurotransmitters Chemical messengers released from nerve cells that transmit information.

neutron A subatomic particle, in the nucleus of an atom, with no electrical charge.

niacin (vitamin B_3) An essential water-soluble vitamin involved in energy metabolism, electron transport chain, synthesis of fatty acids and proteins, metabolism of vitamin C and folate, glucose homeostasis, and cholesterol metabolism.

niacin equivalent (NE) A unit of measure that describes the niacin and/or tryptophan content in food.

nicotinamide adenine dinucleotide (NAD) The oxidized form of the coenzyme that is able to accept two electrons and two hydrogen ions, forming NADH + H⁺.

nicotinamide adenine dinucleotide phosphate (NADP⁺) The oxidized form of the coenzyme that is able to accept two electrons and two hydrogen ions, forming NADPH + H⁺.

night blindness A condition characterized by impaired ability to see in the dark.

night eating syndrome (NES) A disordered eating pattern characterized by a cycle of daytime food restriction, excessive food intake in the evening, and nighttime insomnia.

nitrites Nitrogen-containing compounds that are often added to processed meats to enhance color and flavor.

nitrogen balance The condition in which protein (nitrogen) intake equals protein (nitrogen) loss by the body.

nitrosamines Nitrogen-containing chemical carcinogens, produced from nitrites, that have been shown to cause cancer.

nocturnal sleep-related eating disorder (SRED) A disordered eating pattern characterized by eating while asleep without any recollection of having done so.

nonessential nutrient A substance found in food and used by the body to promote health but not required to be consumed in the diet.

nonexercise activity thermogenesis (NEAT) Energy expended for spontaneous movement such as fidgeting and maintaining posture.

nonheme iron Iron that is not a component of a heme group.

noninfectious agent of foodborne illness An inert (nonliving) substance in food that causes illness.

noninfectious disease An illness that is not contagious.

nonpolar molecule One that does not have differently charged portions.

nonprovitamin A carotenoid A carotenoid that cannot be converted to vitamin A.

norovirus A type of infectious pathogen (virus) that often causes foodborne illness.

nucleus A membrane-enclosed organelle that contains the genetic material DNA.

nutrient absorption The transfer of nutrients from the lumen of the GI tract to the circulatory system.

nutrient content claim FDA-regulated phrase or words that can be included on a food's packaging to describe its nutrient content; based on a 2,000-kcal diet.

nutrient density The relative ratio of nutrients in a food in comparison to total calories.

nutrient requirement The lowest intake level of a nutrient that supports basic physiological functions and promotes optimal health.

nutrient A substance in foods used by the body for energy, maintenance of body structures, or regulation of chemical processes.

nutrients of concern Those nutrients, identified in the 2010 Dietary Guidelines for Americans, that are somewhat lacking in the typical U.S. diet.

nutrigenomics The science of how genetics and nutrition together influence health.

Nutrition Facts panel A required component of food packaging that contains information about the nutrient content of the food.

nutrition transition The shift from undernutrition to overnutrition or unbalanced nutrition that often occurs simultaneously with the industrialization of a society.

nutrition The science of how living organisms obtain and use food to support processes required for life.

nutritional adequacy The situation in which a person consumes the required amount of a nutrient to meet physiological needs.

nutritional sciences A broad spectrum of academic and social disciplines related to nutrition.

nutritional scientist A person who conducts and/or evaluates nutrition-related research.

nutritional status The health of a person as it relates to how well his or her diet meets that person's individual nutrient requirements.

nutritional toxicity Overconsumption of a nutrient resulting in dangerous (toxic) effects.

O

ob gene The gene that codes for the protein leptin.

ob/ob mouse Obese mouse with a mutation in the genes that code for the hormone leptin.

obese Having excess body fat.

oil A lipid that is liquid at room temperature.

oligosaccharide Carbohydrate made of relatively few (3 to 10) monosaccharides.

omega (ω) end The end of a fatty acid, which consists of a methyl (–CH$_3$) group.

omega-3 fatty acid A fatty acid in which the first double bond is located between the third and fourth carbons from the methyl or omega (ω) end.

omega-6 fatty acid A fatty acid in which the first double bond is located between the sixth and seventh carbons from the methyl or omega (ω) end.

oncogene An abnormal gene that transforms normal cells to cancer cells.

oncology The branch of medicine related to cancer and its treatment.

opsin The protein component of rhodopsin.

organ system Organs that work collectively to carry out related functions.

organ A group of tissues that combine to carry out coordinated functions.

organelles Cellular structures that have a particular function.

organic compound A substance that contains carbon–carbon bonds or carbon–hydrogen bonds.

osmosis Movement of water molecules from a region of lower solute concentration to that of a higher solute concentration, until equilibrium is reached.

ossification The process by which minerals are added to cartilage, ultimately resulting in bone formation.

osteoblast A bone cell that promotes bone formation.

osteoclast A bone cell that promotes bone breakdown.

osteomalacia Softening of the bones in adults that can be due to vitamin D deficiency.

osteopenia A condition whereby bone mineral density is lower than normal.

osteoporosis A serious bone disease resulting in weak, porous bones.

overload principle The body's adaptive response when challenged by physical exertion.

overweight Having excess weight for a given height.

oxaloacetate The final product of the citric acid cycle, which becomes the substrate for the first reaction in this pathway.

oxidation The loss of one or more electrons.

oxidative phosphorylation The chemical reactions that link the oxidation of NADH + H$^+$ and FADH$_2$ to the phosphorylation of ADP to form ATP and water.

oxidized LDL A lipoprotein particle (low-density lipoprotein) with lipids or proteins that have been oxidized by free radicals.

oxytocin A hormone, produced by the hypothalamus and stored in the pituitary gland, that stimulates the movement of milk into the mammary ducts.

P

pancreatic α-amylase Enzyme, produced by the pancreas, which digests starch by hydrolyzing α-1,4 glycosidic bonds.

pancreatic juice Pancreatic secretions that contain bicarbonate and enzymes needed for digestion.

pancreatic lipase An enzyme, produced in the pancreas, that hydrolyzes ester linkages between fatty acids and glycerol molecules.

pancreatitis Inflammation of the pancreas.

pantothenic acid (vitamin B$_5$) A water-soluble vitamin involved in energy metabolism, hemoglobin synthesis, and phospholipid synthesis.

parasite An organism that, during part of its life cycle, must live within or on another organism without benefiting its host.

parathyroid hormone (PTH) A hormone, produced in the parathyroid glands, that is released in response to low blood calcium concentration; stimulates the conversion of 25-(OH)D$_3$ to 1,25-(OH)$_2$D$_3$ in the kidneys.

parietal cells Exocrine cells within the gastric mucosa that secrete hydrochloric acid and intrinsic factor.

partial hydrogenation A process by which some carbon–carbon double bonds found in PUFAs are converted to carbon–carbon single bonds, resulting in the production of *trans* fatty acids.

passive transport mechanism Transport mechanism that enables substances to cross cell membranes without expenditure of energy (ATP).

pasteurization A food preservation process that subjects foods to heat to kill bacteria, yeasts, and molds.

peak bone mass The greatest amount of bone mineral that a person has during his or her life.

peer-reviewed journal A publication that requires a group of scientists to read and approve a study before it is published.

pellagra A disease caused by niacin and/or tryptophan deficiency.

pepsin An enzyme needed for protein digestion.

pepsinogen The inactive form (proenzyme) of pepsin.

peptic ulcer An irritation or erosion of the mucosal lining in the stomach, duodenum, or esophagus.

peptide bond A chemical bond that joins amino acids.

periconceptional period Time shortly before and after conception.

perimenopaue Literally, the time "around" the time of menopause.

periosteum The outer covering of bone, consisting of blood vessels, nerves, and connective tissue.

peristalsis Waves of muscular contractions that move materials in the GI tract in a forward direction.

pernicious anemia An autoimmune disease that causes vitamin B$_{12}$ deficiency due to lack of intrinsic factor.

pH scale A scale, ranging from 0 to 14, that signifies the acidity or alkalinity of a solution.

pharnyx Region toward the back of the mouth that is the shared space between the oral and nasal cavities.

phenylketonuria (PKU) An inherited disease in which the body cannot convert phenylalanine into tyrosine.

phosphatidylcholine (also called lecithin) A phospholipid that contains choline as its polar head group; commonly added to foods as an emulsifying agent.

phospholipase A$_2$ An enzyme, produced in the pancreas, that hydrolyzes fatty acids from phospholipids.

phospholipid A lipid composed of a glycerol bonded to two fatty acids and a polar head group.

phosphorus (P) A major mineral needed for cell membranes, bone and tooth structure, DNA, RNA, ATP, lipid transport, and a variety of metabolic reactions.

photosynthesis Process whereby plants trap energy from the sun to produce glucose from carbon dioxide and water.

phylloquinone (vitamin K$_1$) A form of vitamin K found in foods and dietary supplements.

physical activity Bodily movement that uses skeletal muscles and that results in a substantial increase in energy expenditure over resting energy expenditure (REE).

physical fitness The ability to perform moderate to vigorous levels of physical activity without undue fatigue and the capability of maintaining such ability throughout life.

phytates Phosphorus-containing compounds often found in the outer coatings of kernels of grain as well as vegetables and legumes; can bind dietary minerals.

phytochemical (also called phytonutrient) A substance found in plants and thought to benefit human health above and beyond the provision of essential nutrients and energy.

phytoestrogen An estrogen-like phytochemical.

phytostanol Sterol-like compound made by plants.

phytosterol Sterol made by plants.

pica A desire to consume nonfood items such as laundry starch, clay, soil, and burned matches.

placebo effect The phenomenon in which there is an apparent effect of the treatment because the individual expects or believes that it will work.

placebo A "fake" treatment given to the control group that cannot be distinguished from the actual treatment.

placenta An organ, consisting of fetal and maternal tissues, that supplies nutrients and oxygen to the fetus, and aids in the removal of metabolic waste products from the fetal circulation.

plaque A complex of cholesterol, fatty acids, cells, cellular debris, and calcium that can form inside blood vessels and within vessel walls.

plasma The fluid component of blood.

plica circulares Circular folds in the mucosal lining of the small intestine.

polar head group A phosphate-containing charged chemical structure that is a component of a phospholipid.

polar molecule A molecule (such as water) that has both positively and negatively charged portions.

polycyclic aromatic hydrocarbons Cancer-causing compounds that can be formed when meat is grilled.

polymorphism An alternation in a gene that is present in at least 1% of the population.

polypeptide A string of more than 12 amino acids held together via peptide bonds.

polyphenols Organic compounds found in some plant-based foods; can bind dietary minerals.

polysaccharide Complex carbohydrate made of many monosaccharides.

polyunsaturated fatty acid (PUFA) A fatty acid that contains more than one carbon–carbon double bond in its backbone.

positive correlation An association between factors in which a change in one is related to a similar change in the other.

positive energy balance A state in which energy intake is greater than energy expenditure.

positive nitrogen balance The condition in which protein (nitrogen) intake is greater than protein (nitrogen) loss by the body.

postabsorptive state The period of time (4 to 24 hours after a meal) when no dietary nutrients are being absorbed.

postpartum amenorrhea The span of time between the birth of the baby and the return of menses.

post-term infant Baby born with a gestational age greater than 42 weeks.

potassium (K) A major mineral important in fluid balance, muscle and nerve function, and energy metabolism.

pre-embryonic phase The early phase of the embryonic period that begins with fertilization and continues through implantation.

prebiotic food Food that stimulates the growth of bacteria that naturally reside in the large intestine.

preformed toxin Poisonous substance produced by microbes while they are in a food (prior to ingestion).

pregnancy-induced hypertension (also called pre-eclampsia or toxemia of pregnancy) A form of pregnancy-related hypertension characterized by high blood pressure, sudden swelling and weight gain due to fluid retention, and protein in the urine.

preterm infant (premature infant) A baby born with a gestational age less than 37 weeks.

prevalence The total number of people who have a condition in a given period of time.

previtamin D₃ (precalciferol) An intermediate product made during the conversion of 7-dehydrocholesterol to cholecalciferol (vitamin D₃) in the skin.

primary malnutrition Poor nutritional status caused strictly by inadequate diet.

primary structure The sequence of amino acids that make up a single polypeptide chain.

prion A misshapen protein that causes other proteins to also become distorted, damaging nervous tissue.

probiotic Food or dietary supplement that contains beneficial live bacteria.

product A molecule produced in a chemical reaction.

proenzyme An inactive precursor of an enzyme.

progression The third stage of cancer in which tumor cells rapidly divide and invade surrounding tissues.

prolactin A hormone, produced in the pituitary gland, which stimulates the production of milk in alveoli.

promotion The second stage of cancer, in which the initiated cell begins to replicate itself, forming a tumor.

proof A measure of the alcohol content of distilled liquor.

prospective dietary assessment Type of dietary assessment that evaluates adequacy of food and beverage intake.

prostaglandins A group of eicosanoids involved in regulation of blood pressure; there are both ω-6 and ω-3 prostaglandins, having somewhat opposite effects.

prosthetic group A nonprotein component of a protein that is part of the quaternary structure.

protease An enzyme that cleaves peptide bonds.

protein complementation Combining incomplete protein sources to provide all essential amino acids in relatively adequate amounts.

protein turnover The cycle involving both protein degradation and protein synthesis in the body.

protein (peptide) Nitrogen-containing macronutrient made from amino acids.

protein-energy malnutrition (PEM) Protein deficiency accompanied by inadequate intake of protein and often of other essential nutrients as well.

proteolysis The breakdown of proteins into amino acids.

prothrombin A clotting factor (protein) that is converted to the enzyme thrombin.

proton A subatomic particle, in the nucleus of an atom, that carries a positive charge.

protozoa Very small (single-cell) organisms that are sometimes parasites.

provitamin A carotenoid A type of carotenoid that can be converted to vitamin A.

puberty Maturation of the reproductive system.

PubMed A computerized database that allows access to approximately 11 million biomedical journal citations.

pulmonary artery Blood vessel that transports oxygen-poor blood from the right side of the heart to the lungs.

pulmonary circulation The division of the cardiovascular system that circulates deoxygenated blood from the heart to the lungs, and oxygenated blood from the lungs back to the heart.

pulmonary vein Blood vessel that transports oxygen-rich blood from the lungs to the heart.

purging Self-induced vomiting and/or misuse of laxatives, diuretics, and/or enemas.

purine A constituent of DNA and RNA.

pyloric sphincter A circular muscle that regulates the flow of food between the stomach and the duodenum.

pyridoxal phosphate (PLP) The coenzyme form of vitamin B₆.

pyrimidine A constituent of DNA and RNA.

pyruvate An intermediate product formed during metabolism of carbohydrates and some amino acids.

Q

qualified health claim Statement concerning less well established health benefits that have been ascribed to a particular food or food component.

quaternary structure The combining of peptide chains with other peptide chains in a protein.

R

R protein A protein, produced in the stomach, that binds to vitamin B₁₂.

random assignment When study participants have equal chance of being assigned to each experimental group.

rate A measure of the occurrence of a certain type of event within a specific period of time.

reabsorption The return of previously removed materials to the blood.

reactive hypoglycemia (also called idiopathic postprandial hypoglycemia) Low blood glucose that occurs when the pancreas releases too much insulin in response to eating carbohydrate-rich foods.

ready-to-use therapeutic food (RUTF) Packaged, convenient, nutrient-dense food products that require no preparation or refrigeration and have long shelf lives.

rebound weight gain Weight regain that often follows successful weight loss.

Recommended Dietary Allowance (RDA) The average intake of a nutrient thought to meet the nutrient requirements of nearly all (97%) healthy people in a specified life stage and sex.

rectum The lower portion of the large intestine between the sigmoid colon and the anal canal.

red tide Phenomenon in which certain ocean algae grow profusely, causing reddish discoloration of the surrounding water.

reduction The gain of one or more electrons.

reduction–oxidation (redox) reactions Chemical reactions that take place simultaneously whereby one molecule gives up one or more electrons (is oxidized) while the other molecule receives one or more electrons (is reduced).

regular health claim Statement concerning scientifically backed health benefit associated with a food or food component.

renin An enzyme produced in the kidneys in response to low blood pressure; converts angiotensinogen to angiotensin I.

researcher bias When the researcher influences the results of a study.

resistance training Physical activities that, using weights or other methods of physical resistance, overload specific groups of muscles to make them work harder.

resorption The breakdown and assimilation of a substance in the body.

resting energy expenditure (REE) Energy expended for resting metabolism over a 24-hour period.

resting metabolic rate (RMR) Energy expended for resting metabolism per hour that is assessed under less stringent conditions than is BMR.

restrained eaters People who experience cycles of fasting followed by bingeing.

resveratrol An antioxidant found in the skin of red grapes and abundant in red wine and purple grape juice.

retina The inner lining of the back of the eye.

retinoid (preformed vitamin A) A term used to describe all forms of preformed vitamin A.

retinol activity equivalent (RAE) A unit of measure used to describe the combined amount of preformed vitamin A and provitamin A carotenoids in foods.

retinol-binding protein and **transthyretin** Proteins that carry retinoids in the blood.

retrospective dietary assessment Type of dietary assessment that assesses previously consumed foods and beverages.

reverse cholesterol transport Process whereby HDLs remove cholesterol from nonhepatic (non-liver) tissues for transport to the liver.

R-group The portion of an amino acid's structure that distinguishes it from other amino acids.

rhodopsin A compound in the retina that consists of the protein opsin and the vitamin A derivative *cis*-retinal; needed for night vision.

riboflavin (vitamin B₂) An essential water-soluble vitamin involved in energy metabolism, the synthesis of a variety of vitamins, nerve function, and protection of lipids.

ribosome An organelle, associated with the endoplasmic reticulum in the cytoplasm, involved in gene translation.

rickets A condition caused by vitamin D deficiency in young children; characterized by deformed bones, especially in the legs.

risk factor A lifestyle, environmental, or genetic factor related to a person's chances of developing a disease.

rugae Folds that line the inner stomach wall.

S

saliva A secretion released into the mouth by the salivary glands; moistens food and starts the process of digestion.

salivary α-amylase Enzyme released from the salivary glands that digests starch by hydrolyzing α-1,4 glycosidic bonds.

satiety The state in which hunger is satisfied and a person feels he or she has had enough to eat.

saturated fatty acid (SFA) A fatty acid that contains only carbon–carbon single bonds in its backbone.

scientific method Steps used by scientists to explain observations.

scurvy A condition caused by vitamin C deficiency; symptoms include bleeding gums, bruising, poor wound healing, and skin irritations.

secondary diabetes Diabetes that results from other diseases, medical conditions, or medication.

secondary malnutrition Poor nutritional status caused by factors such as illness.

secondary structure Folding of a protein because of weak bonds that form between elements of the amino acid backbone (not R-groups).

secretin A hormone, secreted by the small intestine, that stimulates the release of sodium bicarbonate and enzymes from the pancreas.

segmentation A muscular movement in the gastrointestinal tract that moves the contents back and forth within a small region.

selective estrogen receptor modulator (SERM) A drug that does not contain the hormone estrogen but causes estrogen-like effects in the body.

selenium (Se) An essential trace mineral that is important for reduction–oxidation (redox) reactions, thyroid function, and activation of vitamin C.

selenomethionine The amino acid methionine that has been altered to contain selenium instead of sulfur.

selenoprotein A protein that contains selenomethionine instead of sulfur-containing methionine.

selenosis Selenium toxicity.

semipermeable membrane A barrier that allows passage of some, but not all, molecules across it.

senescence The phase of aging during which function diminishes.

sensory receptors Receptors that monitor conditions and changes in the GI tract.

serosa Connective tissue that encloses the gastrointestinal tract.

serotype A specific strain of a larger class of organism.

serum transferrin saturation A measure of iron status that reflects the percentage of transferrin that contains iron.

set point theory A theory suggesting that a hormonal signaling system regulates body weight by making adjustments in energy intake and energy expenditure.

shellfish poisoning A group of foodborne illnesses caused by consuming shellfish that contain marine toxins.

short-chain fatty acid A fatty acid having <8 carbon atoms in its backbone.

sickle cell anemia A disease in which a small change in the amino acid sequence of hemoglobin causes red blood cells to become misshapen and decreases the ability of the blood to carry oxygen and carbon dioxide.

sign Physical indicator of disease that can be seen by others, such as pale skin and skin rashes.

simple carbohydrate, or simple sugar Category of carbohydrates consisting of mono- and disaccharides.

simple diffusion A passive transport mechanism whereby substances cross cell membranes from a region of higher concentration to a region of lower concentration.

simple relationship A relationship between two factors that is not influenced or modified by another factor.

single-blind study A human experiment in which the participants do not know to which group they have been assigned.

skeletal fluorosis Weakening of the bones caused by excessive fluoride intake.

skinfold caliper An instrument used to measure the thickness of subcutaneous fat.

small for gestational age (SGA) infant A baby that weighs less than the 10th percentile for weight for gestational age.

sodium (Na) A major mineral important for regulating fluid balance, nerve function, and muscle contraction.

soluble fiber Dietary fiber that dissolves in water.

solute A substance that dissolves in a solvent.

solution A mixture of two or more substances that are uniformly dispersed.

Special Supplemental Nutrition Program for Women, Infants, and Children (WIC) A federally funded program that assists families in a targeted, at-risk population in making nutritious food purchases.

specific heat The energy required to raise the temperature of a substance.

sphincter A muscular band that narrows an opening between organs in the GI tract.

spina bifida A form of neural tube defect in which the spine does not properly form.

sports anemia A physiological (adaptive) response to training caused by a disproportionate increase in plasma volume relative to the increase in number of red blood cells.

stable isotope A form of an element that contains additional neutrons.

starvation response Adaptive physiological mechanisms that promote the storage of excess energy during times of food abundance and conserve energy during times of limited food availability.

starvation state Food deprivation for an extended period of time, typically lasting longer than one week.

stent A device made of rigid wire mesh that is threaded into an atherosclerotic blood vessel to expand and provide support for a damaged artery.

steroid hormone A hormone made from cholesterol.

sterol ester A chemical compound consisting of a sterol molecule bonded to a fatty acid via an ester linkage.

sterol A type of lipid with a distinctive multi-ring structure; a common example is cholesterol.

stomatitis Inflammation of the mucous membrane of the mouth.

strength training Athletic workouts that increases muscle growth and strength.

stroke volume The amount of blood pumped out of the heart with each contraction.

stroke A condition that occurs when a portion of the brain does not receive enough blood.

structure/function claim A statement that can be placed on food packaging stating the relationship between a nutrient or other dietary ingredient and health; not FDA approved.

stunted growth Delayed growth resulting from chronic undernutrition.

subcutaneous adipose tissue (SCAT) Adipose tissue found directly under the skin.

submucosa A layer of tissue that lies between the mucosa and muscularis tissue layers.

substrate phosphorylation The transfer of an inorganic phosphate (P_i) group to ADP to form ATP.

substrate (reactant) A molecule that enters a chemical reaction.

sucrase Brush border enzyme that hydrolyzes sucrose into glucose and fructose.

sucrose Disaccharide consisting of glucose and fructose; found primarily in fruits and vegetables.

sulfite A naturally occurring compound that is sometimes used as a food additive to prevent discoloration and bacterial growth.

superoxide dismutase A copper-containing enzyme that reduces the superoxide free radical to form hydrogen peroxide.

Supplemental Nutrition Assistance Program (SNAP) A federally funded program, formerly known as the Food Stamp Program, that helps low-income households pay for food.

symptom Manifestation of disease that cannot be seen by others, such as stomach pain or loss of appetite.

systemic circulation The division of the cardiovascular system that begins and ends at the heart and delivers blood to all the organs except the lungs.

T

talk test A method to judge the intensity level of workouts by assessing a person's ability to converse while exercising.

target heart rate That needed to achieve a desired activity intensity; 50–70% and 70–85%

of maximum heart rate are often used as targets for moderate- and vigorous-level intensities, respectively.

teratogen Environmental agent that can alter normal cell growth and development, causing a birth defect.

tertiary structure Folding of a polypeptide chain because of interactions among the R-groups of the amino acids.

tetany A condition in which muscles tighten and are unable to relax.

tetrahydrofolate (THF) The active form of folate.

Therapeutic Lifestyle Changes (TLC) diet A set of heart-healthy diet recommendations put forth by the National Cholesterol Education Program.

thermic effect of food (TEF) Energy expended for the digestion, absorption, and metabolism of nutrients.

thiamin (vitamin B₁) An essential water-soluble vitamin involved in energy metabolism, synthesis of DNA, RNA, and NADPH + H⁺, and nerve function.

thiamin pyrophosphate (TPP) The coenzyme form of thiamin that has two phosphate groups.

thiamin triphosphate (TTP) A form of thiamin with three phosphate groups.

thrombin The enzyme that catalyzes the conversion of fibrinogen to fibrin.

thyroid-stimulating hormone (TSH) A hormone, produced in the pituitary gland, that stimulates uptake of iodine by the thyroid gland.

thyroxine (T₄) The less-active form of thyroid hormone; contains four atoms of iodine.

tissue An aggregation of specialized cell types that are similar in form and function.

Tolerable Upper Intake Level (UL) The highest level of chronic intake of a nutrient thought to be not detrimental to health.

tolerance Response to high and repeated alcohol exposure that results in reduced alcohol-related effects.

total energy expenditure (TEE) Total energy expended or used by the body.

total fiber The combination of dietary fiber and functional fiber.

total iron-binding capacity (TIBC) A measure of iron status that relates to the total number of free (unbound) iron-binding sites on transferrin.

trabecular bone Inner, less dense layer of bone; contains the bone marrow.

trace mineral An essential mineral that is required in amounts less than 100 mg daily.

***trans* double bond (*trans*, across)** A carbon–carbon double bond in which the hydrogen atoms are positioned on opposite sides of the double bond.

***trans* fatty acid** A fatty acid containing at least one *trans* double bond.

transamination The process whereby an amino group is formed via the transfer of an amino group from one amino acid to another organic compound (an α-keto acid).

transcobalamin The protein that transports vitamin B₁₂ in the blood.

transcription The process whereby mRNA is made using DNA as a template.

transfer ribonucleic acid (tRNA) A form of RNA in the cytoplasm involved in gene translation.

transferrin A protein, produced in the liver, important for iron transport in the blood.

transferrin receptor Protein found on cell membranes that binds transferrin, allowing the cell to take up iron.

transient ischemic attack (TIA) A "ministroke" that is caused by a temporary decrease in blood flow to the brain.

transit time Amount of time between the consumption of food and its elimination as solid waste.

translation The process whereby amino acids are linked together via peptide bonds on ribosomes, using mRNA and tRNA.

triglyceride (also called triacylglycerol) (tri, three) A lipid composed of a glycerol molecule bonded to three fatty acids.

triiodothyronine (T₃) The more-active form of thyroid hormone; contains three atoms of iodine.

trypsin, chymotrypsin, elastase, and **carboxypeptidase** Active enzymes (proteases) involved in protein digestion in the small intestine.

trypsinogen, chymotrypsinogen, proelastase, and **procarboxypeptidase** Inactive proenzymes produced in the pancreas and released into the small intestine in response to CCK.

tumor A growth; sometimes caused by cancer.

type 1 diabetes Previously known as juvenile-onset diabetes and as insulin-dependent diabetes mellitus, this form of diabetes results when the pancreas is no longer able to produce insulin due to a loss of insulin-producing β-cells.

type 1 osteoporosis The form of osteoporosis that occurs mostly in women; caused by hormone-related bone loss.

type 2 diabetes Previously known as adult-onset diabetes and as non–insulin-dependent diabetes mellitus, this form of diabetes results when insulin-requiring cells have difficulty responding to insulin.

type 2 osteoporosis The form of osteoporosis that occurs in men and women; caused by age-related and lifestyle factors.

U

ulcerative colitis A type of inflammatory bowel disease (IBD) that causes chronic inflammation of the colon.

umami A taste, in addition to the four basic taste components, that imparts a savory or meat-like flavor.

undernutrition (or nutritional deficiency) Inadequate intake of one or more nutrients and/or energy.

unsaturated fatty acid A fatty acid that contains at least one carbon–carbon double bond in its backbone.

up-regulation In the context of protein synthesis, increased expression of a gene.

urbanization A shift in a country's population from primarily rural to urban regions.

urea A relatively nontoxic, nitrogen-containing compound that is produced from ammonia.

U.S. Centers for Disease Control and Prevention (CDC) A governmental agency that monitors the nation's health in order to prevent disease outbreaks.

U.S. Peace Corps A federally funded program that sends American volunteers to live and work with people in underdeveloped countries.

USDA Food Guides and Food Patterns Dietary recommendations developed by the USDA based on categorizing foods into "food groups."

V

variant Creutzfeldt-Jakob disease A form of Creutzfeldt-Jakob disease that may be caused by consuming BSE-contaminated foods.

vegan A type of vegetarian who consumes no animal products.

vegetarian A person who does not consume any or selected foods and beverages made from animal products.

vein A blood vessel that carries blood toward the heart.

ventilation rate Rate of breathing (breaths/minute).

venule Small blood vessel that branches off from veins.

very low food security Households that experience multiple indications of disrupted eating patterns due to inadequate resources or access to food, sometimes leading to reduced food intake.

very low density lipoprotein (VLDL) A lipoprotein made by the liver that contains a large amount of triglyceride; its major function is to deliver fatty acids to cells.

vesicular active transport An energy-requiring mechanism whereby large molecules move into or out of cells by an enclosed vesicle.

vigorous-intensity activity Physical exertion that causes profound increases in breathing, sweating, and heart rate.

villi (plural of *villus*) Small, finger-like projections that cover the inner surface of the small intestine.

visceral adipose tissue (VAT) Adipose tissue deposited between the internal organs in the abdominal area.

vitamin A deficiency disorder (VADD) A multifaceted disease resulting from vitamin A deficiency.

vitamin B$_{12}$ (cobalamin) A water-soluble vitamin involved in energy metabolism and the conversion of homocysteine to methionine.

vitamin B$_6$ A water-soluble vitamin involved in the metabolism of proteins and amino acids, the synthesis of neurotransmitters and hemoglobin, glycogenolysis, and regulation of steroid hormone function.

vitamin C (ascorbic acid) A water-soluble vitamin that has antioxidant functions in the body.

vitamin K deficiency bleeding A disease that occurs in newborn infants; characterized by internal bleeding (hemorrhage) from inadequate vitamin K.

W

waist circumference A measure used as an indicator of central adiposity.

wet beriberi A form of thiamin deficiency characterized by severe edema.

whole-grain foods Cereal grains that contain bran, endosperm, and germ in the same relative proportion as they exist naturally.

X

xerophthalmia A condition caused by vitamin A deficiency and characterized by serious damage to the cornea. Xeropthalmia can lead to blindness.

Z

zinc (Zn) An essential trace mineral involved in gene expression, immune function, and cell growth.

zinc finger Zinc containing three-dimensional structure of some proteins that allows them to regulate gene expression.

zoonutrient (also called zoochemical) A substance found in animal foods and thought to benefit human health above and beyond the provision of essential nutrients and energy.

zygote An ovum that has been fertilized by a sperm.

INDEX

Alzheimer's disease, 477
AMDRs. *See* Acceptable Macronutrient Distribution Ranges
American Academy of Pediatrics, 603, 610
American College of Sports Medicine (ACSM), 390 (Table 9.2)
American Diabetes Association, 155 (Table 5), 157
American Dietetic Association (ADA), 644
American Heart Association (AHA), 20 (Table 1.2), 22
American Society for Nutirtion (ASN), 644
Amino acids, 163
    athletes and, 190, 403–404 (Table 9.7), 407, 412
    catabolism, 280–281, 284–286 (Figure7.13)
    composition, 164 (Figure 5.1)
    deamination of, 184, 284–286
    DRIs for, 187
    essential/nonessential/conditionally essential, **5–6**, 163–164 (Table 5.1), 165
    fat converted from, 184–185
    functions of, 179–185
    gluconeogenesis from, 140, 184, 291–292
    glucose from, 117, 140, 184
    ketogenic, 292–294
    limiting, 165
    main components of, 163
    protein diversity from, 163
    protein turnover maintaining, 185
    RDAs for, 188 (Figure 5.10)
    small intestine, absorption of, 178–179
    transamination of, 284–286
    vitamin B for making nonessential, 434–435
Amino group, **163**
Amphibolic pathway, **273**, 283–284 (Figure 7.11)
Amphipathic, **231–232, 273**
Amylopectin, 123–124
    digestion of, 130
Amylose, 123–124
    digestion of, 130
AN. *See* Anorexia nervosa
Anabolic neurotransmitters, **326**, 348–349 (Figure 8.12)
Anabolic pathways, **272–273**, 281–282
    catabolic pathways compared to, 272–273
    energy metabolism, contributions of, 281–282, 295 (Figure 7.19), 400
Anaerobic capacity, 272–273, **401**
Anaerobic pathways, 272–273, 281–282, 395–400
Anal sphincters, **107**
Anaphylaxis, **179**
Anemia. *See also* Sickle cell anemia; Sports anemia
    iron deficiency causing, 562, 618
    folate, vitamin B and, 442
    hemolytic, **480**
    megaloblastic macrocytic, 442
    microcytic hypochromic, **435–436**, 563
    pernicious, **444**
Aneurysm, **254**
Angina pectoris, **256**
Angiogram, **256**
Angioplasty, **257**
Angiotensinogen, **520**
Angiotensinogen I, **520**
Angiotensinogen II, **520**
Animal studies, **19**
Anions, **72**
    formation of, 72–73 (Figure 3.3)
    magnesium stabilizing, 72, 518
Anisakis simplex, **205**
Anorexia nervosa, binge-eating/purging type, 366 (Table 1), 367, 371–372

Anorexia nervosa, restricting type, 366 (Table 1)
Anorexia nervosa (AN), **366**–369
    health concerns with, 369
    routines/rituals associated with, 367–369
    signs/symptoms/consequences of, 368 (Table 3)
    thoughts/behaviors associated with, 366 (Table 2), 369 (Figure 3)
    types of, 367 (Figure 1)
Anorexia of aging, 374, **627**
Anthropometric measurements, **34–35**
Antibiotics, 206
Antibodies, **182**
Antidiuretic hormone (ADH), **530**
Antioxidants
    iron and, 562
    nonprovitamin A carotenoids as, 467–468
    Se as, 571–572
    vitamin C as, 445–448
    vitamin E as, 477, 478–480
Antithiamin factors, 426
Aorta, **101**
Apoproteins, **243**, 244–245
Apoptosis, 489
Appetite, 329–331
Appropriate for gestational age (AGA) infant, **591**
Arachidonic acid, **226**
Ariboflavinosis, **429**
Arsenic, 507 (Figure 12.1), 578
Arteriole, **103**
Artery, **101**
Artesian water, 532
Ascites, 193
Ascorbic acid, **445**. *See also* Vitamin C
Aspartame, 122
Aspergillus, 201 (Table 1), 202
Assignment, random, **18**
Associations. *See* Correlations
Atherosclerosis, **254**
Athletes
    amino acids and, 190, 403–404 (Table 9.7), 407, 412
    calcium and, 409
    carbohydrate loading and, 407
    carbohydrates and, 114
    carnitine and, 287, 403 (Table 9.7)
    creatine supplements and, 389
    diet and, 405–413
    eating disorders and, 377–379
    endurance training for, 386, 390, 401–404
    ergogenic aids enhancing performance for, 397, 402–404 (Table 9.7)
    fluids and electrolytes required for, 409–411
    iron and, 408, 564–565
    lipids and, 407–408
    macronutrients required by, 406–408
    micronutrients required by, 408–409
    nutrition and post-exercise recovery, 411–413
    physical activity, energy consumption needed for, 345, 405–406 (Table 9.8)
    physiological changes during training, 401–404
    protein and, 188, 190
    strength training for, 401–404
    training for, 401–404
    water and, 409–411, 412, 531–532
    zinc and, 409
Atoms
    chemical bonds and, 73 (Figures 3.3, 3.4), 74 (Figures 3.5, 3.6)
    components of, 72 (Figure 3.2)
    defined, **72**
    oxidation and, 72–73 (Figure 3.4), 287–289
    reduction and, 73 (Figure 3.4)

ATP. *See* Adenosine triphosphate
ATP–creatine phosphate (ATP–CP) pathway, **396**–397 (Figure 9.2)
ATP synthesis, 183, 277–278, **279**
ATP–CP pathway. *See* ATP–creatine phosphate pathway
Atrophic gastritis, **627**
Atrophy, **401**
AUDs. *See* Alcohol use disorders
Autoimmune diseases, **24**, 570
Avidin, **437**

B

Babies, 228. *See also* Infancy
"Baby Boom Generation," 622–623 (Figure 14.11)
Baby bottle tooth decay, **612**
BAC. *See* Blood alcohol concentration
Backyard Harvest, 584
Bacteria, intestinal, 121, 125
Balance, 323–325 (Figure 8.2), 352–353. *See also* specific terms
Banting, Frederick, 147
Bariatric surgery, 327–328 (Figure 8.5)
Bariatrics, 327
Barker, David, 593
Basal energy expenditure (BEE), 332, 359
Basal metabolic rate (BMR), 331–333
Basal metabolism, **331**
Basic, **75**
Basolateral membrane, **100**
BCAA. *See* Branched-chain amino acids
B-complex vitamins, **421–422** (Table 10.2). *See also* Biotin; Folate; Niacin; Pantothenic acid; Riboflavin; Thiamin; Vitamin B_6; Vitamin B_{12}
    energy metabolism and, 424 (Figure 10.3)
    names and, 420–421 (Table 10.1)
    vegetarians and, 192
Beauty pageants, 375
BED. *See* Binge-eating disorder
BEE. *See* Basal energy expenditure
Benign tumor, **490**
Beriberi, **427**
Best, Charles, 147
Beta (β) carotene. *See* β-Carotene
Beta (β) glycosidic bond, **120**
Beta (β) oxidation, **287**–288 (Figure 7.15)
Bias, **17**–18
Bicarbonate (HCO_3^-) subunits, 438
Bile, 238–239
    defined, **99**
    digestion, role of, bile and, 86 (Table 3.2), 87 (Table 3.3), 96 (Figure 3.21), 97, 99 (Figure 3.24), 236–240
    lipid emulsification by, 237–239 (Figure 6.14)
Bile salt–dependent cholesteryl ester hydrolase, **240**
Binge drinking, **313**
Binge-eating disorder (BED), 371–372
Bingeing, 313, 366 (Table 1), 367, 371–372
Bioavailability, **101**, 409, 447, 510, 557–558
Biochemical measurements, 35–36
Bioelectrical impedance, 339–340
Biological marker (biomarker), **35**
Bioterrorism Act, **213**
Biotin (vitamin B_7), **436**–438
    carboxylation and, 438
    food sources of, 437
    raw eggs causing deficiency of, 437
    RDAs for, 438
    sources for, 437

Prostaglandins, **227**
Prosthetic groups, **172**–173 (Figure 5.5), 555–556
Protease, **177**
Protein(s), 161–196. *See also* Enzymes; Heme
   proteins; Prions; Retinol-binding protein
   Acceptable Macronutrient Distribution
      Ranges for (AMDRs), 189, 596
   amino acids and diversity of, 163–164
      (Figure 5.1, Table 5.1), 164, 165, 184–185
      (Figure 5.8), 187–188 (Figure 5.10)
   analysis, 195–196
   athletes and, 188, 190, 412, 407
   body protected with, 182
   catabolism, 181, 284–286
   cell signaling and, 167, 182
   complementation, 166
   complete, 165
   deficiency of, 193–194
   defined, **163**
   denaturation altering, **173**–174
   Dietary Guidelines for Americans on, 52
      (Table 2.4), 189
   digestion stages of, 176–179
   Dietary Reference Intakes for, 38–40, 187–188
      (Figure 5.10), 188, 189
   energy metabolism and, 183–184
      (Figure 5.8), 185
   as energy sources, 9, 284–286
   excess of, 194–195
   fats and, 191
   fluid balance regulated by, 182–183
      (Figure 5.7)
   functions of, 9, 166–169, 179–180 (Table 5.2),
      181–185, 173–174
   gene alteration and, 167, 174–176
   as glucose sources, 183–185
   health and, 175–176, 182, 193–195
   hormones as, 182
   in human milk, 604
   incomplete, 165
   infancy and, 193
   movement facilitated by, 181
   muscle, 181
   muscle recovery and, 412
   MyPlate recommendations, 189, 547–548
   nitrogen and, 186–187
   overview, 160
   pH scale and, 183
   primary structure of, 169–170 (Figure 5.3)
   prosthetic groups and, 172–173 (Figure 5.5)
   quality of, 166
   quaternary structure of, 172–173 (Figure 5.5)
   RDAs for, 189
   secondary structure of, 170, 172 (Figure 5.4)
   small intestine digestion of, 178–179
   stages of, 176–177 (Figure 5.6), 178–179
   stomach digestion of, 176–177 (Figure 5.6)
   structural materials provided by, 180
   structure of, 163–165, 169–170 (Figure 5.3),
      171–174, 180
   synthesis, 167 (Figure 5.2)
   tertiary structure of, 170–172
   translation, peptide chains and, 169
   transcription, transfer, 168–169 (Figure 5.2)
   transport service and, 181–182
   turnover, 185
   urea, 186 (Figure 5.9)
   vegetarian diets and, 191–192
Protein complementation, **166**
Protein turnover, **185**
Protein-energy malnutrition (PEM), **193**, 194,
   469–470, 641
Proteolysis, 184 (Figure 5.8), 185, 273 (Table 7.1),
   279_280 (Figure 7.7)

Prothrombin, **482**
Protons, **72**
Protozoa, **204**
Provitamin A carotenoids, **462**, 464, 467–468
   food sources of, 463–464
Psychological disorders, 343, 380
Psyllium, 127 (Table 4.3)
PTH. *See* Parathyroid hormone
Puberty, **619**
Public health organizations, 22
PubMed database, **21**
PUFAs. *See* Polyunsaturated fatty acids
Pulmonary arteries, **104**
Pulmonary circulation
   of blood, 104
   defined, **104**
   diagram of, 102 (Figure 3.26)
   endurance training and, 401
Pulmonary veins, **104**
Purging, **367**, 370
Purified water, 532
Purines, **440**
Pyloric sphincter, **90**
Pyridoxal phosphate (PLP), **434**
Pyrimidines, **440**
Pyruvate, **281**
   Cytochrome P450, **562**
   Acetyl-CoA converted from, 282–283
      (Figure 7.10), 307 (Figure 4)
   oxygen availability and, 282–283

Q

Qualified health claims, **64**
Quaternary structure, **172**–173 (Figure 5.5)

R

R protein, **443**
Race, cancer risk and, 494
Radiation,
   birth defects and, 588
   of foods, 6–7
RAE. *See* Retinol activity equivalent
Random assignment, **18**
Rate, **22**
Rating of perceived exertion, 393 (Table 9.5)
Raw eggs, 437
RDAs. *See* Recommended Dietary Allowances
Reabsorption, **105**
Reactive hypoglycemia, **140**
Rebound weight gain, **333**
Recommended Dietary Allowances (RDAs),
      38–39 (Table 2.1)
   for amino acids, 187–188 (Figure 5.10)
   for biotin, 438
   for calcium, 514–515, 597
   for carbohydrates, 142, 596
   for carotenoids, 469
   for chromium, 447, 573
   defined, **38**, **42**
   for folate, 442, 598
   for iodine, 570
   for iron, 597–598, 564–565
   for lipids, 250
   for magnesium, 518, 574
   for sodium, 264, 522
   for niacin, 432
   for pantothenic acid, 434
   for phosphorus, 516
   for protein, 188, 596
   for riboflavin, 429
   for thiamin, 427
   for vitamin A, 469, 597
   for vitamin B$_6$, 596, 445, 597

   for vitamin B$_{12}$, 445, 597
   for vitamin C, 449, 605
   for vitamin D, 477
   for vitamin E, 480
   standards for, **38**–39 (Table 2.1)
Rectum, **107**
Red blood cells, 104 (Figure 3.27), 105
      (Figure 3.28), 170, 171, 287, 408
Red meat, cancer and, 194–195, 499
Red tide, **206**
Red wine, 309
Redox reactions, **73**
Reduction, **73**
Reduction–oxidation (redox) reactions, 73
      (Figure 3.4)
   copper and, 566
   defined, **73**
   niacin and, 432
   riboflavin assisting, 429
REE. *See* Resting energy expenditure
Regular health claims, **64**
Renin, **520**–521 (Figure 12.8)
Reproduction, 466. *See also* Pregnancy
Researcher
   bias, **17**
   credibility, 20–21
Resistance training, 390
Resorption, **512**
Resting energy expenditure (REE), **332**, 359
Resting metabolic rate (RMR), **332**
Restrained eaters, 372
Resveratrol, **309**
Retina, **466**
Retinoids, 462
Retinol activity equivalent (RAE), **463**
Retinol-binding protein, **464**
Retrospective dietary assessment, **36**–37
Reverse cholesterol transport, **245**
R-group, **163**
Rhodopsin, **466**
Riboflavin (B$_2$), **428**, 600 (Figure 14.6)
   ariboflavinosis from deficiency of, 429
   food with, 428–429
   RDAs for, 429
   reduction–oxidation reactions assisted
      with, 429
   sources of, 428 (Figure 10.4)
Ribonucleic acid (RNA). *See also* Messenger
      ribonucleic acid; Transfer ribonucleic
      acid
   folate and synthesis of, 442, 448, 451
   zinc and, 576
Ribosomes, **169**
Rickets, **475**–476
Risk factors, **26**
   cardiovascular disease, 258–259
   cancer, 492–495, 499
   chronic diseases and, 26
   diabetes and, 151 (Table 3)
   osteoporosis and, 546–547
RMR. *See* Resting metabolic rate
Rods, **466**
Rugae, **90**

S

Saccharin, 122
Saliva, **88**
Salivary α-amylase, **130**
Salmonella, 200 (Table 1)
Salt. *See also* Sodium chloride
   cancer prevention, limiting, 499–500
   hypertension and, 260, 522
   iodine deficiency and fortification of, 568

Chapter Summary

What Do We Mean by "Nutrition"?

- The term nutrition refers to how the body uses substances in foods to promote and sustain life.
- Essential nutrients must be consumed, whereas we can make sufficient amounts of the nonessential nutrients when needed; conditionally essential nutrients are needed by subsets of the population.
- Water, carbohydrates, proteins, and lipids are macronutrients; vitamins and minerals are micronutrients.
- Phytochemicals and zoonutrients are health-promoting compounds found in plant- and animal-derived foods, respectively. Foods that contain enhanced amounts of essential nutrients, phytochemicals, or zoonutrients are called functional foods.

What Are the Major Nutrient Classes?

- Carbohydrates, proteins, and lipids all provide energy. They are referred to as energy-yielding nutrients and have many structural and regulatory functions in the body.
- Water serves as the medium in which all chemical reactions occur, helps eliminate waste products, regulates body temperature, and provides insulation and protection.
- Vitamins are either water or lipid soluble and serve many purposes, mostly having to do with regulation of chemical reactions; essential minerals have many specific functions regarding structure, regulation, and energy metabolism.

How Do Foods Provide Energy?

- Energy is the capacity to perform work, and the body transforms the chemical energy in foods to a usable form called adenosine triphosphate (ATP).
- Carbohydrates, proteins, lipids, and alcohol provide 4, 4, 9, and 7 kcal per gram, respectively.

How Is Nutrition Research Conducted?

- Most research is conducted using a three-step process called the scientific method, which involves making an observation, generating an explanation (or hypothesis), and testing the explanation by conducting a study.
- Epidemiologic studies investigate associations or correlations, whereas intervention studies can test causal relationships; many techniques decrease study bias, including randomization, control groups, placebos, and blinding.
- Sometimes it is not possible or practical to test a hypothesis using humans as participants. In these cases, researchers turn to animal models or cell culture systems.

Are All Nutrition Claims Believable?

- Publication in a peer-reviewed journal indicates the information is probably reliable.
- Although most research is not likely to be influenced by the source of funding, it is possible for the funding agency to have biased the conclusions made by the investigators.
- It is important to consider whether the design of the study was appropriate and whether major public health groups support the study conclusions before applying them to your life.

Nutrition and Health: What Is the Connection?

- Over the past century, the primary public health concerns have shifted from infectious disease and nutritional deficiencies to chronic disease and overnutrition. These shifts are reflected in morbidity and mortality rates of various diseases as well as life expectancy and longevity.
- The shift from undernutrition to overnutrition or unbalanced nutrition often occurs as a society becomes more industrialized; this is called the nutrition transition.

Review Questions

What Do We Mean by "Nutrition"? (pp. 5–8)

1. What is a nutrient, and what is meant by the term *nutrition*? Why is the meaning of the term *nutrient* evolving?

2. Which of the following are NOT considered essential nutrients?
   a. lipids
   b. water-soluble vitamins
   c. phytochemicals
   d. minerals

3. Explain what the difference is between organic compounds and organic foods. Do all experts agree that organic foods are more nutritious than those that are conventionally grown?

What Are the Major Nutrient Classes? (pp. 8–10)

4. Name the six nutrient classes. Which are organic? Which are energy yielding?

5. Which macronutrient class constitutes the largest proportion of body weight?
   a. carbohydrates
   b. proteins
   c. lipids
   d. water

How Do Foods Provide Energy? (pp. 10–13)

6. How many kilocalories would be obtained from a food containing 45 grams fat, 50 grams carbohydrate, and 4 grams protein? If you were to eat one serving of this food, what percentage of calories would you get from the macronutrient classes?

How Is Nutrition Research Conducted? (pp. 13–19)

7. What are the components of the scientific method, and in which order do they proceed?

8. Explain the major differences between epidemiologic and intervention studies in terms of how they are carried out and what kinds of conclusions can be drawn from them. What are two strengths and weaknesses of each type of study design?

9. Hawthorne effect is defined as which of the following?
   a. bias introduced when the researcher knows to which group the subjects have been assigned
   b. an effect that occurs because the subject believes that the treatment will be effective
   c. an effect that occurs because a subject changes his or her behavior due to the study
   d. results are valid because the intervention was implemented correctly

10. Explain two situations in which researchers might choose to conduct a cell culture study instead of a human intervention trial.

Are All Nutrition Claims Believable? (pp. 19–22)

11. Why are peer-reviewed journals considered some of the best sources of scientific findings? What other primary sources are considered credible in this regard?

12. What is a randomized, double-blind, placebo-controlled study, and why is this considered the "gold standard" in nutrition research?

Nutrition and Health: What Is the Connection? (pp. 22–28)

13. The _____ of a disease is the number of new cases in a given year.
    a. mortality rate
    b. prevalence
    c. incidence
    d. longevity

14. Life expectancy has _____ in the United States during the past century.
    a. increased
    b. decreased
    c. not changed

15. Poor nutrition is considered a(n) _____ risk factor for many chronic diseases.
    a. lifestyle
    b. environmental
    c. genetic
    d. all of the above

16. What is meant by the term *nutrition transition*? How has this phenomenon influenced the types of diseases common in the United States? In developing countries?

Chapter Summary

What Do We Mean by "Nutritional Status"?

- A person's nutritional status depends on whether he or she is getting sufficient amounts of nutrients and energy to support optimal health. Nutritional adequacy supports optimal physiological function, whereas under- and over-nutrition (types of malnutrition) do not.
- Primary malnutrition is due to inadequate diet, whereas secondary malnutrition may be caused by other factors such as illness.
- Nutrient requirements vary greatly among individuals, and are influenced by genetic, lifestyle, and environmental factors.

How Is Nutritional Status Assessed?

- Nutritional status can be assessed using anthropometric measurements (body dimensions and composition), chemical analysis of biological samples, clinical assessment, and dietary assessment.

How Much of a Nutrient Is Adequate?

- The Institute of Medicine has formulated a set of standards called the Dietary Reference Intakes (DRIs) that include the Estimated Average Requirements (EARs), Recommended Dietary Allowances (RDAs), Adequate Intake Levels (AIs), and Tolerable Upper Intake Levels (ULs).
- EARs estimate average nutrient requirements in various population groups and should be used to assess dietary intake in a population.
- RDAs, based on EARs, were established to be used as recommended intakes for individuals.
- AIs are used when the Institute of Medicine could not establish EARs and RDAs.
- ULs represent the upper range of nutrient intake values that is considered safe.
- Estimated Energy Requirement (EER) equations and Acceptable Macronutrient Distribution Ranges (AMDRs) provide guidance as to total energy intake and distribution of energy intake from the macronutrients.

How Can You Easily Assess and Plan Your Diet?

- The U.S. government has been providing advice concerning dietary planning and assessment for over a century.
- These recommendations are currently provided as the Dietary Guidelines for Americans, which are released every five years.

2010 Dietary Guidelines for Americans: Our Current Recommendations

- The 2010 Dietary Guidelines for Americans provide science-based nutritional recommendations to promote optimal health and reduce the risk of chronic disease in the United States.
- The Food Patterns form the basis of MyPlate, a graphic representation of its food groups and recommended intakes.
- The MyPlate website includes a wide variety of information, including personal plates and the Food Tracker feature, which allows you to conduct a relatively thorough dietary self-assessment.
- Healthy People 2020 outlines our nation's goals for attaining optimal health in the population.

How Can You Use Food Labels to Plan a Healthy Diet?

- Food labels provide consumers with useful information to help make food choices.
- Using the Daily Values (DVs) and % Daily Values (%DVs) can help you compare foods and choose those that fit your dietary goals.
- Nutrient content claims, structure/function claims, and health claims are valuable tools when planning a healthy diet.

Can You Put These Concepts into Action?

- You should now be able to use a variety of tools (e.g., the Dietary Reference Intakes, 2010 Dietary Guidelines for Americans, Nutrition Facts labels, MyPlate) to both evaluate your diet and begin to make better food choices.
- However, remember that along with conducting your own dietary self-assessment, you should also consult with appropriate health care providers concerning your overall health.

Review Questions

What Do We Mean by "Nutritional Status"? (pp. 33–34)

1. Which of the following situations is an example of *malnutrition*?
   a. when a person consumes too much of a nutrient
   b. when a person consumes spoiled food and becomes ill
   c. when a person does not eat enough of one or more essential nutrients
   d. both a and c

2. Malnutrition caused solely by dietary inadequacy is called _____ malnutrition.
   a. primary
   b. secondary
   c. simple
   d. complex

3. Why might one person's nutrient requirements differ from those of another person? List four factors that can influence a person's nutrient requirements.

How Is Nutritional Status Assessed? (pp. 34–38)

4. List the four basic methods by which nutritional status is assessed, and when each type might be used. What are their benefits and drawbacks?

5. Which of the following represents a challenge in using a 24-hour recall to assess nutritional status?
   a. A person must write down everything he or she eats while consuming it.
   b. A biological sample, such as blood, must be collected.
   c. It is difficult to remember the portion sizes that were consumed.
   d. A person must collect samples of everything he or she eats for 24 hours.

How Much of a Nutrient Is Adequate? (pp. 38–46)

6. Describe the four categories of Dietary Reference Intake (DRI) values. Which of these is/are meant to be used as dietary intake goals? How are these sets of standards related to each other?

7. If an EAR could not be established for a certain nutrient, AIs were put forth instead of RDAs. Why was this so? What is the difference between an AI and an RDA?

8. Which of the following is the Acceptable Macronutrient Distribution Range (AMDR; % of calories) for dietary fat?
   a. 20–35%
   b. 10–35%
   c. 45–50%
   d. 45–65%

9. The Acceptable Macronutrient Distribution Range (AMDR) for carbohydrates is 45 to 65% of calories. If your estimated caloric requirement is 2,000 kcal/day, how many kilocalories should come from carbohydrates? What dietary changes might you make if your carbohydrate intake is too high? Would there be any potential negative outcomes of decreasing carbohydrate intake? What would they be?

10. Which of the following should be used as a nutrient intake goal for *populations*?
    a. EAR
    b. RDA
    c. AI
    d. UL

11. Using the following equation, calculate the Estimated Energy Requirement (EER) for a 21-year-old person who weighs 150 pounds and is 5 feet, 10 inches tall. Assume an "active" physical activity level with a Physical Activity (PA) value of 1.25.
    - $EER = 662 - [9.53 \times age\ (y)] = PA \times [15.91 \times wt\ (kg) + 539.6 \times ht\ (m)]$

12. Consider a relatively healthy woman who consumes 50 mg of vitamin E each day. The Estimated Average Requirement (EAR) for this nutrient is 12 mg/day; the Recommended Dietary Allowance (RDA) is 15 mg/day; and the Tolerable Upper Intake Level (UL) is 1,000 mg/day. What percentage of her RDA is she consuming? Should she be concerned about her vitamin E intake? Explain.

How Can You Easily Assess and Plan Your Diet? (pp. 46–48)

13. Why has the U.S. Department of Agriculture for decades put forth recommendations for optimal food intake. In general, how have these changed over the past century?

2010 Dietary Guidelines for Americans: Our Current Recommendations (pp. 48–58)

14. List four strategies put forth in the 2010 Dietary Guidelines for Americans to help people balance calories to maintain weight. Why is this especially important today?

15. According to the current Dietary Guidelines, which foods (or components therein) should we try to reduce? Which ones should we try to eat more of? Why?

How Can You Use Food Labels to Plan a Healthy Diet? (pp. 58–64)

16. Which of the following represents the type of statement *with the greatest scientific evidence*, and who approves these statements?
    a. regular health claim, approved by the U.S. Department of Agriculture
    b. qualified health claim, approved by the U.S. Department of Agriculture
    c. regular health claim, approved by the U.S. Food and Drug Administration
    d. qualified health claim, approved by the U.S. Food and Drug Administration

17. For a food to be considered "an excellent source" of a nutrient, how much of your daily requirement must be provided by a single serving?
    a. 0–5%
    b. 10–15%
    c. 15–19%
    d. ≥20%

Can You Put These Concepts into Action? (pp. 64–66)

18. Using the food composition table that accompanies this text or other sources, calculate the vitamin B_{12} content of a lunch containing a hamburger on a white bun (with ketchup and mayonnaise), a small order of french fries, and a large cola soft drink. What percentage of your RDA of vitamin B_{12} does this represent? What foods could be added to this lunch to increase its vitamin B_{12} content?

Chapter Summary

How Does Chemistry Apply to the Study of Nutrition?

- Atoms consist of subatomic particles—neutrons, protons, and electrons.
- When the number of protons (particles with positive charges) equals the number of electrons (particles with negative charges), an atom is neutral.
- Atoms that have unequal numbers of protons and electrons are called ions. Ions with positive charges are called cations, and ions with negative charges are called anions.
- Atoms that gain electrons become reduced, whereas those that lose electrons become oxidized.
- Molecules result when chemical bonds join together two or more atoms. Molecules composed of a single type of atom are called elements. Molecules composed of two or more different atoms are called compounds.
- Acidic fluids have more hydrogen ions than hydroxide ions and have a pH less than 7. Basic fluids have more hydroxide ions than hydrogen ions and have a pH greater than 7. Fluids with a pH of 7 are neutral.

How Do Biological Molecules Form Cells, Tissues, Organs, and Organ Systems?

- Transport mechanisms that do not require energy (ATP) (simple diffusion, facilitated diffusion, and osmosis), are called passive transport systems, whereas mechanisms that require energy are called active transport systems. Active transport systems include carrier-mediated active transport and vesicular transport.
- There are four tissue types (epithelial, connective, muscle, and neural), which collectively carry out functions such as movement, communication, protection, and structure.

How Does the Digestive System Break Down Food into Absorbable Components?

- The digestive system consists of the gastrointestinal (GI) tract and accessory organs.
- The three functions of the digestive system are the chemical and physical breakdown of food (digestion), the transfer of nutrients from the GI tract into the blood or lymphatic circulatory systems (absorption), and the formation and elimination of feces from the body (egestion).
- The GI tract consists of four tissue layers—mucosa, submucosa, muscularis, and serosa.
- Secretions needed for digestion include water, hydrochloric acid, electrolytes, mucus, salts, bicarbonate, hormones, and enzymes.

How Do Gastrointestinal Motility and Secretions Facilitate Digestion?

- Sphincters located throughout the GI tract regulate the flow of the luminal contents from one organ to the next.
- Segmentation both mixes and slowly propels food, while peristalsis involves more vigorous propulsive movements that move the food mass through the GI tract.
- Secretions needed for digestion are produced and released from specialized cells in the mucosal lining of the GI tract and from accessory organs.
- GI motility and the release of secretions are regulated by neural and hormonal signals.

How Does the Gastrointestinal Tract Coordinate Functions to Optimize Digestion and Nutrient Absorption?

- The cephalic phase of digestion is initiated by the central nervous system in anticipation of food, and the gastric phase begins with the arrival of food in the stomach. GI motility and the release of GI secretions increase during these phases. The intestinal phase begins when food moves into the small intestine, and the rate of gastric motility and the release of gastric juice decreases.
- After swallowing, food moves down the esophagus and enters the stomach, passing through the gastroesophageal sphincter.
- The hormone gastrin is released when food enters the stomach, stimulating the release of gastric juice. Food mixes with gastric juice and is then referred to as chyme.
- Secretions from the pancreas (pancreatic juice) and gallbladder (bile) are released into the small intestine, and facilitate the process of digestion.
- Nutrient absorption is the process whereby nutrients are transported into and then out of the enterocytes, and taken up into circulation.
- Nutrients cross cell membranes by a variety of passive and active transport mechanisms.
- Water-soluble nutrients absorbed from the digestive tract are circulated in the blood directly to the liver via the hepatic portal vein. Fat-soluble nutrients are taken up into lacteals and circulated by the lymphatic system.

How Does the Body Circulate Nutrients and Eliminate Cellular Waste Products?

- The delivery of nutrients and oxygen to cells is accomplished by the cardiovascular system.
- The cardiovascular system consists of the systemic circulation and the pulmonary circulation. The systemic blood circulates between the heart, the body organs, and back again. The pulmonary route circulates blood between the heart and lungs.
- The process of metabolism generates cellular waste products (e.g., carbon dioxide, urea, and water) that are removed from the body primarily by the lungs, kidneys, and the skin.
- The urinary system is the primary site for eliminating cellular waste products other than carbon dioxide.
- The kidneys remove cellular waste products from the blood as it flows through filtering units called nephrons. Water and other essential substances are returned to the blood by a process called reabsorption.

What Is the Role of the Large Intestine?

- The large intestine, which consists of the cecum, colon, rectum, and anal canal, prepares undigested food residue for elimination as feces by removing water.
- Bacteria in the large intestine help form feces by breaking down undigested food residue.
- The rectum serves as a holding chamber for feces until the body eliminates it.

Review Questions

How Does Chemistry Apply to the Study of Nutrition? (pp. 71–76)

1. Subatomic particles associated with atoms include all of the following *except*
   a. protons.
   b. free radicals.
   c. neutrons.
   d. electrons.

2. The molecular formula for glucose is $C_6H_{12}O_6$. This means that
   a. glucose consists of three different types of atoms.
   b. glucose consists of 6 carbon, 12 hydrogen, and 6 oxygen atoms.
   c. glucose is a molecule.
   d. all of the above are true.

3. When molecule A transfers electrons to molecule B,
   a. molecule A is reduced.
   b. molecule A is oxidized.
   c. molecule B is oxidized.
   d. molecule B becomes a cation and molecule A becomes an anion.

How Do Biological Molecules Form Cells, Tissues, Organs, and Organ Systems? (pp. 76–81)

4. Which of the following is NOT an example of a passive transport mechanism?
   a. simple diffusion
   b. osmosis
   c. facilitated diffusion
   d. endocytosis

5. The net movement of water molecules across cell membranes is determined by the concentration of _____ in a solution.
   a. solutes
   b. protons
   c. electrons
   d. cytoplasm (cytosol)

6. The inner lining of organs, ducts, and cavities is made of
   a. connective tissue.
   b. muscle tissue.
   c. neural tissue.
   d. epithelial tissue.

7. Glands release _____ into the blood, which play a very important role in communication and homeostasis.
   a. enzymes
   b. hormones
   c. ions
   d. buffers

How Does the Digestive System Break Down Food into Absorbable Components? (pp. 81–84)

8. Which of the following organs is NOT a part of the digestive tract?
   a. the stomach
   b. the small intestine
   c. the pancreas
   d. the esophagus

9. List and describe the four tissue layers making up the GI tract.

How Do Gastrointestinal Motility and Secretions Facilitate Digestion? (pp. 84–87)

10. The flow of material through the GI tract is regulated by
    a. the gallbladder.
    b. the pancreas.
    c. the epiglottis.
    d. sphincters.

11. Which of the following is a type of GI motility?
    a. segmentation
    b. condensation
    c. emulsification
    d. egestion

12. List the organs that produce and release secretions needed for digestion. For each secretion, describe its role in the process of digestion.

How Does the Gastrointestinal Tract Coordinate Functions to Optimize Digestion and Nutrient Absorption? (pp. 88–101)

13. Which of the following organs releases the hormone gastrin?
    a. the pancreas
    b. the small intestine
    c. the gallbladder
    d. the stomach

14. The flow of material from the stomach into the small intestine is regulated by the
    a. epiglottis.
    b. ileocecal sphincter.
    c. pyloric sphincter.
    d. pharynx.

15. Structures called villi are found in the
    a. stomach.
    b. small intestine.
    c. large intestine.
    d. all of the above

16. Chyme is neutralized by a secretion released from the
    a. small intestine.
    b. gallbladder.
    c. pancreas.
    d. stomach.

17. Describe the digestive events that take place in the mouth, stomach, and small intestine.

18. Describe the inner surface of the small intestine and how this surface facilitates the process of nutrient absorption.

How Does the Body Circulate Nutrients and Eliminate Cellular Waste Products? (pp. 101–106)

19. Upon absorption, nutrient-rich blood leaves the small intestine and circulates to the
    a. gallbladder.
    b. lungs.
    c. large intestine.
    d. liver.

20. Which of the following is NOT part of the systemic circulation?
    a. lymph
    b. arteries
    c. veins
    d. capillaries

21. Which organ(s) filter the blood and assist in the excretion of metabolic waste products in the urine?
    a. the pancreas
    b. the gallbladder
    c. the kidneys
    d. all of the above

What Is the Role of the Large Intestine? (pp. 106–111)

22. The term used to describe bacteria residing in the large intestine is
    a. chyme.
    b. bolus.
    c. macrophages.
    d. microbiota.

23. Which of the following is NOT part of the large intestine?
    a. colon
    b. cecum
    c. duodenum
    d. rectum

24. Describe the events that take place once undigested food matter enters the large intestine until its elimination through the anal canal.

Chapter Summary

What Are Simple Carbohydrates?

- Monosaccharides (containing one sugar) and disaccharides (containing two sugars) are classified as simple sugars.
- Glucose, fructose, and galactose are monosaccharides, whereas lactose, maltose, and sucrose are disaccharides.
- Plants combine carbon dioxide and water in the presence of sunlight to make glucose. This process is called photosynthesis.
- Fruits, vegetables, and milk contain naturally occurring simple sugars, whereas many processed foods contain added sugars.

What Are Complex Carbohydrates?

- Polysaccharides (starch, glycogen, and fiber) and oligosaccharides are classified as complex carbohydrates.
- The two forms of starch found in plants are amylose and amylopectin.
- Amylose consists of glucose molecules arranged in a linear (unbranched) arrangement, whereas glucose molecules in amylopectin are in a branched arrangement.
- Glycogen, a storage form of glucose found mainly in liver and skeletal muscle, consists of glucose molecules bonded together in a highly branched arrangement.
- The term dietary fiber describes a diverse group of plant substances that are not digested or absorbed in the human small intestine.
- Dietary fiber is found naturally in foods, whereas functional fiber is added to foods.

How Are Carbohydrates Digested, Absorbed, and Circulated in the Body?

- Carbohydrate digestion requires enzymes produced in the salivary glands, pancreas, and small intestine.
- Starch digestion begins in the mouth, but mostly occurs in the small intestine via pancreatic α-amylase. Digestion is completed by the intestinal enzyme maltase, resulting in the release of glucose.
- In addition to pancreatic α-amylase, the digestion of amylopectin also requires the enzyme α-dextrinase to hydrolyze α-1,6 glycosidic bonds.
- Disaccharides are digested by the disaccharidases, which are enzymes produced by enterocytes.

- Glycemic response is the rise in blood glucose that occurs in response to eating.
- Glycemic index and glycemic load are rating systems used to assess the glycemic response to specific foods.

How Do Hormones Regulate Blood Glucose and Energy Storage?

- The pancreatic hormones insulin and glucagon regulate blood glucose and energy storage.
- The release of insulin increases when blood glucose levels are high. Insulin lowers blood glucose levels by (1) enabling glucose to cross cell membranes and (2) stimulating the conversion of glucose to glycogen and fatty acids. It also decreases the use of amino acids for glucose production (gluconeogenesis).
- The release of glucagon increases when blood glucose decreases.
- Glucagon stimulates the breakdown of liver glycogen and the subsequent release of glucose into the blood.
- During periods of stress, the hormones epinephrine and cortisol cause blood glucose levels to increase.
- When glycogen is depleted, the body generates glucose from noncarbohydrate sources via gluconeogenesis.
- The body decreases its need for glucose (and hence gluconeogenesis) by using ketones as an energy source.

How Much Carbohydrate Do We Require?

- The Institute of Medicine's Recommended Dietary Allowances (RDAs) for carbohydrates ensure that the brain has adequate glucose for its energy needs.
- People who exercise regularly are advised to consume enough carbohydrate to prevent fatigue and replenish glycogen stores.
- There is no Tolerable Upper Intake Level (UL) for any individual carbohydrate class or for total carbohydrates.
- The Institute of Medicine recommends that adults consume 14 g of dietary fiber per 1,000 kcal.
- The Institute of Medicine's Acceptable Macronutrient Distribution Range (AMDR) for carbohydrates is 45 to 65% of total energy.

Review Questions

What Are Simple Carbohydrates? (pp. 115–121)

1. Which monosaccharide is present in all three disaccharides (maltose, sucrose, and lactose)?
   a. galactose
   b. fructose
   c. glucose
   d. amylose

2. Which of the following carbohydrates is found in milk?
   a. fructose
   b. sucrose
   c. amylose
   d. lactose

3. The process whereby plants use the energy from the sun to combine carbon dioxide and water to form glucose is called
   a. radiation.
   b. photosynthesis.
   c. hydrolysis.
   d. nitrification.

4. List the two monosaccharides that make up sucrose, lactose, and maltose, respectively. Describe the chemical reaction that takes place when two monosaccharides react to form a disaccharide.

What Are Complex Carbohydrates? (pp. 121–129)

5. Plants store glucose in the form of _____, whereas animals store glucose in the form of _____.
   a. dietary fiber; amylose
   b. starch; maltose
   c. glycogen; amylopectin
   d. starch; glycogen

6. Which of the following carbohydrates consists of a highly branched arrangement of glucose molecules?
   a. lactose
   b. sucrose
   c. amylose
   d. glycogen

7. One difference between dietary fiber and amylose is that
   a. amylose is a monosaccharide, whereas dietary fiber is a polysaccharide.
   b. amylose has α-glycosidic bonds, whereas dietary fiber has β-glycosidic bonds.
   c. amylose is not found in plant foods, whereas dietary fiber is abundant in plants.
   d. amylose is a disaccharide, whereas dietary fiber is a monosaccharide.

8. Which of the following foods provides a good source of dietary fiber?
   a. dried beans
   b. milk
   c. cheese
   d. yogurt

9. Explain why humans are unable to digest dietary fiber.

10. What is the difference between dietary fiber and functional fiber?

How Are Carbohydrates Digested, Absorbed, and Circulated in the Body? (pp. 129–135)

11. Dextrins are formed during the digestion of
    a. lactose.
    b. dietary fiber.
    c. amylose and amylopectin.
    d. sucrose.

12. Foods eliciting a high glycemic response
    a. help lower the level of glucose in the blood.
    b. cause a rapid and large surge in blood glucose levels.
    c. do not contain glucose.
    d. are not easily digested.

13. Enzymes needed for disaccharide digestion are made by the
    a. pancreas.
    b. stomach.
    c. salivary glands.
    d. small intestine.

14. Describe the steps associated with the digestion of amylose and amylopectin to form glucose.

How Do Hormones Regulate Blood Glucose and Energy Storage? (pp. 135–142)

15. The hormone released from the pancreas that helps decrease blood glucose is called
    a. glucagon.
    b. epinephrine.
    c. insulin.
    d. all of the above

16. Cells with insulin-responsive glucose transporters include
    a. nerve tissue.
    b. red blood cells.
    c. skeletal muscle and adipose tissue.
    d. liver cells.

17. Which of the following is NOT true of insulin?
    a. Insulin promotes glycogen storage.
    b. Insulin promotes glycogenolysis.
    c. Insulin promotes fat synthesis.
    d. Insulin promotes muscle synthesis.

18. Explain how insulin lowers blood glucose and how glucagon increases blood glucose.

19. Why does ketone formation increase when glucose is limited?

How Much Carbohydrate Do We Require? (pp. 142–144)

20. The RDA for carbohydrates (130 g/day) is based on the amount of glucose needed by the
    a. liver.
    b. kidneys.
    c. skeletal muscles.
    d. brain.

21. The Dietary Reference Intakes (DRIs) recommend that healthy adults consume approximately 45 to 65% of their total energy as carbohydrates. Calculate the minimum amount of carbohydrates (in grams) a person should consume based on a total caloric intake of 1,800 kcal/day.

Chapter Summary

What Are Proteins?

- Proteins are made from amino acids linked together by peptide bonds.
- Amino acids contain a central carbon, a carboxylic acid group, an amino group, and a side chain, or R-group.
- The body needs 20 amino acids; 9 of these are essential nutrients and 6 are conditionally essential.

Are All Food Proteins Equal?

- Foods that contain all of the essential amino acids in the appropriate proportions are complete protein sources; those lacking or having low amounts of at least one of the essential amino acids are incomplete protein sources.
- Combining incomplete proteins so that all of the essential amino acids are consumed in appropriate amounts is called protein complementation.

How Are Proteins Made?

- Protein synthesis involves cell signaling, transcription, and translation.
- Chromosomes (made from deoxyribonucleic acid, DNA, and proteins) can be subdivided into genes, which contain information used to make proteins.
- Protein synthesis involves DNA, enzymes, messenger ribonucleic acids (mRNAs), ribosomes, transfer ribonucleic acids (tRNAs), and amino acids.

How Do Proteins Get Their Shapes?

- The sequence of amino acids making up a protein is called its primary structure. The primary structure folds into secondary and tertiary structures.
- Peptide chains can join together to form quaternary structures, and these can combine with prosthetic groups to form the protein's final three-dimensional shape.
- Disruption of a protein's shape is called denaturation and can influence the ability of a protein to function.

Genetics, Epigenetics, Nutrition, and Nutrigenomics

- Mutations in DNA (genes) can affect the primary structure of a protein, which in turn can influence the ability of the protein to function.
- Alterations in gene expression not involving changes to the DNA can also impact protein synthesis; these are referred to as *epigenetic* modifications.
- Nutrigenomics is the study of how nutrition interacts with a person's genetic and epigenetic makeup to influence protein synthesis and health.

How Are Dietary Proteins Digested, Absorbed, and Circulated?

- Protein digestion involves GI hormones (gastrin, secretin, and CCK) as well as a variety of proteases produced in the stomach, pancreas, and small intestine.
- Protein-digesting enzymes are released as proenzymes and then converted to their active protease forms in the GI tract.
- Amino acids are absorbed primarily along the duodenum and circulated in the blood to the liver.

What Are the Major Functions of Proteins and Amino Acids in the Body?

- Amino acids are used to synthesize proteins needed for structure, catalysis, movement, transport, communication, and protection.
- Amino acids are also used to produce ATP, regulate metabolic reactions, and synthesize nitrogen-containing nonprotein substances. In addition, amino acids can be transformed into fat.

Protein Turnover, Urea Excretion, and Nitrogen Balance

- Protein turnover represents the balance between protein synthesis and degradation.
- When protein (or nitrogen) intake exceeds loss, the body is in positive nitrogen balance. When nitrogen loss is greater than intake, the body is in negative nitrogen balance.
- Amino acids are degraded via deamination; the resultant ammonia is converted to urea.

How Much Protein Do You Need?

- It is recommended that healthy adults consume approximately 0.8 g of protein for each kilogram of body weight daily.
- The Institute of Medicine's Acceptable Macronutrient Distribution Ranges (AMDRs) recommend that adults obtain 10 to 35% of their energy intake from protein.

Vegetarian Diets: Healthier Than Other Dietary Patterns?

- Consuming adequate amounts of protein is typically not difficult for vegetarians, although they may be at greater risk for calcium, zinc, iron, and vitamin B_{12} deficiencies.

What Are the Consequences of Protein Deficiency?

- In children, protein-energy malnutrition (PEM) takes two forms: kwashiorkor and marasmus. Most adults with PEM have signs and symptoms associated with marasmus.
- Treatment of children and adults with PEM is multifaceted, involving a variety of therapies.

Protein Excess: Is There Cause for Concern?

- For most people, excess protein intake does not result in health complications.
- There is growing evidence that excessive consumption of red meat is associated with increased risk of colorectal cancer, but whether this is a causal relationship has not been established.

Review Questions

What Are Proteins? (pp. 163–165)

1. Which of the following is the basic "building block" of a protein?
   a. amino acid   c. phospholipid
   b. glucose   d. DNA
2. Which structure is "unique" to each amino acid?
   a. central carbon   c. amino group
   b. R-group   d. carboxylic acid group
3. Which amino acid becomes conditionally essential for people with phenylketonuria (PKU)?
   a. phenylalanine   c. tyrosine
   b. alanine   d. glutamine

Are All Food Proteins Equal? (pp. 165–166)

4. Protein complementation is important in some regions of the world where people consume limited amounts of animal products. Define protein complementation, and describe why it is especially important during periods of rapid growth and development.

How Are Proteins Made? (pp. 166–169)

5. Creating a genetically modified organism (GMO) can involve which of the following?
   a. modifying the genetic material of a plant
   b. modifying the genetic material of an animal
   c. altering the amino acid or protein content of a food
   d. all of the above

How Do Proteins Get Their Shapes? (pp. 169–174)

6. Describe the basic cause and complications of sickle cell anemia. Your answer should appropriately use the following terms: *genetic code, translation, malaria, iron, DNA, symptoms,* and *protein*.
7. Which level of protein structure is due to chemical bonding among amino acid side chains (R-groups)?
   a. primary   c. tertiary
   b. secondary   d. quaternary
8. The shape of a protein is very important for its structure. Describe the four levels of protein structure, and explain how mercury can influence the shape of a protein.

Genetics, Epigenetics, Nutrition, and Nutrigenomics (pp. 174–176)

9. Compare and contrast what is meant by the terms *genetics* and *epigenetics*. Does nutrient intake influence genetics? What about epigenetics?

How Are Dietary Proteins Digested, Absorbed, and Circulated? (pp. 176–179)

10. The chemical digestion of protein begins in the _____.
    a. mouth   c. pancreas
    b. stomach   d. large intestine

11. Cholecystokinin (CCK) is released in response to _____.
    a. insulin   c. secretin
    b. amino acids   d. stomach stretching

What Are the Major Functions of Proteins and Amino Acids in the Body? (pp. 179–185)

12. Gluconeogenesis is typically stimulated in connection with which of the following?
    a. high carbohydrate intake
    b. low protein intake
    c. high fat intake
    d. low carbohydrate intake
13. List four functions of proteins, providing specific examples of each.

Protein Turnover, Urea Excretion, and Nitrogen Balance (pp. 185–187)

14. Name two physiologic conditions in which a person would expect to be in positive protein balance?

How Much Protein Do You Need? (pp. 187–191)

15. The Acceptable Macronutrient Distribution Range (AMDR) for protein is 10 to 35% of calories. If your Estimated Energy Requirement (EER) is 2,000 kcal/day, how many calories should come from protein? How many grams of protein would this be? What dietary changes might you make if your protein intake was too high? What about if it was too low?
16. Using the food composition table that accompanies this text or other sources, calculate the protein content of a lunch containing a hamburger on a white bun, a small order of french fries, and a large cola soft drink. What percentage of your Recommended Dietary Allowance (RDA) of protein does this represent? What foods might be added to or substituted for the foods in this lunch to increase the protein content?

Vegetarian Diets: Healthier Than Other Dietary Patterns? (pp. 191–193)

17. Which form of protein-energy malnutrition is characterized by the presence of edema? Why does this occur?

What Are the Consequences of Protein Deficiency? (pp. 193–194)

18. Suboptimal functioning of which system may accompany protein deficiency?
    a. immune   c. circulatory
    b. reproductive   d. all of the above

Protein Excess: Is There Cause for Concern? (pp. 194–195)

19. Excessive consumption of which of the following protein sources is linked to increased risk of colorectal cancer?
    a. poultry   c. fish
    b. red meat   d. legumes

Chapter Summary

What Are Lipids?

- The three main classes of lipids are triglycerides, phospholipids, and sterols.
- Alpha (α) nomenclature designates the number of carbon atoms, the number and placement of double bonds, and the type of double bonds relative to the carboxylic acid (α) end.
- Omega (ω) nomenclature designates the placement of double bonds relative to the methyl (ω) end.
- Saturated fatty acids (SFAs) contain only carbon–carbon single bonds; monounsaturated fatty acids (MUFAs) have one carbon–carbon double bond; and polyunsaturated fatty acids (PUFAs) have two or more carbon–carbon double bonds.
- *Cis* double bonds have hydrogen atoms on the same side of the molecule, whereas *trans* double bonds have hydrogen atoms on opposite sides.

Which Fatty Acids Do We Need, and Where Do They Come From?

- The two essential fatty acids are linoleic acid and linolenic acid, which can be metabolized to longer-chain fatty acids and other compounds such as eicosanoids.
- Arachidonic acid and docosahexaenoic acid (DHA) are conditionally essential fatty acids during infancy.

Mono-, Di-, and Triglycerides: What's the Difference?

- Triglycerides consist of three fatty acids attached to a glycerol molecule via an ester linkage; monoglycerides have one fatty acid; and diglycerides have two fatty acids.
- During periods of energy need, the enzyme called *hormone-sensitive lipase* breaks down stored triglycerides, resulting in the mobilization of free fatty acids.
- During times of energy abundance, fatty acids are stored in adipose tissue as triglycerides.

What Are Phospholipids and Sterols?

- A phospholipid consists of a glycerol, two fatty acids, and a phosphate-containing polar head group.
- Phospholipids are amphipathic, meaning that one portion is hydrophobic and another is hydrophilic.
- Phospholipids make up cell membranes and are needed to circulate lipids in the body.
- Sterols are multi-ring compounds that have many roles in the body.
- Most sterols are bonded to a fatty acid, forming a sterol ester.
- Cholesterol is important as a component of cell membranes and a precursor for the synthesis of steroid hormones such as estrogen and testosterone.
- Cholesterol is made in the body and is found in many animal-based foods.
- Many experts recommend that we limit our consumption of cholesterol to decrease risk for cardiovascular disease.

- Phytosterols are sterols found in plants, some of which may reduce the risk for heart disease.

How Are Dietary Lipids Digested?

- Lipid digestion occurs in the mouth (via lingual lipase), stomach (via gastric lipase), and small intestine (via pancreatic lipase).
- Bile emulsifies large lipid globules into smaller droplets in the small intestine.
- Phospholipids are digested by the pancreatic enzyme *phospholipase A$_2$*.
- Cholesteryl esters are broken down by the pancreatic enzyme bile salt–dependent cholesteryl ester hydrolase.

How Are Dietary Lipids Absorbed and Circulated in the Body?

- Following digestion, lipids must be transported into the enterocytes.
- Small lipid molecules are circulated directly to the liver in the blood; large lipids are packaged into chylomicrons and circulated in the lymph.
- The enzyme *lipoprotein lipase* breaks down triglycerides in chylomicrons, which are then taken up by cells.

What Is the Role of Other Lipoproteins in Lipid Transport and Delivery?

- The liver produces very low density lipoproteins (VLDLs) that deliver fatty acids to cells.
- The loss of fatty acids from a VLDL results in its conversion to an intermediate-density lipoprotein (IDL) and ultimately to a low-density lipoprotein (LDL).
- LDLs deliver cholesterol to cells.
- The liver produces high-density lipoproteins (HDLs) that pick up cholesterol and return it to the liver.
- High levels of LDL and low levels of HDL are associated with increased risk of cardiovascular disease.

What Is the Relationship between Lipid Intake and Health?

- Evidence suggests that certain types of dietary fat (e.g., saturated and *trans* fatty acids) are associated with risk for cardiovascular disease and type 2 diabetes.
- Some types of dietary fat may play a role in development of cancer, although this effect may be due to the influence of dietary fat on obesity.

What Are the Dietary Recommendations for Lipids?

- The 2010 Dietary Guidelines and Institute of Medicine recommend that we consume 20 to 35% of our calories from lipid, limit intake of SFA to less than 7–10% of calories, choose foods low in *trans* fatty acids and cholesterol (200–300 mg/day), and emphasize foods high in PUFAs and MUFAs.

Review Questions

What Are Lipids? (pp. 219–225)

1. Which of the following best describes the physical and chemical characteristics of saturated fatty acids?
   a. They tend to be solids at room temperatures.
   b. They tend to be oils at room temperatures.
   c. They typically have low melting points.
   d. They are lower in calories than other fats.

2. A fatty acid with the nomenclature of *cis*9, *cis*12, *cis*15–18:3 has
   a. 9 carbons.
   b. 18 carbons.
   c. 9 double bonds.
   d. 15 carbons.

Which Fatty Acids Do We Need, and Where Do They Come From? (pp. 225–229)

3. Arachidonic acid and docosahexaenoic acid (DHA) are considered conditionally essential fatty acids for human infants because
   a. these fatty acids are important for infant health but not adult health.
   b. human milk contains very little of these fatty acids.
   c. babies can synthesize adequate amounts of them from other fatty acids.
   d. babies cannot synthesize adequate amounts of them from other fatty acids.

Mono-, Di-, and Triglycerides: What's the Difference? (pp. 229–231)

4. Monoglycerides have _____ fatty acid(s), whereas diglycerides have _____ fatty acid(s).
   a. 2; 3
   b. 1; 2
   c. 2; 1
   d. They have the same number of fatty acids.

5. The major form of lipid stored in adipose tissue is which of the following?
   a. phospholipid
   b. cholesterol
   c. monoglyceride
   d. triglyceride

What Are Phospholipids and Sterols? (pp. 231–236)

6. Which of the following lipids naturally contains cholesterol?
   a. beef tallow
   b. peanut oil
   c. coconut oil
   d. corn oil

7. Describe the basic structure of a phospholipid. Label the hydrophobic and hydrophilic components. Why is this compound said to be amphipathic? What are phospholipids used for in the body?

How Are Dietary Lipids Digested? (pp. 236–240)

8. Digestion of dietary triglycerides occurs in the _____.
   a. mouth
   b. stomach
   c. small intestine
   d. all of the above

9. The arrival of _____ in the small intestine signals the release of _____.
   a. lipids; cholecystokinin (CCK)
   b. VLDLs; lipoprotein lipase
   c. triglycerides; lingual lipase
   d. bile; hormone-sensitive lipase

How Are Dietary Lipids Absorbed and Circulated in the Body? (pp. 240–243)

10. Which of these types of lipids are circulated away from the small intestine within chylomicrons?
    a. phospholipids
    b. cholesterol
    c. long-chain fatty acids
    d. all of the above

What Is the Role of Other Lipoproteins in Lipid Transport and Delivery? (pp. 243–246)

11. Which of the following classes of lipoproteins is referred to as "good cholesterol"?
    a. chylomicron
    b. VLDL
    c. LDL
    d. HDL

12. The enzyme lipoprotein lipase is needed to deliver _____ to cells.
    a. phospholipids
    b. cholesterol
    c. chylomicrons
    d. fatty acids

What Is the Relationship between Lipid Intake and Health? (pp. 246–248)

13. In general, how is lipid intake related to obesity?

What Are the Dietary Recommendations for Lipids? (pp. 248–251)

14. What are the current recommendations for lipid intake for adults? Please refer to both the Acceptable Macronutrient Distribution Ranges (AMDRs) and the 2010 Dietary Guidelines in your answer.

15. The Acceptable Macronutrient Distribution Range (AMDR) for lipids is approximately 20 to 35% of the energy requirement. If you require 2,750 kcal/day, how many kilocalories should you obtain from lipids? How many grams of lipid does this represent?

Chapter Summary

What Is Energy Metabolism?

- Metabolic pathways are made up of interrelated, enzyme-catalyzed chemical reactions that can be categorized as catabolic, anabolic, or amphibolic.
- Anabolic pathways promote the synthesis of new compounds and energy storage, whereas catabolic pathways promote the mobilization of stored energy and the breakdown of energy-yielding nutrients. Amphibolic pathways provide intermediate products that can bridge catabolic and anabolic metabolic pathways.
- Chemical reactions are catalyzed by enzymes, some of which require cofactors or coenzymes to function.
- A cofactor is an inorganic substance that is part of the enzyme's structure, whereas coenzymes are organic substances that assist enzymes.
- The hormone insulin promotes energy storage, whereas the hormone glucagon promotes energy mobilization.

What Is the Role of ATP in Energy Metabolism?

- Cells rely on the energy contained in the chemical bonds of ATP to fuel their activities.
- Some ATP is generated by substrate phosphorylation, a process that adds a phosphate group (P_i) directly to ADP. Most ATP is synthesized by oxidative phosphorylation, which involves a series of chemical reactions that comprise the electron transport chain.
- When NADH + H^+ and $FADH_2$ enter the electron transport chain, their electrons and hydrogen ions are removed. Electrons pass along protein complexes, and the energy released is used to pump the hydrogen ions out of the mitochondrial matrix. The movement of hydrogen ions back into the mitochondrial matrix releases energy that is used by the enzyme ATP synthase to attach a phosphate group to ADP, generating ATP.

How Do Catabolic Pathways Release Stored Energy?

- Liver and muscle cells break down glycogen into glucose by a process called glycogenolysis.
- Glucose catabolism begins with glycolysis, which converts glucose to pyruvate. Oxygen availability determines whether pyruvate will be converted to acetyl-CoA or lactate. If oxygen is available, acetyl-CoA is formed and enters the citric acid cycle, resulting in the formation of NADH + H^+ and $FADH_2$. These coenzymes can enter the electron transport chain and drive ATP formation via oxidative phosphorylation.

- Protein is broken down to amino acids by proteolysis. For amino acids to be used as an energy source, the nitrogen-containing amino group is removed via transamination and deamination. The remaining structure (α-keto acid) can enter the citric acid cycle, ultimately generating ATP via oxidative phosphorylation.
- Lipid catabolism (triglycerides) begins with lipolysis, releasing fatty acids and glycerol. Fatty acids are oxidized via β-oxidation and the citric acid cycle to form NADH + H^+ and $FADH_2$, which then enter the electron transport chain to generate additional ATP.

How Do Anabolic Pathways Contribute to Energy Metabolism?

- Anabolic pathways play important roles in storing excess energy and in synthesizing energy-yielding molecules when glucose availability is limited.
- The hormone insulin stimulates liver and muscle tissues to store excess glucose as glycogen (glycogenesis), promotes the conversion of glucose and amino acids to fatty acids, and signals the subsequent production of triglyceride in adipose and liver tissues (lipogenesis).
- During starvation, anabolic pathways are used to synthesize glucose from noncarbohydrate sources (gluconeogenesis). Substances used for gluconeogenesis include glucogenic amino acids, lactate, and glycerol.
- High rates of gluconeogenesis deplete the amount of oxaloacetate. When this occurs, acetyl-CoA cannot participate in the citric acid cycle and is diverted to ketone production. Some cells are able to use ketones for ATP production.

How Is Energy Metabolism Influenced by Feeding and Fasting?

- Energy metabolism pathways are responsive to intermittent physiologic conditions referred to as the fed state, postabsorptive state, fasting state, and starvation state.
- Most of the major anabolic pathways operate during the absorptive state, including those promoting protein, triglyceride, and glycogen synthesis. During this time, glucose is the major source of energy for all tissues.
- The postabsorptive state, which is the period 4 to 24 hours after the last intake of food, relies heavily on energy supplied by the breakdown of stored energy reserves, especially glycogen.

Review Questions

What Is Energy Metabolism? (pp. 271–276)

1. Which of the following play a role in energy metabolism?
   a. coenzymes
   b. cofactors
   c. enzymes
   d. all of the above

2. _____ promotes the activity of anabolic pathways.
   a. Insulin
   b. Glucagon

What Is the Role of ATP in Energy Metabolism? (pp. 276–279)

3. The majority of ATP is generated by a metabolic process called
   a. substrate phosphorylation.
   b. proteolysis.
   c. oxidative phosphorylation.
   d. glycolysis.

4. When NADH + H⁺ and FADH₂ enter the electron transport chain, their electrons and hydrogen ions are removed. This is called
   a. electrification.
   b. hydrogenation.
   c. reduction.
   d. oxidation.

5. The end products of the electron transport chain are
   a. carbon dioxide and oxygen.
   b. oxygen and glucose.
   c. ATP and water.
   d. ADP and inorganic phosphate.

6. Describe two pathways used by cells to generate ATP.

How Do Catabolic Pathways Release Stored Energy? (pp. 279–289)

7. Which of the following nutrients can be used by cells to generate ATP?
   a. glucose
   b. fatty acids
   c. amino acids
   d. all of the above

8. Which of the following cannot provide cells with a source of glucose?
   a. fatty acids
   b. glycogen
   c. amino acids
   d. glycerol derived from triglycerides

9. Which of the following is associated with protein metabolism?
   a. proteolysis
   b. transamination
   c. deamination
   d. all of the above

10. The metabolic pathway called β-oxidation begins with _____ and ends with multiple molecules of _____.
    a. glucose; pyruvate
    b. fatty acids; pyruvate
    c. amino acids; acetyl-CoA
    d. fatty acids; acetyl-CoA

11. The citric acid cycle begins when _____ combine(s) with oxaloacetate to form citrate.
    a. ketones
    b. pyruvate
    c. acetyl-CoA
    d. water

12. The metabolic pathway that splits glucose to form two molecules of pyruvate is called
    a. gluconeogenesis.
    b. glycogenolysis.
    c. glycolysis.
    d. the citric acid cycle.

13. Provide a brief description of the stages of catabolism for: carbohydrates, proteins, and triglycerides.

How Do Anabolic Pathways Contribute to Energy Metabolism? (pp. 289–294)

14. Which of the following is *true* for gluconeogenesis?
    a. It results in the formation of triglycerides.
    b. It occurs when liver and muscle glycogen stores are full.
    c. It is the process whereby the amino group is removed from amino acids.
    d. It occurs when glucose availability is limited.

15. Glycogenesis results in the
    a. formation of triglycerides.
    b. formation of pyruvate.
    c. breakdown of glycogen to glucose.
    d. formation of glycogen.

16. During conditions of low glucose availability, cells depend on _____ to generate glucose.
    a. gluconeogenesis
    b. glycogenesis
    c. glycolysis
    d. lipogenesis

17. When glucose is limited, _____ provide(s) the brain with an alternative energy source.
    a. fatty acids
    b. ketones
    c. glycogen
    d. amino acids

18. Ketone production increases
    a. after a meal.
    b. when glucose is readily available.
    c. after a period of prolonged fasting.
    d. when glycogen stores are full.

19. Explain how the formation of ketones helps protect the breakdown of lean body mass during starvation.

How Is Energy Metabolism Influenced by Feeding and Fasting? (pp. 294–299)

20. Anabolic pathways are most active during which state?
    a. fed
    b. postabsorptive
    c. fasting
    d. starvation

21. Which of the following metabolic pathways is NOT very active during the postabsorptive state?
    a. glycogenolysis
    b. lipolysis
    c. lipogenesis

22. Describe the metabolic response following a meal and the metabolic response after 24 hours without food (fasting state).

Chapter Summary

What Is Energy Balance?

- Energy balance describes the overall relationship among energy intake, energy expenditure, and energy storage.
- When a person is in positive energy balance, energy intake exceeds energy expenditure, which results in weight gain.
- When a person is in negative energy balance, energy intake is less than energy expenditure, which results in weight loss.
- Adipose tissue serves as the body's primary energy reserve.
- When body fat increases, adipocytes can increase in size (hypertrophy) and/or number (hyperplasia). When body fat decreases, adipocytes decrease in size but not in number.
- Visceral adipose tissue (VAT) refers to adipose tissue deposited between the internal organs in the abdominal cavity, whereas subcutaneous adipose tissue (SCAT) is adipose tissue found directly beneath the skin.

What Determines Energy Intake?

- Hunger is the basic physiological need for food, whereas satiety is the sensation of having eaten enough.
- The brain releases catabolic and anabolic neurotransmitters that promote weight loss and weight gain, respectively.
- Signals from the GI tract play a role in regulation of short-term food intake and include gastric stretching and GI hormones. Circulating concentrations of glucose, fatty acids, and amino acids also influence hunger and satiety.
- Appetite reflects psychological factors that influence hunger and satiety. A food aversion is a strong psychological dislike of a particular food, whereas a food craving is a strong desire for a specific food.
- The majority of GI hormones inhibit food intake, with the exception of ghrelin, which stimulates hunger.

What Determines Energy Expenditure?

- The body expends energy to maintain basal metabolism, to perform physical activity, and to process food. These components make up a person's total energy expenditure (TEE).
- Smaller components of TEE include adaptive thermogenesis and nonexercise activity thermogenesis (NEAT).
- Direct and indirect calorimetry can be used to assess TEE. Other estimates of TEE are based on the use of stable isotopes and mathematical formulas.

How Are Body Weight and Body Composition Assessed?

- A person who is overweight has excess weight for his or her height, whereas an obese person has an abundance of body fat in relation to lean tissue.
- Anthropometric indices such as weight for height and body mass index (BMI) are used to assess body weight.
- Body mass index (BMI) is based on the ratio of weight to height squared and is considered a good indicator of body fat.
- Measures of body composition include hydrostatic weighing, dual-energy X-ray absorptiometry (DEXA), bioelectrical impedance, and skinfold thickness.
- The accumulation of body fat within the abdomen, called central adiposity, increases a person's risk of weight-related health problems.
- Waist circumference is used as an indicator of central adiposity.

How Does Lifestyle Contribute to Obesity?

- Many factors have contributed to the increasing prevalence of obesity in the United States, including increased energy intake and decreased energy expenditure.
- In addition to lifestyle choices, socioeconomic and cultural factors greatly affect a person's susceptibility to obesity.

Can Genetics Influence Body Weight?

- A gene mutation discovered in the *ob/ob* mouse prevents production of a hormone called leptin.
- A gene mutation discovered in the *db/db* mouse causes leptin resistance.
- Leptin (encoded for by the *ob* gene) is a satiety hormone produced mainly by adipose tissue that helps regulate body weight.
- Leptin deficiency is rare in humans, although researchers believe that some obese people may be somewhat unresponsive to leptin.
- In rare cases of leptin-deficiency in humans, leptin injections have resulted in profound weight loss.

How Does the Body Regulate Energy Balance and Body Weight?

- The "set point" theory of energy balance suggests that body weight is relatively stable because of the body's ability to adjust energy intake and expenditure.
- The hormone leptin plays a role in regulating body weight by communicating the body's energy reserve to the brain.
- When leptin concentrations decrease, anabolic neurotransmitters are released, resulting in overall increased energy intake and decreased energy expenditure. When leptin concentrations increase, catabolic neurotransmitters are released, resulting in overall decreased energy intake and increased energy expenditure.
- Most obese people produce adequate amounts of leptin, but defects in leptin signaling may contribute to obesity.

What Are the Best Approaches to Weight Loss?

- Maintaining weight loss requires lasting lifestyle changes, including eating reasonable amounts of healthy foods and regular exercise.
- Healthy weight-loss and weight-maintenance programs set reasonable goals; encourage intake of low-energy-dense, nutrient-dense foods; and promote regular physical activity.
- A realistic weight-loss goal is to decrease body weight by 5 to 10%, not exceeding one to two pounds each week.
- The majority of individuals who lose weight and maintain their weight loss do so by a combination of a healthy diet and physical activity.

Does Macronutrient Distribution Matter?

- Proponents of low-fat diets believe that less fat in the diet leads to consumption of fewer calories and to greater weight loss, whereas advocates of low-carbohydrate diets believe that high carbohydrate intake can cause insulin levels to rise, leading to weight gain.
- Limiting carbohydrates can cause the body to go into a ketotic state, which results in decreased appetite.

Review Questions

What Is Energy Balance? (pp. 323–325)

1. A person in positive energy balance experiences
   a. an increase in body weight.
   b. a decrease in body weight.
   c. a decrease in BMR.
   d. memory loss.

2. Subcutaneous adipose tissue
   a. is found exclusively in males.
   b. is found exclusively in females.
   c. refers to adipose tissue beneath the skin.
   d. is found only in the upper torso of the body.

3. When adipocytes undergo hypertrophy,
   a. their size decreases.
   b. they are embedded directly under the skin.
   c. their number increases.
   d. their size increases.

4. When a person loses body fat,
   a. the number of adipocytes decreases.
   b. adipocytes undergo a process called hypertrophy.
   c. adipocytes undergo a process called hyperplasia.
   d. the size of adipocytes decreases.

What Determines Energy Intake? (pp. 325–331)

5. An anabolic neurotransmitter
   a. stimulates hunger.
   b. causes a decrease in body weight.
   c. is produced by adipose tissue.
   d. causes food cravings and aversions.

6. The release of the gastric hormone _____ increases when there is little food in the stomach.
   a. cholecystokinin (CCK)   c. ghrelin
   b. leptin   d. insulin

7. Hunger is driven primarily by _____ cues.
   a. physiological   c. emotional
   b. psychological   d. social

8. Which of the following is a satiety signal?
   a. increased gastric stretching
   b. low blood glucose
   c. release of ghrelin
   d. low levels of cholecystokinin (CCK)

9. List and describe three signals that originate in the gastrointestinal tract and trigger satiety.

What Determines Energy Expenditure? (pp. 331–335)

10. The majority of energy associated with total energy expenditure (TEE) is expended to support
    a. basal metabolism.   c. the thermic effect of food.
    b. physical activity.   d. adaptive thermogenesis.

11. List and explain the three major components of total energy expenditure (TEE). How are these affected by weight loss?

How Are Body Weight and Body Composition Assessed? (pp. 335–341)

12. Body mass index (BMI) is based on the
    a. ratio of waist to hip circumference.
    b. ratio of weight to height.

c. amount of fat deposited directly under the skin.
    d. self-reported weight and height of people who buy life insurance.

13. A person's waist circumference provides an indication of
    a. leptin concentration.
    b. the amount of lean tissue in the body (muscle mass).
    c. ideal body weight.
    d. central adiposity.

14. List and describe three methods used to assess body composition. What are their strengths and weaknesses?

15. How would you assess whether a person's body fat distribution was associated with increased health risk?

16. Calculate body mass index (BMI; kg/m^2) for a male who is 5 feet, 10 inches tall and weighs 175 pounds. How would you assess whether this person has a healthy body weight?

How Does Lifestyle Contribute to Obesity? (pp. 341–345)

17. In the United States, the occurrence of obesity has increased among
    a. children.   c. adults.
    b. adolescents.   d. all of the above

Can Genetics Influence Body Weight? (pp. 345–346)

18. Leptin is produced primarily by
    a. the brain.   c. adipose tissue.
    b. the liver.   d. skeletal muscle.

How Does the Body Regulate Energy Balance and Body Weight? (pp. 346–350)

19. When leptin levels decrease,
    a. hunger is stimulated.
    b. satiety is stimulated.
    c. glycogen stores increase.
    d. energy expenditure increases.

20. Describe the set point theory of body weight regulation.

21. Describe the proposed role of the hormone leptin in terms of long-term body weight regulation.

What Are the Best Approaches to Weight Loss? (pp. 350–353)

22. A realistic weight-loss goal should be
    a. one to two pounds per week.
    b. three to four pounds per week.
    c. one to two pounds per day.

23. Calculate the recommended initial weight loss goal for a person weighing 200 pounds.

Does Macronutrient Distribution Matter? (pp. 353–359)

24. Low-carbohydrate diets sometimes induce a state of
    a. hyperglycemia.
    b. ketosis.
    c. leptin deficiency.
    d. low levels of calcium in the blood (hypocalcemia).

25. Describe the physiological rationale for why some people believe that low-carbohydrate diets promote weight loss better than high-carbohydrate diets.

Chapter Summary

What Are the Health Benefits of Physical Activity?

- Thirty minutes of sustained physical activity on most days can improve health and reduce a person's risk for certain chronic diseases.
- Physical activity is defined as any bodily movement that substantially increases energy expenditure produced by the contraction of skeletal muscle, which substantially increases energy expenditure.
- Exercise is a type of physical activity that is planned, structured, and results in repetitive bodily movement; exercise is usually done to improve or maintain physical fitness.
- Physical fitness is defined as the ability to carry out daily tasks with vigor and alertness, and to perform leisure-time pursuits without undue fatigue.
- Five outcomes of most successful fitness programs are cardiovascular fitness, muscle strength, endurance, flexibility, and attainment of healthy body composition.
- The four components of the "FITT" principle—frequency, intensity, type, time—form the foundation of many physical fitness plans.

How Does Physical Activity Impact Energy Metabolism?

- During energy metabolism, ATP can be generated both in the presence of oxygen (aerobically) and in its relative absence (anaerobically).
- Aerobic pathways generate large amounts of ATP over an extended period of time, whereas anaerobic pathways generate smaller amounts of ATP quickly.
- Aerobic metabolic pathways include the citric acid cycle, β-oxidation, and oxidative phosphorylation, whereas anaerobic metabolic pathways include the ATP–creatine phosphate (ATP–CP) pathway and glycolysis.
- The ATP–CP pathway serves as an immediate energy system, glycolysis serves as a short-term energy system, and aerobic pathways serve as long-term energy systems.
- The ATP–CP pathway uses creatine phosphate for rapid ATP generation, whereas glycolysis uses glucose.
- Both immediate and short-term energy systems operate during the first few minutes of exercise and during short bursts of high-intensity activity.
- The aerobic pathways use acetyl-CoA, derived from glucose, fatty acids, and—to a lesser extent—amino acids, for low-intensity, long-duration activities.

- Although aerobic pathways are relatively slow in regard to the rate of ATP formation, their ATP yield is rich.

What Physiologic Adaptations Occur in Response to Athletic Training?

- Compared to those of sedentary individuals, muscles of highly active people use glucose more sparingly, and are better able to use fatty acids as an energy source.
- Training increases the number and size of mitochondria, in turn increasing maximal oxygen consumption (called VO_2 max) and delaying fatigue.
- Training strengthens the heart, resulting in more forceful and efficient cardiac function and better nutrient and oxygen delivery to muscles.
- The expansion of capillary blood vessels increases blood flow to muscle cells, whereas the increased production of red blood cells increases the oxygen-carrying capacity of the blood.

How Does Physical Activity Influence Dietary Requirements?

- To meet the demands of physical activity, athletic people may require additional fluids and energy.
- Carbohydrates are needed to maintain glycogen stores, whereas high-quality protein sources are needed to maintain, build, and repair muscle.
- Dietary fat is needed for energy, to provide essential fatty acids, and to facilitate the absorption of fat-soluble substances.
- Iron is necessary for oxygen delivery to tissues, and calcium is needed for building and repairing bone tissue.
- Zinc is needed for energy metabolism and the maintenance, repair, and growth of muscles.
- Excess sweating can disrupt fluid and electrolyte balance, impairing the regulation of body temperature and fluid balance.
- Physically active individuals can prevent dehydration by consuming adequate amounts of fluids before, during, and after exercise.
- To stay hydrated, two to three cups of fluid should be consumed two to three hours before exercise.
- After exercise, body fluids should be replenished by consuming two to three cups of fluids for each pound of weight lost.

Review Questions

What Are the Health Benefits of Physical Activity? (pp. 387–394)

1. To determine your target heart rate, you must first calculate your
   a. body mass index.
   b. maximum heart rate.
   c. basal metabolic rate.
   d. all of the above

2. The Borg Scale can be used to assess
   a. your target heart rate.
   b. glycogen stores in muscle.
   c. maximum heart rate.
   d. intensity level of physical activity.

How Does Physical Activity Impact Energy Metabolism (pp. 394–400)

3. During _____-intensity activities, muscles rely heavily on anaerobic metabolic systems to generate ATP.
   a. low
   b. medium
   c. high

4. Creatine phosphate
   a. is a metabolic waste product excreted in the urine.
   b. can be used to generate ATP during high-intensity activities.
   c. is the end product of glycolysis.
   d. is generated when a phosphate group is split off of ATP.

5. Low oxygen availability in active muscles results in increased production of
   a. glucose.
   b. glycogen.
   c. acetyl-CoA.
   d. lactate.

6. The Cori cycle generates _____ from _____.
   a. glucose; lactate
   b. fatty acids; lactate
   c. glucose; fatty acids
   d. fatty acids; glycogen

7. Explain how each of the body's energy systems (immediate, short term, and long term) is used to generate ATP during physical activity.

What Physiologic Adaptations Occur in Response to Athletic Training? (pp. 401–404)

8. Compared with nonathletes, athletes
   a. have a lower anaerobic threshold.
   b. have a lower VO_2 max.
   c. have a lower resting heart rate.
   d. All of the above are true.

9. Sports anemia is caused by
   a. inadequate intake of iron.
   b. a lack of red meat in the diet.
   c. a disproportionate increase in plasma volume relative to the increase in red blood cell synthesis.
   d. hemolysis (rupture) of red blood cells.

10. Describe how cardiovascular fitness improves in response to regular physical activity.

How Does Physical Activity Influence Dietary Requirements? (pp. 405–412)

11. Which of the following is associated with dehydration?
    a. pale urine
    b. high blood pressure
    c. rapid, weak pulse
    d. all of the above

12. Sports drinks may be advantageous during prolonged exercise because, in addition to water, they also supply
    a. fatty acids.
    b. electrolytes.
    c. dietary fiber.
    d. creatine phosphate.

13. What is the role of glycogen during prolonged physical exertion?

14. Explain why dehydration can contribute to heat exhaustion and heat stroke.

15. The American Dietetic Association recommends that athletes in training consume 5 to 10 g of carbohydrates for each kilogram of body weight. How many grams of carbohydrate would you recommend for an athlete weighing 175 lbs (80 kg)?

16. Protein requirements for athletes remain controversial. While some experts claim that protein requirements for athletes are the same as those of nonathletes (0.8 g/kg/day), other experts recommend that endurance athletes consume 1.2 to 1.4 g/kg/day protein. Calculate the difference in terms of grams of protein using these different recommendations for a 150-lb athlete.

Chapter Summary

The Water-Soluble Vitamins: A Primer

- Water-soluble vitamins play important functions as coenzymes, but many also have non-coenzyme roles.
- Water-soluble vitamins are often added to foods (fortification), and when added in certain amounts, these foods can be labeled as "enriched."

Thiamin (Vitamin B$_1$)—Needed for Production of Acetyl Coenzyme A

- There are three forms of thiamin: free thiamin, thiamin pyrophosphate (TPP), and thiamin triphosphate (TTP). TPP functions as a coenzyme, catalyzing reactions involved in energy metabolism, and is also involved in the synthesis of DNA, RNA, and NADP$^+$.
- Good sources of thiamin include pork, peas, whole grains, fish, and enriched cereal products. Deficiency causes beriberi.

Riboflavin (Vitamin B$_2$)—Coenzyme Required for Reduction–Oxidation Reactions

- There are three forms of riboflavin: free riboflavin, flavin mononucleotide (FMN), and flavin adenine dinucleotide (FAD).
- Riboflavin functions as a coenzyme in a variety of reduction–oxidation (redox) reactions, especially those involved in energy metabolism reactions.
- Good sources of riboflavin include liver, meat, dairy products, whole-grain products, and enriched cereals. Severe deficiency causes ariboflavinosis.

Niacin (Vitamin B$_3$)—Required for Energy Metabolism

- There are two forms of niacin—nicotinic acid and nicotinamide—both of which can be converted to NAD$^+$ and NADP$^+$, coenzymes important in energy metabolism. Niacin is needed for synthesizing fatty acids, cholesterol, steroid hormones, and DNA and for metabolizing vitamin C and folate.
- Niacin can be obtained directly from food or made from tryptophan. Good sources include liver, chicken, fish, pork, mushrooms, and enriched foods. Deficiency causes pellagra.

Pantothenic Acid (Vitamin B$_5$)—A Component of Coenzyme A

- Pantothenic acid is a component of coenzyme A (CoA), needed for ATP production, and for synthesizing heme, cholesterol, bile salts, fatty acids, phospholipids, and steroid hormones.
- Good food sources include mushrooms, organ meats (such as liver), and sunflower seeds. Deficiency causes "burning feet syndrome."

Vitamin B$_6$—Critical for Metabolism of Amino Acids

- There are three forms of vitamin B$_6$: pyridoxine, pyridoxal, and pyridoxamine, all of which are converted to their coenzyme form, pyridoxal phosphate (PLP). Vitamin B$_6$ is involved in metabolism of amino acids and is needed for synthesizing neurotransmitters and heme, converting tryptophan to niacin, and glycogenolysis.
- Good food sources include chickpeas, fish, liver, and potatoes. Deficiency causes microcytic hypochromic anemia.

Biotin (Vitamin B$_7$)—Coenzyme for Carboxylation Reactions

- Biotin acts as a coenzyme for enzymes catalyzing reactions required for energy metabolism, synthesis of fatty acids, breakdown of the amino acid leucine, and cell growth and development.
- Good sources include nuts, eggs, mushrooms, and tomatoes. Biotin deficiency causes neurological problems and can be severe, especially in infants.

Folate—Required for Methylation Reactions

- The active form of folate is tetrahydrofolate acid (THF). Folate is needed for amino acid metabolism, such as the conversion of homocysteine to methionine, and for DNA synthesis.
- Good sources include liver, legumes, mushrooms, green leafy vegetables, and enriched products. Deficiency causes megaloblastic macrocytic anemia and in some cases neural tube defects.

Vitamin B$_{12}$ (Cobalamin)—Vitamin Made Only by Microorganisms

- Vitamin B$_{12}$ (cobalamin) is required as a coenzyme in reactions that (1) allow some amino acids and fatty acids to enter the citric acid cycle and (2) catalyze the conversion of homocysteine to methionine, in turn regenerating active folate (THF).
- Good sources include shellfish, liver, fish, and meat. Deficiency causes megaloblastic macrocytic anemia (due to secondary folate deficiency) and severe neurological complications.

Vitamin C (Ascorbic Acid)—Critical Antioxidant

- Vitamin C provides important antioxidant functions and modulates the synthesis of important compounds such as collagen and neurotransmitters.
- Vitamin C is found in many foods, including a variety of fruits and vegetables. Deficiency causes scurvy.

Is Choline a "New" Conditionally Essential Nutrient?

- Choline is needed for synthesizing several phospholipids (e.g., lecithin), production of the neurotransmitter acetylcholine, muscle control, and a variety of other metabolic reactions.
- Choline is found in many foods but is especially high in eggs, liver, legumes, and pork. Deficiency may cause liver damage in some people.

Carnitine—Needed for Fatty Acid Transport

- Although carnitine is not an essential nutrient for adults, it may be conditionally essential for newborns. It is important for fatty acid transport across biological membranes and is therefore needed to obtain energy from lipids.

Summary of the Water-Soluble Vitamins and Use of Supplements

- Water-soluble vitamins are found in all food groups; therefore, consuming a varied and balanced diet is necessary if one is to consume adequate amounts of each one.
- When a person's diet is limited, dietary supplements may be beneficial.

Review Questions

The Water-Soluble Vitamins: A Primer (pp. 419–424)

1. Which of the following vitamins is readily stored in the body?
   - a. thiamin
   - b. riboflavin
   - c. vitamin C
   - d. none of the above

2. Name at least two good dietary sources of each essential water-soluble vitamin.

Thiamin (Vitamin B₁)—Needed for Production of Acetyl Coenzyme A (pp. 424–428)

3. Beriberi is caused by _____ deficiency.
   - a. riboflavin
   - b. thiamin
   - c. vitamin B₆
   - d. folate

Riboflavin (Vitamin B₂)—Coenzyme Required for Reduction–Oxidation Reactions (pp. 428–430)

4. To help preserve _____, milk is stored in cloudy cartons or paper containers.
   - a. niacin
   - b. biotin
   - c. folate
   - d. riboflavin

Niacin (Vitamin B₃)—Required for Energy Metabolism (pp. 430–433)

5. The amino acid tryptophan is used by the body to make _____.
   - a. vitamin C
   - b. choline
   - c. niacin
   - d. pantothenic acid

Pantothenic Acid (Vitamin B₅)—A Component of Coenzyme A (pp. 433–434)

6. Which of these molecules requires pantothenic acid for its synthesis?
   - a. NAD⁺
   - b. coenzyme A (CoA)
   - c. FADH₂
   - d. collagen

Vitamin B₆—Critical for Metabolism of Amino Acids (pp. 434–436)

7. Microcytic anemia can be caused by _____ deficiency.
   - a. vitamin B₆
   - b. folate
   - c. vitamin B₁₂
   - d. thiamin

Biotin (Vitamin B₇)—Coenzyme for Carboxylation Reactions (pp. 436–438)

8. Biotin deficiency was once called _____.
   - a. egg white injury
   - b. scurvy
   - c. pernicious anemia
   - d. microcytic hypochromic anemia

Folate—Required for Methylation Reactions (pp. 438–442)

9. Folate is needed for which of the following types of reactions in the body?
   - a. carboxylation reactions
   - b. decarboxylation reactions
   - c. reduction–oxidation (redox) reactions
   - d. single-carbon transfer reactions

10. Using the food composition table that accompanies this text or other sources, calculate the folate content of a lunch containing a hamburger on a white bun, a small order of french fries, and a large cola soft drink. What percentage of your folate RDA does this represent? What foods might be added to or substituted for the food in this lunch to increase the folate content?

Vitamin B₁₂ (Cobalamin)—Vitamin Made Only by Microorganisms (pp. 442–445)

11. Which of these vitamins contains cobalt?
    - a. folate
    - b. vitamin B₁₂
    - c. vitamin B₆
    - d. vitamin C

12. Which of these vitamin-related conditions is caused by an autoimmune disease?
    - a. scurvy
    - b. pellagra
    - c. ariboflavinosis
    - d. pernicious anemia

13. Folate deficiency is sometimes said to "mask" vitamin B₁₂ deficiency. What is meant by this?

Vitamin C (Ascorbic Acid)—Critical Antioxidant (pp. 445–449)

14. The main function of vitamin C in the body is
    - a. acting as a coenzyme for energy metabolism reactions.
    - b. involvement as an antioxidant.
    - c. regulation of lipid synthesis.
    - d. inhibition of cell growth.

15. Describe how an antioxidant functions, and provide two examples of how vitamin C functions in this capacity.

Is Choline a "New" Conditionally Essential Nutrient? (pp. 449–450)

16. Which of the following organs can be damaged in response to choline deficiency?
    - a. liver
    - b. kidneys
    - c. pancreas
    - d. heart

Carnitine—Needed for Fatty Acid Transport (pp. 450–451)

17. Carnitine transports _____ across biological membranes.
    - a. monosaccharides
    - b. amino acids
    - c. cholesterol
    - d. fatty acids

Summary of the Water-Soluble Vitamins and Use of Supplements (pp. 451–453)

18. Which of the following situations might warrant dietary supplementation?
    - a. vegetarianism
    - b. lactose intolerance
    - c. limited time for food preparation
    - d. all of the above

Chapter Summary

What Makes the Fat-Soluble Vitamins Unique?

- Fat-soluble vitamins are absorbed in the small intestine, requiring the presence of other lipids and bile.
- Fat-soluble vitamins are circulated away from the small intestine in the lymph via chylomicrons.
- Consuming large amounts of fat-soluble vitamins (other than vitamin K) can result in toxicities, sometimes with serious consequences.

Vitamin A and the Carotenoids—Needed for Eyesight and Much More

- Vitamin A refers to a group of compounds called *retinoids*. Retinol, retinoic acid, and retinal are all forms of vitamin A.
- Vitamin A–like substances found in plants and some animals are called carotenoids. Carotenoids that can be converted to vitamin A in the body are called *provitamin A carotenoids*. Those that cannot be converted are called *nonprovitamin A carotenoids*.
- Vitamin A and the carotenoids are important regulators of growth, reproduction, vision, immune function, gene expression, and bone formation. The carotenoids are also potent antioxidants that help protect proteins, DNA, and cell membranes from free radical damage.
- Good sources of preformed vitamin A are liver, fish, whole milk, and fortified foods. Carotenoids tend to be found in brightly colored fruits and vegetables, especially those that are yellow, orange, or red.
- Vitamin A deficiency can result in blindness, increased risk of infection, and even death.
- Excessive intake of vitamin A can be toxic and may result in fetal malformations. Although high intake of carotenoids from food is not dangerous, it can cause the skin to become orange.

Vitamin D—The "Sunshine Vitamin"

- Vitamin D is found in fatty fish and in some plant foods.
- In the presence of sunlight, the skin can produce vitamin D from a cholesterol metabolite.
- The active form of vitamin D is 1,25-dihydroxyvitamin D (1,25-[OH]$_2$D$_3$), also called calcitriol. Calcitriol stimulates calcium absorption in the intestine, decreases calcium loss in the urine, and increases breakdown of bone. Vitamin D also regulates genes important for cell differentiation and maturation.
- Vitamin D deficiency can cause rickets in children and osteomalacia and osteoporosis in adults.
- Vitamin D toxicity from foods is rare, but excessive intake from supplements can be dangerous.

Vitamin E—Antioxidant That Protects Biological Membranes

- Alpha-tocopherol is the most biologically active form of vitamin E and is found in oils, nuts, seeds, and some fruits and vegetables.
- Vitamin E functions mainly as an antioxidant, protecting biological membranes from free radical damage. Vitamin E may also protect the eyes from cataract formation and influence cancer risk by decreasing DNA damage.
- Vitamin E deficiency is rare but results in neuromuscular problems and hemolytic anemia. Vitamin E toxicity, which is also rare, results in bleeding.

Vitamin K—Critical for Coagulation

- Vitamin K refers to three related compounds: phylloquinone, menaquinone, and menadione.
- Dark green vegetables are often good sources of vitamin K. Light and heat can destroy this vitamin in foods.
- Intestinal bacteria produce vitamin K$_2$.
- Vitamin K is involved in calcium-requiring reactions that are needed for blood clotting and proper bone mineralization.
- Vitamin K deficiency causes bleeding and, perhaps, weak bones. In infants, severe vitamin K deficiency can cause vitamin K deficiency bleeding, which is largely prevented by vitamin K injections.

Fat-Soluble Vitamins: Summary and Overall Recommendations

- Because fat-soluble vitamins are found in many plant- and animal-based foods, it is important to choose a wide variety of foods from each food group daily.

Review Questions

What Makes the Fat-Soluble Vitamins Unique? (pp. 461–462)

1. Fat-soluble vitamins are generally absorbed in the _____ and circulated in the _____ .
   a. stomach; lymph
   b. small intestine; lymph
   c. large intestine; blood
   d. small intestine; blood

Vitamin A and the Carotenoids—Needed for Eyesight and Much More (pp. 462–470)

2. The most biologically active form of vitamin A is _____ .
   a. retinol   c. retinal
   b. retinoic acid   d. lycopene

3. Preformed vitamin A is typically found in what kinds of foods?
   a. carrots
   b. green leafy vegetables
   c. animal products
   d. legumes

4. In general, fat-soluble vitamins are absorbed in the _____ and circulated away from the gastrointestinal tract in the _____ .
   a. large intestine; blood
   b. large intestine; lymph
   c. small intestine; blood
   d. small intestine; lymph

5. Using the food composition table that accompanies this text or other sources, calculate the vitamin A content of a lunch containing a hamburger on a white bun, a small order of french fries, and a large cola soft drink. What percentage of your Recommended Dietary Allowance (RDA) of vitamin A does this represent? What foods might be added to or substituted for the food in this lunch to increase the vitamin A content?

Vitamin D—The "Sunshine Vitamin" (pp. 471–477)

6. Why is vitamin D often called the "sunshine vitamin"? Outline how vitamin D is synthesized in the skin and how it is converted to calcitriol in the body.

7. Parathyroid hormone stimulates the conversion of $25\text{-}(OH)D_3$ to $1,25\text{-}(OH)_2D_3$ when
   a. blood calcium is low.
   b. blood calcium is high.
   c. a person is consuming excessive amounts of calcium.
   d. a person is injured and bleeding.

8. Which of the following can result from vitamin D deficiency in infants or children?
   a. calcification of the kidneys
   b. osteomalacia
   c. rickets
   d. osteoporosis

9. Using the MyPlate food guidance system (http://www.choosemyplate.gov), determine the amount of foods from the dairy group that is recommended for you. Then make up a sample menu of dairy foods you commonly consume that would satisfy these recommendations, and calculate the amount of vitamin D you would get if you consumed this amount of foods.

Vitamin E—Antioxidant That Protects Biological Membranes (pp. 477–480)

10. Which of the following is a major function of vitamin E?
    a. regulation of energy metabolism during periods of fasting
    b. carboxylation reactions
    c. blood clotting
    d. protection of biological membranes from free radical damage

11. Vitamin E deficiency can cause which of the following?
    a. neuromuscular problems
    b. loss of coordination
    c. hemolytic anemia
    d. all of the above

Vitamin K—Critical for Coagulation (pp. 480–483)

12. The functions of vitamin K depend largely on which of the following minerals?
    a. selenium
    b. iron
    c. calcium
    d. magnesium

13. For which fat-soluble vitamin has the Institute of Medicine not established a UL?
    a. vitamin A
    b. vitamin D
    c. vitamin E
    d. vitamin K

Fat-Soluble Vitamins: Summary and Overall Recommendations (pp. 483–485)

14. Name two food sources of each fat-soluble vitamin and state whether food processing (such as excessive heating or exposure to air) can influence the fat-soluble vitamin content of each of these foods.

15. Several vitamins—both water soluble and fat soluble—participate in antioxidant functions in the body. Which vitamins are these, and how do they act as antioxidants? Describe, in general, how these compounds are thought to be important for cancer prevention.

Chapter Summary

What Are Minerals?

- Minerals are inorganic atoms or molecules (other than water) required by the body.
- The major minerals are those needed in amounts greater than 100 mg/day.
- The body requires six major minerals: calcium, phosphorus, magnesium, sodium, chloride, and potassium.
- Although animal foods tend to be the best sources of many major minerals, they are also abundant in myriad plant foods.
- In general, major minerals are absorbed in the small intestine and circulated in the blood; excess amounts are excreted in the urine.
- Major mineral toxicities are rare.

Calcium—The Body's Most Abundant Mineral

- Although both plant and animal foods provide calcium, the most abundant and bioavailable sources are dairy products.
- Calcium is important for the structure of bones and teeth, blood clotting, nerve and muscle function, and energy metabolism.
- Blood calcium concentrations are regulated primarily by the small intestine, bone, and kidneys. This involves parathyroid hormone (PTH), calcitriol (vitamin D), and calcitonin.
- Calcium deficiency in children can cause rickets. In adults, calcium deficiency can be related to osteopenia and osteoporosis.
- Excessive calcium intake can cause the mineralization of soft tissues such as muscle and kidney.

Phosphorus—A Component of Biological Membranes

- Good sources of phosphorus include dairy products, meat, seafood, nuts, and seeds.
- Phosphorus is a component of phospholipids, which make up all biological membranes, and is also found in bone, DNA, RNA, and ATP. Phosphorus can activate or inactivate a variety of enzymes.
- Regulation of blood phosphorus levels occurs via the actions of parathyroid hormone (PTH) and vitamin D (calcitriol) and involves the small intestine and bone.
- Phosphorus deficiency is rare but can be fatal; toxic intakes can damage the kidneys.

Magnesium—Needed for Building Bones and Stabilizing Enzymes

- Magnesium is found in seafood, legumes, nuts, chocolate, and unprocessed or minimally-processed grains.
- Magnesium is required for energy metabolism, is a cofactor for hundreds of enzymes, and is needed for nerve and muscle function.
- Magnesium deficiency is rare but can result in poor nerve and muscle action, especially in heart tissue.
- Magnesium toxicity from supplements and some medications can cause intestinal upset and, when severe, cardiac arrest.

Sodium and Chloride—Regulators of Fluid Balance

- Table salt (NaCl; sodium chloride) is the major source of sodium and chloride in the diet in the United States.
- Sodium chloride, when dissolved in water, dissociates into ions (Na^+ and Cl^-) important in regulating fluid balance.
- Sodium is needed for nerve and muscle function, whereas chloride is required for digestion, elimination of carbon dioxide, and immune function.
- Deficiencies of sodium and chloride are uncommon, except in sick infants and children and endurance athletes. Symptoms include nausea, dizziness, and muscle cramps.
- For some people, excessive salt intake is associated with hypertension and increased risk for heart disease.

Potassium—An Important Intracellular Cation

- Potassium is found in legumes, potatoes, seafood, a variety of fruits and vegetables, dairy products, and meat.
- Potassium, an important cation in the body, helps maintain fluid balance. Potassium is also needed for muscle function (especially in the heart), nerve impulses, and energy metabolism.
- Potassium deficiency is rare except during illness and in some eating disorders. It can cause cardiac disturbances, weakness, and confusion.
- Excessive potassium intake from supplements can cause cardiac arrest.

Water—The Essence of Life

- Water is the most abundant substance in the body and is required for everything from energy metabolism to temperature regulation.
- Water crosses cell membranes via osmosis, often following the active transport of various electrolytes.
- Dehydration can influence cognitive function, ability to engage in aerobic activity, temperature regulation, and risk of urinary tract infections.
- The body responds to dehydration by releasing the hormones antidiuretic hormone (ADH) and aldosterone, which act to conserve water in the kidneys by decreasing urine production.
- People who maintain high levels of physical activity or live in hot environments are advised to consume additional water.

How Are the Functions and Food Sources of the Major Minerals Related?

- It is important to remember that the major minerals often work together in the body; for example, the electrolytes coordinately regulate fluid balance.
- Consuming a balanced, varied diet is important when it comes to obtaining sufficient amounts of all the major minerals.

Review Questions

What Are Minerals? (pp. 507–508)

1. Which of the following statements is *false?*
   a. Minerals are inorganic substances.
   b. Major minerals are those needed in amounts greater than 100 mg/day.
   c. The mineral contents of whole-grain foods tend to be lower than those of refined-grain foods.
   d. Toxicity from excessive intake of the major minerals is usually associated with dietary supplement use.

Calcium—The Body's Most Abundant Mineral (pp. 509–515)

2. Calcitonin is released by the _____ during periods of _____ blood calcium concentration.
   a. thyroid gland; low
   b. thyroid gland; high
   c. parathyroid glands; low
   d. parathyroid glands; high

3. Which of the following can influence calcium bioavailability?
   a. oxalates   c. aging
   b. phytates   d. all of the above

Phosphorus—A Component of Biological Membranes (pp. 515–516)

4. Which of the following major minerals is a component of biological membranes?
   a. potassium   c. calcium
   b. sodium   d. phosphorus

5. Characterize the mechanisms by which phosphorus deficiency may cause weakness and fatigue.

Magnesium—Needed for Building Bones and Stabilizing Enzymes (pp. 517–518)

6. Most of the body's magnesium is found in _____.
   a. muscle   c. bones
   b. liver   d. blood

7. Explain what is somewhat unusual about the UL value for magnesium.

Sodium and Chloride—Regulators of Fluid Balance (pp. 518–523)

8. Hyponatremia stimulates the release of _____.
   a. parathyroid hormone
   b. aldosterone
   c. growth hormone
   d. insulin

9. Positively charged ions, such as Na^+, are called _____.
   a. electrons
   b. cations
   c. anions
   d. both a and c

10. Using the food composition table that accompanies this text or other sources, calculate the sodium content of a lunch containing a hamburger on a white bun, a small order of french fries, and a large cola soft drink. What percentage of your AI of sodium does this represent? How does this value compare with what is recommended in the 2010 Dietary Guidelines for Americans? What foods could be substituted for the food in this lunch to decrease the sodium content?

11. The 2010 Dietary Guidelines for Americans recommend that we "choose and prepare foods with less salt." List five suggestions for daily dietary changes that can help accomplish this goal.

12. Sodium and chloride are important electrolytes in maintaining fluid balance. Describe how they function in the process of osmosis.

Potassium—An Important Intracellular Cation (pp. 523–524)

13. Increased circulating levels of _____ are associated with decreased blood pressure in some people.
    a. potassium
    b. chloride
    c. manganese
    d. sodium

14. Blood concentrations of sodium, chloride, and potassium are regulated by similar hormones. Compare and contrast these mechanisms.

Water—The Essence of Life (pp. 525–532)

15. Most of the fluid within the body is contained within the _____ space.
    a. interstitial   c. circulatory
    b. intravascular   d. intracellular

16. During osmosis, active transport of electrolytes causes the subsequent passive transport of _____.
    a. water
    b. sodium and potassium
    c. plasma proteins
    d. urea

17. Cystic fibrosis is a condition caused by the inability of cells to move _____ across cell membranes.
    a. water
    b. carbohydrates
    c. lipids
    d. chloride

18. Dehydration causes the release of _____ from the _____.
    a. aldosterone; brain
    b. angiotensinogen; kidneys
    c. antidiuretic hormone; pituitary gland
    d. antidiuretic hormone; liver

How Are the Functions and Food Sources of the Major Minerals Related? (pp. 532–534)

19. Describe how various major minerals work together to help keep bones and teeth strong.

Chapter Summary

What Do the Trace Minerals Have in Common?

- Trace minerals are inorganic atoms or compounds needed by the body in very small quantities.
- The bioavailability of trace minerals is influenced by nutritional status and other components of foods.

Iron—Transporter of Oxygen

- Two forms of iron in foods are heme and nonheme iron.
- The absorption of heme iron is primarily determined by iron status, whereas that of nonheme iron is influenced mostly by dietary factors.
- Sources of heme iron include meat and seafood; nonheme iron is found in vegetables and fortified foods.
- Iron from hemoglobin and myoglobin is needed for oxygen transport and storage. Iron associated with the electron transport chain is needed for ATP production.
- Iron deficiency can cause fatigue, decreased work performance, impaired intellectual abilities, and microcytic hypochromic anemia.
- Iron toxicity can be fatal.

Copper—Cofactor in Redox Reactions

- Copper homeostasis occurs via alterations in absorption in the intestine and excretion in the bile.
- Copper is a cofactor for enzymes involved in reduction–oxidation (redox) reactions and is important for energy metabolism, neural function, and antioxidant reactions.
- Good sources of copper include liver, shellfish, mushrooms, and nuts.
- Copper deficiency causes connective tissue and bones to weaken, anemia, and neural problems; copper toxicity results in gastrointestinal distress and liver damage.

Iodine (Iodide)—An Essential Component of the Thyroid Hormones

- Iodine is found in fortified salt (iodized salt), shellfish, and dairy products.
- Iodine bioavailability is high, and excess is excreted in the urine.
- Iodine is a component of thyroid hormones and is important for growth, development, brain function, and energy metabolism.
- Iodine deficiency can cause cretinism and goiter.
- Iodine toxicity causes both hypo- and hyperthyroidism.

Selenium—A Mineral with Antioxidant Functions

- Food sources of selenium include nuts, seafood, meats, and cereal products.
- Selenium absorption is efficient; excess selenium is excreted in the urine.
- Selenium is a component of selenoproteins involved in redox reactions and protecting the body from oxidative damage. Others regulate thyroid metabolism and activation of vitamin C.
- Selenium deficiency can cause Keshan disease; toxicity (selenosis) causes a garlic-like odor of the breath, gastrointestinal upset, and brittle teeth and fingernails.

Chromium—Implicated in Glucose Homeostasis

- Chromium is found in whole grains, fruits, and vegetables.
- Chromium is needed for the function of insulin and for optimal growth and development.
- Chromium deficiency is rare but causes high blood glucose concentrations, decreased insulin sensitivity, and weight loss; toxicity from foods is rare, but is sometimes caused by environmental exposure.

Manganese—Important for Gluconeogenesis and Bone Formation

- Manganese is found in water, whole-grain products, nuts, legumes, and some fruits and vegetables.
- Little manganese is absorbed; excess is excreted in the bile.
- Manganese is involved in energy metabolism, bone formation, gluconeogenesis, and antioxidant function.
- Manganese deficiency causes dry skin, weak bones, and poor growth; toxicity from foods is rare, but environmental exposure can cause serious problems.

Molybdenum—Required in Very Small Quantities

- Legumes, grains, and nuts are good sources of molybdenum.
- Molybdenum is a cofactor for enzymes involved in the metabolism of sulfur-containing amino acids and the purines important for protein, DNA, and RNA structures. It is also needed for protein synthesis and cellular growth.
- Molybdenum deficiency is rare but causes headaches, abnormal cardiac function, and visual difficulties; toxicity has never been documented in humans but causes reproductive failure in laboratory animals.

Zinc—Involved in RNA Synthesis and Gene Expression

- Zinc is found in shellfish, fortified cereals, meat, legumes, and chocolate, and its absorption is influenced by many factors, such as phytates.
- Zinc absorption increases during deficiency and decreases during toxicity.
- Zinc is needed for RNA synthesis and modulation of gene expression. It is also important for cell maturation and immune function.
- Zinc deficiency causes poor appetite, growth failure, skin irritations, diarrhea, delayed physical maturation, and infertility; toxicity is rare but can influence immune function, lipid metabolism, and appetite.

Fluoride—Nonessential Mineral That Strengthens Bones and Teeth

- Fluoride is not an essential nutrient, but increased fluoride intake decreases tooth decay and strengthens bones.
- Fluoride is not found naturally in many foods. Instead, we tend to consume it from the water we drink.
- Fluoride toxicity can discolor teeth (dental fluorosis) and weaken bones (skeletal fluorosis).

Are There Other Important Trace Minerals?

- In addition to the trace minerals known to be essential for life, many others may influence health. These include nickel, aluminum, silicon, vanadium, arsenic, and boron.

Review Questions

What Do the Trace Minerals Have in Common? (pp. 555–556)

1. The trace minerals are found mainly in which of the following food groups?
   a. grains and cereals
   b. meat
   c. nuts and seeds
   d. all of the above

2. The mineral content of foods can vary greatly. For example, wheat produced in one region of the United States can have very high levels of selenium, whereas wheat raised elsewhere can have very low amounts. Why is this?

3. Are there any trace minerals that a vegetarian may have trouble consuming enough of? If so, what foods might she or he need to consume more of?

Iron—Transporter of Oxygen (pp. 556–565)

4. Which of the following would be expected in iron *deficiency?*
   a. increased ferroportin production and decreased ferritin production
   b. decreased ferroportin production and decreased ferritin production
   c. elevated hematocrit
   d. increased transferrin saturation

5. Vitamin C increases the bioavailability of nonheme iron by
   a. binding to ferritin in the intestinal cell.
   b. increasing the activity of the digestive enzymes in the small intestine.
   c. reducing ferric iron to ferrous iron.
   d. helping iron bind to chelators in the intestinal tract.

Copper, Iodine, and Selenium (pp. 565–572)

6. Ceruloplasmin transports _____ in the blood.
   a. cobalt
   b. copper
   c. selenium
   d. molybdenum

7. Thyroid-stimulating hormone (TSH) stimulates
   a. uptake of iodine by the thyroid gland.
   b. excretion of iodine in the urine.
   c. increased absorption of iodine in the small intestine.
   d. release of iodine from the thyroid gland into the blood.

8. What distinguishes a person with cretinism from a person with goiter? What are the reasons that the United States has a low occurrence of these conditions?

9. Glutathione peroxidases require which of the following trace minerals?
   a. manganese
   b. iron
   c. fluoride
   d. selenium

Chromium, Manganese, and Molybdenum (pp. 572–574)

10. _____ requires chromium for optimal function.
    a. Gastrin
    b. Insulin
    c. Thyroid hormone
    d. Glucagon

11. Regulation of the amount of manganese in the body is via its loss in the _____.
    a. urine
    b. blood
    c. sweat
    d. feces

12. Using the MyPlate food guidance system (http://www.choosemyplate.gov), determine the amount of foods from the protein, vegetables, fruits, grains, and dairy groups that are recommended for you. Then, make up a sample menu consistent with what you normally eat that would satisfy these recommendations. Calculate the amount of magnesium you would get if you consumed these amounts of foods. Do you think your diet provides adequate amounts of magnesium?

Zinc and Fluoride (pp. 574–577)

13. Which of the following best describes acrodermatitis enteropathica?
    a. a primary zinc deficiency caused by inadequate zinc in the diet
    b. a genetic disease that inhibits zinc excretion
    c. a genetic disease that results in zinc toxicity
    d. a genetic disease that causes inadequate zinc absorption

14. Using the food composition table that accompanies this book, calculate the zinc content of a lunch containing a hamburger on a non-fortified white bun, a small order of french fries, and a large cola soft drink. What percentage of your RDA for zinc does this represent? What foods might be added to or substituted for the food in this lunch to increase the zinc content?

15. Fluoride is NOT considered an essential nutrient because
    a. the body does not use it for any processes.
    b. it is not important for ATP production.
    c. it is not required for growth, reproduction, or basic body functions.
    d. we do not consume any fluoride in food or water.

Are There Other Important Trace Minerals? (pp. 577–578)

16. Scientists all agree that all of the essential trace minerals have been discovered.
    a. true
    b. false

Chapter Summary

What Physiological Changes Take Place during the Human Life Cycle?

- Periods of growth and development coincide with stages of life (infancy, childhood, adolescence, adulthood, and for women, the special life stages of pregnancy and lactation).
- Changes in body size and composition influence nutrient requirements.
- Growth occurs when the number (hypertrophy) and/or the size (hyperplasia) of cells increases.
- When physical maturity is reached, the rate of cell turnover is in equilibrium. With increasing age, the rate of new cell formation slows, resulting in a decline in physiological function called senescence.

What Are the Major Stages of Prenatal Development?

- The two stages of prenatal development are the embryonic period and the fetal period.
- The placenta transfers nutrients, gases, hormones, and other products between the mother and the fetus.
- Substances and conditions that disrupt normal growth and development are called teratogens.
- Whereas gestation length is determined by counting the weeks since conception, gestational age is determined by counting the number of weeks between the first day of a woman's last normal menstrual period and her infant's birth.
- Infants with a gestational age of 37 to 42 weeks are considered full term; those with a gestational age less than 37 weeks are considered preterm; and those born after 42 weeks are considered post term.
- Although most preterm infants are born with low birth weight (LBW), LBW can also be caused by intrauterine growth retardation (IUGR). Babies who experience IUGR are referred to as being small for gestational age (SGA).

What Are the Recommendations for a Healthy Pregnancy?

- Adequate weight gain, a healthy diet, and refraining from smoking can influence birth weight and gestation length.
- Carbohydrates should remain the primary energy source throughout pregnancy.
- During pregnancy, additional protein (25 g/day) is needed for the formation of fetal and maternal tissues.
- Dietary fat should provide 20 to 35% of total calories during pregnancy.
- Extra calcium is needed for the fetus to grow and develop, although changes in maternal physiology can accommodate these needs without an increase in dietary intake.
- Iron is needed to form hemoglobin and for the growth and development of the fetus and the placenta.
- Folate is needed for cell division and is critical to the development of the nervous system. Women with poor folate status early in the pregnancy are at increased risk of having a baby born with a neural tube defect.

Why Is Breastfeeding Recommended during Infancy?

- Milk production, a process called lactogenesis, takes place in mammary secretory cells.
- Nursing stimulates the hypothalamus, which in turn signals release of prolactin and oxytocin from the pituitary gland.
- Prolactin stimulates milk synthesis, and oxytocin is needed for the release of milk out of the alveoli and into the ducts.

What Are the Nutritional Needs of Infants?

- By 6 months of age, infant weight doubles, and by 12 months, it triples.
- The American Academy of Pediatrics and World Health Organization recommend exclusive breastfeeding during the first 4 to 6 months of life (preferably 6 months).
- Infant formula is derived typically from either cow milk (cow-based formula) or soybeans (soy-based formula).
- Human milk and infant formula should be the primary sources of nutrients and energy throughout the first year of life.
- Nonmilk complementary foods can be introduced sometime between 4 to 6 months of age, depending on an infant's readiness.
- Older infants should be given a wider variety of foods, although these should be chosen carefully to pose minimal risk for choking.

What Are the Nutritional Needs of Toddlers and Young Children?

- Children between the 85th and 95th percentile based on BMI-for-age are considered overweight. Those at or above the 95th percentile are considered obese. Children below the 5th percentile are considered underweight.

How Do Nutritional Requirements Change during Adolescence?

- Hormonal changes begin the transformation from childhood into adolescence, causing changes in height, weight, and body composition.
- Linear growth is largely completed at the end of the adolescent growth spurt, although bone mass continues to increase into early adulthood.
- During adolescence, males experience an increase in percentage of lean mass accompanied by a decrease in percentage of fat mass; females experience the opposite.
- The rapid growth and development associated with adolescence increases the body's need for certain nutrients (such as iron and folate) and energy.

How Do Age-Related Changes in Adults Influence Nutrient and Energy Requirements?

- Adults who remain physically active and maintain a nutritious diet tend to live longer and have fewer health problems than those who eat nutrient-poor diets and lead sedentary lives.
- As individuals grow older, they often experience a relative loss of lean mass and increase in fat mass. For this reason, energy requirements often decrease.
- Age-related bone loss can make bones fragile; thus, it is important for older adults to obtain adequate intakes of protein, calcium, vitamin D, phosphorus, and magnesium.
- Many factors can contribute to inadequate food intake in older adults, including poor oral health, altered taste, and decreased ability to smell.
- A decrease in production of gastric secretions can impair absorption of iron, calcium, biotin, folate, vitamin B_{12}, and zinc.
- Services that provide food to older adults include congregate meal programs, meal delivery programs, and the Supplemental Nutrition Assistance Program (SNAP).

Review Questions

What Physiological Changes Take Place during the Human Life Cycle? (pp. 585–587)

1. Which stage of the lifespan has the lowest rate of hyperplasia?
   a. infancy
   b. adolescence
   c. pregnancy
   d. senescence

What Are the Major Stages of Prenatal Development? (pp. 587–592)

2. Which of the following represents the correct sequence of prenatal development?
   a. embryo, fetus, zygote
   b. fetus, zygote, embryo
   c. zygote, embryo, fetus
   d. blastocyst, fetus, zygote

3. The first trimester for pregnancy corresponds to the
   a. embryonic period.
   b. embryonic period and part of the fetal period.
   c. fetal period.
   d. zygotic period.

What Are the Recommendations for a Healthy Pregnancy? (pp. 592–601)

4. Weight gain recommendations during pregnancy are based on
   a. pre-pregnancy BMI.
   b. the ratio of lean mass to fat mass.
   c. waist circumference.
   d. hip circumference.

5. Which of the following groups of women has the lowest recommended weight gain during pregnancy?
   a. underweight women
   b. overweight and obese women
   c. women over 30 years of age
   d. women between 25 and 30 years of age

6. Impaired _____ status can increase the risk of having a child born with a neural tube defect.
   a. calcium
   b. linoleic acid
   c. iron
   d. folate

7. During pregnancy, a 25-year-old woman's recommended calcium intake (RDA)
   a. increases.
   b. decreases.
   c. stays the same.

Why Is Breastfeeding Recommended during Infancy? (pp. 601–606)

8. Milk production takes place in specialized structures called _____.
   a. alveoli
   b. mammary ducts
   c. the endometrium
   d. milk sinus

9. Colostrum is
   a. a brand of infant formula.
   b. the first milk released from the breast after a baby is born.
   c. mucus-like feces that collect in the large intestine in the fetus.
   d. a brand of nutrient supplements that are given to infants.

10. Describe the hormonal regulation of milk production.

What Are the Nutritional Needs of Infants? (pp. 606–612)

11. A baby weighing 7 pounds at birth should weigh approximately _____ pounds by 6 months of age.
    a. 10
    b. 14
    c. 21
    d. 32

12. Growth charts are used to monitor a baby's _____ during the first 12 months of life.
    a. iron status
    b. weight and height
    c. developmental stages

What Are the Nutritional Needs of Toddlers and Young Children? (pp. 612–618)

13. A young child above the 95th percentile for BMI-for-age would be classified as
    a. appropriate for gestational age.
    b. underweight.
    c. having a healthy weight.
    d. obese.

14. Young children (toddlers) should
    a. be in positive energy balance.
    b. have a faster rate of growth compared to infants.
    c. consume infant formula rather than cow milk.
    d. be in negative energy balance.

How Do Nutritional Requirements Change during Adolescence? (pp. 618–621)

15. During which stage of life does menarche occur?
    a. pregnancy
    b. infancy
    c. adolescence
    d. adulthood

16. Compare and contrast growth patterns (weight, length/ height, and BMI) during infancy, childhood, and adolescence.

How Do Age-Related Changes in Adults Influence Nutrient and Energy Requirements? (pp. 621–629)

17. Which of the following changes in body composition is/are often associated with aging in older adults?
    a. increased bone mass
    b. increased lean mass
    c. increased body fat
    d. all of the above

18. List and describe three physiological changes often associated with aging and how these changes can affect nutrient status and requirements.

19. Explain why older adults are considered at particularly high risk for developing impaired nutritional status.

Answers to Multiple-Choice Questions

1. d; 2. c; 3. b; 4. a; 5. b; 6. d; 7. c; 8. a; 9. b; 11. b; 12. b; 13. d; 14. a; 15. c; 17. c.

Make half your plate fruits and vegetables

» Choose fresh, frozen, canned, or dried fruits and vegetables.
» Eat red, orange, and dark-green vegetables, such as tomatoes, sweet potatoes, and broccoli, in main and side dishes.
» Use fruit as snacks, salads, or desserts.
» Keep raw, cut-up vegetables handy for quick snacks.
» Choose whole or cut-up fruits more often than fruit juice.

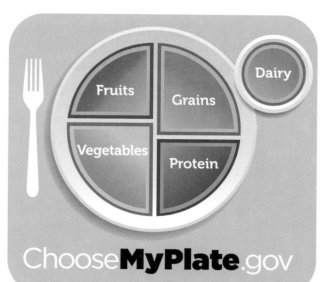

ChooseMyPlate.gov

Switch to skim or 1% milk

» They have the same amount of calcium and other essential nutrients as whole milk, but less fat and calories.

Make at least half your grains whole

» Choose 100% whole-grain cereals, breads, crackers, rice, and pasta.
» Check the ingredients list on food packages to find whole-grain foods.

Vary your protein food choices

» Choose a variety of foods including seafood, beans and peas, nuts, lean meats, poultry, and eggs.
» Keep meat and poultry portions small and lean.
» Try grilling, broiling, poaching, or roasting. These methods do not add extra fat.

Cut back on foods high in solid fats, added sugars, and salt

• Choose foods and drinks with little or no added sugars.
• Look out for salt (sodium) in foods you buy.
• Eat fewer foods that are high in solid fats.

Eat the right amount of calories for you

• Enjoy your food, but eat less.
• Cook more often at home, where you are in control of what's in your food.
• When eating out, choose lower calorie menu options.

Get your personal daily calorie limit at www.ChooseMyPlate.gov and keep that number in mind when deciding what to eat.

Be physically active your way

Pick activities that you like and start by doing what you can, at least 10 minutes at a time. Every bit adds up, and the health benefits increase as you spend more time being active.

MyPlate was created by the U.S. Department of Agriculture, Center for Nutrition Policy and Promotion.   Printed and distributed by Learning ZoneXpress.   1-888-455-7003 • www.learningzonexpress.com • UPC 846742000857

2010 Dietary Guidelines for Americans—Major Concepts and Key Recommendations

Major Concepts	Key Recommendations
Balance Calories to Manage Weight	• Prevent and/or reduce overweight and obesity through improved eating and physical activity behaviors. • Control total calorie intake to manage body weight. For people who are overweight or obese, this will mean consuming fewer calories from foods and beverages. • Increase physical activity and reduce time spent in sedentary behaviors. • Maintain appropriate calorie balance during each stage of life.
Reduce Certain Foods and Food Components	• Reduce daily sodium intake to less than 2,300 mg/day and further reduce intake to 1,500 mg/day among individuals with increased risk for hypertension. • Consume less than 10% of calories from saturated fats by replacing them with monounsaturated and polyunsaturated fats. • Consume less than 300 mg/day of dietary cholesterol. • Keep *trans* fatty acid consumption as low as possible. • Reduce intake of calories from solid fats (saturated and *trans* fats) and added sugars. • Limit consumption of foods that contain refined grains. • If you consume alcohol, consume it in moderation—up to one drink per day for women and two drinks per day for men; only adults of legal drinking age should drink alcoholic beverages.
Increase Certain Foods and Nutrients[1]	• Increase vegetable and fruit intake. • Eat a variety of vegetables, especially peas and beans and vegetables that are dark green, red, or orange. • Consume at least half of all grains as whole grains. • Increase intake of fat-free or low-fat milk and milk products or fortified soy beverages. • Choose a variety of protein foods, including seafood, lean meat and poultry, eggs, beans and peas, soy products, and unsalted nuts and seeds. • Replace protein foods high in solid fats with choices that are lower in solid fats and calories. • Choose foods that provide more potassium, dietary fiber, calcium, and vitamin D, such as vegetables, fruits, whole grains, and dairy products.
Build Healthy Eating Patterns	• Select an eating pattern that meets nutrient needs over time at an appropriate calorie level. • Account for all foods and beverages consumed and assess how they fit within a total healthy eating pattern. • Follow food safety recommendations when preparing and eating foods to reduce the risk of foodborne illness.

[1] Additional recommendations are also made for specific population groups. For example, women capable of becoming pregnant are advised to choose foods that supply heme iron, additional iron sources, and enhancers of iron absorption (e.g., vitamin C). It is also recommended that they consume synthetic folic acid (from foods or supplements) in addition to foods high in folate. Women who are pregnant or breastfeeding are advised to consume 8 to 12 oz seafood per week, being careful to avoid those types known to be high in mercury. Individuals who are 50 years or older should consume foods fortified with vitamin B_{12} or a dietary supplement.

SOURCE: U.S. Department of Agriculture and U.S. Department of Health and Human Services. Dietary guidelines for Americans, 2010. 7th edition, Washington, DC: US Government Printing Office; December 2010.

Diet Analysis Plus
Quick Start Guide

Logging in to *Diet Analysis Plus*

1. Go to the CourseMate for Nutritional Sciences at **www.cengagebrain.com**.
2. If you have logged in to *Diet Analysis Plus* before, or if you have a CengageBrain account, enter your user name and password and click the **Log In** button to open the **My Home** page. If you have not logged in previously, skip to step 4.
3. If you have logged in before, simply click the **Open** button under *Diet Analysis Plus* on the My Home page to log in to the program. If this is your first time, enter the *Diet Analysis Plus* access code from your access card under **Register an Access Code** and click the **Register** button. Skip steps 4–8.
4. ***CengageBrain Registration:*** Click the **Create an Account** button.
5. Enter the *Diet Analysis Plus* access code from your access card under **Enter Code** and click the **Continue** button.
6. On the **Account Information** page, enter your personal information and preferences, accept the license agreement, and click the **Continue** button.
7. On the **Select Institution** page, use the menus and **Search** button to select the name of your school, and then click the **Continue** button.
8. Click the **Open** button under *Diet Analysis Plus* on the **My Home** page to log in to the program.

Creating a Profile

1. The first time you log in to *Diet Analysis Plus*, you will see the **Create Profile** page. Enter the requested information for your primary profile, and click the **Next** button.

Home	
ⓘ Welcome to DA Plus 10. Please create your primary profile.	
ⓘ Welcome	
Create Profile	
Profile Name*	Main Profile
Birth Date	Apr 19, 1990
Gender	Male
Height Feet*	5 ft.
Height Inches*	6 inches
Weight*	100 lbs.
Smoker	☐
Vegetarian	☐
Next	

2. You will see the **Activity Questionnaire** page. Answer the questions on this page and click the **Next** button.

Activity Questionnaire	
Profile Name:	Main Profile
Age:	20 years
Gender:	Male
Height:	5 ft. 6 inches
Weight:	100.0 lbs.
Smoker:	No
Vegetarian:	No
Edit	

1. Do you know your body fat percentage?
   ○ Yes ● No
2. What type of occupation do you have?
   Sedentary desk occupation (I am seated most of the day; ex. student, administr
3. How many hours per week do you perform this occupation?* 0.0 hrs
4. How much time do you spend on leisure time and activities of daily living in an
5. How would you rate your walking pace? Strolling/Casual (<2 mph) ⬍
6. How much time do you spend performing light physical exercise in an average
7. How much time do you spend performing moderate physical exercise in an av
   I consider myself an elite athlete. ☐
8. How much time do you spend performing high-intensity physical exercise in a

Next

3. You will see the **Confirm Profile** page. Review your information. You can click the **Edit** button under your personal or activity information if you need to make corrections. Click the **Save** button.
4. To create an alternate profile, click the **Home** tab, then click the **Create New Profile** link beneath My Profile. Repeat steps 1–3.

Entering Your Course ID

1. Click the **Home** tab.
2. Under My Profile, locate the **Course Identification Number** box and enter the Course Identification Number provided to you by your instructor.
3. Click the **Submit** link. The name of the course for which you are using *Diet Analysis Plus* should now appear beside **Current Course**.

Tracking Your Diet

1. Click the **Track Diet** tab.
2. Click the calendar to the right of the "Select a Date" label and choose the date for which you would like to enter or edit the foods you ate.
3. Next to the "Find Foods" label, enter the name of a food/beverage you ate and click the **Go** button. A food list will appear (with arrows and page numbers if there are more than 15 choices). You may use the Category Filter to narrow your search to a particular group of foods.

4. Click on the food name closest to what you ate/drank to open the **Serving Size** pop-up screen.

5. Enter the amount that you ate/drank, being careful to choose the correct unit of measurement from the first menu. Select the correct meal or snack from the second menu. Click the **Save** button. Repeat steps 3–5 for each food or drink you consumed that day.

Tracking Your Activity

1. Click the **Track Activity** tab.
2. Click the calendar to the right of the "Select a Date" label and then choose the date for which you would like to enter or edit your activities.
3. Next to the "Find Activities" label, enter the name of an activity you performed and click the **Go** button. An activities list will appear (with arrows and page numbers if there are more than 15 choices).

4. Click on the name of the activity that most closely matches what you did to open the **Activity Duration** pop-up screen.

5. Use the menus to enter the amount of time you performed this activity. Click the **Save** button. Repeat steps 3–5 for each activity you completed that day. You must enter activities for all 24 hours of the day (including time spent sleeping, getting dressed, and driving).

Viewing and Printing Reports

1. Click the **Reports** tab.
2. Click on the name of the report you would like to create.
3. Most reports have options for you to adjust at the top of the page. Click the calendars under **Select a Date/Date Range** or **Day 1/2/3** to choose the date(s) you would like to analyze. Under **Choose Meals**, you can uncheck the boxes if you want to exclude meals or snacks from the report. (For the *3 Day Average* report, you will also need to click the **Preview Report** button to view the report.)
4. To generate a version of the report that you can print or e-mail, click the **Print PDF** button to create a PDF of the report, or **Print RTF** to create a Word document.
5. To submit the report to your instructor so that he/she can view it online, click the **Submit Report** button.

Requesting Technical Support

1. Go to **www.cengage.com/support**.
2. You can either click the **Select Product** button under "Self-help" and then choose your version of *Diet Analysis Plus* to access instructions and tutorial videos, or sign in to submit a request for assistance online (click on **create a new case**).
3. Alternatively, you can call technical support at 1-800-354-9706 (option 5, then option 2). Support is available Monday through Thursday, 8:30 a.m.–9:00 p.m. EST, and Friday 8:30 a.m.–6:00 p.m EST.